WIRELESS TRANSCEIVER ARCHITECTURE

WIRELESS TRANSCEIVER ARCHITECTURE

BRIDGING RF AND DIGITAL COMMUNICATIONS

Pierre Baudin

WILEY

This edition first published 2015

Registered office
John Wiley & Sons Ltd, The Atrium, Southern Gate, Chichester, West Sussex, PO19 8SQ, United Kingdom

For details of our global editorial offices, for customer services and for information about how to apply for permission to reuse the copyright material in this book please see our website at www.wiley.com.

Library of Congress Cataloging-in-Publication Data

Baudin, Pierre (Electrical engineer)
Wireless transceiver architecture : bridging RF and digital communications / Pierre Baudin.
pages cm
Includes bibliographical references and index.
ISBN 978-1-118-87482-0 (hardback)
1. Radio–Transmitter-receivers. I. Title.
TK6564.3.B38 2014
621.384–dc23

2014011413

A catalogue record for this book is available from the British Library.

ISBN 9781118874820

Set in 10/12pt Times by Aptara Inc., New Delhi, India

1 2015

To my children, Hugo and Chloé, and in memory of my parents

Contents

Preface

The origins of this book lie in the frequent questions that I have been asked by colleagues in the different companies I have worked for about how to proceed in the dimensioning and optimization of a transceiver line-up. The recurrence of those questions, along with the problem of identifying suitable reference sources, made me think there could be a gap in the literature. There is indeed an abundant literature on the physical implementation of wireless transceivers (e.g. the RF/analog CMOS design), or on digital communications theory itself (e.g. the signal processing required), but little on how to proceed for dimensioning and optimizing a transceiver line-up.

Furthermore, the fact is that those questions were coming from two distinct categories of engineers. On the one hand, RF/analog designers are curious to understand how the specifications of their blocks are derived. On the other hand, the digital signal processing engineers in charge of the baseband algorithms need to understand the mechanisms involved in the degradation of the wanted signal along the line-up for optimizing their processing. Obviously, it is the job of an RFIC architect to make the link between the two communities and to attempt to overcome the communication problems between those two groups. Roughly speaking, you have on the one hand the baseband engineers that process complex envelopes while benchmarking their algorithms using AWGN, and on the other hand the RF/analog designers that optimize their designs based on the use of CW tones for evaluating the degradation of a signal expected to be modulated. Based on my experience, even if the difficulty in the discussion can be related somehow to the different nature of the technical issues addressed by the two communities, it is also closely related to the different formalisms traditionally used in those two domains of knowledge.

I therefore wrote this book with two aims in mind. I first tried to detail the mindset required, at least based on my professional experience, to take care of the system design of a transceiver. Expressed like this, we understand that the purpose is not to be exhaustive about how to perform such dimensioning for all the architectures one can imagine. Rather the goal is to explain the spirit of it, and how to initiate such work in practice. Conversely, in order to be able to react correctly whatever the architecture under consideration, there is a need to understand as far as possible the constraints we have to take into account in order to dimension a transceiver. Practically speaking, this means understanding the system design of transceiver line-up in its various aspects. We can, for instance, mention the need to understand the purpose of a transceiver from the signal processing perspective. Indeed, from the transmitted or received signal point of view, the transceiver implements nothing more than signal processing functions, mainly analog signal processing, but signal processing when all is said and done. Alternatively, we can mention the need to understand the various limitations one can face in the implementation of this analog signal processing using electronic devices. Those limitations can indeed be encountered whatever the architecture implemented.

I then also tried to unify the formalisms used in the various domains of knowledge involved in the field of wireless transceivers. In practice, this means considering the digital communications formalism and the extensive use of the complex envelope concept for modeling modulated RF signals. As discussed above, the first goal was to make easier the link between RF and digital communications people who need to work together in order to optimize a line-up. This approach also happens to have many additional benefits. It allows us to correctly define RF concepts for modulated signals often introduced in a more intuitive way. It also allows us to perform straightforward analytical derivations in many situations of interest, as in nonlinearity for instance, while allowing explicit graphical representations in the complex plane. I am now fully convinced that this formalism is of much interest to RF problems and I would be pleased if this book can help to propagate its use.

As a result, this book consists of three parts. Part I focuses on the explanation of what is expected from a transceiver. This part is composed of three chapters dedicated to the three areas that drive those requirements: (a) the digital communications theory itself, which allows us to define the minimum set of signal processing functions to be embedded in a transceiver, as well as introducing key concepts such as complex envelopes; (b) the electromagnetism theory, as theoretical results in the field of propagation allow us to explain some architectural constraints for transceivers; (c) the practical organization of wireless networks, as it drives most of the performance required from transceivers in practice. By the end of Part I we should thus have an understanding of the functionalities required in a transceiver as well as their associated performance.

Part II is then dedicated to a review of the limitations we face in the physical implementation using electronic devices of the signal processing functions derived in Part I. Those limitations are sorted into three groups, leading to three chapters dedicated to: (a) the noise sources to be considered in a line-up; (b) the nonlinearity in RF/analog components; (c) what are classically labeled RF impairments.

Part III then turns to the transceiver architecture and system design itself. We can now focus on how to dimension a transceiver that fulfills the requirements derived in Part I while taking into account the implementation limitations reviewed in Part II. Practically speaking, this is done through three chapters. The first of these is dedicated to the illustration of a transceiver budget for a given architecture. This shows how a practical line-up budget can be done, i.e. how the constraints linked to the implementation limitations can be balanced between the various blocks of a given line-up in order to achieve the performance. The second chapter reviews different architectures of transceivers. In contrast to what is done in the previous chapter, we can see here how the fundamental limitations of a given line-up can be overcome by changing its architecture. The third chapter then examines some algorithms classically used for improving or optimizing the performance of transceiver line-ups.

At this stage, I need to highlight that, due to the organization of the book, only the reasoning used for the architecture and system design of transceivers is discussed in Part III. All the theoretical results, as well as the description of the elementary phenomena that are involved in this area, are detailed in Parts I and II. As a result, I recommend that the reader should not embark on Part III without sufficient understanding of the phenomena discussed in Parts I and II.

To conclude, I would like to thank all those people who helped in completing this project. First of all, I would like to thank my former colleagues at Renesas who participated in one way or another during this project, i.e. Alexis Bisiaux, Pascal Le Corre, Mikaël Guenais, Stéphane Paquelet, Arnaud Rigollé, Patrick Savelli, and in particular Larbi Azzoug and Anis Latiri. Then, I would like to warmly thank Marc Hélier, who taught me microwave engineering at Supélec some years ago, and who was kind enough to go through Chapter 2. Finally, I would like to thank Fabrice Belvèze, as it was all the good technical discussions we had during the old STMicroelectronics times that first convinced me that it was interesting to using the digital communications formalism for the system design of transceiver line-up. Things have changed since then, but the origins are there.

Rennes, January 2014

List of Abbreviations

ABB	Analog Baseband
ACPR	Adjacent Channel Power Ratio
ACLR	Adjacent Channel Leakage Ratio
ACS	Adjacent Channel Selectivity
ADC	Analog to Digital Converter
ADPLL	All Digital PLL
AFC	Automatic Frequency Correction
AGC	Automatic Gain Control
AM	Amplitude Modulation
AWGN	Additive White Gaussian Noise
BBIC	Baseband IC
BER	Bit Error Rate
BTS	Base Transceiver Station
CALLUM	Combined Analog Locked Loop Universal Modulator
CCP	Cross-Compression Point
CDMA	Code Division Multiple Access
CDE	Code Domain Error
CDF	Cumulative Distribution Function
CCDF	Complementary Cumulative Distribution Function
CDP	Code Domain Power
CF	Crest Factor
CMOS	Complementary Metal Oxide Semiconductor
CORDIC	COordinate Rotation DIgital Computer
CP	Compression Point
CW	Continuous Wave
DAC	Digital to Analog Converter
DC	Direct Current
DNL	Differential NonLinearity
DR	Dynamic Range
DSB	Double SideBand
DtyCy	Duty Cycle
EDGE	Enhanced Data Rates for GSM Evolution
EER	Envelope Elimination and Restoration
EMF	ElectroMotive Force

EMI	ElectroMagnetic Interference
ERR	Even Order Rejection Ratio
ET	Envelope Tracking
EVM	Error Vector Magnitude
FDD	Frequency Division Duplex
FDMA	Frequency Division Multiple Access
FE	Front-End
FEM	Front-End Module
FIR	Finite Impulse Response
FM	Frequency Modulation
FS	Full Scale
GMSK	Gaussian Minimum Shift Keying
GSM	Global System for Mobile communications
HPSK	Hybrid Phase Shift Keying
I/F	InterFace
IC	Integrated Circuit
ICP	Input Compression Point
ICCP	Input Cross-Compression Point
IF	Intermediate Frequency
IIP	Input Intercept Point
IMD	Intermodulation Distortion
INL	Integral NonLinearity
IP	Intercept Point
IPsat	Input Saturated Power
IRR	Image Rejection Ratio
ISI	InterSymbol Interference
ISR	Input Spurious Rejection
LINC	LInear amplification using Nonlinear Component
LNA	Low Noise Amplifier
LO	Local Oscillator
LSB	Least Significant Bit
LTE	Long-Term Evolution
LUT	LookUp Table
NCO	Numerically Controlled Oscillator
NF	Noise Figure
OCP	Output Compression Point
OFDM	Orthogonal Frequency Division Multiplexing
OIMD	Output InterModulation Distortion
OIP	Output Intercept Point
OPsat	Output Saturated Power
OSR	OverSampling Ratio
PA	Power Amplifier
PAPR	Peak to Average Power Ratio
PGA	Programmable Gain Amplifier
PDF	Probability Density Function
PFD	Phase Frequency Detector

PLL	Phase Locked Loop
PM	Phase Modulation
Psat	Saturated Power
PSD	Power Spectral Density
PSK	Phase Shift Keying
QAM	Quadrature Amplitude Modulation
RF	Radio Frequency
RFIC	Radio Frequency Integrated Circuit
RL	Return Loss
RMS	Root Mean Square
RRC	Root Raised Cosine
RX	Receiver
RXFE	RX Front-End
SEM	Spectrum Emission Mask
SFDR	Spurious Free Dynamic Range
SiNAD	Signal to Noise and Distortion ratio
SNR	Signal to Noise Power Ratio
SSB	Single SideBand
TDD	Time Division Duplex
TDMA	Time Division Multiple Access
TE	Transverse Electric
TEM	Transverse Electromagnetic
TM	Transverse Magnetic
THD	Total Harmonic Distortion
TX	Transmitter
TRX	Transceiver
UE	User Equipment
VGA	Variable Gain Amplifier
VSWR	Voltage Standing Wave Ratio
WCDMA	Wideband CDMA
WSS	Wide Sense Stationarity
XM	Cross-Modulation
ZIF	Zero-IF

Nomenclature

t, τ	time variables
f	frequency variable
ω	angular frequency variable ($= 2\pi f$)
$x(t)$	continuous time signal
$x[n]$	discrete time signal
$X(f)$, $X(\omega)$	frequency representations of $x(t)$ or $x[n]$
$\mathscr{F}_{\{x(t)\}}(f)$, $\mathscr{F}_{\{x(t)\}}(\omega)$	Fourier transforms of $x(t)$
v, i, j	voltage, current, current density
P, $p(t)$	in-phase, in-phase component
Q, $q(t)$	quadrature, quadrature component
j	$\sqrt{-1}$
$\mathrm{Re}\{.\}$, $\mathrm{Im}\{.\}$	real part, imaginary part
$\lvert.\rvert$, $\arg\{.\}$	modulus, argument
$.^{*}$	complex conjugate
$\star$	convolution
$\delta(.)$	Dirac delta distribution
$\mathrm{U}(.)$	Heaviside unit step function
$x_{\mathrm{a}}(t)$	analytical signal associated with $x(t)$
$X_{\mathrm{a}}(f)$, $X_{\mathrm{a}}(\omega)$	frequency domain representations of $x_{\mathrm{a}}(t)$
$\hat{x}(t)$	Hilbert transform of $x(t)$
$\hat{X}(f)$, $\hat{X}(\omega)$	frequency domain representations of $\hat{x}(t)$
$\tilde{x}(t)$	complex envelope associated with $x(t)$
$\tilde{X}(f)$, $\tilde{X}(\omega)$	frequency domain representations of $\tilde{x}(t)$
$\overline{(.)}$	time average value
$\mathbb{E}\{.\}$	stochastic expectation value
$\gamma_{x\times y}(t_1, t_2)$	cross-correlation function $\left(= \mathbb{E}\left\{x_{t_1} y_{t_2}^{*}\right\}\right)$
$\gamma_{x\times x}(t_1, t_2)$	autocorrelation function
$\gamma_{x\times x}(\tau)$	autocorrelation function in stationary case $\left(= \mathbb{E}\left\{x_t x_{t-\tau}^{*}\right\}\right)$
$\Gamma_{x\times x}(f)$, $\Gamma_{x\times x}(\omega)$	power spectral densities of $x(t)$
$\mathbf{X}$, $\boldsymbol{X}$	vector, matrix
$.^{\mathrm{T}}$	transpose
$\lVert.\rVert$	Hermitian norm
$\cdot$	dot product
$\times$	cross product

Part I

Between Maxwell and Shannon

1

The Digital Communications Point of View

When detailing how to dimension a transceiver, it can seem natural to first clarify what is expected from such a system. This means understanding both the minimum set of functions that need to be implemented in a transceiver line-up as well as the minimum performance expected from them. In practice, these requirements come from different topics which can be sorted into three groups. We can indeed refer to the signal processing associated with the modulations encountered in digital communications, to the physics of the medium used for the propagation of the information, and to the organization of wireless networks when considering a transceiver that belongs to such system, or alternatively its coexistence with such systems.

The last two topics are discussed in Chapter 2 and Chapter 3 respectively, while this chapter focuses on the consequences for transceiver architectures of the signal processing associated with the digital communications. In that perspective, a first set of functions to be embedded in such a system can be derived from the inspection of the relationship that holds between the modulating waveforms used in this area and the corresponding modulated RF signals to be propagated in the channel medium.

As a side effect, this approach enables us to understand how information that needs a complex baseband modulating signal to be represented can be carried by a simple real valued RF signal, thus leading to the key concept of the complex envelope. It is interesting to see that this concept allows us to define correctly classical quantities used to characterize RF signals and noise, in addition to its usefulness for performing analytical derivations. It is therefore used extensively throughout this book.

Finally, in this chapter we also review some particular modulation schemes that are representative of the different statistics that can be encountered in classical wireless standards. These schemes are then used as examples to illustrate subsequent derivations in this book.

Wireless Transceiver Architecture: Bridging RF and Digital Communications, First Edition. Pierre Baudin.

1.1 Bandpass Signal Representation

1.1.1 RF Signal Complex Modulation

Digital modulating waveforms in their most general form are represented by a complex signal function of time in digital communications books [1]. But, even if we understand that this complex signal allows us to increase the number of bits per second that can be transmitted by working on symbols using this two-dimensional space, a question remains. The final RF signal that carries the information, like the RF current or voltage generated at the transmitter (TX) output, is a real valued signal like any physical quantity that can be measured. Accordingly, we may wonder how the information that needs a complex signal to be represented can be carried by such an RF signal. Any RF engineer would respond by saying that an electromagnetic wave has an amplitude and a phase that can be modulated independently. Nevertheless, we can anticipate the discussion in Chapter 2, and in particular in Section 2.1.2, by saying that there is nothing in the electromagnetic theory that requires this particular structure for the time dependent part of the electromagnetic field. In fact, the right argument remains that this time dependent part, like any real valued signal, can be represented by two independent quantities that can be interpreted as its instantaneous amplitude and its instantaneous phase as long as it is a *bandpass* signal. Here, "bandpass signal" means that the spectral content of the signal has no low frequency component that spreads down to the zero frequency. In other words, the spectrum of the RF signal considered, whose positive and negative sidebands are assumed centered around $\pm\omega_c$, must be non-vanishing only for angular frequencies in $[-\omega_u - \omega_c, -\omega_c + \omega_l] \cup [+\omega_c - \omega_l, +\omega_c + \omega_u]$, with ω_c, ω_l and ω_u defined as positive quantities, and with $\omega_c > \omega_l$.

To understand this behavior, let us consider the complex baseband signal $\tilde{s}(t)$ expressed as

$$\tilde{s}(t) = p(t) + jq(t), \tag{1.1}$$

where $p(t)$ and $q(t)$ are respectively the real and imaginary parts of this complex signal. We can assume that the spectrum of this signal spreads over $[-\omega_l, +\omega_u]$. Such baseband signals with a non-vanishing DC component in their spectrum are called *lowpass* signals in contrast to the bandpass signals as given above. If we now wish to shift the spectrum of this signal around the central carrier angular frequency $+\omega_c$, we have to convolve its spectrum with the Dirac delta distribution $\delta(\omega - \omega_c)$. In the time domain, this means multiplying the signal by the Fourier transform of this Dirac delta distribution, i.e. the complex exponential $e^{+j\omega_c t}$ [2]. This results in the complex signal $s_a(t)$ defined by

$$s_a(t) = \tilde{s}(t)e^{+j\omega_c t} = (p(t) + jq(t))e^{+j\omega_c t}. \tag{1.2}$$

Suppose now that we take the real part of this signal. Using

$$e^{+j\omega t} = \cos(\omega t) + j\sin(\omega t), \tag{1.3}$$

we get the classical form of the resulting RF signal $s(t)$ we are looking for,

$$s(t) = \mathrm{Re}\{s_a(t)\} = p(t)\cos(\omega_c t) - q(t)\sin(\omega_c t). \tag{1.4}$$

But what is interesting to see is that even if we took only the real part of the upconverted initial complex lowpass signal transposed around $+\omega_c$, we have no loss of information compared to the initial complex baseband signal as long as $\omega_c > \omega_l$. Indeed, under that condition, the original complex modulating waveform $\tilde{s}(t)$ can be reconstructed from the bandpass RF real signal $s(t)$.

To understand this, let us first consider the spectral content of the resulting bandpass signal $s(t)$. What would be a good mathematical tool to choose for the spectral analysis? Dealing with digital modulations that are randomly modulated most of the time, the natural choice would be to use the stochastic approach to derive the signal power spectral density. The problem with this approach is that the power spectral density (PSD) of a signal is only linked to the modulus of the Fourier transform of the original signal. It thus leads to a loss of information compared to the time domain signal. As a result, in some cases of interest in this book, we need to keep the simple Fourier transform representation in order to be able to discuss the phase relationship between different sidebands present in the spectrum. Here, by "sideband" we mean a non-vanishing portion of spectrum of finite frequency support. This phase relationship is indeed required to understand the underlying phenomenon involved in concepts as reviewed in this chapter, but also in Chapter 6, for instance, when dealing with frequency conversion and image rejection. The existence of such Fourier transforms can be justified thanks to the practical finite temporal support of the signals of interest that ensures a finite energy. This is indeed the practical use case when dealing with the post-processing of a finite duration measurement or simulation result. The signals we deal with are therefore assumed to have a finite temporal support and a finite energy, i.e. they are assumed to belong to $L^2[0, T]$, the space of square-integrable functions over the bounded interval $[0, T]$. Nevertheless, when dealing with a randomly modulated signal, this approach means that we consider only the spectral properties of a single realization of the process of interest. Thus, even if this direct Fourier analysis is suitable for discussing some signal processing operations involved in transceivers, the power spectral analysis should be considered when possible for taking into account the statistical properties of the modulating process of interest, as done in "Power spectral density" (Section 1.1.3).

Let us therefore derive the Fourier transform of $s(t)$. As the aim is to make the link between the spectral representation of $s(t)$ and that of $\tilde{s}(t)$, we can first expand the relationship between $s(t)$ and the complex signal $s_a(t)$ given by equation (1.4). To do so, we use the general property that for any complex number $\tilde{s}(t)$, we have

$$\mathrm{Re}\{\tilde{s}(t)\} = \frac{1}{2}(\tilde{s}(t) + \tilde{s}^*(t)), \tag{1.5}$$

where $\tilde{s}^*(t)$ stands for the complex conjugate of $\tilde{s}(t)$. This means that we can write

$$s(t) = \mathrm{Re}\{s_a(t)\} = \frac{1}{2}(s_a(t) + s_a^*(t)). \tag{1.6}$$

Using the relationship between $s_a(t)$ and $\tilde{s}(t)$ given by equation (1.2), we finally get that

$$s(t) = \frac{1}{2}\tilde{s}(t)e^{+j\omega_c t} + \frac{1}{2}\tilde{s}^*(t)e^{-j\omega_c t}. \tag{1.7}$$

It now remains to take the Fourier transform of this signal. For that, we can use two properties of the Fourier transform. The first states that for any signal, $\tilde{s}(t)$, the Fourier transform of the complex conjugate, $\tilde{s}^*(t)$, of such a signal can be related to that of $\tilde{s}(t)$ through

$$\begin{aligned}\mathcal{F}_{\{\tilde{s}^*(t)\}}(\omega) &= \int_{-\infty}^{+\infty} \tilde{s}^*(t)\mathrm{e}^{-\mathrm{j}\omega t}\mathrm{d}t \\ &= \left[\int_{-\infty}^{+\infty} \tilde{s}(t)\mathrm{e}^{+\mathrm{j}\omega t}\mathrm{d}t\right]^* \\ &= \left[\mathcal{F}_{\{\tilde{s}(t)\}}(-\omega)\right]^*. \end{aligned} \tag{1.8}$$

We observe that this derivation remains valid when $\tilde{s}(t)$ reduces to a real signal $s(t)$. In that case, having $s^*(t) = s(t)$ leads to having $S^*(-\omega) = S(\omega)$. We then recover the classical property of real signals, i.e. the Hermitian symmetry of their spectrum. Then we can use the property that the Fourier transform of a product of signals is equal to the convolution of the Fourier transforms of each signal. Indeed, we get that

$$\begin{aligned}\mathcal{F}_{\{\tilde{s}_1(t)\tilde{s}_2(t)\}}(\omega) &= \int_{-\infty}^{+\infty} \tilde{s}_1(t)\tilde{s}_2(t)\mathrm{e}^{-\mathrm{j}\omega t}\mathrm{d}t \\ &= \int_{-\infty}^{+\infty} \tilde{S}_1(\omega') \int_{-\infty}^{+\infty} \tilde{s}_2(t)\mathrm{e}^{-\mathrm{j}(\omega-\omega')t}\mathrm{d}t\mathrm{d}\omega' \\ &= \int_{-\infty}^{+\infty} \tilde{S}_1(\omega')\tilde{S}_2(\omega-\omega')\mathrm{d}\omega', \end{aligned} \tag{1.9}$$

i.e. that

$$\mathcal{F}_{\{\tilde{s}_1(t)\tilde{s}_2(t)\}}(\omega) = \mathcal{F}_{\{\tilde{s}_1(t)\}}(\omega) \star \mathcal{F}_{\{\tilde{s}_2(t)\}}(\omega). \tag{1.10}$$

Thus, using the two properties above, we get that the Fourier transform of equation (1.7) reduces to

$$S(\omega) = \frac{1}{2}\tilde{S}(\omega) \star \delta(\omega - \omega_\mathrm{c}) + \frac{1}{2}\tilde{S}^*(-\omega) \star \delta(\omega + \omega_\mathrm{c}), \tag{1.11}$$

where[1] $\tilde{S}(\omega)$ stands for the Fourier transform of $\tilde{s}(t)$ and where the Dirac delta distribution is the Fourier transform of the complex exponential. As this distribution is even, i.e. we have $\delta(\omega + \omega_\mathrm{c}) = \delta(-\omega - \omega_\mathrm{c})$, the spectrum of $s(t)$ can be expressed as the sum of two components as illustrated in Figure 1.1. The first component corresponds to the positive frequencies part, denoted $S_+(\omega)$, and is referred to as the positive sideband of the spectrum of $s(t)$. The second component corresponds to the negative part of the spectrum, $S_-(\omega)$, and is therefore referred to as the negative sideband of the spectrum of $s(t)$. As $s(t)$ is assumed to be bandpass, there is

[1] Recall our convention in this book that $\tilde{S}(\omega)$ stands for the spectral domain representation of the complex envelope $\tilde{s}(t)$ and not for the complex envelope of the signal $S(\omega)$.

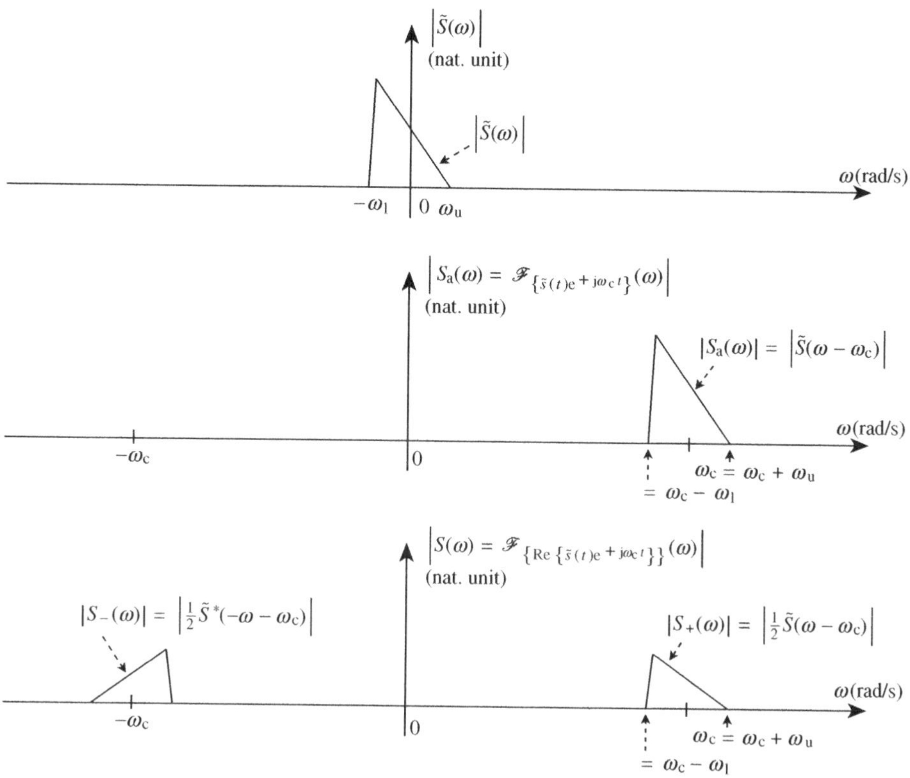

Figure 1.1 Spectral domain representation of the complex modulation of a real bandpass RF signal – The information linked to a complex lowpass signal, whose spectrum does not fulfill Hermitian symmetry (top), can be carried by a real bandpass signal as long as the intermediate complex upconverted signal is such that its spectrum does not spread toward zero frequency (middle), i.e. as long as $\omega_c > \omega_l$. In that case, the two sidebands of the spectrum of the resulting real bandpass signal do not overlap (bottom).

no overlap between those sidebands. They are thus defined without ambiguity from equation (1.11) as

$$\begin{aligned} S(\omega) &= \frac{1}{2}\tilde{S}(\omega - \omega_c) + \frac{1}{2}\tilde{S}^*(-\omega - \omega_c) \\ &= S_+(\omega) + S_-(\omega), \end{aligned} \tag{1.12}$$

where

$$S_+(\omega) = \frac{1}{2}\tilde{S}(\omega - \omega_c), \tag{1.13a}$$

$$S_-(\omega) = \frac{1}{2}\tilde{S}^*(-\omega - \omega_c). \tag{1.13b}$$

As expected for a real signal, the spectrum of $s(t)$ fulfills Hermitian symmetry as we have $S_-(\omega) = S_+^*(-\omega)$. Reconsidering our derivation so far, we thus see that $S_+(\omega)$ represents the spectrum of the initial complex signal $\tilde{s}(t)$ transposed around $+\omega_c$, whereas $S_-(\omega)$ represents the spectrum of the complex conjugate of the initial complex signal, i.e. of $\tilde{s}^*(t)$, transposed around $-\omega_c$. We therefore see that the action of taking the real part of $s_a(t)$ leads to a symmetrization of the spectrum by creating a negative part of the spectrum that is a flipped copy of its positive part. What is interesting to remark is that even if the resulting real RF signal spectrum fulfills Hermitian symmetry, the portion that lies in the positive frequencies part still represents precisely the initial complex lowpass signal spectrum. Consequently, as long as the resulting signal $s(t)$ is bandpass the two sidebands of its spectrum do not overlap. The initial complex lowpass signal can thus be reconstructed, theoretically without distortion, by downconversion to baseband of the positive sideband of the passband signal.

As a consequence of the concepts presented so far, we can see that in a transceiver, as long as we are dealing with a lowpass representation of a complex modulating waveform, $\tilde{s}(t)$, we need to work with two analog real quantities to represent, for instance, its real and imaginary parts $p(t)$ and $q(t)$. However, this is not required when working on the corresponding modulated real bandpass signal for which a representation using a single analog signal is always possible with no loss of information. We also mention that the real and imaginary parts are classically labeled $p(t)$ and $q(t)$, as used in this book, or alternatively as $i(t)$ and $q(t)$. The origin of these labels is that $p(t)$ or $i(t)$ stand for the in-phase component and $q(t)$ stands for the in-quadrature component of the bandpass signal as those lowpass modulating signals are carried by cosine and sine waveforms that are effectively in quadrature.

1.1.2 Complex Envelope Concept

We saw in Section 1.1.1 that a real bandpass RF signal can be modulated in such a way that it carries information that needs a complex lowpass modulating signal, i.e. two independent real baseband signals, to be represented. We can now go through the reverse process and examine how to reconstruct the equivalent complex lowpass modulating signal that represents the modulation of a given bandpass RF signal $s(t)$.

For that purpose, we can retain the approach of Section 1.1.1, i.e. assume that it is the positive sideband of $s(t)$ that is an image of the spectrum of the original complex lowpass modulating waveform. Let us first define the complex signal $s_a(t)$ whose spectral representation $S_a(\omega)$ is equal to twice the positive sideband of $s(t)$, i.e. such that

$$S_a(\omega) = 2\mathrm{U}(\omega)S(\omega), \tag{1.14}$$

where U(.) stands for the Heaviside step function. Here, the factor 2 is used for the sake of consistency with the definition introduced in Section 1.1.1. The signal $s_a(t)$ defined as such is said to be the *analytic* signal associated with $s(t)$. Based on its definition, we clearly get that this signal is unique, which is an important difference compared to the complex envelope concept defined later on. We observe that the term "analytic" is used for a signal whose spectrum is null for negative frequencies. This is thus the equivalent, in the spectral domain, of the causality concept encountered in the time domain. In order to recover the complex modulating waveform we are looking for, it thus remains to downconvert this analytic signal toward the zero frequency. In that perspective, it remains of interest to consider a time domain representation of $s_a(t)$. In doing so, rather than directly taking the inverse Fourier transform of equation (1.14), it is convenient to remark that an alternative expression for $S_a(\omega)$ is

$$\begin{aligned} S_a(\omega) &= S(\omega) + \text{sign}\{\omega\}S(\omega) \\ &= S(\omega) + \text{j}(-\text{jsign}\{\omega\})S(\omega), \end{aligned} \tag{1.15}$$

where sign{.} denotes the sign function. This expression is in fact the direct transcription of the cancellation of the negative part of the spectrum of $s(t)$ for expressing $s_a(t)$. If we now transpose this equation in the time domain, this signal can then be expressed as

$$s_a(t) = s(t) + \text{j}\hat{s}(t), \tag{1.16}$$

where $\hat{s}(t)$ represents the signal resulting from the filtering of $s(t)$ by the filter whose transfer function can be expressed in the frequency domain as

$$-\text{j}\,\text{sign}\{\omega\} = \begin{cases} -\text{j} & \text{when } \omega > 0, \\ 0 & \text{when } \omega = 0, \\ \text{j} & \text{when } \omega < 0. \end{cases} \tag{1.17}$$

At this stage, it may be of interest to remark that this filter simply behaves as a $\pi/2$ phase shifter when acting on a bandpass signal. This can be highlighted by considering, for instance, its effect on a pure sine wave such as $s(t) = \cos(\omega_c t)$. Expanding the cosine function as the sum of two complex exponentials results in

$$\cos(\omega_c t) = \frac{1}{2}\left(e^{\text{j}\omega_c t} + e^{-\text{j}\omega_c t}\right). \tag{1.18}$$

Thus, considering the effect of the transfer function given by equation (1.17), we get that the first complex exponential, $e^{\text{j}\omega_c t}$, changes to $-\text{j}e^{\text{j}\omega_c t}$ and the second one, $e^{-\text{j}\omega_c t}$, changes to $\text{j}e^{-\text{j}\omega_c t}$. The signal recovered at the filter output is therefore the sine wave:

$$\sin(\omega_c t) = \frac{1}{2\text{j}}\left(e^{\text{j}\omega_c t} - e^{-\text{j}\omega_c t}\right). \tag{1.19}$$

Given that $\sin(\omega t) = \cos(\omega t - \pi/2)$, we thus recover the expected behavior of a $\pi/2$ phase shifter. This effect can in fact be confirmed for any bandpass signals by reconsidering equation (1.17). Indeed, given that $\text{j} = e^{\text{j}\pi/2}$, we get that this filter simply subtracts a constant $\pi/2$ phase offset from all the frequency components of the input signal lying at positive frequencies and adds the same $\pi/2$ phase offset to all the components lying at negative frequencies, thus confirming the behavior of an ideal $\pi/2$ phase shifter. The transfer function of this filter can also be transposed in the time domain, thus leading to the implementation of a Hilbert transform corresponding to the convolution of the input signal with the distribution principal value of $1/(\pi t)$ [2]. The filter impulse response can thus be expressed in the distribution sense as

$$\hat{s}(t) = \text{pv}\left(\frac{1}{\pi t}\right) \star s(t) = \lim_{\epsilon \to 0} \frac{1}{\pi} \int_{|t'|>\epsilon} \frac{s(t-t')}{t'} \mathrm{d}t', \tag{1.20}$$

where pv(.) stands for the Cauchy principal value.

Now that we have an expression for $s_a(t)$, it remains to downconvert this signal toward the zero frequency to recover the complex modulating waveform $\tilde{s}(t)$. This can be achieved by multiplying its time domain expression by the negative complex exponential $e^{-j\omega_c t}$ so that [2]

$$\tilde{s}(t) = s_a(t)e^{-j\omega_c t}. \tag{1.21}$$

Using equation (1.16) we then get that

$$\tilde{s}(t) = (s(t) + j\hat{s}(t))e^{-j\omega_c t}. \tag{1.22}$$

At first glance, the signal $\tilde{s}(t)$ defined this way matches the definition of the complex modulating waveform as defined in Section 1.1.1. This signal is called the complex envelope of the bandpass signal $s(t)$. The different steps for its construction are illustrated in the frequency domain in Figure 1.2. Alternatively, its real and imaginary parts, $p(t)$ and $q(t)$ respectively, can also be expressed in the time domain as functions of $s(t)$ and $\hat{s}(t)$ by expanding the above equation. This results in

$$p(t) = s(t)\cos(\omega_c t) + \hat{s}(t)\sin(\omega_c t), \tag{1.23a}$$

$$q(t) = \hat{s}(t)\cos(\omega_c t) - s(t)\sin(\omega_c t). \tag{1.23b}$$

Reconsidering our definitions so far, we observe that we have called $\tilde{s}(t)$ *the* complex envelope of the bandpass signal $s(t)$. But we should in fact say that $\tilde{s}(t)$ is *a* complex envelope of $s(t)$. Indeed, we defined the lowpass signal $\tilde{s}(t)$ by the downconversion of the analytic signal $s_a(t)$ around the zero frequency. This notion of "around" the zero frequency is somewhat imprecise. The point for the definition of this downconversion is only that the resulting signal must have a non-vanishing DC component in order to be effectively lowpass. We can thus define as many complex envelopes as we want for a given bandpass signal, depending on the chosen angular frequency ω_c used for the downconversion of its associated analytic signal. We see that this definition necessarily leads to an interesting relationship between the different complex envelopes that can be defined for a given bandpass signal. Indeed, considering for instance the complex envelopes $\tilde{s}_1(t)$ and $\tilde{s}_2(t)$ as defined through the downconversion of the analytic signal $s_a(t)$ using ω_{c1} and ω_{c2} respectively, we get that

$$\tilde{s}_1(t) = s_a(t)e^{-j\omega_{c1} t}, \tag{1.24a}$$

$$\tilde{s}_2(t) = s_a(t)e^{-j\omega_{c2} t}. \tag{1.24b}$$

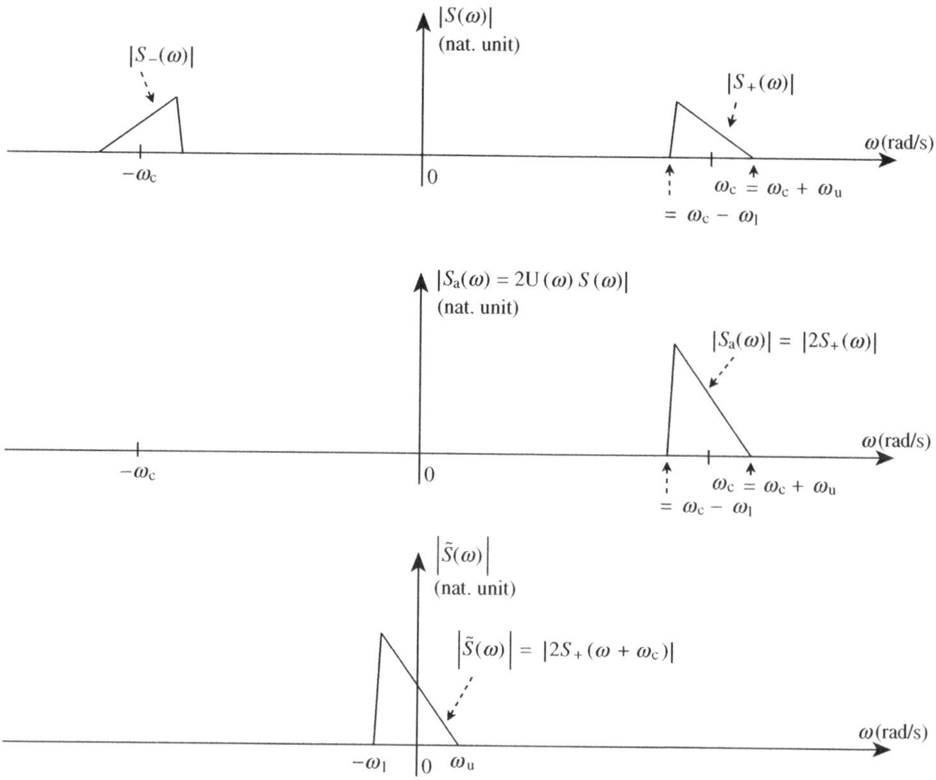

Figure 1.2 Illustration in the frequency domain of the complex envelope derivation of a bandpass RF signal – The analytic signal associated with a bandpass RF signal (top) is defined by considering twice its positive sideband (middle). The complex envelope then results from the downconversion to baseband of this analytic signal (bottom).

We thus have

$$\tilde{s}_1(t) = \tilde{s}_2(t)\mathrm{e}^{-\mathrm{j}(\omega_{c1}-\omega_{c2})t}. \tag{1.25}$$

Consequently, all the complex envelopes representing a given bandpass signal necessarily have the same modulus. There is only a continuous phase rotation difference between them. This property is important, as in transceiver budgets we often use such complex envelopes to perform analytic signal power evaluation for the purpose of signal to noise power ratio (SNR) budgets. And as illustrated for instance in "Average power" (Section 1.1.3) or throughout Chapter 5, it is the modulus of those complex envelopes that is involved in such evaluations. It is therefore good to see that the quantities involved in those derivations are well defined whatever the complex envelope considered. However, in digital communications, we generally do not have this kind of ambiguity in the selection of the center angular frequency used for the definition of the complex envelopes we deal with. Indeed, for reasons discussed in "Impact of spectral symmetry on top of stationarity" (Section 1.1.3), the complex lowpass modulating waveforms that are classically used in wireless systems almost always have a symmetrical PSD around the zero frequency, as illustrated for instance in Section 1.3. Thus, the angular frequency of the corresponding center of symmetry in the spectrum of the modulated RF bandpass signals remains a natural choice for the definition of the corresponding complex envelope.

Reconsidering our derivations so far, we see that we have derived an expression for the complex envelopes in their Cartesian form, i.e. through their real and imaginary parts $p(t)$ and $q(t)$. But, as for any complex number, we can instead use a polar representation. Denoting by $\rho(t)$ and $\phi(t)$ respectively the modulus and phase of the complex number $\tilde{s}(t)$, we can write

$$\tilde{s}(t) = p(t) + jq(t) = \rho(t)e^{j\phi(t)}. \tag{1.26}$$

In this expression and throughout the rest of this book we assume that $\rho(t)$ is a positive quantity. This is indeed required in order to define $\phi(t)$ without ambiguity. Thus by using the Cartesian or polar representation of $\tilde{s}(t)$, we can derive two different (albeit equivalent) analytical expressions for the corresponding bandpass RF signal $s(t)$. Indeed, since by equations (1.2) and (1.4) we have that

$$s(t) = \mathrm{Re}\{s_a(t)\} = \mathrm{Re}\left\{\tilde{s}(t)e^{j\omega_c t}\right\}, \tag{1.27}$$

we can write, using either representation of $\tilde{s}(t)$,

$$s(t) = \mathrm{Re}\left\{(p(t) + jq(t))e^{j\omega_c t}\right\} = p(t)\cos(\omega_c t) - q(t)\sin(\omega_c t), \tag{1.28a}$$

$$= \mathrm{Re}\left\{\rho(t)e^{j(\omega_c t+\phi(t))}\right\} = \rho(t)\cos(\omega_c t + \phi(t)). \tag{1.28b}$$

We can thus see that the modulus of the complex envelope represents the amplitude part of the modulation of the corresponding bandpass RF signal, while its argument represents the phase/frequency part. In that sense, it is often said that an RF signal is complex modulated when it is both amplitude and phase/frequency modulated. We see here that this statement is not directly related to the nature of the corresponding complex envelope. The latter can only be a real signal for a pure amplitude modulation. A phase/frequency modulation is still said to be a real modulation scheme, whereas the corresponding complex envelope is truly complex, even if with a constant modulus. In practice a real modulation corresponds to a modulating signal that can be represented by a single real signal, i.e. either $\rho(t)$ or $\phi(t)$, regardless of the mathematical nature of the complex envelope. In this book we therefore reserve the term "complex modulated RF signals" for signals that are both amplitude and phase modulated.

In conclusion we highlight that, as illustrated in the next sections, the concept of the complex envelope is of particular importance as this lowpass signal embeds all the statistical characteristics of interest of the modulated bandpass RF signal it represents. We can thus say that from the signal processing point of view, the transceiver architecture reduces to how to upconvert or recover a given complex envelope while minimizing its degradation due to implementation limitations. We can also point out that such complex envelopes have additional practical advantages over the use of real bandpass signal representations when performing analytical derivations. The root cause for this comes from its complex nature that can allow for more straightforward analytical derivations. This behavior is illustrated for instance in Chapter 5 where the complex envelope polar notation allows easy analytical derivations when dealing with nonlinearity due to the fact that trigonometric polynomials are naturally linearized compared to power of sine and cosine functions. This complex nature is

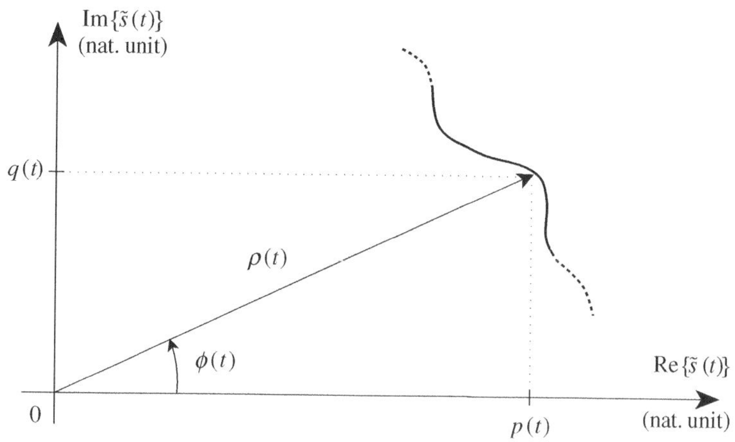

Figure 1.3 Complex envelope and corresponding trajectory representation in the complex plane – The complex envelope as well as the analytic signal concept allow vectorial representations for complex RF modulated signals. It can be seen as a generalization of the Fresnel representation used for continuous waves. The set of points that successively represent, in time, the complex envelope values, i.e. the extremity of its representative vector in the complex plane, define the trajectory of the modulation.

also of interest as it allows useful vectorial representations of signals in the complex plane. As shown in Figure 1.3, we are dealing here with a generalization of the Fresnel representation used for continuous waves. This approach leads to simple vectorial interpretations of analytical derivations that are often not so obvious to interpret directly. Furthermore, considering a simple practical modulating waveform, we get that the corresponding complex envelope describes a curve in the complex plane as a function of time. This curve represents what is called the trajectory associated with the modulating waveform. This trajectory in fact results from the pulse shaping filtering of the original sequence of symbols. If this pulse shaping filter does not introduce intersymbol interference (ISI), the trajectory simply links the symbols represented as a constellation in the complex plane. Some trajectory examples are shown in Section 1.3. Finally, we get that complex envelopes and the associated analytic signal concept allow us to correctly define the quantities of interest for characterizing RF signals, as illustrated in the following sections.

1.1.3 Bandpass Signals vs. Complex Envelopes

Due to the importance of the complex envelope concept introduced in the previous section, it is of interest to consider the characteristics of such complex lowpass signals in more detail, in particular with respect to the characteristics of the real bandpass signals they represent.

Positive vs. Negative Sidebands

Returning the definition of the complex envelope $\tilde{s}(t)$ of a real bandpass signal $s(t)$ given in Section 1.1.2, we see that we considered the positive sideband, i.e. the positive part of the spectrum, of that signal to define the associated analytic signal, $s_a(t)$, the complex envelope being the downconversion of this analytic signal toward baseband. Practically speaking, this definition necessarily leads to the use of the negative complex exponential, i.e. the complex exponential with a negative angular frequency $e^{-j\omega_c t}$, for the ideal implementation of the downconversion processing of $s_a(t)$, as illustrated by equation (1.21). Conversely, it is thus the positive complex exponential, i.e. the complex exponential with a positive angular frequency $e^{+j\omega_c t}$, that was considered for the upconversion of an initial complex lowpass envelope in order to generate the corresponding modulated real bandpass signal according to equation (1.27). We might wonder whether we could consider the opposite choice for the complex exponentials. We observe that the processing corresponding to $\mathrm{Re}\{\tilde{s}(t)e^{-j\omega_c t}\}$ would result in a real bandpass signal that has the same spectral location as that corresponding to our definition so far. This would result in different characteristics for the resulting modulated signals.

By way of illustration, let us consider a first bandpass signal $s_1(t)$ resulting from the upconversion of the complex lowpass signal $\tilde{s}(t)$ following our legacy definition, i.e. using the positive complex exponential $e^{+j\omega_c t}$. According to Section 1.1.2, $s_1(t)$ can be written as

$$s_1(t) = \mathrm{Re}\left\{\tilde{s}(t)e^{+j\omega_c t}\right\}. \tag{1.29}$$

For our analysis, it is of interest to decompose this bandpass signal as the sum of two time domain bandpass signals that correspond to the two sidebands present in its spectrum, i.e. the positive one lying around $+\omega_c$ and the negative one lying around $-\omega_c$. We can use equation (1.5) to expand the real part in the above expression as

$$s_1(t) = \frac{1}{2}\tilde{s}^*(t)e^{-j\omega_c t} + \frac{1}{2}\tilde{s}(t)e^{+j\omega_c t}. \tag{1.30}$$

As expected, we recover a bandpass signal that is the sum of the upconversion of $\tilde{s}(t)$ around $+\omega_c$, and the upconversion of its complex conjugate $\tilde{s}^*(t)$ around $-\omega_c$. Those two bandpass signals centered around symmetric angular frequencies with respect to the zero frequency are obviously complex valued, but also complex conjugates to each other. This is indeed required in order to have the spectrum of the real valued signal $s_1(t)$ that exhibits Hermitian symmetry, as can be seen by taking the Fourier transform of the above expression. Using equation (1.8), we get that

$$S_1(\omega) = \frac{1}{2}\tilde{S}^*(-\omega - \omega_c) + \frac{1}{2}\tilde{S}(\omega - \omega_c). \tag{1.31}$$

We thus see that the sideband centered on $-\omega_c$ is a flipped copy of that corresponding to $\tilde{s}(t)$.

Let us now consider the upconversion of $\tilde{s}(t)$ using the negative complex exponential $e^{-j\omega_c t}$. This results in a second real bandpass signal $s_2(t)$ that can be expressed as

$$\begin{aligned} s_2(t) &= \mathrm{Re}\left\{\tilde{s}(t)e^{-j\omega_c t}\right\} \\ &= \frac{1}{2}\tilde{s}(t)e^{-j\omega_c t} + \frac{1}{2}\tilde{s}^*(t)e^{+j\omega_c t}, \end{aligned} \tag{1.32}$$

and whose representation in the frequency domain is given by

$$S_2(\omega) = \frac{1}{2}\tilde{S}(\omega + \omega_c) + \frac{1}{2}\tilde{S}^*(-\omega + \omega_c). \tag{1.33}$$

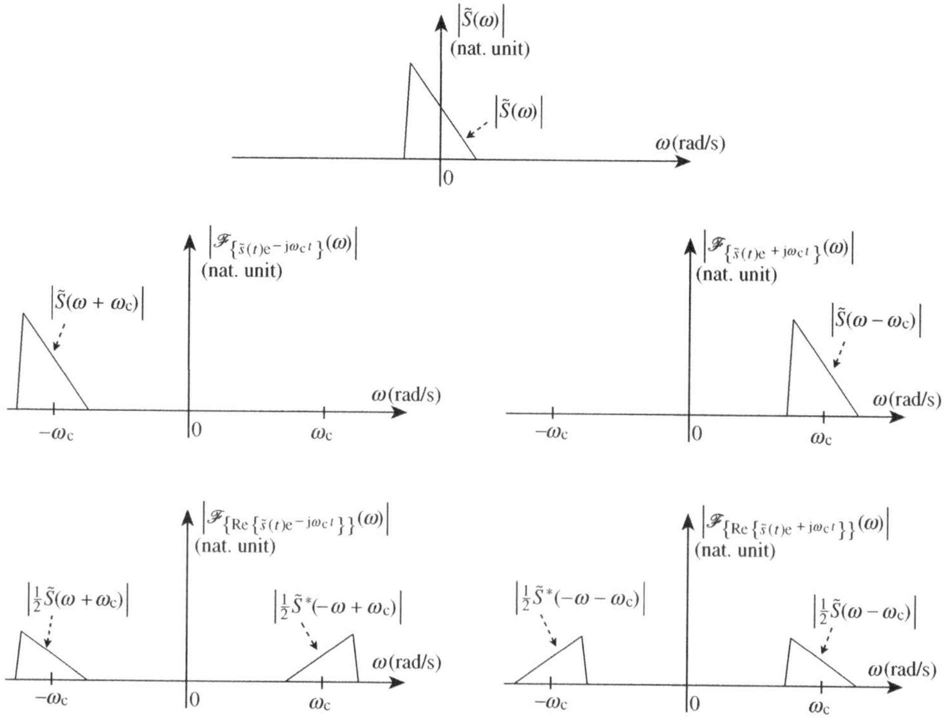

Figure 1.4 Spectral domain representation of the complex envelope frequency upconversion and resulting RF bandpass signal – Depending on the sign of the complex exponential, $e^{+j\omega_c t}$ (middle right) or $e^{-j\omega_c t}$ (middle left), used for the frequency upconversion, the same complex lowpass envelope (top) results in different RF bandpass signals. In the first case, the positive sideband of the resulting spectrum corresponds to the spectrum of the original complex envelope (bottom right) whereas in the second case, it corresponds to a flipped copy of it (bottom left).

Comparing this expression with equation (1.31), we see that we now face an inversion of the positive and negative sidebands of the spectrum of $s_2(t)$ with respect to those of $s_1(t)$, as illustrated in Figure 1.4. Reconsidering this figure and equation (1.32), we can moreover see that $s_2(t)$ can alternatively be interpreted as the result of the upconversion of the complex lowpass signal $\tilde{s}^*(t)$ using the positive complex exponential $e^{+j\omega_c t}$.

The same kind of comment applies when considering the downconversion operation performed to recover the complex envelope of a given bandpass signal. Indeed, depending on whether we use the negative or the positive complex exponential to perform the operation, we obtain through this processing either the positive or the negative sideband of the original bandpass signal. Moreover, depending on whether the bandpass signal considered for such downconversion has been originally generated by the upconversion of an initial complex envelope through the use of the negative or the positive complex exponential, we can finally recover at the output of the downconversion process the sideband that corresponds to the complex conjugate of the initially upconverted complex envelope. In fact, this is not so problematic as it is just a matter of a complex conjugate operation in the complex envelope and hence of managing the sign of the $q(t)$ modulating component. We then get that the choice of sideband transmitted on the positive part of the resulting RF bandpass signal spectrum can be simply made through this $q(t)$ sign selection. We observe that for a given standardized modulation scheme definition, the convention is often that the positive sideband of the resulting bandpass signal corresponds to $\tilde{s}(t) = p(t) + jq(t)$ and the negative sideband to its complex conjugate. Consequently, in this book we mostly assume that theoretical upconversions are implemented using the positive complex exponential $e^{+j\omega_c t}$ and reciprocal downconversions with the corresponding negative complex exponential, $e^{-j\omega_c t}$. Trickier configurations can occur in practical architecture choices. In some transceiver implementations we can indeed consider successive up- or downconversions based on the selection of different sidebands to be transposed each time. This may be the case in order to optimize the frequency planning of the transceiver through the use of supradyne or infradyne conversions. This topic is discussed in Chapter 6 when dealing with more general real and complex frequency conversions.

Power Spectral Density

So far, we have used Fourier transforms directly to examine the spectral representation of a given signal. As discussed in Section 1.1.1, this was done with the aim of keeping the good phase relationship between the signal sidebands and thus interpreting the behavior of a signal processing function in the frequency domain. However, for digital modulation schemes generated through random data bits, the Fourier transform only provides a frequency representation of a given realization of the modulating process over a finite time frame. It is thus of interest to derive the PSD of those randomly modulated waveforms, taking into account their statistical properties. Given that the complex envelope $\tilde{s}(t)$ of an RF bandpass signal $s(t)$ contains all its statistical information, it is of interest to examine the link between the power spectral densities of these two signals.

The simplest way to proceed is to derive the autocorrelation function of $s(t)$ in order to take its Fourier transform [3, 4]. We suppose that we are dealing with stationary bandpass signals, at least up to second order. This is an assumption we can rely on in most cases for processes we have to deal with in wireless transceivers, as discussed in Appendix 2. Assuming the ergodicity of the modulating process, we get that the autocorrelation function of $s(t)$, defined by

$$\gamma_{s\times s}(t_1, t_2) = \lim_{\substack{\tau_1 \to -\infty \\ \tau_2 \to +\infty}} \frac{1}{\tau_2 - \tau_1} \int_{\tau_1}^{\tau_2} s(t - t_1)s(t - t_2)\mathrm{d}t, \tag{1.34}$$

can be evaluated as

$$\gamma_{s\times s}(t_1, t_2) = \mathbb{E}\left\{s_{t_1} s_{t_2}\right\}. \tag{1.35}$$

Here $\mathbb{E}\{.\}$ stands for the stochastic expectation value and s_t for the random variable corresponding to the sample of $s(t)$ at time t. In order to go further, we can then use the fact that $s(t)$ is related to its complex envelope $\tilde{s}(t)$, assumed defined as centered around the angular frequency ω_c, according to equation (1.7). As a result, the above expectation can be written as

$$\gamma_{s \times s}(t_1, t_2) = \frac{1}{4}\mathbb{E}\left\{\left(\tilde{s}_{t_1} e^{j\omega_c t_1} + \tilde{s}_{t_1}^* e^{-j\omega_c t_1}\right)\left(\tilde{s}_{t_2} e^{j\omega_c t_2} + \tilde{s}_{t_2}^* e^{-j\omega_c t_2}\right)\right\}. \tag{1.36}$$

After expansion, we finally get that

$$\gamma_{s \times s}(t_1, t_2) = \frac{1}{4}\mathbb{E}\left\{\tilde{s}_{t_1}\tilde{s}_{t_2}^*\right\} e^{j\omega_c(t_1 - t_2)} + \frac{1}{4}\mathbb{E}\left\{\tilde{s}_{t_1}^*\tilde{s}_{t_2}\right\} e^{-j\omega_c(t_1 - t_2)}$$
$$+ \frac{1}{2}\mathrm{Re}\left\{\mathbb{E}\left\{\tilde{s}_{t_1}\tilde{s}_{t_2}\right\} e^{j\omega_c(t_1 + t_2)}\right\}. \tag{1.37}$$

As discussed in Appendix 2, a side effect of the stationarity of $s(t)$, at least up to second order, is that any of its complex envelopes must fulfill equation (A2.11). We thus get in particular that $\mathbb{E}\left\{\tilde{s}_{t_1}\tilde{s}_{t_2}\right\} = 0$, and that the autocorrelation of $s(t)$ finally reduces to

$$\gamma_{s \times s}(t_1, t_2) = \gamma_{s \times s}(\tau) = \frac{1}{4}\gamma_{\tilde{s} \times \tilde{s}}(\tau) e^{j\omega_c \tau} + \frac{1}{4}\gamma_{\tilde{s} \times \tilde{s}}(-\tau) e^{-j\omega_c \tau}. \tag{1.38}$$

with $\tau = t_1 - t_2$.

We can then derive the PSD of $s(t)$ by taking the Fourier transform of this autocorrelation function.[2] For this, we can use the Fourier transform property, valid for all signals real or complex, that links the Fourier transform of a signal with that of its time reversal copy according to

$$\begin{aligned}\mathcal{F}_{\{\tilde{s}(-t)\}}(\omega) &= \int_{-\infty}^{+\infty} \tilde{s}(-t) e^{-j\omega t} dt\\ &= \int_{-\infty}^{+\infty} \tilde{s}(t) e^{-j(-\omega)t} dt\\ &= \mathcal{F}_{\{\tilde{s}(t)\}}(-\omega).\end{aligned} \tag{1.39}$$

Thus, using this property while taking the Fourier transform of equation (1.38) finally results in

$$\Gamma_{s \times s}(\omega) = \frac{1}{4}\Gamma_{\tilde{s} \times \tilde{s}}(\omega - \omega_c) + \frac{1}{4}\Gamma_{\tilde{s} \times \tilde{s}}(-\omega - \omega_c). \tag{1.40}$$

---

[2] Here and in the rest of this book, we follow the signal processing approach for the definition of the PSD, i.e. we implicitly assume that we are working with normalized impedances when dealing with analog quantities for $s(t)$. Otherwise, as illustrated in Section 2.2.3, a factor $\mathrm{Re}\{1/Z_0\}$ should appear when $s(t)$ represents a voltage over an impedance Z_0. In the same way, a factor $\mathrm{Re}\{Z_0\}$ should appear when $s(t)$ represents a current across Z_0. But in most formulas derived in this book, only relative power quantities are involved so that such normalization factors cancel, as illustrated for instance in Chapter 5, in "Characterization of RF device nonlinearity" (Section 5.1.2). Nevertheless, for the derivation of the absolute PSD of an analog quantity across a given load, the reader needs to add the corresponding impedance term compared to the results derived in this book.

As might be expected for the modulated bandpass RF signal $s(t)$ that corresponds to the frequency upconversion of the lowpass modulating signal $\tilde{s}(t)$, its PSD is simply proportional to that of $\tilde{s}(t)$ once transposed around the angular frequencies $\pm\omega_c$. Here, we recover that the resulting sideband of the spectrum lying at negative frequencies is the flipped version of the original lowpass signal spectrum, i.e. a transposition of $\Gamma_{\tilde{s}\times\tilde{s}}(-\omega)$ around $-\omega_c$. As highlighted in earlier sections, this behavior is related to the Hermitian symmetry that must be retrieved on the Fourier transform of any real signal. We also observe that each sideband of the PSD of the bandpass signal $s(t)$, i.e. both that centered on $+\omega_c$ and that centered on $-\omega_c$, is one fourth that of its complex envelope. This means that the integration of the PSD of such sideband corresponds to one half the original complex envelope signal power. As discussed in "Average power" later in this section, this can be linked to the convention followed in this book for the definition of this complex lowpass signal that is defined as twice the positive sideband of the original bandpass signal it represents.

Impact of Stationarity

In the previous section, we assumed the bandpass signal $s(t)$ to be stationary in order to derive its PSD. However, we did not examine all the resulting characteristics of the spectrum for such stationarity. In particular, in the light of the derivations performed in Appendix 2, we get that the stationarity of $s(t)$ implies that all of its complex envelopes $\tilde{s}(t)$ satisfy equation (A2.11), i.e.

$$\gamma_{\tilde{s}\times\tilde{s}^*}(\tau) = 0. \tag{1.41}$$

We can then expand this equation in terms of the real and imaginary parts of $\tilde{s}(t) = p(t) + jq(t)$ following equation (A1.46b). Referring to equation (A2.13a), we then get that

$$\gamma_{p\times p}(\tau) = \gamma_{q\times q}(\tau). \tag{1.42}$$

Now taking the Fourier transform of this equation leads to

$$\Gamma_{p\times p}(\omega) = \Gamma_{q\times q}(\omega). \tag{1.43}$$

We thus recover the remarkable result that the stationarity of $s(t)$ implies that the real and imaginary parts of any of its complex envelopes have the same PSD. We observe that this result is related to the discussion in Appendix 2 that gives more insight into the conditions required to achieve the generation of an RF bandpass signal that is stationary.

Impact of Spectral Symmetry on Top of Stationarity

Let us now go a step further and suppose that on top of dealing with a stationary bandpass signal $s(t)$, at least up to second order, we also face an even symmetry in the PSD of either of its sidebands when inspected independently of the other one. Referring to equation (1.40), we get that the spectral shape of the sidebands of such bandpass signal is proportional to that of any of its complex envelopes. We thus get for instance that a symmetry regarding the zero frequency in the spectrum of such complex envelope, let say $\tilde{s}(t)$, necessarily corresponds to a symmetry regarding ω_c of the positive sideband of $s(t)$, ω_c being the center angular frequency chosen for the definition of $\tilde{s}(t)$. The point is that such a characteristic leads to interesting properties.

In order to examine this, let us assume that the PSD of $\tilde{s}(t)$ exhibits an even symmetry, i.e. is such that

$$\Gamma_{\tilde{s}\times\tilde{s}}(\omega) = \Gamma_{\tilde{s}\times\tilde{s}}(-\omega). \tag{1.44}$$

At the same time, we know from the results derived in Appendix 1 that the PSD of any process is necessarily a real valued function. Thus the relationship

$$\Gamma_{\tilde{s}\times\tilde{s}}(\omega) = \Gamma^*_{\tilde{s}\times\tilde{s}}(\omega) \tag{1.45}$$

always holds. From the above two equations, we can thus deduce that the complex envelope considered fulfills

$$\Gamma^*_{\tilde{s}\times\tilde{s}}(\omega) = \Gamma_{\tilde{s}\times\tilde{s}}(-\omega). \tag{1.46}$$

This means that the even symmetry in the PSD of $\tilde{s}(t)$ leads to a quantity that exhibits Hermitian symmetry. Thus, recalling the property discussed during the derivation of equation (1.8), we can deduce that the autocorrelation function of $\tilde{s}(t)$, $\gamma_{\tilde{s}\times\tilde{s}}(\tau)$, is necessarily a real valued function. So its imaginary part must be null. We then deduce from equation (A1.46a) that

$$\gamma_{p\times q}(-\tau) - \gamma_{p\times q}(\tau) = 0. \tag{1.47}$$

On the other hand, we get that any complex envelope of a stationary bandpass signal necessarily fulfills equation (1.41). Using equation (A1.46b) to expand $\gamma_{\tilde{s}\times\tilde{s}^*}(\tau)$ in terms of the real and imaginary parts of $\tilde{s}(t) = p(t) + jq(t)$, we then get that

$$\gamma_{p\times q}(-\tau) + \gamma_{p\times q}(\tau) = 0. \tag{1.48}$$

From the above two equations, we thus deduce that

$$\gamma_{p\times q}(\tau) = \gamma_{q\times p}(\tau) = 0. \tag{1.49}$$

The real and imaginary parts of $\tilde{s}(t) = p(t) + jq(t)$ are thus necessarily uncorrelated, even when considered at different times. This is obviously a new property as, according to Appendix 2, the stationarity of $s(t)$ implies that $p(t)$ and $q(t)$ are uncorrelated when considered at the same time t.

Conversely, considering a complex lowpass signal $\tilde{s}(t) = p(t) + jq(t)$ such that equation (1.49) holds, we necessarily get that $\gamma_{\tilde{s}\times\tilde{s}}(\tau)$ is a real valued quantity by equation (A1.46a). It follows that the PSD of $\tilde{s}(t)$, which reduces to the Fourier transform of this autocorrelation function, satisfies Hermitian symmetry. But, given at the same time that this quantity is also real valued, we then recover the even spectral symmetry of this PSD, i.e. that $\Gamma_{\tilde{s}\times\tilde{s}}(\omega) = \Gamma_{\tilde{s}\times\tilde{s}}(-\omega)$. As highlighted at the beginning of the derivation, this symmetry property is thus necessarily recovered on the sidebands of the corresponding modulated bandpass signal $s(t)$ relative to the angular frequency ω_c used for the definition of the complex envelope $\tilde{s}(t)$. At the same time, we get that most of the modulating waveforms used in practical wireless standards are defined such that $\gamma_{p\times q}(\tau) = \gamma_{q\times p}(\tau) = 0$, as discussed in Appendix 2. This explains why this even spectral symmetry in the spectrum of modulated bandpass signals is encountered in most practical cases. In spite of this, we often represent the spectrum as asymmetrical throughout this book. This allows us to clearly highlight which sideband we are dealing with, i.e. whether the complex envelope we are dealing with is $\tilde{s}(t) = p(t) + jq(t)$ or its complex conjugate $\tilde{s}^*(t) = p(t) - jq(t)$ as discussed in "Positive vs. negative sidebands" earlier in this section. Given that those complex envelopes are complex conjugates to each other, it follows that their Fourier transforms are related as in equation (1.8), i.e. their magnitudes are symmetric to each other with respect to the zero frequency. Thus, representing them as asymmetrical allows us to distinguish easily between them, even if, when $p(t)$ and $q(t)$ are uncorrelated, the modulus of their Fourier transform, and thus their power spectral densities, are evenly symmetrical. In this last case, it is only the argument of their Fourier transform that remains an odd function of the frequency, and it is only the phase difference between them that distinguishes these sidebands.

We can even go a step further. Recall that for a stationary bandpass signal $s(t)$, the real and imaginary parts of any of its complex envelopes $\tilde{s}(t) = p(t) + jq(t)$ necessarily have the same autocorrelation function as given by equation (1.42). Thus, if we now assume that we are also dealing with a signal whose spectrum exhibits the even symmetry discussed above, we necessarily have that $p(t)$ and $q(t)$ are also uncorrelated as given by equation (1.49). Based on those two results, equation (A1.46a) reduces to

$$\gamma_{\tilde{s}\times\tilde{s}}(\tau) = \gamma_{p\times p}(\tau) + \gamma_{q\times q}(\tau) = 2\gamma_{p\times p}(\tau) = 2\gamma_{q\times q}(\tau). \tag{1.50}$$

Now taking the Fourier transform of this equation, we finally obtain

$$\Gamma_{\tilde{s}\times\tilde{s}}(\omega) = \Gamma_{p\times p}(\omega) + \Gamma_{q\times q}(\omega) = 2\Gamma_{p\times p}(\omega) = 2\Gamma_{q\times q}(\omega). \tag{1.51}$$

We now see that $\tilde{s}(t)$ and thus $s(t)$ also have the same spectral shape as $p(t)$ and $q(t)$ when this spectrum is evenly symmetric regarding the carrier center angular frequency used to define the complex envelope. Practically speaking, this property is important not only for the study of modulated RF bandpass signals but also for RF bandpass noise, as discussed in Section 1.2. Indeed, the stationarity of an RF bandpass noise source allows us to link the PSD of the bandpass noise term $n(t)$ it delivers to its load, to the PSD of the real and imaginary parts of any of its complex envelopes $\tilde{n}(t)$. It is important to link the properties of the noise terms at the input and output of a complex frequency conversion, for instance.

Average Power

Let us now focus on the average power of a bandpass signal, or more precisely on the link between this quantity and the statistical properties of any of its complex envelopes. Let us consider the expression for the bandpass signal $s(t)$ as a function of its complex envelope $\tilde{s}(t)$, defined as centered around the carrier angular frequency ω_c. According to equation (1.27), this expression reduces to

$$s(t) = \text{Re}\left\{\tilde{s}(t)e^{+j\omega_c t}\right\}. \quad (1.52)$$

In order to derive the average power of this signal, we can thus consider further either the direct time domain average or the stochastic approach. Under ergodicity, the two approaches are expected to give the same result. Here, we initially adopt the time domain average approach as it leads to an intuitive explanation for the result when considering the derivation in the frequency domain.

Let us therefore evaluate the long-term average power P_s of this bandpass signal as its root mean square (RMS) value. Considering the complex envelope $\tilde{s}(t)$ expressed in its polar form as $\tilde{s}(t) = \rho(t)e^{j\phi(t)}$, we can then write:[3]

$$P_s = \lim_{\substack{\tau_1 \to -\infty \\ \tau_2 \to +\infty}} \frac{1}{\tau_2 - \tau_1} \int_{\tau_1}^{\tau_2} [\rho(t)\cos(\omega_c t + \phi(t))]^2 dt. \quad (1.53)$$

In order to evaluate this integral, we can linearize the cosine function using

$$\cos^2(\theta) = \frac{1 + \cos(2\theta)}{2}. \quad (1.54)$$

This leads to

$$\begin{aligned} P_s = & \lim_{\substack{\tau_1 \to -\infty \\ \tau_2 \to +\infty}} \frac{1}{\tau_2 - \tau_1} \int_{\tau_1}^{\tau_2} \frac{\rho^2(t)}{2} dt \\ & + \lim_{\substack{\tau_1 \to -\infty \\ \tau_2 \to +\infty}} \frac{1}{\tau_2 - \tau_1} \int_{\tau_1}^{\tau_2} \frac{\rho^2(t)}{2} \cos(2\omega_c t + 2\phi(t)) dt. \end{aligned} \quad (1.55)$$

The first term on the right-hand side obviously depends only on the amplitude part of the modulating signal, i.e. on the modulus of the complex envelope $\tilde{s}(t)$. More precisely, it is equal to half the RMS value of the modulus of this complex envelope. We remark that it reduces to the classical value $\rho^2/2$ when dealing with a constant amplitude bandpass signal. And given that it directly gives the average power of this particular bandpass signal, we can then guess that the second term in the above expression is null. This is indeed true as long as the spectrum of $\tilde{s}(t)$ can be assumed narrowband with respect to the carrier angular frequency ω_c. In other words, if the spectrum of $\tilde{s}(t)$ lies over the frequency band $[-\Omega/2, \Omega/2]$, we need $\Omega \ll \omega_c$.

We first interpret this condition in the time domain. As can be seen in the left part of Figure 1.5, as long as the time domain variations of $\rho^2(t)$ are weak over a period $2\pi/\omega_c$ of the carrier, the integral of $\rho^2(t)\cos(2\omega_c t + 2\phi(t))$ over this period is necessarily almost null.

[3] As highlighted in "Power spectral density" earlier in this section, we adopt the signal processing approach for this definition as this expression implicitly assumes that we are working with normalized impedances when dealing with analog quantities for $s(t)$.

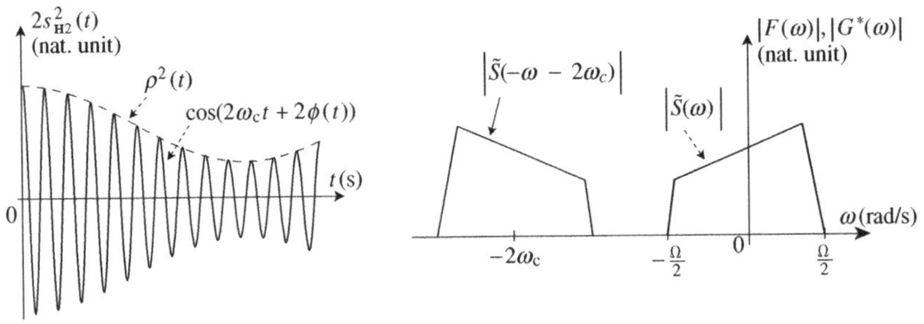

Figure 1.5 Illustration in the frequency and time domain of the bandpass signal average power estimation mechanism – The long-term average power estimation of $s(t) = \rho(t)\cos(\omega_c t + \phi(t))$ involves the time average of the high-frequency term $\rho^2(t)\cos(2\omega_c t + 2\phi(t))$ (left). This contribution is expected to be negligible as can be seen in the spectral domain (right). If we refer to equation (1.60), it indeed involves the averaging of the product of two non-overlapping spectrum, $F(\omega) = \tilde{S}(\omega)$ and $G^*(\omega) = \tilde{S}(-\omega - 2\omega_c)$.

We can thus expect that the overall integral toward infinity is also almost null, or at least negligible regarding the first term of equation (1.55). This condition can be interpreted more clearly in the frequency domain. Now we may need to consider Fourier transforms of the corresponding signals. As discussed in Section 1.1.1, for that to be possible we can initially assume a signal corresponding to an observation over a finite duration, for instance limited to $[0, T]$, of a realization of the modulating process. Under that assumption, we can rewrite the second term of equation (1.55) as

$$I = \frac{1}{2T}\int_0^T \rho^2(t)\cos(2\omega_c t + 2\phi(t))\mathrm{d}t. \tag{1.56}$$

This quantity can then be expressed in terms of the complex envelope $\tilde{s}(t)$ through the use of equation (1.52). This leads to

$$I = \frac{1}{2T}\mathrm{Re}\left\{\int_0^T \tilde{s}^2(t)\mathrm{e}^{\mathrm{j}2\omega_c t}\mathrm{d}t\right\}. \tag{1.57}$$

With $\tilde{s}(t)$ vanishing outside $[0, T]$, this integral can be extended toward infinity:

$$I = \frac{1}{2T}\mathrm{Re}\left\{\int_{-\infty}^{+\infty} \tilde{s}(t)\left[\tilde{s}^*(t)\mathrm{e}^{-\mathrm{j}2\omega_c t}\right]^*\mathrm{d}t\right\}. \tag{1.58}$$

Then, using the Fourier transform property, which states that

$$\int_{-\infty}^{+\infty} f(t)g^*(t)\mathrm{d}t = \frac{1}{2\pi}\int_{-\infty}^{+\infty} F(\omega)G^*(\omega)\mathrm{d}f, \tag{1.59}$$

we finally get that I is proportional to

$$\mathrm{Re}\left\{ \int \tilde{S}(\omega)\tilde{S}(-\omega - 2\omega_c)\mathrm{d}\omega \right\}, \tag{1.60}$$

where $\tilde{S}(\omega)$ stands for the Fourier transform of the complex envelope $\tilde{s}(t)$. As can be understood by inspecting the representation shown in Figure 1.5, we then get that for ω_c high enough with respect to Ω, the spectra involved in that equation are non-overlapping. We can thus expect that their product is null, as is the above integral. We must mention that having assumed a finite temporal support over $[0, T]$ for the time domain signals, their spectra necessarily spread toward infinity. On the other hand, we get that their asymptotic behavior is also necessarily decreasing in order to have an overall power that remains finite. This means that the above integral necessarily tends to zero as ω_c becomes higher and higher compared to Ω. We thus see that the second term of the right-hand side of equation (1.55) necessarily remains negligible compared to the first. As a result, the power of the modulated bandpass signal $s(t)$ can effectively be estimated as half the RMS value of the modulus of its complex envelope according to

$$P_s = \lim_{\substack{\tau_1 \to -\infty \\ \tau_2 \to +\infty}} \frac{1}{\tau_2 - \tau_1} \int_{\tau_1}^{\tau_2} \frac{\rho^2(t)}{2} \mathrm{d}t. \tag{1.61}$$

Then, assuming the ergodicity of the process we are dealing with, we can rewrite this result as

$$P_s = \frac{\mathbb{E}\{\rho_t^2\}}{2}, \tag{1.62}$$

so that under stationarity we finally get

$$P_s = \frac{\mathbb{E}\{\rho^2\}}{2}. \tag{1.63}$$

Alternatively, given that $\tilde{s}(t)$ is a lowpass process, we remark that its power is simply given by $\mathbb{E}\{\tilde{s}\tilde{s}^*\} = \mathbb{E}\{\rho^2\}$. We can therefore write

$$P_s = \frac{\mathbb{E}\{\rho^2\}}{2} = \frac{1}{2} P_{\tilde{s}}. \tag{1.64}$$

The long-term average power of a bandpass signal $s(t)$ is therefore equal to half the long-term average power of its lowpass complex envelope $\tilde{s}(t)$. This factor $1/2$ is in fact related to the definition used in this book for the complex envelope as already discussed in Section 1.1.3. In the same way, as reviewed in Section 1.1.2, we may recall that we can define as many complex envelopes as we want for a given bandpass signal. All those complex envelopes necessarily have the same modulus if we refer to equation (1.25). We thus get that the quantity defined by equation (1.61) or (1.64) necessarily remains well defined whatever the complex envelope considered to perform the derivation.

In conclusion, it is of interest to highlight the link that holds between the average power of the bandpass signal $s(t)$ and the average power of the real and imaginary parts of one of its complex envelopes $\tilde{s}(t) = p(t) + jq(t)$. Using equation (A1.39), we can immediately write under stationarity that

$$P_s = \frac{1}{2}P_{\tilde{s}} = \frac{1}{2}\mathbb{E}\{\tilde{s}\tilde{s}^*\} = \frac{1}{2}\gamma_{\tilde{s}\times\tilde{s}}(0). \tag{1.65}$$

This result can be expanded in terms of the autocorrelation functions of the real and imaginary parts of the complex envelope $\tilde{s}(t)$ by using equation (A1.46a). This finally leads to

$$P_s = \frac{1}{2}\gamma_{\tilde{s}\times\tilde{s}}(0) = \frac{1}{2}[\gamma_{p\times p}(0) + \gamma_{q\times q}(0)]. \tag{1.66}$$

We can thus sum the average power of the in-phase and in-quadrature components, $p(t)$ and $q(t)$ respectively, to get twice the average power of the corresponding bandpass signal $s(t)$. This property remains valid for $p(t)$ and $q(t)$ whether correlated or not. The deep reason for this is that the bandpass signal $s(t)$ is constructed through the sum of orthogonal waveforms that carry the information of $p(t)$ and $q(t)$. Thus, whatever the correlation behavior of $p(t)$ and $q(t)$, the powers of the orthogonal waveforms that carry their information sum together during the derivation of the power of the final bandpass RF signal.

Peak to Average Power Ratio and Crest Factor

Two quantities are of particular interest for the characterization of the waveforms that are processed in a transceiver: the peak to average power ratio (PAPR) and the crest factor (CF). Although the purpose of those two quantities is to characterize the amplitude variations of a given signal, they are up to a point related to different problems in the dimensioning of transceivers.

For instance, the PAPR of a bandpass signal $s(t)$ of the form $\rho(t)\cos(\omega_c t + \phi(t))$ is defined as the ratio between its peak and average power. Although this definition may look clear enough at first glance, it remains of interest to discuss what we call peak power in the wireless transceiver perspective. In so doing, we may anticipate the discussion in Chapter 5 as the analytical derivations performed in this chapter clearly highlight that it is the square of the magnitude of the modulating waveform, i.e. $\rho^2(t)$, that is involved in the formulations related to RF compression. We can then understand that when talking about the instantaneous power variations of a bandpass RF signal, we are in fact talking about the variations linked to the modulation part only of the signal. Consequently, the peak power involved in the PAPR definition reduces to a peak average power of the bandpass modulated signal $s(t)$ over a duration short enough that the characteristics of the modulation remain almost constant, but long enough that many carrier periods have occurred. Classically, this duration can be assumed in the range of a data symbol.

We can now derive an analytical expression for the PAPR of $s(t)$ assuming that its spectrum spreads over a bandwidth of width Ω such that $\Omega \ll \omega_c$; that is, assuming that $s(t)$ is narrowband, we can directly reuse the material derived in the previous section to express its average power P_s through equation (1.61). In order to derive the PAPR of this signal, it thus remains to derive an expression for its peak power. We therefore reconsider the expansion of $s^2(t)$ as

$$s^2(t) = \frac{\rho^2(t)}{2} + \frac{\rho^2(t)}{2}\cos(2\omega_c t + 2\phi(t)). \tag{1.67}$$

The first term is linked to the modulation process only, and the second term is a high frequency component. The latter term thus vanishes when averaged. In the present case, the power variations we are looking for can be estimated through an average of this instantaneous power over a short duration $\tau_2 - \tau_1$ that is in the range of the data symbol, say $2\pi/\Omega$. Thus, given that the modulation scheme is assumed narrowband regarding the carrier frequency, we necessarily get that at least a few carrier periods occur during $\tau_2 - \tau_1$. This means that as long as the averaging period is short enough that the modulation scheme amplitude does not vary much, but long enough to have an averaging of the carrier over few periods, we can write

$$p_1(t) = \frac{1}{\tau_2 - \tau_1}\int_{t+\tau_1}^{t+\tau_2} \frac{\rho^2(t')}{2} dt' \approx \frac{\tau_2 - \tau_1}{\tau_2 - \tau_1}\frac{\rho^2(t)}{2} = \frac{\rho^2(t)}{2}, \tag{1.68a}$$

$$p_2(t) = \frac{1}{\tau_2 - \tau_1}\int_{t+\tau_1}^{t+\tau_2} \frac{\rho^2(t')}{2} \cos\left(2\omega_c t' + 2\phi(t')\right) dt' \approx 0. \tag{1.68b}$$

We then get that the second contribution $p_2(t)$ is almost null, whereas the first term is still proportional to the instantaneous power of the modulating waveform. As a result, the instantaneous power variations $p(t)$ linked only to the envelope contributions can be estimated as

$$p(t) = p_1(t) + p_2(t) \approx \frac{\rho^2(t)}{2}. \tag{1.69}$$

We can thus express the PAPR as the ratio between the peak value of $p(t)$ and the average power P_s given by equation (1.61). This results in

$$PAPR = \frac{\max_t\{\rho^2(t)\}}{\overline{\rho^2(t)}}, \tag{1.70}$$

where $\overline{(.)}$ denotes the time average and $\max\{.\}$ the maximum value. Since we consider $\rho(t)$ to be a positive quantity, and due to the fact that the square function is strictly monotonic for such a positive quantity, the maximum of the square is therefore equal to the square of the maximum. Thus, denoting by ρ_{pk} the peak value of $\rho(t)$, we simply get that

$$PAPR = \frac{\rho_{pk}^2}{\overline{\rho^2(t)}} \tag{1.71}$$

or, in decibels,

$$PAPR|_{\mathrm{dB}} = 10\log_{10}(PAPR) = 10\log_{10}\left(\frac{\rho_{\mathrm{pk}}^2}{\overline{\rho^2(t)}}\right). \tag{1.72}$$

Up to now, we have adopted a deterministic approach to the derivation of the PAPR quantity through time averaging. But, dealing with randomly modulated signals, we can transpose those definitions using stochastic concepts. Under ergodicity and assuming the stationarity of those processes, which is a reasonable assumption in the wireless transceiver area as discussed in Appendix 2, we can write the above expressions as

$$PAPR = \frac{\rho_{\mathrm{pk}}^2}{\mathbb{E}\{\rho^2\}}, \tag{1.73a}$$

$$PAPR|_{\mathrm{dB}} = 10\log_{10}\left(\frac{\rho_{\mathrm{pk}}^2}{\mathbb{E}\{\rho^2\}}\right). \tag{1.73b}$$

The peak value ρ_{pk} of $\rho(t)$ may thus be defined in turn in a probabilistic way. In that case, this peak value can be defined such that a given percentage of the realizations of ρ are lower than ρ_{pk}. In practice the value of 99.9% is often used for the definition. Having that ρ is a positive quantity, we thus get

$$F(\rho_{\mathrm{pk}}) = \int_{-\infty}^{\rho_{\mathrm{pk}}} p(\rho)\mathrm{d}\rho = \int_{0}^{\rho_{\mathrm{pk}}} p(\rho)\mathrm{d}\rho = 0.999. \tag{1.74}$$

Here, $p(\rho)$ stands for the probability density function (PDF) of ρ and thus $F(\rho)$ for its cumulative distribution function (CDF). Obviously, the definition using the complementary cumulative distribution function (CCDF), $F_{\mathrm{c}}(\rho_{\mathrm{pk}})$, can be used instead. This leads to

$$F_{\mathrm{c}}(\rho_{\mathrm{pk}}) = \int_{\rho_{\mathrm{pk}}}^{+\infty} p(\rho)\mathrm{d}\rho = 0.001. \tag{1.75}$$

We observe that effective modulating waveforms necessarily remain bounded in amplitude whatever the pattern of modulating bits that is used. Thus, an upper bound is always well defined for the amplitude of those waveforms. But for some schemes this kind of maximum value can occur with a very low probability. In that case, it may be of interest to adopt a statistical approach in order not to overdimension the data path only for events with no significant contributions to overall system performance. The orthogonal frequency division multiplexing (OFDM) signals introduced in "OFDM" (Section 1.3.3) are a good example of this.

Let us now focus on the CF. This quantity is classically defined as the peak amplitude of a given waveform over its RMS value. As a result, we can write the crest factor CF of the signal $s(t)$ as

$$CF = \frac{s_{\mathrm{pk}}}{\sqrt{\overline{s^2(t)}}}, \tag{1.76}$$

with $s_{\text{pk}} = \max\{s(t)\}$ the peak value of $s(t)$. Alternatively, it can be expressed in decibels as

$$CF|_{\text{dB}} = 20\log_{10}(CF) = 20\log_{10}\left(\frac{s_{\text{pk}}}{\sqrt{\overline{s^2(t)}}}\right). \tag{1.77}$$

As was done for the PAPR, under ergodicity and still assuming the stationarity of the processes we are dealing with, we can transpose those expressions to the stochastic case:

$$CF = \frac{s_{\text{pk}}}{\sqrt{\mathbb{E}\{s^2\}}}, \tag{1.78a}$$

$$CF|_{\text{dB}} = 20\log_{10}\left(\frac{s_{\text{pk}}}{\sqrt{\mathbb{E}\{s^2\}}}\right), \tag{1.78b}$$

where the signal peak value depends on a chosen bound for the CDF of the signal as discussed above. Compared to the definition of the PAPR, the CF is not related to the instantaneous power of the considered signal but directly to its amplitude. Thus, according to our discussion so far, we may initially expect the CF to be more useful for characterizing lowpass signals than RF bandpass ones. As illustrated, for instance in Chapter 7, the CF of baseband modulating waveforms indeed drives the scaling of those signals along the data path in order to avoid any potential saturation. Nevertheless, even if the CF is more suited to addressing lowpass signal problems, there is nothing to prevent its use in deriving bandpass signals. For instance, a continuous wave (CW) signal has both a PAPR of 1, i.e. 0 dB, and a CF of $\sqrt{2}$, i.e. 3 dB. In terms of physical interpretation of those results, it may be of interest to highlight that the deep reason for this ratio of $\sqrt{2}$ between those quantities in the present case comes from the fact that the PAPR focuses on the power variations, i.e. on the variations of the complex envelope only, whereas the CF takes into account the variations of the overall signal, i.e. including those of the sinusoidal carrier. Thus, it is natural to wonder if this ratio of $\sqrt{2}$ cannot be generalized to any bandpass signal $s(t)$. We can consider the following expression for its complex envelope $\tilde{s}(t)$:

$$\tilde{s}(t) = p(t) + \text{j}q(t) = \rho(t)e^{\text{j}\phi(t)}. \tag{1.79}$$

We can then write

$$|\tilde{s}(t)|^2 = \rho^2(t) = p^2(t) + q^2(t). \tag{1.80}$$

Using the stochastic approach, this means that under stationarity we can transpose equation (1.66) as

$$\mathbb{E}\{\rho^2\} = \mathbb{E}\{p^2\} + \mathbb{E}\{q^2\}. \tag{1.81}$$

But, as discussed in "Impact of stationarity" earlier in this section, the stationarity of $s(t)$ also leads to having the same power for $p(t)$ and $q(t)$. We can thus write

$$\mathbb{E}\{\rho^2\} = 2\mathbb{E}\{p^2\} = 2\mathbb{E}\{q^2\}. \tag{1.82}$$

On the other hand, from a statistical point of view we can expect that the maximum value that $p(t)$ or $q(t)$ can reach is of the same order of magnitude as the maximum value of $\rho(t)$. We can indeed assume in most cases that the maximum value of $\rho(t)$ occurs when the contributions of one of the signals $p(t)$ or $q(t)$ is negligible with regard to the other. Thus, comparing equations (1.73a) and (1.78a), we see that we can approximate for most bandpass signals the result that was exact for CW, i.e. that

$$CF_{\mathrm{p}} = CF_{\mathrm{q}} \approx \sqrt{2PAPR_{\rho}}, \tag{1.83}$$

or, in decibels,

$$CF_{\mathrm{p}}|_{\mathrm{dB}} = CF_{\mathrm{q}}|_{\mathrm{dB}} \approx PAPR_{\rho}|_{\mathrm{dB}} + 3, \tag{1.84}$$

In those equations, CF_{p} and CF_{q} stand respectively for the CF of the real and imaginary parts $p(t)$ and $q(t)$ of the complex envelope that describes the modulating waveform considered, and $PAPR_{\rho}$ for the PAPR of the corresponding bandpass RF modulated signal. This behavior is confirmed by the examples detailed in Section 1.3.

Instantaneous Amplitude and Instantaneous Frequency

Let us now clarify two other concepts classically associated with bandpass RF signals, namely their instantaneous amplitude and their instantaneous phase or frequency. At first glance, we might be tempted to say that these quantities are quite straightforward to define. Indeed, considering, for instance, a modulated bandpass signal that can be written using equation (1.28b) in the form $s(t) = \rho(t)\cos(\omega_{\mathrm{c}}t + \phi(t))$, we could obviously call $\rho(t)$ the instantaneous amplitude of $s(t)$ and $\omega_{\mathrm{c}}t + \phi(t)$ as the instantaneous phase of $s(t)$. Nevertheless, we might wonder whether this analytical expression for $s(t)$ is unique and thus whether the resulting expressions for its instantaneous amplitude and phase are also unique. Moreover, considering the situation where $s(t)$ is a bandpass signal for which no particular analytical expression of this form is available, we can guess what is the correct way to define and derive those quantities.

Let us first focus on the problem of uniqueness by considering the definition of the instantaneous frequency. Recall that the concept of frequency in itself is naturally associated with Fourier analysis. Perhaps we can rely on this theoretical tool to formalize the instantaneous frequency concept we are looking for. Unfortunately, in order to correctly evaluate the angular frequency of a CW in the form

$$s_{\mathrm{cw}}(t) = \rho\cos(\omega_{\mathrm{c}}t), \tag{1.85}$$

we require an infinite duration of observation. Practically speaking, this is the cost of having infinite integration bounds in the integral of the Fourier transform. As soon as we try to "locate" the frequency content of a signal in the time domain we get a limitation in terms of frequency resolution related to the Heisenberg uncertainty principle. This limitation shows that Fourier analysis may not be the right tool for the definition of an *instantaneous* frequency. We can look at another approach when considering a generalization of the above CW signal with an arbitrary time dependent phase $\varphi(t)$ in the form

$$s_{\text{PM}}(t) = \rho \cos(\varphi(t)). \tag{1.86}$$

Looking at the analytical expression for this constant amplitude signal, it seems natural to define its instantaneous frequency, $f(t)$, as

$$f(t) = \frac{1}{2\pi}\frac{\mathrm{d}\varphi(t)}{\mathrm{d}t}. \tag{1.87}$$

This definition indeed leads to no ambiguity in the definition of $f(t)$ and gives good results when applied to a CW signal such as $s_{\text{CW}}(t)$. Nevertheless, a problem occurs if we now consider a general bandpass signal,

$$s(t) = \rho(t)\cos(\varphi(t)), \tag{1.88}$$

that is now also amplitude modulated. In this latter case the direct application of equation (1.87) gives a result that is not unique. This can be seen by considering a time dependent function $\xi(t)$, such that $0 < \xi(t) < 1$, that allows $s(t)$ to be rewritten as

$$s(t) = \frac{\rho(t)}{\xi(t)}\xi(t)\cos(\varphi(t)), \tag{1.89}$$

or alternatively, in a form more suited to our purposes, as

$$s(t) = \rho'(t)\cos(\varphi'(t)), \tag{1.90}$$

with

$$\rho'(t) = \rho(t)/\xi(t), \tag{1.91a}$$

$$\varphi'(t) = \arccos(\xi(t)\cos(\varphi(t))). \tag{1.91b}$$

We thus get two different analytical expressions for the same bandpass signal $s(t)$. The direct application of equation (1.87) to those expressions therefore results in two different expressions for the instantaneous frequency of $s(t)$. Obviously, this is unsatisfactory.

Nowadays, the most commonly used definition for the instantaneous frequency of a bandpass signal is that given by Ville in 1948 [5]. This definition involves the analytic signal associated with the considered bandpass signal according to the definition given in Section 1.1.2. Indeed, reconsidering the former CW signal $s_{\text{CW}}(t)$, we can see that by using equations (1.16) and (1.17) we can express its analytic signal, $s_{\text{CW,a}}(t)$, as

$$s_{\text{CW,a}}(t) = \rho e^{j\omega_c t}. \tag{1.92}$$

We can therefore conclude that the angular frequency of $s_{\mathrm{cw}}(t)$ matches the rotation angular speed of the vectorial representation in the complex plane of $s_{\mathrm{cw,a}}(t)$. What is interesting in this way of proceeding is that the analytic signal derived from any real bandpass signal is unique. It can therefore be used for the generalization of the concept of instantaneous frequency without ambiguity. Thus, we can use equations (1.27) and (1.5) to express the bandpass signal $s(t)$ as a function of its analytic signal $s_{\mathrm{a}}(t)$:

$$s(t) = \mathrm{Re}\{s_{\mathrm{a}}(t)\} = \frac{1}{2}\left(s_{\mathrm{a}}(t) + s_{\mathrm{a}}^{*}(t)\right). \tag{1.93}$$

Then, given that $s_{\mathrm{a}}(t)$ and $s_{\mathrm{a}}^{*}(t)$ have the same modulus but opposite arguments, we can represent those quantities in the complex plane as vectors of the same magnitude but symmetrical with regard to the real axis. As expected, and as shown in Figure 1.6, their sum therefore remains a real quantity equal to twice the value of $s(t)$. Practically speaking, we thus get that the instantaneous phase $\varphi(t)$ of $s(t)$ can be defined as the argument of $s_{\mathrm{a}}(t)$ or equivalently its instantaneous frequency, $f(t)$, as the angular velocity of the representation in the complex plane of this analytic signal. In the same way, the instantaneous amplitude $\rho(t)$ of $s(t)$ can be defined without ambiguity as the modulus of $s_{\mathrm{a}}(t)$. In terms of analytical expressions, we obtain

$$f(t) = \frac{1}{2\pi}\frac{\mathrm{d}\varphi(t)}{\mathrm{d}t} = \frac{1}{2\pi}\frac{\mathrm{d}\arg\{s_{\mathrm{a}}(t)\}}{\mathrm{d}t}, \tag{1.94a}$$

$$\rho(t) = |s_{\mathrm{a}}(t)|. \tag{1.94b}$$

For our purposes, it is of interest to link those quantities to the characteristics of the complex envelopes of the bandpass signals we are dealing with. Let us therefore suppose

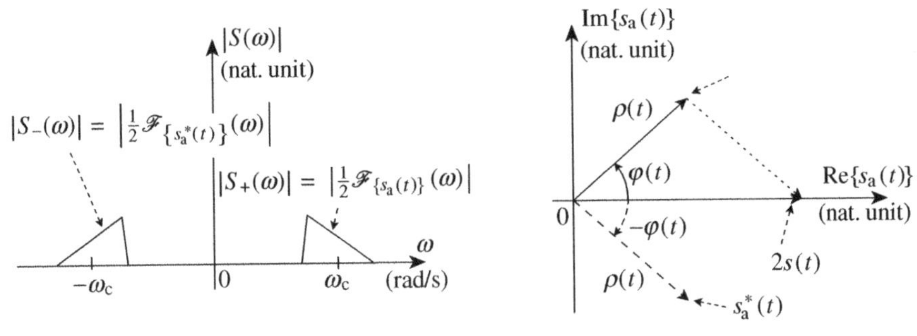

Figure 1.6 Instantaneous amplitude and phase definition of a real bandpass signal through its associated analytic signal – A real bandpass signal, $s(t)$, can be decomposed as half the sum of its analytic signal, $s_{\mathrm{a}}(t)$, and its complex conjugate according to equation (1.93). In this decomposition, the analytic signal is linked to the signal corresponding to the positive sideband of $s(t)$ and its complex conjugate to the signal corresponding to the negative sideband (left). The representation in the complex plane of those signals illustrates the definition of the instantaneous amplitude, $\rho(t)$, and phase, $\varphi(t)$, of $s(t)$ as the modulus and argument of $s_{\mathrm{a}}(t)$, respectively (right).

that $\tilde{s}(t) = \rho(t)\,e^{j\phi(t)}$ is the complex envelope defined as centered around ω_c of $s(t)$. Inverting equation (1.21), we immediately get the following expression for the analytic signal $s_a(t)$ associated with $s(t)$:

$$s_a(t) = \tilde{s}(t)e^{j\omega_c t} = \rho(t)\,e^{j(\omega_c t+\phi(t))}. \tag{1.95}$$

The instantaneous phase of $s(t)$ can then be expressed as

$$\begin{aligned}\varphi(t) = \arg\{s_a(t)\} &= \omega_c t + \arg\{\tilde{s}(t)\} \\ &= \omega_c t + \phi(t),\end{aligned} \tag{1.96}$$

and its instantaneous frequency as

$$\begin{aligned}f(t) = \frac{1}{2\pi}\frac{d\arg\{s_a(t)\}}{dt} &= \frac{1}{2\pi}\left(\omega_c + \frac{d\arg\{\tilde{s}(t)\}}{dt}\right) \\ &= \frac{1}{2\pi}\left(\omega_c + \frac{d\phi(t)}{dt}\right).\end{aligned} \tag{1.97}$$

In the same way, its instantaneous amplitude can then be expressed as

$$\rho(t) = |s_a(t)| = |\tilde{s}(t)|. \tag{1.98}$$

Thus the instantaneous frequency of an RF bandpass signal is precisely the sum of the carrier frequency and the normalized time derivative of the phase of its complex envelope that is defined as centered around this carrier frequency. This instantaneous frequency is thus related to the angular rate at which the complex envelope describes the modulation trajectory in the complex plane. As a side effect, it also allows us to understand how the spectral limitation of the modulating signal impacts the speed at which the vector representing the complex envelope can go from a data symbol representation to another one in the complex plane, and thus impacts the relationship between the data symbol rate and the instantaneous frequency variations of the resulting modulated bandpass signal.

Following those definitions, it may be of interest to recall that an infinity of complex envelopes can be defined for a given bandpass signal, as discussed in Section 1.1.2 "Complex envelope concept" (Section 1.1.2). Nevertheless, all these complex envelopes are derived from the same unique analytic signal, so that they all fulfill equation (1.25). As a result, the above relationships linking the instantaneous amplitude and phase or frequency to the characteristics of the complex envelopes result in well-defined quantities.

In conclusion, it may be of interest to illustrate the concepts discussed so far in this section with an example. In Section 1.3.3 we examine the characteristics of a wideband code division multiple access (WCDMA) modulated bandpass signal, $s(t)$, in two different configurations. In the first, the carrier angular frequency ω_c is low enough that it is of the same order of magnitude as the modulation spectrum width even though still ensuring the bandpass condition. Thus, given that the spectrum of a WCDMA modulation spreads in a frequency bandwidth slightly lower than $[-1/(2T_c), 1/(2T_c)]$, where T_c represents the chip rate as detailed in "Code

Division Multiple Access" in Section 1.3.3, we can select ω_c as equal to $0.8 \times 2\pi/T_c$ for our example. In the second configuration, the carrier angular frequency is high enough that $s(t)$ can be considered as narrowband. We can select, for instance, ω_c as equal to $4 \times 2\pi/T_c$. As illustrated in Figure 1.7, we then get in the first situation that the frequency of occurrence of the instantaneous amplitude variations is of the same order of magnitude as those of the carrier. As a result, the definition of an instantaneous amplitude and phase or frequency of $s(t)$ is not so obvious just from the time domain representation of $s(t)$. In contrast, as soon as ω_c increases compared to the modulating bandwidth, the shape of $s(t)$ clearly matches that of a CW signal whose amplitude and frequency are slowly varying according to its instantaneous amplitude and frequency. Nevertheless, it is interesting to notice that the modulus and phase of the analytic signal of $s(t)$ lead to an unambiguous definition for those quantities in both cases. Moreover, we recover that whatever the chosen carrier angular frequency, all the resulting bandpass signals necessarily have the same instantaneous amplitude. The same holds for their instantaneous frequency variations when considered relative to their carrier frequencies.

1.2 Bandpass Noise Representation

The complex envelope is a valuable concept in performing analytical derivations for transceiver budgets, and we now focus on extending it to bandpass noise. Or rather, we now focus on the additional properties of complex envelopes for bandpass noise as the complex envelope, originally introduced for bandpass deterministic signals in Section 1.1.2, has already been implicitly extended to bandpass stochastic processes when considering modulated bandpass signals. Practically speaking, we consider that each realization of a given bandpass process is also bandpass. We can thus define a complex envelope associated with each of its realizations, thus in turn allowing us to define the complex envelope as a stochastic process.

There are new features of complex envelopes to consider when dealing with bandpass noises as classically encountered in practical wireless transceiver implementations. For instance, we get that the distribution of most such bandpass noises is Gaussian, which leads to interesting characteristics for their complex envelopes. Furthermore, those bandpass noises can often be seen as additive with regard to the signal being processed along the line-up. In that case, the bandpass behavior of both the noise and the signal allows for an interesting decomposition of the noise in terms of amplitude and phase noise, i.e. in terms of noise components that corrupt the instantaneous amplitude or the instantaneous phase respectively of the bandpass signal it adds to. Due to the importance of this decomposition in the field of transceivers, it is important to introduce it correctly based on the complex envelope concept.

Before going any further, however, we discuss the bandpass concept itself for noise signals as classically encountered in practical transceivers. Referring to the discussion at the beginning of Chapter 4, we get that noise components such as thermal noise have a spectrum that spreads up to a very high frequency. Although the bandpass concept requires only that the spectral extent has no power density at the zero frequency, it turns out that the noise processes we need to consider for the derivation of transceiver budgets often have limited bandwidth. Practically speaking, there are various reasons for that:

(i) On the receive side, we always get channel filters that select the desired signal. We thus get that the noise contributions that would lie outside their passband are canceled, or at

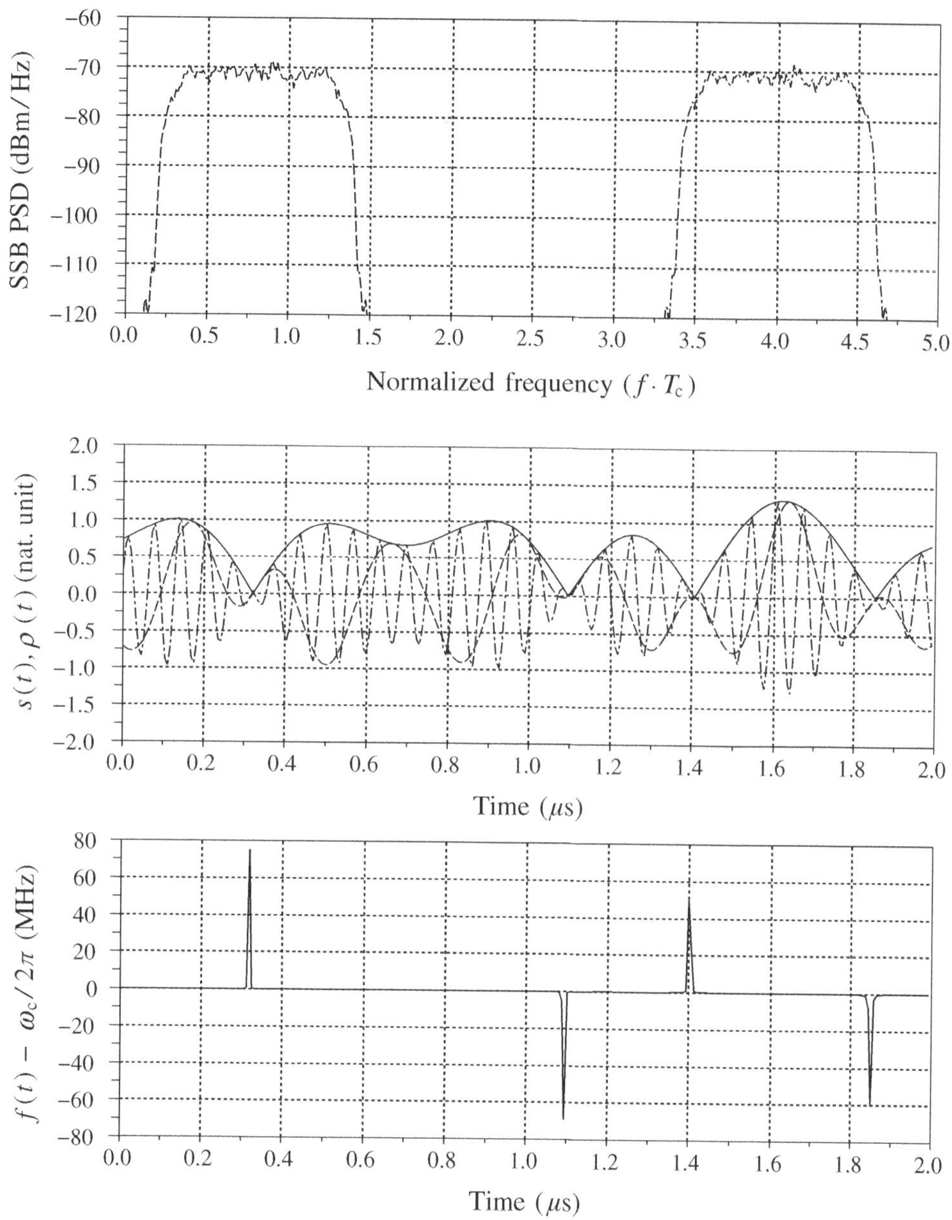

Figure 1.7 Bandpass signal and associated instantaneous amplitude and frequency – When the carrier frequency is set to $0.8/T_c$, where T_c is the chip rate, a WCDMA bandpass signal has a spectrum with low frequency components (top, dashed). This results in a time domain waveform (middle, dashed) that does not highlight clearly the shape of the instantaneous amplitude waveform (middle, solid). This is not the case when the carrier frequency is set to $4/T_c$ (top, dot-dashed), which results in a clear shaping of the time domain waveform by the instantaneous amplitude (middle, dot-dashed). In these two cases, the bandpass signals have the same instantaneous amplitude (middle, solid) and the same instantaneous frequency variations around the carrier frequency (bottom).

least reduced to a negligible level. As a result, even when dealing with noises that are broadband in essence due to their physical origin, we can restrict their description to the fraction that lies within the receiver passband in order to perform the line-up budget. We observe that it leads to the concept of receiver noise bandwidth, which directly matches the receiver passband when dealing with a noise process that has a flat spectral density.

(ii) On the transmit side, we can invoke the same kind of argument. We observe, for instance, that analog reconstruction filters are often required in the line-up as discussed in Section 4.6. Any broadband noise contribution that experiences such filtering can thus be seen as a narrower bandpass noise at the output of the line-up.

We also observe that various noise contributions have a spectral extent naturally in the region of that of the signal being processed. This is the case, for instance, for the distortion terms that are generated from it through linear or nonlinear effects. In any case, the concept of the complex envelope is well suited for performing analytical derivations involving such noise contributions. Among other things, it allows interesting vectorial representations in the complex plane that illustrate underlying phenomena that are sometimes not immediately obvious.

1.2.1 Gaussian Components

Let us now detail some characteristics of complex envelopes of real bandpass noises that exhibit a Gaussian distribution – or, rather, of Gaussian bandpass processes in general. The Gaussian distribution is naturally associated with analog noises, the most famous being thermal noise as discussed in Chapter 4, and is encountered in many situations in practice, among them the following:

(i) The modulating waveforms used in some wideband wireless systems. As discussed in Section 1.3.3, the Gaussian distribution can be seen as a limit case for the distribution of most of the resulting modulated bandpass signal.

(ii) The wireless propagation channel in a mobile environment. As discussed in Chapter 2, the simplest propagation channel is the Rayleigh channel. In this channel model we have a Gaussian distribution for the bandpass signal that goes through it, thus resulting in a Rayleigh distribution for its instantaneous amplitude.

Looking at those various use cases, we can appreciate the importance of going through the statistical properties of the Gaussian real bandpass process and its associated complex envelopes.

Let us suppose that we are dealing with a Gaussian bandpass process $s(t)$ whose PSD spreads over a bandwidth $\delta\omega$ centered around ω_c as illustrated in Figure 1.8. Generally speaking, the term "Gaussian process" means that for any set of N samples of $s(t)$, taken at times $t_1, t_2, \dots, t_N$, the vector $\left(s_{t_1}, s_{t_2}, \dots, s_{t_N}\right)^{\mathrm{T}}$ is a real normal random vector as defined in Appendix 3. Practically speaking, when it exists, the PDF of $\left(s_{t_1}, s_{t_2}, \dots, s_{t_N}\right)^{\mathrm{T}}$, $p(x_1, \dots, x_N)$, has a multivariate normal distribution as given by equation (A3.2). Furthermore, correlations can exist between successive time samples due to the finite spectral bandwidth of the process.

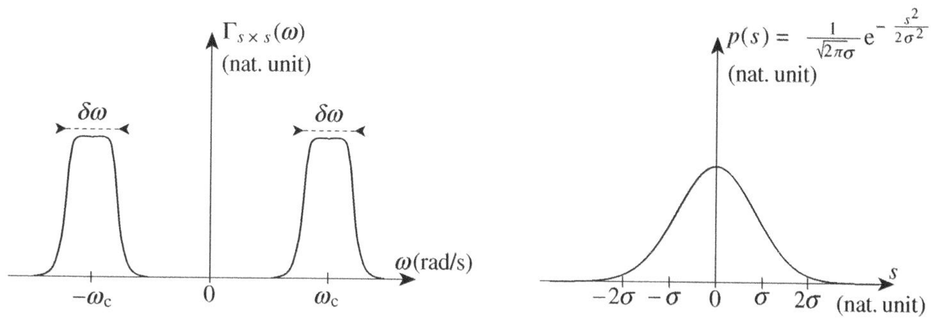

Figure 1.8 Power spectral density and probability density function of a real bandpass Gaussian process – A real Gaussian bandpass process, $s(t)$, has a PSD that is non-vanishing only over a finite frequency band, $\delta\omega$, that does not go down to the zero frequency (left). Due to the finite spectral width of the process, correlation terms can exist between successive time samples of the process. Considered at a given time t, the random variable s_t has a Gaussian distribution (right).

This simply means that the noise process we are considering is not white in the sense that its flat spectrum does not spread toward infinity. As a result, the covariance matrix $\mathbf{\Sigma}$, whose elements are given by equation (A3.6), can be non-diagonal. The PDF of a single sample random variable s_t reduces to a Gaussian distribution.

Let us now suppose that we are dealing with a centered and stationary bandpass process. This is indeed a reasonable assumption in the field of wireless transceivers. Practically speaking, it means we can assume that the PDF $p(s_t)$ of s_t is independent of t and that it takes the simple form

$$p(s_t) = p(s) = \frac{1}{\sqrt{2\pi}\sigma}\mathrm{e}^{-\frac{s^2}{2\sigma^2}}, \tag{1.99}$$

with

$$\mathbb{E}\{s\} = 0, \tag{1.100a}$$

$$\mathbb{E}\{s^2\} = \sigma^2. \tag{1.100b}$$

As the realization $s(t)$ is bandpass, we can consider for it the complex envelope $\tilde{s}(t)$, defined as centered around the angular frequency ω_c. As a first step, let us consider a Cartesian representation of $\tilde{s}(t) = p(t) + \mathrm{j}q(t)$. This means that $s(t)$ can be expressed as

$$\begin{aligned} s(t) &= \mathrm{Re}\left\{\tilde{s}(t)\mathrm{e}^{\mathrm{j}\omega_c t}\right\} \\ &= p(t)\cos(\omega_c t) - q(t)\sin(\omega_c t). \end{aligned} \tag{1.101}$$

On the other hand, we get that the stationarity of $s(t)$ leads to $\gamma_{\tilde{s}\times\tilde{s}^*}(\tau) = 0$ following the arguments in Appendix 2, and in particular equation (A2.11). This property, by which the complex normal random vector $\left(\tilde{s}_{t_1}, \tilde{s}_{t_2}, \ldots, \tilde{s}_{t_N}\right)^{\mathrm{T}}$ is said to be circular, also leads to $p(t)$ and $q(t)$ being uncorrelated when considered at the same time, as given by equation (A2.13b) for

$\tau = 0$. Under our assumptions, the result is that p_t and q_t are independent, with a distribution given by [3, 4]

$$p(p_t) = p(p) = \frac{1}{\sqrt{2\pi}\sigma} e^{-\frac{p^2}{2\sigma^2}}, \tag{1.102a}$$

$$p(q_t) = p(q) = \frac{1}{\sqrt{2\pi}\sigma} e^{-\frac{q^2}{2\sigma^2}}, \tag{1.102b}$$

with

$$\mathbb{E}\{p\} = \mathbb{E}\{q\} = \mathbb{E}\{s\} = 0, \tag{1.103a}$$

$$\mathbb{E}\{p^2\} = \mathbb{E}\{q^2\} = \mathbb{E}\{s^2\} = \sigma^2. \tag{1.103b}$$

We thus get that the real and imaginary parts, $p(t)$ and $q(t)$, of the complex envelopes that represent a stationary and centered bandpass Gaussian process also have a centered Gaussian distribution with the same variance as the original bandpass process.

An additional interesting property for $p(t)$ and $q(t)$ occurs when $s(t)$ also exhibits a spectral symmetry with regard to the angular frequency chosen for the definition of its complex envelope $\tilde{s}(t) = p(t) + \mathrm{j}q(t)$ – or, more precisely, when its PSD is evenly symmetrical with regard to the frequency used to define the complex envelope $\tilde{s}(t)$. In that case, the results in "Impact of spectral symmetry on top of stationarity" (Section 1.1.3) apply and we get from equation (1.51) that

$$\Gamma_{\tilde{s}\times\tilde{s}}(\omega) = \Gamma_{p\times p}(\omega) + \Gamma_{q\times q}(\omega) = 2\Gamma_{p\times p}(\omega) = 2\Gamma_{q\times q}(\omega). \tag{1.104}$$

In that case, then, $p(t)$ and $q(t)$ have the same spectral shape as the original stationary bandpass process they represent. The converse situation is also obviously true. This means that the frequency upconversion of two stationary and centered lowpass processes with Gaussian distributions results in a bandpass process that also exhibits a Gaussian distribution. And when the two initial lowpass components are also uncorrelated, we get that the PSD of the resulting RF bandpass signal is the sum of the input ones.

Let us now focus on the characteristics of the polar form of the complex envelope $\tilde{s}(t) = p(t) + \mathrm{j}q(t) = \rho(t)\mathrm{e}^{\mathrm{j}\phi(t)}$. Assuming that we are dealing with a stationary and centered bandpass process, we have seen that $p(t)$ and $q(t)$ are lowpass processes such that p_t and q_t are also centered and have a Gaussian distribution. Under the same assumption, it can be shown that the magnitude $\rho(t)$ of the complex envelope $\tilde{s}(t)$ is such that ρ_t follows a Rayleigh distribution, i.e. that [3]

$$p(\rho_t) = p(\rho) = \frac{\rho}{\sigma^2} e^{-\frac{\rho^2}{2\sigma^2}}. \tag{1.105}$$

In the same way, as illustrated in Figure 1.9, the phase $\phi(t)$ of $\tilde{s}(t)$ is necessarily uniformly distributed on $[0, 2\pi]$, i.e.

$$p(\phi_t) = p(\phi) = \frac{1}{2\pi}. \tag{1.106}$$

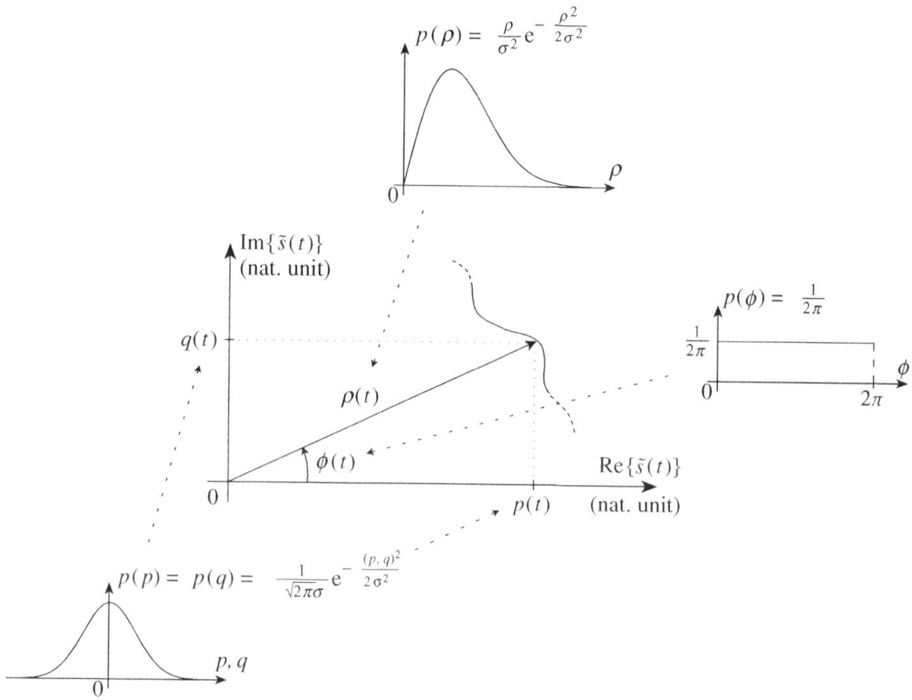

Figure 1.9 Statistical properties of the complex envelope of a stationary and centered bandpass Gaussian process – The complex envelope of a stationary and centered bandpass Gaussian process, whose PSD and PDF are of the form shown in Figure 1.8, has noticeable statistical properties: its real and imaginary parts also have a Gaussian distribution with the same variance as the original bandpass process, its modulus is Rayleigh distributed and its argument uniformly distributed over $[0, 2\pi]$. When considered at the same time, the modulus and argument are mutually independent.

Moreover, we get that the joint distribution of the random variables ρ_t and ϕ_t corresponding to two realizations of $\rho(t)$ and $\phi(t)$, when considered at the *same* time, is such that

$$p(\rho_t, \phi_t) = p(\rho_t)p(\phi_t). \tag{1.107}$$

Practically speaking, this means that the variables ρ_t and ϕ_t are independent when considered at the same time. Nevertheless, we have to keep in mind that this is true only when $p(t)$ and $q(t)$ are centered. Indeed, if

$$\mathbb{E}\{p\} = m_{\text{p}}, \tag{1.108a}$$

$$\mathbb{E}\{q\} = m_{\text{q}}, \tag{1.108b}$$

with

$$m_{\text{p}} + \text{j}m_{\text{q}} = m\text{e}^{\text{j}\theta}, \tag{1.109}$$

it can be shown that [3]

$$p(\rho_t, \phi_t) = \frac{1}{2\pi}\frac{\rho_t}{\sigma^2}e^{-\frac{1}{2\sigma^2}\left(\rho_t^2 - 2\rho_t m\cos(\phi_t - \theta) + m^2\right)}, \tag{1.110}$$

which is a Rayleigh–Rice distribution. And if equation (1.107) no longer holds, then ρ_t and ϕ_t are no longer independent. However, as already highlighted, we can assume that all processes of interest in wireless transceivers are centered.

To conclude this section, we briefly review the moments of the modulus $\rho(t)$ of the complex envelope of such a Gaussian and centered bandpass process. Those moments will be needed to perform analytical derivations, for instance when dealing with nonlinear transfer functions, as is done in Chapter 5. From our discussion so far, we have that those moments are nothing more than those of the Rayleigh distribution. We can thus write that [1]

$$\mathbb{E}\{\rho^n\} = (2\sigma^2)^{n/2}\Gamma\left(1 + \frac{n}{2}\right), \tag{1.111}$$

where $\Gamma(.)$ is the gamma function, which can be expressed for instance in its Euler integral form as [6]

$$\Gamma(z) = \int_0^{+\infty} t^{z-1}e^{-t}\mathrm{d}t. \tag{1.112}$$

One property of this function is that it reduces, for integers z, to a factorial expression:

$$\Gamma(z) = (z-1)! = (z-1)(z-2)\cdots 1. \tag{1.113}$$

Thus, for any even $n = 2k$, the moments of the Rayleigh distribution given by equation (1.111) reduce to

$$\mathbb{E}\{\rho^{2k}\} = (2\sigma^2)^k k!. \tag{1.114}$$

It is these even order moments that will later be of interest.

1.2.2 Phase Noise vs. Amplitude Noise

Let us now take another step forward in the description of how an additive bandpass noise corrupts a bandpass signal. Such bandpass noise can be decomposed into two terms that, under the assumption that the SNR is good enough, corrupt up to first order either the instantaneous amplitude or the instantaneous phase of the bandpass signal it is added to. Such decomposition is of particular interest in practice as there are times when only one of the two terms is involved in some of the mechanisms we deal with in the field of transceivers. One example is the transfer of the phase noise part of the local oscillator (LO) signal during a frequency transposition based on the use of mixers behaving as choppers, as discussed in Section 4.3.2. Another is the cancellation mechanism of the amplitude part of such an additive noise signal in nonlinear devices exhibiting compression behavior, as discussed in Sections 5.1.4 and 5.2.3.

Let us consider a real bandpass signal, $s_{\mathrm{w}}(t)$, resulting from the upconversion of the complex envelope

$$\tilde{s}_{\mathrm{w}}(t) = \rho_{\mathrm{w}}(t)e^{j\phi_{\mathrm{w}}(t)}, \tag{1.115}$$

around the center angular frequency ω_{c}. We can thus express $s_{\mathrm{w}}(t)$ as

$$s_{\mathrm{w}}(t) = \mathrm{Re}\left\{\tilde{s}_{\mathrm{w}}(t)e^{j\omega_{\mathrm{c}}t}\right\}. \tag{1.116}$$

Let us suppose that this signal is corrupted by an additive bandpass noise, $n(t)$, described by the complex envelope

$$\tilde{n}(t) = \rho_{\mathrm{n}}(t)e^{j\phi_{\mathrm{n}}(t)}, \tag{1.117}$$

also defined as centered around the *same* center frequency ω_{c}. We thus assume that $n(t)$ can be written as

$$n(t) = \mathrm{Re}\left\{\tilde{n}(t)e^{j\omega_{\mathrm{c}}t}\right\}. \tag{1.118}$$

Using these expressions, we can express the total signal $s(t)$, defined as the superposition of $s_{\mathrm{w}}(t)$ and $n(t)$, as

$$s(t) = s_{\mathrm{w}}(t) + n(t) = \mathrm{Re}\left\{(\tilde{s}_{\mathrm{w}}(t) + \tilde{n}(t))e^{j\omega_{\mathrm{c}}t}\right\}. \tag{1.119}$$

We then see that the complex envelope $\tilde{s}(t)$ of $s(t)$, when also defined as centered around ω_{c}, can be written as

$$\tilde{s}(t) = \tilde{s}_{\mathrm{w}}(t) + \tilde{n}(t). \tag{1.120}$$

Having defined those signals, we can now focus on the decomposition of $n(t)$. Let us consider the decomposition of its complex envelope, $\tilde{n}(t)$, as the sum of two terms

$$\tilde{n}(t) = \tilde{n}_{\parallel}(t) + \tilde{n}_{\perp}(t). \tag{1.121}$$

As illustrated in Figure 1.10, in this expression $\tilde{n}_{\parallel}(t)$ is defined as collinear to $\tilde{s}_{\mathrm{w}}(t)$ when represented in the complex plane, and $\tilde{n}_{\perp}(t)$ as orthogonal to it. Obviously, this definition requires that $\tilde{s}_{\mathrm{w}}(t)$ is non-vanishing in order to be able to define the direction of its representation in the complex plane. Unfortunately, practical modulating complex envelopes can exhibit such zero crossings, as illustrated in Section 1.3. However, when they occur, we can assume that it is only for a set of discrete instants in practice. We can thus assume that $\tilde{n}_{\parallel}(t)$ and $\tilde{n}_{\perp}(t)$ can still be defined at any time by continuity. The result is that we can in turn interpret $\tilde{n}_{\parallel}(t)$

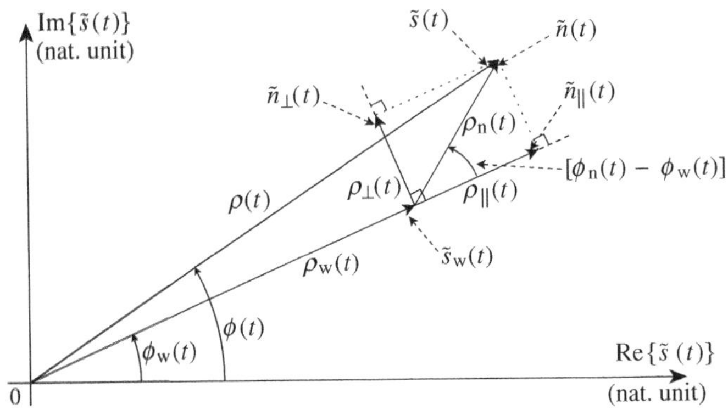

Figure 1.10 Additive bandpass noise decomposition in terms of amplitude and phase noise terms – The complex envelope of a bandpass noise, when defined as centered around the same center frequency as that considered for the definition of the complex envelope of a bandpass signal it is added to, can be decomposed into two components. The first, collinear to the signal complex envelope, represents a bandpass noise that corrupts, at first order, mainly its instantaneous amplitude. The second, orthogonal to it, represents a bandpass noise that corrupts, still at first order, mainly its instantaneous phase.

and $\tilde{n}_{\perp}(t)$ as the complex envelopes, still defined as centered around ω_c, of the bandpass noise processes $n_{||}(t)$ and $n_{\perp}(t)$ that can be written as

$$n_{||}(t) = \mathrm{Re}\left\{\tilde{n}_{||}(t)e^{j\omega_c t}\right\}, \tag{1.122a}$$

$$n_{\perp}(t) = \mathrm{Re}\left\{\tilde{n}_{\perp}(t)e^{j\omega_c t}\right\}. \tag{1.122b}$$

Thus, given that all the complex envelopes we are dealing with are defined as centered around the same angular frequency, by substituting equation (1.121) into equation (1.118) and using the above definition we can write

$$\begin{aligned} n(t) &= \mathrm{Re}\left\{\tilde{n}(t)e^{j\omega_c t}\right\} \\ &= n_{||}(t) + n_{\perp}(t). \end{aligned} \tag{1.123}$$

We have thus decomposed the additive bandpass noise $n(t)$ as the sum of two bandpass terms that have interesting properties. This can be advantageously investigated by expressing the complex envelopes of those bandpass terms as a function of the characteristics of $\tilde{s}_w(t)$ and $\tilde{n}(t)$. This can be done in a straightforward way following geometrical considerations that can be derived from the illustration in Figure 1.10. Based on this figure, we immediately have that

$$\tilde{n}_{||}(t) = \rho_{||}(t)\epsilon_{||}(t)e^{j\phi_w(t)} = \rho_{\epsilon||}(t)e^{j\phi_w(t)}, \tag{1.124a}$$

$$\tilde{n}_{\perp}(t) = \rho_{\perp}(t)\epsilon_{\perp}(t)e^{j(\phi_w(t)+\pi/2)} = j\rho_{\epsilon\perp}(t)e^{j\phi_w(t)}, \tag{1.124b}$$

where we have defined

$$\rho_{\parallel}(t) = \rho_{\rm n}(t)\, |\cos(\phi_{\rm n}(t) - \phi_{\rm w}(t))|, \tag{1.125a}$$

$$\rho_{\perp}(t) = \rho_{\rm n}(t)\, |\sin(\phi_{\rm n}(t) - \phi_{\rm w}(t))|, \tag{1.125b}$$

and

$$\epsilon_{\parallel}(t) = \text{sign}\, \{\cos(\phi_{\rm n}(t) - \phi_{\rm w}(t))\}, \tag{1.126a}$$

$$\epsilon_{\perp}(t) = \text{sign}\, \{\sin(\phi_{\rm n}(t) - \phi_{\rm w}(t))\}. \tag{1.126b}$$

The reason for this definition comes from the fact that, following our conventions, $\rho_{\parallel}(t)$ and $\rho_{\perp}(t)$ are positive quantities that represent moduli of complex envelopes. We thus introduce $\epsilon_{\parallel}(t)$ and $\epsilon_{\perp}(t)$, which can be only ± 1, in order to represent the π phase shift of the arguments of $\tilde{n}_{\parallel}(t)$ and $\tilde{n}_{\perp}(t)$ depending on the relative instantaneous phase between $\tilde{s}_{\rm w}(t)$ and $\tilde{n}(t)$. However, as will be seen shortly, the quantities of interest for our derivations are $\rho_{\epsilon\parallel}(t)$ in equation (1.124a) and $\rho_{\epsilon\perp}(t)$ in equation (1.124b), defined as

$$\rho_{\epsilon\parallel}(t) = \rho_{\rm n}(t) \cos(\phi_{\rm n}(t) - \phi_{\rm w}(t)), \tag{1.127a}$$

$$\rho_{\epsilon\perp}(t) = \rho_{\rm n}(t) \sin(\phi_{\rm n}(t) - \phi_{\rm w}(t)). \tag{1.127b}$$

Based on this decomposition of $n(t)$, we can then go further in the interpretation of the degradation of $s_{\rm w}(t)$. We can express the modulus and argument of the complex envelope $\tilde{s}(t) = \rho(t)e^{j\phi(t)}$ of the total signal $s(t) = s_{\rm w}(t) + n(t)$ as a function of the characteristics of those two bandpass noises. For that purpose, we can consider both the decomposition of $\tilde{n}(t)$ given by equation (1.121), and the expressions for $\tilde{n}_{\parallel}(t)$ and $\tilde{n}_{\perp}(t)$ given by equation (1.124). Following the representation shown in Figure 1.10, we can then immediately write

$$\rho(t) = |\tilde{s}_{\rm w}(t) + \tilde{n}(t)| = \sqrt{\rho_{\rm w}^2(t) + 2\rho_{\epsilon\parallel}(t)\rho_{\rm w}(t) + \rho_{\rm n}^2(t)}, \tag{1.128a}$$

$$\phi(t) = \arg\{\tilde{s}_{\rm w}(t) + \tilde{n}(t)\} = \phi_{\rm w}(t) + \arctan\left\{\frac{\rho_{\epsilon\perp}(t)}{\rho_{\rm w}(t) + \rho_{\epsilon\parallel}(t)}\right\}, \tag{1.128b}$$

with $\rho_{\epsilon\parallel}(t)$ and $\rho_{\epsilon\perp}(t)$ given by equation (1.127). Obviously, this definition requires that $\tilde{s}_{\rm w}(t)$ is non-vanishing so that $\rho_{\rm w}(t) \neq 0$ so as to be able to define at least $\phi(t)$ correctly. Nevertheless, as highlighted previously, we can assume when dealing with practical modulations that this situation occurs only for a set of discrete instants. We can thus assume that we can define such a quantity at any time by continuity. Moreover, we can assume in many practical use cases that $\rho_{\rm n}(t) \ll \rho_{\rm w}(t)$. At first glance, we could consider linking this assumption to the fact of having a good enough SNR, i.e. to the assumption of having $\mathbb{E}\{\rho_{\rm n}^2\} \ll \mathbb{E}\{\rho_{\rm w}^2\}$ according to equation (1.64). This is unfortunately not true in all cases, as we have just recalled that practical modulating complex envelopes can have their instantaneous amplitude going down toward zero. This means that whatever their power, at some instant we can potentially have that the modulus of their complex envelope is lower than that of a bandpass noise it is added to. However, as the SNR increases, we can imagine that the set of instants for which this situation occurs diminishes. Furthermore, there are classical modulating waveforms for which the instantaneous amplitude has a lower bound. This is obviously the case for constant amplitude schemes, but is also the case for some complex modulations such as the modified 8PSK used in the GSM/EDGE standard and detailed in Section 1.3.2. For such a modulation, there is thus a minimum SNR for which the condition $\rho_{\rm n}(t) \ll \rho_{\rm w}(t)$ can be considered as true at any time. Finally, we recall that the classical use case for such decomposition of a bandpass noise corresponds to a signal that is nothing more than the LO waveform flowing from an RF synthesizer and used to drive a mixer for implementing a frequency transposition. In any case, given that $\rho_{\rm n}(t) \ll \rho_{\rm w}(t)$ holds, we can rely on a small angle approximation in our derivations so that the above expressions for $\rho(t)$ and $\phi(t)$ reduce to

$$\rho(t) = \rho_{\rm w}(t) + n_\rho(t), \tag{1.129a}$$

$$\phi(t) = \phi_{\rm w}(t) + n_\phi(t), \tag{1.129b}$$

with

$$n_\rho(t) \approx \rho_{\epsilon\|}(t), \tag{1.130a}$$

$$n_\phi(t) \approx \frac{\rho_{\epsilon\perp}(t)}{\rho_{\rm w}(t)}. \tag{1.130b}$$

Under our present assumption, we thus get that up to first order $\rho_{\epsilon\|}(t)$ corrupts only the magnitude $\rho_{\rm w}(t)$ of the signal complex envelope, i.e. the instantaneous amplitude of $s_{\rm w}(t)$, whereas $\rho_{\epsilon\perp}(t)$ corrupts only the argument $\phi_{\rm w}(t)$ of the signal complex envelope, i.e. the instantaneous frequency of $s_{\rm w}(t)$. We also have an additive behavior for the terms $n_\rho(t)$ and $n_\phi(t)$ that can naturally be labeled as respectively the *amplitude noise* and the *phase noise* which corrupt respectively the instantaneous amplitude and the instantaneous phase of the signal.

We thus have two distinct, albeit equivalent, ways of describing the impact of the bandpass noise $n(t)$ on the bandpass signal $s_{\rm w}(t)$ it is added to. On the one hand, we can keep the *additive* behavior of $n(t)$ regarding $s_{\rm w}(t)$ while achieving its decomposition as the sum of $n_\|(t)$ and $n_\perp(t)$. Using equations (1.119) and (1.123), we can write $s(t)$ as

$$s(t) = s_{\rm w}(t) + n_\|(t) + n_\perp(t). \tag{1.131}$$

Up to first order, i.e. given that $\rho_{\rm n}(t) \ll \rho_{\rm w}(t)$, we then get that $n_\|(t)$ and $n_\perp(t)$ corrupt either the instantaneous amplitude of $s_{\rm w}(t)$ or its instantaneous phase or frequency. On the other hand, we can directly *embed* the noise terms in the components of the complex envelope of $s_{\rm w}(t)$ in order to derive an expression for $s(t)$. In that case, with the small angle approximation, i.e. still assuming that $\rho_{\rm n}(t) \ll \rho_{\rm w}(t)$, we can use the expression for the complex envelope of $s(t)$ given by equation (1.129) to write

$$\tilde{s}(t) = (\rho_{\rm w}(t) + n_\rho(t))e^{j(\phi_{\rm w}(t)+n_\phi(t))}, \tag{1.132}$$

with $n_\rho(t)$ and $n_\phi(t)$ given by equation (1.130). In that case, $s(t)$ can be written as

$$\begin{aligned} s(t) &= \mathrm{Re}\left\{\tilde{s}(t)e^{j\omega_c t}\right\} \\ &= (\rho_w(t) + n_\rho(t))\cos(\omega_c t + \phi_w(t) + n_\phi(t)). \end{aligned} \tag{1.133}$$

Although the latter formulation hides the additive behavior of the original bandpass noise with regard to the signal, it is of interest for some practical analytical derivations. In any case, the two representations are strictly equivalent when they exist. The use of either one thus depends only on how easily derivations can be done with it. Nevertheless, despite this equivalence we need to keep in mind that there is a fundamental difference between the terms involved in the two different representations. On the one hand we are dealing with *bandpass* noise terms, $n_\parallel(t)$ and $n_\perp(t)$, directly related to the decomposition of the original bandpass noise $n(t)$. On the other hand, we have *lowpass* components, $n_\rho(t)$ and $n_\phi(t)$, which are involved in the modulus and argument of the complex envelope of the total signal $s(t)$.

To conclude this section, it is of interest to mention additional considerations that arise when the bandpass noise, implicitly assumed centered, also has a Gaussian distribution. In this case it is easy to derive the fraction of power of the bandpass noise that corrupts either the instantaneous amplitude of the signal or its instantaneous phase or frequency. Assuming we are dealing with stationary processes, we can express the power $P_{n_\parallel}$ and $P_{n_\perp}$ of the bandpass processes $n_\parallel(t)$ and $n_\perp(t)$ based on the expressions we have derived so far for their complex envelopes through the use of equation (1.64). Focusing first on the average power of $n_\parallel(t)$, given that $\epsilon_\parallel(t)$ can be only ±1, from equations (1.124a) and (1.125a) we have that

$$P_{n_\parallel} = \frac{\mathbb{E}\left\{\rho_n^2\cos^2(\phi_n - \phi_w)\right\}}{2}. \tag{1.134}$$

But, following the derivations performed in Section 1.2.1, we also get that the modulus and argument of the complex envelope of a stationary and centered Gaussian process are independent when considered at the same time. Because the wanted signal $s_w(t)$ and the noise component $n(t)$ can also be assumed independent, we can therefore write

$$P_{n_\parallel} = \frac{\mathbb{E}\left\{\rho_n^2\right\}}{2}\mathbb{E}\left\{\cos^2(\phi_n - \phi_w)\right\}. \tag{1.135}$$

Using equation (1.54), we can then expand the last term of the right-hand side of this equation:

$$\mathbb{E}\left\{\cos^2(\phi_n - \phi_w)\right\} = \frac{1}{2} + \frac{1}{2}\mathbb{E}\{\cos[2(\phi_n - \phi_w)]\}. \tag{1.136}$$

As we have the independence of ϕ_n and ϕ_w, we can then evaluate the expectation on the right-hand side of this equation by integrating first the PDF of ϕ_n. But, as we deal with a stationary centered Gaussian bandpass noise, we get that ϕ_n is uniformly distributed over

$[0, 2\pi]$ according to equation (1.106). Due to this even distribution over a period of the cosine function, it follows that

$$\mathbb{E}\{\cos[2(\phi_{\mathrm{n}} - \phi_{\mathrm{w}})]\} = 0, \tag{1.137}$$

and thus that

$$P_{\mathrm{n}_{\parallel}} = \frac{1}{2}\frac{\mathbb{E}\{\rho_{\mathrm{n}}^2\}}{2} = \frac{1}{2}P_{\mathrm{n}}, \tag{1.138}$$

where P_{n} is the power of the original bandpass Gaussian noise. As exactly the same derivation can be performed for the perpendicular component $n_{\perp}(t)$, we finally get that

$$P_{\mathrm{n}_{\parallel}} = P_{\mathrm{n}_{\perp}} = \frac{1}{2}P_{\mathrm{n}}. \tag{1.139}$$

We see that whatever the characteristics of the signal such bandpass noise is added to, we recover an equal noise power over the components defined as perpendicular and parallel to the signal complex envelope. We also observe that this result highlights the derivations performed in Chapter 5 when considering the SNR improvement through nonlinearity in Sections 5.1.4 and 5.2. Indeed, in those sections we find that the fraction of the input bandpass noise corresponding to a complex envelope collinear to that of the constant amplitude signal can be canceled due to RF compression. Only the fraction of noise that corresponds to an orthogonal complex envelope succeeds in going through it. Reconsidering the above derivation, we can then understand why the SNR improvement can asymptotically reach 3 dB as the input SNR increases, as shown for instance in Figure 5.30.

1.3 Digital Modulation Examples

Let us now review some modulation schemes that are classically encountered in wireless standards and that are used as examples throughout this book. From a transceiver dimensioning perspective, we are mainly interested in the statistical properties of the modulating waveform and of the associated modulated bandpass signal we have to process in a line-up, rather than in the way such modulating waveforms are generated in practice. As a result, we review here different schemes that are representative of almost all the statistics of complex envelopes we encounter in practice, from the simple constant envelope modulation, illustrated here through the Gaussian minimum shift keying (GMSK) scheme used in the GSM standard, to the most complex waveforms as used in CDMA or OFDM based systems. However, we still need to detail the way those waveforms are generated in order to be able to derive their statistical properties and the associated constraints for transceivers.

1.3.1 Constant Envelope

Let us focus first on constant envelope modulation schemes. As the name suggests, we expect in this case to keep the instantaneous amplitude of the resulting modulated bandpass signal constant. The modulations we are referring to are thus in fact pure phase or frequency modulation schemes. This means that it is only the instantaneous phase or frequency of the resulting modulated RF bandpass signals that carries the information. To illustrate this kind of modulating waveform we can consider GMSK as defined in the GSM standard [7].

We first need to express the instantaneous phase of the resulting modulated bandpass signal as a function of the data bits. From the discussion in "Instantaneous amplitude and instantaneous frequency" (Section 1.1.3), we get that this instantaneous phase is linked to the argument of any of the complex envelopes of this bandpass signal. And due to the nature of the modulation we are considering, the polar form of such complex signal is of interest. Accordingly, let us consider the bandpass signal $s(t)$ resulting from the upconversion of the modulating complex signal $\tilde{s}(t) = e^{j\phi(t)}$ around the carrier angular frequency ω_c. It can be expressed as

$$\begin{aligned} s(t) &= \mathrm{Re}\left\{\tilde{s}(t)e^{j\omega_c t}\right\} \\ &= \mathrm{Re}\left\{e^{j(\omega_c t+\phi(t))}\right\}. \end{aligned} \tag{1.140}$$

It thus simply means that $\tilde{s}(t)$ can be further interpreted as the complex envelope, defined as centered around ω_c, of the bandpass signal $s(t)$. Here, we observe that we have assumed for the sake of simplicity a normalized modulus for $\tilde{s}(t)$ in the definition of those quantities, i.e. that $\rho = |\tilde{s}(t)| = 1$. In any case, according to equation (1.97) we get that the instantaneous frequency $f(t)$ of $s(t)$ is given by

$$f(t) = \frac{1}{2\pi}\left(\omega_c + \frac{d\phi(t)}{dt}\right). \tag{1.141}$$

In the GSM standard, the modulation is defined so that the instantaneous frequency of the resulting modulated RF bandpass signal can be written as a function of the modulating data according to

$$f(t) = \frac{\omega_c}{2\pi} + \frac{h}{2}\sum_k \alpha_k g(t - kT). \tag{1.142}$$

As a result, the argument $\phi(t)$ of $\tilde{s}(t)$ can in turn be written as

$$\phi(t) = \pi h \sum_k \alpha_k \int_{-\infty}^{t-kT} g(u)du, \tag{1.143}$$

thus leading to the expression for $\tilde{s}(t)$:

$$\tilde{s}(t) = \exp\left(j\pi h \sum_k \alpha_k \int_{-\infty}^{t-kT} g(u)du\right). \tag{1.144}$$

In equations (1.142)–(1.144), the parameter h stands for the modulation index and T for the symbol period. These parameters are respectively equal to $1/2$ and $48/13 = 3.69\ \mu s$ in the standard. In the same way $g(t)$ is the pulse shaping filter that smooths the transitions of the instantaneous frequency of $s(t)$ when going from one representation of input data to the next, and α_k is the sequence of input data, which can be ± 1 here.[4]

With regard to the above expressions, it is of interest to interpret further how this modulating scheme works. For instance, when a single datum $\alpha_k = +1$ enters the modulator, we get that the instantaneous frequency signal $f(t)$ experiences a positive deviation compared to the carrier frequency $\omega_c/2\pi$. This frequency offset stands for a finite duration corresponding to the length of the impulse response of the filter $g(t)$. We thus get that the impact of the datum $\alpha_k = +1$ on the final modulated signal $s(t)$ is simply a finite phase shift on its instantaneous phase compared to the average term $\omega_c t$ linked to the carrier, this shift being equal to the integral of the instantaneous frequency deviation during the duration of the impulse response of $g(t)$. The magnitude of this phase shift is thus linked to the properties of the filter $g(t)$. In the GSM standard, this pulse shaping filter is a Gaussian filter defined as the convolution of the gate function $\Pi(t)$,

$$\Pi\left(\frac{t}{T}\right) = \begin{cases} \dfrac{1}{T} & \text{when } |t| < T/2, \\ 0 & \text{otherwise,} \end{cases} \tag{1.145}$$

with a Gaussian function $h(t)$ defined by

$$h(t) = \frac{1}{\delta T\sqrt{2\pi}} \exp\left(-\frac{t^2}{2\delta^2 T^2}\right). \tag{1.146}$$

Here, δ is defined by

$$\delta = \frac{\sqrt{\ln(2)}}{2\pi BT}, \tag{1.147}$$

with B the filter cut-off frequency at 3 dB, defined from the symbol duration through the relation $BT = 0.3$. After some algebra, we can express the impulse response $g(t)$ of the Gaussian pulse shaping filter as

$$\begin{aligned} g(t) &= h(t) \star \Pi\left(\frac{t}{T}\right) \\ &= \frac{1}{2T}\left\{\mathrm{Erfc}\left[\frac{1}{\sqrt{2}\delta}\left(\frac{t}{T}-\frac{1}{2}\right)\right] - \mathrm{Erfc}\left[\frac{1}{\sqrt{2}\delta}\left(\frac{t}{T}+\frac{1}{2}\right)\right]\right\}, \end{aligned} \tag{1.148}$$

where Erfc(.) stands for the complementary error function defined by

$$\mathrm{Erfc}(t) = \frac{2}{\sqrt{\pi}} \int_t^{+\infty} \mathrm{e}^{-x^2} \mathrm{d}x. \tag{1.149}$$

[4] In the GSM standard, it is in fact the differential encoded input bits that are provided to the modulator as the α_k data.

From these expressions, we can thus see that the GMSK pulse shaping filter is normalized so that $\int_{-\infty}^{+\infty} g(u)\mathrm{d}u = 1$. As a result, in the case of a single input datum α_k equal to $+1$, we get that the phase of the modulating signal given by equation (1.143) varies from 0 at $t = 0$ to $\pi/2$ rad at $t = +\infty$. And for the same reason, for a single input datum of opposite sign, we get that the phase of the signal varies from 0 at $t = 0$ to $-\pi/2$ rad at $t = +\infty$. With this behavior in mind, we can now understand that when considering a stream of constant input data all equal to $+1$, we obtain a constant phase rotation at the modulator output of $\pi/2$ rad for each symbol period, i.e. of 2π after four symbol periods. We can thus expect with such a sequence of constant input data values that the resulting modulated RF bandpass signal behaves as an almost CW signal with a frequency offset of $1/(4T) = 13 \times 10^6/(48 \times 4) = 67708.33$ Hz compared to the carrier frequency. Moreover, we understand that this value corresponds to the maximum we can achieve for the instantaneous frequency of the modulated signal, that is coherent with the plot shown in Figure 1.11. However, this maximum deviation of $\pm h/(2T) = \pm 67.70833$ kHz around the carrier frequency does not mean that the spectrum of the modulated RF signal does not spread over a wider frequency range, as shown in Figure 1.12. Indeed, we get that the final RF modulated signal is the cosine of the phase corresponding to this instantaneous frequency. The relationship between its spectrum and the spectral content of the instantaneous frequency signal is therefore not so straightforward. This topic is related to what is encountered in the derivation of the Carson bandwidth when dealing with an analog frequency modulation.

Let us now focus on the consequences on the architecture of transceivers, and more precisely of transmitters, of using such a pure phase or frequency modulating waveform. The first consequence comes from the fact that the instantaneous amplitude of the resulting RF bandpass

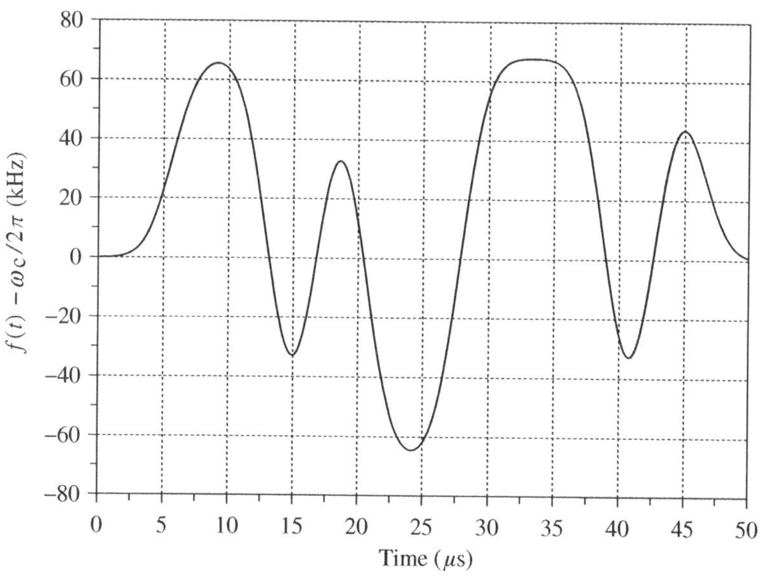

Figure 1.11 Instantaneous frequency of a GMSK modulated bandpass signal as used in the GSM standard – For the instantaneous frequency to reach the maximum theoretical value of ± 67.70833 kHz, we need at least three input data of the same polarity. In the present case, the input data sequence is $1, 1, -1, 1, -1, -1, 1, 1, 1, -1, 1$.

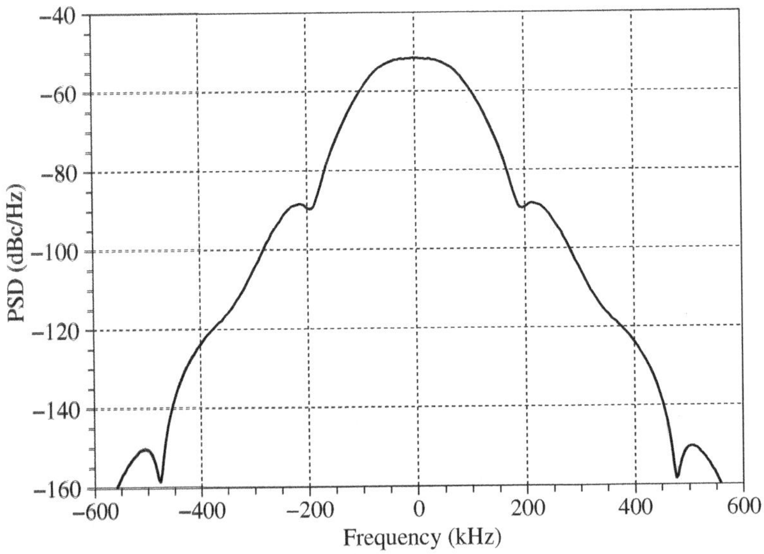

Figure 1.12 Power spectral density of a randomly modulated GMSK complex envelope as used in the GSM standard – The PSD of a randomly modulated GMSK waveform is wider than the instantaneous frequency range due to the nonlinear behavior of the cosine function and to the rate of change of the instantaneous frequency that follows the Gaussian filter response.

signal is constant. Another way to express this is to say that the PAPR of this signal, defined by equation (1.71), reduces to 1, or 0 dB. Practically speaking, this characteristic is of interest as it corresponds to RF bandpass signals that are resistant to compression, at least when considering their sideband centered around the carrier angular frequency. As discussed in Chapter 5, this allows, for instance, the use of saturated power amplifiers (PAs) that exhibit good efficiency, thus leading to an optimization of the power consumption of the transceiver implementation. Nevertheless, the counterpart of using such a constant amplitude modulating waveform is obviously a limited efficiency in terms of transmitted bit rate for a given spectrum bandwidth, as we use only one dimension out of the two available for mapping the modulation scheme.

Another interesting consequence in terms of physical implementation of using such constant instantaneous amplitude behavior can be seen when reconsidering the expression for the resulting RF bandpass signal $s(t)$ derived from the polar form of the complex modulating signal. Using equation (1.144), we get

$$\begin{aligned} s(t) &= \mathrm{Re}\left\{\tilde{s}(t)e^{j\omega_c t}\right\} \\ &= \cos\left(\omega_c t + \pi h \sum_k \alpha_k \int_{-\infty}^{t-kT} g(u)\mathrm{d}u\right). \end{aligned} \quad (1.150)$$

This expression shows that once the modulating waveform is generated, we can directly modulate the phase or frequency of the RF synthesizer that is used to generate the carrier waveform in order to directly recover the modulated RF bandpass signal. This is what allows the use of the very compact and power efficient direct phase locked loop (PLL) modulator architecture for such a transmitter, as discussed in Chapter 8. However, this does not prevent us considering a Cartesian form of the complex envelope $\tilde{s}(t)$ if required. In that case, using equation (1.144), we can write

$$p(t) = \cos\left(\pi h \sum_k \alpha_k \int_{-\infty}^{t-kT} g(u)\mathrm{d}u\right), \tag{1.151a}$$

$$q(t) = \sin\left(\pi h \sum_k \alpha_k \int_{-\infty}^{t-kT} g(u)\mathrm{d}u\right), \tag{1.151b}$$

so that we can express $s(t)$ as

$$\begin{aligned} s(t) = {} & \cos\left(\pi h \sum_k \alpha_k \int_{-\infty}^{t-kT} g(u)\mathrm{d}u\right)\cos(\omega_c t) \\ & - \sin\left(\pi h \sum_k \alpha_k \int_{-\infty}^{t-kT} g(u)\mathrm{d}u\right)\sin(\omega_c t). \end{aligned} \tag{1.152}$$

Alternatively, this last expression for the RF signal can be obtained directly from equation (1.150) using

$$\cos(\alpha + \beta) = \cos(\alpha)\cos(\beta) - \sin(\alpha)\sin(\beta). \tag{1.153}$$

Such Cartesian expansion may be required, for instance, when using a direct conversion transmitter. We then see that the real and imaginary parts of its complex envelope $\tilde{s}(t) = p(t) + jq(t)$ are sinusoidal waveforms that have a CF equal to $\sqrt{2}$, i.e. 3 dB. However, we get that even if we are dealing with a final modulated bandpass signal $s(t)$ that is insensitive to RF compression when considering the sideband centered around the carrier angular frequency as mentioned above, this property is untrue for the $p(t)$ and $q(t)$ signals that are by definition lowpass. Any degradation of those Cartesian components through compression would result in a distorted RF bandpass signal that could exhibit a non-constant amplitude waveform, as originally expected. We also mention that the same behavior would result from any imbalance between those components, as illustrated in Section 6.2.2. In any case, having such non-perfect constant amplitude for $s(t)$ may be an issue if we are faced with AM-PM conversion in the final PA stage, as is frequently the case with saturated devices used for constant amplitude RF waveforms. It would thus lead to some additional distortion of the transmitted signal on top of the former non-perfect constant amplitude behavior. Here, we can thus anticipate the discussion in Chapter 8 and highlight that the generation of $s(t)$ through a direct PLL modulation would allow us to overcome this kind of limitation linked to the implementation of transmitters based on the Cartesian processing of the complex envelope of such constant amplitude signal.

We conclude with a comment on the complexity of the equations considered so far. We understand that it would be quite unrealistic to consider an analog implementation of those expressions for generating the modulating signals. This is in fact a general statement for most of the modulations used in the field of digital communication, as illustrated through the examples discussed in the forthcoming sections. It thus follows that these waveforms are almost always generated in the digital domain and then converted to analog for further processing. This therefore leads to traditional high level system block partitions in transceivers, as discussed in Section 1.4.

1.3.2 Complex Modulation

As highlighted in the previous section, the use of constant envelope modulation schemes is driven by both the potential simplicity in the generation of the corresponding modulated RF bandpass signals and their good behavior with regard to RF compression. Nevertheless, the need for increasing data rates while preserving the frequency bandwidth that is used may lead to the consideration of complex modulations. Here, by "complex" we mean a modulation scheme that effectively uses the two dimensions allowed by the complex envelope representation. Practically speaking, it thus corresponds to a modulated RF bandpass signal that is both amplitude and phase or frequency modulated. It is thus of interest to focus on the modified 8PSK modulation used in the GSM/EDGE standard. Indeed, in addition to illustrating a complex modulation in itself, this scheme gives the opportunity to highlight the way some features considered at the definition level of the modulation may or may not make realizable some particular transmit architectures as discussed at the end of this section.

Let us begin by detailing how to generate the modulating complex envelope as a function of the input data bits according to the standard specification [7]. The first step is to map the data bits into 8PSK (eight-phase shift keying) symbols. Practically speaking, the data bits are Gray mapped by groups of three bits onto eight possible symbols uniformly spread over the unit circle when represented in the complex plane. For a given sequence of input data bits, the sequence of 8PSK symbols, $\tilde{S}_k$, is in fact defined by

$$\tilde{S}_k = e^{j\frac{2l\pi}{8}}, \tag{1.154}$$

with l given in Table 1.1 as a function of the input data bits. Up to now, this symbol mapping corresponds to a classical 8PSK modulation. But, in the GSM/EDGE standard, the symbol mapping is slightly modified so that a continuous phase rotation of $3\pi/8$ from symbol to symbol is added to the former sequence of symbols. This results in the new sequence,

$$\tilde{S}'_k = \tilde{S}_k e^{j\frac{3k\pi}{8}}. \tag{1.155}$$

This continuous phase rotation is the reason for the epithet "modified" in the term "modified 8PSK modulation". Practically speaking, it is this simple added feature that explains the properties of the resulting modulated RF bandpass signal mentioned at the beginning of the section and further discussed at the end of the section. We can see that the final modulating waveform is obtained as the filtered version of this sequence of data symbols in order to shape the spectrum of the resulting modulated signal. Denoting this pulse shaping filter by C_0 and

Table 1.1 Symbol mapping parameter l for the GSM/EDGE modified 8PSK modulation.

Modulating bits $d_{3k}, d_{3k+1}, d_{3k+2}$	Symbol parameter l
(1,1,1)	0
(0,1,1)	1
(0,1,0)	2
(0,0,0)	3
(0,0,1)	4
(1,0,1)	5
(1,0,0)	6
(1,1,0)	7

its impulse response by $C_0(t)$, the resulting complex envelope $\tilde{s}(t) = p(t) + \mathrm{j}q(t)$ can thus be written as

$$\tilde{s}(t) = \sum_k \tilde{S}'_k C_0(t - kT + 2T), \tag{1.156}$$

so that we get

$$p(t) = \sum_k \mathrm{Re}\{\tilde{S}'_k\} C_0(t - kT + 2T), \tag{1.157a}$$

$$q(t) = \sum_k \mathrm{Im}\{\tilde{S}'_k\} C_0(t - kT + 2T). \tag{1.157b}$$

We observe that this is in fact the spectrum of both the real and the imaginary parts of the modulating complex envelope $\tilde{s}(t)$ that the C_0 filter shapes. But, referring to the discussion in "Power spectral density" (Section 1.1.3), we get that under common assumptions, those real and imaginary parts have the same spectral shape as the overall complex envelope, and thus as the final modulated bandpass signal $s(t)$. We can thus understand the importance of defining such complex modulating waveforms in their Cartesian form in order to directly control the characteristics of the transmitted output signal in terms of spectral shape. This explains why most practical complex modulations are defined in such a Cartesian way. In our present case the final modulated RF bandpass signal $s(t)$ resulting from the upconversion of this complex envelope around the carrier angular frequency ω_c can therefore be expressed as

$$s(t) = \mathrm{Re}\{\tilde{s}(t)\mathrm{e}^{\mathrm{j}\omega_\mathrm{c}t}\}, \tag{1.158}$$

or, based on the Cartesian representation of $\tilde{s}(t)$, as

$$s(t) = \sum_k \mathrm{Re}\{\tilde{S}'_k\} C_0(t - kT + 2T)\cos(\omega_\mathrm{c}t) - \sum_k \mathrm{Im}\{\tilde{S}'_k\} C_0(t - kT + 2T)\sin(\omega_\mathrm{c}t). \tag{1.159}$$

In the EDGE standard, the impulse response of this C_0 filter is defined as

$$C_0(t) = \begin{cases} \prod_{l=0}^{3} S(t+lT) & \text{when } 0 \le t \le 5T, \\ 0 & \text{otherwise,} \end{cases} \tag{1.160}$$

where T stands for the symbol period. In the EDGE standard, this period is the same as for the GMSK modulation of the GSM standard, i.e. $48/13 = 3.69\ \mu$s. In this expression, $S(t)$ is defined as[5]

$$S(t) = \begin{cases} \sin\left(\pi \int_0^t g(u)\mathrm{d}u\right) & \text{when } 0 \le t \le 4T, \\ \sin\left(\frac{\pi}{2} - \pi \int_0^{t-4T} g(u)\mathrm{d}u\right) & \text{when } 4T < t \le 8T, \\ 0 & \text{otherwise,} \end{cases} \tag{1.161}$$

with $g(t)$ given by

$$g(t) = \frac{1}{2T}\left\{\mathrm{Q}\left(2\pi 0.3\frac{t-5T/2}{T\sqrt{\ln(2)}}\right) - \mathrm{Q}\left(2\pi 0.3\frac{t-3T/2}{T\sqrt{\ln(2)}}\right)\right\}. \tag{1.162}$$

In this expression, Q(.) stands for

$$\mathrm{Q}(t) = \frac{1}{\sqrt{2\pi}}\int_t^{+\infty} \mathrm{e}^{-\frac{x^2}{2}}\mathrm{d}x. \tag{1.163}$$

Alternatively, we can make the link between this function and the Gaussian filter defined in the previous section. For that purpose, we can express $g(t)$ using both the complementary error function, Erfc(.) defined by equation (1.149), and the δ parameter defined by equation (1.147). We obtain

$$g(t) = \frac{1}{4T}\left\{\mathrm{Erfc}\left[\frac{1}{\sqrt{2\delta}}\left(\frac{t}{T}-\frac{5}{2}\right)\right] - \mathrm{Erfc}\left[\frac{1}{\sqrt{2\delta}}\left(\frac{t}{T}-\frac{3}{2}\right)\right]\right\}. \tag{1.164}$$

[5] In equation (1.161), the summation of the Gaussian filter transfer function $g(t)$ is done over a finite range as defined in the standard, thus resulting in $S(t)$ not being continuous for $t = 4T$ [7]. With this definition, C_0 is not exactly symmetric, whereas the main pulse in the Laurent series expansion of the GMSK modulation should be, as the original Gaussian filter is.

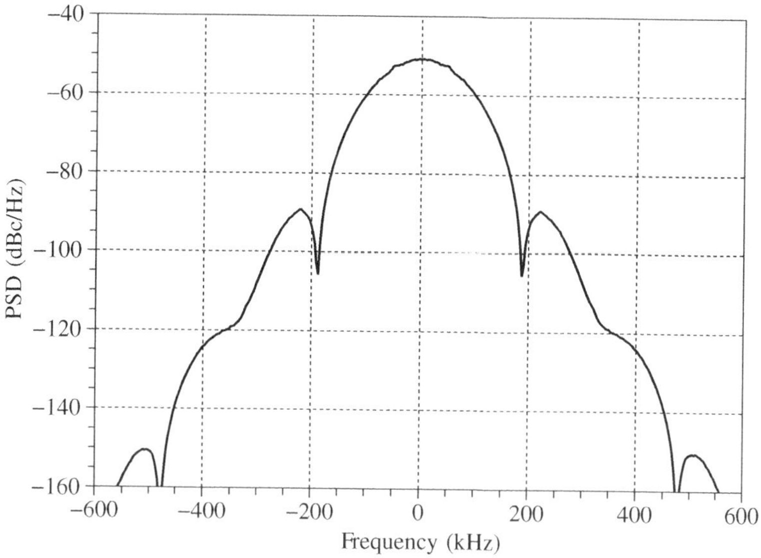

Figure 1.13 Power spectral density of a randomly modulated GSM/EDGE complex envelope – The PSD of a randomly modulated modified 8PSK waveform, as used in the GSM/EDGE standard, is comparable to that of a GMSK modulated waveform used in the GSM standard, as shown in Figure 1.12.

We thus see that $g(t)$ is nothing more than the Gaussian filter of the GMSK modulation described in the previous section, but translated around $t = 2.5T$ and with a different normalization coefficient. The reason for the correspondence is that the C_0 filter used in the present modulation scheme is nothing more than the main pulse in the Laurent series decomposition of the GMSK scheme [8]. It thus follows that the modified 8PSK scheme shaped through the use of this filter results in almost the same spectrum as the original GMSK one, as can be seen by comparing Figures 1.13 and 1.12.

Let us now focus on the statistical properties of the resulting modulating waveform. We begin with the continuous phase rotation of $3\pi/8$ added from symbol to symbol. The direct impact of this feature can be seen in Figure 1.14 where the modulating complex envelope trajectories are plotted in the two cases where the continuous phase rotation is either implemented or not. This figure clearly shows that this simple continuous rotation, easily implemented on the transmit side and compensated on the receive side, allows us to avoid any zero crossing of the modulation trajectory in the complex plane. As summarized in Figure 1.15, we get that this behavior results in interesting statistical properties, in particular the following:

(i) The dynamic range (DR) of the complex envelope magnitude $\rho(t) = |\tilde{s}(t)|$, and thus of the PAPR of the corresponding modulated bandpass signal, is minimized. This parameter can be evaluated to be 3.2 dB when this continuous rotation is used and 3.5 dB when not. This reduction of 0.3 dB can thus be directly transposed in terms of gain on the back-off regarding the saturated power to be used in order to avoid compression. Here we anticipate Section 8.1.7 where it is mentioned that this reduction can be directly transposed in terms

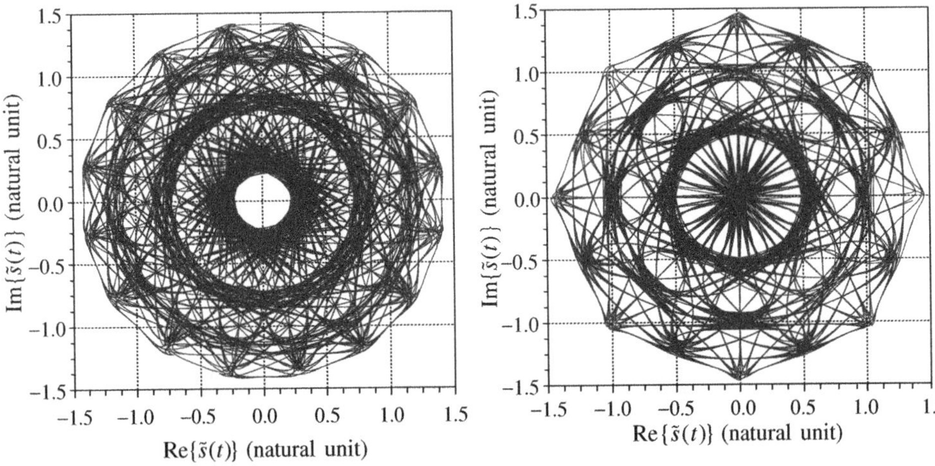

Figure 1.14 Trajectories of a randomly modulated GSM/EDGE complex envelope with and without continuous phase rotation – Compared to the case where no continuous rotation is applied (right), the trajectory of the modified 8PSK as defined in [7] exhibits no zero crossing (left).

of gain on the power consumption of the solution, at least when considering a statically polarized linear amplifier. At first glance, 0.3 dB may seem a small improvement, but we have to keep in mind that in a typical cellular system, the PA consumption can be up to hundreds of milliamperes for mobile equipment. A few tenths of a decibel on this power consumption is therefore not negligible.

(ii) A trajectory that does not vanish leads to other advantages that make simpler (and even make possible) the use of architectures such as the polar transmitter. Indeed, as discussed in Section 8.1.6, having zero crossings in the trajectory leads to important spikes in the instantaneous frequency of the resulting modulated RF bandpass signal. This is obviously due to the π shifts in the argument of the complex envelope at those instants that lead to a derivative theoretically infinite. Such important instantaneous frequency variations require, on the one hand, synthesizers with a high enough bandwidth, and, on the other hand, a very precise timing alignment between the instantaneous amplitude and the instantaneous frequency to achieve a good reconstructed modulated RF bandpass signal. In addition, having such zero crossings leads to an important DR of the modulus of the complex envelope that can theoretically reach infinity when expressed in decibels. Having an analog device that is linear over such a wide DR in order to be able to perform the amplitude modulation of an RF carrier without distortion is almost impossible. However, those problems are not too severe for the modified 8PSK modulation of the GSM/EDGE standard. We get, for instance, that the DR of the instantaneous amplitude remains within 17 dB. We thus see that a simple trick like this continuous phase rotation from symbol to symbol can make the difference between being and not being able to use a given transmit architecture.

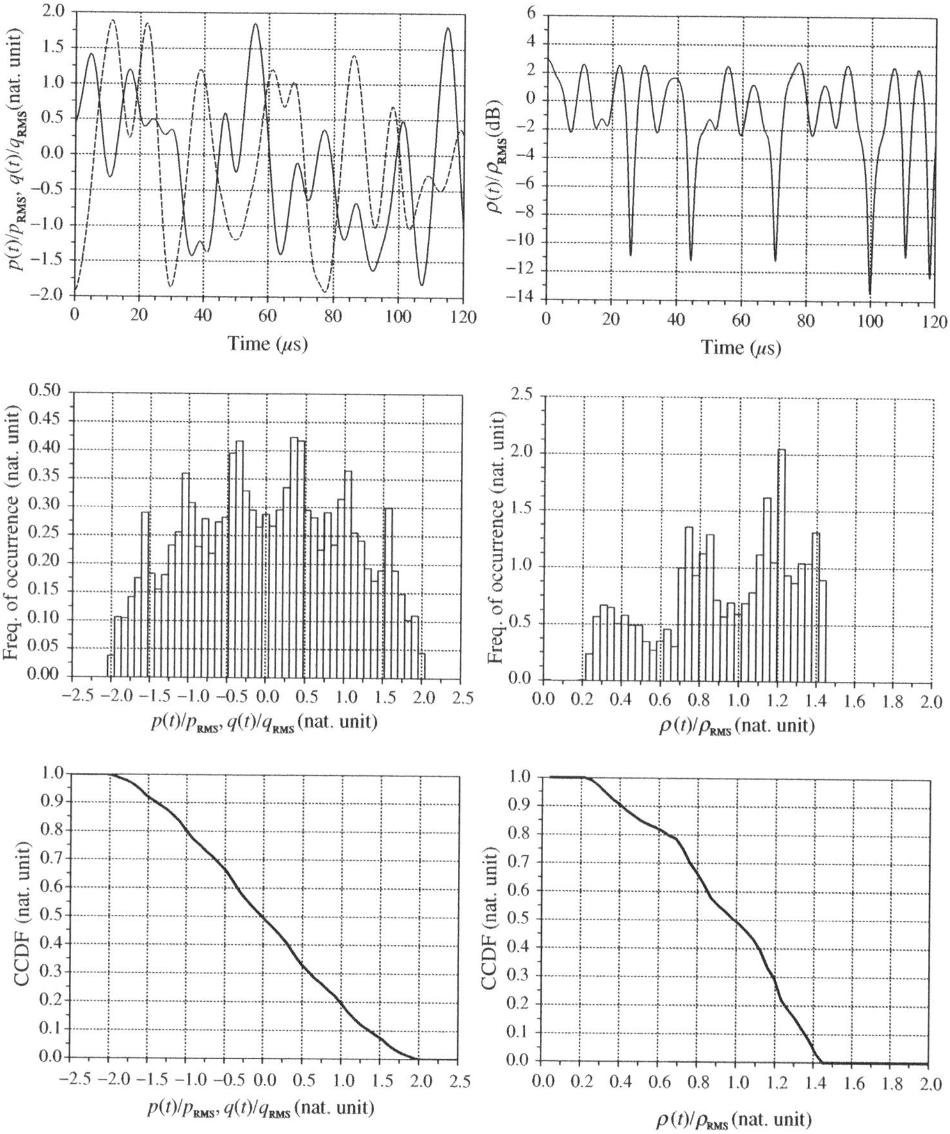

Figure 1.15 Statistical characteristics of a randomly modulated GSM/EDGE complex envelope – The real and imaginary parts $p(t)$ (top left, solid) and $q(t)$ (top left, dashed) of the GSM/EDGE complex envelope have a distribution that is bounded by ±2 times their RMS value (middle left). Therefore their CCDF reaches 0 for $p(t) = 2p_{\text{RMS}}$ or $q(t) = 2q_{\text{RMS}}$ (bottom left). The CF of those waveforms is thus equal to 2, i.e. 6 dB. The corresponding modulus $\rho(t)$ (top right) has a bounded distribution relative to its RMS value (middle right). This upper bound of 1.4, or 3.2 dB, corresponds to the modulation PAPR as recovered on the CCDF of this parameter (bottom right).

To conclude, as can be seen in Figure 1.15, we highlight the fact that the CF of the $p(t)$ and $q(t)$ waveforms is around 6 dB, i.e. about 3 dB higher than the modulating waveform PAPR. This result is consistent with equation (1.84) and the discussion presented in "Peak to average power ratio and crest factor" (Section 1.1.3).

1.3.3 Wideband Modulation

To conclude this overview of digital modulation examples, let us now focus on wideband modulations. The problem is that the criteria used to designate a modulating waveform as wideband may depend on the point of view adopted. For instance, taking the RF implementation point of view, we may be tempted to label as wideband a modulation that leads to a modulated bandpass signal that spreads over a non-negligible frequency band with regard to the carrier frequency. Unfortunately, most of the schemes classically classified as wideband still lead to modulated signals that remain narrowband with regard to the carrier frequency in respect to RF implementation issues. Alternatively, taking the wireless mobile signal processing point of view, we may be tempted to label as wideband a modulating waveform whose spectrum is wider than the coherence bandwidth of the propagation channel according to the concept introduced in Section 2.3, the reason being that this property allows some interesting structures for the channel estimation and equalization. However, with this definition we see that even the modulating schemes reviewed in the previous sections can be considered as wideband as long as the symbol data rate is chosen high enough. In any case, this situation does not really occur in practice as the schemes leading to more efficient channel equalization when there is the possibility of using such high frequency bandwidth rely on different modulation strategies than those former ones. Moreover, in the wireless network perspective it is not necessarily of interest to have the high data rate corresponding to such high bandwidth statically allocated to a single end user. On the other hand, it may be of interest to consider the possibility of multiplexing different users over the same wideband modulating waveform in order to have the flexibility for the data rate to be dynamically allocated to any of them, as introduced in Section 3.1. In both cases, it classically results in two families of modulating waveforms that we can classify as wideband in what follows and that we can examine as such, namely those encountered in CDMA systems and in OFDM systems. As will be seen, this results in waveforms that exhibit interesting statistical properties compared to those seen up to now.

Code Division Multiple Access

Let us begin our review of wideband modulations by examining the behavior of waveforms encountered in CDMA systems. In such systems the multiplexing of the users belonging to the same network cell, as introduced in Section 3.1, or at least of the different data and logical channels of a given user, is done in the same frequency band and same time slot through the use of orthogonal codes. That said, it is of interest to remark that those codes need to have a high frequency content, at least comparable to the overall data rate of the system, in order to allow for this multiplexing. As a side effect, we then get that, from a network perspective, the modulating waveform that needs to be processed by a given receiver (RX) may potentially carry information for all the users belonging to the same cell at the same time, not just its own.[6] This deep difference with respect to the modulations reviewed in previous sections explains the fact that the modulating signals to be processed by the transceivers belonging to such a network have in most cases a much higher bandwidth than their own effective data rate, thus highlighting the label "wideband" for characterizing such modulating waveforms.

Let us consider the WCDMA standard [9, 10] as an example. In this standard, the orthogonal codes used for the multiplexing of the channels are the Walsh–Hadamard codes that can be expressed in a general form as

$$w_k(t) = \sum_{n=0}^{L-1} a_k(n)\delta(t - nT_c). \tag{1.165}$$

In this expression, T_c stands for the duration of the elementary part a_k of the code sequence, usually referred to as the chip duration. Given that this chip duration is expected to be the shortest time base of the toggling of the modulating waveform, we can expect an overall bandwidth for this signal in the region of $1/T_c$. We thus get that the choice for this chip duration is driven by the targeted bandwidth of the final modulating waveform. Then, we get that the a_k values are equal to ± 1 and are such that the scalar product between two different codes is zero. Each data symbol $\tilde{S}$ to be transmitted – which can be a QPSK symbol, a quadrature amplitude modulation (QAM) symbol, or whatever is considered in the standard – is then associated with an entire code signal $w_k(t)$. As a result, given a symbol duration equal to T, we then need to select the length of the code L, also defined as the spreading factor, such that $T = LT_c$. A sequence of data symbols $\tilde{S}_k(m)$ associated with the kth code $w_k(t)$ therefore leads to a sequence of modulating chips $\tilde{c}_k(t)$ given by

$$\tilde{c}_k(t) = \sum_{m} \tilde{S}_k(m) w_k(t - mT). \tag{1.166}$$

Now, supposing that we have K different users, or at least K different channels, to be multiplexed through the use of K different codes, we get that the overall sequence of modulating chips $\tilde{c}_K(t)$ is nothing more than the sum of the sequence associated with each code, i.e. $\sum_{k=1}^{K} \tilde{c}_k(t)$. Assuming for simplicity that the spreading factor L is the same for any of them, i.e. that we have the same data symbol rate for each user or channel, we can then use equations (1.165) and (1.166) to express $\tilde{c}_K(t)$ as

$$\begin{aligned} \tilde{c}_K(t) &= \sum_{k=1}^{K} \beta_k \sum_{m} \tilde{S}_k(m) w_k(t - mT) \\ &= \sum_{k=1}^{K} \beta_k \sum_{m} \tilde{S}_k(m) \sum_{n=0}^{L-1} a_k(n)\delta(t - nT_c - mT). \end{aligned} \tag{1.167}$$

[6] We leave the interested reader to think about the burden, for instance in terms of power consumption, of having a receiver that needs to demodulate and process a wideband signal that carries the information of all the users belonging to the same network cell when its own data represents only a fraction of it.

Here, β_k stands for the relative power of the kth code. This parameter depends on the data rate used with the associated code.

Practically speaking, if the characteristics of the Walsh–Hadamard codes are good in terms of orthogonality, we get that this is obviously not true in terms of autocorrelation. This can be a problem in terms of both the initial synchronization to the network, and the PSD of the resulting modulating waveform. This explains why additional scrambling codes, $\tilde{sc}$, are used in practice for that purpose. Even if not true in the considered standard, we can assume for the sake of simplicity that their length is the same as that of the channelization codes. This then leads to a modulating chip sequence $\tilde{c}(t)$ that finally takes the form

$$\tilde{c}(t) = \sum_{m} \sum_{n=0}^{L-1} \tilde{sc}(n) \sum_{k=1}^{K} \beta_k \tilde{S}_k(m) a_k(n) \delta(t - nT_c - mT). \tag{1.168}$$

We observe that, in contrast to what happens for the Walsh–Hadamard codes, the $\tilde{sc}(n)$ elements of those additional scrambling codes are complex numbers. They thus scramble together the data from the real and imaginary parts of the constellation recovered after channelization. It is then of interest to mention that in the WCDMA standard, at least on the uplink side, this scrambling sequence takes a particular form in order to limit the modulated signal PAPR. This particular scheme, referred to as hybrid phase shift keying (HPSK), avoids, or at least limits, the zero crossings of the modulating waveform trajectory in the complex plane. It can thus be compared to the mechanism discussed in the previous section for minimizing the same parameter in the modified 8PSK scheme of the GSM/EDGE standard. We need not give too much detail in the present case, as the modified 8PSK scheme is already a good example of such optimization. Nevertheless, it highlights that the statistical parameters of the modulating waveform are of particular importance for the efficiency of the implementation of the RF/analog part of a transmitter at least. We can then understand why we need to consider the optimization of such parameters at the early stages of the derivation of a wireless standard.

Having derived the sequence of modulating chips $\tilde{c}(t)$ with both good orthogonality properties to allow recovery of the different data channels, and good autocorrelation properties to allow for white behavior, it remains to apply a pulse shaping filter in order to shape correctly the spectrum of the modulating waveform. In the standard being considered, this filter is a root raised cosine (RRC) filter defined by its impulse response,

$$h_{\mathrm{RRC}}(t) = \frac{\sin\left(\pi \frac{t}{T_c}(1-\alpha)\right) + 4\alpha \frac{t}{T_c} \cos\left(\pi \frac{t}{T_c}(1+\alpha)\right)}{\pi \frac{t}{T_c}\left(1 - \left(4\alpha \frac{t}{T_c}\right)^2\right)}, \tag{1.169}$$

with

$$h_{\mathrm{RRC}}(0) = (1-\alpha) + \frac{4\alpha}{\pi} \tag{1.170}$$

and

$$h_{\mathrm{RRC}}\left(\pm\frac{T_{\mathrm{c}}}{4\alpha}\right)=\frac{1}{2}\left[\frac{4\alpha}{\pi}\sin\left(\frac{1-\alpha}{\alpha}\frac{4}{\pi}\right)\right.$$
$$\left.+(1+\alpha)\sin\left(\frac{1+\alpha}{\alpha}\frac{4}{\pi}\right)-(1-\alpha)\cos\left(\frac{1-\alpha}{\alpha}\frac{4}{\pi}\right)\right]. \tag{1.171}$$

Here, α is the roll-off factor of the filter, set to 0.22 in the present case. Now the complex envelope of the modulating signal, $\tilde{s}(t)$, that results from the RRC filtering of equation (1.168) can be written as

$$\tilde{s}(t)=\sum_{m}\sum_{n=0}^{L-1}\tilde{sc}(n)\sum_{k=1}^{K}\beta_k\tilde{S}_k(m)a_k(n)h_{\mathrm{RRC}}(t-nT_{\mathrm{c}}-mT). \tag{1.172}$$

The corresponding real and imaginary parts of $\tilde{s}(t)=p(t)+\mathrm{j}q(t)$, can therefore be expressed as

$$p(t)=\sum_{m}\sum_{n=0}^{L-1}\sum_{k=1}^{K}\beta_k a_k(n)\mathrm{Re}\{\tilde{S}_k(m)\tilde{sc}(n)\}h_{\mathrm{RRC}}(t-nT_{\mathrm{c}}-mT), \tag{1.173a}$$

$$q(t)=\sum_{m}\sum_{n=0}^{L-1}\sum_{k=1}^{K}\beta_k a_k(n)\mathrm{Im}\{\tilde{S}_k(m)\tilde{sc}(n)\}h_{\mathrm{RRC}}(t-nT_{\mathrm{c}}-mT). \tag{1.173b}$$

This results in the spectral shape for $\tilde{s}(t)$ shown in Figure 1.16. Alternatively, it is of interest to note that the impulse response defined by equation (1.169) has non-zero values at instants $t_s=sT_{\mathrm{c}}$. Hence, ISI exists in the resulting modulating waveform. Nevertheless, this filter being a Nyquist filter, the self-convolution of its transfer function leads to an ISI-free signal, as illustrated in Section 4.4. This explains why this filter, used as a pulse shaping filter on the transmit side, is also implemented on the receive side of transceivers belonging to this system.

Having derived the structure of $\tilde{s}(t)$, we can now focus on the statistical properties of this modulating waveform. From equation (1.173), we can understand that the statistical characteristics of the modulating waveform obviously depend on the number of data codes that are summed together. Indeed, assuming that the data symbols are statistically independent and evenly distributed in the complex plane, we can expect that the more codes are summed, the more the distributions of the real and imaginary parts of the complex envelope $\tilde{s}(t)=p(t)+\mathrm{j}q(t)$ tend toward a centered Gaussian distribution thanks to the central limit theorem. In that case, recalling the discussion in Section 1.2, we can expect that $\rho(t)=|\tilde{s}(t)|$ tends toward a Rayleigh distribution. This behavior is indeed confirmed by comparing the simulation results shown in Figures 1.17 and 1.18 that correspond to uplink WCDMA modulating waveforms based on either one or four data codes. We see that when only one code is used, the CF of $p(t)$ or $q(t)$ is around 2, i.e. 6 dB, whereas the PAPR of the modulation is found to be 1.45, i.e. 3.2 dB. But when four codes are used, the CF on $p(t)$ or $q(t)$ attains a value of 3, i.e. 9.6 dB, whereas the PAPR is simulated to reach 2.2, i.e. 6.8 dB. In addition, we observe that even if the use of the HPSK scheme allows us to reduce the modulation PAPR, we still have

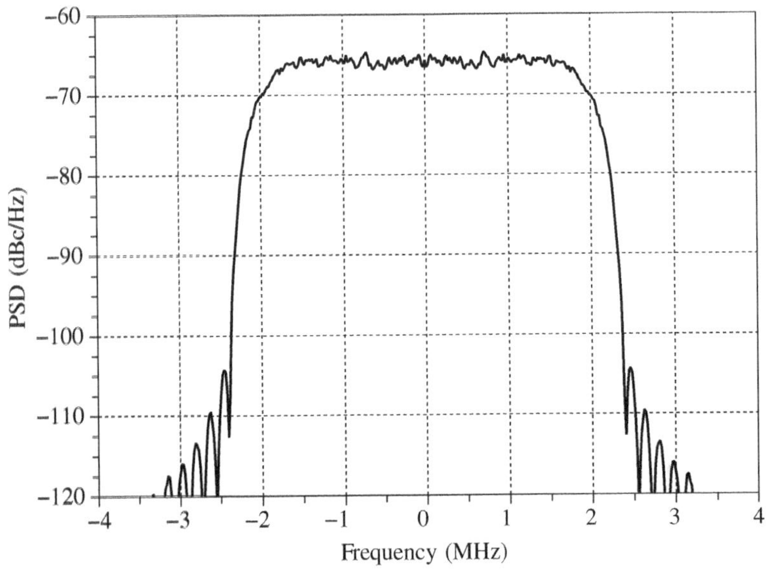

Figure 1.16 Power spectral density of a randomly modulated WCDMA complex envelope – Due to the properties of the scrambling codes defined in the standard, the spectrum of a WCDMA signal remains white within the bandwidth of the RRC filter. The spectral side lobe heights then depend mainly on the RRC filter implementation in terms of the impulse response length.

the minimum value of the modulus $\rho(t)$ of the complex envelope that can go close to 0 when the number of data codes increases. In contrast to what happens with the modified 8PSK modulation discussed in the previous section, we can expect in the present case to be faced with a huge DR for the instantaneous amplitude of the resulting modulated bandpass signal as well as wideband instantaneous frequency variations. We may thus conclude at first glance that this kind of waveform is not well suited to polar transmitter architectures, as discussed in Section 8.1.6.

OFDM

Let us now turn our attention to the modulating waveforms that can be encountered in wireless systems based on the OFDM technique. To understand their statistical properties, we first need to detail the construction of such waveforms. Practically speaking, the idea behind the OFDM is to use different subcarriers, which can be seen as orthogonal in a given sense, to multiplex the information, rather than orthogonal codes as in WCDMA systems. More precisely, the modulating waveform is constructed as the succession of slices of time domain waveforms of duration T_s, each one resulting from the superposition of $2N + 1$ subcarriers of the form $s_k(t) = \cos(2\pi f_k t)$. Here, the number of subcarriers is chosen as odd only for the sake of simplicity in our formulations. Each of those subcarriers is then in turn modulated – either in amplitude, phase, or both – in order to carry one data symbol $\tilde{S}_k$. These data symbols can result from the symbol mapping of any classical modulation scheme as a QPSK, QAM or

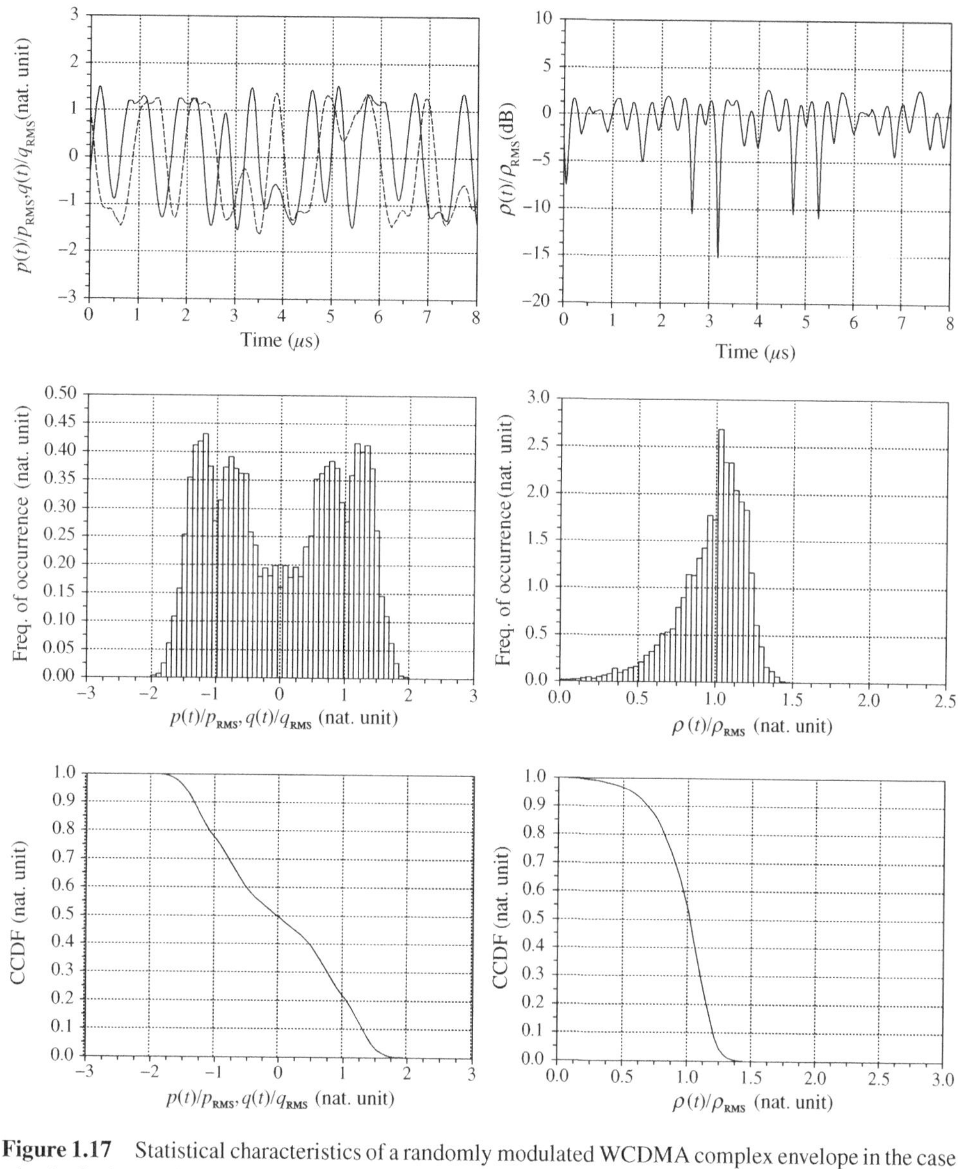

Figure 1.17 Statistical characteristics of a randomly modulated WCDMA complex envelope in the case of a single data code – The real and imaginary parts $p(t)$ (top left, solid) and $q(t)$ (top left, dashed) of the WCDMA uplink complex envelope when one data code is used have a distribution that is bounded by ± 2 times their RMS value (middle left). Therefore their CCDF reaches 0 for $p(t) = 2p_{\text{RMS}}$ or $q(t) = 2q_{\text{RMS}}$ (bottom left). The CF of those waveforms is thus equal to 2, i.e. 6 dB. The corresponding modulus $\rho(t)$ (top right) can reach 1.45 or 3.2 dB above its RMS value (middle right). This upper bound for $\rho(t)$ corresponds to the modulation PAPR as recovered on the CCDF of this parameter (bottom right).

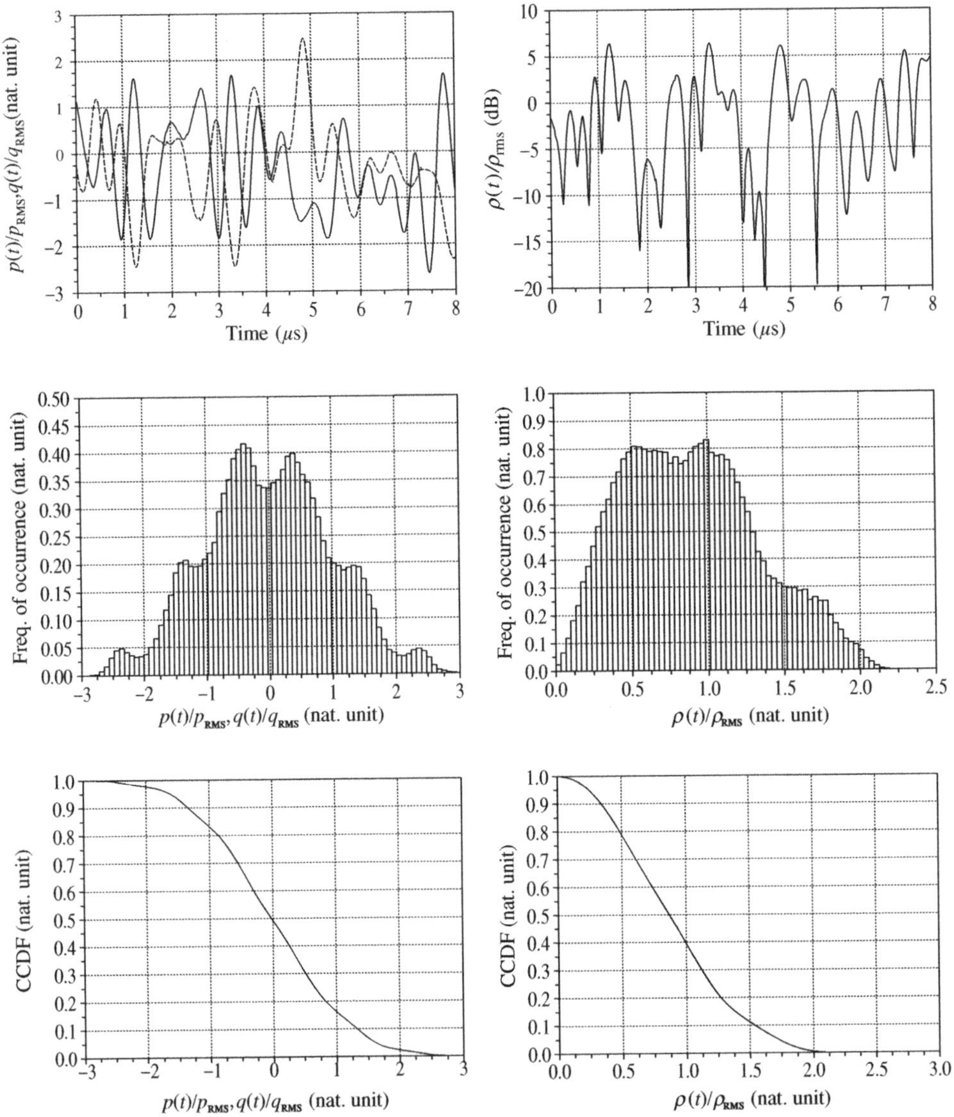

Figure 1.18 Statistical characteristics of a randomly modulated WCDMA complex envelope in the case of four data codes – The real and imaginary parts $p\,(t)$ (top left, solid) and $q\,(t)$ (top left, dashed) of the WCDMA uplink complex envelope when four data codes are used have a distribution that is bounded by ± 3 times their RMS value (middle left). Therefore their CCDF reaches 0 for $p\,(t) = 3p_{\text{RMS}}$ or $q\,(t) = 3q_{\text{RMS}}$ (bottom left). The CF of those waveforms is thus equal to 3, i.e. 9.6 dB. The corresponding modulus $\rho(t)$ (top right) can reach 2.2 or 6.8 dB above its RMS value (middle right). This upper bound for $\rho(t)$ corresponds to the modulation PAPR as recovered on the CCDF of this parameter (bottom right).

whatever. Such a slice of modulating waveform is then called an OFDM symbol, and thus T_s represents the OFDM symbol duration. In addition, the subcarriers $s_k(t)$ are chosen so that their frequencies f_c are equally distributed on each side of the carrier. With our present assumptions, these frequencies are thus of the form $f_k = f_c + k\delta f$ with $k = -N, \ldots, N$. This periodicity in the frequency domain in fact allows us to achieve the orthogonality we were referring to, as can be understood when examining the expression for the complex envelope $\tilde{s}_l(t)$ of the modulated bandpass signal carrying the lth OFDM symbol. Assuming that we are dealing with complex envelopes defined as centered around the carrier frequency, we can refer to the definitions given in Section 1.1.2 to see that the corresponding complex envelope of the kth subcarrier $s_k(t) = \cos(2\pi f_k t)$ can be directly written as $e^{j2\pi k\delta f t}$. Thus, given that all the complex envelopes we are dealing with are defined as centered around the same frequency, we can immediately write $\tilde{s}_l(t)$ as the sum of the complex envelopes of the N subcarriers,

$$\tilde{s}_l(t) = [U((l-1)T_s) - U(lT_s)] \sum_{k=-N}^{N} \tilde{S}_{k,l}\, e^{j2\pi k\delta f t}. \tag{1.174}$$

Here, U(.) stands for the Heaviside step function and $\tilde{S}_{k,l}$ represents the data symbols carried by the kth subcarrier during the lth OFDM symbol. We then observe that due to the periodicity in the subcarrier spacing, the transposition of this expression in the sampled time domain leads to a discrete Fourier transform that links the values of the data symbols carried by each subcarrier to the values of the time domain samples of the corresponding OFDM symbol. In that case we get that the OFDM symbol duration T_s is inversely proportional to the subcarrier frequency spacing δf, thus leading to the orthogonality between vectors composed of the sample of each subcarrier. Finally, the complex envelope $\tilde{s}(t)$ of the overall modulated signal carrying the succession of OFDM symbols can be written as

$$\tilde{s}(t) = \sum_{l} \tilde{s}_l(t). \tag{1.175}$$

Having derived the structure of $\tilde{s}(t)$ for a general OFDM signal, we can now focus on its characteristics. We will use a concrete example by way of illustration. We examine the 10 MHz bandwidth long-term evolution (LTE) downlink signal that corresponds in the configuration considered to an OFDM modulation composed of 600 useful subcarriers separated by a frequency offset of 15 kHz. This results in a complex envelope signal whose spectrum is spread over the band [−4.5, 4.5] MHz. Obviously, the first consequence of summing subcarriers that are evenly distributed in the frequency domain and carry the same power on average is that the PSD of the resulting complex envelope is expected to be almost flat. This is indeed confirmed by inspecting the spectrum shown in Figure 1.19. Nevertheless, we observe that in the present case, even if not visible in the figure, a non-negligible amount of power necessarily leaks in the adjacent channels through the secondary lobes of the sinc function resulting from the lack of time domain windowing of the OFDM symbols. But due to the orthogonality resulting from the discrete Fourier transform approach, this leakage can be made visible only through an oversampling of the corresponding waveforms, which is obviously not done here.

Let us now focus on the statistical characteristics of such modulating waveform. Practically speaking, in most of the standards based on an OFDM technique, the number of subcarriers

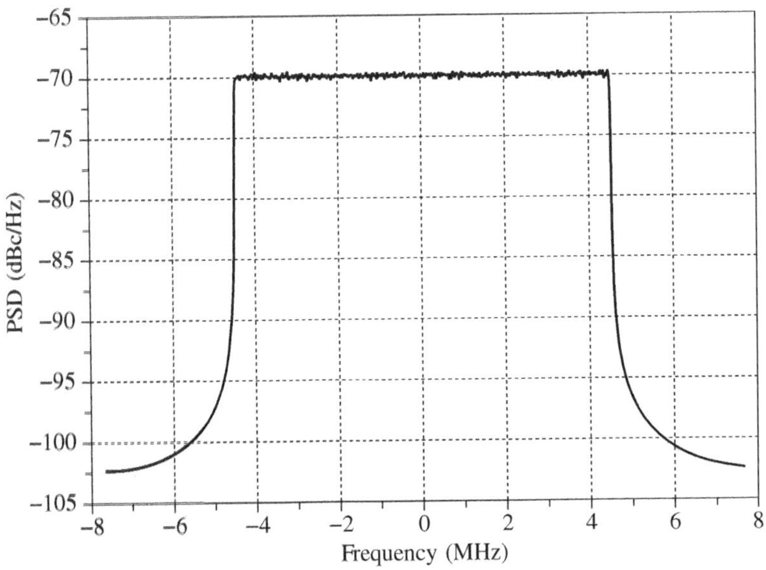

Figure 1.19 Power spectral density of a randomly modulated OFDM complex envelope – An LTE downlink waveform using 600 useful subcarriers separated by a frequency offset of 15 kHz exhibits a PSD that spreads over the band [−4.5, 4.5] MHz.

involved in the generation of the modulating waveforms is important, as illustrated by our LTE example. Thus, referring to equation (1.174), we see that the real and imaginary parts of the complex envelope $\tilde{s}(t) = p(t) + jq(t)$ result from the summation of a great number of data symbols. Thus, as long as they can be considered as evenly distributed in the complex plane and statistically independent, both $p(t)$ and $q(t)$ tend to have a centered Gaussian distribution according to the central limit theorem. In that case, from the discussion in Section 1.2.1, we can also expect that the modulus $\rho(t)$ of $\tilde{s}(t)$ tends to have a Rayleigh distribution. Those behaviors are confirmed by the simulations shown in Figure 1.20. There is indeed an obvious fit between the histograms of those waveforms and the theoretical Gaussian and Rayleigh distributions displayed in the same graphs. Nevertheless, even if there is such a good agreement, the fact is that the summation of the subcarriers remains finite in the definition of $\tilde{s}(t)$. This means that the maximum values of the modulating waveforms remain well defined and cannot reach infinity. Practically speaking, those values can be used to derive the CF and PAPR of the modulating waveforms as defined in "Peak to average power ratio and crest factor" (Section 1.1.3). As an example, we can reconsider our former 10 MHz bandwidth LTE downlink signal. In that case, time domain simulations lead to both a CF for $p(t)$ and $q(t)$ of around 5, i.e. 14 dB, and a PAPR of around 3.6, i.e. 11.1 dB. We observe that these numbers are larger than what could be expected when considering the complementary cumulative distribution function (CCDF) curves shown in Figure 1.20. The reason for this behavior is that the exact maximum values reached by the waveforms during such a time domain simulation correspond to a very low probability of occurrence. This justifies having a statistical definition for the CF and the PAPR that reflects this behavior. Obviously, the exact thresholds to be considered

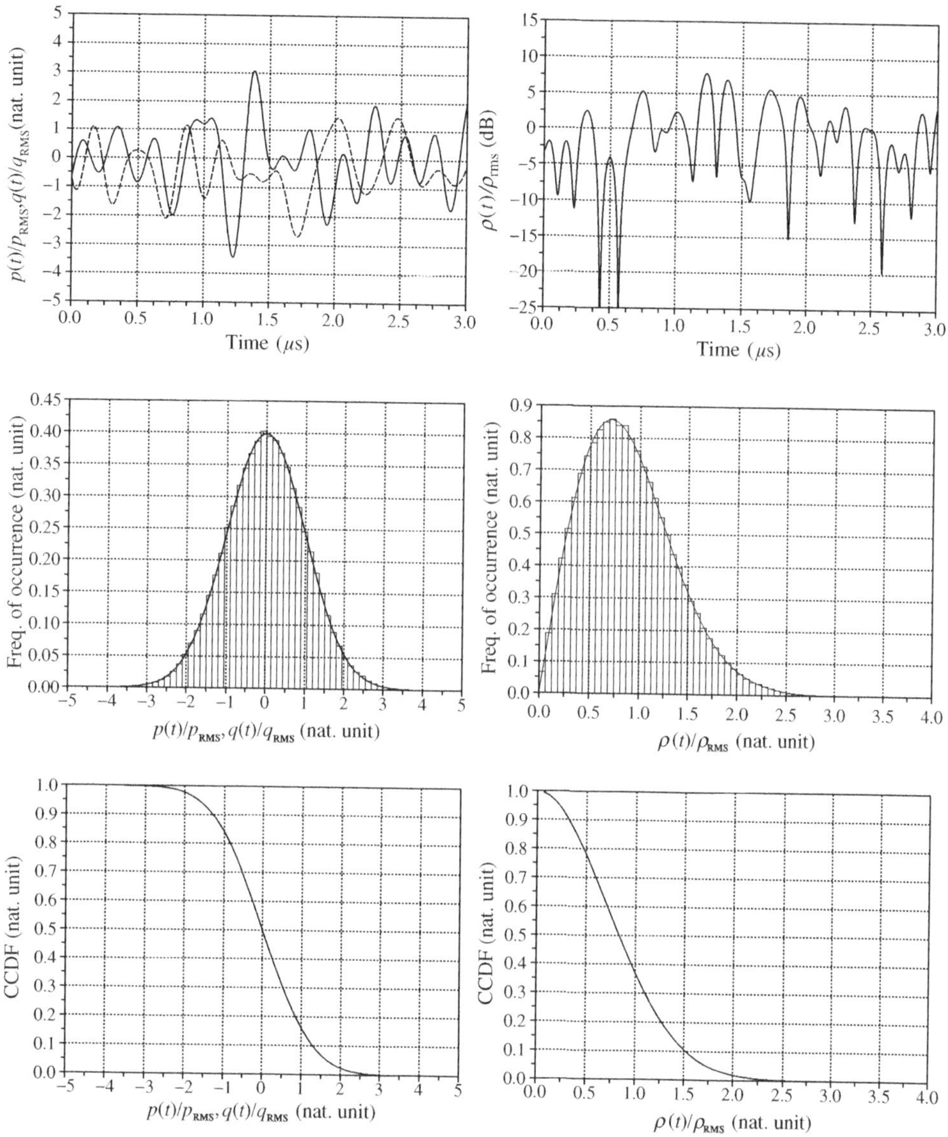

Figure 1.20 Statistical characteristics of a randomly modulated OFDM complex envelope – The real and imaginary parts $p(t)$ (top left, solid) and $q(t)$ (top left, dashed) of an OFDM complex envelope tend to have a Gaussian distribution (middle left), but with a deterministic maximum value due to the finite summation of the subcarriers. Using the statistical definition for the maximum values, their CCDF reaches 0.001 for $p(t) = 3.1p_{RMS}$ or $q(t) = 3.1q_{RMS}$ (bottom left). The CF of those waveforms thus reaches 9.8 dB. The modulus $\rho(t)$ (top right) follows a Rayleigh distribution and can reach 2.2 times its RMS value (middle right). This corresponds to the modulation PAPR as recovered on the CCDF of $\rho(t)$ (bottom right).

to decide whether or not a frequency of occurrence has negligible impact on the average performance of a line-up should be refined. It is realistic to define the peak value of those waveforms as the value for which their CCDF reaches 0.001. In other words, we can define the peak value of $p(t)$ or $q(t)$ as the upper bound below which those waveforms are 99.9% of the time. Applying this definition to $p(t)$, for instance, we get that its peak value p_{pk} is defined from equation (1.75) such that

$$F_{\text{c}}(p_{\text{pk}}) = 0.001. \tag{1.176}$$

Assuming a Gaussian distribution for this parameter, we get that its CCDF is given by

$$F_{\text{c}}(p_{\text{pk}}) = \frac{1}{2}\text{Erfc}\left(\frac{1}{\sqrt{2}}\frac{p_{\text{pk}}}{p_{\text{RMS}}}\right), \tag{1.177}$$

with Erfc(.) the complementary error function defined by equation (1.149). It then follows that the CF of $p(t)$, defined as the ratio $p_{\text{pk}}/p_{\text{RMS}}$, is approximately equal to 3.1, or 9.8 dB, for such a Gaussian waveform. This is in line with the values that can be deduced from the CCDF shown in Figure 1.20 and should be compared to the 14 dB derived previously. We thus see more than a 4 dB reduction in the evaluation of the CF, merely by considering a realistic upper bound that takes into account the frequency of occurrence of the peak values. The same can obviously be done for the modulating waveform PAPR. Recall that, according to equation (1.84), the PAPR must remain around 3 dB lower than the CF of the $p(t)$ or $q(t)$ waveforms. This then results in a direct estimation of this quantity in the region of 7 dB, i.e. 2.2 in natural units. Here again, this is in line with the CCDF of $\rho(t)$ displayed in Figure 1.20.

To conclude this review, we highlight that even when taking into account the low probability of occurrence of the peak values of the waveforms in the dimensioning of a line-up, we get that such OFDM signals still exhibit really high PAPR. This is one of the main reasons for the use of a slightly different modulation scheme for the uplink in the LTE standard. For that scheme, chosen modulation is indeed based on the single carrier frequency division multiple access (FDMA) scheme rather than a pure OFDM one. The idea behind this is to introduce some correlation terms between the symbols used to modulate each subcarrier so that their sum no longer tends toward a normal distribution as would be the case if they were statistically independent. This correlation, introduced through the use of an extra discrete Fourier transform, is enough in practice to quite significantly reduce the PAPR of the modulated signal compared to the case of a pure OFDM scheme. In that sense, this approach can be related to what has already been encountered in both the modified 8PSK scheme of the GSM/EDGE standard and in the HPSK scheme of the WCDMA. This is thus another example of an optimization of a modulation scheme that allows for greater efficiency in the physical implementation of the transmit side.

1.4 First Transceiver Architecture

Based on what we have done so far in this chapter, we can already derive first block diagrams for transceiver architectures based on the signal processing functions involved in the modulation and demodulation of an RF bandpass signal. Here we continue with the general case of a complex modulation. These structures, which can be seen as the minimal ones that enable transceiver functionality, are then further completed in the subsequent chapters by taking into account the constraints linked to other aspects of wireless systems as well as the limitations of their physical implementation. This will lead to the practical architectures discussed in Chapter 8.

When considering this minimum set of functions derived from the communication theory, it is hard to totally get rid of additional constraints linked to their physical implementation if we want to keep the approach realistic. Despite this, we try to stay as much as possible at a theoretical level at this first stage.

1.4.1 Transmit Side

Let us focus for now on the architectures that can be derived from the signal processing involved in the generation of a complex modulated RF bandpass signal as encountered on the transmit side. Based on the foregoing, we understand that the signal processing we are referring to reduces on the one hand to the generation of the modulating waveform in itself, and on the other hand to the modulation of an RF carrier according to this complex lowpass signal in order to obtain the expected modulated RF bandpass signal. From the discussion in Section 1.1.2, we see that there are basically two different approaches for achieving such a modulation process, depending on the chosen representation of the complex lowpass modulating waveform. Here, by "representation", we mean either Cartesian or polar. We can thus expect two different families of transmit architectures, depending on the chosen representation.

Let us first suppose that we are dealing with a Cartesian representation of the complex lowpass modulating waveform $\tilde{s}(t) = p(t) + jq(t)$. By equation (1.28a), the resulting modulated bandpass signal $s(t)$ can be expressed as

$$\begin{aligned} s(t) &= \mathrm{Re}\left\{\tilde{s}(t)e^{j\omega_c t}\right\} \\ &= p(t)\cos(\omega_c t) - q(t)\sin(\omega_c t). \end{aligned} \tag{1.178}$$

Starting from this equation, as highlighted above, we then need to take into account a minimum set of constraints linked to the physical implementation of the corresponding signal processing functions in order to derive a realistic overview of the line-up. For instance, recalling the examples of modulating waveforms reviewed in Section 1.3, we can understand that the complexity associated with the mathematical operations required for their generation leads to having those operations implemented in the digital domain. As a result, we then need to consider the generation of digital samples $p[k] = p(kT_s)$ and $q[k] = q(kT_s)$ of the real and imaginary parts of the complex modulating signal. These samples are then necessarily converted at a given point in the analog domain. In the present case, we can assume this is done through the use of a digital to analog converter (DAC) that delivers analog baseband signals. Anticipating Section 4.6.2, we observe that such conversion requires in most practical implementations the use of a reconstruction filter in order to cancel the copies still present, even if attenuated, at the output of the DAC. At this stage, we then achieve the generation of analog $p(t)$ and $q(t)$ signals. It thus remains to achieve the modulation of the RF carrier through the implementation of the operation corresponding to the above equation.

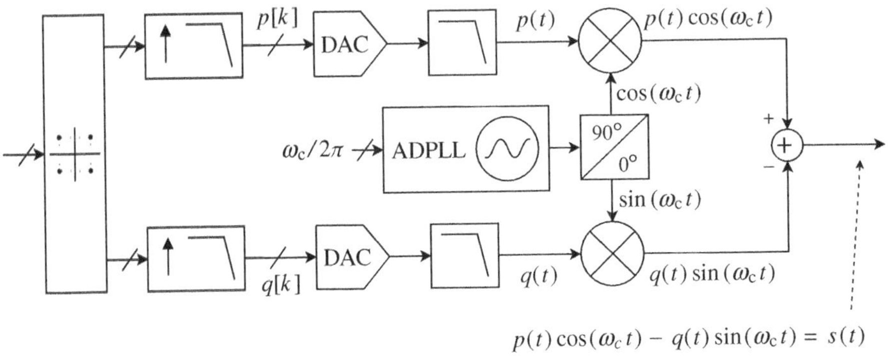

Figure 1.21 Modulated RF bandpass signal generated through the direct upconversion of the complex modulating waveform – Using the Cartesian representation of a complex lowpass modulating waveform $\tilde{s}(t) = p(t) + \mathrm{j}q(t)$, we can derive an implementation for the generation of the corresponding modulated RF bandpass signal through a direct complex frequency upconversion of the modulating signal. This structure is thus referred to as the direct conversion transmitter.

For that purpose, we obviously need to generate the two LO signals in quadrature, $\cos(\omega_c t)$ and $\sin(\omega_c t)$. This can be done through the use of an RF synthesizer, classically implemented as a PLL that locks the oscillating frequency of an RF oscillator to that of an accurate low frequency reference, as discussed in Section 4.3.1. In the present case this reference is not represented in Figure 1.21 for the sake of simplicity, and the PLL is assumed to be an all digital one [11]. Having generated the LO signals, we can then achieve the generation of the modulated bandpass signal we are looking for through their multiplication by the lowpass signals $p(t)$ and $q(t)$, using RF mixing stages for instance, and then subtracting the result. All this results in the structure shown in Figure 1.21. Practically speaking, this structure is a simplified view of a direct conversion, or (zero intermediate frequency, or ZIF), transmitter. This widespread architecture is discussed in more depth in Section 8.1.1, and taken as an example in Chapter 7.

Turning now to a polar representation of the complex lowpass modulating waveform $\tilde{s}(t) = \rho(t)\mathrm{e}^{\mathrm{j}\phi(t)}$, the resulting modulated bandpass signal $s(t)$ can be expressed according to equation (1.28b) as

$$\begin{aligned} s(t) &= \mathrm{Re}\left\{\tilde{s}(t)\mathrm{e}^{\mathrm{j}\omega_c t}\right\} \\ &= \rho(t)\cos(\omega_c t + \phi(t)). \end{aligned} \tag{1.179}$$

However, in the present case we again need to take into account a minimum set of constraints linked to the physical implementation of the signal processing functions we are considering for the derivation of a realistic block diagram. Thinking back to the examples reviewed in Section 1.3, we observe that most of the practical modulating waveforms we are dealing with are defined through their generation in a Cartesian way. As discussed in more depth in Section 8.1.6, there are obviously good reasons for that. In our present polar approach a Cartesian to polar conversion is necessary in order to generate the samples $\rho[k]$ and $\phi[k]$ from the Cartesian $p[k]$ and $q[k]$. We observe that the constant amplitude part of $s(t)$, i.e. $\cos(\omega_c t + \phi(t))$, can be seen as a pure carrier signal, but with a angular frequency that is varying according to the phase modulation term $\phi(t)$. As a result, even if it should obviously be refined, this signal can be generated through the modulation of the PLL that delivers the carrier signal according to the expected instantaneous frequency $f(t)$. Practically speaking, we get from equation (1.97) that $f(t)$ is linked to the carrier frequency and to the argument of the modulating complex envelope through

$$f(t) = \frac{1}{2\pi}\left(\omega_c + \frac{d\phi(t)}{dt}\right). \tag{1.180}$$

As a result, we need to differentiate $\phi(t)$ in order to be able to generate the samples $f[k]$ of $f(t)$ that are further provided to the PLL. Once this is done, it then simply remains to apply the instantaneous amplitude to the constant amplitude signal flowing from the PLL to obtain the expected modulated RF bandpass signal. This can be achieved, for instance, through the use of a variable gain amplifier (VGA), or through a mixing stage that uses this constant amplitude waveform as an LO signal to upconvert the instantaneous amplitude signal $\rho(t)$. As this multiplication is performed in the analog domain, $\rho(t)$ can be reconstructed from the samples $\rho[k]$ through a DAC and a reconstruction filter. The resulting structure shown in Figure 1.22 is a simplified view of the polar transmit architecture that is discussed in more depth in Section 8.1.6.

1.4.2 Receive Side

Let us now focus on the architectures that can be derived from the signal processing corresponding to the recovery of the complex envelope $\tilde{s}(t)$ of a bandpass RF signal $s(t)$ as involved on the receive side. Following the discussion in Section 1.1.2, we get from equation (1.23)

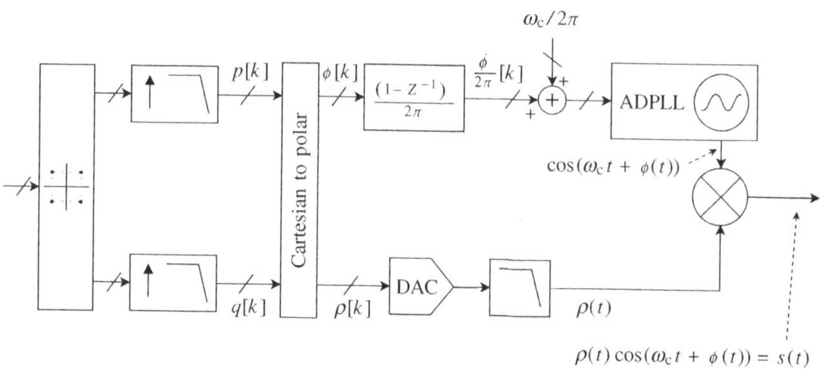

Figure 1.22 Modulated RF bandpass signal generated through a direct instantaneous frequency and amplitude modulation – Using the polar representation of a complex lowpass modulating waveform $\tilde{s}(t) = \rho(t)e^{j\phi(t)}$, we can derive an implementation for the generation of the corresponding modulated RF bandpass signal based on the direct modulation of its instantaneous frequency and amplitude. This structure is referred to as a polar transmitter.

that the real and imaginary parts of $\tilde{s}(t) = p(t) + jq(t)$ can be recovered through the use of a Hilbert transform of $s(t)$ and of dedicated mixing stages using LO signals at the carrier angular frequency. Nevertheless, even if such a Hilbert transform can be implemented for bandpass signals using a pure $\pi/2$ phase shifter as discussed in that section, this approach is not the easiest way forward from the physical implementation perspective. Moreover, it requires a double mixing stage for its correct implementation.

Alternatively, the complex envelope $\tilde{s}(t)$ can be obtained through the use of a complex frequency downconversion followed by a lowpass filtering, as detailed in Chapter 6. With such an approach, we are faced with a situation symmetrical to that encountered on the transmit side when considering the frequency upconversion of the complex modulating waveform expressed in its Cartesian form. Transposed to the present receive case, it results in the simplified view shown in Figure 1.23. Anticipating the discussion in the following chapters, we observe that the signal being processed in a receiver is often composed of both the wanted signal and of unwanted ones linked to the coexistence with other users. As a result, the lowpass filtering dedicated to the cancellation of the unwanted sideband of the wanted signal in the signal processing approach associated with the complex frequency downconversion is also used in practice to cancel, or at least attenuate sufficiently, the unwanted signals. In that sense, it also behaves as an anti-aliasing filter prior to the analog to digital converter (ADC) stage according to the arguments of Section 4.6.1. Such conversion to the digital world is indeed required in practice due to the complexity of the signal processing associated with channel equalization, in particular in a mobile environment, as highlighted in Chapter 2. Here again, we can thus make the link with what was encountered on the transmit side with the complexity of the generation of practical modulating waveforms used in the field of digital communication. Furthermore, we may also mention the interest there is in using the integration capabilities of modern silicon processes for digital logic. The resulting structure shown in Figure 1.23 is a simplified view of a direct conversion, or ZIF receiver. As was the case for its counterpart on the transmit side, this receive architecture is also in widespread use for low cost integrated solutions. As such, it is discussed in more depth in Section 8.2.1, and taken as an example in Chapter 7.

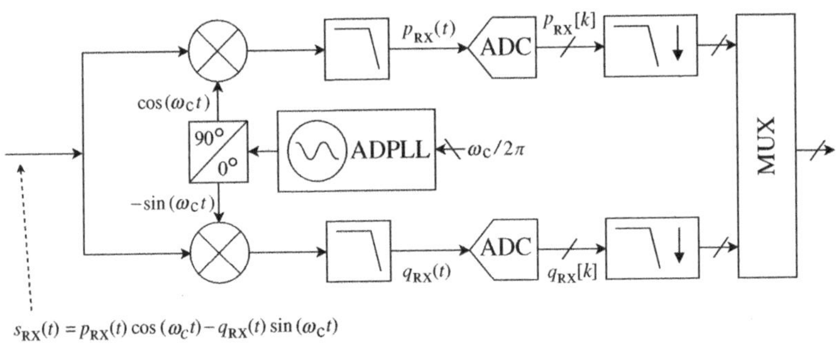

Figure 1.23 Complex envelope recovery using a direct complex frequency downconversion of a modulated bandpass RF signal – The use of a complex frequency downconversion associated with a lowpass filtering on the two branches allows the recovery of the real and imaginary parts of a modulating complex envelope.

To conclude, we observe that we could alternatively imagine recovering the instantaneous frequency and amplitude of the received RF bandpass signal using a sort of polar demodulator that would be the symmetric implementation of the polar transmitter. However, because of various problems this kind of structure is not necessarily considered in practice except in the notable case of a pure frequency modulation. In that case, a PLL can indeed be used to directly recover the instantaneous frequency of the received signal, thus leading to a very compact solution. Such a PLL demodulation scheme is discussed in Section 8.2.4.

2

The Electromagnetism Point of View

Given that a wireless system relies on the propagation of electromagnetic waves for the transmission of information, we discuss the minimum set of functions to be implemented in a transceiver, as derived in Chapter 1, by considering the theory of electromagnetism. This theory highlights why electrical power has to be delivered to the radiating element on the transmit side, whereas the processing of either voltage only or current only waves remains sufficient for recovering the information on the receive side. It also allows us to detail some constraints to be considered for dimensioning a receive line-up due to the dynamic behavior of the propagation channel in a mobile environment. This review of results from the electromagnetic theory also allows us to understand the effective behavior of a wireless link, i.e. how information propagates up to the receive side through the modulation of the electromagnetic field. Our interest in considering such theoretical aspects is thus twofold.

2.1 Free Space Radiation

Let us begin by focusing on free space radiation. Our interest in this comes not only from the guidelines that can be derived for transceivers, but also from the fact that it illustrates how a wireless system behaves in general. We detailed in Chapter 1 the signal processing functions that allow us either to generate an RF bandpass waveform that embeds the information to be transmitted from a given modulating waveform, or to recover this modulating waveform from the received RF bandpass signal. However, we still need to understand how the received RF bandpass signal can have the same modulation, i.e. the same time domain variations, as the RF signal delivered to the radiating element by the active part of a transmitter.

For that purpose, we first need to recall the physical implementation of the signal processing functions detailed in Chapter 1. Indeed, we need to keep in mind that analog voltage or current quantities are used in practice to represent the waveforms we dealt with theoretically in that chapter. This means that on the transmit side, the generation of an RF bandpass waveform from a given lowpass modulating waveform is implemented in such a way that we have the

Wireless Transceiver Architecture: Bridging RF and Digital Communications, First Edition. Pierre Baudin.

generation of an RF current whose time domain variations represent the bandpass modulated waveform. This RF current then feeds a radiating element, the antenna. It then generates an electromagnetic field that in turn, after going through the propagation channel, induces an RF bandpass current on the receive antenna that is further processed by the receiver itself in order to recover its time domain variations. Thus, the key assumption for this to work is that the radiated electromagnetic field embeds in its characteristics the information linked to the time dependent part of the RF bandpass current at the transmitter output in such a way that the time dependent part of the induced current on the receive antenna matches the transmitted modulating waveform. Since we all see that wireless links do indeed work in everyday life, we can surmise that this behavior is exact. Nevertheless, it is of interest to recall the theoretical results that illustrate this, as it also leads to the understanding of interesting constraints for transmitters.

In order to do this, we first recall briefly the theory of electromagnetic far-field radiation assuming a monochromatic excitation. The related results allow straightforward generalizations for narrowband modulated waveforms as in Section 2.1.2.

2.1.1 *Radiated Monochromatic Far-field*

In this first step, let us focus on the far-field radiated by an antenna fed by a monochromatic RF current oscillating at angular frequency ω. Like any electromagnetic topic, this phenomenon can be addressed from a theoretical point of view through Maxwell's equations. In their simple form, those equations are given by

$$\nabla \times \mathbf{E} + \partial_t \mathbf{B} = 0, \tag{2.1a}$$

$$\nabla \times \mathbf{H} - \partial_t \mathbf{D} = \mathbf{J}. \tag{2.1b}$$

Here, by "simple" we mean that we do not consider generalized equations using magnetic currents, as required for instance to derive the radiation from Huygens sources [2, 12]. In equations (2.1), **E** and **D** represent the electric and electric displacement fields respectively, and **H** and **B** the magnetic and magnetic induction fields respectively. **J** denotes the vectorial current density distribution. From our transceiver perspective, this distribution is thus the one that takes place on the transmit antenna in operating conditions. Classically, all the media we deal with, i.e. both the antennas themselves and the media where radiative propagation occurs, are considered as perfect. This means that we can assume linear relationships between **D** and **E** through the medium permittivity ϵ at the angular frequency of interest, i.e. that $\mathbf{D} = \epsilon \mathbf{E}$. In the same way, **B** and **H** are assumed proportional through the medium permeability μ, so that $\mathbf{B} = \mu \mathbf{H}$. In practice, we often deal with free space propagation in the field of wireless, so that μ reduces to $\mu_0 = 4\pi \cdot 10^{-7}$ H/m and ϵ to $\epsilon_0 \approx 8.854 \cdot 10^{-12}$ F/m. These quantities are linked together with the velocity of the corresponding electromagnetic waves in free space c through $\epsilon_0 \mu_0 c^2 = 1$. Nevertheless, there is no additional analytical cost in considering the more general form

$$\nabla \times \mathbf{E} + \mu \partial_t \mathbf{H} = 0, \tag{2.2a}$$

$$\nabla \times \mathbf{H} - \epsilon \partial_t \mathbf{E} = \mathbf{J}, \tag{2.2b}$$

as long as we can assume that both ϵ and μ are constant in time. This holds in most practical media of interest in the field of wireless, so this form of Maxwell's equations is suitable for our purposes. In that case the velocity of the waves is v, with $\epsilon\mu v^2 = 1$. We also observe that all the vectorial quantities involved in the present equations are functions both of the position, through the position vector $\mathbf{r}$, and of time, through the variable t. But, for the sake of simplicity and as usual, this dependency is not explicitly written.

In addition to Maxwell's equations, we also need to consider the electrical charge conservation equation,

$$\nabla \cdot \mathbf{J} + \partial_t \varrho = 0, \tag{2.3}$$

given here in its local form. In this equation, ϱ represents the electrical charge density at a given position and time. The meaning of this equation is that any increase, or decrease, in the electrical charge number in a given volume during a given duration is proportional to the current that flows through its boundary at the same time. Taking the dot product of equation (2.2), and considering on the one hand that $\nabla \cdot (\nabla \times .) = 0$ and on the other hand the above electrical charge conservation law, we thus deduce Gauss's laws. Assuming that ϵ is also constant in position, those laws take the form

$$\nabla \cdot \mathbf{E} = \frac{\varrho}{\epsilon}, \tag{2.4a}$$

$$\nabla \cdot \mathbf{H} = 0. \tag{2.4b}$$

In order to solve the problem of the electromagnetic radiation, a convenient way to proceed is to suppose that we are dealing with a pure monochromatic current excitation. This is convenient because both the form of Maxwell's equations given above and the medium of interest are linear. As a result, all the time dependent quantities we deal with can only be monochromatic with the same angular frequency as the excitation. However, we need to consider a different phase offset at the origin of time that depends only on the position in order to reflect the finite speed of propagation of the information. We thus assume that all the components $s_l(\mathbf{r}, t)$ of the monochromatic field we are dealing with take the general form

$$s_l(\mathbf{r}, t) = \rho_l(\mathbf{r}) \cos(\omega t + \phi_l(\mathbf{r})). \tag{2.5}$$

Here l can stand for x, y or z in the case of a vector field. Considering this general form, we can introduce the classical complex notation and define equivalent complex fields that depend only on the considered position. For instance, we can define the complex electric field, denoted by $\tilde{\mathbf{E}}$, as

$$\tilde{\mathbf{E}} = \begin{pmatrix} \rho_x(\mathbf{r}) e^{j\phi_x(\mathbf{r})} \\ \rho_y(\mathbf{r}) e^{j\phi_y(\mathbf{r})} \\ \rho_z(\mathbf{r}) e^{j\phi_z(\mathbf{r})} \end{pmatrix}. \tag{2.6}$$

This complex field is then related to its real valued physical counterpart through

$$\mathbf{E} = \mathrm{Re}\{\tilde{\mathbf{E}} e^{j\omega t}\}. \tag{2.7}$$

According to the above definition, complex fields are defined in the present case using the convention $+j\omega t$. This definition can be used for fields of any kind, whether vectorial, like the current density distribution field, or scalar, like the charge density field. Note, furthermore, that as the chosen notation suggests, the definition of those complex fields can be seen as the generalization for scalar or vector fields of the concept of complex envelope introduced for signals in Section 1.1.2. This point is discussed in more depth in Section 2.1.2, but we can already observe that the time domain convention followed here, $+j\omega t$, for the definition of the complex fields is consistent with the definition for the complex envelopes of signals chosen in Chapter 1. Returning to our derivation, we see that the complex notation leads to a separation between the space and time variables. This means that we can transpose all the equations derived for the real physical monochromatic field as equations for their complex fields by substituting the time domain derivative $\partial_t \mathbf{E}$ by $\mathrm{j}\omega\tilde{\mathbf{E}}$. Indeed, having $\tilde{\mathbf{E}}$ constant in time leads to

$$\partial_t(\tilde{\mathbf{E}}\mathrm{e}^{\mathrm{j}\omega t}) = \mathrm{j}\omega\tilde{\mathbf{E}}\mathrm{e}^{\mathrm{j}\omega t}. \tag{2.8}$$

As a result, once the equations for the complex fields obtained with this substitution rule are solved, we simply need to multiply the complex fields by $\mathrm{e}^{+\mathrm{j}\omega t}$ and take the real part to recover the real valued physical quantity.

The benefit of using such complex fields is that the equations to be solved are now no longer partial differential equations in time. For instance, the transposition of equations (2.2) and (2.3) leads to

$$\nabla \times \tilde{\mathbf{E}} + \mu \mathrm{j}\omega\tilde{\mathbf{H}} = 0, \tag{2.9a}$$

$$\nabla \times \tilde{\mathbf{H}} - \epsilon \mathrm{j}\omega\tilde{\mathbf{E}} = \tilde{\mathbf{J}}, \tag{2.9b}$$

and

$$\nabla \cdot \tilde{\mathbf{J}} + \mathrm{j}\omega\tilde{\varrho} = 0. \tag{2.10}$$

In the same way, Gauss's laws given by equation (2.4) can be transposed as

$$\nabla \cdot \tilde{\mathbf{E}} = \frac{\tilde{\varrho}}{\epsilon}, \tag{2.11a}$$

$$\nabla \cdot \tilde{\mathbf{H}} = 0. \tag{2.11b}$$

As a result, taking the cross product of equation (2.9) and considering on the one hand that $\nabla \times (\nabla \times .) = \nabla(\nabla \cdot .) - \nabla^2$, and on the other hand the above relationships, we finally recover Helmholtz's equations for complex fields:

$$\nabla^2\tilde{\mathbf{E}} + k^2\tilde{\mathbf{E}} = \frac{\mathrm{j}}{\omega\epsilon}[\nabla(\nabla \cdot \tilde{\mathbf{J}}) + k^2\tilde{\mathbf{J}}], \tag{2.12a}$$

$$\nabla^2\tilde{\mathbf{H}} + k^2\tilde{\mathbf{H}} = -\nabla \times \tilde{\mathbf{J}}. \tag{2.12b}$$

In these equations, k stands for the wavenumber given by

$$k = \omega\sqrt{\epsilon\mu}. \tag{2.13}$$

Practically speaking, Helmholtz's equations link the excitation, i.e. the current density distribution $\tilde{\mathbf{J}}$, and the existing electromagnetic field. We thus need to solve for them in order to get an expression for the electromagnetic field induced by $\tilde{\mathbf{J}}$. Due to the property of the Dirac delta distribution that is the neutral element of the convolution, we get that the general solution for the equations results from the convolution of the elementary solution for the particular equation,

$$\nabla^2 G + k^2 G = \delta(\mathbf{r}), \tag{2.14}$$

and of the right-hand-side terms in the original equations. Here, the solution of this particular equation is nothing more than Green's function:

$$G(r) = -\frac{\mathrm{e}^{-\mathrm{j}kr}}{4\pi r} \tag{2.15}$$

and its complex conjugate. We thus see that $G(r)$ is a function of the norm only of the position vector, $r = \|\mathbf{r}\|$, and not of its direction. However, due to the chosen time domain convention for the definition of the complex fields, $+\mathrm{j}\omega t$, we see that only $G(r)$ can describe outgoing waveforms suited to our radiation problem. We therefore now focus on deriving the complex electric and magnetic fields at position $\mathbf{r}$ as the convolution between $G(r)$ and the right-hand side of equation (2.12). This results in

$$\tilde{\mathbf{E}} = \frac{\mathrm{j}}{\omega\epsilon}\{\nabla[\nabla\cdot(G\star\tilde{\mathbf{J}})] + k^2(G\star\tilde{\mathbf{J}})\}, \tag{2.16a}$$

$$\tilde{\mathbf{H}} = -\nabla\times(G\star\tilde{\mathbf{J}}). \tag{2.16b}$$

In order to go further, we can apply the operators involved in the above equations to Green's function and to the current density distribution. To do so, we need to specify the latter distribution. However, as we do not know a priori the shape of the radiating element we are dealing with, a convenient way to proceed is to perform the derivation with an elementary sinusoidal current density distribution, i.e. a simple sinusoidal dipole, located at the origin of the chosen coordinate system. This means that we consider a current density that takes the form

$$\tilde{\mathbf{J}} = \tilde{\mathcal{J}}\,\delta(\mathbf{r})\mathbf{s}, \tag{2.17}$$

where $\tilde{\mathcal{J}}$ is the complex moment of the dipole and $\mathbf{s}$, defined such that $\|\mathbf{s}\| = 1$, represents its orientation according to the notation shown in Figure 2.1. Our interest in this particular distribution is twofold. We get, on the one hand, that the Dirac delta distribution is the neutral element of the convolution. This thus results in straightforward analytical derivations for solving equation (2.16). On the other hand, this distribution can be seen as an elementary component of any current density distribution. As done later on in this section, the structure of its radiated far-field can thus be used to derive the far-field radiated by any current distribution using the superposition principle linked to the linearity of Maxwell's equations.

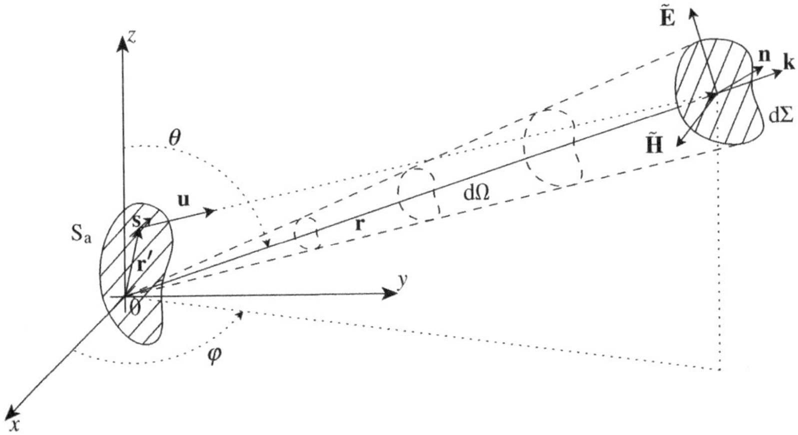

Figure 2.1 Far-field structure of a radiating antenna – The far-field of any radiating antenna has the structure of a plane wave. As a result, the electrical and the magnetic fields are orthogonal, with each other and with the propagation direction. Their magnitude decreases in $1/r$ and thus the radiated power density flowing through a unit area in $1/r^2$.

Consequently, using this particular current density distribution and Green's function given by equation (2.15), one can show that equation (2.16) reduces to [12, 13]

$$\tilde{\mathbf{E}} = j\omega\mu G(r)\left[\left(-1+\frac{3j}{kr}+\frac{3}{k^2r^2}\right)(\tilde{\mathcal{J}}\,\mathbf{s}\times\mathbf{u})\times\mathbf{u}+\left(\frac{2j}{kr}+\frac{2}{k^2r^2}\right)\tilde{\mathcal{J}}\,\mathbf{s}\right],$$

$$\tilde{\mathbf{H}} = -jkG(r)\left(1-\frac{j}{kr}\right)\tilde{\mathcal{J}}\,\mathbf{s}\times\mathbf{u},$$

with $\mathbf{u} = \mathbf{r}/r$, the normalized vector that gives the direction of radiation. In practice, we can suppose for a wireless link that the emitting and receiving antennas are far enough apart so that $kr \gg 1$. This corresponds to what we call the radiated far-field.[1] Applying this condition to the above equations therefore leads to the far-field radiated by the dipole:

$$\tilde{\mathbf{E}} = -j\omega\mu G(r)(\tilde{\mathcal{J}}\,\mathbf{s}\times\mathbf{u})\times\mathbf{u}, \tag{2.18a}$$

$$\tilde{\mathbf{H}} = -jkG(r)\tilde{\mathcal{J}}\,\mathbf{s}\times\mathbf{u}. \tag{2.18b}$$

[1] However, we observe that for a general antenna, this far-field concept has a stricter definition, which involves the physical size of the antenna, and not just the simple condition $kr \gg 1$. The idea is that whatever the elementary radiating element of the antenna, its contribution to the total field at a given position $\mathbf{r}$ is almost in phase with all other contributions so that it sums constructively. There is thus an idea of maximum phase difference in the radiated field by all the elementary contributions of the antenna at the position of observation. This results in the common assumption for the far-field zone of a maximum phase difference of $\pi/8$ [14]. It leads to a minimum distance to reach the far-field that corresponds to $r \geq 2D^2/\lambda$, with D the maximum dimension of the antenna and $\lambda = 2\pi/k$ the wavelength. However, for a simple punctual source, we can hardly talk about the physical size of the radiating element. We then need to consider $kr \gg 1$ as the corresponding assumption in that case.

But in order to go further and derive the structure of the far-field radiated by any current density, we need to consider the case where the dipole is not located at the origin of the coordinate system, but rather at a position defined by $\mathbf{r}'$. This is indeed required in order to be able to sum contributions from elementary elements located all along the radiating antenna. However, in order to use the results derived so far, we need to ensure that the far-field assumption remains valid after the change of coordinates. For that, we assume that the origin of the coordinate system is chosen so that $\|\mathbf{r}'\|$ remains much lower than any distance of interest for the observation of the radiated far-field $\|\mathbf{r}\|$. This is the case, for instance, when this origin is located on the antenna itself, as shown in Figure 2.1. In that case, we can use the above equation but with the position vector between the dipole and the position of radiation that is now given by $\|\mathbf{r}-\mathbf{r}'\|$. However, in the far-field approximation, we get that the direction of radiation, $\mathbf{u}$, theoretically given by $(\mathbf{r}-\mathbf{r}')/\|\mathbf{r}-\mathbf{r}'\|$, remains almost equal to $\mathbf{r}/r$ and thus independent of $\mathbf{r}'$ up to first order. In addition, the assumption $\|\mathbf{r}'\| \ll \|\mathbf{r}\|$ allows us to write

$$\begin{aligned}\|\mathbf{r}-\mathbf{r}'\| &= (\|\mathbf{r}\|^2 + \|\mathbf{r}'\|^2 - 2\mathbf{r}\cdot\mathbf{r}')^{1/2},\\ &\approx \|\mathbf{r}\|\left(1 - 2\frac{\mathbf{r}\cdot\mathbf{r}'}{\|\mathbf{r}\|^2}\right)^{1/2},\\ &\approx \|\mathbf{r}\| - \frac{\mathbf{r}\cdot\mathbf{r}'}{\|\mathbf{r}\|}\end{aligned} \tag{2.19}$$

or, as $\mathbf{u} \approx \mathbf{r}/r$,

$$\|\mathbf{r}-\mathbf{r}'\| \approx \|\mathbf{r}\| - \mathbf{u}\cdot\mathbf{r}'. \tag{2.20}$$

This result can be used in equation (2.15) to derive an expression for $G(\|\mathbf{r}-\mathbf{r}'\|)$:

$$G(\|\mathbf{r}-\mathbf{r}'\|) \approx -\frac{\mathrm{e}^{-\mathrm{j}kr}}{4\pi r}\mathrm{e}^{\mathrm{j}\mathbf{k}\cdot\mathbf{r}'}, \tag{2.21}$$

where

$$\mathbf{k} = k\mathbf{u} \tag{2.22}$$

is the wave vector whose norm is simply the wavenumber defined by equation (2.13), as by construction $\|\mathbf{u}\| = 1$. Thus, using this result in (2.18), we get that the complex field radiated at position $\mathbf{r}$ by an elementary sinusoidal current density located at position $\mathbf{r}'$ is

$$\tilde{\mathbf{E}} = \mathrm{j}\omega\mu\frac{\mathrm{e}^{-\mathrm{j}kr}}{4\pi r}(\tilde{\mathcal{J}}\,\mathbf{s}\times\mathbf{u})\times\mathbf{u}\,\mathrm{e}^{\mathrm{j}\mathbf{k}\cdot\mathbf{r}'}, \tag{2.23a}$$

$$\tilde{\mathbf{H}} = \mathrm{j}k\frac{\mathrm{e}^{-\mathrm{j}kr}}{4\pi r}\tilde{\mathcal{J}}\,\mathbf{s}\times\mathbf{u}\,\mathrm{e}^{\mathrm{j}\mathbf{k}\cdot\mathbf{r}'}. \tag{2.23b}$$

Given this result, we can now focus on the structure of the far-field radiated by a general current density distribution, $\tilde{\mathbf{J}}$, that takes place on an antenna fed by a sinusoidal current generator. Referring to the linearity of Maxwell's equations, we can write the total radiated field as the sum of the contributions of the elementary dipoles that constitute the complete distribution on the antenna. It then follows from the above expression that

$$\tilde{\mathbf{E}} = \mathrm{j}\omega\mu \frac{\mathrm{e}^{-\mathrm{j}kr}}{4\pi r} \int_{\mathrm{S_a}} (\tilde{\mathbf{J}} \times \mathbf{u}) \times \mathbf{u}\, \mathrm{e}^{\mathrm{j}\mathbf{k}\cdot\mathbf{r}'} \mathrm{d}S', \tag{2.24a}$$

$$\tilde{\mathbf{H}} = \mathrm{j}k \frac{\mathrm{e}^{-\mathrm{j}kr}}{4\pi r} \int_{\mathrm{S_a}} \tilde{\mathbf{J}} \times \mathbf{u}\, \mathrm{e}^{\mathrm{j}\mathbf{k}\cdot\mathbf{r}'} \mathrm{d}S', \tag{2.24b}$$

with $\mathrm{S_a}$ the surface of the considered antenna, and $\mathrm{d}S'$ the elementary surface element of $\mathrm{S_a}$ located at position $\mathbf{r}'$. Here, we point out that in order to write this expression, we assumed for the sake of simplicity that the antenna is composed of a perfect conductor material so that the current distribution can be approximated as a surface distribution. In any case, it results that the far-field radiated by any current distribution $\tilde{\mathbf{J}}$ is of the form

$$\tilde{\mathbf{E}} = \frac{\mathrm{e}^{-\mathrm{j}kr}}{r} \mathbf{F}(\mathbf{u}), \tag{2.25a}$$

$$\tilde{\mathbf{H}} = \frac{1}{\eta} \mathbf{u} \times \tilde{\mathbf{E}}, \tag{2.25b}$$

where $\eta = \sqrt{\mu/\epsilon}$ is the wave impedance in the medium considered and

$$\mathbf{F}(\mathbf{u}) = \frac{\mathrm{j}k}{4\pi} \int_{\mathrm{S_a}} (\eta\tilde{\mathbf{J}} \times \mathbf{u}) \times \mathbf{u}\, \mathrm{e}^{\mathrm{j}\mathbf{k}\cdot\mathbf{r}'} \mathrm{d}S' \tag{2.26}$$

is the radiation characteristic vector of the antenna.

What is interesting to see is that whatever the current distribution $\tilde{\mathbf{J}}$, the radiated far-field has a locally plane wave structure with a decreasing amplitude in $1/r$. This result is general and would have been obtained even if we had considered magnetic currents as additional sources. The consequence of this structure is discussed in particular in Section 2.1.3 from a radiated power point of view. For now, we observe that the radiation characteristic vector $\mathbf{F}(\mathbf{u})$ in the present monochromatic case is a function only of the current density distribution on the antenna surface. As this dependency is linear through the integral involved, this simply means that $\mathbf{F}(\mathbf{u})$ is proportional to the complex representation of the current provided to the antenna by our transmitter, at least when the modulation state of the current wave can be assumed as constant along the antenna at a given time. But, on the other hand, it can be shown that a receiving antenna that is illuminated by a plane wave behaves as a generator of internal impedance equal to its transmit impedance and with an internal electromotive force (EMF) proportional to the magnitude of the incoming electrical field [12], as illustrated in Figure 2.2. As in the far-field approximation this electrical field is proportional to the radiation characteristic vector of the emitting antenna, and thus proportional to the transmit current, we thus recover at the receive antenna output a voltage, or current, proportional to the transmit one. Obviously, this behavior is the reason for the possibility of propagating the information from the transmit antenna up to the receive antenna, as discussed in more detail in the next section.

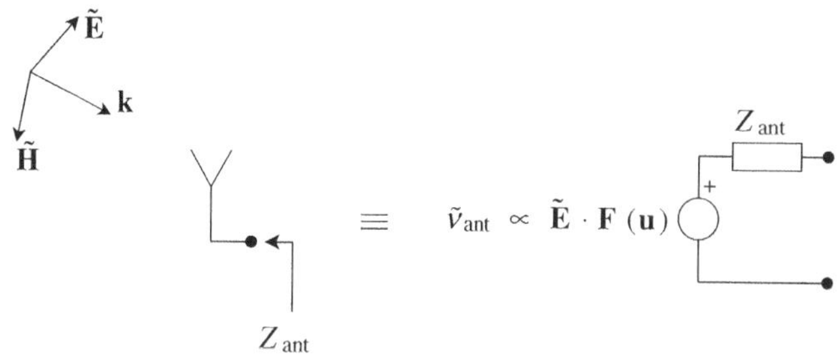

Figure 2.2 Receiving antenna equivalent model when illuminated by a plane wave – When illuminated by a plane wave at angular frequency ω, a reciprocal antenna behaves like a generator with an internal impedance equal in reception and transmission at the same angular frequency and with an electromotive force proportional to the dot product between the incoming electrical field and its radiation characteristic vector.

2.1.2 Narrowband Modulated Fields

As illustrated in Chapter 1, the information propagated in a wireless link is carried by the waveform of the bandpass RF signal flowing from the transmit line-up. Practically speaking, this modulated bandpass RF signal takes the physical form of an RF current that spreads along the transmit antenna, thus resulting in a current density proportional to it. To illustrate the behavior of a wireless link, we need to derive the structure of the electromagnetic field radiated by the bandpass modulated RF current density. The underlying phenomenon we expect is that the same time domain variations would be recovered on this radiated field as well as on the resulting RF current induced on the receive antenna.

Classically, two different approaches can be considered. On the one hand, we can imagine decomposing the modulated bandpass current density excitation that takes place on the transmit antenna in a superposition of monochromatic tones through a Fourier transform on the time variable. Then, using the discussion in the previous section, we can derive the characteristics of the radiated electromagnetic field and thus those of the EMF that is induced at the receiving antenna stage for each of those monochromatic components. Once this is done, we can use inverse Fourier transforms to recover the time domain variations of the induced EMF. On the other hand, we can try to solve the propagation equations directly in the time domain. But in that case the analytical derivations are obviously not straightforward. Furthermore, between these two methodologies, an intermediate way forward may be considered. Reconsidering the modulation scheme detailed in Section 1.3, we see that the widest spectral extent of those waveforms is only as much as a few tens of megahertz for the widest ones. This means that they can be considered as narrowband compared to the carrier frequencies that classically lie in the GHz range. We can thus wonder if the results derived in the previous section in the pure monochromatic case cannot be extended directly, up to a given point, to the narrowband modulation case.

However, in order to be able to perform the corresponding analytical derivations, we first need to generalize a little the concept of the complex field introduced in the pure monochromatic case through equation (2.7) in Section 2.1.1. Indeed, as highlighted in that section, this definition can be linked to the complex envelope concept introduced for real bandpass signals in Section 1.1.2. In order to illustrate this, let us consider as an example a vectorial current density distribution $\mathbf{J}$. If, for a given position where this distribution is non-vanishing, all the components of $\mathbf{J}$ are real bandpass, we can then define complex envelopes for them, and thus in turn a complex envelope vector that is composed of those complex envelopes defined at the position considered. Proceeding like this, we therefore define a complex envelope vector field that we can call $\tilde{\mathbf{J}}$ as it reduces to the classical complex field in the pure monochromatic case. However, in contrast to what happens for the classical complex envelope concept defined for real bandpass signals, we now get that this complex envelope field depends on both space and time. However, it happens that in the present case we can refine the dependence on each of those variables.

On the one hand, the transmitter delivers a bandpass RF current that is a physical representation of the bandpass modulated waveform we want to transmit. Thus, as long as the physical size of the radiating element is such that the modulation state of the current distribution can be considered as almost constant along the antenna at a given time, we can then assume that the system behaves as a quasi-monochromatic system when inspected over a limited duration. As a result, we can assume that during such a time frame the same modulation state applies to all the components of the current density vector field at a given position, as in the pure monochromatic case.

On the other hand, the propagation speed of the information remains finite. This behavior is intrinsically related to the fact that any electromagnetic field that propagates is a solution of a wave equation. In a medium free of sources, for instance, this equation takes the form of d'Alembert's equation, which reduces for the electric field to

$$\nabla^2\mathbf{E} - \frac{1}{v^2}\partial_t^2\mathbf{E} = 0. \tag{2.27}$$

As discussed in Section 2.2.1, the same also applies to current fields, for instance. The result is then that the time dependence of fields that solve the wave equation is of the form $t \pm r/v$ rather than simply t, at least under the spherical symmetry assumption. More precisely, with regard to our present problem, we can assume dependencies of the form $t - r/v$ as this represents outgoing waves (in the direction of increasing r). Thus, at a given distance r from the antenna connector, all behaves as if the information had been delayed by an amount equal to r/v.[2] Rigorously speaking, we should in fact keep in mind that for guided waves (e.g. current waves) we should rather consider a dependency following $t - f(\mathbf{r})/v$ with $f(\mathbf{r})$ that is not necessarily equal to $\|\mathbf{r}\| = r$ and with v that is not necessarily the speed of the wave in free space. Indeed, in this particular case and in the high frequency limit at least, the wave follows the geodesic of the surface on which it propagates. Nevertheless, for the sake of simplicity in our discussion, we can continue to assume what follows a dependency of the form $t - r/v$, as it allows us to illustrate clearly enough the behavior of a wireless link.

[2] This was already the case for the pure monochromatic solution recovered in the previous section. Indeed, considering the complex electromagnetic far-field expression given by equation (2.25), we get that the structure of the corresponding physical electromagnetic field recovered using equation (2.7) has a space-time dependency in $\omega t - kr$. This argument can be expressed as $\omega(t - kr/\omega)$ which is thus a function of $t - r/v$ considering that $k = \omega/v$, as given by equation (2.13).

We can thus assume that the structure of the field component given by equation (2.5) in the pure monochromatic case can be generalized for bandpass modulated fields as

$$s_l(\mathbf{r}, t) = \rho_{S,l}(\mathbf{r})\rho_T(t - r/v)\cos(\omega_c t + \phi_{S,l}(\mathbf{r}) + \phi_T(t - r/v)), \tag{2.28}$$

where l can stand for the spatial coordinate x, y or z when dealing with a vector field.

It then remains to derive the structure of the complex envelope field from this expression. We first recall that an infinity of complex envelopes can be defined for a given bandpass RF waveform, as discussed in Section 1.1.2. Practically speaking, this depends on the center frequency chosen for its definition, and the same holds for fields. For the sake of simplicity, we assume that all the complex envelope fields we deal with are defined as centered around the carrier angular frequency ω_c and that the same holds for the complex envelopes that represent the modulated scalar waveform at the transmitter output. Let us therefore reconsider our current density vector field $\mathbf{J}$ as an example. We can then write from equation (2.28)

$$\mathbf{J} = \mathrm{Re}\left\{\tilde{\mathbf{J}}e^{j\omega_c t}\right\}, \tag{2.29}$$

where

$$\tilde{\mathbf{J}} = \tilde{\mathbf{J}}_S(\mathbf{r})\tilde{j}_T(r, t), \tag{2.30}$$

in which

$$\tilde{\mathbf{J}}_S(\mathbf{r}) = \begin{pmatrix} \rho_{S,x}(\mathbf{r})e^{j\phi_{S,x}(\mathbf{r})} \\ \rho_{S,y}(\mathbf{r})e^{j\phi_{S,y}(\mathbf{r})} \\ \rho_{S,z}(\mathbf{r})e^{j\phi_{S,z}(\mathbf{r})} \end{pmatrix} \quad \text{and} \quad \tilde{j}_T(r, t) = \rho_T(t - r/v)e^{j\phi_T(t-r/v)}. \tag{2.31}$$

In this definition, we thus recover the two kinds of terms we were expecting. On the one hand, we get the term $\tilde{\mathbf{J}}_S(\mathbf{r})$ denoting the spatially dependent part of $\tilde{\mathbf{J}}$. This contribution is thus a vector constant in time that gives the attenuation and polarization of the wave considered; it is nothing more than the classical complex field defined in the pure monochromatic case in the previous section. On the other hand, we get the term $\tilde{j}_T(r, t)$ denoting the time dependent part of $\tilde{\mathbf{J}}$. This is a scalar complex envelope that carries the modulation information while taking into account the finite propagation speed of the information in order to give an overall solution that remains a function of $t - r/v$, as required for an outgoing wave solution of a wave equation.

Obviously, nothing prevents us from going further in the generalization of the concepts introduced for real bandpass signals in Chapter 1. For instance, we can define an analytic field whose components at a given position are the analytic signals of the components of the field at that position. In the present current density vector field example case where ω_c was chosen as the center angular frequency for the definition of the complex envelope vector, this would lead to an analytic vector field $\mathbf{J}_a$ of the form

$$\mathbf{J}_a = \tilde{\mathbf{J}}e^{j\omega_c t}. \tag{2.32}$$

Based on equation (2.29) we can then write

$$\mathbf{J} = \mathrm{Re}\left\{\tilde{\mathbf{J}}e^{j\omega_c t}\right\} = \mathrm{Re}\{\mathbf{J}_a\}, \tag{2.33}$$

which can be compared, for instance, to equation (1.27) in the real bandpass signal case. However, in the present propagating field case we can consider the decomposition of the complex envelope field in terms of spatial and time dependent parts as given by equation (2.30). Using this decomposition we can rewrite equation (2.32) as

$$\mathbf{J}_a = \tilde{\mathbf{J}}_s(\mathbf{r})j_{T,a}(r,t), \tag{2.34}$$

with

$$j_{T,a}(r,t) = \tilde{j}_T(r,t)e^{j\omega_c t}. \tag{2.35}$$

We thus see that the analytic field is nothing more than the product of the spacial dependent part of the complex envelope field, i.e. $\tilde{\mathbf{J}}_s(\mathbf{r})$, and of the analytic signal associated with the time dependent part of the original bandpass field. This analytic field allows us to generalize the concepts of instantaneous amplitude and instantaneous phase or frequency, introduced in "Instantaneous amplitude and instantaneous frequency" (Section 1.1.3), to bandpass modulated scalar or vector fields.

For consistency with the classical complex field definition, we may assume in addition that

$$\mathbb{E}\{|\tilde{j}_{T,t}(r)|^2\} = 1 \tag{2.36}$$

whatever the distance r from the antenna. Indeed, looking back at equation (2.31), we see that the distance r only corresponds to a time shift on $\tilde{j}_T(r,t)$. We can thus assume that it has no impact on the statistical properties of the modulating waveform as long as we assume the stationarity of the modulating process. The benefit of this normalization can be seen when deriving the average power $P_{\mathbf{J}}(\mathbf{r})$ associated with the field $\mathbf{J}$ considered at position $\mathbf{r}$. We get that $P_{\mathbf{J}}(\mathbf{r})$ is given by the time domain average of the instantaneous power $P_{\mathbf{J}}(\mathbf{r},t) = \|\mathbf{J}\|^2$. But referring to the discussion in "Average power" (Section 1.1.3), we can see that in the random modulation perspective the time average of $\|\mathbf{J}\|^2$ can alternatively be estimated as half the expectation of the square norm of any of the associated complex envelope fields $\tilde{\mathbf{J}}$. We can thus write

$$P_{\mathbf{J}}(\mathbf{r}) = \mathbb{E}\left\{\frac{\|\tilde{\mathbf{J}}_t(\mathbf{r})\|^2}{2}\right\}. \tag{2.37}$$

Now using equations (2.30) and (2.36), this expression reduces to

$$P_{\mathbf{J}}(\mathbf{r}) = \|\tilde{\mathbf{J}}_s(\mathbf{r})\|^2 \frac{\mathbb{E}\{|\tilde{j}_{T,t}(r)|^2\}}{2} = \frac{1}{2}\|\tilde{\mathbf{J}}_s(\mathbf{r})\|^2 = \frac{1}{2}P_{\tilde{\mathbf{J}}_s}(\mathbf{r}). \tag{2.38}$$

We then recover the power of $\mathbf{J}$ as half the norm of the spatial part of its complex envelope field. This result matches the classical relationship for complex fields in the pure monochromatic case as in that particular case $\tilde{j}_\mathrm{T}(r,t) = 1$ and thus $\tilde{\mathbf{J}} = \tilde{\mathbf{J}}_\mathrm{s}(\mathbf{r})$.

The interesting point regarding the present definition is that the time domain derivative expression of the form given by equation (2.8) is still consistent in the present case as long as the time dependent part of the complex envelope field, i.e. the modulating waveform, is narrowband enough compared to the carrier angular frequency. In order to illustrate this, let us reconsider our current density field example. Based on the decomposition of its complex envelope vector field given by equation (2.30), we can write

$$\begin{aligned}\partial_t \mathbf{J}_\mathrm{a} = \partial_t(\tilde{\mathbf{J}}\mathrm{e}^{\mathrm{j}\omega_\mathrm{c}t}) &= \tilde{\mathbf{J}}_\mathrm{s}(\mathbf{r})\partial_t\left(\tilde{j}_\mathrm{T}(r,t)\mathrm{e}^{\mathrm{j}\omega_\mathrm{c}t}\right)\\ &= \tilde{\mathbf{J}}_\mathrm{s}(\mathbf{r})(\mathrm{j}\omega_\mathrm{c}\tilde{j}_\mathrm{T}(r,t) + \partial_t\tilde{j}_\mathrm{T}(r,t))\mathrm{e}^{\mathrm{j}\omega_\mathrm{c}t}.\end{aligned} \tag{2.39}$$

It then remains to show that the time domain derivative of $\tilde{j}_\mathrm{T}(r,t)$ is negligible with respect to $\mathrm{j}\omega_\mathrm{c}\tilde{j}_\mathrm{T}(r,t)$ when dealing with a sufficiently narrowband modulation. But as we obviously rely on an argument based on relative frequency bandwidths, we surmise that the result can be achieved more easily in the frequency domain. Thus, assuming that for a given realization of the modulation process the Fourier transform of the current density field exists, as can be the case when considering a finite observation duration for instance, we can write

$$j_\mathrm{T,a}(r,t) = \int_{-\infty}^{+\infty} \mathcal{F}_{\{j_\mathrm{T,a}(r,t)\}}(r,\omega)\mathrm{e}^{\mathrm{j}\omega t}\mathrm{d}t \tag{2.40}$$

and then

$$\partial_t j_\mathrm{T,a}(r,t) = \int_{-\infty}^{+\infty} \mathrm{j}\omega\mathcal{F}_{\{j_\mathrm{T,a}(r,t)\}}(r,\omega)\mathrm{e}^{\mathrm{j}\omega t}\mathrm{d}t. \tag{2.41}$$

We recover the general result that links the Fourier transform of the time domain derivative of a given function with the Fourier transform of that function, i.e. that

$$\mathcal{F}_{\{\partial_t j_\mathrm{T,a}(r,t)\}}(r,\omega) = \mathrm{j}\omega\mathcal{F}_{\{j_\mathrm{T,a}(r,t)\}}(r,\omega). \tag{2.42}$$

Consequently, when the time domain part of $\mathbf{J}_\mathrm{a}$, i.e. $j_\mathrm{T,a}(r,t)$, has a spectrum that spreads over $[\omega_\mathrm{c} - \Omega, \omega_\mathrm{c} + \Omega]$ with $\Omega \ll \omega_\mathrm{c}$, we can write

$$\begin{aligned}\mathcal{F}_{\{\partial_t j_\mathrm{T,a}(r,t)\}}(r,\omega) &= \mathrm{j}\omega_\mathrm{c}\left(1 + \frac{\delta\omega}{\omega_\mathrm{c}}\right)\mathcal{F}_{\{j_\mathrm{T,a}(r,t)\}}(r,\omega)\\ &\approx \mathrm{j}\omega_\mathrm{c}\mathcal{F}_{\{j_\mathrm{T,a}(r,t)\}}(r,\omega),\end{aligned} \tag{2.43}$$

where $\delta\omega = \omega - \omega_\mathrm{c} \ll \omega_\mathrm{c}$ for ω in $[\omega_\mathrm{c} - \Omega, \omega_\mathrm{c} + \Omega]$. But then, taking the inverse Fourier transform of this relationship we get that

$$\partial_t j_\mathrm{T,a}(r,t) = \mathrm{j}\omega_\mathrm{c} j_\mathrm{T,a}(r,t). \tag{2.44}$$

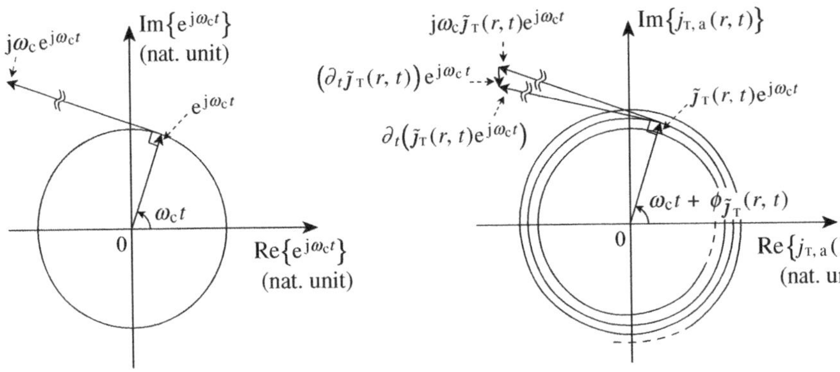

Figure 2.3 Time domain derivative of the analytic field representing a narrowband modulated waveform – For a pure monochromatic field, the time dependent part of its associated analytic field given by equation (2.35) reduces to $e^{j\omega_c t}$ as $\tilde{j}_T(r,t) = 1$ in that case. Its time domain derivative is thus simply equal to $j\omega_c e^{j\omega_c t}$ whose representation remains exactly orthogonal to it (left). For a bandpass modulated field, as long as the modulation remains narrowband with regard to the carrier angular frequency, the variations of the analytic field remain driven by the rotation due to the carrier angular frequency. As a result, its time domain derivative can be approximated as almost orthogonal to it (right).

Referring to equation (2.34), we can now multiply each side of this equation with the spatially dependent part of the analytic vector field, $\tilde{\mathbf{J}}_s(\mathbf{r})$, which is independent of t, to write

$$\partial_t \mathbf{J}_a = j\omega_c \mathbf{J}_a. \tag{2.45}$$

Remembering the relationship between the complex envelope field, assumed defined as centered around ω_c, and the analytic field, given by equation (2.32), we can finally write

$$\partial_t\left(\tilde{\mathbf{J}}e^{j\omega_c t}\right) = j\omega_c \tilde{\mathbf{J}}e^{j\omega_c t}. \tag{2.46}$$

We thus see that in the narrowband modulation case we can neglect the time domain derivative of $\tilde{j}_T(r,t)$ with respect to the term $j\omega_c \tilde{j}_T(r,t)$ in equation (2.39), as illustrated in Figure 2.3. We then see that the low frequency behavior of the modulating waveform leads to a slowly varying time dependent part of the analytic field with regard to the rotation speed linked to the carrier.

Under this narrowband modulation assumption, we can moreover expect the characteristics of the propagation medium, i.e. both ϵ and μ, to remain almost constant in the frequency band $[\omega_c - \Omega, \omega_c + \Omega]$, and thus expect the proportionality between the complex envelope vector fields of **D** and **E**, or between those of **B** and **H**, when defined around the same carrier angular frequency, still to hold as in the pure monochromatic case. Indeed, at a given position **r** we can, for instance, write in the frequency domain that

$$\mathcal{F}_{\{\mathbf{D}\}}(\mathbf{r},\omega) = \epsilon(\omega)\mathcal{F}_{\{\mathbf{E}\}}(\mathbf{r},\omega). \tag{2.47}$$

Assuming that the modulation is sufficiently narrowband so that $\epsilon(\omega) \approx \epsilon$ for all angular frequencies within $[\omega_c - \Omega, \omega_c + \Omega]$, we thus get that

$$\mathscr{F}_{\{\mathbf{D}\}}(\mathbf{r}, \omega) \approx \epsilon \mathscr{F}_{\{\mathbf{E}\}}(\mathbf{r}, \omega). \tag{2.48}$$

In particular, this relationship is valid when considering the positive part only of these frequency domain representations. But from our discussion so far, these positive sidebands are nothing more than the spectra of the analytic vector fields associated with $\mathbf{D}$ and $\mathbf{E}$, i.e. $\mathbf{D}_a$ and $\mathbf{E}_a$ respectively. We can thus write

$$\mathscr{F}_{\{\mathbf{D}_a\}}(\mathbf{r}, \omega) = \epsilon \mathscr{F}_{\{\mathbf{D}_a\}}(\mathbf{r}, \omega). \tag{2.49}$$

Taking the inverse Fourier transform of this relationship results in

$$\mathbf{D}_a = \epsilon \mathbf{E}_a. \tag{2.50}$$

Given that all the complex envelope fields are defined as centered around the same center angular frequency ω_c, we can then write from equation (2.32) that

$$\tilde{\mathbf{D}} e^{j\omega_c t} = \epsilon \tilde{\mathbf{E}} e^{j\omega_c t}, \tag{2.51}$$

so that

$$\tilde{\mathbf{D}} = \epsilon \tilde{\mathbf{E}}. \tag{2.52}$$

The same derivation can obviously be carried out for the magnetic components.

Given that relationships like equations (2.46) and (2.52) still hold when dealing with sufficiently narrowband modulation schemes, we can then directly transpose the electromagnetic equations derived for complex fields in the pure monochromatic case in Section 2.1.1 to complex envelope fields in the present narrowband modulation case. This holds as long as those complex envelope fields are all defined as centered around the same center angular frequency, equal to ω_c here. Consequently, as the analytical expressions for Maxwell's equations remain identical, so does the expression for their solutions. We can thus directly reuse the results from Section 2.1.1 to derive the far-field structure in the narrowband modulation case. For instance, reconsidering the decomposition of the modulated current density complex envelope vector field given by equation (2.30), we can express the radiating characteristic function of the antenna in the narrowband modulation case from equation (2.26) as

$$\mathbf{F}(\mathbf{u}, t) = \frac{jk_c}{4\pi} \int_{S_a} \tilde{\jmath}_T(r' + \|\mathbf{r} - \mathbf{r}'\|, t)(\eta \tilde{\mathbf{J}}_s(\mathbf{r}') \times \mathbf{u}) \times \mathbf{u}\, e^{j\mathbf{k}_c \cdot \mathbf{r}'} dS', \tag{2.53}$$

where we use the notation of Figure 2.1. In particular, $\mathbf{k}_c$ stands for the wave vector corresponding to the carrier angular frequency. Its norm is thus equal to ω_c/v. We can also observe in this expression the dependency in $(r' + \|\mathbf{r} - \mathbf{r}'\|)/v$ that models the finite speed of propagation of the information. This term can be decomposed as the sum of the term r'/v that represents the propagation delay along the antenna, and the term $\|\mathbf{r} - \mathbf{r}'\|/v$ that represents the delay between the radiating element and the observation point. However, these dependencies can be further simplified, remembering that we are focusing here on the radiated far-field. Indeed, using equation (2.20) to expand $\|\mathbf{r}-\mathbf{r}'\|$, we can rewrite the above equation as

$$\mathbf{F}(\mathbf{u},t)=\frac{\mathrm{j}k_{\mathrm{c}}}{4\pi}\int_{\mathrm{S_a}}\tilde{\jmath}_{\mathrm{T}}(r'-\mathbf{u}\cdot\mathbf{r}'+r,t)(\eta\tilde{\mathbf{J}}_{\mathrm{s}}(\mathbf{r}')\times\mathbf{u})\times\mathbf{u}\,\mathrm{e}^{\mathrm{j}\mathbf{k}_{\mathrm{c}}\cdot\mathbf{r}'}\mathrm{d}S'. \tag{2.54}$$

We observe that, still due to the narrowband nature of the modulation, we can assume that the modulation state of the current distribution remains almost constant over the antenna device at a given time. In other words, due to its high propagation speed, the current wave can spread all over the antenna within the characteristic time frame of the modulation scheme. In an analytical form, this means that for antennas of reasonable dimensions we should be able to write that $\tilde{\jmath}_{\mathrm{T}}(r'-\mathbf{u}\cdot\mathbf{r}'+r,t)$ is of the form $\tilde{\jmath}_{\mathrm{T}}(f(\mathbf{r}'_0)+r,t)$ for all t, where $f(\mathbf{r}'_0)$ is a function of $\mathbf{r}'_0$, $\mathbf{r}'_0$ referring to a given point on the surface of the antenna. Here again, given that this behavior results from the narrowband nature of the modulation scheme, we should be able to establish this result more easily in the frequency domain. With that in mind, we continue to assume that the Fourier transform of the current density distribution for our realization of the modulation process exists. Performing a simple change of variable on t, we can thus write

$$\begin{aligned}&\int_{-\infty}^{+\infty}\tilde{\jmath}_{\mathrm{T}}\left(t-\frac{(l(\mathbf{r}')+r)}{v}\right)\mathrm{e}^{-\mathrm{j}\omega t}\mathrm{d}t\\&=\mathrm{e}^{-\mathrm{j}\omega\frac{(l(\mathbf{r}')-l(\mathbf{r}'_0))}{v}}\int_{-\infty}^{+\infty}\tilde{\jmath}_{\mathrm{T}}\left(t'-\frac{(l(\mathbf{r}'_0)+r)}{v}\right)\mathrm{e}^{-\mathrm{j}\omega t'}\mathrm{d}t',\end{aligned} \tag{2.55}$$

where $l(\mathbf{r}')$ stands for the length $r'-\mathbf{u}\cdot\mathbf{r}'$. Thus, given that the carrier wavelength λ_{c} is related to the carrier angular frequency through $v=\lambda_{\mathrm{c}}\omega_{\mathrm{c}}/2\pi$, we can write

$$\begin{aligned}&\mathcal{F}_{\{\tilde{\jmath}_{\mathrm{T}}(l(\mathbf{r}')+r,t)\}}(l(\mathbf{r}')+r,\omega)\\&=\mathrm{e}^{-\mathrm{j}2\pi\frac{\omega}{\omega_{\mathrm{c}}}\frac{(l(\mathbf{r}')-l(\mathbf{r}'_0))}{\lambda_{\mathrm{c}}}}\mathcal{F}_{\{\tilde{\jmath}_{\mathrm{T}}(l(\mathbf{r}'_0)+r,t)\}}(l(\mathbf{r}'_0)+r,\omega).\end{aligned} \tag{2.56}$$

Assuming that $\tilde{\jmath}_{\mathrm{T}}(r,t)$ has a time-related spectrum that spreads over $[-\Omega,+\Omega]$ with $\Omega\ll\omega_{\mathrm{c}}$, we can then write that $\mathrm{e}^{-\mathrm{j}2\pi\frac{\omega}{\omega_{\mathrm{c}}}\frac{l(\mathbf{r}')-l(\mathbf{r}'_0)}{\lambda_{\mathrm{c}}}}\approx 1$ as long as the variations of $\mathbf{r}'-\mathbf{r}'_0$, and thus the size of the antenna in practice, are not much greater than a few wavelengths. Thus, taking the inverse Fourier transform of the above equation leads to

$$\tilde{\jmath}_{\mathrm{T}}(l(\mathbf{r}')+r,t)\approx\tilde{\jmath}_{\mathrm{T}}(l(\mathbf{r}'_0)+r,t) \tag{2.57}$$

for all the possible vectors $\mathbf{r}'$ referring to a position on the antenna. We can thus deduce that under our assumptions the current density has an instantaneous amplitude and phase that are almost constant over the antenna, and in particular equal to the magnitude and phase of the time dependent part of the current density wave flowing from the transmitter through the antenna connector. Practically speaking, given that this time dependent part is nothing more than an image of the RF bandpass modulated waveform we want to transmit, we get that its complex envelope at the antenna connector $\tilde{j}_{\mathrm{T,TX}}(t)$ is proportional to $p(t) + \mathrm{j}q(t)$, with $p(t)$ and $q(t)$ the real and imaginary parts of the complex envelope of the modulating waveform. As a result, we can write that

$$\tilde{j}_{\mathrm{T}}(l(\mathbf{r}'_0) + r, t) \approx \tilde{j}_{\mathrm{T,TX}}(t - r/v)$$

so that the radiation characteristic function of the transmit antenna, given by equation (2.54), reduces to

$$\mathbf{F}(\mathbf{u}, t) = \frac{\mathrm{j}k_{\mathrm{c}}}{4\pi}\tilde{j}_{\mathrm{T,TX}}(t - r/v)\int_{\mathrm{S_a}} (\eta\tilde{\mathbf{J}}_{\mathrm{s}}(\mathbf{r}') \times \mathbf{u}) \times \mathbf{u}\,\mathrm{e}^{\mathrm{j}\mathbf{k}_{\mathrm{c}}\cdot\mathbf{r}'}\mathrm{d}S'. \tag{2.58}$$

We can then rewrite equation (2.25) in the narrowband modulated case as

$$\tilde{\mathbf{E}} = \frac{\mathrm{e}^{-\mathrm{j}k_{\mathrm{c}}r}}{r}\tilde{j}_{\mathrm{T,TX}}(t - r/v)\mathbf{F}(\mathbf{u}), \tag{2.59a}$$

$$\tilde{\mathbf{H}} = \frac{1}{\eta}\mathbf{u} \times \tilde{\mathbf{E}}, \tag{2.59b}$$

with $\mathbf{F}(\mathbf{u})$ still given by equation (2.26). Thus, reconsidering the decomposition of the complex envelope field given by equation (2.30), we can write

$$\tilde{E}_{\mathrm{T}}(r, t) = \tilde{H}_{\mathrm{T}}(r, t) = \tilde{j}_{\mathrm{T,TX}}(t - r/v) \propto p(t - r/v) + \mathrm{j}q(t - r/v). \tag{2.60}$$

We find the behavior we were expecting, i.e. that the far-field radiated by the antenna is also time varying proportionally to the considered modulation.

The same kind of reasoning could be applied to a receiving antenna illuminated by a bandpass electromagnetic field that exhibits the above structure. Referring to the discussion at the end of Section 2.1.1, we would then find that the induced EMF has the same time dependency as the incoming waveform. This thus illustrates in a convenient way the propagation of information through a wireless link in the narrowband modulation case. As a side effect, the present derivation also highlights that the structure of the field as given by equation (2.28) is deeply related to the bandpass behavior of its time domain dependent part. As recalled in Section 1.1.1, it is indeed this bandpass behavior that is the root cause for the possibility of having a physical quantity like the electrical field, which is after all a real quantity, whose time dependency carries information that needs to be represented by two real lowpass waveforms $p(t)$ and $q(t)$.

2.1.3 Radiated Power

Let us now focus on another aspect of a wireless transceiver, namely the fact that a transmitter needs to deliver RF power to the transmit antenna. After all, the information we want to process, or transmit in the present case, is both carried by the RF current wave, and by the

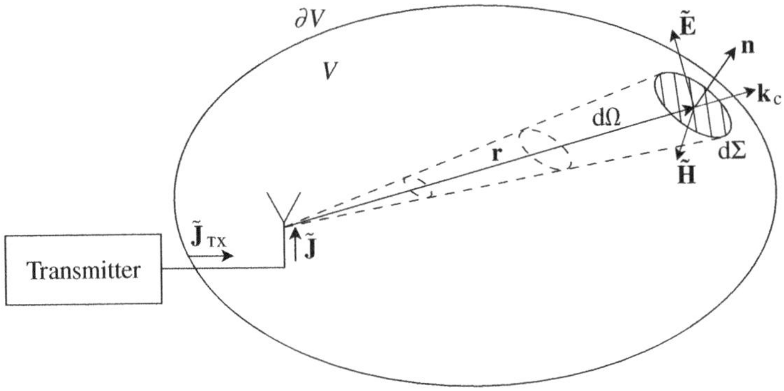

Figure 2.4 Configuration for the derivation of the power budget of a radiating antenna – According to Poynting's theorem in the form given by equation (2.67), the power radiated through the closed surface ∂V is equal to the sum of the power provided by the transmitter through the current component $\tilde{\mathbf{J}}_{TX}$, the power dissipated by Joule effect in the antenna conductor, and the time domain variations of the electromagnetic energy stored in the volume V, bounded by ∂V.

corresponding RF voltage[3] wave flowing from the transmitter. But we may wonder why we need to deliver *both* to the antenna and thus consume power. This is an important difference compared to the receive side where we can imagine configurations where the analog devices process only one of the two quantities in order to recover the information while little, if any, power is delivered to their load. The deep reason for this difference comes from the electromagnetic theory, which requires power for radiation to occur.

In order to illustrate this, let us consider a system composed of a transmit antenna that is fed by a transmitter that delivers a bandpass RF current to it. This antenna is thus expected in turn to radiate a bandpass electromagnetic field. In order to balance the power that is provided and consumed by the system, we assume it to be within a closed surface ∂V that delimits a volume V as illustrated in Figure 2.4. Accordingly, we can write the electromagnetic energy, $T(t)$, that is stored at time t in the volume V as [15]

$$T(t) = T_{\mathbf{E}}(t) + T_{\mathbf{H}}(t). \tag{2.61}$$

Here

$$T_{\mathbf{E}}(t) = \frac{1}{2}\int_V \mathbf{E}\cdot\mathbf{D}\,\mathrm{d}V \tag{2.62}$$

is the stored electrical energy and

$$T_{\mathbf{H}}(t) = \frac{1}{2}\int_V \mathbf{H}\cdot\mathbf{B}\,\mathrm{d}V \tag{2.63}$$

[3] The concept of RF voltage is discussed in Section 2.2.1.

is the stored magnetic energy. If in addition the antenna radiates in a medium that can be assumed perfect, then we have that $\mathbf{D} = \epsilon\mathbf{E}$ and $\mathbf{B} = \mu\mathbf{H}$, with ϵ and μ constant. We can then write

$$T(t) = \frac{1}{2}\int_V \epsilon\|\mathbf{E}\|^2 \mathrm{d}V + \frac{1}{2}\int_V \mu\|\mathbf{H}\|^2 \,\mathrm{d}V. \tag{2.64}$$

Nevertheless, from the discussion in Section 2.1.2 and in particular the transposition of equation (2.52) for the corresponding real bandpass electromagnetic quantities, we observe that this holds even for a medium such that ϵ and μ are frequency dependent, as long as their variations remain weak enough in the frequency band of the modulation we are dealing with. Consequently, under the assumption of sufficiently narrowband modulation we can continue to assume that ϵ and μ are constant in time in this expression for $T(t)$. The energy stored in the volume V, equal to the time derivative of this expression, can thus be simply expressed as

$$\frac{\mathrm{d}T(t)}{\mathrm{d}t} = \int_V \epsilon\mathbf{E}\partial_t\mathbf{E} + \mu\mathbf{H}\partial_t\mathbf{H}\,\mathrm{d}V. \tag{2.65}$$

In order to rewrite this expression in a form suited to our purposes, we can use Maxwell's equations to replace the time domain derivative. As a result, using equation (2.2), we can write

$$\frac{\mathrm{d}T(t)}{\mathrm{d}t} = \int_V \mathbf{E}\cdot(\nabla\times\mathbf{H} - \mathbf{J} - \mathbf{J}_{\mathrm{TX}}) - \mathbf{H}\cdot(\nabla\times\mathbf{E})\,\mathrm{d}V. \tag{2.66}$$

In this expression, we get a clear distinction between the current fed by the transmitter, which we can thus assume located at the antenna connector, for instance, and leading to the term $\mathbf{J}_{\mathrm{TX}}$, and the current density occurring on the antenna itself, $\mathbf{J}$. If we now assume that the antenna is made of a conducting material that follows Ohm's law, i.e. such that $\mathbf{J} = \sigma\mathbf{E}$ with σ its conductivity, we can then expand the above equation in the form

$$W_{\mathrm{TX}} = \frac{\mathrm{d}T(t)}{\mathrm{d}t} + W_{\mathrm{r}} + W_{\mathrm{j}}. \tag{2.67}$$

In this expression,

$$W_{\mathrm{TX}} = -\int_V \mathbf{E}\cdot\mathbf{J}_{\mathrm{TX}}\,\mathrm{d}V \tag{2.68}$$

represents the power delivered by the transmitter to the antenna, and

$$W_{\mathrm{j}} = \int_V \frac{1}{\sigma}\|\mathbf{J}\|^2\,\mathrm{d}V \tag{2.69}$$

the power dissipated by Joule effect in the conductor the antenna is made of. The last term we need to interpret is thus:

$$W_{\mathrm{r}} = -\int_V \mathbf{E}\cdot(\nabla\times\mathbf{H}) - \mathbf{H}\cdot(\nabla\times\mathbf{E})\,\mathrm{d}V. \tag{2.70}$$

We can rewrite this in a more explicit way by using on the one hand the relationship

$$\mathbf{E} \cdot (\nabla \times \mathbf{H}) - \mathbf{H} \cdot (\nabla \times \mathbf{E}) = \nabla \cdot \mathbf{P}, \tag{2.71}$$

where

$$\mathbf{P} = \mathbf{E} \times \mathbf{H} \tag{2.72}$$

is the Poynting vector, and on the other hand Stokes' theorem which states that

$$\int_V \nabla \cdot \mathbf{P} \, \mathrm{d}V = \int_{\partial V} \mathbf{P} \cdot \mathbf{n} \, \mathrm{d}\Sigma, \tag{2.73}$$

where $\mathbf{n}$ represents the outgoing normal to ∂V. Finally, equation (2.70) can be written as

$$W_\mathrm{r} = \int_{\partial V} \mathbf{P} \cdot \mathbf{n} \, \mathrm{d}\Sigma. \tag{2.74}$$

Based on these expressions for W_TX, W_j and W_r, we can construct an interpretation of equation (2.67). We see that the power delivered by the transmitter to the antenna through the term W_TX is distributed between the variation of the electromagnetic energy stored in the volume V through the term $\mathrm{d}T(t)/\mathrm{d}t$, the heating of the conductors by the Joule effect through the term W_j, and the flux of the Poynting vector through the boundary surface of V, as given by the term W_r. This last term therefore represents the power radiated through the electromagnetic field over the surface ∂V. We thus see that the Poynting vector physically represents the power density per unit area carried by the electromagnetic field in the corresponding direction. In fact, we have derived nothing more than Poynting's theorem through equation (2.67).

However, focusing on the illustration of practical wireless links, we can go a step further and use the results derived under the narrowband modulation assumption. Referring to our discussion in Section 2.1.2, we see that when dealing with sufficiently narrowband modulation schemes regarding the carrier angular frequency, we can directly transpose the electromagnetic equations derived for complex fields in the pure monochromatic case to complex envelope fields in the narrowband modulation case. The condition for this to hold is to have the complex envelope fields all defined as centered around the same center angular frequency, chosen as equal to the carrier angular frequency ω_c here. Consequently, using on the one hand the transposition of equation (2.71) for complex envelope fields, and on the other hand the transposition of Maxwell's equations originally given in the pure monochromatic case by equation (2.9), we can write

$$\begin{aligned}\frac{1}{2}\nabla \cdot (\tilde{\mathbf{E}} \times \tilde{\mathbf{H}}^*) &= \frac{1}{2}[\tilde{\mathbf{H}}^* \cdot (\nabla \times \tilde{\mathbf{E}}) - \tilde{\mathbf{E}} \cdot (\nabla \times \tilde{\mathbf{H}}^*)] \\ &= \mathrm{j}\omega_\mathrm{c}\left(\epsilon\frac{\|\tilde{\mathbf{E}}\|^2}{2} - \mu\frac{\|\tilde{\mathbf{H}}\|^2}{2}\right) - \frac{1}{\sigma}\frac{\|\tilde{\mathbf{J}}\|^2}{2} - \frac{\tilde{\mathbf{E}} \cdot \tilde{\mathbf{J}}_\mathrm{TX}}{2}.\end{aligned} \tag{2.75}$$

We can then define the Poynting vector in the narrowband modulation case through its complex envelope vector field as

$$\tilde{\mathbf{P}} = \frac{1}{2}\tilde{\mathbf{E}} \times \tilde{\mathbf{H}}^*. \tag{2.76}$$

This vector is constant in time when dealing with pure monochromatic waveforms, but has a time dependent part when dealing with a narrowband modulation. Indeed, considering the structure of the radiated far-field given by equation (2.59b), we can write

$$\tilde{\mathbf{P}} = \frac{1}{2}\tilde{\mathbf{E}} \times (\mathbf{u} \times \tilde{\mathbf{E}}^*) = \frac{1}{2\eta}\|\tilde{\mathbf{E}}\|^2\mathbf{u}, \tag{2.77}$$

or, considering the general decomposition of the complex envelope field as given by equation (2.30),

$$\tilde{\mathbf{P}} = \tilde{\mathbf{P}}_S(\mathbf{r})\tilde{P}_T(r,t) = \frac{1}{2\eta}|\tilde{E}_T(r,t)|^2\|\tilde{\mathbf{E}}_S\|^2\mathbf{u}. \tag{2.78}$$

Assuming that equation (2.60) holds, we then get that

$$\tilde{P}_T(r,t) = |\tilde{E}_T(r,t)|^2 = |\tilde{H}_T(r,t)|^2. \tag{2.79}$$

We thus see that the time dependent part of the Poynting vector, and thus the time domain variations of the instantaneous power of the electromagnetic field at the position considered, is simply related to the square magnitude of the time dependent part of the electromagnetic complex envelope field. As this time dependent part has by definition unit average power (see equation (2.36)), we then find that even if the instantaneous electromagnetic power flowing through a given surface fluctuates according to the electromagnetic field modulation, the average power flowing through it is in turn related only to the space dependent part of the Poynting vector as given by the pure monochromatic theory.

What is interesting to see for our purposes is that the power density carried by the electromagnetic field radiated by *any* antenna, as given by the norm of the Poynting vector, is not null. This is in fact something that we could have guessed, knowing that wireless links work effectively. This means that the radiated electromagnetic field succeeds in inducing a current on a receive antenna. But, as electrons have a non-vanishing mass, this means that a transfer of energy occurs between the electromagnetic field and the electrons in order to provide them with kinetic energy. And for that, we at least need an electromagnetic field that carries power. From our wireless transceiver perspective, this simply means that we need to supply RF power on the transmit side. And even if the maximum amount of power to be provided can vary from one wireless standard to another, this often leads to implementation problems or at least constraints in addition to the obvious power consumption problems. Technologies are available that can be used to deliver such RF power. This fact often leads to the use of external PAs, designed using suitable technologies. We also observe that such high power RF signals, implemented close to other RF functionalities, can lead to RF pollution. This can happen when coupling occurs between the transmitted RF signal and the RF synthesizers that are used for frequency conversions. It can even lead to the frequency pulling of RF voltage controlled oscillators through injection locking. This problem particularly has to be considered for some transceiver architectures more sensitive to this, as detailed in Section 8.1.1, for instance.

2.1.4 Free Space Path Loss

We can now use the results of the previous sections to derive what is called the free space path loss. This term refers to the loss that occurs in a wireless link between the power provided by the transmitter to the transmit antenna and the power delivered by the receive antenna to its load, i.e. the receiver. It is of interest to understand the mechanism involved in the free space propagation case as an introduction to the discussion in Section 2.3 concerning the characteristics of the wireless propagation channel in a moving environment.

We begin by recalling the definition of the Poynting vector derived in the previous section. This vector is defined in such a way that its flux over a given surface represents the instantaneous electromagnetic power that flows through it. Thus, for narrowband modulated fields, under stationarity and ergodicity we can express the radiated average power, $\delta W_\text{r}(r, \mathbf{u})$, through the elementary surface element $\text{d}\Sigma$ located at distance r as the expectation of equation (2.77) times the projection of this surface element orthogonally in the propagation direction:

$$\delta W_\text{r}(r, \mathbf{u}) = \frac{1}{2\eta}\mathbb{E}\{\|\tilde{\mathbf{E}}_{\mathbf{r},t}\|^2\}\text{d}\Sigma\mathbf{n}\cdot\mathbf{u}. \tag{2.80}$$

Thus, for a position in the far-field of the transmit antenna, i.e. assuming that the structure of the electromagnetic field is given by equation (2.59), and remembering that equation (2.60) holds, we can write the above expression as

$$\delta W_\text{r}(r, \mathbf{u}) = \frac{1}{2\eta}\|\mathbf{F}(\mathbf{u})\|^2\mathbb{E}\{|\tilde{E}_{\text{T},r,t}|^2\}\frac{\text{d}\Sigma\mathbf{n}\cdot\mathbf{u}}{r^2}, \tag{2.81}$$

with $\mathbf{n}$ the outgoing normal to $\text{d}\Sigma$ as shown in Figure 2.1. But, as the time dependent part of the complex envelope field is defined so that it has unit power as given by equation (2.36), we finally get that

$$\delta W_\text{r}(\mathbf{u}) = \frac{1}{2\eta}\|\mathbf{F}(\mathbf{u})\|^2\text{d}\Omega, \tag{2.82}$$

with $\text{d}\Omega = \text{d}\Sigma\mathbf{n}\cdot\mathbf{u}/r^2$ the elementary solid angle under which $\text{d}\Sigma$ is seen by the radiating antenna. We thus see that the power radiated through the elementary surface $\text{d}\Sigma$ only depends on the solid angle $\text{d}\Omega$, and not on the distance r. We also remark that this radiated power density per solid angle unit depends only on the magnitude of the radiation characteristic vector of the antenna. This quantity thus characterizes the ability of the radiating antenna to focus the electromagnetic power in a given direction $\mathbf{u}$.

This behavior allows us to define the concept of directivity of the radiating antenna. Supposing that the transmit antenna is fed by a power P_TX, the directivity is simply defined as the ratio between the power density that is effectively radiated in a given direction and the power density that would be radiated in that direction by the same antenna if it were isotropic. As its name suggests, an isotropic antenna is nothing more than the theoretical radiating element that has no preferred direction, i.e. that radiates the same power density in every direction. It is thus the appropriate element to refer to the directivity of an antenna. We therefore classically define the directivity of a transmit antenna, $D_{\mathrm{TX}}(\mathbf{u})$, as the ratio[4]

$$D_{\mathrm{TX}}(\mathbf{u}) = \frac{\delta W_{\mathrm{r}}(\mathbf{u})}{\delta W_{\mathrm{r,i}}(\mathbf{u})}, \tag{2.83}$$

where $\delta W_{\mathrm{r,i}}(\mathbf{u})$ is the power radiated by the theoretical isotropic antenna through the same elementary surface $\mathrm{d}\Sigma$, when fed by the same power P_{TX}. This quantity can thus be easily derived using the isotropic assumption. Indeed, as the elementary surface $\mathrm{d}\Sigma$ is assumed to be at a distance r from the antenna, the radiated power density per unit area at that distance is simply equal to $P_{\mathrm{TX}}/4\pi r^2$ as $4\pi r^2$ represents the area of the sphere of radius r. As a result, we have

$$\delta W_{\mathrm{r,i}}(\mathbf{u}) = \frac{P_{\mathrm{TX}}}{4\pi r^2}\mathrm{d}\Sigma\mathbf{n}\cdot\mathbf{u} = P_{\mathrm{TX}}\frac{\mathrm{d}\Omega}{4\pi}. \tag{2.84}$$

Using this result and equation (2.82), we can finally express the antenna directivity defined by equation (2.83) as

$$D_{\mathrm{TX}}(\mathbf{u}) = \frac{2\pi}{\eta}\frac{\|\mathbf{F}(\mathbf{u})\|^2}{P_{\mathrm{TX}}}. \tag{2.85}$$

Let us pause for a moment and comment on this expression. Recalling the definition of $\mathbf{F}(\mathbf{u})$ given by equation (2.26), we see that its square norm is in fact proportional to the square magnitude of the current that is provided to the antenna. It follows that the above ratio $\|\mathbf{F}(\mathbf{u})\|^2/P_{\mathrm{TX}}$ is in fact independent of the power supplied to the antenna. We thus obtain that $D_{\mathrm{TX}}(\mathbf{u})$ effectively characterizes only the ability of the transmit antenna to focus the radiated electromagnetic power in the direction $\mathbf{u}$.

However, in the above definition of directivity, we assumed no loss in the antenna itself. But, due either to the finite conductivity of practical conductors or to the losses in the dielectrics, for instance, only a fraction of the power P_{TX} that is provided to the antenna is effectively radiated. Practically speaking, this means that the magnitude of the current distribution that has to be considered in the derivation of $\mathbf{F}(\mathbf{u})$, i.e. in equation (2.26), is only a fraction of the magnitude of the current flowing from the transmitter. This leads to the concept of antenna gain, defined as the antenna directivity times the antenna efficiency. It follows from this definition that the radiated power density per unit area orthogonal to the propagation direction $\mathbf{u}$, $\partial_\Sigma W_{\mathrm{r}}(\mathbf{u})$, when the antenna is fed with a power P_{TX} is simply equal to

$$\partial_\Sigma W_{\mathrm{r}}(\mathbf{u}) = \frac{\delta W_{\mathrm{r}}(\mathbf{u})}{\mathrm{d}\Sigma\mathbf{n}\cdot\mathbf{u}} = D_{\mathrm{TX}}(\mathbf{u})\frac{\xi P_{\mathrm{TX}}}{4\pi r^2}. \tag{2.86}$$

[4] This is the reason why the directivity, and the gain as introduced in the sequel, are often expressed in dBi units. Here, the "i" suffix stands for isotropic. For instance, $D_{\mathrm{TX}}(\mathbf{u}) = 3$ dBi means that the antenna radiates twice as much power as would an isotropic antenna of the same efficiency in the direction $\mathbf{u}$ when fed with the same power.

In this expression, $P_{\rm TX}/4\pi r^2$ represents the power density that would be radiated by an ideal, i.e. lossless, isotropic antenna, and ξ the antenna efficiency, i.e. the ratio between the total power that is effectively radiated in the entire space and the power fed by the transmitter. We therefore have

$$\partial_\Sigma W_{\rm r}(\mathbf{u}) = G_{\rm TX}(\mathbf{u})\frac{P_{\rm TX}}{4\pi r^2}, \tag{2.87}$$

where $G_{\rm TX}(\mathbf{u}) = \xi D_{\rm TX}(\mathbf{u})$ is the transmit antenna gain in the direction $\mathbf{u}$.

Now that we have introduced those concepts, we are ready to derive a simple link budget. In fact, we already have from the above equation the power density per unit area radiated by the transmit antenna, i.e. $\partial_\Sigma W_{\rm r}(\mathbf{u})$. It thus remains to evaluate the fraction of power that is effectively recovered by the receive antenna and delivered to its load, i.e. to the receiver. Assuming that $\mathbf{u}$ effectively represents the direction of the receive antenna and r its distance from the transmit antenna, we simply need to multiply $\partial_\Sigma W_{\rm r}(\mathbf{u})$ by an amount that has the dimension of an area. This quantity is in fact what is called the effective area, $S_{\rm RX}(\mathbf{u})$, of the receiving antenna when illuminated by an incoming plane wave from direction $\mathbf{u}$. As the name suggests, the effective area of an antenna does not necessarily match its physical size. For instance, considering a simple half-wave dipole that can be implemented using wires, we cannot really talk about an area for this one-dimensional device. However, as the line strength of the electromagnetic field is bent at the boundary of the wire, we can talk about the area around the receive antenna, orthogonal to the propagation direction of the incoming wave, which corresponds to the fraction of energy retrieved by the antenna from the electromagnetic field. For some antenna shapes, and in the high frequency approximation, this effective absorption area does match the physical size of the device. This is true, in particular, for aperture antennas like horns. In any case, it can be shown that this effective area is linked to the gain of the receive antenna $G_{\rm RX}(\mathbf{u})$ by [12]

$$S_{\rm RX}(\mathbf{u}) = \frac{\lambda_{\rm c}^2}{4\pi}G_{\rm RX}(\mathbf{u}). \tag{2.88}$$

Here $\lambda_{\rm c}$ represents the carrier wavelength that is related to the carrier angular frequency $\omega_{\rm c}$ according to $\lambda_{\rm c} = 2\pi v/\omega_{\rm c}$, with v the speed of the electromagnetic wave which equals c in free space.

There are other phenomena to consider in deriving the power retrieved at the receive antenna connector. The condition of polarization of the incoming wave compared to that of the receiving antenna should be considered separately as the definition of the effective area only refers to gain and thus to the norm of the radiation characteristic vector. In the same way, we should take into account the impedance matching between the receive antenna and its load. This power matching operation, detailed in Section 2.2.3, directly impacts the amount of power retrieved by the receive antenna as any reflected power is in fact radiated again.

Here, we can assume for the sake of simplicity that each of these two phenomena is compensated so that we get the optimum power reception, thus corresponding to the configuration shown on Figure 2.5. It follows that the power delivered by the receive antenna to its load $P_{\rm RX}$ can be written as the power density per unit area radiated by the transmit antenna,

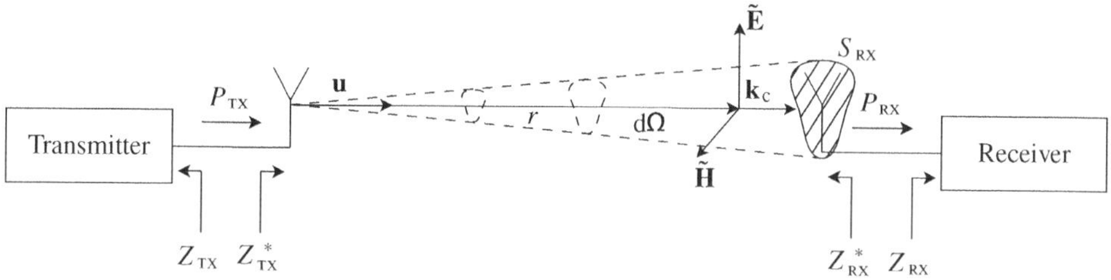

Figure 2.5 Configuration for the derivation of a transmission wireless link budget – Assuming on the one hand the power matching between the transmitter and the transmit antenna and between the receive antenna and the receiver, and on the other hand that the antennas have the same polarization, we can derive the transmission link budget in free space as given by equation (2.90).

$\partial_\Sigma W_r(\mathbf{u})$, times the effective area of the receive antenna $S_{RX}(\mathbf{u})$. It follows from equation (2.87) that

$$P_{RX} = P_{TX}G_{TX}(\mathbf{u})\frac{S_{RX}(\mathbf{u})}{4\pi r^2}. \tag{2.89}$$

This relationship can also be expressed as a function of the receive antenna gain using equation (2.88). This leads to the classical Friis transmission equation in its simplest form [16],

$$P_{RX} = P_{TX}G_{TX}(\mathbf{u})G_{RX}(\mathbf{u})\left(\frac{\lambda_c}{4\pi r}\right)^2. \tag{2.90}$$

In this expression, the term

$$L(r) = \left(\frac{4\pi r}{\lambda_c}\right)^2 \tag{2.91}$$

is called the free space path loss.

Looking at this expression, it appears at first glance that the free space path loss increases as the frequency does. In other words, for a given amount of power provided to the transmit antenna and a given geometrical configuration between the transmit and the receive antenna, it seems that the amount of power delivered to the receiver by the receive antenna decreases as the carrier angular frequency increases. This is indeed true for antennas with a constant gain. But what can be confusing is to say that it is the path loss that increases with the carrier angular frequency. Indeed, free space remains a lossless medium that is independent of the frequency of the electromagnetic field that propagates through it. And, looking again at equation (2.87), we see that the radiated power density per unit area orthogonal to the propagation direction $\mathbf{u}$, $\partial_\Sigma W_r(\mathbf{u})$, when the antenna is fed with a power P_{TX}, is independent of the carrier angular frequency for a given transmit antenna gain. Thus, as long as the physical areas of receive antennas considered at different carrier angular frequencies are identical, i.e. have the same effective areas, we get the same recovered power on the receive side as given by equation (2.89). It is only when we express the received power as a function of the receive antenna gains instead of their effective areas that this dependency on the carrier angular frequency appears. The deep reason for this is that, for a given antenna physical area, we get a higher gain capability when the frequency increases according to equation (2.88). This general behavior means that the larger the antenna in terms of wavelength, the better its ability to focus the electromagnetic power in a given direction. As a result, when considering receive antennas with the same gain whatever the carrier angular frequency, their physical and thus their effective areas are necessarily different. This means that they necessarily retrieve different fractions of power from the incoming electromagnetic field as their different physical sizes, or effective areas, make them intercept a different fraction of the incoming electromagnetic field power. It thus looks like the path loss increases when the carrier angular frequency increases. But it is only the effective area of the receive antenna that necessarily decreases when considering constant gain receive antennas.

This behavior is often encountered in wireless systems as handsets are for the most part defined and designed incorporating almost isotropic antennas. Indeed, unless using smart space filtering based on multi-antenna systems, such an isotropic antenna is a necessity – we can hardly define in advance a preferred direction for the wireless link as we do not know a priori the position of the user. But, in accordance with our discussion, having antennas with the same gain whatever the frequency bands considered means having different effective areas for the antennas and thus different path losses. This effect is one of the main reasons, on top of the multipath propagation conditions in effective operating conditions, for having reduced area cells in high frequency bands compared to what can be achieved in low frequency ones.

To conclude, we can mention that the main reason for deriving a power link budget as we have done here is that most receivers are power matched to the receive antenna. However, we can imagine configurations for which this is untrue, as discussed in Section 2.2.3. In that case, the received power derived above would be interpreted as the available power gain at the receive antenna connector. Equivalent electrical parameters could then be derived using the antenna and receiver impedances.

2.2 Guided Propagation

At this point it is appropriate to say a few words about the constraints linked to electromagnetic guided propagation. This mode of conduction may be used in different parts of a transceiver, for instance between the radiating elements and the active parts of the line-up. Practically speaking, the length of the corresponding lines may be much greater than the carrier wavelength so that the propagation phenomenon cannot be avoided and some related implementation constraints have to be considered. The results below complement our discussion of the free space case in order to illustrate the overall mechanisms involved in the transmission of information.

2.2.1 Transmission Lines

Solving Maxwell's equations for the derivation of the structure of electromagnetic waves able to propagate along conductors, different modes of propagation exist depending on the configuration of the conductors, the carrier angular frequency, and the characteristics of the dielectrics. In the simple situation where only one dielectric is involved, i.e. when dealing with a homogeneous medium, there are three modes of propagation. These modes are classically classified depending on the existence or not of a longitudinal component for the electromagnetic field collinear to the direction of propagation. Practically speaking, when dealing with at least two conductors we may encounter [13, 17]: *transverse electromagnetic* (TEM) waves characterized by electric and magnetic fields both orthogonal to the propagation direction; *transverse electric* (TE) waves which have only their electrical component orthogonal to the propagation direction, and whose magnetic field structure exhibits a longitudinal component; *transverse magnetic* (TM) waves with the reverse configuration compared to the TE mode, i.e. where only the magnetic field has a component orthogonal to the propagation direction.

What is interesting to observe for our purposes is that for a given geometry of the conductors we are dealing with, both the TE and TM modes necessarily have a cut-off frequency. This means that those modes can exist only for a carrier frequency that is set above this cut-off frequency. This is not the case for the TEM mode, which can always occur. However, in practical implementations we may be faced with more than one dielectric as is, for instance, the case in the widely used microstrip lines. This may complicate the picture. Fortunately, when the thickness of the dielectric slab decreases we can expect to observe the behavior of the homogeneous case at some point. This is what occurs in practical implementations where a quasi-TEM mode results. The latter mode exhibits properties quite similar to the TEM mode, except for the relative permeability that now depends on the geometry of the line. Practically speaking, there are two such properties of interest:

(i) The electromagnetic field in this mode has a quasi-plane wave structure, i.e. the same structure as that of the radiated far-field in free space as detailed in Section 2.1.1. In particular, the wavenumber, i.e. the norm of the wave vector $k_c = \|\mathbf{k}_c\|$, is of the same form as it would be for a free space radiation in the same medium. It is given by $k_c = \omega_c\sqrt{\epsilon\mu}$, with ω_c the carrier angular frequency. However, the wave impedance η now depends on the geometry of the propagation structure, which is a main difference with respect to free space radiation.
(ii) Then, the transverse structure of a TEM mode allows us to define a scalar potential from which the transverse electrical field derives. This means that, even for a high frequency propagation phenomenon, we can continue to work with the concept of voltage difference between the different conductors of the line as would be done under quasi-static conditions. It follows that electrical relationships involving RF voltages and current densities can still be written locally for the structure considered.

These two properties, which do not hold for TE or TM modes, allow us to derive the wave equations fulfilled by the RF voltage v and the current density j along the conductors. However, due to the invariance by translation along the propagation direction of the guided structures used in practice, we can alternatively write those equations directly for the total current i flowing in the system rather than in terms of current density. We can thus continue to work with this current in our review of the transmission line theory.

This theory is classically derived for pure monochromatic waves. However, we can adapt those results for narrowband modulated fields, as we have already done for the free space radiation in Section 2.1.2. To do so, we first need to return to the concept of complex envelope field introduced in that section. We need to consider waves that propagate in both directions along the line in order to take into account potential reflections that were not possible in free space. This means that we now consider complex envelope fields with time dependent parts

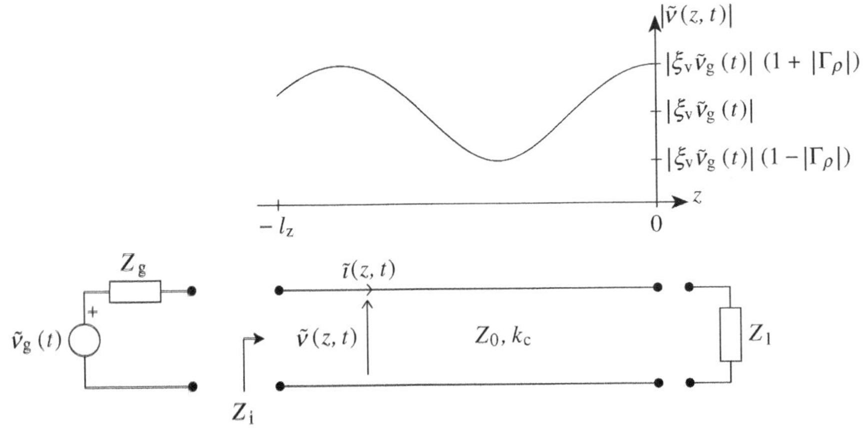

Figure 2.6 Propagation of narrowband modulated voltage and current waves along a transmission line – Assuming a TEM mode propagating along a transmission line, equivalent RF voltage and current can be defined and used for electrical derivations. Both are the sum of forward and reflected waves. As long as the transmission line length is not much than a few carrier wavelengths, the time dependent parts of the electrical quantities can be considered constant along the line.

that may be a function of either $t + z/v$ or $t - z/v$, with z the coordinate along the transmission line as shown in Figure 2.6, and v the speed of the wave. For the sake of clarity, we identify the forward wave, which increases z, by subscript + and the reflected one, which decreases z, by subscript −. Thus, following equation (2.30), we can define the complex envelopes of those scalar fields as the product of a time dependent part and a spatially dependent part:

$$\tilde{v}_\pm(z,t) = \tilde{v}_{s,\pm}(z)\tilde{v}_{T,\pm}(\mp z, t), \tag{2.92a}$$

$$\tilde{\imath}_\pm(z,t) = \tilde{\imath}_{s,\pm}(z)\tilde{\imath}_{T,\pm}(\mp z, t). \tag{2.92b}$$

Here, we must keep in mind that the dependency on t and z of the time dependent part of those complex envelopes means that we are dealing with functions of $t \pm z/v$, and that due to the time domain convention taken for the definition of the complex fields, i.e. $+j\omega t$, it is a function of $-z$ for the forward wave and of $+z$ for the reflected one.

We then need to detail the structure of the spatially dependent part of these complex envelopes. Recall that the average power of these complex envelopes reduces by definition to the magnitude of their spatially dependent parts. Thus, assuming that we are dealing with a lossless medium and, for the sake of simplicity, a propagation structure that is invariant along z, we see that this magnitude must be constant, i.e. independent of z. We can thus write $|\tilde{v}_{s,\pm}(z)| = \tilde{V}_\pm$ and $|\tilde{\imath}_{s,\pm}(z)| = \tilde{\imath}_\pm$. Moreover, under the same assumptions, the theory says that Green's function reduces to a complex exponential with a propagation constant equal to the wavenumber. We can thus finally assume in our narrowband modulated case that

$$\tilde{v}_\pm(z,t) = \tilde{V}_\pm \tilde{v}_{T,\pm}(\mp z, t)e^{\mp jk_c z}, \tag{2.93a}$$

$$\tilde{\imath}_\pm(z,t) = \tilde{I}_\pm \tilde{\imath}_{T,\pm}(\mp z, t)e^{\mp jk_c z}. \tag{2.93b}$$

Also assuming for the sake of simplicity that the complex envelopes we are dealing with are all defined as centered around the carrier angular frequency ω_c, we can express the corresponding real forward and reflected bandpass modulated waves as

$$v_{\pm}(z,t) = \mathrm{Re}\{\tilde{v}_{\pm}(z,t)e^{j\omega_c t}\} = \mathrm{Re}\left\{\tilde{V}_{\pm}\tilde{v}_{T,\pm}(\mp z,t)e^{j(\omega_c t \mp k_c z)}\right\}, \tag{2.94a}$$

$$i_{\pm}(z,t) = \mathrm{Re}\{\tilde{\imath}_{\pm}(z,t)e^{j\omega_c t}\} = \mathrm{Re}\left\{\tilde{I}_{\pm}\tilde{\imath}_{T,\pm}(\mp z,t)e^{j(\omega_c t \mp k_c z)}\right\}. \tag{2.94b}$$

Physically speaking, as the propagation equations are linear, we have that the general expression for the voltage and current along the line is a result of the linear superposition of those forward and reflected waves. We can thus write

$$v(z,t) = v_+(z,t) + v_-(z,t), \tag{2.95a}$$

$$i(z,t) = i_+(z,t) + i_-(z,t). \tag{2.95b}$$

Given that all those complex envelopes are assumed defined as centered around the same center angular frequency, we can even derive the complex envelope of the voltage and current fields from those of the respective forward and reflected waves. Indeed, based on equation (2.94), we can write

$$v(z,t) = \mathrm{Re}\left\{\tilde{v}_+(z,t)e^{j\omega_c t} + \tilde{v}_-(z,t)e^{j\omega_c t}\right\}, \tag{2.96a}$$

$$i(z,t) = \mathrm{Re}\left\{\tilde{\imath}_+(z,t)e^{j\omega_c t} + \tilde{\imath}_-(z,t)e^{j\omega_c t}\right\}. \tag{2.96b}$$

We can thus finally assume that

$$\tilde{v}(z,t) = \tilde{v}_+(z,t) + \tilde{v}_-(z,t), \tag{2.97a}$$

$$\tilde{\imath}(z,t) = \tilde{\imath}_+(z,t) + \tilde{\imath}_-(z,t). \tag{2.97b}$$

In order to go further and reuse the results of the monochromatic case, we need to say more about the narrowband modulation condition than we did in Section 2.1.2. We said in that section that in order to recover the propagation equations in the same form as in the monochromatic case, and thus obtain the same structure for their solution, we only need to ensure that the width of the spectrum of the time domain modulation is negligible with respect to the carrier angular frequency. This condition remains necessary here, but is not sufficient. Indeed, we now need to consider reflected waves due to the potential impedance mismatch at the line termination. There is thus a new phenomenon to take into account compared to the free-space radiation case. For the corresponding derivations, the monochromatic theory uses the proportionality between equivalent complex voltage and current quantities across the terminal impedance stage. This kind of relationship can also be used under the condition that the terminal impedance is assumed almost constant over the spectrum of the modulated wave. In order to understand this, let us consider a particular realization of the modulation process, and let us assume that we can take the Fourier transform of the quantities of interest, which can be assumed observed over a finite duration for instance. In the frequency domain, we can thus write for the real bandpass modulated voltage and current fields across the terminal impedance that

$$V(0,\omega) = Z_l(\omega)I(0,\omega). \tag{2.98}$$

Here, we indeed assume that the load impedance occurs at $z = 0$, as shown in Figure 2.6. Thus, if we now assume that the modulation is sufficiently narrowband so that $Z(\omega) \approx Z(\omega_c)$ for all angular frequencies within the modulated waveform spectrum, we can write

$$V(0,\omega) = Z_l(\omega_c)I(0,\omega). \tag{2.99}$$

In particular, this relationship is valid when considering the positive part only of the spectrum of $V(0,\omega)$ and $I(0,\omega)$. According to Section 2.1.2, these sidebands are in fact nothing more than the spectrum of the analytic fields associated with $v(z,t)$ and $i(z,t)$ when considered at position $z = 0$. Thus, denoting those analytic fields by $v_a(z,t)$ and $i_a(z,t)$, we have

$$V_a(0,\omega) = Z_l(\omega_c)I_a(0,\omega). \tag{2.100}$$

Now taking the inverse Fourier transform of this relationship yields

$$v_a(0,t) = Z_l(\omega_c)i_a(0,t). \tag{2.101}$$

But, as we assumed that the complex envelopes for $v(z,t)$ and $i(z,t)$, i.e. $\tilde{v}(z,t)$ and $\tilde{\imath}(z,t)$, are defined as centered around the same center angular frequency ω_c, according to equation (2.32) we can write that

$$\tilde{v}(0,t)e^{j\omega_c t} = Z_l(\omega_c)\tilde{\imath}(0,t)e^{j\omega_c t} \tag{2.102}$$

and, finally, that

$$\tilde{v}(0,t) = Z_l(\omega_c)\tilde{\imath}(0,t). \tag{2.103}$$

This means that under the narrowband assumption and dealing with complex envelopes defined as centered around the same center angular frequency, we can still write electrical relationships between those complex envelopes that represent voltages and currents as is usually done using the complex notation in the pure monochromatic case.

As a consequence, the transmission line equations can be transposed to the narrowband modulation case as

$$\tilde{v}(z,t) = \tilde{V}\left(\tilde{v}_{T,+}(-z,t)e^{-jk_c z} + \Gamma_\rho \tilde{v}_{T,-}(z,t)e^{jk_c z}\right), \tag{2.104a}$$

$$\tilde{\imath}(z,t) = \frac{\tilde{V}}{Z_0}\left(\tilde{v}_{T,+}(-z,t)e^{-jk_c z} - \Gamma_\rho \tilde{v}_{T,-}(z,t)e^{jk_c z}\right). \tag{2.104b}$$

Here, Γ_ρ stands for the amplitude reflection coefficient between the load impedance $Z_l(\omega_c)$ and the line characteristic impedance Z_0. As we are dealing with electrical quantities that are assumed derived from a TEM propagating mode, Z_0 is in fact equal to the wave impedance η of this mode.[5] An expression for Γ_ρ can then be derived form the comparison of the above transmission equations taken for $z = 0$ and of the electrical relationship between $\tilde{v}(0, t)$ and $\tilde{\imath}(0, t)$ across the load impedance as given by equation (2.103). After some algebra, we recover the classical relationship:

$$\Gamma_\rho = \frac{Z_l - Z_0}{Z_l + Z_0}, \tag{2.105}$$

where we used the notation $Z_l = Z_l(\omega_c)$.

However, we can go a step further in considering practical implementations. Classically, the length of the lines to be considered is not much more than a few wavelengths at the carrier angular frequency. It follows that having a narrowband modulation scheme leads to the time dependent part of the waves involved, i.e. $\tilde{v}_{T,\pm}(\mp z, t)$, being almost constant along the line at a given time. Once again, as this behavior is linked to the relative importance of the modulation scheme bandwidth compared to the carrier angular frequency, we can highlight it in a simple way in the frequency domain. Let us consider a realization of the modulation process over a finite duration of observation so that we assume that we can derive its Fourier transform. Performing a simple change of variables on t, we can then write for any position z along the line that

$$\int_{-\infty}^{+\infty} \tilde{v}_{T,\pm}\left(t \mp \frac{z}{v}\right) e^{-j\omega t} dt = e^{\mp j\omega \frac{(z-z_0)}{v}} \int_{-\infty}^{+\infty} \tilde{v}_{T,\pm}\left(t' \mp \frac{z_0}{v}\right) e^{-j\omega t'} dt', \tag{2.106}$$

for some reference point z_0 along the line. As the carrier wavelength λ_c of the propagating TEM wave is related to the carrier angular frequency through $v = \lambda_c \omega_c / 2\pi$, we can write

$$\mathcal{F}_{\{\tilde{v}_{T,\pm}(\mp z, t)\}}(\mp z, \omega) = e^{\mp j2\pi \frac{\omega}{\omega_c} \frac{(z-z_0)}{\lambda_c}} \mathcal{F}_{\{\tilde{v}_{T,\pm}(\mp z_0, t)\}}(\mp z_0, \omega). \tag{2.107}$$

As a result, assuming that $\tilde{v}_{T,\pm}(\mp z, t)$ has a time-related spectrum that spreads over $[-\Omega, +\Omega]$ with $\Omega \ll \omega_c$, we can write that $e^{\mp j2\pi \frac{\omega}{\omega_c} \frac{z-z_0}{\lambda_c}} \approx 1$ for ω in this frequency band as long as the variations of $z - z_0$, and thus the length of the line considered, are not much greater than a few wavelengths. Thus taking the inverse Fourier transform of the above equation leads to

$$\tilde{v}_{T,\pm}(\mp z, t) \approx \tilde{v}_{T,\pm}(\mp z_0, t) \tag{2.108}$$

for all the possible z referring to a position along the line. We can choose, for example, $z_0 = -l_z$ which refers to the input of the line as shown in Figure 2.6. In this case, we can write that $\tilde{v}_{T,\pm}(\mp z, t)$ is proportional to $\tilde{v}_g(t)$, i.e. to the complex envelope of the internal EMF of the generator connected to the line; this complex envelope being assumed defined as centered

[5] The characteristic impedance of the line thus depends on its geometrical dimensions, as does the TEM wave impedance. The line must thus be designed properly in order to achieve the appropriate target characteristic impedance for matching purposes.

around the carrier angular frequency as we have done for all the complex envelopes we deal with here. As a result, equation (2.104) reduces to

$$\tilde{v}(z,t) = \xi_{\mathrm{v}}\tilde{v}_{\mathrm{g}}(t)\left(\mathrm{e}^{-\mathrm{j}k_{\mathrm{c}}z} + \Gamma_{\rho}\mathrm{e}^{\mathrm{j}k_{\mathrm{c}}z}\right), \tag{2.109a}$$

$$\tilde{\imath}(z,t) = \frac{\xi_{\mathrm{v}}}{Z_0}\tilde{v}_{\mathrm{g}}(t)\left(\mathrm{e}^{-\mathrm{j}k_{\mathrm{c}}z} - \Gamma_{\rho}\mathrm{e}^{\mathrm{j}k_{\mathrm{c}}z}\right). \tag{2.109b}$$

In these equations, the complex quantity ξ_{v} can be determined by the relationship that must hold between the complex envelope of the voltage at the input of the line $\tilde{v}(-l_{\mathrm{z}},t)$ and the generator EMF, $\tilde{v}_{\mathrm{g}}(t)$. Indeed, using electrical relationships of the form given by equation (2.103) in the narrowband modulation case, we can write that

$$\tilde{v}(-l_{\mathrm{z}},t) = \frac{Z_{\mathrm{i}}}{Z_{\mathrm{i}} + Z_{\mathrm{g}}}\tilde{v}_{\mathrm{g}}(t), \tag{2.110}$$

with $Z_{\mathrm{i}} = Z_{\mathrm{i}}(\omega_{\mathrm{c}}) = \tilde{v}(-l_{\mathrm{z}},t)/\tilde{\imath}(-l_{\mathrm{z}},t)$ the input impedance of the line at the carrier angular frequency, and Z_{g} the generator internal impedance at the same frequency. We thus get using equation (2.109) that

$$\xi_{\mathrm{v}} = \frac{Z_{\mathrm{i}}}{Z_{\mathrm{i}} + Z_{\mathrm{g}}}\frac{1}{\left(\mathrm{e}^{\mathrm{j}k_{\mathrm{c}}l_{\mathrm{z}}} + \Gamma_{\rho}\mathrm{e}^{-\mathrm{j}k_{\mathrm{c}}l_{\mathrm{z}}}\right)} \tag{2.111}$$

and

$$Z_{\mathrm{i}} = \frac{\tilde{v}(-l_{\mathrm{z}},t)}{\tilde{\imath}(-l_{\mathrm{z}},t)} = Z_0\frac{Z_{\mathrm{l}} + \mathrm{j}Z_0\tan(k_{\mathrm{c}}l_{\mathrm{z}})}{Z_0 + \mathrm{j}Z_{\mathrm{l}}\tan(k_{\mathrm{c}}l_{\mathrm{z}})}. \tag{2.112}$$

This last expression is of interest as it shows that the impedance seen at the input of the line depends both on its length, l_{z}, and on the relative value of its characteristic impedance compared to the load impedance. In particular, we get the interesting property that Z_{i} is equal to Z_{l}, whatever the length of the line l_{z}, if and only if its characteristic impedance Z_0 is equal to Z_{l}. This condition corresponds in fact to the amplitude matching of the line.

This condition, and more generally the guidelines for transceiver implementations that can be deduced from the above transmission line theory, are discussed in the following section. We conclude the present section by mentioning two side effects that can be illustrated by this theory. First of all, this theory gives new insights that illustrate the transmission of information in a wireless link. Indeed, on the receive side, for instance, the generator considered in the above derivations, and represented in Figure 2.6, can represent the Thévenin equivalent of the receive antenna illuminated by an incoming plane wave. In that case, according to the discussion in Section 2.1.2, the time domain variations of the generator EMF are proportional to the modulation or the received plane wave, which is in turn proportional to the radiated modulation at the transmit antenna level. We then obtain via equation (2.109) that this modulation is transmitted up to the load impedance of the line, which can be the input impedance of the active part of the receiver, for instance. In the same way, on the transmit side, the generator can represent the Thévenin equivalent of the transmitter. In that case, the time domain variations of its EMF represent the modulation to be transmitted. We then see that this modulation is provided to the load, i.e. the transmit antenna in that case. Second, this transmission line theory enables us to understand the limit between the lumped and distributed regimes. Indeed, we can see from equation (2.109) that in the general case the voltage and current along the line depend on the position and thus the impedance of the loaded line seen at that stage. It is only when the length of the line is much less than the carrier wavelength, i.e. when $l_z \ll \lambda_c$ or equivalently when $k_c l_z \ll 2\pi$, that the voltage and current along the line become independent of the position and that Kirchhoff's law can be applied, for instance. We indeed recover in that case that the input impedance of the loaded line, given by equation (2.112), reduces to Z_l and remains independent of the length of the line as long as $l_z \ll \lambda_c$. This last criterion then provides the limit between the lumped and distributed regimes.

2.2.2 *Amplitude Matching*

Let us now return to the condition of amplitude matching encountered at the end of the previous section and discuss its consequences for practical implementation.

First of all, thinking back to the derivations of the previous section, we see that this condition, which corresponds to having $Z_0 = Z_l$, can be interpreted from different points of view. On the one hand, we see from equation (2.112) that this condition leads to having a constant input impedance for the loaded line, i.e. having $Z_i = Z_0 = Z_l$ whatever its length l_z. On the other hand, this condition can also be interpreted in terms of amplitude reflection coefficients: having $Z_0 = Z_l$ is equivalent to having $\Gamma_\rho = 0$ according to equation (2.105). Fortunately, these two behaviors are intimately related. Indeed, having $\Gamma_\rho = 0$ means that we have no reflected waves, as can be seen from equation (2.109). As a result, it is only when $\Gamma_\rho \neq 0$ that we can have a coherent combining of the forward and reflected waves that leads to a local increase or decrease in the instantaneous amplitudes of the voltage and current along the line. But, as the combining of those forward and reflected waves is different for the voltage and current fields due to the negative sign that weights the reflected current wave, we get different variations of the instantaneous amplitudes of those voltage and current. We thus have variations of the wave impedance along the line. This effect cannot exist without reflected waves.

The presence of reflected waves also leads to the existence of voltage nodes and anti-nodes due to the coherent combining of the forward and reflected waves. We therefore get an amount of standing wave that takes place along the line when the amplitude matching is not realized, i.e. when $Z_0 \neq Z_l$. The amount of standing wave is often quantified using the voltage standing wave ratio (VSWR). This quantity is defined simply as the ratio between the maximum value that the amplitude the voltage can reach along the line and its minimum value, all this at a given time. This quantity classically defined in the pure monochromatic case can be generalized in the narrowband modulation case when considering transmission lines no longer than a few wavelengths. Indeed, in that case we saw in the previous section that the time dependent part of the waves remains the same over the entire line. This means that the variations of the voltage along the line at a given time remains only due to the spatial dependency of the forward and reflected waves. In order to illustrate this, we can derive the modulus of the voltage complex envelope field along the line. We get from equation (2.109a) that

$$|\tilde{v}(z,t)|^2 = |\xi_v \tilde{v}_g(t)|^2 (1 + |\Gamma_\rho|^2 + 2|\Gamma_\rho| \cos(2k_c z + \arg\{\Gamma_\rho\})). \tag{2.113}$$

This expression thus confirms that, at a given time, the variations along the line of $|\tilde{v}(z,t)|$ are due to the recombining of the spatially dependent parts of the forward and reflected waves. We observe that the maximum and minimum values occur for a difference of a quarter wavelength between the corresponding positions. In particular, we have

$$\max_z\{|\tilde{v}(z,t)|\} = |\xi_v \tilde{v}_g(t)|(1 + |\Gamma_\rho|), \tag{2.114a}$$

$$\min_z\{|\tilde{v}(z,t)|\} = |\xi_v \tilde{v}_g(t)|(1 - |\Gamma_\rho|). \tag{2.114b}$$

By the definition of the VSWR, we can thus write that

$$VSWR = \frac{\max_z\{|\tilde{v}(z,t)|\}}{\min_z\{|\tilde{v}(z,t)|\}} = \frac{1 + |\Gamma_\rho|}{1 - |\Gamma_\rho|}. \tag{2.115}$$

We therefore see that from an implementation point of view, an impedance mismatch somewhere along the line leads to the existence of reflected and thus standing wave. The resulting wave impedance variations along the line are merely a different expression of this effect. Nevertheless, we may wonder how important such a standing wave is. The problem is that the standing wave result in voltage nodes with a local magnitude that can be up to twice the level that would be required to carry only the active power that is transferred to the load. This factor of 2 can in fact be seen in equation (2.114a), given that Γ_ρ can have a modulus up to 1. This effect can be a problem for active parts of transceiver front-ends (FEs) that have to handle such voltage. Even if not destructive when taken into account at the early stages of conception, it can lead to an oversizing of those parts and thus to an additional cost of the final solution. This explains why amplitude matching is often targeted in implementations. However, power matching also often needs to be considered, as discussed in the next section.

There is also another side effect linked to amplitude mismatch that has to be considered in transceivers. This can be seen by reconsidering the complex envelope field of the voltage that is delivered to the load of the line as a function of the generator EMF complex envelope. In the notation of Figure 2.6, this quantity corresponds to $\tilde{v}(0,t)$. Using equation (2.109a), we can thus write this quantity as

$$\tilde{v}(0,t) = \xi_v \tilde{v}_g(t)(1 + \Gamma_\rho). \tag{2.116}$$

We then see that both the magnitude and the phase of the modulated bandpass voltage delivered to the line load are functions of the amplitude reflection coefficient at this stage. But, due to practical implementation limitations, this coefficient cannot be exactly null in practice even if the amplitude matching is targeted. Thus, the problem is that in some implementations this factor can vary while the transceiver is receiving or transmitting. Considering the receive side, for instance, the load we are talking about is classically the input impedance of the receiver, i.e. of the low noise amplifier (LNA). But, as discussed in Chapter 7, the gain of this stage is often switched to optimize performance during the reception, depending on the power of the received signal. Thus, if the input impedance presented by the device is not exactly the same in its different gain states, which is often the case in real implementations, we have to cope with different values of the amplitude reflection coefficient for those different gain states. The variations of this complex number thus lead to both an amplitude step and a phase shift of the complex envelope of the voltage delivered to this LNA and thus processed by the receiver. The latter effect can be problematic as it directly impacts the carrier phase and can thus possibly perturb the synchronization of the receiver, for instance. Generally speaking, this behavior can occur for any RF device that is dynamically set during a transmission and whose input or output impedances vary depending on its configuration. We can thus have the same kind of problem on the transmit side, for instance when considering gain steps in a PA with the aim of saving current when lower radiated power is required. In that case, we have to ensure that the resulting carrier phase shift is indeed bounded so that the overall wireless link is not degraded.

2.2.3 *Power Matching*

In addition to the amplitude matching detailed in the previous section, it is often what is called the power matching that is targeted in transceiver implementations. For a given amount of power supplied by a generator, we want the maximum proportion of it effectively delivered to its load and not consumed in internal impedance. When this occurs, we say that the load is power matched to the generator.

This behavior can be anticipated in various ways in transceiver implementations. The most obvious situation concerns the transmit side. As discussed in Section 2.1.3, a transmitter needs to deliver RF power to the transmit antenna. This RF power corresponds to an important fraction of the total power that is consumed in a transmitter. We thus have a direct interest in minimizing the fraction of power that is not delivered to the load and thus is lost. Moreover, in this particular case, due to the relatively high power that can be involved, any reflected wave can cause irreversible damage to the transmitter. On the receive side, this kind of behavior is less immediately obvious. Indeed, as detailed in Section 2.1.2, the information we need to further process is carried by the RF currents induced on the receive antenna. It is thus recovered at the EMF stage of the equivalent generator model of this receive antenna. We can therefore imagine a receiver that senses and further processes either the induced voltage or the current at the receive antenna output but does not require both at the same time. The practical limitation for this to be possible often comes from the use of passive FE devices that can be placed between the receive antenna and the active part of the receiver. RF filters, for instance, often use wave reflections at target frequencies to achieve selectivity. As a result, their transfer function is guaranteed only when loaded using the correct impedance as used during their conception. But, this impedance can hardly be chosen different from the characteristic impedances of transmission lines. Indeed, this characteristic impedance is the only predictable load impedance that can be used for the conception of an RF filter in practice. The reason is that when not loaded by its characteristic impedance, the impedance presented by the transmission line to the RF filter depends on its length. In that case, it is difficult to anticipate precisely the impedance that will load the filter in the final board implementation and thus to design it correctly. This often makes it necessary to target amplitude matching and thus power matching for the input active part of receivers when dealing with a real characteristic impedance, as discussed in the following. But we should keep in mind that when a passive filter is not used, other strategies can be considered as long as the signal to noise ratio performance is not compromised [18].

After this not so short introduction, we can derive the implementation guidelines that allow this power matching to be achieved. Let us reconsider the voltage and current fields that can be derived from a propagating TEM mode along a transmission line as depicted in Figure 2.6. As we have done up to now, we assume that we are dealing with narrowband modulated fields and that the length of the line, l_z, is no more than a few wavelengths. Thus, according to the discussion in Section 2.2.1, the complex envelope of the voltage and current fields, assumed defined as centered around the carrier angular frequency ω_c, are given by equation (2.109). Using those expressions, we can write the instantaneous power, $p(z,t)$, consumed by the impedance $Z(z) = \tilde{v}(z,t)/\tilde{\imath}(z,t)$ of the loaded line at position z as

$$p(z,t) = v(z,t)i(z,t). \tag{2.117}$$

As the voltage and current are related to their complex envelope fields through

$$v(z,t) = \mathrm{Re}\left\{\tilde{v}(z,t)e^{j\omega_c t}\right\}, \tag{2.118a}$$

$$i(z,t) = \mathrm{Re}\left\{\tilde{\imath}(z,t)e^{j\omega_c t}\right\}, \tag{2.118b}$$

we can use equation (1.5) to expand equation (2.117) as

$$\begin{aligned} p(z,t) &= \frac{1}{4}\left(\tilde{v}(z,t)e^{j\omega_c t} + \tilde{v}^*(z,t)e^{-j\omega_c t}\right)\left(\tilde{\imath}(z,t)e^{j\omega_c t} + \tilde{\imath}^*(z,t)e^{-j\omega_c t}\right) \\ &= \frac{1}{2}\mathrm{Re}\{\tilde{v}(z,t)\tilde{\imath}^*(z,t)\} + \frac{1}{2}\mathrm{Re}\left\{\tilde{v}(z,t)\tilde{\imath}(z,t)e^{2j\omega_c t}\right\}. \end{aligned}$$

If we look at this equation in more detail, we see that the second term of its right-hand side represents nothing more than what is called the reactive power. Indeed, as we assumed that $\tilde{v}_g(t)$ and thus both $\tilde{v}(z,t)$ and $\tilde{\imath}(z,t)$ are narrowband regarding ω_c during a carrier frequency period, we get an alternation of electrical energy transfer, occurring at angular frequency $2\omega_c$, between each side of the line at position z. This results in an overall energy transfer that is null on average at that frequency. Thus, only the first term of the right-hand side of this equation corresponds to a non-vanishing average power transfer and thus to the active power transferred to the load. We can evaluate this term using the stochastic approach considering the fields as narrowband modulated. This means that with the assumptions of stationarity, which can classically be considered for wireless modulating processes as discussed in Appendix 2, and ergodicity, we can write that

$$P_a(z) = \frac{1}{4}\mathbb{E}\{\tilde{v}(z)\tilde{\imath}^*(z) + \tilde{v}^*(z)\tilde{\imath}(z)\} = \frac{1}{2}\mathbb{E}\{\mathrm{Re}\{\tilde{v}(z)\tilde{\imath}^*(z)\}\}, \tag{2.119}$$

where P_a stands for the average active power.

In order to go further, we can use the electrical relationships that hold between the complex envelope of the narrowband modulated voltage and current fields. Having assumed that we are dealing with complex envelopes defined as centered around the same center frequency, we can use the results of Section 2.2.1 to write

$$\tilde{v}(-l_z, t) = Z_i \tilde{\imath}(-l_z, t), \tag{2.120}$$

with Z_i the input impedance of the loaded line at the carrier angular frequency as given by equation (2.112). It follows that the active power delivered by the generator to the loaded line, $P_{a,ll} = P_a(-l_z)$, reduces to

$$P_{a,ll} = \text{Re}\left\{\frac{1}{Z_i^*}\right\} \frac{\mathbb{E}\{|\tilde{v}(-l_z)|^2\}}{2} = \text{Re}\left\{\frac{1}{Z_i}\right\} \frac{\mathbb{E}\{|\tilde{v}(-l_z)|^2\}}{2}. \tag{2.121}$$

This active power can be expressed as a function of the generator EMF using equation (2.110). This yields

$$P_{a,ll} = \text{Re}\left\{\frac{1}{Z_i}\right\} \frac{|Z_i|^2}{|Z_i + Z_g|^2} \frac{\mathbb{E}\{|\tilde{v}_g|^2\}}{2}. \tag{2.122}$$

This relationship can be reordered using equation (1.5) to expand $\text{Re}\{1/Z_i\}$. We then get

$$\text{Re}\left\{\frac{1}{Z_i}\right\} = \frac{1}{2}\left(\frac{1}{Z_i} + \frac{1}{Z_i^*}\right) = \frac{1}{2}\frac{Z_i + Z_i^*}{Z_i Z_i^*} = \frac{\text{Re}\{Z_i\}}{|Z_i|^2}. \tag{2.123}$$

Finally, we obtain

$$P_{a,ll} = \frac{\text{Re}\{Z_i\}}{|Z_i + Z_g|^2} \frac{\mathbb{E}\{|\tilde{v}_g|^2\}}{2}. \tag{2.124}$$

This active power transferred by the generator to the loaded line is therefore maximum when its input impedance Z_i fulfills

$$\partial_{\text{Re}\{Z_i\}} P_{a,ll} = 0, \tag{2.125a}$$

$$\partial_{\text{Im}\{Z_i\}} P_{a,ll} = 0. \tag{2.125b}$$

The resolution of this system gives the classical result that power matching occurs when

$$Z_i = Z_g^*. \tag{2.126}$$

For a given generator internal impedance Z_g, we get a maximum active power that can be retrieved from it when the above condition holds. This maximum power in fact corresponds by definition to what is called the available power from this generator. This concept is often used to characterize RF devices that deal with RF power. We can derive an expression for this available power, denoted by P_{av}, from equation (2.124):

$$P_{av} = P_{a,ll}|_{Z_i = Z_g^*} = \frac{1}{4\text{Re}\{Z_g\}} \frac{\mathbb{E}\{|\tilde{v}_g|^2\}}{2}. \tag{2.127}$$

We can use this quantity to express the active power effectively delivered by the generator to its load in a form more suitable for a physical interpretation, reordering equation (2.124) as

$$P_{a,ll} = P_{av}\frac{4\mathrm{Re}\{Z_i\}\mathrm{Re}\{Z_g\}}{|Z_i + Z_g|^2}. \tag{2.128}$$

Thus, again using equation (1.5) to expand the real parts, we get that

$$P_{a,ll} = P_{av}\frac{(Z_i + Z_i^*)(Z_g + Z_g^*)}{|Z_i + Z_g|^2}. \tag{2.129}$$

Let us now reorder the numerator as

$$\begin{aligned}(Z_i + Z_i^*)(Z_g + Z_g^*) &= (Z_i + Z_g - Z_g + Z_i^*)(Z_g + Z_g^*)\\ &= (Z_i + Z_g)(Z_g + Z_g^*) + (Z_i^* - Z_g)(Z_g + Z_g^*).\end{aligned}$$

We also observe that the terms $Z_g + Z_g^*$ can be expanded in two complementary ways as

$$Z_g + Z_g^* = Z_g - Z_i^* + Z_i^* + Z_g^* = Z_g - Z_i + Z_i + Z_g^*. \tag{2.130}$$

Using these two expressions, we can then write that

$$\begin{aligned}(Z_i + Z_i^*)(Z_g + Z_g^*) &= (Z_i + Z_g)(Z_i^* + Z_g^*) - (Z_i - Z_g^*)(Z_i^* - Z_g)\\ &= |Z_i + Z_g|^2 - |Z_i - Z_g^*|^2.\end{aligned}$$

Using this result in equation (2.129), we finally get

$$\begin{aligned}P_{a,ll} &= P_{av}\frac{|Z_i + Z_g|^2 - |Z_i - Z_g^*|^2}{|Z_i + Z_g|^2}\\ &= P_{av}(1 - |\Gamma_p|^2),\end{aligned} \tag{2.131}$$

with

$$\Gamma_p = \frac{Z_i - Z_g^*}{Z_i + Z_g}. \tag{2.132}$$

This last term is defined as the power reflection coefficient. Written in that form, the interpretation of the different terms involved in equation (2.131) becomes straightforward. The square of the magnitude of the power reflection coefficient represents nothing more than the fraction of the available power from the generator that is reflected back to it due to the impedance

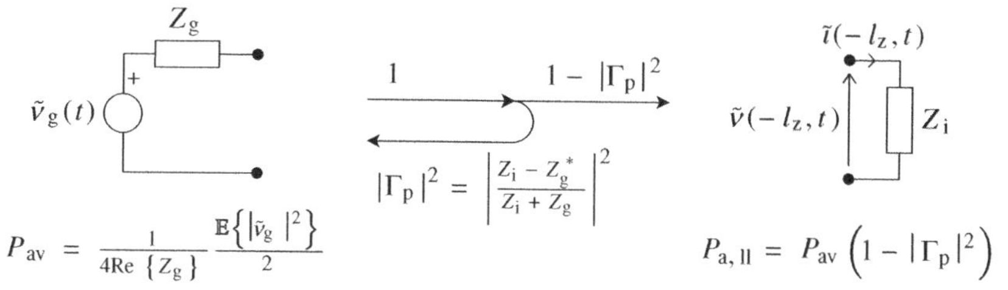

Figure 2.7 Power matching between a loaded line and a generator connected to its input – According to equation (2.131), the active power delivered by the generator to its load can be interpreted as the fraction of its available power that is not reflected back to it. In our case of interest, Z_i can be interpreted as the input impedance of the loaded line at the carrier angular frequency as given by equation (2.112).

mismatch between its internal impedance, Z_g, and its load impedance, Z_i, in the present case. The term $1 - |\Gamma_p|^2$ therefore represents the fraction of active power that is effectively delivered to the load as represented graphically in Figure 2.7. We observe that this power reflection coefficient is classically used to define the return loss, *RL*, which is the ratio between the incident power and the reflected power. Expressed in decibels, this is given by

$$RL|_{dB} = -20\log_{10}(|\Gamma_p|). \tag{2.133}$$

The return loss thus varies from 0 dB when all the incident power is reflected, to ∞ dB when the power matching is realized.

However, it is noteworthy that in general the condition for power matching does not correspond to the condition for the amplitude matching detailed in the previous section. On the one hand, power matching means that $\Gamma_p = 0$ and thus that $Z_g = Z_i^*$. Here, we get a condition between the generator internal impedance and the impedance seen at the line input. Thus, under that condition, the maximum active power is retrieved from the generator and delivered to the loaded line. But it says nothing about the active power that is effectively delivered to the load at the end of the line. On the other hand, to get no standing wave along the line, and thus no voltage node, we need to have Γ_ρ and thus $Z_0 = Z_l$ according to equation (2.105). Here we get a condition between the characteristic impedance of the line and its load impedance. Thus, it seems that we get different constraints to optimize either amplitude matching or power matching at the generator output. This is in fact not true in practical implementations. Indeed, in most cases, we deal with line characteristic impedances that are real, i.e. we can assume that we have $Z_0 = R_0$. As a result, amplitude matching is realized when the load impedance is also equal to R_0. But when $Z_0 = Z_l = R_0$, the input impedance of the loaded line is also equal to R_0. This result holds whatever the length l_z of the line, as can be seen considering equation (2.112). This means that dealing with lines with real characteristic impedances R_0 and having realized the amplitude matching, the load on the generator is also necessarily R_0. This means that the power matching is also realized when the generator input impedance is set to R_0. Moreover, as in the present case the impedance of the loaded line becomes independent of its length, we get that the power matching is realized all along the transmission line, and thus also at the load stage.

We thus see that when dealing with real characteristic impedances for the transmission line, we can realize at the same time amplitude matching and power matching all along the line. A side effect of this is that dealing with real characteristic impedances for the transmission line, we can directly link the generated standing wave ratio and the power loss at the generator output for weak load impedance mismatch. This property is in fact not obvious in the general case comparing, on the one hand, equation (2.115) with Γ_ρ given by equation (2.105) and, on the other hand, (2.131) with Γ_p given by equation (2.132). In order to achieve this link, let us start by considering the case where

$$Z_\mathrm{g} = Z_0 = R_0. \tag{2.134}$$

Suppose that we now are dealing with a small amount of impedance mismatch at the load stage in order to reflect the non-ideal matching. We thus further suppose that

$$Z_\mathrm{l} = R_0 + \delta Z_\mathrm{l} = R_0\left(1 + \frac{\delta Z_\mathrm{l}}{R_0}\right) \tag{2.135}$$

with $|\delta Z_\mathrm{l}| \ll R_0$. Under these assumptions, the amplitude reflection coefficient at the load stage of the line, given by equation (2.105), reduces to

$$\Gamma_\rho = \frac{Z_\mathrm{l} - R_0}{Z_\mathrm{l} + R_0} = \frac{\delta Z_\mathrm{l}}{2R_0 + \delta Z_\mathrm{l}} \approx \frac{\delta Z_\mathrm{l}}{2R_0}. \tag{2.136}$$

In the same way, the loaded line input impedance seen by the generator, Z_i, given by equation (2.112) in the general case, now reduces to

$$Z_\mathrm{i} = R_0\frac{(1 + \delta Z_\mathrm{l}/R_0) + \mathrm{j}\tan(k_\mathrm{c}l_\mathrm{z})}{1 + \mathrm{j}(1 + \delta Z_\mathrm{l}/R_0)\tan(k_\mathrm{c}l_\mathrm{z})} \approx R_0\left(1 + \frac{\delta Z_\mathrm{l}}{R_0}\frac{1 - \mathrm{j}\tan(k_\mathrm{c}l_\mathrm{z})}{1 + \mathrm{j}\tan(k_\mathrm{c}l_\mathrm{z})}\right).$$

But, using equation (2.136), we can express this impedance in terms of the amplitude reflection coefficient at the load stage of the line as

$$Z_\mathrm{i} = R_0\left(1 + 2\Gamma_\rho\frac{1 - \mathrm{j}\tan(k_\mathrm{c}l_\mathrm{z})}{1 + \mathrm{j}\tan(k_\mathrm{c}l_\mathrm{z})}\right). \tag{2.137}$$

It results that the power reflection coefficient given by equation (2.132) now reduces to

$$\Gamma_\mathrm{p} = \frac{Z_\mathrm{i} - R_0}{Z_\mathrm{i} + R_0} = \frac{\Gamma_\rho\dfrac{1 - \mathrm{j}\tan(k_\mathrm{c}l_\mathrm{z})}{1 + \mathrm{j}\tan(k_\mathrm{c}l_\mathrm{z})}}{1 + \Gamma_\rho\dfrac{1 - \mathrm{j}\tan(k_\mathrm{c}l_\mathrm{z})}{1 + \mathrm{j}\tan(k_\mathrm{c}l_\mathrm{z})}}. \tag{2.138}$$

We observe that the term $1 - \mathrm{j}\tan(k_\mathrm{c}l_\mathrm{z})$ is the complex conjugate of $1 + \mathrm{j}\tan(k_\mathrm{c}l_\mathrm{z})$. Those numbers thus have the same modulus. It results that

$$\left|\Gamma_\rho\frac{1 - \mathrm{j}\tan(k_\mathrm{c}l_\mathrm{z})}{1 + \mathrm{j}\tan(k_\mathrm{c}l_\mathrm{z})}\right| = |\Gamma_\rho|\frac{|1 - \mathrm{j}\tan(k_\mathrm{c}l_\mathrm{z})|}{|1 + \mathrm{j}\tan(k_\mathrm{c}l_\mathrm{z})|} = |\Gamma_\rho|. \tag{2.139}$$

This means that assuming a weak load impedance mismatch, i.e. that $|\delta Z_l| \ll R_0$, we get that $|\Gamma_\rho| \ll 1$ according to equation (2.136). Thus,

$$\left|1 + \Gamma_\rho \frac{1 - \mathrm{j}\tan(k_c l_z)}{1 + \mathrm{j}\tan(k_c l_z)}\right| \approx 1. \tag{2.140}$$

As a result, the modulus of the power reflection coefficient given by equation (2.138) reduces in the present case to

$$|\Gamma_\mathrm{p}| \approx |\Gamma_\rho|. \tag{2.141}$$

We thus get that the modulus of the power reflection coefficient of the generator loaded by the loaded line is almost the same as that of the amplitude reflection coefficient of the line loaded by the load impedance. This allows us to express the power effectively delivered by the generator to the loaded line, originally given by equation (2.131), as a function of the modulus of the amplitude reflection coefficient at the line load stage instead of the modulus of the power reflection coefficient at the generator output stage. This results in

$$P_{\mathrm{a,ll}} \approx P_{\mathrm{av}}(1 - |\Gamma_\rho|^2). \tag{2.142}$$

But, as we can link the amplitude reflection coefficient at the load stage and the VSWR along the line using equation (2.115), we can finally write that

$$P_{\mathrm{a,l}} \approx P_{\mathrm{av}} \frac{4VSWR}{(1 + VSWR)^2}. \tag{2.143}$$

In light of the above derivations, we may wonder how ideal the power matching may be in a real implementation. In practice, dealing with transmission lines that have a real characteristic impedance R_0, we need to make both the load impedance and the generator impedance equal to R_0 in order to achieve power matching. But, considering the receive side for instance, the load impedance can be the input impedance of an LNA stage. Unfortunately, for the most part such input impedance does not reduce to a pure resistor due to the capacitive behavior of transistor devices. The result is that an impedance adaptation device, called a matching network, often needs to be used to make Z_l as close as possible to the line characteristic resistance R_0, as shown in Figure 2.8. However, as in that case the impedance we have to

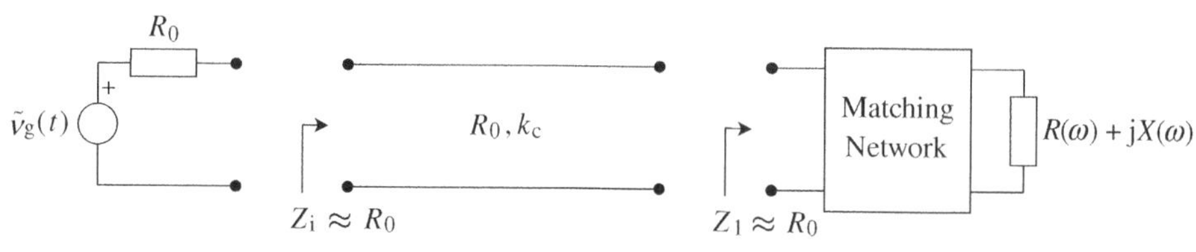

Figure 2.8 Impedance matching using an interstage network – Using a physical network to adapt an impedance $R(\omega) + \mathrm{j}X(\omega)$ that varies in frequency, the performance achieved can only be limited according to equation (2.146). As a side effect, a correct impedance matching in a given frequency band results in an overall passband filtering behavior at the load stage.

match to R_0 experiences variations in frequency, we can expect problems in achieving a load impedance Z_l that is exactly constant and equal to R_0 over a given frequency band.

Indeed, Bode has shown that for a function $F(s)$ of a complex variable $s = \sigma + j\omega$, that is analytic in the right half plane, i.e. for $\sigma \geq 0$, and such that $F(s)$ is real for s real, then we have [19]:

$$\int_0^\infty \frac{X(\omega)}{\omega} \mathrm{d}\omega = \frac{\pi}{2}[R(\infty) - R(0)], \tag{2.144a}$$

$$\int_0^\infty [R(\omega) - R(\infty)] \mathrm{d}\omega = -\frac{\pi}{2} \lim_{\omega \to \infty} \omega X(\omega), \tag{2.144b}$$

with $F(j\omega) = R(\omega) + jX(\omega)$. These relationships are known as the reactance and resistance integral theorems, respectively. Thus, as any physical impedance fulfills those conditions, the above relationships hold for its real and imaginary parts. In that case, it follows that when dealing with an impedance that does not reduce to a pure resistor, the achievable reflection coefficient

$$\Gamma_l = \frac{Z_l - R_0}{Z_l + R_0} \tag{2.145}$$

fulfills a bound of the form [19, 20]

$$\int_0^\infty \ln \left| \frac{1}{\Gamma_l} \right| \mathrm{d}\omega = \Omega, \tag{2.146}$$

where Ω represents a constant that depends on the exact structure of the network that implements the load impedance considered. We observe that here Γ_l represents either the power reflection coefficient or the amplitude reflection coefficient, both at the load stage, as we assume that we are dealing with a transmission line that has a real characteristic impedance R_0. We can interpret the above equation as saying that the area under the curve of $\ln |1/\Gamma_l|$ cannot be greater than Ω. Thus, assuming that we want to achieve impedance matching over the bandwidth $[\omega_1, \omega_2]$, for instance, the best we can do is to make Γ_l equal to 1 outside this frequency band in order to concentrate the available area of $\ln |1/\Gamma_l|$ within $[\omega_1, \omega_2]$ in order to achieve the best possible matching. In that case, the above integral reduces to

$$\int_{\omega_1}^{\omega_2} \ln \left| \frac{1}{\Gamma_l} \right| \mathrm{d}\omega = \Omega. \tag{2.147}$$

Thus, if we assume for the sake of simplicity that Γ_l can be set constant over $[\omega_1, \omega_2]$, we finally obtain that

$$|\Gamma_l| = \mathrm{e}^{-\Omega/\delta\omega}, \tag{2.148}$$

with $\delta\omega = \omega_2 - \omega_1$. We thus derive two main conclusions for impedance matching:

(i) By the above equation, $\Gamma_l = 0$ only when $\delta\omega = 0$. Thus, even if we assumed for the derivation that Γ_l was constant over a given frequency band, it is in fact a general result that perfect matching can be achieved only for a set of discrete frequencies.

(ii) In the same way, in order to achieve the best possible impedance matching, we need to set Γ_l as close as possible to 1 outside the frequency band considered for the adaptation. This is done in order to have all the available area of $\ln|1/\Gamma_l|$ within this frequency band. Thus, achieving the best matching over a frequency band necessarily results in a passband filtering behavior at the achieved load stage.

We conclude this section with some comments on the theoretical tools we used for the analytical derivations. We saw that voltage and current fields are perhaps not so well suited to power derivations as electrical power is a composite of those waves. Moreover, we highlighted at the beginning of Section 2.2.1 that voltage waves cannot be defined in a simple way for non-TEM propagating modes. That has lead to the introduction of *power waves* that can be defined for any propagating mode based on the components of the corresponding electromagnetic field [13]. Of course, in the particular case where a TEM mode is involved, those waves can be related to the voltage and current waves defined above, as originally proposed by Kurokawa [21]. Power derivations then become more straightforward using those incident and reflected power waves and their related scattering parameters, or S parameters. However, we will not pursue that topic further as those concepts have to be considered as specific tools to handle matching problems, but add nothing more to the physical meaning of the phenomena involved.

2.3 The Propagation Channel

Another set of constraints on the dimensioning of transceivers comes from the characteristics of the propagation channel. Here, by "propagation channel" we mean the electromagnetic propagation medium between the transmit and receive antennas.

Up to now in this chapter, when we have considered free space propagation for wireless links, we have assumed that no particular obstacles were present between the transmit and receive antennas, and also that there was a static relative configuration between them. This is unfortunately often not the case. Indeed, the presence of obstacles, or even of the ground itself, leads to multiple reflections and thus to a received signal that is the superposition of echoes of the transmit signal. Due to the coherent sum of the waves corresponding to those echoes, we get a selectivity of the propagation channel in frequency even if the propagation medium is not selective in itself. Moreover, dealing with transceivers embedded in mobile devices, we get that the characteristics of the propagation channel themselves become a function of time.

From the wireless signal processing point of view, time-varying multipath propagation channels are mainly associated with the equalization stage on the receive side. Such apparatus is indeed required to compensate for the distortions experienced by the transmitted signal through the propagation channel. Nevertheless, from the RF transceiver implementation point of view, there are impacts to consider that lead, for instance, to the necessity to have an automatic gain control (AGC) system, or to the margins to consider in a receiver line-up as discussed in Section 2.3.3.

Thus, we illustrate in the following the basic concepts associated with the classical description of the propagation channel in order to introduce their related impacts on receivers.

2.3.1 Static Behavior

Channel Selectivity

Let us begin with the configuration of the elementary wireless link shown in Figure 2.5. In this simple configuration we are dealing with a direct electromagnetic propagation between the transmitter and the receiver, i.e. without any obstacle between them. In addition, we suppose in this first step that both the transmitter and the receiver are in a static relative configuration, i.e. with a null relative speed. As a result, the characteristics of the propagation channel we are dealing with remain constant in time.

Under those assumptions, and still supposing that we are dealing with narrowband modulated waveforms, we can reuse the material of Section 2.1.2. In particular, we get, on the one hand, that the far-field radiated by the transmit antenna in the direction of the receive antenna is proportional to the bandpass modulated RF current flowing from the transmitter connected to it, i.e. $j_{TX}(t)$, but delayed in time by the propagation delay d/v following equation (2.59). In this expression, $d = \|\mathbf{d}\|$ represents the distance between the transmitter and the receiver, and v the speed of the wave in the relevant medium. On the other hand, we have that the EMF of the receiving antenna equivalent generator, $v_{RX}(t)$, generated by a received modulated plane wave, is proportional to the time domain variations of the received electromagnetic field. Thus, in this simple configuration,

$$v_{RX}(t) \propto j_{TX}(t - d/v), \tag{2.149}$$

where $j_{TX}(t)$ is nothing more than an analog image of the bandpass modulated waveform we want to transmit. And in the same way, $v_{RX}(t)$ represents the received bandpass modulated waveform that the receiver is expected to process in order to recover its modulating waveform and thus the transmitted information. Thus, in order to stay at the signal processing level in the following, we can label those waveforms in a general way as $s_{TX}(t)$ and $s_{RX}(t)$, i.e. independently of their physical representation using either voltages or currents. According to the above equation, then, the received RF bandpass waveform can be written in terms of the transmitted one as

$$s_{RX}(t) = cs_{TX}(t - \tau), \tag{2.150}$$

where $\tau = d/v$ stands for the propagation delay and c can be related to the free space path loss as detailed in Section 2.1.4.

To deal with bandpass modulated waveforms, the above relationship can advantageously be transposed to the corresponding complex envelopes. Assuming for the sake of simplicity that we are dealing with complex envelopes defined as centered around the same center angular frequency, ω_c, we can write in particular that

$$s_{TX}(t) = \mathrm{Re}\left\{\tilde{s}_{TX}(t)e^{j\omega_c t}\right\}. \tag{2.151}$$

It follows that

$$\begin{aligned} s_{TX}(t - \tau) &= \mathrm{Re}\left\{\tilde{s}_{TX}(t - \tau)e^{j\omega_c(t-\tau)}\right\} \\ &= \mathrm{Re}\left\{\tilde{s}_{TX}(t - \tau)e^{-j\omega_c \tau}e^{j\omega_c t}\right\}. \end{aligned} \tag{2.152}$$

As we have considered free space propagation in this first step, the coefficient c defined via equation (2.150) representz the path loss. It is thus a real and positive number in this simple approach and we can write

$$cs_{\mathrm{TX}}(t-\tau) = \mathrm{Re}\left\{ c e^{-j\omega_c \tau} \tilde{s}_{\mathrm{TX}}(t-\tau) e^{j\omega_c t} \right\}. \tag{2.153}$$

As $\tilde{s}_{\mathrm{RX}}(t)$ is also assumed defined as centered around ω_c, we can then assume that

$$\tilde{s}_{\mathrm{RX}}(t) = c e^{-j\omega_c \tau} \tilde{s}_{\mathrm{TX}}(t-\tau). \tag{2.154}$$

This means that dealing with a medium that is not dispersive, i.e. that is not selective in frequency, as is the case with free space, and considering a direct path between the transmitter and the receiver, we get that the complex envelope of the received waveform is simply a time shifted version of the transmitted waveform, weighted by a complex number that represents both the path attenuation and the carrier phase shift due to the propagation delay.

Based on the above derivations, we observe that the received signal can be expressed as the filtered version of the transmitted signal. Indeed, due to the property of the Dirac delta distribution that is the identity element of the convolution, we can write equation (2.150) in the form

$$s_{\mathrm{RX}}(t) = h_{\mathrm{CH}}(t) \star s_{\mathrm{TX}}(t), \tag{2.155}$$

with $h_{\mathrm{CH}}(t)$ the impulse response or transfer function of the propagation channel given by

$$h_{\mathrm{CH}}(t) = c\delta(t-\tau). \tag{2.156}$$

Equivalently, we can define a lowpass transfer function that acts on the complex envelopes of those bandpass modulated waveforms assumed defined as centered around the same center angular frequency ω_c. We use the notation $\tilde{h}_{\mathrm{CH}}(t)$ for this equivalent lowpass transfer function even if it is not, strictly speaking, the complex envelope of $h_{\mathrm{CH}}(t)$ in the mathematical sense, but acts on the complex envelopes of the original bandpass signals processed by $h_{\mathrm{CH}}(t)$. Thus, according to equation (2.154), we can write

$$\tilde{s}_{\mathrm{RX}}(t) = \tilde{h}_{\mathrm{CH}}(t) \star \tilde{s}_{\mathrm{TX}}(t), \tag{2.157}$$

with

$$\tilde{h}_{\mathrm{CH}}(t) = c e^{-j\omega_c \tau} \delta(t-\tau) \tag{2.158}$$

We thus get that the propagation channel behaves as a finite impulse response (FIR) filter.

However, in real life the propagation channel often does not reduce to a single path between the transmit and receive antennas. A major reason for this is that, as already highlighted at the end of Section 2.1.4, classical antenna systems used for wireless communications are for the most part isotropic, or at least almost isotropic. Indeed, unless using smart space filtering based on multi-antenna systems, transmitters do not know a priori where receivers are located so that no particular preferred directions can be decided. This means that most of the radiated

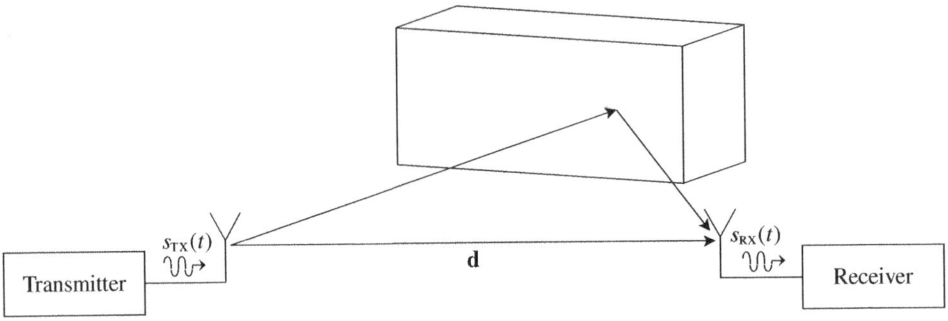

Figure 2.9 Static two-path propagation channel composed of a direct path and a secondary one due to a reflection in the geometric optical approximation – In the presence of a reflecting obstacle, due to the almost isotropic behavior of antennas classically used in wireless transceivers, a secondary path is expected to exist in addition to the direct one. In the geometric optical approximation, i.e. in the high frequency limit, this secondary path results from the specular reflection in the obstacle. The received signal at the antenna connector is then proportional to a linear superposition of the electromagnetic waves coming from the two paths.

power is not radiated in the direction of the receiver. We thus have a large probability that a fraction of it encounters an obstacle and thus that part of it is reflected, diffracted or scattered back toward the receiver again. This results in what we call a multipath channel.

In order to illustrate this, let us consider the situation shown in Figure 2.9 where we assume that only one obstacle exists, a reflecting wall in this case. In this simple situation, a fraction of the electromagnetic field that reaches the obstacle is reflected back to the receiver. Assuming in this first step that we are dealing with an ideal specular reflection, which would give the high frequency optical limit, we now get two received echoes at the receive antenna stage. Due to the linearity of Maxwell's equations, the received bandpass modulated signal can now be written in the form

$$s_{\text{RX}}(t) = c_1 s_{\text{TX}}(t - \tau_1) + c_2 s_{\text{TX}}(t - \tau_2), \tag{2.159}$$

with c_1 and c_2 the respective attenuations experienced by the electromagnetic field along the two paths, and τ_1 and τ_2 the corresponding delays. Here again, we have to keep in mind that to write such a relationship, we need to assume that the propagation along each path occurs in a non-dispersive way. In particular, the reflection in the obstacle must be non-selective in frequency so that the corresponding path tap remains constant over the frequency band of the modulated waveform. This can be justified by considering narrowband modulations leading to the involved electromagnetic phenomenon being non-selective over the corresponding frequency band. Moreover, we assumed implicitly in the above equation that no additional phase shift is induced by the reflexion, i.e. that c_1 and c_2 are real positive numbers. This is done in this first step for the sake of simplicity in the introduction of some concepts linked to the propagation channel and a more complete treatment is given in the forthcoming sections. Nevertheless, under the present assumptions we can write the transfer function of the equivalent filter that models the propagation channel for those real bandpass signals as

$$h_{\text{CH}}(t) = c_1 \delta(t - \tau_1) + c_2 \delta(t - \tau_2), \tag{2.160}$$

or equivalently, for the corresponding complex envelopes assumed defined as centered around the same center angular frequency ω_c,

$$\tilde{h}_{CH}(t) = c_1 e^{-j\omega_c \tau_1} \delta(t - \tau_1) + c_2 e^{-j\omega_c \tau_2} \delta(t - \tau_2). \tag{2.161}$$

Let us now introduce the concept of coherence bandwidth. As described above, the propagation channel behaves as a filter on the transmitted signal. Thus, as with any filter, we can characterize its behavior in the frequency domain rather than in the time domain. But we might wonder what we can gain by looking at the frequency domain. We assumed that we were dealing with a non-dispersive propagation medium, i.e. with no frequency selectivity. This is effectively the case for each path that composes the channel. But, as soon as we get the coherent recombination of at least two such paths at the receive stage, we can have either constructive or destructive recombinations of the electromagnetic components, depending on the relative electrical length experienced along the two paths. As this electrical length difference depends on the carrier angular frequency, we can expect a frequency selectivity of the overall propagation channel even if the propagation medium is not dispersive in itself. The channel coherence bandwidth is then an evaluation of the maximum bandwidth of a signal that can be transmitted over it without experiencing too much distortion.

This concept can be illustrated by considering the above two-tap propagation channel transfer function in the frequency domain. We can either directly take the Fourier transform of equation (2.160), or consider the channel transfer function given by equation (2.161) for pure monochromatic waves for different angular frequencies ω. In this last case, we then consider complex envelopes constant in time, $\tilde{s}_{TX}$ and $\tilde{s}_{RX}$, defined as centered around ω. As expected, we have the same result in the two cases, i.e.

$$H_{CH}(\omega) = \frac{\tilde{s}_{TX}}{\tilde{s}_{RX}} = c_1 e^{-j\omega\tau_1} + c_2 e^{-j\omega\tau_2}. \tag{2.162}$$

If we now focus on the magnitude of this transfer function, assuming that c_1 and c_2 are real positive numbers, we can write

$$\begin{aligned} |H_{CH}(\omega)|^2 &= c_1^2 \left| 1 + \frac{c_2}{c_1} e^{-j\omega\delta\tau} \right|^2 \\ &= c_1^2 \left[1 + \left(\frac{c_2}{c_1} \right)^2 + 2\frac{c_2}{c_1} \cos(\omega\delta\tau) \right], \end{aligned} \tag{2.163}$$

with $\delta\tau = \tau_2 - \tau_1$. As can be seen in Figure 2.10, we effectively recover a propagation channel that is selective in frequency due to the coherent sum of waves from the two paths at the receive antenna stage. The depth of the resulting nulls depends on the relative magnitude of the two paths as an exact cancellation can occur only when they have the same amplitude. We see that we have a maximum attenuation of the received signal at frequencies that correspond to a phase difference of π between the two paths, i.e. to an electrical length difference along the two physical paths that is a half integer multiple of the wavelength. This means that if we denote by δl the physical length difference between the two paths, a maximum attenuation of the received signal occurs when

$$\delta l = n\lambda + \lambda/2, \quad \text{with } n \in \mathbb{Z}. \tag{2.164}$$

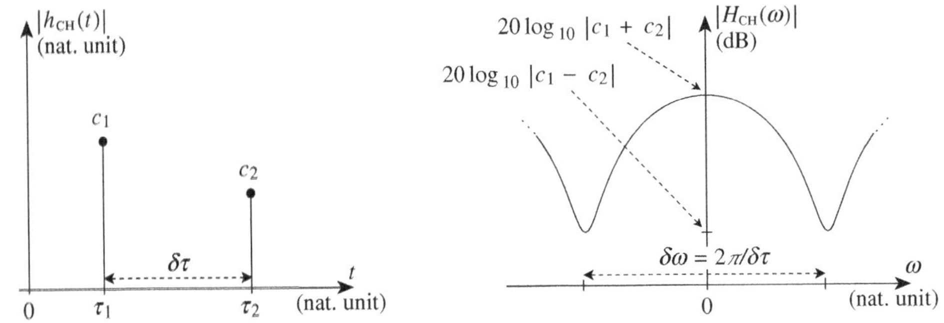

Figure 2.10 Frequency selective behavior of a two-path propagation channel – Due to the coherent sum of the electromagnetic field components coming from the different paths at the receive antenna stage, we get a selectivity in frequency of the propagation channel even when the propagation medium is not dispersive (right). As given by equation (2.166), the coherence bandwidth of the channel, $\delta\omega$, is inversely proportional to the delay spread $\delta\tau$ of its impulse response (left). In the simple case of a two-tap channel, the delay spread $\delta\tau$ is simply equal to the delay difference along the two paths $\tau_2 - \tau_1$.

This relationship allows us to make the link between the delay difference between the two paths, $\delta\tau$, and the angular frequencies at which this maximum attenuation occurs. Indeed, on the one hand we get that the delay difference between the two paths is linked to the physical length difference between them through the speed of the wave v according to $\delta l = v\delta\tau$. On the other hand, we have the wavelength that is linked to the wave angular frequency through $\lambda = 2\pi v/\omega$. We can then rewrite the above relationship as

$$\delta\tau = n\frac{2\pi}{\omega} + \frac{\pi}{\omega}. \tag{2.165}$$

It follows that the presence of the nulls is periodic in the angular frequency of the monochromatic electromagnetic wave. This period, $\delta\omega$, is given by the difference between two successive angular frequencies that fulfill the above equation and thus depends on the delay difference $\delta\tau$. As a result, we have

$$\delta\tau\delta\omega = 2\pi. \tag{2.166}$$

Thus, the period of the propagation channel transfer function in the wave angular frequency domain is inversely proportional to the delay difference $\delta\tau$ between the two paths. This is in fact nothing more than a manifestation of the uncertainty principle linked to the limitation in the resolution of the Fourier transform due to the limitation in the delay spread of the observed signal. This clarifies the concept of the coherence bandwidth of the propagation channel; this coherence bandwidth being no more than the maximum frequency bandwidth over which we can expect not to have much signal distortion. It follows that this coherence bandwidth can be taken as $\delta\omega$ and is thus directly proportional to $1/\delta\tau$.

We can now generalize the above results to the general multipath case. We can write the impulse response, $h_{CH}(t)$, of the channel under the same assumptions as considered up to now, i.e. dealing with sufficiently narrowband modulated waveforms, so that each propagation path can be considered as non-selective in frequency. This results in the generalization of equation (2.156) in the multipath case as

$$h_{\mathrm{CH}}(t) = \sum_{n} c_n \delta(t - \tau_n). \tag{2.167}$$

This impulse transfer function can be transposed as a lowpass transfer function that behaves on the complex envelopes of the transmitted and received bandpass signals, assuming that the complex envelopes are defined as centered around the same center angular frequency. We can then write

$$\tilde{h}_{\mathrm{CH}}(t) = \sum_{n} c_n \mathrm{e}^{-\mathrm{j}\omega_c \tau_n} \delta(t - \tau_n). \tag{2.168}$$

However, we observe that with our present simple assumptions, the waves along each path experience only attenuation, taken into account by the term c_n, and phase shift due to the propagation delay, taken into account by the term $\mathrm{e}^{-\mathrm{j}\omega_c \tau_n}$. With those simple assumptions about the propagation channel, we thus have that the coefficients c_n involved in the above equation are real positive numbers. Nevertheless, we can anticipate the discussion in the next section and highlight that this is no longer true when taking into account more realistic electromagnetic phenomena. In the general case, the taps involved in the expression for the lowpass transfer function $\tilde{h}_{\mathrm{CH}}(t)$ are indeed complex.

The magnitude of the above transfer functions as a function of time then gives the channel intensity profile. It thus corresponds to the magnitude profile of the received signal power as a function of time. The width of this profile gives the delay spread $\delta\tau$ of the channel as illustrated in Figure 2.10. Conversely, the width of the Fourier transform of the impulse response directly gives the coherence bandwidth of the channel and thus the maximum frequency bandwidth that we can expect to use without much distortion due to frequency selectivity. According to the above derivation, this coherence bandwidth is related to the inverse of the delay spread of the channel. Thus, at first glance it can look like we should focus on modulating waveforms that have a spectrum that spreads over a bandwidth smaller than the coherence bandwidth of the propagation channel in order to obtain a reliable wireless link. This is in fact not necessarily the case as having a much wider frequency band gives a time domain resolution that may be thinner than the delay difference between successive echoes. This results in the possibility of processing each received path separately and can lead to efficient equalization stages [1]. This is the case of the rake structure, for instance, often associated with the CDMA waveforms illustrated in Section 1.3.3.

However, we observe that these channel intensity profiles and associated Fourier transforms are only part of the functions that are classically used for the characterization of multipath propagation channels [22, 23]. We also need to characterize the variations of the characteristics detailed here in case we have to deal with mobile transmitters or receivers. This is the purpose of Section 2.3.2. But, prior to that, we can discuss the statistical models classically used for the characterization of the taps in the above channel transfer function. This is done in the following section.

Fading Distribution

For a given geometrical configuration between the transmitter and the receiver, we have given characteristics for the propagation channel. Unfortunately, dealing with the dimensioning of transceivers when we do not know a priori where they are located during their use, we cannot rely on an exact knowledge of the characteristics of the propagation channel they are expected to experience. This is of course important when considering mobile transceivers as in that case the characteristics of the propagation channel can vary in time. Nevertheless, we have to keep in mind that even in static conditions we often cannot have the knowledge of the propagation channel at the conception level. Thus, the best we can do is to consider both c_n and τ_n as random variables and derive realistic characteristics for them.

For that purpose, we need to reconsider the physical behavior of the electromagnetic propagation considered in the previous section for the derivation of the structure of the propagation channel transfer function. During this derivation we supposed that the electromagnetic phenomenon at the origin of the re-radiated waves of the secondary paths, i.e. a reflection in that particular case, was localized at a single point. This assumption was considered for simplicity of the derivations, but is nevertheless accurate enough for the introduction of channel selectivity due to multipath. We have to keep in mind that such a single point spatial localization remains a high frequency, or optical, limit for the electromagnetism theory. In fact, in the frequency range classically used in wireless cellular networks, i.e. up to few gigahertz, the re-radiated electromagnetic field that has a non-negligible contribution to the overall field at the receive antenna stage has been reflected by a non-vanishing surface surrounding the specular reflection point of the high frequency limit. This configuration is depicted in Figure 2.11.

Theoretically speaking, this surface can be evaluated using the stationary phase approximation of the re-radiated field, which leads to the definition of Fresnel zones [24]. These zones are in fact ellipses with their minor and major axis proportional on the one hand to multiples of half the carrier wavelength $\lambda_c/2$, but also on the other hand to geometrical characteristics of the configuration considered. If we take the 10th Fresnel zone, it can be shown that we already take into account more than 85% of the total electromagnetic power recovered at the receive antenna stage and coming from the reflection area. Thus, we typically need to consider

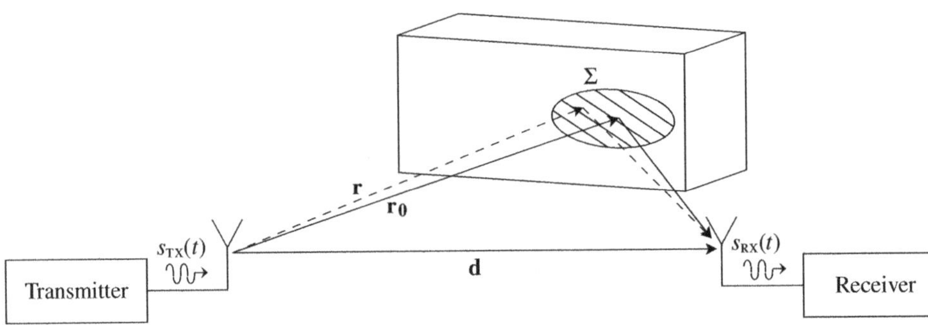

Figure 2.11 Static two-path propagation channel composed of a direct path and of a secondary one due to a reflection – In the frequency range classically used for carriers in wireless standards, an electromagnetic reflection cannot reduce to the simple specular case of the optical limit shown in Figure 2.9. We get a non-vanishing area, Σ, with characteristic dimensions of the order of wavelengths, which is involved in the reflection in the obstacle. If different kinds of reflecting material exist within the area of Σ, and assuming that we are dealing with a narrowband modulation scheme, we can assume that the instantaneous amplitude of the reflected wave corresponds to that of the incoming wave corrupted by a random variable with a Rayleigh PDF.

a surface Σ with characteristic dimensions of some wavelengths around the specular reflection point, as illustrated in Figure 2.11. In the high frequency limit, i.e. as λ_c tends toward zero, this surface reduces to the single point of the specular reflection. But, from our perspective, what we need to understand is that having Σ with a non-vanishing area leads to interesting characteristics for the reflected wave when dealing with a narrowband modulated waveform. According to equation (2.150), we can write in the specular case that the received bandpass signal $s_{RX}(t)$ is proportional to a time delayed version of the transmitted signal $s_{TX}(t)$. We thus have

$$s_{RX}(t) = c s_{TX}(t - \tau), \tag{2.169}$$

where τ is the propagation delay and c the amplitude attenuation experienced along the path considered. Now that we are dealing with a reflection that spreads over a non-vanishing area surface, Σ, we can use the linearity of Maxwell's equations to write the resulting received bandpass signal as the superposition of the re-radiated waves due to the contributions from the entire elementary surface area, referenced by the position vector $\mathbf{r}$, over Σ. We can therefore write[6]

$$s_{RX}(t) = \int_{\Sigma} c(\mathbf{r}) s_{TX}(t - \tau(\mathbf{r})) \mathrm{d}\Sigma, \tag{2.170}$$

with $\mathrm{d}\Sigma$ the elementary surface element centered on $\mathbf{r}$, and $c(\mathbf{r})$ the reflection coefficient density at that position. Still assuming that we are dealing with complex envelopes defined as centered around the carrier angular frequency ω_c, we have

$$s_{TX}(t) = \mathrm{Re}\left\{\tilde{s}_{TX}(t) \mathrm{e}^{\mathrm{j}\omega_c t}\right\}. \tag{2.171}$$

We can thus express equation (2.170) as

$$\begin{aligned} s_{RX}(t) &= \int_{\Sigma} c(\mathbf{r}) \mathrm{Re}\left\{\tilde{s}_{TX}(t - \tau(\mathbf{r})) \mathrm{e}^{\mathrm{j}\omega_c (t-\tau(\mathbf{r}))}\right\} \mathrm{d}\Sigma \\ &= \mathrm{Re}\left\{ \int_{\Sigma} c(\mathbf{r}) \tilde{s}_{TX}(t - \tau(\mathbf{r})) \mathrm{e}^{\mathrm{j}\omega_c (t-\tau(\mathbf{r}))} \mathrm{d}\Sigma \right\}. \end{aligned} \tag{2.172}$$

In order to go further, we need to consider the impact of dealing with a narrowband modulation scheme on the above expression. This assumption indeed leads to this time domain modulation being almost constant over all the positions on Σ at a given time. To understand that, we need to recall that the delay $\tau(\mathbf{r})$ involved in the above equation in fact equals the ratio of the length of the path, denoted by $l(\mathbf{r})$ in what follows, to the wave propagation speed v. As we expect to use an argument based on relative orders of magnitude of characteristics in the frequency domain, we may obtain the result more easily by working on the frequency domain

[6] In fact, we should consider the vector behavior of the electromagnetic field and write this kind of relationship for each of its components. However, due to the linearity of Maxwell's equations and considering narrowband modulated waveforms, we can directly write the complex envelopes of the signals as proportional to the components of the electromagnetic field through both polarization and matching coefficients. We can therefore write this kind of linear relationship between the transmit and received complex envelopes.

representation of the waveforms involved. Thus, assuming that the Fourier transform of the realization of the modulation process exists, which can be justified by considering a finite duration of observation, for instance, we can write, for any position **r** on Σ,

$$\int_{-\infty}^{+\infty} \tilde{s}_{\text{TX}}\left(t - \frac{l(\mathbf{r})}{v}\right) \mathrm{e}^{-\mathrm{j}\omega t} \mathrm{d}t = \mathrm{e}^{-\mathrm{j}\omega \frac{l(\mathbf{r})-l(\mathbf{r}_0)}{v}} \int_{-\infty}^{+\infty} \tilde{s}_{\text{TX}}\left(t' - \frac{l(\mathbf{r}_0)}{v}\right) \mathrm{e}^{-\mathrm{j}\omega t'} \mathrm{d}t', \tag{2.173}$$

where $\mathbf{r}_0$ represents a reference point on Σ, for which the point of the optical specular reflection can be chosen, for instance. But, as the carrier wavelength λ_{c} is related to the carrier angular frequency through the relationship $v = \lambda_{\text{c}}\omega_{\text{c}}/2\pi$, we can write that

$$\mathscr{F}_{\{\tilde{s}_{\text{TX}}(\mathbf{r},t)\}}(\mathbf{r}, \omega) = \mathrm{e}^{-\mathrm{j}2\pi \frac{\omega}{\omega_{\text{c}}} \frac{(l(\mathbf{r})-l(\mathbf{r}_0))}{\lambda_{\text{c}}}} \mathscr{F}_{\{\tilde{s}_{\text{TX}}(\mathbf{r}_0,t)\}}(\mathbf{r}_0, \omega). \tag{2.174}$$

As a result, assuming that $\tilde{s}_{\text{TX}}(t)$ has a spectrum that spreads over $[-\Omega, +\Omega]$ with $\Omega \ll \omega_{\text{c}}$, we can write that $\mathrm{e}^{-\mathrm{j}2\pi \frac{\omega}{\omega_{\text{c}}} \frac{l(\mathbf{r})-l(\mathbf{r}_0)}{\lambda_{\text{c}}}} \approx 1$ for ω in $[-\Omega, +\Omega]$ as long as the variations of $\mathbf{r} - \mathbf{r}_0$, and thus the width of Σ, are not much greater than a few wavelengths. Thus, taking the inverse Fourier transform of the above equation, we get that $\tilde{s}_{\text{TX}}(t - l(\mathbf{r})/v) \approx \tilde{s}_{\text{TX}}(t - l(\mathbf{r}_0)/v)$ and thus that

$$\tilde{s}_{\text{TX}}(t - \tau(\mathbf{r})) \approx \tilde{s}_{\text{TX}}(t - \tau_0) \tag{2.175}$$

for all possible **r** referring to a position on Σ. Here, τ_0 refers to the average delay of the path, which thus matches that of the optical specular case. It thus finally results from all this that equation (2.172) can be written as

$$s_{\text{RX}}(t) = \text{Re}\left\{ \int_{\Sigma} c(\mathbf{r})\mathrm{e}^{\mathrm{j}\omega_{\text{c}}(\tau_0-\tau(\mathbf{r}))}\mathrm{d}\Sigma\, \mathrm{e}^{-\mathrm{j}\omega_{\text{c}}\tau_0}\tilde{s}_{\text{TX}}(t - \tau_0)\mathrm{e}^{\mathrm{j}\omega_{\text{c}}t} \right\}. \tag{2.176}$$

Comparing this equation with equation (2.153) and the subsequent derivation, we see that the tap for the equivalent lowpass channel transfer function acting on the signal complex envelopes can be expressed as

$$c\mathrm{e}^{-\mathrm{j}\omega_{\text{c}}\tau_0} = \mathrm{e}^{-\mathrm{j}\omega_{\text{c}}\tau_0} \int_{\Sigma} c(\mathbf{r})\mathrm{e}^{\mathrm{j}\omega_{\text{c}}(\tau_0-\tau(\mathbf{r}))}\mathrm{d}\Sigma. \tag{2.177}$$

We see in that expression that c now has to be considered as a complex valued quantity when involved in the equivalent lowpass channel transfer function, which is a main difference with respect to the specular reflection case assumed as a first step in the previous section.

What is interesting about this relationship is that it allows us to derive a realistic behavior for the distribution of the tap. Suppose we approximate the above integral as a discrete Riemann sum. This corresponds to what is physically expected, i.e. the reflecting surface Σ is not homogeneous and can be decomposed as the superposition of different reflecting small areas with different characteristics. This approximation leads to

$$c = \sum_m c(\mathbf{r}_m)e^{j\omega_c(\tau_0-\tau(\mathbf{r}_m))}\delta\Sigma(\mathbf{r}_m). \quad (2.178)$$

From this expression we can deduce a first characteristic for the statistics of the taps for the equivalent lowpass channel model, c, i.e. that it can be considered as a centered variable. Indeed, according to the above derivation, we can write the complex exponentials involved in the summation in the form

$$e^{j\omega_c(\tau_0-\tau(\mathbf{r}_m))} = e^{j\omega_c\frac{l(\mathbf{r}_0)-l(\mathbf{r}_m)}{v}} = e^{j2\pi\frac{l(\mathbf{r}_0)-l(\mathbf{r}_m)}{\lambda_c}}. \quad (2.179)$$

Thus, assuming that Σ spreads over several wavelengths as seen previously, we get variations of the length difference between the paths, $l(\mathbf{r}_0) - l(\mathbf{r}_m)$, of several times the carrier wavelength. We thus get a variation of several times 2π of the argument of the above complex exponential for all the possible m involved in the expression for c given by equation (2.178). We thus get that the summation leads to an averaging so that

$$\text{Re}\left\{\sum_m c(\mathbf{r}_m)e^{j\omega_c(\tau_0-\tau(\mathbf{r}_m))}\delta r_m\right\} \approx \text{Im}\left\{\sum_m c(\mathbf{r}_m)e^{j\omega_c(\tau_0-\tau(\mathbf{r}_m))}\delta r_m\right\} \approx 0.$$

As a result, it seems realistic to assume that c can be considered as a centered process as its real and imaginary parts are. We can go a step further by assuming that all the elementary contributions involved in equation (2.178) are independent and identically distributed. This is another assumption that has the great advantage of giving a distribution for c that is representative of what is encountered in real life. Thus, under that assumption and due to the central limit theorem, the above real and imaginary parts follow a Gaussian distribution. And given that those variables are centered, we also deduce that c can be written in polar form as

$$c = \rho_{\text{CH}}e^{j\theta_{\text{CH}}}, \quad (2.180)$$

where ρ_{CH} follows a Rayleigh distribution and θ_{CH} is uniformly distributed over $[0, 2\pi]$. This property for the modulus and argument of c corresponds to results already detailed for Gaussian noise processes as discussed in Chapter 1, and specifically in Section 1.2.1. This distribution for the channel tap is the reason for the term “Rayleigh fading”. Nevertheless, one should keep in mind that it is related to the distribution of the magnitude only of the tap. It thus corresponds to the statistics of the random variable that corrupts the instantaneous amplitude of the wave following this path; its phase being corrupted by a random variable uniformly distributed over $[0, 2\pi]$.

Thus, assuming that we are dealing with a multipath channel, each path being linked to reflection, diffraction or scattering, we can finally write the overall channel transfer function for the bandpass signal complex envelopes according to equation (2.168), with c_n in the form

$$c_n = \rho_{\text{CH},n}e^{j\theta_{\text{CH},n}}, \quad (2.181)$$

$\rho_{\mathrm{CH},n}$ assumed to follow a Rayleigh distribution and $\theta_{\mathrm{CH},n}$ uniformly distributed over $[0, 2\pi]$. We observe, however, that different distributions can be derived for the magnitude and phase of each path in order to refine the simple reasoning considered above. Different hypotheses can lead to Rician fading, Nakagami fading, etc. But, even if a precise model of the propagation channel is required in order to derive the structure and performance of the baseband equalization stage, for instance, this is not necessarily the case for the dimensioning of the RF/analog parts of a receiver. Indeed, as discussed in Section 2.3.3, it is mainly the worst case statistic of the received signal, mainly in terms of PAPR, which needs to be taken into account in order to carry out correct dimensioning of the receive path back-off. Nevertheless, again referring to that section, it happens that the statistics of the modulation itself have very little influence on the overall statistics of the received signal that has gone through the propagation channel. Consequently, we can convince ourselves that the Rayleigh path fading leads to a worst case in terms of the distribution of the received signal instantaneous amplitude. We may recall that when different paths sum together, we have a kind of averaging that limits the variance of the received signal. But then, when more and more paths exist and are summed together, assuming that each path is independent and identically distributed, due to the central limit theorem we get that the sum tends again toward a Gaussian distribution for the real and the imaginary parts of the received complex envelope and then to a Rayleigh distribution for its amplitude. Thus, a single Rayleigh path indeed leads to a worst case in terms of received signal amplitude and thus of PAPR. This is why in the following we keep our focus on Rayleigh fading, and more precisely on single path Rayleigh fading for the derivation of RF/analog receiver dimensioning guidelines.

However, in order to determine the exact impact of such Rayleigh fading on the statistics of the received signal instantaneous amplitude, we need to take into account the relative rate of change of the channel tap amplitude compared to that of the amplitude modulation part of the transmitted signal, as discussed in "Instantaneous amplitude variations" (Section 2.3.3). But, in order to do so, we first need to detail the dynamic behavior of the propagation channel, and it is to this that we now turn.

2.3.2 *Dynamic Behavior*

Small Scale Fading vs. Large Scale Fading

Let us suppose now that either the transmitter, the receiver or both are moving. The geometrical configuration between the transmit antenna, the receive antenna and the obstacles encountered by the propagating electromagnetic wave will now change during the displacement. As a result, the characteristics of the propagation channel are time dependent. In order to reflect this, the transfer function of the multipath propagation channel, derived in the previous section, will now have time dependent parameters. Thus, we henceforth consider the transfer function form as suited for baseband complex envelopes in order to easily take into account the impact of the electromagnetic propagation on both the amplitude and phase of the electromagnetic wave. Assuming that we are dealing with sufficiently narrowband modulated waveforms, we adapt equation (2.168) to the non-static case, obtaining

$$\tilde{h}_{\mathrm{CH}}(t) = \sum_n c_n(t) e^{-j\omega_c \tau_{0,n}(t)} \delta(t - \tau_{0,n}(t)), \tag{2.182}$$

with channel taps $c_n(t)$ that can be derived from equation (2.181) as processes in the form

$$c_n(t) = \rho_{\mathrm{CH},n}(t)e^{j\theta_{\mathrm{CH},n}(t)}. \tag{2.183}$$

Moreover, as also discussed in the previous section, considering a worst case for our transceiver dimensioning in terms of received signal PAPR, we suppose that both the real and imaginary parts of the sample variables $\rho_{\mathrm{CH},n,t}$ follow a Gaussian distribution. We thus assume that their magnitudes are Rayleigh distributed and their arguments uniformly distributed over $[0, 2\pi]$.

In order to go further on the dynamic behavior of the propagation channel, we first need to focus on the difference between what are called large scale fading, or path loss, and small scale fading. From the discussion of the previous section, the relative amplitude of the different taps that compose the channel transfer function is mainly related to the nature of the reflection, diffraction or scattering experienced by the electromagnetic waves propagating along each path. As a result, for a given geometrical configuration for the transmitter, the receiver, and the obstacles encountered by the electromagnetic field during the propagation, we have a given relative amplitude between all the channel taps. Let us now therefore imagine applying a scaling factor to all the dimensions involved in the configuration. This operation impacts the power of the electromagnetic wave along each path due to the free-space propagation path loss, but not the nature of the involved reflection, diffraction or scattering. As a result, we can expect a change in the average power of the received signal, but not necessarily a difference in the relative amplitude of the taps. This simply means that in the derivation of the channel transfer function, it is meaningful to make the distinction between the average path loss experienced by the transmitted signal, and the relative amplitude and phase of the channel taps. We will discuss later on the practical reasons for making this distinction.

In order to perform the analytical derivations, and bearing in mind the physical origins of the phenomena involved, we need to detail some realistic assumptions about the stochastic processes $c_n(t)$ and $\tau_n(t)$:

(i) Even if the characteristics of the taps $c_n(t)$ vary in time, their PDFs can be assumed to be the same at each sample time.
(ii) Two different channel taps, $c_n(t)$ and $c_m(t)$ with $n \neq m$, can be considered independent. This seems realistic as they represent different echoes, i.e. are linked to re-radiated waves generated by different obstacles. We can thus expect no dependency between the resulting attenuations for the two paths. For the same reason, we can assume that two different delays $\tau_n(t)$ and $\tau_m(t)$, with $n \neq m$, are independent.
(iii) The argument of each tap c_n, i.e. $\theta_{\mathrm{CH},n}$, can be assumed independent of the delay τ_n associated with the path. It seems realistic to assume that this random phase offset experienced by the carrier, which is linked to the nature of the reflection, diffraction or scattering only, is independent of the delay and thus the length of this particular path.
(iv) Finally, due to the structure of the tap distribution for the Rayleigh fading, the magnitude and argument of a tap are independent when considered at the same sample time. This is in fact a general property of bandpass Gaussian processes, discussed in Section 1.2.1.

Let us now return to our derivation and make the link between our channel transfer function model and the channel path loss, denoted by $L_{\mathrm{CH}}(d(t))$ in what follows. This quantity is defined as the ratio between the average power of the transmit signal $s_{\mathrm{TX}}(t)$, and the average power of the received signal $s_{\text{RX}}(t)$. Obviously, this path loss is a function of the distance $d(t)$ between the transmit antenna and the receive antenna that is now time dependent. In the particular case of free space radiation, we saw in Section 2.1.4 that this average received power decreases as $1/d(t)^2$. In that particular case, we can thus write that

$$L_{\text{CH}}(d(t)) = L_{\text{CH}}(d_0)\left(\frac{d(t)}{d_0}\right)^2, \tag{2.184}$$

with d_0 a reference point somewhere between the transmitter and the receiver. However, practical measurements in urban areas, for instance, show that due to the density of the obstacles, this average power can experience greater average path loss than is predicted by the simple free space propagation theory. Realistic path loss variations are of the form

$$L_{\text{CH}}(d(t)) = L_{\text{CH}}(d_0)\left(\frac{d(t)}{d_0}\right)^n, \tag{2.185}$$

with n up to 5 or 6. Moreover, experience also shows that the dependency of the path loss does not necessarily vary exactly as an nth power of the distance, and additional correction factors that depend on $d(t)$ are often added to the above equation [22, 25]. However, as already discussed at the end of the previous section, we will not pursue this model and its related accuracy further. It is mainly of interest for the dimensioning of the baseband equalization stage of a receiver, but not necessarily for RF transceiver dimensioning. We simply suppose in the following that the path loss is given by $L_{\text{CH}}(d(t))$, with no further assumptions on it.

We can now return to the analytical expression for the propagation channel transfer function using this path loss quantity in order to highlight the distinction between large scale and small scale fading. Let us suppose for the sake of simplicity that we are dealing with a pure monochromatic waveform, i.e. a CW tone, that is transmitted through the channel. This means that we now assume that the transmit complex envelope is constant in time, i.e. that $\tilde{s}_{\text{TX}}(t) = \tilde{s}_{\text{TX}}$. This assumption, in a static configuration, leads us to expect that the received signal is also CW. However, assuming a propagation channel that now changes in time, we can expect that the received complex envelope changes in time. Thus, still assuming that we are dealing with complex envelopes defined as centered around the same center angular frequency, we can use equations (2.157) to express $\tilde{s}_{\text{RX}}(t)$ as a filtered version of $\tilde{s}_{\text{TX}}$. Using the general form for the propagation channel transfer function for complex envelopes as given by equation (2.182), we can write that

$$\tilde{s}_{\text{RX}}(t) = \left(\sum_n c_n(t)e^{-j\omega_c\tau_{0,n}(t)}\delta(t-\tau_{0,n}(t))\right)\star\tilde{s}_{\text{TX}}. \tag{2.186}$$

But, according to equation (1.64), we can derive the average power of these bandpass processes as half the square modulus of their complex envelopes. Thus, refering to the general form for $c_n(t)$ given by equation (2.183), we can write that

$$\mathbb{E}\{|\tilde{s}_{\text{RX},t}|^2\} = \left(\sum_n \mathbb{E}\{|c_{n,t}|^2\} + \sum_{\substack{n,m\\ n\neq m}} \mathbb{E}\left\{|c_{n,t}||c_{m,t}|e^{-j\omega_c(\tau_{0,n,t}-\theta_{\text{CH},n,t}-\tau_{0,m,t}+\theta_{\text{CH},m,t})}\right\}\right)|\tilde{s}_{\text{TX}}|^2.$$

But, due to the assumed independence of the arguments of the taps $\theta_{\text{CH},n,t}$ with on the one hand their amplitude $|c_{m,t}|$ and on the other hand the associated delays $\tau_{0,m,t}$, we get that

$$\begin{aligned}&\mathbb{E}\left\{|c_{n,t}||c_{m,t}|e^{-j\omega_c(\tau_{0,n,t}-\theta_{\text{CH},n,t}-\tau_{0,m,t}+\theta_{\text{CH},m,t})}\right\}\\ &\quad=\mathbb{E}\left\{|c_{n,t}||c_{m,t}|e^{-j\omega_c(\tau_{0,n,t}-\tau_{0,m,t})}\right\}\mathbb{E}\left\{e^{j\omega_c(\theta_{\text{CH},n,t}-\theta_{\text{CH},m,t})}\right\}.\end{aligned}$$

Thus, given the independence of the arguments of different paths and given the uniform distribution of those arguments over $[0, 2\pi]$ for Rayleigh paths, we get immediately that the above expression is null. It follows from those derivations that

$$L_{\text{CH}}(d(t)) = \frac{|\tilde{s}_{\text{TX}}|^2}{\mathbb{E}\{|\tilde{s}_{\text{RX},t}|^2\}} = \frac{1}{\sum_n \mathbb{E}\{|c_{n,t}|^2\}}. \tag{2.187}$$

Finally, this expression for the path loss can be used to reconsider the expression for the equivalent lowpass channel transfer function for the waveform complex envelopes as given by equation (2.182). It results in

$$\tilde{h}_{\text{CH}}(t) = \frac{1}{\sqrt{L_{\text{CH}}(d(t))}} \sum_n \mathscr{C}_n(t) e^{-j\omega_c \tau_{0,n}(t)} \delta(t - \tau_{0,n}(t)), \tag{2.188}$$

with

$$\mathscr{C}_n(t) = \frac{c_n(t)}{\sqrt{\sum_n \mathbb{E}\{|c_n|^2\}}}. \tag{2.189}$$

In our case of interest, we assume in addition that $c_n(t)$ is given by equation (2.183) with $\rho_{\text{CH},n,t}$ Rayleigh distributed and $\theta_{\text{CH},n,t}$ uniformly distributed over $[0, 2\pi]$. This means that we can still assume that the modulus of the normalized tap, i.e. $|\mathscr{C}_{n,t}|$, follows a Rayleigh distribution and that the argument, $\arg\{\mathscr{C}_{n,t}\}$, remains uniformly distributed over $[0, 2\pi]$. Moreover, as we are now dealing with normalized taps, we can expect that their variance remains constant in time. It follows that we can assume stationarity, at least up to second order, of the $\mathscr{C}_n(t)$ processes.

As discussed in more depth in Section 2.3.3, this distinction between small scale and large scale is not just a matter of formalism. The impacts of those components on the received signal are indeed handled in different ways at the receiver stage. The reason for this comes from the DR of the variations linked to those phenomena as well as their characteristic rates of change, which are really different. Thus, before saying any more about the way such small scale and large scale are handled, we first need to derive and discuss the relative rate of change of those two phenomena.

Relative Rate of Change of Small Scale and Large Scale Fading

Let us first consider small scale fading and try to understand its typical rate of change in a mobile environment. As discussed in the previous section, small scale fading results from the coherent summation of the different electromagnetic paths at the receive antenna stage. Thus, in order to derive its characteristic rate of change, we should consider on the one hand how the characteristics of each path that make up the propagation channel vary in time, and on the other hand how the coherent summation of the waves from those different paths changes in time. Thinking about the physics behind those two effects, we can in fact identify two phenomena that are root causes of these variations during the displacement of either the transceivers or the obstacles:

(i) On the one hand, the Doppler effect leads to a dilatation or contraction of time. It thus results in a distortion of the spectrum of the waveform transmitted over the propagation channel. This means that, even considering a single direct path, we have a distortion of the spectrum of the transmitted waveforms, and thus of its time domain variations.
(ii) On the other hand, the change in the relative electrical length of different paths leads to changes in the resulting coherent summation of the corresponding electromagnetic fields. What is interesting to understand is that the characteristic rate of change for this phenomenon represents not only the rate of change for the selectivity of the channel but also the rate of change for the channel tap processes themselves. Recalling the derivation of their PDF in the simple reflection case in the previous section, we see that the resulting Gaussian distribution for their real and imaginary parts results from the summation in a coherent way of the different elementary radiating parts of the overall reflecting area. We thus see that we can expect the same order of magnitude for the rate of change of the tap processes that result from such coherent sum as for the coherent recombination of the paths at the receive antenna stage.

Let us therefore first consider the impact of the Doppler effect. Suppose that a signal is radiated from a fixed transmit antenna toward a receive antenna that is moving at a relative speed $\mathbf{v}_{RX}$ compared to the transmitter. As depicted in Figure 2.12(left), we assume that the

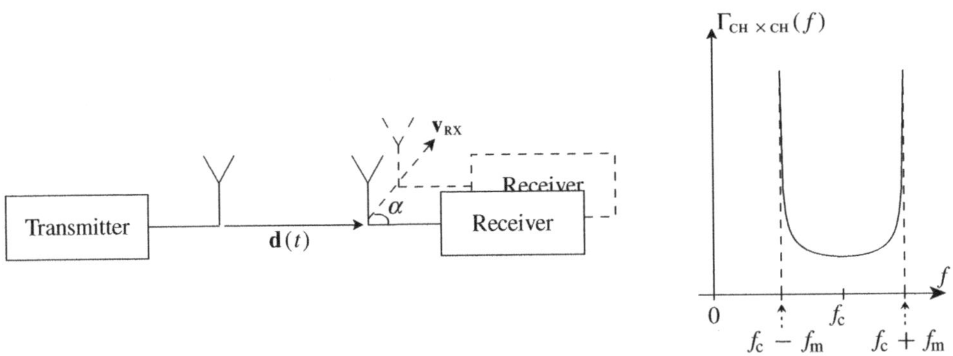

Figure 2.12 Spectrum distortion of a CW signal due to the Doppler effect during a displacement of the receiver with no preferred direction – During the transmission of a CW signal from a transmitter toward a receiver that moves at speed $\mathbf{v}_{RX}$ relative to the transmitter (left), assuming on the one hand no preferred direction in the displacement, i.e. that α is uniformly distributed, and on the other hand that we are dealing with isotropic transmit and receive antennas, so that the PSD of the received signal has the characteristic U-shape due to the Doppler effect, as given by equation (2.196) (right).

vector $\mathbf{v}_{RX}$ forms an angle α with the vector $\mathbf{d}(t)$ that goes from the transmit antenna toward the receive antenna. In order to derive the dynamic behavior of the propagation channel in that case, we can focus on the spectrum of its transfer function. In this simple situation, only a single tap is involved in the channel model as no multipath exists. Thus, in that situation the complex envelope of the received signal is simply the product of the time shifted transmitted complex envelope and the channel tap, according to the definition of $\tilde{h}_{CH}(t)$ given by equation (2.157). As a result, reasoning as above using a transmitted CW tone, i.e. a transmitted waveform with a constant complex envelope, the derivation of the received signal complex envelope spectrum immediately gives the spectrum of the tap we are looking for. Let us therefore suppose that we are dealing with a transmitted CW tone with a frequency f_c. Due to the Doppler shift, it happens that we can receive a fraction of the transmitted power in the frequency band $[f, f + \delta f]$, where f may be different from the transmitted CW frequency f_c. Indeed, by our convention, we can write the received signal carrier frequency, f, taking into account the Doppler shift in the non-relativistic approximation, as [26]

$$f = f_c + f_m \cos(\alpha), \tag{2.190}$$

with

$$f_m = f_c \frac{v_{RX}}{v} \tag{2.191}$$

the maximum Doppler shift, where $v_{RX} = \|\mathbf{v}_{RX}\|$ and v is the speed of the wave in the relevant medium. We then see that there is a direct relationship between the angle α and the location of the received power in the frequency domain. Thus, denoting by $\Gamma_{CH \times CH}(f)$ the PSD of the received signal, i.e. of the channel, at frequency f and by $P_{RX}(\alpha)$ the received power angular density in the configuration where the receiver speed vector makes an angle α with the relative transmitter and receiver positions, we can write

$$\Gamma_{CH \times CH}(f)|df| = P_{RX}(\alpha)|d\alpha|, \tag{2.192}$$

with f and α linked by equation (2.190). Thus, by differentiating this last relationship, we get that

$$|df| = f_m |\sin(\alpha)||d\alpha|. \tag{2.193}$$

But, using equation (2.190) again, we can express $\sin(\alpha)$ as a function of the frequencies of interest so that we can write

$$|df| = \sqrt{f_m^2 - (f - f_c)^2}|d\alpha|. \tag{2.194}$$

Using this result in equation (2.192), we finally get

$$\Gamma_{CH \times CH}(f) = \begin{cases} \dfrac{P_{RX}(\alpha)}{\sqrt{f_m^2 - (f - f_c)^2}} & \text{when } |f - f_c| \le f_m, \\ 0 & \text{when } |f - f_c| > f_m. \end{cases} \tag{2.195}$$

Practically speaking, dealing with mobile transceivers, we can expect that there is no preferred direction for their displacement a priori. This means that the power density $P_{\mathrm{RX}}(\alpha)$ can be assumed uniformly distributed and thus equal to $P_{\mathrm{RX}}/2\pi$, with P_{RX} the average received power. It follows that

$$\Gamma_{\mathrm{CHXCH}}(f) = \begin{cases} \dfrac{P_{\mathrm{RX}}}{2\pi\sqrt{f_{\mathrm{m}}^2 - (f - f_{\mathrm{c}})^2}} & \text{when } |f - f_{\mathrm{c}}| \leq f_{\mathrm{m}}, \\ 0 & \text{when } |f - f_{\mathrm{c}}| > f_{\mathrm{m}}. \end{cases} \tag{2.196}$$

We thus arrive at the classical U-shaped Doppler spectrum depicted in Figure 2.12(right). This final spectrum depends on the precise characteristics of the antennas, for instance through their polarization [25, 27]. What is interesting to see is that whatever the exact shape for those spectrum, we get that they spread only up to $\pm f_{\mathrm{m}}$ at maximum. This is something that we could have guessed as, physically speaking, the Doppler effect effectively leads to a shift in the frequency of the radiated wave by a maximum of $\pm f_{\mathrm{m}}$.

Let us now focus on the rate of change in the coherent summation between electromagnetic waves from different paths. As highlighted at the beginning of the section, this phenomenon also gives the rate of change of the tap processes, $\mathscr{C}_n(t)$, themselves. This is illustrated in Figure 2.13 where we can see that a small physical displacement can lead at the same time to a large change in the characteristics of the paths as well as a change in their electrical lengths. To go from a configuration where two paths sum constructively to one where they do so destructively, we need to add an electrical length of $\lambda_{\mathrm{c}}/2$ to one of the two paths, where λ_{c} is the carrier wavelength. We can assume that the rate of change is linked to the time it takes one of the devices in the system to experience a displacement of the order of $\lambda_{\mathrm{c}}/2$. Thus,

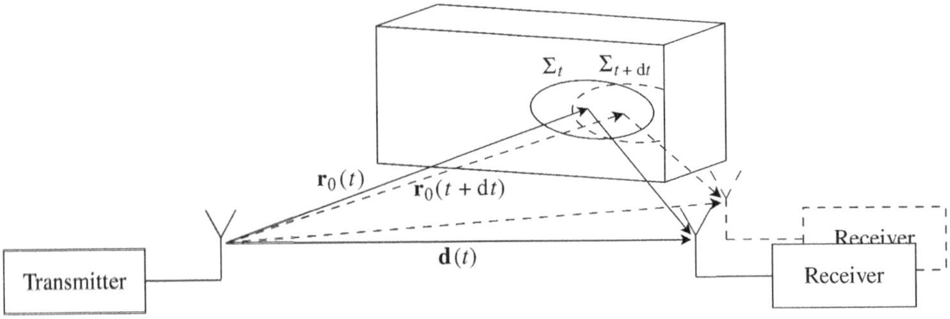

Figure 2.13 Dynamic view of a two-path propagation channel composed of a direct path and a secondary one due to a reflection – When the receiver is moving, considering the snapshot of the configuration of the link at time t (solid) and $t + \mathrm{d}t$ (dashed), we get that the distance $d(t) = \|\mathbf{d}(t)\|$ between the transmitter and the receiver has not changed much up to first order. As a result we expect little variation in the path loss, or large scale fading. But, meanwhile, there may be a large change in the amplitude and phase of the reflected wave as well as in the electrical length difference between the two paths. Thus the rate of change in small scale fading can be much more important than that in the large scale fading.

still considering the situation detailed above with a receiver moving at a speed of v_{RX} relative to the transmitter, we get that a distance $\lambda_c/2$ is covered in a duration $T_c = \lambda_c/2v_{RX}$. But, as the carrier wavelength is related to the speed of the wave in the relevant medium through $\lambda_c = v/f_c$, we can finally write that

$$T_c = \frac{1}{f_c}\frac{v}{2v_{RX}} = \frac{1}{2f_m}, \tag{2.197}$$

with f_m the maximum Doppler frequency given by equation (2.191). We thus see that we have the same order of magnitude for the rate of change of this phenomenon as for the pure Doppler effect.

The characteristic duration T_c as defined above is referred to the coherence time of the channel. Generally speaking, it is defined as the width of the Fourier transform of the channel tap spectrum. It is thus related to the inverse of the maximum Doppler frequency as derived above. This quantity characterizes the rate of change of small scale fading and thus gives an idea of the time during which the characteristics of the channel have not much changed. It gives guidelines for the design of equalization stages that should operate on slices of received samples recovered during a period shorter than this coherence bandwidth. We also observe that this concept is complementary to the coherence bandwidth concept of the channel as introduced in "Channel selectivity" (Section 2.3.1). Indeed, whereas the coherence bandwidth gives an indication of the maximum bandwidth that can be transmitted without much distortion due to selectivity through the multipath channel at a given time, the coherence time gives a time duration during which the characteristics of the channel do not change much. These two quantities are therefore of equal importance in characterizing the behavior of a dynamic multipath channel.

However, what is interesting to see is that all the physical phenomena involved in the dynamic behavior of small scale fading have the same characteristic rate of change, directly given by the maximum Doppler frequency f_m. This explains why no distinction is made in practice between the physical origins of the phenomena that impact this dynamic behavior. In order to model the time domain dependency of such a channel, and thus of the taps involved in its transfer function, we classically associate a PSD for the process as given for instance by the U-shaped function plotted in Figure 2.12(right). This spectrum is then expected to represent the time domain variations of the process, whatever their physical origins.

It is important to keep in mind that when talking about this U-shaped spectrum, we are talking about the spectrum of the bandpass processes $\mathcal{C}_n(t)$, i.e. of the spectrum of the processes that corrupt the transmitted bandpass waveform. But we remark that this spectrum is symmetric with respect to the carrier frequency f_c, which is also the center frequency implicitly chosen to define the complex envelopes of the waveforms we are dealing with in order to use the form of the channel transfer function as used up to now. Moreover, dealing with normalized taps, we can assume the stationarity of these processes in line with the discussion of the previous section. Thus, by the discussion in "Impact of spectral symmetry on top of stationarity" (Section 1.1.3), we get that the spectrum of both the real and imaginary parts of each tap process is the same as that of the tap process itself. As a result, when talking about Rayleigh fading with a U-shaped Doppler spectrum, the Rayleigh distribution refers to the distribution of the modulus of the channel tap whereas the U-shaped spectrum refers to the spectrum of the Gaussian processes that are the real and imaginary parts of the tap process. In that case, the spectrum of the modulus of the tap is no longer U-shaped [25]. This kind of relationship between the spectrum of the modulus of a complex bandpass process and the spectrum of its real and imaginary parts can be related to the derivation carried out for noise-like modulations in nonlinearity, for instance in Chapter 5 (see "Spectral regrowth" (Section 5.1.3)), or to the case of the polar transmitter in Chapter 8. In order to model a normalized Rayleigh channel tap, we can use two white Gaussian processes that are filtered using the square root of the U-shaped filter transfer function given by equation (2.196). Using those filtered processes as the real and imaginary parts of a complex envelope, we thus ensure that the corresponding bandpass process has the good U-shaped PSD with the good Rayleigh distribution for its amplitude. Note that other methods can be used to generate such processes [25, 28].

Returning to our derivation, i.e. to the comparison between small scale and large scale fading, we already know that the rate of change of small scale fading is of the order of $1/f_{\text{m}}$, with f_{m} the maximum Doppler frequency given by equation (2.191). Thus, let us now put numbers on the rate of change for large scale fading for the sake of comparison. Assume a worst case in terms of relative speed between the transmitter and receiver, i.e. that we are dealing with a receiver that is located in a high speed train running at 500 km/h. Then the rate of change of small scale fading, given by f_{m}, is of the order of 1 kHz (for a carrier frequency around 2 GHz). This means that the coherence time of the channel, linked to f_{m} through equation (2.197), is around 0.5 ms. We thus get really short time frames between potential deep changes in the channel characteristics. On the other hand, during the coherence time of the channel, the receiver has experienced a displacement of only about 7 cm. Compared to the size of a cellular cell that can be typically more than a few kilometers, there is little variation in the received average power in such a time frame, even considering a path loss as given by equation (2.185) with n up to 5 or 6. We thus see that the orders of magnitude for the rates of change of small scale and large scale fading are totally different. This confirms what we could have expected looking at Figure 2.13. We can see that a small displacement of the receiver can give large variations of the multipath configuration and thus of the final received signal power after coherent recombination. This behavior is also confirmed by the simulation of a received CW power at a moving receiver stage after going through a single Rayleigh fading path, as shown in Figure 2.14. This figure clearly shows the slow behavior of large scale fading compared to small scale fading. This difference in rate of change between the fading components is one of the reasons why they are classically handled in different ways in receivers, as discussed in the next section.

2.3.3 Impact on Receivers

According to Section 2.3.2, a received signal that has gone through a time-varying propagation channel is corrupted by both large scale and small scale fading. The main differences between the two fading components are due to their different characteristic rates of change, their different DRs and the fact that large scale fading corrupts only the instantaneous amplitude of the received signal. As a result, those fading components and the compensation of their related impacts on the received signal are classically handled in different ways at the receive stage.

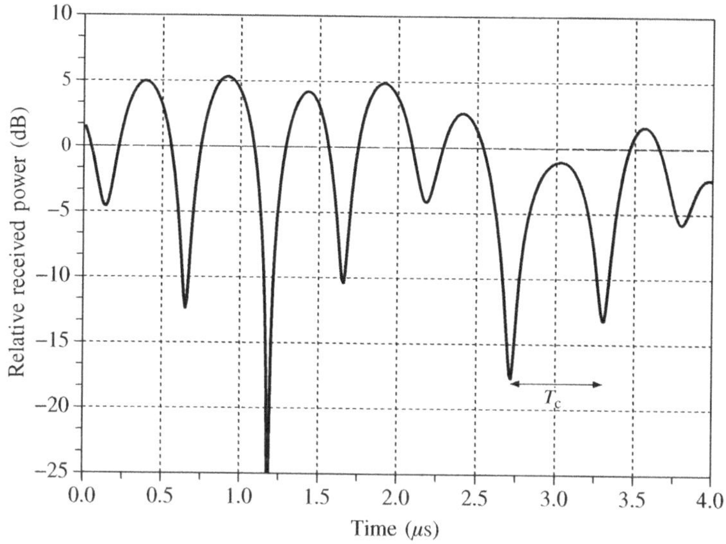

Figure 2.14 Relative time domain variations in small scale and large scale fading for a single Rayleigh fading path – Assuming that at $t = 0$ the receiver is already 500 m from the transmitter and that it moves at a speed of 500 km/h away from it, large scale fading, i.e. path loss, experiences almost no variation over a duration of 4 μs (dashed). In the present case, the dependency of the path loss is assumed to be given by equation (2.185) with $n = 5$. The tap that models the Rayleigh path is assumed with a U-shaped spectrum and a maximum Doppler frequency of 1 kHz, which corresponds to a carrier frequency of 2 GHz. We identify from the resulting amplitude variations that the coherence time, T_c, is around 0.5 μs as given by equation (2.197).

In order to highlight how this is done in practice, let us suppose for the sake of simplicity that we are dealing with a single path Rayleigh fading. Indeed, as already discussed in "Fading distribution" (Section 2.3.1), we can expect that this already leads to a worst case in terms of the received signal amplitude statistic, which is one of our main concerns for the dimensioning of RF/analog receivers. Thus, assuming that we are dealing with complex envelopes for the transmitted signal, $\tilde{s}_{TX}(t)$, and for the received signal, $\tilde{s}_{RX}(t)$, that are defined as centered around the same center angular frequency, we can model the propagation channel as a filter acting on these complex envelopes according to equation (2.157). In that equation, $\tilde{h}_{CH}(t)$ represents the channel transfer function for the considered complex envelopes, given in its general form by equation (2.188). But, under the present single path assumption, this transfer function reduces to

$$\tilde{h}_{CH}(t) = \frac{1}{\sqrt{L_{CH}(d(t))}} \mathscr{C}(t) e^{-j\omega_c \tau_0(t)} \delta(t - \tau_0(t)), \qquad (2.198)$$

with $L_{\text{CH}}(d(t))$ the path loss given by equation (2.185). In addition, as we are assuming that we are dealing with Rayleigh fading, $\mathscr{C}(t)$ can be written in the form

$$\mathscr{C}(t) = \rho_{\text{CH}}(t)e^{j\theta_{\text{CH}}(t)}, \tag{2.199}$$

with $\rho_{\text{CH},t}$ following a Rayleigh distribution and $\theta_{\text{CH},t}$ uniformly distributed over $[0, 2\pi]$. Then, if we write the transmitted signal complex envelope in its polar form as

$$\tilde{s}_{\text{TX}}(t) = \rho_{\text{TX}}(t)e^{j\phi_{\text{TX}}(t)}, \tag{2.200}$$

we can express the received signal complex envelope using equation (2.157) and the three equations above as

$$\tilde{s}_{\text{RX}}(t) = \frac{1}{\sqrt{L_{\text{CH}}(d(t))}}\rho_{\text{CH}}(t)\rho_{\text{TX}}(t-\tau_0(t))e^{j(\phi_{\text{TX}}(t-\tau_0(t))+\theta_{\text{CH}}(t)-\omega_c\tau_0(t))}. \tag{2.201}$$

Thus, now taking $\tilde{s}_{\text{RX}}(t)$ in its polar form, i.e. as $\rho_{\text{RX}}(t)e^{j\phi_{\text{RX}}(t)}$, we can write that

$$\rho_{\text{RX}}(t) = \frac{1}{\sqrt{L_{\text{CH}}(d(t))}}\rho_{\text{CH}}(t)\rho_{\text{TX}}(t-\tau_0(t)), \tag{2.202a}$$

$$\phi_{\text{RX}}(t) = \phi_{\text{TX}}(t-\tau_0(t)) + \theta_{\text{CH}}(t) - \omega_c\tau_0(t). \tag{2.202b}$$

We see that $\rho_{\text{RX}}(t)$ and $\phi_{\text{RX}}(t)$ are effectively different from the expected values, $\rho_{\text{TX}}(t)$ and $\phi_{\text{TX}}(t)$, which represent the transmitted modulation. Using this model, we can now further discuss how those corruptions of the received signal are classically handled at the receive stage.

Instantaneous Amplitude Variations

Equation (2.202a) tells us that the instantaneous amplitude $\rho_{\text{RX}}(t)$ of a bandpass signal that has gone through a single Rayleigh path is linked on the one hand to the path loss $L_{\text{CH}}(d(t))$ and on the other hand to the product of the propagation channel tap amplitude, $\rho_{\text{CH}}(t)$, and the transmitted amplitude modulation delayed in time by the propagation duration, $\rho_{\text{TX}}(t-\tau_0(t))$. As highlighted previously, this split into two categories for the received signal amplitude in fact corresponds to different sets of orders of magnitudes for their main characteristics.

First, as detailed in "Relative rate of change of small scale and large scale fading" (Section 2.3.2), we have different rates of change for these phenomena. Indeed, as already discussed, the rate of change of small scale fading is much higher than that of large scale fading, i.e. of path loss. This is even worse for the rate of change of the modulation scheme that can classically range from hundreds of kilohertz up to tens of megahertz, as illustrated in Section 1.3. Compare this with the small scale fading rate of change that is linked to the maximum Doppler frequency, f_m, given by equation (2.191), and is scarecely higher than a few kilohertz.

Second, the magnitudes of these amplitude variations are also completely different. Indeed, for both the modulating scheme and the channel tap amplitude, we are dealing with constant average processes. This means that we only have to cope with their instantaneous variations around their average value, i.e. with a DR that is classically of the order of a few tens of decibels most of the time. Compare this with the path loss that experiences huge variations when the mobile station goes from one edge of the cell to the other following equation (2.185). This can classically lead to variations of the average received power by more than 80 dB in total.

As a result, large scale fading, or path loss, leads to the largest variations in terms of the average received power but also has the lowest rate of change. It thus seems reasonable to imagine a system able to track these low frequency variations in the averaged received power and compensate for them by adjusting the gain of the receiver. The main use of such a system is to minimize the DR of the received signal at the RF/analog receiver output and thus to minimize the number of bits required to implement the subsequent digital signal processing and equalization stages. Such a system is called the automatic gain control (AGC) of the receiver and is discussed in more depth in Chapter 9, Section 9.2.1.

An interesting question is why we expect to compensate for the variations of the instantaneous amplitude of the received signal due to the path loss only when considering an AGC system and not also for small scale fading. In fact, this is related to the first set of different characteristics recalled above, i.e. the associated rate of change. Indeed, an AGC system, at least a classical one, is basically based on a tracking of the received power or amplitude variations. Thus, if wanted to track small scale fading variations, which have a rate of change that can lie in the kilohertz range, we would need an AGC loop with a passband much higher than the kilohertz range. But in that case the loop would also compensate for the low frequency components of the received signal amplitude modulation that lies within its passband. This is obviously undesirable as it would result in a degradation of the modulation scheme. Thus, even if the exact impact should be investigated on a case by case basis, the best we can do in the receiver is to minimize the DR of the received signal by compensating the path loss and then to provide the scaled signal to the digital equalization stage, which is able to compensate for the remaining channel variations using more sophisticated techniques than just instantaneous amplitude tracking [1].

An overview of how such an AGC system impacts the signal waveform at the RF/analog receiver output is shown in Figure 2.15. It depicts the results of a time domain simulation assuming the same configuration as used for Figure 2.14. This means a transmitted CW waveform that experiences on the one hand a single path Rayleigh fading for small scale fading, with a U-shaped spectrum and a Doppler frequency of 1 kHz, and on the other hand the path loss model given by equation (2.185), with $n = 5$. Here, we have again assumed that at $t = 0$ the receiver is already at a distance of 500 m from the transmitter and that it moves at a speed of 500 km/h away from it. Finally, we have assumed that the AGC system update frequency of the receiver gain is set at 215 Hz, which corresponds to the GSM frame rate and thus to the effective update rate of most GSM AGC system implementations. It is evident that the low frequency update is intended to compensate the path loss but effectively does not succeed in compensating the variations due to small scale fading. However, we have to keep in mind that we have considered a worst case in terms of relative speed between the transmitter and the receiver, and thus in terms of Doppler frequency. For low speed displacements, such AGC systems can effectively track and compensate small scale fading, or at least part of it. But in the case of low speed displacement, the distinction between small scale and large scale fading becomes less obvious.

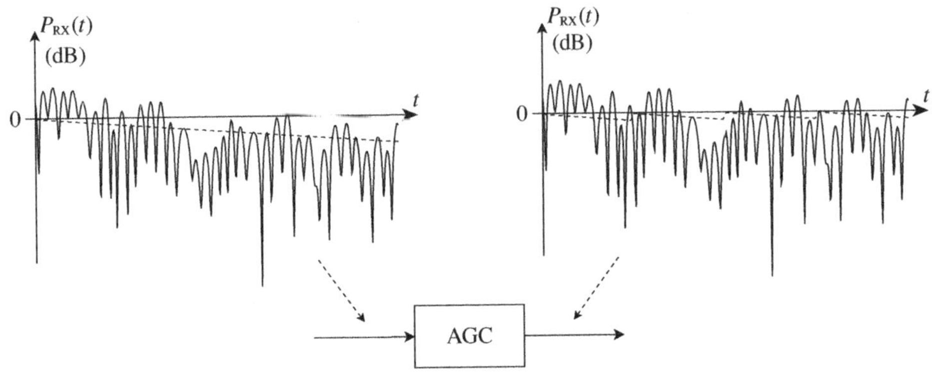

Figure 2.15 Overview of the impact of AGC on the received signal power – Due to its low update frequency, intended to not corrupt the amplitude modulation of the received signal, a classical AGC system compensates mainly for large scale fading, i.e. path loss (dashed). As a result, the instantaneous power of the received signal still exhibits the variations of small scale fading after the gain compensation command from the AGC, at least for high speed channels (solid).

Given that the AGC system scales the average received power to a target level, we wonder what back-off we realistically have to consider in a receiver line-up. Here, by "back-off" we mean the ratio between the available full scale (FS) to represent the received signal and the target RMS level at which the AGC system scales it. Basically, for the RF/analog parts of a receiver, "full scale" means the maximum voltage swing that can be handled or delivered by the devices. Thus, the back-off is something that has to be well defined in order to correctly set the target levels throughout the receive path and avoid saturations while using the maximum available DR. We might surmise that the back-off depends on the modulation of the signal we expect to process. This is indeed true on the transmit side as no other time-varying phenomenon is expected to modify this statistic. But it is not true on the receive side due to the impact of the channel statistics.

This is confirmed by the simulation results displayed in Figure 2.16 that show the statistics of the received signal for different modulation schemes. For these simulations, we use the same single Rayleigh path propagation channel and path loss parameters as for Figure 2.15. The same procedure is also carried out on the signal recovered at the AGC output, after compensation of the path loss at an update rate of 215 Hz. In order to interpret the results, we assume that the instantaneous amplitude of the received signal, originally given by equation (2.202a), can be written once the path loss is compensated according to

$$\rho_{RX}(t) \approx \rho_{CH}(t)\rho_{TX}(t - \tau_0(t)). \tag{2.203}$$

This is in fact only an approximation as between two gain updates of the AGC system, we have the path loss that still impacts the received signal. Nevertheless, the present assumption is accurate enough for our purposes. In addition, we also assume that during the observation time of the received signal we have almost constant $\tau_0(t) \approx \tau$. Indeed, the variations of this delay are proportional to the displacement experienced by the transceivers over the speed of the wave in the relevant medium. This results in delay drifts that can be low with respect to the characteristic rate of change of the modulation scheme. In practice, the equalization processing is indeed performed on slices of received signals that last for durations that allow $\tau_0(t)$ to

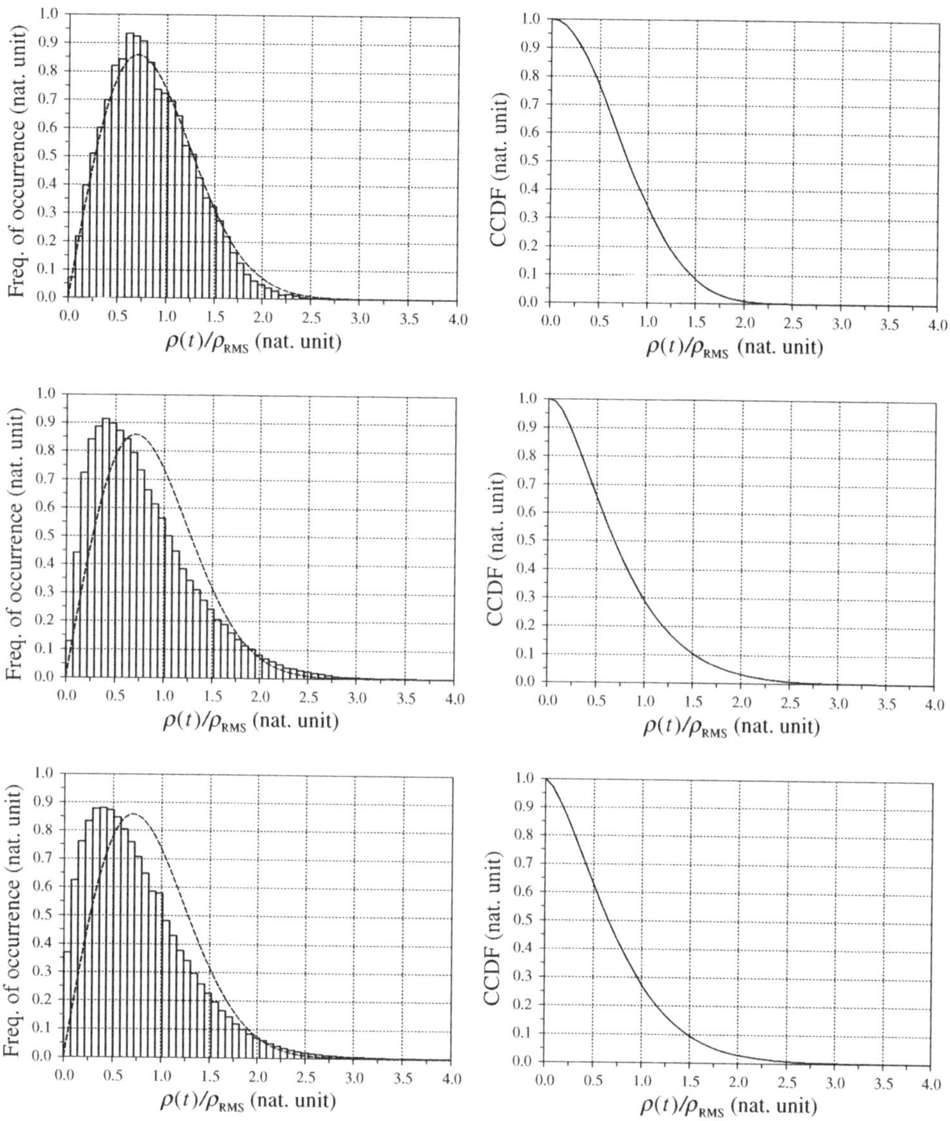

Figure 2.16 Amplitude distributions of randomly modulated waveforms corrupted by a single path Rayleigh fading – With the propagation channel characteristics used in Figure 2.14 and an AGC update frequency of 215 Hz, we get that a constant amplitude GMSK waveform exhibits a Rayleigh distribution for its amplitude part at the AGC system output (top). For an 8PSK scheme (middle) or an OFDM waveform (bottom) the distributions of the received amplitudes are linked to the product of the Rayleigh tap of the channel and the amplitude modulation. However, the waveform PAPR, corresponding to the CCDF of their normalized amplitude which reaches 0.001, is equal to almost $\rho(t) = 2.3\rho_{RMS}$ or 7.2 dB whatever the modulation scheme.

be considered almost constant. Moreover, most wireless transceivers embed an automatic frequency correction (AFC) scheme that compensates for Doppler drifts at the level of the system clock itself. As discussed in Section 9.2.5, we thus get a natural compensation for such delay drift in the samples provided to the equalization stage. Thus we assume that we can write, at the AGC system output,

$$\rho_{\text{RX}}(t) \approx \rho_{\text{CH}}(t)\rho_{\text{TX}}(t-\tau). \tag{2.204}$$

This means that the statistics of the received signal instantaneous amplitude are well approximated by the product of the channel tap amplitude and the amplitude part of the transmit signal modulation scheme. This explains why when considering a constant amplitude modulation scheme, i.e. with $\rho_{\text{TX}}(t) = \rho_{\text{TX}}$ constant, such as the GMSK modulation of the GSM standard, we recover at the receive stage the Rayleigh distribution of the channel tap amplitude, as can be seen in Figure 2.16. We observe, however, that the theoretical Rayleigh PDF, plotted as the dashed curve, is not exactly recovered at the receive signal stage. This is linked to the path loss that is not compensated in practice by the AGC system between two gain updates. It thus slightly impacts the resulting distribution. In contrast, we see that for an OFDM signal, with an instantaneous amplitude that already has a Rayleigh distribution, we recover at the AGC output the distribution of the product of two such Rayleigh variables. In any case, what is interesting to see is that whatever the type of modulation scheme considered, the distribution of the received signal amplitude almost always exhibits the same PAPR. This PAPR, equal to the ratio of the peak instantaneous amplitude to its RMS value, can be evaluated as the value for which the CCDF of the normalized instantaneous amplitude reaches 0.001, as discussed in Section 1.1.3. In the present case, we obtain a value of about 2.3, or 7.2 dB in the three cases considered. This is thus more or less the value of the Rayleigh distribution itself. We thus see that the impact of the modulation scheme on the back-off in the receive path is negligible compared to the channel statistic itself.

We can even go a step further. As discussed in "Peak to average power ratio and crest factor" (Section 1.1.3), the CF of the real and imaginary parts of the received signal complex envelope are estimated to be about 3 dB higher than the bandpass signal PAPR. This thus corresponds to a minimum back-off of about 7.2 dB for the received RF bandpass signal and thus 10.2 dB for the lowpass $p(t)$ and $q(t)$ modulating waveforms along the baseband part of the receive path. This is much more than could be expected from low PAPR modulating waveforms, for instance. In contrast, dealing with noise-like waveforms, we see that we do not have to sum the PAPR linked to the fading with that linked to the modulation. Statistically speaking, this means that there is a low probability of the high PAPR of the two variables occurring at the same time.

To conclude, we observe that the back-off discussed here must be understood as the minimum back-off that we need to consider in a receiver. Other contributors such as the gain spread, the AGC inaccuracy or the residual DC offsets have to be taken into account in order to derive the final back-off in a receive path. Further discussion of this problem is deferred to "Filtering budget vs. ADC dynamic range" (Section 7.3.3).

Instantaneous Phase Variations

Equation (2.202b) tells us that the instantaneous phase, $\phi_{\text{RX}}(t)$, of a bandpass signal that has gone through a single path Rayleigh fading is corrupted on the one hand by a propagation delay that impacts both the modulation scheme and the carrier phase offset, and on the other hand by a random phase offset linked to the nature of the propagation path itself. Rigorously speaking, the propagation delay $\tau_0(t)$ involved in this model is a function of time when considering moving transceivers. Nevertheless, as discussed in the previous section, for the time frame of interest during the observation of the received signal we can assume that $\tau_0(t) \approx \tau$ is constant. This means that we can assume for our discussion that the instantaneous phase of the received signal can be written as

$$\phi_{\mathrm{RX}}(t) \approx \phi_{\mathrm{TX}}(t-\tau) + \phi(t), \tag{2.205}$$

with

$$\phi(t) = \theta_{\mathrm{CH}}(t) - \omega_{\mathrm{c}}\tau. \tag{2.206}$$

Such a phase offset, if not compensated, can be a problem when it comes to recovering the good transmitted modulating waveforms $p_{\mathrm{TX}}(t)$ and $q_{\mathrm{TX}}(t)$ and thus the good data bits. In order to illustrate this, let us suppose that the modulus of the received complex envelope is simply a time shift copy of that of the transmitted complex envelope. This means that in contrast to the approximation at the AGC output derived in the previous section and given by equation (2.204), we now assume that $\rho_{\mathrm{CH}}(t)$ is constant. Even if not realistic in a moving configuration, this assumption simply allows us to focus on the impact of the phase shift. Under that assumption,

$$\tilde{s}_{\mathrm{RX}}(t) = \rho_{\mathrm{RX}}(t)\mathrm{e}^{\mathrm{j}\phi_{\mathrm{RX}}(t)} \approx \rho_{\mathrm{TX}}(t-\tau)\mathrm{e}^{\mathrm{j}(\phi_{\mathrm{TX}}(t-\tau)+\phi(t))}, \tag{2.207}$$

and thus, assuming that we are dealing with complex envelopes defined as centered around the angular frequency ω_{c}, the corresponding received RF bandpass signal can be written as

$$s_{\mathrm{RX}}(t) = \mathrm{Re}\left\{\tilde{s}_{\mathrm{RX}}(t)\mathrm{e}^{\mathrm{j}\omega_{\mathrm{c}}t}\right\} = \rho_{\mathrm{RX}}(t)\mathrm{e}^{\mathrm{j}(\omega_{\mathrm{c}}t+\phi_{\mathrm{RX}}(t))}. \tag{2.208}$$

Basically, the modulating waveforms can be recovered on the receive side using a direct frequency downconversion as detailed in Chapter 1, and in more detail in Chapter 6. This means that we recover the sideband of $s_{\mathrm{RX}}(t)$ that is centered on $+\omega_{\mathrm{c}}$ by multiplying it by the negative complex exponential $\mathrm{e}^{-\mathrm{j}\omega_{\mathrm{c}}t}$ and then lowpass filter the resulting baseband waveforms in order to cancel the remaining unwanted sideband. The resulting $p_{\mathrm{BB}}(t)$ and $q_{\mathrm{BB}}(t)$ baseband waveforms recovered at the output of the P and Q branches of the frequency downconversion are then the real and imaginary part of the complex signal $s_{\mathrm{BB}}(t)$ corresponding to the positive sideband we are looking for. We can write these as

$$s_{\mathrm{BB}}(t) = p_{\mathrm{BB}}(t) + \mathrm{j}q_{\mathrm{BB}}(t) = \tilde{s}_{\mathrm{RX}}(t), \tag{2.209}$$

or, using equation (2.207), as

$$s_{\mathrm{BB}}(t) = \rho_{\mathrm{TX}}(t-\tau)\mathrm{e}^{\mathrm{j}(\phi_{\mathrm{TX}}(t-\tau)+\phi(t))}. \tag{2.210}$$

We thus see that compared to the transmitted waveform, we have to cope with a phase shift that results in a rotation of the constellation used for the modulation as illustrated in Figure 2.17 in

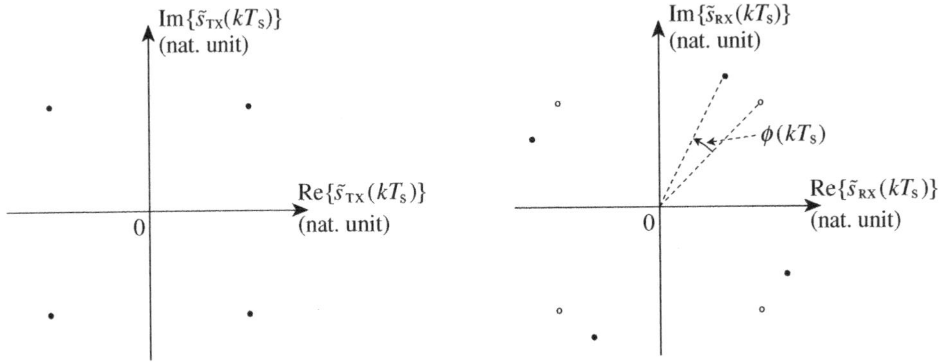

Figure 2.17 Carrier phase shift impact on the recovered symbol constellation – Using an LO signal for the frequency transposition down to baseband that does not have a coherent phase relationship with the received signal carrier, we get a phase shift of the received data symbol constellation (right, solid circles) compared to the transmitted constellation (right, empty circles and left, solid circles), here corresponding to a simple 4QAM modulation. For a Rayleigh fading path, this phase shift $\phi(t)$, given by equation (2.206), depends on the carrier phase offset due to the average propagation delay, on the random phase offset introduced by the involved electromagnetic phenomenon at the origin of the Rayleigh path distribution, and on the phase offset at the origin of the LO used both at the transmit and receive side for the generation and recovery of the waveforms.

the simple case of a 4QAM constellation. This phase shift therefore needs to be compensated in order to correctly decide the transmitted bits.

In fact, some really basic modulation schemes can behave without this phase shift compensation, albeit with poor performance. All modern systems using complex modulation need to compensate for it. In that case, we talk about coherent reception. We might think that such reception can be done by ensuring that the LO embedded in the receiver has the correct phase relationship with the received signal. Indeed, using a complex exponential of the form $e^{-j(\omega_c t+\phi(t))}$ instead of $e^{-j\omega_c t}$ for the frequency transposition leads to a recovered baseband complex signal that can be written as

$$s_{BB}(t) = \rho_{RX}(t)e^{j(\phi_{RX}(t)-\phi(t))} = \rho_{TX}(t-\tau)e^{j\phi_{TX}(t-\tau)}, \tag{2.211}$$

using equation (2.205) to expand $\phi_{RX}(t)$. The problem is then to have the RF synthesizer used for the generation of the LO that has a good phase relationship with the received signal. Moreover, as this phase offset can vary in time as $\phi(t)$ does, we need to consider a loop that tracks those variations. A classical way to handle this has been introduced by Costas through what is now called a Costas loop [29]. A general overview of a loop able to track the received carrier phase offset is shown in Figure 2.18. The underlying assumption for such a loop to work is that the potential phase modulation of the received signal has a high frequency compared to the fluctuations of the phase offset due to the channel variations. This is indeed what can be expected, as already discussed in "Instantaneous amplitude variations" earlier in this section, concerning the amplitude components of the received signal. Under that assumption, the lowpass filtered version of the argument of the baseband complex envelope,

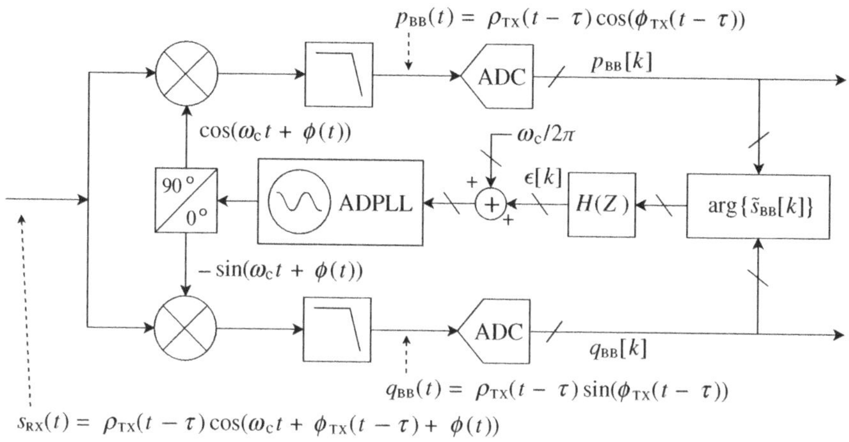

Figure 2.18 Coherent reception by compensation of the received carrier phase offset at the LO stage – The compensation of the low frequency drift of the received carrier phase offset can be done at the LO stage. The residual phase offset can be estimated as the argument of the received baseband complex envelope, $\tilde{s}_{BB}(t) = p_{BB}(t) + jq_{BB}(t)$, lowpass filtered in order to cancel the potential phase modulation. Once the loop is locked, the error signal $\epsilon[k]$ is expected to be null.

i.e. $\epsilon[k]$, represents the error in this phase offset compensation. This error term is therefore null when the compensation is exact.

However, in most cases this compensation is not done at the RF synthesizer stage. Indeed, we saw in the above-mentioned section that we needed an equalization stage to compensate for the amplitude variations of the channel. This means that such an algorithm needs to make an estimate for the channel before compensating for it. Thus, it is simpler from an implementation point of view to deal with a channel estimator that estimates both $\rho_{CH}(t)$ and $\theta_{CH}(t) - \omega_c\tau$ at the same time and performs the overall equalization in the digital domain. This is indeed what is done in practice. At this channel estimator and compensation output, we can then write the received complex signal in the form

$$s_{BB_0}(t) = \rho_{TX}(t - \tau)e^{j\phi_{TX}(t-\tau)}. \tag{2.212}$$

We thus see that the remaining problem is the synchronization of the receiver. Indeed, the correct estimation of τ is required in order to sample correctly the received waveform and correctly decide the data bits. However, all those functions required in a receiver are digital baseband related and we will not pursue this issue further.

In the rest of this book, we do not explicitly write the dependence of the instantaneous amplitude and phase of the received signals on the propagation channel parameters. As those parameters are handled at the digital baseband stage, there is no particular need to write them explicitly for the purpose of discussing RF/analog transceiver topics. A first exception obviously occurred in the previous section for the discussion of the back-off in RF/analog receivers. Another exception occurs during the study of the LO waveform properties in Section 4.3.

3

The Wireless Standards Point of View

A final set of constraints that drive the functionalities to be embedded in a transceiver as well as their performance come from the requirements set by the wireless standard we want to address. Those requirements are in fact mainly driven by two aspects of a wireless system. We need to ensure on the one hand a sufficiently high quality for the wireless link for the reliable transfer of data, and on the other hand that this quality remains sufficiently high for a given user while coexisting with other users. Such other users may or may not belong to the same wireless system.

From that perspective, it is interesting to see that medium access strategies have a deep impact on the transceiver architectures as well as on their ultimate performance. It is also interesting to see that common metrics have been derived to put numbers on the ability of a wireless transceiver to fulfill those two goals. We give a quick overview of the classical metrics in order to understand the underlying network system constraints. These metrics are used throughout the rest of this book, in particular to illustrate the transceiver budgets in Chapter 7.

3.1 Medium Access Strategies

By a "medium access strategy" we mean the strategy that is used by a wireless system in order to share in an optimal way the limited radio resource, i.e. the radio frequency band in practice, between on the one hand the different users and on the other hand the uplink and the downlink. Here, "uplink" means the radio link that goes from the user equipment (UE) toward the base transceiver station (BTS) that is in charge of the cell; and "downlink" means the radio link that goes from the BTS toward the UE. We give a quick overview of the different strategies classically used to achieve these two goals, with the aim of examining the associated implementation constraints for transceivers.

3.1.1 Multiplexing Users

Given that all the users share the same physical resource – the same radio frequency band, whether at the same time or not – we can sort wireless systems into two groups [1]. Into the

Wireless Transceiver Architecture: Bridging RF and Digital Communications, First Edition. Pierre Baudin.

first group we can put systems in which users have a dedicated radio resource, i.e. a dedicated frequency band over a given time. We can split this group into two subgroups. Classically when all users use different carrier frequencies with non-overlapping frequency bands, we talk about a FDMA system. In contrast, when users use the same frequency band but at different times, we talk about a time division multiple access (TDMA) system. These two configurations are illustrated in Figure 3.1(top and middle), where uplink configurations are represented.

Into the second group we can then put systems where all the users are multiplexed over the same frequency band at the same time. In practice, this multiplexing is for the most part realized through the use of orthogonal codes, as in CDMA systems. However, we can also include systems that use orthogonal trigonometric polynomials for this multiplexing, as in OFDM systems. An overview of this is shown in Figure 3.1(bottom), again with uplink configurations represented.

It is of interest from our transceiver perspective that the different strategies for the multiplexing of users directly impact the implementation performance. For instance, as highlighted in Section 1.3.3, a system based on CDMA or OFDM relies on transceivers that process wideband waveforms, thus with high PAPRs compared to what is encountered in systems using TDMA or FDMA. Different implementation constraints result, as discussed in greater depth in "Narrowband and low PAPR vs. wideband and high PAPR waveforms" (Section 3.1.3).

3.1.2 Multiplexing Uplink and Downlink

Once the multiplexing of the users is done, we need to share the radio resource between the uplink and the downlink. Here again there are two classical ways to proceed. On the one hand, we have systems that use two different frequency bands for these two radio links. These are called frequency division duplex (FDD) systems. On the other hand, we have systems that use the same frequency band, but during different time slots. These are called time division duplex (TDD) systems.

However, it is theoretically possible to imagine a system that can transmit and receive at the same time while using the same carrier frequency for both links. This might be possible, for instance, based on the use of directive couplers and isolators to make the separation between the forward and reverse waves corresponding to the transmit and receive paths. Unfortunately, this remains theoretical only as load mismatch, for instance, would lead to poor isolation between the transmit and receive paths and thus to transceivers that are unable to receive and transmit at the same time in practice. We therefore stick to the classical FDD and TDD configurations in what follows.

Let us first focus on FDD systems. The advantage of this way of multiplexing the uplink and downlink is to allow the implementation of transceivers able to transmit and receive at the same time. The use of different frequency bands allows the use of RF filters able to isolate the transmit and receive paths so that transmission and reception can be considered at the same time. This behavior, which would not be easily achieved when solely based on the directivity of microwaves devices, leads to the possibility of implementing a transceiver that is said to be a full duplex device. The interest in such full duplex devices is obvious in terms of data transmission as the simultaneous transmission and reception allows higher potential data rates. However, this is achieved at some implementation cost, as discussed in "Full duplex vs. half duplex" (Section 3.1.3). FDD systems based on the use of full duplex transceivers are often

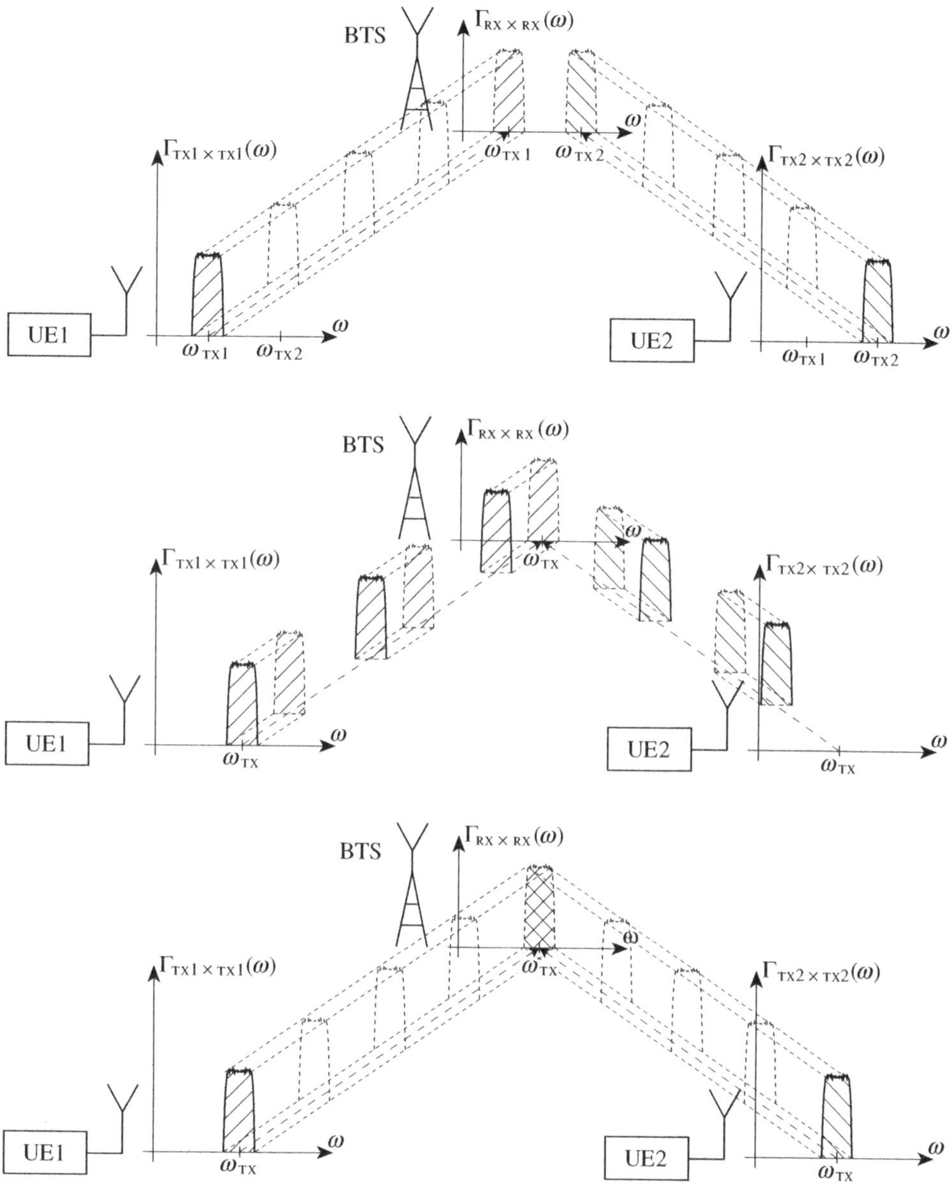

Figure 3.1 Classical ways to multiplex users on the radio resource – Classically, different users of a wireless system can be multiplexed on the radio resource using different frequency bands, as in an FDMA system (top), or using the same frequency band but during different time slots, as in a TDMA system (middle), or using the same frequency band at the same time through the use of orthogonal waveforms as in CDMA or OFDM systems (bottom). Here, uplink configurations are represented.

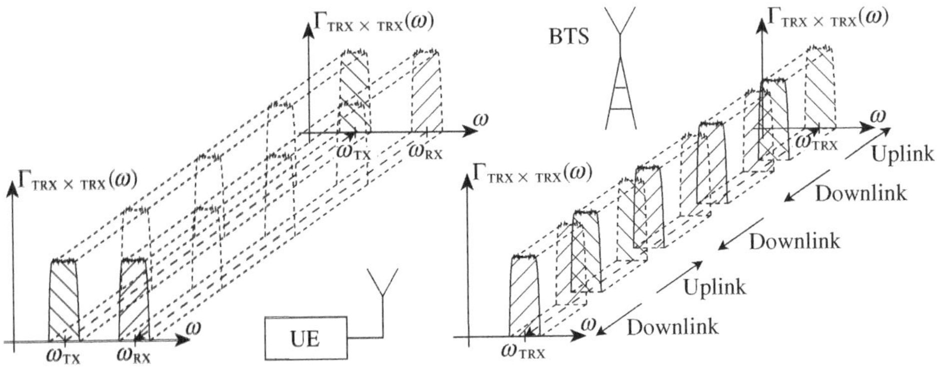

Figure 3.2 Ways to multiplex uplink and downlink in FDD and TDD systems – FDD systems multiplex the uplink and the downlink on different carrier frequencies, thus classically allowing transmission and reception at the same time for the UE (left). In contrast, TDD systems multiplex the uplink and the downlink over the same carrier frequency, thus during different time slots (right).

referred to as full duplex systems. In practice, even if FDD systems can be different from full duplex systems, they are almost always as represented in Figure 3.2(left).

In the same way, TDD systems are associated in practice to half duplex devices, i.e. devices that do not transmit and receive at the same time, as illustrated in Figure 3.2(right). However, although it might seem that half duplex leads to a simpler implementation, there are still particular constraints associated with such an alternative transmit and receive configuration, as also discussed in "Full duplex vs. half duplex" (Section 3.1.3). Even if a TDD system leads to a potential overall reduced data rate for the end user compared to the same system in a FDD version, due to the sharing of the radio resource between the uplink and the downlink, it has the advantage of allowing for a flexible allocation of this radio resource between the two radio links. The network can, for instance, allocate more uplink slots to the user when a higher uplink data rate is necessary and fewer when a higher downlink data rate is required. This kind of flexibility theoretically allows a more efficient use of the radio resource compared to the fixed allocation in the FDD case.

Thus, recalling our discussion in Section 3.1.1, we conclude that wireless systems are based in practice on a couple of access strategies, the first addressing the multiplexing of the users, and the second the multiplexing of the uplink and the downlink. This juxtaposition of strategies is particularly highlighted by wireless standards that exist in both FDD and TDD versions. For instance, both the WCDMA, based on CDMA for the multiplexing of the users, and the LTE, based on OFDM for the same purpose, exist in FDD and TDD versions [30–32]. However, we can have situations where the distinction is less obvious as we get a mix of those different strategies. This is, for instance, the case of the GSM standard which uses both TDMA and FDMA at the same time to multiplex the users, and both TDD and FDD at the same time to multiplex the uplink and downlink. Moreover, we can have some frequency hopping on top of that, meaning that the carrier frequency of the received or transmitted signals changes periodically, basically from one time slot to the next. Thus, even if the strategies used for the multiplexing can be tricky, we can derive some high level guidelines for the implementation of transceivers, depending on the multiplexing strategies adopted in a given wireless standard. This is done in the following sections.

3.1.3 Impact on Transceivers

Narrowband and Low PAPR vs. Wideband and High PAPR Waveforms

From the transceiver implementation point of view, the knowledge of the exact origins of the structure of the signal to be processed is not so relevant up to first order. Indeed, considering for instance a CDMA receiver, the knowledge that the received signal being processed contains the data for all the users in the cell or only different data channels dedicated to a single user does not much change the way the RF/analog receiver has to process the received signal as long as it has the same bandwidth and statistics in the two cases. It is only those characteristics that set the requirements for the RF/analog block performance.

In that sense, the medium access strategy can impact the dimensioning of transceivers. For instance, as highlighted in Section 3.1.1, systems that multiplex the users over the same frequency band at the same time lead to wideband waveforms, often with high PAPR. This statement is obvious with regard to the bandwidth as the waveform we need to process, at least on the receive side, may contain the data for all the users at the same time. This results in a higher bandwidth than would be strictly necessary to carry the data associated with a single user. In the same way, recalling the central limit theorem, we can imagine that the sum of all the waveforms carrying the data of each user can lead to a total waveform with Gaussian-like statistics, i.e. with a high PAPR.

Thus, a potential problem linked to this way of multiplexing users is that even if one of them requires only a low data rate on the downlink, it still needs to process the overall wideband signal that carries the data for all the users in the cell. It is only after digital signal processing that the split between the data for all the users occurs. Thus, in any case, the RF/analog receiver needs to process correctly the total signal. But, as this signal has a higher bandwidth and PAPR than for a waveform that carries only the data for the user, we can expect a higher power consumption of the receiver in order to implement analog devices with higher bandwidth and linearity, ADC and digital signal processing with higher sampling rates, etc. In fact, we see that a UE receiver in such a system needs to have the power consumption corresponding to what is required to process correctly the overall signal in the cell, i.e. the signal corresponding to the maximum data rate that is allowed by the standard in a given cell. This holds whatever the effective data rate assigned to this UE. In that sense, TDMA or FDMA systems potentially lead to more efficient power consumption implementations as only the waveform that carries the data for the user is processed by its receiver in practice.

Full Duplex vs. Half Duplex

The strategy adopted for the multiplexing of the uplink and downlink also has many impacts on the transceiver architecture and performance. Depending on whether we need to transmit and receive at the same time as in a full duplex transceiver, or periodically to turn the transmitter on and off as in a half duplex transceiver, there are different sets of constraints to consider.

For instance, we need to keep in mind that in any physical implementation, due to coupling through either free or guided radiation, only a finite RF isolation can be achieved. Moreover,

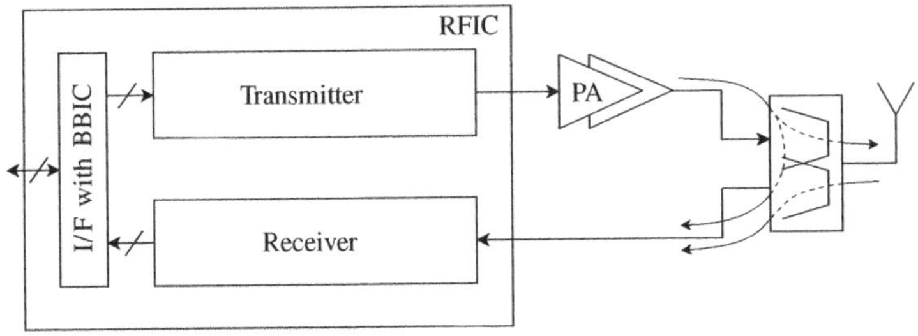

Figure 3.3 Transmitter leakage at the receiver input when using a single antenna through a duplexer – Due to the finite rejection achieved by physical RF filters, and to the RF coupling between paths, a transmitter leakage is recovered at the receiver input of the transceiver. When dealing with a full duplex system, the receiver performance needs to be guaranteed in the presence of this transmitter leakage.

when dealing with a low cost transceiver for the consumer market, for instance, we may want to use a single antenna for both the transmit and the receive path in order to reduce the overall cost. This can lead to problematic coupling from the transmitter output to the receiver input. As a result, we can face a non-negligible level of transmitter leakage at the receiver input, as illustrated in Figure 3.3. In practice, this signal is composed of both the useful part of the transmit signal, which can thus behave as a blocker at the receiver input, and also to the noise contributions recovered at the output of the active part of the transmitter that lies within the receive band. This behavior can therefore be problematic when implementing a full duplex transceiver. In that case, we need to ensure that the receiver still performs well enough while having to cope with this transmitter leakage. This can lead to significant requirements for the receiver, mainly in terms of linearity as illustrated in Section 7.3.

Moreover, when using a classical antenna,[1] the isolation between the transmit and receive paths is classically achieved through the use of either a diplexer, involving highpass and lowpass filters, or a duplexer, involving bandpass filters. The duplexer is thus of additional interest as it also provides some filtering effect on out-of-band blocking signals inherently present in the wireless environment, as discussed in Section 3.3.1. This explains why it is often duplexers that are encountered in practice, as represented in Figure 3.3. However, what is interesting to understand is that this filtering effect is always achieved at the cost of insertion loss. In particular, the sharper the transition between the passband of the filter and the stopband, the higher the passband insertion loss. This is unfortunately normal behavior for passive technologies classically used for such FE filters, as high quality factor devices are hard to achieve in practice. The problem then comes from the fact that in practice full duplex systems use frequency bands that are close to each other for the uplink and the downlink. As a result, in addition to the presence of the transmitter leakage on the receive side, we may anticipate higher insertion loss compared to what might be anticipated for a half duplex transceiver due to the use of such a duplexer with a sharp transfer function.

In contrast, when dealing with half duplex devices dedicated to TDD systems, we obviously no longer need to consider such transmitter leakage as we can turn off the transmitter during

[1] Here, by "classical antenna" we mean a radiating element that does not provide any intrinsic isolation between two ports that would excite modes with orthogonal polarization, for instance.

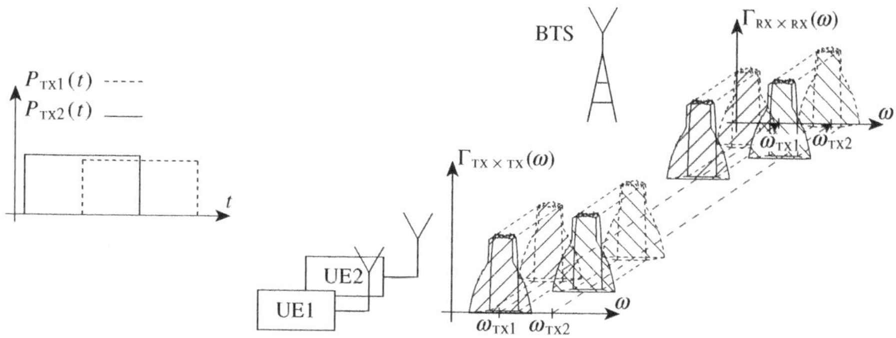

Figure 3.4 Adjacent channel pollution due to turning on and off of transmitters in half duplex systems – Due to the turning on and off of the transmitters (left), RF pollution occurring during the transitions can leak onto adjacent channels and degrade the quality of those channels. A worst case occurs when two UEs using adjacent transmit channels belong to different networks that are not synchronized. In that case, transitions of the bursts of one transmitter can occur during the useful part of the bursts of the other transmitter (right).

reception. This results in fewer linearity constraints for the receiver and fewer rejection constraints for the FE i.e. in less insertion loss for the FE in line with the above discussion. However, this simpler behavior is achieved at the cost of periodically turning the transmitter on and off. And when dealing with non-negligible transmit RF powers, we need to perform those variations for the output power with some caution in order not to pollute adjacent channels. This can be particularly critical when these adjacent channels are, for instance, dedicated to other networks that are not necessarily synchronized with that of the user, as illustrated in Figure 3.4. In that case, the spectrum pollution linked to the shape of the transitions can degrade the quality of the corresponding wireless links. Due to the importance of this pollution, specific systems are for the most part required at the transmitter implementation stage to fulfill the spectrum pollution requirements, as discussed in "Time mask and switching transient" (Section 3.2.1). Specific metrics are also classically used to check the performance achieved.

In conclusion, we mention a final difference between full and half duplex systems. As different carrier frequencies are used in the former, we need for the most part two RF synthesizers to generate them. This is not necessarily the case for transceivers belonging to half duplex systems. Thus, in addition to the potential injection locking problems linked to the potential coupling between the RF oscillators used to generate the carrier frequencies, we also have a direct impact on the cost of the final solution that is necessarily higher for full duplex transceivers due to the two RF synthesizers that are needed.

3.2 Metrics for Transmitters

As highlighted in the introductory part of this chapter, there are two main aspects to the transceiver requirements set by a wireless standard. Obviously, such requirements first need to ensure that, once implemented, a transceiver adhering to the standard performs well enough to be able to reach the expected data rate with a reliable radio link. But, in addition, it also needs to guarantee the ability of the transceiver to coexist with others belonging to the same wireless network, or to another wireless network. These two sets of requirements lead to different metrics classically used to check the characteristics of the transceivers, on both the transmit and receive sides. We focus on the transmit side in this section, before addressing the receive side in Section 3.3.

3.2.1 Respect for the Wireless Environment

When discussing in Section 3.1 the strategies used for the multiplexing of either the users or the uplink and downlink, we highlighted that one of the main objectives of a wireless system is to use the radio resource (i.e. the frequency spectrum) in the most efficient way. This is achieved in practice by an optimized scheduling in the time frequency plan of the signals carrying the data, on the one hand for the different users and on the other hand for their uplink or downlink.

In order to ensure no overlapping or pollution between the users, this scheduling must be carefully checked. Transmitters need to fulfill requirements both in the time domain, through time masks for instance, and in the frequency domain, through spectrum masks. We can thus follow this natural categorization of the requirements to illustrate the metrics classically used to check the ability of transmitters to respect their wireless environment.

Time Domain Requirements

We begin with the time domain requirements classically encountered for transmitters. Such requirements are obviously critical for systems that multiplex either users or the uplink and downlink in the time domain, i.e. for TDMA and TDD systems in practice. This thus explains why we often have recourse to examples from those standards in what follows. However, it should be kept in mind that, even if less stringent for the most part, such time domain requirements exist in any kind of wireless standard.

Timing Advance

Let us first consider a general requirement, related to the finite speed of the electromagnetic waves, which leads to different flight times between a BTS and its related UE depending on their relative positions. Recall the time domain multiplexing used in a TDMA system and shown in Figure 3.1(middle). Given the different time flights involved between the UE and the BTS, a precise scheduling of the bursts transmitted over the same frequency band by the UE toward the same physical BTS is required in order to avoid any collision between those bursts at the receive stage of the BTS.

More than that, some kind of *asymmetry* is required between the scheduling of the transmitter and of the receiver of each UE. Let us consider the more complete situation of a TDD system that uses TDMA to multiplex the users. For our example, we assume two UEs, both using the same single carrier angular frequency ω_{TRX} for their uplink and downlink toward the BTS. If we now assume that the first UE is closer to the BTS and the second is further from it, compared to a reference configuration where all the UE are at the same average distance from the BTS, any burst sent by the BTS arrives first at the first UE and then at the second UE.

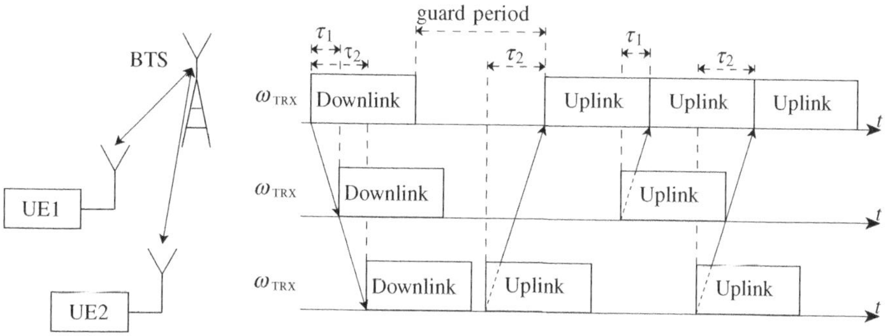

Figure 3.5 Timing advance mechanism in a TDD/TDMA system – Having a first UE close to the BTS and a second one far from it results in a shorter flight time τ_1 for the first UE than for the second UE, τ_2. The downlink bursts then arrive earlier at the first UE and later at the second. In contrast, to avoid any collision at the receiver stage of the BTS, the second UE needs to send its uplink burst sooner than the first UE relative to the BTS time frame. There is thus a need for a guard period in the system frames between the downlink and the uplink bursts in order to avoid simultaneous emission and reception for the UE furthest away from the BTS.

More precisely, it arrives sooner at the first UE than it would in the reference configuration, and later for the second UE. This is obviously due to the difference in the flight time between the different configurations in this downlink direction, as illustrated in Figure 3.5. But, now considering the uplink, as we need to avoid any collision at the input of the BTS receiver, the first UE needs to send its uplink burst later than it would in the reference configuration, whereas the second UE needs to do it earlier. Once again this is done with the aim of compensating the difference in flight time in the uplink direction.

Thus, the first UE receives its downlink bursts sooner and transmits its uplink bursts later than in the reference configuration, while the reverse holds for the second UE. But if the two UEs move and exchange their position, it is now the second UE that receives its downlink bursts sooner and transmits its uplink bursts later than in the reference configuration. This means that such a UE device needs to *adapt* its own transmit scheduling relative to the scheduling of the received bursts from the BTS depending on its distance from it. This for the most part leads to the necessity to implement at the UE transmit level some timing advance system, allowing a time shift of the data sent compared to a reference schedule, usually taken as the received sequence time frame.

Time Mask and Switching Transient

In addition to the precise relative scheduling of the uplink and downlink waveforms, as detailed in the previous section, we may encounter particular time domain requirements linked to the transitions in the transmitted output RF power. Again, such requirements are encountered mainly in wireless standards that use burst waveforms, i.e. in TDD or TDMA systems. However, such time domain requirements can be encountered in any kind of standards with the aim of controlling the transitions in the transmit output power.

Staying with TDD or TDMA as an example, in order to generate the RF bursts used for the transmission with a given nominal RF power, we may need to turn critical parts of a transmitter, such as the amplifiers that deliver the RF power, repeatedly on and off. For that

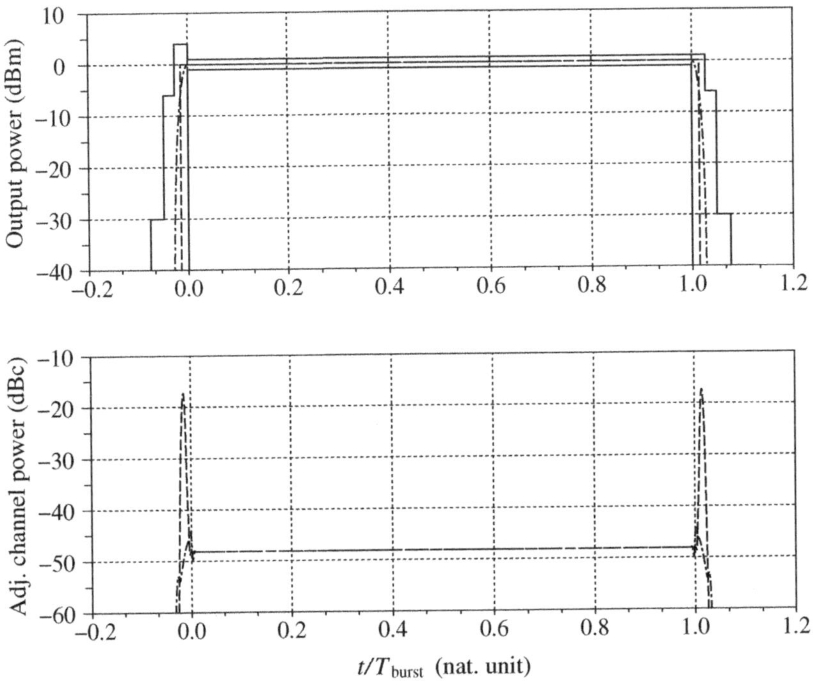

Figure 3.6 GSM time mask and impact of the burst shaping on adjacent channel pollution – RF bursts in TDD or TDMA systems are often required to fulfill a time mask in order to minimize the guard time, as done in the GSM standard [33] (top, solid). In addition, control of the shape of the transitions from off to on and from on to off is also required to minimize the pollution of adjacent channels. For instance, a simple gate function (top and bottom, dashed) leads to a peak power pollution more than 25 dB higher than a sine wave arch (top and bottom, dot-dashed).

to be done properly, some time is required before and after the useful part of the burst that carries the active part of the modulated waveforms, i.e. the data itself. But wireless systems expect to minimize such guard time between two such successive bursts in order to allocate the maximum available time to the effective data transfer. This results in constraints for the ramping up and down of the transmit output power for a duration as short as possible. Such constraints are classically set by a time mask to be fulfilled by the transmit burst. This means that the shape of the transmitted output power as a function of time must remain within the bounds set by such time masks as shown in Figure 3.6(top).

Moreover, as highlighted in “Full duplex vs. half duplex” (Section 3.1.3), such off–on and on–off transitions of the transmitter can lead to pollution in the spectral domain and thus to the degradation of the signal quality in the adjacent channels. This is why, in addition to the short duration requirement, the turning on and off of the transmitter must be done such that the shape of the output power as a function of time leads to minimal pollution in those adjacent channels. This pollution is classically evaluated through a switching transient requirement that simply sets a maximum allowable fraction of RF power radiated within the adjacent channels during the transmission of the burst. Classically, this fraction of power can be evaluated using

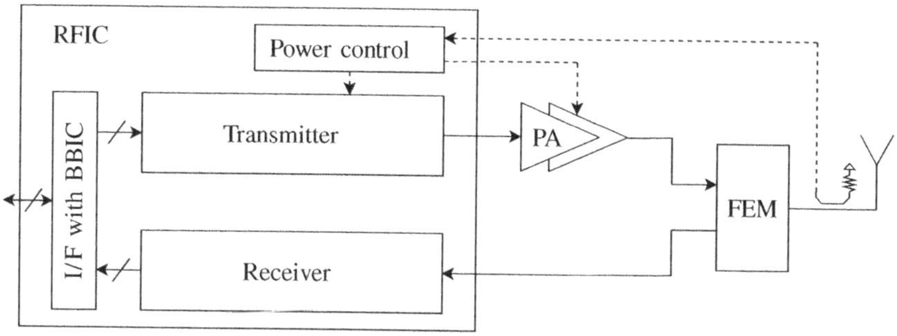

Figure 3.7 Feedback loop as a system to control the transmit output power – A feedback loop may be required to control the output power of a transmitter. This may be either a complete real time control loop with a sufficiently high bandwidth to allow control of the shape of the transmitted bursts used in TDD or TDMA systems, or a simple sense and correct loop that periodically controls the long-term average output power during the transmission.

a passband RF filter centered around the adjacent channel. Thus, when the transmitted burst is passed through this filter, we recover a power profile in time that gives an image of the RF power leaking in this adjacent channel due successively to the power ramp-up, the modulation itself, and the power ramp-down. This sliding power evaluation over the length of the burst leads to a profile with maximum occurring, hopefully, during the transitions between on and off of the transmitter, as shown in Figure 3.6(bottom). Once the measurement filter is specified, a maximum peak value for the resulting power profile can be set in order to ensure low pollution of the other users.

In order to minimize the pollution in these adjacent channels and thus fulfill the switching transient responses, it is often necessary to make the shape of the transition for the RF power smooth enough. Indeed, as shown in Figure 3.6, the switching transient response corresponding to a simple gate function, i.e. to an abrupt transition, leads to a peak power pollution of the adjacent channel more than 25 dB higher than for a sine wave arch. The problem is that the quality of the control that is required to achieve such transition shape with accuracy is often hard to achieve by direct control of the RF devices. Such devices are traditionally sensitive to process, supply and temperature variations. In practice, therefore, control loops may be required in order to achieve this precise transmitted power transition. As shown in Figure 3.7, this is often achieved at the cost of power detectors and baseband control systems that, even if implemented in the analog domain, are less sensitive to variations than RF devices.

Long-Term Average Output Power

Precise control of the output power may be required not only during the turning on and off of the transmitter as discussed in the previous section, but also during the transmission of the data, regardless the nature of the standard being followed.

A good reason for this is linked to the fact that we need to ensure a minimum SNR at the input of the receivers in the system. This is obviously required in order to ensure a sufficiently good data rate. We thus need to have the powers of transmitters set to a level that at least compensate for the path loss of the radio link. But, as detailed in Sections 2.1.4 and 2.3.2, this path loss depends on the distance between the transmitter and the receivers. We might imagine requiring that all the transmitters have the same output power by default, corresponding to the worst case path loss, i.e. to the maximum distance we can have in the network between transmitters and receivers. Recall, however, that a particular wireless link between one transmitter and a receiver is nothing more than a potential adjacent or blocking signal for another wireless link. There is thus an interest in minimizing the level of such unwanted signals in order to minimize the electromagnetic pollution between users. The optimization of the wireless network thus requires some real time adaptation of the transmitted powers.

Furthermore, some systems of multiplexing for the users need to have the output power of the UE controlled with high precision. In a CDMA system, for instance, all the users share the same frequency band at the same time. But, as the orthogonality of the codes used for the multiplexing is up to a given point broken by the propagation channel, there is a maximum discrepancy allowable between the signal strengths of the users at the input of the BTS receiver stage so that the signals, once superposed in time and frequency, can be correctly split by signal processing.

Thus, even when dealing with systems that are not burst in nature, such as most FDD systems, we may need some mechanism to control and guarantee the average transmit power with precision. The main difference with burst systems is that the real time constraint is not so relevant at first glance compared to what is required for a loop that tracks transition shapes over a short duration. It thus simplifies the implementation, but not necessarily the need for a control system, which can for instance still take the form shown in Figure 3.7.

Frequency Domain Requirements

Let us now focus on the frequency domain requirements for transmitters that are classically encountered in wireless standards. In order to introduce the corresponding metrics, it is of interest to first examine a typical spectrum as recovered at the output of a real transmitter. Looking at the WCDMA transmit spectrum shown in Figure 3.8, as a result of imperfections in the functions implemented in the transmit line-up, we obviously obtain a spectrum that we were not expecting.

The reasons for those discrepancies fall within the scope of Part II of this book. Here, we observe that the resulting impacts of these spectral distortions on the other users in the network can be categorized into three groups:

(i) We first get a spectral regrowth of the transmit spectrum, i.e. a close in-band spectrum that is a bit wider than expected. This phenomenon can then lead to a non-negligible fraction of power lying in the adjacent channels.

(ii) Then we get some wideband noise floor that limits the decrease in the transmitted PSD at high frequencies. This leads to potential pollution in frequency bands dedicated to other wireless standards.

(iii) Last, although not present in the spectrum shown in Figure 3.8, we can have some spurious tones, resulting from the RF coupling of clock harmonics for instance, superposed on the transmitted spectrum. Once again, this results in pollution of adjacent channels or other frequency bands.

We thus classically encounter different kinds of metrics and requirements to specify and quantify these behaviors.

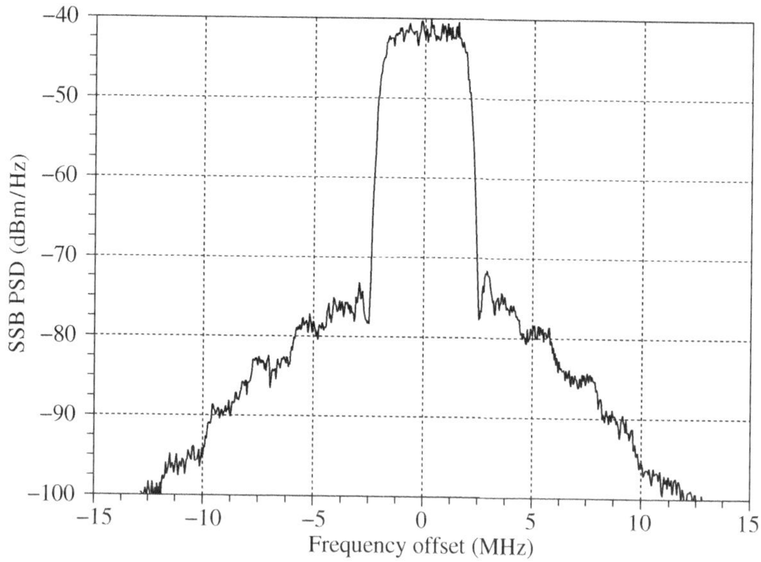

Figure 3.8 Typical close in-band WCDMA transmit spectrum at maximum output power – Classically, due to implementation imperfections, the spectrum recovered at the output of a WCDMA transmitter exhibits distortions compared to the theoretical spectrum discussed in Section 1.3.3. These distortions lead to the existence of unwanted signal components lying outside the transmit channel.

Spectrum Emission Mask and Spurious Emissions

Most of the requirements aimed at control of pollution of either the adjacent channels, following the same standard, or other frequency bands, potentially following other standards, are often summarized using the same formalism, i.e. through the specification of a spectrum emission mask (SEM) in practice.

We can, however, distinguish between close in-band pollution, i.e. the pollution of the adjacent channels, and the wideband pollution that impacts the other RF frequency bands. These different kinds of pollution are often related to different implementation limitations as detailed throughout Part II of this book. It is the spectral regrowth of the modulation spectrum, often due to nonlinearity as detailed in "Spectral regrowth" (Section 5.1.3), that mainly impacts the adjacent channels, whereas it is mainly wideband noise components as well as spurious tones that impact the remaining part of the spectrum located at higher frequency offsets.

The consequence is that we often have differences in the specification of the SEM to reflect these different behaviors. For instance, the close in-band spectrum mask must be shaped to match the physical behavior of the spectral regrowth. In the WCDMA example shown in Figure 3.9 the maximum allowable signal PSD in the adjacent channels is a more complicated function of the frequency offset than a simple staircase function. However, we observe that although the spectrum mask is plotted here in terms of a maximum allowable PSD it is often a maximum power in a given measurement frequency band that is specified at a given frequency offset. The reason for this is that such PSD remains well suited for the specification of requirements for continuous noise-like spectra but not necessarily when spurious tones are

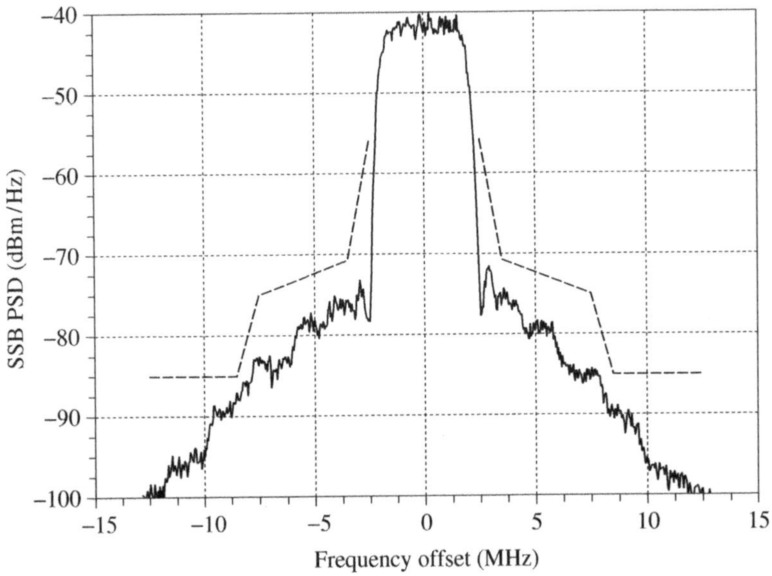

Figure 3.9 Typical close in-band WCDMA transmit spectrum at maximum output power and associated spectrum emission mask – The WCDMA standard specifies a SEM to ensure limited pollution of the adjacent channels [34]. Although plotted here as a PSD limit, such emission mask is often specified in terms of a maximum power in a given measurement bandwidth in order to take into account potential spur tones more efficiently.

present. From the system point of view, even if we locally get an unwanted spur tone with a higher PSD than the effective noise floor, it may remain acceptable if the total noise plus tone power, integrated over a system bandwidth, is lower than a given bound. This total noise plus unwanted tone power is often meaningful from the system point of view.

Focusing now on the transmit spectrum that corresponds to higher frequency offsets, we have already observed that the degradations that are classically encountered are linked to wideband noise components or spurious tone radiations rather than spectral regrowth. As a result, requirements are often set through the use of some staircase-like maximum allowable power in those frequency bands. Again the specification is often expressed in terms of a maximum power measured in a measurement bandwidth in order to take into account the system impact of spurious tones, but it can be converted in terms of a maximum PSD, as shown in Figure 3.10. The figure shows that more stringent requirements can be set in RF frequency bands that are dedicated to the use of other wireless standards. This is obviously done for the sake of coexistence between users of different wireless systems.

There is another major difference between the close in-band distortion of the modulation spectrum due to nonlinearity and the wideband noise floor that impacts the far part of the spectrum. As it is related to the distortion of the modulation spectrum through nonlinearity, we expect the power of the close in-band pollution to reduce more quickly than the transmit power does if we refer to the material in Chapter 5. Thus, for lower output powers, we expect a lower

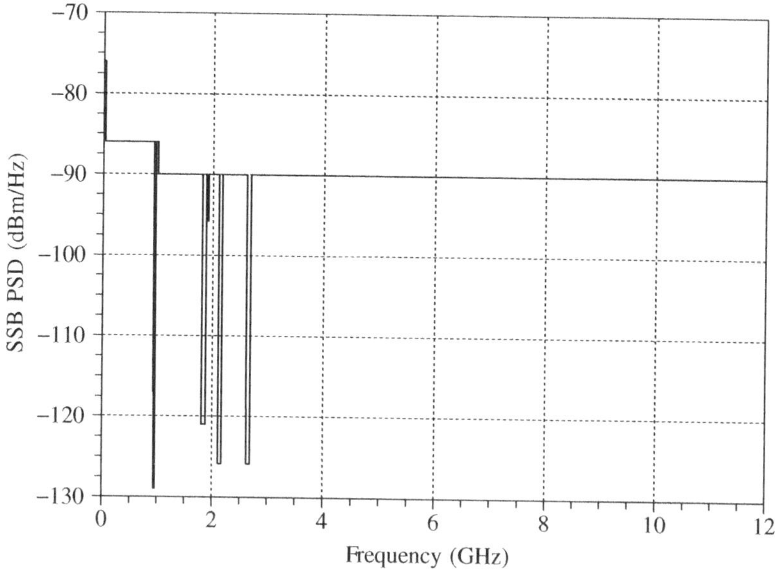

Figure 3.10 WCDMA spurious emission mask – The WCDMA standard specifies a spurious emission mask to ensure limited pollution of the total RF spectrum [34]. As for the SEM, the spurious emission mask is often specified in terms of maximum power in a given measurement bandwidth in order to take the spur tones into account more efficiently. We also get more stringent specifications in RF bands dedicated to other wireless standards in order to ensure a good coexistence.

power for the pollution of the adjacent channels. This is not so obvious for the wideband noise floor that is often linked to an absolute noise power, added in the last stages of the transmitter line-up, and thus more or less independent of the signal power. This difference in the behavior explains why an additional kind of metric is often used for the close in-band distortion on top of the SEM in order to be able to check the performance improvements when the transmit power reduces. This is the purpose of the adjacent channel leakage ratio metric to which we now turn.

Adjacent Channel Leakage Ratio

As highlighted in the previous section, the pollution of adjacent channels is mainly driven by the spectral regrowth of the transmit spectrum through nonlinearity. Due to the dependence of this distortion on the transmit modulated signal power, we expect the pollution of the adjacent channels to improve when the transmit power reduces. This behavior explains why, on top of the SEM that sets only a maximum absolute limit for this pollution that is expected to be reached only for the maximum output power, there is an interest in having a metric that allows more and more stringent requirements to be set on close in-band pollution when the output power reduces.

This metric often used is the adjacent channel leakage ratio (ACLR) or adjacent channel power ratio (ACPR). This quantity, which we denote by $ACLR$, is simply defined as the ratio between the power recovered in an adjacent channel, P_{adj}, and the power in the main transmit channel, which can simply be taken as equal to the transmit output power P_{out} up to first order. We can thus write

$$ACLR = \frac{P_{out}}{P_{adj}}. \tag{3.1}$$

In practice we can have specifications for the first adjacent channel, $ACLR_{1st}$, the second, $ACLR_{2nd}, \ldots$. The more adjacent channels we consider, located at higher frequency offsets from the transmit channel in the frequency domain, the more stringent the requirement can be as we expect the pollution to be decreasing as a function of the frequency offset. This is shown in Figure 3.11 where we can see two different ACLR requirements, one for the first adjacent channel and one for the second. In the figure, the ACLR requirements are represented graphically in the spectral domain as an equivalent maximum allowable PSD in the channel, that holds for the maximum transmit output power.

However, in practice the dependence between the transmit output power and the pollution of the adjacent channels holds only in the upper range of the transmit power. Indeed, in any real life implementation, for sufficiently low transmit power, the distortion of the modulation

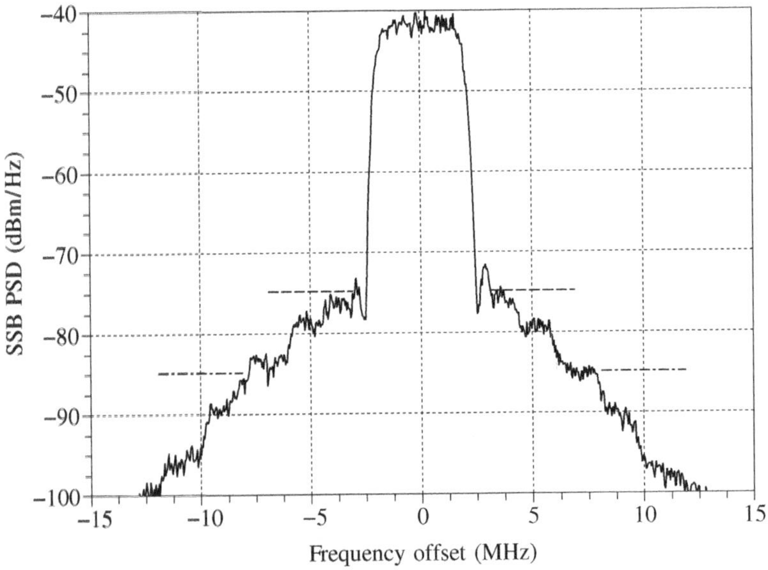

Figure 3.11 Typical close in-band WCDMA transmit spectrum at maximum output power and associated ACLR limit – The WCDMA standard specifies a minimum ACLR of 33 dB for the first adjacent channel (dashed) and 43 dB for the second (dot-dashed) [34]. This specification must be interpreted in terms of the ratio between integrated signal powers in the relevant channel and holds for the higher part of the output power DR, as illustrated in Figure 3.12.

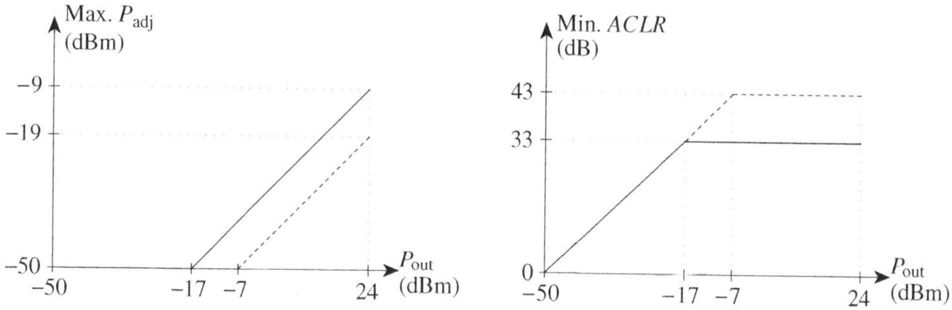

Figure 3.12 ACLR requirement as a function of the transmit output power in the WCDMA standard – The minimum ACLR requirement of 33 dB for the first adjacent channel (solid) and 43 dB for the second (dashed) holds only as long as the power in the adjacent channel remains above −50 dBm [34]. The existence of this floor level is equivalent to a reduction of the required ACLR by 1 dB per decibel in the lower part of the transmit power DR.

spectrum is negligible compared to the unavoidable additive wideband noise that thus sets an absolute lower bound for this pollution, as detailed in Section 7.2.3. This lower bound should match up to a given point that of the spurious requirement detailed in the previous section, as at high frequency offsets from the transmit channel, it is mainly this noise floor itself that exists in addition to the spurs.

This physical behavior explains why this ACLR requirement is often written in two different and complementary ways. On the one hand, we often get a minimum ratio that has to be fulfilled between the transmitted power and the power leakage in the adjacent channel in the higher part of the output power DR. But then, for the lower output transmit power range, only an absolute maximum leakage power specification in the adjacent channel stands. It thus corresponds to a degradation of the ACLR requirement of 1 dB per decibel relative to the output transmit power in the lower part of its DR, as represented graphically in Figure 3.12.

3.2.2 *Transmitted Signal Modulation Quality*

On top of the necessity to respect the wireless environment, i.e. the other wireless users in practice, a transmitter needs to deliver waveforms of sufficiently high quality to achieve a reliable radio link able to sustain high data rates. But in practical implementations a degradation of the modulated waveforms results from the distortions experienced by the signal along the transmit line-up. There are thus classical metrics used in most wireless standards to quantify this degradation.

Although the root causes of this degradation at the implementation level are for the most part the same as those involved in the degradation of the radiated spectrum and discussed in "Frequency domain requirements" (Section 3.2.1), in the present case we are dealing with *in-band* distortions, necessarily requiring different metrics.

In order to introduce those metrics naturally, we can consider a generic form for the complex envelope of a modulated bandpass waveform as recovered at the output of a real transmitter. This obviously falls within the scope of Part II of this book. At the risk of getting ahead of ourselves, we observe that due to classical implementation limitations, the complex envelope $\tilde{s}_{TX}(t)$ of the bandpass modulated waveform recovered at the transmitter output can roughly be expressed as a function of the complex envelope of the signal expected to be transmitted, $\tilde{s}_o(t)$, as

$$\tilde{s}_{TX}(t) = (\tilde{s}_o(t) + \tilde{n}(t) + \tilde{\kappa})e^{j\delta\omega t}. \tag{3.2}$$

In this expression, we assume that $\tilde{s}_{TX}(t)$ is defined as centered around the expected transmit carrier angular frequency ω_{TX}, i.e. around the center of the transmit channel in practice. Looking at this equation, we can identify different terms:

(i) First of all, we have a complex exponential term that leads to a rotation of the transmitted complex envelope in the complex plane at a constant rate of $\delta\omega$ radians per second. This term is often related to the frequency error in the generation of the LO used for the frequency upconversion.
(ii) Then we have a constant term, $\tilde{\kappa}$, that behaves as a complex offset for this transmitted complex envelope. This term can be related to DC offset present in the baseband part of the transmitter, or to the direct LO leakage to the output.
(iii) Finally, we have an additive term, $\tilde{n}(t)$, that also leads to an error compared to the expected trajectory. This term can be related to noise sources present in the line-up, or to the distortion of the transmitted signal itself due to nonlinearity or imbalance.

There are thus classical metrics used to put numbers on those degradations as detailed in the following. We observe, however, that this list is not exhaustive, and there may be additional requirements related to specific behaviors not taken into account in this very simple expression.

Error Vector Magnitude

Let us first focus on the term $\tilde{n}(t)$ in equation (3.2). In practice, this term can be linked either to all the noise contributions detailed throughout Chapter 4, or to the distortion of the signal being processed through some filtering stages as discussed in Section 4.4, or through nonlinearity as discussed in "Nonlinear EVM due to odd order nonlinearity" (Section 5.1.3). In any case, this term leads to an error in the sampled modulating waveform at the symbol rate $1/T_s$, i.e. on the modulation constellation as shown in Figure 3.13, which takes the simple case of a 4QAM constellation.

Assuming that no frequency offset exists, i.e. that $\delta\omega = 0$, and, as a first step, that no LO leakage exists either, i.e. that $\tilde{\kappa} = 0$, we can write the sampled transmitted complex envelope $\tilde{s}_{TX}[k] = \tilde{s}_{TX}(kT_s)$ from equation (3.2) as

$$\tilde{s}_{TX}[k] = \tilde{s}_o[k] + \tilde{n}[k]. \tag{3.3}$$

The characteristics of $\tilde{n}[k]$, which becomes an error vector when represented in the complex plane, can thus be used as a metric. More precisely, due to its origins, we can assume that $\tilde{n}(t)$ represents a noise component in practice, i.e. that $\tilde{n}_k$ is centered and uncorrelated with $\tilde{s}_o(t)$. Thus, having assumed that $\tilde{\kappa} = 0$, i.e. that no additional offset is added in the line-up,

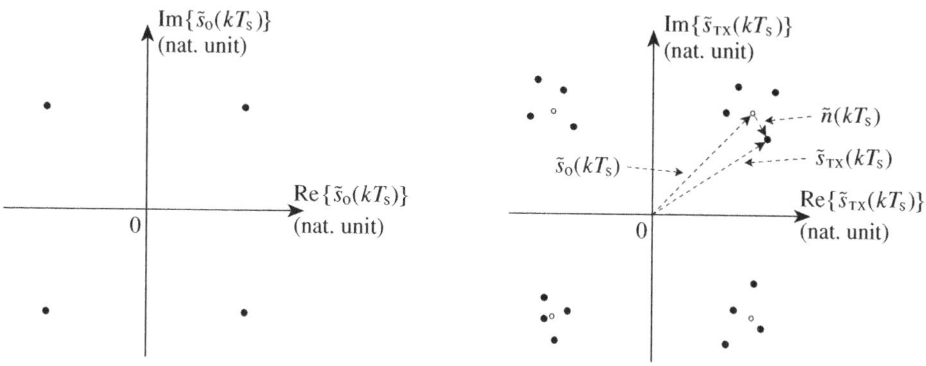

Figure 3.13 Definition of the EVM based on the noise corruption of the transmit constellation – Due to additive noise components, or to linear or nonlinear distortions, the complex envelope of a transmitted signal, sampled at the symbol rate T_s, exhibits an error vector signal (right) compared to the theoretical expected constellation (left), here a simple 4QAM. Once the signal offset is retrieved, the power of the resulting centered error vector process normalized by the constellation power defines the RMS EVM according to equation (3.4).

we can directly define the RMS error vector magnitude (EVM), which we denote by *EVM*, as the ratio between the RMS value of the above defined error vector and the RMS value of the theoretical constellation. We thus get

$$EVM = \sqrt{\frac{\sum_k |\tilde{n}[k]|^2}{\sum_k |\tilde{s}_o[k]|^2}}. \tag{3.4}$$

However, in practice a measurement filter is often specified for the evaluation of the EVM. This means that the distorted waveform $\tilde{s}_{TX}(t)$ is first filtered using this measurement filter before the EVM is evaluated. This is done, on the one hand, with the aim of having a standardized noise bandwidth used for every evaluation and, on the other hand, in order to equalize for the potential linear distortions that can be inherently linked to the generation of the modulating waveform itself. This generation can involve a pulse shaping filter which may induce some ISI for instance.

What is the exact difference between the above definition of the RMS EVM and the more classical SNR metric? We can argue that the SNR is often associated with the degradation of an analog signal rather than with the distortion of a constellation as it is for the EVM according to our present definition. But despite those differences, the correspondence between the SNR of a waveform and the RMS EVM of its sampled version often holds so that in practice we can write

$$EVM = \frac{1}{\sqrt{SNR}}. \tag{3.5}$$

However, this should not be allowed to obscure the fact that the definition of the EVM is subject to interpretation depending on wireless standards. Indeed, in some cases we may have to include in the EVM metrics additional contributions not classically associated with noise terms. This is the case, for instance, for LO leakage, neglected here due to our assumption that $\tilde{\kappa} = 0$.

Hitherto we have only considered RMS EVM. However, recalling that the definition of the EVM includes not only intrinsic noise sources but also signal distortion terms, due to either nonlinearity or filtering, we can have statistics for the sum of the contributions that turn out to be non-Gaussian. This explains why a peak EVM metric is also often used in addition to the RMS EVM in order to reflect the DR of the distribution for the distortion of the signal.

We observe that for this parameter also, wireless standards often set requirements only for the higher part of the output power DR. Indeed, for low output powers, the transmit waveform remains corrupted by noise floors, or even LO leakage when such a component is considered, thus leading to an absolute minimum in-band noise power. This results in an increase of the EVM, or a decrease of the SNR, which can reach 1 dB per decibel relative to the output transmit power for the lower part of its DR.

Frequency Error

Let us now focus on the rotation of the transmitted complex envelope as evaluated through the complex exponential in equation (3.2). Looking at this term, we see that in the present case we are not talking about a constant phase offset as could be caused by the finite propagation duration between a transmitter and a receiver, for instance, following the discussion in Chapter 2. Rather, in the present case, having defined the complex envelope $\tilde{s}_{\mathrm{TX}}(t)$ of the transmitted modulated bandpass signal as centered around the expected carrier angular frequency ω_{TX}, we are faced with a frequency offset of this transmitted signal compared to this expected one. Obviously, such frequency offset leads to a continuous phase rotation of the transmitted constellation from symbol to symbol, as shown in Figure 3.14, where are we still considering the simple case of a 4QAM. As a consequence, when such a modulated RF signal is demodulated on the

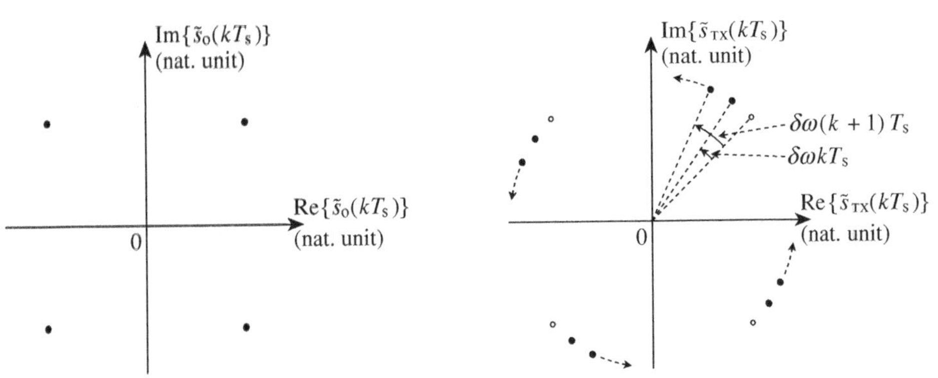

Figure 3.14 Impact of an LO frequency error on the transmit constellation – An angular frequency offset $\delta\omega$ of the transmit LO signal relative to the theoretical expected carrier angular frequency ω_{TX} results in a distortion of the modulated transmit RF signal. We get a signal centered around ω_{TX} that appears to be modulated by a constellation that experiences a continuous rotation (right) compared to the theoretically expected one (left), here a simple 4QAM.

receive side using an LO signal at the expected exact angular frequency, i.e. ω_{TX}, we recover this rotating complex envelope at baseband. We can thus understand that correctly determining the symbols can be a problem if nothing is done to compensate for this continuous rotation.

We observe that in practice baseband algorithms often perform the equalization of the received waveform and thus decide which data bits are working on slices of received samples during a finite time duration T. This characteristic duration T obviously depends on the wireless standard, but always exists. There is thus an interest in setting a requirement for a maximum frequency offset in order to ensure that the resulting maximum phase offset during the duration T is sufficiently low to ensure good data recovery. It can indeed ensure good enough performance even if no particular care is taken over this potential frequency offset at the data bit decision stage. Although other arguments can be put forward, this explains why absolute maximum frequency errors as low as 0.1 parts per million for the carrier angular frequencies can be required in wireless standards.

However, although setting such a maximum frequency error can potentially simplify the digital algorithms for the bit decision, it does not necessarily simplify the overall implementation of the transceiver. Indeed, in practice, such frequency error is related to the imprecise reference that is used to generate the various clocks and LO waveforms. At least for transceivers embedded in low cost UE, the poor performance of classical reference clocks leads to the necessity to embed mechanisms that estimate the absolute frequency error and compensate for it in real time during the operation. Such an AFC algorithm, discussed in more depth in Section 9.1.5, can also lead to the implementation of additional hardware blocks for compensation purposes, on top of the additional complexity of the algorithms.

Origin Offset Suppression

We now turn our attention to the term $\tilde{\kappa}$ in the expression for the complex envelope of the transmitted signal $\tilde{s}_{TX}(t)$ in equation (3.2). Given that this complex envelope is defined as centered around the expected transmit carrier angular frequency ω_{TX}, the term $\tilde{\kappa}$ represents the amount of LO leakage present in the transmitted RF signal. As discussed in Section 6.5, the presence of this LO signal can be linked to either a direct RF feedthrough from the LO generation to the transmit RF output, or the presence of DC offset in the baseband part of the transmit line-up.

Practically speaking, the presence of this leakage can have different system impacts depending on the structure of the signal being processed in the line-up. For instance, if we take a CDMA signal as introduced in Section 1.3.3, we expect that the reverse scrambling and spreading processing on the receive side will make such leakage behave as an additional noise component, as would any DC offset present at the input of such processing [35]. For simpler waveforms resulting from a direct modulation of the data symbol, we get that such LO leakage results in a direct origin offset error for the constellation being transmitted, as shown in Figure 3.15. We may then be interested in guaranteeing a maximum amount for such leakage in the transmitted signal in order to ensure no significant degradations of the performance on the receive side due to this offset. Thus, when not included in the EVM specification, a dedicated origin offset suppression requirement may be put in place to handle this leakage problem.

However, we are again faced with the same limitations as highlighted in the previous sections. The origins of the LO leakage in the line-up implementation may lead to an absolute

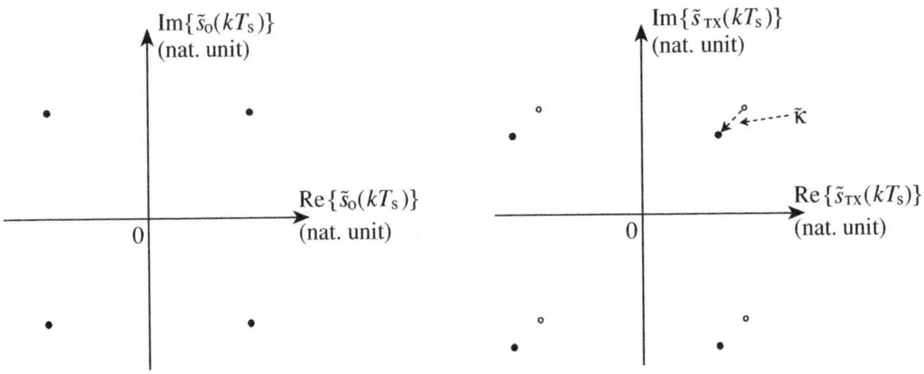

Figure 3.15 Definition of the origin offset in the transmit symbol constellation – Due to either DC offset in the baseband part of the transmitter, or direct feedthrough in the RF stages, we get an LO leakage component superposed on the modulated signal at the transmitter output. We thus get a total RF signal that appears modulated by a constellation that experiences an offset (right) compared to the theoretical expected one (left), here a simple 4QAM.

value for this LO component at the transmitter output. When this occurs, we get the same absolute leakage level whatever the power of the transmitted signal, and thus an origin offset suppression requirement that is more and more difficult to handle as the transmit output power decreases.

Carrier Phase Shift

In addition to the main metrics detailed in the previous sections, we may also encounter various additional requirements linked to different implementation limitations. This should obviously be discussed on a case by case basis. A good example is the phase shift that can corrupt the carrier phase of a modulated RF bandpass signal. This phenomenon, discussed in Section 2.2.2, can occur each time an RF device is switched with a discontinuity in the impedance seen by the RF modulated signal. It can lead to a phase offset $\delta\phi$ on the recovered complex envelope once demodulated using an LO waveform at the carrier angular frequency.

This problem may appear less critical than the frequency offset that leads to a continuous phase rotation of the expected complex envelope, as shown in Figure 3.14. Indeed, in the present case we are talking about a single phase jump occurring each time a given RF device is switched. This is in fact not necessarily the case as the phase drift from symbol to symbol in the frequency error case can be very small and thus allow for controlling the corresponding radio link degradation even if the continuous behavior of the phenomenon may appear problematic. In the present case, we are talking about a phase shift that can be important, i.e. classically in the range of many tens of degrees. Thus, even if only occurring from time to time, this phase discontinuity can lead to problems at the receive stage when dealing with coherent reception, i.e. with a receiver that needs to track the phase of the received carrier. We can then specify some phase shift requirements in order to prevent any problems for the network.

3.3 Metrics for Receivers

Let us now focus on the receive side. As for the transmit side, it is meaningful to make the distinction between, on the one hand, the requirements linked to the coexistence of the receiver with other equipment and, on the other hand, the requirements expected to ensure that sufficiently high performance can be achieved. However, before reviewing the most widely encountered of those requirements and the associated metrics, we should recall the main differences compared to the transmit side.

As far as coexistence with other devices is concerned, the problem is no longer to ensure a satisfactory spectral shape or appropriate scheduling of the events in order to avoid electromagnetic pollution, but rather to check the ability of the receiver to preserve the quality of the radio link while having to cope with RF signals coming from all the users at its antenna connector.

In the same way, some of the requirements set to ensure the quality of the radio link cannot be directly translated by the wireless standard itself as absolute requirements on the modulation accuracy at the output of the RF/analog path of a receiver, as done for the transmit side. For the most part, the RF/analog path of a receiver is expected to be co-designed with the digital baseband algorithms that perform the equalization and the final bit decision. We thus have an additional degree of freedom in the sense that for an overall target radio link performance, often expressed in terms of bit error rate (BER), there is a potential joint optimization between the RF/analog path and the digital baseband algorithms. This explains why, in practice, a non-negligible proportion of requirements for the RF/analog part of a receiver do not come directly from the wireless standard itself but rather from the performance of the digital algorithms that set the minimum quality allowable for the signal delivered by the RF/analog path in order to achieve the overall performance required by the standard. This overall performance, in terms of BER, is the quantity of interest for the standard even for the wireless link that involves the transmitter of a device. But in that case the person in charge of the RF/analog system design of the transmitter can hardly make assumptions on the performance of the receiver that will process the signal its transmitter generates as those line-ups are not part of the same device. Thus, the best that can be done is to require an absolute accuracy for the transmitted modulation in order to ensure no problems for receivers that could process this signal, whatever their architecture. This explains why wireless standards need to set up requirements that check such absolute accuracy at transmitter outputs, whereas mainly overall BER performance is encountered for receivers.

3.3.1 Resistance to the Wireless Environment

We start with the classical metrics used to set the requirements that check the ability of a receiver to correctly demodulate its dedicated signal while having to cope at the same time with unwanted signals linked to other users. Unless using smart antenna array schemes that exhibit some spatial filtering capabilities, we classically recover all the different electromagnetic waves present at the antenna stage at its connector output, and thus at the input of the receive stage. Obviously, the power of unwanted signals depends on many parameters, on top of which we can find the frequency response of the receive antenna. However, although based on some assumptions, wireless standards need to specify a mask in the frequency domain for the power

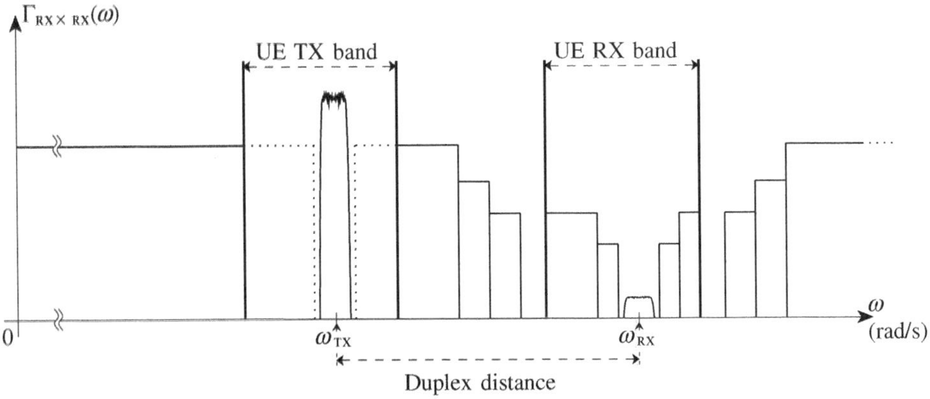

Figure 3.16 Power profile in the frequency domain for the unwanted signals present at the receive antenna connector – As receive antennas used in UE are not necessarily highly selective in frequency or direction, unwanted signals linked to other users or systems are present at the receiver input. Wireless standards often specify a profile in the frequency domain for those unwanted signals at the antenna connector. In full duplex systems the transmitter leakage must also be considered as an additional unwanted signal that occurs at the duplex distance, in the frequency domain, from the received carrier angular frequency.

of the signals that have to be assumed as potentially present at the antenna connector stage of the receiver in addition to the wanted signal. An example of such a mask is shown in Figure 3.16.

Looking at this figure, we see that the unwanted signals can represent either signals in adjacent channels, mostly linked to other users in the same network, or signals at higher frequency offsets, mostly belonging to other networks. In this last case we talk about blocking signals, or blockers. Those blockers are said to be in-band if they remain within the UE receive system band of the wireless standard, or out-of-band if they are outside this band. Due to their different nature, adjacent channels and blockers are for the most part treated differently by wireless standards. Indeed, as adjacent channels are expected to belong to the same network as the wanted signal, we may possibly define their power relative to the power of the wanted signal. This can be done with the aim of reflecting the behavior of networks that optimize the power of their different RF channels in order to achieve the highest overall performance while minimizing the level of interference. As discussed in more depth in "Adjacent channel selectivity" later in this section, this correlation between the power of the different channels belonging to the same network often results in the definition of a profile for the maximum adjacent channel power to be handled by receivers that is a function of the wanted signal power. This is in fact not the same for blockers that are expected to represent signals not belonging to the same network. Thus, the best that can be done for a wireless standard is to define a maximum level to be considered for their power, based for the most part on experience.

Furthermore, in full duplex systems, receivers have to cope with transmitter leakage as discussed in "Full duplex vs. half duplex" (Section 3.1.3). This signal, even if not directly linked to other users, has to be considered carefully as it makes the requirements set by the standard trickier to achieve, as illustrated in the following sections. In particular, we observe that the duplex distance, i.e. the distance in the frequency domain between the transmit and receive channels, impacts the sensitivity of receivers to the presence of other blockers through nonlinear limitations, for instance.

To conclude this introduction, we can say that, as for the transmit side, there are some classical implementation limitations, detailed in Part II of this book, that lead to degradation of the reception when those unwanted signals are present. In the present receive case, it is mainly the selectivity of the filters embedded in the receive line-up and the linearity of the RF/analog parts that are the root causes of these limitations. We now review the metrics classically used to specify and check these behaviors.

Input Signal Dynamic Range

Before going further, we observe that a receiver has to support its wireless environment while receiving a wanted signal whose level can fluctuate over a non-negligible DR. One reason for this comes from the variations of the power radiated by transmitters in the network. As discussed in "Long-term average output power" (Section 3.2.1), transmitters are expected to vary their output power in order to optimize, or rather minimize, the level of interference. This necessarily results in variations in the power of the received signals. We also need to consider the path loss experienced by the received signal. As detailed in Sections 2.1.4 and 2.3.2, this path loss depends on the distance between the transmitter and receiver. There is thus a variation of the signal received by a UE depending on whether or not it is located close to the BTS when assuming that this path loss is not fully compensated by an adaptation of the transmit power of the BTS.

It appears that, in practice, wireless standards define requirements for a DR of the received signals that is linked to the physical area of the cells used in the networks. This DR may be very wide, as illustrated by cellular standards that classically define requirements for a received wanted signal whose power can go from -110 dBm up to -25 dBm. Such a wide DR has many impacts on the design of receivers that often need to deliver a signal with a constant RMS level to the digital algorithms dedicated to equalization and bit decisions. This leads, for instance, to the necessity to use AGC schemes, as discussed in Section 9.2.1. In the same way, it leads to the necessity to consider carefully the linearity of the receiver in order to ensure no degradation of the received wanted signal, whether due to distortion when reaching its maximum input power, or to the presence of blocking signals when reaching its minimum input power.

In practice, there are thus requirements that cover the full DR of the received wanted signal level. As an example of the requirements that cover the lower part of this DR we can mention the classical sensitivity test case. This kind of requirement for the minimum performance while receiving the weakest possible wanted signal mainly sets the trade-off between the insertion loss of the passive FE of a receiver and the noise performance of its active part, as discussed in Chapter 7. We can also mention the test cases involving blocking signals, performed in order to test the linearity of the receiver, as discussed in the following sections. In contrast, we often get requirements that apply to the maximum input level of the received wanted signal in order to check the performance of the device while having potential compression and saturation in the receive path.

Adjacent Channel Selectivity

Looking again at Figure 3.16, we see that the main differences between the characteristics of the various unwanted signals a receiver has to cope with come from their frequency offsets regarding the edge of the receiver channel as well as their powers. The higher the frequency offset, the higher the potential level of those unwanted signals.

This is in fact a general behavior widely encountered in wireless standards, and its root cause comes from the effective capabilities of any physical implementation of receivers. In order to deliver a clean wanted signal to the digital algorithms that perform the channel equalization and then the bit decision, a receiver first needs to filter out the unwanted signals. This is the purpose of what is often called the channel filter in the RF/analog part of the receive line-up, i.e. of the filter that cleans up the received signal in order to let only the wanted signal, centered around the received channel frequency, go toward the digital baseband algorithms. But, in most cases, the physical implementation of lowpass or bandpass filters leads to a rejection that increases as the frequency offset does. This simple behavior explains why in practice wireless systems have to manage the level of the signals in the network so that the closest signals to their wanted signals received by each receiver are also the weakest.

But, given that in practice receivers are able to suppress by a given fixed amount the power of signals in adjacent channels relative to that of the wanted signal, they can support increasing powers for the adjacent channels when the power of the wanted signal power also increases. This holds as long as the ratio between the adjacent channel power and the wanted signal remains within the filtering capability of the receiver; this preserves the wanted to unwanted signal power ratio after the filtering stage. This quite natural capability of receivers, in addition to the DR for the wanted signal at the receiver input, leads to the possibility for the network to optimize the level of the different channels while ensuring that all users are faced with an adjacent channel to wanted signal power ratio within a given bound. This naturally explains why wireless standards often set a direct requirement for the selectivity of the receiver channel filtering through an adjacent channel selectivity (ACS) specification that requires a minimum effective attenuation of the adjacent signal, as illustrated for instance in Figure 3.17.

However, thinking back to the remark in the introductory part of Section 3.3.1, there are alternative ways to set requirements that check the resistance to adjacent channels. For instance, a wireless standard may simply require that receivers achieve a given performance, a minimum BER for instance, while having to cope with a given ratio of adjacent over wanted signal power at their input. With this approach there is no direct requirement for the rejection of the adjacent signal. It is therefore the responsibility of the overall receiver to achieve satisfactory rejection of the adjacent signal, or whatever is considered in terms of signal processing in order to achieve the equalization and bit decision with satisfactory performance.

Intermodulation and Blocking Test Case

As discussed in the previous section, blockers that take place at higher frequency offsets than adjacent signals can have higher levels than that of the adjacent signals. Moreover, as they mostly belong to different networks than the received wanted signal, there is obviously a *non-correlation* between their levels and that of the wanted signal. As a result, receivers need to ensure correct performance whatever the configuration for the wanted signal power relative to that of the blockers. This is a main difference compared to adjacent channels whose power

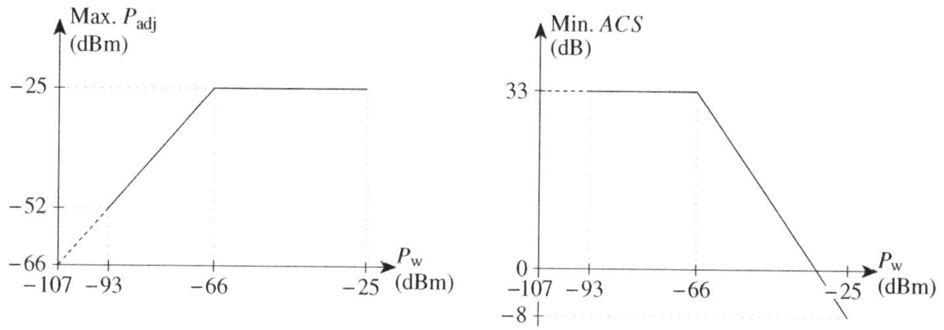

Figure 3.17 ACS requirement as a function of the received wanted signal power in the WCDMA standard – An ability to reject signals in adjacent channels is required for networks to work correctly. For instance, a minimum ACS of 33 dB is required in the WCDMA standard (right) [34]. Equivalently, the signal power in the first adjacent channel, P_{adj}, cannot be higher than that of the wanted signal, P_w, by more than a given amount (left). But, like any signal in the network, the adjacent signal cannot have a level above a given maximum absolute value, here −25 dBm. We thus get a reduction in the ACS requirement of 1 dB per decibel in the higher part of the wanted signal DR.

is managed by the same network and thus remains coherent with the wanted signal level. Thus, a worst case for receivers occurs when the wanted signal is close to sensitivity while the blockers are at their maximum possible level. It is indeed in this configuration that their distortion due to nonlinearity can adversely impact the reception of the wanted signal, as discussed throughout Chapter 5. This explains why blocking signal requirements in wireless standards mostly test the linearity limitations of receivers.

On top of those requirements we often find an intermodulation test case involving two blockers centered around the angular frequencies ω_1 and ω_2. In practice, the angular frequencies are chosen so that the difference $\delta\omega = |\omega_2 - \omega_1|$ is also equal to the difference between the carrier angular frequency of one of the two blockers and that of the wanted signal. This is done in order to directly test the odd order nonlinearity performance of the receiver, as discussed in "Intermodulation" (Section 5.1.2) and in "Revisiting intermodulation and harmonic tones" (Section 5.1.3). With this frequency planning, we get one of the generated third order intermodulation tones that is superposed on the wanted signal in the frequency domain and thus leads to an SNR limitation, as illustrated in Figure 3.18. We thus often get requirements set by wireless standards for a minimum performance to be achieved while having this kind of two-blocker configuration at the receiver input.

In addition to intermodulation test cases, we can have test cases involving a single blocker. Even if it looks simpler than an intermodulation test case, such a blocking signal requirement can be important as it often involves blockers with the maximum allowable power, which is not necessarily the case for intermodulation tests. It allows us to test the desensitization of the receiver due to compression, or the performance in terms of phase noise of the LO signal used for the frequency downconversion, for instance. It can even test some particular behaviors linked to the periodic nature of waveforms used in TDD or TDMA standards. Indeed, in those standards, waveforms are burst in nature, as discussed in Sections 3.1.1 and 3.1.2. On the transmit side, this can thus lead to pollution of the adjacent channels, as discussed in "Time

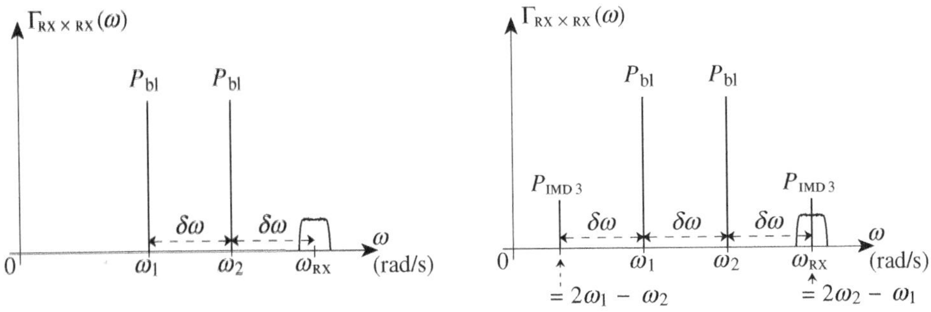

Figure 3.18 Typical intermodulation test case configuration involving two blockers – Classically, intermodulation tests are used to check the odd order nonlinearity behavior of receivers. As discussed in Chapter 5, such nonlinear behavior, often linked to compression effects in the RF/analog devices, leads to the generation of odd order intermodulation tones. Using two blockers lying at selected frequency offsets from the wanted signal (left), one of the generated third order intermodulation tones corrupts the wanted signal and thus limits the performance of the receiver (right).

mask and switching transient" (Section 3.2.1). But it can also lead to trouble on the receive side due, for instance, to transient responses of filters if blockers appear during the reception of the useful part of the wanted signal. The reception degradation can also be linked to nonlinearity problems such as the even order nonlinear behavior of the downmixer used in the receive line-up, in particular in direct conversion receivers, as discussed in Chapter 8. A good illustration of this kind of behavior is the AM-demodulation test case of the GSM standard which involves burst blockers appearing in the middle of the active part of the received wanted signal [33].

For full duplex systems, the single blocker test case can indirectly lead to an intermodulation test due to the presence of the transmitter leakage signal. Fortunately, this can occur only for some particular configurations of the blocking signals relative to the transmitter signal. This can be illustrated with reference to the odd order nonlinearity case, and by recalling that having two blockers lying at angular frequencies ω_1 and ω_2 leads to having the third order intermodulation tones generated at the angular frequencies $2\omega_1 - \omega_2$ and $2\omega_2 - \omega_1$ as shown in Figure 3.18. Thus, assuming that the transmitter signal, lying at the angular frequency ω_{TX}, can play the role of either the first or second blocker, the resulting intermodulation tone can be superposed on the wanted signal, lying at ω_{RX}, if and only if the angular frequency of the other blocker is such that

$$2\omega_{TX} - \omega_1 = \omega_{RX}, \tag{3.6}$$

or

$$2\omega_2 - \omega_{TX} = \omega_{RX}. \tag{3.7}$$

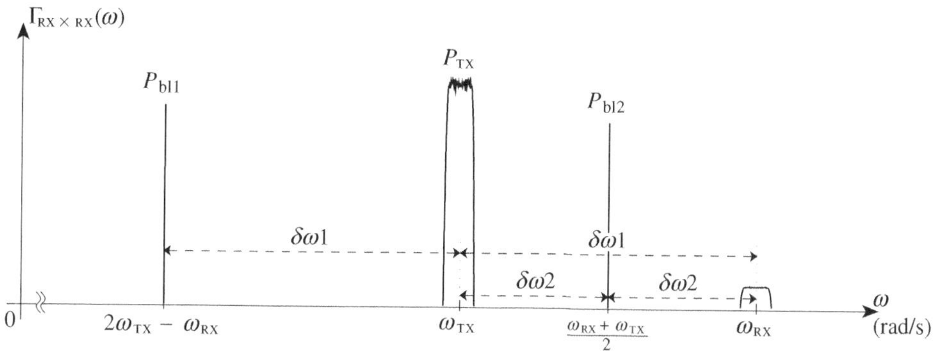

Figure 3.19 Typical blocking test case configuration involving one blocker for a full duplex transceiver – Blocking test cases also exist with only one blocker present at the input of the receiver. But in full duplex transceivers such tests can still lead to a two-blocker intermodulation configuration when considering the transmitter leakage signal. Due to the duplex distance set by the standard, this occurs only for a particular instance of frequency planning, here corresponding to the odd order nonlinear response of the device. But the same kind of configuration can be derived when considering the even nonlinear response of the receiver.

As shown in Figure 3.19, this leads to the two possible angular frequencies for the blocking signal:

$$\omega_1 = 2\omega_{TX} - \omega_{RX}, \tag{3.8a}$$

$$\omega_2 = \frac{\omega_{TX} + \omega_{RX}}{2}. \tag{3.8b}$$

Obviously, this frequency planning is valid only for the generation of the third order intermodulation tones. But the same reasoning would apply to the second order intermodulation tones, lying at angular frequencies $\omega_2 \pm \omega_1$. We would then obtain

$$\omega_1 = \omega_{RX} - \omega_{TX}, \tag{3.9a}$$

$$\omega_2 = \omega_{RX} + \omega_{TX}, \tag{3.9b}$$

so that the second order intermodulation tones involving the transmitter signal indeed lie around the wanted signal carrier angular frequency. In any case, these considerations show that we can cumulate the problems in full duplex systems as we still need to fulfill the requirements set for single blocker test cases, but with the potential additional intermodulation effects. These simple considerations highlight the potential cost, in terms of linearity, of full duplex transceivers.

Spurious Responses and Emissions

Another set of problems can occur due to the presence of unwanted signals at the antenna connector. As discussed throughout Chapter 6, in practice the frequency downconversion of the received signal is done by multiplication with an LO waveform. But, due to various implementation limitations discussed for instance in Section 6.3, this LO waveform is not necessarily the pure sine wave that would be expected when implementing an ideal frequency conversion. We rather get the presence of either harmonics, often due to the implementation of the mixing stage using a chopper-like mixer, or spurious tones, often due to RF coupling with clock signals.

Thus, when a blocker is present at a angular frequency offset $\delta\omega$ from the wanted signal carrier angular frequency and a harmonic or a spurious tone is also present in the LO waveform at the same angular frequency offset $\delta\omega$ from the LO fundamental tone, we get the folding of the blocker on the wanted signal during the frequency conversion, as discussed in greater depth in Section 6.4.1. This is illustrated, for instance, in Figure 6.42 in the harmonic mixing case, i.e. the mixing of the received signal with the harmonics of the LO signal. But whatever the origins of the unwanted tone in the LO waveform spectrum, we are faced with a spurious response of the receiver occurring at the angular frequency of the input blocker for which this folding on the wanted received signal happens. In practice, these spurious responses are very difficult to cancel completely. This is why in practice some exceptions for the level of the blockers for which those spurious responses occur are almost always allowed in wireless standards.

A side concept to this spurious response is input spurious rejection (ISR). This quantity is simply the conversion gain, experienced by the blocking signal from the receiver input up to its folding on the wanted signal, relative to the conversion gain of the wanted signal itself. This quantity therefore directly gives the attenuation experienced by the blocking signal relatively to the wanted signal and is thus a good metric to quantify the ability of a receiver to handle the blocking signals lying at those spurious responses frequencies. This attenuation can be provided by the relative level of the harmonics or spurs in the LO waveform compared to that of the fundamental tone, but also by filtering effects present in the receiver at the blocking signal frequency prior to the frequency downconversion.

Spur tones present at the LO stage of a receiver can be a problem in terms of radiated emissions. Indeed, by electromagnetic coupling or direct LO feedthrough, they can lead to propagating signals at the RF input of the receiver that can eventually reach the antenna. We may therefore have a spur emission problem even for the receive path. This explains why some spurious emission requirements can be set in practice even for receivers. As a result, frequency planning, which involves predicting and possibly selecting the angular frequency of those unwanted spurious tones, can be critical in ensuring good transceiver behavior and thus fulfilling the requirements set by the wireless standard.

3.3.2 *Received Signal Modulation Quality*

In addition to its ability to resist its electromagnetic environment, a receiver needs to be able to deliver the modulating waveform of the received wanted signal to the digital baseband algorithms in charge of the bit detection with sufficiently high quality that the performance required by the wireless standard is achieved. As discussed in the introductory part of Section 3.3, it is mainly the overall performance of a receiver, often expressed in terms of BER, that is of interest to a wireless network. In particular, the minimum quality required for the signal delivered by the RF/analog part of the receiver is driven by the performance of the digital baseband algorithms. In practice, an early task involves the translation of the overall performance required by a wireless standard, expressed in terms of minimum BER achieved for different configurations of propagation channel, input power, etc., in terms of requirements for the quality of the modulating waveform provided by the RF/analog part based on the performance of the digital algorithms.

For that purpose, there are also classical metrics used to quantify the quality of the modulating waveform. These metrics are very close to those used by wireless standards for the transmit side as detailed in Section 3.2.2. This is reasonable as in both cases we talk about quantifying the quality of a modulating waveform. There should be only slight differences as, on the one hand, we talk about the distortion experienced by the modulating waveform whereas, on the other hand, we talk about the same modulation distortion but transposed around a carrier frequency. This explains, for instance, why we talk about DC offset on the receive side but origin offset on the transmit side.

Thus, it is of interest to say a few words on the metrics adapted to the receive side. We can follow the derivations done for the transmit side and introduce them by considering the simplified expression for the lowpass complex envelope $\tilde{s}_{\text{RX}}(t) = p(t) + \text{j}q(t)$ as recovered at the output of the RF/analog part of the receiver. According to the implementation limitations detailed in Part II of this book, at the channel filter output of the receiver, $\tilde{s}_{\text{RX}}(t)$ can be expressed as a function of the complex envelope of the wanted signal, $\tilde{s}_{\text{w}}(t)$:

$$\tilde{s}_{\text{RX}}(t) = (\tilde{s}_{\text{w}}(t) + \tilde{n}(t))\text{e}^{\text{j}\delta\omega t} + \tilde{dc}. \tag{3.10}$$

In this expression, we assume that $\tilde{s}_{\text{w}}(t)$ is defined as centered around the carrier angular frequency of the received wanted signal, ω_{RX}, and that this angular frequency is that of the LO used in the receiver for the downconversion to baseband. Looking at this expression, we first observe the analogy with equation (3.2). This similarity is not surprising as we are dealing with the same kind of limitations for the RF/analog implementation of a receive path as for a transmit path. We can thus identify the same kind of terms:

(i) A complex exponential term that is related to the frequency error in the generation of the LO used for the frequency downconversion compared to the expected theoretical received carrier frequency. This means that, compared to the expected modulating complex envelope, $\tilde{s}_{\text{w}}(t)$, the received one exhibits a rotation in the complex plane at a constant rate of $\delta\omega$ radians per second.

(ii) We then have a constant term, $\tilde{dc}$, which represents the DC offset inherently present in the baseband part of the receiver. As this offset either is directly injected in the line-up after the downconversion to baseband, or results from the self-mixing of the LO signal, it does not depend on the potential frequency offset of the LO.

(iii) Finally, we have the additive term, $\tilde{n}(t)$, representing either noise sources present in the line-up, or the distortion of the received signal due to nonlinearity or imbalance. Rigorously speaking, these sources are partly located in the baseband side of the receiver. This means that a term that does not experience the frequency offset of the LO should be present in the above expression in order to reflect the baseband sources. However, RF devices are hopefully the main contributors to both the noise sources and to the distortion experienced by the signal being processed. The above expression is thus of interest up to first order.

There are thus classical metrics encountered on the receive side to quantify these degradations. However, we observe that for the receive side also this list is not exhaustive and there may be some additional requirements related to particular limitations of the RF/analog design, or of the digital algorithms themselves. This thus needs to be checked on a case by case basis.

Signal to Noise Power Ratio

We focus here on how to set requirements for the noise performance of a receiver, i.e. how to set constraints on the term $\tilde{n}(t)$ in equation (3.10). As discussed in the introductory part of Section 3.3.2, the requirements in terms of quality of the modulating waveform delivered by the RF/analog part of a receiver are driven by the performance of the digital algorithms dedicated to bit decision. This means that we first need to understand the characteristics of the signal being processed that the performance depends on in order to derive the corresponding metrics.

Dealing with an additive noise only, often assumed as an additive white and white Gaussian noise (AWGN), it turns out that the theoretical BER performance that can be achieved by a given modulation scheme is mostly driven by the ratio E_b/N_0 [1]. In this expression, E_b represents the energy per bit in the waveform being processed, and N_0 the PSD of the AWGN. However, before going further, we may first wonder why we consider only AWGN for such theoretical results. First of all, it comes from the fact that this is the situation encountered in most cases in real life implementations, in particular when the received signal is corrupted by the thermal noise inherently present in the receive line-up. As discussed in Chapter 4, on the one hand this noise contribution is additive compared to the received signal, and on the other hand it also has a Gaussian distribution. But, we could also say that nature does things well, as having a Gaussian distribution for this noise term allows us to derive theoretical results in a quite straightforward way, thus explaining the number of available results in the literature.

However, dealing with our BER concerns applied to wireless tranceivers, we need to say a little more about this E_b/N_0 concept. Bear in mind that the BER is a statistical quantity as it represents the probability of obtaining the wrong value for a data bit. This means that, assuming the ergodicity of the modulation process, the long-term ratio between the number of wrong bit decisions and the number of good ones for a given realization of the modulation process should approach the BER value. Thus, saying that this BER value depends on the ratio E_b/N_0 in fact means that, for instance, E_b represents a *long-term average value*. This remark is important as, in fading channels, the instantaneous value of energy per bit fluctuates in the received signal as a function of time. As a result, only the average value of this quantity can be linked to the performance of the modulation scheme. We discuss this point again at the end of the section. Following the above general arguments, although the real life implementation of digital algorithms leads to some discrepancies with the theory, the performance of digital baseband algorithms in terms of BER is mainly driven by the ratio E_b/N_0 in the waveform that is provided to it. And in practice, baseband algorithm designers still benchmark their designs using this kind of metric. The problem is that obviously the quantity E_b/N_0 is not really suited to the characterization of RF/analog devices. For instance, in the analog world, the concept of "bit" is not really meaningful as we are dealing with continuous time modulating waveforms derived from the symbols used by the modulation, the time sequence of those symbols being linked to the data bits. There is thus a need to link the characteristics of the continuous time modulating waveforms with the E_b/N_0 quantity that drives algorithm performance.

This can be done by considering the signal to noise power ratio, *SNR*, of the modulating waveform. On the one hand, we can write that the power of the received signal, assuming that only the wanted signal remains present at the channel filter output, is simply equal to the energy per bit, E_b, times the bit rate, R_b, when expressed in bits per second. On the other hand, we get that the noise signal power is equal to its PSD N_0 times the receiver noise bandwidth BW, this noise bandwidth being no more than the bandwidth of the overall equivalent channel filter experienced by the noise signal. We can therefore write

$$SNR = \frac{E_b R_b}{N_0 BW} = \frac{E_b}{N_0}\frac{R_b}{BW}. \tag{3.11}$$

This kind of relationship then allows us to make the link between the performance of the digital algorithms and the requirements for the RF/analog part of the receiver. As a side comment, we also point out that the quantity

$$G_p = \frac{BW}{R_b} \tag{3.12}$$

represents the processing gain linked to the modulation scheme. The meaning of this term appears when we rewrite equation (3.11) as

$$\frac{E_b}{N_0} = G_p\,SNR. \tag{3.13}$$

We see that the ratio E_b/N_0 experienced by the digital algorithms is equal to G_p times the SNR delivered by the receiver. In practice G_p can have a non-negligible value when we use a high bandwidth signal to transmit a low bit rate. To achieve this high bandwidth when a low bit rate is used, much redundancy is introduced into the modulating waveform, and by careful signal processing the information linked to this redundancy can be used to improve the final performance. This is exactly what happens in CDMA systems as introduced in Section 1.3.3. The processing we are talking about is the correlation with the spreading codes in that case.

From the foregoing discussion, SNR requirements for the RF/analog part of a receiver can be derived from the performance of the digital baseband algorithms by considering the test cases defined by the wireless standard in terms of overall BER performance. Indeed, in practice, BER requirements are set for various configurations of the modulation or coding scheme used, of the propagation channel and of the power of the wanted signal at the antenna connector. This results in an SNR curve as a function of the wanted signal input power, as shown in Figure 3.20. More precisely, we should say that many SNR curves can be provided, depending on whether we talk about test cases involving the wanted signal alone or in the presence of adjacent signals or blockers. In this last case, different noise contributors have to be considered, compared to the case where only the wanted signal is present. Thus, as discussed in more depth in Chapter 7, different budgets have to be derived in those different cases. We also discuss in that chapter why the SNR performance of a receiver is inherently a function of the received signal power and thus why SNR requirements need to be presented in such a way, i.e. as a function of this received signal power.

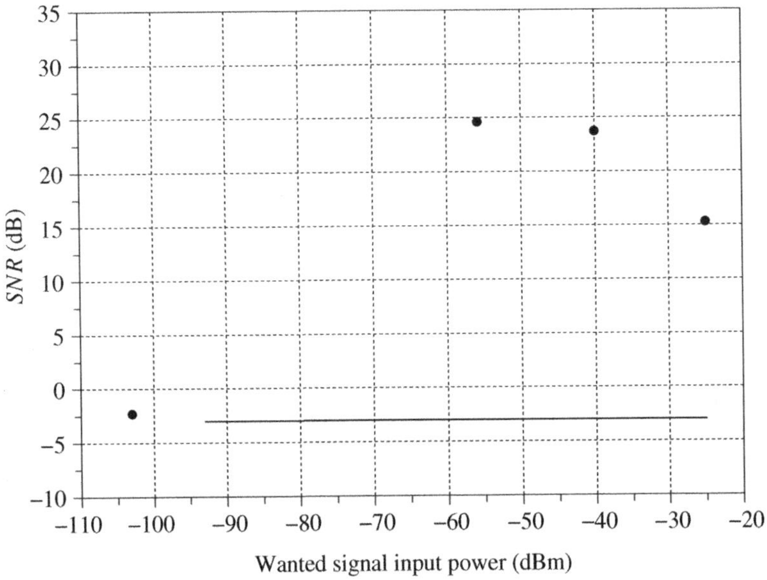

Figure 3.20 Typical SNR requirements for the RF/analog part of a receiver – From the analysis of the performance of the digital baseband algorithms, the BER requirements set by wireless standards for various configurations of received signal can be translated in terms of SNR requirements for the waveform that is delivered to those algorithms. This results in a curve that represents the SNR requirements for the RF/analog part of the receiver as a function of the received wanted signal power. However, due to differences in performance of such RF/analog receive paths, it may be necessary to distinguish between test cases involving the wanted signal alone (dots) or either adjacent channels or blockers (solid line), as detailed in Chapter 7.

At this point it would be appropriate to say a few words about the validity of deriving SNR requirements for the RF/analog part of receivers in the way we have detailed so far. We may wonder, for instance, about the validity of the Gaussian assumption for the distribution of the noise terms we deal with in receivers. Assuming in practice that any unwanted signal uncorrelated with the wanted signal is a noise component, many noise sources other than thermal noise exist in transceivers. And depending on their origins, those additional terms have statistics that can differ from the Gaussian distribution. This is typically the case when dealing with terms generated by distortion of either the wanted signal or potential adjacent or blocking signals through nonlinearity. The statistics of these terms are indeed linked to those of the modulation of the signal and can thus be different from the Gaussian distribution.

Another example of discrepancy between the AWGN model and real life is linked to the additive assumption for the noise. As highlighted previously, this assumption means that even when the received wanted signal experiences some fading, for instance, the noise, which is added to the signal in the receiver, i.e. after the propagation channel, remains at a constant instantaneous power that is therefore equal to its long-term average power. This situation is depicted in Figure 3.21(left). But, as discussed in Chapter 7, some noise components encountered in transceivers are said to be multiplicative. Unlike a pure additive noise, the instantaneous power of a multiplicative noise is *proportional* to that of the wanted signal. As

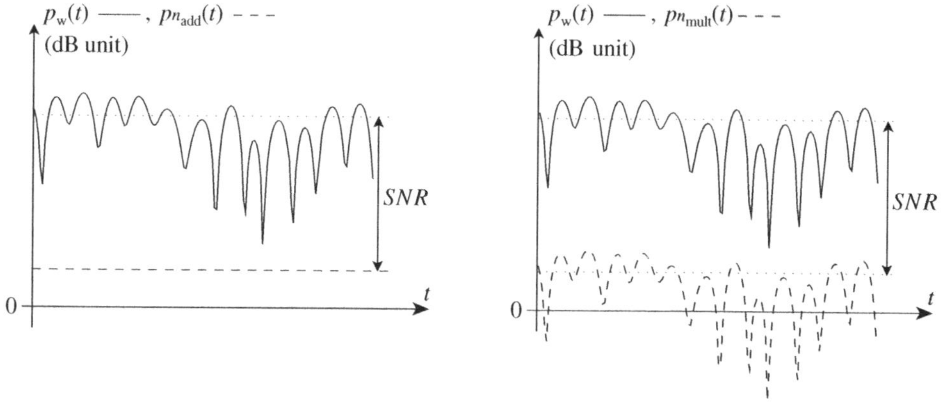

Figure 3.21 Instantaneous vs. long-term average SNR for additive and multiplicative noises – Dealing with a multiplicative noise (right), the instantaneous variations of the noise power (dashed) follow those of the received signal power (solid). This is not the case when dealing with an additive noise (left). However, although we can have different instantaneous SNR values, as illustrated here with a received wanted signal that has experienced fading, for instance, the long-term SNR remains the same in the two cases.

a consequence, the power of such a noise component follows the variations of the propagation channel, as shown in Figure 3.21(right). But, as observed at the beginning of the section, the SNR that is meaningful for evaluating the BER performance is linked to the long-term average power of the signals up to first order. However, although this long-term average SNR remains the same in the two cases, we may still wonder if this different behavior for the instantaneous power variations has an impact on algorithm performance. This obviously has to be checked on a case by case basis depending on the algorithms. The situation is even worse in the case of a distortion noise component due to the compression of the signal being processed. Since in this case the power of the noise term increases more rapidly than the power of the signal, we can expect some discrepancies in the long-term average power when dealing with faded signals.

Although the two previous examples seem to show that the SNR curves derived from the baseband algorithm performance in AWGN are not the ultimate metric for setting the requirements of the RF/analog part of the receiver, they remain a good starting point. Such SNR curves can indeed be interpreted as the lower bound for the required SNR to be delivered by the RF/analog part. This allows the derivation of a budget using this lower bound and some reasonable margins to handle the remaining uncertainties. More precise system evaluations can then be considered in order to tune some parameters more accurately. However, in practice, such SNR curves derived from the performance of the baseband algorithms in AWGN are of particular importance for budgeting a receiver, as illustrated in Chapter 7.

Frequency Error, DC Offset, Carrier Phase Shift

Thinking back to equation (3.10), we see that in addition to the noise term $\tilde{n}(t)$ we get the same kind of degradations as encountered on the transmit side. For instance, we can identify the complex exponential term that represents a continuous phase drift of the received complex envelope compared to that of the expected theoretical modulating waveform. This results in a rotation of the constellation recovered after sampling at the symbol rate of those waveforms. Such frequency offset, already encountered on the transmit side as shown in Figure 3.14, has its origins in an error between the frequency of the LO signal used for the downconversion to baseband of the received wanted signal and the frequency used for its generation on the transmit side. However, the difference is that in the present case we do not necessarily get direct requirements on such frequency error set by the wireless standards on the receive side. Here again we need to check what error on this parameter can be handled by the baseband algorithms before saying what precision we need to achieve for the RF/analog part.

This is exactly the same for the residual DC term, $\tilde{dc}$, which results in an offset for the received constellation. We get the same impact as that linked to the presence of the LO leakage encountered on the transmit side as illustrated in Figure 3.15. However, in the present case the allowable residual DC offset at the output of the RF/analog path is a parameter internal to the receiver and its value must then be driven by the performance of subsequent algorithms.

The list of degradations reviewed for receivers in the present chapter is not exhaustive, though most of the key classical parameters have been discussed. In particular, an analysis based on a case by case basis may be necessary for some additional parameters, depending on the targeted implementation. This can be illustrated by considering the potential carrier phase shifts due to the switching of RF devices that can take place on the receive side also. This behavior classically occurs when switching the gain of the LNA. Such a device can present a different input impedance value in its different gain modes, resulting in variations in the phase of the amplitude transmission coefficient, as discussed in Section 2.2.2. The impact of the carrier phase shift must then be checked at the digital algorithm level in order to set a requirement for a maximum phase shift due to gain switching at the RF implementation level.

Part II

Implementation Limitations

4

Noise

Having derived in Part I of this book the minimum set of functionalities to be implemented in transceivers as well as the minimum performance required from them, we may now review the limitations we face in their physical implementation before being able to perform practical line-up system design as illustrated in Part III.

For an implementation based on the use of electronic devices we necessarily face limitations that have various physical origins. However, from the point of view of the signal processing dedicated to data bit recovery, any signal that does not correlate with the wanted signal can be considered as an additional noise term that degrades the radio link performance. We thus anticipate that whatever its physical origin, we can take into account the degradation of the wanted signal in a SNR or EVM budget by simply considering as an additional noise term the fraction of signal that does not correlate with the wanted signal. By these lights, *any* implementation limitation can be interpreted as a source of noise.

However, in practice some categorization is classically done of the sources of degradation so that the term "noise" is usually associated with specific phenomena. But it is interesting to note that this categorization may depend on the point of view adopted. For instance, when performing a transceiver budget as illustrated in Chapter 7, it is convenient to consider, on the one hand, the noise components that add with a given absolute level to the wanted signal when going through the line-up. In that case, we classically talk about additive noise components that thus result in a given noise floor whatever the wanted signal power. On the other hand, we can distinguish between noise components whose power is proportional to the power of the wanted signal, and those whose power increases more rapidly than the power of the wanted signal. The former, related to RF impairments or to the presence of LO phase noise for instance, are said to be multiplicative. They result in an upper bound for the SNR whatever the wanted signal level. The latter, being related to the compression in nonlinear devices for instance, can be classified as distortion noises. They result in an SNR that degrades as the power of the signal being processed increases. According to this classification, there are additive noises in the line-up before a signal is processed. This explains that their power is necessarily independent of that of the signal being processed. This is not the case for the multiplicative or distortion noises which exist in the line-up only when such a signal is processed. But when discussing the physical origins of the degradations experienced by the wanted signal, it is convenient to

Wireless Transceiver Architecture: Bridging RF and Digital Communications, First Edition. Pierre Baudin.

bear in mind a more traditional categorization in order to more easily understand what kind of degradation must be associated with which block of a line-up. We follow this approach in this part of the book, consisting of three chapters devoted to this categorization. Chapter 5 is devoted to nonlinearity, and Chapter 6 to RF impairments. In the present chapter, we focus on the remaining mechanisms leading to the presence of unwanted terms more traditionally labeled as "noise", such as analog electronic noises, LO phase noise, and quantization noise.

4.1 Analog Electronic Noise

4.1.1 Considerations on Analog Electronic Noise

Noises in analog electronic devices are unfortunately unavoidable. Many physical root causes can be found for these noise components. Some are linked to the technology used to implement the active devices. This is for instance the case for shot noise, burst or popcorn noise, avalanche noise, flicker noise, etc, when dealing with semiconductor devices [36]. But some are independent of the technology. This is the case for the thermal noise present in any physical device, active or passive. The fact that this noise component is unavoidable and for the most part is the main contributor to the overall noise performance of a system is the reason for its particular importance. It explains the origins of some concepts used to characterize, in a general way, the noise performance of a device, such as its effective noise temperature. A brief discussion of this noise term will serve to introduce some interesting quantities classically used from a system point of view.

4.1.2 Thermal Noise

As its name suggests, thermal noise originates in the random motion of electrons due to thermal agitation. The classical picture of the phenomenon is that in a finite piece of conductor, large enough to contain a significant number of electrons to get a reliable statistic but small enough to have macroscopic quantities such as the temperature well defined and constant, the electrons have a finite velocity linked to the conductor thermodynamic temperature. This means that an electric current, proportional to the sum of the velocities times the elementary charge, occurs in the conductor. This random current therefore has the same characteristics as the sum of the velocities, i.e. a vanishing mean but a non-vanishing RMS value. The interesting point is that this classical description of the phenomenon allows us to understand the statistical distribution of the resulting random current: the independence of the elementary currents associated with the motion of individual electrons means that their sum tends to a Gaussian distribution due to the central limit theorem.

However, this classical picture fails to explain the PSD that can be associated with this random process. We need to evaluate the average energy of the electromagnetic modes generated in the conductor by this random current. The exact derivation must be done using the result from quantum mechanics that gives the energy of each of those modes. It can be shown [37] that the energy, E_{m}, carried by the mode of angular frequency ω_{m}, can be written as

$$E_{\mathrm{m}} = \left(n_{\mathrm{m}} + \frac{1}{2}\right)\hbar\omega_{\mathrm{m}}, \tag{4.1}$$

where $\hbar = h/2\pi$ with h the Planck constant, and n_{m} stands for the number of photons excited into that mode. The resulting average energy, $\mathbb{E}\{E_{\text{m}}\}$, that can be expected in that mode is therefore linked to the expected number of photons that reach the corresponding energy level. Invoking the Bose–Einstein partition function which holds for photons, we thus finally get that

$$\begin{aligned}\mathbb{E}\{E_{\text{m}}\} &= \left(\mathbb{E}\{n_{\text{m}}\} + \frac{1}{2}\right)\hbar\omega_{\text{m}}\\ &= \left(\frac{1}{e^{\hbar\omega_{\text{m}}/kT} - 1} + \frac{1}{2}\right)\hbar\omega_{\text{m}},\end{aligned} \tag{4.2}$$

where k is the Boltzmann constant, and T the temperature in kelvin. The term $\hbar\omega_{\text{m}}/2$ in the above relationship is independent of the temperature. It therefore still exists when $T = 0$, i.e. when there is no thermal motion in the material, and it is linked to the vacuum energy. Its physical meaning is outside the scope of this book. However, for normal temperature and frequency ranges, it is negligible compared to the temperature dependent term. Indeed, with $k = 1.381 \times 10^{-23}$ J/K and $h = 6.626 \times 10^{-34}$ J·s, for normal temperatures and frequencies up to terahertz, i.e. far above the usual RF frequencies, we have $\hbar\omega_{\text{m}} \ll kT$. This allows us to approximate the PSD of the thermal noise, $n_{\text{th}}(T, t)$, as

$$\Gamma_{n_{\text{th}} \times n_{\text{th}}}(T, \omega) = kT, \tag{4.3}$$

which means that, for practical RF transceivers, the thermal noise can be considered as white over the frequency range of interest. However, in this formulation all the noise PSD is concentrated on the positive frequencies only. This is therefore a single sideband (SSB) PSD. As a result, the total available noise power, when measured in a frequency band δf centered around a given frequency f, is given by

$$P_{\text{av},n_{\text{th}}} = kT\,\delta f. \tag{4.4}$$

In this expression we have dropped the temperature dependence of the noise power, as is usually done. Note that the subscript av means that we are dealing with available power. Indeed, as long as the material is in thermal equilibrium, i.e. as long as a thermostat provides thermal energy that compensates for the electrical energy that is extracted from the device through electrical thermal noise, the above equation effectively represents the maximum power that can be extracted in a given frequency band. This quantity therefore represents the *available* electrical power.

A few comments are in order regarding the above derivation:

(i) Due to the macroscopic physical dimension of devices in practice, the angular frequencies of the electromagnetic modes ω_{m}, which are inversely proportional to the length of the conductor, are close enough that the resulting noise spectrum looks continuous in frequency.

(ii) As mentioned above, the noise PSD given by equation (4.3) is SSB, i.e. it spreads from DC toward infinity but is null for negative frequencies. This comes from the physical meaning of the energy of the electromagnetic modes given by equation (4.1) that were defined for the positive angular frequencies ω_{m} only. Thus, equation (4.4) represents the total power as effectively measured at the output of an ideal real bandpass filter with a passband δf centered around the frequency f, i.e. with two symmetrical sidebands centered around $\pm f$ in the total double sideband (DSB) frequency spectrum.

(iii) Finally, as the thermal noise is white for RF frequencies of interest, the noise power retrieved at the output of practical RF bandpass filters depends only on its bandpass δf and not on its center frequency.

We conclude that thermal noise can thus be modeled up to RF frequencies of interest by an additive white Gaussian process with a SSB PSD equal to kT.

4.2 Characterization of Noisy Devices

4.2.1 Noise Temperatures

We saw in the previous section (equation (4.4)) that the electrical power, $P_{\text{av},n}$, available at the output of a dipole device in a frequency band δf due to thermal motion is directly related to the physical temperature of the device, T. But, in the general case where the device is active, there are other noise sources (see Section 4.1.1). That said, due to the historical importance of thermal noise, the noise performance of devices is often characterized by an *effective* noise temperature T_e. This temperature is defined so that the available electrical noise power retrieved in a frequency band δf is such that

$$P_{\text{av},n} = kT_e\,\delta f. \tag{4.5}$$

We immediately observe that the available electrical noise power effectively measured at the output of a device must be greater than or equal to that linked to the thermal noise only, and thus the effective noise temperature is greater than or equal to the device physical temperature. Equality obviously holds when only thermal noise exists, i.e. for passive dipoles.

Practically speaking, this quantity is used to characterize the noise performance of devices in a straightforward way, whatever the origins of the noise sources. For instance, it can apply to antennas for which noise terms are linked not only to the analog electronic noises as discussed up to now, but also to received signals that act as noises. However, we can anticipate a potential problem with that definition when dealing with noise contributors that do not have a flat PSD. This major difference with respect to thermal noise results in an available noise power that depends not only on the width of the frequency band δf, but also on the central frequency. As a result, it becomes convenient to define an effective noise temperature that is a function of the frequency, $T_e(f)$, in order to represent the shape of the device noise PSD. Thus, the available noise power on the frequency band δf at the dipole output becomes

$$P_{\text{av},n} = \int_{\delta f} kT_e(f)\,\mathrm{d}f. \tag{4.6}$$

Nevertheless, in order to keep derivations simple, it is often assumed that the noise frequency band is sufficiently narrow to allow the quantities involved to be considered as constant. In that case, we talk about the spot noise characteristics of the device at a given frequency f_0. For a sufficiently small noise bandwidth δf centered on this frequency, we get that $T_e(f) \approx T_e(f_0)$ for $f \in \delta f$. Moreover, for the sake of simplicity, we can keep implicit the dependence of those spot noise quantities on the frequency. Under those assumptions, it follows that the available noise power on the frequency band δf at the dipole output can be written as

$$P_{\mathrm{av},n} \approx kT_e(f_0)\,\delta f = kT_e\,\delta f. \tag{4.7}$$

This approach is particularly fruitful when dealing with noise voltages or currents as it allows us to work with simple electrical equations, as illustrated in Section 4.2.3.

Obviously, these concepts can be extended to quadrupoles. Consider as a first step the configuration where a noisy generator is connected at the input of a noiseless quadrupole. This generator is a dipole. It can therefore be characterized by its internal impedance Z_g and its spot noise temperature T_g in the sufficiently narrow frequency band δf. Let us now assume that the quadrupole acts as a linear amplifier with an available power gain equal to G_{av}. Generally speaking, this available power gain is the ratio of the available power at the quadrupole output to the available power at the generator output. As illustrated for instance in Section 4.2.3, those powers depend on the output impedance of the device. Moreover, in the general case the output impedance of a quadrupole also depends on the internal impedance of the generator used to characterize it. This means that G_{av} depends in general on Z_g. In the present context it is interesting to observe that quantities that appear to be intrinsic to a quadrupole in fact depend on their environment. We return to this in Section 4.2.4 when dealing with cascades of noisy devices. Denote by $G_{\mathrm{av}(Z_g)}$ the available power gain of the considered quadrupole when operated with an input generator of internal impedance Z_g. Then the available noise power retrieved at the output of the system can be written as

$$P_{\mathrm{av},n} = G_{\mathrm{av}(Z_g)}\,kT_g\,\delta f. \tag{4.8}$$

Now supposing that the same quadrupole is noisy, we obviously have a noise term, $P_{\mathrm{av,q}}$, that *adds* to the output available noise power of the noiseless case according to

$$P_{\mathrm{av},n} = G_{\mathrm{av}(Z_g)}\,kT_g\,\delta f + P_{\mathrm{av,q}}. \tag{4.9}$$

This new term linked to the internal noise sources of the quadrupole thus characterizes the device intrinsic noise performance. We can thus define the effective input spot noise temperature of the quadrupole at the relevant frequency in δf by

$$P_{\mathrm{av,q}} = G_{\mathrm{av}(Z_g)}\,kT_e\,\delta f. \tag{4.10}$$

Although this term characterizes the noise sources of the quadrupole only, its expression depends on the input generator connected at its input through its impedance which affects $G_{\mathrm{av}(Z_g)}$. This means in particular that although the effective noise temperature of a quadrupole effectively characterizes its intrinsic noise sources, its value depends on the internal impedance of the generator connected to its input. There are different reasons for this.

The main reason is that the noise performance of the device is an intrinsic parameter of it, whereas we defined its effective temperature as an additional contribution coming from the input generator. Hence, in order to achieve the same noise contribution at the quadrupole output whatever the input generator impedance Z_g, the effective noise temperature must depend on Z_g to compensate potential power reflection variations due to impedance mismatch at the device input. This behavior is discussed in more depth in "RF and baseband devices" (Section 4.2.5).

Another reason comes from the fact that the intrinsic noise performance of a quadrupole also depends in general on the generator internal impedance connected at its input, as illustrated in "Noise factor dependency on generator impedance" (Section 4.2.5).

Let us therefore keep this dependence explicit in what follows, as we did for the available power gain. We denote this effective temperature by $T_{e(Z_g)}$, and will do the same in the next section for noise factors or noise figures. All this leads to the following simple expression for the available noise power at the quadrupole output:

$$P_{av,n} = G_{av(Z_g)} k\left(T_g + T_{e(Z_g)}\right)\delta f. \tag{4.11}$$

We thus see that the effective input spot noise temperature of the quadrupole is defined so that, from the output noise power point of view, everything behaves as if the quadrupole were noiseless but with the noise source temperature increased by $T_{e(Z_g)}$. The interesting point is that although its value depends on the input generator impedance, for a given topological configuration this effective noise temperature remains a characteristic of the intrinsic noise performance of the quadrupole. It does not depend on the source noise temperature, i.e. on the intrinsic noise power flowing from the generator. This is a main difference with respect to the noise factor, or noise figure (NF), both defined in Section 4.2.2, which depend not only on the noise performance of the considered quadrupole but also on that of the generator.

Practically speaking, other noise temperatures derived from the effective noise temperature can be used to characterize a system. For example, the operational temperature, $T_{op(Z_g)}$, of a system is defined by

$$T_{op(Z_g)} = T_g + T_{e(Z_g)}. \tag{4.12}$$

Here, the label "operational" indicates that this quantity characterizes the noise delivered by the quadrupole in operating conditions, i.e. taking into account the noise coming from the generator. From this definition, it is obvious that the operational spot noise temperature value also depends on the noise source impedance Z_g at the relevant frequency in δf. Up to a point, this concept can be related to the noise factor concept, described in Section 4.2.2, in the sense that it is related not only to the noise performance of the quadrupole itself but also to the source noise performance.

Another concept is that of apparent noise temperature. As for the operational temperature, the apparent noise temperature characterizes the complete system composed of the quadrupole and the noise source, but now as seen from the output. And from this output, the entire system is equivalent to a noisy dipole with an internal impedance equal to the output impedance of the system, and with an apparent temperature $T_{a(Z_g)}$ defined in such a way that the output available noise power in the frequency band δf is equal to

$$P_{av,n} = kT_{a(Z_g)}\,\delta f. \tag{4.13}$$

This concept can easily be linked to the operational temperature concept introduced above. We have from equations (4.11) and (4.12) that

$$P_{av,n} = G_{av(Z_g)}\,kT_{op(Z_g)}\,\delta f = kT_{a(Z_g)}\,\delta f. \tag{4.14}$$

As a result,

$$T_{a(Z_g)} = G_{av(Z_g)}\,T_{op(Z_g)} = G_{av(Z_g)}\left(T_g + T_{e(Z_g)}\right). \tag{4.15}$$

The equivalence between those noise definitions is shown in Figure 4.1.

In order to illustrate those concepts, we now derive as an example the noise temperatures of a passive quadrupole device. In this particular case, only the thermal noise is involved so that the different noise temperatures can be derived in a straightforward way from the physical

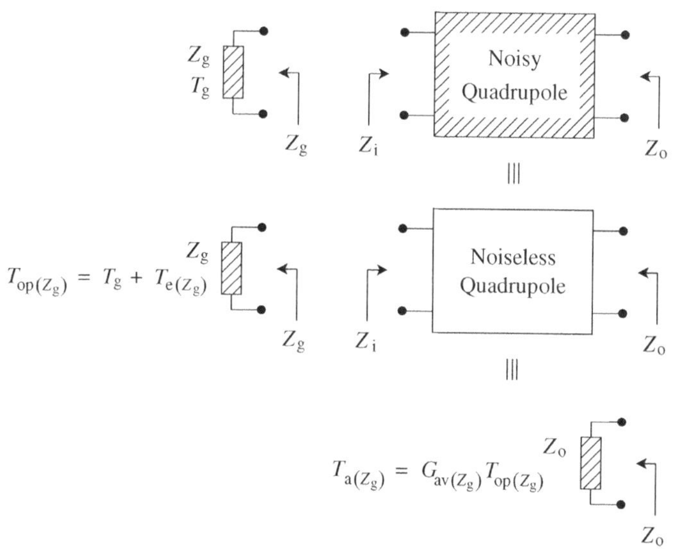

Figure 4.1 Noise temperature characterizations of a noisy quadrupole – The effective input spot noise temperature $T_{e(Z_g)}$ of a noisy quadrupole characterizes the fraction of the output noise that is linked to the quadrupole's own noise, as defined by equations (4.10) and (4.11), at the relevant frequency in δf. In that case, the noisy system (top) behaves as if the quadrupole were noiseless but with the noise source temperature increased by $T_{e(Z_g)}$. This also allows us to define the operational noise temperature, $T_{op(Z_g)}$, of the system according to equation (4.12) (middle). This system also behaves as a noise source with an internal impedance equal to the system output impedance, Z_o, and with a noise temperature equal to the system apparent noise temperature $T_{a(Z_g)}$ (bottom).

thermodynamic temperature of the device T. This is an important example as passive devices are often required in practical implementations, for instance between the antenna and the active part of the transceiver. Consider the configuration shown in Figure 4.1(top), when composed of a passive quadrupole device with a passive noisy dipole connected at its input. For the sake of simplicity in our derivations we can assume that the power matching of the device input is realized. We thus assume that $Z_g = Z_i^*$, as discussed in Section 4.2.3. As a result, we can simply write that the available power gain of the quadrupole, $G_{av(Z_i^*)}$, reduces to $1/L$, where L is the loss of the device under power matching conditions.

To determine the effective input noise temperature of the quadrupole, we recall that this quantity does not depend on the noise temperature of the source. Thus, we can consider a simple configuration that makes it particularly simple. Suppose that both the passive noise generator dipole and the passive quadrupole are at the same physical thermodynamic temperature T. Then we can see the entire system as a unique passive dipole with a physical temperature T. As only thermal noise is involved, the output available noise power, $P_{av,n}$, in a frequency band δf, is therefore simply derived from equation (4.4) as

$$P_{av,n} = kT\,\delta f. \tag{4.16}$$

If we now consider the definition of the effective input noise temperature of the quadrupole as given by equation (4.11), the output available noise power $P_{av,n}$ recovered at the system output can be written as

$$P_{av,n} = G_{av(Z_i^*)}\,k\left(T + T_{e(Z_i^*)}\right)\delta f, \tag{4.17}$$

or, in terms of the quadrupole losses under matching conditions,

$$P_{av,n} = \frac{1}{L}k\left(T + T_{e(Z_i^*)}\right)\delta f. \tag{4.18}$$

Using equation (4.16), we thus obtain that

$$T_{e(Z_i^*)} = T(L-1). \tag{4.19}$$

Physically, this means that the effective noise temperature of the quadrupole exactly compensates for the attenuation of the noise source power through the attenuation stage so that the output noise power of the system is equal to that linked to its physical thermodynamic temperature. Given that the effective noise temperature of a device is an intrinsic characteristic of it and is independent of the source noise temperature used for its derivation, the above expression remains valid when the noise source temperature is T_g. As a result, using this expression in equation (4.15) we can deduce the apparent temperature of the system composed of the noisy source and the noisy quadrupole:

$$\begin{aligned} T_{a(Z_i^*)} &= \frac{1}{L}T_g + \frac{1}{L}T(L-1) \\ &= T + \frac{1}{L}(T_g - T). \end{aligned} \tag{4.20}$$

In the above derivations we have used the fact that the available power gain $G_{\mathrm{av}(Z_i^*)}$ of the passive quadrupole is equal to $1/L$ under power matching conditions. However, it can be generalized under unmatched conditions, taking into account the relationship between the available power gain of the passive device and its insertion loss [17]. It follows in all cases that all the noise contributions and hence the equivalent temperature definitions can be expressed directly in terms of the physical temperature of the device T as only thermal noise is involved in such passive devices.

In conclusion, the important feature of these noise temperatures is that they are quantities that preserve the additive behavior of the physical noise terms they represent. This is of particular interest as this property drives most of the reasoning about additive noise budgeting, as illustrated for instance in Section 4.2.4. This is obviously one drawback of the noise factor concept introduced in the next section.

4.2.2 *Noise Factor*

Another alternative concept used to characterize the noise performance of RF devices is the noise factor or noise figure. The difference between the two terms is simply that the NF is the noise factor in decibel units. Thus, denoting the noise factor of a device by F and the noise figure by NF, the following simple relationship holds:

$$NF = F|_{\mathrm{dB}} = 10\log_{10}(F). \tag{4.21}$$

In that expression, the $10\log_{10}$ dependency comes from the fact that the noise factor is linked to the noise power delivered by the device as detailed below.

Recall the configuration discussed in the previous section and shown in Figure 4.1, i.e. a noisy generator, of internal impedance Z_g and noise temperature T_g, connected at the input of a noisy quadrupole. Assuming in a first step that the quadrupole is noiseless, the output available noise power $P_{\mathrm{av},n}$ in the frequency band δf is simply the input available power times the quadrupole available power gain $G_{\mathrm{av}(Z_\mathrm{g})}$. The implicit assumption for that to be possible is that δf is sufficiently narrow that the quantities $G_{\mathrm{av}(Z_\mathrm{g})}$ and T_g can be considered constant on the frequency band δf. Here again, this means we expect to work on spot noise quantities. Thus, we can write that

$$P_{\mathrm{av},n} = G_{\mathrm{av}(Z_\mathrm{g})}\, kT_\mathrm{g}\, \delta f. \tag{4.22}$$

This is nothing more than equation (4.11) in the particular case where $T_\mathrm{e} = 0$ as we consider the quadrupole as noiseless in this first step.

Now assuming that the quadrupole is noisy, there is necessarily an increase in the available noise power experienced at the output of the system. To characterize this phenomenon, the concept of effective noise temperature of the quadrupole was introduced in the previous section to represent the *addition* of an amount of noise power linked to the quadrupole contribution. This approach was consistent with the behavior of the physical phenomenon experienced, i.e. addition of a given amount of noise power to the noise coming from the source. However, we can alternatively consider a *proportionality* factor to represent this noise power increase. Labeling this factor as the spot noise factor of the device at the relevant frequency in δf, we can write the output available noise power $P_{av,n}$ as F times that obtained in the noiseless case,

$$P_{av,n} = F\, G_{av(Z_g)}\, kT_g\, \delta f. \tag{4.23}$$

But, as discussed for the effective temperature and the available power gain in the previous section, the noise factor also depends on the source impedance. This phenomenon is illustrated in more depth in "RF and baseband devices" (Section 4.2.5) and in "Noise factor dependency on generator impedance" (Section 4.2.5). As a result, it is of interest to make this dependence apparent, as illustrated for instance when dealing with cascades of noisy devices as discussed in Section 4.2.4. As a result, the above equation can be rewritten as

$$P_{av,n} = F_{(Z_g)}\, G_{av(Z_g)}\, kT_g\, \delta f. \tag{4.24}$$

From the output available noise power point of view, we thus get that everything behaves as if the quadrupole were noiseless but fed by a noise source at temperature $F_{(Z_g)}\, T_g$.

However, reconsidering this definition of the noise factor, this quantity also necessarily depends on the source temperature. In that sense this is not an intrinsic characteristic of the quadrupole itself, as were the noise temperatures. This obviously comes from the multiplicative behavior of this quantity, which does not match the physics of the phenomenon we are trying to model, i.e. its *additive* behavior. To better understand this, suppose that the noise source temperature T_g is very high so that the available noise power delivered by this source is much higher than that added by the quadrupole. In other words, assume that the intrinsic noise contribution of the quadrupole is negligible compared to that from the input source. In that case, the available noise power at the quadrupole output is almost equal to the input available power times the quadrupole available power gain. The quadrupole noise factor is therefore almost equal to 1. But suppose now that the noise contribution from the source is almost equal to that of the quadrupole. In that case, the output available noise power is almost twice that which would result from the source contribution only. This means that the quadrupole noise factor is around 2. And finally, if now the noise contribution from the source is negligible compared to that of the quadrupole, the output available noise power remains finite and almost equal to the quadrupole contribution only, whereas the input available power from the source can be considered as almost null. We end up with a quadrupole noise factor that tends toward infinity. This discussion explicitly shows that although the noise factor is expected to characterize the intrinsic performance of the quadrupole, its value depends on the noise source temperature used for its characterization. We thus see that there is a potential problem for comparing noise factors of different devices. Practically speaking, this has been resolved by adopting a standardized noise temperature, $T_0 = 290$ K, for the source to be used to reference all noise factors.

However, we observe that in practice other source noise temperatures may be used. In particular, we may select the physical thermodynamic temperature T of the device to refer its noise factor to. This is useful when the circuit board that embeds the quadrupole is operating at the same temperature T and if the input noise source can also be considered at temperature T. Then the source noise temperature seen by the device is also equal to T and the noise factor referred to the temperature T directly represents the SNR degradation through the device as discussed in Section 4.2.6. This is why we are interested in the noise factor concept in comparison to the effective noise temperature in this particular case where the temperature used for its reference corresponds to the source noise temperature seen by the device. It explains why, although less close to the physical reality of the noise phenomenon, the noise factor is interesting and convenient for system purposes as the SNR degradation is the main concern for system dimensioning.

In the following we can thus use the notation $F_{(Z_g,T)}$ to show explicitly that the device noise factor is referred to a source noise temperature T. Hence, $F_{(Z_g,T_0)}$ corresponds to the standardized definition of the noise factor, assuming that the noise source impedance is Z_g. However, in the following, $F_{(Z_g,T_0)}$ is used to express that the considered noise factor is referred to a given reference source noise temperature T_0, whatever the effective value of this reference temperature. However, this notation must not be allowed to obscure the fact that the intrinsic noise performance of a device depends on many parameters, such as its own physical temperature or its power supply in the case of active devices. This dependency is true for both the effective noise temperature and the noise factor and is here left implicit. The difference is that for a given intrinsic configuration and physical temperature, only the noise factor value depends also on the input source noise temperature.

From the above discussion, the available noise power retrieved at the output of a noisy quadrupole characterized by the spot noise factor $F_{(Z_g,T_0)}$ can be written as

$$P_{\mathrm{av},n} = G_{\mathrm{av}(Z_g)} kF_{(Z_g,T_0)} T_0 \,\delta f \tag{4.25}$$

if and only if the noise source temperature is effectively equal to T_0 and its internal impedance equal to Z_g. As a consequence, we can hazard a guess at the available noise power we effectively retrieve at the output of the system when the noise temperature of the source now becomes different from T_0 while the only characterization of the quadrupole we have for our derivation remains its noise factor referred to a noise source temperature equal to T_0. To proceed, we first need to convert the noise factor in terms of effective noise temperature. This noise temperature characteristic is indeed *the* intrinsic quantity linked to the device noise performance and not to the source noise temperature. We thus expect to be able to derive the output available noise power for any source noise temperature in that way. We observe that the available noise power $P_{\mathrm{av},n}$ given by equation (4.25) can also be directly expressed in terms of the quadrupole effective noise temperature $T_{\mathrm{e}(Z_g)}$ by using equation (4.11) with $T_g = T_0$. This leads to

$$P_{\mathrm{av},n} = G_{\mathrm{av}(Z_g)} k\left(T_0 + T_{\mathrm{e}(Z_g)}\right)\delta f. \tag{4.26}$$

Direct comparison of this expression with equation (4.25) thus leads to the relationship between the noise factor and effective noise temperature illustrated in Figure 4.2, i.e. to[1]

$$F_{(Z_g,T_0)} = 1 + \frac{T_{\mathrm{e}(Z_g)}}{T_0}. \tag{4.27}$$

[1] This relationship justifies the above discussion on the dependence of the noise factor on the input noise source temperature used for its definition. If the input source noise temperature is already very high, i.e. if $T_0 \gg T_{\mathrm{e}(Z_g)}$, then the resulting noise factor almost reduces to 1 as the noise power degradation through the device is almost null. On the other hand, if the input noise power is negligible compared to that added by the device, i.e. if $T_0 \ll T_{\mathrm{e}(Z_g)}$, the resulting noise factor tends toward infinity to give an output noise power that remains finite.

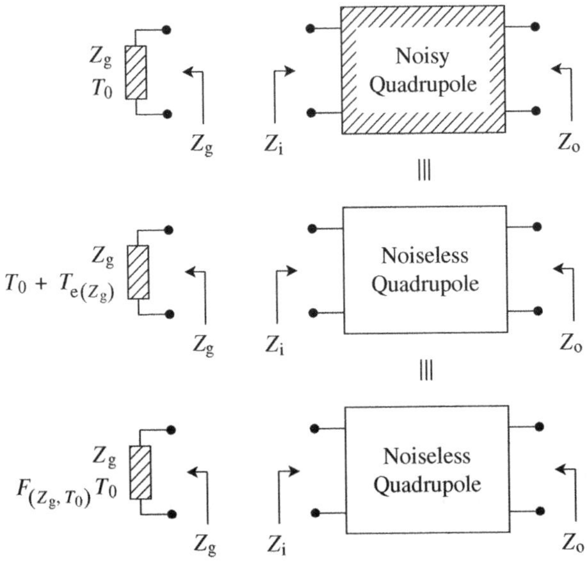

Figure 4.2 Equivalent noise representations of a noisy quadrupole using its effective noise temperature or its noise factor – RF noisy quadrupoles (top) are often characterized using two kinds of quantities: the effective input noise temperature $T_{e(Z_g)}$, linked to the quadrupole's own noise performance as defined by equation (4.11); and the noise factor $F_{(Z_g,T_0)}$ as defined via equation (4.25) that depends not only on the quadrupole's own noise contribution but also on the input source noise level through its noise temperature T_0. As a result, the noisy system fed by a noise source at temperature T_0 behaves, from the noise point of view, as if the quadrupole were noiseless but with the noise source temperature increased by $T_{e(Z_g)}$ (middle) or multiplied by $F_{(Z_g,T_0)}$ (bottom).

If we now suppose that the input noise source temperature $T_g \neq T_0$, we can use equation (4.11) to express the available noise power recovered at the output of the system in the narrow frequency band δf as a function of the quadrupole effective noise temperature $T_{e(Z_g)}$. Then inverting equation (4.27) to substitute the quadrupole noise factor into the noise temperature, we finally get that

$$P_{av,n} = G_{av(Z_g)} k\left[T_g + \left(F_{(Z_g,T_0)} - 1\right)T_0\right]\delta f. \tag{4.28}$$

This expression obviously reduces to equation (4.25) when $T_g = T_0$. But as illustrated in Figure 4.3, in the case where the noise factor is referred to an input noise source temperature different from that effectively used in the application, the expression for the output available noise power is less straightforward than the initial definition of the noise factor might suggest.

Another way to highlight this behavior is to make explicit the dependence of the noise factor on the temperature of the noisy source used as the reference. We can express the noise factor as

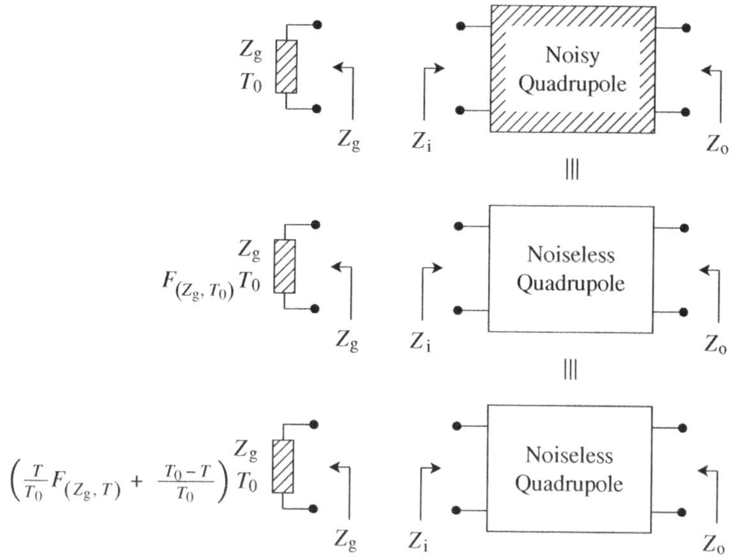

Figure 4.3 Equivalent noise representations of a noisy quadrupole using its noise factor defined for different source noise temperatures – The noise behavior of a noisy quadrupole (top) can be characterized using its noise factor as defined via equation (4.25). If the source noise temperature T_0 is equal to the temperature used for the noise factor characterization, the system behaves as if the quadrupole were noiseless but with the noise source temperature multiplied by $F_{(Z_g,T_0)}$ (middle). If the quadrupole noise factor has been characterized using a noise source temperature T different from the present source temperature T_0, then $F_{(Z_g,T_0)}$ has to be reconstructed by inverting equation (4.30) to get the correct noise level at the system output (bottom).

a function of the intrinsic noise temperature of the quadrupole when considering two distinct source noise temperatures, T and T_0. Using equation (4.27), this leads to

$$F_{(Z_g,T)} = 1 + \frac{T_{e(Z_g)}}{T}, \tag{4.29a}$$

$$F_{(Z_g,T_0)} = 1 + \frac{T_{e(Z_g)}}{T_0}. \tag{4.29b}$$

Canceling the effective temperature term yields

$$F_{(Z_g,T)} = \frac{T_0}{T} F_{(Z_g,T_0)} + \frac{T - T_0}{T}. \tag{4.30}$$

This relationship can be used in an alternative derivation of the available noise power $P_{av,n}$ seen at the output of the system characterized by its noise factor $F_{(Z_g,T_0)}$ referred to T_0, while fed by a source exhibiting a noise temperature T_g different from T_0. Substituting $F_{(Z_g,T_0)}$ into $F_{(Z_g,T_g)}$ in equation (4.25) we can immediately write

$$\begin{aligned} P_{av,n} &= G_{av(Z_g)}\, k \left(\frac{T_0}{T_g} F_{(Z_g,T_0)} + \frac{T_g - T_0}{T_g} \right) T_g\, \delta f \\ &= G_{av(Z_g)}\, k \left[T_g + \left(F_{(Z_g,T_0)} - 1 \right) T_0 \right] \delta f. \end{aligned} \tag{4.31}$$

As expected, we recover the results derived through equation (4.28).

In conclusion, it is of interest to complement the discussion about the lossy passive quadrupole at the end of the previous section with the noise factor concept. Given in this particular case that only thermal noise is involved, straightforward derivations can be done involving only the thermodynamic temperature T of the device. Practically speaking, in that former section we already derived an expression, equation (4.19), for the effective temperature of the system which remains valid under power matching conditions. Thus, using equation (4.27) directly, we can now give in turn an expression for the noise factor under the same matching condition:

$$F_{(Z_i^*,T_0)} = 1 + \frac{T}{T_0}(L - 1), \tag{4.32}$$

where L is the insertion loss of the device. We observe that $F_{(Z_i^*,T_0)}$ effectively reduces to L if and only if the thermodynamic temperature of the device, T, is equal to the noise source temperature T_0 used to characterize its noise factor. More generally, under the same temperature conditions but when unmatched, we would get that the passive device noise factor reduces to the inverse of its available power gain instead of its insertion loss [17].

Noise Factor for Mixing Stage

Noisy devices also need to be characterized in the particular case of mixing stages. This is due to an additional mechanism that we have to deal with during the operation of such devices, namely the folding of different noise sidebands [38]. Anticipating the discussion in Chapter 6, we observe that this folding mechanism is linked to the presence of unwanted tones in the waveform of the LO signal that physically drives the mixing stage. Practically speaking, these unwanted tones are either harmonics of the targeted LO frequency (this occurs, for instance, when the mixing stage is implemented as a chopper) or image tones that are necessarily present when considering a real frequency conversion, but that still exist during a complex frequency conversion in the presence of gain and phase imbalance. For the sake of simplicity and convenience, we consider the folding mechanism linked to the presence of the image tone as discussed in Section 6.1.2.

Let us consider a single mixing device that we want to characterize in operating conditions from a noise performance perspective. As we are considering a single device, we necessarily talk about its characterization during a real frequency conversion. Suppose that a bandpass RF signal $s_{RF}(t)$, centered on the carrier angular frequency ω_{RF}, as well as two bandpass noise terms are present at the input of the mixing stage. The first noise term, $n_{RF}(t)$, is also assumed to be centered around ω_{RF}, while the second, $n_{IM}(t)$, is assumed to be centered around the input image angular frequency ω_{IM}. In practice, these bandpass signals can be decomposed in terms of positive and negative sidebands that are involved in the frequency conversion processing. This decomposition can be achieved by expressing them as a function of their complex envelopes, assumed defined here as centered around the respective carrier angular frequencies. Using equation (1.5), we have

$$s_{RF}(t) = \mathrm{Re}\left\{\tilde{s}_{RF}(t)e^{j\omega_{RF}t}\right\} = \frac{1}{2}\left(\tilde{s}_{RF}(t)e^{j\omega_{RF}t} + \tilde{s}^*_{RF}(t)e^{-j\omega_{RF}t}\right), \tag{4.33a}$$

$$n_{RF}(t) = \mathrm{Re}\left\{\tilde{n}_{RF}(t)e^{j\omega_{RF}t}\right\} = \frac{1}{2}\left(\tilde{n}_{RF}(t)e^{j\omega_{RF}t} + \tilde{n}^*_{RF}(t)e^{-j\omega_{RF}t}\right), \tag{4.33b}$$

$$n_{IM}(t) = \mathrm{Re}\left\{\tilde{n}_{IM}(t)e^{j\omega_{IM}t}\right\} = \frac{1}{2}\left(\tilde{n}_{IM}(t)e^{j\omega_{IM}t} + \tilde{n}^*_{IM}(t)e^{-j\omega_{IM}t}\right). \tag{4.33c}$$

The important fact to understand is that, depending on the frequency planning used for the frequency conversion, different SNR degradations can be predicted, as illustrated in Figure 4.4. For instance, we can see on the left-hand side that in an infradyne downconversion the negative

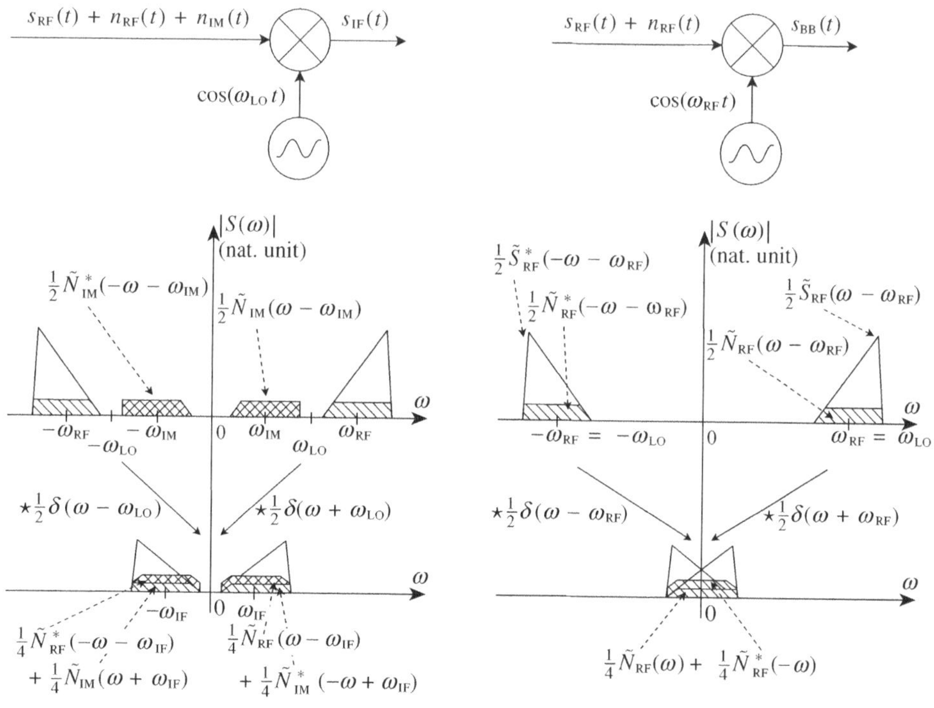

Figure 4.4 Noise sideband folding during a real frequency conversion – Depending on the relative LO angular frequency compared to that of the carrier at the input of the mixer, a folding on the wanted signal of an additional noise term lying at the image angular frequency occurs, as in the heterodyne case (left), or not, as in the homodyne case (right). As a result, in the heterodyne case we get an increase of the equivalent noise contribution of the mixer relative to the wanted signal power due to the noise sideband folding (bottom left). This is not true in the homodyne case (bottom right).

sideband of the image noise component is folded on the positive sideband of the main noise component originally centered on ω_{RF}. This means that we have a SNR degradation when going through the frequency conversion process compared to the case where no frequency conversion occurs. In contrast, we may preserve this SNR in the homodyne conversion case by having a folding of the wanted signal too, as illustrated in Figure 4.4(right). We thus see through this simple example that the apparent noise performance of a mixing stage depends on its operating conditions. This behavior has led to a first refinement in the definition of the noise quantities used to characterize such mixing stage:

(i) In an operational condition where two noise sidebands are folded on one single signal sideband, the noise performance of the overall processing is referred to as the SSB noise performance of the mixing stage. In that case we can thus define an SSB effective noise temperature for this mixer and an SSB noise factor or noise figure for it.
(ii) In a configuration where the same number of signal and noise sidebands are folded together, we talk about the DSB noise performance of the mixer. In that case, we recover the intrinsic noise performance of the mixing stage through the DSB quantities.

Practically speaking, we thus see that the term SSB refers to the performance recovered during the use of the mixing stage in a heterodyne configuration while DSB refers to the homodyne configuration.

We can even go a step further and give some relationships between the DSB and SSB quantities when assuming that the noise contributions we are dealing with have a flat PSD. This assumption seems realistic as it is representative of thermal noise behavior. In that case, we can assume the same magnitude for the different noise sidebands involved in our derivation. The relationship between the DSB and SSB noise characteristics of the mixing stage then seems straightforward as it is just a matter of doubling the noise contributions lying around ω_{RF} at the input of the mixer in order to reflect the SSB behavior compared to the DSB behavior. Given that the effective noise temperature of a device is directly proportional to the noise power of its intrinsic contributors, as given by equation (4.5), we can deduce that the SSB effective noise temperature of the mixer, $T_{e,SSB}$, can be estimated as twice its DSB effective noise temperature, $T_{e,DSB}$. As a result, for a given configuration where the internal impedance of the source generator connected at the mixer input is Z_g, we can write

$$T_{e,SSB(Z_g)} = 2T_{e,DSB(Z_g)}. \tag{4.34}$$

In the same way, everything behaves in the SSB configuration as if the contribution of the source noise to the total noise power recovered at the device output were also doubled due to the folding process. Assuming that the source noise temperature is T_0 and based on the relationship between the noise factor and the effective temperature of the device given by equation (4.27), we may write for the SSB noise factor that

$$F_{SSB(Z_g,T_0)} = \frac{2T_0 + T_{e,SSB(Z_g)}}{T_0} = \frac{2T_0 + 2T_{e,DSB(Z_g)}}{T_0} = 2F_{DSB(Z_g,T_0)}. \tag{4.35}$$

As might be expected, we recover the 3 dB degradation due to the noise folding.

However, this statement may be refined by reconsidering what really is a realistic operational condition for a mixer in a transceiver. On the receive side, for instance, we get that the image signal necessarily needs to be filtered out before performing a real frequency conversion, as discussed in Section 6.1.2. Obviously, such an image reject filter necessarily also acts on the source noise flowing from the early stage of the line-up. One can thus assume that only the input-referred noise contribution of the mixing stage is doubled during the frequency conversion process, and not the source noise contribution. In case the source noise temperature seen by the mixer remains equal to T_0 in the presence of the filter, that would result in a relationship of the form

$$F_{\text{SSB}(Z_g,T_0)} = \frac{T_0 + T_{e,\text{SSB}(Z_g)}}{T_0} = 1 + \frac{2T_{e,\text{DSB}(Z_g)}}{T_0} < 2\left(1 + \frac{T_{e,\text{DSB}(Z_g)}}{T_0}\right), \tag{4.36}$$

or

$$F_{\text{SSB}(Z_g,T_0)} < 2F_{\text{DSB}(Z_g,T_0)}. \tag{4.37}$$

We thus get with this new definition that the SSB noise factor of the mixer is less than twice the DSB noise factor. Only if the contribution of the source noise remains negligible compared to the intrinsic noise performance of the mixer, i.e. if $T_0 \ll T_{e,\text{SSB}(Z_g)}$, do we have $F_{\text{SSB}(Z_g,T_0)} \approx 2F_{\text{DSB}(Z_g,T_0)}$. But in that case, the mixer we are dealing with has very poor intrinsic noise performance compared to the performance of the earlier stages of the line-up in the targeted application.

According to our discussion so far, the noise characterization of mixing stages needs to match the exact usage of the device in operating conditions. Particular care must be taken to use the correct definition in terms of noise sideband folding, both for the intrinsic contributions of the mixer and for those of the reference source used for the characterization. This holds as long as the problem is to work on reliable quantities to budget a line-up. In contrast, to compare the performance of two different devices, we need to ensure that the same definition has been used for the two characterizations.

4.2.3 *Noise Voltage and Current Sources*

For some topics, we may wish to work on equivalent noise voltage or current sources rather than with available noise powers as hitherto. A typical application is the characterization of devices that process baseband signals. As discussed in Section 2.2.3, in that case the power matching is often not required in order to obtain the transfer function of interest. The transfer function of such devices is often targeted on either the voltages or the currents that carry the information, but not on both at once. We can thus understand the interest in working with noise voltages or current sources to characterize such devices. From a system perspective there is a need to link these new noise quantities with those we reviewed earlier.

As a first step, let us briefly discuss how to model these noise voltages and currents themselves. The efficient way to perform electrical derivations between currents, voltages and impedances is to work on pure sine waves using the complex notation. This has the

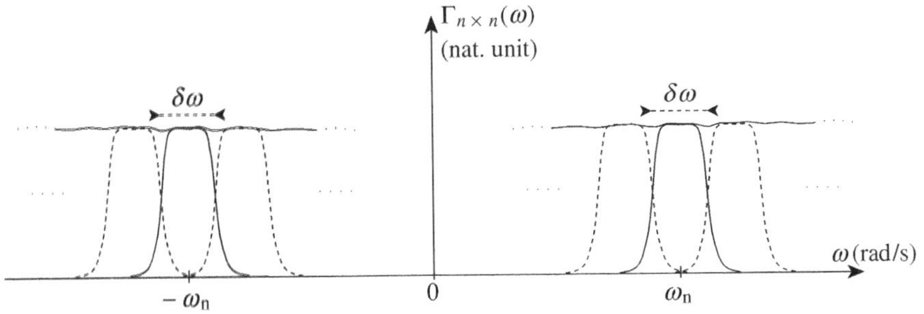

Figure 4.5 Spectral representation of the decomposition of a wideband noise process as a sum of narrowband terms – The spectrum of a wideband noise process, $n(t)$, can be decomposed in terms of adjacent narrowband portions. In the time domain, this means that $n(t)$ can be expressed as the sum of narrowband processes to which can be associated a complex envelope. When the bands $\delta\omega$ are sufficiently narrow, straightforward electrical equations as equation (4.43) can be derived using these complex envelopes.

great advantage of changing differential equations into linear ones. The problem is that such CW signals cannot correctly represent the statistical characteristics of noise waveforms. As discussed in Appendix 2, this is particularly the case for the stationarity of the noise sources and the related properties of interest for analytical derivations such as equation (A1.31). One way to handle this is to decompose the total spectrum, possibly wideband, of a given noise process in terms of a sum of adjacent narrowbands, each assumed of width $\delta f = \delta\omega/2\pi$ as shown in Figure 4.5 [39]. For each of those narrowband slices, we can associate a sine wave signal with an amplitude such that its RMS value matches the power of the noise process in the band of width δf. An independent random phase offset uniformly distributed over $[0, 2\pi]$ is then added in order to achieve the stationarity of the process, as discussed in Appendix 2. This approach can even be generalized using a general complex envelope, not necessarily with a constant amplitude, as long as it fulfills the conditions that lead to stationarity. This gives the possibility of matching the statistics for the amplitude and phase of the complex envelope to those of the noise that would be recovered at the output of a real bandpass filter of bandwidth δf and centered around $f_n = \omega_n/2\pi$.

This approach is of interest because as long as the frequency band δf can be considered as sufficiently narrowband that the electrical characteristics of the devices under consideration can be taken as almost constant, we can still think of writing electrical relationships as in the pure CW case, but now with complex envelopes. Indeed, considering for instance a bandpass voltage noise process $v(t)$, bandlimited to δf and centered around $f_n = \omega_n/2\pi$, applied across a dipole of impedance $Z(\omega)$, we get a current $i(t)$ generated through it such that it can be expressed in the frequency domain as

$$V(\omega) = Z(\omega)I(\omega). \tag{4.38}$$

Assuming that $Z(\omega) \approx Z(\omega_n)$ for all angular frequencies in $\delta\omega$, we get

$$V(\omega) \approx Z(\omega_n)I(\omega). \tag{4.39}$$

This relationship is valid for the positive part of the spectrum only. According to Section 1.1.2, this positive part of the spectrum is nothing more than the Fourier transform of the analytic signals of $v(t)$ and $i(t)$, i.e. the Fourier transform of $v_a(t)$ and $i_a(t)$ respectively. We can thus write

$$V_a(\omega) = Z(\omega_n)I_a(\omega) \tag{4.40}$$

or, taking the inverse Fourier transform of this relationship,

$$v_a(t) = Z(\omega_n)i_a(t). \tag{4.41}$$

Thus, if we assume that the complex envelopes for $v(t)$ and $i(t)$, $\tilde{v}(t)$ and $\tilde{i}(t)$ respectively are defined as centered around the same center angular frequency, for instance ω_n for the sake of simplicity, we get from equation (1.21) that

$$\tilde{v}(t)e^{j\omega_n t} = Z(\omega_n)\tilde{i}(t)e^{j\omega_n t}. \tag{4.42}$$

Canceling terms,

$$\tilde{v}(t) = Z(\omega_n)\tilde{i}(t). \tag{4.43}$$

Thus for sufficiently narrowband processes, we can write electrical relationships between complex envelopes that represent voltages and currents as is usually done using complex notation for pure CWs. The condition for this to be possible is that our complex envelopes are defined as centered *around the same center frequency*. This is not so restrictive, however, as this condition is also required for many analytical derivations. Examples are the possibility of writing the complex envelope of the sum of signals as the sum of the complex envelopes of those signals, and being able to derive the non-correlation between bandpass signals from that of their complex envelopes, as detailed in Appendix 1. Thus, we will henceforth always assume that we are dealing with complex envelopes defined as centered around the same center frequency. However, for the sake of simplicity, the dependence on this center frequency is not necessarily explicitly written in subsequent derivations, both for complex envelopes, impedances and more generally for all noise characteristics of the devices. But this should not obscure the fact that, defined in this way, the corresponding bandpass noise voltages and currents we deal with represent the noise behavior of the device at a given frequency. This means that we are working with the spot noise characteristics of the device.

Dipole Case

Let us now focus on the equivalent Thévenin model for the electrical noise delivered by a dipole of internal impedance Z_g and noise temperature T_g to a load of impedance Z_l. We can proceed by expressing the noise power available at the dipole output in the frequency band δf, $P_{av,n}$, in terms of both noise temperature and equivalent voltage or current quantities.

Considering first a model for the dipole based on its noise temperature, we can use equation (4.5) to write

$$P_{\mathrm{av},n} = kT_{\mathrm{g}}\,\delta f. \tag{4.44}$$

This equation is valid as long as the device noise temperature is almost constant over the frequency band δf. This in fact matches our assumption of working on spot noise quantities to be able to use simple electrical equations. In order to go further and determine the source noise voltage of the equivalent Thévenin model of the dipole, $v_{\mathrm{g}}(t)$, such that its available power corresponds to the above expression, we can introduce various complex envelopes for the quantities of interest. As discussed in the introductory part of this section, all these complex envelopes, which represent the fraction of noise present in the frequency band δf, are defined as centered around the same angular frequency $\omega_{\mathrm{n}} = 2\pi f_{\mathrm{n}}$, with $f_{\mathrm{n}} \in \delta f$, in order to allow simple analytical derivations. In the notation of Figure 4.6, the source noise voltage can thus be expressed as

$$v_{\mathrm{g}}(t) = \mathrm{Re}\left\{\tilde{v}_{\mathrm{g}}(t)\mathrm{e}^{\mathrm{j}\omega_{\mathrm{n}}t}\right\}, \quad \text{with } \tilde{v}_{\mathrm{g}}(t) = \rho_{\mathrm{v,g}}(t)\mathrm{e}^{\mathrm{j}\phi_{\mathrm{v,g}}(t)}. \tag{4.45}$$

In the same way, we can write the voltage and current noise signals across the load as

$$v_{\mathrm{l}}(t) = \mathrm{Re}\left\{\tilde{v}_{\mathrm{l}}(t)\mathrm{e}^{\mathrm{j}\omega_{\mathrm{n}}t}\right\}, \quad \text{with } \tilde{v}_{\mathrm{l}}(t) = \rho_{\mathrm{v,l}}(t)\mathrm{e}^{\mathrm{j}\phi_{\mathrm{v,l}}(t)}, \tag{4.46a}$$

$$i_{\mathrm{l}}(t) = \mathrm{Re}\left\{\tilde{\imath}_{\mathrm{l}}(t)\mathrm{e}^{\mathrm{j}\omega_{\mathrm{n}}t}\right\}, \quad \text{with } \tilde{\imath}_{\mathrm{l}}(t) = \rho_{\mathrm{i,l}}(t)\mathrm{e}^{\mathrm{j}\phi_{\mathrm{i,l}}(t)}. \tag{4.46b}$$

Following our assumptions on the definition of the complex envelopes we are dealing with, we can then directly reuse the results derived for narrowband modulated waveforms in

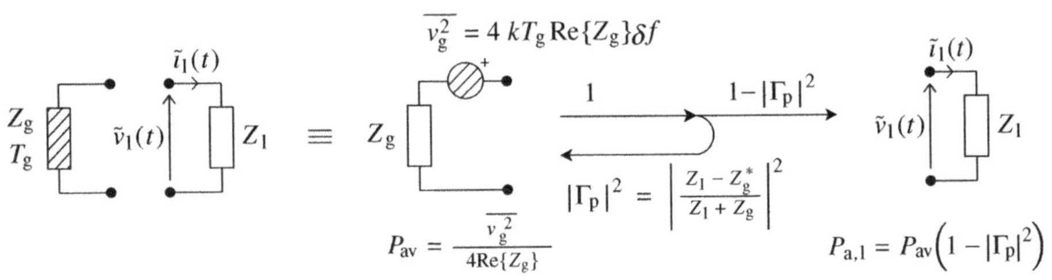

Figure 4.6 Noisy dipole model as a voltage generator – A noisy dipole with an internal impedance Z_{g} and a noise temperature T_{g} behaves as noise source characterized by the available noise power that can be transfered to a load Z_{l}. As a result, an equivalent voltage generator model of the noise source can be derived just with knowledge of the RMS value of this source in the noise frequency band δf according to equation (4.50). In the general case where the source and load impedances are not matched, the active power that is effectively delivered to the load corresponds to the fraction that is not reflected back to the source according to equation (4.52).

Section 2.2.3. In particular, based on equation (2.127), we can write the available power at the output of the equivalent Thévenin generator as

$$P_{\text{av}} = \frac{1}{4\text{Re}\{Z_{\text{g}}\}} \frac{\mathbb{E}\{|\tilde{v}_{\text{g}}|^2\}}{2}. \tag{4.47}$$

The last term on the right-hand side of this equation is nothing more than the power of the bandpass process $v_{\text{g}}(t)$ as given by equation (1.64). Under ergodicity, this term thus directly represents the RMS value of any realization of $v_{\text{g}}(t)$. As this RMS value is the quantity of interest for power transfer purposes, the specific notation $\overline{v_{\text{g}}^2}$ is often used for it. It thus corresponds to the definition

$$\overline{v_{\text{g}}^2} \stackrel{\Delta}{=} \overline{v_{\text{g}}^2(t)} = \frac{\mathbb{E}\{|\tilde{v}_{\text{g}}|^2\}}{2}. \tag{4.48}$$

We can therefore rewrite the available power from the generator given by equation (4.47) as

$$P_{\text{av}} = \frac{1}{4\text{Re}\{Z_{\text{g}}\}} \overline{v_{\text{g}}^2}. \tag{4.49}$$

Finally, direct comparison of this equation with equation (4.44) leads to

$$\overline{v_{\text{g}}^2} = 4kT_{\text{g}}\text{Re}\{Z_{\text{g}}\}\delta f. \tag{4.50}$$

This noise RMS voltage depends only on the real part of the source impedance at the relevant frequency in δf. Referring to the derivation in Section 2.2.3, we can say that the reason for this is that the available power, given here by equation (4.47), represents an average power, i.e. an active power only. A consequence of this dependence on the real part only is that the noise power delivered can be null if and only if there are no resistors in the dipole. However, when such resistive devices exist, it does not mean that reactive components cannot be involved in the value of this real part through some resonant behavior at the relevant frequency. Another way to see this is that when looking for such an equivalent model, two strategies can be considered. We can first replace each individual component in the dipole by its equivalent noise voltage source model; the above equation show that only resistors need be considered for this. But then the value of other components can be involved in the derivation of the equivalent Thévenin voltage of the dipole when solving the electrical equations. Alternatively, we can directly derive the overall model of the dipole by applying the above equation with the real part of its internal impedance. Components other than resistors can then be involved in this real part when resonance occurs. Whatever the method, we must end up with the same result.

Up to now we have considered the equivalent Thévenin generator for our model. But, the same could obviously be done with an equivalent Norton representation. This would lead to the dual representation as a current generator with an RMS value given by

$$\overline{i_{\mathrm{g}}^2} = 4kT_{\mathrm{g}}\mathrm{Re}\left\{\frac{1}{Z_{\mathrm{g}}}\right\}\delta f, \tag{4.51}$$

in parallel with the impedance Z_{g}. The use of one representation or the other should be driven by how easy the analytical derivations are. But whatever the representation considered, we observe we can only derive RMS values of voltages or current noise sources. The deep reason for that comes from the physics of noise which gives knowledge of available powers only, as we have seen so far. But in practice, noise voltages and currents are required in order to write electrical relationships. This is no big deal, bearing in mind that only RMS values can be taken and linked to physical quantities. However, the concept of independence or correlation between different noise sources must be taken into account as their contributions to the overall noise power can add or not, depending on the case. This point is discussed in more depth shortly when we come to the quadrupole case.

It may be instructive to conclude our present discussion by focusing on the noise power that is effectively delivered by the dipole to its load. Practically speaking, this quantity, $P_{\mathrm{a,l}}$, can be expressed from equation (2.131) as

$$P_{\mathrm{a,l}} = P_{\mathrm{av}}(1 - |\Gamma_{\mathrm{p}}|^2), \tag{4.52}$$

where

$$\Gamma_{\mathrm{p}} = \frac{Z_{\mathrm{l}} - Z_{\mathrm{g}}^*}{Z_{\mathrm{l}} + Z_{\mathrm{g}}} \tag{4.53}$$

is the power reflection coefficient. As illustrated in Figure 4.6, the square of the magnitude of this factor represents the fraction of the available power from the source that is reflected back to it due to the impedance mismatch. The term $1 - |\Gamma_{\mathrm{p}}|^2$ therefore represents the fraction of the total available power that is effectively delivered to the load. What is interesting for our discussion is that the load impedance value only changes the amount of noise that is effectively delivered by the dipole to the load and not its *available* noise power, which remains an intrinsic characteristic of it. The same statement therefore holds for the equivalent noise voltage or current generator.

Quadrupole Case

Generally speaking, an electronic quadrupole device embeds active parts that require either voltage or current sources to model their intrinsic noise behavior. But when searching for an equivalent model for such a system, the resolution of the electrical equations makes it necessary to use at the same time a voltage and a current noise source to model correctly the quadrupole behavior. This is due to the two degrees of freedom associated with the load and the generator impedances which can both be variable in the present case. This is the main

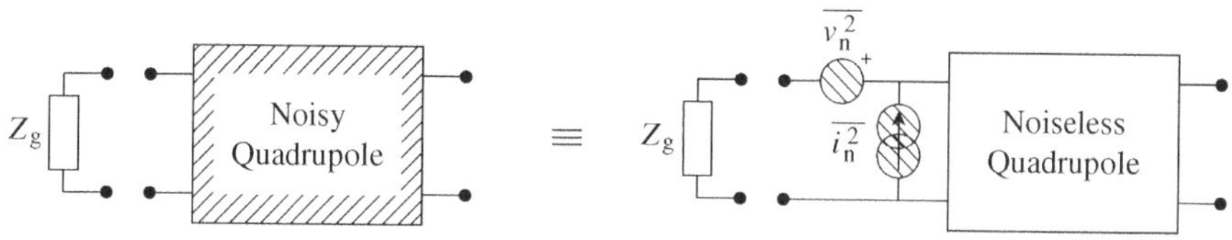

Figure 4.7 Noisy quadrupole model using voltage and current sources – A noisy quadrupole composed of passive and active solid state devices can be modeled by a noiseless quadrupole with both input voltage and current noise sources. In most cases, the two noise sources are correlated so that the overall noise power delivered to the quadrupole is not equal to the sum of the power from each source.

difference with respect to the dipole configuration, which can be interpreted as a quadrupole but seen from the output with a fixed input generator impedance. This fixed impedance results in an additional electrical relationship that links the two generators required in the quadrupole case, thus allowing for a single equivalent source through the equivalent Thévenin or Norton model. Practically speaking, assuming that we are working as usual with equivalent input-referred quantities, we can then rely on a topology as shown on the right-hand side of Figure 4.7 to model our noisy quadrupole.

There is more to say about this topology as the relative importance of the two generators represented in the figure depends on the relative magnitude of the input impedance of the quadrupole, Z_i, in respect to the internal impedance of the generator, Z_g [40]. For instance, if $|Z_g| \ll |Z_i|$ the equivalent current noise source can be considered as almost in short-circuit so that its impact can be taken to be almost negligible compared to the voltage source contribution. Practically speaking, this situation occurs for an interface (I/F) in voltage between the generator and the quadrupole. Thus, when dealing with a cascade of voltage gain amplifiers, for instance, a single noise voltage source is often accurate enough to model the noise behavior of the devices. It is also obviously true that when $|Z_g| \gg |Z_i|$, the noise voltage source contribution can be considered as almost null and a single noise current source is accurate enough. As a result, it is in matched conditions, i.e. for devices that operate on RF signals in practice, that the two noise sources are necessary in order to correctly model the quadrupole noise behavior.

It may be worth dwelling briefly on the general case where the voltage and current noise sources in this equivalent model are correlated, as its extension to bandpass signals and their complex envelopes is not completely straightforward. Practically speaking, the property of interest related to this correlation behavior is that the RMS value, i.e. the average power, of the sum of two correlated time domain signals is not necessarily equal to the sum of the power of each individual signal. This simple behavior can be illustrated by considering two voltage processes $v_1(t)$ and $v_2(t)$ that sum together to give

$$v(t) = v_1(t) + v_2(t). \tag{4.54}$$

With no assumptions on the processes involved except stationarity and ergodicity, the average power P_v of the real valued signal $v(t)$ can be written as

$$P_v = \mathbb{E}\{v^2\} = \mathbb{E}\{(v_1 + v_2)^2\}. \tag{4.55}$$

Expanding, we get

$$\begin{aligned} P_v &= \mathbb{E}\{v_1^2\} + \mathbb{E}\{v_2^2\} + 2\mathbb{E}\{v_1 v_2\} \\ &= P_{v_1} + P_{v_2} + 2\mathbb{E}\{v_1 v_2\}. \end{aligned} \tag{4.56}$$

The last term of this equation is just the correlation between the two signals with no time offset. Thus, we can define a correlation coefficient, $C_{(v_1,v_2)}$, between $v_1(t)$ and $v_2(t)$ given by

$$C_{(v_1,v_2)} = \frac{\mathbb{E}\{v_1 v_2\}}{\sqrt{P_{v_1} P_{v_2}}} = \frac{\mathbb{E}\{v_1 v_2\}}{\sqrt{\mathbb{E}\{v_1^2\}\mathbb{E}\{v_2^2\}}}. \tag{4.57}$$

This factor can be used to rewrite equation (4.56) as

$$P_v = P_{v_1} + P_{v_2} + 2C_{(v_1,v_2)}\sqrt{P_{v_1} P_{v_2}}. \tag{4.58}$$

In practice, $C_{(v_1,v_2)}$ varies between -1 and $+1$, the extreme values corresponding to perfect correlation between the processes [3]. For example, when we have perfect correlation and in-phase processes, i.e. when $C_{(v_1,v_2)} = 1$, the above relationship reduces to

$$\sqrt{P_v} = \sqrt{P_{v_1}} + \sqrt{P_{v_2}}. \tag{4.59}$$

Thus, the standard deviations of the initial voltage processes sum together. In the particular case where $P_{v_1} = P_{v_2}$, the power of the sum is therefore equal to four times that of each process. In contrast, when the two voltages are perfectly correlated but in opposite phase, we get $C_{(v_1,v_2)} = -1$, which leads to

$$\sqrt{P_v} = \sqrt{P_{v_1}} - \sqrt{P_{v_2}}. \tag{4.60}$$

If $P_{v_1} = P_{v_2}$, the amplitudes of the processes involved thus cancel each other so that the power of the sum is null. Finally, in the situation where the voltages are uncorrelated, i.e. $C_{(v_1,v_2)} = 0$, we have

$$P_v = P_{v_1} + P_{v_2}. \tag{4.61}$$

This is now the power of the two signals that sum together so that when $P_{v_1} = P_{v_2}$, the power of the sum of the voltage processes is now twice the initial power.

Supposing now that we are dealing with bandpass noise processes, we may want to transpose this correlation concept to their associated complex envelopes in order to be able to perform simple electrical derivations for spot quantities, as discussed in the previous section. Further insights can be found in Appendix 1, but here we follow the same simple approach as above in order to discuss the transposition of the correlation coefficient to complex envelopes. Assume for now that the two spot noise voltages considered above are described by their complex envelopes defined around the same center frequency, $f_{\rm n} = \omega_{\rm n}/2\pi$, assumed to be in the narrow frequency band δf. We can thus write

$$v_1(t) = \mathrm{Re}\left\{\tilde{v}_1(t)e^{j\omega_{\rm n}t}\right\}, \quad \text{with } \tilde{v}_1(t) = \rho_1(t)e^{j\phi_1(t)}, \tag{4.62a}$$

$$v_2(t) = \mathrm{Re}\left\{\tilde{v}_2(t)e^{j\omega_{\rm n}t}\right\}, \quad \text{with } \tilde{v}_2(t) = \rho_2(t)e^{j\phi_2(t)}. \tag{4.62b}$$

As a result,

$$v(t) = v_1(t) + v_2(t) = \mathrm{Re}\left\{(\tilde{v}_1(t) + \tilde{v}_2(t))e^{j\omega_{\rm n}t}\right\}. \tag{4.63}$$

Assuming the same center frequency for all the complex envelopes of interest, we can thus write the complex envelope of the sum as the sum of the complex envelopes:

$$\tilde{v}(t) = \tilde{v}_1(t) + \tilde{v}_2(t). \tag{4.64}$$

Now considering the stationarity of the considered processes, we can directly write from equation (1.64) that

$$\begin{aligned} P_v &= \frac{\mathbb{E}\{|\tilde{v}_1 + \tilde{v}_2|^2\}}{2} \\ &= \frac{\mathbb{E}\{|\tilde{v}_1|^2\}}{2} + \frac{\mathbb{E}\{|\tilde{v}_2|^2\}}{2} + \frac{\mathbb{E}\{\tilde{v}_1\tilde{v}_2^* + \tilde{v}_1^*\tilde{v}_2\}}{2} \\ &= P_{v_1} + P_{v_2} + \mathrm{Re}\{\mathbb{E}\{\tilde{v}_1\tilde{v}_2^*\}\}. \end{aligned} \tag{4.65}$$

We thus see that the powers of the two processes sum together if and only if $\mathrm{Re}\{\mathbb{E}\{\tilde{v}_1\tilde{v}_2^*\}\} = 0$. By equation (A1.27), this is indeed the condition that complex envelopes, assumed defined as centered around the same center frequency, must fulfill for the bandpass processes they represent to be uncorrelated. Thus, a logical extension of the correlation coefficient using complex envelopes as defined by equation (4.57) would be to have it proportional to $\mathrm{Re}\{\mathbb{E}\{\tilde{v}_1\tilde{v}_2^*\}\}$. This is in fact not what is classically done in noise derivations. The reason is that when dealing with noise sources that are stationary, as is the case in practice in transceivers as discussed in Appendix 2, we get that equation (A1.31) holds. There is thus an equivalence between the non-correlation of the bandpass processes, and $\mathbb{E}\{\tilde{v}_1\tilde{v}_2^*\}$ being null. Thus, what is defined as the correlation coefficient between the bandpass processes $v_1(t)$ and $v_2(t)$ is classically transposed for their complex envelopes, defined as centered around the same carrier frequency, as

$$C_{(\tilde{v}_1,\tilde{v}_2)} = \frac{\mathbb{E}\{\tilde{v}_1\tilde{v}_2^*\}}{\sqrt{\mathbb{E}\{|\tilde{v}_1|^2\}\mathbb{E}\{|\tilde{v}_2|^2\}}} \tag{4.66}$$

In order to illustrate the use of this correlation coefficient, we can now apply it to our present case of interest, i.e. to the equivalent input-referred noise sources for a quadrupole. We retain the notation of Figure 4.7 and denote by $v_{\mathrm{n}}(t)$ the equivalent input spot noise voltage and by $i_{\mathrm{n}}(t)$ the equivalent input spot noise current of the device. Given that these two noise sources are correlated in the general case, $v_{\mathrm{n}}(t)$ is classically decomposed into two terms, $v_{\mathrm{c}}(t)$ and $v_{\mathrm{u}}(t)$, such that $v_{\mathrm{u}}(t)$ is uncorrelated with $i_{\mathrm{n}}(t)$ while $v_{\mathrm{c}}(t)$ is perfectly correlated with it. Assuming as usual that we are dealing with complex envelopes defined as centered around the same center frequency, we can then transpose

$$v_{\mathrm{n}}(t) = v_{\mathrm{u}}(t) + v_{\mathrm{c}}(t) \tag{4.67}$$

as

$$\tilde{v}_{\mathrm{n}}(t) = \tilde{v}_{\mathrm{u}}(t) + \tilde{v}_{\mathrm{c}}(t). \tag{4.68}$$

Due to the equivalence in the correlation behavior between stationary bandpass signals and their complex envelopes as recalled above, $\tilde{v}_{\mathrm{u}}(t)$ is then defined as uncorrelated with $\tilde{\imath}_{\mathrm{n}}(t)$ while $\tilde{v}_{\mathrm{c}}(t)$ is perfectly correlated with it. As a result, the random variable $\tilde{v}_{\mathrm{c},t}$ can be written as proportional to $\tilde{\imath}_{\mathrm{n},t}$ with probability one. As justified below, under stationarity we can in fact go a step further and suppose in our decomposition that we can express $\tilde{v}_{\mathrm{c}}(t)$ as proportional to $\tilde{\imath}_{\mathrm{n}}(t)$ through the complex factor $Z_{\mathrm{c}} = R_{\mathrm{c}} + \mathrm{j}X_{\mathrm{c}}$, i.e. that

$$\tilde{v}_{\mathrm{c}}(t) = Z_{\mathrm{c}}\tilde{\imath}_{\mathrm{n}}(t). \tag{4.69}$$

The term Z_{c} is called the correlation impedance of the noise sources $v_{\mathrm{n}}(t)$ and $i_{\mathrm{n}}(t)$. By our assumptions so far, we can finally express $\tilde{v}_{\mathrm{n}}(t)$ as

$$\tilde{v}_{\mathrm{n}}(t) = \tilde{v}_{\mathrm{u}}(t) + Z_{\mathrm{c}}\tilde{\imath}_{\mathrm{n}}(t). \tag{4.70}$$

However, due to its definition we can suspect that Z_{c} can be expressed as a function of the statistical properties of the noise sources. This can indeed be done in a straightforward way by multiplying each side of the above equation by $\tilde{\imath}_{\mathrm{n}}^{*}$ and taking the expectation of the result. Due to the assumed non-correlation between $\tilde{v}_{\mathrm{u}}(t)$ and $\tilde{\imath}_{\mathrm{n}}(t)$, this leads to

$$Z_{\mathrm{c}} = \frac{\mathbb{E}\{\tilde{v}_{\mathrm{n}}\tilde{\imath}_{\mathrm{n}}^{*}\}}{\mathbb{E}\{|\tilde{\imath}_{\mathrm{n}}|^2\}}. \tag{4.71}$$

With this expression, it is easy to check that $\tilde{v}_{\mathrm{u}}(t) = \tilde{v}_{\mathrm{n}}(t) - Z_{\mathrm{c}}\tilde{\imath}_{\mathrm{n}}(t)$ is uncorrelated with $\tilde{\imath}_{\mathrm{n}}(t)$. Indeed, multiplying each side of this equation by $\tilde{\imath}_{\mathrm{n}}^{*}(t)$ and taking expectations results in $\mathbb{E}\{\tilde{v}_{\mathrm{u}}\tilde{\imath}_{\mathrm{n}}^{*}\} = \mathbb{E}\{\tilde{v}_{\mathrm{n}}\tilde{\imath}_{\mathrm{n}}^{*}\} - Z_{\mathrm{c}}\mathbb{E}\{|\tilde{\imath}_{\mathrm{n}}|^2\}$. Using equation (4.71), we then immediately see that $\mathbb{E}\{\tilde{v}_{\mathrm{u}}\tilde{\imath}_{\mathrm{n}}^{*}\} = 0$. This obviously justifies our decomposition and the introduction of Z_{c}, constant in time under our assumptions.

It is also interesting to observe that all the quantities introduced up to now can be fully expressed as a function of only the RMS values of the two original spot noise sources $v_{\mathrm{n}}(t)$ and

$i_n(t)$, which is the main metric of importance for them, as well as their correlation coefficient. For instance, using equation (4.66), we can rewrite the expression for Z_c as

$$Z_c = C_{(\tilde{v}_n,\tilde{\imath}_n)}\sqrt{\frac{\mathbb{E}\{|\tilde{v}_n|^2\}}{\mathbb{E}\{|\tilde{\imath}_n|^2\}}} = C_{(\tilde{v}_n,\tilde{\imath}_n)}\sqrt{\frac{\overline{v_n^2}}{\overline{i_n^2}}}. \tag{4.72}$$

Based on this expression, the real part R_c of this impedance reduces to

$$R_c = \mathrm{Re}\left\{C_{(\tilde{v}_n,\tilde{\imath}_n)}\right\}\sqrt{\frac{\overline{v_n^2}}{\overline{i_n^2}}}. \tag{4.73}$$

Written like this, we observe that the sign of R_c depends on the original sign of the real part of the correlation coefficient $C_{(\tilde{v}_n,\tilde{\imath}_n)}$. It can thus be either positive or negative. The physical meaning is that if R_c were a physical resistor, the voltage $v_c(t)$ applied to it and the current $i_n(t)$ flowing through it would lead to electrical noise power delivered or retrieved from this load depending on its sign. In the same way, the imaginary part X_c of the correlation impedance takes the form

$$X_c = \mathrm{Im}\left\{C_{(\tilde{v}_n,\tilde{\imath}_n)}\right\}\sqrt{\frac{\overline{v_n^2}}{\overline{i_n^2}}}. \tag{4.74}$$

Here again, if Z_c were a physical impedance the sign of its imaginary part, and thus of the imaginary part of the correlation coefficient between the complex envelopes, would indicate whether we were dealing with an inductive or capacitive behavior. We can also express in the same way the RMS values of $\tilde{v}_u(t)$ and $\tilde{v}_c(t)$, still based on the assumption of ergodicity and stationarity. Indeed, based on equation (4.69) we can write that

$$|\tilde{v}_c(t)|^2 = |Z_c|^2|\tilde{\imath}_n(t)|^2. \tag{4.75}$$

Hence,

$$\mathbb{E}\{|\tilde{v}_c|^2\} = |Z_c|^2\mathbb{E}\{|\tilde{\imath}_n|^2\}. \tag{4.76}$$

Then, using equation (4.72) we finally get

$$\overline{v_c^2} = \left|C_{(\tilde{v}_n,\tilde{\imath}_n)}\right|^2\overline{v_n^2}. \tag{4.77}$$

Due to the non-correlation between $\tilde{v}_u(t)$ and $\tilde{v}_c(t)$, we can also write from equation (4.68) that

$$\mathbb{E}\{|\tilde{v}_n|^2\} = \mathbb{E}\{|\tilde{v}_u|^2\} + \mathbb{E}\{|\tilde{v}_c|^2\}. \tag{4.78}$$

That finally yields

$$\overline{v_u^2} = \left(1 - \left|C_{(\tilde{v}_n,\tilde{i}_n)}\right|^2\right)\overline{v_n^2}. \tag{4.79}$$

In conclusion, the spot noise voltage and current sources, $v_n(t)$ and $i_n(t)$, are classically associated with equivalent noisy impedances. Referring to the derivations performed in the previous section, we can write the RMS noise voltage $\overline{v_n^2}$ as that delivered by a serial noisy resistor considered at a given thermodynamic temperature T, following equation (4.50). In the present case, we can therefore introduce a formal resistor R_n defined by

$$\overline{v_n^2} = 4kTR_n\,\delta f. \tag{4.80}$$

In the same way we can define from equation (4.51) an equivalent shunt conductance G_n for the spot current noise source by

$$\overline{i_n^2} = 4kTG_n\,\delta f. \tag{4.81}$$

The introduction of these resistors is useful for analytical derivations. An illustration of their use is given for instance in "Noise factor dependency on generator impedance" (Section 4.2.5). However, as $\overline{v_n^2}$ and $\overline{i_n^2}$ are intrinsic characteristics of the quadrupole, the value of R_n or G_n then necessarily depends on the noise temperature T used for their definition. We thus need to keep in mind that those quantities are in turn not intrinsic characteristics of the device, but rather a means of characterization suitable for analytical derivations.

4.2.4 *Cascade of Noisy Devices*

Let us now focus on how to predict the overall noise performance of a cascade of noisy devices. As done since the beginning of the chapter, here again only the intrinsic noise contributions of devices are considered. This means that we are dealing with *additive* noise contributions only, in contrast to the behavior of multiplicative and distortion noises linked to the presence of the wanted signal in the device. This additive behavior is of great importance as it is the root cause for the possibility of minimizing the impact of the noise contributions of subsequent stages with respect to that of earlier stages in a line-up as long as the gain of those earlier stages is sufficiently high. It is important to understand the following reasoning as it forms the basis of most of the guidelines for the dimensioning of transceivers from the SNR optimization perspective. However, it should be kept in mind that additional guidelines are needed when considering multiplicative and distortion noise contributions, as illustrated in Chapter 7.

Let us suppose that the devices under consideration are characterized from a noise perspective by their effective noise temperature. As discussed earlier in this section, this noise temperature preserves the additive behavior of the intrinsic noise contributions, which is not the case for the noise factor, for instance. As a result, we suppose that we are dealing with a cascade of m noisy devices, each one characterized by

(i) its available power gain, $G_{\mathrm{av}(Z_{\mathrm{o},k-1}),k}$;
(ii) its effective input noise temperature $T_{\mathrm{e}(Z_{\mathrm{o},k-1}),k}$.

In the present case, we assume that we are dealing with spot noise quantities in a frequency band δf sufficiently narrow that each quantity can be assumed constant. As highlighted by the notation used, we still need to clarify the impedances used to refer those characteristic quantities to. Indeed, as discussed in Section 4.2.1, both the available power gain and the effective noise temperature of a device depend on the internal impedance of the generator that is used for its characterization. In the present case, it is of interest to assume that the impedance used to refer the characteristic quantities of the kth quadrupole effectively corresponds to the output impedance $Z_{\mathrm{o},k-1}$ seen from the earlier part of the line-up connected at its input. Under those assumptions, the cascade of m noisy devices shown in Figure 4.8 can be considered as a single equivalent system with an overall available power gain G_{av} equal to the product $G_{\mathrm{av}(Z_{\mathrm{g}}),1}G_{\mathrm{av}(Z_{\mathrm{o},1}),2}\cdots G_{\mathrm{av}(Z_{\mathrm{o},m-1}),m}$ and with an effective input noise temperature $T_{\mathrm{e}(Z_{\mathrm{g}})}$. We can then derive an expression for $T_{\mathrm{e}(Z_{\mathrm{g}})}$ as a function of the characteristics of the devices involved in the line-up.

From that perspective, let us give an expression for the available noise power $P_{\mathrm{av,o},1}$ recovered in the frequency band δf at the output of the first stage in the line-up. Assuming that the

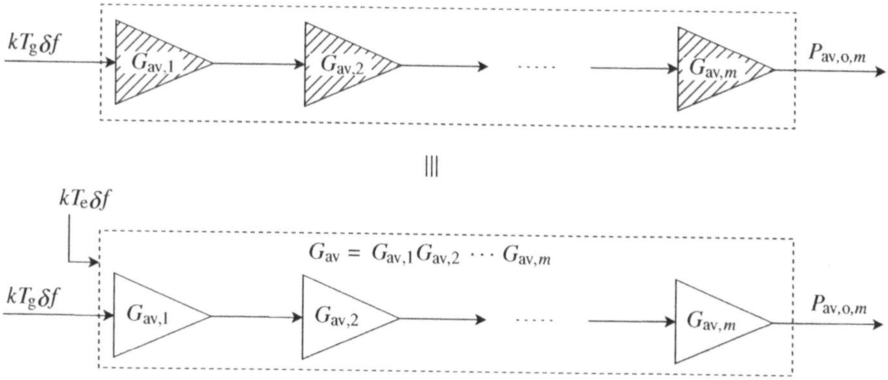

Figure 4.8 Input effective noise temperature of a cascade of noisy quadrupoles – A cascade of noisy quadrupoles, each one characterized at a given frequency by its available power gain and effective noise temperature defined as referred to an impedance equal to the output impedance of the preceding stage, behaves like a single noisy device with an available power gain equal to the product of the individual one and with an effective noise temperature given by equation (4.86). Here, the dependency of the available power gains and noise temperature on the input stage output impedance is not explicitly written.

source noise temperature connected to the system input is T_g, we can immediately write from equation (4.11) that

$$P_{av,o,1} = G_{av(Z_g),1}\, k\left(T_g + T_{e(Z_g),1}\right)\delta f. \tag{4.82}$$

Now taking into account the noise contribution of the second stage, we can express the available noise power delivered by the system composed of the two first devices. Dealing with input-referred quantities for the noise temperatures, we can first express this total noise power as referred to the input of this second stage. This leads to the quantity

$$\begin{aligned} P_{av,i,2} &= P_{av,o,1} + kT_{e(Z_{o,1}),2}\,\delta f \\ &= G_{av(Z_g),1}\, k\left(T_g + T_{e(Z_g),1}\right)\delta f + kT_{e(Z_{o,1}),2}\,\delta f. \end{aligned} \tag{4.83}$$

This expression already illustrates the behavior we are looking for. It can be seen even more clearly when assuming that $T_{e(Z_g),1} \approx T_{e(Z_{o,1}),2}$, i.e. that the two devices have comparable intrinsic noise performance. Then the contribution of the first stage in the overall noise power, $G_{av(Z_g),1}\, kT_{e(Z_g),1}\delta f$, is higher than the contribution of the second stage by an amount equal to $G_{av(Z_g),1}$. The contribution of the second stage can be made as negligible as we want by amplifying sufficiently the contribution of the first stage. This behavior, illustrated in Figure 4.9, obviously holds because we are dealing with additive noise contributions. Their absolute level is therefore totally independent of the level of the signal entering the device.

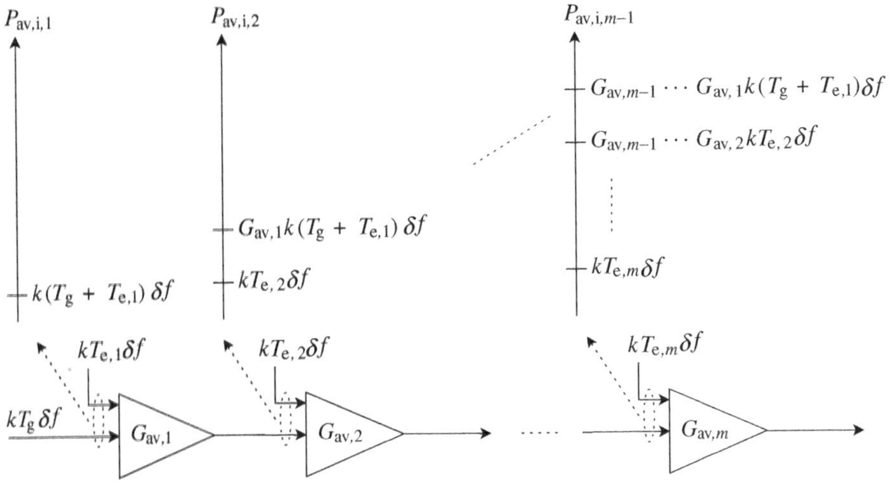

Figure 4.9 Relative noise contributions of each stage of a cascade of noisy quadrupoles – The fact that the noise coming from earlier stages has been amplified before the addition of the noise component of a given device in the line-up makes the overall noise performance mainly linked to the noise characteristics of the first stage according to equation (4.86). Here, the dependency of the available power gains and noise temperature on the input stage output impedance is not explicitly written.

This signal level can thus be amplified sufficiently in earlier stages of the line-up in order to have a much higher power than the intrinsic noise contribution of the device they are entering. As highlighted at the beginning of the section, the reasoning would of course be different for multiplicative distortion noises.

Now iterating that process, we can express the available noise power, $P_{\text{av,o},m}$, recovered at the output of the system as

$$\begin{aligned} P_{\text{av,o},m} = {} & G_{\text{av}(Z_\text{g}),1} G_{\text{av}(Z_{\text{o},1}),2} \cdots G_{\text{av}(Z_{\text{o},m-1}),m}\, k\left(T_\text{g} + T_{\text{e}(Z_\text{g}),1}\right) \delta f \\ & + G_{\text{av}(Z_{\text{o},1}),2} \cdots G_{\text{av}(Z_{\text{o},m-1}),m}\, kT_{\text{e}(Z_{\text{o},1}),2}\, \delta f \\ & + \cdots + G_{\text{av}(Z_{\text{o},m-1}),m}\, kT_{\text{e}(Z_{\text{o},m-1}),m}\, \delta f. \end{aligned} \tag{4.84}$$

Alternatively, considering the overall system as a single device with an available power gain of $G_{\text{av}(Z_\text{g}),1} G_{\text{av}(Z_{\text{o},1}),2} \cdots G_{\text{av}(Z_{\text{o},m-1}),m}$ and an effective noise temperature of $T_{\text{e}(Z_\text{g})}$, we can use equation (4.11) to directly express $P_{\text{av,o},m}$ as

$$P_{\text{av,o},m} = G_{\text{av}(Z_\text{g}),1} G_{\text{av}(Z_{\text{o},1}),2} \cdots G_{\text{av}(Z_{\text{o},m-1}),m}\, k\left(T_\text{g} + T_{\text{e}(Z_\text{g})}\right) \delta f. \tag{4.85}$$

Direct comparison of the two expressions above therefore leads to

$$\begin{aligned} T_{\text{e}(Z_\text{g})} = {} & T_{\text{e}(Z_\text{g}),1} + \frac{T_{\text{e}(Z_{\text{o},1}),2}}{G_{\text{av}(Z_\text{g}),1}} + \frac{T_{\text{e}(Z_{\text{o},2}),3}}{G_{\text{av}(Z_\text{g}),1} G_{\text{av}(Z_{\text{o},1}),2}} \\ & + \cdots + \frac{T_{\text{e}(Z_{\text{o},m-1}),m}}{G_{\text{av}(Z_\text{g}),1} G_{\text{av}(Z_{\text{o},1}),2} \cdots G_{\text{av}(Z_{\text{o},m-2}),m-1}}. \end{aligned} \tag{4.86}$$

Written like this, we see that each term on the right-hand side of the above equation represents the noise contribution of the successive stages in the line-up, but expressed as an equivalent input quantity. We thus recover in another way that the contribution of a given stage can be made negligible when having enough amplification in the earlier stages. As a consequence, if the available power gain of the first device, which is the only one involved in all terms of this expression, is sufficiently high, the effective input noise temperature of the overall system almost reduces to that of the first stage, i.e.

$$T_{\text{e}(Z_\text{g})} \approx T_{\text{e}(Z_\text{g}),1}. \tag{4.87}$$

Alternatively, we may want to find an expression for the noise factor of the system. Although it hides the additive behavior of intrinsic noise contributions of devices, the concept of noise factor is often used for convenience, as discussed in Section 4.2.6. In that perspective, we can in fact directly use the relationship between this quantity and the effective noise temperature given by equation (4.27). However, as usual when dealing with noise factors, we need to be careful in the source temperature that is used to refer the noise factor to, on top of the source impedance problem common with the effective noise temperature. In a theoretical case where the kth device in the cascade is characterized by a noise factor $F_{(Z_{o,k-1},T_k),k}$ referred to a source noise temperature T_k, the equivalent noise factor of the system when referred to the noise temperature T_i, $F_{(Z_g,T_i)}$, can be expressed from equation (4.86) as

$$T_i\left(F_{(Z_g,T_i)}-1\right) = T_1\left(F_{(Z_g,T_1),1}-1\right) + \frac{T_2\left(F_{(Z_{o,1},T_2),2}-1\right)}{G_{av(Z_g),1}} + \cdots + \frac{T_m\left(F_{(Z_{o,m-1},T_m),m}-1\right)}{G_{av(Z_g),1}G_{av(Z_{o,1}),2}\cdots G_{av(Z_{o,m-2}),m-1}}. \tag{4.88}$$

In the particular case where all the noise factors in the line-up are referred to the same standardized temperature T_0, this relationship reduces to the classical Friis formula for noise [41]:

$$F_{(Z_g,T_0)} = F_{(Z_g,T_0),1} + \frac{F_{(Z_{o,1},T_0),2}-1}{G_{av(Z_g),1}} + \cdots + \frac{F_{(Z_{o,m-1},T_0),m}-1}{G_{av(Z_g),1}G_{av(Z_{o,1}),2}\cdots G_{av(Z_{o,m-2}),m-1}}. \tag{4.89}$$

We therefore recover the fact that if the available power gain of the first device is sufficiently high, the contribution of the subsequent stages can be made negligible so that the noise factor of the system almost reduces to that of the first stage, i.e.

$$F_{(Z_g,T_0)} \approx F_{(Z_g,T_0),1}. \tag{4.90}$$

In conclusion, we have considered here effective noise temperatures or noise factors for all the devices involved. But, in practice, this kind of quantity is used mainly for RF devices, whereas in transceivers the devices act partly on baseband signals. In this last case, equivalent voltage or current noise sources are often used, as introduced in Section 4.2.3. Practically speaking, we thus often deal with line-ups in which part of the devices are characterized by RF quantities, i.e. noise temperatures and noise factors, and part by baseband quantities, i.e. noise voltage and currents. Thus, although it does not change the conclusions on the possibility of minimizing some contributions using the gain of previous stages, analytical derivations have to be performed in a different way. An example of this is given in "RF and baseband devices" Section 4.2.5.

4.2.5 *Illustration*

Passive Front-end

In order to illustrate the concepts introduced so far, let us consider first the case depicted in Figure 4.10, a receiver whose active part, here depicted as a radio frequency integrated circuit

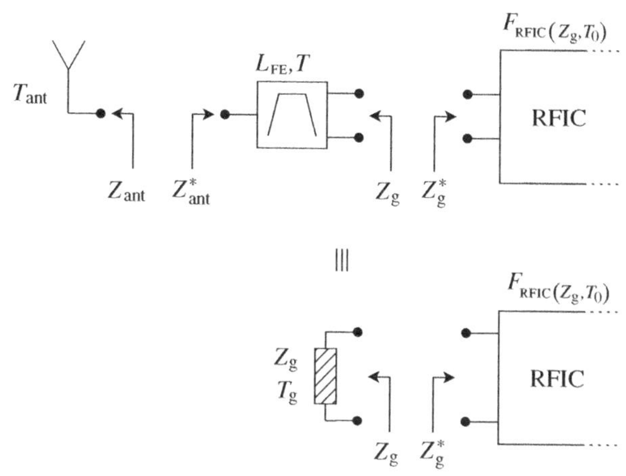

Figure 4.10 Receiver configuration with a passive front-end between the antenna and the RFIC – Assuming the impedance matching in power of the devices, the equivalent noise source temperature T_g seen by the RFIC can easily be derived as a function of the passive FE insertion loss L_{FE}, its physical thermodynamic temperature T, and the source noise temperature at the antenna connector T_{ant}. The resulting noise source temperature T_g is given by equation (4.100).

(RFIC), is directly connected to the antenna through a passive FE. In this simple configuration, the passive FE can be, for instance, a RF filter dedicated to the attenuation of out-of-band blocking signals. We can then focus on the effective noise temperature and on the noise factor of this system as seen from the antenna connector.

We first use equation (4.86) to derive an expression for the effective noise temperature of the receive path, $T_{e,RX(Z_{ant})}$, referred here to the impedance seen from the antenna connector. We obtain

$$T_{e,RX(Z_{ant})} = T_{e(Z_{ant}),1} + \frac{T_{e(Z_g),2}}{G_{av(Z_{ant}),1}}. \tag{4.91}$$

In this expression, $T_{e(Z_{ant}),1}$ stands for the effective input noise temperature of the passive FE, $G_{av(Z_{ant}),1}$ for its available power gain, and $T_{e(Z_g),2}$ for the effective input noise temperature of the RFIC. Let us assume that the devices under consideration are matched in power. This is indeed what mostly happens in practice, as discussed in Section 2.2.3. This allows us to express the effective noise temperature of the passive FE as a function of its insertion loss L_{FE} and physical thermodynamic temperature T. From equation (4.19) we can write that

$$T_{e(Z_{ant}),1} = T(L_{FE} - 1). \tag{4.92}$$

In the same way, $G_{av(Z_{ant}),1}$ reduces to the inverse of its insertion loss $1/L_{FE}$. Finally, assuming that the RFIC is characterized by its noise factor $F_{RFIC(Z_g,T_0)}$, defined as referred to the impedance physically connected to its input Z_g and to the standard source temperature T_0 as classically done, we can use equation (4.27) to express its effective input noise temperature as

$$T_{e(Z_g),2} = T_0\left(F_{RFIC(Z_g,T_0)} - 1\right). \tag{4.93}$$

Substituting these expressions into equation (4.91), we can thus now express the effective input noise temperature of the overall receiver as seen from the antenna connector as

$$T_{e,RX(Z_{ant})} = T(L_{FE} - 1) + L_{FE}T_0\left(F_{RFIC(Z_g,T_0)} - 1\right). \tag{4.94}$$

Obviously, referring to the relationship between noise factor and effective noise temperature as given by equation (4.27), we could directly derive an expression for the noise factor of the overall receiver $F_{RX(Z_{ant},T_0)}$ based on the above equation. Following our notation, we assume that this noise factor is defined as referred to the impedance seen from the antenna connector and to the standard source temperature T_0. However, this quantity can also be derived using the transposition for noise factor of equation (4.91), i.e. the Friis formula given by equation (4.89). Direct application of this expression leads to

$$F_{RX(Z_{ant},T_0)} = F_{(Z_{ant},T_0),1} + \frac{F_{(Z_g,T_0),2} - 1}{G_{av(Z_{ant}),1}}. \tag{4.95}$$

Here $F_{(Z_{ant},T_0),1}$ stands for the noise factor of the passive FE under matched conditions and can be derived from its insertion loss and physical temperature through equation (4.32) as

$$F_{(Z_{ant},T_0),1} = 1 + \frac{T}{T_0}(L_{FE} - 1). \tag{4.96}$$

As $F_{(Z_g,T_0),2}$ is no more than the RFIC noise factor, it follows that

$$F_{RX(Z_{ant},T_0)} = 1 + \frac{T}{T_0}(L_{FE} - 1) + L_{FE}\left(F_{RFIC(Z_g,T_0)} - 1\right). \tag{4.97}$$

As expected, we effectively recover that the receiver noise factor and its effective noise temperature are linked to each other by equation (4.27). But what is interesting to see is how the above expressions for both the system effective temperature and noise factor simplify when the thermodynamic temperature of the passive device T is equal to the noise temperature T_0 used to refer the RFIC noise factor to. Indeed, setting $T = T_0$ in equations (4.94) and (4.97) leads to

$$T_{e,RX(Z_{ant})} = T_0\left(L_{FE}F_{RFIC(Z_g,T_0)} - 1\right), \tag{4.98a}$$

$$F_{RX(Z_{ant},T_0)} = L_{FE}F_{RFIC(Z_g,T_0)}. \tag{4.98b}$$

Only in this particular case do we get that the noise factor of the receiver reduces to the product of the passive device insertion loss and the noise factor of the active part of the line-up. This point can be related to the discussion in Section 4.2.6 about the use of the noise factor as a metric for SNR degradation.

A final quantity of interest for deriving RFIC budgets is the source noise temperature T_g seen at its input. Indeed, it is often convenient to work with signal levels referred to the RFIC input in order to derive link budgets. As also discussed in Section 4.2.6, this requires knowledge of the noise power delivered by the generator, i.e. the source noise temperature seen by the RFIC. In the present case, this derivation is straightforward as this source noise power is simply equal to the sum of that delivered by the antenna and the equivalent input noise power of the passive FE times its available power gain. Once transposed to equivalent noise temperatures, we get that

$$T_g = G_{av(Z_{ant}),1}\left(T_{ant} + T_{e(Z_{ant}),1}\right). \tag{4.99}$$

Given on the one hand that $G_{av(Z_{ant}),1}$ is equal to the inverse of the FE insertion loss $1/L_{FE}$ with our matching assumption, and on the other hand that equation (4.92) holds, we finally get that

$$T_g = T + \frac{T_{ant} - T}{L_{FE}}. \tag{4.100}$$

We remark that it is only when $T = T_{ant}$ that the RFIC sees an input source noise temperature that is equal to the physical temperature of the application board.

Active Front-end

Let us now consider the case of a receiver in which a passive device is used in between a first active device and the rest of the active part of the line-up. A shown in Figure 4.11, this active device can be an external LNA used to optimize the overall noise performance of the receiver, whereas the remainder of the active part of the line-up can be an RFIC in itself, for instance.

In order to work out noise budgets for the RFIC part, it is necessary to have access to the generator noise temperature T_g it sees. For this purpose, we need to cascade the contributions of all the devices from the antenna connector up to its input. We start by determining the generator noise temperature seen by the LNA, $T_{g,LNA}$. This noise temperature is in fact equal to the sum of the noise temperature of the antenna and that of the first passive FE component, this sum being multiplied by the available power gain of this passive component. This is the same as the situation discussed in the previous section. As a result, still assuming that the devices under consideration are matched in power, we can refer to equation (4.100) to express $T_{g,LNA}$ as

$$T_{g,LNA} = T + \frac{T_{ant} - T}{L_{FE\,I}}. \tag{4.101}$$

We can now add the intrinsic contribution of the LNA. Given that this stage is classically characterized by its noise factor, $F_{LNA(Z_i^*,T_0)}$, referred to the standard temperature T_0 and to its

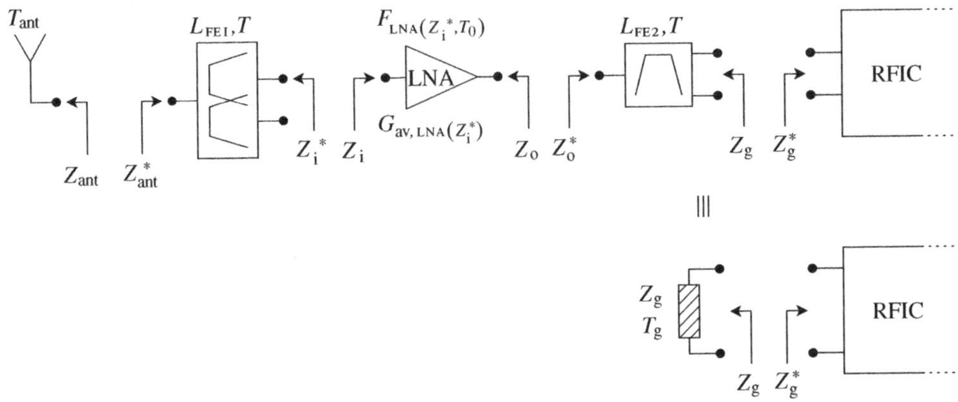

Figure 4.11 Receiver configuration with an active front-end between the antenna and the RFIC – Assuming the impedance matching in power of the devices, the equivalent noise source temperature T_g seen by the RFIC in the case of an active FE can be derived as a function of the FE characteristics and of the physical temperature of the devices. The resulting noise source temperature T_g is given by equation (4.103).

power matched input impedance Z_i^*, we can use equation (4.27) to derive its effective input noise temperature. The generator noise temperature $T_{g,FE2}$ seen by the second passive device can then be written as the sum of the generator noise temperature seen by the LNA, $T_{g,LNA}$, and the effective input noise temperature of the LNA, this sum being multiplied by the available power gain of this stage, $G_{av,LNA(Z_i^*)}$. We obtain

$$T_{g,FE2} = G_{av,LNA(Z_i^*)} \left[T + \frac{T_{ant} - T}{L_{FE1}} + T_0 \left(F_{LNA(Z_i^*,T_0)} - 1 \right) \right]. \tag{4.102}$$

It then remains to take into account the effective input noise temperature of the second passive device. This is in fact the same derivation as carried out above with the first passive device, but now with $T_{g,FE2}$ as the generator noise temperature. We finally get that the source noise temperature T_g seen by the RFIC can be expressed as

$$T_g = \frac{G_{av,LNA(Z_i^*)}}{L_{FE2}} \left[T + \frac{T_{ant} - T}{L_{FE1}} + T_0 \left(F_{LNA(Z_i^*,T_0)} - 1 \right) \right] + T \left(1 - \frac{1}{L_{FE2}} \right). \tag{4.103}$$

In contrast to what occurs in the pure passive FE case, we now see that even with $T_{ant} = T = T_0$, the source noise temperature seen by the RFIC is a complex function of the FE parameters. We can thus understand that there is little chance that $T_g = T_0$ in practice. It follows that the RFIC noise factor referred to T_0 cannot be interpreted as directly representative of the SNR degradation when going through this device, as discussed in Section 4.2.6.

RF and Baseband Devices

Let us now take an example that illustrates the management of blocks of different type in the active part of a receiver. Referring to the discussion in Section 4.2.3, we get that the noise performance of active devices is classically not characterized in the same way when working on RF signals and when not. Indeed, whereas power matching is often required in the RF world, it is often sufficient in terms of propagation of the information to process either only voltage or only current waves in baseband. As a result, noise temperatures or noise factors are mainly encountered in practice to characterize matched RF devices, whereas noise voltage or noise current sources are often preferred for unmatched baseband devices.

Let us consider the configuration shown in Figure 4.12 where the active part of a receiver, here taken as an RFIC, is considered as the cascade of an active receiver front-end (RXFE) and of an analog baseband (ABB). The noise performance of the RXFE block is supposed to be characterized by its effective input noise temperature $T_{e,\mathrm{RXFE}(Z_g)}$, or equivalently by its noise factor $F_{\mathrm{RXFE}(Z_g,T_0)}$, here defined as referred to the standard source noise temperature T_0. These two quantities are also both referred to the impedance Z_g that effectively corresponds to the internal impedance of the dipole connected at the RXFE input. Conversely, the ABB is supposed to work as a voltage gain amplifier. Its input impedance is therefore assumed much higher than the output impedance of the RXFE block. To simplify, we can assume that this ABB input impedance is infinite and that the output impedance of the RXFE block is null. Referring to the discussion in Section 4.2.3, we can then assume that the noise performance of this block is characterized using only an equivalent input noise voltage generator defined thought its spot noise RMS value in the narrow frequency band δf, $\overline{v^2_{\mathrm{ABB}}}$. However, as discussed in that former section, it is of interest to represent this ABB noise voltage source by its complex envelope, $\tilde{v}_{\mathrm{ABB}}(t)$, assumed defined as centered around a given frequency f_n within δf. This complex envelope, which can be related to $\overline{v^2_{\mathrm{ABB}}}$ through equation (4.48), allows straightforward electrical derivations under the narrowband assumption.

As we may require noise budgets for the total receiver, we may need an equivalent input quantity that characterizes the overall line-up. We thus need to derive an equivalent input-referred quantity for the voltage noise contribution of the ABB. In that perspective, let us suppose that the voltage gain of the RXFE block in its operational configuration, i.e. when an impedance Z_g connected to its input and an open circuit at its output, is G_v. Seen from the receiver output, all behaves as if the ABB were noiseless but with an input voltage present at the RXFE input such that its complex envelope $\tilde{v}_{i,\mathrm{ABB}}(t)$, defined as centered around f_n, is given by

$$\tilde{v}_{i,\mathrm{ABB}}(t) = \frac{\tilde{v}_{\mathrm{ABB}}(t)}{G_v}. \tag{4.104}$$

It remains to derive equivalent input noise quantities referred to the input source impedance Z_g. This means deriving equivalent noise contributions as coming from the source generator

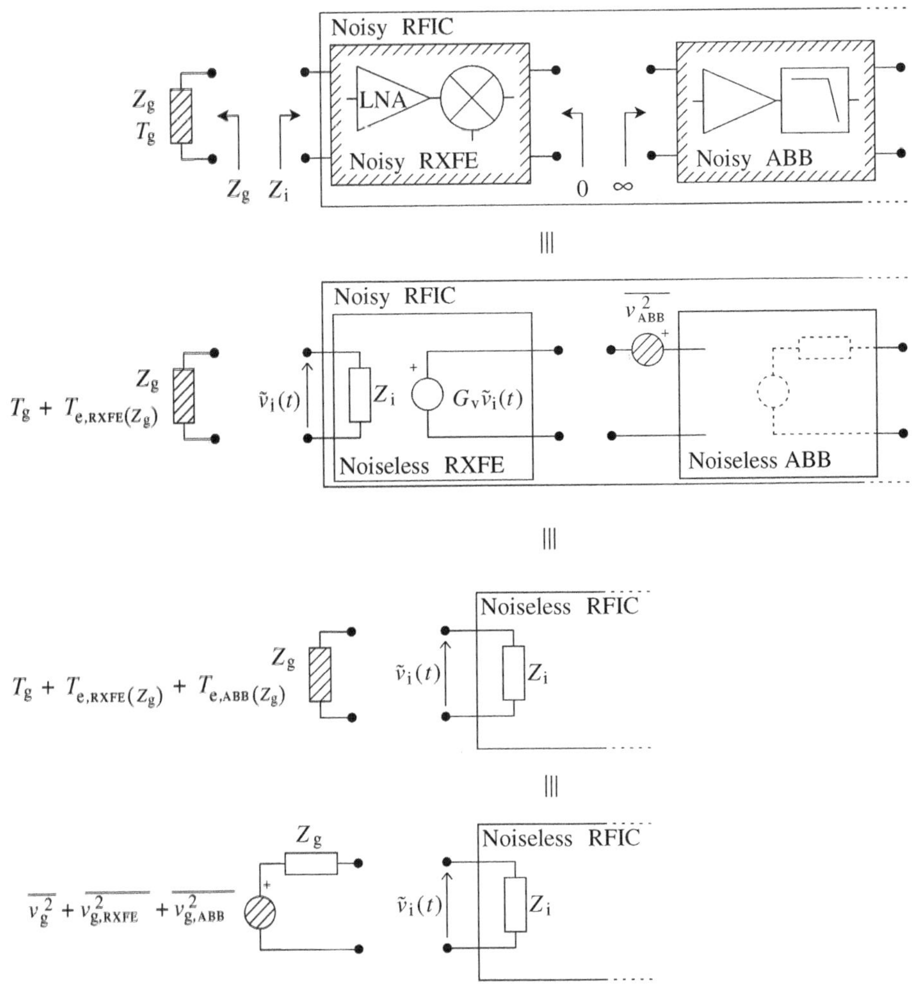

Figure 4.12 RF and baseband noise model and contributions to the overall noise performance of a receiver – As described in Section 4.2.3, the noise performance of active devices is classically not characterized in the same way when working with RF signals and when not. For a general receiver, its RXFE operates on RF signals whereas its analog baseband (ABB) part operates on baseband voltages or currents (top). This means that these blocks exhibit noise performance that is classically characterized in different ways. Noise temperatures or noise factors are used for the RXFE block whereas noise voltage can be used for the ABB block if this stage processes input voltage (middle top). However, equivalent RF input noise quantities can be derived for the ABB noise contributions and thus for the overall receiver (bottom and middle bottom).

connected at the input of the receiver. Thus, referring to the topology shown in Figure 4.12(bottom), we can write that the complex envelope, $\tilde{v}_{\mathrm{g,ABB}}(t)$, of the equivalent source noise voltage that represents the ABB contribution is related to $\tilde{v}_{\mathrm{i,ABB}}(t)$ according to

$$\tilde{v}_{\mathrm{i,ABB}}(t) = \frac{Z_\mathrm{i}}{Z_\mathrm{i} + Z_\mathrm{g}} \tilde{v}_{\mathrm{g,ABB}}(t). \tag{4.105}$$

From the above two equations, we then have that

$$\tilde{v}_{\mathrm{g,ABB}}(t) = \frac{Z_\mathrm{i} + Z_\mathrm{g}}{Z_\mathrm{i}} \frac{\tilde{v}_{\mathrm{ABB}}(t)}{G_\mathrm{v}}. \tag{4.106}$$

Under our assumption, we can then express the RMS value of this equivalent source noise contribution as

$$\overline{v^2_{\mathrm{g,ABB}}} = \frac{|Z_\mathrm{i} + Z_\mathrm{g}|^2}{|Z_\mathrm{i}|^2} \frac{\overline{v^2_{\mathrm{ABB}}}}{G_\mathrm{v}^2}. \tag{4.107}$$

Using equation (4.50), we can therefore define an equivalent noise temperature for the ABB contribution, $T_{\mathrm{e,ABB}(Z_\mathrm{g})}$, as

$$\begin{aligned} kT_{\mathrm{e,ABB}(Z_\mathrm{g})}\,\delta f &= \frac{\overline{v^2_{\mathrm{g,ABB}}}}{4\mathrm{Re}\{Z_\mathrm{g}\}} \\ &= \frac{1}{4\mathrm{Re}\{Z_\mathrm{g}\}} \frac{|Z_\mathrm{i} + Z_\mathrm{g}|^2}{|Z_\mathrm{i}|^2} \frac{\overline{v^2_{\mathrm{ABB}}}}{G_\mathrm{v}^2}. \end{aligned} \tag{4.108}$$

This equation can be reordered to highlight the physical meaning of the different terms involved. For instance, we can write that

$$kT_{\mathrm{e,ABB}(Z_\mathrm{g})}\,\delta f = \frac{\mathrm{Re}\{Z_\mathrm{i}\}}{|Z_\mathrm{i}|^2} \frac{|Z_\mathrm{i} + Z_\mathrm{g}|^2}{4\mathrm{Re}\{Z_\mathrm{i}\}\mathrm{Re}\{Z_\mathrm{g}\}} \frac{\overline{v^2_{\mathrm{ABB}}}}{G_\mathrm{v}^2}. \tag{4.109}$$

Using equation (1.5), the first term of the right-hand side of this equation can be written as

$$\frac{\mathrm{Re}\{Z_\mathrm{i}\}}{|Z_\mathrm{i}|^2} = \frac{1}{2} \frac{Z_\mathrm{i} + Z_\mathrm{i}^*}{Z_\mathrm{i} Z_\mathrm{i}^*} = \frac{1}{2}\left(\frac{1}{Z_\mathrm{i}} + \frac{1}{Z_\mathrm{i}^*}\right) = \mathrm{Re}\left\{\frac{1}{Z_\mathrm{i}}\right\}. \tag{4.110}$$

Then, recalling Section 2.2.3 – more precisely, comparing equations (2.128) and (2.131) – we can write

$$\frac{4\mathrm{Re}\{Z_\mathrm{i}\}\mathrm{Re}\{Z_\mathrm{g}\}}{|Z_\mathrm{i}+Z_\mathrm{g}|^2} = 1 - |\Gamma_\mathrm{p}|^2, \tag{4.111}$$

with

$$\Gamma_\mathrm{p} = \frac{Z_\mathrm{i} - Z_\mathrm{g}^*}{Z_\mathrm{i} + Z_\mathrm{g}} \tag{4.112}$$

the power reflection coefficient at the input of the RFIC. Using these results, equation (4.109) therefore reduces to

$$kT_{\mathrm{e,ABB}(Z_\mathrm{g})}\,\delta f(1 - |\Gamma_\mathrm{p}|^2) = \mathrm{Re}\left\{\frac{1}{Z_\mathrm{i}}\right\}\frac{\overline{v_\mathrm{ABB}^2}}{G_\mathrm{v}^2}. \tag{4.113}$$

Written in this way, the interpretation of the terms is straightforward. $kT_{\mathrm{e,ABB}(Z_\mathrm{g})}$ is the equivalent average noise power available from the source in the frequency band δf that represents the ABB noise contribution. As $|\Gamma_\mathrm{p}|^2$ is the fraction of power that is reflected back to the generator due to the impedance mismatch, $1 - |\Gamma_\mathrm{p}|^2$ represents the fraction of power that is effectively transferred to the load. Thus, the left-hand term of the above equation is nothing more than the average noise power, P_a, effectively delivered from the generator to the load Z_i, i.e. to the RFIC. Alternatively this equivalent noise power can be directly derived as the average electrical power of $\tilde{v}_{\mathrm{i,ABB}}(t)$ across the input impedance of the RFIC Z_i. Using equation (4.104), we obtain

$$P_\mathrm{a} = \mathbb{E}\left\{\mathrm{Re}\left\{\frac{|\tilde{v}_{\mathrm{i,ABB}}|^2}{2Z_\mathrm{i}}\right\}\right\} = \frac{\mathbb{E}\left\{|\tilde{v}_{\mathrm{i,ABB}}|^2\right\}}{2}\mathrm{Re}\left\{\frac{1}{Z_\mathrm{i}}\right\} = \frac{\overline{v_\mathrm{ABB}^2}}{G_\mathrm{v}^2}\mathrm{Re}\left\{\frac{1}{Z_\mathrm{i}}\right\},$$

which is just the right-hand side of equation (4.113).

From equation (4.113) we can then see that the equivalent noise temperature $T_{\mathrm{e,ABB}(Z_\mathrm{g})}$, referred to the source generator internal impedance Z_g, that represents the noise power of the ABB block in the frequency band δf, is related to the RMS spot noise voltage at the ABB input through

$$T_{\mathrm{e,ABB}(Z_\mathrm{g})} = \frac{1}{k}\frac{1}{(1 - |\Gamma_\mathrm{p}|^2)}\mathrm{Re}\left\{\frac{1}{Z_\mathrm{i}}\right\}\frac{1}{G_\mathrm{v}^2}\frac{\overline{v_\mathrm{ABB}^2}}{\delta f}. \tag{4.114}$$

As the noise contribution from the RXFE and the ABB can be assumed independent, the total effective input noise temperature of the total RFIC, $T_{\mathrm{e,RFIC}(Z_\mathrm{g})}$, is simply the sum of that of the

RXFE, $T_{e,RXFE(Z_g)}$, and that of the ABB, $T_{e,ABB(Z_g)}$. Using equation (4.27), we can thus derive the equivalent noise factor of the RFIC $F_{RFIC(Z_g,T_0)}$ as

$$\begin{aligned} F_{RFIC(Z_g,T_0)} &= 1 + \frac{T_{e,RFIC(Z_g)}}{T_0} \\ &= 1 + \frac{T_{e,RXFE(Z_g)} + T_{e,ABB(Z_g)}}{T_0}. \end{aligned} \tag{4.115}$$

However, if we get that the RXFE is also characterized by its noise factor $F_{RXFE(Z_g,T_0)}$ instead of its noise temperature, we can still use equation (4.27) to rewrite the above expression as

$$\begin{aligned} F_{RFIC(Z_g,T_0)} &= 1 + \frac{T_0\left(F_{RXFE(Z_g,T_0)} - 1\right)}{T_0} + \frac{T_{e,ABB(Z_g)}}{T_0} \\ &= F_{RXFE(z_g,T_0)} + \frac{T_{e,ABB(z_g)}}{T_0}. \end{aligned} \tag{4.116}$$

Using equation (4.114), we can then finally write

$$F_{RFIC(Z_g,T_0)} = F_{RXFE(Z_g,T_0)} + \frac{1}{kT_0}\frac{1}{(1-|\Gamma_p|^2)}\mathrm{Re}\left\{\frac{1}{Z_i}\right\}\frac{1}{G_v^2}\frac{\overline{v_{ABB}^2}}{\delta f}. \tag{4.117}$$

Here, we see that the power reflection coefficient at the input of the RFIC appears in the term linked to the ABB noise contribution only. As discussed above, this term makes the link between the voltage spot noise quantity as effectively used in the RFIC for the ABB characterization and its equivalent noise source from the generator. This term does not appear in the contribution of the RXFE as the effective noise temperature and noise factor assumed for its characterization are already equivalent noise sources from the generator, thus referred to the correct impedance. This example illustrates the importance of referring the noise temperature and noise factor of a device to the correct quantities as used in operating conditions. This is necessary in order to obtain reliable quantities that can be easily interpreted from a system point of view. One can link this remark to the derivations carried out in Section 4.2.4.

Noise Factor Dependency on Generator Impedance

As highlighted in Section 4.2.1 or 4.2.2 for instance, the input-referred quantities classically used to quantify the noise performance of electronic devices depend on the internal impedance of the source generator used for their characterization. For instance, in the example discussed in the previous section we get that the analytical expression of quantities such as the effective noise temperature or the noise factor need to embed a factor linked to the power reflection coefficient that models the mismatch between the generator internal impedance and that of the device input. This term is necessary in order to correctly represent the intrinsic noise performance of the device, whereas quantities such as the effective noise temperature or the noise factor correspond to the source noise increase that is required to achieve the same available noise power as effectively retrieved at the device output. However, in this previous example the device under consideration was a baseband amplifier that needed only a single equivalent input voltage noise source to model its behavior. But when a quadrupole operates under matched conditions it generally requires at the same time an equivalent voltage noise source and an equivalent current noise source to correctly represent its behavior. As detailed in Section 4.2.3, those sources can be correlated. This results in a dependence of quantities like the device effective noise temperature or noise factor on the generator impedance used for the characterization that is more complex than in the discussion so far.

Let us reconsider the equivalent noise model of a quadrupole based on input voltage and current spot noise quantities shown in Figure 4.13 (middle). The noise sources involved in that model thus represent the device noise performance on the frequency band δf assumed sufficiently narrowband that all the electrical characteristics involved can be assumed constant on this frequency band. As discussed in Section 4.2.3, these sources are quantities intrinsic to the quadrupole and are therefore independent of the generator impedance used for the characterization. With the aim of deriving the device noise factor or effective noise temperature, suppose that the generator internal impedance is Z_g and its noise temperature is T_0. In order

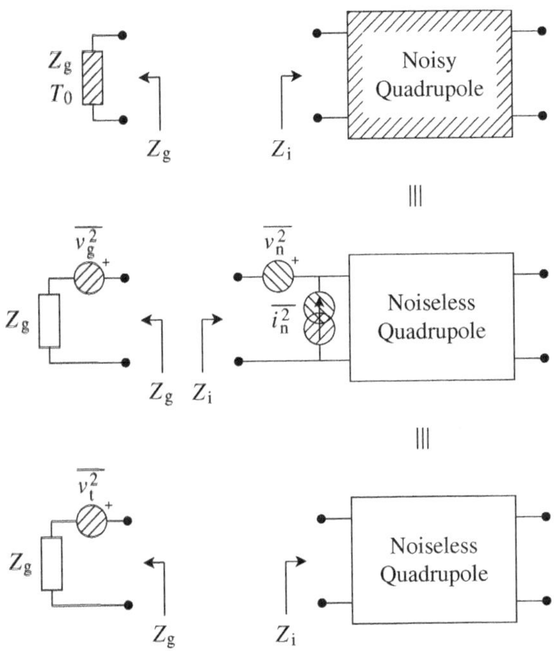

Figure 4.13 Equivalent model for a noisy quadrupole fed by a noisy source – The noise behavior of both the quadrupole and the source (top) can be represented using both voltage and current noise sources as detailed in Section 4.2.3 (middle). These equivalent noise sources can be considered together as a single voltage generator through an equivalent Thévenin model (bottom).

to discuss the dependence of these noise characteristics on Z_g, it is of interest to model the intrinsic noise behavior of this dipole generator using a voltage noise source proportional to T_0 and Z_g. Based on equation (4.50), the RMS value of this voltage source, $\overline{v_g^2}$, can thus be expressed as

$$\overline{v_g^2} = 4kT_0\mathrm{Re}\{Z_g\}\delta f. \tag{4.118}$$

To derive the device noise factor or noise temperature, we then need to derive the amount by which we need to increase the source noise in order to correctly represent the noise power as recovered at the output of the system composed of the generator and the noisy quadrupole. We can derive the equivalent Thévenin model for the system composed of a noisy source generator plus the equivalent input-referred voltage and current noise sources of the quadrupole. As shown in Figure 4.13(bottom), we obtain an equivalent noisy generator that embeds all the noise contributions of the system. Obviously, the internal impedance of this equivalent generator remains Z_g and the RMS value of its voltage source, $\overline{v_t^2}$, can easily be derived from $\overline{v_g^2}$, $\overline{v_n^2}$ and $\overline{i_n^2}$ when dealing with spot noise quantities. As discussed in Section 4.2.3 and illustrated in the previous section, in that case we can rely on simple electrical equations when considering complex envelopes defined as centered around the same center frequency. In the present case we can thus write under those assumptions that

$$\tilde{v}_t(t) = \tilde{v}_g(t) + \tilde{v}_n(t) + Z_g\tilde{i}_n(t). \tag{4.119}$$

We see that although the noise voltage and current sources that characterize the quadrupole are intrinsic quantities, the total equivalent voltage source depends on the generator source impedance in the equivalent Thévenin model. Now focusing on the RMS values of $v_t(t)$, by Appendix 2 we can rely on the stationarity and ergodicity of its complex envelope. We can thus use equation (4.48) to write that

$$\overline{v_t^2} = \frac{\mathbb{E}\{|\tilde{v}_t|^2\}}{2}. \tag{4.120}$$

In order to go further it is of interest to use equation (4.67) to decompose the quadrupole voltage noise source $v_n(t)$ as the sum of two components, $v_u(t)$ and $v_c(t)$. In this classical decomposition, $v_u(t)$, is defined as uncorrelated with $i_n(t)$, while $v_c(t)$ is perfectly correlated with it. Due to our assumptions, the same behavior then holds for the complex envelopes of the spot noise quantities. As a result, we can write $\tilde{v}_c(t)$ as proportional to $\tilde{i}_n(t)$ through the correlation impedance defined by equation (4.69). We obtain

$$\tilde{v}_n(t) = \tilde{v}_u(t) + Z_c\tilde{i}_n(t). \tag{4.121}$$

This correlation between noise sources is in fact the main difference compared to the analog baseband example discussed in the previous section. Using this decomposition in equation (4.119), we can then express $\tilde{v}_t(t)$ as

$$\tilde{v}_t(t) = \tilde{v}_g(t) + \tilde{v}_u(t) + (Z_g + Z_c)\tilde{i}_n(t). \tag{4.122}$$

Multiplying each side of this equation by its complex conjugate then gives

$$|\tilde{v}_t(t)|^2 = |\tilde{v}_g(t)|^2 + |\tilde{v}_u(t)|^2 + |Z_g + Z_c|^2 |\tilde{\imath}_n(t)|^2 + 2\mathrm{Re}\{\tilde{v}_g(t)\tilde{v}_u^*(t)\}$$
$$+ 2\mathrm{Re}\{(Z_g + Z_c)\tilde{\imath}_n(t)\tilde{v}_g^*(t)\} + 2\mathrm{Re}\{(Z_g + Z_c)\tilde{\imath}_n(t)\tilde{v}_u^*(t)\}. \tag{4.123}$$

Finally, taking expectations yields

$$\overline{v_t^2} = \overline{v_g^2} + \overline{v_u^2} + |Z_g + Z_c|^2\, \overline{i_n^2} + \mathrm{Re}\{\mathbb{E}\{\tilde{v}_g\tilde{v}_u^*\}\}$$
$$+ \mathrm{Re}\{(Z_g + Z_c)\mathbb{E}\{\tilde{\imath}_n\tilde{v}_g^*\}\} + \mathrm{Re}\{(Z_g + Z_c)\mathbb{E}\{\tilde{\imath}_n\tilde{v}_u^*\}\}. \tag{4.124}$$

This expression involves different correlation terms between noise voltage and current complex envelopes. However, it seems realistic to assume independence between the noise components coming from the generator and those coming from the quadrupole. In addition, these noise components can be assumed centered. As the complex envelopes are defined around the same center frequency, it follows from the discussion in Appendix 1 that these complex envelopes can be assumed uncorrelated. We can thus write that

$$\mathbb{E}\{\tilde{v}_g\tilde{v}_u^*\} = \mathbb{E}\{\tilde{\imath}_n\tilde{v}_g^*\} = \mathbb{E}\{\tilde{\imath}_n\tilde{v}_u^*\} = 0. \tag{4.125}$$

It finally follows that

$$\overline{v_t^2} = \overline{v_g^2} + \overline{v_u^2} + |Z_g + Z_c|^2\, \overline{i_n^2}. \tag{4.126}$$

By Section 4.2.2, we can express the device noise factor, $F_{(Z_g,T_0)}$, referred to the source generator internal impedance Z_g and noise temperature T_0, as the ratio of the total available noise power from the source generator plus the input-referred contribution of the quadrupole to the contribution of the source generator. Given that those available noise powers can be derived from the RMS value of the spot noise voltages according to equation (4.49), we can then write that

$$F_{(Z_g,T_0)} = \frac{\overline{v_t^2}}{4\mathrm{Re}\{Z_g\}} \frac{4\mathrm{Re}\{Z_g\}}{\overline{v_g^2}}$$
$$= 1 + \frac{\overline{v_u^2} + |Z_g + Z_c|^2\, \overline{i_n^2}}{\overline{v_g^2}}. \tag{4.127}$$

As our task is to discuss the dependence of this noise factor on the generator internal impedance Z_g, we can express $\overline{v_g^2}$ as a function of the real part of this impedance using equation (4.118). In order to continue to work with homogeneous quantities in the equations, we can then formally define a noise resistance R_u and a noise conductance G_n for $\overline{v_u^2}$ and $\overline{i_n^2}$ respectively as done for $\overline{v_g^2}$. We can then rewrite the above expression as

$$F_{(Z_g,T_0)} = 1 + \frac{R_u + [(R_g + R_c)^2 + (X_g + X_c)^2]G_n}{R_g}, \tag{4.128}$$

where $Z_g = R_g + jX_g$ and $Z_c = R_c + jX_c$. The structure of this expression involves the sum of positive quantities. We therefore get that $F_{(Z_g,T_0)}$ has an absolute minimum value F_0 that is obtained when

$$\partial_{R_g} F_{(Z_g,T_0)} = 0, \tag{4.129a}$$

$$\partial_{X_g} F_{(Z_g,T_0)} = 0. \tag{4.129b}$$

Solving the corresponding equations leads to $F_{(Z_g,T_0)} = F_0$ when

$$R_g = R_{g_0} = \sqrt{\frac{R_u}{G_n} + R_c^2}, \tag{4.130a}$$

$$X_g = X_{g_0} = -X_c. \tag{4.130b}$$

As a result, writing $Z_{g_0} = R_{g_0} + jX_{g_0}$, we get that the spot noise factor of the quadrupole at the relevant frequency takes the simple form

$$F_{(Z_g,T_0)} = F_0 + G_n \frac{|Z_g - Z_{g_0}|^2}{R_g}. \tag{4.131}$$

In order to discuss further the dependence of the noise factor on the generator impedance and allow physical interpretations, we need to bring in the power reflection coefficient at the input of the quadrupole. Denoting by Z_i the input impedance of the quadrupole at the relevant frequency, this power reflection coefficient Γ_p is defined by equation (2.132). However, it remains more convenient for our present derivations to introduce the modified factor Γ'_p defined by

$$\Gamma'_p = \frac{Z_g - Z_i^*}{Z_g + Z_i}. \tag{4.132}$$

We then remark that $|\Gamma'_p|^2 = |\Gamma_p|^2$, which is the quantity of interest when discussing power reflection. In order to introduce this factor Γ'_p into the expression for the device noise factor, we make Z_g the subject of the above equation:

$$Z_g = \frac{Z_i^* + Z_i\Gamma'_p}{1 - \Gamma'_p}. \tag{4.133}$$

If we denote by Γ'_{p_0} the value of Γ'_p when $Z_g = Z_{g_0}$, we therefore have

$$Z_g - Z_{g_0} = \frac{(Z_i^* + Z_i\Gamma'_p)(1 - \Gamma'_{p_0}) - (Z_i^* + Z_i\Gamma'_{p_0})(1 - \Gamma'_p)}{(1 - \Gamma'_p)(1 - \Gamma'_{p_0})}$$

$$= 2\mathrm{Re}\{Z_i\}\frac{\Gamma'_p - \Gamma'_{p_0}}{(1 - \Gamma'_p)(1 - \Gamma'_{p_0})}. \tag{4.134}$$

In the same way, we can express R_g as half the sum of Z_g and its complex conjugate:

$$2R_g = \frac{(Z_i^* + Z_i\Gamma'_p)(1 - \Gamma'^*_p) + (Z_i + Z_i^*\Gamma'^*_p)(1 - \Gamma'_p)}{(1 - \Gamma'_p)(1 - \Gamma'^*_p)}$$

$$= 2\mathrm{Re}\{Z_i\}\frac{1 - |\Gamma'_p|^2}{|1 - \Gamma'_p|^2} = 2\mathrm{Re}\{Zi\}\frac{1 - |\Gamma_p|^2}{|1 - \Gamma'_p|^2}. \tag{4.135}$$

As a result,

$$\frac{|Z_g - Z_{g_0}|^2}{R_g} = \frac{4\mathrm{Re}\{Z_i\}}{|1 - \Gamma'_{p_0}|^2}\frac{|\Gamma'_p - \Gamma'_{p_0}|^2}{(1 - |\Gamma_p|^2)}. \tag{4.136}$$

Finally, equation (4.131) can be expressed as

$$F_{(Z_g,T_0)} = F_0 + \frac{4G_n\mathrm{Re}\{Z_i\}}{|1 - \Gamma'_{p_0}|^2}\frac{|\Gamma'_p - \Gamma'_{p_0}|^2}{(1 - |\Gamma_p|^2)}. \tag{4.137}$$

Looking at this equation, we see that only the last fraction on the right-hand side exhibits a dependency of the quadrupole noise factor on the generator impedance. In the denominator of this fraction, we recover the same term as found during the review of the example in the previous section, i.e. $1 - |\Gamma_p|^2$. The interpretation of this term remains the same. As $|\Gamma_p|^2$ is the fraction of power delivered by the generator that is reflected back to it, $1 - |\Gamma_p|^2$ is the fraction of power effectively delivered to the load (the quadrupole here) by the input generator. This means that under impedance mismatch conditions, the quadrupole noise factor referred to the generator impedance increases in order to effectively get the correct noise power of the quadrupole intrinsic contribution recovered at its output whatever the input impedance mismatch. But, compared to the previous example, we get in addition the term $|\Gamma'_p - \Gamma'_{p_0}|^2$, which shows the noise factor optimization when the generator impedance gets closer to the optimum impedance Z_{g_0}. This is an additional phenomenon that exists here as we are dealing with correlated voltage and current noise sources that model the behavior of the quadrupole. This example thus shows that the input generator impedance not only influences the quantities that characterize the device under study in order to recover the correct level of intrinsic noise contribution at the device output whatever this impedance, but also the intrinsic noise performance of the device. We can also consider an optimization of such intrinsic noise performance due to the generator impedance tuning.

4.2.6 *SNR Degradation*

Up to now, we have focused on how to characterize the noise performance of electronic devices. However, from the system design point of view, our main interest is the possibility of predicting the SNR degradation experienced by the wanted signal when going through a line-up. It is useful to clarify the exact relationship between this degradation and the quantities introduced so far.

We first observe that when talking about SNR we are obviously talking about noise *power*, i.e. the long-term average active power of the noise process. But by the discussion so far in this chapter, concepts such as the noise temperature and the noise factor of a device are related to the noise *available power* that can be retrieved from it. The reason for this is the physics of thermal noise, as discussed in Section 4.1.2. We then wonder if there are practical ways to use these concepts to finally quantify the SNR degradation. This is in fact no big deal if we refer to the relationship between the available power from a source, P_{av}, and the average active power, $P_{\text{a,l}}$, effectively delivered by this source to a given load. Under the narrowband assumption, this relationship is given by equation (2.131):

$$P_{\text{a,l}} = P_{\text{av}}(1 - |\Gamma_{\text{p}}|^2), \tag{4.138}$$

where Γ_{p} is the power reflection coefficient given by equation (2.132). This expression therefore shows that the fraction of power that is reflected back to the generator depends only on the relative impedances of the devices we are dealing with, and not on the statistical properties of the signal under consideration. As a result, the proportionality factor between the available power and the average power necessarily remains the same for both the wanted signal and the noise signal flowing from a given device. We thus get that the signal to noise power ratio $SNR = P_{\text{w}}/P_{\text{n}}$ is equal to the ratio of the available powers,

$$SNR = \frac{P_{\text{w}}}{P_{\text{n}}} = \frac{P_{\text{av,w}}}{P_{\text{av},n}}. \tag{4.139}$$

Let us now reconsider the case of a noisy quadrupole characterized by its effective input noise temperature, $T_{\text{e}(Z_{\text{g}})}$, referred to an input generator internal impedance Z_{g}. This corresponds to the configuration shown in Figure 4.1. In order to derive the SNR degradation through the device, we first need to give an expression for the SNR as it is at the input of the noisy quadrupole, SNR_i. We can refer to Section 4.2.1 to derive the available noise power from the noisy source generator in the narrow frequency bandwidth δf. We obtain

$$SNR_i = \frac{P_{w,i}}{P_{n,i}} = \frac{P_{av,w,i}}{P_{av,n,i}} = \frac{P_{av,w,i}}{kT_g\,\delta f}. \tag{4.140}$$

At the output of the device, the available power of the wanted signal is simply that of its input times the available power gain of the quadrupole, $G_{av(Z_g)}$. Here, this available power gain needs to correspond to the operating conditions, i.e. the gain effectively experienced when the impedance Z_g is connected at the quadrupole input. Concerning the available power of the noise, we also need to take into account the noise contribution of the quadrupole through its effective noise temperature $T_{e(Z_g)}$. Thus, referring to equation (4.11), we can express the SNR recovered at the output of the noisy quadrupole, SNR_o, as

$$SNR_o = \frac{P_{w,o}}{P_{n,o}} = \frac{P_{av,w,o}}{P_{av,n,o}} = \frac{G_{av(Z_g)}\,P_{av,w,i}}{G_{av(Z_g)}\,k\left(T_g + T_{e(Z_g)}\right)\delta f}. \tag{4.141}$$

From the two above equations we can finally derive the SNR degradation, $D = SNR_i/SNR_o$ as

$$D = \frac{SNR_i}{SNR_o} = \frac{T_g + T_{e(Z_g)}}{T_g} = 1 + \frac{T_{e(Z_g)}}{T_g}. \tag{4.142}$$

This is a natural result as the noisy device adds a given amount of noise power proportional to its effective noise temperature $T_{e(Z_g)}$. As the noise power provided by the generator is proportional to its own noise temperature T_g, the SNR degradation is simply related to the sum of the contributions referred to the contribution of the generator only. This shows the importance of characterizing the quadrupole through an effective noise temperature that is referred to the generator internal impedance Z_g. This is indeed this assumption that allows us to get the same proportionality factor linked to the impedance mismatch at the quadrupole input for all the noise contributions, and hence this simple formulation for D.

Nevertheless, the SNR degradation D takes this simple form because the noise temperatures used to characterize the device preserve the additive behavior of the noise contributions. Using quantities that are close to the physics of the phenomenon, the model leads to straightforward expressions. As this is not the case for noise factors, we expect a trickier relationship when using the latter. Practically speaking, we can directly express D as a function of the device noise factor from its expression as a function of the effective noise temperature by using equation (4.27). Assuming that we are dealing with a noise factor $F_{(Z_g,T_0)}$ referred to an input source noise temperature T_0, we obtain

$$D = \frac{SNR_i}{SNR_o} = 1 + \frac{T_{e(Z_g)}}{T_g} = 1 + \frac{T_0}{T_g}\left(F_{(Z_g,T_0)} - 1\right). \tag{4.143}$$

We see that the SNR degradation is equal to the device noise factor $F_{(Z_g,T_0)}$ if and only if the source noise temperature is equal to the noise temperature used to refer the noise factor to. As detailed in Section 4.2.2, this behavior is related to the definition of the noise factor that embeds the source noise contribution at the same time as that of the device it is supposed to characterize. The interpretation of the above equation is thus straightforward. Given this source noise contribution T_0 that is not necessarily equal to that of the generator, T_g, we then first need to reconstruct the fraction of noise power that is linked only to the noisy device before adding to it the contribution of the generator. This operation simply corresponds to the reconstruction of the device effective noise temperature from its noise factor, i.e. the term $T_0\left(F_{(Z_g,T_0)} - 1\right)$ in the above expression if we refer to equation (4.27), before adding the generator noise temperature T_g to it.

To conclude our discussion, we highlight one potential reason for referring the noise factor of a receiver to the physical temperature T of the application board. We can assume that all the passive devices and the antenna have the same physical temperature. As illustrated in Section 4.2.5, the source noise temperature seen by the active part of the receiver is then also equal to T. As a result, the noise factor referred to the same temperature directly gives the SNR degradation when going through the line-up. However, the examples given in the previous sections should convince the reader that the source noise temperature seen by the active part of a receiver is often not the thermodynamic temperature. As a result, the use of the noise factor must be considered carefully depending on whether or not the noise temperature this noise factor is referred to is equal to the source noise temperature effectively experienced in operating conditions. In general, working with noise temperatures remains more straightforward and less confusing for system derivations.

4.3 LO Phase Noise

Let us now focus on another source of degradation in transceivers, the LO signal quality. As detailed extensively in Chapter 6, the frequency conversion in the analog domain is classically achieved through the multiplication of the signal to be processed by a LO waveform. Obviously, the generation of a LO signal in the RF/analog domain requires the use of electronic devices that inherently generate noise. The LO signal then generally takes the form of the ideal waveform we are looking for, but with an additive bandpass noise centered around it. There are different system impacts linked to this additive noise term, for example:

(i) the SNR degradation due to the fraction of the noise component of the LO that lies within the bandwidth of the signal being processed;
(ii) the degradation of the spectrum of the signal being processed;
(iii) the reciprocal mixing issue that occurs when an additional blocking signal is present on top of the wanted signal during the processing.

To further investigate these problems, we first need to detail on the one hand the characteristics of the additional noise term that corrupts the LO waveform in practical implementations, and on the other hand how this noise term propagates toward the signal being processed in a practical analog mixing stage. We discuss the common characteristics of RF synthesizers used

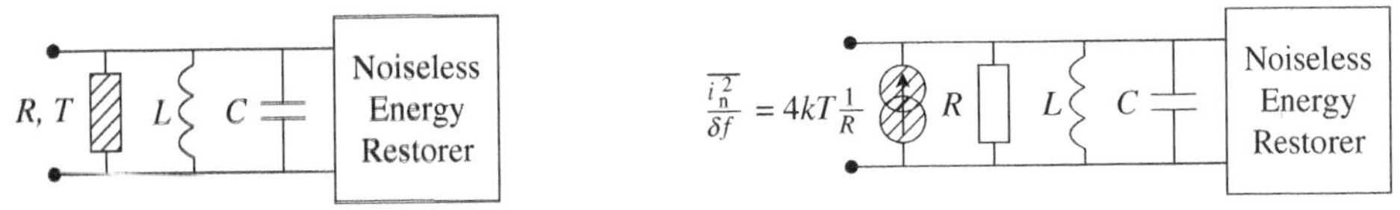

Figure 4.14 LC tank oscillator for the generation of RF sinusoidal waveform and equivalent noise model – Due to inherent lossy devices, an equivalent parasitic resistor exists in an LC tank oscillator (left). This parasitic resistor leads to a noise source with a non-vanishing available active power due to thermal noise in the equivalent noise model of the oscillator (right). The filtering effect of the resonant LC circuit on this white noise source leads to the spectral shape of the noise delivered by the oscillator.

for the generation of LO signals in Section 4.3.1 before addressing practical mixing stages in Section 4.3.2. System impacts are then discussed in Section 4.3.3.

4.3.1 RF Synthesizers

LO signals at RF frequencies are classically generated through the use of RF synthesizers. From the system point of view we thus need to understand the characteristics of the noise term that necessarily corrupts the RF wanted signal flowing from such systems. Practically speaking, there are numerous structures for RF synthesizers and it is not the purpose of this book to discuss them extensively. Nevertheless, a general discussion of the characteristics of this noise term is warranted, based on the inspection of simple structures for the RF oscillators and the PLLs that are the main blocks that make up practical RF synthesizers.

RF Oscillators

In silicon integrated transceivers, oscillators are classically implemented as either ring oscillators or LC tanks. The use of either structure is application dependent. But for RF applications that require high noise performance, LC oscillators are almost always used despite their size and the potential coupling issues. For the sake of simplicity, we focus on this structure in the following as the conclusions are sufficiently general for our purposes.

The goal here is not to derive the general theory of noise in such oscillators as can be found in the literature [42, 43]. We focus on the phenomenon involved in the generation of the noise term that adds to the wanted signal delivered by a RF oscillator in order to derive its characteristics. Consider the model shown in Figure 4.14(left) for LC oscillators. This model is composed of a resonator and of an energy restorer that compensates for both the energy loss due to the dissipation in the parasitic resistor R and for the energy delivered by the oscillator to the external world. But by Section 4.2.3, a resistor R with a thermodynamic temperature T plays the role of a noise generator able to deliver an active power to a load. It can be modeled by a noise current source, $i_n(t)$, whose RMS value in the frequency band δf is given by equation (4.51). The PSD of the noise signal delivered by this source at RF frequencies therefore takes the form

$$\frac{\overline{i_n^2}}{\delta f} = 4kT\frac{1}{R}. \tag{4.144}$$

Now assuming that the energy restorer of the resonator is noiseless, we can derive the equivalent noise model shown in Figure 4.14(right) for our oscillator. Assuming in addition that this restorer presents a sufficiently high impedance to the oscillator, the noise voltage v_{n} across the reactive elements is simply the noise current flowing from the noise current source times the impedance of the oscillator. In the Laplace domain, this impedance $Z(s)$ is such that

$$\frac{1}{Z(s)} = \frac{1}{R} + \frac{1}{Ls} + Cs, \tag{4.145}$$

so that

$$Z(s) = \frac{R}{Q}\frac{\frac{s}{\omega_{\mathrm{osc}}}}{\left(\frac{s}{\omega_{\mathrm{osc}}}\right)^2 + \frac{1}{Q}\frac{s}{\omega_{\mathrm{osc}}} + 1}, \tag{4.146}$$

with

$$\omega_{\mathrm{osc}} = \frac{1}{\sqrt{LC}},$$

$$Q = \frac{R}{L\omega_{\mathrm{osc}}} = RC\omega_{\mathrm{osc}}. \tag{4.147}$$

Here, ω_{osc} represents the resonance frequency of the resonator and Q its quality factor. For $s = \mathrm{j}2\pi f$ purely imaginary, this expression for $Z(s)$ therefore gives the value of the oscillator impedance at the frequency f. Now focusing on the RMS values of $v_{\mathrm{n}}(t)$, by Appendix 2 we can rely on the stationarity and ergodicity of its complex envelopes in order to use equation (4.48). Now assuming that we are dealing with spot noise quantities, we can rely on simple electrical equations between complex envelopes defined as centered around the same center frequency as discussed in Section 4.2.3. We can then finally write that

$$\frac{\overline{v_{\mathrm{n}}^2}}{\delta f}(f) = |Z(2\pi f)|^2 \frac{\overline{i_{\mathrm{n}}^2}}{\delta f}. \tag{4.148}$$

We thus see that this PSD of the noise voltage recovered across the oscillator at the frequency f has a shape that follows the frequency response of the oscillator impedance. As we are dealing with a second order cell, we obtain a spectrum with slope −2, i.e. decreasing by an amount of −20 dB per decade when going away from the center angular frequency ω_{osc}. At the same time, the width of the spectrum is driven by the quality factor Q. Moreover, for practical RF frequencies of interest and realistic Q factors, we can assume the even symmetry of this PSD regarding ω_{osc}. This holds at least up to first order, i.e. in the in-band part close to ω_{osc}, which is the part of interest where the noise power is concentrated. We also observe that in practical implementations the noise spectrum does not decrease to zero as the frequency offset increases more and more. The thermal noise sources distributed in the materials present in between the core oscillator and the outer world necessarily result in a noise floor that must be added to our simple model.

Finally, we see that the signal $s(t)$ delivered by the RF oscillator can be written as the sum of a sine wave at the resonance angular frequency ω_{osc}, and an additive bandpass noise $n_{\text{osc}}(t)$. In its general form, we can write this sum as

$$s(t) = \rho \cos(\omega_{\text{osc}} t + \theta) + n_{\text{osc}}(t), \tag{4.149}$$

where θ represents a potential phase offset at the origin of time. Although this represents the physical behavior of the oscillator, it is not necessarily considered in all our subsequent derivations, for the sake of simplicity. We then get that $n_{\text{osc}}(t)$ exhibits an almost evenly symmetric PSD that is decreasing with slope -2 away from ω_{osc}, but with an absolute floor level at RF frequencies of interest. Moreover, as thermal noise is at the origin of this process, we can assume that we are dealing with a stationary noise source with a Gaussian amplitude distribution according to the discussion in Section 4.1.2.

Revisiting Amplitude and Phase Noise

We can now go a step further in the examination of the properties of the signal $s(t)$ delivered by the RF oscillator considered in the previous section. Practically speaking, this means going through the decomposition of the term $n_{\text{osc}}(t)$ into two components that corrupt both the instantaneous amplitude and the instantaneous phase of the wanted signal $\rho \cos(\omega_{\text{osc}} t + \theta)$ in equation (4.149). Anticipating the discussion in Section 4.3.2, we recall that only the phase noise part of the additive bandpass noise term that corrupts the LO signal succeeds in going through mixing stages implemented as choppers. It is therefore useful to discuss the characteristics of this phase noise term from the perspective of investigating the system impacts related to it.

This decomposition was introduced in Section 1.2.2. However, in the present case we have additional assumptions to consider as we know the statistics of both the wanted signal, which is CW, and of $n_{\text{osc}}(t)$, which can be considered as a stationary Gaussian bandpass process with a PSD that exhibits an even symmetry regarding ω_{osc}, at least up to first order. We thus expect to be able to derive more characteristics of interest for the phase and amplitude noise terms in the practical LO context than in the general case.

Recall the decomposition in Section 1.2.2 of the additive bandpass noise $n_{\text{osc}}(t)$ as the sum of two bandpass terms of interest, $n_{\parallel}(t)$ and $n_{\perp}(t)$. Practically speaking, the definition of these bandpass terms requires us to work on the complex envelope of both the wanted signal, $\rho e^{j\theta}$, and the additive bandpass noises we are dealing with, i.e. $\tilde{n}_{\text{osc}}(t)$, $\tilde{n}_{\parallel}(t)$ and $\tilde{n}_{\perp}(t)$. Assuming that we are dealing with complex envelopes that are defined as centered around the same angular frequency ω_{osc}, we can start by writing

$$\tilde{n}_{\text{osc}}(t) = \tilde{n}_{\parallel}(t) + \tilde{n}_{\perp}(t). \tag{4.150}$$

In the same way, the complex envelope $\tilde{s}(t)$ of $s(t) = \rho \cos(\omega_{\text{osc}} t + \theta) + n_{\text{osc}}(t)$ can be written as

$$\begin{aligned} \tilde{s}(t) = \rho_s(t) e^{j\phi_s(t)} &= \rho e^{j\theta} + \tilde{n}_{\text{osc}}(t) \\ &= \rho e^{j\theta} + \tilde{n}_{\parallel}(t) + \tilde{n}_{\perp}(t). \end{aligned} \tag{4.151}$$

As illustrated in Figure 1.10, $n_{\|}(t)$ is then defined so that its complex envelope $\tilde{n}_{\|}(t)$ is the projection of $\tilde{n}_{\mathrm{osc}}(t)$ on $\rho e^{j\theta}$, and $n_{\perp}(t)$ so that $\tilde{n}_{\perp}(t)$ is the orthogonal projection of $\tilde{n}_{\mathrm{osc}}(t)$ on $\rho e^{j\theta}$.

Alternatively, this decomposition allows us to introduce new lowpass quantities, $n_{\rho_{\mathrm{osc}}}(t)$ and $n_{\phi_{\mathrm{osc}}}(t)$, that corrupt respectively the instantaneous amplitude of the wanted signal flowing from the RF oscillator and its instantaneous phase. By equations (1.127), (1.129) and (1.130), we can write

$$\rho_s(t) = \rho + n_{\rho_{\mathrm{osc}}}(t), \tag{4.152a}$$

$$\phi_s(t) = \theta + n_{\phi_{\mathrm{osc}}}(t), \tag{4.152b}$$

with

$$n_{\rho_{\mathrm{osc}}}(t) \approx \rho_{\epsilon\|}(t), \tag{4.153a}$$

$$n_{\phi_{\mathrm{osc}}}(t) \approx \frac{\rho_{\epsilon\perp}(t)}{\rho}, \tag{4.153b}$$

and

$$\rho_{\epsilon\|}(t) = \rho_{n_{\mathrm{osc}}}(t)\cos(\phi_{n_{\mathrm{osc}}}(t) - \theta), \tag{4.154a}$$

$$\rho_{\epsilon\perp}(t) = \rho_{n_{\mathrm{osc}}}(t)\sin(\phi_{n_{\mathrm{osc}}}(t) - \theta). \tag{4.154b}$$

Here $\rho_{n_{\mathrm{osc}}}(t)$ and $\phi_{n_{\mathrm{osc}}}(t)$ stand for the magnitude and phase of the complex envelope $\tilde{n}_{\mathrm{osc}}(t)$ of the additive noise term flowing from the RF oscillator $n_{\mathrm{osc}}(t)$:

$$\tilde{n}_{\mathrm{osc}}(t) = p_{n_{\mathrm{osc}}}(t) + jq_{n_{\mathrm{osc}}}(t) = \rho_{n_{\mathrm{osc}}}(t)e^{j\phi_{n_{\mathrm{osc}}}(t)}. \tag{4.155}$$

The approximations in those expressions hold as long as the SNR remains good enough at the output of the RF oscillator. By substituting equation (4.154) into equation (4.153) we can then express the amplitude and phase noise components $n_{\rho_{\mathrm{osc}}}(t)$ and $n_{\phi_{\mathrm{osc}}}(t)$ as

$$n_{\rho_{\mathrm{osc}}}(t) \approx \rho_{n_{\mathrm{osc}}}(t)\cos(\phi_{n_{\mathrm{osc}}}(t) - \theta), \tag{4.156a}$$

$$n_{\phi_{\mathrm{osc}}}(t) \approx \frac{\rho_{n_{\mathrm{osc}}}(t)}{\rho}\sin(\phi_{n_{\mathrm{osc}}}(t) - \theta). \tag{4.156b}$$

We finally get that the complex envelope $\tilde{s}(t)$ can be expressed as

$$\tilde{s}(t) = \rho_s(t)e^{j\phi_s(t)} = (\rho + n_{\rho_{\mathrm{osc}}}(t))e^{j(\theta + n_{\phi_{\mathrm{osc}}}(t))}. \tag{4.157}$$

Referring to the original expression in equation (4.151), we see that $s(t)$ can be expressed in two different but equivalent ways:

$$\begin{aligned} s(t) = \mathrm{Re}\left\{\tilde{s}(t)e^{j\omega_{osc}t}\right\} &= \rho\cos(\omega_{osc}t+\theta)+n_{osc}(t) \\ &= (\rho+n_{\rho_{osc}}(t))\cos(\omega_{osc}t+\theta+n_{\phi_{osc}}(t)). \end{aligned} \quad (4.158)$$

As observed at the beginning of this section, given knowledge of the statistics of both the wanted signal, which is CW, and of $n_{osc}(t)$, which is Gaussian, we can now go a step forward in the investigation of the statistical properties of the phase and amplitude noise terms $n_{\phi_{osc}}(t)$ and $n_{\rho_{osc}}(t)$. Given that the wanted signal is CW, we immediately see from equation (4.153) that $n_{\rho_{osc}}(t)$ and $n_{\phi_{osc}}(t)$ are proportional to $\rho_{\epsilon\|}(t)$ and $\rho_{\epsilon\perp}(t)$, respectively. But the latter quantities are in turn linked to the characteristics of the complex envelope $\tilde{n}_{osc}(t)$ of the original additive bandpass noise $n_{osc}(t)$ through equation (4.154). Based on this relationship and referring to the expression for $\tilde{n}_{osc}(t)$ given by equation (4.155), we can then write that

$$\begin{aligned} \rho_{\epsilon\|}(t)+j\rho_{\epsilon\perp}(t) &= \rho_{n_{osc}}(t)e^{j(\phi_{n_{osc}}(t)-\theta)} \\ &= (p_{n_{osc}}(t)+jq_{n_{osc}}(t))e^{-j\theta}. \end{aligned} \quad (4.159)$$

Thus,

$$\begin{pmatrix} \rho_{\epsilon\|}(t) \\ \rho_{\epsilon\perp}(t) \end{pmatrix} = \begin{pmatrix} \cos(\theta) & \sin(\theta) \\ -\sin(\theta) & \cos(\theta) \end{pmatrix} \begin{pmatrix} p_{n_{osc}}(t) \\ q_{n_{osc}}(t) \end{pmatrix}. \quad (4.160)$$

Recalling Section 1.2.1, we see that $n_{osc}(t)$ being Gaussian and stationary leads to the real and imaginary parts of its complex envelope $\tilde{n}_{osc}(t)$ also being Gaussian and stationary. But it can then be shown that any linear transformation of Gaussian random variables yields Gaussian random variables [3, 4]. As a result, $\rho_{\epsilon\|}(t)$ and $\rho_{\epsilon\perp}(t)$ are necessarily Gaussian, and so are the amplitude and the phase noise terms $n_{\rho_{osc}}(t)$ and $n_{\phi_{osc}}(t)$ that are proportional to them according to equation (4.153).

It is also of interest examine the spectral shape of the phase and amplitude noise terms. We can follow the same approach as above and observe that those quantities are proportional to $\rho_{\epsilon\|}(t)$ and $\rho_{\epsilon\perp}(t)$. They thus necessarily have the same spectral shape as the latter processes. We observe further that a simple linear transformation holds between $(\rho_{\epsilon\|}(t),\rho_{\epsilon\perp}(t))^T$ and $(p_{n_{osc}}(t),q_{n_{osc}}(t))^T$. We thus expect to be able to link quite easily the spectral shape of the real and imaginary parts of the complex envelope of the bandpass noise term $n_{osc}(t)$ to that of $\rho_{\epsilon\|}(t)$ and $\rho_{\epsilon\perp}(t)$. We can therefore consider the expansion of $\gamma_{\rho_{\epsilon\|}\times\rho_{\epsilon\|}}(\tau)$. Using equation (4.160) leads to

$$\begin{aligned} \gamma_{\rho_{\epsilon\|}\times\rho_{\epsilon\|}}(\tau) &= \cos^2(\theta)\mathbb{E}\left\{p_{n_{osc},t}p_{n_{osc},t-\tau}\right\}+\sin^2(\theta)\mathbb{E}\left\{q_{n_{osc},t}q_{n_{osc},t-\tau}\right\} \\ &\quad + \cos(\theta)\sin(\theta)\left(\mathbb{E}\left\{p_{n_{osc},t}q_{n_{osc},t-\tau}\right\}+\mathbb{E}\left\{q_{n_{osc},t}p_{n_{osc},t-\tau}\right\}\right). \end{aligned} \quad (4.161)$$

Now referring to the stationarity of $n_{osc}(t)$ and the symmetry characteristics of its PSD around the angular frequency ω_{osc} used to define its complex envelope, we have that the real and imaginary parts of this complex envelope can be considered uncorrelated whatever the time difference τ, as discussed in "Impact of spectral symmetry on top of stationarity" (Section 1.1.3). Given that $\tilde{n}_{\text{osc}}(t) = p_{\text{n}_{\text{osc}}}(t) + \text{j}q_{\text{n}_{\text{osc}}}(t)$ is effectively defined as centered around ω_{osc}, we can therefore write that

$$\mathbb{E}\left\{p_{\text{n}_{\text{osc}},t} q_{\text{n}_{\text{osc}},t-\tau}\right\} = \mathbb{E}\left\{q_{\text{n}_{\text{osc}},t} p_{\text{n}_{\text{osc}},t-\tau}\right\} \approx 0. \tag{4.162}$$

Then equation (4.161) reduces to

$$\begin{aligned}\gamma_{\rho_{\epsilon\|}\times\rho_{\epsilon\|}}(\tau) &= \cos^2(\theta)\mathbb{E}\left\{p_{\text{n}_{\text{osc}},t} p_{\text{n}_{\text{osc}},t-\tau}\right\} + \sin^2(\theta)\mathbb{E}\left\{q_{\text{n}_{\text{osc}},t} q_{\text{n}_{\text{osc}},t-\tau}\right\}\\ &= \cos^2(\theta)\gamma_{p_{\text{n}_{\text{osc}}}\times p_{\text{n}_{\text{osc}}}}(\tau) + \sin^2(\theta)\gamma_{q_{\text{n}_{\text{osc}}}\times q_{\text{n}_{\text{osc}}}}(\tau).\end{aligned}$$

As highlighted in "Impact of stationarity" (Section 1.1.3), the stationarity of $n_{\text{osc}}(t)$ also leads to the real and imaginary parts of its complex envelope having the same autocorrelation function,

$$\gamma_{p_{\text{n}_{\text{osc}}}\times p_{\text{n}_{\text{osc}}}}(\tau) = \gamma_{q_{\text{n}_{\text{osc}}}\times q_{\text{n}_{\text{osc}}}}(\tau). \tag{4.163}$$

We thus get that

$$\gamma_{\rho_{\epsilon\|}\times\rho_{\epsilon\|}}(\tau) = \gamma_{p_{\text{n}_{\text{osc}}}\times p_{\text{n}_{\text{osc}}}}(\tau) = \gamma_{q_{\text{n}_{\text{osc}}}\times q_{\text{n}_{\text{osc}}}}(\tau). \tag{4.164}$$

Taking the Fourier transform of this equation, we finally obtain

$$\Gamma_{\rho_{\epsilon\|}\times\rho_{\epsilon\|}}(\omega) = \Gamma_{p_{\text{n}_{\text{osc}}}\times p_{\text{n}_{\text{osc}}}}(\omega) = \Gamma_{q_{\text{n}_{\text{osc}}}\times q_{\text{n}_{\text{osc}}}}(\omega). \tag{4.165}$$

An identical derivation could be done to get an expression for $\Gamma_{\rho_{\perp}\times\rho_{\perp}}(\omega)$. This would result in:

$$\Gamma_{\rho_{\epsilon\perp}\times\rho_{\epsilon\perp}}(\omega) = \Gamma_{p_{\text{n}_{\text{osc}}}\times p_{\text{n}_{\text{osc}}}}(\omega) = \Gamma_{q_{\text{n}_{\text{osc}}}\times q_{\text{n}_{\text{osc}}}}(\omega). \tag{4.166}$$

At the same time, we get that the stationarity and spectral symmetry of $n_{\text{osc}}(t)$ yields that the PSD of $p_{\text{n}_{\text{osc}}}$ and $q_{\text{n}_{\text{osc}}}$ is half that of the complex envelope of $n_{\text{osc}}(t)$ by equation (1.51). We therefore get that

$$\Gamma_{\rho_{\epsilon\|}\times\rho_{\epsilon\|}}(\omega) = \Gamma_{\rho_{\epsilon\perp}\times\rho_{\epsilon\perp}}(\omega) = \frac{1}{2}\Gamma_{\tilde{n}_{\text{osc}}\times\tilde{n}_{\text{osc}}}(\omega). \tag{4.167}$$

Finally, we can derive the PSD of $n_{\rho_{\text{osc}}}(t)$ and $n_{\phi_{\text{osc}}}(t)$ by using their relationship with $\rho_{\epsilon\|}(t)$ and $\rho_{\epsilon\perp}(t)$ (equation (4.153)):

$$\Gamma_{n_{\rho_{\text{osc}}}\times n_{\rho_{\text{osc}}}}(\omega) = \frac{1}{2}\Gamma_{\tilde{n}_{\text{osc}}\times\tilde{n}_{\text{osc}}}(\omega), \tag{4.168a}$$

$$\Gamma_{n_{\phi_{\text{osc}}}\times n_{\phi_{\text{osc}}}}(\omega) = \frac{1}{2\rho^2}\Gamma_{\tilde{n}_{\text{osc}}\times\tilde{n}_{\text{osc}}}(\omega). \tag{4.168b}$$

As a result, we get the interesting result that both the amplitude noise $n_{\rho_{osc}}(t)$ and the phase noise $n_{\phi_{osc}}(t)$ have the same spectral shape as the original RF bandpass noise $n_{osc}(t)$ delivered by the free-running RF oscillator.

We conclude with a comment on the relative magnitude of the two noise components under discussion. Integrating equation (4.168) while taking into account the relationship between the power of a bandpass process and its complex envelope as given by equation (1.64), we can write that

$$P_{\mathrm{n}_{\rho_{osc}}} = \frac{1}{2} P_{\tilde{\mathrm{n}}_{osc}} = P_{\mathrm{n}_{osc}}, \tag{4.169a}$$

$$P_{\mathrm{n}_{\phi_{osc}}} = \frac{1}{2} \frac{2}{\rho^2} \frac{P_{\tilde{\mathrm{n}}_{osc}}}{2} = \frac{1}{2} \frac{2}{\rho^2} P_{\mathrm{n}_{osc}} = \frac{1}{2} \frac{P_{\mathrm{n}_{osc}}}{P_{CW}}. \tag{4.169b}$$

Here P_{CW} represents the power of the wanted signal at the oscillator output, i.e. the power of $\rho\cos(\omega_{osc}t + \theta)$. Looking at this equation, we see that the power $P_{\mathrm{n}_{\phi_{osc}}}$ of the *lowpass* term $n_{\phi_{osc}}(t)$ that corrupts the instantaneous phase of the wanted signal is negligible regarding the power $P_{\mathrm{n}_{\rho_{osc}}}$ of the *lowpass* term $n_{\rho_{osc}}(t)$ that corrupts the instantaneous amplitude of this wanted signal. It thus looks like all the power of the bandpass noise $n_{osc}(t)$ is transferred to the amplitude part of the noise only. This is in fact only a way of representing the phenomenon. $n_{\phi_{osc}}(t)$ represents a normalized version of the modulus of $\tilde{n}_{\perp}(t)$ by the amplitude of the CW wanted signal in order to get a phase term. But, under the small angle approximation, this phase noise term can be expanded as an additive bandpass noise term that adds to the wanted signal. This additive term is nothing more than $n_{\perp}(t)$ in practice. And in the same way $n_{\rho_{osc}}(t)$ can be expanded back to $n_{\parallel}(t)$. Both of those terms are additive *bandpass* to the wanted signal and then carry half of the power of the original bandpass noise component $n_{osc}(t)$, as discussed at the end of Section 1.2.2.

Phase Locked Loop

In practical implementations, the LO signal is not directly derived from the signal flowing from a free-running RF oscillator. There are numerous reasons for this. One reason is the spread in the devices in mass production. This can result in huge variations for the resonant frequencies of the oscillator. In the same way LC tanks are inherently sensitive to coupling with other RF signals [44–47]. Both phenomena could result in unacceptable inaccuracies in the frequency of oscillation, depending on the environment. We also mention that in real implementations we expect the reuse of the same oscillator to generate many RF carrier frequencies, as can be required in many applications. All those reasons lead to the necessity on the one hand to make the resonant frequency of the RF oscillator variable and on the other hand to find a way to tune it toward the exact expected frequency. In an LC tank resonator, the first point is classically addressed by making the capacitor C variable. For ring oscillators, this can be achieved through the modification of the supply voltage, for instance. Depending on whether this variation is continuous or discrete, it results in a voltage controlled oscillator or a digitally controlled oscillator. But whatever the structure of the RF oscillator, the accurate tuning of its resonance frequency is always achieved through the use of a PLL that locks it to an accurate low frequency reference [48]. The overall system composed of the RF oscillator and the PLL thus behaves as a programmable frequency multiplier of this accurate low frequency reference.

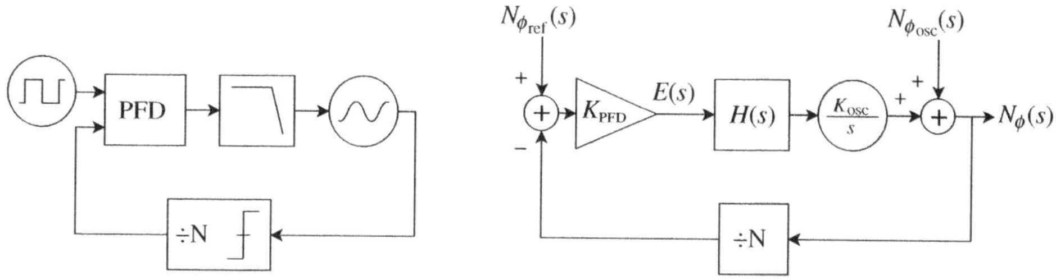

Figure 4.15 Simple integer phase locked loop (PLL) and equivalent phase noise model – An integer PLL locks the instantaneous phase/frequency of the RF oscillator, once it is divided by N, to the instantaneous phase/frequency of a stable reference. Usually, this is done through the use of a phase frequency detector (PFD) which provides an error signal that is further filtered before being provided to the oscillator (left). Even if this PFD is often implemented as working on the edges of the incoming signals, the equivalent noise model of the PLL applies to the fundamental tones of both the oscillator and the reference. Here, only the noise contributions of the reference and the oscillator are taken into account (right).

In practical transceivers, the exact setting of this multiplication value is the responsibility of the AFC function presented in Sections 9.1.5 and 9.2.5.

Reconsidering our discussion so far, we can understand the interest in investigating further the characteristics of the phase noise term flowing from a PLL in order to derive the properties of the LO signal. Although it is not very realistic for most practical products, for the sake of simplicity we can focus on the integer PLL structure shown in Figure 4.15(left). We then need to find an equivalent phase noise model for it. To do so we must identify each noise source in the system and derive the transfer function experienced by its contribution up to the PLL output. Although all the blocks in the system add some noise in practice, we can restrict our simple analysis to the two main contributors in order to highlight the behavior we are looking for: the bandpass noise term flowing from the oscillator, $n_{\rm osc}(t)$; and the bandpass contribution from the reference source, $n_{\rm ref}(t)$. As we are focusing here on the phase noise that is added to the wanted signal flowing from the oscillator, we can restrict ourselves to the phase noise contributions $n_{\phi_{\rm osc}}(t)$ and $n_{\phi_{\rm ref}}(t)$ linked to those bandpass noise terms.

It then remains to derive the transfer function experienced by those phase noise terms. And for that, we need an equivalent model that applies to the instantaneous phase of the bandpass signals in our system. We may thus need to anticipate the discussion in "Frequency division" (Section 4.3.2) and observe that frequency dividers are often implemented as edge counters. Such devices thus work on a squared copy of the CW signal in the RF oscillator. The same often holds for a phase frequency detector (PFD) device. We also recall from "Phase noise vs. timing jitter" (Section 4.3.2) the equivalence between the jitter of this square wave and the phase noise of its fundamental tone. In the present case, this fundamental tone is nothing more than the CW in the RF oscillator, i.e. the tone of interest for the generation of the LO signal in practice. Based on the discussion in those sections, we then rely on the equivalent phase noise model shown in Figure 4.15(right) for this tone.

Based on this model, we first focus on the transfer functions for the phase noise term $n_{\phi_{\rm ref}}(t)$ associated with the noisy reference. Assuming that $n_{\phi_{\rm osc}}(t) = 0$, we can then write in the Laplace domain the error signal of the loop $E(s)$ as

$$E(s) = K_{\rm PFD}\left(N_{\phi_{\rm ref}}(s) - \frac{1}{N}N_{\phi,{\rm LP}}(s)\right). \tag{4.170}$$

Here $N_{\phi,\mathrm{LP}}(s)$ represents the Laplace transform of the phase noise term recovered at the output of the oscillator in the present case. At the same time, the signal present at the input of the oscillator is converted into a variation of the instantaneous frequency of the bandpass signal delivered by this device through a conversion gain K_{osc}. The instantaneous phase of this signal is therefore the integral of its instantaneous frequency. We can thus write

$$N_{\phi,\mathrm{LP}}(s) = K_{\mathrm{osc}}\frac{H(s)}{s}E(s). \tag{4.171}$$

The two relationships above then lead to a first transfer function $H_{\mathrm{LP}}(s)$ experienced by the phase noise flowing from the reference:

$$H_{\mathrm{LP}}(s) = \frac{N_{\phi,\mathrm{LP}}(s)}{N_{\phi_{\mathrm{ref}}}(s)} = N\frac{H(s)}{H(s) + sN/K}, \tag{4.172}$$

with

$$K = K_{\mathrm{PFD}}K_{\mathrm{osc}}. \tag{4.173}$$

In the same way, we can derive the transfer function $H_{\mathrm{HP}}(s)$ experienced by the phase noise flowing from the oscillator when $n_{\phi_{\mathrm{ref}}}(t) = 0$. Denoting by $N_{\phi,\mathrm{HP}}(s)$ the Laplace transform of this term, we get that

$$H_{\mathrm{HP}}(s) = \frac{N_{\phi,\mathrm{HP}}(s)}{N_{\phi_{\mathrm{osc}}}(s)} = \frac{sN/K}{H(s) + sN/K}. \tag{4.174}$$

We assume now that the loop filter is a simple integrator, i.e. that $H(s) = 1/s$. This assumption obviously does not make our example more realistic, for instance in terms of loop stability, but it remains meaningful for illustrating our problem. We can now write $H_{\mathrm{LP}}(s)$ and $H_{\mathrm{HP}}(s)$ as

$$H_{\mathrm{LP}}(s) = N\frac{1}{1 + (s/\omega_{\mathrm{c}})^2}, \tag{4.175a}$$

$$H_{\mathrm{HP}}(s) = \frac{(s/\omega_{\mathrm{c}})^2}{1 + (s/\omega_{\mathrm{c}})^2}, \tag{4.175b}$$

with

$$\omega_{\mathrm{c}} = \sqrt{K/N}. \tag{4.176}$$

We thus see in that case that the transfer function experienced by the phase noise contribution from the reference, $H_{\mathrm{LP}}(s)$, is a second order lowpass transfer function with a 3 dB cut-off angular frequency equal to ω_{c}. But what is interesting to see is that the DC gain of this transfer function is equal to the division ratio in the feedback path of the PLL, N. This behavior comes from the fact that the instantaneous phase of the signal recovered at the output of the oscillator is first divided by a factor N before being compared to the reference phase/frequency. When the PLL is locked, i.e. when the error signal ϵ is equal to 0, we necessarily have that $N_\phi(s)/N$ is equal to $N_{\phi_{\text{ref}}}(s)$. In other words, we then get that $N_\phi(s) = NN_{\phi_{\text{ref}}}(s)$. We thus recover here that the PLL behaves as a frequency multiplier by a factor N. However, this behavior holds only up to the passband edge of the loop. Higher frequency components from the reference and from the feedback path are necessarily filtered out by the loop filter so that no correction on the oscillator can occur. Only the low part of the reference phase noise spectrum can be replicated on the output signal of the PLL, and with a gain N. Conversely, the signal of the oscillator is corrected by the loop, at least for its components that lie within its passband. In that case the loop corrects the phase of the signal flowing from the oscillator in order to make it equal to N times the reference phase as just discussed. In the passband of the loop the phase noise term $n_{\phi_{\text{osc}}}(t)$ from the oscillator is thus canceled, whereas outside it this term is directly recovered at the PLL output. This results in an equivalent highpass filter for the transfer function $H_{\text{HP}}(s)$ experienced by the phase noise from the oscillator. Moreover, the 3 dB cut-off frequency for this equivalent highpass filter is the same as for the lowpass filter experienced by the phase noise from the reference.

We can now give an expression for the total phase noise $n_\phi(t)$ that corrupts the fundamental tone of the bandpass signal flowing from the PLL. By superposition, it can be written as the sum

$$\begin{aligned} n_\phi(t) &= n_{\phi,\text{LP}}(t) + n_{\phi,\text{HP}}(t) \\ &= h_{\text{LP}}(t) \star n_{\phi_{\text{ref}}}(t) + h_{\text{HP}}(t) \star n_{\phi_{\text{osc}}}(t). \end{aligned} \tag{4.177}$$

Given the independence of $n_{\phi_{\text{ref}}}(t)$ and $n_{\phi_{\text{osc}}}(t)$, and assuming the stationarity of those processes, the PSD of $n_\phi(t)$ can be evaluated using the interference formula given by equation (A1.21). We finally obtain

$$\Gamma_{n_\phi \times n_\phi}(\omega) = |H_{\text{LP}}(\omega)|^2 \Gamma_{n_{\phi_{\text{ref}}} \times n_{\phi_{\text{ref}}}}(\omega) + |H_{\text{HP}}(\omega)|^2 \Gamma_{n_{\phi_{\text{osc}}} \times n_{\phi_{\text{osc}}}}(\omega). \tag{4.178}$$

Here $H_{\text{LP}}(\omega)$ and $H_{\text{HP}}(\omega)$ can be evaluated from equation (4.175) using the fact that for s pure imaginary, $s = \text{j}\omega$. An example of those contributions is shown in Figure 4.16 using a realistic PLL cut-off frequency of 50 kHz. This simulation result confirms that in the passband of the PLL, the phase noise from the oscillator is corrected by the loop so that the output spectrum corresponds to that of the reference signal. Out of the passband, the phase noise spectrum is simply that of the free-running oscillator. Practically speaking, even if only two sources of noise in the PLL are considered, the conclusions remain valid in any case. The fact that the PLL can track and correct the instantaneous phase of the output signal only in its passband is general. Conversely, the contributions coming from blocks that occur prior to the loop filter are filtered by it before reaching the oscillator. As a result, it is the phase noise characteristics of the free-running oscillator that we recover outside the loop passband at the PLL output.

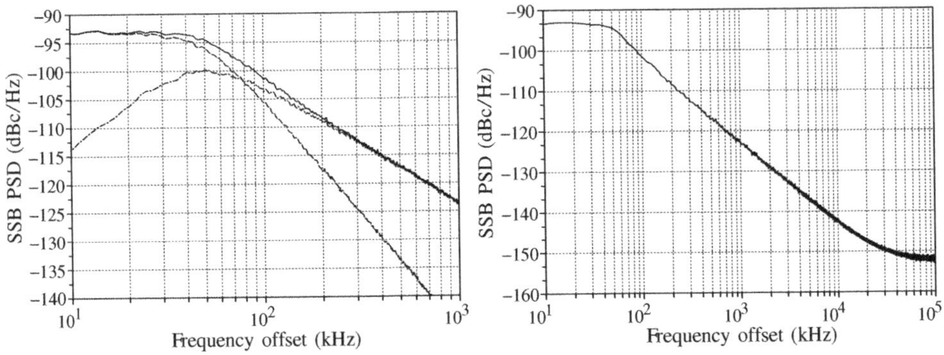

Figure 4.16 Phase noise spectrum at the PLL output – In the passband of the loop, the phase noise recovered at the PLL output is set proportional to that of the reference. This results in an equivalent highpass filtering of the free-running oscillator contribution (left, dot-dashed) and an equivalent lowpass filtering on the contribution from the reference (left, dashed). Both transfer functions have the same cut-off frequency at 3 dB, here set to 50 kHz. Due to the independence between the sources, the phase noise spectrum at the PLL output (left, solid) is simply the sum of those for the elementary contributors. As a result, the wideband phase noise spectrum at the PLL output (right) reduces to that of the free-running oscillator as soon as we consider frequency offsets greater than the PLL closed loop cut-off frequency, plus a potential noise floor due to unavoidable internal contributions.

Based on our discussion so far, we can summarize the characteristics of the LO signal $lo(t)$ as recovered at the output of the PLL system. We can proceed as in the previous section and decompose the additive bandpass noise $n_{\text{pll}}(t)$ that corrupts the wanted signal $\rho\cos(\omega_{\text{LO}}t+\theta)$ delivered by the RF oscillator in terms of amplitude and phase noise, $n_\rho(t)$ and $n_\phi(t)$, respectively. This leads to

$$\begin{aligned} lo(t) &= \rho\cos(\omega_{\text{LO}}t+\theta)+n_{\text{pll}}(t) \\ &= (\rho+n_\rho(t))\cos(\omega_{\text{LO}}t+\theta+n_\phi(t)). \end{aligned} \tag{4.179}$$

Given that the PLL system reacts only on the instantaneous phase of the bandpass signal delivered by the RF oscillator, $n_\rho(t)$ remains exactly the same as in the free-run case. All the characteristics discussed in the previous section thus hold for this term in the present case. This is obviously not so straightforward for the phase noise term $n_\phi(t)$ as the free-run contribution from the oscillator is compensated in the passband of the PLL. At the same time contributions from constituent parts of the PLL are recovered as components of $n_\phi(t)$ in the low frequency part of its spectrum. This results in a typical spectral shape for $n_\phi(t)$ where the phase noise spectrum of the free-running oscillator is recovered above the PLL cut-off frequency, as illustrated in Figure 4.16. On top of that, if all the noise contributions in the system can be considered bandpass Gaussian, $n_{\text{pll}}(t)$ can be approximated as bandpass Gaussian. This property then holds for $n_\phi(t)$ by the previous section. Finally, an illustration

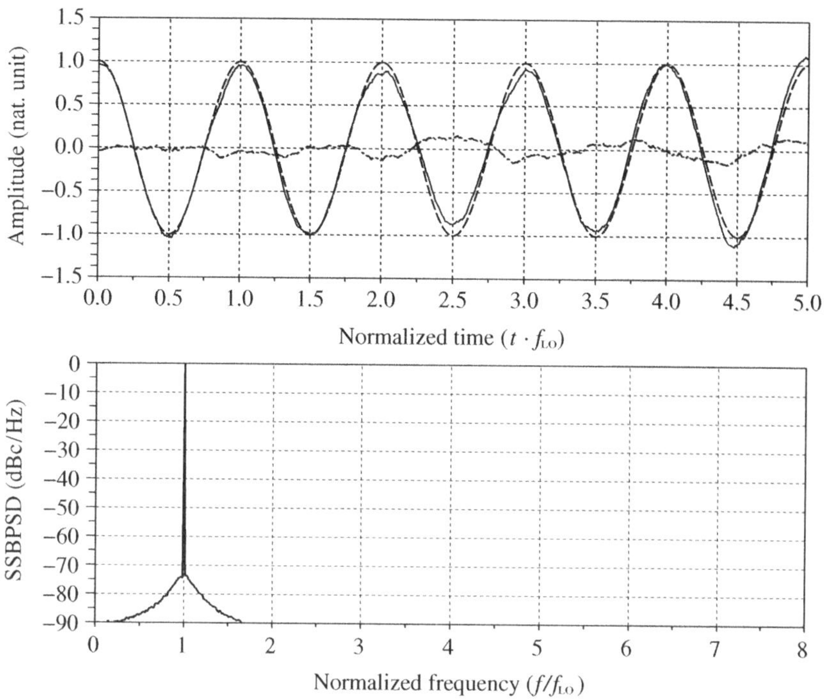

Figure 4.17 Noisy LO sinusoidal waveform in the time and frequency domains – A noisy LO sinusoidal waveform composed of the sum of a pure sine wave at the angular frequency ω_{LO} (top, dashed) and a bandpass noise centered around the same angular frequency (top, dot-dashed) results in a pseudo-sinusoidal waveform that exhibits both amplitude and phase noise (top, solid). In this simulation, the bandpass noise power is set 20 dB below the sine wave power and its bandwidth to a tenth of ω_{LO} as illustrated on its PSD (bottom).

of the bandpass signal $lo(t)$ as effectively recovered at the output of such system is shown in Figure 4.17.

4.3.2 *Square LO Waveform for Chopper-like Mixers*

Noisy Square LO Waveform Characteristics

As discussed in Chapter 6, practical RF/analog mixers often behave as choppers. From the signal processing point of view, the operation of such a device can be seen as the multiplication of the signal to be frequency transposed by a square LO waveform, as illustrated in Section 6.3.1.

In practical implementations, such an LO waveform is obtained mostly using inverter-like buffers that transform the sine wave delivered by the resonant part of the oscillator embedded in an RF synthesizer in a square waveform. From the signal processing point of view, all thus behaves as if this sine wave had gone through a hard limiter as illustrated in Figure 4.18. Based on the characteristics for the sinusoidal waveform $lo(t)$ derived in the previous section, we can

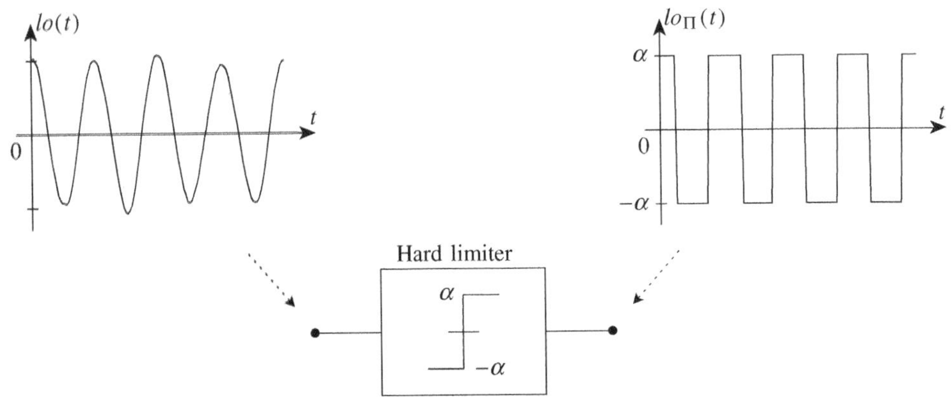

Figure 4.18 Square LO waveform seen as the result of a sine wave that has gone through a hard limiter – The characteristics of the square LO waveform as used for chopper-like mixers can be derived from the hard limiter theory and the characteristics of the sine wave as generated by RF synthesizers.

thus expect to be able to derive the characteristics of the square LO waveform, $lo_{\Pi}(t)$, using the hard limiter theory. Referring to the expression for $lo(t)$ given by equation (4.179), it can be shown that $lo_{\Pi}(t)$ takes the form [4]

$$\begin{aligned} lo_{\Pi}(t) &= \sum_{\substack{m=1\\ m\ \text{odd}}}^{\infty} lo_m(t) \\ &= \sum_{\substack{m=1\\ m\ \text{odd}}}^{\infty} 2C(0,m)\cos(m\omega_{\text{LO}}t + m\theta + mn_{\phi}(t)), \end{aligned} \tag{4.180}$$

with

$$2C(0,m) = \frac{\epsilon_m \alpha}{\Gamma\left(1-\frac{m}{2}\right)\Gamma\left(1+\frac{m}{2}\right)}. \tag{4.181}$$

Here, ϵ_m is the Neumann factor defined as $\epsilon_0 = 1$ and $\epsilon_m = 2$ for all $m \neq 0$, $\Gamma(.)$ denotes the gamma function as defined by equation (1.112) and α is the amplitude of the hard limiter. Looking at this expression, we observe that the amplitude noise component of $lo(t)$ has effectively been canceled by the hard limiter. All that remains is the phase noise part of the original bandpass noise that corrupted the sine wave at the RF synthesizer output. This explains why only the phase noise of the LO waveform is of importance in terms of system impacts when dealing with chopper-like mixers.

In almost all cases of interest, the frequency planning of the frequency conversion is such that this is the fundamental tone of the decomposition of $lo_{\Pi}(t)$ that is used for the frequency conversion of the wanted signal. One obvious reason for that is that this is the tone of highest amplitude, thus offering the highest conversion gain. From now we can thus focus on the characteristics of this particular component of interest for our system design perspective, $lo_1(t)$. According to equations (4.180) and (4.181), this term takes the form

$$\begin{aligned} lo_1(t) &= 2C(0,1)\cos(\omega_{\mathrm{LO}}t + \theta + n_\phi(t)) \\ &= \frac{4\alpha}{\pi}\cos(\omega_{\mathrm{LO}}t + \theta + n_\phi(t)). \end{aligned} \tag{4.182}$$

In order to perform analytical derivations, we can consider the complex envelope of this bandpass waveform, $\tilde{lo}_1(t)$, defined as centered around ω_{LO}. Given that

$$lo_1(t) = \mathrm{Re}\left\{\tilde{lo}_1(t)e^{j\omega_{\mathrm{LO}}t}\right\}, \tag{4.183}$$

we can thus write

$$\tilde{lo}_1(t) = \frac{4\alpha}{\pi}e^{j(\theta + n_\phi(t))}. \tag{4.184}$$

At this stage, it is realistic to assume that the signal delivered by the RF synthesizer exhibits a good SNR. By the discussion in "Revisiting amplitude and phase noise" (Section 4.3.1) we can then write that $|n_\phi(t)| \ll 2\pi$ for all t. This last result allows us to use the small angle approximation in the above equation, leading to

$$\tilde{lo}_1(t) = \frac{4\alpha}{\pi}e^{j\theta}e^{jn_\phi(t)} \approx \frac{4\alpha}{\pi}e^{j\theta}(1 + jn_\phi(t)). \tag{4.185}$$

We thus see that $\tilde{lo}_1(t)$ can be expressed as the sum of the complex envelope of the expected pure LO sine wave at ω_{LO}, and a time-varying noise term that is nothing more than the original phase noise term $n_\phi(t)$.

Based on this decomposition, we can now derive the characteristics of the spectrum of $lo_1(t)$. We first derive the autocorrelation of $\tilde{lo}_1(t)$. From the above equation, we have

$$\gamma_{\tilde{lo}_1 \times \tilde{lo}_1}(\tau) = \mathbb{E}\{\tilde{lo}_{1,t}\tilde{lo}^*_{1,t-\tau}\} = \left(\frac{4\alpha}{\pi}\right)^2 \mathbb{E}\{(1 + jn_{\phi,t})(1 - jn_{\phi,t-\tau})\}. \tag{4.186}$$

Now referring to the stationarity of $n_\phi(t)$, we can write that $\mathbb{E}\{n_{\phi,t}\} = \mathbb{E}\{n_{\phi,t-\tau}\}$. After expansion, the expression for $\gamma_{\tilde{lo}_1 \times \tilde{lo}_1}(\tau)$ then becomes

$$\begin{aligned} \gamma_{\tilde{lo}_1 \times \tilde{lo}_1}(\tau) &= \left(\frac{4\alpha}{\pi}\right)^2 (1 + \mathbb{E}\{n_{\phi,t}n_{\phi,t-\tau}\}) \\ &= \left(\frac{4\alpha}{\pi}\right)^2 (1 + \gamma_{n_\phi \times n_\phi}(\tau)). \end{aligned} \tag{4.187}$$

Now taking the Fourier transform of this equation, we get

$$\Gamma_{\tilde{lo}_1 \times \tilde{lo}_1}(\omega) = \left(\frac{4\alpha}{\pi}\right)^2 \left(\delta(\omega) + \Gamma_{n_\phi \times n_\phi}(\omega)\right). \tag{4.188}$$

We then have the remarkable result that the spectrum centered around the fundamental tone of the square LO waveform $lo_\Pi(t)$ is nothing more than that of the phase noise term of the original sinusoidal noisy LO waveform flowing from the resonant part of the RF oscillator embedded in the synthesizer. But, by the section "Phase locked loop" (Section 4.3.1), we get that outside the passband of the RF synthesizer, the spectrum of the phase noise that corrupts $lo(t)$, i.e. $\Gamma_{n_\phi \times n_\phi}(\omega)$, matches the spectrum of the phase noise contribution of the free-running oscillator, i.e. $\Gamma_{n_{\phi_{osc}} \times n_{\phi_{osc}}}(\omega)$. Now by the section "Revisiting amplitude and phase noise" (Section 4.3.1), we get in turn that $\Gamma_{n_{\phi_{osc}} \times n_{\phi_{osc}}}(\omega)$ is proportional to $\Gamma_{\tilde{n}_{osc} \times \tilde{n}_{osc}}(\omega)$ if we refer to equation (4.168b). We thus see that the spectral shape of the additive bandpass noise corrupting $lo_1(t)$ matches the spectral shape of the additive bandpass noise flowing from the free-running oscillator, at least outside the passband of the synthesizer. At the same time we have that the amplitude noise was canceled when going through the hard limiter so that only the phase noise part of $n_{osc}(t)$ is recovered in $lo_1(t)$. From the discussion in Section 1.2.2, we then expect that although they have the same spectral shape, the bandpass noise that corrupts $lo_1(t)$ is 3 dB weaker than $n_{osc}(t)$. This is indeed confirmed by simulations, as illustrated in Figure 4.19.

In conclusion, this correspondence in the spectral shape of the different bandpass noises we are dealing with does not hold when considering the bandpass components centered around the higher order harmonics of $lo_\Pi(t)$. It can be shown that for a phase noise term $n_\phi(t)$ with a Gaussian distribution, the spectrum sidebands located on the mth harmonics involve mth order self-convolutions of the spectrum of $n_\phi(t)$ [4]. In that case, the self-convolution leads to a spectral regrowth according to the mechanism discussed in "Spectral regrowth" (Section 5.1.3). This should be considered in terms of system impacts if one expects to use frequency planning such that the wanted signal is converted by a higher-order tone instead of the fundamental one as is usually done.

Phase Noise vs. Timing Jitter

When dealing with square LO signals, or more generally with any signal that is used to drive events through its edges, for example clock signals in digital implementations or in analog to digital conversions as detailed in Section 4.6, the amount of noise that corrupts the waveform is classically characterized through the uncertainty in the location of those edges in the time domain. Given that the amplitude of such theoretical waveform is fixed, the only uncertainty that remains is obviously the time at which the switching occurs. This timing accuracy is classically measured through the concept of timing jitter that is nothing more for such square wave than the time difference between the actual transitions and those we would have in the case of a noiseless waveform. However, the quantity of interest from the transceiver system design point of view remains the phase noise term recovered as centered around the fundamental tone in the series expansion of this waveform, as discussed in the previous section. There is thus an interest in detailing the relationship between the timing jitter of the square waveform and the characteristics of this phase noise.

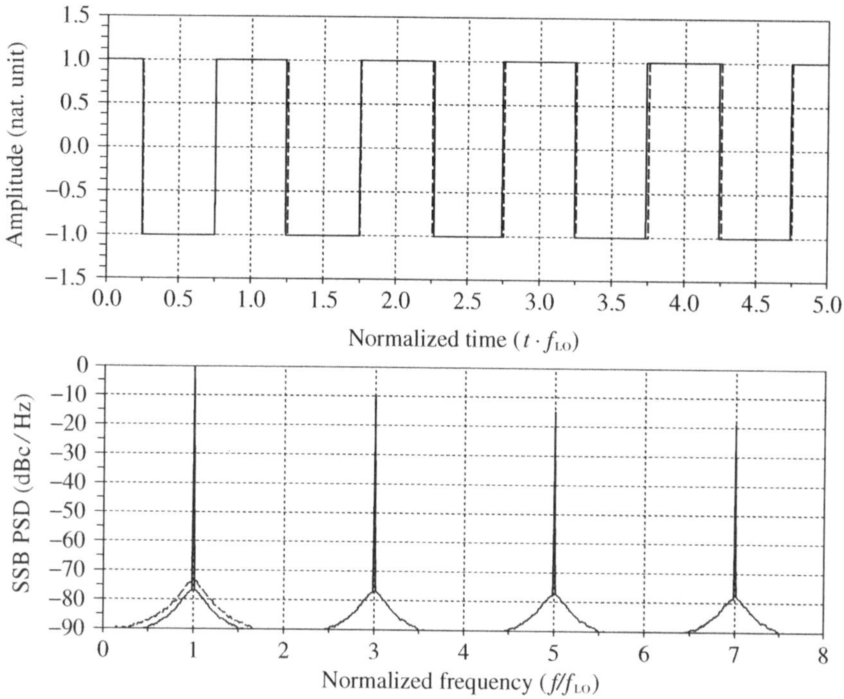

Figure 4.19 Noisy square LO waveform in the time and frequency domains – The square LO waveform can be interpreted as the sine waveform shown in Figure 4.17 seen at the hard limiter output. We get a jitter error on the transitions of the noisy waveform (top, solid) compared to the ideal waveform (top, dashed). The bandpass noise term centered on the fundamental tone corresponds to the phase noise term of the sinusoidal waveform. Thus, outside the passband of the PLL used for the synthesis, the PSD of this output noise sideband (bottom, solid) is 3 dB lower than the original one centered on the sinusoidal LO waveform (bottom, dashed).

In that perspective, we first observe that the switching instants of such square waveform corresponds to the zeros of all the tones involved in its series expansion. This can be understood by reconsidering the expansion of the signal $lo_{\Pi}(t)$ discussed in the previous section. Based on equation (4.180), and supposing for the sake of clarity that $\theta = 0$, we can write

$$lo_{\Pi}(t) = \sum_{\substack{m=1 \\ m \text{ odd}}}^{\infty} lo_m(t) = \sum_{\substack{m=1 \\ m \text{ odd}}}^{\infty} 2C(0, m)\cos(m\omega_{\text{LO}}t + mn_{\phi}(t)), \tag{4.189}$$

with $2C(0, m)$ given by equation (4.181). We then see that the instantaneous phase, $\phi_m(t)$, of the mth tone, $lo_m(t)$, can be written as

$$\phi_m(t) = m\omega_{\text{LO}}t + mn_{\phi}(t) = m(\omega_{\text{LO}}t + n_{\phi}(t)) = m\phi_1(t). \tag{4.190}$$

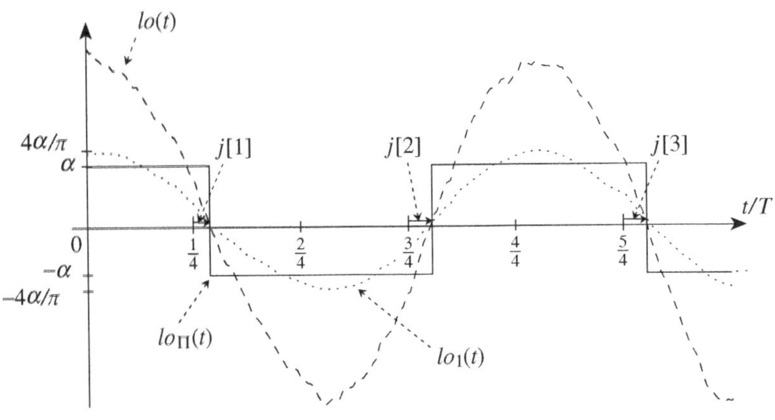

Figure 4.20 Jitter sequence of a noisy LO square wave and relationships with corresponding sinusoidal waveforms – A square LO signal, $lo_{\Pi}(t)$ (solid), generated from a noisy sinusoidal waveform, $lo(t)$ (dashed), having gone through a hard limiter of amplitude α, exhibits a timing uncertainty on its switching points compared to the noiseless case. The sequence of timing errors defines the jitter sequence $j[k]$. The fundamental tone $lo_1(t)$ (dotted) in the decomposition of $lo_{\Pi}(t)$ is corrupted by the same phase noise term as the original sinusoidal waveform $lo(t)$.

As a consequence, when $\phi_1(t)$ is of the form $\pi/2 + k\pi$, $\phi_m(t) = m\phi_1(t)$ is necessarily also of the form $\pi/2 + k'\pi$ as m is an odd number. All the harmonics involved in the decomposition of $lo_{\Pi}(t)$ thus vanish at the same time. Consequently, the timing jitter sequence corresponding to the square waveform $lo_{\Pi}(t)$ can be evaluated through the timing jitter function associated with each of those tones when sampled at the instants corresponding to their zeros.

Due to this equivalence, illustrated in Figure 4.20, we can now focus on the timing jitter function of the fundamental tone $lo_1(t)$ in order to make the link with its phase noise term $n_{\phi}(t)$. We may recall that this jitter function is defined for a continuous sinusoidal waveform as the time difference such that the instantaneous phase of the noisy and ideal waveforms are equal. This time difference is thus defined as a function of the phase φ_1 of this ideal waveform. Thus, the jitter function $j(\varphi_1)$ of $lo_1(t)$ is such that

$$\phi_1[t + j(\varphi_1)] = \varphi_1. \tag{4.191}$$

Practically speaking, by the expression for $lo_1(t)$ that can be deduced from equation (4.189), the ideal noiseless waveform whose φ_1 is the instantaneous phase can simply be written as $2C(0,1)\cos(\omega_{\text{LO}}t)$. We then get for such CW that

$$\varphi_1(t) = \omega_{\text{LO}}t. \tag{4.192}$$

As a consequence, we can use this linear relationship between φ_1 and t to define the timing jitter function j as a function of time instead of φ_1. This is more suitable for practical derivations.

On top of that, using this relationship in equation (4.191) and expanding the expression of $\phi_1(t)$ using equation (4.190), we get that

$$\omega_{\rm LO} j(t) + n_\phi[t + j(t)] = 0. \tag{4.193}$$

Or alternatively, as a function of the signal period T using the fact that $\omega_{\rm LO} = 2\pi/T$:

$$\frac{j(t)}{T} + \frac{n_\phi}{2\pi}[t + j(t)] = 0. \tag{4.194}$$

Based on a Taylor series expansion, it can then be shown that that the second term in this equation can be approximated so that finally [49]

$$\frac{j(t)}{T} \approx -\frac{n_\phi(t)}{2\pi}. \tag{4.195}$$

This approximation holds as long as

$$\Delta\omega j(t) \ll \frac{1}{2\pi}, \tag{4.196}$$

with $\Delta\omega$ the spectral extent of $j(t)$. Under this assumption, we then get that $j(t)$ is proportional to $n_\phi(t)$. We see that the two representations necessarily have the same spectral shape.

Finally, the timing jitter sequence $j[k]$ measured on the square waveform $lo_\Pi(t)$ at its switching instants is simply the sampled version of $j(t)$. Practically speaking, we get that the bandwidth $\Delta\omega$ of the noise term recovered at the output of realistic synthesizers is much lower than the carrier frequency $\omega_{\rm LO}$. There is thus no aliasing problem to deal with during such a sampling operation. As a result, the spectrum and the characteristics of the discrete timing jitter that applies to $lo_\Pi(t)$ can still be considered proportional to those of the LO phase noise term recovered at the synthesizer output. However, as discussed in Section 4.3.1, the noise spectrum at the output of such an RF synthesizer may also exhibit a noise floor that is wideband by construction. Thus, the sampled version $j[k]$ of $j(t)$ can exhibit some divergence with it for this wideband noise floor part. This behavior is illustrated in the next section when dealing with the frequency division operation.

Frequency Division

Let us now examine a final kind of device that is classically used to generate LO signals and that can impact their phase noise performance, namely the frequency divider. Such a device, already encountered in the feedback path of PLL systems in "Phase locked loop" (Section 4.3.1), is indeed often used to obtain a LO signal at an angular frequency $\omega_{\rm LO}$ different from that of the RF oscillator used for its generation. There are different reasons for that, all related to the frequency planning problem of the line-up. Among others, we can mention the reuse of the same oscillator to address the generation of various RF carriers in different frequency bands, or the minimization of the potential frequency pulling or injection locking of the RF oscillator in some architectures as illustrated in Section 8.1.1.

In order to explore the impact of this kind of device, it is of interest to recall the theoretical behavior expected from it. From the signal processing point of view, the obvious target is to have the instantaneous frequency of a given bandpass signal that is divided by a factor N. Recall from the discussion in "Instantaneous amplitude and instantaneous frequency" in Chapter 1 that the instantaneous frequency $f(t)$ of a bandpass signal is nothing more than the time domain derivative of its instantaneous phase. Therefore dividing the instantaneous frequency by a factor N results in a division of the instantaneous phase by the same amount. Application of this process to the fundamental tone $lo_1(t)$ of the decomposition of the noisy square LO signal $lo_\Pi(t)$, as given by equation (4.189), would thus lead to the generation of the signal:

$$lo_{1,\mathrm{o}}(t) = 2C(0,1)\cos\left(\frac{\omega_{\mathrm{LO}}}{N}t + \frac{n_\phi(t)}{N}\right). \tag{4.197}$$

We then see that the division of the instantaneous phase of $lo_1(t)$ leads to the division of both the carrier angular frequency ω_{LO}, and the phase noise term. As a result, the amplitude of this phase noise is reduced by a factor N when going through such a theoretical device, which is good news.[2] For instance, a 6 dB reduction in the phase noise power can be expected with a frequency division by a factor 2.

Nevertheless, the frequency division we have considered so far is a theoretical one. We can guess how this function might be implemented in practice and what the resulting effective behavior might be. Let us focus on the most widely used kind of divider, those that work as edge counters. Such a device directly processes a waveform with edges – a square waveform such as $lo_\Pi(t)$. Practically speaking, it simply counts the edges of the input waveform, using flip-flops for instance, and delivers a waveform in which only one transition over N is kept. As a result, the period T_o of the output waveform is increased by a factor N compared to that of the input, leading to a division by N of the signal angular frequency. This behavior is illustrated in Figure 4.21 in the case of $N = 2$. However, it remains to clarify the practical impact of this process on the phase noise that corrupts the tones involved in the series expansion of this square waveform.

For that purpose, we make the obvious remark that a counter can count an edge only when this edge reaches the counter. The time locations of the edges on the output waveform are thus directly related to the time locations of the corresponding edges on the input waveform. As a result, the jitter samples, $j_\mathrm{o}[k]$, of the waveform $lo_{\Pi,\mathrm{o}}(t)$ recovered at the output of the divider are simply equal to the jitter samples, $j[Nk]$, of the input waveform. But writing $j_\mathrm{o}[k] = j[Nk]$ shows immediately that the output jitter sequence is a downsampled version by a factor N of the input jitter sequence. Obviously, when referring to a downsampling process, we may consider the bandwidth of the signal being processed in order to see if we may face any aliasing issues. By the discussion in the previous section, in practical implementations the jitter can be considered as a narrowband process regarding the LO angular frequency, as is the phase noise of the fundamental tone $lo_1(t)$ in the series expansion of $lo_\Pi(t)$. We thus expect

---

[2] If $n_\phi(t)$ were a phase/frequency modulating signal instead of a noise component, the use of this kind of device on the modulated signal would be a problem as it would lead to a reduction in the range of the modulation. This is a phenomenon to take into account when considering the direct modulation of a PLL, for instance.

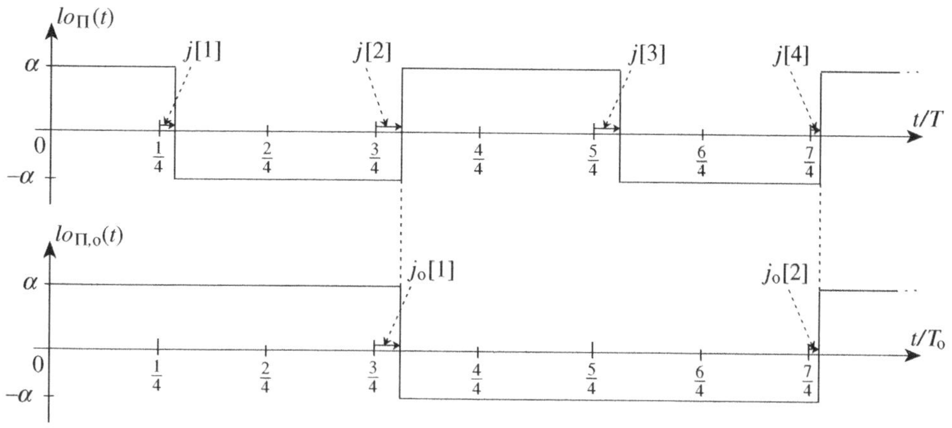

Figure 4.21 Jitter propagation through a frequency division of a square waveform – When a frequency divider is implemented as an edge counter operating on square waveforms, the edges of the output waveform occur at the same instant as the input waveform. We therefore get a propagation of the jitter of the input waveform to the output waveform.

no such aliasing issue and thus assume for the timing jitter functions of $lo_{1,\mathrm{o}}(t)$ and $lo_1(t)$ that $j_\mathrm{o}(t) = j(t)$. Then, given that equation (4.195) holds, we can write

$$\frac{j(t)}{T} = -\frac{n_\phi(t)}{2\pi}, \tag{4.198a}$$

$$\frac{j_\mathrm{o}(t)}{T_\mathrm{o}} = -\frac{n_{\phi,\mathrm{o}}(t)}{2\pi}, \tag{4.198b}$$

where $n_{\phi,\mathrm{o}}(t)$ and $n_\phi(t)$ are the phase noise terms that corrupt $lo_{1,\mathrm{o}}(t)$ and $lo_1(t)$, respectively. Then, given that the frequency division ratio is N, the output period T_o is simply N times the input period T. It then follows that $n_{\phi,\mathrm{o}}(t)$ and $n_\phi(t)$ are linked by

$$n_{\phi,\mathrm{o}}(t) = \frac{T}{T_\mathrm{o}} n_\phi(t) = \frac{n_\phi(t)}{N}. \tag{4.199}$$

We therefore recover the expected behavior of the ideal frequency division, i.e. a reduction of the input phase noise by a factor N. However, it may seem strange that although the jitter remains the same on the input and output waveforms, the phase noise decreases. In fact, although the two quantities represent a measure of the same noise component that corrupts the LO waveform, the phase noise is a normalized version of this noise term relative to the signal period. The difference in the behavior is thus only a matter of representation. However, we need to keep in mind that the additive bandpass noise processes that are recovered as centered around the harmonics of the LO waveform, and in particular around the fundamental tone, are proportional to the phase noise process in the small angle approximation, as discussed in "Noisy square LO waveform characteristics" earlier in this section. We thus get that this additive bandpass noise effectively experiences the noise reduction due to the frequency division.

However, thinking back to the discussion hold in the previous section, the narrowband assumption for the timing jitter does not hold for the unavoidable wideband noise floor. We then suspect a potential folding issue for this part of the spectrum during the frequency division process. In order to illustrate this, let us reconsider the example of the frequency division by a factor $N = 2$, shown in Figure 4.21. In this particular case, the PSD of $j_o(t)$ is a folded version of the PSD of $j(t)$ so that its spectral extent is now divided by 2. As a result, the noise floor power for $j_o(t)$ may be evaluated as twice that of $j(t)$:

$$\Gamma_{j_o \times j_o}(\omega) = 2\Gamma_{j \times j}(\omega), \quad \text{for } \omega > \omega_{\text{floor}}. \tag{4.200}$$

Here ω_{floor} represents the angular frequency above which the spectrum corresponds to the noise floor. But then, we can write from equation (4.198) that

$$\frac{\Gamma_{j \times j}(\omega)}{T^2} = \frac{\Gamma_{n_\phi \times n_\phi}(\omega)}{(2\pi)^2}, \tag{4.201a}$$

$$\frac{\Gamma_{j_o \times j_o}(\omega)}{T_o^2} = \frac{\Gamma_{n_{\phi,o} \times n_{\phi,o}}(\omega)}{(2\pi)^2}. \tag{4.201b}$$

Using equation (4.200), we then get

$$\Gamma_{n_{\phi,o} \times n_{\phi,o}}(\omega) = 2\left(\frac{T}{T_o}\right)^2 \Gamma_{n_\phi \times n_\phi}(\omega), \quad \text{for } \omega > \omega_{\text{floor}}. \tag{4.202}$$

Having in the present case that $T_o = 2T$, we finally can write that

$$\Gamma_{n_{\phi,o} \times n_{\phi,o}}(\omega) = \frac{1}{2}\Gamma_{n_\phi \times n_\phi}(\omega) \text{ for } \omega > \omega_{\text{floor}}. \tag{4.203}$$

This results in only a 3 dB improvement in the phase noise floor instead of the 6 dB expected when $N = 2$. In fact, this 3 dB improvement can be decomposed as the expected 6 dB linked to the frequency division but with an additional 3 dB degradation coming from the downsampling effect of the jitter process. This second effect, negligible in the frequency band where the PSD of the phase noise is non-negligible, i.e. close to the carrier frequency, is in fact an important limitation on the noise floor improvement during the frequency division process.

4.3.3 System Impact

Revisiting the Frequency Conversion

In order to examine the impact of the LO phase noise on the performance of a frequency conversion, we can focus on the elementary operation involved in a such process, i.e. the multiplication of the signal being transposed by a LO signal (see Chapter 6). It would be advantageous to revisit this operation in terms of the transformation of the complex envelope of the signal being processed. This is the most straightforward way to take into account the results derived in the previous sections about the structure of the noisy LO.

Let us suppose that we are dealing with the bandpass signal $s_\mathrm{i}(t)$ centered around the carrier angular frequency ω_i at the input of the frequency conversion process. We can then define its complex envelope $\tilde{s}_\mathrm{i}(t)$, centered around ω_i, such that

$$s_\mathrm{i}(t) = \mathrm{Re}\left\{\tilde{s}_\mathrm{i}(t)\mathrm{e}^{\mathrm{j}\omega_\mathrm{i}t}\right\}. \tag{4.204}$$

For a LO signal $lo(t)$ at the angular frequency ω_LO, its complex envelope $\tilde{lo}(t)$ defined as centered around this angular frequency is such that

$$lo(t) = \mathrm{Re}\left\{\tilde{lo}(t)\mathrm{e}^{\mathrm{j}\omega_\mathrm{LO}t}\right\}. \tag{4.205}$$

Using equation (1.5), we can now expand the real parts in these expressions so that the signal $s_\mathrm{i}(t)lo(t)$ recovered at the output of the mixing stage can be written as

$$\begin{aligned} s_\mathrm{i}(t)lo(t) &= \frac{1}{4}\left(\tilde{s}_\mathrm{i}(t)\mathrm{e}^{\mathrm{j}\omega_\mathrm{i}t} + \tilde{s}_\mathrm{i}^*(t)\mathrm{e}^{-\mathrm{j}\omega_\mathrm{i}t}\right)\left(\tilde{lo}(t)\mathrm{e}^{\mathrm{j}\omega_\mathrm{LO}t} + \tilde{lo}^*(t)\mathrm{e}^{-\mathrm{j}\omega_\mathrm{LO}t}\right) \\ &= \frac{1}{4}\left(\tilde{s}_\mathrm{i}(t)\tilde{lo}^*(t)\mathrm{e}^{\mathrm{j}(\omega_\mathrm{i}-\omega_\mathrm{LO})t} + \tilde{s}_\mathrm{i}^*(t)\tilde{lo}(t)\mathrm{e}^{-\mathrm{j}(\omega_\mathrm{i}-\omega_\mathrm{LO})t}\right. \\ &\quad \left. + \tilde{s}_\mathrm{i}(t)\tilde{lo}(t)\mathrm{e}^{\mathrm{j}(\omega_\mathrm{i}+\omega_\mathrm{LO})t} + \tilde{s}_\mathrm{i}^*(t)\tilde{lo}^*(t)\mathrm{e}^{-\mathrm{j}(\omega_\mathrm{i}+\omega_\mathrm{LO})t}\right). \end{aligned} \tag{4.206}$$

As might be expected from the discussion in Section 6.1.2, the signal recovered at the output of the processing is composed of a collection of sidebands. This is due to the presence of complex exponentials that are complex conjugates to each other in the expansions of the real bandpass signals involved here. Depending on the frequency planning that is under consideration for the frequency conversion of the wanted signal, one or other of these sidebands is of interest. In that respect we can assume that in practical implementations the unwanted sidebands are canceled using a zonal bandpass filter acting as an image reject filter. As a result, if for instance the sideband of interest is that centered around $\omega_\mathrm{i} + \omega_\mathrm{LO}$, the corresponding bandpass signal $s_\mathrm{o}(t)$ takes the form

$$s_\mathrm{o}(t) = \frac{1}{2}\mathrm{Re}\left\{\tilde{s}_\mathrm{i}(t)\tilde{lo}(t)\mathrm{e}^{\mathrm{j}(\omega_\mathrm{i}+\omega_\mathrm{LO})t}\right\}. \tag{4.207}$$

Selecting $\omega_\mathrm{o} = \omega_\mathrm{i} + \omega_\mathrm{LO}$ to define its complex envelope $\tilde{s}_\mathrm{o}(t)$, we can then write that

$$\tilde{s}_\mathrm{o}(t) = \frac{1}{2}\tilde{s}_\mathrm{i}(t)\tilde{lo}(t). \tag{4.208}$$

We thus recover the same behavior for complex envelopes as for the bandpass signals they represent, i.e. that $\tilde{s}_\mathrm{o}(t)$ is proportional to the product of the complex envelopes of the input and LO signals. We also observe that since by convention ω_i and ω_LO are positive quantities, the sideband centered on $\omega_\mathrm{i} + \omega_\mathrm{LO}$ necessarily corresponds to an upconversion case. Conversely, the other sideband centered on $|\omega_i - \omega_{LO}|$ corresponds to a downconversion. In that case, the complex envelope $\tilde{s}_o(t)$, defined as centered around $|\omega_i - \omega_{LO}|$ takes the form

$$\tilde{s}_o(t) = \frac{1}{2}\tilde{s}_i(t)\tilde{lo}^*(t) \tag{4.209}$$

or

$$\tilde{s}_o(t) = \frac{1}{2}\tilde{s}_i^*(t)\tilde{lo}(t), \tag{4.210}$$

depending on whether we are dealing with an infradyne or supradyne frequency conversion, i.e. whether $\omega_i - \omega_{LO}$ is positive or negative. In the following sections we can thus rely on one or other of these expressions, depending on the configuration considered. But we anticipate that the conclusions deduced in these particular cases remain general. We indeed see that the only difference is a complex conjugate on the complex envelopes involved in these expressions. This thus does not change the power of the corresponding processes. Only a potential flip in their PSD can be predicted in accordance with equation (1.8), for instance. But given that LO spectra are mostly symmetric according to the discussion so far, we see that the same conclusions hold whatever the structure considered for $\tilde{s}_o(t)$.

However, in order to be able to perform analytical derivations we also need to assume a structure for $\tilde{lo}(t)$. Obviously, focusing on a given sideband of interest at the output of the frequency conversion implicitly means that we can reduce the structure of $lo(t)$ to a sinusoidal form. By Section 4.3.2, this signal can advantageously be chosen as the fundamental tone in the series expansion of the square LO signal used for chopper-like mixers. We can then rely on equation (4.184) for the expression for $\tilde{lo}(t)$. However, to investigate the impact of the phase noise term, it would be preferable to have an expression in which the noise component is additive to the wanted signal. This is fortunately possible when the SNR of this bandpass signal is good enough. We can rely in that case to the small angle approximation to simplify this expression as per equation (4.185), thus yielding

$$\tilde{lo}(t) = \frac{4\alpha}{\pi}e^{j\theta}(1 + jn_\phi(t)). \tag{4.211}$$

SNR Limitation

A first limitation that can be analyzed when using a noisy LO waveform is the SNR degradation. Given that the additive bandpass noise component that corrupts the LO signal is centered around its carrier angular frequency ω_{LO}, we expect to recover it as centered around the wanted signal after the frequency conversion, thus leading to an in-band degradation.

To illustrate this behavior, suppose here that we are dealing with an upconversion. Referring to the previous section, we can then assume that the complex envelope $\tilde{s}_o(t)$ of the bandpass signal of interest recovered at the output of the process is given by equation (4.208). Assuming that the LO complex envelope $\tilde{lo}(t)$ is given by equation (4.211), we can thus write that

$$\tilde{s}_o(t) = \frac{1}{2}\frac{4\alpha}{\pi}e^{j\theta}(1 + jn_\phi(t))\tilde{s}_i(t)$$

$$= \frac{2\alpha}{\pi} e^{j\theta} \tilde{s}_i(t) + j\frac{2\alpha}{\pi} e^{j\theta} \tilde{s}_i(t) n_\phi(t). \quad (4.212)$$

Since the first term of the right-hand side of this equation is proportional to the complex envelope of the wanted signal, $\tilde{s}_i(t)$, we need to check that the second term can be interpreted as the complex envelope of a noise component in respect of the wanted signal. The criterion for that is obviously the non-correlation between the corresponding bandpass signals. However, it is of interest to continue to work with complex envelopes in order to carry out analytical derivations. We observe that $n_\phi(t)$ is a real lowpass signal. As a result, $\tilde{s}_i(t)n_\phi(t)$ can be interpreted as the complex envelope, still defined as centered around the same carrier frequency as that chosen for the definition of $\tilde{s}_i(t)$, of the bandpass signal expected to be categorized as a noise. By Appendix 1, this commonality in the center angular frequencies ensures that checking the non-correlation between $\tilde{s}_i(t)$ and $\tilde{s}_i(t)n_\phi(t)$ implies the non-correlation of the corresponding bandpass signals. Then, assuming the phase noise process $n_\phi(t)$ and the modulation process of the input signal $s_i(t)$ are independent, we can immediately write that

$$\mathbb{E}\left\{\tilde{s}_{i,t}^* n_{\phi,t} \tilde{s}_{i,t}\right\} = \mathbb{E}\{n_{\phi,t} |\tilde{s}_{i,t}|^2\} = \mathbb{E}\{n_{\phi,t}\}\mathbb{E}\{|\tilde{s}_{i,t}|^2\}, \quad (4.213)$$

and assuming in turn that $n_\phi(t)$ is centered, that

$$\mathbb{E}\left\{\tilde{s}_{i,t}^* n_{\phi,t} \tilde{s}_{i,t}\right\} = 0. \quad (4.214)$$

As a result, the bandpass signals corresponding to the complex envelopes can effectively be assumed to be uncorrelated when considered at the same time.

We can thus now focus on an expression for the SNR, recovered at the output of the frequency conversion. Based on our discussion so far and given the proportionality between the power of a bandpass signal and of its complex envelope as per equation (1.64), this quantity can be expressed as the power ratio of $\tilde{s}_i(t)$ and $\tilde{s}_i(t)n_\phi(t)$. Given the stationarity of the processes involved, we can write

$$SNR = \frac{\mathbb{E}\{|\tilde{s}_i|^2\}}{\mathbb{E}\{|\tilde{s}_i n_\phi|^2\}} = \frac{\mathbb{E}\{|\tilde{s}_i|^2\}}{\mathbb{E}\left\{|\tilde{s}_i|^2 n_\phi^2\right\}}. \quad (4.215)$$

Still assuming the independence of the phase noise process $n_\phi(t)$ and the input wanted signal $s_i(t)$, we get

$$SNR = \frac{\mathbb{E}\{|\tilde{s}_i|^2\}}{\mathbb{E}\{|\tilde{s}_i|^2\}\mathbb{E}\left\{n_\phi^2\right\}} = \frac{1}{\mathbb{E}\left\{n_\phi^2\right\}}. \quad (4.216)$$

As $n_\phi(t)$ is a lowpass process, the term $\mathbb{E}\left\{n_\phi^2\right\}$ directly represents its power. We thus see that the SNR recovered at the output of the frequency conversion is inversely proportional to the

LO phase noise power.[3] What is interesting to see is that this SNR is independent of the level of the input signal $s_i(t)$. From the system design point of view, we thus have here what we call a multiplicative noise component. As illustrated in Chapter 7, this particular behavior prevents us from using the same strategies in terms of optimization of a line-up as when dealing with additive noise components.

To conclude this section, we point out that the result may be different if one considers using alternative frequency planning for the frequency conversion. Indeed, using a chopper-like mixer, one can consider using the kth harmonic of the LO signal instead of the fundamental tone to carry out the frequency transposition of the wanted signal. But, by equation (4.180), the phase noise of the kth harmonic is k times the phase noise of the fundamental tone. The SNR resulting from such a frequency conversion would thus be accordingly degraded compared to the present case. This degradation, moreover, cumulates with the conversion gain of the processing that is also lower on such harmonics, proportionally to their amplitude. These two reasons illustrate the benefit of considering classical frequency planning based on the fundamental tone of the LO square waveform.

Spectrum Degradation

Another impact that results from the use of a noisy LO waveform is the degradation of the spectrum of the signal being processed. To illustrate this phenomenon, we can continue to consider that we are dealing with a frequency upconversion as in the previous section. Although this degradation exists in any configuration involving a noisy LO, it is often associated with the transmit side where a spectrum mask has to be fulfilled most of the time, as discussed in Chapter 3.

Thus, we can still assume that the complex envelope $\tilde{s}_o(t)$ of the bandpass signal of interest recovered at the output of the processing is given by equation (4.208). And due to the equivalence in the spectral shape of a bandpass signal and its complex envelope as given by equation (1.40), it is useful to examine the PSD of $\tilde{s}_o(t)$ for our discussion on the spectral degradation of $s_o(t)$. We first need an expression for the autocorrelation function of $\tilde{s}_o(t)$:

$$\gamma_{\tilde{s}_o \times \tilde{s}_o}(t_1, t_2) = \mathbb{E}\{\tilde{s}_{o,t_1} \tilde{s}^*_{o,t_2}\} = \frac{1}{4}\mathbb{E}\{\tilde{s}_{i,t_1} \tilde{lo}_{t_1} \tilde{s}^*_{i,t_2} \tilde{lo}^*_{t_2}\}. \tag{4.217}$$

At this stage, it seems reasonable to assume the independence of the LO signal $lo(t)$ and the input wanted signal $s_i(t)$. As a result, we have

$$\gamma_{\tilde{s}_o \times \tilde{s}_o}(t_1, t_2) = \frac{1}{4}\mathbb{E}\{\tilde{s}_{i,t_1} \tilde{s}^*_{i,t_2}\}\mathbb{E}\{\tilde{lo}_{t_1} \tilde{lo}^*_{t_2}\}. \tag{4.218}$$

[3] This phase noise power is classically expressed through its RMS value. But, as we are dealing with a phase term, it is usually expressed in degrees. This is simply $180/\pi$ times the RMS value of the signal defined here in natural units, i.e. in radians.

Referring to Appendix 2, it is also reasonable to assume the stationarity of those processes. We thus get that

$$\gamma_{\tilde{s}_o \times \tilde{s}_o}(\tau) = \frac{1}{4}\gamma_{\tilde{s}_i \times \tilde{s}_i}(\tau)\gamma_{\tilde{lo} \times \tilde{lo}}(\tau), \tag{4.219}$$

with $\tau = t_1 - t_2$. Taking the Fourier transform of this equation and using equation (1.10), we finally have that

$$\Gamma_{\tilde{s}_o \times \tilde{s}_o}(\omega) = \frac{1}{4}\Gamma_{\tilde{s}_i \times \tilde{s}_i}(\omega) \star \Gamma_{\tilde{lo} \times \tilde{lo}}(\omega). \tag{4.220}$$

We thus see that the PSD of the output signal is simply proportional to the convolution of the PSD of the LO and input signals.

This convolution product leads to interesting properties. To investigate them we continue to assume that the LO complex envelope $\tilde{lo}(t)$ is given by equation (4.211). The PSD of this signal is given by equation (4.188). Thus, substituting this result into the above equation leads to

$$\Gamma_{\tilde{s}_o \times \tilde{s}_o}(\omega) = 4\left(\frac{\alpha}{\pi}\right)^2 \left(\Gamma_{\tilde{s}_i \times \tilde{s}_i}(\omega) + \Gamma_{\tilde{s}_i \times \tilde{s}_i}(\omega) \star \Gamma_{n_\phi \times n_\phi}(\omega)\right). \tag{4.221}$$

The first term on the right-hand side of this equation is obviously related to the PSD of the wanted signal. The second term corresponds to the PSD of the complex envelope of the noise term identified as such in the previous section. We can thus focus on the analysis of the PSD

$$\begin{aligned}\Gamma_{\tilde{n}_d \times \tilde{n}_d}(\omega) &= \Gamma_{\tilde{s}_i \times \tilde{s}_i}(\omega) \star \Gamma_{n_\phi \times n_\phi}(\omega)\\ &= \int_{-\infty}^{\infty} \Gamma_{\tilde{s}_i \times \tilde{s}_i}(\omega')\Gamma_{n_\phi \times n_\phi}(\omega - \omega')\mathrm{d}\omega',\end{aligned} \tag{4.222}$$

of this contribution. Due to the convolution product that is involved in this expression, the spectrum of this distortion term is not directly proportional to that of the phase noise of the LO waveform. This is indeed the case if and only if the PSD of the input signal reduces to the identity element of the convolution product, i.e. to a Dirac delta distribution. This occurs for an input CW signal only. But in the general case of a complex modulated input wanted signal, the PSD of this additive distortion term is more complicated to derive. However, it can be detailed using a decomposition of $\Gamma_{\tilde{s}_i \times \tilde{s}_i}(\omega)$ as a sum of slices of width $\delta\omega$, centered on the angular frequencies $\omega_k = k\delta\omega$, $k \in \mathbb{Z}$. These slices can be chosen sufficiently narrow that the power of $\tilde{s}_i(t)$ in the relevant frequency band can be evaluated as $\delta\omega\Gamma_{\tilde{s}_i \times \tilde{s}_i}(\omega_k)$. It follows that $\Gamma_{\tilde{s}_i \times \tilde{s}_i}(\omega)$ can be approximated as

$$\Gamma_{\tilde{s}_i \times \tilde{s}_i}(\omega) \approx \sum_{k \in \mathbb{Z}} \delta\omega\Gamma_{\tilde{s}_i \times \tilde{s}_i}(\omega_k)\delta(\omega - \omega_k), \tag{4.223}$$

i.e. as a Dirac comb. As highlighted above, this approximation is of interest because the Dirac delta distribution is the identity element of the convolution product. Thus, using this decomposition in equation (4.222) directly leads to

$$\Gamma_{\tilde{n}_d \times \tilde{n}_d}(\omega) = \sum_{k \in \mathbb{Z}} \delta\omega \Gamma_{\tilde{s}_i \times \tilde{s}_i}(\omega_k) \Gamma_{n_\phi \times n_\phi}(\omega - \omega_k). \tag{4.224}$$

As a result, the spectrum of the distortion term can be interpreted as the sum of the power spectral densities of the LO noise transposed around each angular frequency ω_k and weighted by the fraction of power of the input signal in that particular frequency band.

Based on this decomposition, we can now interpret the distortion experienced by the spectrum of the wanted signal during the frequency conversion. We focus first on the close in-band part of $\Gamma_{\tilde{s}_o \times \tilde{s}_o}(\omega)$. Having that the spectrum of the LO phase noise term is recovered as centered around each of the slices of the input spectrum, we observe that it is recovered in particular around the slices located at the extreme parts of it. The bandwidth of the output spectrum is thus enhanced by the spectral extent of the phase noise term. This behavior thus leads to a spectral regrowth of the input signal, as illustrated in Figure 4.22. However, the behavior is slightly different in the present case from that encountered in nonlinear blocks as detailed in “Spectral regrowth” in Chapter 5. In the latter case, we are dealing with the

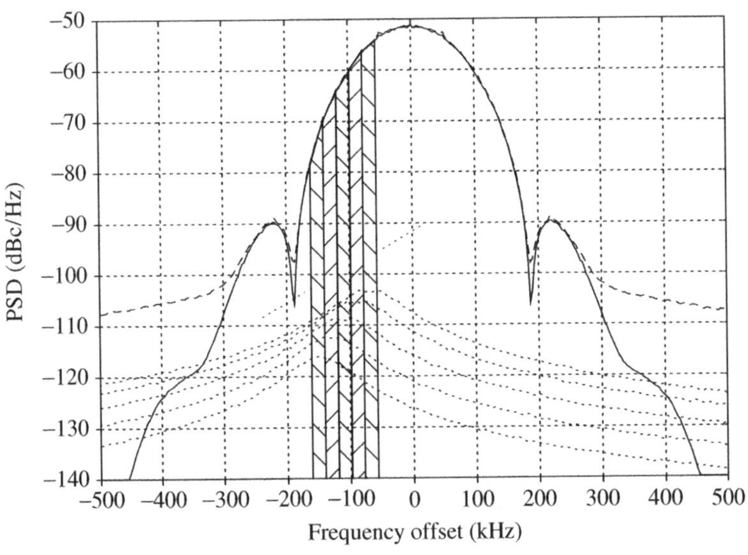

Figure 4.22 Spectral degradation due to LO noise during a frequency conversion – During a frequency conversion, the spectrum of the output signal (dashed) is the convolution of the spectrum of the input signal (solid) with the spectrum of the LO waveform. The impact of this convolution can be understood using a decomposition of the spectrum of the input signal in terms of adjacent narrowband slices (five of those slices are shown here, but the decomposition of all the input spectrum has to be considered). The LO noise spectrum is therefore transposed around each slice (dotted). Thus, the output spectrum, which is the sum of all those contributions, exhibits a spectral degradation compared to the input spectrum. Here, the case of a GSM/EDGE modified 8PSK modulated signal is shown. The LO noise power is set 33 dB below the carrier power and its bandwidth to ±50 kHz around the carrier, as compared to the ±135 kHz of the modulation bandwidth.

self-convolution of the wanted signal, thus with the convolution of signals of the same spectral extent. This necessarily leads to a non-negligible spectral spread. However, in the present case the relative width of the spectrum involved in the convolution processing can be very different. The narrower the phase noise spectrum, the weaker the spread of the wanted signal. It is thus only on a case by case basis that we can draw conclusions on the impact of such phenomena on some performance measure such as the ACLR of a transmitter, for instance.

Let us now focus on the far away part of the output spectrum. We recall that in real implementations the LO noise spectrum exhibits an absolute noise floor as discussed in Section 4.3.1. Assuming that ω_{floor} represents the angular frequency above which the spectrum of $n_\phi(t)$ reaches this noise floor, we can write that $\Gamma_{n_\phi \times n_\phi}(\omega) \approx \Gamma_{n_\phi \times n_\phi}(\omega_{\text{floor}})$ for $|\omega| \geq \omega_{\text{floor}}$. Referring to equation (4.224), we can thus write that

$$\Gamma_{\tilde{n}_\text{d} \times \tilde{n}_\text{d}}(\omega) \approx \Gamma_{n_\phi \times n_\phi}(\omega_{\text{floor}}) \sum_{k \in \mathbb{Z}} \delta\omega \Gamma_{\tilde{s}_\text{i} \times \tilde{s}_\text{i}}(\omega_k) \tag{4.225}$$

for ω far enough away from the in-band part of the wanted signal spectrum. But in this equation the summation represents nothing more than the power $P_{\tilde{s}_\text{i}}$ of the input signal complex envelope. We can then write that

$$\Gamma_{\tilde{n}_\text{d} \times \tilde{n}_\text{d}}(\omega) \approx \Gamma_{n_\phi \times n_\phi}(\omega_{\text{floor}}) P_{\tilde{s}_\text{i}} \tag{4.226}$$

for the far away part of the wanted signal spectrum. This result can be interpreted in a straightforward way if we recall that the noise floor derived above directly sums with the PSD of the input signal complex envelope. Referring to equation (4.221), we can write for this part of the spectrum that

$$\Gamma_{\tilde{s}_\text{o} \times \tilde{s}_\text{o}}(\omega) = 4 \left(\frac{\alpha}{\pi} \right)^2 \left(\Gamma_{\tilde{s}_\text{i} \times \tilde{s}_\text{i}}(\omega) + \Gamma_{n_\phi \times n_\phi}(\omega_{\text{floor}}) P_{\tilde{s}_\text{i}} \right). \tag{4.227}$$

Comparing this expression with equation (4.188), we thus see that we have the same power ratio for the noise floor regarding the input signal power as for the LO phase noise floor compared to the LO carrier. As a result, when the phase noise floor of the LO waveform is expressed relative to the LO carrier power, i.e. in dBc/Hz for instance, this ratio can be directly used to derive the noise floor recovered at the frequency conversion output relative to the wanted signal power.

Reciprocal Mixing

A final problem related to the presence of the phase noise term in the LO signal can be investigated by considering the presence of two bandpass signals at the input of the frequency conversion. This configuration is classically encountered in the reception of a wanted signal $s_\text{w}(t)$ while a blocking signal $s_\text{bl}(t)$ is also present, for instance. Although the mechanism remains the same whatever the frequency planning, let us for illustrative purposes consider an infradyne downconversion. The complex envelope of the signal recovered at the output of the processing is given by equation (4.209). However, given that the input bandpass signal $s_\text{i}(t) = s_\text{w}(t) + s_\text{bl}(t)$ is composed of the superposition of two bandpass signals, we first need to give an expression for its complex envelope in order to use this equation.

Suppose that the wanted signal $s_{\rm w}(t)$ is centered around the angular frequency $\omega_{\rm w}$, and that the blocker $s_{\rm bl}(t)$ is centered around the angular frequency $\omega_{\rm bl}$. We can then define the complex envelopes $\tilde{s}_{\rm w}(t)$ and $\tilde{s}_{\rm bl}(t)$ of the bandpass signals as centered around these angular frequencies according to

$$s_{\rm w}(t) = \mathrm{Re}\left\{\tilde{s}_{\rm w}(t)e^{j\omega_{\rm w}t}\right\}, \tag{4.228a}$$

$$s_{\rm bl}(t) = \mathrm{Re}\left\{\tilde{s}_{\rm bl}(t)e^{j\omega_{\rm bl}t}\right\}. \tag{4.228b}$$

We can then write that

$$\begin{aligned} s_{\rm i}(t) &= s_{\rm w}(t) + s_{\rm bl}(t) \\ &= \mathrm{Re}\left\{(\tilde{s}_{\rm w}(t) + \tilde{s}_{\rm bl}(t)e^{j(\omega_{\rm bl}-\omega_{\rm w})t})e^{j\omega_{\rm w}t}\right\}. \end{aligned} \tag{4.229}$$

Assuming for the sake of simplicity that we are dealing with the complex envelope $\tilde{s}_{\rm i}(t)$ of $s_{\rm i}(t)$, defined as centered around $\omega_{\rm i} = \omega_{\rm w}$, and that $\omega_{\rm bl} > \omega_{\rm w}$ for instance, we can immediately write

$$\tilde{s}_{\rm i}(t) = \tilde{s}_{\rm w}(t) + \tilde{s}_{\rm bl}(t)e^{j(\omega_{\rm bl}-\omega_{\rm w})t}. \tag{4.230}$$

Substituting this expression into equation (4.209) yields

$$\tilde{s}_{\rm o}(t) = \frac{1}{2}\left(\tilde{s}_{\rm w}(t) + \tilde{s}_{\rm bl}(t)e^{j(\omega_{\rm bl}-\omega_{\rm w})t}\right)\tilde{lo}^*(t), \tag{4.231}$$

where $\tilde{s}_{\rm o}(t)$ is defined as centered around $\omega_{\rm o} = \omega_{\rm i} - \omega_{\rm LO} = \omega_{\rm w} - \omega_{\rm LO}$.

The phenomenon of reciprocal mixing can then be highlighted when looking at the PSD of $s_{\rm o}(t)$. Here again, due to the equivalence in the spectral shape of a bandpass signal and its complex envelope as given by equation (1.40), we can continue to work on the PSD of $\tilde{s}_{\rm o}(t)$ for the sake of simplicity. We first need an expression for the autocorrelation function of $\tilde{s}_{\rm o}(t)$:

$$\begin{aligned} \gamma_{\tilde{s}_{\rm o}\times\tilde{s}_{\rm o}}(t_1,t_2) = \frac{1}{4}\mathrm{E}\Big\{\Big(&\tilde{s}_{{\rm w},t_1}\tilde{s}^*_{{\rm w},t_2} + \tilde{s}_{{\rm bl},t_1}\tilde{s}^*_{{\rm bl},t_2}e^{j(\omega_{\rm bl}-\omega_{\rm w})(t_1-t_2)} \\ &+ \tilde{s}_{{\rm w},t_1}\tilde{s}^*_{{\rm bl},t_2}e^{-j(\omega_{\rm bl}-\omega_{\rm w})t_2} + \tilde{s}^*_{{\rm w},t_2}\tilde{s}_{{\rm bl},t_1}e^{j(\omega_{\rm bl}-\omega_{\rm w})t_1}\Big)\tilde{lo}^*_{t_1}\tilde{lo}_{t_2}\Big\}. \end{aligned} \tag{4.232}$$

At this stage, it seems reasonable to assume that the wanted signal, the blocking signal and the LO signal are independent processes. We can thus write

$$\begin{aligned} \gamma_{\tilde{s}_{\rm o}\times\tilde{s}_{\rm o}}(t_1,t_2) = \frac{1}{4}\mathrm{E}\Big\{&\tilde{s}_{{\rm w},t_1}\tilde{s}^*_{{\rm w},t_2} + \tilde{s}_{{\rm bl},t_1}\tilde{s}^*_{{\rm bl},t_2}e^{j(\omega_{\rm bl}-\omega_{\rm w})(t_1-t_2)} \\ &+ \tilde{s}_{{\rm w},t_1}\tilde{s}^*_{{\rm bl},t_2}e^{-j(\omega_{\rm bl}-\omega_{\rm w})t_2} + \tilde{s}^*_{{\rm w},t_2}\tilde{s}_{{\rm bl},t_1}e^{j(\omega_{\rm bl}-\omega_{\rm w})t_1}\Big\}\mathrm{E}\left\{\tilde{lo}^*_{t_1}\tilde{lo}_{t_2}\right\} \end{aligned} \tag{4.233}$$

or

$$\gamma_{\tilde{s}_o \times \tilde{s}_o}(t_1, t_2) = \frac{1}{4}\Big(\mathbb{E}\big\{\tilde{s}_{w,t_1}\tilde{s}^*_{w,t_2}\big\} + \mathbb{E}\big\{\tilde{s}_{bl,t_1}\tilde{s}^*_{bl,t_2}\big\}e^{j(\omega_{bl}-\omega_w)(t_1-t_2)}$$

$$+\mathbb{E}\big\{\tilde{s}_{w,t_1}\tilde{s}^*_{bl,t_2}\big\}e^{-j(\omega_{bl}-\omega_w)t_2} + \mathbb{E}\big\{\tilde{s}^*_{w,t_2}\tilde{s}_{bl,t_1}\big\}e^{j(\omega_{bl}-\omega_w)t_1}\Big)\mathbb{E}\big\{\tilde{lo}^*_{t_1}\tilde{lo}_{t_2}\big\}.$$

Assuming now that the modulation processes of the wanted signal and blocking signals are also centered, we get that

$$\mathbb{E}\big\{\tilde{s}_{w,t_1}\tilde{s}^*_{bl,t_2}\big\} = \mathbb{E}\big\{\tilde{s}_{w,t_1}\big\}\mathbb{E}\big\{\tilde{s}^*_{bl,t_2}\big\} = 0, \tag{4.234a}$$

$$\mathbb{E}\big\{\tilde{s}^*_{w,t_2}\tilde{s}_{bl,t_1}\big\} = \mathbb{E}\big\{\tilde{s}^*_{w,t_2}\big\}\mathbb{E}\big\{\tilde{s}_{bl,t_1}\big\} = 0. \tag{4.234b}$$

Given in addition the stationarity of those processes as discussed in Appendix 2, we can finally write

$$\gamma_{\tilde{s}_o \times \tilde{s}_o}(\tau) = \frac{1}{4}\big(\gamma_{\tilde{s}_w \times \tilde{s}_w}(\tau) + \gamma_{\tilde{s}_{bl} \times \tilde{s}_{bl}}(\tau)e^{j(\omega_{bl}-\omega_w)\tau}\big)\gamma_{\tilde{lo} \times \tilde{lo}}(-\tau), \tag{4.235}$$

with $\tau = t_1 - t_2$. The PSD of the output signal can now be estimated by taking the Fourier transform of this equation. Using equations (1.39) and (1.10) yields

$$\Gamma_{\tilde{s}_o \times \tilde{s}_o}(\omega) = \frac{1}{4}\big[\Gamma_{\tilde{s}_w \times \tilde{s}_w}(\omega) \star \Gamma_{\tilde{lo} \times \tilde{lo}}(-\omega)$$

$$+\Gamma_{\tilde{s}_{bl} \times \tilde{s}_{bl}}(\omega - (\omega_{bl} - \omega_w)) \star \Gamma_{\tilde{lo} \times \tilde{lo}}(-\omega)\big]. \tag{4.236}$$

In order to go further, we need to consider a particular expression for the spectrum of the LO complex envelope. As in the previous section, we can continue to assume that $\tilde{lo}(t)$ is given by equation (4.211). The PSD of this signal is given by equation (4.188). Substituting this result into the above equation results in

$$\Gamma_{\tilde{s}_o \times \tilde{s}_o}(\omega) = 4\left(\frac{\alpha}{\pi}\right)^2 \big[\Gamma_{\tilde{s}_w \times \tilde{s}_w}(\omega) + \Gamma_{\tilde{s}_{bl} \times \tilde{s}_{bl}}(\omega - (\omega_{bl} - \omega_w))$$

$$+\Gamma_{\tilde{s}_w \times \tilde{s}_w}(\omega) \star \Gamma_{n_\phi \times n_\phi}(-\omega) + \Gamma_{\tilde{s}_{bl} \times \tilde{s}_{bl}}(\omega - (\omega_{bl} - \omega_w)) \star \Gamma_{n_\phi \times n_\phi}(-\omega)\big].$$

Looking at this expression, we see that we recover at the output of the processing the two signals present at the input of the frequency conversion, i.e. the wanted signal and the blocker, but with two additive bandpass noises centered around them. In fact, due to the distributivity of the multiplication operation that models the frequency transposition, we simply recover in the present case the same noise structure as discussed in "Spectrum degradation" earlier in this section, but now centered around each of the bandpass signals present at the input of the processing. Recalling the discussion in that section, there is a fraction of the noise recovered as centered around one signal that may lie within the frequency band of the other signal, as

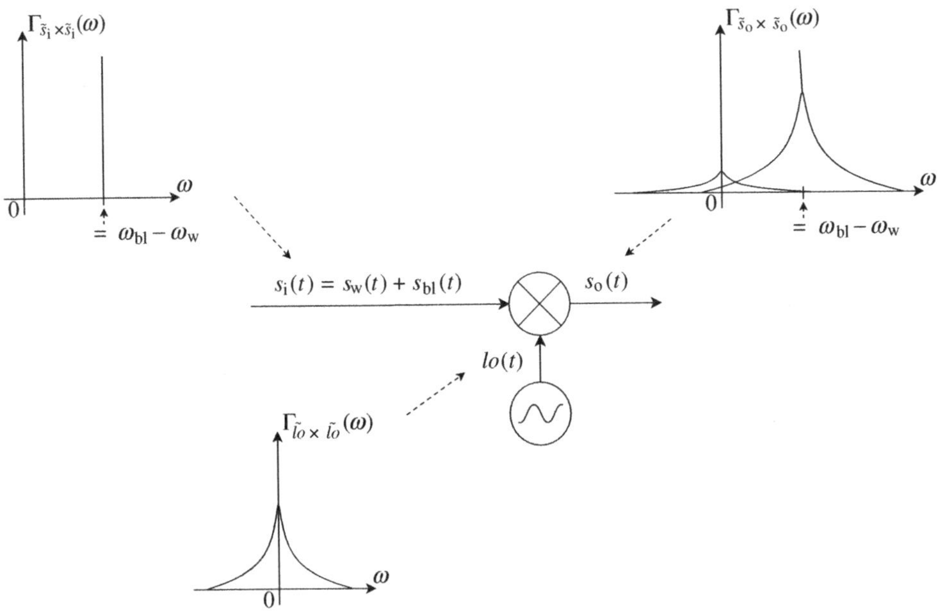

Figure 4.23 Reciprocal mixing phenomena using a noisy LO waveform – When two signals are present simultaneously at the input of a frequency conversion (top left) that uses a noisy LO waveform (bottom), the noise spectrum of the LO waveform is recovered around each signal at the frequency conversion output (top right). As a result, the noise spectrum of each output signal corrupts that of other signals, and due to the symmetry in the LO noise spectrum, it is the same spectral part of this noise component that corrupts the two signals.

illustrated in Figure 4.23. As this behavior holds for the two signals, they thus degrade each other through this mechanism. This explains the term "reciprocal mixing".

From the system design point of view, it is of interest to estimate the SNR degradation resulting from this phenomenon in order to be able to carry out line-up budgets as illustrated in Chapter 7. As it is obviously not easy to handle convolution products, first orders of magnitude are often derived using the approximation for the power of the noise complex envelope given by equation (4.226). In the present case, this equation can be adapted so that the PSD of the complex envelope of the bandpass noise centered around the blocker can be approximated as

$$\Gamma_{\tilde{n}_d \times \tilde{n}_d}(\omega) = \Gamma_{n_\phi \times n_\phi}(-\omega + (\omega_{bl} - \omega_w))P_{\tilde{s}_{bl}}. \tag{4.237}$$

Thus, when the PSD of the LO phase noise $\Gamma_{n_\phi \times n_\phi}(\omega)$ is expressed in dBc/Hz, we simply need to add the power of the blocker, to derive the PSD of the noise component that corrupts the wanted signal. However, we need to keep in mind that this equation remains an approximation. It is exact only when the blocker is CW or when $\Gamma_{n_\phi \times n_\phi}(-\omega + (\omega_{bl} - \omega_w))$ is constant. The latter case occurs when it is the floor part of the phase noise spectrum that lies within the wanted signal frequency band. This is fortunately a common situation when dealing with far away blockers. Assuming that the wanted signal spreads toward the frequency band $[-\delta\omega/2, \delta\omega/2]$ around ω_w, we can then derive the amount of power of $\tilde{n}_\mathrm{d}(t)$ that lies within the wanted signal band as

$$P_{\tilde{n}_\mathrm{d}} = P_{\tilde{s}_\mathrm{bl}} \int_{-\delta\omega/2}^{\delta\omega/2} \Gamma_{n_\phi \times n_\phi}(-\omega + (\omega_\mathrm{bl} - \omega_\mathrm{w}))\mathrm{d}\omega. \tag{4.238}$$

But referring again to the case where it is the floor part of the phase noise spectrum that is involved, we can approximate this expression by using the value of its PSD at the frequency offset corresponding to the wanted signal times the noise bandwidth. This results in the relationship

$$P_{\tilde{n}_\mathrm{d}} \approx P_{\tilde{s}_\mathrm{bl}} \Gamma_{n_\phi \times n_\phi}(\omega_\mathrm{bl} - \omega_\mathrm{w})\delta\omega. \tag{4.239}$$

Referring to equation (1.64), the same relationship obviously holds for original bandpass signals:

$$P_{n_\mathrm{d}} \approx P_{s_\mathrm{bl}} \Gamma_{n_\phi \times n_\phi}(\omega_\mathrm{bl} - \omega_\mathrm{w})\delta\omega. \tag{4.240}$$

To conclude, we observe that in practical implementations the LO noise spectrum can be assumed symmetrical with regard to the LO angular frequency, as highlighted in Section 4.3.1. We can thus write that $\Gamma_{n_\phi \times n_\phi}(\omega_\mathrm{bl} - \omega_\mathrm{w}) = \Gamma_{n_\phi \times n_\phi}(\omega_\mathrm{w} - \omega_\mathrm{bl})$. The above equation can alternatively be written as

$$P_{n_\mathrm{d}} = P_{s_\mathrm{bl}} \Gamma_{n_\phi \times n_\phi}(\omega_\mathrm{w} - \omega_\mathrm{bl})\delta\omega. \tag{4.241}$$

This means that only the relative frequency distance between the wanted signal and blocking signal is of importance in the reciprocal mixing phenomenon and not the exact frequency location of those signals.

4.4 Linear Error Vector Magnitude

Another source of degradation classically encountered in transceiver line-ups is the distortion experienced by the wanted signal when going through filtering stages. The cause can be both the in-band group delay variations of those filters, and their amplitude ripple. Referring to practical filtering stages, this behavior is emphasized close to their cut-off frequency. This phenomenon is thus of particular importance when the passband of the filter is of the same order of magnitude as the wanted signal bandwidth. It is therefore classically encountered at the channel filtering stage of a line-up. When such a distortion occurs on the P and Q paths of a transceiver, it then results in an error on the waveforms $p(t)$ and $q(t)$ compared to what would be theoretically expected. This error therefore propagates on the constellation corresponding to the time samples of those waveforms. Referring to the discussion in "Error vector magnitude" in Chapter 3, we are then dealing with the generation of an EVM term.

In order to illustrate this behavior, we can reconsider the effect of the RRC filter used in the WCDMA standard as detailed in Section 1.3.3. Assume for the sake of simplicity that we are dealing with two sequences of chips, regardless of how they have been generated through potential scrambling processes. Consider a first sequence, $\{c_{\mathrm{p},k}\}$, used to generate the P waveform, and a second, $\{c_{\mathrm{q},k}\}$, used to generate the Q waveform. Both take their values in the set $\{-1, +1\}$. The signals $p(t)$ and $q(t)$ are then derived from $\{c_{\mathrm{p},k}\}$ and $\{c_{\mathrm{q},k}\}$ according to

$$p(t) = \sum_k c_{\mathrm{p},k} h_{\mathrm{RRC}}(t - kT_{\mathrm{c}}), \tag{4.242a}$$

$$q(t) = \sum_k c_{\mathrm{q},k} h_{\mathrm{RRC}}(t - kT_{\mathrm{c}}), \tag{4.242b}$$

with $h_{\mathrm{RRC}}(t)$ the impulse response of the RRC filter given by equation (1.169), and T_{c} the chip duration. We thus see that with our simple assumption that $c_{\mathrm{p},k}$ and $c_{\mathrm{q},k}$ take their values in $\{-1, +1\}$, the waveform we are dealing with can simply be seen as an RRC filtered version of a 4QAM symbol mapping.

The impact of the RRC filtering on the generated waveforms is shown in Figure 4.24. Although the average group delay and amplitude of the RRC filter are compensated in order to best fit the time samples of the $p(t)$ and $q(t)$ signals to the original chip values, we obviously have a residual error. As this error is linked to the impulse response of the RRC filter that is not null at the integer multiples of T_{c}, we then refer this error to an intersymbol interference. We get that the error experienced by a data chip at a given sample time is linked to the value of the other chips in the sequence. As this error is necessarily retrieved on the symbol constellation that represents the time samples of the complex envelope $\tilde{s}(t) = p(t) + \mathrm{j}q(t)$ in the complex plane, we then face an EVM as illustrated in Figure 4.25(left). However, what is interesting is that if we again apply the same RRC filter to the $p(t)$ and $q(t)$ signals, we cancel this EVM. This behavior is linked to the property that the self-convolution of the impulse response of the RRC filter corresponds to a Nyquist filter. As a result, there exist ideal sampling times on the filtered waveforms $h_{\mathrm{RRC}}(t) \star p(t)$ and $h_{\mathrm{RRC}}(t) \star q(t)$ that return the correct values of the input chip sequences, as illustrated in Figures 4.24 and 4.25.

Based on this simple example, we see that a linear distortion as experienced when going through a filtering stage can possibly be equalized using another linear operation. This is an important difference with the EVM generated by nonlinearities, as introduced in "Nonlinear EVM due to odd order nonlinearity" (Section 5.1.3). Indeed, even if it results in EVM in both cases, in the latter case only a nonlinear operation can possibly compensate for it (see Section 9.1.4). Consequently, the impact on the overall link quality is often different in practical implementations for these two kinds of EVM. The root cause of this lies in the receive side of a wireless link where almost all the adaptive equalization schemes present in a baseband are defined for compensating a propagation channel that behaves as a linear filter, as discussed in Chapter 2. As a result, such an equalization scheme can often compensate linear distortions as it does for the propagation channel. A classical example of this is the OFDM systems for which the channel equalization is mostly done in the frequency domain. Any linear distortion related to the selective behavior of filters is therefore also naturally compensated in that case. When dealing with such systems, we can then neglect, up to a point, the linear EVM in the overall link budget. This explains the distinction made between the linear EVM discussed

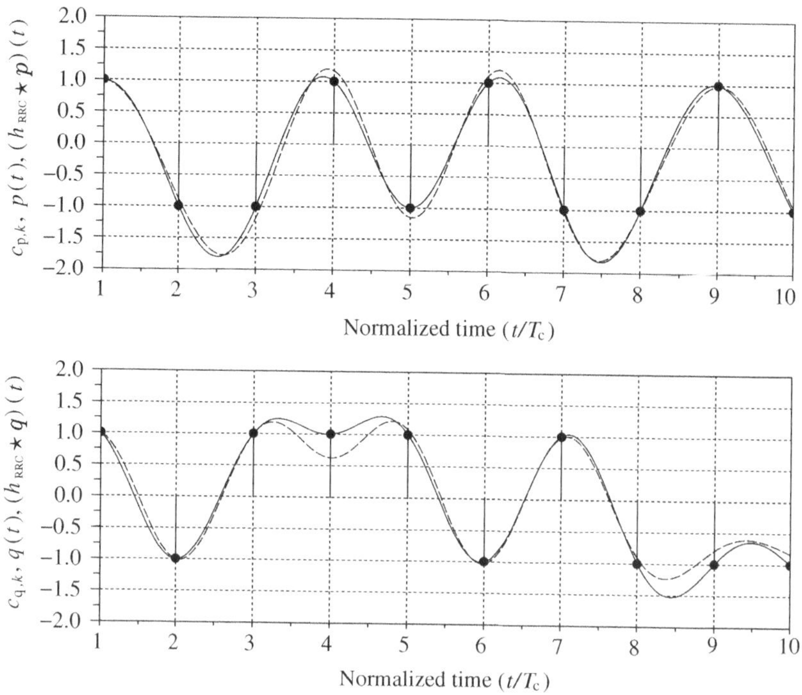

Figure 4.24 Linear EVM generation and equalization using RRC filters – modulating waveforms – The RRC filtering of two chip sequences (bars) leads to the generation of $p(t)$ and $q(t)$ waveforms as given by equation (4.242). The waveforms exhibit an error in their sampled values compared to the original chip amplitude even if the average group delay and amplitude of the filter are compensated (dashed). This EVM can be canceled by again applying the RRC filter on the waveforms (solid). The corresponding constellations are shown in Figure 4.25.

here and the nonlinear one introduced in Chapter 5 in practical budgets. However, the final word on this topic really depends on the modulation scheme and how the equalization stage is implemented.

In conclusion, we can examine whether the EVM signal introduced by such linear distortion really can be considered as an additional noise term with respect to the wanted signal. One might fear that the linear dependency that holds between the two signals leads to correlations between them. This is in fact not the case, as can be understood by reconsidering the above example. We can write the error term $e_{p,l}$ that exists between the lth sample of $p(t)$ and the lth data chip $c_{p,l}$ from equation (4.242a) as

$$e_{p,l} = p(lT_c) - h_{RRC}(0)c_{p,l} = \sum_{k \neq l} c_{p,k} h_{RRC}(lT_c - kT_c). \tag{4.243}$$

Thus the error term on the lth chip value is a linear combination of the sequence of the input chips different from it. As a result, the correlation between $e_{p,l}$ and $c_{p,l}$ can be written as

$$\mathbb{E}\{e_{p,l}c_{p,l}\} = \sum_{k \neq l} \mathbb{E}\{c_{p,k}c_{p,l}\} h_{RRC}(lT_c - kT_c). \tag{4.244}$$

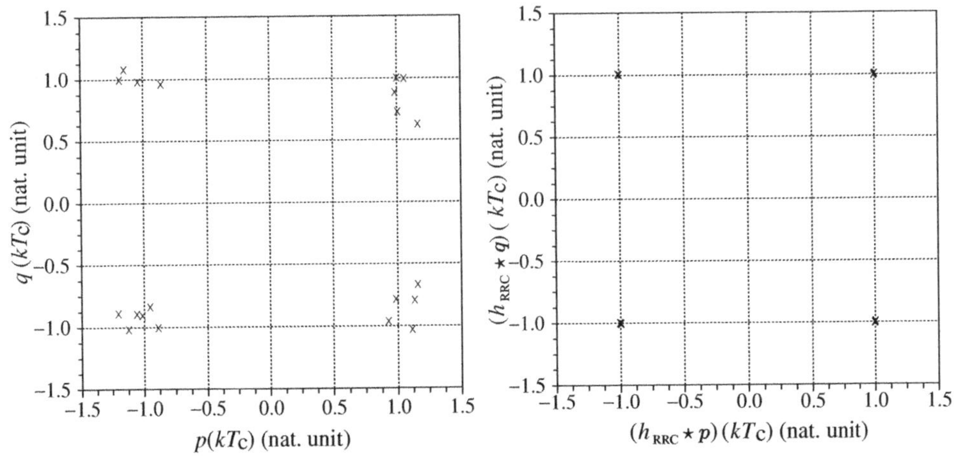

Figure 4.25 Linear EVM generation and equalization using RRC filters – constellations – Even with the compensation of the average group delay and amplitude of the RRC filter used for the generation of the waveforms $p(t)$ and $q(t)$ shown in Figure 4.24, we get an error in the sampled amplitudes compared to the initial chip values. This results in an EVM on the corresponding constellation (left) that can be canceled by filtering the waveforms with the RRC filter again (right).

But, if we assume on the one hand the independence of the data chips of the input sequence, and on the other hand that we are dealing with a centered process, i.e. that we have an equal probability of getting a +1 and a −1, we get for all $k \neq l$ that

$$\mathbb{E}\{c_{p,k}c_{p,l}\} = \mathbb{E}\{c_{p,k}\}\mathbb{E}\{c_{p,l}\} = 0. \tag{4.245}$$

As a result, $\mathbb{E}\{e_{p,l}c_{p,l}\} = 0$, i.e. the sample considered and the EVM term are uncorrelated. This obviously holds in the general case as long as the successive data that are used to construct the modulating waveforms can be considered as independent and evenly distributed. Finally, we observe that if the input data can effectively be considered as independent, the sum in equation (4.243) tends to be Gaussian distributed, as does the EVM term, in light of the central limit theorem [3].

4.5 Quantization Noise

Among all the noise contributions that we have to deal with in transceivers, there is also what is called the quantization noise. This refers to the distortion of the amplitude of a signal linked to its representation using a finite set of integer numbers. Such approximation is often required for implementing realistic digital signal processing functions based on the use of a finite number of bits. This distortion has to be considered carefully as the natural trend, for various motivations, is to implement as much as possible of the signal processing functions required in a transceiver line-up in the digital domain.

There is some benefit in reviewing the conditions that allow us to consider this quantization distortion as an additional noise contribution in a transceiver budget, enabling us to understand the direct impact of quantization on the performance of a transceiver and to derive guidelines for the fixed point implementations of digital functionalities. This review also highlights that the properties of this quantization noise also depend on the time sampling operation that is always associated with quantization in practical implementations. In order to clearly differentiate the effects, we first focus in the next section on the quantization operation itself. The impact of time sampling on this quantization noise is then addressed in Section 4.5.2.

4.5.1 Quantization Error as a Noise

Quantizer Model

In order to highlight the main results about quantization noise, we restrict our review to uniform quantization. This is the most common use case and it allows us to carry out simple analytical derivations necessary for understanding the underlying phenomena.

We begin by considering a device whose operation on an input signal is to deliver output values within a finite set, evenly spread in a given DR. If the input signal amplitude lies somewhere in a quantization bin of width q, the quantizer output is the value equal to the center of that bin according to the transfer function shown in Figure 4.26(top).

Looking at this figure, note that we are assuming for the sake of simplicity that the conversion gain of the quantizer, i.e. the average slope of its transfer function, is unity. This assumption allows a direct comparison of the input and output quantities. However, it should be kept in mind that, depending on the interpretation of the output binary words, a (non-unit) conversion gain may need to be included. This can for instance be the case for ADC devices in order to have the maximum output digital word to effectively correspond to the maximum input analog value, as discussed in Section 4.6. In that case, a conversion gain should be applied to transpose the results derived in what follows as equivalent input quantities.

We also observe that the quantization error is a periodic function of the input signal amplitude as long as this quantity remains within the outer bounds of the extreme bins. This condition defines the no overload region of the quantizer. The no overload condition is of particular importance in carrying out analytical derivations. Moreover, this is a condition that must be satisfied to ensure the correct functionality of some practical structures such as $\Sigma\Delta$ converters (as discussed later in Section 4.6 under "Noise shaping"). We thus stay with this assumption in what follows.

As a side effect, we notice that the limits of this no overload region can be used to define the quantizer FS as the maximum input amplitude that guarantees the error signal to be bounded within $[-q/2, q/2]$. This FS quantity can then be defined both as an input and output quantity, taking into account a potential conversion gain. Alternatively, the conversion gain of the quantizer can be seen as the ratio between the device output and input FS.

Moments of the Quantization Error

The derivation of the properties of the quantization error is far from obvious if we consider the nonlinearity of the transfer function of the quantizer shown in Figure 4.26. There are various ways to overcome this difficulty. Widrow's approach has the great advantage of giving intuitive results as it uses the classical tools of the time sampling theory [50, 51].

Figure 4.27 shows that the probability $P[Q(s) = kq]$ of the sample $Q(s)$ of the signal recovered at the output of the quantizer being equal to kq is simply the probability of the sample s

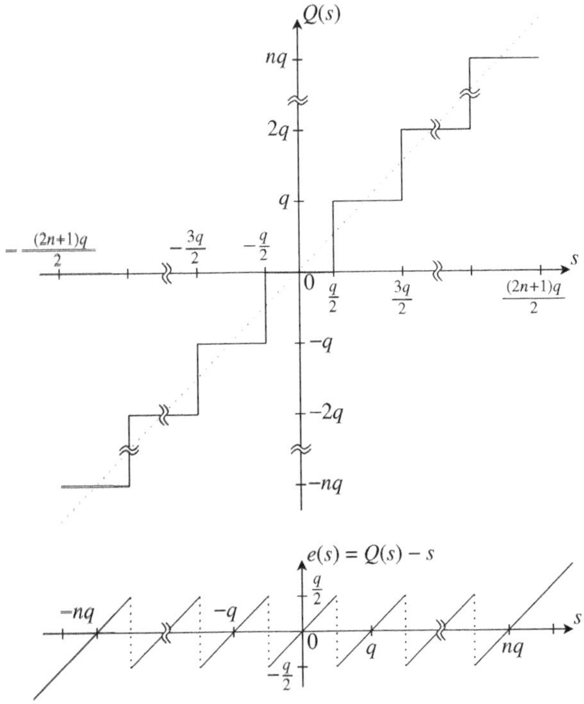

Figure 4.26 Ideal uniform quantizer model – The purpose of the ideal quantizer is to approximate the amplitude of an input signal by a discrete set of possible values. Hence for an input amplitude s lying somewhere within a quantization bin of width q, we get an output value $Q(s)$ corresponding to the center of that bin (top). This means that when no overload occurs, i.e. when the input signal amplitude remains lower than the input FS, the quantization error $e(s) = Q(s) - s$ remains within $\pm q/2$ (bottom). As we are considering uniform quantizers only, the bin width q is the same for all bins.

of the signal present at the input of the quantizer lying between $kq - q/2$ and $kq + q/2$. Thus

$$P[Q(s) = kq] = \int_{kq-q/2}^{kq+q/2} p(s)\mathrm{d}s, \tag{4.246}$$

with $p(s)$ the PDF of s. Alternatively, this integral can be rewritten as a convolution product:

$$\begin{aligned} P[Q(s) = kq] &= \int p(s)q\Pi\left(\frac{s-kq}{q}\right)\mathrm{d}s \\ &= q\left(p(s) \star \Pi\left(\frac{s}{q}\right)\right)\Bigg|_{s=kq}. \end{aligned} \tag{4.247}$$

Here, $\Pi(s)$ is the unitary gate function of width q and height $1/q$ as defined by equation (1.145). We then remark that the even property of this function allows for the use of the

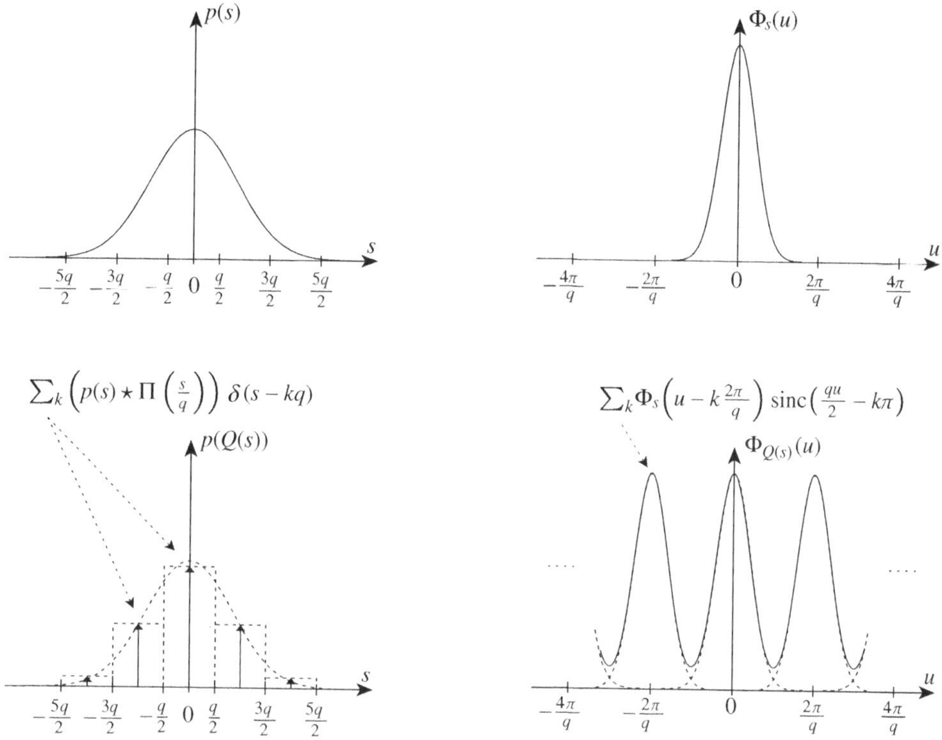

Figure 4.27 The ideal uniform quantizer acting as a sampler for the input amplitude probability density function – The nonlinear operation realized by the uniform quantizer can be analyzed using traditional tools of sampling theory when reasoning on the PDF of the quantized signal [50, 51]. The output signal PDF (bottom left) is a Dirac comb whose amplitudes are the result of the convolution of the input signal PDF (top left) with the gate function corresponding to width q and height $1/q$. The output characteristic function (bottom right) is therefore the sum of the copies of the input one (top right).

convolution form in the above equation. Consequently, the PDF $p(Q(s))$ of $Q(s)$ can simply be interpreted as the Dirac comb distribution:

$$\begin{aligned} p(Q(s)) &= \sum_k P[Q(s) = kq]\delta(Q(s) - kq) \\ &= \left(p(s) \star \Pi\left(\frac{s}{q}\right)\right) q \sum_k \delta(s - kq). \end{aligned} \tag{4.248}$$

We thus see that the PDF of $Q(s)$ can be interpreted as the result of sampling the PDF of s. The only refinement in respect of this interpretation is the presence of the convolution operation with the gate function that reflects the width of the bins used for the quantization.

As classically done in the time sampling theory, we expect to derive interesting conclusions when examining the sampled quantities in the spectral domain. Let us take the Fourier transform of the above equation. Given that the Fourier transform of the Dirac comb is itself a Dirac comb [2], we obtain the following expression for the characteristic function $\Phi_{Q(s)}(u)$ of $Q(s)$:

$$\Phi_{Q(s)}(u) = \sum_k \Phi_s\left(u - k\frac{2\pi}{q}\right)\text{sinc}\left(\frac{qu}{2} - k\pi\right). \tag{4.249}$$

Here,

$$\Phi_s(u) = \mathbb{E}\{e^{jus}\} \tag{4.250}$$

is the characteristic function of s, and $\text{sinc}(u) = \sin(u)/u$. We thus recover that the sampling process performed by the quantizer on the PDF of s leads to a periodization of its Fourier transform, i.e. of its characteristic function. The period is $2\pi/q$ in the characteristic function domain, to be related to the sampling period q in the PDF domain. As can be understood when looking at Figure 4.27, we then need to have at least

$$\Phi_s(u) = 0, \quad \text{when } |u| > \pi/q, \tag{4.251}$$

in order to be able to reconstruct the statistics of the sample s of the input signal from those of $Q(s)$. We thus recover here a bandlimited condition equivalent to the Nyquist criterion that applies to time sampling as discussed in "Aliasing" (Section 4.6.1). However, we see in the present case that even under such a bandlimited condition, the PDF that can be reconstructed from the main copy in equation (4.249) is not directly that of s. This main component $\Phi_{Q(s),0}(u)$ takes the form

$$\Phi_{Q(s),0}(u) = \Phi_s(u)\text{sinc}\left(\frac{qu}{2}\right). \tag{4.252}$$

The right-hand side of this equation can obviously be interpreted as the product of two characteristic functions – more precisely as the product of the characteristic function of s times that of a new random variable e, defined such that

$$Q(s) = s + e, \tag{4.253}$$

with e and s independent. The latter property is required in order to be able to write

$$\begin{aligned}\Phi_{Q(s)}(u) = \mathbb{E}\{e^{juQ(s)}\} = \mathbb{E}\{e^{ju(s+e)}\} &= \mathbb{E}\{e^{jus}\}\mathbb{E}\{e^{jue}\}\\ &= \Phi_s(u)\Phi_e(u).\end{aligned} \tag{4.254}$$

Obviously, e can only be interpreted as the quantization error necessarily recovered at the output of the quantization process if we refer to Figure 4.26. Consequently, when dealing with a stationary process $s(t)$ at the input of the quantizer, we recover at its output a stationary noise process $n_q(t) = e(s(t)) = Q(s(t)) - s(t)$ whose characteristic function of any of its sample is equal to $\mathrm{sinc}(qu/2)$ when equation (4.251) is fulfilled. The PDF of any sample of this quantization noise is therefore given by the Fourier transform of this function. It is thus equal to the gate function $\Pi(s/q)$ of width q and height $1/q$ introduced earlier. In other words, the PDF of the quantization noise is uniform on $[-q/2, q/2]$ in that case. We can thus evaluate its first and second moments as

$$\mathbb{E}\{n_\mathrm{q}\} = \int_{-q/2}^{q/2} \frac{s}{q} \mathrm{d}s = 0, \tag{4.255a}$$

$$\mathbb{E}\{n_\mathrm{q}^2\} = \int_{-q/2}^{q/2} \frac{s^2}{q} \mathrm{d}s = \frac{q^2}{12}. \tag{4.255b}$$

Here the expectation is null because the quantizer transfer function is centered, as can be seen in Figure 4.26(top).

The advantage of this approach is that it gives intuitive results as it refers to widely known sampling theory. Its disadvantage is that it results in a bandlimited condition on the characteristic function of the input signal samples that is only a sufficient condition, but not a necessary one. It is therefore of interest to recall the exact conditions originally derived by Sripad and Snyder for this to be true [52, 53]. It can be shown that the PDF $p(n_\mathrm{q})$ of n_q can be expressed as a function of the characteristic function $\Phi_s(u)$ of s:

$$p\left(n_\mathrm{q}\right) = \begin{cases} \dfrac{1}{q} + \dfrac{1}{q} \displaystyle\sum_{k \neq 0} \Phi_s\left(\frac{2\pi}{q}k\right) \mathrm{e}^{-\mathrm{j}\frac{2\pi}{q}kn_\mathrm{q}} & \text{for } -q/2 \leq n_\mathrm{q} < q/2, \\ 0 & \text{otherwise.} \end{cases} \tag{4.256}$$

In the same way, a general expression for the first and second moments of n_q can be derived based on a Fourier series expansion of the error signal $e(s)$ and its square $e^2(s)$. Assuming no overload of the quantizer, we can interpret $e(s) = Q(s) - s$ as a periodic sawtooth function over a finite support width, as illustrated in Figure 4.26. By periodizing the pattern $e(s) = s$ that holds over $[-q/2, q/2]$, we can then write, for input ranges of interest for s, that

$$e(s) = \sum_{k \neq 0} (-1)^k \frac{\mathrm{j}q}{2\pi k} \mathrm{e}^{\mathrm{j}\frac{2\pi}{q}ks}. \tag{4.257}$$

In the same way, periodizing the pattern $e^2(s) = s^2$ that holds over $[-q/2, q/2]$, we can write

$$e^2(s) = \frac{q^2}{12} + \sum_{k \neq 0} (-1)^k \frac{q^2}{2(\pi k)^2} \mathrm{e}^{\mathrm{j}\frac{2\pi}{q}ks}. \tag{4.258}$$

The interest of this decomposition lies in the presence of the complex exponentials that may allow a link to be made with the characteristic functions of the quantities involved. We can indeed derive an expression for the first and second moments of the quantization noise $n_\mathrm{q}(t) = e(s(t))$ by taking the expectations of those equations. Referring to the definition of the characteristic function $\Phi_s(u)$ of the input signal given by equation (4.250), we can then write

$$\mathbb{E}\{n_\mathrm{q}\} = \sum_{k\neq 0} (-1)^k \frac{jq}{2\pi k}\Phi_s\left(\frac{2\pi}{q}k\right), \tag{4.259a}$$

$$\mathbb{E}\left\{n_\mathrm{q}^2\right\} = \frac{q^2}{12} + \sum_{k\neq 0}(-1)^k \frac{q^2}{2(\pi k)^2}\Phi_s\left(\frac{2\pi}{q}k\right). \tag{4.259b}$$

We thus see that the necessary and sufficient condition for the quantization noise to be uniformly distributed, and consequently with its moments given by equation (4.255), is to have $\Phi_s(u)$ null for all the multiples of the sampling angular frequency $2\pi/q$, i.e. that

$$\Phi_s\left(\frac{2\pi}{q}k\right) = 0, \quad \forall k \neq 0. \tag{4.260}$$

This is obviously a milder condition than the bandlimited condition derived previously through equation (4.251).

Nevertheless, we may wonder if there are many signals that fulfill this condition in practice. For instance, we observe that a random variable that has a uniform distribution over $[-q/2, q/2]$ has a characteristic function in sinc$(qu/2)$ that exhibits zeros at multiples of $2\pi/q$. But, in general, there are few reasons for the characteristic function of the input signal samples to have such periodic nulls. However, it happens that the bandlimited condition can be easily fulfilled in practice. The PDF $p(s)$ of any sample of the stationary process present at the input of the quantizer and its characteristic function $\Phi_s(u)$ are Fourier transforms of each other. Their respective support widths, Δs and Δu, are therefore linked by the Heisenberg uncertainty principle through $\Delta s \Delta u \geq 2\pi$ [37]. As a result, if Δs is wider than many bin widths q, we expect Δu to be less than $2\pi/q$. We can therefore assume that the bandlimited condition given by equation (4.251) is well approximated in that case. The utility of this approach is that this situation can be easily guaranteed in practice, as ensuring that Δs is wider than many bin widths reduces to ensuring a correct scaling of the signal to be quantized at the input of the device. The assumption that the quantization noise is uniformly distributed can then be considered valid in many practical transceiver line-up use cases as long as we ensure that many bits remain active during the conversion process. However, this does not mean we should forget the potential limitations in this approach. As an example we can mention the $\Sigma\Delta$ converters that rely on quantizers with a very small number of levels, as discussed in Section 4.6. In that case, the input signal cannot be scaled so that the associated PDF spreads over many bin widths. This behavior indeed leads to classical spur problems in those devices [53, 54].

In our discussion so far we have used the term "noise" for the process corresponding to the quantization error resulting from the quantization of a given input process. However, to ensure that this error can effectively be considered as an additional noise term, we need it to be uncorrelated with the process being quantized. This property is not so obvious at first glance as the quantization error term is a deterministic function of this input process. We thus need to examine the higher order statistics of the quantization error, and it is to this that we now turn.

Higher Order Statistics

To examine the higher order statistics of the quantization error, we begin with the non-correlation between this error and the process being converted. This is necessary in order to be able to consider the quantization error as an effective additional noise term. Then we focus on the characteristics of the PSD of the resulting quantization noise process.

Consider the correlation $\gamma_{s \times n_q}(0) = \mathbb{E}\{n_q s\}$ between the stationary process $s(t)$ present at the input of the quantizer and the quantization noise process $n_q(t) = e(s(t))$ retrieved at its output when considered at the same sample time. Recall the Fourier series expansion of $e(s)$ given by equation (4.257). The presence of complex exponentials in that equation may allow us to make the link with the characteristic functions of the quantities involved. Multiplying this equation by s and taking the expectation yields

$$\begin{aligned}\gamma_{s \times n_q}(0) &= \sum_{k \neq 0} (-1)^k \frac{jq}{2\pi k} \mathbb{E}\left\{ s e^{j\frac{2\pi}{q}ks} \right\} \\ &= \sum_{k \neq 0} (-1)^k \frac{q}{2\pi k} \dot{\Phi}_s\left(\frac{2\pi}{q}k\right),\end{aligned} \tag{4.261}$$

where $\dot{\Phi}_s(u)$ is the derivative of $\Phi_s(u) = \mathbb{E}\{e^{jus}\}$. As a result, a sufficient condition for this non-correlation to hold is that [52]

$$\dot{\Phi}_s\left(\frac{2\pi}{q}k\right) = 0, \quad \forall k \neq 0. \tag{4.262}$$

Now focusing on the PSD of $n_q(t)$, we first need to derive the autocorrelation function $\gamma_{n_q \times n_q}(\tau)$ of this stationary process. For that, we could rely on the derivations performed in the next sub-section, for instance on equation (4.271). However, an alternative approach is of interest for our present perspective, still based on the work of Sripad and Snyder. They showed that the joint PDF of the two variables $n_{q,1}$ and $n_{q,2}$ corresponding to two time samples of $n_q(t) = Q(s(t)) - s(t)$ can be written as [52]

$$p(n_{q,1}, n_{q,2}) = \begin{cases} \dfrac{1}{q^2} + \dfrac{1}{q^2} \displaystyle\sum_{k \neq 0} \sum_{l \neq 0} \Phi_{s_1, s_2}\left(\frac{2\pi}{q}k, \frac{2\pi}{q}l\right) e^{-j\frac{2\pi}{q}(k n_{q,1} + l n_{q,2})} \\ \quad \text{for } -q/2 \leq n_{q,1} < q/2, \ -q/2 \leq n_{q,2} < q/2, \\ 0 \quad \text{otherwise.} \end{cases}$$

We thus see that

$$p(n_{q,1}, n_{q,2}) = \begin{cases} \dfrac{1}{q^2} & \text{for } -q/2 \leq n_{q,1} < q/2, \ -q/2 \leq n_{q,2} < q/2, \\ 0 & \text{otherwise,} \end{cases} \tag{4.263}$$

if and only if

$$\Phi_{s_1,s_2}\left(\frac{2\pi}{q}k, \frac{2\pi}{q}l\right) = 0, \quad \forall (k,l) \neq (0,0). \tag{4.264}$$

Looking at equation (4.271), we see that this condition implies the non-correlation we were looking for between $n_{q,1}$ and $n_{q,2}$. However, referring to equation (4.256), we can remark now that having equation (4.260) also fulfilled leads to $p\left(n_{q,1}\right) = 1/q$. As a result, we get in that case that

$$p(n_{q,1}, n_{q,2}) = p(n_{q,1})p(n_{q,2}). \tag{4.265}$$

The joint PDF of two samples of $n_q(t)$ is then simply the product of the PDFs of the two samples. They are therefore statistically independent and not only uncorrelated. However, the latter condition remains sufficient for having the autocorrelation $\gamma_{n_q \times n_q}(\tau)$ that reduces to a Dirac distribution so that the PSD of $n_q(t)$ is necessarily flat.

At first glance, it may not be obvious that a process can simultaneously fulfill all the various conditions reviewed up to now, i.e. equations (4.260), (4.262) and (4.264). However, we observe that the narrowband condition for the characteristic function of s, as per equation (4.251), or its equivalent for the joint characteristic function, is sufficient to ensure this situation. And by the discussion at the end of the previous section, appropriate scaling of the signal at the input of the quantizer can ensure a sufficiently good approximation of this bandlimited condition. We thus expect in most cases to have the quantization noise to be both uncorrelated with the wanted signal, uniformly distributed and white. However, this remains only a quantitative approach, and problems could be encountered in practice. This prompts the introduction of an additional phenomenon discussed in the next section, dithering, which is inherently present in transceiver line-ups and can compensate for those potential problems.

Dithering

Let us now focus on an additional phenomenon of importance in transceiver line-ups, namely dithering. In order to examine this phenomenon, which results in desirable behavior of the quantization noise whatever the statistics of the wanted signal, $s_w(t)$, being converted, we suppose that the process $s(t)$ present at the input of the quantizer is the sum of $s_w(t)$ and a dither signal $s_d(t)$:

$$s(t) = s_w(t) + s_d(t). \tag{4.266}$$

We assume that these processes are stationary and statistically independent. A first consequence of this independence can be seen in the characteristic function $\Phi_s(u) = \mathbb{E}\{e^{jus}\}$ of a sample $s = s_w + s_d$. We have

$$\begin{aligned}\Phi_s(u) = \mathbb{E}\left\{e^{ju(s_w+s_d)}\right\} &= \mathbb{E}\left\{e^{jus_w}\right\}\mathbb{E}\left\{e^{jus_d}\right\}\\ &= \Phi_{s_w}(u)\Phi_{s_d}(u).\end{aligned} \tag{4.267}$$

If the characteristic function $\Phi_{s_d}(u)$ of the dither signal is sufficiently narrowband with regard to the amplitude sampling frequency $2\pi/q$, then $\Phi_s(u)$ can also be made narrowband in that respect. By the discussions in the two previous sections, we then expect to be able to recover the statistical characteristics of the wanted signal from those of the quantized signal. We also expect to make the quantization noise uniformly distributed, white and uncorrelated with $s_w(t)$ whatever the statistical properties of the latter signal.

Nevertheless, the conditions on the dither signal for having those characteristics of the quantization noise to be true can be refined a little. We first reconsider the PDF of a sample n_q of the quantization noise process $n_q(t)$. Substituting equation (4.267) into the original expression given by equation (4.256) when no dither signal is present, we get that

$$p(n_q) = \begin{cases} \frac{1}{q} + \frac{1}{q} \sum_{k \neq 0} \Phi_{s_w}\left(\frac{2\pi}{q}k\right) \Phi_{s_d}\left(\frac{2\pi}{q}k\right) e^{-j\frac{2\pi}{q}kn_q} & \text{for } -q/2 \leq n_q < q/2, \\ 0 & \text{otherwise.} \end{cases}$$

Thus whatever the characteristics of $s_w(t)$, the distribution of the quantization noise can be made uniform as long as the dither process fulfills Schuchman's condition [53, 55]:

$$\Phi_{s_d}\left(\frac{2\pi}{q}k\right) = 0, \quad \forall k \neq 0. \tag{4.268}$$

We then recover that the variance of the process is equal to $q^2/12$.

In the same way, we can reconsider the correlation between the wanted signal and $n_q(t) = e(s(t))$ when considered at the same time sample. We proceed as in the previous section when we derived equation (4.261) and consider the Fourier series expansion of the quantization error transfer function given by equation (4.257). This allows us to make the link with the characteristic functions of the quantities involved through the presence of the complex exponentials. Thus, given that $n_q(t) = e(s(t)) = e(s_w(t) + s_d(t))$, we can write, under the no overload assumption, that

$$e(s_w + s_d) = \sum_{k \neq 0} (-1)^k \frac{jq}{2\pi k} e^{j\frac{2\pi}{q}k(s_w+s_d)}. \tag{4.269}$$

As a result, $\gamma_{s_w \times n_q}(0) = \mathbb{E}\{n_q s_w\}$ can be written as

$$\begin{aligned} \gamma_{s_w \times n_q}(0) &= \sum_{k \neq 0} (-1)^k \frac{jq}{2\pi k} \mathbb{E}\left\{s_w e^{j\frac{2\pi}{q}k(s_w+s_d)}\right\} \\ &= \sum_{k \neq 0} (-1)^k \frac{jq}{2\pi k} \Phi_{s_d}\left(\frac{2\pi}{q}k\right) \mathbb{E}\left\{s_w e^{j\frac{2\pi}{q}ks_w}\right\}, \end{aligned} \tag{4.270}$$

We thus see again that Schuchman's condition allows the quantization noise to be uncorrelated with the wanted signal.

Turning to the spectrum of the quantization noise, we first derive the autocorrelation function $\gamma_{n_q \times n_q}(\tau)$ of this stationary process. We again consider the Fourier series expansion of the quantization error transfer function, as given by equation (4.257). Under the non-overload assumption, we can thus write that, for $\tau \neq 0$,

$$\begin{aligned}\gamma_{n_q \times n_q}(\tau) &= \mathbb{E}\{n_{q,t} n_{q,t-\tau}\}\\ &= -\sum_{k\neq 0}\sum_{l\neq 0}(-1)^{k+l}\frac{q}{2\pi k}\frac{q}{2\pi l}\Phi_s^{(\tau)}\left(\frac{2\pi}{q}k, \frac{2\pi}{q}l\right),\end{aligned} \tag{4.271}$$

with

$$\Phi_s^{(\tau)}(u,v) = \mathbb{E}\left\{e^{jus_t + vs_{t-\tau}}\right\}. \tag{4.272}$$

We can now use the fact that the signal $s(t)$ is the sum of two independent processes, $s_w(t)$ and $s_d(t)$, to expand $\Phi_s^{(\tau)}(u,v)$ as was done in order to derive equation (4.267). If we assume in addition that $s_d(t)$ yields independent random variables when considered at different time samples, it results in

$$\Phi_s^{(\tau)}(u,v) = \Phi_{s_d}(u)\Phi_{s_d}(v)\Phi_{s_w}^{(\tau)}(u,v). \tag{4.273}$$

Substituting this into equation (4.271) yields

$$\gamma_{n_q \times n_q}(\tau) = -\sum_{k\neq 0}\sum_{l\neq 0}(-1)^{k+l}\frac{q}{2\pi k}\frac{q}{2\pi l}\Phi_{s_d}\left(\frac{2\pi}{q}k\right)\Phi_{s_d}\left(\frac{2\pi}{q}l\right)\Phi_{s_w}^{(\tau)}\left(\frac{2\pi}{q}k, \frac{2\pi}{q}l\right).$$

Once again, under Schuchman's condition given by equation (4.268), we get that the autocorrelation of the quantization noise reduces to a Dirac delta distribution. The quantization noise is therefore white.

Finally, we see that a dither signal that fulfills equation (4.268) and that yields independent random variables when considered at different time samples ensures that the quantization noise effectively has a uniform PDF, thus with a variance equal to $q^2/12$, is white and uncorrelated with $s_w(t)$. This holds whatever the statistical properties of the latter signal as long as it is independent of the dither signal. Obviously, a parallel can be made between this condition on the dither signal and the conditions derived so far for the wanted signal when alone at the input of the quantizer, as in equation (4.260). We furthermore see that the bandlimited condition discussed in that previous case can also be used for the dither signal to fulfill equation (4.268). We may thus wonder what the real added value of this signal is, compared to the previous situation. In fact, the interest comes from the statistics of the signals encountered in practice in transceivers. They can indeed be various, or even unknown on the receive side when considering blocking signals for instance. Conversely, we get a known signal, always present in the analog stages, and that can act as a dither signal, i.e. the thermal noise. This may ensure good behavior of the quantization error as an additional noise component treated as such in noise budgets. This behavior is illustrated in the next two sections.

Dithering on the Receive Side

In order to illustrate the dithering effect, we first consider the situation classically encountered on the receive side. The signal present at the input of the P or Q ADC stage is composed of the wanted signal, potential residual unwanted signals, and the unavoidable electronic noises present in the line-up. As illustrated in Chapter 7, for a standard dimensioning of such a line-up, we can assume that the thermal noise is amplified so that its power is 20 dB above that of the ADC quantization noise. As a result, the standard deviation of the thermal noise is set far above the equivalent input quantization step q of the quantizer. But, by the discussion in Section 4.1.2, the PDF $p(n_{\text{th}})$ of any sample of such thermal noise can be considered as Gaussian. We thus have

$$p(n_{\text{th}}) = \frac{1}{\sqrt{2\pi}\sigma} e^{-\frac{n_{\text{th}}^2}{2\sigma^2}}. \tag{4.274}$$

Now referring in turn to the definition of the characteristic function in equation (4.250), one can see that

$$\Phi_{n_{\text{th}}}(u) = e^{-\frac{\sigma^2 u^2}{2}}. \tag{4.275}$$

Thus, even when assuming a worst case where σ is nothing more than q, we get that $\Phi_{n_{\text{th}}}(u)$ can be considered as narrowband with respect to the amplitude sampling frequency $2\pi/q$. We can thus assume that Schuchman's condition given by equation (4.268) is well approximated in that case. We, moreover, get that two time samples of a thermal noise process can be considered as independent random variables. This gives the result that the thermal noise behaves as a dither signal with respect to the quantization processing.

This behavior can be illustrated by considering the quantization of a simple sine wave. However, for a sine wave, the quantization noise is not white and is not uncorrelated with the signal [53]. As illustrated in Figure 4.28, the PSD of the quantized sine wave clearly exhibits spur tones. Furthermore, the addition of a Gaussian noise, even if coded on a few bits only, effectively whitens the quantization noise. Such behavior illustrates the dithering properties of quantities like the thermal noise in a practical line-up. It justifies in many cases of interest treating the quantization noise as an additional noise component among others in a receiver budget.

However, we should keep in mind that this statement relies on the examination of the quantization processing only. For the final word on this, we refer to the discussion in Section 4.5.2 to highlight the fact that the effect of time sampling must also be considered. The two phenomena are always implemented together for effective digital signal processing.

Dithering on the Transmit Side

The typical situation on the transmit side is slightly different from that on the receive side as the generation of practical modulating signals is mostly done in the digital domain. Unless something special is done, we thus cannot rely on any dither effect in the earliest stages of the generation of the modulating signals. Only the statistics of the signal being processed drive the properties of the quantization noise.

To illustrate the potential impact of this, we can reconsider the example of the GMSK modulation as used in the GSM standard. As discussed in Section 1.3.1, the instantaneous frequency of a GMSK modulated bandpass signal is expected to be constant when a constant input data stream enters the modulator. The corresponding modulating frequency word is

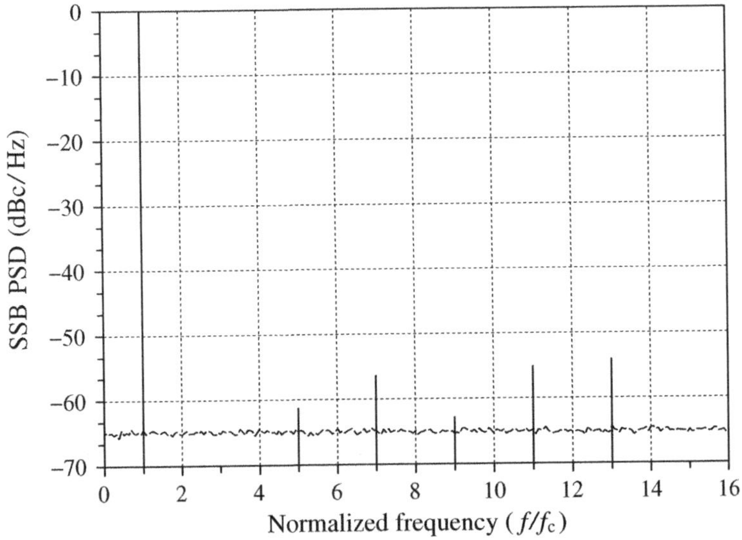

Figure 4.28 Power spectral density of a quantized sine wave with and without dithering – The quantization of a sine wave can lead the quantization noise not being white (solid). In the present case, the sine wave amplitude is set to the quantizer input full scale *FS*. The signal is then quantized on 8 bits including the sign. The quantization step is thus $q \approx FS/2^7$ as given by equation (4.290). The presence of an additive Gaussian noise whitens the quantization noise (dashed). Here, the standard deviation of the Gaussian noise that is added to the analog input sine wave is set equal to the quantizer step.

therefore also constant and equal to $\pm f_{symb}/4 = \pm 67\,708.33$ Hz. We thus expect only the least significant bit (LSB) to be active in the digital words representing this modulating frequency signal. As a result, we surmise that the characteristic function associated with this process can hardly fulfill the narrowband condition discussed up to now. We consequently expect problems in the characteristics of the associated quantization noise.

As illustrated in Figure 4.29, we indeed recover spur tones in the spectrum of this quantized signal. One could argue that this situation of a constant signal is extreme as in general we are dealing with modulations randomly modulated. Nevertheless, this shows that some problems can occur more easily on the transmit side than on the receive side due to this lack of natural dithering.

4.5.2 *Sampling Effect on Quantization Noise*

Up to now we have focused on the amplitude quantization only. However, all practical implementations that work on quantized signals also necessarily work on time samples of those signals. We thus need to say a few words on the impact of the sampling frequency f_s on the characteristics of the quantization noise $n_q(t)$.

Recalling the discussion in Section 4.5.1, all the first order characteristics of the quantization noise depend only on the statistics of a single time sample s of the signal $s(t)$ present at the input of the quantizer. Those characteristics are, moreover, independent of time as long as we can assume the stationarity of $s(t)$, as is common in wireless transceivers (see Appendix 2).

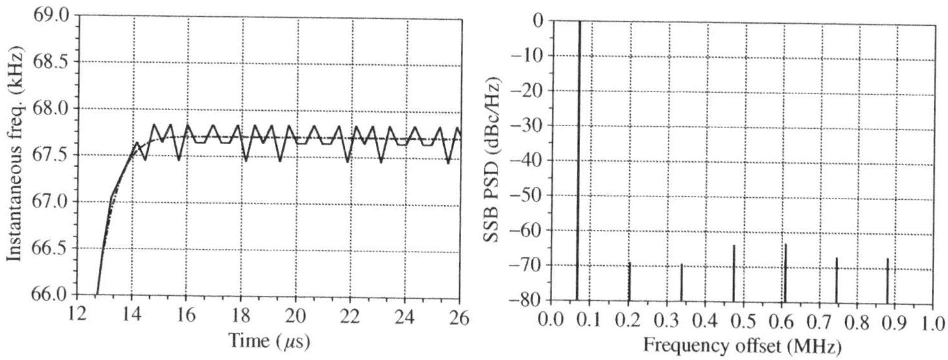

Figure 4.29 Effect of quantization on the GMSK modulation of the GSM standard in the constant input data case – In the case of constant input data bits, the ideal GSM/GMSK modulator provides a constant instantaneous frequency word that is reached after the Gaussian filter rise time (left). Only the LSB remains active on the quantized signal. We therefore assume that the corresponding characteristic function cannot be narrowband, and thus that the quantization noise PSD is not flat (right).

Thus the time difference, $1/f_s$, between two consecutive samples has no impact on the first order characteristics, i.e. on the quantization noise PDF and its moments. The same obviously holds for the correlation between $n_q(t)$ and $s(t)$ by equation (4.261). Conversely, a joint PDF appears as soon as we are dealing with the autocorrelation $\gamma_{n_q \times n_q}(\tau)$ of $n_q(t)$ according to equation (4.271). But this joint PDF can in turn depend on the sampling period $1/f_s$. This is obviously related to the limited spectral extent of $s(t)$ that impacts its rate of change in the time domain. Consequently, the PSD of $n_q(t)$, which is nothing more than the Fourier transform of $\gamma_{n_q \times n_q}(\tau)$, also depends on the sampling rate used to derive the sequence of samples being considered [56].

Expressed in this way, this appears to be nothing more than an aliasing problem related to the sampling of a signal with a finite bandwidth. However, in the present case the problem is trickier as it involves amplitude sampling in addition to time domain sampling. This can be seen, for instance, in equation (4.264) where the amplitude sampling frequency $2\pi/q$ directly drives the condition for having a uniform joint PDF.

The impact of this imbrication in the amplitude and time sampling on the statistics of the quantization noise can be illustrated by reconsidering the quantization of a sine wave. From Figure 4.30, the smaller the quantization step q with respect to the signal amplitude, the higher the number of transitions per unit of time. As a result, the PSD of the continuous time quantization noise signal $n_q(t)$ necessarily spreads over a wider frequency range as q reduces. However, whatever this spectral extent, it can only be finite. Considering now the time sampling of this signal, we anticipate a trade-off between the sampling rate f_s and the spectral behavior of the quantization noise sequence $n_q[k] = n_q(k/f_s)$. If f_s remains lower than the spectral extent Δf_q of $n_q(t)$, we necessarily have spectral aliasing in accordance with Section 4.6.1. The folding of these slices of spectrum can make the result almost flat. Conversely, as f_s goes higher and higher, the aliasing effect vanishes and the spectrum of $n_q(t)$ is recovered without distortion on its sampled version.

This behavior is illustrated in Figure 4.31, taking as an example an LTE 5 MHz signal as introduced in Section 1.3.3. In these simulations, the signal $s(t)$ is scaled at the input of

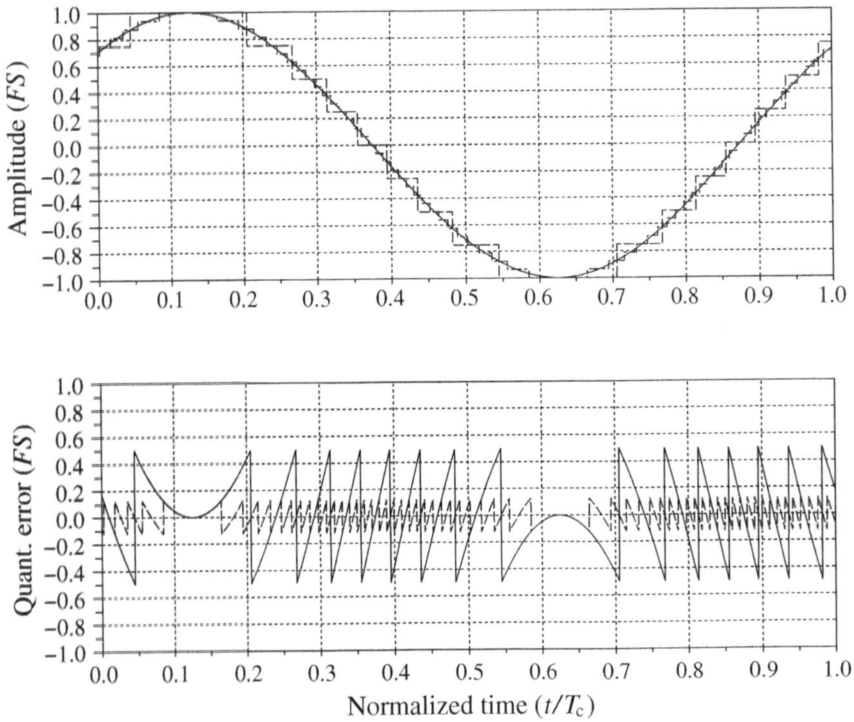

Figure 4.30 Impact of the quantization step level on the quantization of a sine wave – The quantization of a sine wave (top, solid) using a rough quantization step (top, dashed) leads to a huge quantization error (bottom, solid). The use of a smaller quantization step (top, dot-dashed) leads to a smaller quantization error (bottom, dashed) but also to an error signal that exhibits many more transitions per time unit, here the input sine wave period T_c. As a result, the quantization noise PSD spreads over a wider frequency band.

the quantizer in such a way that the bandlimited assumption discussed in Section 4.5.1 can be assumed valid. The quantization noise can thus be assumed as effectively uncorrelated with the wanted signal, and uniformly distributed over $[-q/2, q/2]$. However, it appears that the spectral partition of this noise power indeed depends on the sampling frequency f_s. For sampling frequencies lower than the quantization noise bandwidth Δf_q, the PSD of this quantization noise remains flat. The level of this PSD thus decreases linearly with the increase in the sampling rate, each doubling of f_s resulting in a 3 dB decrease in this level. But, as soon as f_s is sufficiently high that the quantization noise spectrum is correctly represented, this effect vanishes.

The behavior illustrated in Figure 4.31 can alternatively be highlighted by considering the fraction of quantization noise power that is retrieved in the wanted signal bandwidth as a function of the sampling frequency f_s for a given quantization step q. As shown in Figure 4.32, for low sampling rates the in-band quantization noise power decreases linearly with f_s. But once this sampling rate becomes much higher than the quantization noise bandwidth, this phenomenon stops as the quantization noise spectrum is already correctly represented. We

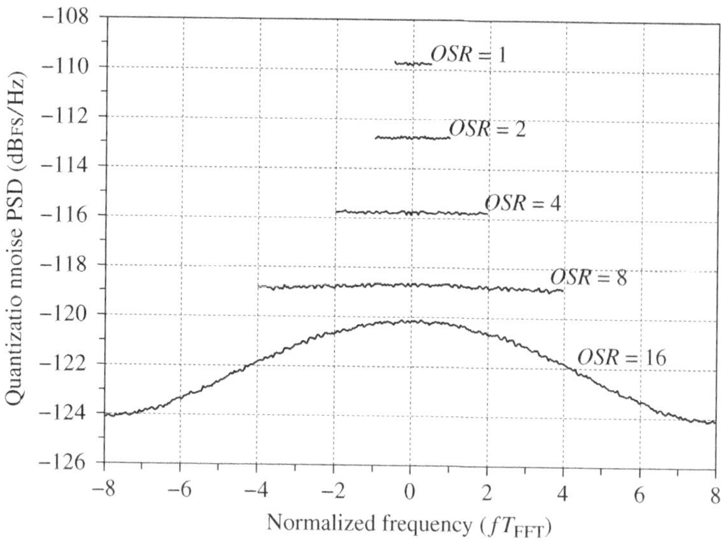

Figure 4.31 Quantization noise PSD for an OFDM signal with different oversampling ratios – The quantization noise PSD level depends both on the quantizer step value q and on the sampling frequency f_s. Here the OFDM signal, an LTE $BW = 5$ MHz signal, is set such that its RMS value is $FS/8$, i.e. 18 dB under the quantizer full scale FS. The quantizer step is $q \approx FS/2^5$ as given by equation (4.290), i.e. the number of bits used to represent the samples is 6 (including the sign). The PSD is derived on $\pm f_s/2$ for different OSRs, i.e. for different sampling periods compared to the original OFDM FFT sampling period. For low OSRs, the PSD level decreases linearly with the sampling frequency according to equation (4.276). This behavior stops when the sampling frequency is high enough to represent correctly the spectrum of the continuous time quantization noise.

also find that the thinner the quantization step with respect to the signal amplitude scaling, the wider the spectral extent of the quantization noise.

However, we observe that for realistic quantization steps as used in practical digital line-ups, the sampling rate for which the white quantization noise approximation fails corresponds to very high oversampling ratios (OSRs) compared to the wanted signal bandwidth. This behavior can also be enhanced by the dithering effect, at least on the receive side as discussed in the previous section, and often related to wideband thermal noise components. As a result, the quantization noise can be considered as white in most practical cases. On top of that, the wanted signal scaling at the input of ADC stages is often such that the bandlimited assumption discussed in Section 4.5.1 is valid. Consequently, the quantization noise can be considered uniformly distributed, with its RMS value given by equation (4.255b). Under those assumptions, the PSD of the quantization noise, $\Gamma_{n_q \times n_q}(f)$, can be written as the ratio of this RMS value to the sampling frequency f_s:

$$\Gamma_{n_q \times n_q}(f) = \frac{q^2}{12 f_s}. \tag{4.276}$$

This expression is widely used for the dimensioning of line-ups as illustrated in the next section, for instance.

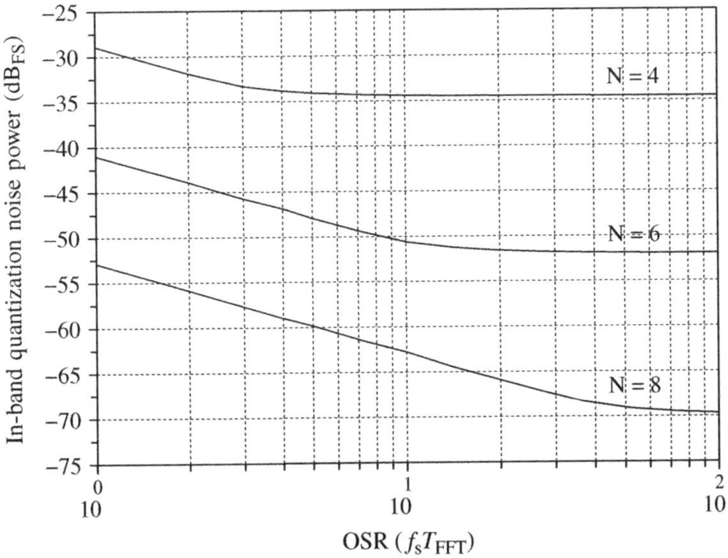

Figure 4.32 In-band quantization noise power for an OFDM signal for different quantization step and oversampling ratios – The in-band quantization noise depends both on the quantizer step value q and on the OSR, as can be seen in Figure 4.31. Here the LTE $BW = 5$ MHz signal is also scaled so that its RMS value is 18 dB under the quantizer full scale FS. The three curves correspond to different quantization steps values $q \approx FS/2^{N-1}$ as given by equation (4.290), where N is the number of bits (including the sign). For $OSR = 1$, the in-band quantization noise power is equal to the overall power and is given by equation (4.255b). This power then decreases in accordance with equation (4.276) as long as the white noise assumption remains valid. The validity of this assumption vanishes when the sampling frequency is high enough to represent correctly the spectrum of the continuous time quantization noise.

4.5.3 *Illustration*

Filtering, Decimation and Quantization Noise Floor

In order to illustrate the concepts introduced so far, let us examine as a first example how to manage the quantization along a digital data path consisting of a cascade of filtering and decimation stages.

Recall that the goal in such dimensioning is the minimization of the SNR degradation when going through the data path. But at the same time, it is often required to minimize the number of bits used to represent the quantities of interest, with the aim of minimizing the area and power consumption of the solution. In practice, this minimization takes the form of a reduction in the number of bits used to represent the signals compared to the number that would be required to achieve the full resolution. This may require implementing truncation operations after some arithmetic operations, as can be understood by considering a simple multiplication between an input sample $s_i[k]$, quantized on N_i bits, and a fixed number, say a filter tap, coded on N_t bits. Assuming that all quantities are signed, the full precision for the result of the multiplication would require the use of $N_i + N_t - 1$ bits for its fixed point representation. But, as only N_i bits were necessary to correctly represent the signal before the processing, we can expect that a similar number of bits will be enough at its output as long as the signal remains correctly scaled. There is thus a temptation to dump LSBs on the output word through a truncation operation. The problem is then to determine the number of bits that are really required to represent the output signal with a minimum loss of information.

In order to go further, we must convince ourselves that the truncation operation can also be interpreted as a quantization processing as long as the LSBs we dump are effectively active. In that case, the obvious resolution loss on the signal can still be interpreted in terms of amplitude sampling as in any other quantization operation. By the discussion in Section 4.5.1, as long as the signal is correctly scaled at the input of the truncation operation, its characteristic function can be assumed narrowband compared to the amplitude sampling frequency $2\pi/q_o$, with q_o the quantization step recovered at the output of the operation. This results in the possibility of interpreting the error resulting from this truncation as an additional quantization noise term. This noise can furthermore be assumed as white, with an RMS value given by equation (4.255b), and with PSD given by equation (4.276). As a result, assuming that the signal is coded on N_o bits after truncation, and that it is sampled at the sampling rate f_s, we can consider an equivalent model for the truncation stage as shown in Figure 4.33.

Thus, let us now focus on a single filtering and decimation stage, as illustrated in Figure 4.34. The filter is here supposed to exhibit a DC gain of G. Due to the arithmetic operations in its processing, we then get a natural increase of the number of bits used to represent the signal. The quantization step recovered at the filter output, q_f, is thus much thinner than the input step q_i. The decimation stage is then supposed to dump a given amount of samples so that the sampling rate goes from $f_{s,i}$ to $f_{s,o}$. Finally, the truncation stage dumps LSBs so that the quantization step increases from q_f to q_o. In order to evaluate the resulting SNR degradation when going through such a line-up, we simply need to consider the degradation of the PSD $\Gamma_{n_q \times n_q}(f)$ of the quantization noise at a given frequency within the passband of the filters, let's say the zero frequency. For a white noise, this is a direct image of the total noise recovered in the wanted signal frequency band while allowing us to get rid of the exact bandwidth of this signal. Thus, we can derive the PSD of the quantization noise recovered at the output of such a line-up. This is quite straightforward if we assume on the one hand that the truncation behaves as an additive quantization noise source as illustrated in Figure 4.33, and on the other hand that the filter effectively behaves as a decimation filter, i.e. that it cancels signal components

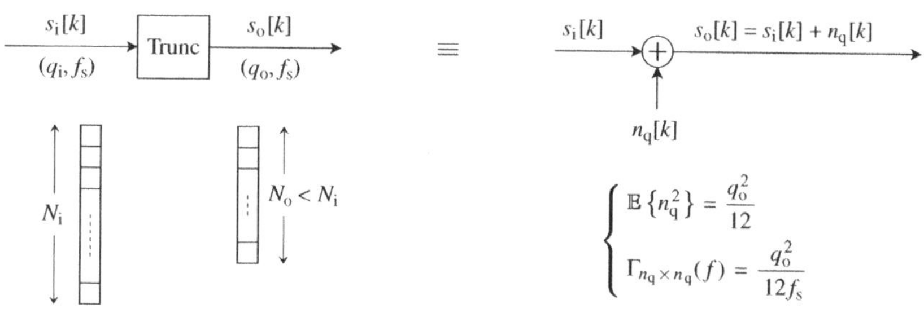

Figure 4.33 Truncation operation as an additive quantization noise source – From the equivalent noise model point of view, the truncation operation, i.e. the dumping of a significant number of active LSBs, can be modeled by an additive quantization noise source. Under assumptions derived in Section 4.5.1, the RMS value of this noise source is given by equation (4.255b) and its PSD by equation (4.276).

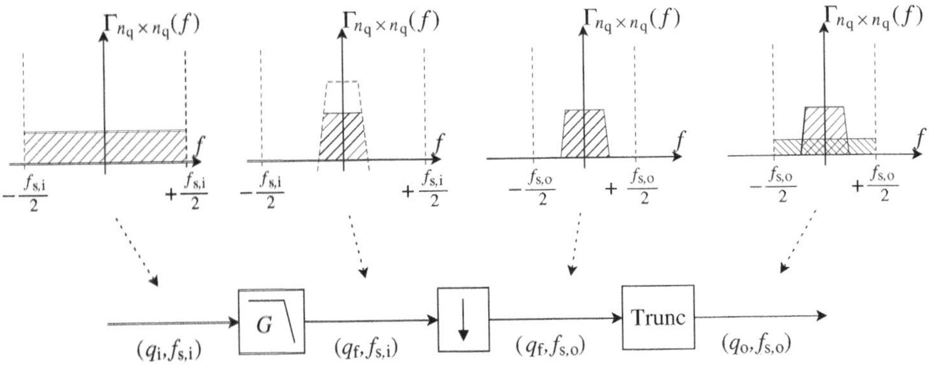

Figure 4.34 Quantization noise floor management during a decimation stage – The ideal implementation of a filtering stage results in the filtering of the input quantization noise. If the filter is well defined, no residual out-of-band noise component remains and is thus folded in-band during the decimation processing. The truncation of the resulting signal leads to an addition of a new quantization noise that can be approximated as white. Its PSD is therefore spreads in a frequency band equal to the output sampling frequency.

that could be folded in-band during the downsampling. Assuming that the PSD $\Gamma_{n_{q,i} \times n_{q,i}}(f)$ of the quantization noise present at the input of the line-up can be written as

$$\Gamma_{n_{q,i} \times n_{q,i}}(f) = \frac{q_i^2}{12 f_{s,i}}, \tag{4.277}$$

it immediately follows that

$$\begin{aligned}\Gamma_{n_{q,o} \times n_{q,o}}(0) &= G^2 \Gamma_{n_{q,i} \times n_{q,i}}(0) + \frac{q_o^2}{12 f_{s,o}} \\ &= G^2 \frac{q_i^2}{12 f_{s,i}} + \frac{q_o^2}{12 f_{s,o}}.\end{aligned} \tag{4.278}$$

Looking at this expression, we see that in order to keep the additional quantization noise contribution resulting from the truncation negligible compared to input, we need to ensure that

$$\frac{q_o^2}{f_{s,o}} \ll G^2 \frac{q_i^2}{f_{s,i}}. \tag{4.279}$$

In most applications, the scaling of the signal remains consistent in terms of back-off along the data path so that $G \approx 1$. We then see that it is the ratio q^2/f_s that has to be managed and controlled in each digital block. The weaker the sampling rate, the thinner the quantization step, and thus the higher the number of bits required to preserve the noise performance.

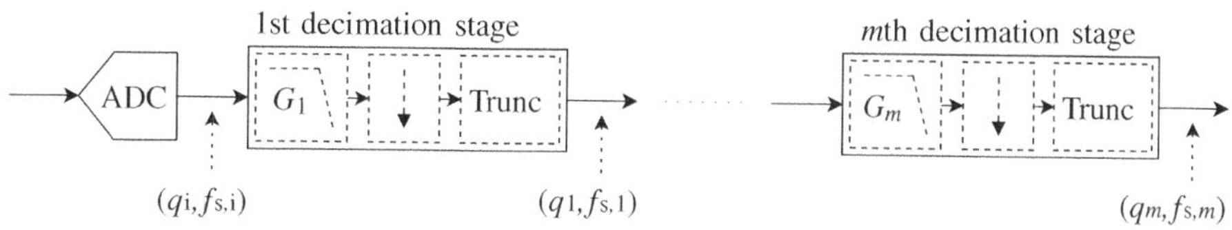

Figure 4.35 Quantization noise floor management during a cascade of decimation stages – The quantization step has to be managed all along a cascade of filtering and decimation stages in order to preserve the overall noise performance in accordance with equation (4.280).

This behavior can be generalized by considering the cascade of decimation blocks illustrated in Figure 4.35. This configuration is classically encountered in receivers to convert the samples recovered at a high data rate at the output of an ADC to a reasonable sampling rate that enables the baseband processing dedicated to synchronization, equalization, etc. The PSD $\Gamma_{n_{q,o}\times n_{q,o}}(f)$ recovered at the output of the line-up can then be derived in the passband of the filters by the recursive application of equation (4.278). Assuming that the quantization step and sampling frequency recovered at the output of the lth stage are respectively q_l and $f_{s,l}$, we can indeed write that

$$\Gamma_{n_{q,o}\times n_{q,o}}(0) = G_1^2 G_2^2 \cdots G_m^2 \Gamma_{n_{q,i}\times n_{q,i}}(0) + G_2^2 \cdots G_m^2 \frac{q_1^2}{12 f_{s,1}} + \cdots + G_m^2 \frac{q_{m-1}^2}{12 f_{s,m-1}} + \frac{q_m^2}{12 f_{s,m}}, \tag{4.280}$$

where G_l is the amplitude gain of the lth filter stage. As already highlighted, when considering a consistent scaling of the signal along the data path, we get that the contribution of each stage is proportional to the square of its output quantization step, and inversely proportional to its output sampling rate. As a result, when assuming for instance a decrease in the sampling frequency after each decimation block, the number of bits used to represent the signal must be increased after each of those blocks in proportion to the square root of the ratio between its input and output sampling rates.

To conclude, we observe that the above arguments are based on the white noise assumption for the quantization noise. However, as discussed in Section 4.6, some ADCs are based on noise shaping. In that case, the quantization noise at the converter output is not white. But even in that case the correct scaling of the signal in the data path allows us to assume that the subsequent additive quantization noise sources linked to the truncations present in the line-up are white. The criterion of minimizing the degradation of the input noise PSD along the digital receive path, as per equation (4.280), thus remains valid.

Loss Below the Quantization Step

Another interesting aspect of the concepts introduced so far lies in the possibility of recovering the statistics of the signal present at the input of the quantizer by processing its output samples.

This is of interest because this behavior holds for *any* of statistical characteristics of the input signal, even for those that may look inconsistent with the quantization step q used for the conversion.

To illustrate, let us suppose that the signal $s(t)$ present at the input of the quantizer results from the superposition of a simple DC component, DC, and a thermal noise signal $n_{\text{th}}(t)$. Then $s(t)$ can be written as

$$s(t) = n_{\text{th}}(t) + DC. \tag{4.281}$$

Assume now that the signal is scaled in the line-up so that the standard deviation σ of the thermal noise, assumed to be Gaussian, is equal to the quantization step q. By the discussion in "Dithering" (Section 4.5.1), we can then assume that this noise component behaves as a dithering signal. The conclusion on the characteristics of the quantization noise derived in that section can thus be considered as true as well as the possibility of recovering the statistical characteristics of $s(t)$ by examination of the quantized signal. The interesting point is that this holds whatever the value of those characteristics. For instance, if we assume that the DC term is equal to $q/3$, we may at first think that this term can only be lost by the rounding operation associated with the quantization operation. However, this DC term is nothing more than a component of the first moment of $s(t)$. It is even exactly the expectation of this process if we consider the thermal noise to be centered. According to our discussion so far, we should thus be able to recover this DC term by examining the statistics of the quantized signal.

This behavior is confirmed by the simulations shown in Figure 4.36. The histogram of the quantized signal is biased in the presence of the DC component. Furthermore, an averaging of this quantized signal over 50,000 samples leads to an estimate of 0.3323σ, which corresponds to the DC value considered here. This confirms that the DC term can be estimated as the first moment of the quantized signal. This example is a precise illustration of what can be encountered in practice, for instance in a WCDMA receiver. In this standard introduced in Section 1.3.3, the dedicated data channel can indeed be below the receiver thermal noise in a sensitivity configuration – classically, up to 20 dB below. Expressed like this, we might think

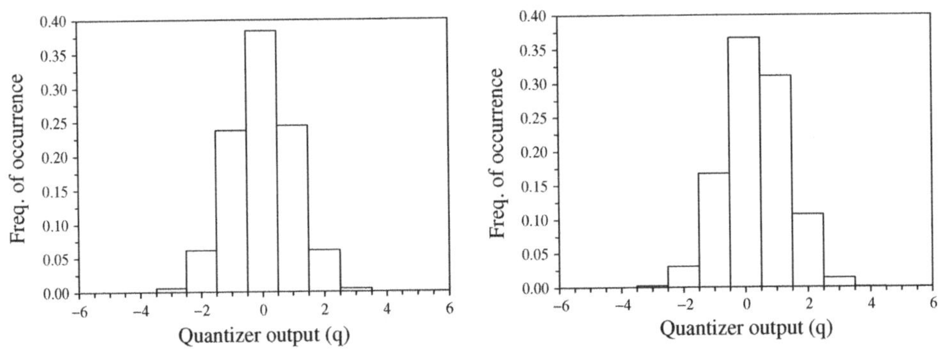

Figure 4.36 Distribution of a quantized Gaussian noise plus DC when the DC level is below the quantization step – The presence of a DC component equal to $q/3$, with q the quantization step, biases the distribution of the quantized signal (right) compared to the DC free distribution (left). Although the DC component is below the quantization step, its value can be recovered from the statistics of the quantized signal. Here the standard deviation σ of the Gaussian noise is set to q.

that we need to scale the received signal so that the power of the quantization error remains negligible enough with regard to the power of the dedicated data channel. We might indeed think this is required in order to be able to process this dedicated channel correctly and recover the data. This is obviously not a good criterion in terms of our discussion so far. Our focus should only be on the receiver thermal noise. Indeed, as long as we can ensure that this noise component is scaled so that the quantization error behaves as the expected noise component, the statistics of the data channel remain present in the output quantized signal as in its input. They can thus be recovered by appropriate signal processing. Another way to see this is that, as long as the quantization error behaves as an additive noise term, we can write the samples $s_q[k]$ recovered at the output of the quantizer as

$$s_q[k] = s_w[k] + n_{th}[k] + n_q[k]. \tag{4.282}$$

The distinction cannot be made between the quantization noise and any other noise components like the thermal noise. Any processing that could be predicted before quantization for recovering the data on the wanted signal $s_w(t)$ can still be considered on the quantized signal. The only valid criterion for scaling the signal at the input of the quantizer is that the quantization noise remains negligible compared to the thermal noise in the perspective of minimizing the SNR degradation. In practice, analog noises of the order of 20 dB above the quantization noise are sufficient to obtain a dithering effect and minimal SNR degradation, as illustrated in Chapter 7.

4.6 Conversion Between Analog and Digital Worlds

Throughout Section 4.5 we discussed the characteristics of the quantization noise in order to derive guidelines useful for the dimensioning of transceiver data paths. However, rather than theoretical quantizers and time samplers, the conversion between the analog and digital worlds is achieved in practical line-ups using effective ADC and DAC blocks that may already be complex systems in themselves. We thus expect to have additional constraints to consider for the correct dimensioning of such line-ups in addition to considerations about the simple additive quantization noise contribution. These constraints can be related for instance to the principles used by the converters, as for noise shaping based devices, or to the limitations in the implementation that can impact the linearity of the quantizer, or even to the quality of the clock that is used to drive the conversion.

Even if not directly related to "noise" in itself, this chapter is the appropriate place to say a few words on ADC and DAC from the system design point of view. This means discussing the phenomena we may be faced with in practice and how to manage them in a line-up.

4.6.1 Analog to Digital Conversion

Aliasing

When considering the implementation of signal processing functions in the digital domain, we obviously consider a system that processes quantized samples of a continuous time signal. As a result, any ADC implements both the quantization *and* the time sampling operation. Having discussed the quantization operation extensively in Section 4.5, we can now go deeper into the system impact of the time sampling operation.

Suppose that we sample the signal $s(t)$ at the constant[4] sampling rate $f_s = 1/T_s$. We can then usefully interpret the sequence of output samples $s_o[k] = s_o\left(kT_s\right)$ in terms of the Dirac comb distribution:

$$s_o(t) = s(t) \sum_k \delta(t - kT_s). \tag{4.283}$$

As the Fourier transform of a Dirac comb is itself a Dirac comb we can write the Fourier transform of $s_o(t)$ as [2]

$$\begin{aligned} S_o(f) &= S(f) \star \frac{1}{T_s} \sum_k \delta\left(f - \frac{k}{T_s}\right) \\ &= f_s \sum_k S(f - kf_s). \end{aligned} \tag{4.284}$$

This result can be interpreted in terms of functions as long as the sum converges, which is the case in particular when the translated spectra do not overlap. We then recover the classical result of the sampling theorem which states that the spectrum of the sampled signal is a superposition of the copies of the spectrum of the input signal, each copy being centered around multiples of the sampling frequency f_s.

As a first consequence, if $s(t)$ is a lowpass signal whose spectrum spreads over the frequency range $[-B/2, B/2]$, we need to have at least $f_s > B$ in order to have a correct representation of $S(f)$ in any sideband of the series expansion of $S_o(f)$. Here, we recover nothing more than the Nyquist criterion for the correct sampling of a continuous time signal. However, in classical implementations of receivers, we almost always have that $f_s \gg B$. The main reason for this is that a receiver collects a large number of signals at its input. By Section 3.3.1, the total spectrum of the received signal can be assumed almost infinite with respect to practical ADC sampling rates. Obviously, even if f_s can be made greater than B, filtering needs to be carried out in order to limit the frequency extent of the unwanted signals present at the ADC input before doing the time sampling. As discussed in Section 7.3.3, such filtering can hardly be fully implemented in the RF world. We need to have the channel selection done so that a baseband filter can be used. This filtering classically takes place prior to the ADC to ensure that the total bandwidth of the signal present at its input remains lower than f_s. Due to its function, this filter is often referred to as an anti-aliasing filter. We can then understand a posteriori the interest in having $f_s \gg B$. The higher the sampling rate, the lower the filtering required in respect of the aliasing problem, but then the higher the power consumption of the converter.

All this results in the classical line-up structure shown in Figure 4.37 whose behavior in the frequency domain is illustrated in Figure 4.38. We remark that this filter is required prior to the time sampler itself rather than prior to the ADC block. The distinction may be worthwhile as the input of what is called the ADC is not necessarily the sampler itself. Some ADCs embed in their own structure some analog filtering that can act as an anti-aliasing filter before carrying

[4] Non-uniform sampling is beyond the scope of this book.

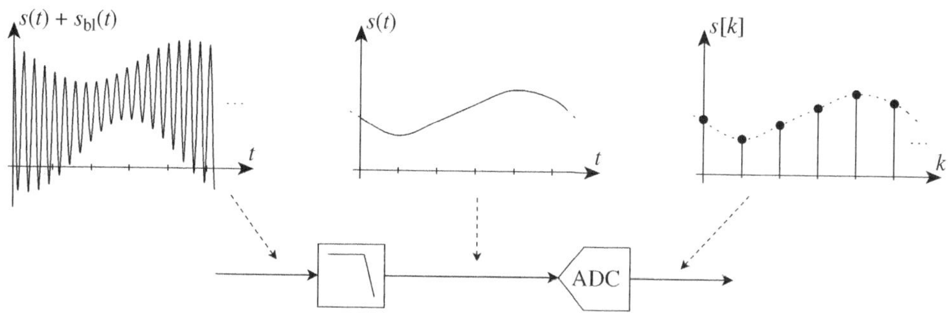

Figure 4.37 Anti-aliasing filter required prior to analog to digital conversion – As illustrated in the frequency domain in Figure 4.38, when a high frequency blocker, $s_{bl}(t)$, is superposed on the wanted signal $s(t)$, an anti-aliasing filter may be needed to prevent aliasing during the analog to digital conversion.

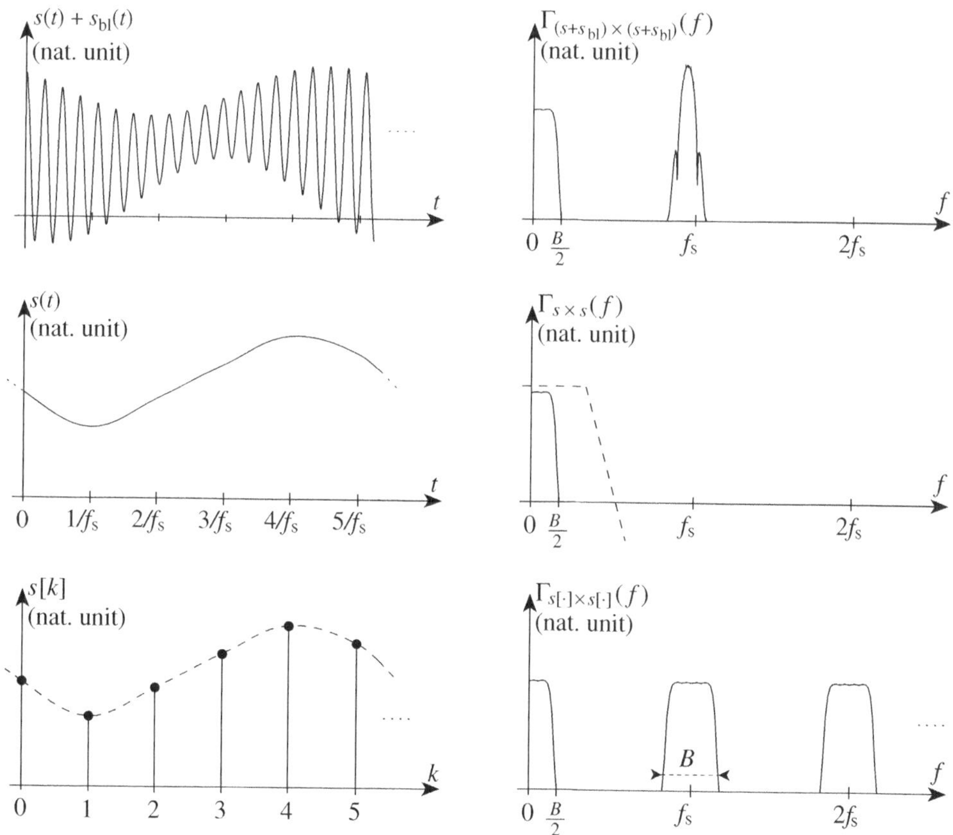

Figure 4.38 Illustration of anti-aliasing and analog to digital conversion in the spectral domain – When a high frequency blocker, $s_{bl}(t)$, lies around multiples of the sampling rate f_s (top), an anti-aliasing filter needs to cancel them (middle) in order to prevent a folding on the wanted signal $s(t)$ during the sampling operation (bottom).

out time sampling as illustrated in "Noise shaping" later in this section. In that case, the need for an additional external filter must be investigated based on the characteristics of the ADC block.

To conclude, we observe that this filtering stage is also often needed for another purpose. It allows minimization of the DR, or the effective number of bits as introduced in the next section, required on the ADC block. This is an important additional topic for this filtering stage and is discussed in Section 7.3.3.

Dynamic Range and Effective Number of Bits

As discussed extensively in Section 4.5, the quantization operation results in the generation of a quantization error that can be considered as an additive noise term in most practical cases. The fact is that the power of this quantization noise is a quantity intrinsic to the quantizer and is thus independent of the scaling of the signal present at the input of the device. But at the same time, dealing with samples coded over a finite number of bits N, we are obviously faced with an upper bound in the quantities we can represent at the output of the quantizer. This maximum value is nothing more than the ADC FS (see the quantity introduced in "Quantizer model" (Section 4.5.1)). There is thus naturally a limitation on the SNR that can be recovered at the output of such an ADC block. The capability of the device in that respect is often quantized through its dynamic range, DR, defined as

$$DR = \frac{\mathbb{E}\{s_{\text{max}}^2\}}{\mathbb{E}\{n_{\text{q}}^2\}} \tag{4.285}$$

or, in decibel units,

$$DR|_{\text{dB}} = 10\log_{10}\left(\frac{\mathbb{E}\{s_{\text{max}}^2\}}{\mathbb{E}\{n_{\text{q}}^2\}}\right). \tag{4.286}$$

Here, $s_{\text{max}}(t)$ stands for the signal of maximum amplitude that can be set at the input of the device without experiencing clipping during the conversion, and $n_{\text{q}}(t)$ for the equivalent input quantization noise. It is convenient to work on equivalent input quantities in order to allow straightforward comparisons with other analog contributions during the derivation of budgets, as illustrated in Chapter 7. The relationship between equivalent input and output quantities is straightforward, referring to the device conversion gain as illustrated in Figure 4.39.

As the purpose of an ADC is to deliver samples coded on N bits, it is often convenient to link the DR of the device and this bit number. For that purpose, we need to remark that $s_{\text{max}}(t)$ is defined such that its peak value can be assumed, up to first order, equal to the ADC full scale FS as illustrated in Figure 4.39. Dealing with a waveform that exhibits a CF, i.e. a ratio between its peak value over its RMS value, equal to CF, we thus need to scale $s_{\text{max}}(t)$ such that

$$\mathbb{E}\{s_{\text{max}}^2\} = \frac{FS^2}{CF^2}. \tag{4.287}$$

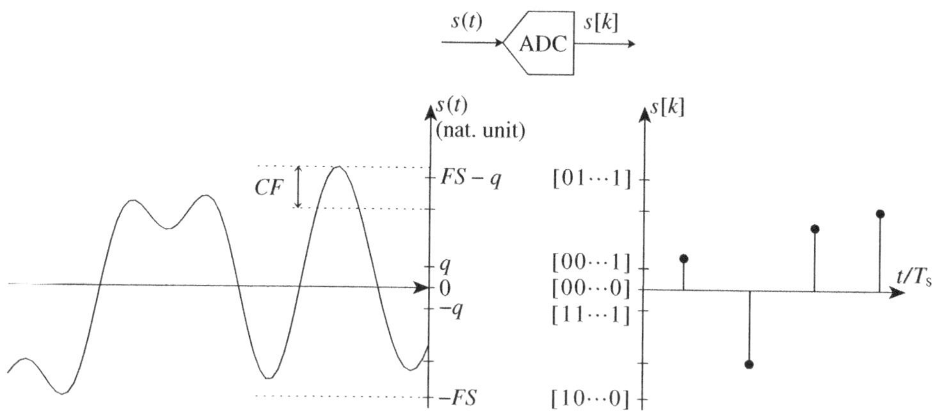

Figure 4.39 ADC equivalent input quantities and input signal maximum level scaling – The characteristics of an ADC can be derived as equivalent input quantities in order to facilitate comparisons with analog input contributions. This is true, for instance, for its full scale FS and its quantization step q. To optimize the ADC DR while avoiding overload, the best that can be done is to scale the peak value of the input signal to the ADC full scale. In that case, the input signal RMS value has to be set accordingly to the waveform crest factor CF.

At the same time, under common assumptions, we get that the second moment of the quantization noise can be estimated using equation (4.255b). Substituting those two results into equation (4.285), we can write

$$DR = \frac{FS^2}{CF^2}\frac{12}{q^2}. \tag{4.288}$$

It then remains to find an expression for the equivalent input quantizer step q based on the ADC full scale FS and its number of bits N. Assuming for instance that we are dealing with signed quantities encoded in two's complement, the ADC can deliver 2^N words that represent input signal amplitudes from $-FS$ to $+FS - q$. We can therefore write

$$q = \frac{2FS - q}{2^N - 1}. \tag{4.289}$$

As in almost all cases we have $N \gg 1$ so that $FS \gg q$, this expression is often approximated as

$$q \approx \frac{FS}{2^{N-1}}. \tag{4.290}$$

Using this expression in equation (4.288), we finally get that

$$DR = \frac{3 \cdot 2^{2N}}{CF^2} \tag{4.291}$$

or, in decibel units,

$$DR|_{\text{dB}} = 4.76 - CF|_{\text{dB}} + 6.02N. \tag{4.292}$$

In the particular case where the input signal is a sine wave, we get that $CF = \sqrt{2}$ so that the above equation reduces to

$$DR|_{\text{dB}} = 1.76 + 6.02N. \tag{4.293}$$

What remains important to keep in mind is that due to its definition, the DR depends on the statistics of the signal used for its evaluation through its CF. In that respect, equation (4.292) is obviously a less confusing formulation.

However, in practical ADC implementations, the RMS value of the noise that is delivered by the device does not match perfectly the ideal value linked only to the characteristics of the quantizer as given by equation (4.255b). As in any active electronic device, there may be additional noise sources that may degrade the overall noise power delivered by the device. In order to reflect this, the concept of effective number of bits is often used. The idea behind this is nothing more than the reverse of what we have done so far. For a given DR effectively measured at the output of a converter, we can define an effective number of bits by inverting equation (4.292). And given that additional noise sources are always present on top of the quantization noise, the effective number of bits is necessarily lower than the number of bits physically delivered by the quantizer.

In the same way, we observe that the noise power considered so far in our definitions is the *total* noise delivered by the converter. It is therefore the noise lying in the total sampling band $[-f_s/2, f_s/2]$. But, by the discussion in the previous section, the sampling rate f_s used for practical ADC stages can be much higher than the spectral extent of the wanted signal being converted. We can thus understand that the fraction of noise lying in the wanted signal band can be much lower than the total noise power considered so far. Given that it is the SNR in the wanted signal frequency band that is usually of interest, we can often define an effective number of bits in that frequency band. This concept is well illustrated by the $\Sigma\Delta$ converters introduced in "Noise shaping" later in this section. For such devices, the total quantization noise recovered at the output of the converter corresponds to poor overall DR performance. Conversely, considering only the fraction of power that lies in the wanted signal frequency band, we can achieve very good performance and a high effective number of bits.

Clock Jitter

As discussed in the previous section, different internal noise contributions can be found at the origin of the limitations of ADC performance in terms of DR and thus effective number of bits. On top of the inherent electronic noise sources related to the physical implementation of the ADC, the timing jitter that corrupts the clock used for the sampling is of particular importance when dealing with uniform sampling.

In order to investigate this, we can reconsider the sampling of the signal $s(t)$ at the constant rate f_s. In the noiseless clock case, the kth sample of $s(t)$, $s[k]$, can be directly written as $s(k/f_s)$. But, in the presence of timing jitter, the kth sample of $s(t)$, $s_j[k]$, takes the form

$$s_j[k] = s\left(\frac{k}{f_s} + j\left(\frac{k}{f_s}\right)\right), \tag{4.294}$$

with $j(k/f_s)$ the error on the kth sampling time. Obviously, the sampling clock signal is designed to minimize the jitter term. We can thus assume that $|j(k/f_s)| \ll 1/f_s$. Consequently, the above equation can be approximated as

$$\begin{aligned} s_j[k] &\approx s\left(\frac{k}{f_s}\right) + j\left(\frac{k}{f_s}\right)\dot{s}\left(\frac{k}{f_s}\right) \\ &= s[k] + j[k]\dot{s}[k], \end{aligned} \tag{4.295}$$

where $\dot{s}(t)$ is the derivative of $s(t)$. Finally, the error on the value of the kth sample resulting from the clock jitter, $e[k]$, can be written as

$$e[k] = s_j[k] - s[k] = j[k]\dot{s}[k]. \tag{4.296}$$

As illustrated in Figure 4.40, we thus see that this error is linked up to first order to the slope of the signal at the ideal sampling instant.

In order to evaluate the resulting SNR limitation, we first need to check that this error term is uncorrelated with the input signal when considered at the same time. This can be done in a straightforward way assuming the independence of the timing jitter process with $s(t)$. We can write in that case that

$$\mathbb{E}\{e_k s_k\} = \mathbb{E}\{j_k \dot{s}_k s_k\} = \mathbb{E}\{j_k\}\mathbb{E}\{\dot{s}_k s_k\}. \tag{4.297}$$

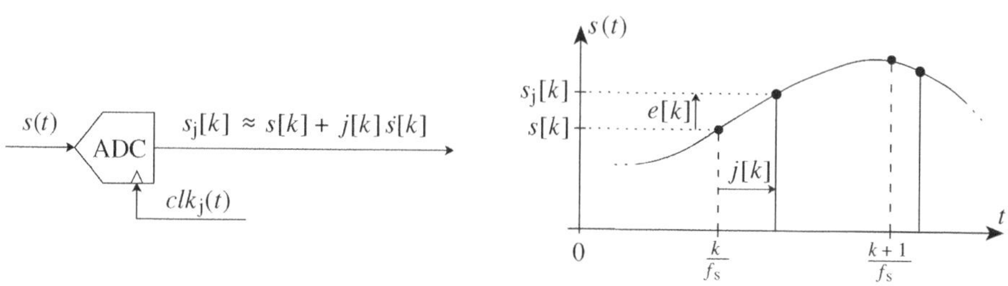

Figure 4.40 Analog to digital conversion driven by a jittered clock signal – During the sampling of an analog signal (left), the use of a jittered clock signal $clk_j(t)$ leads to the generation of an error signal in addition to the ideal samples taken at exact multiples of the sampling period. Up to first order, the error term generated is linked to the slope of the sampled signal at the sampling time (right).

Given that the jitter process is naturally centered, $\mathbb{E}\{j_k\}$ is thus null. The non-correlation between the error signal and $s(t)$ therefore holds. To derive an expression for the SNR, we can thus now focus on the power of the error process. We can assume for the sake of simplicity that we are dealing with stationary processes as classically encountered in the field of wireless (see Appendix 2). Thus, based on equation (4.296), we can write that

$$\mathbb{E}\{e^2\} = \mathbb{E}\{j^2\}\mathbb{E}\{\dot{s}^2\}. \tag{4.298}$$

In this equation, $\mathbb{E}\{j^2\}$ simply represents the power of the jitter process. As we are dealing with a centered process, this power reduces to its standard deviation σ_{j}. We can then write the power of the error process as

$$\mathbb{E}\{e^2\} = \sigma_{\mathrm{j}}^2\mathbb{E}\{\dot{s}^2\}. \tag{4.299}$$

The SNR resulting from the timing jitter, SNR_{j}, can thus be written as

$$SNR_{\mathrm{j}} = \frac{\mathbb{E}\{s^2\}}{\mathbb{E}\{e^2\}} = \frac{1}{\sigma_{\mathrm{j}}^2}\frac{\mathbb{E}\{s^2\}}{\mathbb{E}\{\dot{s}^2\}}. \tag{4.300}$$

Considering this expression, it is hard to go further without any assumptions on the statistics of the signal $s(t)$. Suppose that we are dealing with a pure sine wave, i.e. that

$$s(t) = \rho\sin(2\pi f t). \tag{4.301}$$

Given that the derivative of this signal is still a sine wave at the same angular frequency, the power of those two bandpass signals can be directly written as

$$\mathbb{E}\{s^2\} = \rho^2/2, \tag{4.302a}$$

$$\mathbb{E}\{\dot{s}^2\} = (2\pi f\rho)^2/2. \tag{4.302b}$$

Then equation (4.300) reduces to

$$SNR_{\mathrm{j}} = \frac{1}{(2\pi f)^2\sigma_{\mathrm{j}}^2}. \tag{4.303}$$

This expression thus gives an indication of the maximum jitter that is allowed on the clock signal to achieve a given target SNR, and thus a target DR during the conversion of a sine wave whose angular frequency is $\omega = 2\pi f$. Alternatively, we expect to be able to use this relationship to give preliminary orders of magnitude during the sampling of any signal whose spectral content lies within $[-f, f]$.

However, the fact is that the magnitude of the error term, and hence its power, is related to the slope of the input signal at the sampling instants. The assumption that we are dealing with a pure CW is thus necessarily restrictive. However, we can refine our approach slightly by using Bernstein's theorem, which links the maximum value of the derivative of a signal to the maximum value of its amplitude and to its spectral extent. More precisely, assuming that the Fourier transform of $s(t)$ vanishes outside the frequency domain $[-F, +F]$, and that $|s(t)| \leq M$ for all t, we get that [2]

$$|\dot{s}(t)| \leq 2\pi FM. \tag{4.304}$$

This result can then be used to derive an upper bound for the power of $\dot{s}(t)$. But in that perspective, we can usefully make the distinction between the cases where $s(t)$ is a lowpass and a bandpass signal. In the former case we have

$$\mathbb{E}\{\dot{s}^2\} \leq (2\pi FM)^2. \tag{4.305}$$

But in the latter case, $\dot{s}(t)$ is a bandpass signal as $s(t)$ is. This can be seen, for instance, by differentiating a general bandpass expression such as $\rho_s(t)\cos(\omega t + \phi_s(t))$. As a result, $\dot{s}(t)$ can still be written in the form $\rho_{\dot{s}}(t)\cos(\omega t + \phi_{\dot{s}}(t))$. Assuming that we are dealing with sufficiently narrowband waveforms, we then get that the maximum values of $|\dot{s}(t)|$ almost correspond to the maximum values of $\rho_{\dot{s}}(t)$. We can then write from equation (4.304) that $\rho_{\dot{s}}(t) \leq 2\pi FM$. Given that the power of the bandpass signal $\dot{s}(t)$ is equal to $\mathbb{E}\{\rho^2/2\}$ by equation (1.64), we finally get that

$$\mathbb{E}\{\dot{s}^2\} \leq (2\pi FM)^2/2. \tag{4.306}$$

Based on these upper bounds for $\mathbb{E}\{\dot{s}^2\}$, we can then derive lower bounds for the SNR that can be achieved during the sampling process. Considering first the lowpass case, we can bound the power of the error in the sample values by substituting equation (4.305) into equation (4.299). This results in

$$\mathbb{E}\{e^2\} \leq (2\pi FM)^2\sigma_{\mathrm{j}}^2. \tag{4.307}$$

We can thus derive a lower bound for the SNR achieved during the sampling:

$$SNR_{\mathrm{j}} = \frac{\mathbb{E}\{s^2\}}{\mathbb{E}\{e^2\}} \geq \frac{\mathbb{E}\{s^2\}}{M^2}\frac{1}{(2\pi F)^2\sigma_{\mathrm{j}}^2}. \tag{4.308}$$

We observe that the term $M^2/\mathbb{E}\{s^2\}$ is nothing more than the square of the ratio between the peak amplitude of $s(t)$ and its RMS value. This is therefore the square of the crest factor CF_{LP} of the signal being sampled according to the definition given in "Peak to average power ratio and crest factor" (Section 1.1.3). As a result, the lower bound for SNR_{j} can be written as

$$SNR_{\mathrm{j}} \geq \frac{1}{CF_{\mathrm{LP}}^2}\frac{1}{(2\pi F)^2\sigma_{\mathrm{j}}^2}. \tag{4.309}$$

Obviously, the same derivation can be done in the bandpass case by using equation (4.306) instead of equation (4.305). In that case, the lower bound for SNR_j takes the form

$$SNR_j \geq \frac{2}{CF_{BP}^2} \frac{1}{(2\pi F)^2 \sigma_j^2}, \tag{4.310}$$

with CF_{BP} the CF of the bandpass signal. Obviously, when $s(t)$ is a sine wave, $CF_{BP}^2 = 2$. The right-hand side of this equation then reduces to that of equation (4.303). We see that the above lower bound matches the SNR achieved during the sampling process in this particular case. Conversely, we see that when dealing with a complex modulated signal, the impact of the modulation on the jitter requirement can be non-negligible. Indeed, referring to the examples given in Section 1.3.3, CFs higher than 10 dB can classically be encountered. This means that to ensure a given SNR during a sampling operation, the quality of the clock must be strengthened by the same amount compared to the sampling of a simple CW signal.

However, from the system point of view the two approaches can be seen as complementary. Indeed, referring for instance to Chapter 7, for the most part the blocking signals are defined as CW. In that case, equation (4.303) can be used to determine the maximum jitter allowed in order to achieve the dynamic range required to pass those blocking test cases. Conversely, it may be necessary to take the CF of the waveform into account when dealing with the sampling of the wanted signal only, for instance. In that case, depending on whether or not the wanted signal is sampled in baseband, equations (4.309) and (4.310) can be used to determine the jitter required to ensure the maximum radio link performance.

In conclusion, we observe that in our derivations so far we have considered the total power of the error sequence $e[k]$ for deriving the lower bounds in the achievable SNR. But the fact is that this error is proportional to the product of the time derivative of the signal being converted, $\dot{s}(t)$, times the timing jitter process, $j(t)$, as given by equation (4.296). By the Fourier transform property given by equation (1.10), we then expect the spectrum of this error signal to be proportional to the convolution product between the spectrum of $\dot{s}(t)$ and that of $j(t)$. We may then be faced with spectral regrowth according to the mechanism discussed in "Spectral regrowth" in Chapter 5. This would result in part of the error signal power lying outside the wanted signal bandwidth. The estimations given above should then be refined accordingly and can be seen as pessimistic compared to what may be experienced in practice.

Linearity

In addition to the pure SNR limitations discussed up to now, there may also be some nonlinearity in the transfer function of an ADC, as illustrated in Figure 4.41. Compared to the ideal case discussed throughout Section 4.5 where the device transfer function takes the simple form

$$s_o[k] = G_e s[k] + n[k], \tag{4.311}$$

with $n[k]$ the quantization noise and G_e the constant conversion gain, we now need to consider the form

$$s_o[k] = G_e(s[k])s[k] + n[k]. \tag{4.312}$$

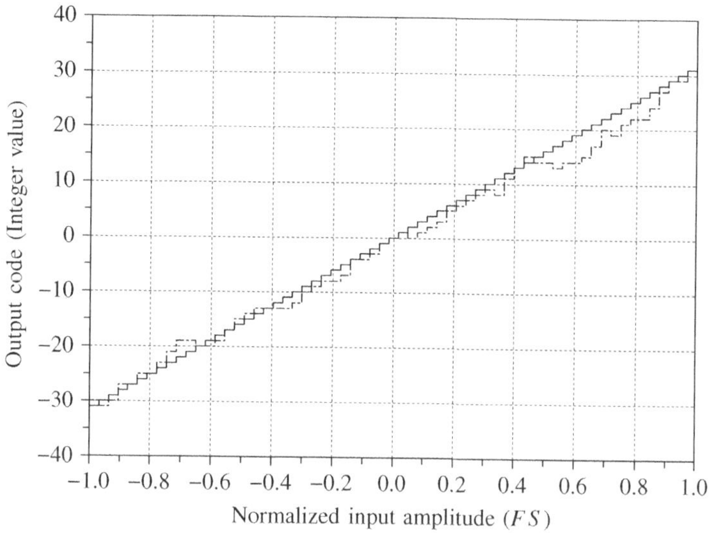

Figure 4.41 Nonlinear transfer function of an analog to digital converter – In the presence of implementation mismatch, the ideal staircase transfer function of the ADC (solid) experiences distortions that result in differential nonlinearity and integral nonlinearity (dot-dashed). In the present case, the differential and integral nonlinearity of the 6-bit ADC considered (including the sign) are respectively equal to 3 and 5 LSBs.

And given that the conversion gain depends on the magnitude of the signal being converted, we can anticipate that we are dealing here with the same behavior as discussed in Chapter 5. However, dealing with a discrete output signal leads to variations in the quantities used to characterize the nonlinearity of the device compared to what is classically done in the pure RF/analog domain. In fact, the definitions for these quantities may even vary from one author to another, depending on whether they are defined with respect to the end point conversion, the best fit conversion slope or any other point of reference [57]. We can thus simply examine the spirit of those quantities by illustrating the potential system impacts of such nonlinear behavior.

We first observe that the performance of the converter with respect to linearity can often be sorted into two categories. We often talk about static or dynamic performance. Here by "static" we mean the performance experienced when only a few bits of the quantizer remain active. This occurs when a quasi-static signal is present at the input of the device. On the other hand, dynamic performance is experienced when the input signal is scaled so that most of the bits are varying. This distinction can seem slightly artificial at first as it is the same nonlinear transfer function of the converter that is involved. However, the difference between the maximum and minimum values of the transfer function in each case justifies the use of different metrics.

Thus, let us focus first on the static performance of the device. Practically speaking, this static performance is characterized by quantities that are direct transcriptions of the shape of the device transfer function. Here, we mention the following:

(i) Differential nonlinearity (DNL) classically gives the largest LSB error that we can expect when increasing by one equivalent input quantization step the amplitude of the analog signal being converted. This definition involves only the difference between successive conversions. As a side effect, when the DNL is greater than 1, we can have some missing codes, i.e. some codes are never delivered by the device. From the system point of view, DNL can be a problem in terms of steady state precision, for instance. This can be understood by considering the case where the converter is embedded in a closed loop system. If the steady state of the system corresponds to a conversion area where the DNL is significant, a weak variation at the input of the converter can lead to a non-negligible error signal in the loop and thus to a perturbation of the internal state of the system compared to what was expected.

(ii) Integral nonlinearity (INL) gives an overview of the maximum overall deviation of the conversion transfer function compared to the theoretical deviation. This quantity is simply the integral or cumulative sum of the DNL. It therefore gives an overview of how much the transfer function deviates from a straight line. This can be a problem if the converter is used in a calibration scheme, for instance. A straight conversion slope allows us to rely on just two calibration points as the overall transfer function of the system can then be easily extrapolated. Conversely, a significant INL could compromise such a strategy.

(iii) The monotonicity of the ADC is a Boolean quantity that indicates the monotonicity of the device transfer function. The impact of a non-monotonic transfer function, like the one shown in Figure 4.41, can here again be illustrated by considering the closed loop example. If the steady state of the loop corresponds, for instance, to the set point where the inversion of the slope occurs, an oscillation of the system may occur around this point instead of a convergence toward a defined value.

Conversely, we get that the dynamic performance of an ADC is linked to its use over its full range. Such performance is often characterized by quantities that are not directly related to the device transfer function but rather to the distortion experienced by the signal being converted. In that respect, when dealing with the nonlinearity of RF/analog stages we expect to recover the system impacts discussed in Chapter 5. Practically speaking, quantities that characterize the dynamic behavior of ADCs are often derived using a sine wave whose amplitude is set to the maximum allowable input as considered for the simulation shown in Figure 4.42. This allows us to define, for instance, the spurious free dynamic range (SFDR). This quantity is simply equal to the maximum spurious tone level relative to the wanted signal amplitude recovered at the output of the converter. It therefore gives the available DR that is spurious-free and thus an amount of distortion on the processed signal that results of the conversion.

Alternatively, we often encounter the signal to noise and distortion ratio (SiNAD). This quantity first takes into account the total power of all the spurs recovered at the converter output, and as such it can be related to the total harmonic distortion (TSD) used for analog baseband devices as discussed in Section 5.4. But the SiNAD also takes into account the quantization noise power. Thus, it also gives an evaluation of the SNR limitation that we can expect from the conversion. However, despite its popularity, a metric that combines quantization noise and distortion is not necessarily meaningful from a system design point of view. The quantization noise is additive with respect to the signal being processed whereas the distortion components are obviously distortion noises. As a consequence, the way to manage those noise components as well as their system requirements may be different when dimensioning a line-up, as illustrated in Chapter 7. It may thus be worth having separate metrics to address this situation.

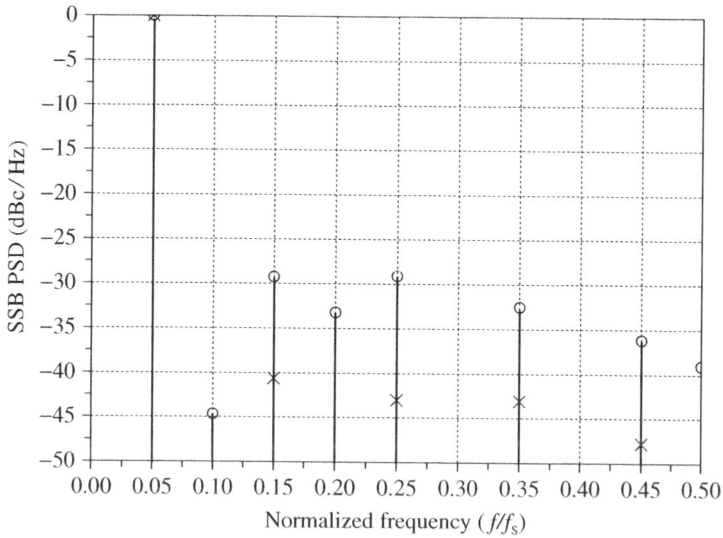

Figure 4.42 Distortion of a sine wave during a conversion using a nonlinear ADC – Although the theoretical spectrum of an ideally quantized sine wave already exhibits spur tones as illustrated in Figure 4.28(crosses), the presence of nonlinearity leads to an increase in the number and level of those spur tones (circles). Here, the 6-bit ADC transfer function shown in Figure 4.41 is used for the time domain simulation, and the amplitude of the input sine wave is set to the ADC input full scale, *FS*.

In conclusion, we observe that it is difficult in practice to specify the linearity of such devices without performing system simulations. Although it should be refined depending on the structure of the converter that is used, the deviation of the transfer function from the ideal straight line is not regular in general, as illustrated in Figure 4.41. We see a different behavior than classically encountered for RF/analog devices for which a simple polynomial expansion often succeeds in modeling the nonlinearity. Moreover, the areas of input signal amplitude for which the discrepancies appear may vary in the present case from part to part due to mismatch problems in mass production. As a result, even in the case where the static performance remains the same for all the parts, we may have to deal with varying dynamic performance depending on the location where those discrepancies appear on the transfer function. Statistical simulations are thus often required to anticipate the exact system performance one can expect.

Noise Shaping

In Section 4.5, we derived that the quantization noise resulting from uniform quantization can be assumed white in most practical cases. However, a quantizer may be embedded in a closed loop system to improve the DR and thus the effective number of bits, of the converter in the frequency band of the wanted signal of interest. This improvement in fact relies on the shaping of the PSD of the quantization noise delivered by the quantizer. But, given that the overall system results in the production of a quantization noise that is no longer white, additional system constraints may need to be considered in order to dimension a line-up. It may thus be worth saying a few words about this.

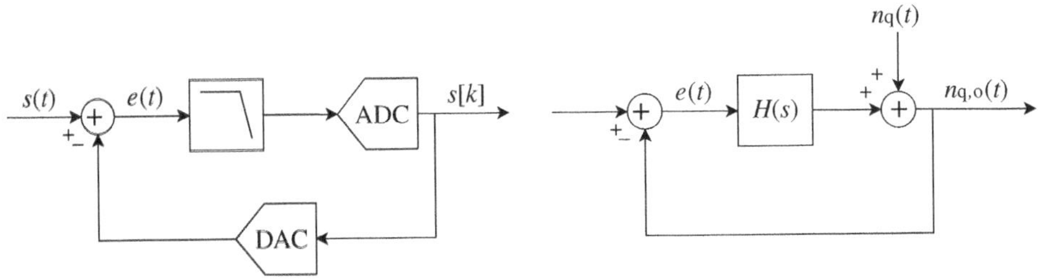

Figure 4.43 Basic continuous time ΣΔ ADC and equivalent quantization noise model – A ΣΔ converter improves the effective number of bits of a low resolution quantizer using a negative feedback loop to cancel the fraction of quantization noise components that lie within its passband. To do so, it uses a DAC in the feedback path in order to get an image of the quantization noise through the comparison of the resulting signal with the input one (left). This behavior is similar to that experienced by the phase noise contribution of an oscillator embedded in a PLL as discussed in "Phase locked loop" (Section 4.3.1). However, in the present case, the only source of quantization noise is the quantizer in the direct path (right).

In order to understand the noise shaping mechanism, we can refer to the discussion in "Phase locked loop" (Section 4.3.1) about the phase noise flowing from a PLL system. A PLL behaves as a noise shaping system with respect to the phase noise contribution of the oscillator. More precisely, the fraction of the oscillator phase noise that lies within the passband of the loop is canceled by comparison with the high precision reference clock. We might then imagine using the same kind of feedback loop to track and cancel the quantization noise flowing from a quantizer by comparing the quantized signal with the input analog signal that is by definition free of quantization noise. This would result in a structure like the one shown in Figure 4.43(left). Based on the equivalent quantization noise model shown on the right in this figure, we can express the transfer function $H_{HP}(s)$ experienced by the quantization noise flowing from the quantizer up to the output of the system as

$$H_{HP}(s) = \frac{N_{q,o}(s)}{N_q(s)} = \frac{1}{1 + H(s)}. \tag{4.313}$$

In order to get the quantization noise efficiently corrected in the passband of the loop, we need a sufficiently high DC gain This is classically achieved through the use of simple integrators as loop filters. For instance, if we assume that $H(s)$ reduces to a first order integrator, i.e. that $H(s) = 1/s$, we get that

$$H_{HP}(s) = \frac{s}{s + 1}. \tag{4.314}$$

The transfer function experienced by the quantization noise is thus a first order high pass filter in that case. But more generally, if we consider a kth order integrator for $H(s)$, we get that $H_{HP}(s)$ reduces to a kth order highpass filter. The corresponding quantization noise spectrum recovered at the output of the loop is then shown in Figure 4.44 for different values of k. We thus see that the higher the order of the lowpass filter used in the loop, the lower the quantization noise power lying in the low frequency part of the spectrum. Moreover, given that a simple 1-bit converter is used in the simulation, we can achieve a good SNR in the frequency band of interest even by using a quantizer with very poor capabilities. The condition for this

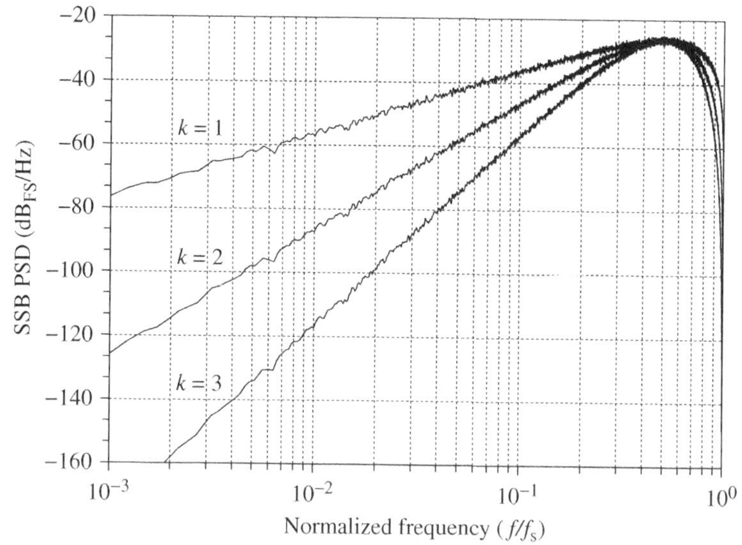

Figure 4.44 Quantization noise shaping using a feedback loop on an ADC – The use of a feedback loop around an ADC as shown in Figure 4.43 leads to the spectrum of its quantization noise being shaped by the transfer function of the loop that tracks the quantization noise to cancel it in its passband. Here, a simple 1-bit quantizer is considered in the loop so that the quantizer equivalent input step is simply $q = 2FS$, i.e. the output states correspond to an equivalent input amplitude equal to $\pm FS$. The quantization noise PSD of the quantizer is assumed flat, as given by equation (4.276). It is then shaped by the loop transfer function in which a kth order integrator is assumed.

is that the system operates at a sample rate f_s sufficiently high compared to the bandwidth of the signal being converted. Nevertheless, downstream from the converter we may need to downsample the signal toward a reasonable sampling rate in relation to the bandwidth of the wanted signal. This may be necessary in order to carry out further signal processing at a reasonable cost. For this purpose, we must first cancel the quantization noise lying in the higher part of the spectrum to avoid any folding issues and preserve the high SNR present in the frequency band of interest. For a kth order system, we get a slope k for the quantization noise spectrum shape. We thus need at least a digital filter with an equivalent lowpass slope $k + 1$ before downsampling the signal. At the output of such a filter and decimation stage we then recover the DR corresponding to the effective number of bits of the converter in its passband. From a system point of view, the overall system composed of the noise shaping converter plus its filtering and decimation stage can be considered as an equivalent classical converter with a DR and thus a quantization step, corresponding to its passband effective number of bits.

The converters based on this noise shaping effect are called $\Sigma\Delta$ converters: the delta represents the difference at the comparison stage between the input analog signal and the feedback signal, while the sigma corresponds to the integrators that are often implemented as accumulators. There are numerous ways to implement such converters [54]. For example, we may mention the existence of continuous time and discrete time $\Sigma\Delta$ converters depending on whether the loop filter takes place before or after the effective time sampling operation. As a result, the loop filter is implemented either as a continuous time filter, i.e. as an analog filer, or as a sampled time filter. This results in different system constraints that need to be considered at the system design phase. A continuous time loop filter can also help in the attenuation of the blocking signals before the time sampling. It can thus behave as an additional anti-aliasing filter, which is not the case if it is implemented as a discrete time filter.

In conclusion, we have considered here an equivalent continuous time model for the converter as illustrated in Figure 4.43. This is done for the sake of simplicity of presentation and to make the link with the PLL behavior discussed previously. This can also be justified in the passband of the signal of interest as long as the sample rate is sufficiently high compared to its bandwidth. However, at the output of the converter the signal remains discrete. Its spectrum is thus periodic with a period equal to the sampling rate f_s. It can simply be reconstructed as the sum of the copies of the main spectrum shown in Figure 4.44 centered on the multiples of the sampling frequency f_s.

4.6.2 Digital to Analog Conversion

Aperture Effect

From the signal processing point of view, the purpose of digital to analog conversion is to reconstruct the continuous time analog signal $s(t)$ that corresponds to the sequence of digital samples $s[k] = s(kT_s)$, where $T_s = 1/f_s$ is the sampling period, assumed constant here. Given that the spectrum of $s(t)$ lies within the frequency band $[-B/2, B/2]$, and that the sampling rate f_s fulfills the Nyquist criterion, i.e. is higher than B, the expression for $s(t)$ is given by the reconstruction formula as [2]

$$s(t) = \sum_k s[k] \operatorname{sinc}\left(\pi \frac{t - kT_s}{T_s}\right). \tag{4.315}$$

Here the sinc(.) function is defined as

$$\operatorname{sinc}(x) = \begin{cases} \dfrac{\sin(x)}{x} & \text{when } x \neq 0, \\ 1 & \text{when } x = 0. \end{cases} \tag{4.316}$$

Obviously, we may wonder how to implement such an expression in the analog world by using physical electronic devices. In practice, the main limitation comes from the fact that we can hardly imagine how to process and convert more than one sample of the sequence $s[k]$ at a time. The best we can expect is thus that what is called a DAC device delivers during the sampling period an analog quantity, voltage or current, proportional to the sample value present at its input at that time. In terms of signal processing, the behavior of most DACs thus reduces to a simple sample and hold functionality. At the output of such a device we therefore recover a staircase analog signal, $s_0(t)$, that is an approximation of $s(t)$ as illustrated in Figure 4.45(left and middle).

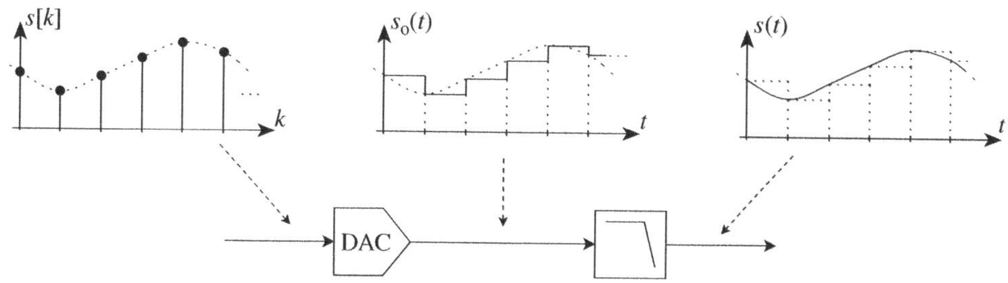

Figure 4.45 Digital to analog conversion as a sample and hold followed by a reconstruction filter – Classically, a DAC maintains a constant voltage or current proportional to the value of the input sample during a sampling period $T_s = 1/f_s$. From the signal processing point of view, it therefore behaves like a sample and hold. As a result, the analog staircase signal (top middle), which represents the input samples (top left), needs to be lowpass filtered to recover the exact analog signal (top right).

Such behavior is an important limitation compared to what is required by equation (4.315). However, the signal $s(t)$ can still be reconstructed by adequate processing of $s_o(t)$. Indeed, $s_o(t)$ can be written as

$$s_o(t) = T_s \sum_k s[k] \Pi\left(\frac{t - (k + 1/2)T_s}{T_s}\right), \tag{4.317}$$

where $\Pi(.)$ is the gate function defined by equation (1.145). But each term of the form $\Pi(t - \tau)$ can in turn be written as the convolution product of the original gate function $\Pi(t)$ with a Dirac delta distribution $\delta(t - \tau)$. Therefore,

$$\begin{aligned} s_o(t) &= T_s \Pi\left(\frac{t}{T_s} - \frac{1}{2}\right) \star \sum_k s[k]\delta(t - kT_s) \\ &= T_s \Pi\left(\frac{t}{T_s} - \frac{1}{2}\right) \star \left(s(t) \sum_k \delta(t - kT_s)\right). \end{aligned} \tag{4.318}$$

The fact is that $s(t) \sum_k \delta(t - kT_s)$ is nothing more than the distribution signal $s_i(t)$ which represents the sequence of samples present at the input of the DAC device. We can thus interpret $s_o(t)$ as the filtered version of $s_i(t)$ through the filter whose impulse response $h(t)$ is the gate function. As classically done in the filtering theory, we can then expect to derive interesting conclusions when examining these quantities in the spectral domain. For that purpose, we can use equation (1.10) to write the Fourier transform of $s_o(t)$ as the product of the Fourier transform of the gate function and that of $s_i(t)$. Given on the one hand that the Fourier transform of the latter distribution signal is given by equation (4.284), and on the other hand that the Fourier transform of the gate function is nothing more than a sinc(.) function, we can immediately write that

$$S_o(f) = e^{-j\pi f/f_s} \text{sinc}(\pi f/f_s) \sum_k S(f - kf_s). \tag{4.319}$$

The shape of the PSD of $s_o(t)$ can then be linked to the square of the modulus of $S_o(f)$:

$$|S_o(f)|^2 = |\mathrm{sinc}(\pi f/f_s)|^2 \left| \sum_k S(f - kf_s) \right|^2. \tag{4.320}$$

But having the sampling criteria $f_s > B$ fulfilled leads to the fact that the spectrum copies in this summation are non-overlapping. When expanding the square modulus of this sum, every product involving two different copies is therefore null for all f. It thus follows that

$$|S_o(f)|^2 = |\mathrm{sinc}(\pi f/f_s)|^2 \sum_k |S(f - kf_s)|^2. \tag{4.321}$$

Finally, we see that the spectrum of the analog signal $s_o(t)$ recovered at the output of the DAC is that of the input sequence but now weighted by the square of the sinc(.) function corresponding to the sample and hold transfer function. This filtering effect linked to the shape of the elementary waveform delivered by the converter is classically referred as the aperture effect.

As illustrated in Figure 4.46, the sinc(.) function has nulls at multiples of the sampling frequency, f_s in the present case. We thus expect these zeros to cancel the copies in the spectrum of $s_o(t)$. However, we see that residual sidebands remain present due to the finite rejection of the sinc(.) function over the frequency band of the wanted signal spectrum. In fact, these residual copies represent nothing more than the spectrum of the error signal $e(t) = s(t) - s_o(t)$ at the DAC output. In order to reconstruct the targeted signal $s(t)$, we thus need to cancel this error signal and filter out these residual copies. This is done in practice by cascading an analog filter with the DAC, resulting in the classical configuration shown in Figure 4.45. Due to its function, this filter is called a reconstruction filter. However, this reconstruction can only be an approximation as the attenuation provided by any physical filter on the signal copies can only be finite. We thus need to consider a tolerance on the residual error and thus a target attenuation on the residual copies in order to derive a specification for the reconstruction filter. For that purpose, we obviously need to take into account the attenuation already provided by the sinc (.) function. It is convenient to focus on the first copy, where the attenuation of the sinc(.) is smallest. This often corresponds to the worst case for the specification of the reconstruction filter. Assuming that the wanted signal bandwidth B is much lower than the sampling frequency f_s, we can use a series expansion of the sinc(.) function up to first order. As illustrated in Figure 4.46, the minimum attenuation of the amplitude of the first copy, A_ρ, is found to be

$$A_\rho = \mathrm{sinc}\left(\pi\left(1 - \frac{B}{2f_s}\right)\right) \approx \frac{B}{2f_s}. \tag{4.322}$$

We thus get the square of this for the power attenuation $A_{\rho 2} = (A_\rho)^2$. The reconstruction filter only needs to provide the residual required attenuation.

According to our discussion so far, it is only the concatenation of a DAC device and a reconstruction filter that implements the digital to analog conversion as expected from the signal processing point of view. We could even add a third functionality that often needs to be implemented, namely an equalization stage. Indeed, the sinc(.) function of the aperture

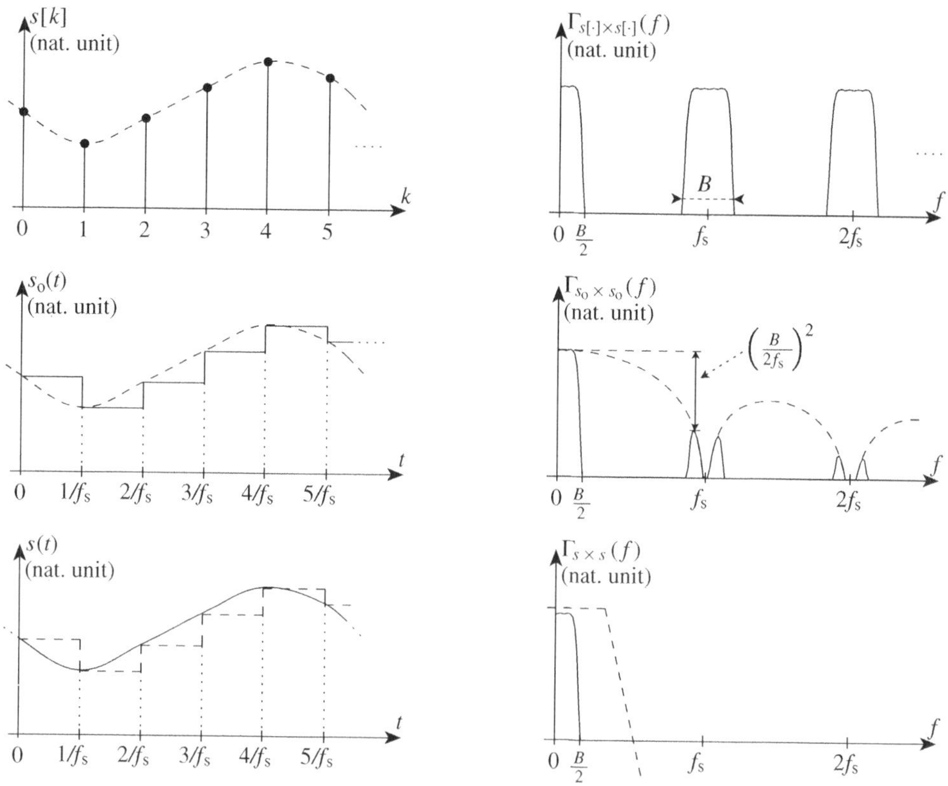

Figure 4.46 Digital to analog conversion behavior in the spectral domain – The copies of a sampled signal (top) are naturally attenuated by the sample and hold behavior of zero-order interpolating DAC devices (middle). However, as illustrated in Figure 4.45, an analog reconstruction filter is almost always required in order to cancel the residual copies and recover the continuous time signal corresponding to the DAC input samples (bottom).

effect can also impact the amplitude of the wanted signal, mainly at the edge of its spectrum, depending on the ratio B/f_s. We may, moreover, be faced with additional distortions due to the presence of the analog reconstruction filter. All this may make it necessary to use an equalization stage, and this can usefully be done prior to the DAC block in the digital domain, in order to preserve the quality of the modulation for instance. In that case, it is the concatenation of this predistortion block, the DAC and the reconstruction filter that implements the expected digital to analog conversion functionality.

Dynamic Range, Linearity, and Noise Shaping

A DAC block is obviously subject to the same kind of performance limitations as encountered for an ADC, due to the similarities in their physical implementation. For instance, the thermal or electronic noises necessarily add to the contribution of the quantization noise also in the DAC case. This can bring about a degradation of the output DR or SNR, compared to what would be theoretically expected when considering only the number of bits used to code the data at the input. We can therefore still talk about an effective number of bits for DAC devices.

In the same way, the mismatch in the implementation of the converter can impact the linearity of its transfer function. In that case, we would expect to use mostly the same quantities as introduced for the ADC in "Linearity" (Section 4.6.1), to characterize the DAC linearity. In particular, we can continue to make the distinction between the static performance, characterized by the DNL, INL or monotonicity, and the dynamic performance, characterized by the SFDR or SiNAD.

Finally, quantization noise shaping can also be used, thus resulting in a ΣΔ DAC [54]. In that case, we recover the behavior detailed in "Noise shaping" (Section 4.6.1). However, we observe that having a quantization noise that is not white at the output of the converter may tighten the requirements for the reconstruction filter compared to the case where only the copy of the converted signal needs to be filtered out. This holds despite the fact that the quantization noise also experiences the aperture effect as well as the signal being converted.

5

Nonlinearity

The nonlinearity of the functions implemented in a transceiver is traditionally seen as an additional source of distortion for the signal present in the line-up. In most cases, we need those functions to be as linear as possible in order to preserve the modulation characteristics of the signal being processed. But in practical electronic implementations, such linearity is always achieved at some cost. For instance, improving the behavior of an RF amplifier with respect to compression always results in an increase in the power consumption. In this particular example, there is thus necessarily a trade-off between the power consumption of the solution and a reasonable degradation of the radio link.

We thus need to review and quantify the impact on the wanted signal of the different kinds of nonlinear behavior that may occur in a practical implementation. This requires us to be able on the one hand to characterize the nonlinearity of a device in a suitable way for analytical derivations, and on the other hand to quantify its impacts, mainly in terms of SNR or EVM degradation. Only this kind of characterization can enable us to minimize the implementation cost by not over-dimensioning the requirements in terms of linearity.

However, it is also interesting to see that the general statement that the functions implemented in a transceiver need to be linear may be wrong in some particular cases. One such case is when dealing with constant amplitude bandpass RF signals. Such phase or frequency only RF modulated signals are insensitive to hard nonlinearity, or at least they can be reconstructed without distortion after having gone through a nonlinear device. This behavior enables, for instance, the use of high efficiency saturated PAs, or even SNR improvements using compression, as illustrated in Section 5.1.4.

Finally, we observe that there are linearity problems in any RF/analog block. However, differences in the practical implementation of RF and analog baseband devices mean that linearity issues are for the most part much more critical on the RF side.[1] Thus, although we say a few words on baseband nonlinearity at the end of the chapter, we focus mainly on the characterization of RF nonlinearity and its impact on transceiver performance.

[1] This is mainly due to the extensive use of operational amplifier based circuits when processing baseband signals. These feedback systems lead to more linear devices than do traditional RF circuits based on single transistor stages.

Wireless Transceiver Architecture: Bridging RF and Digital Communications, First Edition. Pierre Baudin.

5.1 Smooth AM-AM Conversion

As a first step, let us focus on what is called smooth AM-AM conversion. As the name suggests, this kind of behavior corresponds to nonlinearity that remains weak. Although this concept of weakness needs to be explained in detail, we observe that we are talking here about devices that are expected to exhibit linear behavior a priori. Due to the corresponding reduced range in the electronic device characteristics, it often results that it is mainly the amplitude of the signal going through such a device that is corrupted by the nonlinearity. This fact explains why we focus in this first part on AM-AM conversion for smooth nonlinearity. The AM-PM effect, mainly encountered on the transmit side when dealing with high amplitude signals, is addressed in Section 5.3.

We start our discussion with smooth AM-AM conversion as it allows us to introduce most of the concepts classically used for the characterization of nonlinear RF devices. We first show how to model such a device. We can then focus on the system impacts resulting from the distortion of general complex modulated bandpass RF signals as classically encountered in wireless standards.

5.1.1 Smooth AM-AM Conversion Model

In practical implementations, smooth AM-AM conversion is encountered with devices that are expected to be linear. For instance, the output bandpass RF signal $s_o(t)$ flowing from a simple RF amplifier is expected to be proportional to the input signal $s_i(t)$ according to

$$s_o(t) = G s_i(t - \tau), \tag{5.1}$$

with G the small-signal gain of the device and τ its group delay. Dealing with smooth AM-AM conversion, the fact is that the group delay is often of no importance for the description of the phenomenon. This is not true for devices that exhibit memory, as detailed in Section 5.3, but for the time being we can assume that we are dealing with an instantaneous nonlinearity so that

$$s_o(t) = G s_i(t). \tag{5.2}$$

Unfortunately, such a simple transfer function can only be valid over a finite DR of the input signal $s_i(t)$. For instance, an amplifier designed with a finite power supply, as illustrated in Figure 5.1, is not able to deliver more than a given amount of signal level to its load. This means that beyond a given input signal amplitude, the output signal starts to become a clipped version of the input signal. The transfer function of such a device thus cannot reduce to equation (5.2), but rather takes the general form

$$s_o(t) = F[s_i(t)], \tag{5.3}$$

with F a real valued function as illustrated in Figure 5.2. However, by writing this simple relationship we implicitly make some additional assumptions for the model of our device. The most important one is that the transfer function is independent of the spectral content of the input signal $s_i(t)$. One could argue that this is not very realistic for RF devices that often have a limited passband and thus have characteristics that depend on the carrier frequency

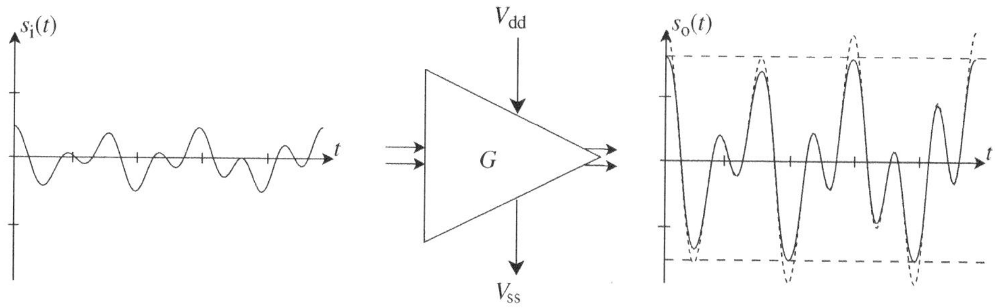

Figure 5.1 Effect of limited power supply on signal amplification – Due to the limited voltage supply $V_{dd} - V_{ss}$ of an amplifier, the input signal $s_i(t)$ (left) does not necessarily experience an ideal linear amplification. Whereas we expect an output signal $s_o(t)$ proportional to $s_i(t)$ (right, dashed), we may rather recover a clipped version of it (right, solid).

of the input signal that goes through it. In fact, this is not so important for our purposes as we expect to use such a model to link the characteristics of the device with the system quantities of interest. Once those relationships are derived, nothing prevents us from using them to consider different characteristics for the device depending on the input signal carrier frequency in order to represent the device selectivity in frequency.

In order to carry out analytical derivations, we now need to consider an explicit expression for F. Although particular formulations have been developed [58], dealing with a series expansion for F remains of particular interest for the derivation of system related relationships. The first reason for this is that it allows us to introduce in a straightforward way the concepts classically used to characterize the nonlinearity of RF devices like the intercept points (IPs). Then, such a series expansion leads to the output signal being expressed as proportional to the input signal plus a sum of unexpected terms. Dealing with randomly modulated signals, those additional terms can often be approximated as additive noise terms, as illustrated in Section 5.1.3. This series expansion approach is therefore well suited to deriving formulations useful for

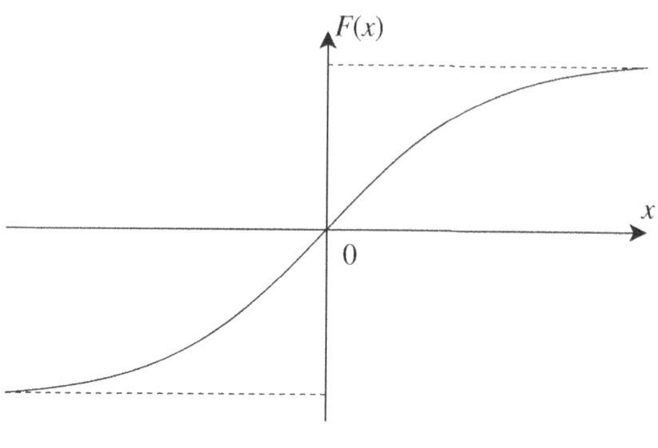

Figure 5.2 Typical transfer function of a power supply limited linear device – Due to power supply limitations, the transfer function of a linear device exhibits saturation. Above a given threshold, increasing the input signal level does not lead to an increase in the output signal level.

performing SNR budgets, as illustrated in Chapter 7. Thus, we can consider in what follows the Taylor series expansion of the device transfer function F given by

$$F(x) = \alpha_0 + Gx + \alpha_2 x^2 + \alpha_3 x^3 + \cdots . \tag{5.4}$$

In this expression, we write the linear proportionality factor as the device small-signal gain G, which is a real valued quantity linking RF bandpass signals in the present AM-AM conversion case. It can be either negative or positive depending on whether or not we are dealing with an inverting device. We can thus use this relationship in equation (5.3) to express the output RF bandpass signal $s_o(t)$ as

$$s_o(t) = \alpha_0 + G s_i(t) + \alpha_2 s_i^2(t) + \alpha_3 s_i^3(t) + \cdots . \tag{5.5}$$

Odd vs. Even Order Nonlinearity

Classically, nonlinearity behaviors are classified into two categories: we talk about odd order and even order nonlinearity. As the names suggest, this classification refers to the parity of the nonlinear terms in the series expansion of the device transfer function $F(x)$. Generally speaking, $F(x)$ can be decomposed as the sum of an odd and an even function, $F_o(x)$ and $F_e(x)$ respectively:

$$\begin{aligned} F(x) &= \overbrace{\frac{F(x) - F(-x)}{2}}^{\text{odd}} + \overbrace{\frac{F(x) + F(-x)}{2}}^{\text{even}} \\ &= F_o(x) + F_e(x). \end{aligned} \tag{5.6}$$

Given the convergence of the respective series expansions, we can then write

$$F_o(x) = \sum_{l=0}^{\infty} \alpha_{2l+1} x^{2l+1}, \tag{5.7a}$$

$$F_e(x) = \sum_{l=0}^{\infty} \alpha_{2l} x^{2l}, \tag{5.7b}$$

with α_n terms corresponding to the coefficients of the series expansion of $F(x)$ given by equation (5.4), e.g. with $\alpha_1 = G$. The interesting point is that this split into two categories of the nonlinear terms is not just a matter of formalism for analytical derivations. It also corresponds to different physical origins for the nonlinear behavior of a device as well as different system impacts in most cases.

To understand this, let us examine the impact of the odd order part of the transfer function F on the input signal $s_i(t) = \cos(\phi(t))$. Considering first the impact of the first nonlinear term in the series expansion of $F_o(x)$, we can approximate the signal $s_o(t)$ recovered at the output of the nonlinear device as

$$s_o(t) = G \cos(\phi(t)) + \alpha_3 \cos^3(\phi(t)). \tag{5.8}$$

Thus, using the relationship

$$\cos^3(\theta) = \frac{3\cos(\theta) + \cos(3\theta)}{4}, \tag{5.9}$$

we get that

$$s_o(t) = \left(G + \frac{3\alpha_3}{4}\right)\cos(\phi(t)) + \frac{\alpha_3}{4}\cos(3\phi(t)). \tag{5.10}$$

As expected, we see that $s_o(t)$ is composed of the superposition of the input sine wave and its third harmonic. But, as illustrated in Figure 5.3(top), it is interesting to observe that

(i) We have exactly one period and a half of the third harmonic tone during half a period of the fundamental tone.
(ii) Due to the phase relationship between the fundamental tone and its third harmonic, for each extremum of the input signal, we have a corresponding extremum of the third harmonic tone. Consequently, if for one extremum of the input signal the corresponding extremum of the third harmonic has an opposite sign, the next extrema of the two signals also have opposite signs.

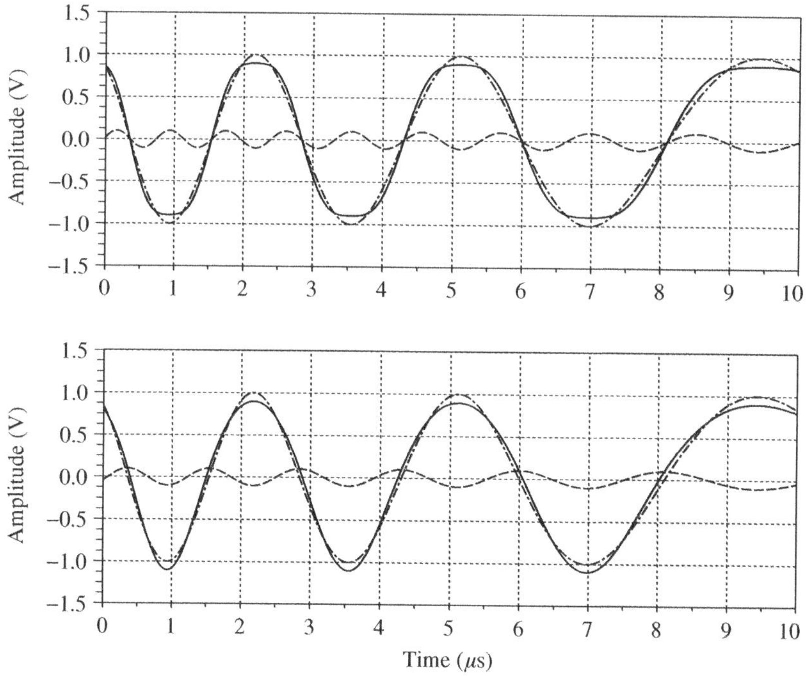

Figure 5.3 Impacts of the nonlinear terms x^3 and x^2 on a sinusoidal input signal – On the one hand, the sum of an input sine wave $s_i(t) = \cos(\phi(t))$ (top, dot-dashed) and the third harmonic related signal $-0.1\cos(3\phi(t))$ (top, dashed) represents a clipped version of $s_i(t)$ (top, solid). On the other hand, the sum of the same sine wave $s_i(t)$ (bottom, dot-dashed) and the second harmonic related signal $-0.1\cos(2\phi(t))$ (bottom, dashed) results in an offset or an imbalance distortion of $s_i(t)$ (bottom, solid).

We thus see that, depending on the relative sign of G and α_3, we get either amplification or compression of the input signal. In most physical implementations, we are faced with a compression of the signal linked to a limited power supply of the device under consideration. We then get that G and α_3 have different signs in practice. But what is interesting to see is that this behavior is in fact general for any odd order terms. This can be confirmed by considering the impact of a term of the form x^{2l+1} on the same input signal $s_i(t) = \cos(\phi(t))$. We can write that [6]

$$\cos^{2l+1}(\phi(t)) = \frac{1}{4^l}\sum_{k=0}^{l}\binom{2l+1}{l-k}\cos((2k+1)\phi(t)). \tag{5.11}$$

As expected, we are faced with only odd harmonics of the fundamental input signal. All the extrema of the fundamental tone and of its harmonics therefore add either constructively or destructively as in the third order case discussed above. In practical physical implementations, we thus get that odd order nonlinearity necessarily represents a compression behavior. However, this is something that we could have guessed by considering a typical transfer function that represents such compression behavior as shown in Figure 5.2. As this transfer function remains an odd function, its series expansion can only be composed of odd order terms.

Let us now focus on the impact of the even order part $F_e(x)$ on the same input signal $s_i(t) = \cos(\phi(t))$. Considering initially only the first nonlinear term in the series expansion of $F_e(x)$, we can write the output signal $s_o(t)$ as

$$s_o(t) = G\cos(\phi(t)) + \alpha_2\cos^2(\phi(t)). \tag{5.12}$$

Thus, using equation (1.54), we have

$$s_o(t) = \frac{\alpha_2}{2} + G\cos(\phi(t)) + \frac{\alpha_2}{2}\cos(2\phi(t)). \tag{5.13}$$

Looking at the time domain curves shown in Figure 5.3(bottom), we now see that

(i) We have exactly one period of the second harmonic tone during half a period of the fundamental tone.
(ii) Still due to the phase relationship between the fundamental tone and its second harmonic, for each extremum of the input signal we have an extremum of the second harmonic. But now, in contrast to the third order case, we get that if for one extremum of the input signal the corresponding extremum of the second harmonic has an opposite sign, the next common extrema of the two signals necessarily have the same sign.

This alternation of addition and subtraction of the extrema does not lead to compression of the input signal but rather to a signal offset or asymmetry between positive and negative alternations of the sine wave. Furthermore, this behavior is enhanced by the presence of the DC term $\alpha_2/2$ related to the non-vanishing zero mean of the always positive signal $\cos^2(\phi(t))$.

Here again, this conclusion remains general for any even order terms of the form x^{2l}. We can write that [6]

$$\cos^{2l}(\phi(t)) = \frac{1}{4^l}\left[\binom{2l}{l} + 2\sum_{k=0}^{l-1}\binom{2l}{l-1-k}\cos(2(k+1)\phi(t))\right]. \tag{5.14}$$

Due to the phase relationship between the fundamental tone and its even order harmonics, we thus see that even transfer functions can only model offsets or asymmetry distortions.

Practical implementations often exhibit both odd and even order nonlinearity. But we see, based on our discussion so far, that each kind of nonlinearity is related to different problems in the physical implementation. Odd order nonlinearity is for the most part related to compression and thus to the limited power supply of devices [59]. On the other hand, even order nonlinearity is mainly linked to offsets and thus to implementation mismatches [60]. Thus, to some extent, the behavior of a device with respect to each kind of nonlinearity can be optimized independently. Moreover, we will see in the subsequent sections that the system impacts of even and odd order nonlinearity are also for the most part different. This justifies the fact that they are treated and specified separately in the practical system design phase of a line-up.

5.1.2 *Phase/Frequency Only Modulated RF Signals*

In investigating the system impacts of both even and odd order nonlinearity, we first need to link the parameters of the device transfer function $F(x)$, i.e. the α_n coefficients involved in equation (5.7), with the nonlinear characteristics of this device.

Classically the nonlinearity of RF devices is not quantized through the direct values of the α_n parameters, but rather through their IPs, compression points (CPs) and other saturated powers (Psat). These quantities are defined in the forthcoming sections, but for the time being we observe that they are derived from the characterization of the device based on the use of pure CWs, i.e. pure sine waves. This is obviously linked to the relatively easy setup for the corresponding evaluation at both laboratory and simulation levels. Consequently, we focus first on the behavior of phase/frequency only modulated bandpass RF signals in smooth AM-AM nonlinearity in order to introduce these RF concepts. This also allows us to derive a first set of system impacts for a line-up involving signals with such characteristics. We can then focus in Section 5.1.3 on the impact of nonlinearity on more realistic complex modulated signals as encountered in modern wireless systems.

Characterization of RF Device Nonlinearity

Even Order Nonlinearity and IP2

Let us focus first on a device that exhibits an even order smooth AM-AM conversion behavior as may be the case for a device implemented with mismatch, for instance. We assume that the even part of its transfer function, $F_e(x)$, could be expanded according to equation (5.7b). However, we can assume that the contribution linked to the first nonlinear term remains the most relevant for the derivation of relationships valid for high level system budgets. Thus, we can assume for now that the device transfer function $F(x)$ can be approximated as

$$F(x) = Gx + F_e(x) \approx Gx + \alpha_2 x^2. \tag{5.15}$$

We also observe that we do not consider in this expression the α_0 term as for the time being we have no interest in an intrinsic DC offset at the output of the device.

The even order nonlinearity of such a device is classically characterized by its even order IPs. The derivation of such IPs, and in particular of the IP2 in the present case, requires the use of two CW test tones at the input of the nonlinear device and the inspection of the signals recovered at its output. We can therefore consider an input signal $s_i(t)$ of the form

$$s_i(t) = \rho_1 \cos(\omega_1 t) + \rho_2 \cos(\omega_2 t). \tag{5.16}$$

Using equation (5.15), we can then write the signal $s_o(t)$ recovered at the output of the device as

$$s_o(t) = F[s_i(t)] = G\,s_i(t) + \alpha_2 s_i^2(t). \tag{5.17}$$

Using equation (5.16) in turn, we get

$$\begin{aligned} s_o(t) &= G(\rho_1 \cos(\omega_1 t) + \rho_2 \cos(\omega_2 t)) \\ &\quad + \alpha_2\left(\rho_1^2 \cos^2(\omega_1 t) + \rho_2^2 \cos^2(\omega_2 t) + 2\rho_1\rho_2 \cos(\omega_1 t)\cos(\omega_2 t)\right). \end{aligned} \tag{5.18}$$

Now using equation (1.54) and

$$\cos(\theta_1)\cos(\theta_2) = \frac{\cos(\theta_1 + \theta_2) + \cos(\theta_1 - \theta_2)}{2} \tag{5.19}$$

yields

$$\begin{aligned} s_o(t) &= \frac{\alpha_2}{2}\left(\rho_1^2 + \rho_2^2\right) \\ &\quad + G(\rho_1 \cos(\omega_1 t) + \rho_2 \cos(\omega_2 t)) \\ &\quad + \frac{\alpha_2}{2}\left(\rho_1^2 \cos(2\omega_1 t) + \rho_2^2 \cos(2\omega_2 t)\right) \\ &\quad + \alpha_2\rho_1\rho_2[\cos((\omega_1 + \omega_2)t) + \cos((\omega_1 - \omega_2)t)]. \end{aligned} \tag{5.20}$$

The presence of the second order nonlinear term thus leads to the generation of new tones as illustrated in Figure 5.4. More precisely:

(i) We now have a DC term. This could be expected as, due to the square operation, the second order term of the transfer function produces a signal with a constant sign, either always positive or always negative depending on the sign of α_2. This signal thus necessarily has a

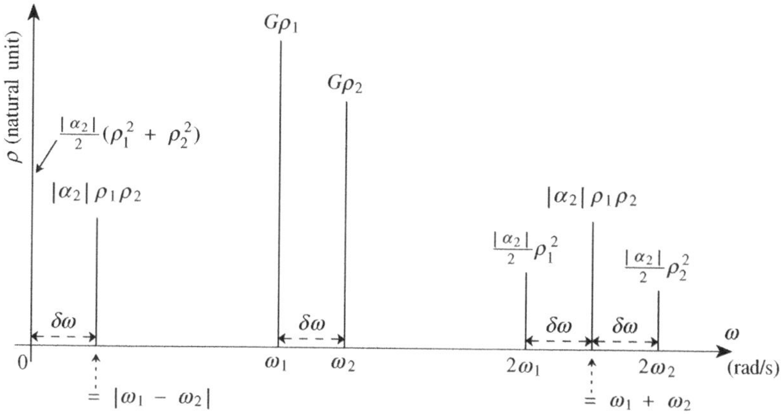

Figure 5.4 CW tone amplitude partition in the frequency domain at the output of a second order smooth AM-AM nonlinear model with two input CW signals – Considering two CW test tones at the input of a transfer function of the form $F(x) = Gx + \alpha_2 x^2$, we recover at the output both the expected test tones and their harmonics but also intermodulation tones, as well as a DC term.

non-zero mean and exhibits DC. This phenomenon is responsible for one of the practical limitations we face in the performance of final downconversion stages of some receive architectures, as discussed in "DC offset generation due to even order nonlinearity" later in this section.

(ii) We then recover the terms that correspond to the expected input test tones lying at angular frequencies ω_1 and ω_2. We observe that those signals are amplified by the device small-signal gain G, as would be expected if the device were linear. This is an important difference compared to the odd order nonlinearity case which leads to compression and thus to a distortion in the gain experienced by the input bandpass signals, as discussed in the next section.

(iii) A third group of terms represents the second harmonics of the input tones.

(iv) A final group of terms represent tones of particular interest for our present study. The output amplitude and angular frequency of each one, $(\rho_{\text{OIMD2}}, \omega_{\text{IMD2}})$, takes the form[2]

$$(\alpha_2 \rho_1 \rho_2, |\omega_1 \pm \omega_2|). \tag{5.21}$$

These two tones lying at angular frequencies $|\omega_1 + \omega_2|$ and $|\omega_1 - \omega_2|$ are called the second order intermodulation distortion (IMD) tones, or IMD2 tones, and can lead to tricky system constraints as discussed in "Intermodulation" later in this section. Here, it may be useful to

[2] We consider here the absolute value for the definition of the tone angular frequency. This is a convention that we maintain throughout Section 5.1.2 as we are dealing with real valued bandpass signals in the present case. We thus do not need to differentiate the positive and negative sidebands of their spectrum that fulfils Hermitian symmetry, in contrast to what is done extensively in Chapter 6 when discussing for instance the complex frequency conversion.

highlight that in this definition the factor 2 refers to the sum of the weighting factors of the angular frequencies ω_1 and ω_2 in the expression for the IMD2 tone angular frequency and not to the order of the nonlinear term involved in the device transfer function. For instance, an $\alpha_4 x^4$ term in the series expansion of $F_e(x)$ would also lead to the generation of IMD2 tones on top of some IMD4 tones as discussed in "Higher order terms and IP*k*" later in this section. We thus must keep in mind that the order of the intermodulation tones refers to their composite angular frequency.

These IMD tones are also involved in the definition of the device IPs. However, this definition is based on the use of two input tones with the same amplitude, i.e. such that

$$\rho_1 = \rho_2 = \rho_i. \tag{5.22}$$

According to equation (5.21), we then get that the amplitude of the two generated output second order intermodulation tones (OIMD), or OIMD2, which is identical[3] for the two tones, now reduces to

$$\rho_{\text{OIMD2}} = \alpha_2 \rho_i^2. \tag{5.23}$$

Considering this expression, we see that the level of the OIMD2 tones increases by 2 dB per decibel of the input tone level. If we linearly extrapolate this behavior, there is an amplitude of the input tones for which the output intermodulation tones would have the *same* amplitude as the theoretical amplitude of the fundamental tones if the device were linear. This behavior is shown in Figure 5.7 for a general OIMD*k* tone amplitude, but remains of course valid for $k = 2$. This theoretical input tone amplitude then defines the input intercept point (IIP) amplitude. In the present case, the IIP2 amplitude ρ_{IIP2} then fulfills

$$|G|\rho_i|_{\rho_i=\rho_{\text{IIP2}}} = |G|\rho_{\text{IIP2}} \stackrel{\Delta}{=} \rho_{\text{OIMD2}}|_{\rho_i=\rho_{\text{IIP2}}} = \alpha_2 \rho_{\text{IIP2}}^2. \tag{5.24}$$

Remembering that the amplitude of such bandpass signals is taken as positive, we thus get that

$$\rho_{\text{IIP2}} = \left|\frac{G}{\alpha_2}\right|. \tag{5.25}$$

By this definition, we see that the IP amplitude physically represents the amplitude of a CW. Moreover, as it is the extrapolated linear curve for the tone amplitude that is involved in their definition, we can link equivalent input and output quantities for these theoretical CW tone

[3] This behavior is not so obvious in real life implementations. We mention here the device impedance rotation for passband devices that can lead to different attenuations at the various frequencies of interest of the tones involved. However, nothing prevents us from using a different device characterization, even if based on this simple model, depending on the frequency planning of the use case under investigation, as highlighted in Section 5.1.1.

amplitudes using the device small-signal gain G.[4] In the present case, we can thus define the amplitude of the device output IP2, ρ_{OIP2}, as

$$\rho_{\text{OIP2}} \stackrel{\Delta}{=} |G|\rho_{\text{IIP2}} = \left|\frac{G^2}{\alpha_2}\right|. \tag{5.26}$$

We can also express these IP characteristics in terms of CW tone power rather than in terms of their amplitude. But in that case, we need to take into account the impedance seen by the signals we are dealing with. Moreover, the derivation of the tone power also depends on whether we are dealing with voltage waves or current waves. For instance, assuming that the CW tones we are dealing with at the input of the device represent a voltage across an impedance Z_0, we can derive the input IP2 power, based on Section 2.2.3, as

$$\text{IIP2} = \text{Re}\left\{\frac{1}{Z_0}\right\}\overline{(\rho_{\text{IIP2}}\cos(\omega t))^2} = \text{Re}\left\{\frac{1}{Z_0}\right\}\left(\frac{\rho_{\text{IIP2}}}{\sqrt{2}}\right)^2, \tag{5.27}$$

where $\overline{(.)}$ denotes time averaging. Thus, using equation (5.25), we get that

$$\text{IIP2} = \text{Re}\left\{\frac{1}{Z_0}\right\}\left(\frac{\rho_{\text{IIP2}}}{\sqrt{2}}\right)^2 = \text{Re}\left\{\frac{1}{2Z_0}\right\}\left|\frac{G}{\alpha_2}\right|^2. \tag{5.28}$$

But, if we now assume that the input CW tones we are dealing with represent currents across the same impedance, we would obtain a term Z_0 instead of $1/Z_0$ in the above relationship:

$$\text{IIP2} = \text{Re}\left\{Z_0\right\}\left(\frac{\rho_{\text{IIP2}}}{\sqrt{2}}\right)^2 = \text{Re}\left\{\frac{Z_0}{2}\right\}\left|\frac{G}{\alpha_2}\right|^2. \tag{5.29}$$

The corresponding output quantities are then obtained by applying the power gain G^2 to these input quantities. We thus see that dealing with power terms is not so easy as we need to manage two different cases depending on whether we are dealing with voltage or current waves. However, this can be overcome easily if we consider working with normalized impedances, i.e. with $Z_0 = 1$ ohm. In that case, whatever the nature of the wave we are dealing with, we can formally write that $\text{Re}\{1/Z_0\} = \text{Re}\{Z_0\} = 1$ and thus that

$$\text{IIP2} = \left(\frac{\rho_{\text{IIP2}}}{\sqrt{2}}\right)^2 = \frac{1}{2}\left|\frac{G}{\alpha_2}\right|^2, \tag{5.30a}$$

$$\text{OIP2} = \left(\frac{\rho_{\text{OIP2}}}{\sqrt{2}}\right)^2 = \frac{1}{2}\left|\frac{G^2}{\alpha_2}\right|^2. \tag{5.30b}$$

[4] However, as discussed in "Input vs. output IP" later in this section, when some filtering effect is involved in the device transfer function, the relationship between input and output quantities is less straightforward.

However, we might wonder if we can really work with such normalized impedances in practical derivations. Indeed, we cannot if we want to fit a device transfer function to time domain simulations, for instance. In that case, we need to link the power of the signals to their amplitude and thus take into account the impedances involved. But we can here as the relationships we expect to derive for system budgets involve the *ratio* between such signals amplitudes or power. This means that the performance or degradation we expect to evaluate is always expressed in terms of the amplitude of the signal we are dealing with relative to an amplitude that characterizes the device, such as its ρ_{IIP2} in the present case. As a result, normalization factors like impedances always cancel in such relationships.

This behavior can be illustrated in a straightforward way by expressing the power of the IMD products recovered at the output of the device. However, we observe that even if these IMD tones appear at the device output only, it is often more efficient to talk about their theoretical equivalent input amplitudes in order to make their comparison with the input test tone level more straightforward. This is very convenient when performing system budgets, as illustrated in Chapter 7. Thus, using the device small-signal gain G and equation (5.25), we can express the collection of generated tones present in equation (5.20) in terms of equivalent input tones of amplitude and angular frequencies, (ρ, ω), of the form

$$\left\{ \left(\frac{(\rho_1^2 + \rho_2^2)}{2\rho_{\text{IIP2}}}, 0 \right), (\rho_1, \omega_1), (\rho_2, \omega_2), \left(\frac{\rho_1 \rho_2}{\rho_{\text{IIP2}}}, |\omega_1 \pm \omega_2| \right), \right.$$

$$\left. \left(\frac{1}{2} \frac{\rho_1^2}{\rho_{\text{IIP2}}}, 2\omega_1 \right), \left(\frac{1}{2} \frac{\rho_2^2}{\rho_{\text{IIP2}}}, 2\omega_2 \right) \right\}. \tag{5.31}$$

Assuming, for instance, that we are dealing with voltage waves across an impedance Z_0, we can thus write the IIMD2 tone power as

$$\text{IIMD2} = \text{Re}\left\{ \frac{1}{Z_0} \right\} \left(\frac{\rho_{\text{IIMD2}}}{\sqrt{2}} \right)^2 = \text{Re}\left\{ \frac{1}{Z_0} \right\} \left(\frac{\rho_1 \rho_2}{\rho_{\text{IIP2}} \sqrt{2}} \right)^2. \tag{5.32}$$

Expanding this last term, we get that

$$\text{IIMD2} = \frac{\text{Re}\left\{ \frac{1}{Z_0} \right\} \left(\frac{\rho_1}{\sqrt{2}} \right)^2 \text{Re}\left\{ \frac{1}{Z_0} \right\} \left(\frac{\rho_2}{\sqrt{2}} \right)^2}{\text{Re}\left\{ \frac{1}{Z_0} \right\} \left(\frac{\rho_{\text{IIP2}}}{\sqrt{2}} \right)^2} = \frac{P_1 P_2}{\text{IIP2}}, \tag{5.33}$$

with P_1 and P_2 the power of the two input tones. Thus the impedance we are dealing with is no longer involved in the final result, which involves only power ratios. We can thus stay with this approach and continue to work with normalized impedances in the following sections.

In conclusion, we can express the set of generated tones in terms of equivalent input power, in dBm for instance, as

$$\begin{cases} \left(2\left[10\log_{10}\left(\dfrac{\rho_1^2+\rho_2^2}{2}\right)+30\right]-\mathrm{IIP2}|_{\mathrm{dBm}}-3, 0\right), \\ (P_1|_{\mathrm{dBm}}, \omega_1), \\ (P_2|_{\mathrm{dBm}}, \omega_2), \\ (P_1|_{\mathrm{dBm}}+P_2|_{\mathrm{dBm}}-\mathrm{IIP2}|_{\mathrm{dBm}}, |\omega_1 \pm \omega_2|), \\ (2P_1|_{\mathrm{dBm}}-\mathrm{IIP2}|_{\mathrm{dBm}}-6, 2\omega_1), \\ (2P_2|_{\mathrm{dBm}}-\mathrm{IIP2}|_{\mathrm{dBm}}-6, 2\omega_2). \end{cases} \tag{5.34}$$

These relationships are of practical importance for performing practical system budgets as they allow us to predict the spur tone levels due to second order nonlinearity assuming two CW tones at the device input. We notice that if the two input tones have the same power, i.e. if $P_1 = P_2 = P_i$, then the DC term reduces to

$$(2P_i|_{\mathrm{dBm}} - \mathrm{IIP2}|_{\mathrm{dBm}} + 3, 0). \tag{5.35}$$

And if one input tone has negligible power compared to the other, i.e. if $\rho_2 \ll \rho_1$ for instance, we get

$$(2P_1|_{\mathrm{dBm}} - \mathrm{IIP2}|_{\mathrm{dBm}} - 3, 0). \tag{5.36}$$

This last configuration corresponds to an important use case when dealing with the DC offset generated by a strong interferer through an even order nonlinearity, for instance.

Odd Order Nonlinearity and IP3, CP1 or Psat

Let us now focus on odd order smooth AM-AM conversion as widely encountered in devices that cope with limited power supply, for instance. As we did for the even order case, we initially assume that the first nonlinear term of the odd part of the device transfer function is of most interest for high level system budgets. We consider the terms up to third order in the series expansion of the transfer function $F_o(x)$ originally given by equation (5.7a), i.e. a device transfer function $F(x)$ given by

$$F(x) = F_o(x) \approx Gx + \alpha_3 x^3, \tag{5.37}$$

where G and α_3 have opposite signs so as to physically represent a compression behavior as discussed in Section 5.1.1. For such compression behavior, simple third order models often allow us to describe most of the important system impacts we see in practice, even dealing with an input signal close to saturation. This is illustrated, for instance, when discussing the potential SNR improvements due to compression as presented in Section 5.1.4.

The odd order nonlinearity of such a device is then characterized by its odd order IPs, which reduce to the IP3 due to the presence of the single nonlinear term x^3 in its transfer function.

To derive an expression for this IP3, we can return to the test configuration detailed in the previous section. This means that we consider the signal

$$s_i(t) = \rho_1 \cos(\omega_1 t) + \rho_2 \cos(\omega_2 t) \tag{5.38}$$

present at the device input. Using equation (5.37), we can then write the signal $s_o(t)$ recovered at its output as

$$s_o(t) = F[s_i(t)] = G\, s_i(t) + \alpha_3 s_i^3(t). \tag{5.39}$$

Substituting equation (5.38) into this expression, we then get that

$$\begin{aligned} s_i^3(t) = {} & \rho_1^3 \cos^3(\omega_1 t) + \rho_2^3 \cos^3(\omega_2 t) + 3\rho_1^2 \rho_2 \cos^2(\omega_1 t)\cos(\omega_2 t) \\ & + 3\rho_1 \rho_2^2 \cos(\omega_1 t)\cos^2(\omega_2 t). \end{aligned} \tag{5.40}$$

Using equations (5.9) and (1.54) leads to

$$\begin{aligned} s_i^3(t) = {} & \left[\frac{3}{4}\rho_1^2 + \frac{3}{2}\rho_2^2\right] \rho_1 \cos(\omega_1 t) + \left[\frac{3}{4}\rho_2^2 + \frac{3}{2}\rho_1^2\right] \rho_2 \cos(\omega_2 t) \\ & + \frac{\rho_1^3}{4}\cos(3\omega_1 t) + \frac{\rho_2^3}{4}\cos(3\omega_2 t) \\ & + \frac{3}{2}\rho_1^2 \rho_2 \cos(2\omega_1 t)\cos(\omega_2 t) + \frac{3}{2}\rho_1 \rho_2^2 \cos(\omega_1 t)\cos(2\omega_2 t). \end{aligned} \tag{5.41}$$

And recalling equation (5.19), we can write

$$\begin{aligned} s_i^3(t) = {} & \left[\frac{3}{4}\rho_1^2 + \frac{3}{2}\rho_2^2\right] \rho_1 \cos(\omega_1 t) + \left[\frac{3}{4}\rho_2^2 + \frac{3}{2}\rho_1^2\right] \rho_2 \cos(\omega_2 t) \\ & + \frac{\rho_1^3}{4}\cos(3\omega_1 t) + \frac{\rho_2^3}{4}\cos(3\omega_2 t) \\ & + \frac{3}{4}\rho_1^2 \rho_2 \cos((2\omega_1 + \omega_2)t) + \frac{3}{4}\rho_1 \rho_2^2 \cos((2\omega_2 + \omega_1)t) \\ & + \frac{3}{4}\rho_1^2 \rho_2 \cos((2\omega_1 - \omega_2)t) + \frac{3}{4}\rho_1 \rho_2^2 \cos((2\omega_2 - \omega_1)t). \end{aligned} \tag{5.42}$$

Finally, using this result in equation (5.39), we can express $s_o(t)$ as

$$\begin{aligned} s_o(t) = {} & \left[G + \frac{3\alpha_3}{4}\left(\rho_1^2 + 2\rho_2^2\right)\right] \rho_1 \cos(\omega_1 t) \\ & + \left[G + \frac{3\alpha_3}{4}\left(\rho_2^2 + 2\rho_1^2\right)\right] \rho_2 \cos(\omega_2 t) \\ & + \frac{\alpha_3}{4}\left(\rho_1^3 \cos(3\omega_1 t) + \rho_2^3 \cos(3\omega_2 t)\right) \\ & + \frac{3\alpha_3}{4}\rho_1\rho_2[\rho_1 \cos((2\omega_1 + \omega_2)t) + \rho_2 \cos((2\omega_2 + \omega_1)t)] \\ & + \frac{3\alpha_3}{4}\rho_1\rho_2[\rho_1 \cos((2\omega_1 - \omega_2)t) + \rho_2 \cos((2\omega_2 - \omega_1)t)]. \end{aligned} \tag{5.43}$$

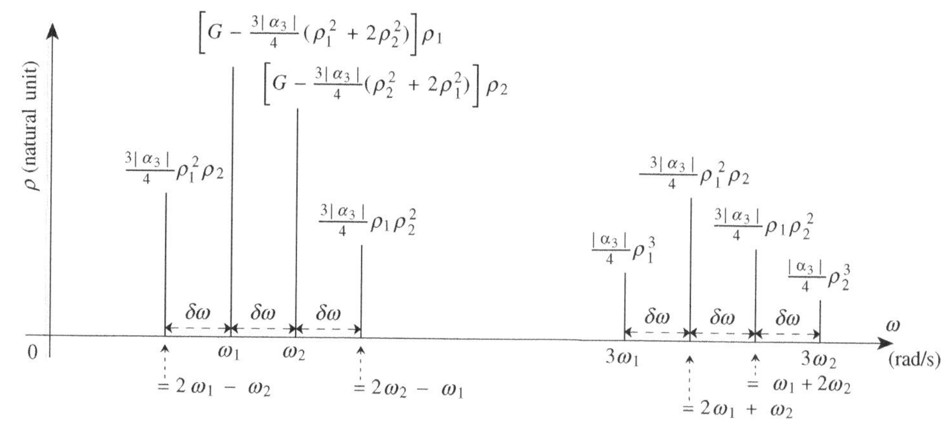

Figure 5.5 CW tone amplitude partition in the frequency domain at the output of a third order smooth AM-AM nonlinear model with two input CW signals – Considering two CW test tones at the input of a transfer function of the form $F(x) = Gx + \alpha_3 x^3$, we recover at the output both the expected test tones and their harmonics, but also intermodulation tones. Unlike in the second order case, the intermodulation tones generated by third order nonlinearity can lie close to the initial input tones in the frequency domain.

We thus see that the presence of the third order nonlinear term in the device transfer function leads to the generation of additional tones at the device output, as illustrated in Figure 5.5. These tones can be sorted into different groups:

(i) We first recover the terms corresponding to the input tones lying at angular frequencies ω_1 and ω_2. But we now see that due to compression, these input tones have experienced an effective gain that is a function of their input amplitude. This phenomenon is in fact at the origin of the terminology of AM-AM conversion that leads to different system issues such as the desensitization of receivers discussed in "Desensitization due to odd order nonlinearity and CCP1" later in this section.
(ii) We then recover the third harmonics of the input tones.
(iii) The last group of output signals represents the intermodulation tones as defined in the previous section in the second order case. These tones are characterized by output amplitudes and angular frequencies, $(\rho_{\text{OIMD3}}, \omega_{\text{IMD3}})$, of the form

$$\left(\frac{3\alpha_3}{4}\rho_1^{|l|}\rho_2^{|m|}, |l\omega_1 + m\omega_2|\right), \quad \text{with } |l| + |m| = 3 \text{ and } l, m \in \mathbb{Z}^*. \tag{5.44}$$

In the present case, we are dealing with the third order intermodulation tones, or IMD3, this term referring to the sum of the absolute values of the weights of ω_1 and ω_2 in the expression for the IMD3 angular frequencies. We also remark that these IMD3 tones can be grouped into two subsets depending on their angular frequencies, as can be seen in Figure 5.5. One of those subsets is composed of tones lying in the area of the third harmonics of the input signals. Thus, considering that we are dealing with RF signals with carrier angular frequencies classically in the gigahertz range, those IMD tones often lie far from the fundamental tones of interest in terms of spectral location. As a result, they can often be handled more easily than those of the other subset that lie close to the fundamental tones of interest, as discussed in "Intermodulation" later in this section.

In order to go further and link the parameters of the transfer function $F(x)$ to the device IP3, we again need to consider input test tones of the same amplitude, i.e. such that $\rho_1 = \rho_2 = \rho_i$. Under this assumption, we get that the amplitude of the IMD3 tones is the same whatever their frequency location as given by equation (5.44). And in that case, the amplitude ρ_i for which the output IMD3 amplitude is equal to the theoretical output amplitude of the test tones if the device were linear, i.e. $|G|\rho_i$, defines the device input IP3 amplitude ρ_{IIP3}. According to this definition, we can write that

$$|G|\rho_i|_{\rho_i=\rho_{IIP3}} = |G|\rho_{IIP3} \stackrel{\Delta}{=} \rho_{OIMD3}|_{\rho_i=\rho_{IIP3}} = \left|\frac{3\alpha_3}{4}\right|\rho_{IIP3}^3. \tag{5.45}$$

We then see the importance of the definition of the IP as the intersection of the extrapolated linear small-signal curves shown in Figure 5.7. It leads to the use of the device small-signal gain G for the amplification of the input tones in the above equation and thus to neglect of the additional terms representing the gain compression. As a result,

$$\rho_{IIP3} = \sqrt{\frac{4}{3}\left|\frac{G}{\alpha_3}\right|}. \tag{5.46}$$

We can then define an equivalent output quantity for this CW tone amplitude. Due to its definition linked to the extrapolated small-signal curves, we can write the output IP3 amplitude as G times that of the input.[5] We thus simply get that

$$\rho_{OIP3} \stackrel{\Delta}{=} |G|\rho_{IIP3} = \sqrt{\frac{4}{3}\left|\frac{G^3}{\alpha_3}\right|}. \tag{5.47}$$

We can then express the IP3 characteristics in terms of power for the CW we are dealing with. As discussed in the previous section, we can continue to assume that we are working on normalized impedances, i.e. that $\text{Re}\{1/Z_0\} = \text{Re}\{Z_0\} = 1$. Consequently, whatever the type of waves we are dealing with, we can write

$$\text{IIP3} = \left(\frac{\rho_{IIP3}}{\sqrt{2}}\right)^2 = \frac{2}{3}\left|\frac{G}{\alpha_3}\right|, \tag{5.48a}$$

$$\text{OIP3} = \left(\frac{\rho_{OIP3}}{\sqrt{2}}\right)^2 = \frac{2}{3}\left|\frac{G^3}{\alpha_3}\right|, \tag{5.48b}$$

[5] Still assuming that no filtering effect is involved, as discussed in "Input vs. output IP" later in this section.

We can then use these results to express the amplitude, or power, of the various IMD3 and harmonic tones recovered at the device output. However, it is convenient here again to express these quantities as equivalent input ones using the device small-signal gain G. We can also use equation (5.46), while remembering that G and α_3 have opposite signs to correctly describe the compression behavior, to write their amplitudes as a function of the device ρ_{IIP3} rather than α_3. We can express the collection of generated tones present in equation (5.43) as equivalent input tones of amplitude and angular frequencies, (ρ, ω), of the form:

$$\left\{\left(\left[1-\frac{(\rho_1^2+2\rho_2^2)}{\rho_{\text{IIP3}}^2}\right]\rho_1, \omega_1\right), \left(\left[1-\frac{(\rho_2^2+2\rho_1^2)}{\rho_{\text{IIP3}}^2}\right]\rho_2, \omega_2\right),\right.$$
$$\left.\left(\frac{\rho_1^2\rho_2}{\rho_{\text{IIP3}}^2}, |2\omega_1 \pm \omega_2|\right), \left(\frac{\rho_1\rho_2^2}{\rho_{\text{IIP3}}^2}, |\pm\omega_1 + 2\omega_2|\right), \left(\frac{1}{3}\frac{\rho_1^3}{\rho_{\text{IIP3}}^2}, 3\omega_1\right), \left(\frac{1}{3}\frac{\rho_2^3}{\rho_{\text{IIP3}}^2}, 3\omega_2\right)\right\}. \tag{5.49}$$

The same can be done in terms of tone power – for instance, in dBm:

$$\left\{\begin{aligned}
&\left(P_1|_{\text{dBm}} + 20\log_{10}\left[1-\frac{(\rho_1^2+2\rho_2^2)}{\rho_{\text{IIP3}}^2}\right], \omega_1\right),\\
&\left(P_2|_{\text{dBm}} + 20\log_{10}\left[1-\frac{(\rho_2^2+2\rho_1^2)}{\rho_{\text{IIP3}}^2}\right], \omega_2\right),\\
&(2P_1|_{\text{dBm}} + P_1|_{\text{dBm}} - 2\text{IIP3}|_{\text{dBm}}, |2\omega_1 \pm \omega_2|),\\
&(P_1|_{\text{dBm}} + 2P_2|_{\text{dBm}} - 2\text{IIP3}|_{\text{dBm}}, |\pm\omega_1 + 2\omega_2|),\\
&(3P_1|_{\text{dBm}} - 2\text{IIP3}|_{\text{dBm}} - 9.6, 3\omega_1),\\
&(3P_2|_{\text{dBm}} - 2\text{IIP3}|_{\text{dBm}} - 9.6, 3\omega_2).
\end{aligned}\right. \tag{5.50}$$

These relationships are of importance for deriving practical nonlinearity system budgets, as illustrated in Chapter 7.

However, we observe that although the IP3 concept seems well suited to intermodulation derivations, it is not really explicit with respect to the compression behavior associated with the odd order nonlinearity. Consequently, there is an interest in having metrics other than the IP3 to give a more explicit idea of the margins we have on signal levels before reaching the saturation of the device. This is the purpose of concepts such as CPs or saturated power (Psat) that are both the direct expression of the limited power, or amplitude, that can be delivered by the device to its load. In order to illustrate their definition, let us suppose now that a single CW test signal is present at the input of the nonlinear device, i.e. that $s_\text{i}(t)$ reduces to

$$s_\text{i}(t) = \rho_\text{i}\cos(\omega t). \tag{5.51}$$

In that case, $s_o(t)$ can still be derived from equation (5.43), but assuming now that one of the two signals used in this former derivation is null. We can thus immediately write that

$$s_o(t) = \left(G + \frac{3\alpha_3}{4}\rho_i^2\right)\rho_i\cos(\omega t) + \frac{\alpha_3}{4}\rho_i^3\cos(3\omega t). \tag{5.52}$$

In order to go further, we can focus on the output fundamental tone, $s_{o,H1}(t)$, which takes the form

$$s_{o,H1}(t) = \left(G + \frac{3\alpha_3}{4}\rho_i^2\right)\rho_i\cos(\omega t). \tag{5.53}$$

In order to link the new concepts we are introducing with the device IP3, we can use equation (5.46), remembering that G and α_3 have opposite signs to correctly describe the compression behavior, to write the above equation as

$$s_{o,H1}(t) = G\left(1 - \frac{\rho_i^2}{\rho_{IIP3}^2}\right)\rho_i\cos(\omega t). \tag{5.54}$$

We thus see that all behaves as if the input tone of interest had experienced an effective gain $G_e(\rho_i)$ given by

$$G_e(\rho_i) = G\left(1 - \frac{\rho_i^2}{\rho_{IIP3}^2}\right). \tag{5.55}$$

This effective gain can also be expressed as a function of the signal powers instead of amplitude. With those powers proportional to the square of the CW tone amplitudes, we get

$$G_e(P_i) = G\left(1 - \frac{P_i}{\text{IIP3}}\right), \tag{5.56}$$

with P_i the power of the input tone, and IIP3 the power of the device input IP3. Looking at this equation, we see that for low input power, i.e. for $P_i \ll \text{IIP3}$, we have that $G_e \approx G$. And as the input signal power increases we get that the effective gain experienced by the input signal decreases, which is in line with the compression behavior we are considering. Consequently, we see that there is an input amplitude for which the gain loss compared to the device small-signal gain G is equal to 1 dB. This input level therefore defines the 1 dB CP,[6] classically denoted by CP1. Based on this definition, we see that the CP corresponds to a particular amplitude of a CW tone. As for the IP, we can therefore either talk about an input CP1 amplitude, ρ_{ICP1}, or an input CP1 power, ICP1, equal to $\rho_{ICP1}^2/2$ when working on a normalized impedance. We can also derive equivalent output quantities, ρ_{OCP1} and the output

[6] Generally speaking, we can have CPs at x dB, i.e. CPx, using a gain loss of x dB for their definition, with x not necessarily reducing to 1. However, the 1 dB CP remains the most commonly used metric in practice.

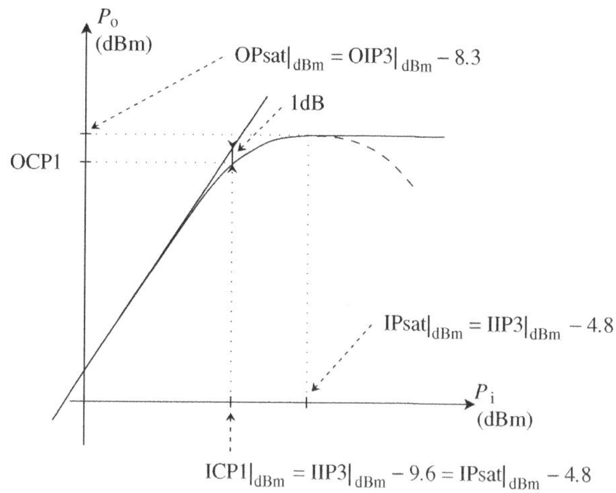

Figure 5.6 Transfer function characterization of a nonlinear device that exhibits odd order smooth AM-AM nonlinearity – The use of a third order series expansion for a nonlinear device transfer function as given by equation (5.37) leads to straightforward relationships between the device CP1, IP3 and Psat. However, the third order polynomial transfer function (dashed) remains valid only up to the device saturated power as it represents a decreasing transfer function above this level (bold solid).

compression point (OCP) $\text{OCP1} = \rho_{\text{OCP1}}^2/2$ respectively. However, there is a major difference in the present case compared to the IP case. Due to their definition the input and output CP1 quantities are linked by the *effective* gain of the device and not by the small-signal one as illustrated in Figure 5.6. In the IP case, this property is indeed linked to their definition corresponding to linear extrapolations of low power behaviors of the device. This is not the case for the CPs whose definitions rely on the power or amplitude of CW signals that can be *effectively* handled and delivered by the device. The correspondence between their equivalent input and output quantities therefore involves the effective gain of the device.

Considering the device transfer function shown in Figure 5.6, we also observe that the third order model can be used only up to a given input power which corresponds to the cancellation of the derivative of this transfer function. For higher input powers, the third order polynomial model leads to an output power that decreases when the input power increases. This is not realistic, and to be able to use such a model at a system simulation level, for instance, a clamping of the output power to the maximum value has to be added above this point. We can link this behavior to the concept of maximum power that can deliver a given RF amplifier to its load, i.e. the saturated power (Psat). For this quantity also we can define either output CW amplitude or power, ρ_{OPsat} and output saturated power (OPsat) respectively, or equivalent input quantities, ρ_{IPsat} and input saturated power (IPsat) respectively. Here, IPsat is defined as the input power for which we effectively have the output power to be equal to OPsat. This means that the IPsat and the OPsat are linked by the device effective gain of the device and not its small-signal gain. A related concept is therefore the back-off used to scale the wanted signal at the input or output of such device. In that case, the back-off has to be understood as the ratio between the maximum allowable power, the Psat in the present case, and the average power of the considered RF signal. This back-off is suitable for scaling the signal level at the input or output of a device that exhibits odd order nonlinearity and is thus extensively used as such in the subsequent sections.

Using these definitions, we can then link the device Psat and its CP1 to its IP3 using our present third order transfer function expansion. For the Psat case, we can use the fact that this quantity corresponds to the input signal level that leads to the maximum level for the output signal that can be delivered by the device. We can thus write that $\rho_\mathrm{i} = \rho_\mathrm{IPsat}$ when the output fundamental tone amplitude $\rho_\mathrm{i}\, G_\mathrm{e}(\rho_\mathrm{i})$ is maximum, i.e. when

$$\partial_{\rho_\mathrm{i}} [\rho_\mathrm{i}\, G_\mathrm{e}(\rho_\mathrm{i})]|_{\rho_\mathrm{i}=\rho_\mathrm{IPsat}} = 0. \tag{5.57}$$

Using the expression for $G_\mathrm{e}(\rho_\mathrm{i})$ given by equation (5.55), we get that

$$\rho_\mathrm{IPsat} = \frac{\rho_\mathrm{IIP3}}{\sqrt{3}} \tag{5.58}$$

or, in terms of signal power,

$$\mathrm{IPsat} = \frac{\mathrm{IIP3}}{3}. \tag{5.59}$$

Conveniently, this last relationship can be expressed in dBm as classically used in practice:

$$\begin{aligned}\mathrm{IPsat}|_\mathrm{dBm} &= \mathrm{IIP3}|_\mathrm{dBm} - 10\log_{10}(3)\\ &= \mathrm{IIP3}|_\mathrm{dBm} - 4.8.\end{aligned} \tag{5.60}$$

We can then do the same for the OPsat value, using the fact that it is related to the IPsat through the effective gain of the device. For $\rho = \rho_\mathrm{IPsat}$,

$$\rho_\mathrm{OPsat} = \rho_\mathrm{IPsat}\, G_\mathrm{e}(\rho_\mathrm{IPsat}). \tag{5.61}$$

Thus, using equations (5.55) and (5.58), we get that

$$\rho_\mathrm{OPsat} = \frac{\rho_\mathrm{IIP3}}{\sqrt{3}} G\left(1 - \frac{1}{3}\right) = G\, \rho_\mathrm{IIP3} \frac{2}{3\sqrt{3}} = \rho_\mathrm{OIP3} \frac{2}{3\sqrt{3}} \tag{5.62}$$

or, in dBm units,

$$\begin{aligned}\mathrm{OPsat}|_\mathrm{dBm} &= \mathrm{OIP3}|_\mathrm{dBm} + 20\log_{10}\left(\frac{2}{3\sqrt{3}}\right)\\ &= \mathrm{OIP3}|_\mathrm{dBm} - 8.3.\end{aligned} \tag{5.63}$$

Comparing this relationship with equation (5.60), we do not have the same offset coefficients between the IPsat and the IIP3 on the one hand and between the OPsat and the OIP3 on the other hand. We thus recover that the IPsat and the OPsat are linked by the effective gain of the device rather than by the small-signal gain as in the case of the IPs.

Let us now focus on the link between CP1 and IP3. For an amplitude of the input CW signal corresponding to $\rho_i = \rho_{ICP1}$, we have that the difference between the effective gain and the linear gain of the device is equal to 1 dB. Thus,

$$20 \log_{10}\left(\frac{G}{G_e(\rho_{ICP1})}\right) = 1. \tag{5.64}$$

Consequently, using equation (5.55),

$$\rho_{ICP1} \approx 0.33\,\rho_{IIP3} = \frac{\rho_{IIP3}}{3}. \tag{5.65}$$

This relationship can be transposed in terms of signal power as

$$\text{ICP1} = \frac{\text{IIP3}}{9} \tag{5.66}$$

or, in dBm units, as

$$\text{ICP1}|_{\text{dBm}} = \text{IIP3}|_{\text{dBm}} - 9.6. \tag{5.67}$$

Using both equations (5.59) and (5.66), we can also express the CP1 as a function of the device Psat as

$$\text{ICP1} = \frac{\text{IPsat}}{3}, \tag{5.68}$$

or, in dBm units,

$$\text{ICP1}|_{\text{dBm}} = \text{IPsat}|_{\text{dBm}} - 4.8. \tag{5.69}$$

Finally, it may be of interest to transpose the CP1 as an output quantity. We can use the fact that the OCP1 is related to the ICP1 through the effective gain of the device:

$$\rho_{OCP1} = \rho_{ICP1}\, G_e(\rho_{ICP1}). \tag{5.70}$$

Using equations (5.55) and (5.65), we then get that

$$\rho_{OCP1} = \frac{\rho_{IIP3}}{3} G\left(1 - \frac{1}{9}\right) = G\,\rho_{IIP3}\frac{8}{27} = \rho_{OIP3}\frac{8}{27} \tag{5.71}$$

or, in dBm units,

$$\begin{aligned}\text{OCP1}|_{\text{dBm}} &= \text{OIP3}|_{\text{dBm}} + 20\log_{10}\left(\frac{8}{27}\right)\\ &= \text{OIP3}|_{\text{dBm}} - 10.6.\end{aligned} \tag{5.72}$$

And in the same way, using equations (5.63), we have

$$\mathrm{OCP1}|_{\mathrm{dBm}} = \mathrm{OPsat}|_{\mathrm{dBm}} - 2.3. \tag{5.73}$$

Higher Order Terms and IPk

Experience shows that considering the main nonlinear term in the series expansion of the device transfer function allows us to derive good enough orders of magnitude for transceiver budgets in most practical use cases. It can even succeed in describing some phenomena involved in the compression range of a device that exhibits odd order nonlinearity (see Section 5.1.4). However, there may be some configurations where investigating higher order harmonics or intermodulation products may be of interest. We should therefore say a few words about such higher order terms, at least with regard to the related IP*k* concept.

By way of illustration, let us reconsider the even order nonlinearity case discussed in "Even order nonlinearity and IP2" earlier in this section. We can extend the series expansion of the even transfer function $F_e(x)$ (equation (5.15)), so that the device transfer function $F(x)$ becomes

$$F(x) \approx Gx + \alpha_2 x^2 + \alpha_4 x^4. \tag{5.74}$$

In order to examine the impact of this fourth order term on the characteristics of the IMD2 tones, we again consider the configuration used for the derivation of the device IP, i.e. an input signal $s_i(t)$ composed of the sum of two CW tones with the same amplitude ρ_i. We thus assume in this first step that

$$s_i(t) = \rho_i(\cos(\omega_1 t) + \cos(\omega_2 t)). \tag{5.75}$$

Given that the contribution of the second order term $\alpha_2 s_i^2(t)$ to the IMD2 tone amplitude has already been derived in "Even order nonlinearity and IP2", we can now focus on the contribution of the fourth order term:

$$\alpha_4 s_i^4(t) = \rho_i^4(\cos(\omega_1 t) + \cos(\omega_2 t))^4. \tag{5.76}$$

After expansion, we get that

$$\begin{aligned}\alpha_4 s_i^4(t) = \alpha_4\rho_i^4(&\cos^4(\omega_1 t) + 4\cos^3(\omega_1 t)\cos(\omega_2 t)\\ &+6\cos^2(\omega_1 t)\cos^2(\omega_2 t) + 4\cos(\omega_1 t)\cos^3(\omega_2 t) + \cos^4(\omega_2 t)).\end{aligned} \tag{5.77}$$

Consequently, using equations (5.11), (5.14) and (5.19) to linearize the power and the product of cosine functions, only the terms of the form $\cos^3(\omega_1 t)\cos(\omega_2 t)$ and $\cos(\omega_1 t)\cos^3(\omega_2 t)$ can contribute to the IMD2 tone lying at angular frequency $\omega_1 + \omega_2$ or $|\omega_1 - \omega_2|$. This results in a contribution of the form $3\alpha_4\rho_i^4 \cos((\omega_1 \pm \omega_2)t)$ to the IMD2 tones. Taking into account this result on top of the contribution linked to the second order term $\alpha_2 x^2$ given by equation (5.23), we can finally write that

$$\rho_{\mathrm{OIMD2}} = \alpha_2\rho_i^2 + 3\alpha_4\rho_i^4. \tag{5.78}$$

A side effect of this result is that the parameter α_2 no longer depends only on the IP2 characteristics of the device but also on its IP4. This means that if we want to fit a fourth order series expansion to the device transfer function, we need to take into account also the relationship between ρ_{IIP4} and α_4. This would result in a system of two equations whose resolution would give both α_2 and α_4 as a function of ρ_{IIP2} and ρ_{IIP4}. However, we can imagine that such a method for fitting the series expansion would soon become tricky as the series order increases. When such a higher order model is effectively required, e.g. on the transmit side, other methods can be considered (see Section 5.3). However, what is interesting to observe about equation (5.78) is that the contribution of the second order term $\alpha_2 x^2$ on the IMD2 tone amplitude has a dependency on ρ_{i}^2 whereas the fourth order term $\alpha_4 x^4$ leads to a contribution that has a dependency on ρ_{i}^4. We thus see that for sufficiently low input levels, we have $\alpha_4 \rho_{\text{i}}^4 \ll \alpha_2 \rho_{\text{i}}^2$. This low input level condition, which corresponds to the configuration for the derivation of the IP of the device based on the linear extrapolation of this low input power behavior, leads to

$$\rho_{\text{OIMD2}} \approx \alpha_2 \rho_{\text{i}}^2. \tag{5.79}$$

We thus recover equation (5.23) and the fact that the IMD2 tone power has a dependence on ρ_{i}^2 for sufficiently low values of ρ_{i} whatever the structure of the device transfer function $F(x)$.

This behavior is in fact general whatever the order and parity of the IMD tones considered. This can be seen by inspecting the contribution to the amplitude of the kth order intermodulation tones of the term $\alpha_n x^n$ in the series expansion of $F(x)$. Consider the input signal $s_{\text{i}}(t)$ given by equation (5.75) and use, on the one hand, the binomial formula [6]

$$(x+y)^n = \sum_{l=0}^{n} \binom{n}{l} x^{n-l} y^l \tag{5.80}$$

and, on the other hand, the trigonometric relationships given by equations (5.11), (5.14), and (5.19). After some algebra, it is evident that the kth order intermodulation tones lying at angular frequencies $\omega_k = |l\omega_1 + m\omega_2|$, with $|l| + |m| = k$ and $l, m \in \mathbb{Z}^*$, can only be generated by terms of the form $\alpha_{k+2n} x^{k+2n}$, with $n \in \mathbb{N}$, in the series expansion of $F(x)$. Consequently, the amplitude of the output kth order intermodulation tone is necessarily of the form

$$\rho_{\text{OIMD}k} = \lambda_k \rho_{\text{i}}^k + \lambda_{k+2} \rho_{\text{i}}^{k+2} + \cdots. \tag{5.81}$$

We thus recover that for sufficiently low input power,

$$\rho_{\text{OIMD}k} \approx \lambda_k \rho_{\text{i}}^k, \tag{5.82}$$

and thus that the IMDk amplitude increases by k dB per decibel of the input tone amplitude, as illustrated in Figure 5.7. The linear part of the curves can thus be extrapolated in order to define the IPk amplitude as their intersection. We thus get that for $\rho_{\text{i}} = \rho_{\text{IIP}k}$ the value of $\rho_{\text{OIMD}k}$ given by the above expression is nothing more than $\rho_{\text{OIP}k}$. We can then write

$$\rho_{\text{OIP}k} = \lambda_k \rho_{\text{IIP}k}^k. \tag{5.83}$$

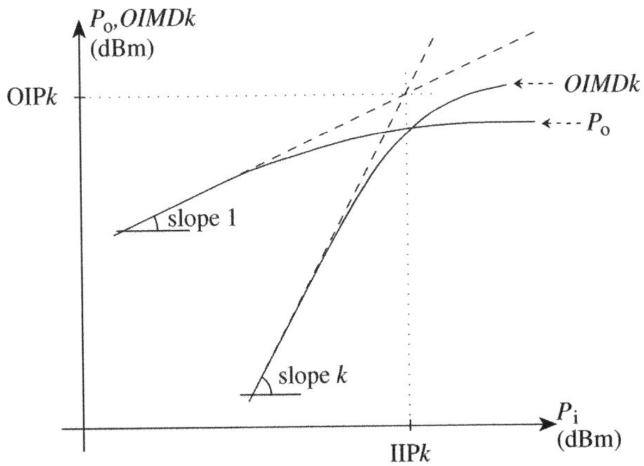

Figure 5.7 Generated IMDk tone power as a function of the input tone power – On a log–log scale, the IMDk tone level increases by k dB per decibel of the input fundamental tones. This occurs as long as the main contribution to the IMDk tone amplitude comes from the term $\alpha_k x^k$ in the series expansion of the device transfer function, i.e. as long as the input tone power remains sufficiently low enough that the contributions of the higher order terms are negligible. The intersection of the extrapolated small-signal linear curves defines the IIPk and OIPk values.

Given by definition of the IP that $\rho_{\text{OIP}k} = G\rho_{\text{IIP}k}$, we get that

$$\lambda_k = \frac{\rho_{\text{OIP}k}}{\rho^k_{\text{IIP}k}} = \frac{G}{\rho^{k-1}_{\text{IIP}k}}. \tag{5.84}$$

We can then use this result in equation (5.82) to derive an expression for the amplitude of the OIMDk tones effectively recovered at the device output in the low input power approximation. This results in

$$\rho_{\text{OIMD}k} = \frac{G\rho^k_{\text{i}}}{\rho^{k-1}_{\text{IIP}k}}. \tag{5.85}$$

Assuming that the same device small-signal gain applies to all the intermodulation tone amplitudes,[7] we can then derive the equivalent input IMDk amplitude as the output amplitude divided by the small-signal gain G. We thus get

$$\rho_{\text{IIMD}k} = \frac{\rho^k_{\text{i}}}{\rho^{k-1}_{\text{IIP}k}}. \tag{5.86}$$

This relationship can also be written in terms of tone power, for instance in dBm units, as

$$(k-1)\text{IIP}k|_{\text{dBm}} = kP_{\text{i}}|_{\text{dBm}} - IIMDk|_{\text{dBm}}. \tag{5.87}$$

[7] The particular case where this device gain depends on the frequency, i.e. when a filtering effect exists, is discussed in "input vs. output IP" later in this section.

As already done in the IP2 and the IP3 case, it is also of interest to derive the characteristics of the generated IMDk tones in the case where the input tones have different amplitudes, i.e. when the input signal is of the form

$$s_{\mathrm{i}}(t) = \rho_1 \cos(\omega_1 t) + \rho_2 \cos(\omega_2 t). \tag{5.88}$$

The IMDk tones of the same order now evidently have different amplitudes, depending on their angular frequency. More precisely, we have output amplitudes and angular frequencies, $(\rho_{\mathrm{OIMD}k_{l,m}}, \omega_{\mathrm{IMD}k_{l,m}})$, of the form

$$\left(\lambda_k \rho_1^{|l|} \rho_2^{|m|}, |l\omega_1 + m\omega_2|\right), \quad \text{with } |l| + |m| = k \text{ and } l, m \in \mathbb{Z}^*. \tag{5.89}$$

Using the expression for λ_k given by equation (5.84), we can derive the device IIPk amplitude as a function of those IMDk tone equivalent input amplitudes by inverting

$$\rho_{\mathrm{OIMD}k_{l,m}} = G \frac{\rho_1^{|l|} \rho_2^{|m|}}{\rho_{\mathrm{IIP}k}^{k-1}}. \tag{5.90}$$

This results in

$$\rho_{\mathrm{IIP}k} = \frac{\left(\rho_1^{|l|} \rho_2^{|m|}\right)^{1/(k-1)}}{\left(\rho_{\mathrm{IIMD}k_{l,m}}\right)^{1/(k-1)}} \tag{5.91}$$

or, in terms of CW tone power, expressed in dBm units for instance,

$$(k-1)\mathrm{IIP}k|_{\mathrm{dBm}} = |l|P_1|_{\mathrm{dBm}} + |m|P_2|_{\mathrm{dBm}} - \mathit{IIMD}k_{l,m}|_{\mathrm{dBm}}. \tag{5.92}$$

These relationships can be transposed at the device output in a straightforward way assuming the device small-signal gain independent of the frequency. Given that $|l| + |m| = k$, we can then write from equation (5.91) that

$$\rho_{\mathrm{OIP}k} = G\,\rho_{\mathrm{IIP}k} = G \frac{\left(\rho_1^{|l|} \rho_2^{|m|}\right)^{1/(k-1)}}{\left(\rho_{\mathrm{IIMD}k_{l,m}}\right)^{1/(k-1)}} = \frac{\left((G\,\rho_1)^{|l|} (G\,\rho_2)^{|m|}\right)^{1/(k-1)}}{\left(G\,\rho_{\mathrm{IIMD}k_{l,m}}\right)^{1/(k-1)}}; \tag{5.93}$$

and in terms of output tone power,

$$(k-1)\mathrm{OIP}k|_{\mathrm{dBm}} = |l|P_{\mathrm{o},1}|_{\mathrm{dBm}} + |m|P_{\mathrm{o},2}|_{\mathrm{dBm}} - \mathit{OIMD}k_{l,m}|_{\mathrm{dBm}}. \tag{5.94}$$

Of course, by taking $k = 2$ or $k = 3$, we recover the results of the second and third order cases discussed in the two previous sections. However, we have to keep in mind that the relationships derived so far remain valid only when the main contribution to the IMDk tone amplitude comes from the nonlinear term $\alpha_k x^k$ in the device series expansion, i.e. for sufficiently low input power.

Otherwise, higher order terms can impact the IMDk tone level effectively recovered at the device output.

Input vs. Output IP

By the definition of the IP as the intersection of the linear extrapolation of low power curves, the relationship between the input and output IP quantities involves the small-signal gain G of the device under consideration. However, this simple property may fail to hold in some cases, for instance when characterizing a device that exhibits some frequency selectivity. In that case, using on the one hand equation (5.92) for deriving the input IPk of a device based on the characteristic of both the test tones and the equivalent input IMDk tones, and on the other hand equation (5.94) for deriving the output IPk based on the collection of tones recovered at the device output, may lead to some discrepancies. Indeed, in order to have $\rho_{\text{OIP}k} = G\rho_{\text{IIP}k}$ we need the same small-signal gain of the device G to apply to all the tones involved in the derivation of the IPk characteristics (see equation (5.93)).

This condition is in fact intrinsically linked to the device model we considered in order to perform our derivations. This model, given by equation (5.3), assumes that the device behaves in the same way whatever the spectral content of the input signal. It thus comes as no surprise that the relationships derived so far based on this model are coherent only when assuming that the same small-signal gain applies to all the tones considered. However, there is a limitation with this approach when dealing with a device whose transfer function (more precisely, small-signal gain) is frequency dependent. From a theoretical point of view, this frequency dependence could be handled through the use of Volterra series, for instance [61]. However, analytical derivations then soon become tricky. It thus becomes desirable to continue to use the relationships derived up to now for the purpose of high level system budgeting. Fortunately, this can be done at some marginal cost.

In order to investigate this, let us consider as an example a nonlinear amplifier that embeds a filtering stage, as illustrated in Figure 5.8. In practice, it can happen that the intermodulation tones of interest lie in the passband of the filter, while the input test tones are outside it. In that case, due to the attenuation experienced by the test tones, the direct application of equation (5.94) using the effective level of the output tones necessarily leads to an estimation of the device OIPk worse than the direct transposition using the device in-band small-signal gain G of the IIPk estimated using equation (5.92) and the equivalent input IMDk tones estimated from the effective OIMk tones using this in-band small-signal gain G. Based on this observation, two strategies can be considered in practice:

(i) We can first consider working independently on input or output quantities, using either equation (5.92) or equation (5.94) to derive the device IIPk or OIPk respectively. In that case, these quantities are linked by a function of the small-signal gain experienced by each tone. More precisely, if we denote by G_1, G_2 and $G_{l,m}$ the small-signal gain of the device at the tone angular frequencies ω_1, ω_2 and $\omega_{\text{IMD}k_{l,m}}$ respectively, equation (5.93) can be rewritten as

$$\rho_{\text{OIP}k} = \frac{\left((G_1\rho_1)^{|l|}(G_2\rho_2)^{|m|}\right)^{1/(k-1)}}{\left(G_{l,m}\rho_{\text{IIMD}k_{l,m}}\right)^{1/(k-1)}} = G_{1,2,l,m}\,\rho_{\text{IIP}k}, \tag{5.95}$$

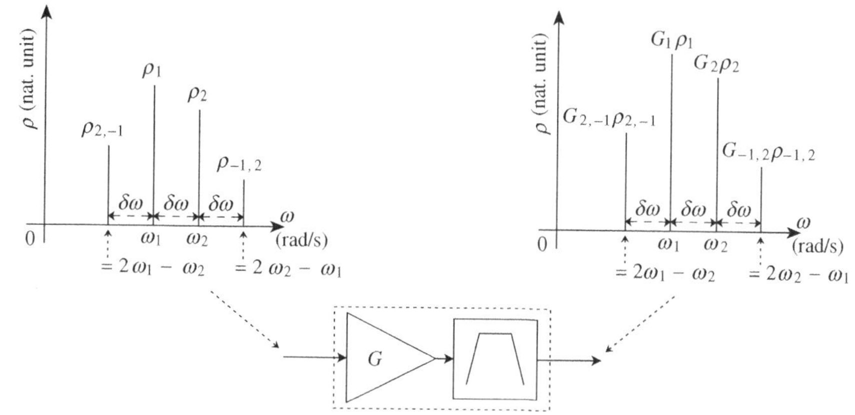

Figure 5.8 Input vs. output tone amplitudes for a device that exhibits both a third order nonlinearity and a frequency dependent gain – Performing an intermodulation test with a frequency selective device leads to tone amplitudes that depend not only on the device linearity, but also on the spectral partition of the tones. If the filtering effect is not compensated, the device IIPk or OIPk characteristics derived by the direct application of equation (5.92) or (5.94) can be linked by a more complicated relationship than the simple device small-signal gain as given by equation (5.96).

with

$$G_{1,2,l,m} = \left(\frac{G_1^{|l|} G_2^{|m|}}{G_{l,m}} \right)^{1/(k-1)} \tag{5.96}$$

and $|l| + |m| = k$. Of course, in the case of a small-signal gain constant in frequency, i.e. when $G_1 = G_2 = G_{l,m} = G$, we recover the fact that the gain experienced by the IPk amplitudes reduces to G. The drawback with this method is that the gain experienced by the IPk amplitudes depends on the frequency configuration of the input tones, as we do not make the distinction between different phenomena, i.e. the nonlinearity itself and the filtering effect that affects the tone amplitude.

(ii) Another strategy is to compensate the output tone levels by the filter attenuation at their corresponding frequency. Once this filtering effect is compensated, the same in-band small-signal gain of the device is valid for all the tones involved and thus between the evaluated device IIPk and OIPk. The method is of interest in that we have a separation between the nonlinearity on the one hand and the filtering effect on the other hand. We thus recover that the model we have used up to now is fully coherent with the description of the nonlinearity part of the phenomenon. When adopting this strategy, given a set of input blocking signals, the equivalent input level of the generated IMD tones can be estimated using equation (5.92) and the device IIPk derived assuming that the filtering effect is compensated. Then, when needed, these input tones can be propagated through the device using the gain corresponding to their frequency locations, thus taking into account the potential filtering effects.

Whatever the strategy used, the point is to keep the derivations coherent throughout the transceiver budgeting. However, experience shows that such system budgets require the ability to analyze separately the impact of each of the phenomena involved in order to optimally balance the constraints. Thus, the second approach discussed here is often better suited in that respect.

Impact on Transceivers

DC Offset Generation Due to Even Order Nonlinearity

A first system impact we can focus on is the DC offset generated by even order nonlinearity.

By the discussion in "Even order nonlinearity and IP2" earlier in this section, and in particular by reference to equation (5.31), this DC component is proportional to the square of the amplitude of the CW signals fed to the device. In practical implementations, this proportionality factor is small enough that the DC component generated by the wanted signal remains negligible compared to its own amplitude. However, problems can arise when a strong blocker is processed at the same time as a weak wanted signal. In that case, the DC term generated by the strong blocker can be non-negligible with regard to the wanted signal. This potential huge difference in the amplitude of the signals being processed at the same time makes receivers more sensitive to this DC generation problem than transmitters. In practical receiver implementations, this situation can be particularly problematic at the final downconversion stage of the line-up when implemented in the RF/analog domain. We indeed recover in that case that the received wanted signal and the DC component generated by the potential blockers are superposed in the frequency domain at the output of the downconversion, as illustrated in Figure 5.9. Anticipating the discussion in Chapter 8, we thus observe that some receiver architectures are more sensitive to this problem than others. This is in particular the case for the direct conversion receiver as no particular filtering takes place before the downconversion stage to attenuate blocking signals.

However, we may wonder whether this DC component is really a new system issue, as DC offset is inherently present in RF/analog device implementations, as discussed in Section 6.5. There is nevertheless an important difference in respect of this former situation as in real life

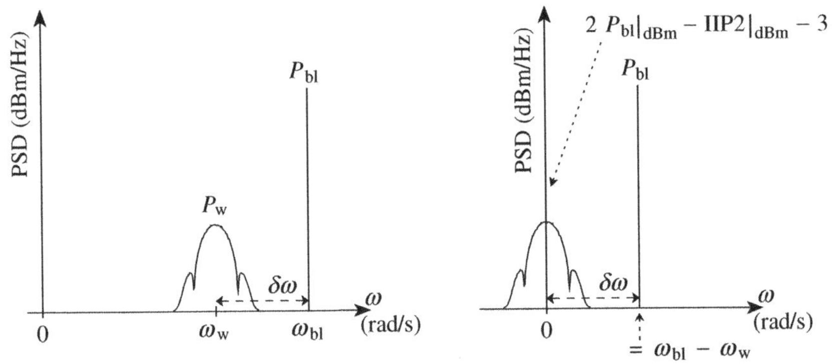

Figure 5.9 DC offset generation resulting from a constant amplitude blocking signal going through an even order nonlinearity – A second order AM-AM nonlinearity generates DC offset from a constant amplitude interferer. Considering a weak wanted signal in addition to a strong blocker at the input of the device (left), we get at the output the downconverted wanted signal and the DC offset pollution superposed both in the time and frequency domain (right). Here, equivalent input quantities are represented.

environment blocking signals can appear or disappear at any time during the reception. We may thus be faced with a DC level that varies abruptly from time to time, which we might call dynamic DC offset, in contrast the DC offset inherently present in the RF/analog part of the line-up which we might call static DC even if it can drift slowly with the temperature of supply of the device. As a first consequence, when considering using a DC offset compensation system as discussed in Section 9.2.2, one needs to ensure that the system can work properly in respect to both DC variations. We will also see in "AM-demodulation due to even order nonlinearity" (Section 5.1.3), that this phenomenon is emphasized when dealing with amplitude modulated blockers. In that case, it becomes mandatory to ensure that the even order linearity of the receiver block is sufficiently good that the degradation of the wanted signal modulation remains negligible.

Desensitization Due to Odd Order Nonlinearity and CCP1

Another system impact we can discuss is the cross-compression that can occur when dealing with an odd order nonlinearity. As already discussed in "Odd vs. even order nonlinearity" (Section 5.1.1), in practical physical implementations an odd order nonlinearity in a transfer function is equivalent to compression behavior of the corresponding device. Such compression corresponds to a gain loss experienced by the signal being processed when its amplitude gets closer to the maximum allowable level. However, it is interesting to see that this gain loss is still experienced by a weak signal, i.e. by a signal whose amplitude is within the linear range of the device transfer function, when a strong blocking signal is in turn within the compression range of the device.

In order to illustrate this behavior, consider an input signal $s_i(t)$ that is the superposition of a weak input wanted signal $s_w(t) = \rho_w \cos(\omega_w t)$ and a strong blocking signal $s_{bl}(t) = \rho_{bl} \cos(\omega_{bl} t)$, i.e. such that

$$s_i(t) = \rho_w \cos(\omega_w t) + \rho_{bl} \cos(\omega_{bl} t). \tag{5.97}$$

This input signal thus has the same structure as that given by equation (5.38) and used for the derivations in "Odd order nonlinearity and IP3, CP1 or Psat" earlier in this section. Still assuming that we can approximate the device transfer function by equation (5.37), we can thus directly use equation (5.43) to derive the expression of the output wanted signal $s_{o,w}(t)$ lying at angular frequency ω_w. Assuming that $\rho_w \ll \rho_{bl}$, we can write that

$$s_{o,w}(t) = \left(G + \frac{3\alpha_3}{2} \rho_{bl}^2 \right) \rho_w \cos(\omega_w t). \tag{5.98}$$

It is then convenient to express this relationship in terms of the device parameters classically used to characterize the nonlinearity instead of the α_3 parameters. Using equation (5.46) and remembering that G and α_3 have opposite signs to correctly reflect the compression behavior, we can then express the effective gain experienced by the wanted signal, G_e, as

$$G_e(\rho_{bl}) = G \left(1 - 2\frac{\rho_{bl}^2}{\rho_{IIP3}^2} \right). \tag{5.99}$$

We can also express this gain in terms of the signal power ratio instead of the amplitude ratio. Given that these powers are proportional to the square of those CW tone amplitude, we have

$$G_e(P_{bl}) = G\left(1 - 2\frac{P_{bl}}{IIP3}\right), \tag{5.100}$$

with P_{bl} the input power of the CW blocking signal and IIP3 the device IIP3 power. However, in discussing compression behavior it is also of interest to use either the device Psat or the CP rather than the IP3 to allow more straightforward interpretations of the results in terms of blocking signal back-off. Using equation (5.59), we can thus write

$$G_e(P_{bl}) = G\left(1 - \frac{2}{3}\frac{P_{bl}}{IPsat}\right). \tag{5.101}$$

Alternatively, as a function of the device CP1 using equation (5.68), we have

$$G_e(P_{bl}) = G\left(1 - \frac{2}{9}\frac{P_{bl}}{ICP1}\right). \tag{5.102}$$

But in the present case the gain loss on the wanted signal is linked to the compression of the blocking signal. And it is interesting to see that the input power of the blocking signal for which we have 1 dB of gain loss on the weak signal is not equal to the device ICP1. It is therefore useful in practice to define a cross-compression point (CCP) at x dB that corresponds to the level of the CW blocker that leads to a gain loss of x dB on the weak signal. As done for the CP, usually x is chosen equal to 1 and the corresponding CCP1 quantity can be expressed either as an equivalent CW tone amplitude or power. This CCP1 quantity can then be linked to the parameters used up to now to characterize the device nonlinearity. For that, we can use the fact that $P_{bl} = \text{ICCP1}$ (where ICCP stands for input compression point) when the difference between the effective gain experienced by the weak signal and the small-signal gain G is equal to 1 dB. We can thus write

$$20\log_{10}\left(\frac{G}{G_e(\text{ICCP1})}\right) = 1. \tag{5.103}$$

Using equation (5.102), we then get that

$$\text{ICCP1} = \frac{\text{ICP1}}{2} \tag{5.104}$$

or, in dBm units,

$$\text{ICCP1}|_{\text{dBm}} = \text{ICP1}|_{\text{dBm}} - 3. \tag{5.105}$$

Thus, for a given device, the CCP1 is theoretically 3 dB lower than the CP1. This means that the effective gain experienced by a signal going through a device exhibiting compression is more sensitive to the amplitude of a potential blocking signal than to the amplitude of the signal itself. This behavior can be understood by reconsidering equation (5.43). We observe that in the expansion of the x^3 term, the coefficients that weight the wanted signal and the blocking signal in the term that represents the effective gain experienced by the wanted signal are not the same. This difference is the reason for a different dependency of this gain on the signal amplitudes.

From the system point of view, the gain loss associated with cross-compression can lead to an overall degradation of the receiver noise performance (recall the mechanism discussed in Section 4.2.4). This behavior explains the use of both the term "desensitization", as we are faced with a loss of sensitivity, and the term "blocker", as such a strong signal can indeed block the reception of the weak signal. In practice, our third order model remains valid only for input powers less than or equal to the device input saturated power, as discussed in "Odd order nonlinearity and IP3, CP1 or Psat" earlier in this section. We can thus evaluate an order of magnitude for the maximum gain loss that can experience a weak signal by considering a blocking signal amplitude equal to ρ_{IPsat} in equation (5.101). This leads to a maximum gain loss of $20\log_{10}(3) = 9.6$ dB for $P_{\text{bl}} = \text{IPsat}$, as shown in Figure 5.10.

We also recall that our derivations so far are based on a model that is not selective in frequency, as discussed in "Input vs. output IP" earlier in this section. This is unfortunately often not the case in practical implementations as out-of-band blockers can cause desensitization to occur. In that case the small-signal gain experienced by these blockers can be different from that of the wanted signal. This often results from the presence of frequency selective blocks

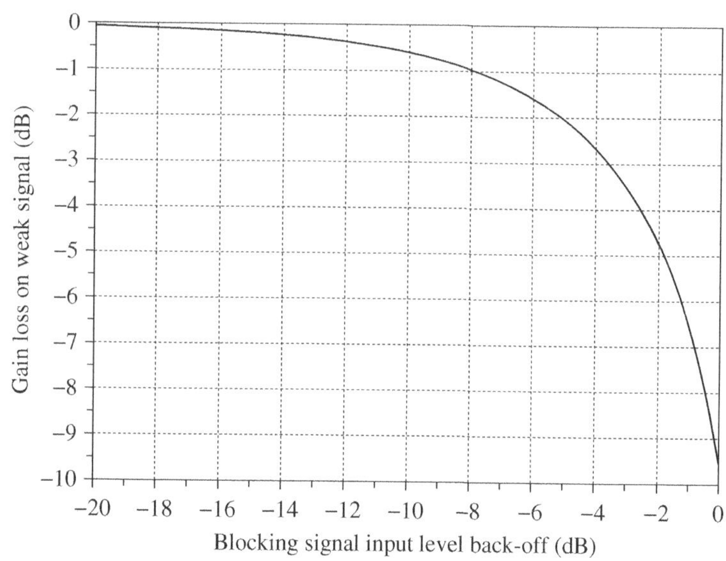

Figure 5.10 Gain loss on the wanted signal due to the compression of a strong CW blocker – When a strong CW interferer is superposed on a weak signal at the input of an amplifier that exhibits saturation, the weak signal experiences a gain loss with respect to the theoretical small-signal gain. Here, the abscissa represents the back-off of the blocking signal regarding the saturated input power of the amplifier, i.e. $P_{\text{bl}}/\text{IPsat}$ expressed in decibels, and the ordinate represents the gain loss with respect to the device small-signal gain using a third order series expansion of the device transfer function as given by equation (5.37).

like matching networks or resonant loads along the data path. In order to take into account such frequency behavior, we can imagine using different Psat or CP1 values depending on whether we are discussing in-band or out-of-band behaviors. However, such quantities are classically often in-band characteristics that apply to the wanted signal. There is thus an interest in expressing the effective gain experienced by the weak signal as a function of the device ICCP1 directly, as this parameter can be defined as a function of the blocker carrier angular frequencies. Using equations (5.102) and (5.104), we can thus write that

$$G_e(P_{bl}) = G\left(1 - \frac{1}{9}\frac{P_{bl}}{\mathrm{ICCP1}}\right), \tag{5.106}$$

with different ICCP1 values depending on the frequency offset of the blocking signal in order to reflect the behavior of the line-up.

In conclusion, we observe that this cross-compression phenomenon is also emphasized when dealing with an amplitude modulated blocker. Given that the effective gain experienced by the weak signal is a function of the blocker amplitude, we expect additional phenomena when the latter amplitude is time varying, i.e. when the blocker is amplitude modulated. In that case, there may be a transfer of amplitude modulation from the blocker to the weak signal rather than a simple gain loss. We can then talk about cross-modulation (XM), as discussed in "Cross-modulation due to odd order nonlinearity" (Section 5.1.3).

Intermodulation

A third and final phenomenon we can focus on is the intermodulation due to either even or odd order nonlinearity. As introduced in former sections for the definition of the IPs, intermodulation tones are generated when a superposition of tones is fed to a nonlinear device. We derived, for instance, the powers and frequency locations of the second order intermodulation tones in the small input power approximation as equation (5.31), and for the third order ones as equation (5.49).

From the system point of view, problems arise when the input blocking signals lie at carrier frequencies such that one of the resulting intermodulation tones folds in the wanted signal band. This occurs, for instance, for the third order intermodulation tones when $|\omega_1 - \omega_2| \approx |\omega_1 - \omega_w|$, as illustrated in Figure 5.11. This kind of configuration is particularly difficult to handle when the input blocking signals lie close to the wanted signal in the frequency domain. There can even be in-band blockers lying within the considered wireless standard receive system band. In that case, it is very unlikely that they can be attenuated before reaching the first downconversion stage of a receiver that allows a channel selection. Thus, all we can do is to ensure that the linearity of the first stages of the line-up is good enough to cope with the unattenuated blockers.

However, although such intermodulation problems are more traditionally associated with the receive side, we have to keep in mind that problems can also be encountered on the transmit side. This is particularly true with architectures based on a frequency planning that uses successive upconversion stages like the heterodyne or variable-IF architectures discussed in Section 8.1.2 and 8.1.3, respectively. In that case, intermodulation tones of the harmonics of the different LOs can indeed fold into frequency bands where specifications are strict in terms of transmitted spectrum mask.

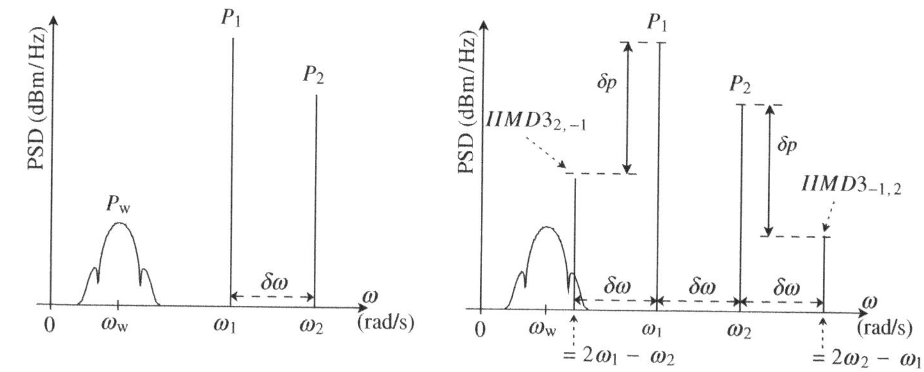

Figure 5.11 Example of unwanted tone generation in the receive band linked to intermodulation between blocking signals – When the frequency offset between two blocking signals at the input of a device that exhibits compression is in the range of the frequency offset between one of the blockers and the wanted signal (left), the resulting third order intermodulation tones can lie in the receive band (right).

5.1.3 Complex Modulated RF Signals

Throughout Section 5.1.2, we focused on the system impacts of AM-AM conversion linked to smooth nonlinearity assuming that only constant amplitude RF signals were present at the input of the nonlinear device. This was in order to introduce in a straightforward way the concepts classically used to characterize the nonlinearity of RF devices, such as IPs and CPs. But we can imagine that potential variations in the instantaneous amplitude of those RF signals can result in additional system impacts. There is thus a need to revisit the derivations of that section for the case of complex modulated bandpass signals.

We first describe the transformation of complex envelopes of RF bandpass signals when going through such nonlinearity. This allows us to derive the analytical expressions required to discuss the corresponding system impacts (AM-demodulation, XM, etc.), which are nothing more than the generalization of those introduced above for constant amplitude signals.

Complex Envelope Transformation

Let us first derive some general results about the transformation of complex envelopes through nonlinear transfer functions of the form as used up to now to model the smooth AM-AM nonlinearity. However, in order to be able to make the link with the derivations in Section 5.1.2 for constant amplitude RF signals, we can consider an input signal $s_i(t)$ that is the sum of two RF bandpass signals, $s_1(t)$ and $s_2(t)$, lying respectively around the carrier angular frequencies ω_1 and ω_2. Such a signal is simply a generalization of the input signal composed of the sum two CW tones as considered for the definition of the IP throughout that previous section.

Recall from the discussion in Chapter 1 that $s_1(t)$ and $s_2(t)$ can be written as a function of their complex envelopes $\tilde{s}_1(t)$ and $\tilde{s}_2(t)$, defined as centered around ω_1 and ω_2 respectively:

$$s_1(t) = \mathrm{Re}\left\{\tilde{s}_1(t)e^{j\omega_1 t}\right\}, \tag{5.107a}$$

$$s_2(t) = \mathrm{Re}\left\{\tilde{s}_2(t)e^{j\omega_2 t}\right\}. \tag{5.107b}$$

The total signal present at the input of the nonlinear device, $s_\mathrm{i}(t)$, can therefore be written as

$$s_\mathrm{i}(t) = \mathrm{Re}\left\{\tilde{s}_1(t)e^{j\omega_1 t} + \tilde{s}_2(t)e^{j\omega_2 t}\right\} \tag{5.108}$$

where $\tilde{s}_1(t)$ and $\tilde{s}_2(t)$ can be written in terms of their real and imaginary parts, $(p_1(t), q_1(t))$ and $(p_2(t), q_2(t))$, or in terms of their instantaneous amplitude and phase, $(\rho_1(t), \phi_1(t))$ and $(\rho_2(t), \phi_2(t))$, as

$$\tilde{s}_1(t) = p_1(t) + jq_1(t) = \rho_1(t)e^{j\phi_1(t)}, \tag{5.109a}$$

$$\tilde{s}_2(t) = p_2(t) + jq_2(t) = \rho_2(t)e^{j\phi_2(t)}. \tag{5.109b}$$

We can thus examine the impact on these expressions of the first nonlinear terms present in the series expansion of the transfer function of a nonlinear device.

Even Order Nonlinearity Case

Let us first consider the even order nonlinearity case. In the smooth AM-AM conversion approximation, we can assume that the device transfer function $F(x)$ can be approximated by its series expansion up to second order given by equation (5.15).

In order to derive the structure of the output signal $s_\mathrm{o}(t)$, we first concentrate on the derivation of the contribution of the term $\alpha_2 x^2$ present in this series expansion. Based on equation (5.108), we can write

$$s_\mathrm{i}^2(t) = \mathrm{Re}\left\{\tilde{s}_1(t)e^{j\omega_1 t} + \tilde{s}_2(t)e^{j\omega_2 t}\right\}\mathrm{Re}\left\{\tilde{s}_1(t)e^{j\omega_1 t} + \tilde{s}_2(t)e^{j\omega_2 t}\right\}. \tag{5.110}$$

We can then use the fact that

$$\forall a \in \mathbb{R}, \forall \tilde{x} \in \mathbb{C}, \quad \mathrm{Re}\{a\tilde{x}\} = a\mathrm{Re}\{\tilde{x}\}, \tag{5.111}$$

to rewrite the expression for $s_\mathrm{i}^2(t)$ as

$$s_\mathrm{i}^2(t) = \mathrm{Re}\left\{\left(\tilde{s}_1(t)e^{j\omega_1 t} + \tilde{s}_2(t)e^{j\omega_2 t}\right)\mathrm{Re}\left\{\tilde{s}_1(t)e^{j\omega_1 t} + \tilde{s}_2(t)e^{j\omega_2 t}\right\}\right\}. \tag{5.112}$$

Using equation (1.5) leads to

$$\begin{aligned} s_\mathrm{i}^2(t) &= \frac{1}{2}\mathrm{Re}\Big\{\left(\tilde{s}_1(t)e^{j\omega_1 t} + \tilde{s}_2(t)e^{j\omega_2 t}\right) \cdot \\ &\quad \left(\tilde{s}_1(t)e^{j\omega_1 t} + \tilde{s}_2(t)e^{j\omega_2 t} + \tilde{s}_1^*(t)e^{-j\omega_1 t} + \tilde{s}_2^*(t)e^{-j\omega_2 t}\right)\Big\} \\ &= \frac{1}{2}\mathrm{Re}\Big\{|\tilde{s}_1(t)|^2 + |\tilde{s}_2(t)|^2 + \tilde{s}_1^2(t)e^{2j\omega_1 t} + \tilde{s}_2^2(t)e^{2j\omega_2 t} \\ &\quad +2\tilde{s}_1(t)\tilde{s}_2(t)e^{j(\omega_1+\omega_2)t} + 2\tilde{s}_1(t)\tilde{s}_2^*(t)e^{j(\omega_1-\omega_2)t}\Big\}. \end{aligned} \tag{5.113}$$

We can thus deduce an expression for the signal $s_o(t) = G\, s_i(t) + \alpha_2 s_i^2(t)$ recovered at the output of our nonlinear device:

$$\begin{aligned} s_o(t) = \mathrm{Re}\Big\{ & \frac{\alpha_2}{2}(|\tilde{s}_1(t)|^2 + |\tilde{s}_2(t)|^2) \\ & + G\big(\tilde{s}_1(t)\mathrm{e}^{\mathrm{j}\omega_1 t} + \tilde{s}_2(t)\mathrm{e}^{\mathrm{j}\omega_2 t}\big) \\ & + \frac{\alpha_2}{2}\big(\tilde{s}_1^2(t)\mathrm{e}^{2\mathrm{j}\omega_1 t} + \tilde{s}_2^2(t)\mathrm{e}^{2\mathrm{j}\omega_2 t}\big) \\ & + \alpha_2\big(\tilde{s}_1(t)\tilde{s}_2(t)\mathrm{e}^{\mathrm{j}(\omega_1+\omega_2)t} + \tilde{s}_1(t)\tilde{s}_2^*(t)\mathrm{e}^{\mathrm{j}(\omega_1-\omega_2)t}\big)\Big\}. \end{aligned} \tag{5.114}$$

This equation is thus simply the generalization of equation (5.20) for complex modulated signals.

We then recover that the output signal is the superposition of a lowpass component and different bandpass signals. We first see that the lowpass signal centered around DC, $s_{\mathrm{o,DC}}(t)$, takes the form

$$s_{\mathrm{o,DC}}(t) = \frac{\alpha_2}{2}(|\tilde{s}_1(t)|^2 + |\tilde{s}_2(t)|^2) = \frac{\alpha_2}{2}\big(\rho_1^2(t) + \rho_2^2(t)\big). \tag{5.115}$$

Then, the RF bandpass part of the output signal is the superposition of signals lying around the angular frequencies of the input signals, their harmonics, or the angular frequencies corresponding to their intermodulation products. Assuming that the complex envelopes $\tilde{s}_{\mathrm{o,1,H1}}(t)$ and $\tilde{s}_{\mathrm{o,2,H1}}(t)$ of the output bandpass signals lying around the fundamental angular frequencies, ω_1 and ω_2, are defined as centered around those angular frequencies, we get that

$$\tilde{s}_{\mathrm{o,1,H1}}(t) = G\,\tilde{s}_1(t), \tag{5.116a}$$

$$\tilde{s}_{\mathrm{o,2,H1}}(t) = G\,\tilde{s}_2(t). \tag{5.116b}$$

In the same way, we can write the complex envelopes of the bandpass signals lying around the harmonic angular frequencies $2\omega_1$ and $2\omega_2$ as

$$\tilde{s}_{\mathrm{o,1,H2}}(t) = \frac{\alpha_2}{2}\tilde{s}_1^2(t), \tag{5.117a}$$

$$\tilde{s}_{\mathrm{o,2,H2}}(t) = \frac{\alpha_2}{2}\tilde{s}_2^2(t). \tag{5.117b}$$

And for the intermodulation bandpass signals $\mathrm{OIMD2}_{l,m}$ lying around the angular frequencies of the form $l\omega_1 + m\omega_2$ with $l, m \in \mathbb{Z}^*$ and $|l| + |m| = 2$ we have

$$\tilde{s}_{\mathrm{OIMD2}_{1,1}}(t) = \alpha_2\tilde{s}_1(t)\tilde{s}_2(t), \tag{5.118a}$$

$$\tilde{s}_{\mathrm{OIMD2}_{1,-1}}(t) = \alpha_2\tilde{s}_1(t)\tilde{s}_2^*(t). \tag{5.118b}$$

The latter expression for $\tilde{s}_{\mathrm{OIMD2}_{1,-1}}(t)$ is valid as long as $\omega_1 > \omega_2$, by equation (5.114). This condition is required in order to match the definition of the complex envelope based on the positive sideband of the corresponding real bandpass signal spectrum as defined in Chapter 1. For $\omega_1 < \omega_2$, we should thus consider the complex conjugate expression for $\tilde{s}_{\text{OIMD}2_{1,-1}}(t)$. However, we can continue to assume the present expression in what follows. Having just a complex conjugate in between the two expressions does not change the characteristics of the corresponding bandpass signal, either in terms of power or spectral shape.

Odd Order Nonlinearity Case

We now turn to the odd order nonlinearity case. Under the smooth AM-AM conversion assumption, we can approximate the device transfer function $F(x)$ by its series expansion up to third order as given by equation (5.37).

In order to derive the structure of the output signal $s_o(t)$ in that case, we first focus on the derivation of the contribution of the term $\alpha_3 x^3$ present in this series expansion. For that, we can simply write that

$$s_i^3(t) = s_i^2(t) s_i(t) \tag{5.119}$$

and use equation (5.113) to expand $s_i^2(t)$. We obtain

$$\begin{aligned} s_i^3(t) = \frac{1}{2}\text{Re}\big\{ &|\tilde{s}_1(t)|^2 + |\tilde{s}_2(t)|^2 + \tilde{s}_1^2(t)e^{2j\omega_1 t} + \tilde{s}_2^2(t)e^{2j\omega_2 t} \\ &+ 2\tilde{s}_1(t)\tilde{s}_2(t)e^{j(\omega_1+\omega_2)t} + 2\tilde{s}_1(t)\tilde{s}_2^*(t)e^{j(\omega_1-\omega_2)t}\big\} \cdot s_i(t). \end{aligned} \tag{5.120}$$

Using in turn equations (5.108), (5.111) and (1.5) eventually yields

$$\begin{aligned} s_o(t) = \text{Re}\Bigg\{ &\left[G + \frac{3\alpha_3}{4}(|\tilde{s}_1(t)|^2 + 2|\tilde{s}_2(t)|^2)\right] \tilde{s}_1(t)e^{j\omega_1 t} \\ &+ \left[G + \frac{3\alpha_3}{4}(|\tilde{s}_2(t)|^2 + 2|\tilde{s}_1(t)|^2)\right] \tilde{s}_2(t)e^{j\omega_2 t} \\ &+ \frac{\alpha_3}{4}\tilde{s}_1^3(t)e^{3j\omega_1 t} + \frac{\alpha_3}{4}\tilde{s}_2^3(t)e^{3j\omega_2 t} \\ &+ \frac{3\alpha_3}{4}\left(\tilde{s}_1^2(t)\tilde{s}_2(t)e^{j(2\omega_1+\omega_2)t} + \tilde{s}_1(t)\tilde{s}_2^2(t)e^{j(2\omega_2+\omega_1)t}\right) \\ &+ \frac{3\alpha_3}{4}\left(\tilde{s}_1^2(t)\tilde{s}_2^*(t)e^{j(2\omega_1-\omega_2)t} + \tilde{s}_1^*(t)\tilde{s}_2^2(t)e^{j(2\omega_2-\omega_1)t}\right)\Bigg\}, \end{aligned} \tag{5.121}$$

which is the generalization of equation (5.43).

We then recover that the output signal is the superposition of RF bandpass signals lying around either the angular frequencies of the input signals, their harmonics, or the angular frequencies corresponding to their intermodulation products. Thus, assuming that the complex envelopes $\tilde{s}_{o,1,\text{H1}}(t)$ and $\tilde{s}_{o,2,\text{H1}}(t)$ of the output bandpass signals lying around the fundamental angular frequencies ω_1 and ω_2 are defined as centered around those angular frequencies, we get that

$$\tilde{s}_{o,1,\text{H1}}(t) = \left[G + \frac{3\alpha_3}{4}\left(\rho_1^2(t) + 2\rho_2^2(t)\right)\right]\tilde{s}_1(t), \tag{5.122a}$$

$$\tilde{s}_{o,2,\text{H1}}(t) = \left[G + \frac{3\alpha_3}{4}\left(\rho_2^2(t) + 2\rho_1^2(t)\right)\right]\tilde{s}_2(t). \tag{5.122b}$$

In the same way, we can write the complex envelopes of the bandpass signals lying around the third harmonic angular frequencies, $3\omega_1$ and $3\omega_2$ respectively, as

$$\tilde{s}_{o,1,\text{H3}}(t) = \frac{\alpha_3}{4}\tilde{s}_1^3(t), \tag{5.123a}$$

$$\tilde{s}_{o,2,\text{H3}}(t) = \frac{\alpha_3}{4}\tilde{s}_2^3(t). \tag{5.123b}$$

And for the intermodulation bandpass signals $\text{OIMD3}_{l,m}$ lying around the angular frequencies of the form $l\omega_1 + m\omega_2$, with $l, m \in \mathbb{Z}^*$ and $|l| + |m| = 3$, we have

$$\tilde{s}_{\text{OIMD3}_{2,1}}(t) = \frac{3\alpha_3}{4}\tilde{s}_1^2(t)\tilde{s}_2(t), \tag{5.124a}$$

$$\tilde{s}_{\text{OIMD3}_{1,2}}(t) = \frac{3\alpha_3}{4}\tilde{s}_1(t)\tilde{s}_2^2(t), \tag{5.124b}$$

$$\tilde{s}_{\text{OIMD3}_{2,-1}}(t) = \frac{3\alpha_3}{4}\tilde{s}_1^2(t)\tilde{s}_2^*(t), \tag{5.124c}$$

$$\tilde{s}_{\text{OIMD3}_{-1,2}}(t) = \frac{3\alpha_3}{4}\tilde{s}_1^*(t)\tilde{s}_2^2(t). \tag{5.124d}$$

The last two expressions for $\tilde{s}_{\text{OIMD3}_{2,-1}}(t)$ and $\tilde{s}_{\text{OIMD3}_{-1,2}}(t)$ are valid as long as $2\omega_1 > \omega_2$ and $2\omega_2 > \omega_1$ respectively (see equation (5.121)). As highlighted for the even order case, this condition is required in order to match the definition of the complex envelope based on the positive sideband of the corresponding real bandpass signal spectrum as defined in Chapter 1. For $2\omega_1 < \omega_2$ and $2\omega_2 < \omega_1$, we should thus consider the complex conjugate expression for $\tilde{s}_{\text{OIMD3}_{2,-1}}(t)$ and $\tilde{s}_{\text{OIMD3}_{-1,2}}(t)$, respectively. However, we can retain the present expressions a complex conjugate if the expression does not change the characteristics of the corresponding bandpass signal, either in terms of power or spectral shape.

Moments of the Instantaneous Amplitude

We now introduce some concepts related to the moments of the instantaneous amplitude of RF bandpass signals. In what follows we often focus on the derivation of average powers of these bandpass signals, mostly with the aim of deriving expressions suitable for carrying out line-up SNR budgets. But assuming the ergodicity and stationarity of the processes we are dealing with, as is classically the case in the field of wireless (see Appendix 2), such long-term average power can be usefully estimated as the expectation of the square of the modulus of the complex envelope of the bandpass signals according to equation (1.64). Referring to the analytical expressions derived so far, we thus see that we necessarily have to cope with the higher order moments of the instantaneous amplitude of the various bandpass signals present at the input of the nonlinear device when performing such an operation. More precisely, we have to cope with their even order moments due to the squaring operation. Thus, it is convenient to introduce the constant Γ_{2k}, defined by

$$\mathbb{E}\{\rho^{2k}\} = (\Gamma_{2k} + 1)\mathbb{E}^k\{\rho^2\}, \tag{5.125}$$

with $k = 1, 2, 3, \ldots$. Equivalently, we can express this relationship as a function of the bandpass signal power P. Given that $P = \mathbb{E}\{\rho^2\}/2$ when working on normalized impedances, we have

$$\mathbb{E}\{\rho^{2k}\} = (\Gamma_{2k} + 1)2^k P^k. \tag{5.126}$$

This definition will be very useful when discussing the system impacts related to the amplitude modulation of bandpass signals. Here, however, we can derive some typical orders of magnitude for Γ_{2k} constants corresponding to classical modulation schemes.

First of all, we observe that for constant amplitude signals, i.e. when $\rho(t) = \rho$ is constant,

$$\mathbb{E}\{\rho^{2k}\} = \rho^{2k} = \mathbb{E}^k\{\rho^2\}. \tag{5.127}$$

Thus, referring to equation (5.125), $\Gamma_{2k} = 0$ for all k. As a second example we can consider a modulation scheme that exhibits a reasonable DR for the variations of the instantaneous amplitude. Obviously the word "reasonable" should be defined, but let us assume that the modified 8PSK modulation of the GSM/EDGE described in Section 1.3.2 fits the bill. In that case, simulations give that

$$\Gamma_{4,8\text{PSK}} = 0.38, \tag{5.128a}$$

$$\Gamma_{6,8\text{PSK}} = 1.20. \tag{5.128b}$$

Finally, let us take the case where the largest DR for the instantaneous amplitude is encountered. This occurs in practice for noise-like bandpass signals that have a Gaussian distribution. By the discussion in Chapter 1, the instantaneous amplitude of those bandpass signals follows a Rayleigh distribution. This is, for instance, the case for most wideband systems such as those based on OFDM. Thus, we can use the moments of the Rayleigh distribution given in Section 1.2.1 to derive the corresponding $\Gamma_{2k,\text{R}}$ constants. Substituting equation (1.114) into equation (5.125), we then get that

$$\Gamma_{2k,\text{R}} = k! - 1. \tag{5.129}$$

So, for instance,

$$\Gamma_{4,\text{R}} = 1, \tag{5.130a}$$

$$\Gamma_{6,\text{R}} = 5. \tag{5.130b}$$

We thus see from these examples that for modulated RF bandpass signals classically encountered in the field of wireless, Γ_{2k} varies from 0 for a constant envelope signal up to $\Gamma_{2k,\mathrm{R}}$ for the extremal case of a noise or noise-like bandpass signal. But given that Γ_{2k} is bounded in practice by $\Gamma_{2k,\mathrm{R}} = k! - 1$, we can write from equation (5.125) that

$$\mathbb{E}\{\rho^{2k}\} \leq k!\mathbb{E}^k\{\rho^2\}. \tag{5.131}$$

On the other hand, due to the convexity of the function $f(x) = x^k$ for $x > 0$, we get from Jensen's inequality that [62]

$$\mathbb{E}^k\{\rho^2\} \leq \mathbb{E}\{\rho^{2k}\}. \tag{5.132}$$

We thus deduce for bandpass signals as encountered in practice in wireless transceivers that

$$\mathbb{E}^k\{\rho^2\} \leq \mathbb{E}\{\rho^{2k}\} \leq k!\mathbb{E}^k\{\rho^2\}. \tag{5.133}$$

This relationship is of particular interest when dealing with two bandpass signals whose powers have different orders of magnitude. Indeed, assuming for instance that $P_1 \ll P_2$ with $P_1 = \mathbb{E}\left\{\rho_1^2\right\}/2$ and $P_2 = \mathbb{E}\left\{\rho_2^2\right\}/2$ when working with normalized impedances, we can write that

$$\mathbb{E}\left\{\rho_1^2\right\} \ll \mathbb{E}\left\{\rho_2^2\right\}. \tag{5.134}$$

But referring to equation (5.133), we get that the above relationship can then be extended to higher order moments of even order of those instantaneous amplitudes. This holds at least as long as this order, and thus $k!$, remains of reasonable value. For instance, for fourth order moments,

$$\mathbb{E}\left\{\rho_1^4\right\} \ll \mathbb{E}\left\{\rho_2^4\right\}. \tag{5.135}$$

As illustrated in subsequent sections, this kind of relationship is useful for simplifying analytical derivations assuming that one of the signals has a much lower power than the other, as classically encountered when faced with a weak wanted signal with a blocking signal.

Impact on Transceivers

Let us now focus on the system impacts of the smooth AM-AM nonlinearity when dealing with bandpass signals that are also amplitude modulated. Most of these system impacts can be seen as generalizations of those detailed above when dealing with constant amplitude RF signals. However, some new phenomena have to be considered in the present case, such as spectral regrowth.

AM-Demodulation Due to Even Order Nonlinearity

Let us first focus on what is classically called the AM-demodulation due to even order nonlinearity. Looking at equation (5.114), the square of the instantaneous amplitudes of the

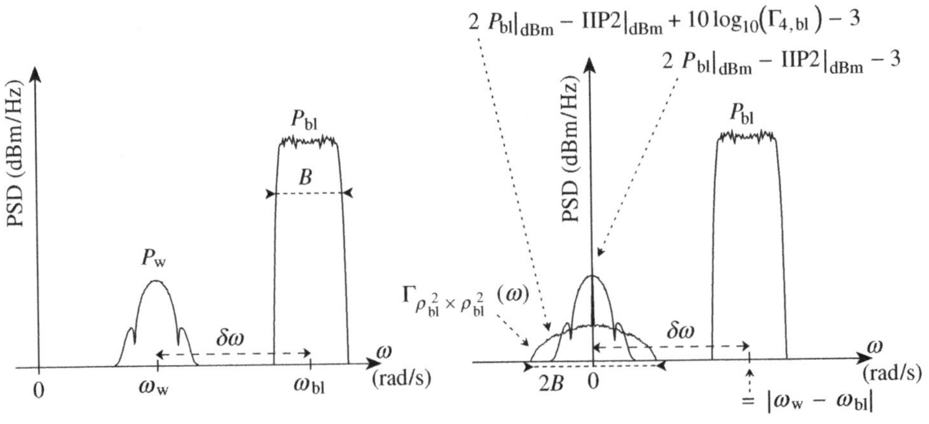

Figure 5.12 PSD of an AM-demodulated baseband signal due to second order AM-AM nonlinearity – Considering the downconversion to baseband of a wanted signal in the presence of a non-constant amplitude modulated blocking signal (left), second order AM-AM nonlinearity at the mixing stage makes the wanted signal superposed both in time and frequency with the AM-demodulated part of the blocking signal (right – equivalent input quantities).

bandpass signals present at the input of a device that exhibits a second order nonlinearity is recovered as centered around DC at its output. This is simply the generalization of the phenomenon of DC offset generation discussed in "DC offset generation due to even order nonlinearity" (Section 5.1.2). But dealing with non-constant amplitude signals, we recover a time varying signal at baseband. This leads to a potential limitation in the SNR when this disturbance is superposed on the wanted signal being processed. This occurs for instance when such even order nonlinearity takes place in the final downconversion stage to baseband of a receiver, as illustrated in Figure 5.12. This problem can be critical in receiver architectures such as direct conversion, as discussed in Section 8.2.1. In that case we traditionally get no filtering of the potential in-band blockers before the direct downconversion to baseband. The receiver performance is thus highly sensitive to the even order nonlinearity of this mixing stage.

In order to evaluate the corresponding SNR limitation, we can consider the situation where a wanted signal is superposed on a blocking signal at the input of such a nonlinear downconversion stage. We can assume that the bandpass signals are represented by their complex envelopes $\tilde{s}_{\text{w}}(t)$ and $\tilde{s}_{\text{bl}}(t)$, defined as centered around the angular frequencies ω_{w} and ω_{bl} respectively, and expressed in polar form as

$$\tilde{s}_{\text{w}}(t) = \rho_{\text{w}}(t)e^{j\phi_{\text{w}}(t)}, \tag{5.136a}$$

$$\tilde{s}_{\text{bl}}(t) = \rho_{\text{bl}}(t)e^{j\phi_{\text{bl}}(t)}. \tag{5.136b}$$

We can thus express the total signal present at the input of the nonlinear device as

$$s_{\text{i}}(t) = \text{Re}\left\{\tilde{s}_{\text{w}}(t)e^{j\omega_{\text{w}}t} + \tilde{s}_{\text{bl}}(t)e^{j\omega_{\text{bl}}t}\right\}. \tag{5.137}$$

Thus, the unwanted signal recovered as centered around DC at the output of the nonlinear device, $s_{o,DC}(t)$, is given by equation (5.115) as

$$s_{o,DC}(t) = \frac{\alpha_2}{2}\left(\rho_w^2(t) + \rho_{bl}^2(t)\right). \quad (5.138)$$

For an easier direct comparison with the input signals, it is convenient to work with equivalent input quantities. We can thus use the small-signal gain G of the device to express the equivalent input disturbance $s_{i,DC}(t)$ as

$$s_{i,DC}(t) = \frac{1}{2}\frac{\alpha_2}{G}\left(\rho_w^2(t) + \rho_{bl}^2(t)\right). \quad (5.139)$$

As usual (see Appendix 2), we can assume the ergodicity and stationarity of the processes we are dealing with. Assuming in addition that we are working with normalized impedances, we can then write the power $P_{s_{i,DC}}$ of $s_{i,DC}(t)$ as

$$P_{s_{i,DC}} = \mathbb{E}\left\{s_{i,DC}^2\right\}. \quad (5.140)$$

As it seems reasonable to also assume independence between the wanted signal and blocking signal modulation schemes, this power can be evaluated from equation (5.139) as

$$P_{s_{i,DC}} = \left(\frac{\alpha_2}{G}\right)^2 \frac{\left(\mathbb{E}\left\{\rho_w^4\right\} + 2\mathbb{E}\left\{\rho_w^2\right\}\mathbb{E}\left\{\rho_{bl}^2\right\} + \mathbb{E}\left\{\rho_{bl}^4\right\}\right)}{4}. \quad (5.141)$$

It is preferable to use the device IIP2 instead of the parameter α_2 in order to get a direct expression in terms of a characteristic of the device. Working with normalized impedances, we can then use equation (5.30a) to write

$$P_{s_{i,DC}} = \frac{1}{2}\frac{1}{\mathrm{IIP2}} \frac{\left(\mathbb{E}\left\{\rho_w^4\right\} + 2\mathbb{E}\left\{\rho_w^2\right\}\mathbb{E}\left\{\rho_{bl}^2\right\} + \mathbb{E}\left\{\rho_{bl}^4\right\}\right)}{4}. \quad (5.142)$$

In order to go further, we recall that the second order nonlinearity of practical implementations is often good enough so that the disturbance generated by the wanted signal itself remains negligible compared to the magnitude of this signal. Problems arise when facing a strong interferer in addition to a weak wanted signal. In that case, the disturbance generated by the interferer can be non-negligible compared to the wanted signal, thus potentially degrading the SNR. Thus, we suppose from now on that the wanted signal power P_w is negligible compared to the blocking signal power P_{bl}. By equation (1.64), we thus assume from now on that

$$\mathbb{E}\left\{\rho_w^2\right\} \ll \mathbb{E}\left\{\rho_{bl}^2\right\}. \quad (5.143)$$

Then by equation (5.133) this relationship can be extended to higher order moments for signals classically encountered in the field of wireless. We can for instance write

$$\mathbb{E}\{\rho_w^4\} \ll \mathbb{E}\{\rho_{bl}^4\}, \tag{5.144}$$

so that equation (5.142) reduces in that case to [63]

$$P_{s_{i,DC}} = \frac{1}{2}\frac{1}{\mathrm{IIP2}}\frac{\mathbb{E}\{\rho_{bl}^4\}}{4}. \tag{5.145}$$

We can then introduce the $\Gamma_{4,bl}$ constant representing the blocking signal fourth power amplitude as defined in equation (5.126). This finally results in

$$P_{s_{i,DC}} = \frac{1}{2}\frac{P_{bl}^2}{\mathrm{IIP2}}(\Gamma_{4,bl} + 1) \tag{5.146}$$

or, expressed in dBm units,

$$P_{s_{i,DC}}|_{dBm} = 2P_{bl}|_{dBm} - \mathrm{IIP2}|_{dBm} + 10\log_{10}(\Gamma_{4,bl} + 1) - 3. \tag{5.147}$$

This last equation is simply the generalization of equation (5.36) for a non-constant envelope input blocking signal. Equation (5.147) reduces to equation (5.36) when dealing with a constant amplitude blocker, i.e. when $\Gamma_4 = 0$.

However, we can go a step further if we observe that the disturbance we are faced with is proportional to the square of the instantaneous amplitude of the blocking signal. This is thus a signal with a constant sign, either always positive or always negative depending on the sign of the proportionality factor, that thus necessarily exhibits a non-vanishing DC term. From a system point of view, we can handle the DC term by using a possible DC offset compensation apparatus (see Section 9.2.2). It is thus of interest to derive the power of the time varying part of this disturbance which needs to be effectively included in an SNR budget. We can estimate the long-term average value of the AM-demodulated signal $\overline{s_{i,DC}}$. Taking the expectation of equation (5.139) and assuming that $\mathbb{E}\{\rho_w^2\} \ll \mathbb{E}\{\rho_{bl}^2\}$ yields

$$\overline{s_{i,DC}} = \mathbb{E}\{s_{i,DC}\} = \frac{\alpha_2}{G}\frac{\mathbb{E}\{\rho_{bl}^2\}}{2} = \frac{\alpha_2}{G}P_{bl}. \tag{5.148}$$

Using again equation (5.30a) to introduce the device IIP2, we finally get that

$$P_{\overline{s_{i,DC}}} = \overline{s_{i,DC}}^2 = \frac{1}{2}\frac{P_{bl}^2}{\mathrm{IIP2}}, \tag{5.149}$$

or, in dBm units,

$$P_{\overline{s_{i,DC}}}|_{dBm} = 2P_{bl}|_{dBm} - \mathrm{IIP2}|_{dBm} - 3. \tag{5.150}$$

This last relationship is in fact precisely equation (5.36). This means that although the blocking signal is amplitude modulated, the DC part of the generated baseband disturbance has the same level as would have been the case if the blocker were a constant amplitude signal. This means in particular that this DC level depends only on the blocking signal power and not on its amplitude statistics. Consequently, the instantaneous amplitude statistics of the input bandpass signals impact only the time varying part of the generated disturbance. To evaluate the power of this noise term part, we can simply write $s_{\text{i,DC}}(t)$ as the sum of an equivalent input DC component, $\overline{s_{\text{i,DC}}}$, and a centered process that represents the equivalent input noise part we are looking for, denoted by $n_{\text{i,DC}}(t)$ below. Given that

$$s_{\text{i,DC}}(t) = \overline{s_{\text{i,DC}}} + n_{\text{i,DC}}(t), \tag{5.151}$$

we can then estimate the power of $n_{\text{i,DC}}(t)$ as

$$\begin{aligned} P_{n_{\text{i,DC}}} &= \mathbb{E}\{(s_{\text{i,DC}} - \overline{s_{\text{i,DC}}})^2\} \\ &= P_{s_{\text{i,DC}}} - 2\,\overline{s_{\text{i,DC}}}\,\mathbb{E}\{s_{\text{i,DC}}\} + \overline{s_{\text{i,DC}}}^2 \\ &= P_{s_{\text{i,DC}}} - \overline{s_{\text{i,DC}}}^2. \end{aligned} \tag{5.152}$$

Using equations (5.146) and (5.149), we then get

$$P_{n_{\text{i,DC}}} = \frac{1}{2}\frac{P_{\text{bl}}^2}{\text{IIP2}}\Gamma_{4,\text{bl}}. \tag{5.153}$$

As expected, the power of this centered process vanishes when the blocking signal is constant amplitude modulated, i.e. when $\Gamma_{4,\text{bl}} = 0$. When this component is not null, we can express its power in dBm units, for instance, as

$$P_{n_{\text{i,DC}}}|_{\text{dBm}} = 2P_{\text{bl}}|_{\text{dBm}} - \text{IIP2}|_{\text{dBm}} + 10\log_{10}(\Gamma_{4,\text{bl}}) - 3. \tag{5.154}$$

The latter relationships allow us to predict the remaining noise power as recovered at the output of a DC offset cancellation scheme, for instance. Nevertheless, other parameters can be considered in order to achieve greater accuracy with regard to the system impacts of this additional disturbance. We can mention, for instance, the spectral shape of this noise term. As detailed in more depth in "Spectral regrowth" later in this section, having this noise term proportional to a power of the input blocker instantaneous amplitude leads to a spectral regrowth of this signal with respect to the initial blocker modulation bandwidth. This can lead to a fraction of this additional noise power lying outside the received signal bandwidth so that it can be partly filtered out by a channel or matched receive filter. However, this can only be evaluated on a case by case basis. We can also mention that statistics of this noise term depend on those of the blocking signal modulation considered. These may deviate from the AWGN traditionally used to benchmark the performance of the baseband algorithms. That said, the present derivations often allow us to obtain good order of magnitude predictions in order to initiate the dimensioning of a line-up.

Revisiting Intermodulation and Harmonic Tones

Let us now examine the characteristics of the intermodulation tones generated when at least two amplitude modulated bandpass signals centered on different carrier angular frequencies go through a nonlinear device. Indeed, as can be seen by comparing equations (5.114) and (5.121) with their counterparts derived in Section 5.1.2 in the constant amplitude case, we now have intermodulation tones that are amplitude modulated.

To evaluate the impact of this new behavior, let us first focus on the even order nonlinearity case and reconsider the second order intermodulation tones. We suppose that we are dealing with a bandpass signal composed of the superposition of two blockers characterized by their complex envelopes, $\tilde{s}_1(t)$ and $\tilde{s}_2(t)$, defined as centered around the angular frequencies ω_1 and ω_2, at the input of the nonlinear device. Recalling that under the smooth AM-AM nonlinearity assumption the device transfer function $F(x)$ can be approximated by equation (5.15), the complex envelopes of the second order intermodulation tones recovered at the output of the device are given by equation (5.118) when assumed defined as centered around $\omega_1 + \omega_2$, or $\omega_1 - \omega_2$ when $\omega_1 > \omega_2$. However, the consequences on the structure of the complex envelopes of the latter condition for the angular frequency order are of no importance for the derivation of the IMD tone power. The forthcoming results are thus general whatever the frequency configuration of the input signals. As usual, it is then convenient to work with equivalent input quantities. We can thus apply the small-signal gain G of the device to derive the complex envelopes of the equivalent input IMD tones as

$$\tilde{s}_{\text{IIMD2}_{1,1}}(t) = \frac{\alpha_2}{G}\tilde{s}_1(t)\tilde{s}_2(t), \tag{5.155a}$$

$$\tilde{s}_{\text{IIMD2}_{1,-1}}(t) = \frac{\alpha_2}{G}\tilde{s}_1(t)\tilde{s}_2^*(t), \tag{5.155b}$$

where the subscript $\text{IIMD2}_{l,m}$ refers to the tone lying as centered around the angular frequency $l\omega_1 + m\omega_2$, with $l, m \in \mathbb{Z}^*$ and $|l| + |m| = 2$. As we have done throughout this book so far, we can assume the ergodicity and stationarity of the processes we are dealing with (see Appendix 2). Assuming in addition that we are working with normalized impedances, we can then express the power of these intermodulation tones as half the expectation of the square modulus of their complex envelopes:

$$IIMD2_{1,1} = \frac{1}{2}\left(\frac{\alpha_2}{G}\right)^2 \mathbb{E}\{\rho_1^2\rho_2^2\}, \tag{5.156a}$$

$$IIMD2_{1,-1} = \frac{1}{2}\left(\frac{\alpha_2}{G}\right)^2 \mathbb{E}\{\rho_1^2\rho_2^2\}. \tag{5.156b}$$

The two second order intermodulation tones thus have the same power. Moreover, assuming independence between the statistics of the two input blockers, we can write that

$$\begin{aligned} IIMD2_{1,1} = IIMD2_{1,-1} &= 2\left(\frac{\alpha_2}{G}\right)^2 \frac{\mathbb{E}\{\rho_1^2\}}{2}\frac{\mathbb{E}\{\rho_2^2\}}{2} \\ &= 2\left(\frac{\alpha_2}{G}\right)^2 P_1 P_2, \end{aligned} \tag{5.157}$$

with P_1 and P_2 the power of the corresponding bandpass signals at the nonlinear device input. Finally, we can use equation (5.30a) to express this relationship in terms of the device IIP2 as

$$IIMD2_{1,1} = IIMD2_{1,-1} = \frac{P_1 P_2}{\text{IIP2}}. \tag{5.158}$$

Thus the power of the second order intermodulation tones expressed in dBm units and their corresponding central angular frequencies takes the simple form

$$(P_1|_{\text{dBm}} + P_2|_{\text{dBm}} - \text{IIP2}|_{\text{dBm}}, |\omega_1 \pm \omega_2|). \tag{5.159}$$

This is precisely the same expression as obtained when dealing with constant amplitude blockers. Thus the statistics of the instantaneous amplitude of the input blockers do not impact the power of the resulting second order intermodulation tones. This behavior is in fact normal. The instantaneous amplitude of the intermodulation tones is indeed linked to the product of the instantaneous amplitude of the two input blockers (see equation (5.155)). It therefore remains linearly dependent on each of those amplitudes, so that no distortion of their statistical characteristics occurs when they are multiplied together under the assumption of statistical independence. We obtain the same second order intermodulation tone power as in the CW input tones case.

This conclusion is not valid in the odd order case. Taking the same superposition of two bandpass blockers at the input of a device that exhibits compression, i.e. whose transfer function $F(x)$ is given by equation (5.37) in the smooth AM-AM approximation, we now expect that the amplitude of the generated third order intermodulation tones depends on the square of the amplitude of one of the input blockers while remaining linearly dependent on the amplitude of the other. This is indeed what can be seen from the expression for the complex envelopes given by equation (5.124) for the tones lying around $2\omega_1 - \omega_2$ and $2\omega_2 - \omega_1$ when $2\omega_1 > \omega_2$ and $2\omega_2 > \omega_1$, respectively. However, in the present case again the consequences on the structure of the complex envelopes of the latter condition for the angular frequency order are of no importance for the derivation of the IMD tone power, as discussed above. We can thus directly apply the small-signal gain G of the device to those expressions to derive an expression valid in any configuration for the power of the equivalent input IMD3 tones:

$$IIMD3_{2,-1} = \frac{1}{2}\left(\frac{3}{4}\frac{\alpha_3}{G}\right)^2 \mathbb{E}\{\rho_1^4 \rho_2^2\}, \tag{5.160a}$$

$$IIMD3_{-1,2} = \frac{1}{2}\left(\frac{3}{4}\frac{\alpha_3}{G}\right)^2 \mathbb{E}\{\rho_1^2 \rho_2^4\}. \tag{5.160b}$$

Assuming independence of the modulating schemes of the two signals, we get

$$IIMD3_{2,-1} = \left(\frac{3}{2}\frac{\alpha_3}{G}\right)^2 \frac{\mathbb{E}\{\rho_1^4\}}{4}\frac{\mathbb{E}\{\rho_2^2\}}{2}, \tag{5.161a}$$

$$IIMD3_{-1,2} = \left(\frac{3}{2}\frac{\alpha_3}{G}\right)^2 \frac{\mathbb{E}\{\rho_1^2\}}{2}\frac{\mathbb{E}\{\rho_2^4\}}{4}; \tag{5.161b}$$

or in terms of the device third order IP power using equation (5.48a),

$$IIMD3_{2,-1} = \frac{1}{\mathrm{IIP3}^2}\frac{\mathbb{E}\{\rho_1^4\}}{4}\frac{\mathbb{E}\{\rho_2^2\}}{2}, \tag{5.162a}$$

$$IIMD3_{-1,2} = \frac{1}{\mathrm{IIP3}^2}\frac{\mathbb{E}\{\rho_1^2\}}{2}\frac{\mathbb{E}\{\rho_2^4\}}{4}. \tag{5.162b}$$

As expected, we now have in each of those expressions the fourth order moment of the instantaneous amplitude of one of the input blockers. To take this into account, we can use the $\Gamma_{4,1}$ and $\Gamma_{4,2}$ constants defined by equation (5.125). The power of the intermodulation tones can then be expressed as

$$IIMD3_{2,-1} = \frac{P_1^2 P_2}{\mathrm{IIP3}^2}(\Gamma_{4,1}+1), \tag{5.163a}$$

$$IIMD3_{-1,2} = \frac{P_1 P_2^2}{\mathrm{IIP3}^2}(\Gamma_{4,2}+1), \tag{5.163b}$$

or, in dBm units at their corresponding central angular frequencies, as

$$(2P_1|_{\mathrm{dBm}} + P_2|_{\mathrm{dBm}} - 2\mathrm{IIP3}|_{\mathrm{dBm}} + 10\log_{10}(\Gamma_{4,1}+1), |2\omega_1 - \omega_2|), \tag{5.164a}$$

$$(P_1|_{\mathrm{dBm}} + 2P_2|_{\mathrm{dBm}} - 2\mathrm{IIP3}|_{\mathrm{dBm}} + 10\log_{10}(\Gamma_{4,2}+1), |2\omega_2 - \omega_1|). \tag{5.164b}$$

As expected, we recover equation (5.50) when $\Gamma_{4,1} = \Gamma_{4,2} = 0$, i.e. when the input blockers are CW. But, if these input signals are effectively amplitude modulated, we observe that the power of the resulting intermodulation tones increases compared to the CW case. Indeed, by the discussion in "Moments of the instantaneous amplitude" earlier in this section, the Γ_4 constants can vary from 0 for a CW signal up to 1 for a signal with an instantaneous amplitude that has a Rayleigh distribution. This means that in practice, the power of the intermodulation tone can increase by 3 dB compared to the prediction of the CW formulation. However, for the sake of accuracy in this regard, we need to take into account the spectral shape of the resulting intermodulation tones. This spectrum is indeed modified compared to that of the input blockers due to the impact of the spectral regrowth, as detailed in "Spectral regrowth" later in this section. This means that we can have a non-negligible fraction of the intermodulation tone power lying outside the bandwidth of the received wanted signal as illustrated in Figure 5.13. In practice, this effect can compensate for most of the power increase linked to the presence of the Γ_4 constant in the expression for the intermodulation tone power. All in all, the CW formulation is often a good starting point for deriving orders of magnitude.

Obviously, this kind of behavior also remains valid for the harmonic tones (see equations (5.117) and (5.123)). Looking at those expressions, the derivation of the power of the bandpass signal lying around the second harmonic angular frequency involves the Γ_4 constant characterizing the corresponding input bandpass signal and the Γ_6 constant for the power of the bandpass signal lying around the third harmonic angular frequency. By equation (5.34) and (5.50), we expect in the present case again a deviation in the power of those signals

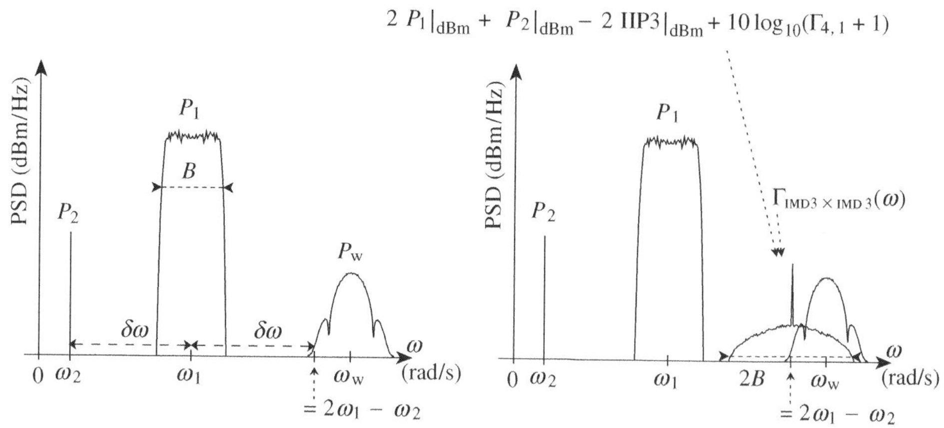

Figure 5.13 IMD3 tone generation due to third order AM-AM nonlinearity with non-constant amplitude input signals – Compared to the case where the input blockers are CW, the amplitude modulation of the input blockers (left) leads to both a potential increase in and a spectral regrowth of the IMD3 tone power compared to the bandwidth of the input blockers (right – equivalent input quantities).

compared to the pure CW case. However, this must be balanced against the spectral regrowth phenomenon which may compensate for this increase when focusing on the power lying in an appropriate frequency band, for instance of the same width as the original signal spectrum.

Cross-Modulation Due to Odd Order Nonlinearity

Another problem that is emphasized when dealing with amplitude modulated blocking signals is that of desensitization due to compression. By the discussion in "Desensitization due to odd order nonlinearity and CCP1" (Section 5.1.2), the small-signal gain experienced by a weak wanted signal depends on the instantaneous amplitude of a strong blocker that is superposed on it at the input of a device that exhibits compression. In the case of a constant amplitude blocker, this behavior leads to a constant gain loss. But when the instantaneous amplitude of the blocker is time varying, the gain experienced by the weak signal varies accordingly. Here, we can thus clearly talk about AM-AM conversion. More precisely, as the blocker amplitude modulation is somehow *transferred* to the wanted signal, we talk about cross-modulation (XM).

In order to put numbers on this phenomenon, we can assume that the wanted and blocking bandpass signals are characterized by their complex envelopes, $\tilde{s}_{\text{w}}(t)$ and $\tilde{s}_{\text{bl}}(t)$ respectively, of the form

$$\tilde{s}_{\text{w}}(t) = \rho_{\text{w}}(t)e^{j\phi_{\text{w}}(t)}, \tag{5.165a}$$

$$\tilde{s}_{\text{bl}}(t) = \rho_{\text{bl}}(t)e^{j\phi_{\text{bl}}(t)}, \tag{5.165b}$$

and defined as centered around the angular frequencies ω_{w} and ω_{bl} respectively. Supposing that in the smooth AM-AM conversion approximation the device transfer function $F(x)$ can be approximated by equation (5.37), the complex envelope of the wanted signal at the device output, $\tilde{s}_{o,w}(t)$, i.e. of the bandpass signal centered around ω_w, can be written from equation (5.122a) as

$$\tilde{s}_{o,w}(t) = G\left[1 + \frac{3}{4}\frac{\alpha_3}{G}\left(\rho_w^2(t) + 2\rho_{bl}^2(t)\right)\right]\tilde{s}_w(t). \tag{5.166}$$

It would be preferable to express this relationship in terms of the device IIP3 instead of the α_3 parameter. Assuming that we are working on normalized impedances and remembering that G and α_3 have opposite signs to physically describe the compression behavior, we can then directly use equation (5.48a) to write that

$$\tilde{s}_o(t) = G\left(1 - \frac{\rho_{bl}^2(t)}{\text{IIP3}} - \frac{\rho_w^2(t)}{2\text{IIP3}}\right)\tilde{s}_w(t). \tag{5.167}$$

In order to derive the resulting SNR limitation, we have to express this output signal as the sum of a term proportional to the wanted signal and a term that can be considered as additive noise. The additive term needs to be at least *uncorrelated* with the wanted signal when the signals are considered at the same time. Here, we observe that the non-correlation has to be interpreted as between the wanted signal and the bandpass RF signals we are dealing with. However, as discussed in Appendix 1, the non-correlation between the complex envelopes defined around the same center frequency, as here, is a sufficient condition for the corresponding bandpass signals to be uncorrelated. There is even an equivalence when dealing with stationary bandpass processes, at least up to second order, that is the case for signals we classically deal with in wireless transceivers (see Appendix 2). Thus, we expect to be able to decompose $\tilde{s}_o(t)$ as

$$\tilde{s}_o(t) = G_e\tilde{s}_w(t) + \tilde{n}(t), \tag{5.168}$$

with G_e the effective gain experienced by the wanted signal while flowing through the device and $\tilde{n}(t)$ the complex envelope of the additional bandpass term generated by the nonlinearity. As by definition this last term is assumed uncorrelated with $\tilde{s}_w(t)$, we can determine G_e by taking the correlation of both members of the above equation with $\tilde{s}_w(t)$. We can thus write that

$$G_e = \frac{\mathbb{E}\{\tilde{s}_{o,t}\tilde{s}_{w,t}^*\}}{\mathbb{E}\{\tilde{s}_{w,t}\tilde{s}_{w,t}^*\}} = G\frac{\mathbb{E}\left\{\left(1 - \frac{\rho_{bl,t}^2}{\text{IIP3}} - \frac{\rho_{w,t}^2}{2\text{IIP3}}\right)\rho_{w,t}^2\right\}}{\mathbb{E}\left\{\rho_{w,t}^2\right\}}. \tag{5.169}$$

But, assuming the independence of the modulating complex envelope processes of the wanted and blocking signals, we get

$$G_e = G\frac{\left(\mathbb{E}\left\{\rho_{w,t}^2\right\} - \frac{\mathbb{E}\left\{\rho_{bl,t}^2\right\}\mathbb{E}\left\{\rho_{w,t}^2\right\}}{\text{IIP3}} - \frac{\mathbb{E}\left\{\rho_{w,t}^4\right\}}{2\text{IIP3}}\right)}{\mathbb{E}\left\{\rho_{w,t}^2\right\}}, \tag{5.170}$$

and thus

$$G_{\mathrm{e}} = G\left(1 - \frac{\mathbb{E}\{\rho_{\mathrm{bl},t}^2\}}{\mathrm{IIP3}} - \frac{1}{2\mathrm{IIP3}}\frac{\mathbb{E}\{\rho_{\mathrm{w},t}^4\}}{\mathbb{E}\{\rho_{\mathrm{w},t}^2\}}\right). \tag{5.171}$$

Using the $\Gamma_{4,\mathrm{w}}$ constant, defined by equation (5.125), which characterizes the statistics of the wanted bandpass signal instantaneous amplitude, we get

$$G_{\mathrm{e}} = G\left(1 - \frac{\mathbb{E}\{\rho_{\mathrm{bl},t}^2\}}{\mathrm{IIP3}} - \frac{\mathbb{E}\{\rho_{\mathrm{w},t}^2\}}{2\mathrm{IIP3}}(\Gamma_{4,\mathrm{w}} + 1)\right). \tag{5.172}$$

In order to go further, we assume that we are dealing in practice with a wanted signal that is weak compared to the blocker. We thus assume that $P_{\mathrm{w}} \ll P_{\mathrm{bl}}$, or that $\mathbb{E}\{\rho_{\mathrm{w}}^2\} \ll \mathbb{E}\{\rho_{\mathrm{bl}}^2\}$ according to equation (1.64). This means that the above equation finally reduces to

$$G_{\mathrm{e}} = G\left(1 - \frac{\mathbb{E}\{\rho_{\mathrm{bl},t}^2\}}{\mathrm{IIP3}}\right) \tag{5.173}$$

or, assuming stationarity of the modulating process,

$$G_{\mathrm{e}} = G\left(1 - \frac{\mathbb{E}\{\rho_{\mathrm{bl}}^2\}}{\mathrm{IIP3}}\right) = G\left(1 - 2\frac{P_{\mathrm{bl}}}{\mathrm{IIP3}}\right) \tag{5.174}$$

as we are working on normalized impedances. Thus the effective gain experienced by the wanted signal exactly matches the gain loss of the desensitization case given by equation (5.100). This means that with our amplitude modulated blocking signal we add one problem to another. We have the same desensitization effect as if the blocker had a constant amplitude and the same power, but we are now also faced with the generation of an additional noise signal that adds to the wanted signal. Substituting equations (5.167) and (5.174) into equation (5.168), we can then derive the complex envelope of this noise component, when defined as centered around ω_{w}, as

$$\tilde{n}(t) = G\left(\frac{\mathbb{E}\{\rho_{\mathrm{bl}}^2\} - \rho_{\mathrm{bl}}^2(t)}{\mathrm{IIP3}} - \frac{\rho_{\mathrm{w}}^2(t)}{2\mathrm{IIP3}}\right)\tilde{s}_{\mathrm{w}}(t). \tag{5.175}$$

Looking at this equation, this complex envelope is the sum of two terms. The first, proportional to $\left(\rho_{\mathrm{bl}}^2(t) - \mathbb{E}\{\rho_{\mathrm{bl}}^2\}\right)\tilde{s}_{\mathrm{w}}(t)$, thus depends on the blocker statistic. This is not the case for the second, which is proportional to $\rho_{\mathrm{w}}^2(t)\tilde{s}_{\mathrm{w}}(t)$. In fact, anticipating the discussion in "Nonlinear EVM due to odd order nonlinearity" later in this section, we can say that this second term is linked to the nonlinear EVM that is generated by the compression of the wanted signal itself. But, assuming for now that the wanted signal remains in the linear range of the device while the blocker enters the compression area, we expect that this last term will remain of negligible magnitude compared to the pure XM term involving the blocking signal when deriving the power of $\tilde{n}(t)$.

This is indeed the behavior we will encounter when evaluating the power $P_n = \mathbb{E}\{|\tilde{n}|^2\}/2$ of this centered process as required to estimate the SNR limitation resulting from the XM phenomenon. Assuming independence of the modulating processes of both the wanted and blocking signals as well as their stationarity, we can write from equation (5.175) that

$$\begin{aligned} P_{\mathrm{n}} &= \frac{1}{2}\frac{G^2}{\mathrm{IIP3}^2}\mathbb{E}\left\{\left[\left(\mathbb{E}\{\rho_{\mathrm{bl}}^2\} - \rho_{\mathrm{bl}}^2\right) - \frac{1}{2}\rho_{\mathrm{w}}^2\right]^2 \rho_{\mathrm{w}}^2\right\} \\ &= \frac{1}{2}\frac{G^2}{\mathrm{IIP3}^2}\left[\mathbb{E}\{\rho_{\mathrm{w}}^2\}\mathbb{E}\left\{\left(\mathbb{E}\{\rho_{\mathrm{bl}}^2\} - \rho_{\mathrm{bl}}^2\right)^2\right\}\right. \\ &\left.+\frac{1}{4}\mathbb{E}\{\rho_{\mathrm{w}}^6\} - \mathbb{E}\{\rho_{\mathrm{w}}^4\}\mathbb{E}\left\{\mathbb{E}\{\rho_{\mathrm{bl}}^2\} - \rho_{\mathrm{bl}}^2\right\}\right]. \end{aligned} \tag{5.176}$$

But recalling the discussion in "Moments of the instantaneous amplitude" earlier in this section, and more precisely equation (5.133), the assumption that $\mathbb{E}\{\rho_{\mathrm{w}}^2\} \ll \mathbb{E}\{\rho_{\mathrm{bl}}^2\}$ can be generalized up to the sixth order moment for practical modulation schemes. Thus, only the first term in the above derivation remains significant and the nonlinear EVM term linked to the compression of the wanted signal can be neglected. The pure XM noise power can thus be expressed as

$$P_{\mathrm{n}} = \frac{1}{2}G^2\frac{\mathbb{E}\{\rho_{\mathrm{w}}^2\}}{\mathrm{IIP3}^2}\left(\mathbb{E}\{\rho_{\mathrm{bl}}^4\} - \mathbb{E}^2\{\rho_{\mathrm{bl}}^2\}\right). \tag{5.177}$$

Now using the constant that characterizes the blocking signal instantaneous amplitude fourth order statistics, $\Gamma_{4,\mathrm{bl}}$, as defined by equation (5.125), we finally get that

$$P_{\mathrm{n}} = 4G^2\frac{P_{\mathrm{w}}P_{\mathrm{bl}}^2}{\mathrm{IIP3}^2}\Gamma_{4,\mathrm{bl}}. \tag{5.178}$$

Looking at this equation, we confirm that the XM noise power vanishes when the blocking signal has a constant instantaneous amplitude, i.e. when $\Gamma_{4,\mathrm{bl}} = 0$. When this is not the case, we can evaluate the resulting SNR at the device output as

$$SNR = \frac{G_{\mathrm{e}}^2 P_{\mathrm{w}}}{P_{\mathrm{n}}} = \frac{1}{\Gamma_{4,\mathrm{bl}}}\left(\frac{\mathrm{IIP3}}{2P_{\mathrm{bl}}} - 1\right)^2. \tag{5.179}$$

We can in fact rearrange this equation to get a more straightforward relationship between the device IIP3, P_{bl} and the resulting SNR. As the IIP3 is a tone power and thus a positive quantity, for reasonable SNR values, we have to keep the positive square root for the above equation so that we get [63]

$$\mathrm{IIP3} = 2P_{\mathrm{bl}}\left(1 + \sqrt{SNR\,\Gamma_{4,\mathrm{bl}}}\right). \tag{5.180}$$

This relationship gives a limitation in the achievable SNR due to the IP3 of the device in the presence of a strong blocking signal that is amplitude modulated. As expected, the resulting SNR does not depend on the wanted signal power as long as the assumption $P_w \ll P_{bl}$ we used for the derivation is valid. However, we observe that as long as this assumption is valid, the SNR achieved remains constant *whatever* the input level of the wanted signal. This is an important difference compared to the desensitization phenomenon that impacts mainly the receiver noise figure and is thus critical for low input powers for the wanted signal only. This is not the case with the XM that induces a distortion component whose power follows the power of the wanted signal, and that can thus be considered as a multiplicative noise component.

It is interesting to observe that the same statistical description of the instantaneous amplitude of the blocking signal through its Γ_4 constant can describe the impact of this blocker in both the AM-demodulation case and the XM case, even though one derivation is based on a second order series expansion of the device transfer function and the other on a third order expansion. This is in fact due to the term of interest involved in the description of the cross-modulation phenomenon that is proportional to the square of the blocking signal amplitude, and that is linear in the wanted signal amplitude. The sum of the exponents is 3, but it involves only the second order statistics of the blocking signal amplitude, as in the AM-demodulation case. This is a difference compared to the nonlinear EVM generation case described in the next section which involves the third order statistics of the input signal amplitude.

It is also of interest to express the SNR limitation in terms of the device Psat. This allows us to express the result in terms of the blocking signal back-off at the device input. Using equation (5.59), we can rewrite equation (5.179) as

$$SNR = \frac{1}{\Gamma_{4,\text{bl}}} \left(\frac{3}{2} \frac{\text{IPsat}}{P_{\text{bl}}} - 1 \right)^2. \tag{5.181}$$

Thus the limit cases predicted by our model are:

$$SNR = \begin{cases} \infty & \text{when } P_{\text{bl}} = 0, \\ \dfrac{1}{4} \dfrac{1}{\Gamma_{4,\text{bl}}} & \text{when } P_{\text{bl}} = \text{IPsat}. \end{cases} \tag{5.182}$$

Two plots of this achieved SNR due to XM are shown in Figure 5.14 for the modified 8PSK modulation of the GSM/EDGE standard and for the limit case of a blocking signal with an amplitude modulation scheme that can be approximated by the Rayleigh distribution as encountered for instance with OFDM signals. The difference between the plots is due to the Γ_4 constants: in the 8PSK case, we have $\Gamma_{4,8\text{PSK}} = 0.38$ as given by equation (5.128a); and in the noise-like case, we have $\Gamma_{4,\text{R}} = 1$ as given by equation (5.130a). As expected, the wider the amplitude variations of the blocker, the higher the power of the generated XM noise term, and thus the lower the SNR achieved.

However, in the present case also we should consider the spectral shape of this generated noise term. Indeed, as discussed in more depth in "Spectral regrowth" later in this section, some spectral regrowth can be considered so that part of the XM noise power lies outside the overall wanted signal bandwidth. This has to be evaluated on a case by case basis as well as

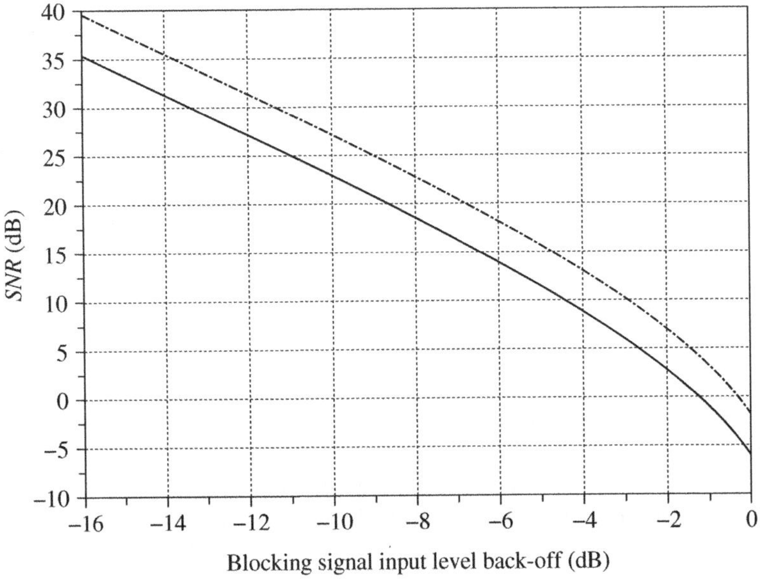

Figure 5.14 SNR limitation due to cross-modulation in the smooth AM-AM nonlinearity approximation – Due to compression, the amplitude modulation of a blocking signal is transfered to the wanted signal. Such XM results in a SNR degradation given by equation (5.181). The SNR limitation, which gets worse when the blocker reaches the compression area of the device, i.e. when its back-off P_{bl}/IPsat tends to 0 dB, is higher when the blocker has a noise-like behavior, i.e. an amplitude that has a Rayleigh distribution with a Γ_4 constant given by equation (5.130a) (solid), than when it has a smoother amplitude modulation like the modified 8PSK of the GSM/EDGE which has a Γ_4 constant given by equation (5.128a) (dot-dashed).

for the exact impact on the performance of the baseband algorithms of the noise term statistics linked to those of the blocking signal modulation scheme. We also need to mention that due to this spectral regrowth phenomenon, there are other configurations in addition to the ones discussed here that lead to a receiver SNR limitation due to such XM. This is the configuration where we have *two* blockers in addition to the wanted signal, when one of those blockers is close to the wanted signal in the frequency domain. Indeed, due to spectral regrowth, we can have that the noise component generated by the XM of the second blocker on the closest one has part of its spectrum overlapping with the wanted signal one. Hence, a SNR limitation occurs as illustrated in due course.

To conclude, we observe that the Psat may be not the most suited metric to refer the blocker back-off to. Indeed, as discussed in "Desensitization due to odd order nonlinearity and CCP1" (Section 5.1.2), the gain experienced by out-of-band blockers, for instance, may be different from that experienced by the wanted signal due to the presence of frequency selective blocks like matching networks or resonant loads along the data path. Given that the Psat is traditionally a metric related to the in-band behavior of the device, it may be preferable to use the device CCP1 corresponding to the frequency offset of the blocker instead of the in-band Psat value. The above relationships can then be expressed in terms of the device CCP1 using equations (5.68) and (5.104), with different ICCP1 values depending on the frequency offset of the blocking signal in order to reflect the behavior of the line-up.

Nonlinear EVM Due to Odd Order Nonlinearity

A final phenomenon introduced with constant amplitude signals and emphasized when dealing with non-constant envelope bandpass signals is the gain loss experienced by such signals when going through a nonlinear device that exhibits compression. We saw in "Odd order nonlinearity and IP3, CP1 or Psat" (Section 5.1.2) that the gain loss experienced by the input signal is proportional to the square of its instantaneous amplitude. We thus get that this gain is constant when dealing with a constant amplitude signal. We therefore see why such constant amplitude bandpass signals are said to be insensitive to odd order nonlinearity, i.e. to compression. However, at this stage we observe that in that case the time domain signal recovered at the output of the nonlinear device looks like the *square wave* version of the constant amplitude sinusoidal signal fed at its input. This behavior, if we refer to the structure of equation (5.52), is seen to be linked to the rise in harmonic tones at the odd multiple of the carrier angular frequency. But if we filter out those harmonics using a zonal bandpass filter centered around the fundamental angular frequency, we can theoretically recover exactly the input constant amplitude sinusoidal signal of interest without any distortion. This is a great theoretical advantage of this kind of phase/frequency only modulated RF signal as we can use for instance saturated PAs with high efficiency even if they behave like hard limiters from a signal processing point of view, as discussed in Section 5.2. But, when considering amplitude modulated bandpass signals, we face the same kind of problem as in the XM case, i.e. a degradation of the amplitude part of the modulation of the input signal due to the time domain variations of the effective gain experienced by the bandpass signal centered around the carrier angular frequency. The difference now is that it is not the modulation carried by a potential blocker that corrupts that of the wanted signal. We have, rather, a self-corruption of the quality of the input signal.

We can thus follow the same strategy as in the XM case and try to derive a limitation of the SNR due to this self AM-AM conversion. Thus, let us consider a single wanted signal at the input of a nonlinear device that exhibits compression. We assume that this bandpass signal is described by the instantaneous amplitude and phase $(\rho_{\mathrm{w}}(t), \phi_{\mathrm{w}}(t))$ of its complex envelope, $\tilde{s}_{\mathrm{w}}(t)$, defined as centered around the carrier angular frequency ω_{w}:

$$\tilde{s}_{\mathrm{w}}(t) = \rho_{\mathrm{w}}(t)\mathrm{e}^{\mathrm{j}\phi_{\mathrm{w}}(t)}. \tag{5.183}$$

We can then continue to assume that in the smooth AM-AM nonlinearity approximation the device transfer function $F(x)$ can be approximated by the third order series expansion given by equation (5.37). Thus, the complex envelope of the wanted signal at the device output, $\tilde{s}_{\mathrm{o,w}}(t)$, i.e. the output complex envelope centered around ω_{w}, can be written from equation (5.122a), supposing that the second input signal considered in that derivation is now null. We obtain

$$\tilde{s}_{\mathrm{o,w}}(t) = G\left(1 + \frac{3}{4}\frac{\alpha_3}{G}\rho_{\mathrm{w}}^2(t)\right)\tilde{s}_{\mathrm{w}}(t). \tag{5.184}$$

It would be preferable to express this relationship in terms of the device IIP3 instead of the α_3 parameter. Assuming that we are working on normalized impedances and remembering that

G and α_3 have opposite signs to physically describe the compression behavior, we can then directly use equation (5.48a) to write that

$$\tilde{s}_{o,w}(t) = G\left(1 - \frac{\rho_w^2(t)}{2IIP3}\right)\tilde{s}_w(t). \tag{5.185}$$

In order to derive the resulting SNR, we need to express the output signal as the sum of a term proportional to the wanted signal and a term that can be considered as an additive noise. The latter term can be considered as a noise component with respect to the wanted signal if it is at least *uncorrelated* with it when considered at the same sample time. Here, we can recall that the non-correlation has to be interpreted as between the wanted signal and the bandpass RF signals we are dealing with. However, as discussed in Appendix 1, the non-correlation between the complex envelopes defined around the same center frequency is equivalent to the non-correlation between the bandpass signals they represent when dealing with stationary processes, at least up to second order, as is classically the case in wireless transceivers as discussed in Appendix 2. Thus, we can achieve our goal with the decomposition of $\tilde{s}_{o,w}(t)$ as

$$\tilde{s}_{o,w}(t) = G_e\tilde{s}_w(t) + \tilde{n}(t), \tag{5.186}$$

where G_e is the effective gain experienced by the wanted signal when flowing through the nonlinear device and $\tilde{n}(t)$ the complex envelope, defined as centered around the same center angular frequency ω_w, of the noise term generated by the nonlinearity. This term is assumed uncorrelated with $\tilde{s}_w(t)$ so that the determination of G_e can be done by taking the correlation of both terms of the above equation with $\tilde{s}_w(t)$. We can thus write that

$$G_e = \frac{\mathbb{E}\left\{\tilde{s}_{o,t}\tilde{s}^*_{w,t}\right\}}{\mathbb{E}\left\{\tilde{s}_{w,t}\tilde{s}^*_{w,t}\right\}} = G\frac{\mathbb{E}\left\{\left(1 - \frac{\rho_{w,t}^2}{2IIP3}\right)\rho_{w,t}^2\right\}}{\mathbb{E}\left\{\rho_{w,t}^2\right\}}. \tag{5.187}$$

Still assuming the stationarity of the modulating process, we get

$$G_e = G\left(1 - \frac{1}{2IIP3}\frac{\mathbb{E}\left\{\rho_w^4\right\}}{\mathbb{E}\left\{\rho_w^2\right\}}\right). \tag{5.188}$$

Then, using the constant that characterizes the wanted signal instantaneous amplitude fourth order statistics, $\Gamma_{4,w}$, as defined by equation (5.125), we finally get that

$$G_e = G\left[1 - \frac{\mathbb{E}\left\{\rho_w^2\right\}}{2IIP3}(\Gamma_{4,w} + 1)\right]. \tag{5.189}$$

This effective gain is in fact nothing more than equation (5.172), assuming that the blocker power is null. Moreover, given that $\mathbb{E}\left\{\rho_w^2\right\}/2$ is nothing more than the power of the input wanted signal P_w when working on normalized impedances, we recover the expression for the effective gain experienced in the constant amplitude case, i.e. equation (5.56), when $\Gamma_{4,\mathrm{w}} = 0$. But, given in general that $\Gamma_{4,\mathrm{w}} \geq 0$, the gain loss is *greater* than in the constant amplitude case if we are dealing with an amplitude modulation. This behavior is in fact related to the creation of an additional noise term whose power can only by retrieved from the input wanted signal in the present case. The resulting output wanted signal power is then necessarily lower in this non-constant amplitude case.

Having derived this effective gain, we can now focus on the expression for the generated noise term to derive the resulting SNR. Indeed, using equations (5.185) and (5.186), we get from the above expression for G_e that

$$\tilde{n}(t) = G\frac{\mathbb{E}\{\rho_\mathrm{w}^2\}}{2\mathrm{IIP3}}\left[(\Gamma_{4,\mathrm{w}}+1) - \frac{\rho_\mathrm{w}^2(t)}{\mathbb{E}\{\rho_\mathrm{w}^2\}}\right]\tilde{s}_\mathrm{w}(t). \tag{5.190}$$

We see that this noise component vanishes for a constant amplitude wanted signal. In that case $\Gamma_{4,\mathrm{w}} = 0$ and $\mathbb{E}\{\rho_\mathrm{w}^2\} = \rho_\mathrm{w}^2$. We thus recover in an alternative way that a constant amplitude bandpass signal fed to a nonlinear device that exhibits compression can theoretically be reconstructed without distortion when considering the sideband centered around the fundamental angular frequency at the device output. Conversely, when $\tilde{n}(t) \neq 0$ we can derive the resulting SNR as

$$SNR = \frac{G_\mathrm{e}^2 P_\mathrm{w}}{P_\mathrm{n}} = G_\mathrm{e}^2\frac{\mathbb{E}\{|\tilde{s}_\mathrm{w}|^2\}}{\mathbb{E}\{|\tilde{n}|^2\}}. \tag{5.191}$$

For this, we can first focus on the denominator and derive the power of the generated noise term. Using equation (5.190), we can write that

$$\mathbb{E}\{|\tilde{n}|^2\} = G^2\left(\frac{\mathbb{E}\{\rho_\mathrm{w}^2\}}{2\mathrm{IIP3}}\right)^2 \mathbb{E}\left\{\left[(\Gamma_{4,\mathrm{w}}+1) - \frac{\rho_\mathrm{w}^2}{\mathbb{E}\{\rho_\mathrm{w}^2\}}\right]^2 \rho_\mathrm{w}^2\right\}. \tag{5.192}$$

We can expand the last expectation on the right-hand side of this equation as

$$\begin{aligned}
&\mathbb{E}\left\{\left[(\Gamma_{4,\mathrm{w}}+1) - \frac{\rho_\mathrm{w}^2}{\mathbb{E}\{\rho_\mathrm{w}^2\}}\right]^2 \rho_\mathrm{w}^2\right\} \\
&= \mathbb{E}\left\{(\Gamma_{4,\mathrm{w}}+1)^2\rho_\mathrm{w}^2 - 2(\Gamma_{4,\mathrm{w}}+1)\frac{\rho_\mathrm{w}^4}{\mathbb{E}\{\rho_\mathrm{w}^2\}} + \frac{\rho_\mathrm{w}^6}{\mathbb{E}^2\{\rho_\mathrm{w}^2\}}\right\} \\
&= \mathbb{E}\{\rho_\mathrm{w}^2\}\left[(\Gamma_{4,\mathrm{w}}+1)^2 - 2(\Gamma_{4,\mathrm{w}}+1)\frac{\mathbb{E}\{\rho_\mathrm{w}^4\}}{\mathbb{E}^2\{\rho_\mathrm{w}^2\}} + \frac{\mathbb{E}\{\rho_\mathrm{w}^6\}}{\mathbb{E}^3\{\rho_\mathrm{w}^2\}}\right].
\end{aligned} \tag{5.193}$$

Then, using both the $\Gamma_{4,\mathrm{w}}$ and $\Gamma_{6,\mathrm{w}}$ constants that characterize the fourth and sixth order moments of the wanted signal instantaneous amplitude as defined by equation (5.125), we finally get

$$\mathbb{E}\left\{\left[(\Gamma_{4,\mathrm{w}}+1)-\frac{\rho_\mathrm{w}^2}{\mathbb{E}\{\rho_\mathrm{w}^2\}}\right]^2\rho_\mathrm{w}^2\right\}=\mathbb{E}\{\rho_\mathrm{w}^2\}[(\Gamma_{6,\mathrm{w}}+1)-(\Gamma_{4,\mathrm{w}}+1)^2].$$

Using this result in equation (5.192), we finally have for the noise power,

$$\mathbb{E}\{|\tilde{n}|^2\}=G^2\left(\frac{\mathbb{E}\{\rho_\mathrm{w}^2\}}{2\mathrm{IIP3}}\right)^2\mathbb{E}\{\rho_\mathrm{w}^2\}[(\Gamma_{6,\mathrm{w}}+1)-(\Gamma_{4,\mathrm{w}}+1)^2]. \tag{5.194}$$

We can then use this expression in conjunction with that of the effective gain given by equation (5.189) to determine the achieved SNR at the output of the nonlinear device. Given that the input wanted signal power, P_w, is equal to $\mathbb{E}\{\rho_\mathrm{w}^2\}/2$ when working on normalized impedances, we can write from equation (5.191) that

$$SNR=\frac{\left[1-\frac{P_\mathrm{w}}{\mathrm{IIP3}}(\Gamma_{4,\mathrm{w}}+1)\right]^2}{\left(\frac{P_\mathrm{w}}{\mathrm{IIP3}}\right)^2[(\Gamma_{6,\mathrm{w}}+1)-(\Gamma_{4,\mathrm{w}}+1)^2]}. \tag{5.195}$$

However, as we did during previous derivations, we can express this achieved SNR in terms of the device Psat. We obtain a convenient expression in terms of the input wanted signal back-off regarding the device Psat. Thus, using equation (5.59), we finally get

$$SNR=\frac{\left[3-\frac{P_\mathrm{w}}{\mathrm{IPsat}}(\Gamma_{4,\mathrm{w}}+1)\right]^2}{\left(\frac{P_\mathrm{w}}{\mathrm{IPsat}}\right)^2[(\Gamma_{6,\mathrm{w}}+1)-(\Gamma_{4,\mathrm{w}}+1)^2]}. \tag{5.196}$$

We observe that the SNR degrades in proportion to the square of the power of the input wanted signal, at least for low input power in respect of the device Psat. The deep reason for this is the power of the generated distortion noise, which increases as the third power of the input signal power in accordance with equation (5.194). Unlike the XM noise component examined in the previous section which behaves as a multiplicative noise, we have here a distortion noise that leads to a quick degradation of the SNR when the input signal level increases. This behavior leads to different management strategies when budgeting a line-up, as illustrated in Chapter 7.

In fact, dealing with the degradation of the quality of modulation of a single wanted signal, it is often the EVM metric, as defined in Section 3.2.2, that is used to express the result. When dealing with an EVM that is generated by a nonlinear phenomenon as in the present case, we talk about nonlinear EVM to distinguish it from the linear EVM discussed in Section 4.4. Here, we recall that this distinction is justified by the different impact these two kinds of distortion have on transceivers. For instance, given that equalization stages are classically implemented as linear processing functions, we can imagine that the linear EVM can be compensated, up to a point, by such baseband algorithms. On the other hand, there is no chance of compensating the nonlinear EVM term by such a linear equalization. This latter term can only be compensated by taking into account the nonlinear behavior of the device. This is for instance done in predistortion schemes as discussed in Chapter 9. In any case, we can give a direct expression for the distortion of the wanted signal in terms of EVM rather than SNR using equation (3.5):

$$EVM = \frac{1}{\sqrt{SNR}} = \frac{\frac{P_{\mathrm{w}}}{\mathrm{IPsat}}\sqrt{(\Gamma_{6,\mathrm{w}}+1)-(\Gamma_{4,\mathrm{w}}+1)^2}}{3-\frac{P_{\mathrm{w}}}{\mathrm{IPsat}}(\Gamma_{4,\mathrm{w}}+1)}. \tag{5.197}$$

Thus the limit cases predicted by our model are:

$$EVM = \begin{cases} 0 & \text{when } P_{\mathrm{w}} = 0, \\ \frac{\sqrt{(\Gamma_{6,\mathrm{w}}+1)-(\Gamma_{4,\mathrm{w}}+1)^2}}{(2-\Gamma_{4,\mathrm{w}})} & \text{when } P_{\mathrm{w}} = \mathrm{IPsat}. \end{cases} \tag{5.198}$$

Figure 5.15 shows two EVM curves for the modified 8PSK modulation of the GSM/EDGE standard and a noise-like signal with an amplitude modulation scheme that can be approximated by the Rayleigh distribution as encountered for instance with OFDM signals. We can clearly see the influence of the signal amplitude modulation statistics on the generated noise component. However, for the sake of accuracy we should also take into account the spectral partition of the generated noise term that we considered for this EVM derivation. Indeed, as detailed in "Spectral regrowth" later in this section, some spectral regrowth can be predicted in the present case so that part of the noise power we consider here lies outside the bandwidth of the wanted signal. Considering the fraction of noise power that is effectively within this bandwidth could improve the final EVM results regarding the prediction of equation (5.197). However, this can only be done on a case by case basis. Finally, we also mention that up to now when deriving equivalent SNR quantities, we have always had some uncertainty regarding the exact impact of the noise component statistics on the demodulation performance. This is not the case for practical EVM measurements, at least on the transmit side. Indeed, transmit EVM is often defined taking into account only the error vector power and does not make any assumption on its statistics. Thus, unless a measurement filter is used as observed above, equation (5.197) really gives the EVM as can be measured by equipment as long as the nonlinear model we used effectively describes the device under test with sufficient accuracy.

Code Domain Error Generation for CDMA Systems

Up to now, we have taken into account the impact of compression due to odd order nonlinearity through the derivation of an equivalent resulting SNR or EVM. However, for CDMA systems we can go a step further and use an alternative metric to quantify the distortion of the signal.

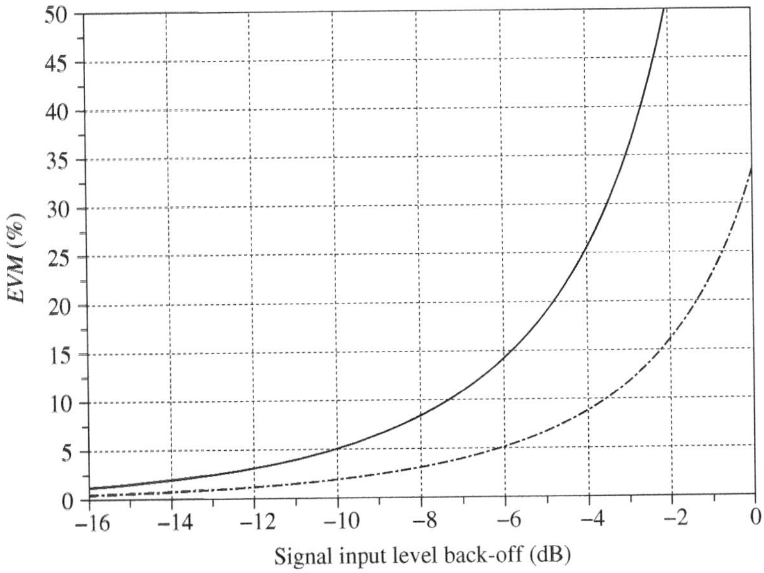

Figure 5.15 EVM generation due to compression in the smooth AM-AM nonlinearity approximation – Due to compression, an amplitude modulated RF bandpass signal experiences distortion. This results in an EVM term given by equation (5.197). This EVM, which increases when the input bandpass signal reaches the compression area of the device, i.e. when its back-off P_w/IPsat tends to 0 dB, is higher when the signal has a noise-like behavior, i.e. an amplitude that has a Rayleigh distribution with the Γ_4 and Γ_6 constants given by equation (5.130) (solid), than when it has a smoother amplitude modulation like the modified 8PSK of the GSM/EDGE which has Γ_4 and Γ_6 constants given by equation (5.128) (dot-dashed).

The combination of the particular structure of the complex envelope of a CDMA signal and the nonlinear transfer function leads to a phenomenon of intermodulation in the code domain.

By way of illustration, we can reconsider the example of the signals encountered in the WCDMA standard, as detailed in Chapter 1. The typical structure for the complex envelope of such signals is given by equation (1.172). However, in order to focus on the intermodulation in the code domain, we make some simplifications to the expression for such complex envelopes. For instance, we can neglect the intersymbol interference introduced by the RRC filter. We simply replace this filter by an ideal gate function. This is obviously a rough approximation. But it has the great advantage of making clear the terms introduced by the nonlinearity in the code domain. Thus, we can assume that the complex envelope of the wanted signal, $\tilde{s}_w(t)$, considered at the input of the nonlinear device can be written as

$$\tilde{s}_w(t) = \sum_m \sum_{n=0}^{L-1} \sum_{k=1}^{K} \beta_k a_k(n) \tilde{sc}(n) \tilde{S}_k(m)\, T_c \Pi\left(\frac{t - nT_c - mT}{T_c}\right). \tag{5.199}$$

In this expression, T represents the symbol period linked to the chip duration T_c by $T = LT_c$, and $\Pi(t)$ represents the impulse response of the ideal gate function as given by equation (1.145). We also recall that in this equation $\tilde{S}_k(m)$ represents the mth data symbol carried by the kth channel code and that $a_k(n)$ and $\tilde{sc}(n)$ are the elements of the spreading and scrambling codes, respectively. According to this expression, at a given sample time lT_c, assumed as an integer multiple of the chip period, we recover the ISI free samples:

$$\tilde{s}_w[l] \triangleq \tilde{s}_w(lT_c) = \sum_{k=1}^{K} \beta_k a_k(n)\tilde{sc}(n)\tilde{S}_k(m). \tag{5.200}$$

In this relationship, n and m stand respectively for the remainder and integer parts of l/LT_c. We can then use this expression to derive the samples of the complex envelope of the wanted signal recovered at the output of the device. Dealing with a device that exhibits smooth AM-AM compression, the complex envelope $\tilde{s}_{o,w}(t)$ of the output wanted signal is given by equation (5.184). Looking at this expression, we thus see that $\tilde{s}_{o,w}(t)$ is the superposition of a term proportional to the wanted complex envelope $\tilde{s}_w(t)$ and an additional term proportional to this complex envelope times its square magnitude, i.e. to $\rho_w^2(t)\tilde{s}_w(t)$. In order to go further, it is thus of interest to take a look at the structure of the latter term. However, for the sake of clarity we first suppose that we are dealing with only two channel codes. This allows simple derivations that are sufficient for our purposes. We then get that equation (5.200) reduces when $K = 2$ to

$$\tilde{s}_w[l] = \tilde{sc}(n)[\beta_1 a_1(n)\tilde{S}_1(m) + \beta_2 a_2(n)\tilde{S}_2(m)]. \tag{5.201}$$

As in the WCDMA case the a_k terms represent the components of the Walsh–Hadamard codes; they are real valued quantities that can only be $+1$ or -1. We can then write that

$$\begin{aligned}\rho_w^2[l] = |\tilde{sc}(n)|^2 \big[&\beta_1^2 a_1^2(n)|\tilde{S}_1(m)|^2 + \beta_2^2 a_2^2(n)|\tilde{S}_2(m)|^2 \\ &+ 2\beta_1\beta_2 a_1(n)a_2(n)\mathrm{Re}\{\tilde{S}_1(m)\tilde{S}_2(m)\}\big].\end{aligned} \tag{5.202}$$

We are thus dealing with two kinds of terms in this expression. The first is proportional to the square of the Walsh–Hadamard code elements, i.e. to either a_1^2 or a_2^2, equal to 1 in both cases. The second is proportional to the product of two elements belonging to two different Walsh–Hadamard codes, i.e. to $a_1(n)a_2(n)$. Thus, looking at the sequence composed of the successive samples $\rho_w^2[l]$, we see that all behaves as if the signals $\rho_w^2(t)$ were composed of the superposition of data symbols carried by new channel codes, either constant and equal to 1, or equal to the product[8] of a_1 and a_2. However, what is interesting is that the product of two Walsh–Hadamard codes results again in a Walsh–Hadamard code. This is indeed one of the properties of such codes.

Using this result, we can now interpret the impact of the nonlinear term $\rho_w^2(t)\tilde{s}_w(t)$ on the structure of the output wanted signal complex envelope. The terms just discussed that compose $\rho_w^2(t)$ are weighting in this expression the complex envelope of the input wanted signal given by equation (5.199). Thus the initial Walsh–Hadamard codes used to multiplex the data channels are now multiplied by the new additional codes generated in the expression for $\rho_w^2(t)$. Referring

[8] In the sense of a term by term product.

again to the property that the product of Walsh–Hadamard codes are Walsh–Hadamard codes, we see that we are necessarily dealing with two kinds of codes present in the expression for $\tilde{s}_{o,w}(t)$.

The first is the result of the multiplication of the channel code composed of constant components equal to 1, as recovered in equation (5.202), with the channel codes originally present in $\tilde{s}_w(t)$. As the code with constant components equal to 1 is the identity for this term by term multiplication, we recover the initial channel codes in the output complex envelope. However, the data symbols carried by these codes are now corrupted by those carried by the other input channel codes. Thus, even if we do not have new generated channel codes, we still have the creation of nonlinear EVM in that case.

The second is the result of the multiplication of the new Walsh–Hadamard codes generated by the nonlinearity in equation (5.202), with the channel codes originally present in $\tilde{s}_w(t)$. The channel codes thus generated involve the product of three codes from the input signal. If the set of codes used for the data channelization in the input signal reduces to two codes as considered above, all the output intermodulation codes are derived from at least the square of one of those input codes. But, as we saw, the square of such Walsh–Hadamard codes is the identity code, i.e. the code with constant components equal to 1. Thus, the final intermodulation code can only be one of the input codes in that case and we recover the situation detailed in the previous paragraph. In fact, for the generation of an additional error code to be different from the input ones, we need to have at least three different codes present in the input signal. In that case the product of those three different Walsh–Hadamard codes may generate an additional code different from the initial ones. This configuration is shown in Figure 5.16 in the code domain power (CDP), i.e. using the representation of the power carried by each channelization code present in the signal. This kind of representation allows us to derive the code domain error (CDE), which is a metric used in CDMA standards that represents the maximum allowable power on any error code referred to the overall signal power.

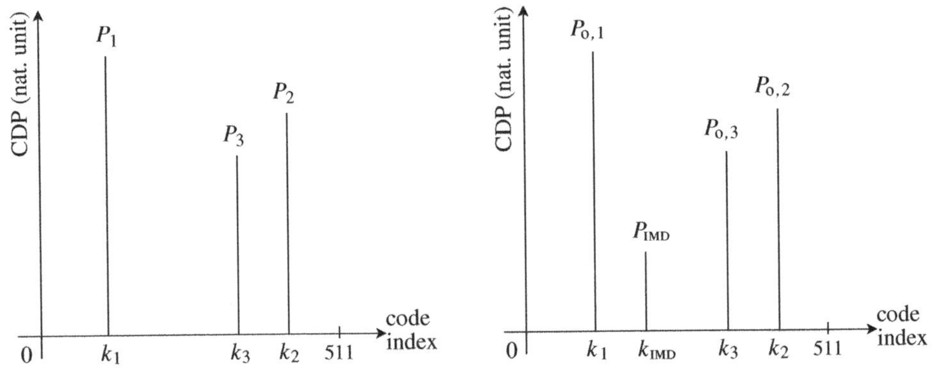

Figure 5.16 Intermodulation code generation in a bandpass WCDMA signal due to odd order nonlinearity – Considering a WCDMA signal composed of three channel codes (left), we can have the generation of an additional IMD code at the output of a device that exhibits compression (right). Here, the signals are represented in the CDP, using on the abscissa the index of the corresponding channel codes, i.e. of the Walsh–Hadamard codes, as defined in the WCDMA standard [9, 10].

To conclude, we observe that by equation (5.184), the power of the generated code is directly linked to the α_3 parameter of the device transfer function, and thus to its IIP3, as if we had to deal with tones in the frequency domain. However, the relationship between the input code power, the resulting output intermodulation code power and the IIP3 is less straightforward than in the tone intermodulation case as it involves the number of codes in the present case, as can be seen by looking at equation (5.202).

Spectral Regrowth

As mentioned throughout the previous sections, spectral regrowth occurs when amplitude modulated bandpass signals pass through a nonlinear device. As the name suggests, spectral regrowth simply means that the signal recovered at the output of a nonlinear device can have a spectrum that spreads over a wider bandwidth than the input spectrum. As detailed in subsequent sections, this phenomenon is deeply related to the presence of terms in the form of powers of input signals in the analytical expression for the output signal.

However, depending on the parity of the nonlinearity we are dealing with, the complex envelopes centered on the various angular frequencies generated at the output of the nonlinear device are not necessarily impacted in the same way by those nonlinear terms and thus by the spectral regrowth:

(i) Considering equation (5.114) and more precisely the bandpass signals centered around the input carriers angular frequencies, we see that those signals are not directly impacted by an even order nonlinearity due to mismatch in the device implementation. We thus do not have to deal with any spectral regrowth for those bandpass signals. Reversely, looking at the component generated around DC, we see that this AM-demodulated component is proportional to the square of the input signal instantaneous amplitude. We then surmise that the spectrum of this component is wider than the initial spectrum of the input signal.

(ii) But considering equation (5.121), we see that we do not have the same behavior when considering odd order nonlinearity linked to compression. In this case, the expression for the signal recovered at the device output around the carrier angular frequency of the input signals involves higher order terms.

In what follows we therefore review the different situations in which spectral regrowth is involved, with an eye on its various system impacts.

Spectral Regrowth for AM-Demodulation Due to Even Order Nonlinearity

Let us consider for now the AM-demodulation due to even order nonlinearity. As discussed throughout "AM-demodulation due to even order nonlinearity" earlier in this section, this phenomenon has a system impact mainly at the final downconversion stage of a receiver, in particular when dealing with a strong blocking signal that is superposed on a weak wanted signal at the input of this downconversion stage. In that case, the noise component related to the presence of the blocking signal that is retrieved folded on the wanted signal is proportional to the square of the instantaneous amplitude of the blocker. However, in order to refine the resulting SNR limitation, it is of interest to examine the spectral shape of the AM-demodulation noise component in order to derive the fraction of equivalent noise power lying in the wanted signal bandwidth.

In order to carry out such derivation, suppose we were to consider using a simple Fourier transform of the square of the instantaneous amplitude of the blocking signal. But that would lead to the evaluation of the spectral content of only one particular realization of the modulating process. Moreover, this analysis would have to be reduced to a finite duration of observation in order to ensure the convergence of the Fourier transform. Thus the preferred choice would be to continue working with the PSD of the signal in order to take into account the statistical properties of the modulating process. But here again a limitation occurs for the derivation as it cannot be performed analytically for any arbitrary distribution of the bandpass signal instantaneous amplitude. However, we observe that it can be done when the blocking signal can be considered as a bandpass Gaussian signal. This is, moreover, a practical case of interest as this configuration represents most of the wideband modulations, as encountered in OFDM systems for instance (see our discussion in Section 1.3.3).

Let us suppose that the blocking signal we are dealing with, $s_{\text{bl}}(t)$, has a complex envelope $\tilde{s}_{\text{bl}}(t) = p_{\text{bl}}(t) + jq_{\text{bl}}(t)$ such that both $p_{\text{bl}}(t)$ and $q_{\text{bl}}(t)$ have a Gaussian distribution. Our goal is then to derive the PSD of $\rho^2_{\text{bl}}(t) = \tilde{s}_{\text{bl}}(t)\tilde{s}^*_{\text{bl}}(t) = p^2_{\text{bl}}(t) + q^2_{\text{bl}}(t)$ as a function of the PSD of the original signal, $\tilde{s}_{\text{bl}}(t)$. For that purpose, we first need to derive the autocorrelation function of $\rho^2_{\text{bl}}(t)$. Assuming the stationarity of the modulating process, we can write for two samples of $\rho^2_{\text{bl}}(t)$ considered at time t_1 and t_2 that

$$\begin{aligned}\gamma_{\rho^2_{\text{bl}}\times\rho^2_{\text{bl}}}(\tau) &= \mathbb{E}\{\rho^2_{\text{bl},t_1}\rho^2_{\text{bl},t_2}\}\\ &= \mathbb{E}\{(\tilde{s}_{\text{bl},t_1}\tilde{s}^*_{\text{bl},t_1})(\tilde{s}_{\text{bl},t_2}\tilde{s}^*_{\text{bl},t_2})\}\end{aligned} \tag{5.203}$$

with $\tau = t_1 - t_2$. But, as discussed in Appendix 2, for modulating signals classically encountered in wireless systems during the transmission of data, the resulting modulated RF bandpass signal, i.e. $s_{\text{bl}}(t)$ here, can also be considered as stationary. Dealing with a Gaussian signal, this means that we can directly use the results derived in Appendix 3, and more precisely equation (A3.23), to write

$$\begin{aligned}\gamma_{\rho^2_{\text{bl}}\times\rho^2_{\text{bl}}}(\tau) &= \mathbb{E}\{\tilde{s}_{\text{bl},t_1}\tilde{s}^*_{\text{bl},t_1}\}\mathbb{E}\{\tilde{s}_{\text{bl},t_2}\tilde{s}^*_{\text{bl},t_2}\} + \mathbb{E}\{\tilde{s}_{\text{bl},t_1}\tilde{s}^*_{\text{bl},t_2}\}\mathbb{E}\{\tilde{s}^*_{\text{bl},t_1}\tilde{s}_{\text{bl},t_2}\}\\ &= \gamma^2_{\tilde{s}_{\text{bl}}\times\tilde{s}_{\text{bl}}}(0) + \gamma_{\tilde{s}_{\text{bl}}\times\tilde{s}_{\text{bl}}}(\tau)\gamma^*_{\tilde{s}_{\text{bl}}\times\tilde{s}_{\text{bl}}}(\tau).\end{aligned}$$

This last equation shows in particular that the autocorrelation of $\rho^2_{\text{bl}}(t)$ can be decomposed into two terms. The first one is constant. Given that the PSD we are looking for is directly proportional to the Fourier transform of this autocorrelation function, this constant term thus leads to a non-vanishing component centered on DC. This is indeed obvious as $\rho^2_{\text{bl}}(t)$ is an always positive signal with thus a non-vanishing average value, as discussed in "AM-demodulation due to even order nonlinearity" earlier in this section. We can then express this DC level in terms of the power P_{bl} of the blocking signal $s_{\text{bl}}(t)$. Using equation (1.66), we can then write that

$$\gamma_{\rho^2_{\text{bl}}\times\rho^2_{\text{bl}}}(\tau) = 4P^2_{\text{bl}} + \gamma_{\tilde{s}_{\text{bl}}\times\tilde{s}_{\text{bl}}}(\tau)\gamma^*_{\tilde{s}_{\text{bl}}\times\tilde{s}_{\text{bl}}}(\tau), \tag{5.204}$$

thus resulting in a PDS for $\rho_{\text{bl}}^2(t)$ of the form

$$\Gamma_{\rho_{\text{bl}}^2\times\rho_{\text{bl}}^2}(\omega) = 4P_{\text{bl}}^2\delta(\omega) + \int_{-\infty}^{\infty} \gamma_{\tilde{s}_{\text{bl}}\times\tilde{s}_{\text{bl}}}(\tau)\gamma^*_{\tilde{s}_{\text{bl}}\times\tilde{s}_{\text{bl}}}(\tau)e^{-j\omega\tau}d\tau. \tag{5.205}$$

To evaluate the integral, we observe that it represents the Fourier transform of the product of two functions. Using equation (1.10), we can then write

$$\Gamma_{\rho_{\text{bl}}^2\times\rho_{\text{bl}}^2}(\omega) = 4P_{\text{bl}}^2\delta(\omega) + \mathscr{F}\left\{\gamma_{\tilde{s}_{\text{bl}}\times\tilde{s}_{\text{bl}}}(\tau)\right\}(\omega) \star \mathscr{F}\left\{\gamma^*_{\tilde{s}_{\text{bl}}\times\tilde{s}_{\text{bl}}}(\tau)\right\}(\omega). \tag{5.206}$$

Given the general property that any autocorrelation function has Hermitian symmetry, we can use the property of the Fourier transform given by equation (A1.41) to evaluate the last term on the right-hand side of this equation. We then get that

$$\Gamma_{\rho_{\text{bl}}^2\times\rho_{\text{bl}}^2}(\omega) = 4P_{\text{bl}}^2\delta(\omega) + \Gamma_{\tilde{s}_{\text{bl}}\times\tilde{s}_{\text{bl}}}(\omega) \star \Gamma_{\tilde{s}_{\text{bl}}\times\tilde{s}_{\text{bl}}}(-\omega). \tag{5.207}$$

We thus see that in addition to the DC term, the spectral shape of ρ_{bl}^2 is proportional to the self-convolution of the PSD of the input blocking signal.

In order to illustrate this behavior, we can assume that this spectral shape can be approximated by a simple gate function of the form

$$\Pi(\omega) = \begin{cases} A & \text{when}|\omega| < B/2, \\ 0 & \text{otherwise.} \end{cases} \tag{5.208}$$

where B is the bandwidth of the signal. This is indeed a common shape for OFDM signals for instance (recall the discussion in Section 1.3.3). In that case, the self-convolution of the gate function,

$$\Pi(\omega) \star \Pi(\omega) = \int_{-\infty}^{\infty} \Pi(\omega')\Pi(\omega - \omega')d\omega', \tag{5.209}$$

reduces to

$$\Pi(\omega) \star \Pi(\omega) = \begin{cases} A^2(B - |\omega|) & \text{when}|\omega| < B. \\ 0 & \text{otherwise,} \end{cases} \tag{5.210}$$

i.e. to a sawtooth function that thus spreads over $[-B, B]$ as illustrated in Figure 5.17. This means that the spectrum of $\rho_{\text{bl}}^2(t)$ spreads over twice the original bandwidth of the signal $s_{\text{bl}}(t)$. This enlargement of the original spectrum is what we refer to as the spectral regrowth phenomenon. It occurs each time a power of the initial modulating signal appears in the analytical expression for the output signal, thus resulting in a self-convolution product in the corresponding PSD expression. We therefore remark that this spectral regrowth is deeply related to the properties of the self-convolution, and can thus be alternatively illustrated without making any assumption on the shape of the initial spectrum. It suffices to follow the

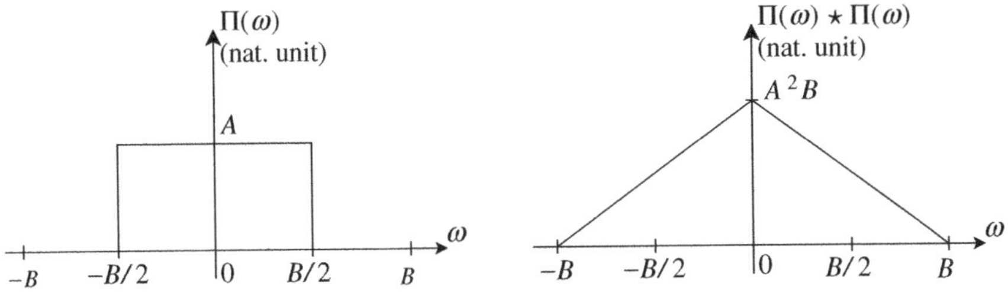

Figure 5.17 Gate function and self-convolution – The self-convolution of a gate function as defined by equation (5.208) (left) leads to a sawtooth waveform as given by equation (5.210) (right). The support of the resulting signal is $2B$, i.e. twice the support of the initial gate function B.

same approach as in "Spectrum degradation" (Section 4.3.3) in Chapter 4, i.e. to consider a decomposition of this input spectrum in terms of narrowband adjacent slices of width $\delta\omega$. Based on such decomposition, we then get that the input spectrum is transposed around each of the slices of the decomposed spectrum due to the convolution process. The result of the self-convolution then necessarily spreads towards a bandwidth that is almost twice the bandwidth of the original spectrum.

This behavior can then be confirmed by comparing our present analytical derivations with simulation results. For that purpose, we can consider an LTE 10 MHz bandwidth signal. As detailed in Section 1.3.3, such a signal has a PSD that spreads over $[-4.5, 4.5]$ MHz. We can then derive the PSD of $\rho_{\text{bl}}^2(t)$ either by direct evaluation from time domain simulations of this signal, or by applying equation (5.207) to the PSD of the initial OFDM signal. The two results are shown in Figure 5.18.

Spectral Regrowth for Cross-Modulation Due to Odd Order Nonlinearity

We now turn to the XM due to odd order nonlinearity. For that purpose, we recall the situation detailed in "Cross-modulation due to odd order nonlinearity" earlier in this section, which corresponds to a wanted signal that remains in the linear range of a nonlinear device, and that is superposed on a blocker that is in turn in the compression range of the device. In that case the XM leads to an SNR limitation through the generation of a noise term centered around the wanted signal carrier angular frequency. We focus on the spectral content of this noise term in order to get a more accurate view of the fraction of its power that effectively lies within the wanted signal bandwidth.

As previously discussed, only the first term in equation (5.175), proportional to $\left(\rho_{\text{bl}}^2(t) - \mathbb{E}\left\{\rho_{\text{bl}}^2\right\}\right)\tilde{s}_{\text{w}}(t)$, represents the complex envelope of the XM noise term of interest. The second term, proportional to $\rho_{\text{w}}^2(t)\tilde{s}_{\text{w}}(t) = \tilde{s}_{\text{w}}^*(t)\tilde{s}_{\text{w}}(t)\tilde{s}_{\text{w}}(t)$, is related to the nonlinear EVM noise term generated by the compression of the wanted signal itself. It is therefore investigated in the next section. Here we derive the PSD of the XM noise complex envelope,

$$\tilde{n}(t) = \rho_{\text{bl,c}}^2(t)\tilde{s}_{\text{w}}(t), \tag{5.211}$$

where

$$\rho_{\text{bl,c}}^2(t) = \rho_{\text{bl}}^2(t) - \mathbb{E}\left\{\rho_{\text{bl}}^2\right\} \tag{5.212}$$

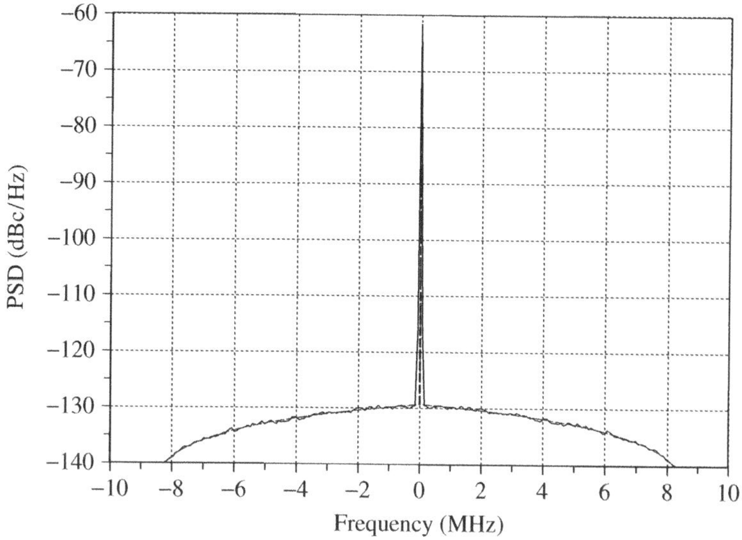

Figure 5.18 PSD of $\tilde{s}_{\text{bl}}(t)\tilde{s}^*_{\text{bl}}(t)$ when $\tilde{s}_{\text{bl}}(t)$ represents a stationary Gaussian bandpass signal – When $\tilde{s}_{\text{bl}}(t)$ represents a stationary bandpass signal that can be assumed Gaussian, the theoretical PSD of $\rho^2_{\text{bl}}(t) = |\tilde{s}_{\text{bl}}(t)|^2$ is given by equation (5.207) (dashed). This result matches the direct evaluation of the PSD using time domain simulations (solid). Here, an LTE 10 MHz bandwidth signal is used for the simulations.

is the centered version of the square instantaneous amplitude of the blocking signal $\rho^2_{\text{bl}}(t)$. In order to derive the PSD of this XM noise component, we first derive its autocorrelation function. Assuming both independence between the modulating processes of the blocker and of the wanted signal, and stationarity, we can write

$$\begin{aligned}\gamma_{\tilde{n}\times\tilde{n}}(\tau) &= \mathbb{E}\left\{\left(\rho^2_{\text{bl,c},t_1}\tilde{s}_{\text{w},t_1}\right)\left(\rho^2_{\text{bl,c},t_2}\tilde{s}_{\text{w},t_2}\right)^*\right\}\\ &= \mathbb{E}\left\{\rho^2_{\text{bl,c},t_1}\rho^2_{\text{bl,c},t_2}\right\}\mathbb{E}\left\{\tilde{s}_{\text{w},t_1}\tilde{s}^*_{\text{w},t_2}\right\}\\ &= \gamma_{\rho^2_{\text{bl,c}}\times\rho^2_{\text{bl,c}}}(\tau)\gamma_{\tilde{s}_{\text{w}}\times\tilde{s}_{\text{w}}}(\tau),\end{aligned} \tag{5.213}$$

with $\tau = t_1 - t_2$. It then remains to take the Fourier transform of this autocorrelation function to derive the PSD of $\tilde{n}(t)$. For that purpose, we can use the property that the Fourier transform of a product of signals is the convolution of the Fourier transform of each term, as given by equation (1.10). We obtain

$$\Gamma_{\tilde{n}\times\tilde{n}}(\omega) = \Gamma_{\rho^2_{\text{bl,c}}\times\rho^2_{\text{bl,c}}}(\omega) \star \Gamma_{\tilde{s}_{\text{w}}\times\tilde{s}_{\text{w}}}(\omega). \tag{5.214}$$

We thus recover that the PSD of the XM noise is the convolution of the PSD of the wanted signal with that of the centered version of the squared instantaneous amplitude of the blocking signal.

We may now consider suitable PDFs for the modulation schemes involved. As detailed in "Spectral regrowth for AM-demodulation due to even order nonlinearity" earlier in this section, this can be done by considering noise-like bandpass signals. Such signals can correctly model wideband signals like those encountered in OFDM systems and allow straightforward analytical derivations. The derivations performed in that section can be reused directly to give an expression for $\Gamma_{\rho^2_{\text{bl,c}} \times \rho^2_{\text{bl,c}}}(\omega)$ in the Gaussian approximation for stationary processes. The only difference in the present case is that we now consider the centered version of the square of the instantaneous amplitude of the blocking signal so we can anticipate that the DC part of $\Gamma_{\rho^2_{\text{bl,c}} \times \rho^2_{\text{bl,c}}}(\omega)$ is necessarily null. Given that $P_{\text{bl}} = \mathbb{E}\{\rho^2_{\text{bl}}\}/2$, we have

$$
\begin{aligned}
\gamma_{\rho^2_{\text{bl,c}} \times \rho^2_{\text{bl,c}}}(\tau) &= \mathbb{E}\{(\rho^2_{\text{bl},t_1} - \mathbb{E}\{\rho^2_{\text{bl}}\})(\rho^2_{\text{bl},t_2} - \mathbb{E}\{\rho^2_{\text{bl}}\})\} \\
&= \mathbb{E}\{\rho^2_{\text{bl},t_1} \rho^2_{\text{bl},t_2}\} - \mathbb{E}^2\{\rho^2_{\text{bl}}\} \\
&= \mathbb{E}\{\rho^2_{\text{bl},t_1} \rho^2_{\text{bl},t_2}\} - 4P^2_{\text{bl}}.
\end{aligned}
\tag{5.215}
$$

Based on the expression for $\gamma_{\rho^2_{\text{bl}} \times \rho^2_{\text{bl}}}(\tau)$ given by equation (5.204), we can write

$$
\gamma_{\rho^2_{\text{bl,c}} \times \rho^2_{\text{bl,c}}}(\tau) = \gamma_{\tilde{s}_{\text{bl}} \times \tilde{s}_{\text{bl}}}(\tau) \gamma^*_{\tilde{s}_{\text{bl}} \times \tilde{s}_{\text{bl}}}(\tau). \tag{5.216}
$$

Taking the Fourier transform of this expression then leads to

$$
\Gamma_{\rho^2_{\text{bl,c}} \times \rho^2_{\text{bl,c}}}(\omega) = \Gamma_{\tilde{s}_{\text{bl}} \times \tilde{s}_{\text{bl}}}(\omega) \star \Gamma_{\tilde{s}_{\text{bl}} \times \tilde{s}_{\text{bl}}}(-\omega). \tag{5.217}
$$

We thus finally see that equation (5.214) reduces to

$$
\Gamma_{\tilde{n} \times \tilde{n}}(\omega) = \Gamma_{\tilde{s}_{\text{bl}} \times \tilde{s}_{\text{bl}}}(\omega) \star \Gamma_{\tilde{s}_{\text{bl}} \times \tilde{s}_{\text{bl}}}(-\omega) \star \Gamma_{\tilde{s}_{\text{w}} \times \tilde{s}_{\text{w}}}(\omega). \tag{5.218}
$$

Referring to the mechanism examined when discussing the AM-demodulation case, we see that the PSD of the XM noise term spreads over a wider bandwidth than that of the blocking signal due to the convolution products present in this equation. More precisely, due to the self-convolution of the spectrum of the blocker, we first get that the resulting spectrum bandwidth is about twice the original one. This bandwidth is then enlarged again by the convolution with the wanted signal spectrum. This shows that correction factors could be considered in respect of the derivation performed in "Cross-modulation due to odd order nonlinearity" earlier in this section to take into account the exact fraction of the generated noise power that remains in the receive bandwidth of the wanted signal. However, this can only be done on a case by case basis.

We can now consider another useful case of interest that is commonly encountered in the field of wireless and where the XM phenomenon is involved. Due to spectral regrowth, the generated noise component can corrupt the wanted signal even if the cross-modulation occurs between blocking signals only. This can happen when a strong interferer cross-modulates an adjacent channel to the wanted signal or a close in-and blocker, for instance. In that case,

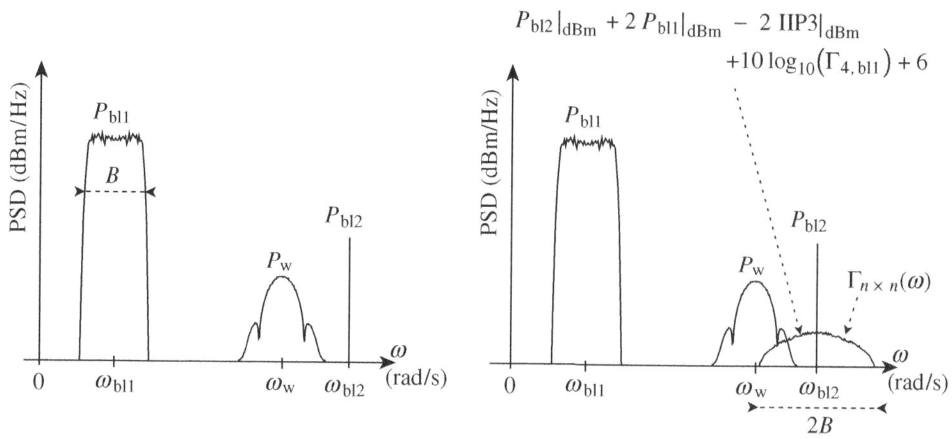

Figure 5.19 Wanted signal degradation due to XM between an adjacent blocker and a strong non-constant amplitude blocking signal – When a blocker lies close to the wanted signal in the frequency domain (left), a fraction of the XM signal generated around this blocker by a strong amplitude modulated blocker can lie within the wanted signal bandwidth and thus degrades the reception (right – equivalent input quantities).

the generated noise term can spread over a wide bandwidth such that a fraction of it lies within the wanted signal frequency band. Such a configuration is shown in Figure 5.19 in the simplified case where the close in-band blocker is CW. From an analytical point of view, the derivation performed up to now can be adapted in a straightforward way as long as we assume that the first blocking signal modulation statistics can be approximated as Gaussian and that $P_w \ll P_{bl2} \ll P_{bl1}$, in the notation of that figure. In that case, we can write the complex envelope of the XM component generated by the first blocker around the second one using equations (5.211) and (5.212). This leads to a term of the form

$$\tilde{n}(t) = \left(\rho_{bl1}^2(t) - \mathbb{E}\left\{\rho_{bl1}^2\right\}\right)\tilde{s}_{bl2}(t), \tag{5.219}$$

and thus to a PSD derived from equation (5.218) as

$$\Gamma_{\tilde{n}\times\tilde{n}}(\omega) = \Gamma_{\tilde{s}_{bl1}\times\tilde{s}_{bl1}}(\omega) \star \Gamma_{\tilde{s}_{bl1}\times\tilde{s}_{bl1}}(-\omega) \star \Gamma_{\tilde{s}_{bl2}\times\tilde{s}_{bl2}}(\omega). \tag{5.220}$$

However, in the simple situation shown in Figure 5.19 where the second blocker is CW, we have that $\Gamma_{\tilde{s}_{bl2}\times\tilde{s}_{bl2}}(\omega) = \delta(\omega)$. It follows that

$$\Gamma_{\tilde{n}\times\tilde{n}}(\omega) = \Gamma_{\tilde{s}_{bl1}\times\tilde{s}_{bl1}}(\omega) \star \Gamma_{\tilde{s}_{bl1}\times\tilde{s}_{bl1}}(-\omega). \tag{5.221}$$

As this expression involves only the two blocking signals which are expected to be of much higher power than that of the wanted signal we understand that the generated noise component can have a non-negligible power lying in the wanted signal frequency band in respect of the power of this latter signal. This can result in a non-negligible SNR degradation. It is in fact a practical case of interest in many full-duplex standards on the receive side where transmitter leakage plays the role of the strong blocking signals, as illustrated in "Close in-band blocker case" in Chapter 7.

Spectral Regrowth for Nonlinear EVM Due to Odd Order Nonlinearity

To conclude our derivations on spectral regrowth, we can now look at the case of a single wanted signal that is distorted due to compression. We studied this distortion in "Nonlinear EVM due to odd order nonlinearity" earlier in this section, in terms of the resulting SNR degradation or nonlinear EVM generation. We can now consider this phenomenon again to take into account the spectral shape of the generated noise component, part of which may lie outside the wanted signal bandwidth.

We first focus on the structure of the complex envelope of the nonlinear EVM noise term given by equation (5.190). This complex envelope is the linear sum of a term proportional to the input wanted signal complex envelope, $\tilde{s}_{\mathrm{w}}(t)$, and a term proportional to this wanted complex envelope times its squared magnitude. Although both terms are important from the SNR point of view, only the second leads to a different spectrum compared to that of the input signal. This is thus the term we focus on in order to study the spectral regrowth. Thus, let us consider the complex envelope

$$\tilde{n}(t) = \rho_{\mathrm{w}}^2(t)\tilde{s}_{\mathrm{w}}(t) = \tilde{s}_{\mathrm{w}}^*(t)\tilde{s}_{\mathrm{w}}(t)\tilde{s}_{\mathrm{w}}(t). \tag{5.222}$$

In order to derive its PSD, we first derive its autocorrelation function. Assuming stationarity of the modulating process we are dealing with, we can write

$$\gamma_{\tilde{n}\times\tilde{n}}(\tau) = \mathbb{E}\left\{\left(\tilde{s}_{\mathrm{w},t_1}^2\tilde{s}_{\mathrm{w},t_1}^*\right)\left(\tilde{s}_{\mathrm{w},t_2}^2\tilde{s}_{\mathrm{w},t_2}^*\right)^*\right\} \tag{5.223}$$

with $\tau = t_1 - t_2$.

We can assume as in the two previous sections that we are dealing with an input stationary Gaussian signal, as is the case for an OFDM signal, for instance. We can then directly use the results derived in Appendix 3, and in particular equation (A3.24), to expand the above relationship as

$$\gamma_{\tilde{n}\times\tilde{n}}(\tau) = 2\mathbb{E}\left\{\tilde{s}_{\mathrm{w},t_1}\tilde{s}_{\mathrm{w},t_2}^*\right\}\left[2\mathbb{E}\left\{\tilde{s}_{\mathrm{w},t_1}\tilde{s}_{\mathrm{w},t_1}^*\right\}\mathbb{E}\left\{\tilde{s}_{\mathrm{w},t_2}\tilde{s}_{\mathrm{w},t_2}^*\right\} + \left|\mathbb{E}\left\{\tilde{s}_{\mathrm{w},t_1}\tilde{s}_{\mathrm{w},t_2}^*\right\}\right|^2\right]. \tag{5.224}$$

We obtain

$$\begin{aligned}\gamma_{\tilde{n}\times\tilde{n}}(\tau) &= 2\gamma_{\tilde{s}_{\mathrm{w}}\times\tilde{s}_{\mathrm{w}}}(\tau)\left[2\gamma_{\tilde{s}_{\mathrm{w}}\times\tilde{s}_{\mathrm{w}}}^2(0) + \left|\gamma_{\tilde{s}_{\mathrm{w}}\times\tilde{s}_{\mathrm{w}}}(\tau)\right|^2\right]\\ &= 4\gamma_{\tilde{s}_{\mathrm{w}}\times\tilde{s}_{\mathrm{w}}}^2(0)\gamma_{\tilde{s}_{\mathrm{w}}\times\tilde{s}_{\mathrm{w}}}(\tau) + 2\gamma_{\tilde{s}_{\mathrm{w}}\times\tilde{s}_{\mathrm{w}}}^2(\tau)\gamma_{\tilde{s}_{\mathrm{w}}\times\tilde{s}_{\mathrm{w}}}^*(\tau).\end{aligned} \tag{5.225}$$

In order to derive the PSD of $\tilde{n}(t)$, it then remains to take the Fourier transform of this autocorrelation function. Given the general property that any autocorrelation function has Hermitian symmetry, we can then use the property of the Fourier transform given by equation (A1.41) to evaluate the Fourier transform of the last term on the right-hand side of this equation. This leads to

$$\Gamma_{\tilde{n}\times\tilde{n}}(\omega) = 4\gamma^2_{\tilde{s}_w\times\tilde{s}_w}(0)\Gamma_{\tilde{s}_w\times\tilde{s}_w}(\omega) + 2\Gamma_{\tilde{s}_w\times\tilde{s}_w}(\omega) \star \Gamma_{\tilde{s}_w\times\tilde{s}_w}(-\omega) \star \Gamma_{\tilde{s}_w\times\tilde{s}_w}(\omega).$$

In the same way, we observe that $\gamma_{\tilde{s}_w\times\tilde{s}_w}(0)$ is equal to twice the power P_w of the input wanted bandpass signal $s_w(t)$ (see equation (1.65)). Thus, finally,

$$\Gamma_{\tilde{n}\times\tilde{n}}(\omega) = 16P_w^2\Gamma_{\tilde{s}\times\tilde{s}}(\omega) + 2\Gamma_{\tilde{s}_w\times\tilde{s}_w}(\omega) \star \Gamma_{\tilde{s}_w\times\tilde{s}_w}(-\omega) \star \Gamma_{\tilde{s}_w\times\tilde{s}_w}(\omega), \tag{5.226}$$

We see that the spectrum of the generated noise component contains a term that is proportional to three times the convolution of the PSD of the original signal. Thus, recalling for instance that the PSD of $s_w(t)$ can be approximated by the simple gate function given by equation (5.208), the spectrum of the noise component is linked to the convolution of the sawtooth waveform resulting from the self-convolution of the gate function, as displayed in Figure 5.17, with the initial gate function again. We thus now expect the result to spread over three times the bandwidth of the initial function.

This behavior can be confirmed by comparing our present analytical derivations with simulations results. For that purpose, we can reconsider the LTE 10 MHz bandwidth signal. As detailed in Section 1.3.3, this leads to a signal with a PSD that spreads over [−4.5, 4.5] MHz. We can then derive the PSD of $\tilde{s}_w^*(t)\tilde{s}_w(t)\tilde{s}_w(t)$ either by direct evaluation from time domain simulations of this signal, or directly by applying equation (5.226) to the power spectral density of the input OFDM signal. The two results are shown in Figure 5.20.

We conclude by observing that spectral regrowth of a signal is also of particular importance on the transmit side. In that case, more than the in-band SNR limitation or EVM generation, it is often the direct degradation of the transmit spectrum that is of importance. The main difference in that case is that higher order terms involved in the series expansion of the device transfer function can effectively lead to non-negligible contributions in terms of leakage in adjacent channels, whereas their contribution to the in-band SNR or EVM can be assumed negligible compared to that of the third order term. This can be illustrated by considering the general form of the expression for the complex envelope that is centered around the wanted signal carrier angular frequency ω_w at the output of a nonlinear device when taking into account the higher order terms in the series expansion of its transfer function. For that purpose, we must anticipate the discussion in Section 5.3.1 on AM-PM conversion and consider the series expansion given by equation (5.307). The structure of this series expansion is valid in any case even if only real valued coefficients are involved as in the pure AM-AM conversion case. Thus, we can see that the next higher order nonlinear term above the third order term to be considered in this expression is the fifth order term of the form $\rho_w^4(t)\tilde{s}_w(t) = \tilde{s}_w^*(t)\tilde{s}_w^*(t)\tilde{s}_w(t)\tilde{s}_w(t)\tilde{s}_w(t)$. By the convolution mechanism as discussed so far, we thus expect the spectrum of this term to spread over about five times the spectral extent of the original wanted signal. This corresponds to the simulation results shown in Figure 5.21 for the LTE 10 MHz example. Consequently, even if in the smooth nonlinearity case we expect the magnitude of the coefficients in the series expansion of the device transfer function to be rapidly decreasing, such a mechanism

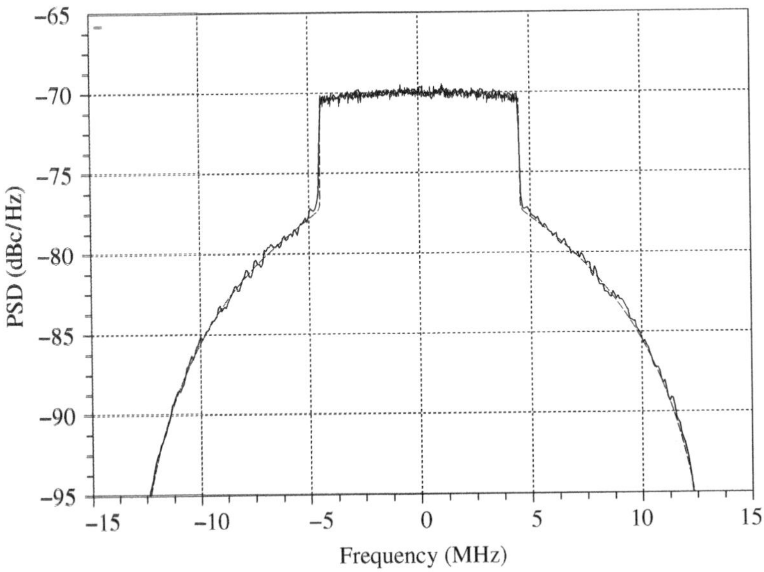

Figure 5.20 Power spectral density of $\rho_w^2(t)\tilde{s}_w(t)$ when $s_w(t)$ can be approximated as a stationary Gaussian bandpass signal – When $\tilde{s}_w(t)$ represents a stationary Gaussian bandpass signal, the theoretical PSD of $\rho_w^2(t)\tilde{s}_w(t)$ is given by equation (5.226) (dashed). This result matches the direct evaluation of the PSD of this term based on time domain simulations (solid). Here, an LTE 10 MHz bandwidth signal is used for the simulations.

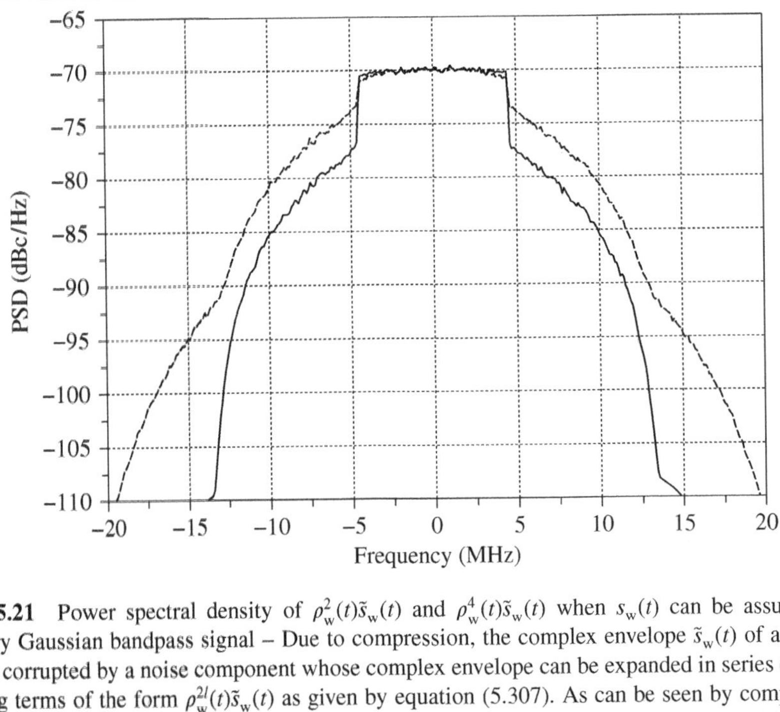

Figure 5.21 Power spectral density of $\rho_w^2(t)\tilde{s}_w(t)$ and $\rho_w^4(t)\tilde{s}_w(t)$ when $s_w(t)$ can be assumed as a stationary Gaussian bandpass signal – Due to compression, the complex envelope $\tilde{s}_w(t)$ of a bandpass signal is corrupted by a noise component whose complex envelope can be expanded in series expansion involving terms of the form $\rho_w^{2l}(t)\tilde{s}_w(t)$ as given by equation (5.307). As can be seen by comparing the PSD of $\rho_w^2(t)\tilde{s}_w(t)$ (solid) with that of $\rho_w^4(t)\tilde{s}_w(t)$ (dashed), the higher the order of the considered term, the wider its spectrum. Here, an LTE 10 MHz bandwidth signal is used for the simulations.

can lead to a non-negligible pollution of higher order adjacent channels in respect of stringent SEM requirements that may be encountered on the transmit side.

5.1.4 SNR Improvement Due to RF Compression

Within certain limits, it is possible for a SNR improvement to occur through devices that exhibit compression. Recall from the discussion in "Nonlinear EVM due to odd order nonlinearity" (Section 5.1.3) that only bandpass RF signals that have a constant instantaneous amplitude can flow without distortion through devices that exhibit compression. We thus anticipate that such SNR improvement can only occur when processing such constant instantaneous amplitude bandpass signals. However, it is still of interest to perform the analytical derivations in the general case. This is an alternative way to show the sensitivity of amplitude modulated bandpass signals to compression at no particular additional analytical cost.

In this section we base these analytical derivations on the use of the series expansion up to third order of the odd order part of the transfer function of a device that exhibits compression. It is interesting that such a simple model succeeds in predicting the phenomenon while allowing simple illustrations of its cause (compare the results derived using this model with those of the hard limiter theory examined in Section 5.2.3).

The phenomenon of SNR improvement is not only of theoretical interest. It is used effectively in practice, for instance in many GSM transmitters, to improve the transmit far noise floor through the use of saturated PAs that behave as hard limiters.

Amplitude Noise Cancellation

Let us suppose for the time being that we are dealing with a wanted bandpass signal $s_\text{w}(t)$ that is superposed on a Gaussian bandpass noise $n(t)$ at the input of a nonlinear device that exhibits a smooth AM-AM compression behavior. Here, the Gaussian assumption is not restrictive in respect of practical use cases encountered in the field of wireless, and allows us to perform derivations at a limited analytical cost, as can be seen in what follows. We describe the two bandpass signals by their complex envelopes $\tilde{s}_\text{w}(t)$ and $\tilde{n}(t)$ respectively, defined as centered around the angular frequencies ω_w and ω_n, respectively. Although our analytical derivations require these complex envelopes to be defined as centered around the same center angular frequency, we make a point of distinguishing between them at first in order to identify easily the different noise sidebands recovered at the nonlinear device output. In the same way, we suppose at first that the wanted bandpass signal we are dealing with can also be amplitude modulated. As will be seen below, this allows us to maintain the generality of the derivation at a limited additional analytical cost. Thus, we assume that the complex envelopes we are dealing with can be expressed in polar form as

$$\tilde{s}_\text{w}(t) = \rho_\text{w}(t)e^{j\phi_\text{w}(t)}, \tag{5.227a}$$

$$\tilde{n}(t) = \rho_\text{n}(t)e^{j\phi_\text{n}(t)}. \tag{5.227b}$$

As a result, the bandpass signal at the input of the nonlinear device can be written as

$$s_\text{i}(t) = \text{Re}\left\{\tilde{s}_\text{w}(t)e^{j\omega_\text{w}t} + \tilde{n}(t)e^{j\omega_\text{n}t}\right\}. \tag{5.228}$$

Thus, assuming that we can use the series expansion of the odd order part of the device transfer function up to third order, the components of interest in the bandpass signal recovered at the output of the nonlinear device, $s_o(t)$, can be derived from equation (5.121) as

$$\begin{aligned} s_o(t) = \mathrm{Re}\Bigg\{ & \left[G + \frac{3\alpha_3}{4}\left(\rho_w^2(t) + 2\rho_n^2(t)\right)\right] \tilde{s}_w(t)e^{j\omega_w t} \\ & + \left[G + \frac{3\alpha_3}{4}\left(\rho_n^2(t) + 2\rho_w^2(t)\right)\right] \tilde{n}(t)e^{j\omega_n t} \\ & + \frac{3\alpha_3}{4}\left(\tilde{s}_w^2(t)\tilde{n}^*(t)e^{j(2\omega_w-\omega_n)t} + \tilde{s}_w^*(t)\tilde{n}^2(t)e^{j(2\omega_n-\omega_w)t}\right) \Bigg\}. \end{aligned} \tag{5.229}$$

For our present purposes, we need to consider only the output sidebands whose center angular frequencies are close to that of the wanted signal, i.e. only the intermodulation terms in addition to the terms lying around the center angular frequencies of the input bandpass signals $s_w(t)$ and $n(t)$. This means that we do not consider the bandpass signals corresponding to the sidebands lying around the third harmonics or around the angular frequencies $2\omega_w + \omega_n$ or $\omega_w + 2\omega_n$. Those terms are expected to occur far away from the wanted signal carrier angular frequency in the frequency domain and thus have no impact on the in-band noise power.

However, although considering different center angular frequencies for the definition of the wanted signal and noise component complex envelopes remains convenient in order to clearly identify the sidebands generated by the nonlinear device, it is necessary to deal with complex envelopes assumed as centered around the same carrier angular frequency. This is the case, for instance, when we check the non-correlation between the bandpass signals corresponding to the different sidebands we are dealing with, by checking the non-correlation between the corresponding complex envelopes, as discussed in Appendix 1. Thus, let us consider from now on that $\omega_w = \omega_n$ so that the above equation directly gives the expression for the bandpass signal centered around the fundamental angular frequency as

$$\begin{aligned} s_{o,H1}(t) = \mathrm{Re}\Bigg\{ & \left[G - \frac{G}{2\mathrm{IIP3}}\left(\rho_w^2(t) + 2\rho_n^2(t)\right)\right] \tilde{s}_w(t)e^{j\omega_w t} \\ & + \left[G - \frac{G}{2\mathrm{IIP3}}\left(\rho_n^2(t) + 2\rho_w^2(t)\right)\right] \tilde{n}(t)e^{j\omega_w t} \\ & - \frac{G}{2\mathrm{IIP3}}\left(\tilde{s}_w^2(t)\tilde{n}^*(t) + \tilde{s}_w^*(t)\tilde{n}^2(t)\right)e^{j\omega_w t} \Bigg\}. \end{aligned} \tag{5.230}$$

In this expression, we have used equation (5.48a) to get a more suitable expression for $s_{o,H1}(t)$ in terms of the device IIP3 instead of α_3, still remembering that G and α_3 have opposite signs to physically describe compression, and that we are working on normalized impedances. Thus, despite our assumption that all the complex envelopes we are dealing with are defined as centered around the same center angular frequency ω_w, we can still clearly identify the origin of the various terms involved in the expression for $s_{o,H1}(t)$. We can thus write its complex envelopes, assumed as defined as centered around $\omega_w = \omega_n$ as the sum

$$\tilde{s}_{o,H1}(t) = \tilde{s}_{o,w}(t) + \tilde{n}_o(t) + \tilde{n}_{IMD}(t). \tag{5.231}$$

The first term on the right-hand side,

$$\tilde{s}_{o,w}(t) = G\left[1 - \frac{\left(\rho_w^2(t) + 2\rho_n^2(t)\right)}{2\text{IIP3}}\right]\tilde{s}_w(t), \tag{5.232}$$

represents the complex envelope of the distorted wanted signal when going through the nonlinear device. The second term,

$$\tilde{n}_o(t) = G\left[1 - \frac{\left(\rho_n^2(t) + 2\rho_w^2(t)\right)}{2\text{IIP3}}\right]\tilde{n}(t), \tag{5.233}$$

represents the complex envelope of the distorted noise when going through the nonlinear device. Finally, having that $2\omega_w > \omega_n$ and that $2\omega_n > \omega_w$,

$$\tilde{n}_{\text{IMD}}(t) = -\frac{G}{2\text{IIP3}}\left(\tilde{s}_w^2(t)\tilde{n}^*(t) + \tilde{s}_w^*(t)\tilde{n}^2(t)\right), \tag{5.234}$$

directly represents the complex envelope of the signal resulting from the intermodulation between the wanted signal and the input noise.

In order to derive the resulting SNR, we have to decompose the complex envelope $\tilde{s}_{o,\text{H1}}(t)$ as the sum of a term proportional to the complex envelope of the wanted signal through a constant factor that represents the effective gain G_e of the device, and an additional complex envelope, $\tilde{n}_\Sigma(t)$, that represents the overall bandpass noise term. For that purpose, the bandpass signal $n_\Sigma(t)$ represented by this complex envelope must be uncorrelated with the wanted bandpass RF signal when considered at the same sample time. Dealing with stationary processes, at least up to second order, as classically occur in wireless transceivers as discussed in Appendix 2, we can use the non-correlation between the complex envelopes defined as centered around the same carrier frequency as discussed in Appendix 1. Thus, we need to express $\tilde{s}_{o,\text{H1}}(t)$ in the form

$$\tilde{s}_{o,\text{H1}}(t) = G_e\tilde{s}_w(t) + \tilde{n}_\Sigma(t), \tag{5.235}$$

where $\tilde{n}_\Sigma(t)$ is uncorrelated with $\tilde{s}_w(t)$. Under this assumption, we can derive G_e by taking the correlation of both terms in the above equation with $\tilde{s}_w(t)$. We obtain

$$G_e = \frac{\mathbb{E}\{\tilde{s}_{o,\text{H1},t}\tilde{s}_{w,t}^*\}}{\mathbb{E}\{\tilde{s}_{w,t}\tilde{s}_{w,t}^*\}}. \tag{5.236}$$

Thus, in order to derive an expression for this effective gain, we need to evaluate the correlation of the complex envelope of the input wanted signal with the three terms that make up the complex envelope $\tilde{s}_{o,\text{H1}}(t)$ of the output signal as given by equation (5.231).

We start with $\tilde{n}_{\mathrm{IMD}}(t)$, given by equation (5.234). Assuming statistical independence between the wanted signal and the input noise term, the correlation between $\tilde{n}_{\mathrm{IMD}}(t)$ and $\tilde{s}_{\mathrm{w}}(t)$ reduces to

$$\mathbb{E}\{\tilde{n}_{\mathrm{IMD},t}\tilde{s}^*_{\mathrm{w},t}\} = -\frac{G}{2\mathrm{IIP3}}\left(\mathbb{E}\left\{\tilde{s}^2_{\mathrm{w},t}\tilde{s}^*_{\mathrm{w},t}\right\}\mathbb{E}\{\tilde{n}^*_t\} + \mathbb{E}\left\{\tilde{s}^{*2}_{\mathrm{w},t}\right\}\mathbb{E}\left\{\tilde{n}^2_t\right\}\right),$$

the sum of two terms proportional to $\mathbb{E}\{\tilde{n}^*_t\}$ and to $\mathbb{E}\{\tilde{n}^2_t\}$, respectively. But, dealing with a bandpass noise, we can assume that its complex envelope is centered so that $\mathbb{E}\{\tilde{n}^*_t\} = 0$. Moreover, as discussed in Appendix 2, the stationarity of this bandpass noise means that any of its complex envelopes fulfills equation (A2.11). Thus, $\mathbb{E}\{\tilde{n}^2_t\} = 0$ and, finally, $\mathbb{E}\{\tilde{n}_{\mathrm{IMD},t}\tilde{s}^*_{\mathrm{w},t}\} = 0$.

By similar reasoning, the correlation between $\tilde{s}_{\mathrm{w}}(t)$ and $\tilde{n}_{\mathrm{o}}(t)$ is the sum of terms proportional to either $\mathbb{E}\{\tilde{n}(t)\}$ or $\mathbb{E}\{\tilde{s}^*_{\mathrm{w}}(t)\}$. In both cases, if the processes involved are centered then the correlation is also null.

By these results, the effective gain G_{e} defined through equation (5.236) reduces to

$$G_{\mathrm{e}} = \frac{\mathbb{E}\{\tilde{s}_{\mathrm{o,w},t}\tilde{s}^*_{\mathrm{w},t}\}}{\mathbb{E}\{\tilde{s}_{\mathrm{w},t}\tilde{s}^*_{\mathrm{w},t}\}}, \tag{5.237}$$

or, using equation (5.232), to

$$G_{\mathrm{e}} = G\frac{\mathbb{E}\left\{\left(1 - \frac{\rho^2_{\mathrm{n},t}}{\mathrm{IIP3}} - \frac{\rho^2_{\mathrm{w},t}}{2\mathrm{IIP3}}\right)\rho^2_{\mathrm{w},t}\right\}}{\mathbb{E}\left\{\rho^2_{\mathrm{w},t}\right\}}. \tag{5.238}$$

This equation can be directly compared to equation (5.169). As the device we consider here exhibits compression, we are dealing in fact with the same kind of configuration as described in "Cross-modulation due to odd order nonlinearity" (Section 5.1.3). We can thus directly reuse the results of that section, and in particular equation (5.172), to give an expression for G_{e}. However, in the present case it is reasonable to assume a sufficiently good SNR at the input of the nonlinear device. Consequently, assuming that $P_{\mathrm{n}} \ll P_{\mathrm{w}}$, or that $\mathbb{E}\{\rho^2_{\mathrm{n}}\} \ll \mathbb{E}\{\rho^2_{\mathrm{w}}\}$ by equation (1.64), we can finally write that

$$G_{\mathrm{e}} = G\left[1 - \frac{\mathbb{E}\{\rho^2_{\mathrm{w}}\}}{2\mathrm{IIP3}}(\Gamma_{4,\mathrm{w}} + 1)\right], \tag{5.239}$$

where $\Gamma_{4,\mathrm{w}}$ is the constant defined by equation (5.125). This expression for G_{e} exactly matches the one derived during the study of the EVM generation due to compression in "Nonlinear EVM due to odd order nonlinearity" (Section 5.1.3). This is not surprising as the effective gain is driven by the signal that *effectively* enters the compression area of the device, i.e. the wanted signal only under our present high SNR assumption. Finally, we can decompose $\tilde{s}_{\mathrm{o,w}}(t)$ as

$$\tilde{s}_{\mathrm{o,w}}(t) = G_{\mathrm{e}}\tilde{s}_{\mathrm{w}}(t) + \tilde{n}_{\mathrm{EVM}}(t), \tag{5.240}$$

where $\tilde{n}_{\mathrm{EVM}}(t)$ is the EVM noise term driven by the compression of the wanted signal in the presence of the weak bandpass noise. Using equations (5.232) and (5.239), this term reduces to

$$\tilde{n}_{\mathrm{EVM}}(t) = G\frac{\mathbb{E}\{\rho_{\mathrm{w}}^2\}}{2\mathrm{IIP3}}\left[(\Gamma_{4,\mathrm{w}}+1) - \frac{\left(\rho_{\mathrm{w}}^2(t) + 2\rho_{\mathrm{n}}^2(t)\right)}{\mathbb{E}\{\rho_{\mathrm{w}}^2\}}\right]\tilde{s}_{\mathrm{w}}(t), \tag{5.241}$$

which can be compared to equation (5.190), but now with the presence of the additive noise component.

Thus, having checked on the one hand that $\tilde{n}_{\mathrm{IMD}}(t)$ and $\tilde{n}_{\mathrm{o}}(t)$ are uncorrelated with the complex envelope of the wanted signal, and on the other hand that $\tilde{s}_{\mathrm{o,w}}(t)$ can be decomposed as the superposition of a term proportional to the complex envelope of the wanted signal and a term $\tilde{n}_{\mathrm{EVM}}(t)$ uncorrelated with it, we can express the complex envelope of the total output noise $\tilde{n}_{\Sigma}(t)$ (equation (5.235)) as

$$\tilde{n}_{\Sigma}(t) = \tilde{n}_{\mathrm{EVM}}(t) + \tilde{n}_{\mathrm{o}}(t) + \tilde{n}_{\mathrm{IMD}}(t). \tag{5.242}$$

In this expression, the complex envelopes of the three noise components are given by equations (5.241), (5.233) and (5.234) respectively. Thus, assuming stationarity of the involved processes, we can now express the output SNR we are looking for as

$$SNR_{\mathrm{o}} = \frac{G_{\mathrm{e}}^2 P_{\mathrm{w}}}{P_{n_{\Sigma}}} = \frac{G_{\mathrm{e}}^2\mathbb{E}\{\tilde{s}_{\mathrm{w}}\tilde{s}_{\mathrm{w}}^*\}}{\mathbb{E}\{(\tilde{n}_{\mathrm{EVM}} + \tilde{n}_{\mathrm{o}} + \tilde{n}_{\mathrm{IMD}})(\tilde{n}_{\mathrm{EVM}} + \tilde{n}_{\mathrm{o}} + \tilde{n}_{\mathrm{IMD}})^*\}}. \tag{5.243}$$

Here, we observe that we are considering the *total* noise power in this SNR derivation. This means that we do not deal with any spectral shape of the power density, which would lead to considering only a fraction of the total noise power in a frequency band of interest, for instance. Considering the total noise power enables us to understand that the SNR improvement we derive in the present case is not related to any spread of the noise density over a wider frequency band due to nonlinearity. However, we will see in the next section through the example of the SSB noise case that this spreading effect can cumulate with the amplitude noise cancellation at the origin of the present overall improvement. But this effect is not considered in the present case. By expanding the above denominator, we see that we have to deal with the power of each output noise component and the three cross-correlation terms. We first focus on the cross products to check the non-correlation between these noise terms.

Let us first consider the correlation between $\tilde{n}_{\mathrm{EVM}}(t)$ and $\tilde{n}_{\mathrm{o}}(t)$. Referring to equations (5.241) and (5.233), we see that the expansion of $\mathbb{E}\{\tilde{n}_{\mathrm{EVM},t}\tilde{n}_{\mathrm{o},t}^*\}$ involves four different terms that are proportional to

$$\mathbb{E}\{\tilde{s}_{\text{w},t}\tilde{n}_t^*\}, \tag{5.244a}$$

$$\mathbb{E}\left\{\tilde{s}_{\text{w},t}^*\tilde{n}_t^2\tilde{n}_t^*\right\}, \tag{5.244b}$$

$$\mathbb{E}\left\{\tilde{s}_{\text{w},t}\left(\rho_{\text{n},t}^2+2\rho_{\text{w},t}^2\right)\tilde{n}_t^*\right\}, \tag{5.244c}$$

$$\mathbb{E}\left\{\tilde{s}_{\text{w},t}\left(\rho_{\text{w},t}^2+2\rho_{\text{n},t}^2\right)\left(\rho_{\text{n},t}^2+2\rho_{\text{w},t}^2\right)\tilde{n}_t^*\right\}. \tag{5.244b}$$

Assuming on the one hand the statistical independence of the wanted signal modulation and the noise processes, and on the other hand that we are dealing with centered processes, we get that equation (5.244a) is null. For the same reasons, equations (5.244b) and (5.244c) also reduce to zero. It then remains to consider equation (5.244d). Expansion of this expression leads to two kinds of terms. On the one hand, we have terms of the same form as treated just above, i.e. proportional to the expectations of either the wanted signal complex envelope or the noise process complex envelope. Those terms are therefore null. On the other hand, we have a term proportional to $\mathbb{E}\{\tilde{n}_t^*\tilde{n}_t\tilde{n}_t^*\}$. But, assuming that we are dealing with a Gaussian bandpass noise, we get from the results derived in Appendix 3 that the odd order moments of such a complex normal random process are null (see equation (A3.19)). Thus $\mathbb{E}\{\tilde{n}_t^*\tilde{n}_t\tilde{n}_t^*\}=0$, so $\tilde{n}_{\text{EVM}}(t)$ and $\tilde{n}_{\text{o}}(t)$ are indeed uncorrelated.

Let us now turn to the correlation between $\tilde{n}_{\text{IMD}}(t)$ and $\tilde{n}_{\text{o}}(t)$. By equations (5.233) and (5.234), expanding $\mathbb{E}\{\tilde{n}_{\text{IMD},t}\tilde{n}_{\text{o},t}^*\}$ leads to two different kinds of terms proportional to

$$\mathbb{E}\left\{\tilde{s}_{\text{w},t}^2\tilde{n}_t^{*2}\right\}, \tag{5.245a}$$

$$\mathbb{E}\left\{\tilde{s}_{\text{w},t}^*\tilde{n}_t^2\tilde{n}_t^*\right\}, \tag{5.245b}$$

$$\mathbb{E}\left\{\tilde{s}_{\text{w},t}^2\tilde{n}_t^{*2}\left(\rho_{\text{n},t}^2+2\rho_{\text{w},t}^2\right)\right\}, \tag{5.245c}$$

$$\mathbb{E}\left\{\tilde{s}_{\text{w},t}^*\tilde{n}_t^2\tilde{n}_t^*\left(\rho_{\text{n},t}^2+2\rho_{\text{w},t}^2\right)\right\}. \tag{5.245d}$$

Due to the independence between the complex envelope processes, the first of the above terms is in fact proportional to either $\mathbb{E}\left\{\tilde{s}_{\text{w},t}^2\right\}$ or $\mathbb{E}\left\{\tilde{n}_t^{*2}\right\}$. But, assuming that we are dealing with stationary bandpass signals, those terms are null by Appendix 2 (see equation (A2.11)), so that $\mathbb{E}\left\{\tilde{s}_{\text{w},t}^2\tilde{n}_t^{*2}\right\}=0$. The same argument can be used for the third term as equation (5.245c) is proportional to either $\mathbb{E}\left\{\tilde{s}_{\text{w},t}^2\right\}$ or $\mathbb{E}\left\{\tilde{n}_t^{*2}\right\}$, so that again $\mathbb{E}\left\{\tilde{s}_{\text{w},t}^2\tilde{n}_t^{*2}\left(\rho_{\text{n},t}^2+2\rho_{\text{w},t}^2\right)\right\}=0$. That leaves the second and fourth terms, given by equations (5.245b) and (5.245d), which again due to the independence between the processes we are dealing with are proportional to, or combination of terms proportional to, $\mathbb{E}\{\tilde{s}_{\text{w},t}^*\}$ and to $\mathbb{E}\{\tilde{n}_t\tilde{n}_t\tilde{n}_t^*\}$. Thus, both terms are also null assuming that we are dealing with centered processes and with a Gaussian bandpass noise. So

$$\mathbb{E}\{\tilde{n}_{\text{IMD},t}\tilde{n}_{\text{o},t}^*\}=0, \tag{5.246}$$

i.e. the two processes $\tilde{n}_{\text{IMD}}(t)$ and $\tilde{n}_{\text{o}}(t)$ can be considered uncorrelated.

We proceed similarly to deal with the correlation between $\tilde{n}_{\text{EVM}}(t)$ and $\tilde{n}_{\text{IMD}}(t)$. The expressions for $\tilde{n}_{\text{EVM}}(t)$, given by equation (5.241), and $\tilde{n}_{\text{o}}(t)$, given by equation (5.233), have the same structure. Thus, the same arguments as used above for the derivation of the non-correlation between $\tilde{n}_{\text{IMD}}(t)$ and $\tilde{n}_{\text{o}}(t)$ lead to the conclusion that

$$\mathbb{E}\{\tilde{n}_{\text{EVM},t}\tilde{n}^*_{\text{IMD},t}\} = 0, \tag{5.247}$$

i.e. that $\tilde{n}_{\text{EVM}}(t)$ and $\tilde{n}_{\text{IMD}}(t)$ are uncorrelated.

Thus, all the noise terms that make up $\tilde{n}_{\Sigma}(t)$ are uncorrelated. Their powers can therefore be summed to derive the total noise power at the nonlinear device output. The resulting SNR expression given by equation (5.243) then reduces to

$$SNR_{\text{o}} = \frac{G_{\text{e}}^2\mathbb{E}\{|\tilde{s}_{\text{w}}|^2\}}{\mathbb{E}\{|\tilde{n}_{\text{EVM}}|^2\} + \mathbb{E}\{|\tilde{n}_{\text{o}}|^2\} + \mathbb{E}\{|\tilde{n}_{\text{IMD}}|^2\}}. \tag{5.248}$$

It then remains to derive the power of the three noise components we are dealing with. For that purpose, let us first consider the EVM noise term. From equation (5.241), and due to the stationarity of the processes considered, we can write that

$$\mathbb{E}\{|\tilde{n}_{\text{EVM}}|^2\} = G^2\left(\frac{\mathbb{E}\{\rho_{\text{w}}^2\}}{2\text{IIP3}}\right)^2 \mathbb{E}\left\{\left[(\Gamma_{4,\text{w}} + 1) - \frac{(\rho_{\text{w}}^2 + 2\rho_{\text{n}}^2)}{\mathbb{E}\{\rho_{\text{w}}^2\}}\right]^2 \rho_{\text{w}}^2\right\}.$$

This expression can be further simplified. We have already assumed that the SNR is good enough at the input of the nonlinear device so that $\mathbb{E}\{\rho_{\text{n}}^2\} \ll \mathbb{E}\{\rho_{\text{w}}^2\}$. By equation (5.133) we can also neglect the higher order moments of ρ_{n} in comparison to those of ρ_{w}. Thus, when expanding the right-hand expectation in the above equation, the remaining terms correspond to the contribution of the wanted signal only. We obtain exactly the same expression as in equation (5.194):

$$\mathbb{E}\{|\tilde{n}_{\text{EVM}}|^2\} = G^2\left(\frac{\mathbb{E}\{\rho_{\text{w}}^2\}}{2\text{IIP3}}\right)^2 \mathbb{E}\{\rho_{\text{w}}^2\}[(\Gamma_{6,\text{w}} + 1) - (\Gamma_{4,\text{w}} + 1)^2]. \tag{5.249}$$

Let us now derive $\mathbb{E}\{|\tilde{n}_{\text{o}}|^2\}$. From equation (5.233) we can write that

$$\mathbb{E}\{|\tilde{n}_{\text{o}}|^2\} = G^2\mathbb{E}\left\{\left[1 - \frac{(\rho_{\text{n}}^2 + 2\rho_{\text{w}}^2)}{2\text{IIP3}}\right]^2 \rho_{\text{n}}^2\right\}. \tag{5.250}$$

Thus, using on the one hand the statistical independence between the processes involved, and on the other hand the possibility of neglecting the moments of the same order of ρ_{n} compared to those of ρ_{w}, we have

$$\mathbb{E}\{|\tilde{n}_{\text{o}}|^2\} = G^2\left(1 - 2\frac{\mathbb{E}\{\rho_{\text{w}}^2\}}{\text{IIP3}} + \frac{\mathbb{E}\{\rho_{\text{w}}^4\}}{\text{IIP3}^2}\right)\mathbb{E}\{\rho_{\text{n}}^2\}. \quad (5.251)$$

Finally, using the $\Gamma_{4,\text{w}}$ constant, we can write for this output noise term that

$$\mathbb{E}\{|\tilde{n}_{\text{o}}|^2\} = G^2\left[1 - 2\frac{\mathbb{E}\{\rho_{\text{w}}^2\}}{\text{IIP3}} + (\Gamma_{4,\text{w}} + 1)\left(\frac{\mathbb{E}\{\rho_{\text{w}}^2\}}{\text{IIP3}}\right)^2\right]\mathbb{E}\{\rho_{\text{n}}^2\}. \quad (5.252)$$

All that remains is to derive the power of the output intermodulation components. Using equation (5.234), we can write that

$$\mathbb{E}\{|\tilde{n}_{\text{IMD}}|^2\} = \frac{G^2}{4\text{IIP3}^2}\mathbb{E}\left\{|\tilde{s}_{\text{w},t}^2\tilde{n}_t^*|^2 + |\tilde{s}_{\text{w},t}^*\tilde{n}_t^2|^2 + \tilde{s}_{\text{w},t}^3\tilde{n}_t^{*3} + \tilde{s}_{\text{w},t}^{*3}\tilde{n}_t^3\right\}.$$

We assume that the noise component we are dealing with can be modeled as a Gaussian process. Thus, as highlighted previously, the odd order moments of such a Gaussian process are null (see Appendix 3). We can then write in the present case that

$$\mathbb{E}\{|\tilde{n}_{\text{IMD}}|^2\} = \frac{G^2}{4\text{IIP3}^2}\mathbb{E}\left\{|\tilde{s}_{\text{w},t}^2\tilde{n}_t^*|^2 + |\tilde{s}_{\text{w},t}^*\tilde{n}_t^2|^2\right\}. \quad (5.253)$$

Expanding the modulus and due to the independence between the modulating complex envelopes, we get that

$$\begin{aligned}\mathbb{E}\{|\tilde{n}_{\text{IMD}}|^2\} &= \frac{G^2}{4\text{IIP3}^2}\mathbb{E}\left\{\rho_{\text{w}}^4\rho_{\text{n}}^2 + \rho_{\text{n}}^4\rho_{\text{w}}^2\right\} \\ &= \frac{G^2}{4\text{IIP3}^2}\left(\mathbb{E}\{\rho_{\text{w}}^4\}\mathbb{E}\{\rho_{\text{n}}^2\} + \mathbb{E}\{\rho_{\text{n}}^4\}\mathbb{E}\{\rho_{\text{w}}^2\}\right). \quad (5.254)\end{aligned}$$

Using the Γ_4 constants for both the wanted signal and the noise components, we can rewrite this equation as

$$\mathbb{E}\{|\tilde{n}_{\text{IMD}}|^2\} = \frac{G^2}{4\text{IIP3}^2}\mathbb{E}\{\rho_{\text{w}}^2\}\mathbb{E}\{\rho_{\text{n}}^2\}\left[(\Gamma_{4,\text{w}} + 1)\mathbb{E}\{\rho_{\text{w}}^2\} + (\Gamma_{4,\text{n}} + 1)\mathbb{E}\{\rho_{\text{n}}^2\}\right].$$

But, as we assumed on the one hand that the SNR at the input of the nonlinear device is good enough so that $\mathbb{E}\{\rho_{\text{n}}^2\} \ll \mathbb{E}\{\rho_{\text{w}}^2\}$, and on the other hand that the Γ_4 constants are bounded in practice by 1 (see "Moments of the instantaneous amplitude" (Section 5.1.3)), we finally have

$$\mathbb{E}\{|\tilde{n}_{\text{IMD}}|^2\} = G^2 \left(\frac{\mathbb{E}\{\rho_{\text{w}}^2\}}{2\text{IIP3}} \right)^2 (\Gamma_{4,\text{w}} + 1)\mathbb{E}\{\rho_{\text{n}}^2\}. \tag{5.255}$$

We can now derive an expression for the change in the SNR due to compression. Using equations (5.239), (5.249), (5.252) and (5.255) we can rewrite equation (5.248) as

$$SNR_{\text{o}} = SNR_{\text{i}} \frac{\left[1 - \frac{P_{\text{w}}}{\text{IIP3}}(\Gamma_{4,\text{w}} + 1)\right]^2}{D(P_{\text{w}})}, \tag{5.256}$$

with $SNR_{\text{i}} = \mathbb{E}\{\rho_{\text{w}}^2\}/\mathbb{E}\{\rho_{\text{n}}^2\}$ and the denominator $D(P_{\text{w}})$ given by

$$\begin{aligned} D(P_{\text{w}}) = {} & SNR_{\text{i}} \left(\frac{P_{\text{w}}}{\text{IIP3}} \right)^2 [(\Gamma_{6,\text{w}} + 1) - (\Gamma_{4,\text{w}} + 1)^2] \\ & + \left[1 - 4\frac{P_{\text{w}}}{\text{IIP3}} + 5(\Gamma_{4,\text{w}} + 1) \left(\frac{P_{\text{w}}}{\text{IIP3}} \right)^2 \right]. \end{aligned} \tag{5.257}$$

The first term in this denominator represents nothing more than the fraction of the overall noise linked to the EVM term resulting from the compression of the non-constant amplitude wanted signal. The second term represents the output noise component that is linked either directly to the input bandpass noise or to its intermodulation with the wanted signal. As might be expected, the ratio between the EVM term and the other output noise terms is proportional to the input SNR. Indeed, as soon as the input noise floor is low enough compared to the input signal power, the overall output noise is dominated by the EVM term generated by the input wanted signal compression. Thus, in order to be able to obtain an SNR improvement, we need to cancel this EVM term, which increases with the input wanted signal power. We thus need to deal with a constant amplitude wanted signal. This assumption corresponds to all the constants $\Gamma_{2k,\text{w}}$ being null, and thus the power of the EVM noise term also, by equation (5.249). Under this assumption we can reconsider equations (5.256) and (5.257) so that the SNR improvement through the nonlinear device, evaluated through the ratio

$$K_{\text{DSB}} = \frac{SNR_{\text{o}}}{SNR_{\text{i}}}, \tag{5.258}$$

now reduces to

$$K_{\text{DSB}} = \frac{\left(1 - \frac{P_{\text{w}}}{\text{IIP3}}\right)^2}{\left(1 - 2\frac{P_{\text{w}}}{\text{IIP3}}\right)^2 + \left(\frac{P_{\text{w}}}{\text{IIP3}}\right)^2}. \tag{5.259}$$

Note the subscript DSB and remember that we have not so far made any assumption on the spectral configuration of the input signals. Indeed, the above result remains valid in all cases as long as we consider the overall noise power at both the input and output of the nonlinear device. However, as discussed in the next section, a symmetrization of the noise spectrum with regard to the wanted signal carrier angular frequency also occurs. It follows that, considering the fraction of the noise that lies within the bandwidth of the input bandpass noise, the ratio K_{DSB} indeed represents the SNR improvement only in the DSB input noise case. To put it another way, K_{DSB} represents the noise floor improvement only when the input bandpass noise has a DSB PSD regarding the wanted signal carrier angular frequency.

As this phenomenon is of interest mainly when reaching the nonlinear device saturation, we can express the K_{DSB} factor using the device Psat. This allows us to express the resulting SNR in terms of the ratio of the back-off of the wanted signal average power to the device Psat. Using equation (5.59), the above equation reduces to

$$K_{DSB} = \frac{\left(3 - \frac{P_W}{\text{IPsat}}\right)^2}{\left(3 - 2\frac{P_W}{\text{IPsat}}\right)^2 + \left(\frac{P_W}{\text{IPsat}}\right)^2}. \tag{5.260}$$

Here we recall that this result has been derived using the third order series expansion of the device odd order transfer function. It thus remains valid only for the wanted signal input power up to IPsat as detailed in "Odd order nonlinearity and IP3, CP1 or Psat" (Section 5.1.2). The limit values of K_{DSB} predicted by our model are then:

$$K_{DSB} = \begin{cases} 1 & \text{when } P_W = 0, \\ 2 & \text{when } P_W = \text{IPsat}. \end{cases} \tag{5.261}$$

As illustrated in Figure 5.22, the SNR improvement can reach 3 dB in the present case.

This SNR improvement is confirmed by the power spectral densities shown in Figure 5.23. To derive these, a white Gaussian noise, bandpass filtered, has been generated and added to a CW signal whose carrier frequency is also the center frequency of the noise PSD. The overall input SNR is set to 15 dB. This signal plus noise then experiences a compression modeled by a transfer function series expansion up to third order as used in the present derivation. The resulting power spectral densities are shown for different ratios between the input CW signal power back-off and the device Psat. The output power of the wanted signal tone is normalized to compensate for the gain loss effect. This allows us to see the SNR improvement on the output noise floor level. It is clear that when the input signal power reaches the device Psat, the SNR improvement reaches 3 dB, as stated by our derivation. Looking at those simulation results, we remark that we do not see any spectral regrowth of the noise component, whereas it is not a constant amplitude signal. This is surprising in light of the discussion in "Spectral regrowth" (Section 5.1.3). But, as we have used a large enough input SNR for the simulation, the noise component has a low enough amplitude so that the additional output terms related to its amplitude have very weak power so that it has a negligible impact on the power spectral densities displayed in this figure. This behavior in fact corresponds to the assumptions used

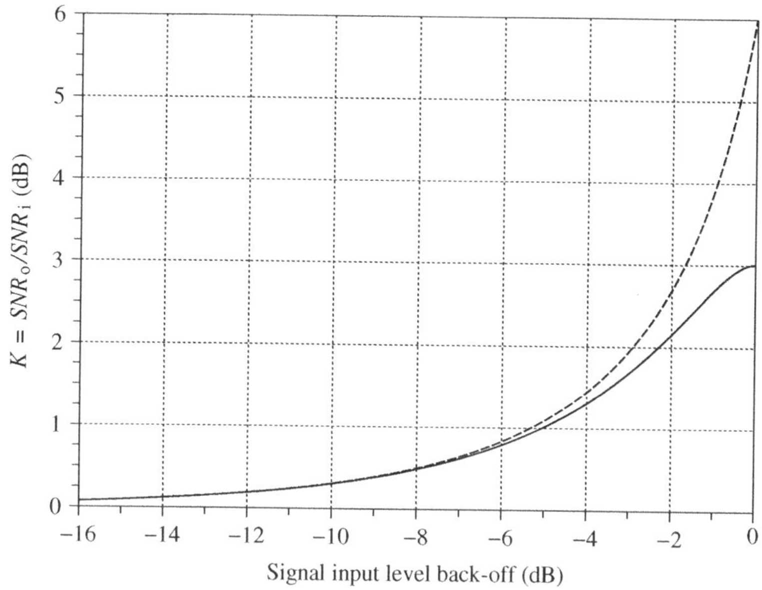

Figure 5.22 SNR improvement due to the compression of an input constant amplitude signal plus a bandpass Gaussian noise – For a DSB noise density, the noise floor improvement, K_{DSB}, through a nonlinear device that exhibits third order compression behavior, can reach 3 dB as given by equation (5.260) (solid). For an SSB noise density, K_{SSB} can reach 6 dB, as given by equation (5.270) (dashed). Here, the abscissa represents the ratio of the back-off of the wanted signal to the saturated input power of the amplifier, i.e. P_w/IPsat, expressed in decibels.

for our derivation. The terms relating to the noise spectral regrowth indeed correspond to those that we have been neglecting in our derivations so far. Therefore, the fact that we do not see any spectral regrowth in Figure 5.23 confirms that this assumption is valid.

Let us now try to understand the phenomenon underlying this SNR improvement. Dealing with a constant amplitude wanted signal, i.e. with $\rho_w(t) = \rho_w$, we have found that the EVM noise part is null. It then remains to consider the noise component corresponding to the input one, i.e. $n_o(t)$, and the intermodulation noise component $n_{IMD}(t)$ whose complex envelopes are given in the general case by equations (5.233) and (5.234), respectively. However, having ρ_w constant also impacts the structure of those complex envelopes. Indeed, we can assume that for realistic physical realizations of the noise process, its instantaneous amplitude remains within a given range above its RMS value, linked to the PAPR of the bandpass signal. But, assuming that we have a good input SNR, i.e. that $\mathbb{E}\{\rho_n^2\} \ll \rho_w^2$, we can now expect $\rho_n^2(t) \ll \rho_w^2$ always to hold. Thus, we can simplify equations (5.233) and (5.234) in the present case:

$$\tilde{n}_o(t) = G\left(1 - \frac{\rho_w^2}{3\mathrm{IPsat}}\right)\tilde{n}(t), \tag{5.262a}$$

$$\tilde{n}_{IMD}(t) = -G\frac{\rho_w^2}{6\mathrm{IPsat}}\tilde{n}^*(t)e^{j2\phi_w(t)}. \tag{5.262b}$$

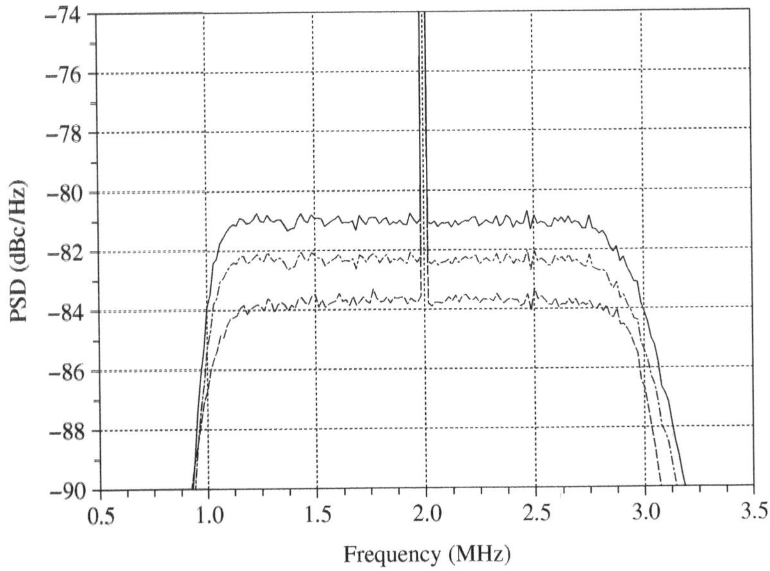

Figure 5.23 Noise floor improvement due to the compression of a CW signal plus a DSB bandpass Gaussian noise – When a CW signal plus a DSB bandpass Gaussian noise go through a nonlinear device that exhibits third order compression behavior, the noise floor improvement can reach 3 dB. Here, the PSD has been estimated from time domain simulations and the power of the output tone is normalized so that the noise floor level variation directly reflects the SNR improvements. Compared to the input situation (solid), the noise floor improvement can reach 1.3 dB for an input back-off, i.e. for an input value of P_w/IPsat, of 4 dB (dot-dashed), and 3 dB for an input back-off of 0 dB (dashed). This is in line with the K_{DSB} factor expression given by equation (5.260) and represented in Figure 5.22.

Here we have used equation (5.59) to write the expressions in terms of the device Psat instead of the IIP3. When reaching full compression, i.e. when $\rho_w = \rho_{IPsat}$, these complex envelopes reduce to

$$\tilde{n}_o(t) = \frac{G}{3}\tilde{n}(t) = \frac{G}{3}\rho_n(t)e^{j\phi_n(t)}, \tag{5.263a}$$

$$\tilde{n}_{IMD}(t) = -\frac{G}{3}\tilde{n}^*(t)e^{j2\phi_w(t)} = \frac{G}{3}\rho_n(t)e^{j(\pi-\phi_n(t)+2\phi_w(t))}. \tag{5.263b}$$

As illustrated in Figure 5.24, in that case the two output noise components have the same instantaneous amplitude and, due to their phase relationship, the sum of their complex envelopes, $\tilde{n}_\Sigma(t)$, remains *orthogonal* to the complex envelope $\tilde{s}_{o,w}(t)$ of the wanted signal. Thus, up to first order, $\tilde{n}_\Sigma(t)$ corrupts only the argument of $\tilde{s}_{o,w}(t)$. It thus represents a phase noise term only. All thus behaves as if the amplitude noise part of the input bandpass noise $n(t)$ has been canceled by the compression. Only the phase noise part succeeds in going through the device. But, by the discussion in Section 1.2.2, the power of an additive Gaussian bandpass noise is equally split between the bandpass term whose complex envelope is orthogonal to that of the wanted signal, i.e. the phase noise part, and the term with its complex envelope parallel to

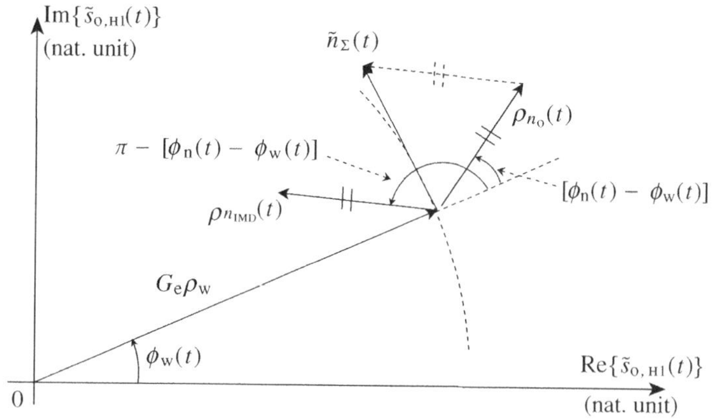

Figure 5.24 Noise amplitude cancellation mechanism due to the compression of a CW wanted signal plus a bandpass Gaussian noise – When the input wanted signal power reaches the Psat of the device, the complex envelope of the output noise component, given by equation (5.263a), has amplitude and phase relationships with the intermodulation noise complex envelope, given by (5.263b), such that their sum $\tilde{n}_\Sigma(t)$ remains orthogonal to the wanted signal complex envelope when plotted in the complex plane. Consequently, the total output noise only impacts the phase of the wanted bandpass signal up to first order. The amplitude part of the additive bandpass noise has thus been canceled.

it, i.e. the amplitude noise part. We can thus understand that canceling this latter component leads to a 3 dB reduction in the noise power, and thus in a 3 dB SNR improvement in the particular case where the wanted signal is insensitive to compression.

Noise Spectrum Symmetrization

On top of the effect of canceling the amplitude part of an additive bandpass noise, compression leads to an interesting property for the spectrum of the generated noise components. In the present case, the phenomenon underlying this spectral behavior can be illustrated by assuming that the wanted signal reduces to a pure CW tone. Recalling the discussion in the previous section, this means that we now consider that $\phi_w(t) = 0$ on top of having a constant instantaneous amplitude for this signal. The complex envelopes of the two non-vanishing noise components recovered at the output of the nonlinear device, i.e. $n_o(t)$ and $n_{IMD}(t)$, originally given by equation (5.262), now reduce to

$$\tilde{n}_o(t) = G\left(1 - \frac{\rho_w^2}{3\text{IPsat}}\right)\tilde{n}(t), \tag{5.264a}$$

$$\tilde{n}_{IMD}(t) = -G\frac{\rho_w^2}{6\text{IPsat}}\tilde{n}^*(t). \tag{5.264b}$$

We thus have that

$$\tilde{n}_{IMD}^*(t) \propto \tilde{n}_o(t), \tag{5.265}$$

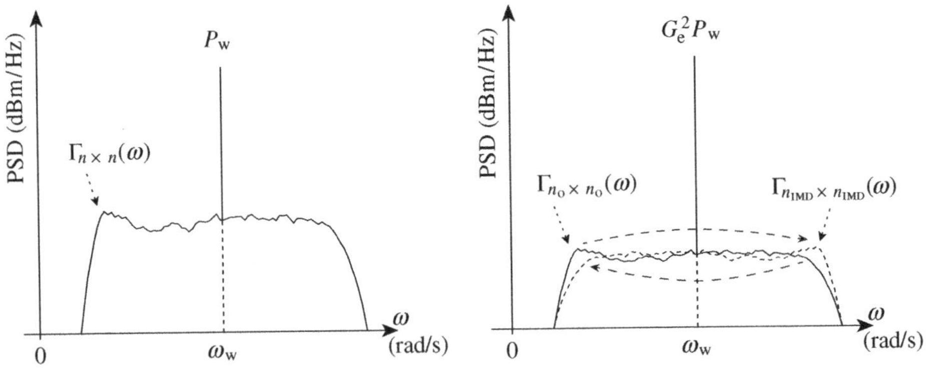

Figure 5.25 Power spectral representation of the compression of a CW signal plus a DSB bandpass Gaussian noise – The RF compression of a CW signal plus a bandpass Gaussian noise (left) generates an intermodulation noise component whose spectrum is *symmetric* to the input one regarding the wanted signal carrier frequency (right). Here, the input wanted signal power is assumed to be equal to the device Psat so that the two output noise components have same power.

i.e. that the complex envelopes of the two output noise terms are complex conjugates to each other. Recalling the Fourier transform property given by equation (1.8), this means that the PSD of these noise components are flipped copies of each other. As the complex envelopes we are dealing with are defined as centered around the same carrier frequency ω_w, the center of symmetry for this flip phenomenon is this carrier angular frequency itself. This symmetrization is in fact something we might have imagined from the simple intermodulation mechanism presented in "Odd order nonlinearity and IP3, CP1 or Psat" (Section 5.1.2). Indeed, dealing with two input tones, one lying at ω_w with a power P_w, the other one lying at ω_n with a power P_n, and such that $P_n \ll P_w$, the IMD3 tone of non-negligible power is the one lying at the angular frequency $2\omega_w - \omega_n$ (ω_w is the center of symmetry between ω_n and $2\omega_w - \omega_n$). Applying this behavior to all the "slices" of the spectrum of $\tilde{n}(t)$, we see that the spectrum of $\tilde{n}_{IMD}(t)$ is necessarily a flipped copy of that of $\tilde{n}_o(t)$ around ω_w when the complex envelopes are defined as centered around this particular angular frequency. This behavior is illustrated in Figure 5.25 in the particular case where $\rho_w = \rho_{IPsat}$, i.e. when the above complex envelopes reduce to

$$\tilde{n}_o(t) = \frac{G}{3}\tilde{n}(t), \tag{5.266a}$$

$$\tilde{n}_{IMD}(t) = -\frac{G}{3}\tilde{n}^*(t). \tag{5.266b}$$

In that case the two noise components also have the same instantaneous amplitude, and thus power.

This phenomenon of noise spectrum symmetrization around the carrier angular frequency is of particular interest when this input noise can be assumed SSB. Such a configuration corresponds for instance to the wanted RF signal lying at one extreme end of the bandpass noise spectrum. As shown in Figure 5.26, the resulting intermodulation component at the device output now lies outside the transmit band. We can thus derive the noise floor improvement in

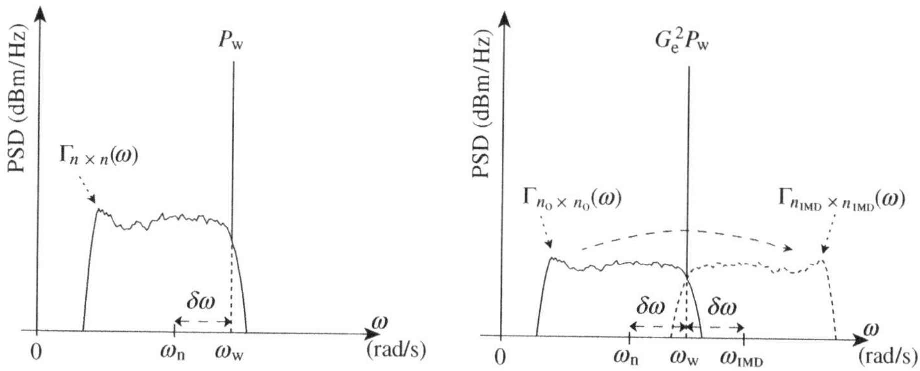

Figure 5.26 Power spectral representation of the compression of a CW signal plus a SSB bandpass Gaussian noise – Compared to the DSB case represented in Figure 5.25, we now have a spreading of the total noise power over a wider bandwidth due to the non-overlapping of the two output noise sidebands.

terms of the original noise sideband residual power relative to the wanted signal power. In fact, this noise power P_{n_o} can be derived from equation (5.264a) as a function of both the device Psat and the input wanted signal average power P_w. Given that $P_w = \rho_w^2/2$ when working on normalized impedances, we then get that

$$P_{n_o} = G^2\left(1 - \frac{2}{3}\frac{P_w}{\text{IPsat}}\right)^2 P_n. \tag{5.267}$$

In the same way, we can express the effective gain experienced by our CW wanted signal from equation (5.239), but now supposing that $\Gamma_{4,w} = 0$. Using equation (5.59) to express the result in terms of the device IPsat, we get

$$G_e = G\left[1 - \frac{1}{3}\frac{P_w}{\text{IPsat}}\right], \tag{5.268}$$

(compare the numerator of equation (5.260)). From these results, the output in-band SNR is equal to

$$SNR_o = \frac{G_e^2 P_w}{P_{n_o}} = K_{SSB}\, SNR_i, \tag{5.269}$$

with

$$K_{SSB} = \frac{\left(3 - \dfrac{P_w}{\text{IPsat}}\right)^2}{\left(3 - 2\dfrac{P_w}{\text{IPsat}}\right)^2}. \tag{5.270}$$

Recalling that so far we have been using the third order series expansion of the device odd order transfer function, the results are valid only for the wanted signal input power up to IPsat. As can be seen in Figure 5.22, the limit values of K_{SSB} predicted by our model are then

$$K_{\text{SSB}} = \begin{cases} 1 & \text{when } P_{\text{w}} = 0, \\ 4 & \text{when } P_{\text{w}} = \text{IPsat}. \end{cases} \tag{5.271}$$

These derivations confirm the two effects we are investigating. First, we have the cancellation of the amplitude part of the noise as described in the previous section. This phenomenon leads to the first 3 dB improvement when the wanted signal input power reaches the device Psat. Second, we now have a spreading of the noise PSD over twice the initial bandwidth. But recalling the discussion in the previous section, and in particular equation (5.246), $n_o(t)$ and $n_{\text{IMD}}(t)$ are uncorrelated. We can thus deduce that the power of the generated intermodulation noise term is directly retrieved from the input noise. In the same way, $n_o(t)$ and $n_{\text{IMD}}(t)$ have the same power when the wanted signal power reaches the device IPsat. As a result, the power of $n_o(t)$ is 3 dB lower than the power of the total bandpass noise, whichs explains the additional 3 dB improvement in the noise floor considering only the initial input noise sideband. Nevertheless, it is important to keep in mind that the phenomenon we are talking about is really a symmetrization of the noise spectrum with regard to the wanted signal carrier frequency and it cannot be related to a potential spectral regrowth effect of the input noise term, as discussed throughout "Spectral regrowth" (Section 5.1.3). Indeed, the present phenomenon does not exist if the wanted signal is not present.

The present behavior is confirmed in another way by the time domain simulations shown in Figure 5.27. For this, we use the same generated white Gaussian bandpass noise added to a CW signal as we used in the DSB case. The difference is that now the wanted CW signal frequency is at the edge of the bandpass noise. This signal plus noise then experiences a compression modeled by the third order expansion of an odd transfer function. The resulting power spectral densities are shown for different input signal back-off compared to the device Psat. We thus see that, as expected, when the input signal power reaches the device Psat, on the one hand the SNR improvement considering the initial noise sideband reaches 6 dB and on the other hand the intermodulation noise power reaches the same level as the power of the noise lying in the initial sideband.

5.2 Hard AM-AM Conversion

As discussed in the previous sections the phenomenon of SNR improvement due to compression can be used only with constant instantaneous amplitude bandpass RF signals. But, in practice, we take advantage of the robustness of such constant instantaneous amplitude signals through the use of RF amplifiers that truly operate in the deep nonlinear area of their transfer function. This is obviously done with the aim of minimizing their power consumption and thus optimizing the efficiency of the overall implementation.

From the signal processing point of view, such devices behave more like hard limiters than like devices exhibiting smooth nonlinearity as considered up to now. We therefore review some of the results related to the hard limiter theory in order to make the link with the arguments rehearsed throughout Section 5.1.

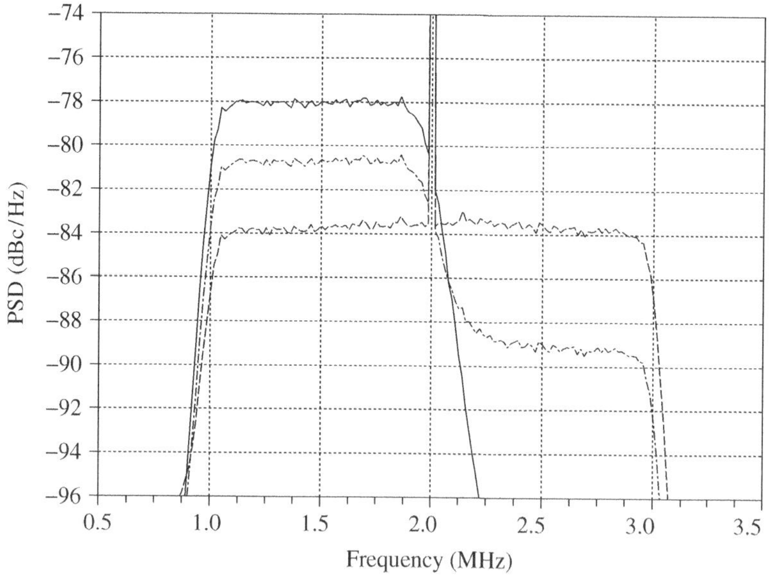

Figure 5.27 Noise floor improvement due to the compression of a CW signal plus a SSB bandpass Gaussian noise – When a CW signal plus a SSB bandpass Gaussian noise experience compression, the noise floor improvement can reach 6 dB. Here, the PSD has been estimated from time domain simulations and the power of the output tone is normalized so that the noise floor level variations directly reflect the SNR improvements. Compared to the input situation (solid), the noise floor improvement can reach 2.7 dB for an input back-off, i.e. for an input value of P_w/IPsat, of 2 dB (dot-dashed) and 6 dB for an input back-off of 0 dB (dashed). This is in line with the K_{SSB} factor expression given by equation (5.270) and shown in Figure 5.22.

5.2.1 Hard Limiter Model

But first, let us say a few words on what we mean by “hard limiter”. Here we are dealing with a device whose transfer function can be approximated by a simple sign function, defined by

$$\text{sign}\{x\} = \begin{cases} 1 & \text{when } x > 0, \\ 0 & \text{when } x = 0, \\ -1 & \text{when } x < 0, \end{cases} \tag{5.272}$$

and weighted by a scaling parameter α representing the maximum output amplitude the device can deliver in the time domain. In other words, the bandpass signal $y(t)$ we recover at the output of the hard limiter when the bandpass signal $x(t)$ is applied to its input takes the simple form

$$y(t) = \alpha\, \text{sign}\{x(t)\}. \tag{5.273}$$

The corresponding transfer function and associated signals are shown in Figure 5.28.

The transfer function we are considering is a pure *odd* function. Practically speaking, this means that we exclude from our present discussion devices such as half wave rectifiers.

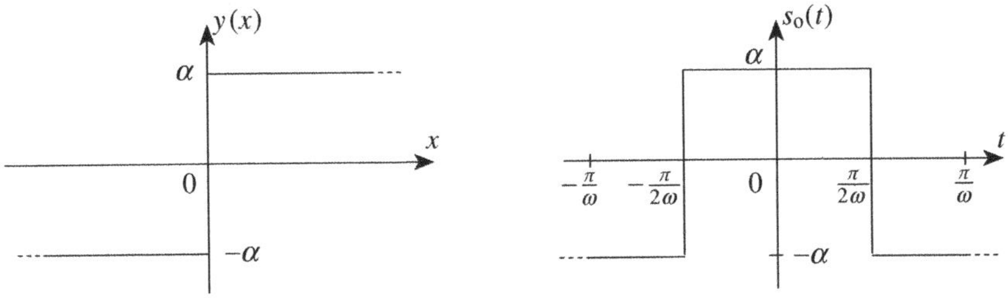

Figure 5.28 Hard limiter transfer function and corresponding output waveform for an input sine wave – The hard limiter transfer function for bandpass signals can be modeled as $\alpha\,\text{sign}\{x\}$ (left). The resulting output waveform in response to an input CW signal of the form $\rho\cos(\omega t)$ is therefore a square wave with the same angular frequency (right).

Although the signal going through such devices experiences hard clipping, it nevertheless results in an output signal that is in essence non-symmetric in terms of positive–negative alternations. This means that the transfer function of such half wave rectifiers contains a non-vanishing even order part (recall the discussion in Section 5.1.1). From the perspective of discussing the SNR improvement linked to RF compression, there is no particular need to worry about this even order part, nor such devices with an unbalanced transfer function.

We also remark that the transfer function we consider remains theoretical in the sense that the output signal amplitude can reach the limit value α only when the input signal amplitude is already sufficiently high, i.e. above a given input level. This is related to the finite gain that we can expect from any realistic physical implementation. In practice, this condition on the input signal amplitude indeed corresponds to the practical operating conditions for devices such as saturated PAs to work correctly. As long as this condition is fulfilled, our model is consistent with behaviors that may be encountered.

5.2.2 Hard Limiter Intercept Points

As for any RF device, the linearity of a device that behaves like a hard limiter can be characterized through its IPs. In the present case where the device transfer function is given by equation (5.273), it is of interest to see that the theoretical expression for the IP of any order can be derived in a straightforward way as a function of the single parameter of this transfer function, i.e. α. The structure for those IPs can then be used to check that a given device indeed behaves like a hard limiter, for instance.

Thus, let us reconsider the configuration used to derive the IP of a device as presented in "Characterization of RF device nonlinearity" (Section 5.1.2). Such derivation involves considering an input signal that is the superposition of two sine waves with the same amplitude ρ_{cw}, and with angular frequencies ω_1 and ω_2. We then assume the input signal

$$s_{\text{i}}(t) = \rho_{\text{cw}}(\cos(\omega_1 t) + \cos(\omega_2 t)). \tag{5.274}$$

Thus, using equation (5.273), we can then write the output signal $s_{\text{o}}(t)$ as

$$s_{\text{o}}(t) = \alpha\,\text{sign}\{\cos(\omega_1 t) + \cos(\omega_2 t)\}. \tag{5.275}$$

We can express this signal as a function of the product of sine waves instead of the sum of such sine waves (recalling that the sign of a product is the product of the signs). Using equation (5.19) we can write that

$$s_o(t) = \alpha\, \text{sign}\{\cos(\omega_\sigma t)\}\text{sign}\{\cos(\omega_\delta t)\}, \tag{5.276}$$

with

$$\omega_\sigma = \frac{\omega_2 + \omega_1}{2}, \tag{5.277a}$$

$$\omega_\delta = \frac{\omega_2 - \omega_1}{2}. \tag{5.277b}$$

We thus see that the output signal $s_o(t)$ is proportional to the product of two periodic square waves, $\Pi_\sigma(t) = \text{sign}\{\cos(\omega_\sigma t)\}$ and $\Pi_\delta(t) = \text{sign}\{\cos(\omega_\delta t)\}$. In order to derive the amplitudes of the intermodulation tones, we can expand each of these signals as their Fourier series.

Taking $\Pi_\sigma(t)$ first, we can write

$$\Pi_\sigma(t) = \text{sign}\{\cos(\omega_\sigma t)\} = a_0 + \sum_{k=1}^{\infty} a_k \cos(k\omega t) + b_k \sin(k\omega t). \tag{5.278}$$

Due to the fact that $\cos(\omega_\sigma t)$ is an even function, we have only even terms in the Fourier series expansion of $\Pi_\sigma(t)$. Thus $b_k = 0$ for all k. In the same way, $a_0 = 0$ as a_0 represents the average value of $\Pi_\sigma(t)$ over a period. The only remaining non-vanishing terms are therefore

$$a_k = \frac{\omega_\sigma}{\pi} \int_{-\frac{\pi}{\omega_\sigma}}^{\frac{\pi}{\omega_\sigma}} \Pi_\sigma(t)\cos(k\omega_\sigma t)\text{d}t = \frac{4}{k\pi} \sin\left(k\frac{\pi}{2}\right), \tag{5.279}$$

for odd values of k. Thus, assuming that $k = 2l + 1$ with $l \in \mathbb{N}$, we get that

$$a_{2l+1} = \frac{4}{\pi} \frac{(-1)^l}{2l+1}. \tag{5.280}$$

Therefore,

$$\Pi_\sigma(t) = \frac{4}{\pi} \sum_{l=0}^{\infty} \frac{(-1)^l}{2l+1} \cos((2l+1)\omega_\sigma t). \tag{5.281}$$

As expected, due to the odd behavior of the transfer function assumed for the device, we have only odd order harmonics of the input signal present at the device output. We also remark that the amplitude of the fundamental tone of this series is $4/\pi$, which is greater than the amplitude of the signal $\Pi_\sigma(t)$. This may appear awkward, but in fact allows the sum of all of the tones involved in the series expansion to be equal to 1 in the time domain, i.e. to the amplitude of $\Pi_\sigma(t)$. As the same expansion holds for $\Pi_\delta(t)$ when substituting ω_δ into ω_σ, we can finally write the series expansion of $s_o(t)$ from equation (5.276) as

$$s_o(t) = \alpha \left(\frac{4}{\pi}\right)^2 \sum_{l=0}^{\infty} \sum_{m=0}^{\infty} \frac{(-1)^{l+m}}{(2l+1)(2m+1)} \cdot \cos((2l+1)\omega_\sigma t)\cos((2m+1)\omega_\delta t). \qquad (5.282)$$

In order to identify the intermodulation tones more easily, it would be preferable to use equation (5.19) to express $s_o(t)$ as a sum of sine waves. We obtain

$$s_o(t) = \alpha \frac{8}{\pi^2} \sum_{l=0}^{\infty} \sum_{m=0}^{\infty} \frac{(-1)^{l+m}}{(2l+1)(2m+1)}$$
$$\cdot \{\cos[((l-m)\omega_1 + (l+m+1)\omega_2)t] + \cos[((l+m+1)\omega_1 + (l-m)\omega_2)t]\}.$$

Finally, to be able easily to identify the order of the intermodulation tones, we can change the summation indexes:

$$u_{l,m} = l - m, \qquad (5.283a)$$

$$v_{l,m} = l + m + 1. \qquad (5.283b)$$

We now have

$$s_o(t) = \alpha \frac{8}{\pi^2} \sum_{l=0}^{\infty} \sum_{m=0}^{\infty} \frac{(-1)^{v_{l,m}-1}}{(v_{l,m} + u_{l,m})(v_{l,m} - u_{l,m})}$$
$$\cdot \{\cos[(u_{l,m}\omega_1 + v_{l,m}\omega_2)t] + \cos[(v_{l,m}\omega_1 + u_{l,m}\omega_2)t]\}. \qquad (5.284)$$

Usually, the intermodulation tone used for the evaluation of the kth order IP is the IMDk closest in the frequency domain to the two initial input tones. In the present case, we have to consider the tone at $3\omega_1 - 2\omega_2$ or at $3\omega_2 - 2\omega_1$ to evaluate the IMD5 tone amplitude instead of the tone at $4\omega_1 - 1\omega_2$ or $4\omega_2 - 1\omega_1$. Thus, to derive the amplitude of the IMDk tone from the previous equation, we have to take $u_{l,m}$ and $v_{l,m}$ as the closest integers to $k/2$. As k can only be an odd integer here, we thus have to take

$$u_{l,m} = \frac{1-k}{2}, \qquad (5.285a)$$

$$v_{l,m} = \frac{1+k}{2}. \qquad (5.285b)$$

Referring back to equation (5.283), we have

$$v_{l,m} + u_{l,m} = 2l + 1 = 1, \qquad (5.286a)$$

$$v_{l,m} - u_{l,m} = 2m + 1 = k. \qquad (5.286b)$$

Using these results in equation (5.284), the IMDk amplitude at the hard limiter output is

$$\rho_{\text{OIMD}k} = \alpha \frac{8}{\pi^2} \frac{1}{k}. \tag{5.287}$$

We observe that this behavior in $1/k$ of the IMDk amplitude is characteristic of the hard limiter in the sense that it is obviously a property of the amplitudes of the harmonics in the decomposition of the square wave.

Using these results, we can derive the hard limiter IP values. For this purpose, we can use for instance the transposition of equation (5.86) for output quantities. Given that all the quantities involved in this equation are linked to their output counterparts through the device small signal gain, we can write that

$$\rho_{\text{OIP}k} = \frac{(\rho_{\text{o}})^{k/(k-1)}}{(\rho_{\text{OIMD}k})^{1/(k-1)}}, \tag{5.288}$$

with ρ_{o} the output amplitude of the two input tones used for the test. This output amplitude in fact corresponds to the amplitude of the tones with index $l = m = 0$ in equation (5.282). We thus get that

$$\rho_{\text{o}} = \alpha \frac{8}{\pi^2}. \tag{5.289}$$

Finally, substituting this result and the values of $\rho_{\text{OIMD}k}$ given by equation (5.287) into equation (5.288), we find that

$$\rho_{\text{OIP}k} = \alpha \frac{8}{\pi^2} k^{1/(k-1)}. \tag{5.290}$$

In practice, it may be easier to have an expression for the IP as a function of the device Psat. This last quantity is of particular interest for PAs, for instance. For this, we need to derive the hard limiter Psat, and thus the amplitude of the fundamental tone when an input sine wave is applied to it. This is straightforward in light of the series expansions carried out so far. We can deduce for instance from equation (5.281) that the amplitude of the fundamental tone in the series expansion of $\alpha\ \text{sign}\{\cos(\omega t)\}$ is $\alpha 4/\pi$. This value is in fact the maximum output amplitude of a fundamental tone the device can deliver. This is therefore the OPsat amplitude of the hard limiter as defined in "Odd order nonlinearity and IP3, CP1 or Psat" (Section 5.1.2). We thus get that

$$\rho_{\text{OPsat}} = \alpha \frac{4}{\pi}. \tag{5.291}$$

Thus, we can express the IPs of the hard limiter in terms of Psat as

$$\rho_{\text{OIP}k} = \rho_{\text{OPsat}} \frac{2}{\pi} k^{1/(k-1)} \tag{5.292}$$

Table 5.1 Measured and calculated IPk for a saturated power amplifier designed for the GSM/GMSK standard.

k	$OIMDk\|_{dBm}$ (measured)	$OIPk\|_{dBm}$ (equation (5.294))	$OIPk\|_{dBm}$ (equation (5.293))
3	18.1	35.9	35.9
5	11.1	34.7	34.6
7	5.5	34.1	33.9
9	0.3	33.7	33.5

or, if preferred, in terms of power, here in dBm units,

$$\begin{aligned} \mathrm{OIP}k|_{\mathrm{dBm}} &= \mathrm{OPsat}|_{\mathrm{dBm}} + \frac{20}{k-1}\log_{10}(k) + 20\log_{10}\left(\frac{2}{\pi}\right) \\ &= \mathrm{OPsat}|_{\mathrm{dBm}} + \frac{20}{k-1}\log_{10}(k) - 3.9. \end{aligned} \tag{5.293}$$

These relationships have been applied for instance to a saturated PA dedicated to the GSM standard. In GMSK mode, this device is given for an OPsat as +35 dBm. Thanks to equation (5.293), we can predict the expected IPk values. Measurements were performed with two input CW tones with power −3 dBm each so that their output power, P_o, was +30 dBm each. Using the measured IMDk tone output power, *OIMDk*, the OIPk values can be determined using equation (5.87), but transposed for output quantities as

$$\mathrm{OIP}k|_{\mathrm{dBm}} = \frac{k}{k-1}P_o|_{\mathrm{dBm}} - \frac{1}{k-1}OIMDk|_{\mathrm{dBm}}. \tag{5.294}$$

The OIPk thus derived that way as well as the theoretical expected values derived from the device Psat are given in Table 5.1. The good agreement observed confirms that the device can indeed be considered to behave like a hard limiter.

5.2.3 SNR Improvement in the Hard Limiter

Let us now consider the SNR improvement due to compression in the case where the device under consideration behaves as a hard limiter. Suppose we begin by again using a series expansion for the device transfer function. Simply looking at the series expansion given by equation (5.281), it is evident that the level of the harmonics is slowly decreasing as their order increases. Thus clearly such a series expansion would have slowly decreasing coefficients so that the derivation would be difficult to handle due to the number of significant terms to consider. Meanwhile, we observe that the hard limiter transfer function given by equation (5.273) can be approximated, for $x > 0$ at least, by a polynomial with a single term αx^{ν}, when ν tends toward 0. We can thus expect that if we can perform the derivation for a transfer function of the form αx^{ν}, we can then get a result valid for the hard limiter by letting $\nu \to 0$. This is in fact what Davenport did when considering a CW input signal in addition to a bandpass Gaussian noise at the hard limiter input [4, 64]. Here, we do not detail the full derivation, but rather review the final result for comparison with the derivation performed in Section 5.1.4.

Before doing so, we observe by inspecting equation (5.273) that the amplitude of the signal recovered at the output of the hard limiter is independent of the input signal amplitude as it only depends on its sign. It is thus totally meaningless in the present case to try to derive the output SNR, SNR_o, as a function of the input signal level. This is an important difference with respect to our previous derivation using third order polynomials. For the hard limiter, results are in fact obtained as a function of the input signal to noise ratio, SNR_i.

That said, considering the superposition of a CW wanted signal and of a bandpass noise, we can first focus on the wanted signal power, $P_{o,w}$, at the hard limiter output. Following Davenport's derivations, it can be shown that [64]

$$P_{o,w} = \frac{2\alpha^2}{\pi} SNR_i \, {}_1F_1^2 \left[\frac{1}{2}; 2; -SNR_i \right], \tag{5.295}$$

with ${}_1F_1[\cdot;\cdot;\cdot]$ the confluent hypergeometric function of the first kind that can be defined for instance through the series [6]

$${}_1F_1[a;b;z] = 1 + \frac{a}{b} z + \frac{a(a+1)}{b(b+1)} \frac{z^2}{2!} + \cdots . \tag{5.296}$$

The resulting signal power is plotted in Figure 5.29. It happens that the wanted signal power limit value for high input SNR is $8\alpha^2/\pi^2$. Working on normalized impedances, this is nothing

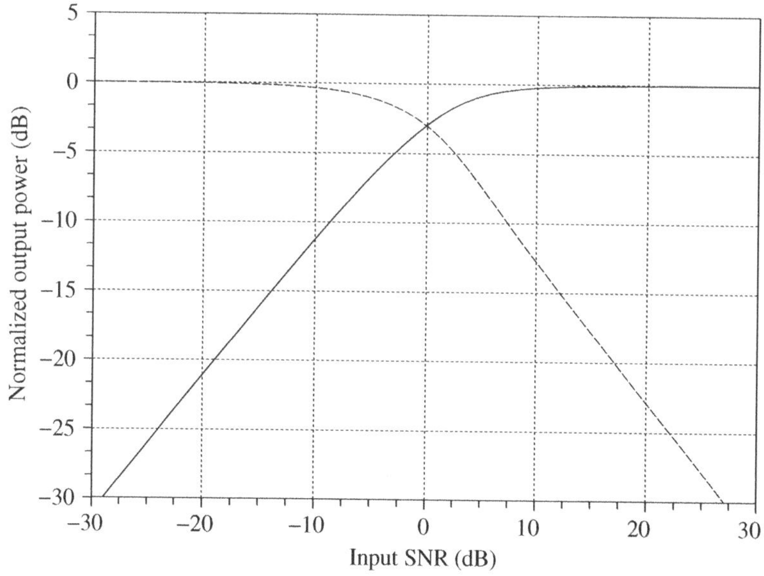

Figure 5.29 Normalized signal and noise power at the hard limiter output – Considering a CW signal superposed with a bandpass Gaussian noise at the input of a hard limiter, both the output wanted signal power, P_o (solid), and the noise power, $P_{n_o} + P_{n_{IMD}}$ (dashed), have the same limit value matching the power of the fundamental tone at the hard limiter output, i.e. $8\alpha^2/\pi^2$.

more than the power of the fundamental tone at the hard limiter output, i.e. the OPsat power we can derive from the OPsat amplitude given by equation (5.291).

When deriving the output noise power, we recover the situation encountered in Section 5.1.4. This means that we are dealing with two noise components. The first, $n_o(t)$, corresponds directly to the input noise component but is now distorted by the nonlinearity. The second, $n_{IMD}(t)$, corresponds to the intermodulation between the wanted signal and the input noise.

The power of the first noise term, P_{n_o} can be expressed as

$$P_{n_o} = \sum_{\substack{k=1 \\ k \text{ odd}}}^{\infty} \frac{\beta_k}{k!} h_{0k}^2 R_n^k(0), \tag{5.297}$$

with

$$h_{mk}^2 R_n^k(0) = \begin{cases} 0 & \text{for } m+k \text{ even,} \\ \alpha^2 2^k \, SNR_i^m \dfrac{{}_1F_1^2\left[\frac{m+k}{2}; m+1; -SNR_i\right]}{\Gamma^2(m+1)\Gamma^2\left(1-\frac{m+k}{2}\right)} & \text{for } m+k \text{ odd,} \end{cases} \tag{5.298}$$

and with $\Gamma(.)$ the gamma function as defined by equation (1.112). The derivation of the β_k factors is a little trickier. However, we observe that they represent the fraction of power of the fundamental tone relative to the total signal power after the kth self-convolution of the PSD of the input signal plus noise. Thus, we can derive those factors by symbolically representing in a vector the power of the components of this input signal centered around the fundamental tone and each harmonic after the kth convolution. For that purpose, let us start with the input signal with normalized power equal to 1 for each sideband centered around $\pm\omega_w$. We can represent this symbolically by the vector $[1, 0, 1]$, with 0 for the power of the component centered around the zero frequency and 1 for the power of the input wanted signal located around the carrier frequency $\pm\omega_w$. After the second convolution with itself, we get the vector $[1, 0, 3, 0, 3, 0, 1]$. The middle 0 always represents the power located at zero frequency. The power of the wanted signal located at $\pm\omega_w$ is now $3+3$ for a total power of $1+3+3+1=8$. The corresponding β_3 factor is thus given by the ratio $6/8 = 3/4$. This iterative process gives the β_k factors series as

$$\{\beta_1, \beta_3, \beta_5, \beta_7, \beta_9, \cdots\} = \left\{1, \frac{3}{4}, \frac{5}{8}, \frac{35}{64}, \frac{63}{128}, \cdots\right\}. \tag{5.299}$$

In the same way, the intermodulation noise component power, $P_{n_{IMD}}$, is given by

$$P_{n_{IMD}} = 2\sum_{k=1}^{\infty} \sum_{\substack{m=1 \\ m+k \text{ odd}}}^{k+1} \frac{\gamma_{mk}}{k!} h_{mk}^2 R_n^k(0), \tag{5.300}$$

with the factors $h_{mk}^2 R_n^k(0)$ given as previously by equation (5.298). Here, the factors γ_{mk} are given by the same kind of derivation as for β_k. The difference is that we are now dealing with the power of the intermodulation products centered around $(1 \pm m)\omega_w$ instead of that of the wanted signal centered around $\pm\omega_w$. So, starting from the same vector $[1, 0, 1]$, after the self-convolution of this vector we get $[1, 0, 2, 0, 1]$. As the wanted signal carrier frequency corresponds to the second zero index, for $m = 1$, the power of the tones to consider for γ_{12} is $2 + 1$, whereas for $m = 3$, i.e. for γ_{32}, we get only one tone with power 1. We thus obtain $\gamma_{12} = 3/4$ and $\gamma_{32} = 1/4$. Successive application of this process gives the γ_{mk} series

$$\{\gamma_{21}, \gamma_{12}, \gamma_{32}, \gamma_{23}, \gamma_{43}, \cdots\} = \left\{\frac{1}{2}, \frac{3}{4}, \frac{1}{4}, \frac{1}{2}, \frac{1}{8}, \cdots\right\}. \tag{5.301}$$

These derivations allow us to determine the resulting overall noise power at the hard limiter output, i.e. $P_{n_o} + P_{n_{IMD}}$ as these two noise components are uncorrelated. This noise power as a function of the input signal to noise ratio is also shown in Figure 5.29. We can see that the behavior for the noise signal is in fact symmetrical regarding that of the wanted signal. This is in fact normal as the roles of the two signals are exchanged depending on whether the SNR is very good or very poor.

The resulting SNR improvement as a function of the input SNR is shown in Figure 5.30. Here, as we consider the total noise power in the derivation of the output SNR, the present result has to be compared with the DSB noise case discussed in Section 5.1.4. We can thus see in the figure that the limit value on the SNR gain is equal to 2, or 3 dB, for high input SNR,

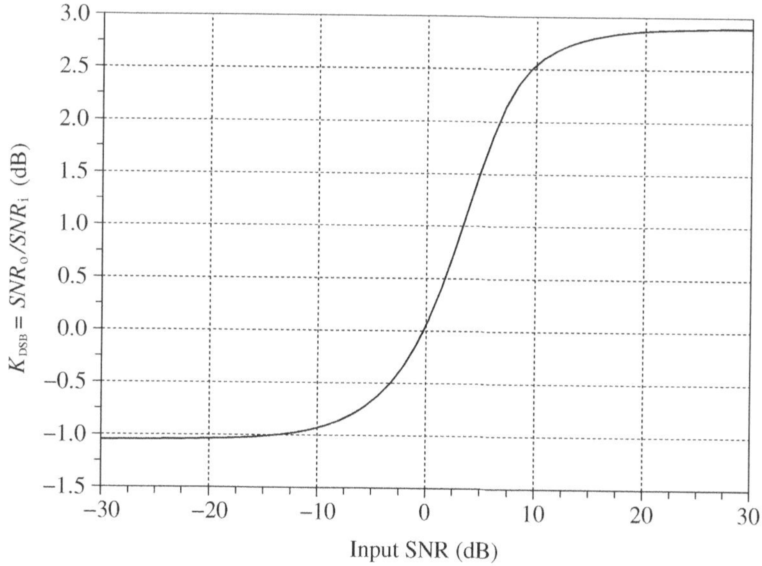

Figure 5.30 SNR improvement through the hard limiter – Considering a CW signal superposed with a bandpass Gaussian noise at the input of a hard limiter, the SNR improvement can reach 3 dB. For low input SNR, the limit value of K_{DSB} is negative and is equal to $\pi/4$.

and $\pi/4$, or -1.05 dB, for low input SNR. What is interesting is that we recover the result derived by our simple third order model when we took the limit of this model for an input power equal to the device IPsat. This simple model enables us to understand the mechanism involved in the cancellation of the term that corresponds to the amplitude component of the input bandpass noise (recall the decomposition presented in Section 1.2.2). Only the phase noise succeeds in going through such a hard limiter device.

5.3 AM-PM Conversion and the Memory Effect

Up to now, in dealing with smooth or hard nonlinearity (Sections 5.1 and 5.2, respectively), we have considered only the consequences of AM-AM conversion. But other kinds of degradations can be linked to nonlinearity in RF devices, more precisely due to the potential existence of AM-PM conversion and memory effect. However, those effects do not exist alone, in the sense that the AM-AM conversion is always present. Thus the AM-PM conversion and the memory effects lead to additional degradations on top of those already linked to AM-AM conversion. Moreover, these additional effects become visible for a given signal only when it enters the nonlinear area of the device transfer function. There is little to be gained by considering these new phenomena in configurations where we have a weak signal in the presence of strong blocking signals, for instance. Conversely, their impact can be non-negligible when dealing with the degradation of the wanted signal alone when the latter enters the truly nonlinear area of the device. This explains why these phenomena are mainly considered in transmitters for which designs are for the most part realized using the minimum allowable back-off for the signal being processed in order to minimize the power consumption of those devices.

Thus, we can review the system impacts of these new phenomena keeping in mind this transmitter perspective, i.e. considering a wanted signal alone in the line-up. But first we need to detail the device model we are considering for these derivations.

5.3.1 Device Model

Model for AM-PM Conversion

We should first say a few words about the physical origins of AM-PM conversion. If we refer to the discussion of AM-AM conversion in Section 5.1.1, we can easily understand the mechanism behind the clipping issue of a signal going through a nonlinear device that has a limited power supply. Conversely, it may seem strange that the instantaneous phase of the bandpass signal being processed can be impacted by the level of its own instantaneous amplitude. In fact, considering the RF current or voltage waves processed in the physical implementation of such a nonlinear device, we can surmise that a rotation of the instantaneous phase comes from a change in the wave impedance. And indeed, considering for instance a transistor device, the associated parasitic capacitors can depend in a nonlinear way on the currents or voltages applied to it. In the case of large input signals, important variations in the signal amplitude result in non-negligible changes in those capacitor values, and thus in the impedance experienced by the signal going through the device. These result in a rotation of the instantaneous phase of the current or voltage wave going through the device, depending on their instantaneous amplitude and thus in AM-PM conversion.

For taking into account such new phenomena at the system level, many models have been developed [61, 65–67]. Obviously, each has its advantages [68], but they share a common characteristic, i.e. that they are suited for performing system simulations once fitted to a particular device transfer function, but not really for highlighting in a didactic way the system impacts corresponding to such phenomena. For this, it is of interest to stay with the series expansion approach as done in the pure AM-AM conversion case. This will allow us to make the link with the derivations performed in that former case as well as highlighting the impact of the new phenomena we are dealing with in terms of additional unwanted terms that corrupt the wanted signal being processed. It moreover allows us to derive in the present case again simple analytical formulations suited for practical system budgets as illustrated in Chapter 7. But first we need to clarify the transfer function we consider in order to derive its series expansion. Assuming that the input bandpass signal that goes through the nonlinear device, $s_{\mathrm{i}}(t)$, is centered around ω_{c}, we observe that:

(i) The signal $s_{\mathrm{o}}(t)$ recovered at the output of the device is composed of a collection of bandpass signals lying around the harmonic angular frequencies of the form $k\omega_{\mathrm{c}}$. However, recalling the discussion in the introductory part of Section 5.3, only the wanted bandpass signal centered round the carrier angular frequency ω_{c}, i.e. $s_{\mathrm{o,H1}}(t)$, is of interest for our present investigation.

(ii) A simple nonlinear transfer function that links the real bandpass signals $s_{\mathrm{i}}(t)$ and $s_{\mathrm{o,H1}}(t)$ cannot reflect the AM-PM conversion. Referring for instance to equation (5.122), we see that only the instantaneous amplitude of the output bandpass signal centered around the carrier angular frequency can be corrupted by that of the input signal using such a model, and not its instantaneous phase.

To solve this issue, one possibility is to consider a series expansion for the transfer function that links the complex envelopes of $s_{\mathrm{i}}(t)$ and $s_{\mathrm{o,H1}}(t)$. However, the structure for this relationship cannot be arbitrary. Indeed, as AM-PM conversion can be seen as an additional distortion on top of AM-AM conversion, we expect such a relationship still to represent correctly the impact of the pure AM-AM conversion when we cancel the impact of the AM-PM conversion. Thus, a realistic way to proceed for the derivation of this structure is to reconsider the relationship that we get between those complex envelopes, i.e. $\tilde{s}_{\mathrm{o,H1}}(t)$ and $\tilde{s}_{\mathrm{i}}(t)$ assumed defined as centered around ω_{c}, when $s_{\mathrm{o}}(t)$ and $s_{\mathrm{i}}(t)$ are linked by the transfer function F considered in the pure AM-AM conversion case. Once this is done, we can expect to be able to generalize this relationship to take AM-PM conversion into account.

Thus, assuming that $s_{\mathrm{o}}(t) = F[s_{\mathrm{i}}(t)]$ with the series expansion of F given by equation (5.4), we can write

$$s_{\mathrm{o}}(t) = \sum_{n=0}^{\infty} \alpha_n s_{\mathrm{i}}^n(t), \tag{5.302}$$

with the real valued coefficients α_n; this relationship makes the link between real bandpass signals. With the aim of deriving a relationship between $\tilde{s}_{\mathrm{i}}(t)$ and $\tilde{s}_{\mathrm{o,H1}}(t)$, we first focus on the contribution to the output signal of the term of order n in this series expansion. For that purpose, given that the complex envelope of $s_\mathrm{i}(t)$, i.e. $\tilde{s}_\mathrm{i}(t)$, is defined as centered around ω_c, using equation (1.5) we can write that

$$s_\mathrm{i}^n(t) = \frac{1}{2^n}\left(\tilde{s}_\mathrm{i}(t)e^{\mathrm{j}\omega_\mathrm{c}t} + \tilde{s}_\mathrm{i}^*(t)e^{-\mathrm{j}\omega_\mathrm{c}t}\right)^n. \tag{5.303}$$

Then, using the binomial formula given by equation (5.80), we can expand this relationship as

$$s_\mathrm{i}^n(t) = \frac{1}{2^n}\sum_{k=0}^{n}\binom{n}{k}\tilde{s}_\mathrm{i}^k(t)\tilde{s}_\mathrm{i}^{*(n-k)}(t)e^{\mathrm{j}(2k-n)\omega_\mathrm{c}t}. \tag{5.304}$$

Looking at the sum on the right-hand side of this equation, we see that the term $s_\mathrm{i}^n(t)$ leads to a contribution to the bandpass signal centered around the carrier angular frequency ω_c if and only if $2k - n = \pm 1$, i.e. if $2k = n \pm 1$. This means that we need n to be an odd number. This is in fact not surprising if we think back to the discussion throughout Section 5.1. Only odd order nonlinearity, i.e. a compression effect, can impact the characteristics of the bandpass signals centered around their fundamental carrier angular frequency. Thus only the odd part of F, i.e. F_o, can lead to a non-vanishing contribution to the signal centered around ω_c. More precisely, we have that the term of order $n = 2l + 1$ leads to a contribution of the form

$$\frac{1}{2^{2l}}\binom{2l+1}{l+1}|\tilde{s}_\mathrm{i}(t)|^{2l}\frac{1}{2}\left(\tilde{s}_\mathrm{i}(t)e^{\mathrm{j}\omega_\mathrm{c}t} + \tilde{s}_\mathrm{i}^*(t)e^{-\mathrm{j}\omega_\mathrm{c}t}\right) \tag{5.305}$$

corresponding to $2k = n \pm 1$, i.e. $k = l + 1$ or $k = l$ in equation (5.304). Finally, considering all the contributions of the odd order terms in the series expansion of F, we can write the complex envelope $\tilde{s}_{\mathrm{o,H1}}(t)$ of the signal corresponding to the output sideband centered around ω_c as

$$\tilde{s}_{\mathrm{o,H1}}(t) = \sum_{l=0}^{\infty}\frac{\alpha_{2l+1}}{2^{2l}}\binom{2l+1}{l+1}|\tilde{s}_\mathrm{i}(t)|^{2l}\tilde{s}_\mathrm{i}(t). \tag{5.306}$$

We recover that having real valued α_n coefficients means that only the instantaneous amplitude of the bandpass signal being processed is impacted by the nonlinearity. However, what is interesting to understand for our purposes is that the above structure for the relationship between $\tilde{s}_\mathrm{i}(t)$ and $\tilde{s}_{\mathrm{o,H1}}(t)$ can still be assumed valid when dealing with the AM-PM conversion case. We simply need to consider that in the general case we can write that

$$\tilde{s}_{\mathrm{o,H1}}(t) = \sum_{l=0}^{\infty}\tilde{\alpha}_{2l+1}|\tilde{s}_\mathrm{i}(t)|^{2l}\tilde{s}_\mathrm{i}(t), \tag{5.307}$$

with complex valued $\tilde{\alpha}_{2l+1}$ coefficients in the general case where both AM-AM and AM-PM conversion are involved. Such a series expansion is used in practice to model the behavior of nonlinear RF devices around the fundamental carrier frequency of interest. This is why this model is often referred to as a model for bandpass nonlinearity. Fitting techniques can then be used to derive the $\tilde{\alpha}_{2l+1}$ sequence in order to correctly represent the effective transfer function of a device. However, the truncation of such a series expansion at the first nonlinear term is important in illustrating the system impacts we are investigating.

In order to take our investigation further, we rewrite the above relationship to express $\tilde{s}_{\text{o,H1}}(t)$ as the product of $\tilde{s}_{\text{i}}(t)$ and a complex gain, denoted by $\tilde{G}$ in the sequel, that depends on $\rho_{\text{i}}^2(t)$, the square of the instantaneous amplitude of $s_{\text{i}}(t)$. This is done in order to make the link with the derivations performed so far in the pure AM-AM case. The transfer function of the nonlinear device takes the form

$$\tilde{s}_{\text{o,H1}}(t) = \tilde{G}\big(\rho_{\text{i}}^2(t)\big)\tilde{s}_{\text{i}}(t). \tag{5.308}$$

Considering only the first nonlinear term in equation (5.307), we see that the magnitude of $\tilde{G}$ reduces to the effective gain of the pure AM-AM conversion case. Thus, retaining the notation used throughout Section 5.1 for the sake of consistency, we use equation (5.184) to write

$$\left|\tilde{G}\big(\rho_{\text{i}}^2(t)\big)\right| = G\left(1 + \frac{3}{4}\frac{\alpha_3}{G}\rho_{\text{i}}^2(t)\right), \tag{5.309}$$

with α_3 real valued. But now dealing also with AM-PM conversion, we can then assume that the argument of $\tilde{G}$ can be written as

$$\arg\left\{\tilde{G}\big(\rho_{\text{i}}^2(t)\big)\right\} = \Phi + \Theta\rho_{\text{i}}^2(t) \tag{5.310}$$

where Θ represents the AM-PM conversion gain, and Φ a constant phase offset that does not carry any information for our system derivations. Let us cancel this last term so that the final model we consider in the sequel reduces to

$$\tilde{s}_{\text{o,H1}}(t) = G\left(1 + \frac{3}{4}\frac{\alpha_3}{G}\rho_{\text{i}}^2(t)\right)e^{j\Theta\rho_{\text{i}}^2(t)}\tilde{s}_{\text{i}}(t). \tag{5.311}$$

In practice, the AM-PM conversion gain is often expressed in degrees per decibel. We thus need to link this practical quantity, denoted by $\Theta_{\circ/\text{dB}}$, with the factor Θ used in our model. For that purpose, we can simply reconsider the effect of the AM-PM conversion on a given bandpass signal and apply the definition of $\Theta_{\circ/\text{dB}}$. Indeed, supposing for instance that the instantaneous amplitude $\rho(t)$ of this bandpass signal varies so that its square value goes from ρ^2 up to $\rho^2 + \delta\rho^2$, we can estimate at first order that its instantaneous phase goes from ϕ to $\phi + \delta\phi$ according to

$$(\phi + \delta\phi) - (\phi) = \Theta_{\circ/\text{dB}}\frac{\pi}{180}[10\log_{10}(\rho^2 + \delta\rho^2) - 10\log_{10}(\rho^2)]. \tag{5.312}$$

Consequently,

$$\delta\phi = \Theta_{\circ/\text{dB}}\frac{\pi}{180}10\log_{10}\left(1 + \frac{\delta\rho^2}{\rho^2}\right). \tag{5.313}$$

Looking at this equation, we first observe that there are obviously some difficulties when dealing with a modulation scheme such that we can sometimes have $\rho^2 = 0$. However, as in practice AM-PM conversion occurs only in the highest part of the device transfer function, the model we are discussing is of interest only for high values of ρ^2. And indeed, in practice clipping can be incorporated into the model so that the AM-PM effect is canceled when the instantaneous amplitude of the bandpass signal being processed goes below a given threshold. Consequently, we get in practical use cases that $\rho^2 > 0$ and $\delta\rho^2 \ll \rho^2$. This means that we can consider an expansion of the ln function up to first order in the above equation so that

$$\Theta \stackrel{\Delta}{=} \frac{\delta\phi}{\delta\rho^2} \approx \Theta_{\circ/\text{dB}} \frac{\pi}{180} \frac{10}{\ln(10)} \frac{1}{\rho^2}. \tag{5.314}$$

We thus see that for a given value of deviation in terms of degrees per decibel, the parameter Θ scales as a function of the signal square amplitude ρ^2. This is obviously due to the decibel definition that quantizes the variations of the amplitude relative to some reference value. It explains that in practice the quantity $\Theta_{\circ/\text{dB}}$ is always associated with a given signal power for which the quantity is defined. Thus in practice the quantity ρ^2 in the above equation must be understood as the average quantity representing this power. As our simple model represented by equation (5.311) involves the characteristics of the bandpass signal $s_\text{i}(t)$ present at the *input* of the nonlinear device, the average power used for the definition of $\Theta_{\circ/\text{dB}}$ must correspond to an input power. Assuming that we are dealing with a random modulation scheme, we can use the stochastic notation so that we can finally write that

$$\Theta = \Theta_{\circ/\text{dB}} \frac{\pi}{180} \frac{10}{\ln(10)} \frac{1}{\mathbb{E}\{\rho_\text{i}^2\}}. \tag{5.315}$$

As a side effect, when $\Theta_{\circ/\text{dB}}$ is originally given for a given output power, $\mathbb{E}\{\rho_\text{i}^2\}$ must then be derived from this output power using the effective gain of the device G_e. An expression for this effective gain in the presence of both AM-AM and AM-PM conversion is derived later on through equation (5.322).

In conclusion, we can go a step forward based on practical orders of magnitude for $\Theta_{\circ/\text{dB}}$. In most integrated transceiver implementations this quantity does not go far beyond 1 degree per decibel. Applying equation (5.315), we then see that

$$\Theta_{\circ/\text{dB}} = 1^\circ/\text{dB} \Leftrightarrow \Theta\mathbb{E}\{\rho_\text{i}^2\} \approx 0.08. \tag{5.316}$$

But recalling the PAPR of realistic modulation schemes examined in Chapter 1, the maximum value of $\rho_\text{i}^2(t)$ does not go far beyond the average quantity $\mathbb{E}\{\rho_\text{i}^2\}$. We can write for modulation schemes classically encountered in the field of wireless that

$$\left|\Theta\rho_\text{i}^2(t)\right| \ll 2\pi. \tag{5.317}$$

This means that the small angle approximation holds for the complex exponential involved in equation (5.311). Keeping the most significant terms in this expression then results in

$$\begin{aligned}\tilde{s}_{\text{o,H1}}(t) &\approx G\left(1 + \frac{3}{4}\frac{\alpha_3}{G}\rho_\text{i}^2(t)\right)\left(1 + \text{j}\Theta\rho_\text{i}^2(t)\right)\tilde{s}_\text{i}(t) \\ &\approx G\left[1 + \left(\frac{3}{4}\frac{\alpha_3}{G} + \text{j}\Theta\right)\rho_\text{i}^2(t)\right]\tilde{s}_\text{i}(t).\end{aligned} \tag{5.318}$$

This simplification is often used in practice. In particular, given that $\frac{3}{4}\frac{\alpha_3}{G}$ is real valued and that $j\Theta$ is pure imaginary, we recover that the AM-AM conversion term, which corrupts the magnitude of the wanted complex envelope $\tilde{s}_i(t)$, remains parallel to it up to first order, whereas the AM-PM term, which corrupts its argument, remains orthogonal to it.

Model for the Memory Effect

It then remains to take into account the memory effect. The term "memory" relates to the fact that the phase shift of the output signal when considered at a given time depends not just on the input signal considered at a single sample time, but on the values of this signal over a finite time interval. We thus have a filtering behavior with a finite impulse response to be added on top of the nonlinear behavior. In practice, a Volterra series is often used to model such behavior [61,68,69]. However, a simplified model can be used for the discussion of the system impacts of this memory effect at a limited analytical cost. An extreme simplification of the phenomenon is that the modulus and argument of the complex gain that represent the device depend on the instantaneous amplitude of the input signal when considered at different times. The nonlinear device thus behaves as if the group delay of the transfer function experienced by the bandpass signal being processed were different for the amplitude part and phase part. If we denote this difference by τ, a simplified device model can then be written from equation (5.311) as

$$\tilde{s}_{o,H1}(t) = G\left(1 + \frac{3}{4}\frac{\alpha_3}{G}\rho_i^2(t)\right) e^{j\Theta\rho_i^2(t-\tau)}\tilde{s}_i(t). \tag{5.319}$$

This formulation involving AM-AM conversion, AM-PM conversion and the memory effect is sufficient to illustrate the system impacts of those phenomenon when considered together. Moreover, as highlighted in the pure AM-PM case through equation (5.317), the small angle approximation can be used to expand the complex exponential in the above equation. Keeping the most significant terms thus results in

$$\begin{aligned}\tilde{s}_{o,H1}(t) &\approx G\left(1 + \frac{3}{4}\frac{\alpha_3}{G}\rho_i^2(t)\right)\left(1 + j\Theta\rho_i^2(t-\tau)\right)\tilde{s}_i(t)\\ &\approx G\left(1 + \frac{3}{4}\frac{\alpha_3}{G}\rho_i^2(t) + j\Theta\rho_i^2(t-\tau)\right)\tilde{s}_i(t).\end{aligned} \tag{5.320}$$

5.3.2 System Impacts

Nonlinear EVM and Spectral Regrowth Due to AM-PM Conversion

The first straightforward system impact linked to the existence of AM-PM conversion is the degradation of the quality of the phase/frequency part of a modulation scheme when dealing with a non-constant amplitude bandpass signal. This results in a direct degradation in the quality of the modulation at the output of the nonlinear device.

However, before going any further, it is interesting to remark that such behavior also potentially needs to be taken into account when considering the system design of a transmitter dedicated to the generation of a constant amplitude RF bandpass signal depending on its architecture. Indeed, if this constant instantaneous amplitude bandpass signal is derived from the direct upconversion of the lowpass modulating complex envelope when represented in

Cartesian form, as is the case in a direct conversion transmitter for instance, there may be a degradation of the modulation quality when dealing with RF impairments. As illustrated in Section 6.2.2, a gain and phase imbalance at the upconversion stage can result in a weak amplitude modulation component of the modulated bandpass signal. But, at the same time, we were expecting a constant instantaneous amplitude bandpass signal. As a result, it is often a saturated PA, i.e. a truly nonlinear device, that is expected to be used to deliver the RF power. This is obviously done with the aim of minimizing the power consumption of the solution while taking advantage of this constant amplitude behavior. But a side effect is that this kind of device often exhibits a non-zero AM-PM conversion factor. Consequently, even if the amplitude modulation component resulting from the RF impairments remains weak, a non-negligible degradation of the phase/frequency modulation part may occur. This has to be taken into account at the system level. However, in this particular case the phenomenon results from the degradation of the amplitude modulation part of the modulated signal due to the way it has been generated.

Whatever the origins of the non-constant behavior of the instantaneous amplitude of the modulated bandpass signal $s_{\mathrm{i}}(t)$ we are dealing with at the input of the nonlinear device, we can move on to analyze the structure of the distortion term resulting from AM-PM conversion. In particular, we can show that it can still be considered as an additional noise component, as in the pure AM-AM conversion case. For that purpose, we can express the complex envelope $\tilde{s}_{\mathrm{o,H1}}(t)$ of the output bandpass signal centered around the carrier fundamental angular frequency as the sum of a term proportional to the expected ideal complex envelope, i.e. $\tilde{s}_{\mathrm{i}}(t)$, through an effective gain G_{e}, and a term $\tilde{n}(t)$ that is expected to represent a bandpass noise centered around the carrier angular frequency. We expect $\tilde{n}(t)$ to be uncorrelated with $\tilde{s}_{\mathrm{i}}(t)$. Here, we recall that the non-correlation we are talking about is understood as the non-correlation between the corresponding bandpass signals. However, as discussed in Appendix 1, the non-correlation between the complex envelopes defined around the same center frequency is equivalent to the non-correlation between the bandpass signals they represent when dealing with stationary processes, at least up to the second order, as is classically the case in wireless transceivers as discussed in Appendix 2. Thus, we expect to be able to write that

$$\tilde{s}_{\mathrm{o,H1}}(t) = G_{\mathrm{e}}\tilde{s}_{\mathrm{i}}(t) + \tilde{n}(t), \tag{5.321}$$

with $\tilde{n}(t)$ uncorrelated with $\tilde{s}_{\mathrm{i}}(t)$. For this derivation, and assuming that we are dealing with an AM-PM conversion of a classical order of magnitude, we can use the expression for $\tilde{s}_{\mathrm{o,H1}}(t)$ in the small angle approximation as given by equation (5.318). This equation has in fact exactly the same structure as that given by equation (5.184) in the pure AM-AM conversion case, except for the additional term jΘ. As this additional term is constant in our model, we can thus directly reuse the results derived in "Nonlinear EVM due to odd order nonlinearity" (Section 5.1.3), to express both G_{e} and $\tilde{n}(t)$. For instance, remembering that $-1/(2\mathrm{IIP3})$ is nothing more than $\frac{3\alpha_3}{4G}$ in the pure AM-AM case – recall equation (5.48a) and the fact that G and α_3 have opposite signs to physically describe compression behavior – we can directly write from equation (5.189) that, in the presence of AM-PM conversion,

$$G_{\mathrm{e}} = G\left[1 + \left(\frac{3\alpha_3}{4G} + \mathrm{j}\Theta\right)(\Gamma_4 + 1)\mathbb{E}\{\rho_{\mathrm{i}}^2\}\right]. \tag{5.322}$$

In this expression, the Γ_4 constant characterizes the fourth order statistics of the instantaneous amplitude of $s_i(t)$, as defined by equation (5.125). Compared to the pure AM-AM conversion case, we thus see that the AM-PM term $j\Theta$ makes the effective gain experienced by the complex envelope of the input signal a complex number. We see that the AM-PM conversion leads to a constant average phase offset proportional to the average power of the input signal, exactly as the average gain loss on the signal amplitude due to the AM-AM conversion was proportional to the same average power. Now proceeding the same way with equation (5.190), we can express $\tilde{n}(t)$ as

$$\tilde{n}(t) = -G\left(\frac{3\alpha_3}{4G} + j\Theta\right)\mathbb{E}\{\rho_i^2\}\left[(\Gamma_4 + 1) - \frac{\rho_i^2(t)}{\mathbb{E}\{\rho_i^2\}}\right]\tilde{s}_i(t). \tag{5.323}$$

We see that we can decompose $\tilde{n}(t)$ as the sum of the contribution of the AM-AM conversion only, originally given by equation (5.190) and now taking the form

$$\tilde{n}_{\text{AM-AM}}(t) = -\frac{3\alpha_3}{4}\mathbb{E}\{\rho_i^2\}\left[(\Gamma_4 + 1) - \frac{\rho_i^2(t)}{\mathbb{E}\{\rho_i^2\}}\right]\tilde{s}_i(t), \tag{5.324}$$

and the contribution of the AM-PM conversion, which can be written as

$$\tilde{n}_{\text{AM-PM}}(t) = -Gj\Theta\mathbb{E}\{\rho_i^2\}\left[(\Gamma_4 + 1) - \frac{\rho_i^2(t)}{\mathbb{E}\{\rho_i^2\}}\right]\tilde{s}_i(t), \tag{5.325}$$

Given that α_3 and Θ are real valued, $\tilde{n}_{\text{AM-AM}}(t)$ is, as expected, collinear to $\tilde{s}_i(t)$ so that it impacts only its magnitude, whereas $\tilde{n}_{\text{AM-PM}}(t)$ is orthogonal to it so that it impacts only its argument up to first order. We can thus associate $\tilde{n}_{\text{AM-AM}}(t)$ with an amplitude noise, whereas $\tilde{n}_{\text{AM-PM}}(t)$ behaves as a phase noise component. Moreover, given the orthogonality between these two complex envelopes, their powers sum together in an SNR or EVM budget. In the same way, we get exactly the same analytical structure for the two noise components. This means that up to first order, AM-PM conversion does not lead to any particular new system issues on top of those detailed in the pure AM-AM case. This holds both in terms of nonlinear EVM generation as detailed in "Nonlinear EVM due to odd order nonlinearity" (Section 5.1.3), and in terms of spectral regrowth as detailed in "Spectral regrowth for nonlinear EVM due to odd order nonlinearity" (Section 5.1.3). This is not the same when the memory effect adds to the AM-PM conversion, as discussed in the next section.

However, before turning to this memory topic we can say a few words on the normalization factor encountered in the expression for the complex envelope of the AM-AM noise component given by equation (5.324). In the pure AM-AM conversion case the α_3 parameter and the device IIP3 are linked by equation (5.48a). Then the expression for the complex envelope of the AM-AM noise term reduces to equation (5.190). But in the presence of AM-PM conversion, we can surmise that the amplitude of the IMD3 tones, and thus of the device IP3, also depends on the Θ parameter. To illustrate this, we need to anticipate the discussion on the memory effect in the next section. Figure 5.31 shows that the configuration for $s_i(t)$ considered in that section directly corresponds to the two CW tone setup used for the characterization of the device IP3. Consequently, reconsidering on the one hand the derivation in "Odd order nonlinearity and IP3, CP1 or Psat" (Section 5.1.2), for such characterization, and on the other

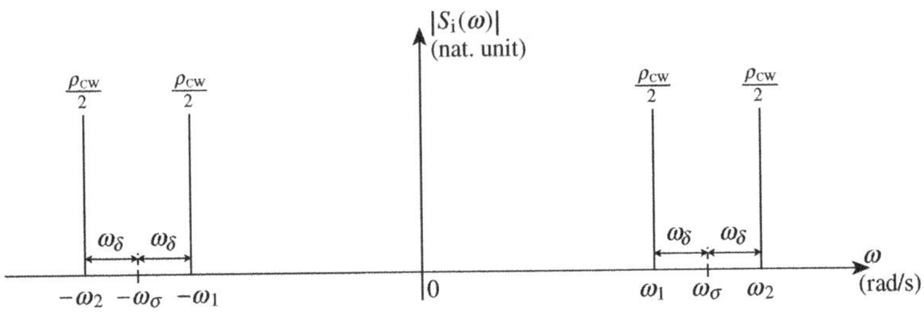

Figure 5.31 Two CW tones configuration considered for the illustration of the impact of the AM-PM conversion and memory effect.

hand the expression for the IMD3 tone given by equation (5.338), but assuming $\tau = 0$ in the present memoryless case, the IIP3 amplitude now takes the form

$$\rho_{\text{IIP3}} = \frac{1}{\sqrt{\left(\frac{3\alpha_3}{4G}\right)^2 + \Theta^2}}. \tag{5.326}$$

This expression must be compared to equation (5.46) in the pure AM-AM conversion case. However, recalling equation (5.316), we get that the parameter Θ, that is null for low values of the input power P_i, remains of small magnitude in most practical implementations even for the highest values of P_i. This behavior has allowed us to use the small angle approximation so far. But in the pure AM-AM conversion case we can write from equation (5.46) that

$$\left(\frac{3\alpha_3}{4G}\right)^2 \mathbb{E}\{\rho_i^2\} = \frac{\mathbb{E}\{\rho_i^2\}}{\rho_{\text{IIP3}}^2} = \frac{P_i}{\text{IIP3}}. \tag{5.327}$$

Thus, given that the highest values for P_i that can be considered for our model are nothing more than the device IPsat, and that IIP3 $\approx$ 3IPsat by equation (5.59), we finally get, for the highest input powers,

$$\left(\frac{3\alpha_3}{4G}\right)^2 \mathbb{E}\{\rho_i^2\} \approx 0.33. \tag{5.328}$$

Comparing this result with equation (5.316) shows that the IMD3 tone amplitude is still driven by the AM-AM conversion term, at least in most practical integrated transceiver implementations. Thus, equation (5.326) can still be approximated by equation (5.46) and the complex envelope of the AM-AM noise term by equation (5.190) as

$$\tilde{n}_{\text{AM-AM}}(t) = G\frac{\mathbb{E}\{\rho_i^2\}}{2\text{IIP3}}\left[(\Gamma_4 + 1) - \frac{\rho_i^2(t)}{\mathbb{E}\{\rho_i^2\}}\right]\tilde{s}_i(t). \tag{5.329}$$

Spectrum Asymmetry Due to Memory Effect on Top of AM-PM Conversion

Let us now focus on the system impacts of the memory effect on top of the AM-PM conversion. For that purpose, we can consider the simple expression for the complex envelope $\tilde{s}_{o,H1}(t)$ of the output bandpass signal of interest, when defined as centered around the carrier fundamental angular frequency, and given by equation (5.320) in the small angle approximation. This means that we continue to assume here that $|\Theta\rho_i^2(t-\tau)| \ll 2\pi$, with $\rho_i(t)$ the instantaneous amplitude of the input bandpass signal $s_i(t)$, as valid in most practical integrated implementations.

However, to highlight the impact of the memory effect, we need to consider a particular form for $s_i(t)$ in order to be able to perform the analytical derivations. We first consider a constant amplitude input signal, $\rho_i(t) = \rho_i$. In that case, equation (5.320) reduces to

$$\tilde{s}_{o,H1}(t) = G\left[1 + \left(\frac{3}{4}\frac{\alpha_3}{G} + j\Theta\right)\rho_i^2\right]\tilde{s}_i(t). \tag{5.330}$$

We thus see that having a constant amplitude input signal cancels the impact of the memory effect. This is in fact obvious as having $\rho_i(t) = \rho_i(t-\tau) = \rho_i$ cancels the impact of having a different group delay on the AM-PM conversion term compared to the AM-AM conversion term. We then recover that the constant amplitude bandpass signals are insensitive to the memory effect in addition to being insensitive to AM-AM and AM-PM conversion. In order to illustrate the impact of the memory effect we then need to consider a non-constant instantaneous amplitude input signal. This can be achieved at a minimal analytical cost by considering an input signal that is the superposition of two CW tones. Such superposition can indeed be interpreted as a single amplitude modulated signal when the correct phase relationship obtains between the two tones. Thus, let us consider the input signal $s_i(t)$ defined as

$$s_i(t) = \rho_{CW}(\cos(\omega_1 t) + \cos(\omega_2 t)). \tag{5.331}$$

We can then define the complex envelope $\tilde{s}_i(t)$ of this input signal as that centered around the angular frequency ω_σ equal to the mean value of ω_1 and ω_2, as illustrated in Figure 5.31. Thus,

$$\tilde{s}_i(t) = \rho_{CW}\left(e^{j\omega_\delta t} + e^{-j\omega_\delta t}\right) = 2\rho_{CW}\cos(\omega_\delta t), \tag{5.332}$$

with

$$\omega_\sigma = \frac{\omega_2 + \omega_1}{2}, \tag{5.333}$$

and

$$\omega_\delta = \frac{\omega_2 - \omega_1}{2}. \tag{5.334}$$

Here, we observe that due to the symmetry of the two input tones regarding the angular frequency used to define the complex envelope, this last one is in fact real. However, we can check that even if each input tone has a constant amplitude, the resulting sum when considered as a single bandpass signal is described by a complex envelope with a non-constant amplitude that represents the expected low frequency beat of the amplitude modulation.

It then remains to derive the complex envelope of the bandpass signal recovered as centered around ω_σ at the output of the nonlinear device, i.e. $\tilde{s}_{o,H1}(t)$. For this, looking at equation (5.320), we see that we have to derive both $\rho^2(t)\tilde{s}(t)$ and $\rho_i^2(t-\tau)\tilde{s}_i(t)$. The latter term can be evaluated using equation (5.332) as

$$\begin{aligned}\rho_i^2(t-\tau)\tilde{s}_i(t) &= \tilde{s}_i(t-\tau)\tilde{s}_i^*(t-\tau)\tilde{s}_i(t)\\ &= \rho_{\text{CW}}^2\left(2+e^{2j\omega_\delta(t-\tau)}+e^{-2j\omega_\delta(t-\tau)}\right)\tilde{s}(t)\\ &= \rho_{\text{CW}}^3\left[\left(2+e^{-2j\omega_\delta\tau}\right)e^{j\omega_\delta t}+\left(2+e^{2j\omega_\delta\tau}\right)e^{-j\omega_\delta t}\right.\\ &\quad\left.+e^{-2j\omega_\delta\tau}e^{3j\omega_\delta t}+e^{2j\omega_\delta\tau}e^{-3j\omega_\delta t}\right].\end{aligned} \tag{5.335}$$

And considering this result for $\tau = 0$, we directly recover the second term we are looking for:

$$\rho_i^2(t)\tilde{s}_i(t) = \rho_{\text{CW}}^3\left(3e^{j\omega_\delta t}+3e^{-j\omega_\delta t}+e^{3j\omega_\delta t}+e^{-3j\omega_\delta t}\right). \tag{5.336}$$

In fact, in order to discuss the impact of the memory effect on the spectral content of $\tilde{s}_{o,H1}(t)$, it is of interest to focus on the elementary complex envelopes corresponding to the different tones involved in this output signal. For that purpose, we can start with those lying at the angular frequencies of the two input tones, ω_1 and ω_2. Assuming that we continue working with complex envelopes defined as centered around the angular frequency ω_σ, these tones correspond to the angular frequency offsets $+\omega_\delta$ and $-\omega_\delta$. Thus, if we denote the output complex envelopes by $\tilde{s}_{o,+\omega_\delta}(t)$ and $\tilde{s}_{o,-\omega_\delta}(t)$, using equations (5.335) and (5.336) in equation (5.320), we can write that

$$\tilde{s}_{o,+\omega_\delta}(t) = G\left[1+\frac{9}{4}\frac{\alpha_3}{G}\rho_{\text{CW}}^2+j\Theta\rho_{\text{CW}}^2\left(2+e^{-2j\omega_\delta\tau}\right)\right]\rho_{\text{CW}}e^{+j\omega_\delta t}, \tag{5.337a}$$

$$\tilde{s}_{o,-\omega_\delta}(t) = G\left[1+\frac{9}{4}\frac{\alpha_3}{G}\rho_{\text{CW}}^2+j\Theta\rho_{\text{CW}}^2\left(2+e^{+2j\omega_\delta\tau}\right)\right]\rho_{\text{CW}}e^{-j\omega_\delta t}. \tag{5.337b}$$

In the same way, the output tones lying at a angular frequency offset of $+3\omega_\delta$ and $-3\omega_\delta$ from ω_σ have complex envelopes, $\tilde{s}_{o,+3\omega_\delta}(t)$ and $\tilde{s}_{o,-3\omega_\delta}(t)$ respectively, given by

$$\tilde{s}_{o,+3\omega_\delta}(t) = G\left(\frac{3}{4}\frac{\alpha_3}{G}+j\Theta e^{-2j\omega_\delta\tau}\right)\rho_{\text{CW}}^3 e^{+3j\omega_\delta t}, \tag{5.338a}$$

$$\tilde{s}_{o,-3\omega_\delta}(t) = G\left(\frac{3}{4}\frac{\alpha_3}{G}+j\Theta e^{+2j\omega_\delta\tau}\right)\rho_{\text{CW}}^3 e^{-3j\omega_\delta t}. \tag{5.338b}$$

These different complex envelopes are represented in the complex plane in Figure 5.32, taking into account that G and α_3 have opposite signs to physically represent a compression behavior. Looking at this figure, we see that due to the non-vanishing delay τ, the two upper and lower sidebands of the output signal now have different amplitudes. Here we refer to the upper (lower) sideband the spectral components that lie at frequencies above (below) the carrier frequency when considering the positive sideband of a bandpass signal. Obviously, when dealing with complex envelopes that already represent the downconverted positive sideband of

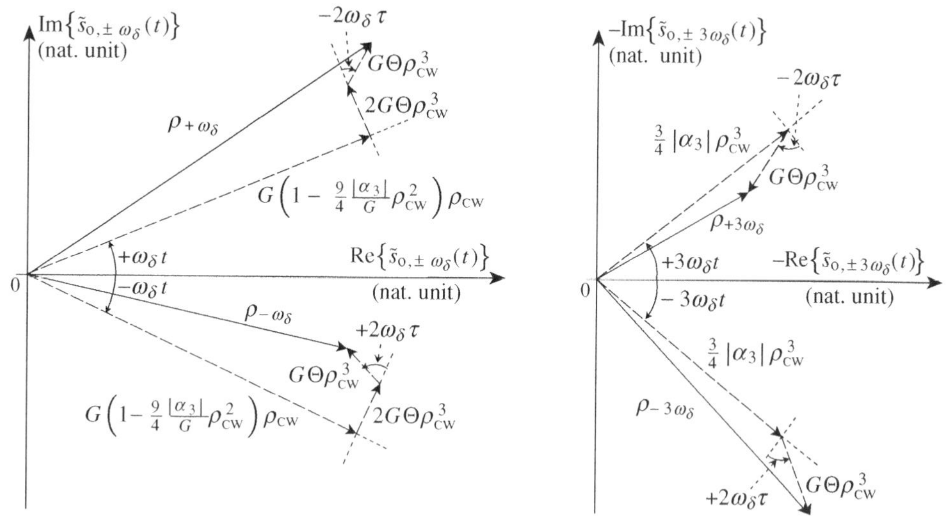

Figure 5.32 Interpretation of the spectral asymmetry resulting from the memory effect on top of the AM-PM conversion in a two CW tone test case – Referring to equations (5.337) and (5.338), the joint effect of the memory effect and the AM-PM conversion results in an asymmetry in the upper and the lower sidebands of the bandpass signal recovered at the output of the nonlinear device. This is caused by the presence of the non-vanishing delay τ that models the memory effect. This term leads to a different phase relationship in the frequency domain between the AM-AM and AM-PM components, depending on whether their frequency offset is positive or negative relative to the center frequency used for the definition of the complex envelope.

the bandpass signal, the upper and lower sidebands of this bandpass signal directly correspond to the positive and negative sidebands of the complex envelope.

Reconsidering the above derivations, we see that only a pure time domain delay can produce the asymmetry in the upper and lower sidebands as only such a delay can lead to a phase offset that increases linearly in the frequency domain. This dependency of the phase term is the reason for the different phase relationships between the AM-PM and the AM-AM terms that compose the positive and the negative sidebands of the complex envelope: it thus results in different amplitudes of the sum of those terms. In particular, when $\tau = 0$ this output spectrum remains symmetric, confirming that it is only when the memory effect is present in addition to the AM-PM conversion that we obtain such asymmetry. This behavior is also illustrated in Figure 5.33 where the PSD is derived from time domain simulations based on an input complex envelope as given by equation (5.332). As a side effect, we observe that having G and α_3 with opposite signs results in $\rho_{o,+\omega_\delta} \geq \rho_{o,-\omega_\delta}$ whereas $\rho_{o,+3\omega_\delta} \leq \rho_{o,-3\omega_\delta}$. For higher order intermodulation tones, this relationship between amplitudes depends on the relative signs of the corresponding terms in the series expansion of the device transfer function.

5.4 Baseband Devices

Baseband devices are devices which process lowpass analog signals. In transceivers, it is often baseband devices that process the modulating signals we are dealing with. In the case of a complex modulation, these signals can be either the real or imaginary part of such a

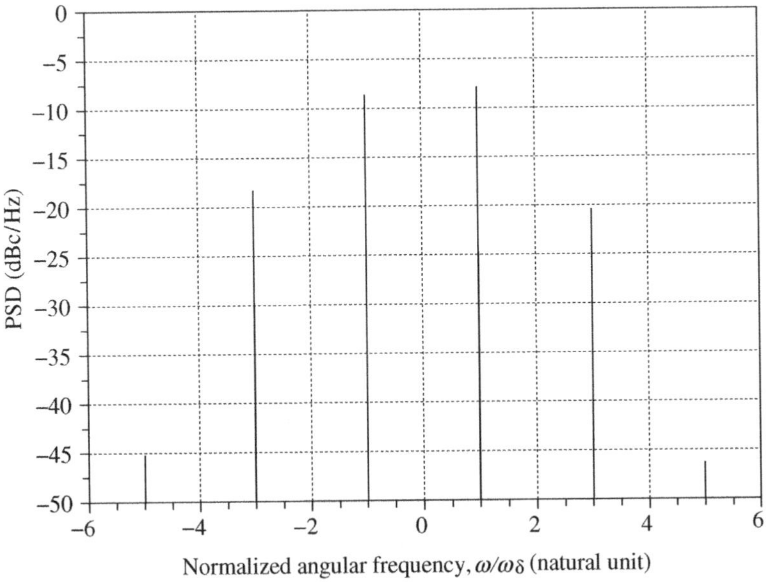

Figure 5.33 Impact of the memory effect on top of the AM-PM conversion on a two CW tone test case – The PSD of $s_{o,H1}(t)$ derived from time domain simulations when the input complex envelope represents a superposition of two CW tones as given by equation (5.332), and when the device transfer function is given by equation (5.319) with $|\Theta\mathbb{E}\{\rho_i^2\}| = |\Theta 2\rho_{CW}^2| = 0.1$, $|\frac{3}{4}\frac{\alpha_3}{G}\mathbb{E}\{\rho_i^2\}| = |\frac{3}{4}\frac{\alpha_3}{G}2\rho_{CW}^2| = 0.3$ and $2\omega_\delta\tau = \pi/4$, confirms the asymmetry resulting from the memory effect on top of the AM-PM conversion. The mechanism for this asymmetry is illustrated in Figure 5.32

modulating complex envelope, or its instantaneous amplitude and phase according to the discussion in Chapter 1. The nonlinearity of such devices should be considered separately from our discussion of RF devices so far. This is due on the one hand to the way such baseband analog devices are traditionally implemented, and on the other hand to the nature of the signal being processed.

As highlighted in the introductory part of this chapter, baseband analog devices are often implemented as feedback systems based on the extensive use of operational amplifiers. As a result, such baseband devices generally exhibit very good linearity behavior compared to RF devices that cannot be embedded easily in such feedback systems. This is a first argument to explain why, in practice, transceiver linearity is mainly driven by the performance of RF devices. However, as for any effective design, the implementation of baseband analog devices cannot be perfect and nonlinearity will occur, whether even or odd nonlinearity. But, due to the above argument, the compression behavior of such devices is often closer to hard clipping than to the smoother third order series expansion approach. This means that as long as full saturation does not occur, the distortion due to odd order nonlinearity can be expected to be very small.

The second difference compared to the RF bandpass case lies in the nature of the signals being processed. For example, we can assume that the device under consideration is expected to process the lowpass signal $p(t)$, which represents for instance the real part of a modulating complex envelope. Assuming that the spectrum of $p(t)$ spreads over the angular frequency bandwidth $[-\Omega;\Omega]$, we can imagine that it is possible to characterize the behavior of the device using a signal $s_i(t)$ that is the superposition of two sine waves with angular frequencies ω_1 and ω_2 according to

$$s_i(t) = \rho_i(\cos(\omega_1 t) + \cos(\omega_2 t)). \tag{5.339}$$

This configuration can in fact be seen as the transposition for baseband devices of the two CW tone test used to characterize RF devices. However, we could argue that pure sine waves are not, rigorously speaking, lowpass signals as they do not exhibit a DC term. This is indeed true, but by choosing that both $\omega_1 \ll \Omega$ and $\omega_2 \ll \Omega$ we can ensure that the impact of the device nonlinearity on this signal represents the distortion effectively experienced by a lowpass signal. Indeed, reconsidering the discussion in Section 5.1.2, we see that with such an input signal, the most significant harmonics and intermodulation tones generated by the nonlinearity remain within the passband of the device, i.e. within $[-\Omega;\Omega]$ as long as ω_1 and ω_2 are low enough compared to Ω. Thus, unlike what happens when dealing with RF bandpass signals, we now have almost all the induced distortion due to nonlinearity that remains within the lowpass signal bandwidth.

Thus, given on the one hand that baseband analog devices behave more like hard clippers and on the other hand that all the induced distortion remains within the passband of the device, concepts previously introduced such as the CP and the even and odd order IPs, whose purpose is to handle the even and odd order nonlinearity separately, are of little relevance here. This explains why in practice it is often the THD metric that is used to characterize baseband devices. The definition of this quantity in fact involves for the most part an input signal composed of only one sine wave with a low enough angular frequency ω, i.e.

$$s_i(t) = \rho_i \cos(\omega t), \tag{5.340}$$

with $\omega \ll \Omega$. Thus, whatever the nonlinear transfer function of the device, it results in an output signal $s_o(t) = F[s_i(t)]$ that is periodic. It can then be decomposed into Fourier series as

$$s_o(t) = a_0 + \sum_{k=1}^{\infty} a_k \cos(k\omega t + \phi_k). \tag{5.341}$$

The THD is then defined as the ratio of the quadratic sum of the harmonic amplitude to the wanted signal carrier amplitude:

$$THD = \frac{1}{a_1}\sqrt{\sum_{k=2}^{\infty} a_k^2}. \tag{5.342}$$

We observe that the DC term is not included in this definition as it is a component of the signal that is often managed separately from the rest of the signal, as discussed, for instance, in Section 6.5.2.

This definition using only a pure sine wave for the measurement is of interest because we can approximate the SNR limitations, or the generated EVM, linked to the baseband nonlinearity as

$$THD = EVM = \frac{1}{\sqrt{SNR}}. \tag{5.343}$$

This means that the power of the harmonics of the test tone is expected to reflect the power of the distortion term that would be generated with a real modulating signal like $p(t)$. However, we cannot make any assumption a priori on the statistics of the modulation. This means that a given amount of THD can indeed be interpreted as a corresponding SNR limitation, but it gives no indication as to how the corresponding noise component effectively impacts the baseband algorithms on the receive side, for instance. It is not the same problem for the EVM on the transmit side, as this metric often does not make any assumption on the statistics of the error signal.

To conclude, we recall that constant amplitude bandpass signals are said to be insensitive to odd order nonlinearity, as highlighted for instance in "Nonlinear EVM due to odd order nonlinearity" (Section 5.1.3), i.e. to RF compression. Considering the structure of $\tilde{s}_{o,1,\mathrm{H1}}(t)$ as given by equation (5.122a), we can see that the output spectrum centered on the carrier frequency is not affected by compression when the instantaneous amplitude of the input signal is constant. This means that, if we filter out the harmonics of this output RF signal using a zonal bandpass filter, we can theoretically recover exactly the input constant amplitude modulated wave with no distortion. However, we observe that the lowpass signals representing the real and imaginary parts of the complex envelope of such constant amplitude modulated bandpass signals are in turn sensitive to baseband nonlinearity as their distortion remains within their passband as discussed above.

6

RF Impairments

A final set of limitations classically encountered in the implementation of wireless transceivers is related to what are called RF impairments. What do we mean by this term? We have in fact already encountered phenomena caused by what could be called RF impairments. A good example of this is even order nonlinearity, detailed in Chapter 5, which is for the most part related to the mismatch between, or the impairment of, elements expected to be identical in RF/analog devices.

In wireless transceivers, the term "RF impairments" is classically associated with degradations linked to the implementation of complex frequency conversions in the analog domain. Given that the implementation of such functionality requires the use of two LO waveforms in quadrature, any imbalance between those LO signals indeed results in a degradation of the processing. The corresponding system limitations are of particular importance as such complex frequency conversion is widely used, for various reasons discussed in this chapter. It even necessarily occurs at some point in a line-up when dealing with complex modulating waveforms as introduced in Chapter 1. However, before going any further, it is of interest to review the frequency conversion processing itself. This allows us to understand in greater depth the mechanisms involved in what are called real and complex frequency conversions and thus the associated system limitations we can expect in the presence of such impairments. It also allows us to introduce some interesting signal processing functions that can be used in the analog domain even if based on the complex signal processing approach.

Finally, we can also discuss in this chapter additional phenomena caused by RF impairments, such as the LO leakage or the DC offset that can be recovered at unexpected ports of a line-up, which have not been covered in previous chapters.

6.1 Frequency Conversion

6.1.1 From Complex to Real Frequency Conversions

When talking about frequency conversion, we simply mean that we want to shift the spectrum of a given signal in the frequency domain. Mathematically, this operation corresponds to the convolution of the spectrum of the initial signal with a Dirac delta distribution centered on the target frequency offset. In the time domain it therefore corresponds to the multiplication

Wireless Transceiver Architecture: Bridging RF and Digital Communications, First Edition. Pierre Baudin.

of the initial signal by the Fourier transform of the Dirac delta distribution, i.e. the complex exponential of the form $e^{\pm j\delta\omega t}$ [2]. In this expression, $\delta\omega$ represents the targeted frequency shift and the sign gives the direction for the shift. The complex nature of this exponential explains why a frequency conversion that implements this operation is called a complex conversion. In such frequency conversions we obviously shift *all* the spectrum of the initial signal in one direction only. But a potential problem comes from the fact that the resulting signal is no longer real valued. This can be easily identified, as the magnitude of its spectrum is no longer symmetric with regard to DC. But given that any available physical quantity in real life is real valued, the physical analog implementation of such complex frequency conversion requires at least two distinct electrical waveforms in order to represent the real and imaginary parts of complex signals.

However, we know from the discussion in Chapter 1 that a RF bandpass signal can be represented using a single real valued signal, even when complex modulated. We may thus wonder whether we cannot find a means to convert such a RF bandpass signal expressed in its real form into another RF bandpass signal still expressed in its real form, but centered around another carrier angular frequency. In order to investigate this, we can consider the frequency transposition of the RF bandpass signal $s_{RF}(t)$. By equation (1.27), we can write this signal as

$$s_{RF}(t) = \text{Re}\left\{\tilde{s}_{RF}(t)e^{j\omega_{RF}t}\right\}, \tag{6.1}$$

where $\tilde{s}_{RF}(t)$ stands for its complex envelope defined as centered around ω_{RF}. Assuming that we want to transpose this signal around the intermediate frequency (IF) f_{IF}, we need to derive a signal process such that the output bandpass signal $s_{IF}(t)$ can be represented by the *same* complex envelope $\tilde{s}_{RF}(t)$ but now defined as centered around the angular frequency $\omega_{IF} = 2\pi f_{IF}$. This means that we want $s_{IF}(t)$ expressed as

$$s_{IF}(t) = \text{Re}\left\{\tilde{s}_{RF}(t)\, e^{j\omega_{IF}t}\right\}. \tag{6.2}$$

In order to go further, we can use equation (1.5) to express $s_{RF}(t)$ and $s_{IF}(t)$ as a function of their positive and negative sidebands. This results in

$$s_{RF}(t) = \frac{1}{2}\left(\tilde{s}_{RF}(t)\, e^{j\omega_{RF}t} + \tilde{s}^*_{RF}(t)\, e^{-j\omega_{RF}t}\right), \tag{6.3}$$

and

$$s_{IF}(t) = \frac{1}{2}\left(\tilde{s}_{RF}(t)\, e^{j\omega_{IF}t} + \tilde{s}^*_{RF}(t)\, e^{-j\omega_{IF}t}\right). \tag{6.4}$$

Comparing those two expressions, we see that to obtain $s_{IF}(t)$ from $s_{RF}(t)$, we need to multiply this input signal by at least *two* complex exponentials of the form $e^{+j\omega_{LO}t}$ and $e^{-j\omega_{LO}t}$, with $\omega_{LO} = |\omega_{RF} - \omega_{IF}|$. Here, the subscript LO stands for "local oscillator" as for the most part this periodic signal is delivered by an oscillator that is local to the transceiver. But, given that the two complex exponentials are complex conjugates, we get that the frequency conversion processing can be achieved by multiplying $s_{RF}(t)$ by simple sine or cosine functions of the form

$$\sin(\omega_{LO}t) = \frac{1}{2j}\left(e^{j\omega_{LO}t} - e^{-j\omega_{LO}t}\right), \tag{6.5a}$$

$$\cos(\omega_{LO}t) = \frac{1}{2}\left(e^{j\omega_{LO}t} + e^{-j\omega_{LO}t}\right). \tag{6.5b}$$

As we are dealing here with only real valued signals, this processing is referred to as a real frequency conversion.

However, our analysis so far is not complete. The simultaneous presence of the two complex exponentials in the LO waveform leads to the generation of an additional unwanted signal. In order to illustrate this, we can suppose for instance that the signal $s_{RF}(t)$ is converted using the cosine function $\cos(\omega_{LO}t)$ with $\omega_{LO} \leq \omega_{RF}$. Using equations (6.3) and (6.5b), we can write

$$s_{RF}(t)\cos(\omega_{LO}t) = \frac{1}{4}\left(\tilde{s}_{RF}(t)\,e^{j\omega_{RF}t} + \tilde{s}^{*}_{RF}(t)\,e^{-j\omega_{RF}t}\right)\left(e^{j\omega_{LO}t} + e^{-j\omega_{LO}t}\right). \tag{6.6}$$

Expanding and reordering, we obtain

$$\begin{aligned} s_{RF}(t)\cos(\omega_{LO}t) &= \frac{1}{2}\left(\mathrm{Re}\left\{\tilde{s}_{RF}(t)\,e^{j(\omega_{RF}-\omega_{LO})t}\right\} + \mathrm{Re}\left\{\tilde{s}_{RF}(t)\,e^{j(\omega_{RF}+\omega_{LO})t}\right\}\right) \\ &= \frac{1}{2}\left(\mathrm{Re}\left\{\tilde{s}_{RF}(t)\,e^{j\omega_{IF}t}\right\} + \mathrm{Re}\left\{\tilde{s}_{RF}(t)\,e^{j\omega_{IM}t}\right\}\right), \end{aligned} \tag{6.7}$$

with $\omega_{IF} = \omega_{RF} - \omega_{LO}$ and $\omega_{IM} = \omega_{RF} + \omega_{LO}$. Comparing with equation (6.4), we see that the first term on the right-hand side of equation (6.7) is nothing more than the expected bandpass signal $s_{IF}(t)$. But, as illustrated in the frequency domain[1] by Figure 6.1, the presence of the two complex exponentials in the LO waveform leads to the generation of an additional unwanted bandpass signal that represents the frequency transposition of $s_{RF}(t)$ around the angular frequency ω_{IM}. This signal is in fact nothing more than the image signal that is retrieved during any real frequency conversion.

The system impacts associated with this signal are discussed in the next section. But for the time being we conclude our discussion on the real frequency conversion by highlighting that, depending on the frequency planning of the conversion, i.e. depending on the value of the LO angular frequency ω_{LO} relative to the center angular frequency ω_{RF} of the input signal $s_{RF}(t)$, it is either the positive or negative complex exponential, $e^{+j\omega_{LO}t}$ or $e^{-j\omega_{LO}t}$, that

[1] In this chapter, the spectral analysis is performed based on Fourier transforms to preserve the phase relationships between sidebands in the spectral domain. This is done in order to explain the corresponding time domain signal processing behavior. When dealing with a randomly modulated waveform, this assumes that the Fourier transform of the realization under consideration exists. As discussed in Chapter 1, this can be justified by considering the corresponding waveform over a finite time duration for instance. Also in this chapter, only the magnitude of complex Fourier transforms is plotted on figures for convenience of representation, as in Chapter 1. However, in contrast to Chapter 1, the full analytical expressions for the sidebands are displayed in the graphs in order to highlight the phase relationship that is involved in the various recombination mechanisms. Finally, recall that our convention in this book is that $\tilde{S}(\omega)$ stands for the spectral domain representation of the complex envelope $\tilde{s}(t)$ and not for the complex envelope of the signal $S(\omega)$.

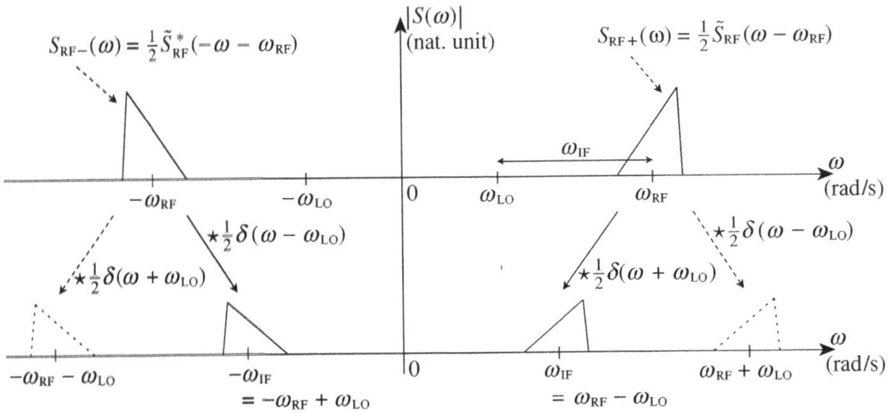

Figure 6.1 Spectral representation of a real frequency conversion – The real frequency conversion around the center angular frequency ω_{IF} of the bandpass signal $s_{RF}(t)$, initially centered around ω_{RF}, can be done by multiplying it by a cosine function at the the angular frequency $\omega_{LO} = |\omega_{RF} - \omega_{IF}|$. In the spectral domain, both sidebands of $s_{RF}(t)$, i.e. $S_{RF-}(\omega)$ and $S_{RF+}(\omega)$ (top), are convolved by the two Dirac delta distributions corresponding to the two complex exponential functions that compose the cosine function. As a result, we get the superposition of the spectrum of the expected transposed real signal, with its sidebands centered around $\pm\omega_{IF}$ (bottom, solid), and an additional unwanted signal, with its sidebands centered around $\pm(\omega_{RF} + \omega_{LO})$ (bottom, dashed line).

composes the LO signal that can be used for the frequency conversion of $s_{RF}(t)$ toward ω_{IF}. However, this choice impacts the sidebands that are recovered in the positive or negative parts of the IF spectrum. In that respect, this behavior can be related to the discussion in "Positive vs. negative sidebands" (Section 1.1.3). Indeed, referring to Figure 6.1, we can understand that when selecting $\omega_{LO} = \omega_{RF} + \omega_{IF}$ rather than $\omega_{LO} = |\omega_{RF} - \omega_{IF}|$, we still recover a transposition of $s_{RF}(t)$ around the angular frequency of interest ω_{IF}, but now with a flip in the sidebands in respect to this former case. In that case, the IF signal can indeed be written as

$$\begin{aligned} s_{IF}(t) &= \frac{1}{2}\left(\tilde{s}^*_{RF}(t)\,e^{j\omega_{IF}t} + \tilde{s}_{RF}(t)\,e^{-j\omega_{IF}t}\right) \\ &= \mathrm{Re}\left\{\tilde{s}^*_{RF}(t)\,e^{j\omega_{IF}t}\right\}. \end{aligned} \tag{6.8}$$

However, if this expression differs from equation (6.4) by a complex conjugate, it is only a matter of managing the sign of the imaginary part of the complex envelope of interest. This explains why such alternative frequency planning can be used in practice if it helps to overcome potential RF coupling issues. As a side effect, specific terminology is used to distinguish between the different possible combinations of a frequency conversion. When $\omega_{LO} < \omega_{RF}$ as in the former case, we say that we are dealing with an infradyne frequency conversion, whereas if $\omega_{LO} > \omega_{RF}$, the frequency conversion is said to be supradyne. In the same way, if $\omega_{LO} = \omega_{RF}$, the frequency conversion is said to be homodyne, whereas in the other case it is said to be heterodyne.

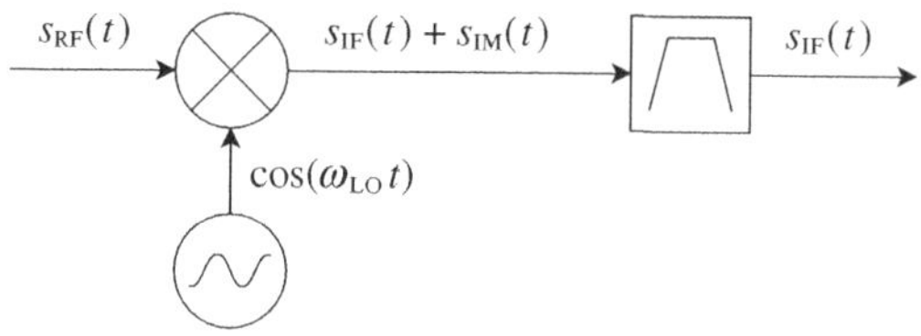

Figure 6.2 Cancellation of the generated unwanted image signal during a real frequency conversion – Recalling the frequency planning shown in Figure 6.1, the real frequency conversion of a bandpass signal $s_{RF}(t)$ at the input of the mixing stage leads to the generation of both a wanted signal $s_{IF}(t)$, here assumed at the IF, and an image signal $s_{IM}(t)$. The suppression of the latter unwanted signal requires the use of a filter after the frequency transposition.

6.1.2 Image Signal

As discussed in the previous section and illustrated in Figure 6.1, an unwanted bandpass signal $s_{IM}(t)$ is necessarily generated during a real frequency conversion. This image signal lies around the image angular frequency ω_{IM} such that the center angular frequency ω_{RF} of the input signal lies midway between ω_{IM} and ω_{IF}. We can thus anticipate potential system problems, in particular on the transmit side. Such an image signal can degrade the quality of the transmit spectrum, and stringent requirements can be set by wireless standards in that respect as discussed in Chapter 3. This unwanted signal can be eliminated by means of a filter located at the output of the real frequency conversion stage; indeed, it is hard to imagine any other solution. We thus need to consider a line-up implementation like the one shown in Figure 6.2. This structure is classically encountered in a transmit line-up relying on a real frequency conversion stage as in the heterodyne architecture discussed in Section 8.1.2. However, we observe that the possibility of filtering out this unwanted image signal in an efficient way depends very much on the frequency planning of the frequency conversion. For instance, if either ω_{RF} or ω_{LO} is too low, the resulting IF and image angular frequencies are too close to allow for an efficient filtering. This highlights the importance of the joint optimization of the frequency planning and the filtering present in a line-up that uses such real frequency conversion.

However, there is a symmetric configuration that leads to potential problems when using an LO waveform composed of two complex conjugate complex exponentials. When an additional unwanted signal is present at the input of the real frequency conversion stage, as classically encountered on the receive side, a folding of an unwanted component may occur toward the wanted signal. As illustrated in Figure 6.3, this may be the case if we assume that an additional unwanted signal, $s_{IM}(t)$, centered around the angular frequency $\omega_{IM} = 2\omega_{LO} - \omega_{RF}$, is present during the frequency conversion of the signal $s_{RF}(t)$ centered around ω_{RF}. In that case again, this signal is also classically referred to as the image signal.[2] This is related here to the fact that ω_{IM} is symmetric to ω_{RF} relative to the LO angular frequency ω_{LO}.

[2] Obviously, this definition is different from that encountered just above during the examination of the unwanted signal generated during the real frequency conversion process. However, due to the symmetries in the corresponding frequency planning, the same term is often used in both cases.

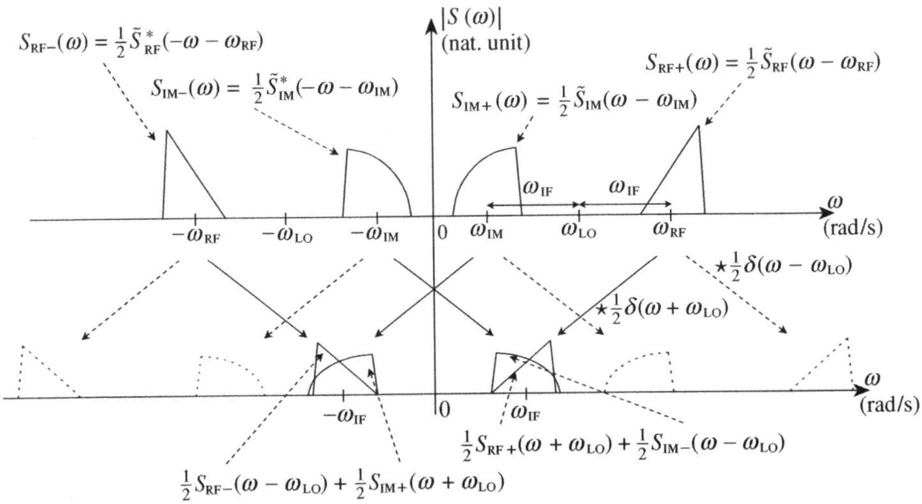

Figure 6.3 Representation in the spectral domain of the image signal folding during a real frequency conversion – When performing a real conversion, the input signal is multiplied by the two complex exponentials that compose the LO waveform, ideally a sine or cosine function at the angular frequency ω_{LO}. As a result, an input signal lying at the image frequency $\omega_{IM} = 2\omega_{LO} - \omega_{RF}$ is folded onto the wanted signal centered around ω_{RF}. Here the downconversion case with $\omega_{RF}/2 < \omega_{LO} < \omega_{RF}$ is considered, otherwise it is sidebands of the same nature, i.e. both positive or both negative, for the input wanted and image signals that get superposed at the output of the frequency conversion stage.

We thus see that if we want to avoid the superposition of the wanted signal with the input image signal during a real frequency conversion, all we can do is to filter out this unwanted signal *before* the frequency conversion occurs. Here again we necessarily face the same kind of frequency planning problem as discussed above for the handling of the generated image signal. When the LO angular frequency ω_{LO} is such that $\omega_{LO} \approx \omega_{RF}$, the input image signal lies very close to the wanted signal in the frequency domain. In that case, it would be costly to have an RF filter that filters out such an adjacent channel in a receiver prior to the frequency downconversion. The image signal can even be the wanted signal itself when $\omega_{LO} = \omega_{RF}$. In this second case, it is exactly the negative (positive) sideband of the wanted signal that is folded on the positive (negative) sideband during the frequency conversion. Obviously, in this configuration the unwanted image signal sideband cannot be filtered out before the downconversion stage as it lies in the passband of any real filter centered on the wanted signal. This configuration corresponds in fact to the direct conversion receive case. However, as these positive and negative sidebands correspond to complex modulating waveforms that are complex conjugates of each other, this superposition leads to problems only when dealing with a modulation that is effectively complex. But, in that case, we necessarily need to use a complex frequency conversion to recover the modulating waveform. This frequency conversion structure in turn solves part of the image signal problem, as detailed in the forthcoming sections.

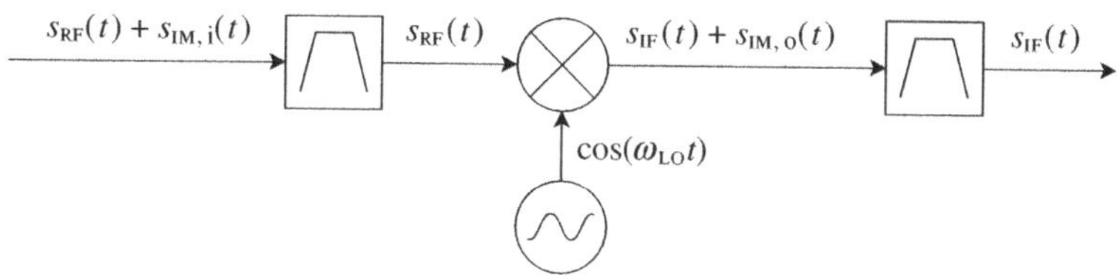

Figure 6.4 Image reject filters used during a real frequency conversion – As the wanted signal $s_{RF}(t)$ and the input image signal $s_{IM,i}(t)$ are folded on each other during a real frequency conversion, an image reject filter needs to be used to reject the image signal before the transposition occurs. Although the cancellation of $s_{IM,i}(t)$ is total, the behavior of the real frequency conversion leads to the generation of an additional unwanted image signal $s_{IM,o}(t)$ at the mixing stage output. In order to cancel it, we may use an additional filtering stage as illustrated in Figure 6.2.

However, when considering the use of real frequency conversions, filtering is still the straightforward way to cancel the image signal. And by our discussion so far, we may need to cancel the image signals present at both the input and the output of the mixing stage. As a result, we may need to consider using two filters in the line-up for a complete clean-up of the signal, as illustrated in Figure 6.4. Moreover, in practical implementations at least one of the two filters needs to be an RF device acting on high frequency signals. Obviously, such a line-up is a costly solution. This explains why we often try to get rid off this kind of filter, and thus to avoid the real frequency conversion itself. The complex frequency conversion advocated above for the direct conversion can be seen as a good alternative even if it also suffers from image problems, as discussed in Section 6.2.1. Alternatively, there is the possibility of using image reject mixers that allow cancellation of the input image signal as discussed in "Image reject mixers" (Section 6.1.4).

There are thus many possible configurations for the implementation of a single real frequency conversion, depending on the frequency planning chosen. This is obviously trickier when considering a cascade of different frequency conversion stages, as is the case in heterodyne architecture, for instance. However, for the sake of simplicity in the rest of this chapter, we often consider the configuration of the infradyne frequency downconversion with $\omega_{RF}/2 < \omega_{LO} < \omega_{RF}$ as an illustrative example for receivers and a symmetric configuration for the transmit side, i.e. an LO angular frequency higher than that of the input signal to be upconverted.

6.1.3 *Reconsidering the Complex Frequency Conversion*

As detailed in the previous sections, the potential folding of an input image signal as well as the generation of an output one during a real frequency conversion is closely related to the presence of the two complex exponentials in the LO signal. Obviously, the complex frequency conversion introduced in Section 6.1.1 can be seen as a solution in that respect. However, like most practical solutions to a problem, it also leads to some implementation constraints as a side effect.

In order to illustrate this, we can consider as an example the downconversion of the RF bandpass signal $s_{RF}(t)$ using the negative complex exponential $e^{-j\omega_{LO}t}$. The conclusions, or at least the guidelines, remain the same, for the most part, i.e. even for a frequency upconversion

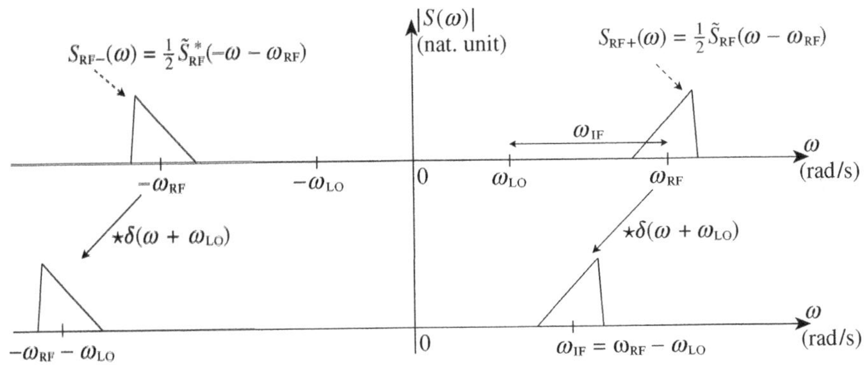

Figure 6.5 Spectral representation of a complex frequency conversion – The complex frequency downconversion of the real bandpass signal $s_{RF}(t)$ is done by multiplying it by the complex exponential $e^{-j\omega_{LO}t}$ in the present example. In the spectral domain, both sidebands of the input spectrum, $S_{RF-}(\omega)$ and $S_{RF+}(\omega)$ (top), are thus convolved with the Dirac delta distribution corresponding to the Fourier transform of the complex exponential. As a result, the input spectrum is shifted in the frequency domain in a single direction (bottom).

or using the positive complex exponential. By our assumptions, the IF signal $s_{IF}(t)$ recovered at the output of the complex frequency conversion can be written as

$$s_{IF}(t) = s_{RF}(t)\,e^{-j\omega_{LO}t}. \tag{6.9}$$

As in Section 6.1.1 when we examined the real frequency conversion, it is of interest to make apparent the sidebands that compose $s_{IF}(t)$. We can use the decomposition of $s_{RF}(t)$ based on its complex envelope $\tilde{s}_{RF}(t)$ defined as centered around ω_{RF}. Using equation (6.3), we can then write $s_{IF}(t)$ as

$$\begin{aligned} s_{IF}(t) &= \frac{1}{2}\left(\tilde{s}_{RF}(t)\,e^{j\omega_{RF}t} + \tilde{s}^*_{RF}(t)\,e^{-j\omega_{RF}t}\right)e^{-j\omega_{LO}t} \\ &= \frac{1}{2}\left(\tilde{s}_{RF}(t)\,e^{j(\omega_{RF}-\omega_{LO})t} + \tilde{s}^*_{RF}(t)\,e^{-j(\omega_{RF}+\omega_{LO})t}\right). \end{aligned} \tag{6.10}$$

As illustrated in Figure 6.5, we recover here that the two sidebands of $s_{RF}(t)$ are shifted in the same direction in the frequency domain. This also means that the signal $s_{IF}(t)$ can only be complex, as its spectrum no longer has Hermitian symmetry. Consequently, this signal can only be represented in an analogous way by using two different physical signals. When based on a Cartesian representation, these signals represent the real and imaginary parts, $s_{IF,p}(t)$ and $s_{IF,q}(t)$ respectively, of $s_{IF}(t)$. As $s_{RF}(t)$ is real valued, we can write

$$\begin{aligned} s_{IF}(t) &= s_{RF}(t)\,e^{-j\omega_{LO}t} \\ &= s_{RF}(t)\cos(\omega_{LO}t) - js_{RF}(t)\sin(\omega_{LO}t). \end{aligned} \tag{6.11}$$

As a result,

$$s_{\mathrm{IF,p}}(t) = s_{\mathrm{RF}}(t)\cos(\omega_{\mathrm{LO}}t), \tag{6.12a}$$

$$s_{\mathrm{IF,q}}(t) = -s_{\mathrm{RF}}(t)\sin(\omega_{\mathrm{LO}}t). \tag{6.12b}$$

Based on those expressions, we can deduce the physical implementation shown in Figure 6.6 for the complex frequency downconversion considered so far. Looking at this figure, we can understand the alternative name for this structure, quadrature downconversion, due to the use of two quadrature LO signals representing the real and imaginary parts of the complex exponential used for the conversion. We also remark that the use of the subscripts p and q to denote the real and imaginary parts of $s_{\mathrm{IF}}(t)$ is no accident. More generally, the corresponding branches of the complex downconverter are often referred to as the P and Q branches. This is because when the LO angular frequency ω_{LO} is equal to ω_{RF}, $s_{\mathrm{IF,p}}(t)$ and $s_{\mathrm{IF,q}}(t)$ match the real and imaginary parts of $\tilde{s}_{\mathrm{RF}}(t)$ after going through a lowpass filter that cancels the unwanted sideband lying around $-(\omega_{\mathrm{RF}} + \omega_{\mathrm{LO}})$. We then recover the in-phase and the in-quadrature components of the input RF bandpass signal $s_{\mathrm{RF}}(t)$ as introduced in Chapter 1. Another way to see this is that such additional lowpass filtering acts as an integration stage on the two P and Q signals. The overall processing that can thus be interpreted as the projection of the input RF bandpass signal on the two signals in quadrature which are the sine and cosine LO signals.

Thinking about our derivation so far, we can effectively manage any potential image signal through the use of a complex frequency conversion. However, this is at the cost of dealing with a complex signal at some point of the conversion processing. Moreover, this requires the use of two physical mixing stages that need to be driven by two LO signals in quadrature. This approach thus necessarily has an impact in terms of area and current consumption of the final

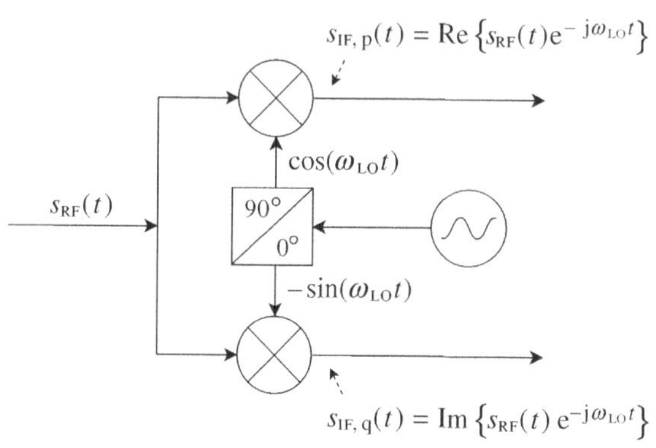

Figure 6.6 Implementation of a complex frequency downconversion using two mixers driven by LO components in quadrature – A complex frequency conversion, corresponding here to the use of the complex exponential $e^{-j\omega_{\mathrm{LO}}t}$, i.e. to the processing depicted in Figure 6.5, can be implemented using two mixers driven by sine waves in quadrature representing the real and imaginary parts of the complex exponential. The resulting output signal being complex, two real valued signals are then required in the analog domain to represent its real and imaginary parts.

solution. However, the gain in terms of integration when avoiding an RF filter almost always makes it attractive. Moreover, from the signal processing point of view there are cases in which the complex frequency conversion is mandatory. This is obviously the case when the wanted signal we are dealing with is complex modulated and located at a low enough frequency at some port of the frequency conversion so that it cannot be represented as bandpass. This simply means that the first upconversion stage on the transmit side and the last downconversion stage on the receive side can only be implemented as a complex frequency conversion as derived in Chapter 1.

In conclusion, it is interesting to observe that the image signal is indeed managed in this way, but only when considering the reconstructed complex signal $s_{\mathrm{IF}}(t) = s_{\mathrm{IF,p}}(t) + \mathrm{j}s_{\mathrm{IF,q}}(t)$. Indeed, when focusing on only one branch of the complex mixer, we recover nothing more than a real frequency conversion. We thus necessarily recover on each of the P or Q branches of the mixer the behavior discussed in Section 6.1.2. For instance, the wanted signal is effectively recovered as superposed with any image signal present at the input of the mixer. This obviously leads to some system constraints for dimensioning each of the P or Q paths. This is well illustrated in the low-IF receive architecture case discussed in Section 8.2.3. We may indeed need to overdimension some parts of those paths to handle the image signal until the complex signal is effectively reconstructed, and thus the image signal canceled. This cancellation mechanism is closely linked to the sign relationship between the sidebands of the wanted and image signal along the P and Q paths. As can be understood from Figure 6.3, the sidebands of the image signal and of the wanted signal that are recovered as superposed have necessarily been converted by complex exponentials of different sign, i.e. $e^{+\mathrm{j}\omega_{\mathrm{LO}}t}$ vs. $e^{-\mathrm{j}\omega_{\mathrm{LO}}t}$. On top of that, the decomposition given by equation (6.5) for the sine and cosine functions that represent the LO signals shows that the two negative complex exponentials that compose those functions are weighted by factors with *different signs* whereas the factors that weight the positive ones have the *same sign*. This results in the possibility of a destructive recombination for one of the signals, while it remains constructive for the other one when performing the operation $s_{\mathrm{IF,p}}(t) + \mathrm{j}s_{\mathrm{IF,q}}(t)$. This mechanism, also involved in image reject mixers, is illustrated further in "Image reject mixers" (Section 6.1.4). It is also illustrated by a complementary approach in Section 6.2 when discussing the limitation in the image rejection that is necessarily experienced due to the practical gain and phase imbalance between the P and Q branches of the complex mixer.

6.1.4 Complex Signal Processing Approach

Signal processing functions based on complex valued operations can be implemented in the RF/analog domain as long as this processing is broken down in terms of real and imaginary parts. Thus it is also of interest to look at the RF/analog implementation of other complex signal processing functions. We can therefore say a few words about the most common of these functions encountered in wireless transceivers: analog complex filters, image reject mixers and some possible alternative implementations of complex frequency conversions. In addition to increasing our understanding of associated signal processing, detailing these functions also allows us to introduce the common limitations in performance that are experienced by all complex signal processing functions when implemented in the RF/analog domain.

These limitations are then studied in more depth for the complex frequency conversion in Section 6.2.

Complex Filters

We focus first on the implementation of complex filters in the analog domain. By "complex filter" we mean a filter whose impulse response is complex valued. Its transfer function in the spectral domain can therefore be made asymmetric with respect to the zero frequency. This behavior can be of interest if we refer to the spectral location of the image signal involved in a frequency downconversion, even complex. Indeed, in the latter case the signal referred to as the image signal in Section 6.1.2 is still present even if no longer folded on the wanted signal during the processing. More precisely (see Figure 6.3), this image signal is centered around the angular frequency symmetric to that of the wanted signal relative to the LO angular frequency before the frequency conversion. After the conversion, the image signal and the wanted signal are thus centered around angular frequencies that are necessarily symmetric relative to DC when considering the reconstructed complex signal. This means that if the sideband of interest of the wanted signal is centered around ω_{IF} (respectively $-\omega_{\mathrm{IF}}$), one sideband of the image signal necessarily falls on $-\omega_{\mathrm{IF}}$ (respectively ω_{IF}). But, in a receiver application the image signal may be a strong interferer of much higher power than the wanted signal. There is therefore a potential need to attenuate this image signal in the analog domain in order to reduce the DR required to pass both the wanted signal and the image signal along the data path. This is often required for the DR of the P and Q ADC stages, as illustrated in the low-IF receive architecture discussed in Section 8.2.3.

The problem is that due to the particular spectral location of those signals, $-\omega_{\mathrm{IF}}$ vs. $+\omega_{\mathrm{IF}}$, the attenuation of the image signal cannot be achieved using two identical real bandpass filters on the P and Q branches of the line-up as illustrated in Figure 6.7. The frequency response of a real filter is necessarily symmetric with respect to DC. A real bandpass filter designed to have its passband centered around $+\omega_{\mathrm{IF}}$ is thus also necessarily bandpass for the negative part of the spectrum centered around $-\omega_{\mathrm{IF}}$. The only way to provide a different rejection for the positive and negative frequencies is to use a complex bandpass filter as illustrated in Figure 6.8(bottom).

The corresponding physical implementation shown in Figure 6.9 can be simply derived by expanding the filtering operation $(h \star s_{\mathrm{IF}})(t) = s_{\mathrm{IF}}(t) \star h(t)$. Decomposing the filter's impulse response $h(t)$ in terms of its real and imaginary parts, $h_{\mathrm{p}}(t)$ and $h_{\mathrm{q}}(t)$ respectively, we can express the real and imaginary parts of the reconstructed filtered signal $(h \star s_{\mathrm{IF}})(t) = (h \star s_{\mathrm{IF}})_{\mathrm{p}}(t) + \mathrm{j}(h \star s_{\mathrm{IF}})_{\mathrm{q}}(t)$ as a function of the real and imaginary parts of the reconstructed IF signal $s_{\mathrm{IF}}(t) = s_{\mathrm{IF,p}} + \mathrm{j}s_{\mathrm{IF,q}}$ present at the output of the frequency conversion. This leads to

$$(h \star s_{\mathrm{IF}})_{\mathrm{p}}(t) = s_{\mathrm{IF,p}}(t) \star h_{\mathrm{p}}(t) - s_{\mathrm{IF,q}}(t) \star h_{\mathrm{q}}(t), \tag{6.13a}$$

$$(h \star s_{\mathrm{IF}})_{\mathrm{q}}(t) = s_{\mathrm{IF,q}}(t) \star h_{\mathrm{p}}(t) + s_{\mathrm{IF,p}}(t) \star h_{\mathrm{q}}(t). \tag{6.13b}$$

Assuming for the sake of simplicity that the complex filter we are dealing with provides an ideal rejection of the signals other than the wanted one, the reconstructed complex signal

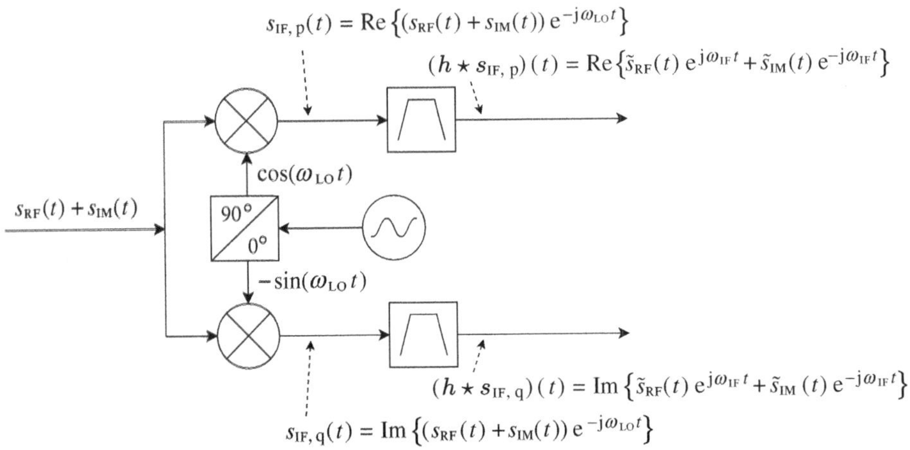

Figure 6.7 Real filters implemented at a complex frequency downconversion output – The implementation of real bandpass filters on each branch of a complex mixer can be useful for the attenuation of adjacent signals or blockers but not for handling the image signal sideband that lies around the angular frequency symmetric to that of the wanted signal relative to DC at the output of the frequency conversion.

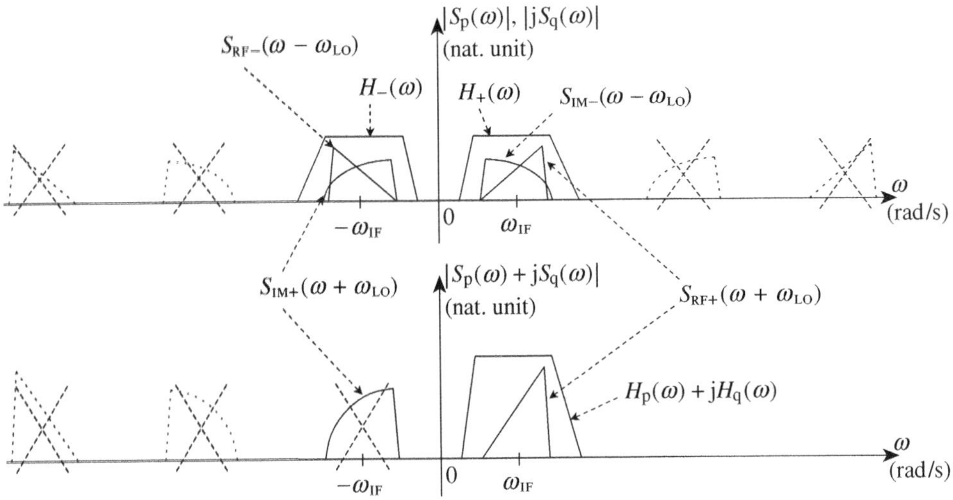

Figure 6.8 Spectral representation of the effects of real and complex filters – Using real bandpass filters on each branch of a complex downmixer, as illustrated in Figure 6.7, cannot provide attenuation at the angular frequency $-\omega_{IF}$ due to its symmetrical frequency response (top). In contrast, the complex filter implementation illustrated in Figure 6.9 can do this when considering the reconstructed complex signal $s_{IF,p}(t) + js_{IF,q}(t)$ (bottom).

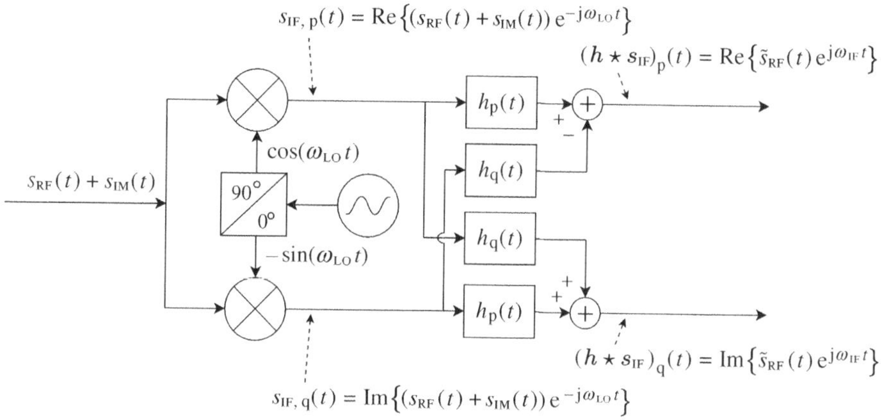

Figure 6.9 Complex filter implementation in the analog domain – The complex filtering of the complex signal $s_{IF}(t)$ is implemented as the filtering of the two signals representing its real and imaginary parts $s_{IF,p}(t)$ and $s_{IF,q}(t)$ by the real and imaginary parts of the filter transfer function $h(t)$. This implementation is a direct transposition of equation (6.13).

recovered at the filter output exactly matches the frequency conversion of the sideband of interest of the input RF bandpass wanted signal $s_{RF}(t)$. In that case, we can thus simply write that

$$(h \star s_{IF})(t) = (h \star s_{IF})_p(t) + j(h \star s_{IF})_q(t) = \tilde{s}_{RF}(t)\, e^{j\omega_{IF} t}, \tag{6.14}$$

with $\tilde{s}_{RF}(t)$ the complex envelope of $s_{RF}(t)$ defined as centered around ω_{RF}.

However, the real life analog domain implementation suffers from limitations that degrade its performance due to the presence of RF impairments. In order to illustrate this, we can assume that there is a gain difference equal to $2\delta G$ between the physical implementations of the real and imaginary parts of the complex filter $h(t)$. Denoting by $h_{p,m}(t)$ and $h_{q,m}(t)$ the transfer functions of the filter in the presence of mismatch, we can then assume for instance that

$$h_{p,m}(t) = (1 + \delta G) h_p(t) \tag{6.15}$$

and

$$h_{q,m}(t) = (1 - \delta G) h_q(t). \tag{6.16}$$

Consequently, we can use equations (6.13a) and (6.15) to express the signal recovered at the output of the P branch of the filter as

$$(h \star s_{IF})_p(t) = (s_{IF,p}(t) \star h_p(t) - s_{IF,q}(t) \star h_q(t)) + \delta G(s_{IF,p}(t) \star h_p(t) + s_{IF,q}(t) \star h_q(t)). \tag{6.17}$$

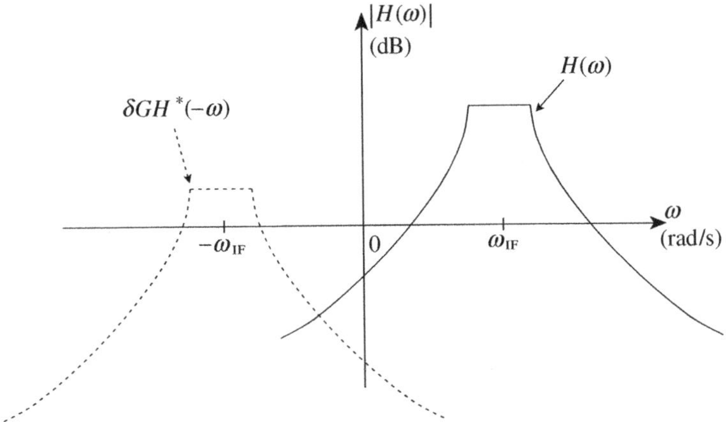

Figure 6.10 Distortion of the transfer function of a complex filter due to mismatch in its analog implementation – Due to practical implementation limitations in its analog implementation, the two instances required to represent both the real and the imaginary parts of the complex filter can exhibit, amongst other things, some gain imbalance, here δG. This leads to a rise in a symmetric copy (dashed line) of the ideal filter transfer function (solid line) relative to DC. The resulting effective transfer function of the filter is then the sum of those two plotted transfer functions as given by equation (6.19).

In the same way, using equations (6.13b) and (6.16), we have for the Q branch

$$(h \star s_{IF})_q(t) = (s_{IF,q}(t) \star h_p(t) + s_{IF,p}(t) \star h_q(t)) + \delta G(s_{IF,q}(t) \star h_p(t) - s_{IF,p}(t) \star h_q(t)). \tag{6.18}$$

The reconstructed complex signal at the filter output can therefore be written as

$$\begin{aligned}(h \star s_{IF})(t) &= (h \star s_{IF})_p(t) + j(h \star s_{IF})_q(t) \\ &= s_{IF}(t) \star h(t) + \delta G\, s_{IF}(t) \star h^*(t) \\ &= s_{IF}(t) \star (h(t) + \delta G\, h^*(t)).\end{aligned} \tag{6.19}$$

All thus behaves as if the impulse response $h(t)$ of the ideal filter were now corrupted by an additive term that is nothing more than its complex conjugate, i.e. $h^*(t)$, but weighted by the gain imbalance δG. In the spectral domain, the complex conjugate of a signal leads to a flipped copy of the original spectrum due to the Fourier transform property given by equation (1.8). When dealing with a complex filter whose passband is centered around ω_{IF}, the term $h^*(t)$ thus leads to a transfer function whose passband is centered around $-\omega_{IF}$, as illustrated in Figure 6.10. By our discussion so far, this results in a limited attenuation of the image signal we were trying to attenuate.

For practical dimensioning, there is a need for consistency between the rejection that is required for such a filter and what can be achieved in practice when taking into account the rejection limitation resulting from the impairments. There is no need to over-specify a filter compared to the rejection allowed by the mismatch we are faced with in practice. At this stage, we can anticipate the discussion in Section 6.2 and observe that the rejection that can be achieved by an analog bandpass complex filter at the image angular frequency is generally nothing more than the equivalent achievable image rejection ratio (IRR) when considering the same gain and phase imbalance. This comes as no great surprise as the same underlying cause is involved; it is the destructive recombination of sidebands when performing the summation of signals that provides the rejection in both cases. Any imbalance thus results in the same amount of rejection limitation. However, from the system point of view the distinction must be made between the latter image rejection limitation problem and the present distortion of the transfer function of a complex filter. Here we are talking about the limitation of an attenuation only. The remaining attenuation can always be handled by using an additional complex filter stage that exhibits no imbalance in its transfer function, as may be the case with a digital implementation for instance. This is not so simple with the image rejection limitation that results in the folding of the image signal on the wanted one. This phenomenon is trickier to handle and needs proper signal processing functions to compensate for it, as discussed in Sections 9.1.3 and 9.2.3.

Image Reject Mixers

As discussed in Section 6.1.3, one great advantage of complex frequency conversion is the possibility of converting an input bandpass signal while avoiding the folding issue of the image signal potentially present at the input of the mixing stage. The drawback is that we need to deal with a complex signal at the output of the system. However, the image cancellation mechanism in the complex frequency transposition is linked to the sign relationship that exists between the sidebands of the wanted and image signals along the P and Q paths of the system. As discussed in Section 6.1.3, the origins of this behavior are twofold. On the one hand, the sidebands recovered as superposed have necessarily been converted by complex exponentials of different sign, i.e. $e^{+j\omega_{LO}t}$ vs. $e^{-j\omega_{LO}t}$. On the other hand, the two negative complex exponentials that compose the sine and cosine LO signals are weighted by factors with *different signs* whereas the factors that weight the positive ones have the *same sign* (see equation (6.5)). The resulting phase relationship leads to cancellation of the image sideband superposed with the wanted one during the reconstruction of the complex signal through the recombination of the P and Q components. But when the output wanted signal is still bandpass, we can continue working with its representation in a real form, as given by equation (1.4), while achieving the same image cancellation by proper processing of the P and Q components before recombination. In that case, we say that we have implemented an image reject mixer.

To investigate this, we can first write analytical expressions for the signals recovered on the P and Q branches of a complex downmixer. More precisely, we can focus for the sake of clarity on the sidebands of interest for our discussion: those centered around $\pm\omega_{IF}$. We can assume for instance that in order to cancel the signal sidebands lying at angular frequencies other than $\pm\omega_{IF}$, a real zonal bandpass filter is duplicated on each branch of the system, as illustrated in Figure 6.7. We can also reconsider here the infradyne complex frequency downconversion using the negative complex exponential $e^{-j\omega_{LO}t}$, with $\omega_{RF}/2 < \omega_{LO} < \omega_{RF}$. This entails the same frequency planning for the downconversion as represented in Figure 6.3 in terms of wanted and image sideband folding. Denoting by $h(t)$ the impulse response of the real zonal bandpass filters, we can express the positive and negative sidebands, $(HS_{\text{IF},\text{p}+})(\omega)$ and $(HS_{\text{IF},\text{p}-})(\omega)$ respectively, of the signal recovered at the output of the filter located on the P branch of the downmixer as

$$(HS_{\text{IF},\text{p}+})(\omega) = \frac{1}{2}(+S_{\text{RF}+}(\omega + \omega_{\text{LO}}) + S_{\text{IM}-}(\omega - \omega_{\text{LO}})), \tag{6.20a}$$

$$(HS_{\text{IF},\text{p}-})(\omega) = \frac{1}{2}(+S_{\text{RF}-}(\omega - \omega_{\text{LO}}) + S_{\text{IM}+}(\omega + \omega_{\text{LO}})), \tag{6.20b}$$

and likewise for the signal recovered on the Q branch,

$$(HS_{\text{IF},\text{q}+})(\omega) = \frac{1}{2\text{j}}(+S_{\text{RF}+}(\omega + \omega_{\text{LO}}) - S_{\text{IM}-}(\omega - \omega_{\text{LO}})), \tag{6.21a}$$

$$(HS_{\text{IF},\text{q}-})(\omega) = \frac{1}{2\text{j}}(-S_{\text{RF}-}(\omega - \omega_{\text{LO}}) + S_{\text{IM}+}(\omega + \omega_{\text{LO}})). \tag{6.21b}$$

As illustrated in Figure 6.11(center), we then recover by inspection of these expressions that the image sideband folded on the wanted sideband is canceled during the reconstruction of the complex signal $(h \star s_{\text{IF},\text{p}})(t) + \text{j}(h \star s_{\text{IF},\text{q}})(t)$. But we also observe that if we succeed in applying a different sign to the positive and negative parts of the spectrum of $(h \star s_{\text{IF},\text{q}})(t)$ before recombining this signal with $(h \star s_{\text{IF},\text{p}})(t)$ we can cancel the two sidebands of the image signal lying around $\pm\omega_{\text{IF}}$ at the same time. Moreover, we recover the two sidebands of the wanted signal as centered around those angular frequencies, and with the correct Hermitian symmetry. This means that the corresponding time domain bandpass signal centered on ω_{IF} is in fact *real* valued. We thus get the result we were looking for.

To derive a corresponding implementation, we need to recall that this application of a different sign to the positive and negative parts of the spectrum is precisely what implements the Hilbert transform (recall the discussion in Section 1.1.2). Considering the transfer function of this transform in the spectral domain, as given by equation (1.17), we can thus write from equations (6.20) and (6.21) that[3]

$$(HS_{\text{IF},\text{p}})(\omega) - (\widehat{HS_{\text{IF},\text{q}}})(\omega) = S_{\text{RF}+}(\omega + \omega_{\text{LO}}) + S_{\text{RF}-}(\omega - \omega_{\text{LO}}). \tag{6.22}$$

Consequently, we can write in the time domain that

$$(h \star s_{\text{IF},\text{p}})(t) - (\widehat{h \star s_{\text{IF},\text{q}}})(t) = \text{Re}\{\tilde{s}_{\text{RF}}(t)\, e^{+\text{j}\omega_{\text{IF}}t}\}, \tag{6.23}$$

with $\tilde{s}_{\text{RF}}(t)$ and $\tilde{s}_{\text{IM}}(t)$ the complex envelopes of the input wanted signal and of the input image signal defined as centered around ω_{RF} and ω_{IM}, respectively. This is effectively the behavior we were expecting in terms of real frequency conversion. Alternatively, one can find that

$$(h \star s_{\text{IF},\text{p}})(t) + (\widehat{h \star s_{\text{IF},\text{q}}})(t) = \text{Re}\{\tilde{s}^*_{\text{IM}}(t)\, e^{+\text{j}\omega_{\text{IF}}t}\}. \tag{6.24}$$

[3] Recall our convention that $\hat{S}(\omega)$ stands for the spectral domain representation of the Hilbert transform $\hat{s}(t)$ and not for the Hilbert transform of the signal $S(\omega)$.

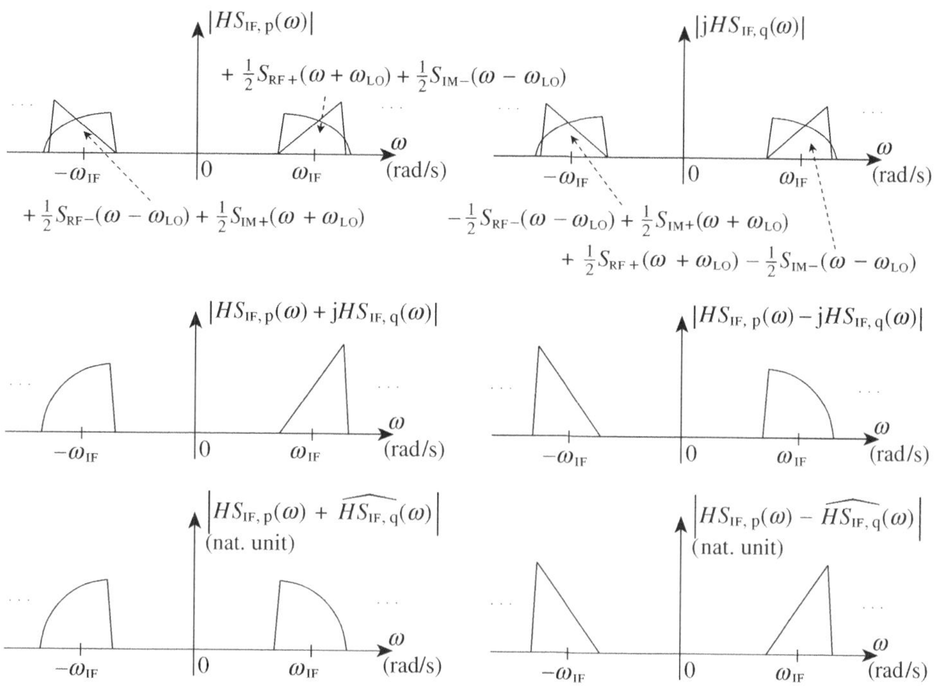

Figure 6.11 Spectral representation of the mechanism involved in an image reject mixer – Considering the implementation shown in Figure 6.12, we get different signs applied to the sidebands of the wanted and image signals on each branch of the complex mixer (top). The wanted signal transposed around $\pm\omega_{\text{IF}}$ can thus be reconstructed in its real form using the Hilbert transform (bottom). This processing is different from that involved in the reconstruction of the complex signal that leads to a simple shift of the input spectrum (center). Here, the case $\omega_{\text{RF}}/2 < \omega_{\text{LO}} < \omega_{\text{RF}}$ is considered.

In practice, the Hilbert transform can be implemented as a simple $\pi/2$ phase shifter when acting on bandpass signals. This leads to the final implementation, known as the Hartley architecture, shown in Figure 6.12. Other architectures can be considered for the image reject mixer stage, but always based on the same root principles [70].

In conclusion, we observe that the image rejection is achieved thanks to the use of an intermediate complex mixer stage. Anticipating the discussion in Section 6.2.1, the use of this quadrature frequency conversion makes the image reject mixer sensitive to any gain and phase imbalance between the P and Q branches of this intermediate stage. We thus necessarily have a limitation of the performance of such a system in real life. Moreover, as in the present case the two branches of the complex IF stage are recombined in order to recover the IF bandpass wanted signal in its real form, no further processing can compensate for this imbalance. This is a major limitation compared to the case of a pure complex mixing stage for which such compensation is possible by proper processing of the P and Q signals (see Sections 9.1.3 and 9.2.3).

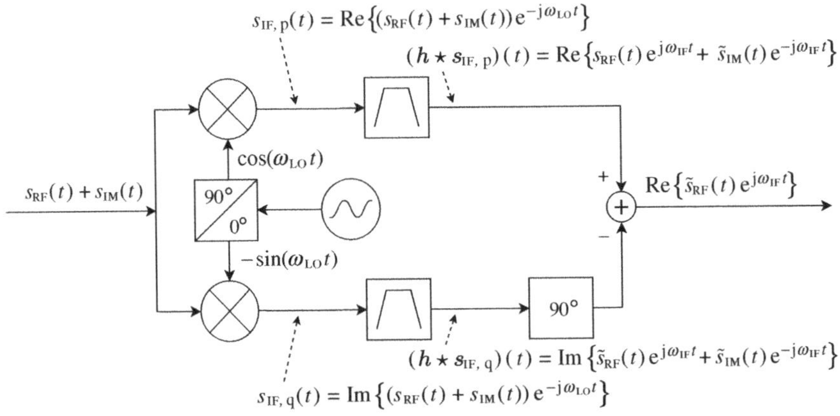

Figure 6.12 Structure for an image reject mixer (Hartley type) – Using a complex mixer and the property of the $\pi/2$ phase shifter that acts differently on the positive and negative sidebands of bandpass signals, an image reject mixer can be implemented. This means we can carry out a *real* frequency conversion of a bandpass signal represented in its real form both at the input and output of the conversion, without using an RF filter before the mixer to cancel the image signal. In the present representation, zonal bandpass filters are assumed on each branch of the complex mixer in order to take into account only the signal sidebands centered around $\pm\omega_{\mathrm{IF}}$ in the analytical expressions.

Alternative Implementation for the Complex Frequency Conversion

Now that we have a good understanding of the root causes for the image signal cancellation mechanism associated with the complex frequency conversion, it is of interest to see that we can achieve the same result, albeit with some restrictions, in an alternative way. Let us recall the infradyne downconversion using the negative complex exponential $\mathrm{e}^{-\mathrm{j}\omega_{\mathrm{LO}}t}$, with $\omega_{\mathrm{RF}}/2 < \omega_{\mathrm{LO}} < \omega_{\mathrm{RF}}$. The underlying phenomenon associated with this frequency planning is illustrated in Figure 6.3 in terms of wanted and image sideband folding. We know for instance from the discussion in Section 6.1.2 that it is the negative (positive) sideband of the image signal that is recovered as superposed on the positive (negative) sideband of the wanted signal lying around the angular frequency $+\omega_{\mathrm{IF}}$ $(-\omega_{\mathrm{IF}})$. This holds along each of the P or Q paths of the complex mixer. And due to the phase relationship between those sidebands resulting from the differences in the structure of the sine and cosine LO signals, the image sideband superposed on the wanted sideband can be canceled during the reconstruction of the complex IF signal. However, with this frequency planning we can obviously achieve the same result by applying different signs directly to the positive and negative parts of the input RF signal spectrum and then using the same real frequency conversion on each branch of the complex mixer. By our discussion in the previous section, this distinction between the positive and negative part of the spectrum can be achieved using a simple $\pi/2$ phase shifter that implements the Hilbert transform.

This behavior can be confirmed by writing analytical expressions for the corresponding sidebands of interest, i.e. those centered around $\pm\omega_{\mathrm{IF}}$. We can assume that a real zonal bandpass filter is duplicated on each of the branches of the line-up in order to cancel the signal sidebands lying at angular frequencies other than $\pm\omega_{\mathrm{IF}}$. This would result in analytical expressions valid

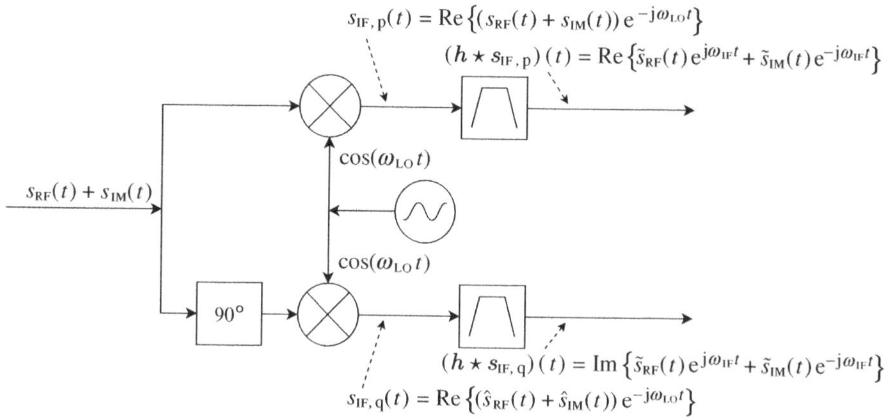

Figure 6.13 Alternative implementation for the complex frequency conversion – The complex frequency conversion can be approximated for some types of frequency planning using the input RF bandpass signal set in two quadrature signals that feed two real mixers that use the same LO waveform. This means that, compared to the ideal implementation shown in Figure 6.6, the $\pi/2$ phase shifter can be used to generate the quadrature LO components *or* the quadrature input RF signals. The corresponding underlying mechanism is shown in Figure 6.14.

for a line-up as represented in Figure 6.13. Assuming that the impulse response of the filters is $h(t)$, we can follow the spectral representation of the signal processing shown in Figure 6.14 to express the positive and negative sidebands of the IF signals recovered at the output of the zonal filters. Denoting by $(HS_{IF,p+})(\omega)$ and $(HS_{IF,p-})(\omega)$ the positive and negative sidebands recovered on the P branch of the mixing stage, we can write

$$(HS_{IF,p+})(\omega) = \frac{1}{2}(+S_{RF+}(\omega + \omega_{LO}) + S_{IM-}(\omega - \omega_{LO})), \tag{6.25a}$$

$$(HS_{IF,p-})(\omega) = \frac{1}{2}(+S_{RF-}(\omega - \omega_{LO}) + S_{IM+}(\omega + \omega_{LO})). \tag{6.25b}$$

Taking into account the transfer function of the $\pi/2$ phase shifter given by equation (1.17), we then get for the Q branch that

$$(HS_{IF,q+})(\omega) = \frac{1}{2j}(+S_{RF+}(\omega + \omega_{LO}) - S_{IM-}(\omega - \omega_{LO})), \tag{6.26a}$$

$$(HS_{IF,q-})(\omega) = \frac{1}{2j}(-S_{RF-}(\omega - \omega_{LO}) + S_{IM+}(\omega + \omega_{LO})). \tag{6.26b}$$

We thus see that we can reconstruct the complex IF signal $(h \star s_{IF,p})(t) + j(h \star s_{IF,q})(t)$ with the same sidebands centered around $\pm\omega_{IF}$ as if we had implemented an exact complex frequency downconversion. We get in the frequency domain that

$$(HS_{IF,p})(\omega) + j(HS_{IF,q})(\omega) = S_{RF+}(\omega + \omega_{LO}) + S_{IM+}(\omega + \omega_{LO}), \tag{6.27}$$

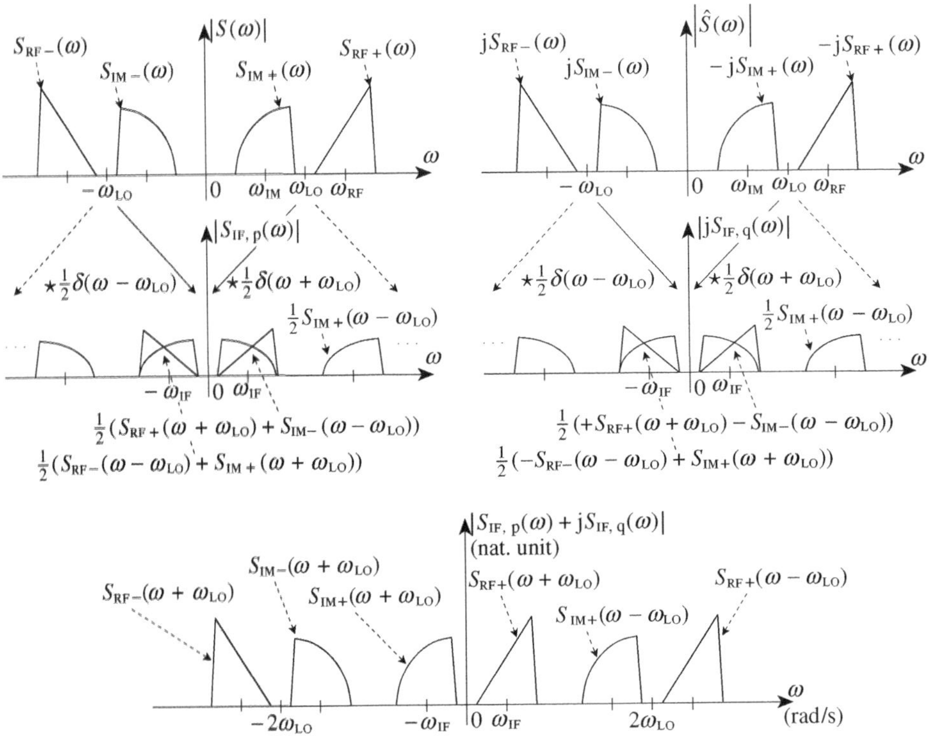

Figure 6.14 Spectral representation of the alternative implementation principle of the complex frequency conversion – The frequency conversion using the multiplication of the input RF signal (top left) and its $\pi/2$ phase shifted version (top right) with a cosine function gives the same spectral partition of the sidebands but with different signs (center left and right). The resulting reconstructed complex IF signal thus gives the same result as the ideal complex frequency conversion when considering the resulting sidebands centered around $\pm\omega_{\text{IF}}$ (bottom). The corresponding implementation is shown in Figure 6.13.

and thus in the time domain that

$$(h \star s_{\text{IF,p}})(t) + \text{j}(h \star s_{\text{IF,q}})(t) = \tilde{s}_{\text{RF}}(t)e^{+\text{j}\omega_{\text{IF}}t} + \tilde{s}_{\text{IM}}(t)e^{-\text{j}\omega_{\text{IF}}t}, \tag{6.28}$$

with $\tilde{s}_{\text{RF}}(t)$ and $\tilde{s}_{\text{IM}}(t)$ the complex envelopes of the input wanted signal and of the input image signal defined as centered around ω_{RF} and ω_{IM}, respectively.

However, we need to keep in mind that the present implementation is not an ideal complex frequency conversion implementation, as it works only for some types of frequency planning, i.e. when the image and wanted signal sidebands that get superposed come from the positive vs. negative part of the input spectrum. Moreover, among the sidebands recovered at the output of the processing, only those centered around $\pm\omega_{\text{IF}}$ match those expected from an ideal complex frequency conversion. The overall downconverted spectrum does not correspond to the input one shifted in frequency, as can be seen in Figure 6.14(bottom).

6.2 Gain and Phase Imbalance

6.2.1 *Image Rejection Limitation*

We know from the discussion in Chapter 1 that complex frequency conversion is mandatory when the signal to be converted is complex modulated and lies at some port of the mixing stage at a sufficiently low frequency that it cannot be represented in its real bandpass form. This holds, for instance, for the first upconversion stage on the transmit side and for the last downconversion stage on the receive side. However, by the discussion so far in this chapter, a great advantage of complex frequency conversion is that it theoretically allows us to get rid of the image problem when considering the reconstructed complex output signal. The possibility of avoiding the use of costly RF filters for the image rejection makes this approach very attractive in many situations for a low cost integrated transceiver solution. However, by Section 6.1.3 the cancellation of the image signal during such processing relies on the exact opposite phase relationship that holds between the sidebands present on the P and Q branches of a complex mixer. Consequently, we may fear that any imbalance between those branches, as may be expected from any RF/analog physical implementation, will result in limited cancellation of the unwanted sidebands during the reconstruction of the complex signal, thus degrading the performance of the system.

In practice, the imbalance we are referring to reduces to both an amplitude and a phase imbalance between the quadrature LO signals used to drive the mixing stages on the two branches of the system. However, the fact is that any gain imbalance between the P and Q branches of the line-up can still be interpreted as an amplitude imbalance between the P and Q LO signals when examined from the signal processing point of view. For the sake of consistency and simplicity in our analytical derivations, we can thus assume that all the imbalance we are faced with, i.e. both phase and amplitude or gain, is embedded in the expression for the P and Q LO signals. Consequently, we can define the gain imbalance g experienced by the signal when going through the P and Q paths of the complex frequency converter as the ratio

$$g = G_q/G_p. \tag{6.29}$$

Here, G_p and G_q represent the equivalent amplitude of the P and Q LO signals, respectively. In the same way, we can define

$$\delta\phi = \phi_q(t) - \phi_p(t) \tag{6.30}$$

as the difference between the instantaneous phase of the quadrature LO signals expected to represent the real and imaginary parts of the targeted complex exponential. We now need to select a practical configuration for a frequency conversion. Let us recall the complex frequency downconversion based on the use of the negative complex exponential $e^{-j\omega_{LO}t}$ already extensively discussed in previous sections. As discussed at the end of the section, the conclusions remain general. Thus, the two quadrature LO signals, $lo_p(t)$ and $lo_q(t)$ expected to represent the real and imaginary parts respectively of this complex exponential, can be written in the presence of impairments as

$$lo_p(t) = G_p \cos(\phi_p(t)), \tag{6.31a}$$

$$lo_q(t) = -gG_p \sin(\phi_p(t) + \delta\phi). \tag{6.31b}$$

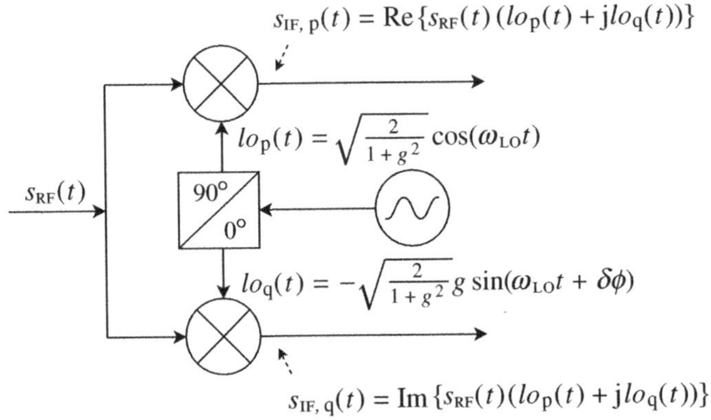

Figure 6.15 Complex frequency downconversion in the presence of gain and phase imbalance – Due to implementation limitations in the RF/analog world, the sine and cosine signals that represent the real and imaginary parts of the complex exponential required for a complex conversion, here $e^{-j\omega_{LO}t}$, are not exactly in quadrature and do not have the same amplitude. Here, g and $\delta\phi$ respectively represent the receiver gain and phase imbalance as defined by equations (6.29) and (6.30). Here, the LO waveforms are normalized by the factor $\sqrt{2/(1+g^2)}$ so that $\overline{(lo_p^2(t)+lo_q^2(t))} = 1$.

We might also assume that we are dealing with an LO signal that has a normalized power whatever the level of imbalance, e.g. such that

$$\overline{\left(lo_p^2(t)+lo_q^2(t)\right)} = 1, \tag{6.32}$$

where $\overline{(.)}$ denotes time averaging. This allows more straightforward interpretations of the results, as can be seen in what follows. It results in $G_p = \sqrt{2/(1+g^2)}$. On top of that, assuming that the angular frequency of those LO signals is ω_{LO}, we can simply assume that $\phi_p(t) = \omega_{LO}t$. Finally, we can suppose that

$$lo_p(t) = \sqrt{\frac{2}{1+g^2}}\cos(\omega_{LO}t), \tag{6.33a}$$

$$lo_q(t) = -\sqrt{\frac{2}{1+g^2}}g\sin(\omega_{LO}t+\delta\phi). \tag{6.33b}$$

The corresponding implementation for the complex downmixer we consider here is represented in Figure 6.15, which should be compared to the ideal structure shown in Figure 6.6.

In order to go further in the derivation of the system impacts of this imbalance, we may investigate the expression for the reconstructed complex LO signal $lo_{-\omega_{LO}}(t) = lo_p(t) + jlo_q(t)$

in order to examine its differences with the targeted complex exponential $e^{-j\omega_{LO}t}$. From the above expressions, we can write that

$$lo_{-\omega_{LO}}(t) = \sqrt{\frac{2}{1+g^2}}[\cos(\omega_{LO}t) - jg\sin(\omega_{LO}t + \delta\phi)]. \tag{6.34}$$

We may now expand the sine and cosine functions involved in this expression using equation (6.5) in order to include complex exponentials. Given that

$$\cos(\omega_{LO}t) = \frac{1}{2}\left(e^{-j\omega_{LO}t} + e^{+j\omega_{LO}t}\right), \tag{6.35a}$$

$$-jg\sin(\omega_{LO}t + \delta\phi) = \frac{g}{2}\left(e^{-j\delta\phi}e^{-j\omega_{LO}t} - e^{+j\delta\phi}e^{+j\omega_{LO}t}\right), \tag{6.35b}$$

we can write

$$lo_{-\omega_{LO}}(t) = (\alpha_-^-)e^{-j\omega_{LO}t} + (\alpha_-^+)e^{+j\omega_{LO}t}, \tag{6.36}$$

with (α_-^-) and (α_-^+) given by

$$(\alpha_-^-) = \frac{1 + ge^{-j\delta\phi}}{\sqrt{2(1+g^2)}}, \tag{6.37a}$$

$$(\alpha_-^+) = \frac{1 - ge^{+j\delta\phi}}{\sqrt{2(1+g^2)}}. \tag{6.37b}$$

As illustrated in Figure 6.16, due to the gain and phase imbalance the tones that compose both the P and Q LO signals no longer have the exact amplitude and phase relationship as could be expected from pure sine and cosine functions. The recombination of those signals then no longer results in an exact cancellation of either of the complex conjugate complex exponentials. Consequently, the reconstructed complex LO signal is now the linear combination of the expected complex exponential, $e^{-j\omega_{LO}t}$ here, and its complex conjugate $e^{+j\omega_{LO}t}$. These complex exponentials are weighted by factors that depend only on the gain and phase imbalance resulting from the physical implementation of the line-up. From the signal processing point of view, we are thus in an intermediate situation between the real frequency conversion, discussed in Section 6.1.1, in which the two complex exponentials have the same amplitude, and the ideal complex frequency conversion, discussed in Section 6.1.3, in which the unwanted complex exponential is completely canceled when considering the reconstructed complex signal. In the present case, the unwanted exponential is only attenuated, not completely canceled. This leads to the folding of an attenuated image signal sideband on the wanted sideband during the frequency conversion processing. We are then faced with what is called a limitation in the image rejection. This image rejection is classically evaluated through the image rejection ratio

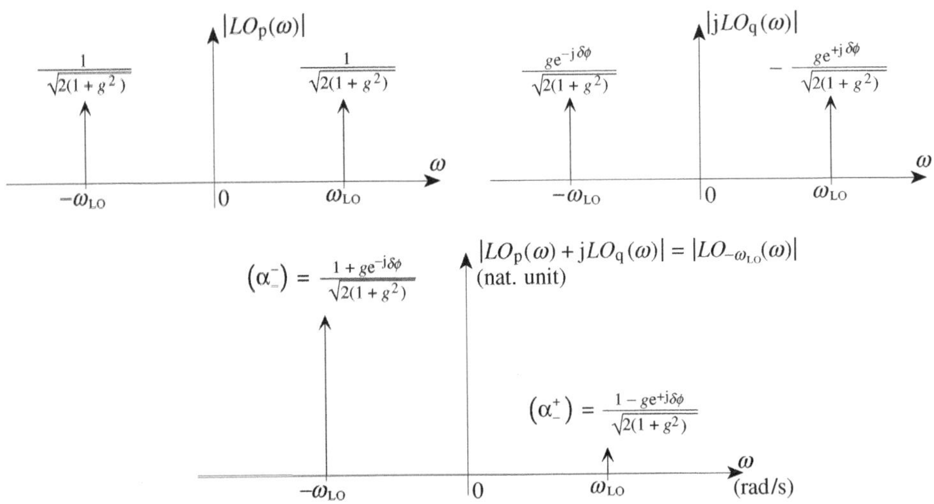

Figure 6.16 Spectral content of the LO waveforms used for the complex frequency conversion in the presence of gain and phase imbalance – In the presence of gain and phase imbalance, the complex conjugate tones that compose both the cosine part (top left) or the sine part (top right) of the LO signal have non-ideal amplitudes and phases. As a result, one of those tones does not exactly vanish when reconstructing the complex LO waveform (bottom).

(IRR), defined as the ratio of the power of the unwanted complex exponential to the power of the wanted complex exponential. Thus, recalling equation (6.36), we deduce that

$$IRR = \frac{|(\alpha_-^+)|^2}{|(\alpha_-^-)|^2} = \frac{(\alpha_-^+)(\alpha_-^+)^*}{(\alpha_-^-)(\alpha_-^-)^*} = \frac{1 - 2g\cos(\delta\phi) + g^2}{1 + 2g\cos(\delta\phi) + g^2}. \tag{6.38}$$

By its definition, this IRR, plotted in Figure 6.17, gives the amount by which the image sideband is attenuated before folding on the wanted sideband.

Here, we observe that for frequency conversions that use different frequency planning, the very same derivation can be done for the reconstructed complex LO signal $lo_{+\omega_{\mathrm{LO}}}(t)$ that approximates the positive complex exponential $\mathrm{e}^{+\mathrm{j}\omega_{\mathrm{LO}}t}$. In that case, the alpha factors can be defined by

$$lo_{+\omega_{\mathrm{LO}}}(t) = (\alpha_+^+)\mathrm{e}^{+\mathrm{j}\omega_{\mathrm{LO}}t} + (\alpha_+^-)\mathrm{e}^{-\mathrm{j}\omega_{\mathrm{LO}}t}. \tag{6.39}$$

However, we observe that the approximation of $\mathrm{e}^{+\mathrm{j}\omega_{\mathrm{LO}}t}$ is obtained by simply taking the opposite sign in the Q component of the LO signal $lo_{-\omega_{\mathrm{LO}}}(t)$ considered so far. We can then directly write that

$$lo_{+\omega_{\mathrm{LO}}}(t) = lo^*_{-\omega_{\mathrm{LO}}}(t). \tag{6.40}$$

Referring to equation (6.36), we then get that

$$lo_{+\omega_{\mathrm{LO}}}(t) = (\alpha_-^-)^*\mathrm{e}^{+\mathrm{j}\omega_{\mathrm{LO}}t} + (\alpha_-^+)^*\mathrm{e}^{-\mathrm{j}\omega_{\mathrm{LO}}t}. \tag{6.41}$$

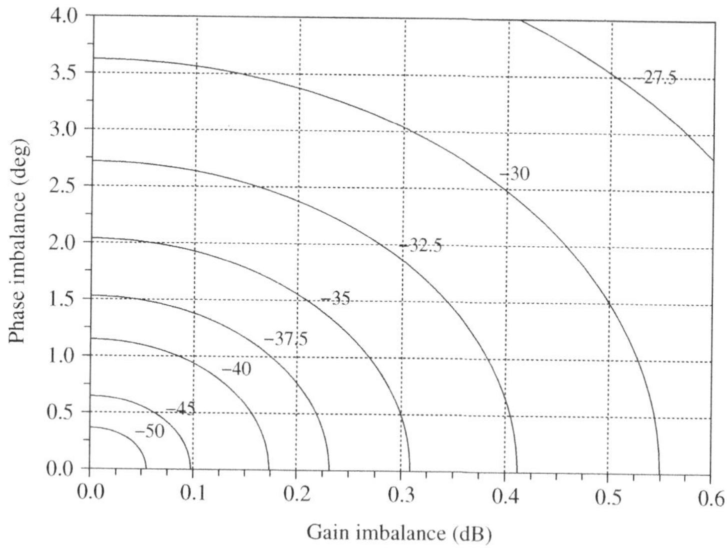

Figure 6.17 IRR as a function of gain and phase imbalance – The IRR gives the maximum achievable power attenuation of the image signal during a complex frequency conversion. This attenuation, given by equation (6.38), is limited by the gain and phase imbalance, g and $\delta\phi$ respectively, between the two branches of the complex mixer as defined by equations (6.29) and (6.30), respectively.

Comparing this equation with equation (6.39) and using the expressions for (α_-^-) and (α_-^+) given by equation (6.37), we can finally write that

$$(\alpha_+^+) = (\alpha_-^-)^* = \frac{1 + g\mathrm{e}^{+\mathrm{j}\delta\phi}}{\sqrt{2(1+g^2)}}, \tag{6.42a}$$

$$(\alpha_+^-) = (\alpha_-^+)^* = \frac{1 - g\mathrm{e}^{-\mathrm{j}\delta\phi}}{\sqrt{2(1+g^2)}}. \tag{6.42b}$$

In particular, we see that the magnitude of the alpha factors remains the same when considering the decomposition of $lo_{+\omega_{\mathrm{LO}}}(t)$ or $lo_{-\omega_{\mathrm{LO}}}(t)$. As could be expected due to the symmetry of the problem, the definition of the IRR given by equation (6.38) remains valid whatever the complex exponential used for a complex frequency conversion. We indeed have the same physical limitations in both cases.

To conclude this section, we observe that the gain imbalance is sometimes expressed as a percentage rather than in decibels. In this case, the gain imbalance, denoted by δG, represents the gain difference between the two branches relative to the average gain, i.e.

$$\delta G = 2\,\frac{G_{\mathrm{q}} - G_{\mathrm{p}}}{G_{\mathrm{q}} + G_{\mathrm{p}}}. \tag{6.43}$$

We thus get that g, defined by equation (6.29), is related to δG by

$$g = \frac{1 + \delta G/2}{1 - \delta G/2} \quad \Longleftrightarrow \quad \frac{\delta G}{2} = \frac{g-1}{g+1}. \tag{6.44}$$

Table 6.1 Correspondence given by equation (6.44) between the gain imbalance expressed as a ratio, i.e. g defined by equation (6.29), and its expression as a normalized difference, i.e. δG defined by equation (6.43).

g (dB)	0	0.1	0.2	0.3	0.4	0.5	0.6	0.7	0.8	0.9
δG (%)	0	1.15	2.3	3.45	4.6	5.75	6.91	8.05	9.2	10.35

Some examples of correspondence are given in Table 6.1.

6.2.2 *Signal Degradation*

According to the discussion in Section 6.2.1, the gain and phase imbalance between the two LO signals used for the implementation of a complex frequency conversion leads to a finite image rejection. We can go a step further and refine the consequences of such limitation in terms of system impacts. So far we have considered mainly frequency downconversions. This may therefore be a good place to review both upconversions and downconversions in order to address in greater depth the system impacts that may be seen in both transmitters and receivers.

Transmit Side

To discuss the system impacts encountered in transmitters, let us consider as an example the complex frequency upconversion of the reconstructed complex signal $\tilde{s}(t) = p(t) + \mathrm{j}q(t)$ based on the use of the positive complex exponential $\mathrm{e}^{+\mathrm{j}\omega_{\mathrm{LO}}t}$. By the discussion in Section 6.2.1, in the presence of gain and phase imbalance, this theoretical LO signal is degraded down to the complex signal $lo_{+\omega_{\mathrm{LO}}}(t) = lo_{\mathrm{p}}(t) + \mathrm{j}lo_{\mathrm{q}}(t)$, still centered on $+\omega_{\mathrm{LO}}$, but now exhibiting a rise in the complex conjugate tone $\mathrm{e}^{-\mathrm{j}\omega_{\mathrm{LO}}t}$. Based on equation (6.40), we can express its real and imaginary parts from equation (6.33) as

$$lo_{\mathrm{p}}(t) = \sqrt{\frac{2}{1+g^2}}\cos(\omega_{\mathrm{LO}}t), \tag{6.45a}$$

$$lo_{\mathrm{q}}(t) = \sqrt{\frac{2}{1+g^2}}g\sin(\omega_{\mathrm{LO}}t + \delta\phi). \tag{6.45b}$$

Assuming that the signal $s_{\mathrm{RF}}(t)$ recovered at the output of the frequency upconversion can be represented in its real bandpass form, we can then write

$$\begin{aligned} s_{\mathrm{RF}}(t) &= \mathrm{Re}\{\tilde{s}(t)(lo_{+\omega_{\mathrm{LO}}}(t))\} \\ &= \mathrm{Re}\{(p(t) + \mathrm{j}q(t))(lo_{\mathrm{p}}(t) + \mathrm{j}lo_{\mathrm{q}}(t))\} \\ &= p(t)lo_{\mathrm{p}}(t) - q(t)lo_{\mathrm{q}}(t), \end{aligned} \tag{6.46}$$

resulting in the implementation shown in Figure 6.18.

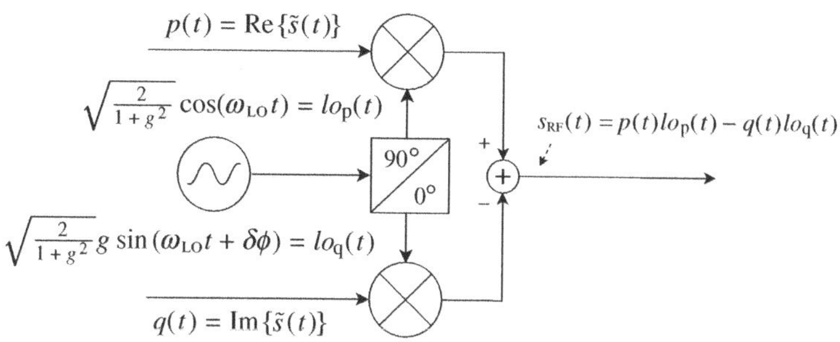

Figure 6.18 Complex frequency upconversion in the presence of gain and phase imbalance – This scheme models the complex frequency upconversion using the positive complex exponential $e^{+j\omega_{LO}t}$, but in the presence of gain and phase imbalance. Here, g and $\delta\phi$ represent the receiver gain and phase imbalance respectively, as defined by equations (6.29) and (6.30). The LO waveforms are normalized by the factor $\sqrt{2/(1+g^2)}$ so that $\overline{(lo_p^2(t) + lo_q^2(t))} = 1$.

In order to take our investigation further, we can expand the reconstructed complex LO signal $lo_{+\omega_{LO}}(t)$ as the linear superposition of the two complex conjugate complex exponentials, $e^{+j\omega_{LO}t}$ and $e^{-j\omega_{LO}t}$. This is required in order to make apparent the mechanism at the origin of the generation of the image signal in this processing. Using equation (6.39), we have

$$\begin{aligned} s_{RF}(t) &= \frac{1}{2}\left(\tilde{s}(t) lo_{+\omega_{LO}}(t) + \tilde{s}^*(t) lo^*_{+\omega_{LO}}(t)\right) \\ &= \frac{1}{2}\Big[((\alpha_+^+)^* \tilde{s}^*(t) + (\alpha_+^-)\tilde{s}(t))e^{-j\omega_{LO}t} \\ &\quad + ((\alpha_+^+)\tilde{s}(t) + (\alpha_+^-)^* \tilde{s}^*(t))e^{+j\omega_{LO}t}\Big]. \end{aligned} \tag{6.47}$$

This expression can be interpreted in the frequency domain in order to discuss the sidebands present in the spectrum of $s_{RF}(t)$. For that purpose, we can take the Fourier transform of a given realization of those processes. Using equation (1.8), we can then directly write that

$$\begin{aligned} S_{RF}(\omega) &= \frac{1}{2}\left((\alpha_+^+)^* \tilde{S}^*(-\omega - \omega_{LO}) + (\alpha_+^-)\tilde{S}(\omega + \omega_{LO})\right) \\ &\quad + \frac{1}{2}\left((\alpha_+^+)\tilde{S}(\omega - \omega_{LO}) + (\alpha_+^-)^* \tilde{S}^*(-\omega + \omega_{LO})\right). \end{aligned} \tag{6.48}$$

Looking at this expression, the spectrum of $s_{RF}(t)$ is composed of positive and negative sidebands that are complex conjugates and flipped copies of each other. This in fact corresponds to the Hermitian symmetry of the spectrum that guaranties that the corresponding time domain signal is a real valued signal. But if we now look in more depth at the sidebands represented in Figure 6.19(bottom), we see that they are composed on the one hand of the sidebands we would ideally have if no mismatch were present, i.e. $\tilde{S}(\omega)$ transposed around $+\omega_{RF}$ and $\tilde{S}^*(-\omega)$ transposed around $-\omega_{RF}$, but on the other hand of unwanted components lying around the angular frequencies $\pm\omega_{IM} = \pm|\omega_{LO} - \omega_{IF}|$. Recalling Section 6.1.2, the unwanted sidebands

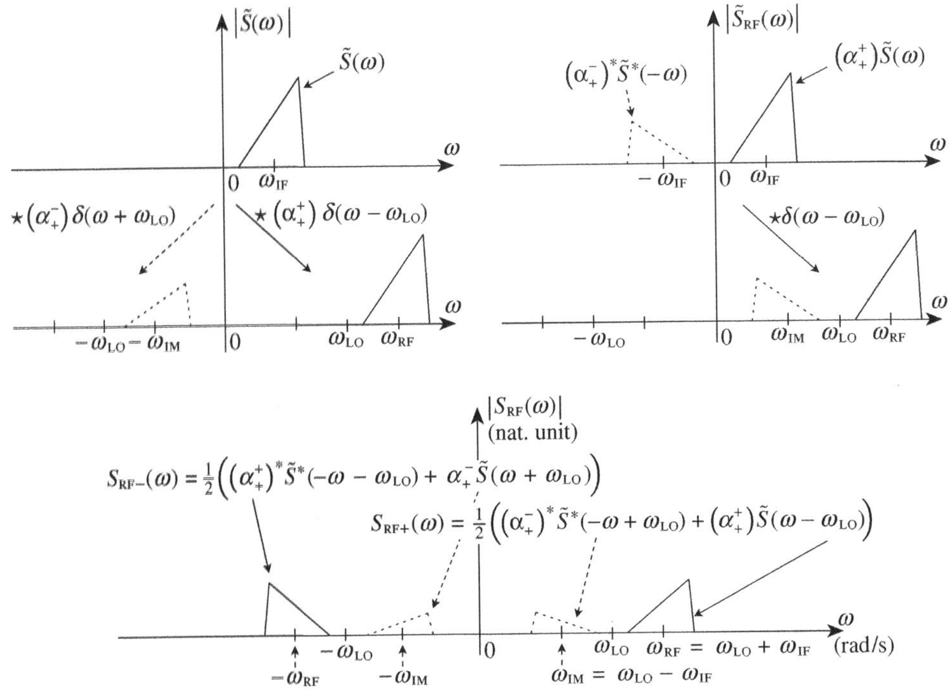

Figure 6.19 Spectral representation of a complex signal frequency upconversion in the presence of gain and phase imbalance – The complex frequency upconversion in the presence of imbalance can be interpreted in two ways. We suppose either that the ideal input complex signal $\tilde{s}(t)$ (top left) is upconverted using an LO waveform corrupted by the attenuated unwanted complex exponential, here $(\alpha_+^-)e^{-j\omega_{LO}t}$ (center left), or that this input complex signal is corrupted by its own image, i.e. its complex conjugate weighted by $(\alpha_+^-)^*$ (top right), but upconverted by the ideal LO waveform, here $e^{+j\omega_{LO}t}$ (center right). Both interpretations lead to the same output real bandpass RF signal (bottom).

match the image signal generated during a real frequency conversion. However, we recover here an intermediate situation between the real frequency conversion case, in which the image signal is generated with the same amplitude as the wanted signal, and the ideal complex frequency conversion, in which this signal is simply not generated at all. Due to the residual unwanted complex exponential present in the reconstructed complex LO signal $lo_{+\omega_{LO}}(t)$, we get a generation of this image signal, but now attenuated with respect to the wanted signal. Dealing with the same alpha factors in equation (6.48) as introduced in Section 6.2.1, we then get that the IRR defined by equation (6.38) directly gives the relative power of the generated image signal relative to the wanted signal. Depending on the application, we may need to filter out the residual unwanted image sidebands at the output of the frequency upconversion in order to fulfill any spectrum emission requirement. In that case, we need to consider carefully the frequency planning of the conversion, as discussed in Section 6.1.2. Depending on the value of ω_{IF}, it may be impossible to consider such filtering in practice, in particular in the RF domain. However, we have to keep in mind that the performance of such a complex mixer in respect of this image rise problem can be improved by using a calibration scheme for the gain and phase imbalance, as discussed for instance in Section 9.1.3.

We can also interpret equation (6.47) in the time domain. We can use equation (1.5) to rewrite this expression as

$$s_{\mathrm{RF}}(t) = \mathrm{Re}\left\{((\alpha_+^+)\tilde{s}(t) + (\alpha_+^-)^*\tilde{s}^*(t))e^{+j\omega_{\mathrm{LO}}t}\right\}. \tag{6.49}$$

We can thus express the complex envelope $\tilde{s}_{\mathrm{RF}}(t)$ of this bandpass signal, when defined as centered around ω_{LO}, as

$$\tilde{s}_{\mathrm{RF}}(t) = (\alpha_+^+)\tilde{s}(t) + (\alpha_+^-)^*\tilde{s}^*(t). \tag{6.50}$$

Consequently, we can interpret the upconversion processing in the presence of gain and phase imbalance in two different but equivalent ways. Up to now we have considered the consequences of this gain and phase imbalance in terms of distortion of the reconstructed complex LO signal used for the conversion. However, we can interpret this processing equivalently as the frequency upconversion still using the pure complex exponential $e^{+j\omega_{\mathrm{LO}}t}$, but of a distorted version of the input wanted signal. In that case, the complex envelope $\tilde{s}_{\mathrm{RF}}(t)$ of the upconverted signal is the linear combination of the complex envelope of the wanted signal $\tilde{s}(t)$, and of $\tilde{s}^*(t)$. The latter signal is therefore nothing more than the complex envelope, defined as centered around ω_{LO}, of the generated image signal. These two interpretations are illustrated in the spectral domain in Figure 6.19.

The latter point of view is useful for interpreting the structure of the distorted upconverted signal $s_{\mathrm{RF}}(t)$. This is particularly true in the case of the direct upconversion of a complex lowpass modulating waveform, i.e. when $\omega_{\mathrm{IF}} = 0$. As Figure 6.19 shows, the image signal is then directly superposed on the wanted signal in the frequency domain. There is thus no way to filter out this unwanted component, and the interpretation of the resulting modulation distortion based on the decomposition given in equation (6.50) makes sense. By way of illustration, we can consider a simple constant amplitude lowpass modulating waveform, for instance a GMSK modulated signal according to the modulation process introduced in Section 1.3.1. The complex lowpass signal representing the modulating waveform, $\tilde{s}(t)$, can be written in terms of the instantaneous phase of the modulation $\phi(t)$ as

$$\tilde{s}(t) = p(t) + jq(t) = \rho e^{j\phi(t)}. \tag{6.51}$$

We thus get that

$$p(t) = \rho\cos(\phi(t)), \tag{6.52a}$$

$$q(t) = \rho\sin(\phi(t)). \tag{6.52b}$$

In order to discuss further the consequences on the modulation of a simple gain imbalance $g = G_{\mathrm{q}}/G_{\mathrm{p}}$ between the branches of the complex frequency upconverter, we can use the first point of view discussed above to derive an expression for $s_{\mathrm{RF}}(t)$. For that purpose, we observe that the real and imaginary parts of the complex LO signal $lo_{+\omega_{\text{LO}}}(t)$, given by equation (6.45) in the general imbalance case, now reduce to

$$lo_{\text{p}}(t) = \sqrt{\frac{2}{1+g^2}}\cos(\omega_{\text{LO}}t), \tag{6.53a}$$

$$lo_{\text{q}}(t) = \sqrt{\frac{2}{1+g^2}}g\sin(\omega_{\text{LO}}t). \tag{6.53b}$$

We can thus use those expressions to write

$$\begin{aligned} s_{\text{RF}}(t) &= p(t)lo_{\text{p}}(t) - q(t)lo_{\text{q}}(t) \\ &= \rho\sqrt{\frac{2}{1+g^2}}\text{Re}\left\{[\cos(\phi(t)) + \text{j}g\sin(\phi(t))]\text{e}^{+\text{j}\omega_{\text{LO}}t}\right\}. \end{aligned} \tag{6.54}$$

The complex envelope of this signal, when defined as centered around ω_{LO}, can thus be written as

$$\tilde{s}_{\text{RF}}(t) = \rho\sqrt{\frac{2}{1+g^2}}[\cos(\phi(t)) + \text{j}g\sin(\phi(t))]. \tag{6.55}$$

As illustrated in Figure 6.20, for $g \neq 1$, $\tilde{s}_{\text{RF}}(t)$ no longer has a constant amplitude. More precisely, the modulation trajectory is now an ellipse due to the fact that the imaginary part of $\tilde{s}_{\text{RF}}(t)$ is stretched by the factor g compared to its real part. We can also use the second point of view discussed above to examine the consequences of this gain imbalance on the structure of $s_{\text{RF}}(t)$. We need to express the complex envelope of $s_{\text{RF}}(t)$ as the linear sum of the initial undistorted modulating waveform $\tilde{s}(t)$ and its complex conjugate by using equation (6.50):

$$\tilde{s}_{\text{RF}}(t) = (\alpha_{+}^{+})\rho\text{e}^{\text{j}\phi(t)} + (\alpha_{+}^{-})^{*}\rho\text{e}^{-\text{j}\phi(t)}. \tag{6.56}$$

In the particular case where we consider only a gain imbalance, (α_{+}^{+}) and (α_{+}^{-}) can be deduced from equation (6.42) as

$$(\alpha_{+}^{+}) = \frac{1+g}{\sqrt{2(1+g^2)}}, \tag{6.57a}$$

$$(\alpha_{+}^{-}) = \frac{1-g}{\sqrt{2(1+g^2)}}. \tag{6.57b}$$

These alpha factors are now real valued. Consequently, although the modulating signal $\tilde{s}(t) = \rho\text{e}^{\text{j}\phi(t)}$ has a constant modulus, its phase relationship with its complex conjugate is such that the sum of those signals weighted by (α_{+}^{+}) and (α_{+}^{-}) results in the effective ellipse trajectory illustrated in Figure 6.20. Even if the resulting RF bandpass signal no longer has a constant

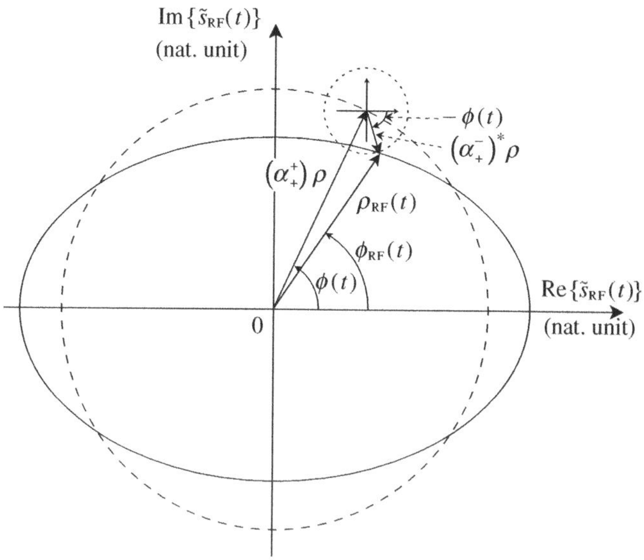

Figure 6.20 Interpretation of the distortion of a constant envelope modulating waveform due to gain imbalance – The complex envelope, $\tilde{s}(t) = \rho e^{j\phi(t)}$, of a phase/frequency only modulated signal has a constant magnitude so that its trajectory is a circle (dashed). The same holds for the trajectory of the complex conjugate complex envelope (dotted). In the presence of gain imbalance between the branches of the complex mixer used to upconvert $\tilde{s}(t)$, the trajectory of the complex envelope of the resulting modulated RF bandpass signal, $\tilde{s}_{RF}(t) = \rho_{RF} e^{j\phi_{RF}(t)}$, becomes an ellipse (solid). But this non-constant amplitude complex envelope can still be decomposed as the linear sum of $\tilde{s}(t)$ and its complex conjugate $\tilde{s}^*(t)$, i.e. as the sum of two constant amplitude complex envelopes.

instantaneous amplitude, it can be expressed as the linear sum of two constant amplitude bandpass signals, the expected ideal signal and its image.

In the particular case where $\omega_{IF} = 0$, we have just shown that we are faced with an in-band distortion of the wanted signal. We can thus alternatively interpret the resulting distortion of the modulated RF bandpass signal in terms of SNR limitation or, equivalently, EVM generation. However, the underlying assumption for this to be true is that the image signal can effectively be considered as a noise component regarding the wanted signal. For that we need the bandpass signals to be uncorrelated. However, we can deduce from equations (6.49) and (6.50) that the complex envelopes of the two bandpass signals, when defined as centered around ω_{LO}, are simply $\tilde{s}(t)$ and $\tilde{s}^*(t)$. We know from the discussion in Appendix 1 that a sufficient condition for non-correlation between two bandpass signals is non-correlation between their complex envelopes when defined as centered around the same carrier angular frequency. But from the discussion in Appendix 2, for signals classically encountered in the wireless transceiver area, equation (A2.11) holds. Thus

$$\gamma_{\tilde{s}\times\tilde{s}^*}(\tau) = 0. \tag{6.58}$$

Consequently, the generated RF bandpass image signal can be considered as an additional noise term that degrades the SNR or EVM performance of the line-up. Moreover, we recover in equation (6.50) the same alpha factors as encountered during the derivation of the IRR in Section 6.2.1. Thus the resulting relative power ratio between the wanted signal and its image is directly given by the IRR factor defined by equation (6.38). The noise term generated this way can thus be classified as a multiplicative noise as its power is directly proportional to the power of the wanted signal. This behavior is important when performing practical line-up budgets, as illustrated in Chapter 7.

We can even go a step further. The image signal appears to be created from scratch during the frequency conversion processing. So where does its power come from? To investigate this, we first derive the power of the overall RF bandpass signal $s_{\mathrm{RF}}(t)$ generated during the upconversion. Based on equation (1.64), this power is proportional to $\mathbb{E}\{\tilde{s}_{\mathrm{RF},t}\tilde{s}^*_{\mathrm{RF},t}\}$. Using equation (6.50), we can then write that

$$\begin{aligned}\mathbb{E}\{\tilde{s}_{\mathrm{RF},t}\tilde{s}^*_{\mathrm{RF},t}\} &= \mathbb{E}\left\{\left((\alpha^+_+)\tilde{s}_t + (\alpha^-_+)^*\tilde{s}^*_t\right)\left((\alpha^+_+)^*\tilde{s}^*_t + (\alpha^-_+)\tilde{s}_t\right)\right\}\\ &= \left(|(\alpha^+_+)|^2 + |(\alpha^-_+)|^2\right)\mathbb{E}\{\tilde{s}_t\tilde{s}^*_t\} + \mathbb{E}\left\{(\alpha^+_+)(\alpha^-_+)\tilde{s}^2_t + (\alpha^+_+)^*(\alpha^-_+)^*\tilde{s}^{*2}_t\right\}.\end{aligned}$$

But, having chosen to work with a normalized LO signal, we see from equation (6.42) that

$$|(\alpha^+_+)|^2 + |(\alpha^-_+)|^2 = (\alpha^+_+)(\alpha^+_+)^* + (\alpha^-_+)(\alpha^-_+)^* = 1. \tag{6.59}$$

Consequently,

$$\begin{aligned}\mathbb{E}\{\tilde{s}_{\mathrm{RF},t}\tilde{s}^*_{\mathrm{RF},t}\} &= \mathbb{E}\{\tilde{s}_t\tilde{s}^*_t\} + 2\mathrm{Re}\left\{(\alpha^+_+)(\alpha^-_+)\mathbb{E}\left\{\tilde{s}^2_t\right\}\right\}\\ &= \gamma_{\tilde{s}\times\tilde{s}}(0) + 2\mathrm{Re}\left\{(\alpha^+_+)(\alpha^-_+)\gamma_{\tilde{s}\times\tilde{s}^*}(0)\right\}.\end{aligned} \tag{6.60}$$

Now referring to the non-correlation between the complex envelopes of the wanted and image signals given by equation (6.58), we finally get that

$$\mathbb{E}\{\tilde{s}_{\mathrm{RF},t}\tilde{s}^*_{\mathrm{RF},t}\} = \gamma_{\tilde{s}\times\tilde{s}}(0). \tag{6.61}$$

The power of $s_{\mathrm{RF}}(t)$ thus remains constant whatever the gain and phase imbalance experienced during the frequency upconversion. But, given at the same time the non-correlation between the wanted signal and its image, we have that the power of the RF bandpass signal is simply the sum of the power of the wanted signal and its image. We thus get that the power of the image signal generated during the upconversion processing is necessarily retrieved from the power of the wanted signal. This could be considered as another drawback on the TX side. However, for realistic IRR values the image power remains far below the power of the wanted signal, thus leading to a negligible power loss on the latter signal.

In conclusion, we point out that the alpha factors are complex numbers in the general case. This means that the instantaneous phase of the transmitted wanted signal is necessarily corrupted by a phase offset linked to the gain and phase imbalance. This is not necessarily a problem in practice as this offset remains either constant or slowly varying in operating conditions as the gain and phase imbalance values do. As long as those variations are slow enough in respect of the variations of the propagation channel characteristics, we expect no additional degradations of the radio link.

Receive Side

Let us now focus on the receive side. We can reconsider the configuration extensively discussed in the previous sections to illustrate the complex frequency downconversion, i.e. the infradyne case based on the use of the negative complex exponential $e^{-j\omega_{LO}t}$, with $\omega_{RF}/2 < \omega_{LO} < \omega_{RF}$. We know from the discussion in Section 6.2.1 that in the presence of gain and phase imbalance this theoretical LO signal is degraded down to the complex signal $lo_{-\omega_{LO}}(t) = lo_p(t) + jlo_q(t)$, still centered on $-\omega_{LO}$, but now exhibiting a rise in the complex conjugate tone $e^{+j\omega_{LO}t}$. The real and imaginary parts of this complex LO signal are given by equation (6.33), thus corresponding to an equivalent model as illustrated in Figure 6.15. Assuming that we are dealing with the complex frequency downconversion of the real bandpass signal $s(t)$, we then recover at the output of the system the reconstructed complex IF signal,

$$s_{IF}(t) = s(t)lo_{-\omega_{LO}}(t). \tag{6.62}$$

Using equation (6.36) to expand $lo_{-\omega_{LO}}(t)$ as the linear sum of the two complex conjugate complex exponentials, $e^{-j\omega_{LO}t}$ and $e^{+j\omega_{LO}t}$, we can then write that

$$s_{IF}(t) = (\alpha_-^-)s(t)e^{-j\omega_{LO}t} + (\alpha_-^+)s(t)e^{+j\omega_{LO}t}, \tag{6.63}$$

with the alpha factors given by equation (6.37). As previously discussed, the presence of the unwanted complex exponential in the LO signal is the reason for the image sideband folding on the wanted sideband. Based on this formalism, this folding mechanism is related to the dual shift of the input RF spectrum in two opposite directions due to the multiplication of $s(t)$ by the two complex exponentials $e^{-j\omega_{LO}t}$ and $e^{+j\omega_{LO}t}$.

There is, however, an alternative interpretation of this behavior, as there was for the frequency upconversion discussed in the previous section. We observe that the first term on the right-hand side of equation (6.63) is proportional to the reconstructed complex IF signal $s_{IF_0}(t)$ we would have expected in the case of an ideal complex frequency conversion, i.e. without gain and phase imbalance. Given that $s(t)$ is a real quantity, we can then rewrite this equation as

$$s_{IF}(t) = (\alpha_-^-)s_{IF_0}(t) + (\alpha_-^+)s_{IF_0}^*(t). \tag{6.64}$$

Using the Fourier transform property given by equation (1.8), we thus get for a given realization of these processes that

$$S_{IF}(\omega) = (\alpha_-^-)S_{IF_0}(\omega) + (\alpha_-^+)S_{IF_0}^*(-\omega). \tag{6.65}$$

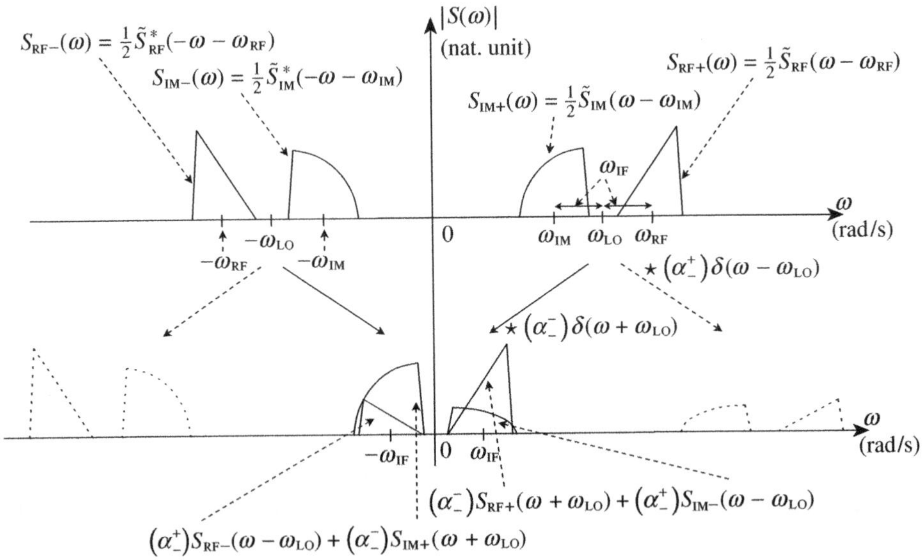

Figure 6.21 Spectral representation of a complex frequency downconversion in the presence of gain and phase imbalance – When executing a complex frequency downconversion of a real bandpass RF signal in the presence of gain and phase imbalance, the output spectrum of the reconstructed output complex signal can be interpreted as the superposition of the expected spectrum as if no imbalance were present, i.e. an exact shift of the input spectrum in a single direction (top), and a flipped copy of it. The two corresponding spectra are weighted by the alpha factors defined by equation (6.37), before summation (bottom).

We can interpret this equation as saying that the spectrum of $s_{IF}(t)$ is effectively composed of the expected term, $S_{IF_0}(\omega)$, but now superposed with an unwanted term that is nothing more than a flipped copy of it, $S^*_{IF_0}(-\omega)$. The reason for this new interpretation is that $s(t)$ is a real quantity and thus has a spectrum $S(\omega)$ that exhibits Hermitian symmetry. The dual shift of $S(\omega)$ in opposite directions is then equivalent to a shift in a single direction followed by the addition of the flipped copy of this shifted version as illustrated in Figure 6.21. As a side effect, we recover that when dealing with complex exponentials of the same amplitude in the expression of the LO signal, thus corresponding to $(\alpha^-_-) = (\alpha^+_-) = 1$, the spectrum of $s_{IF}(t)$ has Hermitian symmetry. We recover in that case the situation of a real frequency conversion, i.e. that $s_{IF}(t)$ is a real quantity, as can be directly seen by factorizing equation (6.63) in that case.

From the system design point of view, we expect the folding of such an unwanted sideband necessarily to result in a degradation of the in-band SNR. However, for this to be true, we need the wanted and unwanted signals to be uncorrelated in order to be able to consider this latter term as a noise component. But, in contrast to the above discussion on the transmit side, we get on the receive side that the folded image signal can be of a different nature, depending on the frequency planning of the conversion. By the discussion in Section 6.1.2, as soon as ω_{IF} is greater than the modulation bandwidth of the wanted signal, the image signal is necessarily either an adjacent or a blocking signal. In that case, classically encountered in the low-IF architecture discussed in Section 8.2.3, the image signal has a statistic that can obviously be assumed independent of that of the wanted signal. Conversely, if ω_{IF} is lower than the modulation bandwidth, the image signal is composed of a fraction of the unwanted sideband of the wanted signal itself. It can even be exactly this unwanted sideband in the case of a direct downconversion to baseband, as encountered in ZIF receivers discussed in Section 8.2.1. However, looking at equation (6.64) we can see that whatever the frequency planning of the conversion, the reconstructed complex image signal is nothing more than the complex conjugate of the reconstructed wanted IF signal. At the same time, we know from the discussion in Appendix 2 that for signals classically encountered in the field of wireless transceivers, $\gamma_{s_{IF0} \times s^*_{IF0}}(\tau) = 0$ (see equation (A2.11)). Finally, we thus see that in any situation of interest the reconstructed complex image signal can effectively be considered as an additional noise component in the SNR budget of a receive line-up.

Looking again at equation (6.64), we remark that the alpha factors it contains are those introduced in Section 6.2.1. We can then rely on the IRR defined by equation (6.38) to evaluate the power attenuation of the image signal before its folding on the wanted signal. Consequently, depending on the achievable IRR, the target SNR, and the power of the image signal $s_{IM}(t)$ relative to that of the wanted signal $s_{RF}(t)$, we may need to filter out this unwanted image signal prior to the frequency downconversion. However, this possibility also highly depends on the frequency planning of the conversion. Moreover, one reason for the considerable interest in the complex frequency conversion lies in the possibility of avoiding the use of an input image reject filter, thus allowing for highly integrated solutions. We thus often rely on the IRR capabilities of such a system to reject the image signal in practice. In the case of a direct frequency downconversion, i.e. when $\omega_{IF} = 0$, the image sideband is the unwanted sideband of the wanted signal. As the two sidebands of an RF bandpass signal have the same power, the IRR then directly gives the maximum SNR that can be achieved during the direct complex frequency downconversion to baseband. In that case, the noise term generated this way can be categorized as a pure multiplicative noise as its power is directly proportional to the power of the wanted signal. This behavior is of importance when performing practical line-up budgets, as illustrated in Chapter 7. But if the image signal is an adjacent or blocking signal, the ratio of the image signal power to the wanted signal power at the input of the downmixing stage needs to be retrieved from the IRR value to get the achieved SNR. This may lead to stringent limitations on the choice of frequency planning. However, we can always consider using a calibration scheme to improve system performance, as discussed in Section 9.2.3.

To conclude, it is of interest to highlight the energy loss on the wanted signal during its downconversion to baseband, i.e. when $\omega_{IF} = 0$. In this particular case, we have already observed that the reconstructed complex image signal is nothing more than the residual unwanted sideband of the wanted signal $s_{RF}(t)$. By the above discussion on the transmit side, we can then anticipate that the residual power in this unwanted sideband is retrieved from the power of the wanted sideband. This can be confirmed by performing the derivation in the receive case. We can assume for the sake of simplicity that only the wanted signal $s_{RF}(t)$ is present at the input of the complex frequency downconversion stage. This signal can be expressed as a function of its complex envelope $\tilde{s}_{\text{RF}}(t)$, defined as centered around ω_{RF}:

$$\begin{aligned} s_{\text{RF}}(t) &= \text{Re}\left\{\tilde{s}_{\text{RF}}(t)\,\text{e}^{\text{j}\omega_{\text{RF}}t}\right\} \\ &= \frac{1}{2}\left(\tilde{s}_{\text{RF}}(t)\,\text{e}^{\text{j}\omega_{\text{RF}}t} + \tilde{s}^*_{\text{RF}}(t)\,\text{e}^{-\text{j}\omega_{\text{RF}}t}\right). \end{aligned} \tag{6.66}$$

We can then substitute this expression into equation (6.63) with $\omega_{\text{IF}} = \omega_{\text{RF}}$ to express the reconstructed complex signal, $s_{\text{BB}}(t)$, that is recovered at the output of the frequency downconversion. Assuming that a lowpass filter is used in order to cancel all the residual bandpass sidebands, we can express the remaining lowpass component as

$$s_{\text{BB}}(t) = (\alpha^-_-)\tilde{s}_{\text{RF}}(t) + (\alpha^+_-)\tilde{s}^*_{\text{RF}}(t). \tag{6.67}$$

In a classical receive line-up, this signal is often provided as such to the channel estimator that compensates at least for the instantaneous phase offset of $s_{\text{BB}}(t)$ in order to achieve a coherent reception. In practical implementations, the IRR is expected to be sufficiently low that $|(\alpha^+_-)| \ll |(\alpha^-_-)|$. We can thus assume that the channel estimator estimates and compensates mainly for the argument φ of (α^-_-). After compensation of this offset, we then recover at the output of the channel estimator the signal

$$s_{\text{BB}_\text{o}}(t) = s_{\text{BB}}(t)\text{e}^{-\text{j}\varphi} = |(\alpha^-_-)|\tilde{s}_{\text{RF}}(t) + (\alpha^+_-)\tilde{s}^*_{\text{RF}}(t)\text{e}^{-\text{j}\varphi}. \tag{6.68}$$

The power of this complex lowpass process can be evaluated as

$$\begin{aligned} \mathbb{E}\left\{s_{\text{BB}_\text{o},t}s^*_{\text{BB}_\text{o},t}\right\} = \mathbb{E}\Big\{&\left(|(\alpha^-_-)|\tilde{s}_{\text{RF},t} + (\alpha^+_-)\tilde{s}^*_{\text{RF},t}\text{e}^{-\text{j}\varphi}\right) \\ &\times\left(|(\alpha^-_-)|\tilde{s}^*_{\text{RF},t} + (\alpha^+_-)^*\tilde{s}_{\text{RF},t}\text{e}^{\text{j}\varphi}\right)\Big\}. \end{aligned} \tag{6.69}$$

Thus,

$$\begin{aligned} \mathbb{E}\left\{s_{\text{BB}_\text{o},t}s^*_{\text{BB}_\text{o},t}\right\} &= \left(|(\alpha^-_-)|^2 + |(\alpha^+_-)|^2\right)\mathbb{E}\left\{\tilde{s}_{\text{RF},t}\tilde{s}^*_{\text{RF},t}\right\} \\ &\quad + \mathbb{E}\left\{(\alpha^-_-)(\alpha^+_-)^*\tilde{s}^2_{\text{RF},t} + (\alpha^-_-)^*(\alpha^+_-)\tilde{s}^{*2}_{\text{RF},t}\right\}. \end{aligned} \tag{6.70}$$

Given that we are dealing with a normalized LO signal, equation (6.59) holds. We thus obtain

$$\begin{aligned} \mathbb{E}\left\{s_{\text{BB}_\text{o},t}s^*_{\text{BB}_\text{o},t}\right\} &= \mathbb{E}\left\{\tilde{s}_{\text{RF},t}\tilde{s}^*_{\text{RF},t}\right\} + 2\text{Re}\left\{(\alpha^-_-)(\alpha^+_-)^*\mathbb{E}\left\{\tilde{s}^2_{\text{RF},t}\right\}\right\} \\ &= \gamma_{\tilde{s}_{\text{RF}}\times\tilde{s}_{\text{RF}}}(0) + 2\text{Re}\left\{(\alpha^-_-)(\alpha^+_-)^*\gamma_{\tilde{s}_{\text{RF}}\times\tilde{s}^*_{\text{RF}}}(0)\right\}, \end{aligned} \tag{6.71}$$

which can be compared to equation (6.60). Assuming in turn that the processes we are dealing with are stationary, we can reuse the results derived in Appendix 2 and in particular that $\gamma_{\tilde{s}_{\mathrm{RF}} \times \tilde{s}^*_{\mathrm{RF}}}(0) = 0$. The above equation then reduces to

$$\mathbb{E}\left\{s_{\mathrm{BB_0},t} s^*_{\mathrm{BB_0},t}\right\} = \gamma_{\tilde{s}_{\mathrm{RF}} \times \tilde{s}_{\mathrm{RF}}}(0), \tag{6.72}$$

which is the transposition of equation (6.61) for our receive case. This relationship shows that the power of the received signal at the channel estimator output is constant and independent of the imbalance value. But we have at the same time the non-correlation between the wanted and image sidebands. The power of $s_{\mathrm{BB}}(t)$ is then equal to the sum of those powers. As a result, the power of the residual image sideband is effectively retrieved from the wanted sideband. We thus have not only a rise in noise power, but also a degradation of the available energy per bit [35]. Both phenomena lead to the degradation of the E_{b}/N_0 ratio. However, in practical implementations the image power remains weak compared to the power of the wanted signal. The energy loss on the wanted signal can thus be considered negligible. Up to first order, the degradation linked to the imbalance between the P and Q branches of the downmixer can be reduced to the creation of an additional noise term to be considered in the SNR budget.

6.3 Mixer Implementation

6.3.1 Mixers as Choppers

For the most part, electronic devices dedicated to the physical implementation of RF/analog frequency conversions do not implement the ideal multiplication of the input signal by a sinusoidal LO waveform as considered so far in our discussion. The reason is that such multiplication requires a device fully linear regarding both its input and LO ports. But this property is hard to obtain over a wide DR at a reasonable cost. Instead, we can take advantage of the switching capability of transistors to implement a device that behaves as a chopper with respect to the signal being processed [40, 70, 71]. From the signal processing point of view, this operation can be interpreted as the multiplication of the signal being converted by an LO signal that takes the form of a periodic square wave whose period $T = 2\pi/\omega_{\mathrm{LO}}$ corresponds to the expected LO angular frequency. If we are dealing with a complex frequency conversion, we can still rely on the structure shown in Figure 6.15 for a downconversion or the one shown in Figure 6.18 for an upconversion, but with P and Q LO signals that are now periodic square waves with a delay offset of $T/4$ between them, as illustrated in Figure 6.22.

To understand how such processing works, we need to expand these square waveforms in order to make apparent the complex exponentials expected to be present in the LO signals for realizing the frequency transposition. We observe that the LO components shown in Figure 6.22 can be interpreted as the signs of sine and cosine functions, but scaled with the amplitude α in order to stay general. Thus, we can write the periodic square LO signal $lo_{\Pi,\mathrm{p}}(t)$ as

$$lo_{\Pi,\mathrm{p}}(t) = \alpha\,\mathrm{sign}\{\cos(\omega_{\mathrm{LO}}t)\}, \tag{6.73}$$

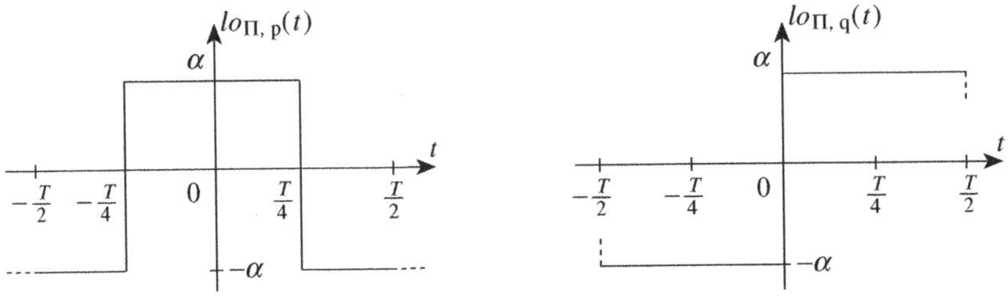

Figure 6.22 Periodic square waves in quadrature as LO waveforms used for trigging RF/analog mixers implemented as choppers – The implementation of a frequency conversion stage in the RF/analog world often uses mixers that behave as choppers. From the signal processing point of view, the signal being converted can be seen as multiplied by a periodic square wave whose period is equal to that of the expected ideal LO signal. In the case of a complex frequency conversion, we need to use two such periodic signals with a square waveform, but in quadrature for both the P (left) and Q (right) branches of the complex mixer.

with the sign function defined by equation (5.272). We can then use equation (5.281) to write the Fourier series expansion of this periodic signal as

$$lo_{\Pi,\mathrm{p}}(t) = \alpha\frac{4}{\pi}\sum_{l=1}^{\infty}\frac{(-1)^{l-1}}{2l-1}\cos((2l-1)\omega_{\mathrm{LO}}t). \tag{6.74}$$

The fundamental tone of $lo_{\Pi,\mathrm{p}}(t)$ is thus nothing more than the cosine function that represents the real part of the complex exponentials expected to be used for an ideal complex frequency conversion. In the same way, we can write for the Q LO signal displayed in Figure 6.22 that

$$lo_{\Pi,\mathrm{q}}(t) = \alpha\,\mathrm{sign}\{\sin(\omega_{\mathrm{LO}}t)\}. \tag{6.75}$$

We then remark that

$$lo_{\Pi,\mathrm{q}}(t) = \alpha\,\mathrm{sign}\{\sin(\omega_{\mathrm{LO}}t)\} = lo_{\Pi,\mathrm{p}}(t - T/4). \tag{6.76}$$

Using the series expansion of $lo_{\Pi,\mathrm{p}}(t)$ given by equation (6.74), we can then write that

$$lo_{\Pi,\mathrm{q}}(t) = \alpha\frac{4}{\pi}\sum_{l=1}^{\infty}\frac{(-1)^{l-1}}{2l-1}\cos\left((2l-1)\omega_{\mathrm{LO}}t - (2l-1)\frac{\pi}{2}\right). \tag{6.77}$$

Given that

$$\cos(\theta - l\pi) = (-1)^l\cos(\theta), \tag{6.78a}$$

$$\cos(\theta + \pi/2) = -\sin(\theta), \tag{6.78b}$$

we finally get that

$$lo_{\Pi,q}(t) = \alpha\, \mathrm{sign}\{\sin(\omega_{LO} t)\} = \alpha \frac{4}{\pi} \sum_{l=1}^{\infty} \frac{1}{2l-1} \sin((2l-1)\omega_{LO} t). \tag{6.79}$$

Based on this series expansion, we can then express the reconstructed complex LO waveform, $lo_{+\omega_{LO}}(t) = lo_{\Pi,p}(t) + j lo_{\Pi,q}(t)$, as

$$lo_{+\omega_{LO}}(t) = \alpha \frac{4}{\pi} \sum_{l=1}^{\infty} \frac{(-1)^{l-1}}{2l-1} e^{j(-1)^{l-1}(2l-1)\omega_{LO} t}. \tag{6.80}$$

Looking at this expression, the fundamental tone in the series expansion of $lo_{+\omega_{LO}}(t)$ is the complex exponential $e^{+j\omega_{LO}t}$. In the same way, considering the opposite sign for the Q LO signal would have resulted in a reconstructed complex LO signal of the form $lo_{-\omega_{LO}}(t) = lo^{*}_{+\omega_{LO}}(t)$, thus approximating the negative complex exponential $e^{-j\omega_{LO}t}$. We thus see how such an implementation can approximate the ideal complex frequency conversion.

However, these expressions also show that a comb of harmonics is also present in the series expansion of the LO signals. We anticipate some folding issues related to their presence, as discussed in more depth in Section 6.4. Here we remark that only odd order harmonics lying alternatively around the angular frequencies $+(2l-1)\omega_{LO}$ or $-(2l-1)\omega_{LO}$ for successive values of l are present in the series expansion of $lo_{+\omega_{LO}}(t)$. As illustrated in Figure 6.23, this behavior results from the exact cancellation of the other harmonic tones present in the $lo_{\Pi,p}(t)$ and $lo_{\Pi,q}(t)$ signals during the reconstruction of the equivalent complex LO signal. By the discussion in Section 6.2, we anticipate that this cancellation cannot be perfect in the presence of imbalance between the P and Q branches of the complex mixer. We also anticipate the presence of even order harmonics when faced with unbalanced P and Q LO signals. Thus we need to investigate the impact of such imbalance before further discussing the system impacts related to the presence of these unwanted harmonic tones.

6.3.2 Impairments in the LO Generation

As long as a chopper is implemented using perfect switches, its performance depends only on the zero crossing of the LO signal that drives it. However, for a complex mixer implemented using two such devices, we can still be faced with an overall gain and delay imbalance between its P and Q branches so that we can expect additional degradations on top of those linked to timing inaccuracies.

We examine the impact of all the sources of impairments with the aim of deriving a general structure for the spectrum of the reconstructed complex LO signal experienced by the signal being converted. We then address the corresponding system impacts in Section 6.4.1.

Duty Cycle

Let us focus first on the impact of the inaccuracies in the switching time of chopper-like mixers. Referring to practical physical implementations, this behavior is for the most part related to a distortion in the duty cycle *DtyCy* of the periodic square LO signal that drives

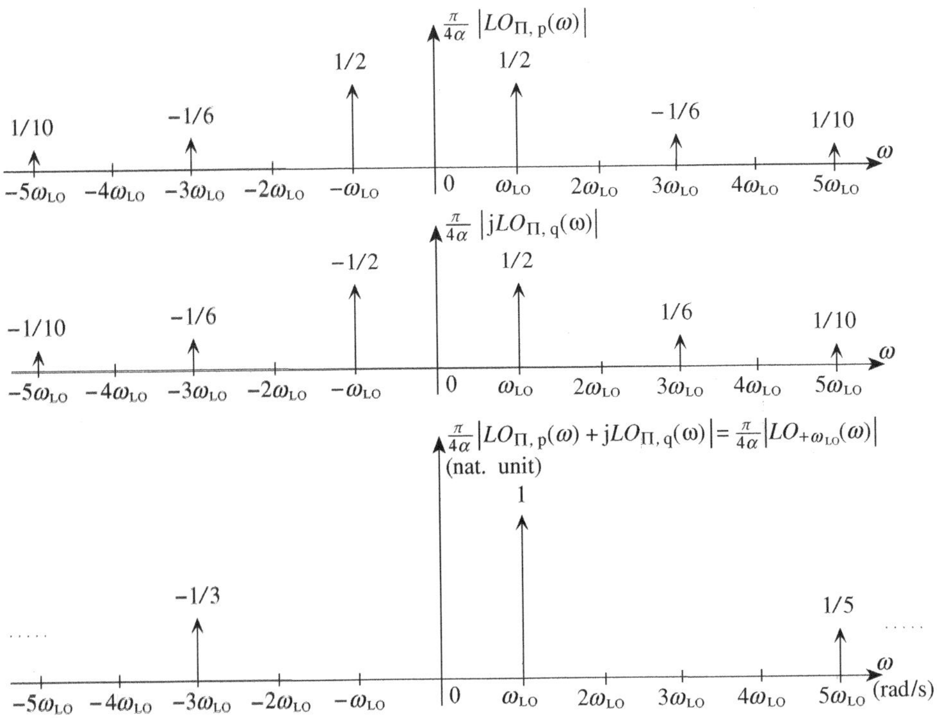

Figure 6.23 Spectral content of the quadrature LO waveforms used for RF/analog mixers implemented as choppers – The spectral content of the $lo_{\Pi,p}(t)$ and $lo_{\Pi,q}(t)$ signals shown in Figure 6.22 exhibits odd order harmonics of the fundamental tone. When no imbalance is present, the amplitude and phase relationships of the tones of the same order on the P (top) and Q (middle) branches are such that they sum or cancel when the complex LO signal is reconstructed (bottom). This complex LO signal is therefore the targeted complex exponential with, in addition, odd order harmonics, lying alternatively around $+(2l-1)\omega_{LO}$ or $-(2l-1)\omega_{LO}$ for successive values of l.

such a device. Classically, this duty cycle (DtyCy) is defined as the ratio of the duration T^+ (T^-) during which the signal remains positive (negative), to the signal period $T = T^+ + T^-$. In what follows we consider the definition based on T^+, i.e. that

$$DtyCy = \frac{T^+}{T} = \frac{T^+}{T^+ + T^-}. \tag{6.81}$$

This quantity is also often expressed as a percentage.

Looking at the waveforms shown in Figure 6.24, we can see that the P and Q LO signals exhibit an imbalance between their positive and negative alternations as soon as their DtyCy is not equal to 0.5. Referring back to the discussion in "Odd vs. even order nonlinearity" (Section 5.1.1), such asymmetry in a signal can be related to an even order distortion of it. We can then expect a rise in the even order harmonics in the series expansion of those signals, whereas only odd order harmonics were present in the case of a DtyCy equal to 0.5, as illustrated in

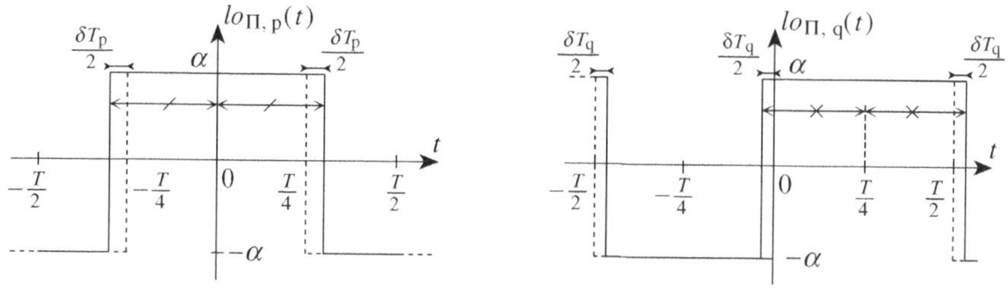

Figure 6.24 Periodic square waves in quadrature as LO waveforms with duty cycles different from 0.5 – A DtyCy different from 0.5 leads to an unsymmetrical behavior of the P and Q LO components. This kind of imbalance leads at least to a rise in the even order tones in the spectrum of the reconstructed complex LO signal and even to a rise in the complex conjugate tones of odd order when $\delta T_p \neq \delta T_q$.

Figure 6.23. This behavior can be checked by updating the series expansion performed in the former case, now taking this distortion in the DtyCy into account. However, in that perspective we can continue to assume a constant delay of $T/4$ modulo T between the respective centers of symmetry of the signals, as illustrated in Figure 6.24. This allows us to distinguish between the impacts of the inaccuracies in the duty cycles discussed in this section and of the delay imbalance developed in "Delay imbalance" later in this section.

But prior to that we observe that the forthcoming comments on the impact of the DtyCy in this section stand for a mixing stage that is effectively expected to be driven by a pure square LO signal as defined so far. But we need to be aware that other implementations can take advantage of alternative equivalent LO waveforms. We can even be faced with structures that combine the signal processing associated with the frequency conversion with other functionalities. This can for instance result in devices that behave more as sample and hold than choppers [18]. However, although they are interesting in practice, we will not discuss such structures here. We merely stress the importance of the spectral content of the equivalent LO waveform experienced by a signal during its frequency conversion. The results discussed here should therefore be adapted on a case by case basis when considering different physical implementations of frequency conversion stages. However, the spirit of the approach is the same.

Balanced Duty Cycles

Let us consider first the case where the DtyCy distortion is the same for the two square LO components $lo_{\Pi,p}(t)$ and $lo_{\Pi,q}(t)$. We thus suppose that $\delta T_p = \delta T_q = \delta T$ in the notation of Figure 6.24.

Based on the fact that these LO signals remain periodic whatever their duty cycles, we can continue to consider their Fourier series expansion here. For $lo_{\Pi,p}(t)$, for instance, this takes the general form

$$lo_{\Pi,p}(t) = a_{p,0} + \sum_{k=1}^{\infty} [a_{p,k} \cos(k\omega_{LO}t) + b_{p,k} \sin(k\omega_{LO}t)], \qquad (6.82)$$

with $\omega_{\mathrm{LO}} = 2\pi/T$. We remark that, based on our assumptions, $lo_{\Pi,\mathrm{p}}(t)$ remains an even function whatever the value of δT. Consequently, the terms

$$b_{\mathrm{p},k} = \frac{2}{T}\int_{-\frac{T}{2}}^{\frac{T}{2}} lo_{\Pi,\mathrm{p}}(t)\sin(k\omega_{\mathrm{LO}}t)\mathrm{d}t \tag{6.83}$$

are necessarily null since the sine is an odd function. For the remaining coefficients we can first evaluate $a_{\mathrm{p},0}$ as

$$a_{\mathrm{p},0} = \frac{1}{T}\int_{-\frac{T}{2}}^{\frac{T}{2}} lo_{\Pi,\mathrm{p}}(t)\mathrm{d}t = 2\alpha\frac{\delta T}{T}. \tag{6.84}$$

The terms $a_{\mathrm{p},k}$ then take the form

$$\begin{aligned} a_{\mathrm{p},k} &= \frac{2}{T}\int_{-\frac{T}{2}}^{\frac{T}{2}} lo_{\Pi,\mathrm{p}}(t)\cos(k\omega_{\mathrm{LO}}t)\mathrm{d}t \\ &= \alpha\frac{4}{T}\left(\int_{0}^{\frac{T}{4}+\frac{\delta T}{2}} \cos(k\omega_{\mathrm{LO}}t)\mathrm{d}t - \int_{\frac{T}{4}+\frac{\delta T}{2}}^{\frac{T_{\mathrm{p}}}{2}} \cos(k\omega_{\mathrm{LO}}t)\mathrm{d}t\right). \end{aligned} \tag{6.85}$$

This results in

$$a_{\mathrm{p},k} = \alpha\frac{4}{\pi k}\sin\left(k\frac{\pi}{2} + k\pi\frac{\delta T}{T}\right). \tag{6.86}$$

A distinction can then be made depending on the parity of k. For $l \in \mathbb{N}^*$ we have

$$a_{\mathrm{p},2l-1} = \alpha\frac{4}{\pi}\frac{1}{2l-1}\sin\left(l\pi - \frac{\pi}{2} + (2l-1)\pi\frac{\delta T}{T}\right), \tag{6.87a}$$

$$a_{\mathrm{p},2l} = \alpha\frac{4}{\pi}\frac{1}{2l}\sin\left(l\pi + 2l\pi\frac{\delta T}{T}\right). \tag{6.87b}$$

Noting that

$$\sin(\theta + l\pi) = (-1)^l\sin(\theta), \tag{6.88a}$$

$$\sin(\theta - \pi/2) = -\cos(\theta), \tag{6.88b}$$

we finally have

$$lo_{\Pi,\mathrm{p}}(t) = 2\alpha\frac{\delta T}{T} + \sum_{l=1}^{\infty}[a_{\mathrm{p},2l-1}\cos((2l-1)\omega_{\mathrm{LO}}t) + a_{\mathrm{p},2l}\cos(2l\omega_{\mathrm{LO}}t)], \tag{6.89}$$

with

$$a_{\text{p},2l-1} = \alpha\frac{4}{\pi}\frac{(-1)^{l-1}}{2l-1}\cos\left((2l-1)\pi\frac{\delta T}{T}\right), \tag{6.90a}$$

$$a_{\text{p},2l} = \alpha\frac{4}{\pi}\frac{(-1)^{l}}{2l}\sin\left(2l\pi\frac{\delta T}{T}\right). \tag{6.90b}$$

Based on this result, we can directly deduce the series expansion of $lo_{\Pi,\text{q}}(t)$, remembering our assumption of, on the one hand, a constant delay of $T/4$ modulo T between the centers of symmetry of the signals, and, on the other hand, the same DtyCy for them. We get that $lo_{\Pi,\text{q}}(t) = lo_{\Pi,\text{p}}(t - T/4)$. Given that $T = 2\pi/\omega_{\text{LO}}$, we can then write from equation (6.89) that

$$lo_{\Pi,\text{q}}(t) = 2\alpha\frac{\delta T}{T} + \sum_{l=1}^{\infty}\left[a_{\text{p},2l-1}\cos\left((2l-1)\omega_{\text{LO}}t - l\pi + \frac{\pi}{2}\right) + a_{\text{p},2l}\cos(2l\omega_{\text{LO}}t - l\pi)\right]. \tag{6.91}$$

Using equations (6.78) and (6.90), we then finally get that

$$lo_{\Pi,\text{q}}(t) = 2\alpha\frac{\delta T}{T} + \sum_{l=1}^{\infty}[b_{\text{q},2l-1}\sin((2l-1)\omega_{\text{LO}}t) + a_{\text{q},2l}\cos(2l\omega_{\text{LO}}t)], \tag{6.92}$$

with

$$b_{\text{q},2l-1} = (-1)^{l-1}a_{\text{p},2l-1} = \alpha\frac{4}{\pi}\frac{1}{2l-1}\cos\left((2l-1)\pi\frac{\delta T}{T}\right), \tag{6.93a}$$

$$a_{\text{q},2l} = (-1)^{l-1}a_{\text{p},2l} = \alpha\frac{4}{\pi}\frac{1}{2l}\sin\left(2l\pi\frac{\delta T}{T}\right). \tag{6.93b}$$

In contrast to what occurs for $lo_{\Pi,\text{p}}(t)$, we now have both even and odd order terms in the series expansion of $lo_{\Pi,\text{q}}(t)$, at least when $\delta T \neq 0$. This is no surprise since $lo_{\Pi,\text{q}}(t)$ is no longer an odd function. There is thus no reason for the $a_{\text{q},2l}$ terms to vanish.

Using the foregoing results, we can now derive an expression for the series expansion of a reconstructed complex LO waveform, for instance $lo_{+\omega_{\text{LO}}}(t) = lo_{\Pi,\text{p}}(t) + \text{j}lo_{\Pi,\text{q}}(t)$. It is of more interest to write the series expansion for this complex signal in terms of complex exponentials than in terms of the real sine and cosine functions used so far in the expression for $lo_{\Pi,\text{p}}(t)$ and $lo_{\Pi,\text{q}}(t)$. We can thus use equation (6.5) to expand the latter function. Consequently, considering on the one hand equations (6.89) and (6.92), and on the other hand equations (6.90) and (6.93), we find that

$$lo_{+\omega_{\text{LO}}}(t) = 2\alpha\frac{\delta T}{T}\sqrt{2}\text{e}^{\text{j}\frac{\pi}{4}} + \sum_{l=1}^{\infty}\left[a_{2l-1}\,\text{e}^{\text{j}(-1)^{l-1}(2l-1)\omega_{\text{LO}}t} + a_{2l}\left(\text{e}^{\text{j}2l\omega_{\text{LO}}t} + \text{e}^{-\text{j}2l\omega_{\text{LO}}t}\right)\right], \tag{6.94}$$

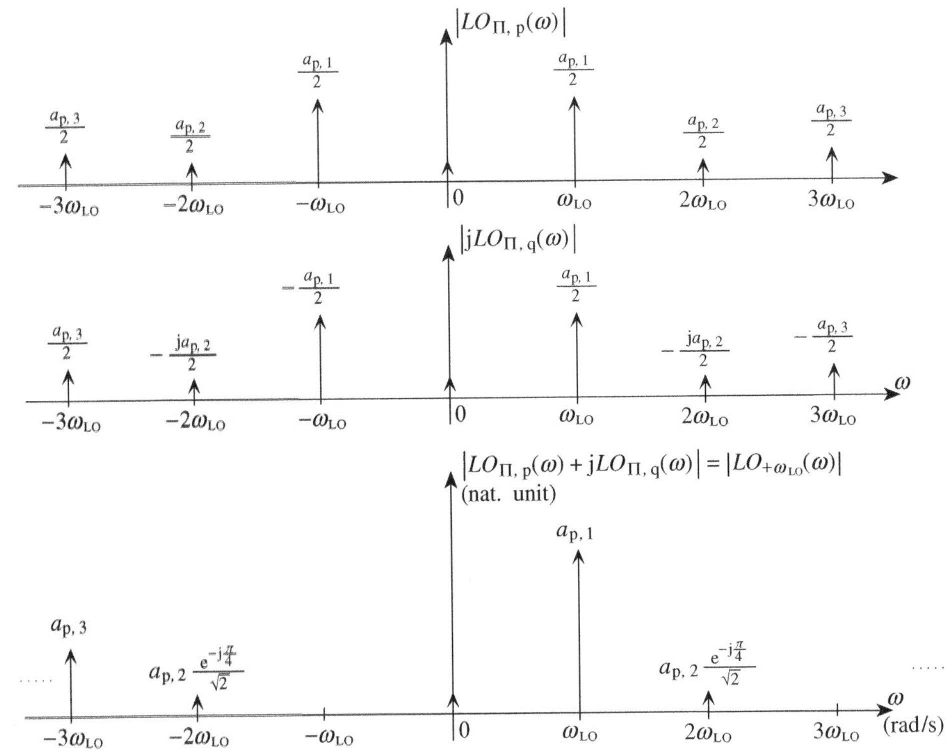

Figure 6.25 Spectral content of the periodic square LO waveforms in the presence of balanced duty cycles – The spectral content of the periodic square LO waveforms shown in Figure 6.24 exhibits a rise in the even order harmonics of the fundamental tone when $\delta T_p = \delta T_q \neq 0.5$. This phenomenon exists for both the P (top) and Q (middle) LO components. The phase relationship between the tones is then such that they are still present in the spectrum of the reconstructed complex LO signal after summation (bottom).

with

$$a_{2l-1} = a_{p,2l-1} = \alpha \frac{4}{\pi} \frac{(-1)^{l-1}}{2l-1} \cos\left((2l-1)\pi \frac{\delta T}{T}\right), \tag{6.95a}$$

$$a_{2l} = a_{p,2l} \frac{1 + j(-1)^l}{2} = \alpha \frac{4}{\pi} \frac{(-1)^l}{2l} \sin\left(2l\pi \frac{\delta T}{T}\right) \frac{e^{j(-1)^l \frac{\pi}{4}}}{\sqrt{2}}. \tag{6.95b}$$

The spectral content of the LO components $lo_{\Pi,p}(t)$ and $lo_{\Pi,q}(t)$ as well as that of the reconstructed complex waveform $lo_{+\omega_{LO}}(t)$ are then shown in Figure 6.25.

Looking at this figure, a couple of observations are in order:

(i) The imbalance between the positive and negative alternations of the LO components when their DtyCy is not equal to 0.5 effectively leads to a rise in the even order harmonics of the fundamental LO angular frequency ω_{LO}. Moreover, we see in Figure 6.26 that the amplitude of those tones quickly rises when this DtyCy deviates from 0.5. For instance, the

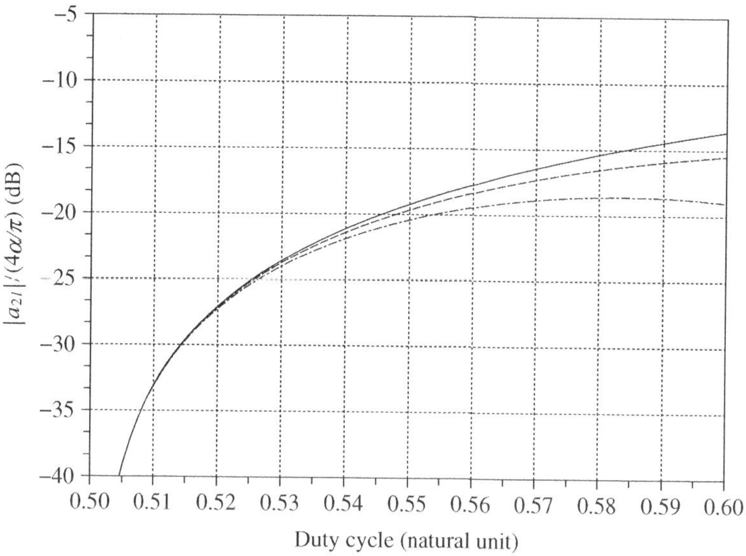

Figure 6.26 Amplitude of the even order harmonics in the reconstructed complex LO waveform as a function of the duty cycle – The even order harmonics of the reconstructed complex LO waveform rise quickly when the LO DtyCy deviates from 0.5 when $\delta T_p = \delta T_q = \delta T$. For low deviations of the DtyCy from 0.5, all even order harmonics have the same amplitude. This can be seen in equation (6.97b) by approximating the sine function by its argument under the small angle approximation. Here the cases $2l = 2$ (solid), $2l = 4$ (dashed) and $2l = 6$ (dot-dashed) are displayed.

level of the even order harmonics rises above −25 dBc for duty cycles as low as 0.53. We also see in that figure that these tones have almost the same amplitude for low duty cycle deviations and low harmonic orders. This behavior can be confirmed by approximating the sine function by its argument under the small angle approximation in equation (6.95b). This results in amplitudes that are almost independent of the harmonic order and that reduce to $|a_{2l}| = 2\sqrt{2}\alpha\delta T/T$. This explains why in practical implementations we often have a comb of even order harmonics with almost the same level whatever their order.

(ii) As illustrated in Figure 6.27, the amplitude of the fundamental tone also decreases when the DtyCy of the LO signals deviates from 0.5. This behavior is in fact a direct consequence of the rise in the even order harmonics. We get that the power of the overall square LO components remains the same whatever its DtyCy. The power of the harmonic tones is thus necessarily retrieved from the power of the fundamental tone when those harmonics exist. From the system point of view, this amplitude loss directly results in a gain loss of the mixing stage, at least in the usual case where it is this fundamental tone that is used to convert the wanted signal. However, we see that this gain loss remains weak for low DtyCy variations around 0.5.

Unbalanced Duty Cycles

Let us suppose now that we have a different DtyCy on the P and Q branches of the considered complex mixer, i.e. that $\delta T_p \neq \delta T_q$. We continue to assume that the delay between the centers of symmetry of the two LO components is equal to $T/4$ modulo T, as illustrated in Figure 6.24.

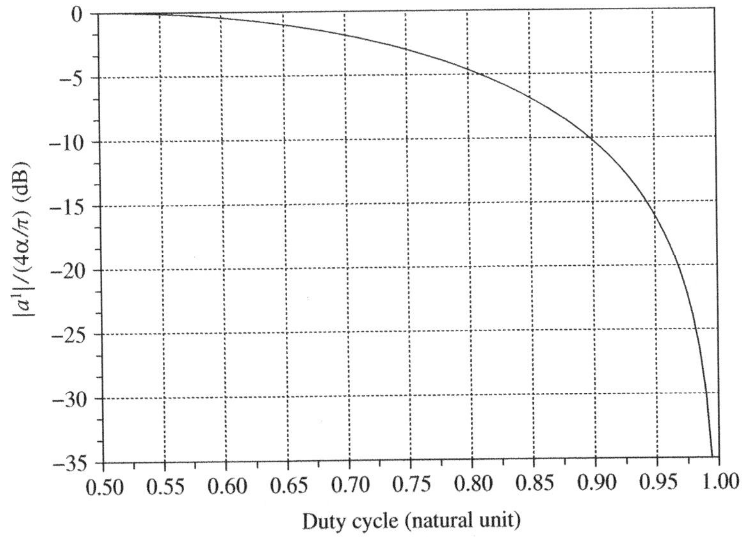

Figure 6.27 Amplitude loss on the fundamental tone of the reconstructed complex LO waveform as a function of the duty cycle – The amplitude loss on the fundamental tone of the reconstructed complex LO waveform relative to the theoretical value $4\alpha/\pi$ remains weak for low variations of the DtyCy around 0.5 when $\delta T_{\mathrm{p}} = \delta T_{\mathrm{q}} = \delta T$. The limit case occurs when the DtyCy reaches 1.0: the LO waveforms are constant in the time domain, thus leading to cancellation of the fundamental tone.

This allows us to distinguish between the impacts of the inaccuracies in the duty cycles discussed in this section and of the delay imbalance developed in "Delay imbalance" later in this section.

Under those assumptions, we can reuse the results derived in the case where $\delta T_{\mathrm{p}} = \delta T_{\mathrm{q}} = \delta T$ to express the Fourier series expansion of both $lo_{\Pi,\mathrm{p}}(t)$ and $lo_{\Pi,\mathrm{q}}(t)$. Substituting δT_{p} for δT in equations (6.89) and (6.90), we can write

$$lo_{\Pi,\mathrm{p}}(t) = 2\alpha\frac{\delta T_{\mathrm{p}}}{T} + \sum_{l=1}^{\infty}[a_{\mathrm{p},2l-1}\cos((2l-1)\omega_{\mathrm{LO}}t) + a_{\mathrm{p},2l}\cos(2l\omega_{\mathrm{LO}}t)], \qquad (6.96)$$

with $\omega_{\mathrm{LO}} = 2\pi/T$ and

$$a_{\mathrm{p},2l-1} = \alpha\frac{4}{\pi}\frac{(-1)^{l-1}}{2l-1}\cos\left((2l-1)\pi\frac{\delta T_{\mathrm{p}}}{T}\right), \qquad (6.97a)$$

$$a_{\mathrm{p},2l} = \alpha\frac{4}{\pi}\frac{(-1)^{l}}{2l}\sin\left(2l\pi\frac{\delta T_{\mathrm{p}}}{T}\right). \qquad (6.97b)$$

In the same way, substituting δT_{q} for δT in equations (6.92) and (6.93), we can write

$$lo_{\Pi,\mathrm{q}}(t) = 2\alpha\frac{\delta T_{\mathrm{q}}}{T} + \sum_{l=1}^{\infty}[b_{\mathrm{q},2l-1}\sin((2l-1)\omega_{\mathrm{LO}}t) + a_{\mathrm{q},2l}\cos(2l\omega_{\mathrm{LO}}t)], \qquad (6.98)$$

with

$$b_{\mathrm{q},2l-1} = \alpha \frac{4}{\pi} \frac{1}{2l-1} \cos\left((2l-1)\pi \frac{\delta T_{\mathrm{q}}}{T}\right), \tag{6.99a}$$

$$a_{\mathrm{q},2l} = \alpha \frac{4}{\pi} \frac{1}{2l} \sin\left(2l\pi \frac{\delta T_{\mathrm{q}}}{T}\right). \tag{6.99b}$$

If we compare the series expansions of $lo_{\Pi,\mathrm{p}}(t)$ and $lo_{\Pi,\mathrm{q}}(t)$, we see that in addition to the rise in the even order harmonics already experienced when $\delta T_{\mathrm{p}} = \delta T_{\mathrm{q}}$, we now have a different amplitude for the tones of the same order in those two expressions. We can thus say that the imbalance in the duty cycles leads to an equivalent gain imbalance between the P and Q branches of the mixing stage. However, there is a major difference with the exact gain imbalance as considered in Section 6.2.1 in the pure sinusoidal case. We now have a gain imbalance that depends on the harmonic order. Thus, it is of interest to generalize the gain imbalance defined by equation (6.29). For that, we can remark from equation (6.93) that in the balanced case, for $l \in \mathbb{N}^*$,

$$b_{\mathrm{q},2l-1} = (-1)^{l-1} a_{\mathrm{p},2l-1}, \tag{6.100a}$$

$$a_{\mathrm{q},2l} = (-1)^{l-1} a_{\mathrm{p},2l}. \tag{6.100b}$$

We can thus define in the present unbalanced case the ratio g_k between the amplitudes of the kth order harmonics such that

$$b_{\mathrm{q},2l-1} = (-1)^{l-1} g_{2l-1} a_{\mathrm{p},2l-1}, \tag{6.101a}$$

$$a_{\mathrm{q},2l} = (-1)^{l} g_{2l} a_{\mathrm{p},2l}. \tag{6.101b}$$

This results in

$$g_{2l-1} = (-1)^{l-1} \frac{b_{\mathrm{q},2l-1}}{a_{\mathrm{p},2l-1}}, \tag{6.102a}$$

$$g_{2l} = (-1)^{l} \frac{a_{\mathrm{q},2l}}{a_{\mathrm{p},2l}}. \tag{6.102b}$$

Using equations (6.97) and (6.99), we can then give an expression for the g_k terms as

$$g_{2l-1} = \frac{\cos\left((2l-1)\pi \frac{\delta T_{\mathrm{q}}}{T}\right)}{\cos\left((2l-1)\pi \frac{\delta T_{\mathrm{p}}}{T}\right)}, \tag{6.103a}$$

$$g_{2l} = \frac{\sin\left(2l\pi \frac{\delta T_{\mathrm{q}}}{T}\right)}{\sin\left(2l\pi \frac{\delta T_{\mathrm{p}}}{T}\right)}. \tag{6.103b}$$

We observe that the definition of g_k is always valid for k odd and reasonable values of δT_p and δT_q. This holds as the corresponding harmonics always exist in the decomposition of the square LO waveforms. But, for k even, the definition remains valid only if the duty cycles of the signals are not equal to 0.5, i.e. when $\delta T_\mathrm{p} \neq 0$ or $\delta T_\mathrm{q} \neq 0$. Otherwise, the even order harmonics simply do not exist in their series expansions.

Based on the results derived so far, we can now derive an expression for the series expansion of a reconstructed complex LO signal, for instance $lo_{+\omega_\mathrm{LO}}(t) = lo_{\Pi,\mathrm{p}}(t) + \mathrm{j}lo_{\Pi,\mathrm{q}}(t)$ as considered in the balanced case. We can introduce the g_k terms in the series expansion of $lo_{\Pi,\mathrm{q}}(t)$. Substituting equation (6.101) into equation (6.98), we then get that

$$lo_{\Pi,\mathrm{q}}(t) = 2\alpha\frac{\delta T_\mathrm{q}}{T} + \sum_{l=1}^{\infty}\big[(-1)^{l-1}g_{2l-1}a_{\mathrm{p},2l-1}\sin((2l-1)\omega_\mathrm{LO}t) + (-1)^l g_{2l}a_{\mathrm{p},2l}\cos(2l\omega_\mathrm{LO}t)\big]. \tag{6.104}$$

Using this expression in conjunction with equation (6.96) while expanding the cosine and sine functions using equation (6.5) to get the result as a function of complex exponentials, we find that

$$lo_{+\omega_\mathrm{LO}}(t) = 2\alpha\left(\frac{\delta T_\mathrm{p}}{T} + \mathrm{j}\frac{\delta T_\mathrm{q}}{T}\right) + \sum_{l=1}^{\infty}\big[(\alpha_+^+)_{2l-1}e^{+\mathrm{j}(2l-1)\omega_\mathrm{LO}t} + (\alpha_+^-)_{2l-1}e^{-\mathrm{j}(2l-1)\omega_\mathrm{LO}t} + (\beta_+^+)_{2l}e^{+\mathrm{j}2l\omega_\mathrm{LO}t} + (\beta_+^-)_{2l}e^{-\mathrm{j}2l\omega_\mathrm{LO}t}\big], \tag{6.105}$$

with

$$(\alpha_+^+)_{2l-1} = a_{\mathrm{p},2l-1}\frac{(1 + (-1)^{l-1}g_{2l-1})}{2}, \tag{6.106a}$$

$$(\alpha_+^-)_{2l-1} = a_{\mathrm{p},2l-1}\frac{(1 - (-1)^{l-1}g_{2l-1})}{2}, \tag{6.106b}$$

and

$$(\beta_+^+)_{2l} = (\beta_+^-)_{2l} = a_{\mathrm{p},2l}\frac{(1 + \mathrm{j}(-1)^l g_{2l})}{2}. \tag{6.107}$$

Here, the $(\alpha_+^+)_{2l-1}$ and $(\alpha_+^-)_{2l-1}$ factors can be seen as a generalization of the alpha factors defined in Section 6.2.1 in the pure sinusoidal LO case. However, we observe that the present expressions correspond to tone powers that are no longer normalized. In the present case, it is the power of the overall time domain square LO signal that is independent of the DtyCy. But the power of the various tones present in its decomposition necessarily depends on the DtyCy.

Here, we have also defined the $(\beta_+^+)_{2l}$ and $(\beta_+^-)_{2l}$ factors that play the role for the even order harmonics corresponding to that played by $(\alpha_+^+)_{2l-1}$ and $(\alpha_+^-)_{2l-1}$ for the odd order harmonics.

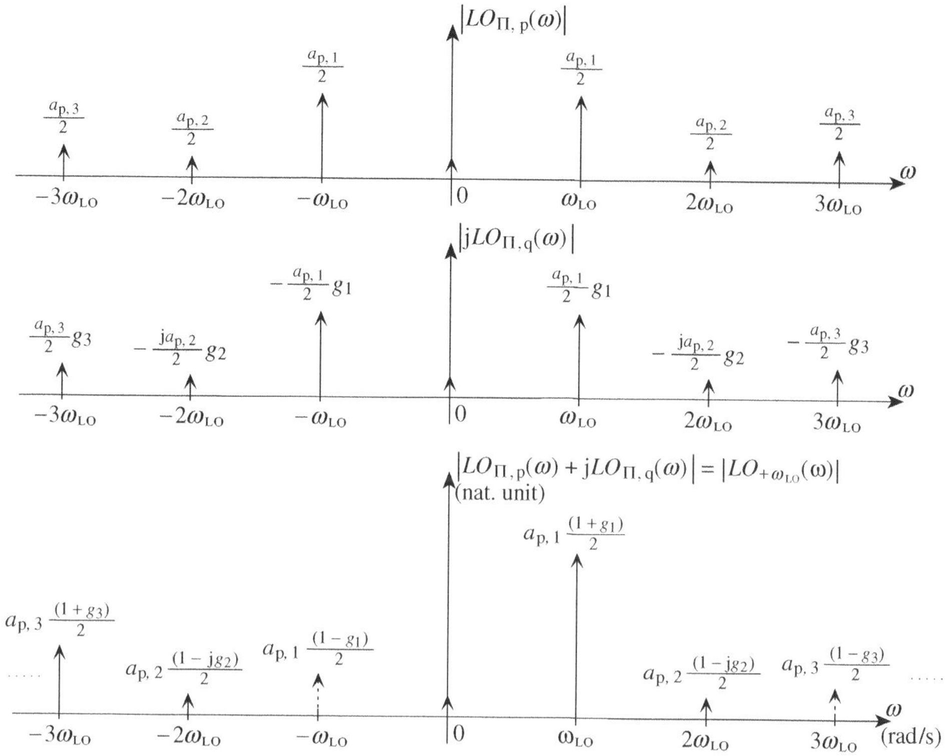

Figure 6.28 Spectral content of the periodic square LO waveforms in the presence of unbalanced duty cycles – The spectral content of the LO waveforms shown in Figure 6.24, when $\delta T_p \neq \delta T_q$, exhibits both a rise in the even order harmonics of the fundamental tone, as already experienced when $\delta T_p = \delta T_q$ and illustrated in Figure 6.25, but also different amplitudes for the odd order harmonic tones in the spectrum of P (top) and Q (middle) LO components. The latter phenomenon results in a rise in the complex conjugate tones of odd order in the spectrum of the reconstructed complex LO signal (bottom).

However, we observe some fundamental differences in the behavior of these even and odd order harmonics. The paired complex conjugate tones of even order have necessarily the same amplitude as illustrated in Figure 6.28. In that sense, we recover the same behavior as encountered in the previous section when $\delta T_p = \delta T_q$ even if with different amplitudes in the two situations. When $\delta T_p \neq \delta T_q$, however, we have a rise in the paired complex conjugate tones of odd order. The DtyCy imbalance between the P and Q LO signals results in an overall gain imbalance between the tones of the same order that compose those signals. This results in an inexact cancellation of the paired complex conjugate tones of odd order when considering the reconstruction of the complex LO signal. We can thus clearly make the link with the derivations performed throughout Section 6.2, and identify the presence of paired complex conjugate tones of odd order as the root cause for the finite image rejection that can be achieved during a complex frequency transposition using such a square LO waveform. However, we need to recall that only the complex conjugate of the tone expected to be used for the frequency conversion of the wanted signal sideband is involved in the folding or in the generation mechanism of the unwanted image sideband during such processing. Assuming that it is the fundamental tone of $lo_{+\omega_{LO}}(t)$ that is expected to be used in that respect, the IRR achieved during such complex frequency conversion is simply linked to the power ratio between this fundamental tone and its complex conjugate. Based on equation (6.105), we can then write that

$$IRR = \left| \frac{(\alpha_{+}^{-})_1}{(\alpha_{+}^{+})_1} \right|^2. \tag{6.108}$$

Thus, using equation (6.106), we have

$$IRR = \left(\frac{1 - g_1}{1 + g_1} \right)^2, \tag{6.109}$$

with g_1 the gain imbalance that exists between the two fundamental tones of $lo_{\Pi,\mathrm{p}}(t)$ and $lo_{\Pi,\mathrm{q}}(t)$ when $\delta T_{\mathrm{p}} \neq \delta T_{\mathrm{q}}$. We observe that this expression matches equation (6.38) in the particular case where $\delta\phi = 0$. This condition corresponds to the present assumption of an exact delay of $T/4$ modulo T between the centers of symmetry of $lo_{\Pi,\mathrm{p}}(t)$ and $lo_{\Pi,\mathrm{q}}(t)$ (see the discussion in the next section). We thus recover that the gain imbalance existing between the quadrature tones of the P and Q LO signals used for the complex frequency conversion leads to a finite image rejection. It is interesting to note that the results derived up to now for the periodic square LO waveform are consistent with the results given in Section 6.2 in the pure sinusoidal case.

We can even go a step further and express the resulting IRR as a function just of the duty cycles on both the P and Q branches. From equation (6.103a) we have

$$g_1 = \frac{b_{\mathrm{q},1}}{a_{\mathrm{p},1}} = \frac{\cos\left(\pi \frac{\delta T_{\mathrm{q}}}{T}\right)}{\cos\left(\pi \frac{\delta T_{\mathrm{p}}}{T}\right)}. \tag{6.110}$$

Using this expression in equation (6.109), we get

$$IRR = \left[\frac{\cos\left(\pi \frac{\delta T_{\mathrm{p}}}{T}\right) - \cos\left(\pi \frac{\delta T_{\mathrm{q}}}{T}\right)}{\cos\left(\pi \frac{\delta T_{\mathrm{p}}}{T}\right) + \cos\left(\pi \frac{\delta T_{\mathrm{q}}}{T}\right)} \right]^2. \tag{6.111}$$

Using the trigonometric identities

$$\cos(\theta_1) - \cos(\theta_2) = -2 \sin\left(\frac{\theta_1 + \theta_2}{2}\right) \sin\left(\frac{\theta_1 - \theta_2}{2}\right), \tag{6.112a}$$

$$\cos(\theta_1) + \cos(\theta_2) = 2 \cos\left(\frac{\theta_1 + \theta_2}{2}\right) \cos\left(\frac{\theta_1 - \theta_2}{2}\right), \tag{6.112b}$$

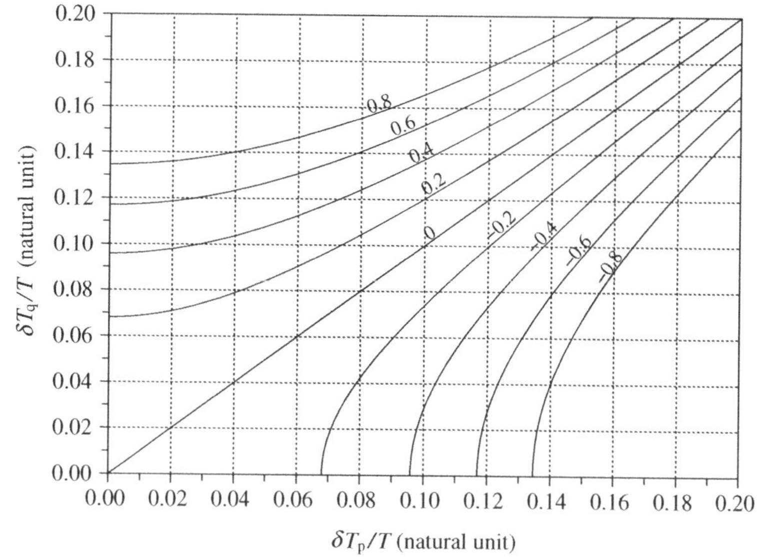

Figure 6.29 Gain imbalance between the fundamental tones of $lo_{\Pi,p}(t)$ and $lo_{\Pi,q}(t)$ due to duty cycles imbalance – The difference in duty cycles between the $lo_{\Pi,p}(t)$ and $lo_{\Pi,q}(t)$ periodic square LO waveforms leads in particular to an amplitude imbalance of the fundamental tones involved in their series expansion as given by equation (6.110). Here, $\delta T_p/T$ and $\delta T_q/T$ represent the signal DtyCy imbalance in the notation of Figure 6.24.

we finally get that

$$IRR = \left[\tan\left(\frac{\pi}{2}\frac{\delta T_p + \delta T_q}{T}\right)\tan\left(\frac{\pi}{2}\frac{\delta T_p - \delta T_q}{T}\right)\right]^2. \tag{6.113}$$

The gain imbalance g_1 as well as the corresponding IRR are plotted as a function of δT_p and δT_q in Figures 6.29 and 6.30, respectively. We thus see that the image power effectively vanishes when the DtyCy is the same on the two branches, i.e. when $\delta T_p = \delta T_q$. This confirms that in this case the amplitude of the fundamental tones involved in the series expansion of both $lo_{\Pi,p}(t)$ and $lo_{\Pi,q}(t)$ are the same. We thus get an exact cancellation of the complex conjugate of the complex exponential corresponding to this fundamental tone when reconstructing the complex LO signal from the periodic square LO components.

Delay Imbalance

Let us now suppose that we have a delay imbalance between the P and Q components $lo_{\Pi,p}(t)$ and $lo_{\Pi,q}(t)$ used to drive a complex mixer implemented using chopper devices. We thus assume now an inexact delay of $T/4$ modulo T, with T the period of the waveform, between the centers of symmetry of those signals, as illustrated in Figure 6.31. However, in order to get rid of the gain imbalance problem discussed in the previous section when $lo_{\Pi,p}(t)$ and $lo_{\Pi,q}(t)$ exhibit different duty cycles, we also assume here that we are dealing with the same DtyCy

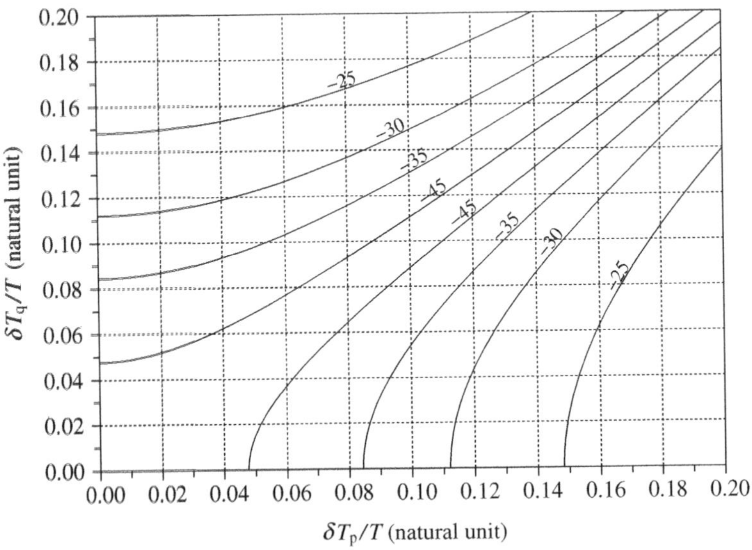

Figure 6.30 Finite IRR achieved in the presence of duty cycle imbalance between $lo_{\Pi,p}(t)$ and $lo_{\Pi,q}(t)$ – The difference in duty cycles between the $lo_{\Pi,p}(t)$ and $lo_{\Pi,q}(t)$ periodic square LO waveforms leads to an amplitude imbalance between their fundamental tones, as illustrated in Figure 6.29. This leads in turn to a finite image rejection at the complex frequency conversion stage output according to equation (6.113). Here, an exact delay of $T/4$ modulo T is assumed between the centers of symmetry of both the $lo_{\Pi,p}(t)$ and $lo_{\Pi,q}(t)$ signals, as shown in Figure 6.24, to ensure a perfect quadrature between the considered fundamental tones.

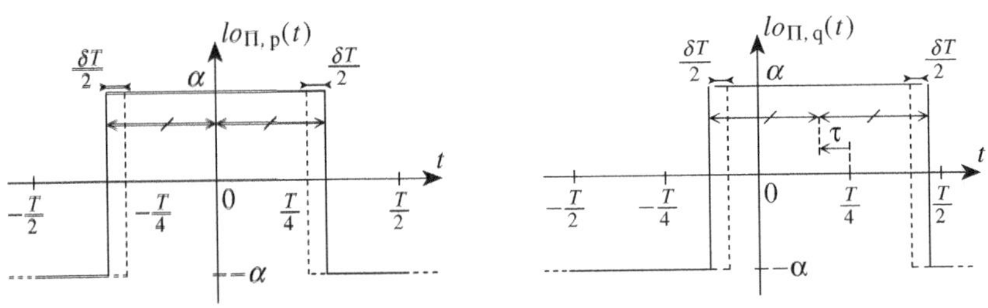

Figure 6.31 Periodic square waves in quadrature as LO waveforms with identical duty cycles, even if different from 0.5, and a delay imbalance – Having an inexact delay of $T/4$ modulo T between the centers of symmetry of the P and Q LO components leads to a phase imbalance between the tones involved in the series expansion of $lo_{\Pi,p}(t)$ and $lo_{\Pi,q}(t)$. The result is a rise in the complex conjugate tones of odd order as well as an imbalance in the amplitude of the paired complex conjugate tones of even order in the spectrum of the reconstructed complex LO signal as shown in Figure 6.32.

for those two signals, i.e. that $\delta T_{\mathrm{p}} = \delta T_{\mathrm{q}} = \delta T$. This is done with the aim of distinguishing between the effects of those two kinds of impairments.

Consequently, our assumptions remain compliant with those in "Balanced duty cycles" earlier in this section, at least for the $lo_{\Pi,\mathrm{p}}(t)$ component. We can thus use equation (6.89) to write the Fourier series expansion of $lo_{\Pi,\mathrm{p}}(t)$ as

$$lo_{\Pi,\mathrm{p}}(t) = 2\alpha\frac{\delta T}{T} + \sum_{l=1}^{\infty}[a_{\mathrm{p},2l-1}\cos((2l-1)\omega_{\mathrm{LO}}t) + a_{\mathrm{p},2l}\cos(2l\omega_{\mathrm{LO}}t)], \tag{6.114}$$

with $\omega_{\mathrm{LO}} = 2\pi/T$ and the coefficients $a_{\mathrm{p},k}$ given by equation (6.90) as

$$a_{\mathrm{p},2l-1} = \alpha\frac{4}{\pi}\frac{(-1)^{l-1}}{2l-1}\cos\left((2l-1)\pi\frac{\delta T}{T}\right), \tag{6.115a}$$

$$a_{\mathrm{p},2l} = \alpha\frac{4}{\pi}\frac{(-1)^{l}}{2l}\sin\left(2l\pi\frac{\delta T}{T}\right). \tag{6.115b}$$

We can then use this result to derive the series expansion of $lo_{\Pi,\mathrm{q}}(t)$, using the fact that by our assumptions $lo_{\Pi,\mathrm{q}}(t) = lo_{\Pi,\mathrm{p}}(t - T/4 + \tau)$. We obtain from equation (6.114) that

$$\begin{aligned} lo_{\Pi,\mathrm{q}}(t) = 2\alpha\frac{\delta T}{T} + \sum_{l=1}^{\infty}\Big[a_{\mathrm{p},2l-1}\cos\left((2l-1)\omega_{\mathrm{LO}}(t+\tau) - l\pi + \frac{\pi}{2}\right) \\ + a_{\mathrm{p},2l}\cos(2l\omega_{\mathrm{LO}}(t+\tau) - l\pi)\Big]. \end{aligned} \tag{6.116}$$

Using equation (6.78) we can finally write that

$$lo_{\Pi,\mathrm{q}}(t) = 2\alpha\frac{\delta T}{T} + \sum_{l=1}^{\infty}[b_{\mathrm{q},2l-1}\sin((2l-1)\omega_{\mathrm{LO}}t + \delta\phi_{2l-1}) + a_{\mathrm{q},2l}\cos(2l\omega_{\mathrm{LO}}t + \delta\phi_{2l})], \tag{6.117}$$

with

$$b_{\mathrm{q},2l-1} = (-1)^{l-1}a_{\mathrm{p},2l-1} = \alpha\frac{4}{\pi}\frac{1}{2l-1}\cos\left((2l-1)\pi\frac{\delta T}{T}\right), \tag{6.118a}$$

$$a_{\mathrm{q},2l} = (-1)^{l}a_{\mathrm{p},2l} = \alpha\frac{4}{\pi}\frac{1}{2l}\sin\left(2l\pi\frac{\delta T}{T}\right), \tag{6.118b}$$

and phase error

$$\delta\phi_k = k\omega_{\mathrm{LO}}\tau = 2k\pi\frac{\tau}{T} \tag{6.119}$$

between the tones of the same order that compose $lo_{\Pi,\mathrm{p}}(t)$ and $lo_{\Pi,\mathrm{q}}(t)$. As expected, the phase error vanishes only when we have $\tau = 0$. In that case only we get an exact quadrature between the fundamental tones of those LO components and thus a chance that they exactly represent the real and imaginary parts of the complex exponential expected for implementing a complex frequency conversion.

We are now in a position to derive an expression for the series expansion of a reconstructed complex LO signal. We consider the signal $lo_{+\omega_{\mathrm{LO}}}(t) = lo_{\Pi,\mathrm{p}}(t) + \mathrm{j}lo_{\Pi,\mathrm{q}}(t)$ already discussed in the previous sections. Using on the one hand equations (6.114) and (6.117) while expanding the cosine and sine functions using equation (6.5) to get the result as a function of complex exponentials, and on the other hand equations (6.115) and (6.118), we obtain

$$\begin{aligned} lo_{+\omega_{\mathrm{LO}}}(t) = 2\alpha\frac{\delta T}{T}(1+\mathrm{j}) + \sum_{l=1}^{\infty} \Big[(\alpha_+^+)_{2l-1}\mathrm{e}^{+\mathrm{j}(2l-1)\omega_{\mathrm{LO}}t} + (\alpha_+^-)_{2l-1}\mathrm{e}^{-\mathrm{j}(2l-1)\omega_{\mathrm{LO}}t} \\ + (\beta_+^+)_{2l}\mathrm{e}^{+\mathrm{j}2l\omega_{\mathrm{LO}}t} + (\beta_+^-)_{2l}\mathrm{e}^{-\mathrm{j}2l\omega_{\mathrm{LO}}t}\Big], \end{aligned} \tag{6.120}$$

with

$$(\alpha_+^+)_{2l-1} = a_{\mathrm{p},2l-1}\frac{\left(1+(-1)^{l-1}\mathrm{e}^{+\mathrm{j}\delta\phi_{2l-1}}\right)}{2}, \tag{6.121a}$$

$$(\alpha_+^-)_{2l-1} = a_{\mathrm{p},2l-1}\frac{\left(1-(-1)^{l-1}\mathrm{e}^{-\mathrm{j}\delta\phi_{2l-1}}\right)}{2}, \tag{6.121b}$$

and

$$(\beta_+^+)_{2l} = a_{\mathrm{p},2l}\frac{\left(1+\mathrm{j}(-1)^{l}\mathrm{e}^{+\mathrm{j}\delta\phi_{2l}}\right)}{2}, \tag{6.122a}$$

$$(\beta_+^-)_{2l} = a_{\mathrm{p},2l}\frac{\left(1+\mathrm{j}(-1)^{l}\mathrm{e}^{-\mathrm{j}\delta\phi_{2l}}\right)}{2}. \tag{6.122b}$$

As already pointed out in "Unbalanced duty cycles" earlier in this section, the $(\alpha_+^+)_{2l-1}$ and $(\alpha_+^-)_{2l-1}$ factors can then be seen as a generalization of the alpha factors defined in Section 6.2.1 in the pure sinusoidal case. The $(\beta_+^+)_{2l}$ and $(\beta_+^-)_{2l}$ factors are introduced to play the same role for the even order terms in the series expansion of $lo_{+\omega_{\mathrm{LO}}}(t)$. However, we observe a different behavior for those even order tones in the present case than encountered when faced with different duty cycles for the P and Q LO components. We now get that the amplitudes of the paired complex conjugate tones of the same order are different, as illustrated in Figure 6.32. This behavior is closely related to the inexact quadrature between $lo_{\Pi,\mathrm{p}}(t)$ and $lo_{\Pi,\mathrm{q}}(t)$ due to the delay imbalance τ. Only such a time offset can lead to a phase offset, here given by equation (6.119), which has an opposite sign for the paired complex conjugate tones lying in the positive and negative parts of the spectrum of $lo_{\Pi,\mathrm{q}}(t)$. This behavior thus leads to a different magnitude for the corresponding tones when carrying out the coherent recombination of $lo_{\Pi,\mathrm{p}}(t)$ and $lo_{\Pi,\mathrm{q}}(t)$.

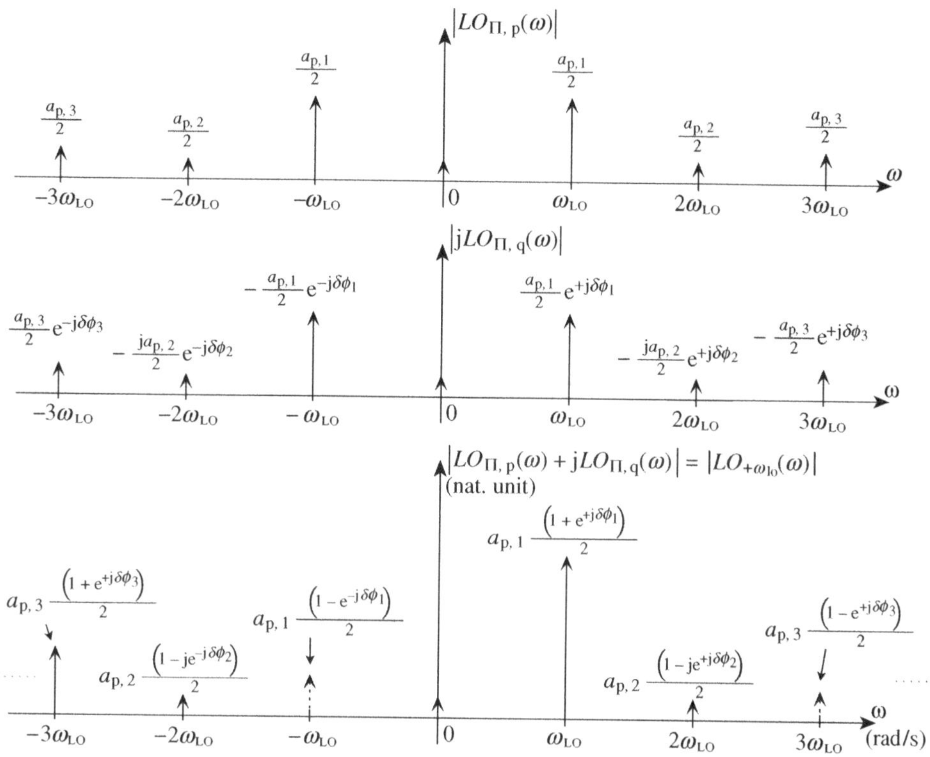

Figure 6.32 Spectral content of the periodic square LO waveforms in the presence of delay imbalance – Considering the waveforms shown in Figure 6.31, a non-vanishing delay imbalance τ leads to a phase imbalance between the tones of the same order in the spectrum of the P (top) and Q (middle) LO components. Even when $\delta T_p = \delta T_q$, we thus have a rise in the complex conjugate tones of odd order and an amplitude imbalance for the paired even order tones in the spectrum of the reconstructed complex LO signal (bottom). This spectrum can be compared to the theoretical one shown in Figure 6.23.

In the same way, this phase imbalance leads to an inexact cancellation of the complex conjugate tones of odd order in the reconstructed complex LO signal $lo_{+\omega_{LO}}(t)$. Here again, we can make the link with the derivations performed throughout Section 6.2, and thus identify the presence of paired complex conjugate tones of odd order as the root cause for the finite image rejection that can be achieved during a complex frequency conversion. Assuming that it is the fundamental tone of $lo_{+\omega_{LO}}(t)$ that is expected to be used for the frequency conversion of the wanted signal sideband, the IRR achieved during such processing is simply linked to the power ratio between this fundamental tone and its complex conjugate. Based on equation (6.120), we can then write that

$$IRR = \left| \frac{(\alpha_+^-)_1}{(\alpha_+^+)_1} \right|^2. \tag{6.123}$$

Using equation (6.121), we then have

$$IRR = \frac{1 - \cos(\delta\phi_1)}{1 + \cos(\delta\phi_1)}, \tag{6.124}$$

with ϕ_1 the phase imbalance between the two fundamental tones of $lo_{\Pi,p}(t)$ and $lo_{\Pi,q}(t)$ when $\tau \neq 0$. We remark that the IRR given by equation (6.38) in the pure sinusoidal LO case reduces to this expression when $g = 1$, i.e. no gain imbalance exists between the P and Q branches of the system. This corresponds to our present assumption of the same DtyCy for the P and Q LO signals, i.e. that $\delta T_p = \delta T_q = \delta T$. This equivalent behavior again confirms the approximation of the ideal complex frequency conversion by using such square LO waveforms.

Finally, we can also express this IRR as a function of the delay imbalance τ. We use equation (6.119) to write

$$IRR = \frac{1 - \cos(2\pi\tau/T)}{1 + \cos(2\pi\tau/T)}. \tag{6.125}$$

The IRR is plotted in Figure 6.33 as a function of the delay imbalance. Looking at equation (6.125), we can remark that the IRR increases to identically equal 1 for $\tau = T/4$. Indeed, referring to Figure 6.31, we can see that the two LO components $lo_{\Pi,p}(t)$ and $lo_{\Pi,q}(t)$ are identical in that case. The resulting reconstructed complex LO signal is then directly proportional to this real valued signal which thus necessarily exhibits complex conjugate tones of equal amplitude in its spectrum. For $\tau > T/4$, the signal $lo_{\Pi,q}(t)$ can be interpreted as the negative of what it is when $\tau < T/4$. In that case, the reconstructed complex LO signal can be interpreted

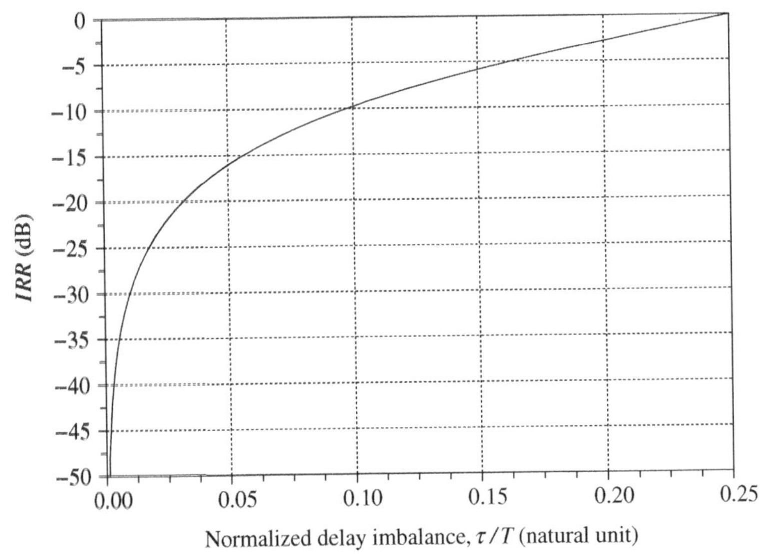

Figure 6.33 Finite IRR achieved in the presence of delay imbalance between $lo_{\Pi,p}(t)$ and $lo_{\Pi,q}(t)$ – An inexact delay of $T/4$ modulo T between the centers of symmetry of the $lo_{\Pi,p}(t)$ and $lo_{\Pi,q}(t)$ signals leads to a phase imbalance between their fundamental tones, as shown in Figure 6.32. This leads to a finite IRR given by equation (6.125) when using this fundamental tone for a frequency conversion.

as $lo_{-\omega_{LO}}(t)$ instead of $lo_{+\omega_{LO}}(t)$ and is thus centered around $-\omega_{LO}$. The complex exponentials considered for the definition of the IRR should then be exchanged, resulting in the inversion of the numerator and denominator of equation (6.125).

Practical LO Spectral Content

So far we have detailed the different impairments involved in the degradation of LO signals used for the physical implementation of frequency conversions based on the use of chopper devices. We have seen that these impairments fall into two groups:

(i) We first get the impairments that impact either the single LO signal used for a real frequency conversion or, identically, the two P and Q LO components used for a complex frequency conversion. When dealing with a mixing stage expected to be driven by an LO signal that exhibits a DtyCy of 0.5, such impairments reduce to the possibility of the waveform DtyCy deviating from this value. As discussed in "Balanced duty cycles" earlier in this section, such imbalance results mainly in a rise in the even order tones.

(ii) When dealing with a complex frequency conversion, we then need to consider the potential imbalance between the P and Q LO components. In practice, this imbalance can be linked to either a difference in the duty cycles of those P and Q components or an inexact quadrature linked to a delay offset between them. As discussed in "Unbalanced duty cycles" and "Delay imbalance" earlier in this section, this results in a gain and phase offset between paired complex conjugate tones of the same order. As a main consequence, we then have a rise in the complex conjugate tones of odd order, thus leading to a finite image rejection during the complex frequency conversion.

Based on this understanding of the phenomenon we are dealing with, we can now give an expression for the LO signals in the general case with the aim of discussing in Section 6.4 the practical consequences to be considered for the system design of a transceiver line-up.

Let us consider the structure of the P and Q LO components shown in Figure 6.34 in the general impairment case. In addition to the impairments taken into account so far, we add a general gain imbalance g between the P and Q LO components. Obviously, a chopper is

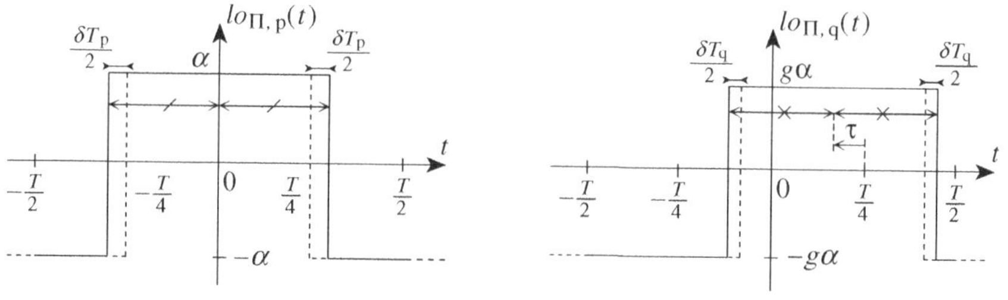

Figure 6.34 Periodic square waves in quadrature as LO waveforms with general impairments – Due to classical limitations in the physical implementation, we have an overall amplitude imbalance ($g\alpha$ vs. α), a DtyCy imbalance (δT_q vs. δT_p) or a delay imbalance ($\tau \neq 0$) between the P and Q LO signals used for trigging mixers implemented as choppers. The consequences for the LO spectrum are shown in Figure 6.35.

sensitive only to the switching time of the LO signal that drives it. Dealing with such imbalance in the amplitude of the LO signals can thus seem pointless at first glance. However, any gain imbalance that is experienced by the signal being converted along the P and Q branches of the line-up leads to a degradation of performance, as illustrated in the pure sinusoidal case in Section 6.2.1. From the signal processing point of view, all behaves as if the LO waveforms that multiply the signal being converted were carrying this gain imbalance. We can retain this formalism for the present discussion.

Our present assumptions remain consistent with the discussion in "Unbalanced duty cycles" earlier in this section, at least for the expression of the P component $lo_{\Pi,\mathrm{p}}(t)$. The Fourier series expansion of this signal can thus be written from equation (6.96) as

$$lo_{\Pi,\mathrm{p}}(t) = 2\alpha\frac{\delta T_\mathrm{p}}{T} + \sum_{l=1}^{\infty}[a_{\mathrm{p},2l-1}\cos((2l-1)\omega_{\mathrm{LO}}t) + a_{\mathrm{p},2l}\cos(2l\omega_{\mathrm{LO}}t)], \qquad (6.126)$$

with $\omega_{\mathrm{LO}} = 2\pi/T$ and the coefficients $a_{\mathrm{p},k}$ given by equation (6.97) as

$$a_{\mathrm{p},2l-1} = \alpha\frac{4}{\pi}\frac{(-1)^{l-1}}{2l-1}\cos\left((2l-1)\pi\frac{\delta T_\mathrm{p}}{T}\right), \qquad (6.127a)$$

$$a_{\mathrm{p},2l} = \alpha\frac{4}{\pi}\frac{(-1)^{l}}{2l}\sin\left(2l\pi\frac{\delta T_\mathrm{p}}{T}\right). \qquad (6.127b)$$

The series expansion of the Q component $lo_{\Pi,\mathrm{q}}(t)$ can then be derived in a straightforward way using the fact that from a purely formal perspective, $lo_{\Pi,\mathrm{q}}(t)$ can be written as $lo_{\Pi,\mathrm{p}}(t - T/4 + \tau)$ when substituting δT_q for δT_p and $g\alpha$ for α. We can thus write from equation (6.126) that

$$\begin{aligned} lo_{\Pi,\mathrm{q}}(t) = 2g\alpha\frac{\delta T_\mathrm{q}}{T} + \alpha\frac{4}{\pi}\sum_{l=1}^{\infty}\Big[a_{\mathrm{p},2l-1}\cos\left((2l-1)\omega_{\mathrm{LO}}(t+\tau) - l\pi + \frac{\pi}{2}\right) \\ + a_{\mathrm{p},2l}\cos(2l\omega_{\mathrm{LO}}(t+\tau) - l\pi)\Big]. \end{aligned} \qquad (6.128)$$

Using equation (6.78), we finally obtain

$$\begin{aligned} lo_{\Pi,\mathrm{q}}(t) = 2g\alpha\frac{\delta T_\mathrm{q}}{T} + \sum_{l=1}^{\infty}[b_{\mathrm{q},2l-1}\sin((2l-1)\omega_{\mathrm{LO}}t + \delta\phi_{2l-1}) \\ + a_{\mathrm{q},2l}\cos(2l\omega_{\mathrm{LO}}t + \delta\phi_{2l})]. \end{aligned} \qquad (6.129)$$

Here, the phase imbalance term $\delta\phi_k$ between tones of the same order corresponds to that introduced through equation (6.119) in the pure delay imbalance case in "Unbalanced duty cycles" earlier in this section, i.e.

$$\delta\phi_k = k\omega_{\rm LO}\tau = 2k\pi\frac{\tau}{T}. \tag{6.130}$$

In the same way, the coefficients

$$b_{{\rm q},2l-1} = g\alpha\frac{4}{\pi}\frac{1}{2l-1}\cos\left((2l-1)\pi\frac{\delta T_{\rm q}}{T}\right), \tag{6.131a}$$

$$a_{{\rm q},2l} = g\alpha\frac{4}{\pi}\frac{1}{2l}\sin\left(2l\pi\frac{\delta T_{\rm q}}{T}\right) \tag{6.131b}$$

can be related to those given by equation (6.118) in the pure unbalanced duty cycles case, but now taking into account the overall gain imbalance g. We can thus update the definition of the gain imbalance factor between harmonics of the same order defined in that section through equation (6.103) to take this overall gain imbalance into account. We obtain

$$g_{2l-1} = (-1)^{l-1}\frac{b_{{\rm q},2l-1}}{a_{{\rm p},2l-1}} = g\frac{\cos\left((2l-1)\pi\frac{\delta T_{\rm q}}{T}\right)}{\cos\left((2l-1)\pi\frac{\delta T_{\rm p}}{T}\right)}, \tag{6.132a}$$

$$g_{2l} = (-1)^{l}\frac{a_{{\rm q},2l}}{a_{{\rm p},2l}} = g\frac{\sin\left(2l\pi\frac{\delta T_{\rm q}}{T}\right)}{\sin\left(2l\pi\frac{\delta T_{\rm p}}{T}\right)}. \tag{6.132b}$$

As discussed in "Unbalanced duty cycles" earlier in this section, g_{2l} exists only when $\delta T_{\rm p} \neq 0$ or $\delta T_{\rm q} \neq 0$, i.e. when at least one of those duty cycles is different from 0.5. This is the condition for an even order harmonic to exist in the spectrum of the corresponding LO signal.

Based on the results derived so far, we can derive an expression for the series expansion of a reconstructed complex LO signal to discuss the consequences of these impairments on the complex frequency conversion. We focus on the complex LO signal $lo_{+\omega_{\rm LO}}(t) = lo_{\Pi,{\rm p}}(t) + jlo_{\Pi,{\rm q}}(t)$, already considered as an example in the previous sections. We can follow the same approach as in "Unbalanced duty cycles" earlier in this section, and introduce the terms g_k in the series expansion of $lo_{\Pi,{\rm q}}(t)$. This results in the same formal expression as given by equation (6.104), except that the g_k terms are now given by equation (6.132). Using this expression in conjunction with equation (6.126) while expanding the cosine and sine functions using equation (6.5) to get the result as a function of complex exponentials, we obtain

$$\begin{aligned} lo_{+\omega_{\rm LO}}(t) = 2\alpha\left(\frac{\delta T_{\rm p}}{T} + {\rm j}g\frac{\delta T_{\rm q}}{T}\right) + \sum_{l=1}^{\infty}\Big[(\alpha_+^+)_{2l-1}{\rm e}^{+{\rm j}(2l-1)\omega_{\rm LO}t} + (\alpha_+^-)_{2l-1}{\rm e}^{-{\rm j}(2l-1)\omega_{\rm LO}t} \\ + (\beta_+^+)_{2l}{\rm e}^{+{\rm j}2l\omega_{\rm LO}t} + (\beta_+^-)_{2l}{\rm e}^{-{\rm j}2l\omega_{\rm LO}t}\Big], \end{aligned} \tag{6.133}$$

with

$$(\alpha_{+}^{+})_{2l-1} = a_{\mathrm{p},2l-1}\frac{\left(1+(-1)^{l-1}g_{2l-1}\mathrm{e}^{+\mathrm{j}\delta\phi_{2l-1}}\right)}{2}, \tag{6.134a}$$

$$(\alpha_{+}^{-})_{2l-1} = a_{\mathrm{p},2l-1}\frac{\left(1-(-1)^{l-1}g_{2l-1}\mathrm{e}^{-\mathrm{j}\delta\phi_{2l-1}}\right)}{2}, \tag{6.134b}$$

and

$$(\beta_{+}^{+})_{2l} = a_{\mathrm{p},2l}\frac{\left(1+\mathrm{j}(-1)^{l}g_{2l}\mathrm{e}^{+\mathrm{j}\delta\phi_{2l}}\right)}{2}, \tag{6.135a}$$

$$(\beta_{+}^{-})_{2l} = a_{\mathrm{p},2l}\frac{\left(1+\mathrm{j}(-1)^{l}g_{2l}\mathrm{e}^{-\mathrm{j}\delta\phi_{2l}}\right)}{2}. \tag{6.135b}$$

The factors expressed in general form here can be compared to the expressions derived earlier in this section in "Duty cycle", when only a gain imbalance exists, or "Delay imbalance", when only a phase imbalance is present. The mechanism involved in the reconstruction of the complex LO waveform in the general case is shown graphically in Figure 6.35. We see that the gain and phase imbalance between harmonics of the same order in the series expansion of the P and Q LO components now depends on their order. The same behavior thus necessarily holds for the magnitude of the resulting tones in the reconstructed complex LO signal $lo_{+\omega_{\mathrm{LO}}}(t)$. However, the consequences remain different for the odd or even order harmonics, as discussed extensively in the previous sections.

In the ideal case, no paired complex conjugate tones of odd order exist in the series expansion of the complex LO signal given by equation (6.80). More precisely, dealing here with a reconstructed complex LO signal centered around the positive complex exponential $\mathrm{e}^{+\mathrm{j}\omega_{\mathrm{LO}}t}$, the tones present in this series are alternately centered around $+\omega_{\mathrm{LO}}, -3\omega_{\mathrm{LO}}, +5\omega_{\mathrm{LO}}, \ldots$ But by equation (6.133), the imbalance between the P and Q LO components leads to a rise in the symmetrical tones that lie alternately around the angular frequencies $-\omega_{\mathrm{LO}}, +3\omega_{\mathrm{LO}}, -5\omega_{\mathrm{LO}}, \ldots$ In the time domain, these unwanted tones correspond to complex exponentials that are proportional to the complex conjugate of those present in the ideal case. Based on the discussion in Section 6.2, we can thus interpret the unwanted complex exponentials as the root cause for the folding, or the generation, of the unwanted sideband labeled as the image signal during a complex frequency conversion. More precisely, considering the frequency conversion of the wanted signal sideband using the complex exponential of order $2l-1$, it is its complex conjugate that is involved in this folding or generation process. But, given now that the relative amplitudes of the paired complex conjugate tones depend on their order, we may thus generalize the IRR defined previously for the fundamental tone through equation (6.38). Consequently, although it is often this fundamental tone that is used in practice in order to achieve the highest conversion gain on the wanted signal, we may define the quantity IRR_{2l-1} that gives the ratio of the power of the tone of order $2l-1$ relative to its complex conjugate. Given that the wanted tone present in the ideal periodic square LO waveform is alternately the

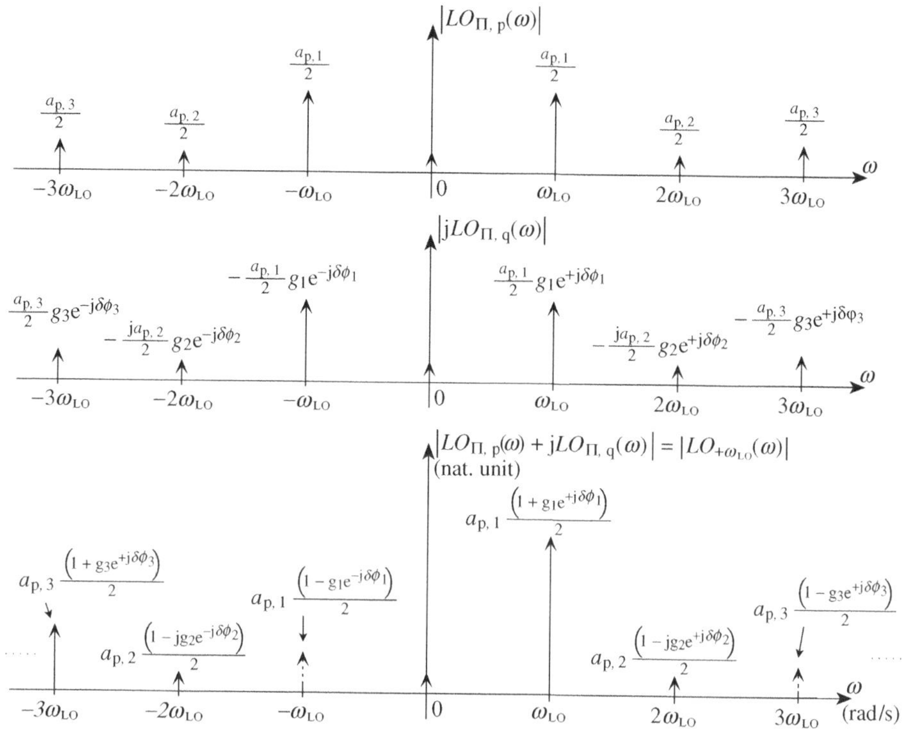

Figure 6.35 Spectral content of the periodic square LO waveforms in the presence of general impairments – The general impairments between the P (top) and Q (middle) LO components shown in Figure 6.34 lead to a rise in both complex conjugate tones of odd order and imbalanced paired tones of even order in the spectrum of the reconstructed complex LO signal (bottom). This spectrum can be compared to the theoretical one shown in Figure 6.23 where no imbalance is present.

one lying at $+(2l-1)\omega_{\mathrm{LO}}$ and the one lying at $-(2l-1)\omega_{\mathrm{LO}}$ depending on the harmonic order, we may define IRR_{2l-1} as

$$IRR_{2l-1} = \left(\frac{|(\alpha_+^-)_{2l-1}|^2}{|(\alpha_+^+)_{2l-1}|^2}\right)^{(-1)^{l-1}}. \tag{6.136}$$

At the same time, we can write from equation (6.134) that

$$|(\alpha_+^+)_{2l-1}|^2 = \frac{|a_{\mathrm{p},2l-1}|^2}{4}\left(1+2(-1)^{l-1}g_{2l-1}\cos(\delta\phi_{2l-1})+g_{2l-1}^2\right),$$
$$|(\alpha_+^-)_{2l-1}|^2 = \frac{|a_{\mathrm{p},2l-1}|^2}{4}\left(1-2(-1)^{l-1}g_{2l-1}\cos(\delta\phi_{2l-1})+g_{2l-1}^2\right).$$

We thus see that only the sign on the cosine function differs between the two terms. As those signs toggle as l does, we always recover the term with the negative sign in the numerator of IRR_{2l-1}. Equation (6.136) thus reduces to

$$IRR_{2l-1} = \frac{1 - 2g_{2l-1}\cos(\delta\phi_{2l-1}) + g_{2l-1}^2}{1 + 2g_{2l-1}\cos(\delta\phi_{2l-1}) + g_{2l-1}^2}, \tag{6.137}$$

with the phase imbalance term $\delta\phi_{2l-1}$ and the gain imbalance term g_{2l-1} given by equations (6.130) and (6.132a), respectively. These expressions clearly show the dependency of the relative magnitude of the paired complex conjugate tones on the harmonic order. This is especially obvious considering the phase imbalance that has a linear dependency on this order. We then have a rapid rise in the complex conjugate tone power when considering increasing harmonic orders. As an example, with a simple delay imbalance of 1% between $lo_{\Pi,\mathrm{p}}(t)$ and $lo_{\Pi,\mathrm{q}}(t)$, i.e. with $\tau/T = 0.01$, we get only 3.6° of phase imbalance between their fundamental tones (see (6.130)). Assuming for the sake of simplicity no gain imbalance, we then get in that case that the IRR achieved when using the fundamental tone for the frequency conversion of the wanted signal sideband, i.e. IRR_1, is of the order of −30 dB. Such a value, in line with the curve shown in Figure 6.33, is a very common order of magnitude for practical RF/analog implementations. But if we now consider using the fifth harmonic of the LO signal to carry out the frequency conversion of the wanted sideband, we get that the phase imbalance reaches 18°. Such imbalance leads to an IRR_5 value of around −15 dB. On top of the conversion gain loss experienced by the wanted signal, this IRR degradation is an additional argument for considering frequency planning relying on the use of the fundamental tone of the LO signal to implement a frequency conversion.

Conversely, the even order tones exist only if at least one of the duty cycles of the P and Q LO components is different from 0.5. Referring to the discussion in "Duty cycle" earlier in this section, we get that in the particular case where no DtyCy imbalance occurs between the P and Q branches, i.e. when $\delta T_\mathrm{p} = \delta T_\mathrm{q} \neq 0.5$, the paired complex conjugate tones lying at $\pm 2l\omega_\mathrm{LO}$ have the same amplitudes. But when a DtyCy imbalance exists, we get an amplitude imbalance between those tones. This is the opposite behavior to that of the odd order harmonics discussed previously. In order to quantize this imbalance between paired complex conjugate tones of even order, we can define an even order rejection ratio (ERR) quantity, denoted by ERR_{2l}, that corresponds to the power ratio between them. This quantity can thus be seen as the transposition for the even order terms of IRR_{2l-1} defined for the odd order ones. Reconsidering equation (6.133), we see that if we want to keep a definition with the weakest tone power in the numerator, we need to consider the tone lying alternately at $+2l\omega_\mathrm{LO}$ and the one lying at $-2l\omega_\mathrm{LO}$ depending on the harmonic order. We may then define ERR_{2l} as

$$ERR_{2l} = \left(\frac{|(\beta_+^-)_{2l}|^2}{|(\beta_+^+)_{2l}|^2} \right)^{(-1)^{l-1}}. \tag{6.138}$$

Based on equation (6.135), we can write

$$(\beta_+^+)_{2l} = a_{\mathrm{p},2l}\frac{\left(1+(-1)^l g_{2l}\mathrm{e}^{+\mathrm{j}(\delta\phi_{2l}+\pi/2)}\right)}{2}, \tag{6.139a}$$

$$(\beta_+^-)_{2l} = a_{\mathrm{p},2l}\frac{\left(1-(-1)^l g_{2l}\mathrm{e}^{-\mathrm{j}(\delta\phi_{2l}+\pi/2)}\right)}{2}. \tag{6.139b}$$

We therefore obtain

$$ERR_{2l} = \frac{1+2g_{2l}\sin(\delta\phi_{2l})+g_{2l}^2}{1-2g_{2l}\sin(\delta\phi_{2l})+g_{2l}^2}, \tag{6.140}$$

with $\delta\phi_{2l}$ and g_{2l} still given by equations (6.130) and (6.132b), respectively. The corresponding achieved ERR is shown in Figure 6.36 as a function of the gain and phase imbalance. In the present case, the gain and phase imbalance to consider depends on the order of the harmonic under consideration. However, we see from this figure that even for high imbalance values we have an ERR that remains weak, of just a few decibels for practical implementations. This is a major difference compared to the odd order terms discussed previously. The root cause for this behavior is that the cancellation of the paired complex conjugate tones in the

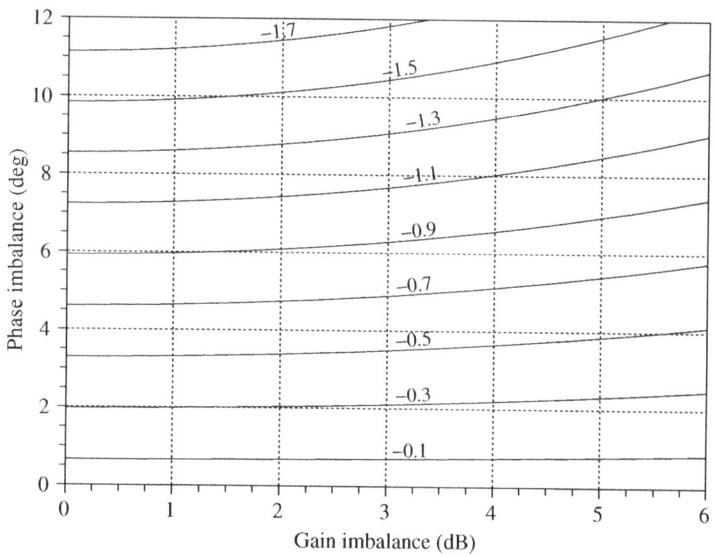

Figure 6.36 ERR due to gain and phase imbalance – The ERR defined by equation (6.140) gives the imbalance between the amplitudes of the paired complex conjugate tones of even order, i.e. lying at $\pm 2l\omega_{\mathrm{LO}}$, present in the series expansion of the reconstructed complex LO waveform. This amplitude difference remains weak, even for large gain and phase imbalance between tones of the same order in the P and Q LO series.

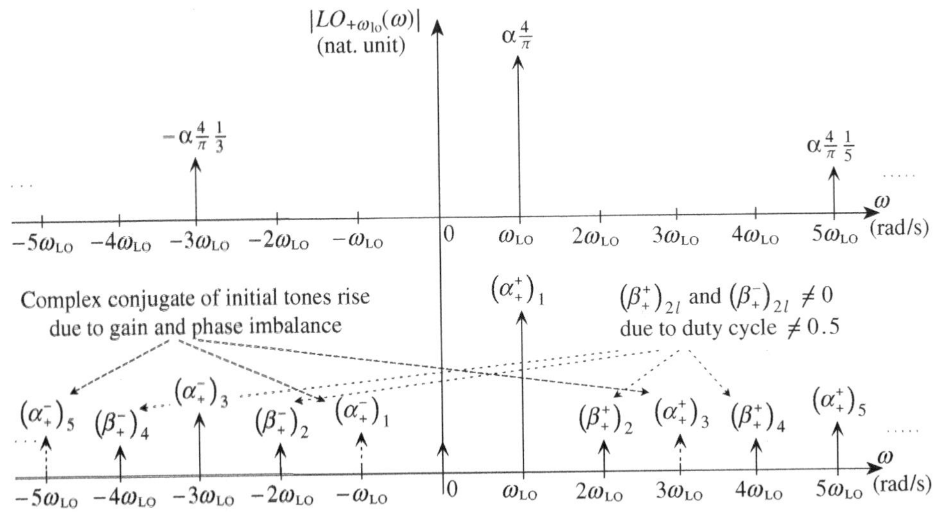

Figure 6.37 Degradation of the spectral content of the reconstructed complex periodic square LO waveform in the presence of general RF impairments – Compared to the ideal spectrum of the reconstructed complex LO signal used for chopper-like mixers (top, corresponding to equation (6.80)), the presence of RF impairments leads to a rise in additional tones (bottom, corresponding to equation (6.133)). Having duty cycles different from 0.5 leads to a rise in the even order harmonics, whereas the presence of imbalance between the P and Q LO components mainly leads to a rise in the complex conjugate of the odd order tones.

odd order case requires that the corresponding tones present on the P and Q LO signals have exactly the same amplitude. Any imbalance between the branches necessarily leads to an inexact cancellation. The amplitude of the resulting tone then increases rapidly. In the even order case we are not dealing with a destructive recombination but rather with a constructive summation. Any gain and phase imbalance between the P and Q branches only leads to weak amplitude variations for those even order tones during the reconstruction of the complex signal in practice.

In conclusion, Figure 6.37 obviously shows that the reconstructed complex LO signal deviates considerably from the pure complex exponential expected to be used to implement a complex frequency conversion. We have a collection of unwanted harmonic tones whose system impacts are discussed in more depth in Section 6.4.1. We have focused here on the mechanisms involved in the reconstruction of the complex LO signal $lo_{+\omega_{LO}}(t)$, i.e. centered around the positive complex exponential $e^{+j\omega_{LO}t}$. However, the conclusions remain exactly the same for the alternative reconstructed LO signal $lo_{-\omega_{LO}}(t)$ centered around $e^{-j\omega_{LO}t}$. Based on the fact that $lo_{-\omega_{LO}}(t) = lo^*_{+\omega_{LO}}(t)$, the spectra of these two signals are simply symmetric to each other with respect to the zero frequency due to the Fourier transform property given by equation (1.8). We also necessarily get that the coefficients involved in the series expansion of these two complex signals are complex conjugate to each other, as the complex exponentials they are weighting are. By equation (6.133), we have

$$lo_{-\omega_{\text{LO}}}(t) = 2\alpha\left(\frac{\delta T_{\text{p}}}{T} - \text{j}g\frac{\delta T_{\text{q}}}{T}\right) + \sum_{l=1}^{\infty}\Big[(\alpha_{-}^{-})_{2l-1}\text{e}^{-\text{j}(2l-1)\omega_{\text{LO}}t} + (\alpha_{-}^{+})_{2l-1}\text{e}^{+\text{j}(2l-1)\omega_{\text{LO}}t}$$
$$+ (\beta_{-}^{-})_{2l}\text{e}^{-\text{j}2l\omega_{\text{LO}}t} + (\beta_{-}^{+})_{2l}\text{e}^{+\text{j}2l\omega_{\text{LO}}t}\Big], \tag{6.141}$$

with $(\alpha_{-}^{-})_{2l-1} = (\alpha_{+}^{+})_{2l-1}^{*}$, $(\alpha_{-}^{+})_{2l-1} = (\alpha_{+}^{-})_{2l-1}^{*}$, $(\beta_{-}^{-})_{2l} = (\beta_{+}^{+})_{2l}^{*}$, and $(\beta_{-}^{+})_{2l} = (\beta_{+}^{-})_{2l}^{*}$. With the IRR and ERR defined through the magnitude of these coefficients, we thus necessarily recover the same metrics in both cases.

Slew Rate

As we have seen, the signal processing associated with mixing stages implemented as choppers corresponds to the multiplication of the signal being converted by a periodic LO signal with a square waveform. But from the various series expansions derived so far we see that the spectral content of such LO signal spreads toward infinity. Obviously, this behavior remains purely theoretical as any physical analog implementation necessarily exhibits a finite passband.

In the present case of interest, this finite passband is related to the capacitive load seen by the LO driver in practical electronic physical implementations. The impact of this load can in turn be interpreted in terms of lowpass filtering of the ideal square wave LO signal. Alternatively, we can also face a slew rate effect resulting from the finite drive capability of the LO driver. In that case, the LO waveform results from the non-vanishing rise and fall time linked to this finite drive capability. However, although we are dealing here with a different phenomenon, the consequences are the same for the characteristics of the LO signal. As illustrated in Figure 6.38, there is indeed an equivalence between the reduction in the level of the harmonics present in the LO signal spectrum and its finite slope in the time domain. This intuitive result is related to Bernstein's theorem, which states that any signal $s(t)$ that is bandlimited, i.e. whose Fourier transform in the frequency domain vanishes outside $[-F, +F]$, and that satisfies $|s(t)| \leq M$ for all t, fulfills [2]

$$\left|\frac{\text{d}s(t)}{\text{d}t}\right| \leq 2\pi FM. \tag{6.142}$$

Consequently, in practical implementations the operation experienced by the signal being processed during a frequency conversion consists in its multiplication with a lowpass filtered version of the theoretical square wave LO signal considered so far.

Obviously, this behavior leads to a reduction of the harmonic tones present in the decomposition of the LO signals. From the system point of view, we can anticipate the discussion in Section 6.4.1 and say that we may consider this reduction in the harmonics level as good news. It allows us to better approximate the operation of the ideal frequency conversion that relies on the use of pure sinusoidal LO signals. Unfortunately, considering the physical implementation of choppers using transistor devices, such degradation in the slopes of the LO signal often results in a degradation of the performance of the mixing stage, for instance in terms of

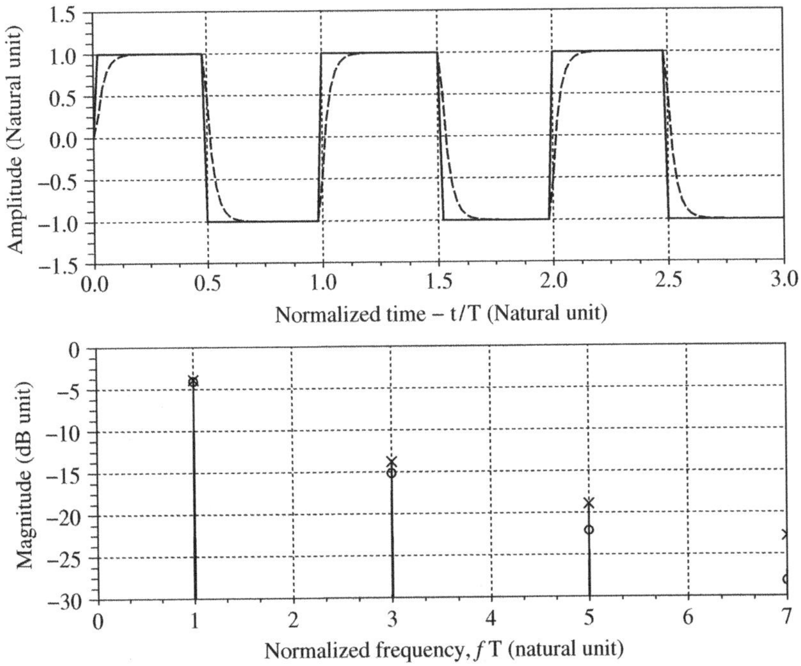

Figure 6.38 Periodic square LO waveform and its lowpass filtered version represented both in the time and frequency domains – The lowpass filtered version of a periodic square signal with period T exhibits a finite rise and fall time (top, dashed vs. solid). This is the result in the time domain of the attenuation in the frequency domain of its high frequency components (bottom, circles vs. crosses). In the present case, the lowpass filter used is a first order one – which explains why we experience only rise and fall times and not overshoot for the filtered time domain waveform – with a cut-off frequency of $5/T$ at 3 dB.

propagation of the LO noise component or in terms of linearity. There is thus a willingness in practice to work with LO signals that approximate the square waveform.

For other applications, however, it may be desirable to reduce the spectral content of such a square wave periodic signal where possible. This holds for instance for digital or data clocks where some tolerance is allowed in the physical implementation of the corresponding digital blocks. As highlighted in Section 6.4.2, this can minimize the electromagnetic interference (EMI) resulting from the clock harmonics coupling toward the RF paths of a transceiver line-up.

6.4 Frequency Planning

Frequency planning is concerned with how to decide on the period of all the periodic signals that are required for the physical implementation of a transceiver. From our discussion in this chapter so far, the LO signals used for the implementation of RF/analog frequency conversions are obviously among the signals to be considered. But in practical transceivers we often need to consider many other periodic signals even if of lower angular frequency in most cases. We can for instance mention the signals used as references by RF synthesizers, the digital clocks used to drive synchronous logic blocks, or even the data clocks that can drain a non-negligible amount of current when used for inter-chip data transfer.

These periodic signals are for the most part implemented as square waves. Recalling the discussion in Section 6.3, harmonics are necessarily present in their spectra. From the system point of view, the presence of these harmonics can lead to different problems that we can discuss now. This highlights the need to select the period of those signals so as to avoid, as far as possible, unwanted harmonic tones lying in critical frequency bands.

6.4.1 Impact of the LO Spectral Content

Let us focus first on the impact of the harmonic tones present in the square wave LO signals used to trig mixers implemented as choppers. Staying with the complex frequency conversion case, we look first at the transmit side and then the receive side.

Transmit Side

We again consider the upconversion of the complex lowpass signal $\tilde{s}(t) = p(t) + jq(t)$ using on the one hand mixers implemented as choppers, and on the other hand the reconstructed complex LO signal $lo_{+\omega_{\mathrm{LO}}}(t) = lo_{\Pi,\mathrm{p}}(t) + jlo_{\Pi,\mathrm{q}}(t)$ centered on the positive complex exponential $e^{+j\omega_{\mathrm{LO}}t}$. In the perspective of illustrating the system impacts of the LO signal waveform, we can examine the structure of the spectrum of the RF bandpass signal $s_{\mathrm{RF}}(t)$ recovered at the output of the frequency upconversion.

Based on the discussion so far in this chapter, the structure of the frequency upconversion reduces to that shown in Figure 6.39. We can thus express $s_{\mathrm{RF}}(t)$ as

$$\begin{aligned} s_{\mathrm{RF}}(t) &= \mathrm{Re}\left\{\tilde{s}(t) lo_{+\omega_{\mathrm{LO}}}(t)\right\} \\ &= p(t) lo_{\Pi,\mathrm{p}}(t) - q(t) lo_{\Pi,\mathrm{q}}(t). \end{aligned} \tag{6.143}$$

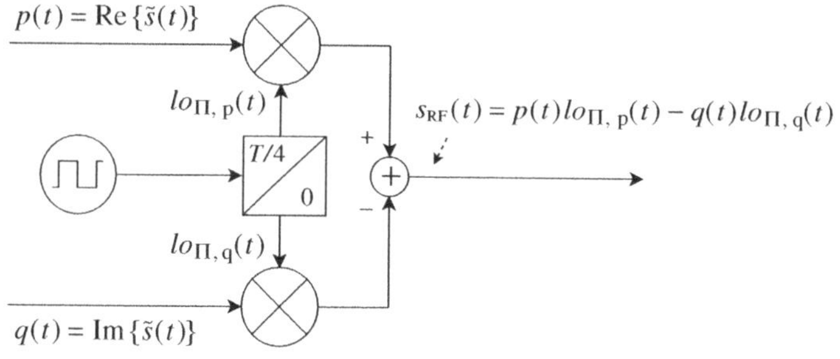

Figure 6.39 Implementation of a complex frequency upconversion using chopper-like mixers – Using mixers implemented as choppers, the effective LO waveforms experienced by the signal being processed are periodic square waves with a delay offset of $T/4$ between the P and Q branches. In the presence of RF impairments, the typical spectrum recovered at the output of the upconversion is shown in Figure 6.40(bottom).

To discuss the sidebands present in the spectrum of this bandpass signal, we can first use (1.5) to expand the real part involved in this expression:

$$\begin{aligned} s_{\mathrm{RF}}(t) &= \mathrm{Re}\left\{\tilde{s}(t) lo_{+\omega_{\mathrm{LO}}}(t)\right\} \\ &= \frac{1}{2}\left\{\tilde{s}(t) lo_{+\omega_{\mathrm{LO}}}(t) + \tilde{s}^*(t) lo^*_{+\omega_{\mathrm{LO}}}(t)\right\}. \end{aligned} \tag{6.144}$$

Now using equations (1.10) and (1.8) in turn, we can derive the Fourier transform of a given realization of $s_{\mathrm{RF}}(t)$ as

$$S_{\mathrm{RF}}(\omega) = \frac{1}{2}\left\{\tilde{S}(\omega) \star LO_{+\omega_{\mathrm{LO}}}(\omega) + \tilde{S}^*(-\omega) \star LO^*_{+\omega_{\mathrm{LO}}}(-\omega)\right\}. \tag{6.145}$$

In our case of interest, the structure of $lo_{+\omega_{\mathrm{LO}}}(t)$ has been derived in "Practical LO spectral content" (Section 6.3.2). In particular, its series expansion in terms of complex exponentials is given by equation (6.133). Given that the Fourier transform of a complex exponential of a given angular frequency is the Dirac delta distribution centered around that angular frequency, we can interpret $LO_{+\omega_{\mathrm{LO}}}(\omega)$ as the distribution signal [2]

$$\begin{aligned} LO_{+\omega_{\mathrm{LO}}}(\omega) = 2\alpha &\left(\frac{T_{\mathrm{p}}}{T} + jg\frac{T_{\mathrm{q}}}{T}\right) \\ &+ \sum_{l=1}^{\infty}[(\alpha^+_+)_{2l-1}\delta(\omega - (2l-1)\omega_{\mathrm{LO}}) + (\alpha^-_+)_{2l-1}\delta(\omega + (2l-1)\omega_{\mathrm{LO}}) \\ &+ (\beta^+_+)_{2l}\delta(\omega - 2l\omega_{\mathrm{LO}}) + (\beta^-_+)_{2l}\delta(\omega + 2l\omega_{\mathrm{LO}})] \end{aligned}$$

plotted in Figure 6.37. The structure of the spectrum of $s_{\mathrm{RF}}(t)$ can then be interpreted in a straightforward way. Given that the convolution with a Dirac delta distribution of the form $\delta(\omega - \omega_0)$ results in a shift of magnitude ω_0, the spectrum of $\tilde{s}(t) lo_{+\omega_{\mathrm{LO}}}(t)$ then reduces to a collection of copies of the expected sideband, but shifted in frequency by all the complex exponentials present in the decomposition of $lo_{+\omega_{\mathrm{LO}}}(t)$. As illustrated in Figure 6.40, these sidebands are then directly centered around the angular frequencies of the complex exponentials when considering an input lowpass signal $\tilde{s}(t)$ centered around DC. We then get that the output RF bandpass signal $s_{\mathrm{RF}}(t)$ is equal to the real part of this upconverted complex signal. Consequently, its spectrum results from the superposition of that of $\tilde{s}(t) lo_{+\omega_{\mathrm{LO}}}(t)$ and a flipped copy of it, according to equation (6.145). The symmetrical structure resulting from this superposition indeed leads to the Hermitian symmetry required for the spectrum of this real bandpass signal.

We thus see that the presence of the sidebands lying at multiples of the LO fundamental tone angular frequency ω_{LO} can be responsible for unwanted spurious emissions. From the system point of view, how we handle these sidebands depends on the frequency planning used for the frequency conversion. More precisely, if ω_{LO} is sufficiently high, we get that they are far away enough from the wanted sideband in the frequency domain so that their attenuation through an RF filtering stage, even if costly in terms of implementation, remains possible. This is obviously not possible when ω_{LO} is low. At first glance, it thus appears that the frequency planning is straightforward in that case, as selecting ω_{LO} sufficiently high prevents any problems. There is, however, another potential issue, i.e. the frequency pulling or injection locking of the RF

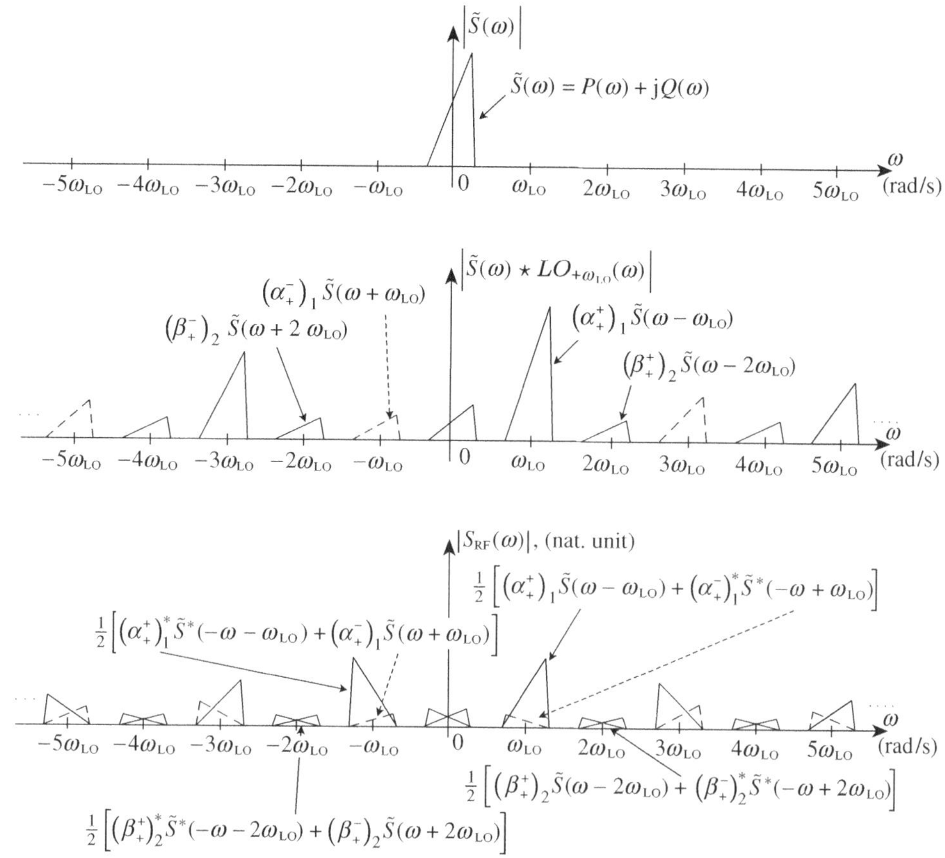

Figure 6.40 Typical spectrum recovered at the output of a complex frequency upconversion implemented using chopper-like mixers – The upconversion of a lowpass complex signal (top) using choppers and in the presence of impairments involves a complex LO signal that exhibits unwanted tones, as shown in Figure 6.37. This results in a copy of the spectrum of the input signal (middle), and then a symmetrization of the spectrum to achieve a real RF bandpass modulated signal (bottom).

oscillator used to generate the LO signal. Indeed, as discussed in Section 8.1.1, for the direct conversion transmitter architecture, the oscillation angular frequency of this RF oscillator must be an integer multiple of ω_{LO}. Consequently, when considering an input lowpass signal $\tilde{s}(t)$ centered around DC, there is necessarily an unwanted sideband present in the spectrum of $s_{RF}(t)$ that lies at this particular oscillation angular frequency. Dealing with high power signals on the transmit side, pollution of the RF oscillator may occur that can potentially destroy the radio link.

In order to avoid this issue, we thus may want to change the frequency planning of the upconversion. We might perhaps think of carrying out this processing in two steps, as in the classical heterodyne transmit architecture discussed in Section 8.1.2. This should indeed solve this integer relationship, at least for the second upconversion stage, and thus the associated pulling issue. But we can anticipate that all the sidebands generated by the first upconversion stage are in turn converted by all the harmonics present in the LO waveform used in the second stage. Consequently, having a first LO angular frequency equal to ω_{LO1} and a second one equal to ω_{LO2}, we can expect the presence of sidebands lying at the output of the line-up at frequency offsets equal to $|l\omega_{LO1} + m\omega_{LO2}|$, with l and m in $\mathbb{Z}^*$, from the wanted sideband. Some of them can thus lie so close to the wanted signal that they become problematic to handle in terms of filtering. The selection of the successive LO angular frequencies can be tricky in practice, thus highlighting the importance of the frequency planning in that case [72].

Receive Side

Let us now consider the receive side through the complex frequency downconversion of the RF bandpass signal $s_{RF}(t)$ centered around ω_{RF}, using the reconstructed complex LO signal $lo_{-\omega_{LO}}(t) = lo_{\Pi,p}(t) + jlo_{\Pi,q}(t)$ centered on the negative complex exponential $e^{-j\omega_{LO}t}$. Dealing with mixers implemented as choppers, both $lo_{\Pi,p}(t)$ and $lo_{\Pi,q}(t)$ are periodic square wave signals in quadrature so that the equivalent model from the signal processing point of view of the complex mixing stage we consider is shown in Figure 6.41.

Thus, we can write the reconstructed complex IF signal $s_{IF}(t)$ recovered at the output of the mixing stage as

$$s_{IF}(t) = s_{RF}(t)lo_{-\omega_{LO}}(t). \tag{6.146}$$

The spectral content of this signal can then be derived in a straightforward way by considering the Fourier transform of a given realization of the modulation process associated with $s_{RF}(t)$. Using equation (1.10), we can write

$$S_{IF}(\omega) = S_{RF}(\omega) \star LO_{-\omega_{LO}}(\omega). \tag{6.147}$$

In our case of interest, the structure of $lo_{-\omega_{LO}}(t)$ has been derived in "Practical LO spectral content" (Section 6.3.2). Its series expansion, given by equation (6.141), is then simply the complex conjugate of that of $lo_{+\omega_{LO}}(t)$ used in the previous section as an example to illustrate

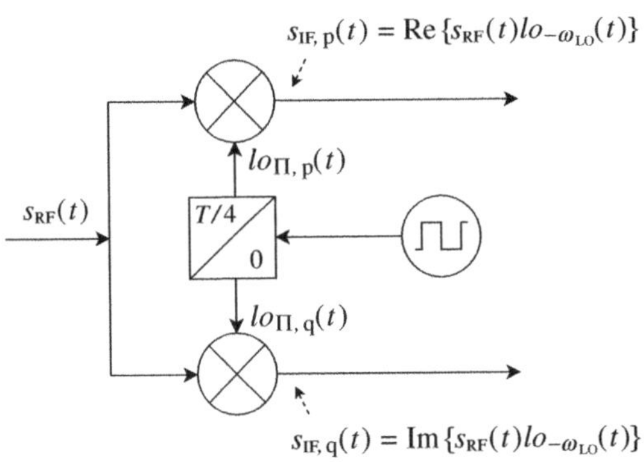

Figure 6.41 Implementation of a complex frequency downconversion using chopper-like mixers – Using mixers implemented as choppers, the effective LO waveforms experienced by the signal being processed are periodic square waves with a delay offset of $T/4$ between the P and Q branches. In the presence of RF impairments, the typical spectrum recovered at the downconversion output is shown in Figure 6.42(bottom).

the transmit side. Using the same approach, we can thus interpret $LO_{-\omega_{LO}}(\omega)$ as the distribution signal

$$LO_{-\omega_{LO}}(\omega) = 2\alpha\left(\frac{T_p}{T} - jg\frac{T_q}{T}\right)$$
$$+ \sum_{l=1}^{\infty}[(\alpha_-^-)_{2l-1}\delta(\omega + (2l-1)\omega_{LO}) + (\alpha_-^+)_{2l-1}\delta(\omega - (2l-1)\omega_{LO})$$
$$+ (\beta_-^-)_{2l}\delta(\omega + 2l\omega_{LO}) + (\beta_-^+)_{2l}\delta(\omega - 2l\omega_{LO})].$$

The structure of the spectrum of the reconstructed complex IF signal $s_{IF}(t)$ can then be interpreted in a straightforward way. Indeed, the convolution with a Dirac delta distribution of the form $\delta(\omega - \omega_0)$ results in a shift of magnitude ω_0. The spectrum of $s_{IF}(t)$ is then the superposition of the copies of the input RF spectrum $S_{RF}(\omega)$ generated by the multiplication of $s_{RF}(t)$ with all the complex exponentials present in the decomposition of $lo_{-\omega_{LO}}(t)$. But based on the discussion in Section 3.3.1, many unwanted signals can be present on top of the wanted signal at the antenna connector, thus resulting in classical power profiles in the frequency domain, as illustrated in Figure 3.16. But assuming such a profile at the input of the conversion stage, there is necessarily an unwanted signal folded on the wanted signal by each of the complex exponentials present in the decomposition of $lo_{-\omega_{LO}}(t)$. In the present case where these unwanted foldings result from the presence of the harmonic tones in the LO signal, we often talk about an harmonic mixing issue. This behavior is illustrated in Figure 6.42 in the particular case where $\omega_{LO} = \omega_{RF}$. However, it holds whatever the frequency planning of the conversion.

Obviously, the folding of the unwanted signal sideband results in a degradation of the SNR and thus of the radio link. For a given frequency planning, i.e. for a given choice of ω_{LO}, such degradation occurs for particular carrier frequencies of the unwanted signals. We classically say that there are spurious responses of the receiver at those particular RF frequencies. By our discussion so far, the ability to minimize this degradation depends on the relative magnitude of the complex exponentials involved in the series expansion of $lo_{-\omega_{LO}}(t)$. However, depending on the frequency planning done, we may also rely on additional RF filtering to attenuate those unwanted signals prior to the downconversion. The higher the LO angular frequency ω_{LO}, the wider the distance between the unwanted signals and the wanted signal in the frequency domain, and thus the easier this filtering. In practical implementations, such attenuation can result from the use of a dedicated RF filter, but also from the use of resonant systems as matching networks or inductive loads along the data path, for instance. We thus see that the spectral content of the LO waveform is not sufficient information for determining the receiver behavior in respect to those spurious responses. This leads to the use of dedicated metrics. This is, for instance, the case for the input spurious rejection (ISR) defined as the gain experienced by the unwanted signal lying at a spurious response frequency, relative to the gain experienced by the wanted signal itself.

6.4.2 *Clock Spurs*

Let us now say a few words about the potential impacts of the harmonic tones present in the clock signals used for the physical implementation of digital or mixed signal blocks. In most

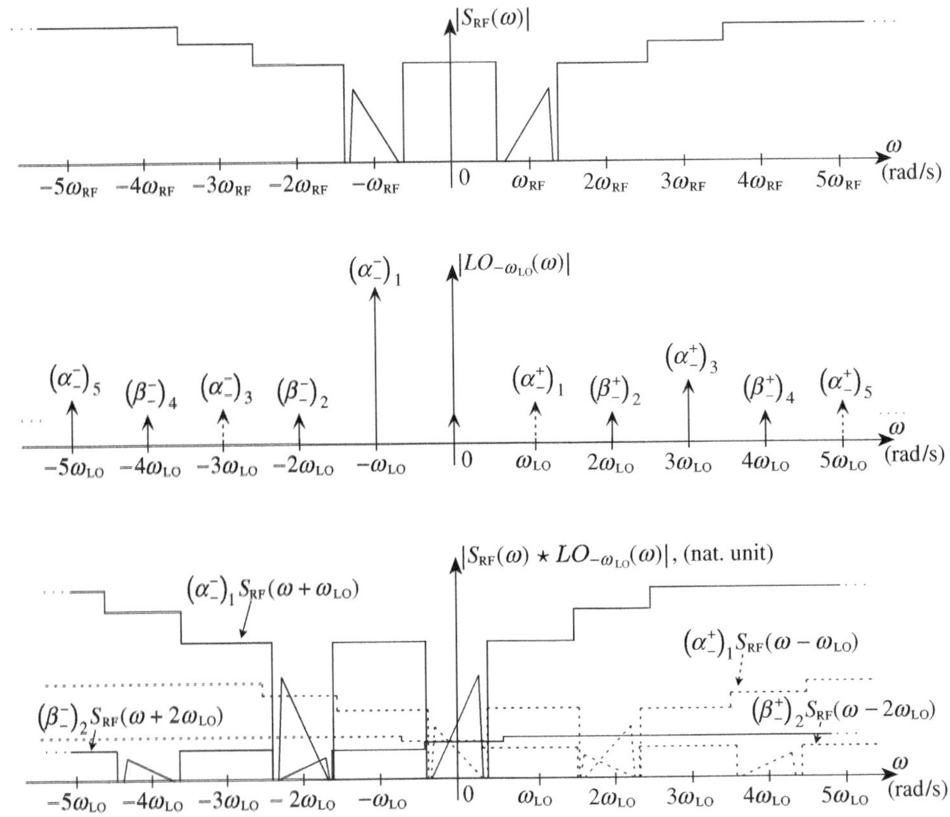

Figure 6.42 Typical spectrum recovered at the output of a complex frequency downconversion implemented using chopper-like mixers – The complex frequency downconversion of a real RF bandpass signal (top) when using choppers and in the presence of impairments involves a complex LO waveform that exhibits unwanted tones (middle). This results in the folding of potential blockers on the wanted signal by each of those unwanted tones (bottom). Here, only the folding due to the first two harmonics is shown for the sake of clarity.

cases there is a major difference compared to the LO signals discussed in the previous section as the angular frequency of those clock signals is often much lower. In practice, the difference may be an order of magnitude or more. The consequences on the spurs potentially recovered on the RF paths of the line-up due to electromagnetic coupling or direct feedthrough are then twofold.

Firstly, only the high order clock harmonics can lie in the RF system bands of interest. Since the power of those harmonics decreases as their order increases, we might at first think that the power of the clock spurs of interest can be sufficiently low that the system impacts remain weak. This is unfortunately not the case in practice, for instance during the reception of a wanted signal near sensitivity. In that case, the received signal is sufficiently weak that a desensitization can take place when a harmonic lies in the wanted signal bandwidth, even with such high order. Alternatively, these clock spurs can also be recovered on the data path of transmitters. As stringent requirements often exist in terms of SEMS, we can also be

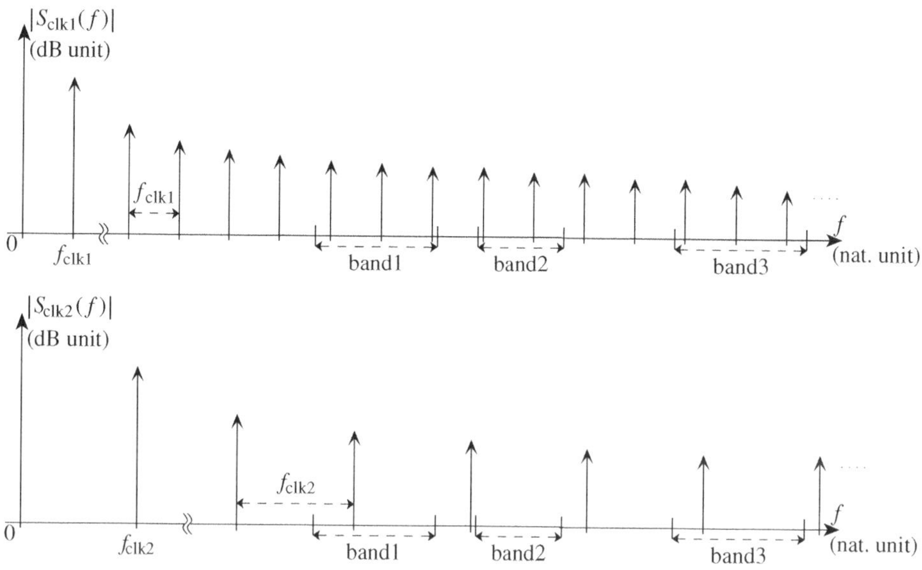

Figure 6.43 Harmonics of clock signals as spurs lying in RF system bands – Clock signals are for the most part implemented with a square waveform. They thus contain harmonics with a slowly decreasing amplitude as the order increases. More harmonics of a low frequency clock signal are recovered in a given RF system band (top) than of a high frequency one (bottom).

faced with problematic spurious emissions in that case. Finally, we can also imagine that the harmonics couple toward the RF oscillator used for the generation of the LO signal. We can then deal with the generation of a spurious tone in the spectrum of this signal. This can result in either a potential degradation of the transmit spectrum when this LO is used on the transmit side, or additional spurious responses when it is used on the receive side.

Many harmonics of these clocks can fall within the overall receiver or transmitter system bands of interest. Figure 6.43 shows that the higher a clock frequency, the fewer potential problems we may foresee. However, for various reasons, such as the power consumption of the solution, there is often a willingness to keep clock frequencies low enough in practice. We can thus understand the importance of frequency planning in that case. Unfortunately, for a transceiver that is expected to support many RF system bands, as is often the case in modern wireless system implementations, we can expect some difficulties in finding one common frequency that is simultaneously suitable for all the possible RF system bands. Strategies thus need to be found to overcome potential clock spur issues in practice, from the handling of different clock frequencies at the implementation level, to the possibility of varying the DtyCy or the rise and fall time of the clock signals in order to tune their spectral content. More systematic strategies can also be considered.

6.5 DC Offset and LO Leakage

The final aspect linked to RF impairments that we discuss in this chapter covers the possible existence of either a DC offset or an LO leakage along the RF/analog paths of a transceiver.

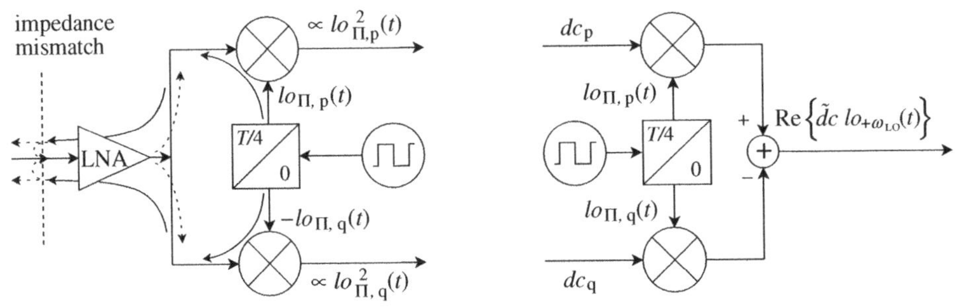

Figure 6.44 Relationship between LO leakage and DC offset in receive and transmit line-ups – On the receive side (left), the LO leakage can reach the input of the receiver by various coupling mechanisms. Any reflected fraction of this signal is therefore self-mixed and produces a DC offset at the downmixing stage output. On the transmit side (right), any DC offset present at the upmixer input leads to an LO leakage component at its output.

We point out here that the DC offset is a baseband quantity, while the LO leakage belongs to the RF world. So why consider those two phenomena together, since they necessarily corrupt the wanted signal at different stages of a line-up? The reason is that from the signal processing point of view, there is up to a point an equivalence between the two phenomena, as illustrated in Figure 6.44. The frequency upconversion of a DC offset results in a signal proportional to the LO signal that is used to drive the mixing stage. Conversely, a self-mixing of the LO signal leads to a DC offset component.

This relationship between the DC offset and LO leakage is important. For example, it allows some strategies to cancel the LO leakage on the transmit side, as discussed in more depth in Chapter 9. In what follows we discuss the causes of DC offset and LO leakage and the corresponding system impacts, giving separate consideration to the transmit and receive sides, and referring to the complex frequency conversion operations shown in Figure 6.44.

6.5.1 LO Leakage on the Transmit Side

We first consider the LO leakage on the transmit side. In most physical implementations, the main root causes for its existence are twofold. We get on the one hand the DC offset that is inherently present at the input of the mixing stage, and on the other hand the direct coupling of the LO signal to the output of the mixing stage due either to some electromagnetic coupling between RF paths or to a direct feedthrough at the device parasitic level. Based on these two phenomena, we can derive an analytical expression for the resulting LO leakage.

Let us first take into account the two DC offsets, dc_p and dc_q, often present respectively on the P and Q branches of the complex upconverter. As these offsets are linked to mismatch in the physical implementation of different baseband devices, they can take different values in practice. Then, we need to take into account the fraction of the $lo_{\Pi,\mathrm{p}}(t)$ and $lo_{\Pi,\mathrm{q}}(t)$ signals directly coupled to the mixer output. But for mixers implemented as choppers, these signals are necessarily periodic square waves (see Section 6.3.1). The spectral extent of the signals is therefore very wide. In practical implementations, the coupling factors experienced by the LO signals can depend on the frequency. This may lead to different weights for the harmonics present in the series expansion of the signals depending on their order. Consequently, we may be faced with a distortion of the waveform of the LO leakage recovered at the output of the mixing stage compared to that of the original LO signals. Let us focus on the tone of interest used for the frequency conversion of the wanted signal. As done in most implementations, we can assume that it is the fundamental tone of the LO signals that is used. As a result, denoting by $lo_{\mathrm{p}}(t)$ and $lo_{\mathrm{q}}(t)$ the P and Q components of this fundamental tone and assuming a structure for the complex upconverter like that shown in Figure 6.44(right), we can write the LO leakage $s_{\mathrm{leak}}(t)$ recovered at the output of the complex upconverter as

$$\begin{aligned} s_{\mathrm{leak}}(t) &= (dc_{\mathrm{p}} + \kappa_{\mathrm{p}})lo_{\mathrm{p}}(t) - (dc_{\mathrm{q}} + \kappa_{\mathrm{q}})lo_{\mathrm{q}}(t) \\ &= \mathrm{Re}\left\{\tilde{s}_{\mathrm{leak}}(t)e^{+j\omega_{\mathrm{LO}}t}\right\}, \end{aligned} \tag{6.148}$$

with κ_{p} and κ_{q} the coupling factors existing on the P and Q branches of the complex mixer at the LO fundamental angular frequency ω_{LO}. Based on this expression, we can then write the complex envelope of $s_{\mathrm{leak}}(t)$, defined as centered around the LO angular frequency ω_{LO}, as

$$\tilde{s}_{\mathrm{leak}}(t) = dc_{\mathrm{p}} + \kappa_{\mathrm{p}} + j(dc_{\mathrm{q}} + \kappa_{\mathrm{q}}), \tag{6.149}$$

or in polar form as

$$\tilde{s}_{\mathrm{leak}}(t) = \rho_{\mathrm{leak}} e^{j\phi_{\mathrm{leak}}}, \tag{6.150}$$

with

$$\rho_{\mathrm{leak}} = \sqrt{(dc_{\mathrm{p}} + \kappa_{\mathrm{p}})^2 + (dc_{\mathrm{q}} + \kappa_{\mathrm{q}})^2}, \tag{6.151a}$$

$$\phi_{\mathrm{leak}} = \arctan\left(\frac{dc_{\mathrm{q}} + \kappa_{\mathrm{q}}}{dc_{\mathrm{p}} + \kappa_{\mathrm{p}}}\right). \tag{6.151b}$$

Using this expression in equation (6.148), we can express $s_{\mathrm{leak}}(t)$ as

$$s_{\mathrm{leak}}(t) = \rho_{\mathrm{leak}} \cos(\omega_{\mathrm{LO}}t + \phi_{\mathrm{leak}}). \tag{6.152}$$

We thus see that whatever the mechanisms involved in the generation of $s_{\mathrm{leak}}(t)$, this signal remains a pure CW tone at the angular frequency ω_{LO} in any case. The DC offsets and coupling factors only impact the characteristics of this leakage, and not its structure.

Suppose now that we are dealing with the frequency upconversion of a baseband complex modulating waveform $\tilde{s}(t)$ using such a non-ideal upconverter. The RF bandpass signal $s_{\mathrm{RF}}(t)$ effectively recovered at the output of the system is the sum of the wanted modulated bandpass signal $s(t)$ and the LO leakage $s_{\mathrm{leak}}(t)$. This result is related to the distributive behavior of the mixing operation. The complex envelope of $s_{\mathrm{RF}}(t)$, when defined as centered around ω_{LO}, is the sum of $\tilde{s}(t)$ and $\tilde{s}_{\mathrm{leak}}(t)$. In the case where $\tilde{s}(t)$ is centered around DC, as encountered for instance in the direct conversion transmit architecture discussed in Section 8.1.1, the LO leakage and the wanted modulated RF bandpass signal are necessarily superposed in the frequency domain. The presence of this LO leakage thus corresponds to a constant offset on

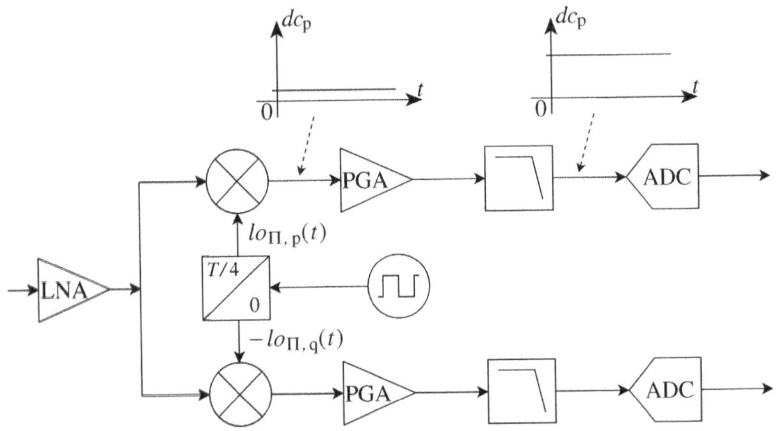

Figure 6.45 Saturation issue along the analog baseband paths of receivers due to DC offset – Classical values for both amplifier offsets and analog baseband gains can lead to saturation along the line-up if nothing is done.

both the P and Q components of $\tilde{s}(t)$. Obviously, the corresponding system impacts depend on the nature of the P and Q signals. But assuming that we are dealing with simple modulating waveforms resulting from the direct mapping of data bits on symbols, we can expect an origin offset error on the modulation scheme as illustrated in "Origin offset suppression" (Section 3.2.2). In practice, we often need to fulfill an origin offset suppression requirement in order to achieve the overall performance in respect of this problem. There are algorithms that can be used to minimize the LO leakage, as discussed in Section 9.1.2.

6.5.2 DC Offset on the Receive Side

Let us now consider the DC offset inherently present along the analog baseband paths of a receiver. In practical physical implementations, this offset mainly results from the mismatches in the implementation of the baseband amplifiers. For instance, an imbalance between the positive and negative branches of a differential amplifier results in a non-vanishing DC component on the differential signal it delivers. However, there are additional causes for this component, e.g. the even order nonlinearities present in the implementation, as discussed in Chapter 5, or even the self-mixing of the LO leakage evoked previously and illustrated in Figure 6.44.

In terms of system impacts, the reconstructed baseband complex signal recovered in the presence of DC offset necessarily exhibits an origin offset in the modulation trajectory. We thus recover the same behavior as in the direct upconversion case in the presence of LO leakage discussed in the previous section. However, unlike what is experienced on the transmit side, saturation problems may also occur in receive line-ups, as illustrated in Figure 6.45. In most practical implementations a non-negligible amount of gain is required in the analog baseband part of receivers in order to correctly scale the received signal at the input of the P and Q ADC stages. This is particularly true in architectures such as the direct conversion one which exhibits a limited amount of gain in RF, as discussed in Section 8.2.1. Assuming for instance that we are dealing with a receiver having 30 dB of analog baseband gain, and that we have around 30 mV of DC offset present at the input of the analog baseband amplifier, we get almost 950 mV of DC offset after amplification. This is already almost the maximum swing that some integrated solutions can support in practice. Compensation systems therefore need to be considered in order to minimize such offset. These can take the form of dedicated algorithms, as discussed in Section 9.2.2.

Part III

Transceiver Dimensioning

7

Transceiver Budgets

Having derived the minimum set of functionalities to be implemented in a transceiver in Part I of this book, as well as the degradations resulting from their physical implementations in Part II, we can now illustrate how to carry out the system design of a transceiver in order to fulfill a given set of requirements while taking into consideration the limitations in its physical implementation.

Two approaches can be followed. We can decide a priori on an architecture that is theoretically able to fulfill the different requirements we have for a transceiver. We can then carry out the budget of the line-up to refine the balance of the constraints between the different blocks of the line-up, and then check that we can achieve the desired performance. Alternatively, we can also imagine tuning the architecture itself in order to overcome the limitations of some constituent blocks that could prevent the achievement of some of the requirements.

In practice, these two approaches can be seen as the two sequences of an iterative process that leads to the optimization of the final implementation of a transceiver. However, for the sake of clarity we first illustrate how to budget a transceiver for a selected architecture. This is done in this chapter for the simplest architecture we can think of in order to carry out the signal processing required by most of the complex modulation schemes used in digital communications, i.e. the direct conversion scheme. Then, we can use the limitations encountered with this architecture to introduce in the next chapter alternative architectures that can overcome those limitations. Finally, in Chapter 9 we discuss classical algorithms that are often implemented in transceivers to improve their behavior.

7.1 Architecture of a Simple Transceiver

The example we consider is a transceiver composed of a direct conversion transmitter and a direct conversion receiver. This architecture is attractive for its simplicity, composed as it is of the minimum set of functions we need to consider to implement the signal processing required to up- or downconvert a modulating waveform, at least a complex one, as detailed in

Wireless Transceiver Architecture: Bridging RF and Digital Communications, First Edition. Pierre Baudin.

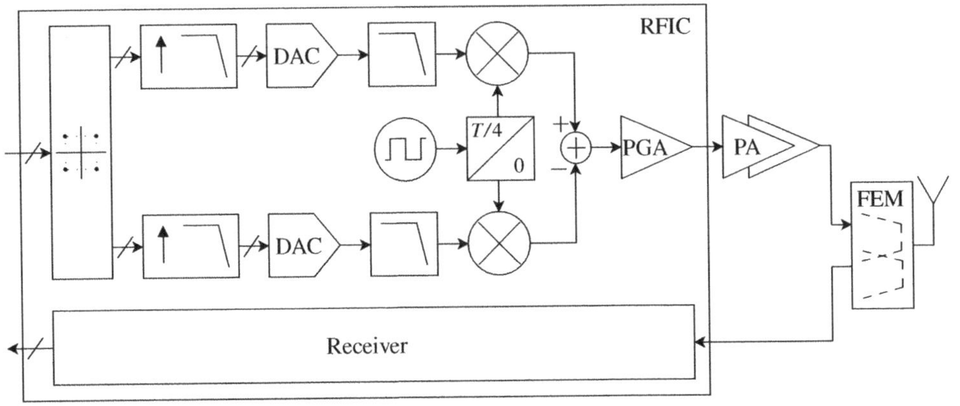

Figure 7.1 Direct conversion transmit line-up considered in this chapter – To illustrate how to budget a transmitter, we consider a simple direct conversion transmitter embedded in a full duplex transceiver. The receive path corresponds to the line-up shown in Figure 7.2.

Chapter 1. It thus allows us to highlight in a simple way the problems of system design and how to carry out the corresponding budgets.[1]

Recalling the architecture derived at the end of Chapter 1, we now consider some updates related not only to the required signal processing itself but also to either the theory of electromagnetism or the medium access strategy as discussed respectively in Chapters 2 and 3. For instance, given the need to deliver RF power to the antenna on the transmit side, we consider RF blocks able to perform this functionality along that line-up. In the same way, we also suppose that we are dealing with a full duplex transceiver, allowing simultaneous transmission and reception. From the system point of view, the main impact of this assumption is that we need to achieve a good enough performance for the receiver while having to deal with transmitter leakage at its input, as discussed in Section 3.1.3. Despite this interaction, we separately consider the transmit and receive line-up to carry out our budgets. The underlying assumption for this to be possible is that we can still budget the receive side by considering only a finite set of parameters that represent the impact of the transmitter on the quality of the reception. Obviously, this statement is not necessarily true from the point of view of programming and real time control of the transceiver as conflicts can occur if not carefully considered from the beginning. However, here we focus on the performance aspect of the transceiver, for which it is still meaningful to consider the two line-ups separately.

As a side effect, we do not have to spend time on the detail of the real time algorithms required to make the whole line-up work properly. Such algorithms, such as the AGC on the receive side or power control on the transmit side, are only mentioned on a case by case basis when necessary and addressed in greater depth in Chapter 9.

In Section 7.2 we consider the problem of budgeting a direct conversion transmit line-up embedded in a full duplex transceiver as illustrated in Figure 7.1. Then in Section 7.3 we

[1] This architecture is widely used in low cost solutions as it potentially allows a high level of integration in silicon while avoiding costly discrete or RF filters.

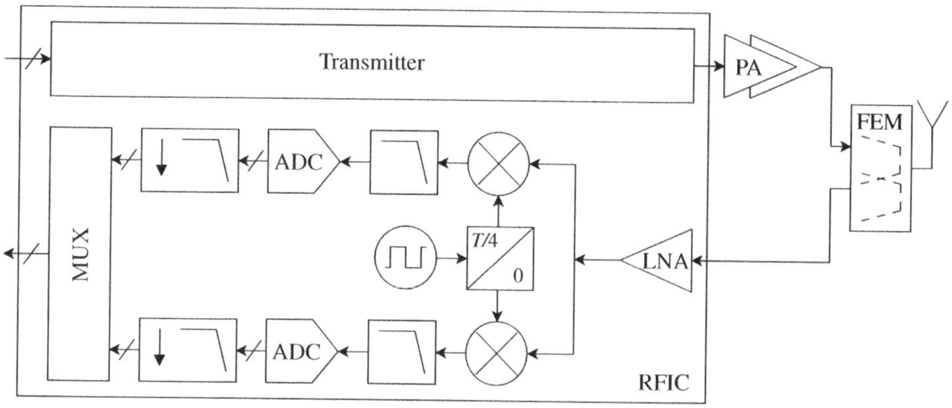

Figure 7.2 Direct conversion receive line-up considered in this chapter – To illustrate how to budget a receiver, we consider a simple direct conversion transmitter embedded in a full duplex transceiver. The transmit path corresponds to the line-up shown in Figure 7.1.

consider the receive side composed of a direct conversion receiver line-up embedded in the same full duplex transceiver as illustrated in Figure 7.2.

7.2 Budgeting a Transmitter

Let us first consider the transmit side. In order to understand how budgets need to be handled for the direct conversion transmitter considered, we first review the functions implemented in such a line-up with the aim of understanding what kind of degradations we have to deal with when trying to achieve the different requirements we have to fulfill. Having done so, we go through the derivation of practical budgets in Sections 7.2.3 and 7.2.4.

7.2.1 Review of the ZIF TX Problem

Even when ideally implemented, the signal processing functions used in the implementation of the direct conversion transmitter shown in Figure 7.1 may result in pollution of the other users in the wireless network. This is obviously exacerbated by the unavoidable limitations in their implementation as we then have to deal with a degradation of the modulation quality on top of the degradation of the out-of-band spectrum.

It is therefore useful to review the consequences of the use of the theoretical signal processing implemented in the line-up, as done in the next section, and then those linked to the limitations in their implementation as done in “Impact of implementation limitations” in the section after that. This approach clearly highlights the root causes in the degradations we are faced with in practice in the performance of the transmitter, and thus what parameters we need to consider for the derivation of our budgets.

The Signal Processing Approach

Let us review the signal processing functions embedded in the transmit line-up shown in Figure 7.1. Assuming that the functions we discuss are implemented in an ideal way, we can detail their impact on the bandpass modulated RF signal recovered at the transmitter output. We remark that even when dealing with functions implemented in an ideal way there is the potential for pollution of the other users in the network. As a result, an efficient way to examine the corresponding impacts is to look at the spectrum of the waveform generated. More precisely, we can look at the spectrum of the signals involved all along the transmit line-up in order to distinguish between the impacts of the different blocks involved in the line-up on the output waveform.

Looking at Figure 7.3, a number of comments are in order:

(i) We first of all get the generation in the digital domain of the real and imaginary parts of the complex modulating waveform. This can hardly be done differently due to the complexity of the functions required to generate those waveforms from the input data bits, as is the case in modern wireless systems. Thus, at the output of the symbol mapping and pulse shaping filtering, we recover the spectrum of this reconstructed complex lowpass modulating waveform, but in a periodic way as we are dealing with a sequence of digital samples, $p_{\text{low}}[k]$ and $q_{\text{low}}[k]$. The period of this spectrum is therefore given by the sampling frequency, $f_{s_{\text{low}}}$, used for the implementation of this processing. In practice, this sampling rate is chosen as low as possible in order to minimize the power consumption of the corresponding digital blocks.

(ii) Then, we often deal with an upsampling stage whose purpose is to deliver the samples $p[k]$ and $q[k]$ to the DAC at a higher sampling rate f_s than that used for the generation of the modulation samples. This means that $f_s > f_{s_{\text{low}}}$. The interest of such functionality appears when considering the constraints on the analog reconstruction filtering stage, i.e. the analog filter that necessarily follows the DAC, as discussed in "Filtering budgets" (Section 7.2.3).

(iii) Then the sequences of digital samples are provided to the P and Q DAC blocks that are driven at the same rate f_s as the sample rate used to compute the digital samples fed to them. As discussed in Section 4.6.2, from the signal processing point of view, for the most part such DACs behave as a simple sample and hold function with regard to the signal being processed. This results in a sinc transfer function experienced by the signal in the spectral domain through the aperture effect. As the zeros of this transfer function are located at the multiples of the sampling frequency f_s, we then get a natural attenuation of the copies in the spectrum of the output analog P and Q waveforms.

(iv) However, the attenuation of the digital copies provided by the DAC sinc transfer function is for the most part not efficient enough. As a result, we still need to consider an analog reconstruction filter that provides the residual required attenuation of the digital copies centered around the multiples of f_s. This results in a sufficiently clean spectrum for the waveforms $p(t)$ and $q(t)$ to be fed to the RF modulator.

(v) We then face the upconversion stage of the complex lowpass modulating waveform $p(t) + jq(t)$ through a complex mixer, which thus uses two physical mixers driven by LO waveforms in quadrature, as introduced in Chapter 1. In practice such mixing stages are implemented using devices that behave as choppers, as discussed in Section 6.3.1. Even

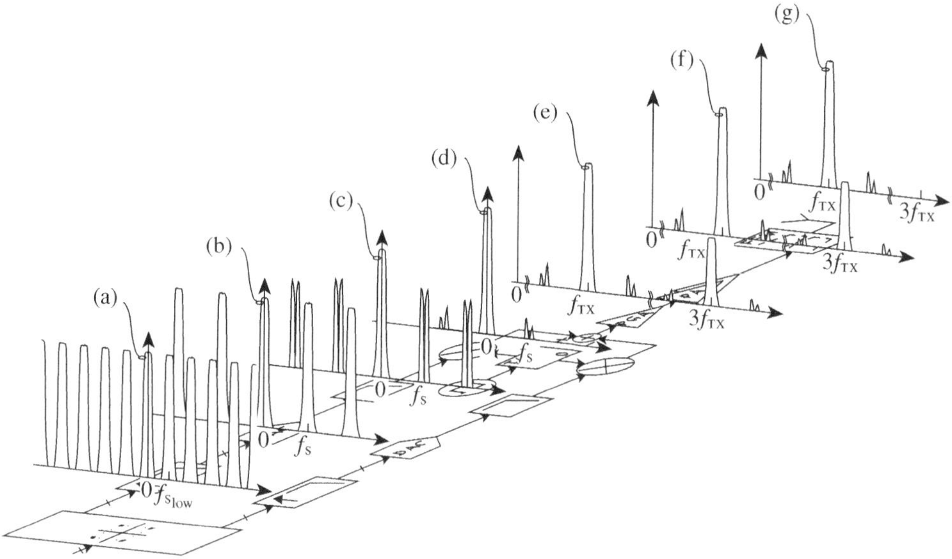

Figure 7.3 Spectrum of the signals recovered at the different stages of a direct conversion transmitter: ideal implementation case – At the output of the symbol mapping and pulse shaping filter, we recover samples of the real and imaginary parts of the modulating waveform, $p_{low}[k]$ and $q_{low}[k]$ respectively, sampled at a relatively low sampling rate $f_{s_{low}}$ in order to limit the power consumption of the corresponding digital blocks. This results in a periodic spectrum that exhibits a relatively low period equal to the sampling rate (a). In order to relax the analog filtering required to cancel the copies present in the periodic spectrum of this sampled waveform, an upsampling toward $f_s > f_{s_{low}}$ can be performed in order to increase the period of the spectrum (b). The DAC used to perform the generation of the $p(t)$ and $q(t)$ waveforms from the $p[k]$ and $q[k]$ samples for the most part behaves as sample and hold. It results in a sinc transfer function whose zeros are located at kf_s and thus provide a natural attenuation of the theoretical modulating waveform copies (c). An analog reconstruction filter is then used to provide the residual required attenuation of those copies (d). The modulating waveform is then upconverted using a complex mixer often implemented using chopper-like devices. This results in the upconversion of the expected sideband around the transmit frequency f_{TX}, but also of unexpected ones centered around the odd order harmonics $(2l+1)f_{TX}$ (e). RF amplifiers can then deliver the required RF power to the radiating element (f). However, a harmonic filter is often required to cancel the unwanted sidebands centered around the odd order harmonics $(2l+1)f_{TX}$ in order to deliver the expected RF signal (g).

if implemented in an ideal way, the use of square wave LO signals to drive such choppers for the direct upconversion toward the carrier frequency f_{TX} results in the generation of unwanted copies of the wanted sideband lying at the odd order harmonics of this frequency, i.e. around the frequencies of the form $(2l+1)f_{TX}$.

(vi) Finally, in order to deliver a modulated RF signal that has the expected spectral content, we often need to add some RF filtering in order to cancel those additional unwanted sidebands lying at the harmonics of the carrier frequency at the transmitter output. Such RF filtering is often located in an external front-end module (FEM), just before the antenna.

This signal processing shows that even when dealing with theoretical signal processing functions, we need to consider at least filtering blocks that allow us to ensure that we are fulfilling the requirements in terms of spectral pollution at the transmitter output. These effects thus have to be considered mainly in the filtering budget, as illustrated in "Filtering budgets" (Section 7.2.3). But on top of that, we now need to take into consideration the limitations in the physical implementation of these signal processing functions, as discussed in the next section.

Impact of Implementation Limitations

Let us now examine the additional degradations we have to take into account due to the limitations in the physical implementation of the signal processing functions detailed in the previous section. We can follow the same approach as in the ideal case and look at the spectral shape of the waveforms recovered along the line-up.

Comparing the spectra displayed in Figures 7.3 and 7.4, we can see that:

(i) Even on the digital side the waveforms we are dealing with are not the pure theoretical samples of the expected waveforms as illustrated in the previous section. Indeed, most of the realistic physical implementation of the digital signal processing is performed in a fixed point way, i.e. using a finite number of bits to represent the samples we are dealing with. As discussed in Section 4.5, under common assumptions, this finite precision leads to a quantization noise that is superposed on the signal being processed. Thus on top of the problematic periodic structure of the sampled waveform, we need to consider a noise contribution that impacts the overall spectrum of the sampled signal being processed. This results in both a wideband noise floor and an in-band SNR or EVM limitation.

(ii) This degradation is enhanced by the P and Q DAC entities that necessarily also have a finite resolution. Classically, this effective resolution is worse than that of the digital data path itself. This is because, for a given sampling frequency and a given DR, the design of a high precision DAC often remains harder to achieve and requires a higher power consumption than the high resolution digital signal processing functions required for the generation of the samples fed to it. It is therefore reasonable to assume that the bottleneck remains the DAC rather than the digital signal processing itself. From the signal processing point of view, the DAC adds an additional noise and distortion contribution in the line-up. But then both the converted signal and the noise contribution from the DAC experience a sample and hold transfer function in the time domain, i.e. a sinc transfer function on their spectrum.

(iii) This noise degradation is emphasized by the contributions of the forthcoming analog reconstruction filter that necessarily adds its own noise contribution. However, in most cases the main noise contribution of this filter often comes from its early stages. As a result, we can expect the noise contribution of this stage to experience the filtering transfer function, as does the incoming signal. We can thus say that in practice the noise contributions of the early stages of the line-up, at least up to the analog reconstruction filter, impact mainly the close in-band part of the spectrum, i.e. either the adjacent channels or the in-band SNR or EVM.

(iv) At the frequency upconversion stage, we can face different additional sources of signal degradation. First of all, we can face an increase in the image signal due to RF impairments

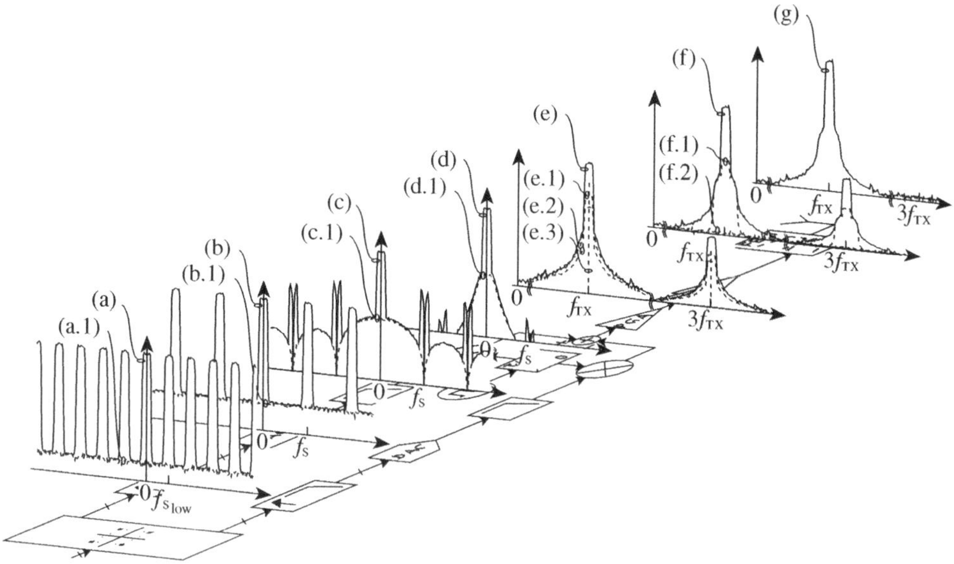

Figure 7.4 Spectrum of the signals recovered at the different stages of a direct conversion transmitter: realistic implementation case – Due to the limitations we face in the physical implementation of the signal processing functions embedded in a direct conversion transmitter, we have to cope with several degradations experienced by the signals being processed along the data path. Those degradations can be examined in the spectral domain by a direct comparison with the ideal case shown in Figure 7.3. In the digital domain we first get the addition of a quantization noise due to the finite number of bits used for the numerical representation of the samples. This results in a noise floor degradation each time a truncation occurs (a.1) and (b.1). Then, the intrinsic noise and distortion contribution of the DAC leads to an overall noise increase. But the total noise recovered at its output is no longer white due to the sample and hold transfer function of the DAC that applies to it (c.1), as it does to the signal (c). The resulting noise spectrum is then further filtered by the analog reconstruction filter (d.1) that is expected to sufficiently attenuate the residual digital copies (d). During the frequency upconversion, we then face both the rise in the copies of the wanted sideband due to the chopper-like behavior of the devices (e), the transposition of the LO signal phase noise (e.1), the rise in the image signal due to gain and phase imbalance (e.2), and the presence of some LO leakage due to potential RF coupling or direct feedthrough (e.3). In order to save current, smooth compression can also occur in the RF amplifiers so that an AM-AM and also potentially an AM-PM noise component (f.1) can add to the signal (f) on top of those blocks' own thermal noise contribution (f.2). At the output of the harmonic filter, we then recover the sideband of interest, but with spectral degradations occurring both in-band, close in-band, and out-of-band (g).

between the P and Q branches of the line-up, as discussed in Chapter 6. Due to the frequency planning of the direct conversion, this phenomenon results mainly in an in-band SNR or EVM limitation, as discussed in Section 6.2. On top of that, such RF impairments can also lead to the rise in the even order harmonics $2lf_{TX}$ of the square wave LO signals used to drive chopper-like mixers, as summarized in "Practical LO spectral content" (Section 6.3.2). However, for the sake of clarity such even order harmonica are not displayed in Figure 7.4. Then we necessarily also face the degradation linked to the use of a noisy LO signal. As discussed in Section 4.3, the use of an LO signal that exhibits a non-vanishing phase noise results in the transposition of the spectrum of this phase noise around the signal being processed. This then leads on the one hand to a limitation in the in-band SNR or EVM performance, and on the other hand to a degradation of the transmit spectrum. Last, we often face a leakage of the LO signal due to electromagnetic coupling or direct feedthrough at the physical device level, as discussed in Section 6.5.1. This leakage results in an origin offset that impacts the quality of the modulation.

(v) Finally, we get the RF stages that mainly provide amplification and gain control in order to deliver the correct RF power to the radiating element. First of all, such blocks also add their own noise contribution. Due to the wideband behavior of those devices and the lack of aggressive filtering that takes place after them, this classically results in a wideband noise contribution and thus to some kind of noise floor in the transmit spectrum. Second, at least when dealing with an amplitude modulated bandpass signal, such a device can also lead to a compression of the bandpass modulated signal so that an AM-AM and also potentially an AM-PM noise component can be generated. This results in both an in-band distortion and a close in-band pollution through the spectral regrowth.

We thus see that degradations linked to the implementation limitations lead to both in-band SNR or EVM limitations that impact the quality of the modulation delivered by the transmitter, but also to degradation either in the adjacent channels or in the wideband part of the spectrum.

Having in mind the impact of the theoretical functions we expect to implement as well as the impact of the degradations linked to their physical implementation, we now review the budgets we have to consider for our transmitter.

Budgets to be Considered

Recalling the requirements set for the performance of transmitters in Chapter 3, we can surmise that we need to perform at least two kinds of budgets in practice. We need to check on the one hand the ability of the transmitter to respect its wireless environment, and on the other hand the quality of the modulation that the transmitter delivers.

From the discussion in the previous sections, different sources of degradations or pollution are involved depending on whether we consider the out-of-band part of the radiated spectrum, its close in-band part, or its in-band part. For instance, the frequency spread of the distortion terms due to the smooth compression of the transmitted signal is proportional to its own frequency extent through the spectral regrowth effect. Such a distortion component can thus only impact either the in-band or close in-band part of the radiated spectrum in practice, but not its far out-of-band part.

We thus need to carry out different budgets depending on which part of the transmitted spectrum is involved. We need at least one budget that involves the in-band noise or distortion components present at the transmitter output in order to check the quality of the modulation; and at least two budgets, one involving the close in-band part of the spectrum and the other one its far out-of-band part, in order to check for respect for the wireless environment.

Before we illustrate these budgets, it is worth going through what are classically called the level diagrams associated with the transmit line-up. By this, we mean the necessary tracking of the level of the signal and of the noise components along the transmit path. This kind of tracking is important as the correct scaling of the signal is essential for the optimization of the performance of the transmitter. It also has didactic value as it naturally explains why overall performance is a function of the output power, and thus why the requirements are set that way in practice, as illustrated in Chapter 3.

7.2.2 *Level Diagrams and Transmitter High Level Parameters*

Let us now focus on how best to scale the signal along the transmit path. We first observe that, unlike what occurs in a receiver, the level of the signal processed in a transmitter is deterministic. We know the target average power at which we need to deliver it to the antenna. We then have all the freedom to generate and scale it in the most suitable way along the transmit path in order to optimize the performance.

Practically speaking, optimizing the performance means that we may minimize both the corruption of the modulation quality and the impact of the transmitter on the other users. But, reconsidering the degradations that we face with such direct conversion architecture, as detailed in Section 7.2.1, the scaling of the signal along the data path can impact the overall performance mainly through two kinds of phenomenon. On the one hand, this scaling directly impacts the margin we have regarding the additive noise contributions in the line-up. Indeed, as the name suggests, those contributions have a constant level whatever the level of the signal being processed. We thus get that the signal should be kept in the highest part of the available DR in order to optimize the SNR with respect to this contribution. But on the other hand, nonlinear distortion terms can be generated when the compression of the signal occurs. As a result, we may need to maintain enough margin for the signal amplitude relative to the available FS to try to minimize such distortion. Thus, a trade-off must be found between those two opposite behaviors in order to optimize the performance.

When reaching the antenna, the highest part of the DR in the line-up obviously corresponds to the maximum possible RF voltage or current swing, and thus to the maximum output power. But, as in many implementations we need to deliver the signal to the radiating element over a wide RF power DR, we necessarily need a variable gain somewhere in the line-up. And in that case, this stage necessarily behaves as a variable attenuator in order to be able to reduce the power delivered to the antenna. As a result, the signal can remain in the highest part of the available DR only up to this variable gain stage. We conclude that in order to optimize the performance, we should have this variable gain as close as possible to the output of the transmitter in order to preserve as much as possible the signal to additive noise power ratio in the line-up.[2] This is indeed what is often done in practice by using an RF programmable gain amplifier (PGA) after the upmixer, as considered in our present line-up illustrated in Figure 7.1.

By these remarks, the typical shape for the level diagrams of our transmit line-up can be expected to correspond to those shown in Figure 7.5 when considering the transmission at a maximum output power. Before commenting further on such diagrams, some preliminary remarks are in order. Figure 7.5 shows that we refer the level of the signals we are dealing with to some maximum available level. In the digital domain, this level FS_{dig} is obviously related to the maximum number of bits used to represent the samples of the modulating waveforms.

[2] The preservation of the signal to LO leakage ratio is another good reason for this.

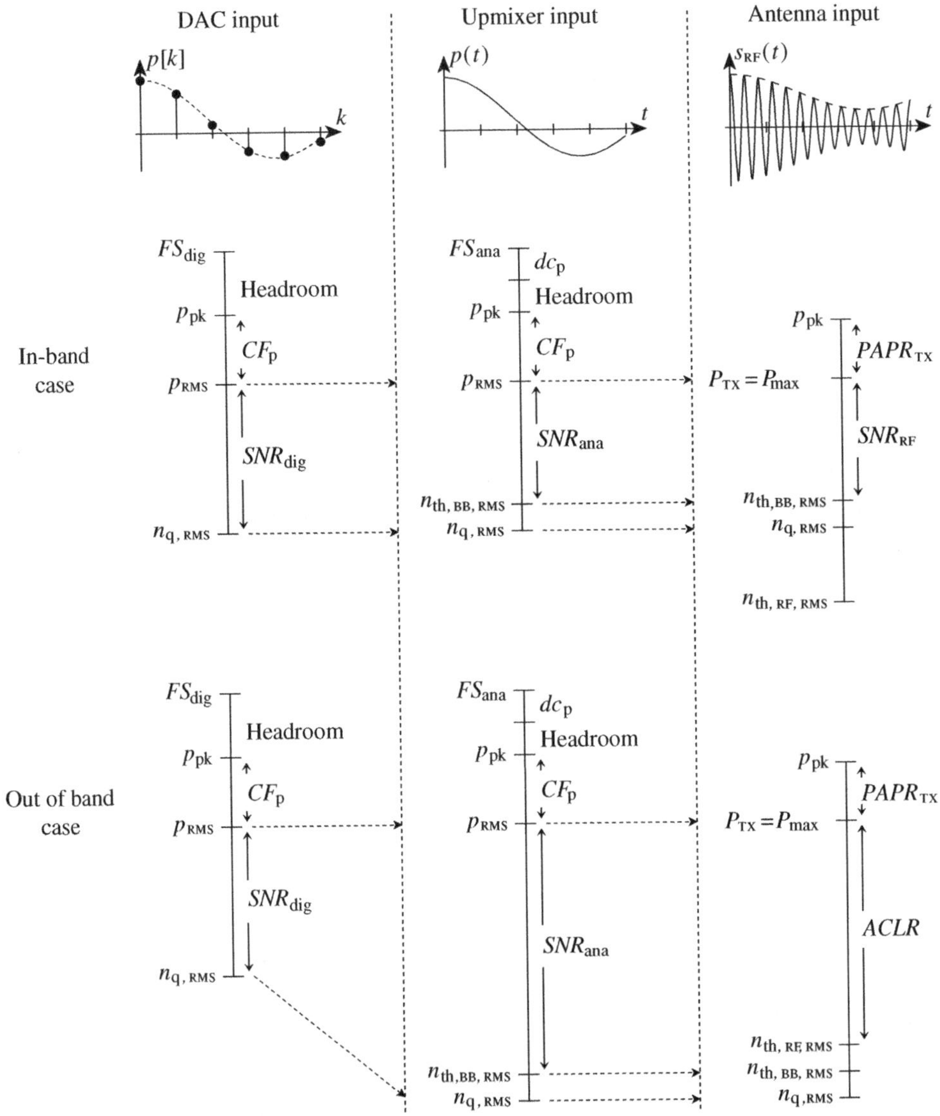

Figure 7.5 Simplified level diagrams along the transmit path considering only additive noise components: maximum output power case – To optimize the SNR along the line-up shown in Figure 7.1, we may generate the samples of the modulating waveform in the highest part of the digital DR (left). On top of the CF, the back-off in the analog data path requires some headroom to pass RF impairments like the DC offset (center). As the in-band baseband noise components (top) and the out-of-band ones (bottom) experience different attenuations by the reconstruction filter, the in-band RF SNR is limited by the baseband noise component, and the ACLR by the RF noise components at the maximum output level.

In contrast, in the RF world we mainly process electromagnetic waves with a given power. As a result, the maximum power to be handled by the transmitter is therefore the natural FS to be used in that domain to refer the signals to. However, in between those two domains, we have the baseband analog domain, in which the use of transistor devices for the implementation of the functions we are dealing with leads to processing either voltages or current waves, but not necessarily both simultaneously. As a result, what we call the FS in that domain, FS_{ana}, should perhaps be refined a little, depending on the nature of the physical waves we are dealing with. This is in fact not necessary in the present case as we do not further discuss the partitioning between the different blocks involved in the analog baseband part of the transmitter. Thus, in contrast to what is discussed on the receive side in Section 7.3.2, we can just assume here that we have a maximum FS available for the analog waves we are dealing with, whatever their physical nature, current or voltage. Second, we observe that, for the sake of simplicity, in Figure 7.5 we have considered only the additive noise contributions on top of the wanted signal. Although such diagrams can be enhanced in practice by adding other noise contributions, the present situation already provides quite a good understanding of how to scale the signal along the data path.

Thus, looking at those level diagrams, we observe that we tend to use the highest part of the available DR all along the data path, at least when transmitting at the maximum output power. All that remains to consider is the back-off, i.e. the level at which we need to regulate the RMS level of the signal regarding the FS in order to avoid compression. On the baseband side the compression we may face is mainly hard clipping, in both the digital and analog worlds. As a result, we may scale the RMS value of the signals in the data path so that their peak value, cannot reach the available FS. Thus the margins required to avoid such clipping must take into account at least the CF of the considered P or Q modulating waveform. But on top of that, we also need to consider additional margins linked to RF impairments in the analog path. This is for instance the case for the DC offset that is necessarily present in the line-up. Such offset leads to an offset in the RMS value of the signal so that it reduces the margins we have with regard to the clipping. This obviously needs to be refined on a case by case basis as many other implementation limitations can burn up such headroom, such as the gain spread of the analog baseband blocks, but this highlights the kind of phenomenon one needs to consider in order confidently to perform such signal scaling.

In Figure 7.5 we also see the difference in the behavior depending on whether we consider the in-band noise components, the close in-band or the out-of-band ones. Even if we consider the maximum output power in each case, we get that due to the filtering effect of the analog lowpass reconstruction filter, the noise power recovered in the close in-band part of the output spectrum remains driven by the noise floor of the RF stages. This filtering effect thus allows a reduction of the noise requirements for the baseband blocks compared to the stringent requirements we can have in the out-of-band part of the spectrum in order not to pollute other users. Such differences in the impact of the intrinsic noise performance of the constituent blocks of the line-up are particularly highlighted in the ACLR and EVM budget examples in "ACLR budget" (Section 7.2.3) and "EVM budget" (Section 7.2.4), respectively.

This partitioning also leads to an interesting behavior that can be highlighted by looking again at the level diagrams, but in the case of a transmission at the minimum output power. As illustrated in Figure 7.6, having an in-band noise performance dominated by the baseband components while the variable gain acts on the RF signal results in an output in-band noise power that scales with the output power of the signal, up to a point at least. Thus, whereas

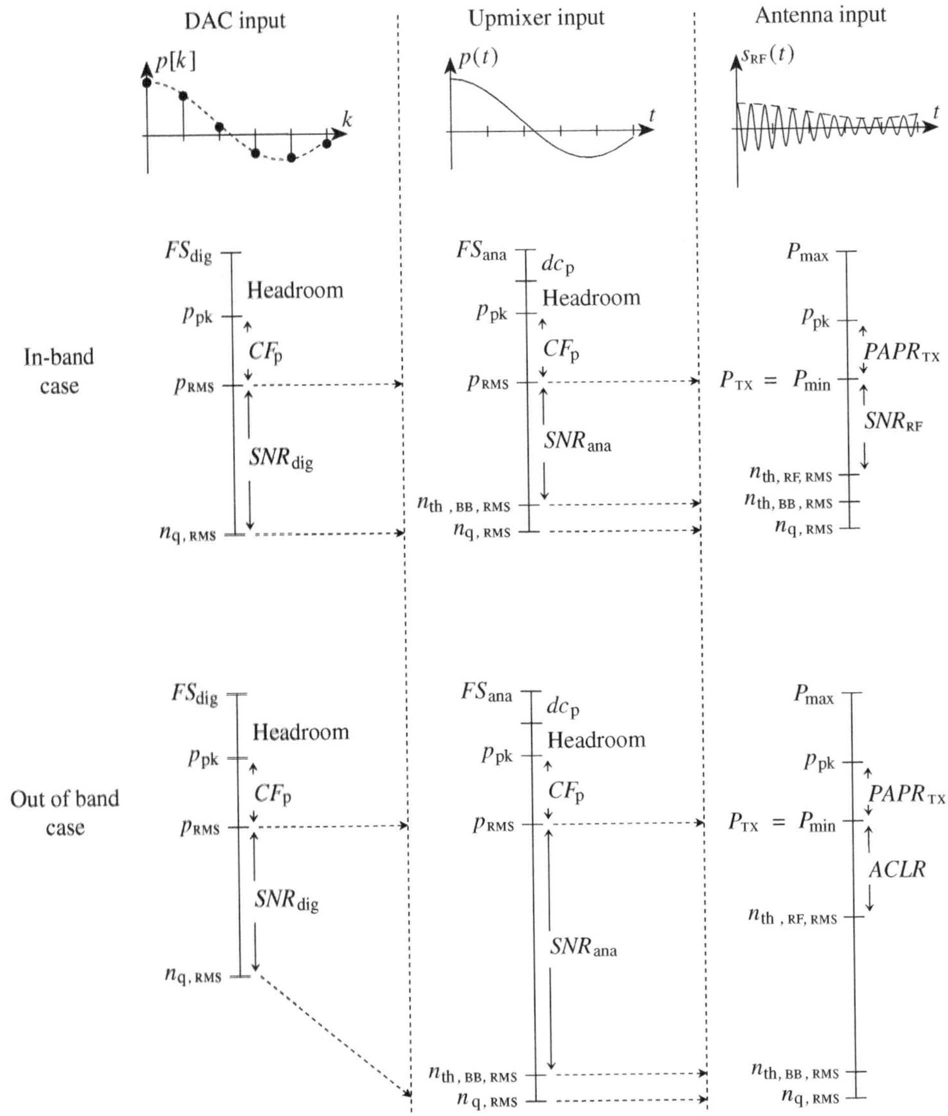

Figure 7.6 Simplified level diagrams along the transmit path considering only additive noise components: minimum output power case – Having all the variable gain in an RF PGA leads to a scaling of both the modulating waveforms and the noise components coming from baseband as a function of the output power. As the opposite behavior holds for the RF noise components injected after the variable gains, we get a constant ratio between the maximum power that can be delivered by the transmitter, P_{max}, and the power of the RF noise components, $n_{th,RF,RMS}$, whatever the gain set for the RF PGA. Thus, compared to the diagrams shown in Figure 7.5, we get a limitation in the achieved in-band SNR and close in-band ACLR due to the RF noise at the minimum output power.

we are considering only additive noise components in these level diagrams, we presently get a multiplicative like behavior when inspected from the output of the transmitter. It is only when looking from the input of the line-up that this additive behavior appears. Obviously, the opposite occurs for the out-of-band components as the noise floor in that part of the output spectrum is linked to the RF blocks that mostly occur after the scaling of the signal in the line-up. This explains the linear reduction in the ACLR as a function of the output power. However, as discussed in "ACLR budget" (Section 7.2.3), this occurs only for the additive noise components in this kind of budget, and many other contributors need to be considered in practice.

In light of the above comments, as a side effect the intrinsic performance of the transmitter necessarily depends on the target output power, as do the settings of its constituent blocks. For instance, for each target output power we have a different gain used in the line-up in order to scale correctly the output RF voltage or current. Those settings can then impact the overall performance of the transmitter in two different ways. They can lead on the one hand to different recombinations of the noise contributions coming from different parts of the line-up as detailed above, but also on the other hand to a direct change in the intrinsic performance of the blocks whose settings have changed. By way of illustration, we can mention the classical behavior associated with a specific block of the transmitter, the PA. This block is classically a power hungry device in the line-up, in particular when set as linear, as required for the processing of amplitude modulated waveforms (see Section 8.1.7). Thus in practical implementations a reduction in its power consumption, either through a direct change in its biasing or through the switching between optimized blocks in the device itself, is often predicted for lower output power, i.e. for output power where its Psat can be scaled down according to the signal power. But as a side effect this power reduction often leads to drawbacks, such as a reduction in the gain and noise performance of the device on top of its linearity. When this kind of strategy is used, we thus need to ensure that the power consumption reduction does not prevent the performance from remaining within acceptable bounds consistent with the requirements.

Thus, given that the performance of the transmitter is necessarily mainly a function of the output power, we may anticipate that all the budgets of the transmitter have to be considered as a function of the output power. As a side effect, this also explains why in practice the requirements are set as a function of this output power, as illustrated in Chapter 3. However, in order to derive practical examples for the various budgets in the following sections, we need to assign numbers to the high level characteristics of our transmitter. And as a first step, we need to refine our gain and power reduction strategy as it drives the final performance of the line-up. For that purpose, we can assume that we are dealing with a transmitter that delivers an RF signal with an average power that ranges between −50 dBm and 25 dBm, as can be encountered in some cellular wireless standards. Furthermore, we can assume that the PA provides a power gain of 25 dB in its nominal mode and that we implement only two switching points for the reduction of its power consumption when the signal scales down. This is purely for for the sake of simplicity; nothing prevents us from having more than two in practice if the implementation of the device permits it. We can then assume that the first switching occurs for an output power equal to 15 dBm for instance, resulting in a power gain of 20 dB for the PA, and the second one for an output power equal to 5 dBm, resulting in a power gain of 15 dB. Assuming that all the variable gain is implemented in the RF PGA, and that the upmixer delivers a constant output average power of 0 dBm, we end up with the gain settings in the transmitter as a function of the output power shown on the top part of Figure 7.7.

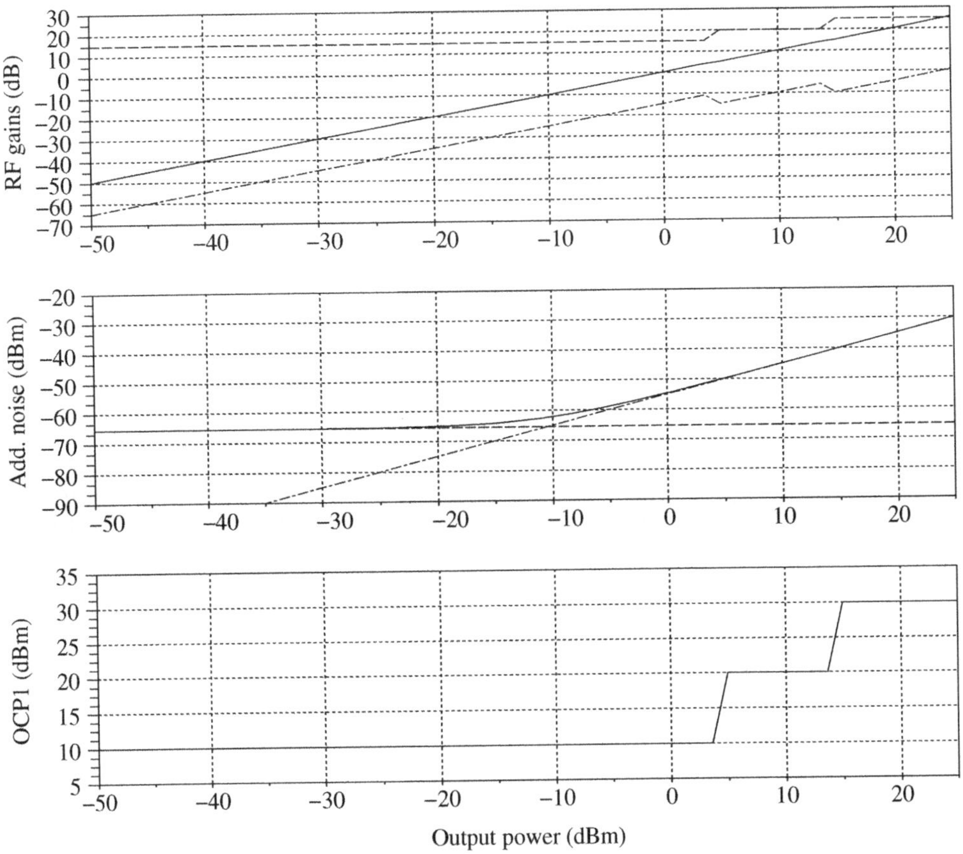

Figure 7.7 Typical transmitter configuration and characteristics considered for the derivation of the budgets – Having three switching points for the PA biasing results in three gains for this stage (top, dashed). The remaining variable gain, set in the RF PGA here, thus needs to compensate for those PA gain variations (top, dot-dashed) in order to achieve the overall target gain (top, solid). Here this target gain is derived assuming a constant power of 0 dBm at the output of the upmixer. This switching in the biasing also results in a drop in the odd order linearity of the transmitter as does its gain (bottom, solid). This results in three different OCP1 values for the transmitter. Moreover, as the signal and noise components from the baseband scale in the same way due to the variable gain set in RF, the baseband noise contribution to the total output in-band noise power increases by 1 dB per decibel of the transmit output power (middle, dot-dashed). The output in-band noise thus seems to behave as a multiplicative noise in the highest part of the output DR (middle, solid). In contrast, the noise component added after the variable gain effectively leads to a noise floor at the transmitter output (middle, dashed). Here the RF noise component corresponds to a noise floor of -135 dBm/Hz at the transmitter output, whereas the baseband components, including the DAC DR, correspond to an equivalent in-band SNR of 55 dB for a sine wave that reaches the ADC FS. The noise bandwidth δf is equal to $[-4.5, 4.5]$ MHz, i.e. corresponds to an LTE 10 MHz waveform.

Having derived the gain strategy, we can then derive the corresponding high level characteristics as a function of the output power. Let us focus for the sake of simplicity on the characteristics that have a significant dependence on the settings of both the RF PGA and the PA. Up to first order, we can assume that these characteristics reduce to the additive noise contributions as discussed above, and to the odd order linearity of the line-up, driven by the performance of the PA itself. Thus, we consider in what follows a transmit line-up with the intrinsic noise and odd order linearity performance shown in Figure 7.7 (middle and bottom). However, looking at those characteristics, one may object that at least the AM-PM conversion performance of the line-up, mainly driven by that of the PA, also depends on its biasing. But, as we assume here that these switchings are set in such a way that the ratio between the resulting OCP1 and the output powers for which they occur remains constant, we can assume for the sake of simplicity that the resulting parameter $\Theta_{\circ/\mathrm{dB}}$ is also constant. This parameter is thus set to $0.2°/\mathrm{dB}$ in what follows. However, as discussed in "Model for AM-PM conversion" (Section 5.3.1), this constant value holds for the output powers at which the switchings occur and should not mask the fact that the parameter Θ used in the model discussed in that section then depends on the relevant powers through equation (5.315). We thus use three distinct values for Θ derived through this equation in the forthcoming budgets. Furthermore, for the sake of simplicity we ignore any memory effects.

In conclusion, we note here that in order to illustrate our budgets we process an OFDM-like signal in our transmitter – more precisely, a waveform corresponding to the LTE 10 MHz case, at least from the frequency bandwidth point of view. This means, as detailed in Section 1.3.3 and shown in Figure 1.19, that we consider a signal with a PSD almost flat over the frequency band $[-4.5, 4.5]$ MHz around the carrier frequency. We also assume that this frequency band can be taken as our line-up noise bandwidth δf. For the sake of simplicity, we assume that the signal we consider here is a pure OFDM signal, i.e. with a statistical distribution for the real and imaginary parts of the modulating complex envelope that can be approximated as Gaussian. The choice for such an amplitude modulated RF bandpass signal also explains why we take a linear line-up as our baseline, i.e. a line-up that exhibits a CP set higher than the output power.

7.2.3 Budgets Linked to Respect for the Wireless Environment

As discussed in "Budgets to be considered" (Section 7.2.1), in order to check that the wireless environment is being respected, we need to carry out at least two kinds of budget, depending on which out-of-band part of the spectrum we are looking at. We get different sources of degradation in the close in-band and far out-of-band parts of the radiated spectrum.

Having said that, there is a common global function embedded in the transmitter that may impact both parts of the spectrum, i.e. filtering. As illustrated in Section 7.2.1, the spectrum recovered at the output of the transmitter can be corrupted by the residual digital copies still present at the DAC output, or by the unwanted sidebands generated at the harmonic of the carrier frequency through the chopper effect of the upmixer. Due to the frequency planning classically used in a direct conversion transmitter, the unwanted copies and sidebands can pollute almost all the areas of the transmitted spectrum. We examine how to budget such filtering in the line-up.

Then we illustrate the budgets for both the close in-band and the out-of-band parts of the radiated spectrum. Here we observe that only the wideband noise floor at the output of the transmitter is of significance for the pollution of the far out-of-band part of the output spectrum. This explains why we comment only briefly on this aspect when discussing the filtering constraints for the far out-of-band part of the transmit spectrum. We focus rather on the budget for the close in-band part of the output spectrum, i.e. the ACLR budget, as it involves many imperfections in addition to the impact of the filtering embedded in the line-up. Finally, we briefly discuss frequency planning and the potential impact of spurs if not well anticipated.

Filtering Budgets

Filtering Problem

We focus here on the filtering problem in a direct-conversion transmitter. Looking the spectrum displayed in Figures 7.3 and 7.4, we can see that due to the signal processing embedded in the transmitter, two kinds of unwanted sideband can pollute the output spectrum: the copies linked to the generation of the modulating waveform in the digital domain; and the unwanted RF sidebands linked to the implementation of the upmixers as choppers. In addition, the modulating waveforms used in some wireless standards lead by construction to power leakage in the adjacent channels. In that case, the filtering embedded in the line-up is also involved in the attenuation of this unwanted contribution lying in the adjacent channel. We say a few words about this contribution to the ACLR budget in due course, but for the time being we focus on the management of the unwanted sidebands, which is one of the first problems to be addressed by the transmit filtering.

There is a major difference between two kinds of copy mentioned above due to the frequency planning classically associated with the direct conversion transmitter. On the one hand, the distance between two copies present in the digital waveform provided to the DAC is equal to the sampling rate of this waveform, i.e. f_s. On the other hand, the distance between two copies generated by the choppers is equal to the carrier frequency, i.e. f_{TX}. In practice, there is often a large difference between the two quantities as f_s is classically in the megahertz range, while the RF carrier frequencies of interest lie in the gigahertz range. As a consequence, the two kinds of copy must be handled in different ways in practice.

The deep reason for this comes from the total TX system frequency band, Δf_{TX}, the transmitter has to address. Here, by TX system band we mean the total frequency band that belongs to the same wireless system and in which all the users have their transmit signals multiplexed in frequency. A transmitter that claims to be compliant with a given wireless system must be able to generate the carrier frequency anywhere in this TX frequency band. As a result, looking at Figure 7.8, any RF filter that occurs after the channel selection, i.e. after the frequency upconversion up to the carrier frequency in the present case, can only be either wider than the overall transmit system band in order not to attenuate the transmit signal whatever the carrier frequency within all the possible system frequencies, or tunable to remain centered around the transmit carrier frequency. But in practical implementations the sampling rate f_s used for the generation of the samples of the modulating waveform is often much lower than the overall transmit band Δf_{TX} of the wireless system. As a result, we may need some attenuation of the digital copies to occur within the overall transmit system band. This can only be done with a tunable RF filter. Moreover, given that $f_s \ll f_{TX}$, the required bandpass RF filter would need a quite impressive quality factor. And an RF filter that has both a high quality factor and whose center frequency is tunable is a solution that is not really cost effective in practice. This explains why, in order to keep the implementation reasonable,

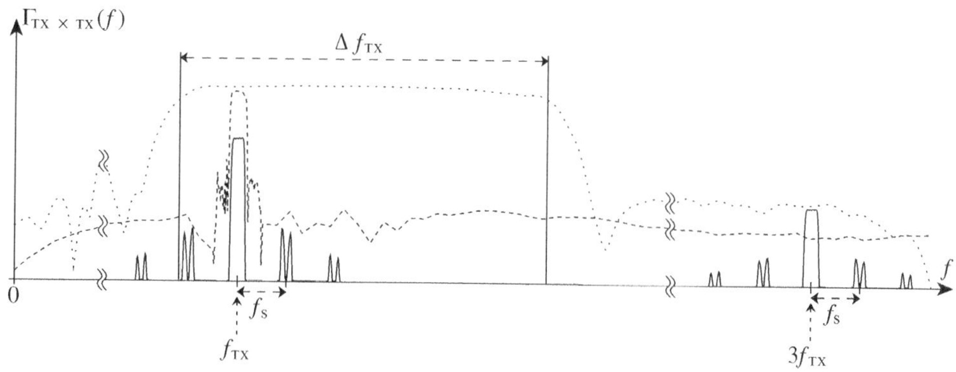

Figure 7.8 Unwanted sidebands in a direct conversion transmitter – Classically, two kinds of unwanted sidebands have to be handled in a transmit line-up like the one shown in Figure 7.2. On the one hand, we have the copies linked to the generation of the waveform in the digital domain, and on the other hand the unwanted RF sidebands generated around the harmonics kf_{TX} of the carrier frequency through the use of mixers implemented as choppers. Assuming that the RF impairments are sufficiently low, these harmonics reduce mainly to the odd order ones $(2l + 1)f_{TX}$ (solid). Thus, due to the frequency planning of the direct conversion transmitter, the harmonics can be handled using a fixed RF filter that is bandpass for the entire transmitter system band Δf_{TX} (dotted). Matters are not so straightforward for the digital copies that can lie within the system band. Any RF filtering in this case would require a filter both tunable in order to remain centered on the used transmit channel, and with a high quality factor due to the low sampling rate f_s with respect to the carrier frequency f_{TX} (dashed).

we need to provide the filtering of the digital copies prior to the RF channel selection, i.e. at baseband. Such lowpass filtering on the real and imaginary parts of the complex modulating waveform allows us to have a filter stage that operates at a fixed frequency as no RF channel selection has been performed yet. But it also allows easier integration, on silicon for instance, compared to what can be achieved when using a tunable RF filter with a high quality factor.

The problem is obviously different when considering the unwanted RF sidebands, centered around the harmonics of the carrier frequency in our direct conversion case, and generated by chopper-like mixers. In this case the distance between the unwanted sidebands allows for the use of a simple wideband fixed RF filter for their attenuation. Such a filter is classically embedded in the FE of the transceiver.

These simple considerations explain the structure of the direct conversion transmitter we consider here, at least from the filtering point of view. And following this split between the baseband and the RF filtering, we can now discuss the corresponding dimensioning.

Cancellation of the Digital Copies

As detailed above, the cancellation, or at least the attenuation, of the digital copies is classically handled through the baseband analog filtering of the $p(t)$ and $q(t)$ modulating waveforms. Recall from "Impact of spectral symmetry on top of stationarity" (Section 1.1.3) that, under the assumption that $p(t)$ and $q(t)$ are uncorrelated and have the same power spectral densities, the spectrum of the RF modulated bandpass signal at the transmitter output matches that of $p(t)$ and $q(t)$. Thus, we can directly apply the constraints we have for the spectrum of the final RF modulated signal at the output of the transmitter to those lowpass modulating processes.

We can then deal with $p(t)$ and $q(t)$ exactly as if we were working on the modulated waveform at the transmitter output.

To continue with the budgeting of the digital copies attenuation, we observe that at least two phenomena come together to provide this attenuation, as illustrated in Figures 7.3 and 7.4. The first is the transfer function of the DAC, which we assume behaves as a sample and hold. As detailed in Section 4.6.2, such an aperture effect results in a sinc transfer function in the spectral domain. As the zeros of this transfer function are located at the multiples of the sampling frequency f_s, we thus get a natural attenuation of the digital copies. The second phenomenon is the additional filtering provided by the analog reconstruction filter. In practice, the reconstruction filter is expected to provide the residual attenuation of the digital copies, as required to achieve the SEM, and that is not already provided by the aperture effect of the DAC. We then see that the first step in our budget is to evaluate the attenuation of the copies that is provided by the natural filtering of the aperture affect.

For that purpose, we first observe that the spectrum of the digital sequence that is fed to the DAC is periodic. This means in particular that all the copies in the spectrum have the same magnitude at the DAC input. Thus, looking at a typical sinc transfer function, we see that it is the first copy, i.e. the one lying around f_s, that necessarily has the highest level at the DAC output. In order to illustrate the attenuation achieved by the aperture effect, it is useful to focus on this first copy. As discussed in Section 7.2.2, we consider an OFDM signal with a frequency bandwidth δf of 9 MHz. As the PSD of this signal can be assumed flat, we thus get that the average power attenuation, $\overline{A}_{\rho^2,\mathrm{DAC}}$, over the frequency band of the first copy is given by

$$\overline{A}_{\rho^2,\mathrm{DAC}} = \frac{1}{\delta f} \int_{f_s-\frac{\delta f}{2}}^{f_s+\frac{\delta f}{2}} |\mathrm{sinc}(\pi f/f_s)|^2 \mathrm{d}f, \tag{7.1}$$

where the sinc(.) function is defined in equation (4.316). The resulting average attenuation is shown in Figure 7.9 in addition to the simplified lower bound for this attenuation as given by equation (4.322). Looking at this figure, we see that the higher the sampling rate used for the samples provided to the DAC, the higher the natural rejection of the copy on the analog side. This seems reasonable as the DAC theoretically approximates the analog waveform with a staircase signal in the time domain. And as the width of the stairs is equal to the sampling period, the higher the sampling rate, the closer to the expected waveform we are at the DAC output. The same must then necessarily occur in the spectral domain. However, this is achieved at the cost of running the DAC at a high rate. And on top of the problem of the physical limit for this rate linked to a given implementation technology, we soon run into a power consumption problem that also necessarily increases with the sampling rate. The trade-off is thus often to maintain a reasonable rate in order to limit the power consumption and then rely on an analog reconstruction filter to cancel unwanted copies and reconstruct correctly the expected waveform in the time domain.

As a result, with a given attenuation already provided by the aperture effect of the DAC, it remains to provide the residual attenuation required to achieve the SEM. But when using an analog reconstruction filter, there is a trade-off to consider. Indeed, if we ignore the refinement related to the prototype of the filters, there is necessarily a balance to find between the cut-off frequency and the order of the filter. In order to illustrate this, we continue to consider our OFDM signal with a frequency bandwidth δf of 9 MHz. We also assume that we are dealing with a reconstruction filter implemented as a lowpass Butterworth filter. This assumption is

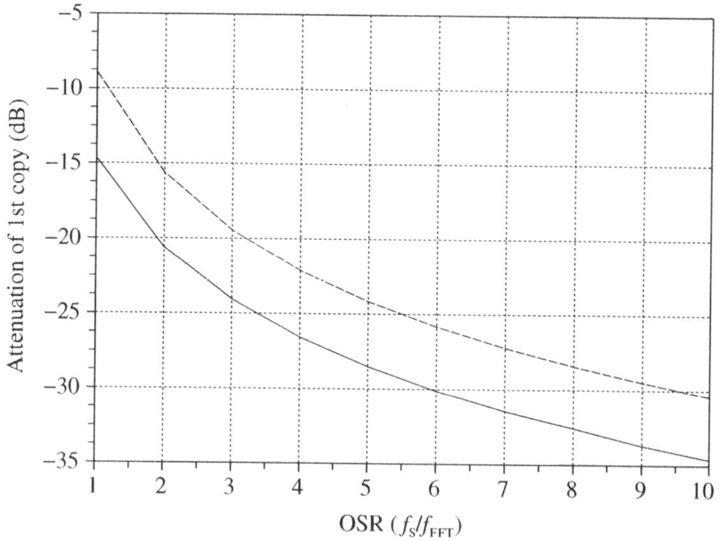

Figure 7.9 Illustration of the attenuation achieved by the aperture effect of the DAC – Considering the digital to analog conversion of a 10 MHz LTE waveform, i.e. of a signal that exhibits an almost flat PSD over the band [−4.5, 4.5] MHz, we get a natural attenuation of the first copy due to the sinc transfer function of the DAC in the spectral domain. The resulting average power attenuation, $\overline{A}_{\rho^2,\text{DAC}}$, given by equation (7.1), increases as a function of the sampling frequency f_s, here expressed as relative to the FFT rate, $f_{\text{FFT}} = 7.68$ MHz in the present case (solid). The same holds for the minimum attenuation achieved at the edge of the signal as given by equation (4.322) (dashed).

convenient due to the simplicity of the power transfer function of such a filter in the frequency domain, which reduces to

$$A_{\rho^2,\text{FILT}}(f) = \frac{1}{(1+f/f_c)^{2n}}, \tag{7.2}$$

with n the order of the filter and f_c its cut-off frequency at 3 dB. Thus the average attenuation over the bandwidth of the first copy, $\overline{A}_{\rho^2,\text{FILT}}$, reduces in the present case to

$$\overline{A}_{\rho^2,\text{FILT}} = \frac{1}{\delta f}\int_{f_s-\frac{\delta f}{2}}^{f_s+\frac{\delta f}{2}} \frac{\mathrm{d}f}{(1+f/f_c)^{2n}}. \tag{7.3}$$

As illustrated in Figure 7.10, a given target attenuation of the first copy can be achieved using different combinations (n, f_c) of orders and cut-off frequencies. For instance, an average attenuation of −70 dB can be achieved either by a third order filter with a cut-off frequency set around $0.35 f_{\text{FFT}}$ or by a fifth order filter with a cut-off frequency set around f_{FFT}. Which solution we choose should be driven by implementation limitations. The cut-off frequency of the fifth order filter is much higher than that of the third order filter. As a result, we expect a smaller capacitor area and thus a more efficient implementation in terms of area. But at the same time we get two more poles, and thus we need to introduce additional capacitors to generate

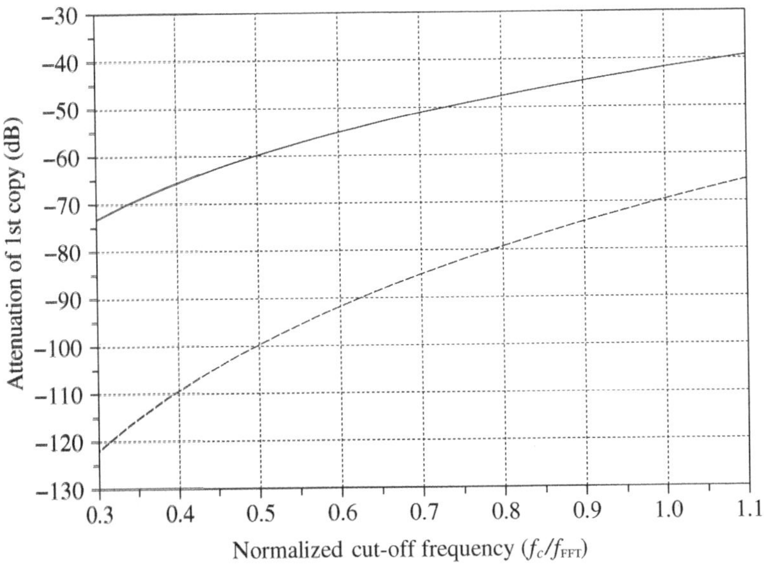

Figure 7.10 Example of the attenuation of the first copy achieved by an analog reconstruction filter – For the same waveform as in Figure 7.9, a given target average power attenuation of the first digital copy, $\overline{A}_{\rho^2,\mathrm{FILT}}$ given by equation (7.3), can be achieved by a Butterworth filter when using different orders n and cut-off frequencies f_c. Here $n = 3$ (solid) and $n = 5$ (dashed) are considered.

them. Moreover, two more poles often means two more operational amplifiers for an active implementation. That means more current in the final solution. For a low power consumption approach the third order filter appears more appropriate. In contrast, a low cut-off frequency like $0.35 f_{\mathrm{FFT}}$ may often lead to in-band distortions of the signal being filtered, which can be interpreted in terms of linear EVM, as discussed in Section 4.4. Equalization of such an analog reconstruction filter, as well as of the DAC sinc transfer function, has to be considered in the digital domain, prior to the digital to analog conversion.

Thus, we see that all is a matter of trade-off between the power consumption and the implementation area. Depending on what are the most important criteria, we can target an efficient implementation for that parameter that still fulfills the required attenuation, and thus the SEM.

Cancellation of the Chopper Copies

Let us now focus on the cancellation of the unwanted sidebands recovered at the output of the transmitter due to the fact that we use upmixers implemented as choppers. As discussed in "Filtering problem" earlier in this section, in a direct conversion transmitter these sidebands lie at the harmonic frequencies kf_{TX} of the TX carrier frequency – more precisely, around the odd order harmonic frequencies $(2l + 1)f_{\mathrm{TX}}$ of this TX carrier frequency as long as the RF impairments in the complex mixing stage remain low enough, as discussed in "Practical LO spectral content" (Section 6.3.2). Thus, considering that f_{TX} is classically in the gigahertz range while the width of the TX system band, Δf_{TX}, is an order of magnitude below, we get that the unwanted sidebands systematically lie outside this frequency band, as illustrated in Figure 7.8. In that case we can thus rely on a simple fixed RF filter, which is bandpass with respect to the full uplink system band, and which provides enough attenuation for the unwanted sidebands. In that case we talk about a simple harmonic filter as it is expected to act on the harmonics of the RF bandpass signal carrier frequencies. In practice, such a filter is often embedded in the FE of the transceiver, as illustrated in our transceiver.

Recall that our primary goal for the cancellation of unwanted sidebands is to fulfill the far part of the SEM. As observed above, there is another contribution that can impact the far out-of-band part of the SEM, the wideband noise floor. In practice, two situations can occur, depending on the noise performance of the active part of the transmitter. Either the noise floor is already good enough to satisfy the requirements; this is obviously the best solution even if at the cost of an additional current consumption in the transmitter. Or it is not, in which case we need to consider the use of a passive RF filter to attenuate the noise component flowing from the transmitter and that remains above the unavoidable absolute thermal noise floor level. In this case, we need to understand that we may require the attenuation to be provided already quite close to the transmit system band. This means in practice a sharper bandpass filter than required just for the suppression of the unwanted sidebands lying at the harmonic frequencies kf_{TX}. Obviously, having this sharper bandpass filter in the transmit line-up leads to additional insertion loss compared to a simple harmonic filter. Such insertion loss means more current consumption in the RF amplifiers in order to achieve the correct transmit power at the radiating element level. But, as the RF amplifiers are often the most power hungry parts of the transmitter, it may be worth using more current in the early stage of the transmitter to achieve a good enough noise floor and then avoid the use of a costly RF bandpass filter. Once again, this is a question of compromise.

ACLR Budget

Let us now focus on the close in-band part of the transmit spectrum. As is evident from comparing the spectra displayed in Figures 7.3 and 7.4, many additional phenomena are involved in the pollution of this part of the transmit spectrum. This is particularly true with regard to the adjacent channels, whose pollution involves for instance the nonlinearity of the line-up on top of the additive noise components and potential signal copies. We illustrate this below in deriving the ACLR budget for the first adjacent channel. We also review the behavior of the noise and distortion terms involved and illustrate why specific requirements are often set in practical wireless standards to check the pollution of adjacent channels (see also Chapter 3).

We begin by listing the noise contributors that we need to consider in our budget. This can be done by thinking about the different blocks involved in the line-up and their potential impact on the first adjacent channel through the limitations in their implementation. For that purpose, we can consider the three high level families of blocks involved in the line-up: the baseband blocks, the blocks dedicated to the frequency upconversion, and the RF blocks. Looking at Figure 7.4, we can make the following observations:

(i) The baseband blocks can be expected to have a negligible impact on the overall noise power in the adjacent channel. As discussed in Section 7.2.2 and illustrated in Figures 7.5 and 7.6, we can assume at first order that the noise and distortion components lying out-of-band as well as the unwanted digital copies are sufficiently attenuated by the analog reconstruction filter so that a negligible noise power contribution in the adjacent channel results. And in the same way, as highlighted in "Filtering problem" earlier in this section, we can assume for the sake of simplicity that this reconstruction filter is efficient enough

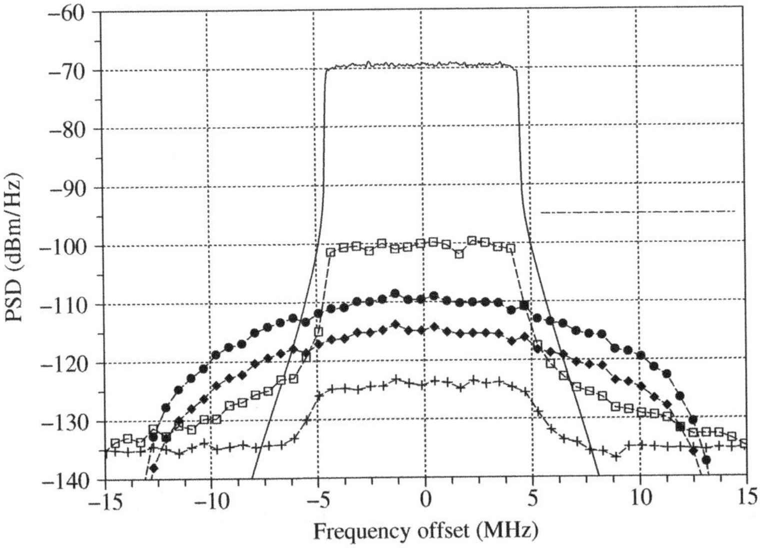

Figure 7.11 Output PSD of the noise components involved in the ACLR budget for the first adjacent channel – In order to derive the ACLR budget for the first adjacent channel, we need to integrate the fraction of PSD of the noise components that lies within this channel. In the present LTE 10 MHz case, this means the fraction of power that lies in the band $[-4.5, 4.5]$ MHz centered at a frequency offset of 10 MHz (dot-dashed). These components reduce here to the additive RF noise floor, given that the baseband contribution is expected to be sufficiently attenuated by the reconstruction filter so that it remains in-band (pluses), the multiplicative LO phase noise (squares), and the distortion noise composed of AM-AM (dots) and AM-PM terms (filled diamonds).

so that the potential sidelobes of the OFDM signal are sufficiently attenuated to also make a negligible contribution to our budget, as displayed in Figure 7.11.

(ii) The frequency upconversion stage can then impact the adjacent channel mainly through the phase noise of the LO signal used to drive the mixers. As discussed in "Spectrum degradation" (Section 4.3.3), even if the spectrum of this noise component has a smaller extent compared to the signal bandwidth in the present case, its convolution with the spectrum of the signal being frequency upconverted necessarily results in a leak in the close in-band part of the transmit spectrum. At this stage we might also consider the rise in the image signal at the complex upconversion stage when dealing with potential RF impairments between the P and Q branches. As detailed in Sections 6.2.2 and 6.3.2, the spectrum of the image signal generated through a complex frequency conversion in the presence of such RF impairments is simply the flipped copy around the LO frequency of the input reconstructed complex signal. Thus, assuming in our present case that the reconstructed complex lowpass modulating waveform is centered around DC at the upconverter input and has a symmetrical spectrum, which is the case for modulating waveforms of interest as discussed in "Impact of spectral symmetry on top of stationarity" (Section 1.1.3), we thus get that the spectra of the signal being processed and its image have exactly the same shape, up to a scaling factor equal to the IRR. Thus, assuming a reasonable value for this IRR, we can expect no additional significant degradations on the adjacent channel. This explains why at the first order we do not consider the contribution of this image signal in our ACLR budget.

(iii) The RF devices contribute to the noise power in the adjacent channel through both their intrinsic thermal noise floor and the direct degradation of the signal being processed due to the odd order nonlinearity. As discussed in "Spectral regrowth" (Section 5.1.3) and "Nonlinear EVM and spectral regrowth due to AM-PM conversion" (Section 5.3.2), such nonlinearity can lead to the generation of a bandpass distortion term that is centered around the signal that experiences the nonlinearity. Due to the associated spectral regrowth effect, we thus recover some unwanted power leakage in the adjacent channel through AM-AM or AM-PM conversion. Thus, assuming that we are dealing with an amplitude modulated RF bandpass waveform, which is the case here, we need to consider such additional contributions in our budget.

It is of interest to classify the various noise components that are involved in our budget depending on whether they can be considered as additive, multiplicative, or distortion noise terms. This will turn out to be useful later on in the analysis of the resulting performance as a function of the output power. Based on the above review, we can expect that the noise contributions in the adjacent channel that we need to consider reduce, for additive noise, to the RF thermal noise existing in the line-up, independently of the presence or absence of the signal being processed; for multiplicative noise, to the LO phase noise; and, for distortion noise, to terms linked to the AM-AM and AM-PM conversion. These last terms are linked to the compression of the amplitude modulated signal occurring at the RF stages.

Then, we need to determine the fraction of these noise terms that leaks into the adjacent channel. Obviously, this cannot be done analytically for all the above-mentioned components. For instance, the distortion noise involves the signal being processed itself. Thus, unless considering a waveform with statistics suited to such analytical derivation (e.g. Gaussian), it is hard to derive the fraction of power that lies in the adjacent channel. This often needs to be done using time domain simulations to derive samples of the noise component of interest. Having a realization of such a process allows first the derivation of its PSD, and second the fraction of power lying in the adjacent channel, as illustrated in Figure 7.11 when using the analytical derivations below. Thus, we can go through the various noise terms listed above and see how to estimate their contributions in the adjacent channel.

For the additive thermal noise, this is quite simple as we are only dealing with a noise floor in the adjacent channel. As discussed above, we assume for the sake of simplicity that the analog reconstruction filtering is efficient enough so that the baseband noise contribution in the adjacent channel remains negligible. Thus, using the values given in Section 7.2.2, we recover in the adjacent channel an RF thermal noise contribution with a constant PSD equal to -135 dBm/Hz. As this level is assumed constant whatever the settings of the transmitter, we thus see that we get a contribution that is independent of the transmitter output power, i.e. that is exactly additive with respect to the output transmitted signal. The power of this noise contribution can then be derived as the level of its PSD, assumed flat, times the noise bandwidth δf, equal to 9 MHz here.

Turning to the phase noise contribution, recall that we consider on the one hand mixers implemented as choppers and on the other hand that the frequency upconversion is based on the use of the fundamental tone of the LO signal that drives those choppers. Thus, by the discussion in "Noisy square LO waveform characteristics" (Section 4.3.2), the phase noise spectrum to be considered as centered around this fundamental tone is the one as recovered at the RF synthesizer output. Thus, we can recover the fraction of phase noise that lies within the adjacent channel in two ways. One way is to use an approximation based on the phase noise specification at the corresponding frequency offset. This approach is illustrated for instance in "Reciprocal mixing" (Section 4.3.3), through equation (4.241). But in the present case we can object that the convolution process between the spectrum of the signal being frequency transposed and the spectrum of the LO phase noise leads to inaccuracies in the application of this kind of approximation, in particular when considering the close in-band part of the spectrum. As a result, it is of interest to continue to use the time domain simulation approach to achieve a spectral domain evaluation. For that purpose, we can assume a simple but realistic model for the RF synthesizer used for the frequency transposition in order to derive a realization of the noise term that corrupts the signal being processed, as given by equation (4.185). In the present case, for the sake of simplicity we assume that $n_\phi(t)$ in that expression is generated using a Gaussian process filtered through a first order lowpass filter with a cut-off frequency of 50 kHz. We also assume that the integrated phase noise is equal to 1.5°. These values are representative of what is encountered in practical implementations.

It remains to derive an expression for the distortion noise terms, $n_{\text{o,AM-AM}}(t)$ and $n_{\text{o,AM-PM}}(t)$, generated respectively through AM-AM and AM-PM conversion and recovered at the output of the transmitter. As we consider the first adjacent channel in the present example, we get that the expression for these distortion terms is dominated by the contributions of the third order term in the series expansion of the device transfer function. Thus, having that we neglect memory effects as highlighted in Section 7.2.2, the complex envelope of these bandpass distortion terms, when defined as centered around the carrier frequency f_{TX}, can be approximated based on our assumptions using the arguments in Chapter 5, more precisely by equations (5.325) and (5.329). Looking at those equations, it is the equivalent input characteristics of the signal being processed, through its equivalent input complex envelope and related characteristics, that are involved. Thus, in order to derive our budget as a function of the actual power at the transmitter output, we can think of two approaches. On the one hand, we can keep such analytical expressions and sweep the equivalent input power. We can make the link between the equivalent input and the output power using the effective gain of the device, G_e, which is in turn a function of the equivalent input power, as can be seen from equation (5.322). We can then finally express the power of the distortion terms derived that way as a function of the output power using this methodology. On the other hand, we can also directly transpose those equations as a function of the output characteristics of the signal being processed. For that purpose, we observe that the equivalent output complex envelope of the expected signal, i.e. when undistorted, $\tilde{s}_\text{o}(t)$, is related to its input complex envelope through the effective gain of the line-up according to $\tilde{s}_\text{o}(t) = G_\text{e}\tilde{s}_\text{i}(t)$. As a result, we can write that $\rho_\text{o}^2(t) = |\tilde{s}_\text{o}(t)|^2 = |G_\text{e}|^2\rho_\text{i}^2(t)$, and that $P_\text{o} = |G_\text{e}|^2 P_\text{i}$ as the average power of such bandpass signal can be derived as half the expectation of any of its complex envelopes, as given by equation (1.64). As a result, we can transpose equation (5.329) as

$$\tilde{n}_{\text{o,AM-AM}}(t) = \frac{G^2}{|G_\text{e}|^2}\frac{G}{G_\text{e}}\frac{P_\text{o}}{G^2\text{IIP3}}\left[(\Gamma_4 + 1) - \frac{\rho_\text{o}^2(t)}{2P_\text{o}}\right]\tilde{s}_\text{o}(t). \tag{7.4}$$

Whatever the selected approach, we also need to express the linearity characteristics of the line-up in terms of the OCP1 as this is the quantity we decided to use in Section 7.2.2. This can be done quite simply, recalling that G^2IIP3 is the device OIP3. Those quantities are indeed linked by the line-up small signal gain G as discussed in Chapter 5. In turn, the device OIP3 can be expressed as a function of the device OCP1 using equation (5.71). In the above equation, the Γ_4 constant characterizes the statistics of the instantaneous amplitude of the bandpass signal being processed in the line-up. Here, as we assume that we are dealing with a pure OFDM signal, we then have the possibility of approximating the distribution of its instantaneous amplitude as a Rayleigh distribution, as illustrated in "OFDM" (Section 1.3.3). We can thus take this constant to be equal to 1 as given by equation (5.130a).

The same kind of derivation can obviously be done for the transposition of equation (5.325). However, we can also get the result directly if we simply remark that the structure of the expressions for $n_{\text{o,AM-AM}}(t)$ and $n_{\text{o,AM-PM}}(t)$ as given by those equations is exactly the same except for the substitution of $1/(2\text{IIP3})$ by $-\text{j}\Theta$. However, we also need to keep in mind that in this substitution, Θ refers to the equivalent input instantaneous amplitude of the signal through equation (5.315). If we want to transpose this equation to the device output, we need to use the equivalent output quantity Θ_o. Due to the structure of equation (5.315), we get that this output quantity can be expressed as $\Theta/|G_\text{e}|^2$. We thus need to simply substitute $1/(2\text{IIP3})$ by $-\text{j}|G_\text{e}|^2\Theta_\text{o}$ to derive an expression for $n_{\text{o,AM-PM}}(t)$ from that for $n_{\text{o,AM-AM}}(t)$. We obtain

$$\tilde{n}_{\text{o,AM-PM}}(t) = -2\text{j}\Theta_\text{o}\frac{G}{G_\text{e}}P_\text{o}\left[(\Gamma_4 + 1) - \frac{\rho_\text{o}^2(t)}{2P_\text{o}}\right]\tilde{s}_\text{o}(t). \tag{7.5}$$

Now equations (7.4) and (7.5) are both obviously functions of the effective gain G_e, the only expression for which is given by equation (5.322). As a result, we still have a dependency of these equations in the equivalent input power of the signal P_i through the term $\mathbb{E}\{\rho_\text{i}^2\}$. In order to get rid of this, we might express G_e as a function of P_o using the fact that $P_\text{o} = |G_\text{e}|^2 P_\text{i}$. But then we would need to invert the equation to solve for G_e. For numerical simulations, it would be more straightforward to sweep the input power to derive an equivalent power transfer function for the line-up. Then for a given output power, we can derive the effective gain to be used in the above equations by considering the appropriate equivalent input power. In practice, we thus see that there are commonalities between the two approaches discussed above.

Thus, using the OCP1 and the AM-PM conversion factor given in Section 7.2.2, we can now perform the simulations and derive the fraction of power retrieved in the adjacent channel as a function of the output power. We obtain the various contributions shown in Figure 7.12. This figure gives a clear illustration of the different impacts of the noise terms depending on whether they are additive, multiplicative, or distortion noises:

(i) The additive noises, which reduce in the present case to the RF thermal noise floor assumed constant at the output of the line-up, lead to an output noise power contribution that is independent of the transmit output power. This necessarily results in an increase in the ACLR achieved of 1 dB per decibel of this output power.

(ii) The multiplicative noises, which reduce with our assumptions to the contribution of the LO phase noise only, lead to an output noise power that exactly scales with the output power. The multiplicative noise contribution results in an upper bound for the maximum achievable ACLR, whatever the considered transmit output power.

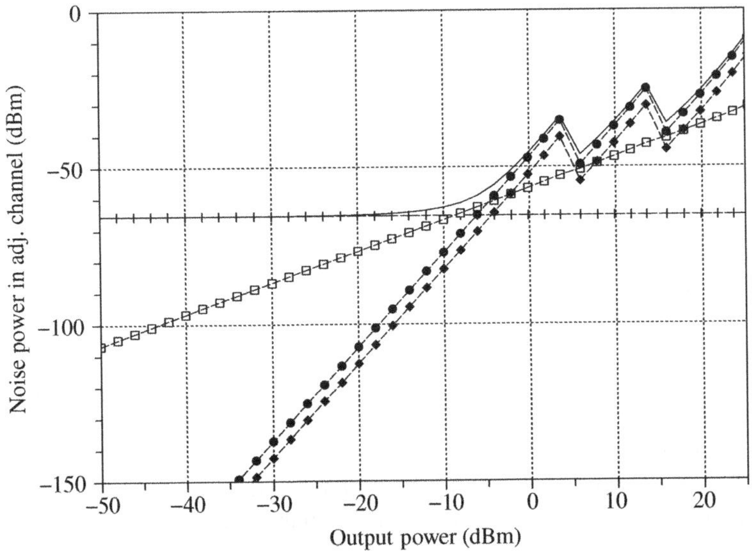

Figure 7.12 Output power of the noise components involved in the ACLR budget for the first adjacent channel as a function of the transmit output power – Due to its additive behavior, the RF thermal noise contribution is constant as a function of the transmit output power (plus). We thus get an increase of the ACLR of 1 dB per decibel of the transmit output power for the lowest part of its DR. Then, the LO phase noise contribution scales with this output power due to its multiplicative behavior (squares). This leads to a given upper bound for the ACLR whatever the transmit output power. Finally, the distortion noise composed of the AM-AM term (dots) and of the AM-PM term (filled diamonds) has a power that rises more rapidly than the output power. This leads to a decrease in the ACLR of 2 dB per decibel of the transmit output power for the highest part of its DR. The limitations associated with each kind of noise term can be retrieved on the total noise curve (solid).

(iii) The distortion noises, composed of the terms linked to AM-AM and AM-PM conversion, lead to an output noise power that increases more rapidly than the transmit output power. Based on a simple third order expansion of the device nonlinear transfer function, the power of such distortion terms rises three times more quickly than that of the output power. This means an ACLR term linked to this component that degrades by 2 dB per decibel of the transmit output power.

It is thus necessarily the additive noises that limit the ACLR for low enough output power. In the same way, it is necessarily the distortion noises, whose powers increase more quickly than for any other noise term, that limit the ACLR for the highest part of the output power. In between those extreme cases, i.e. in the range where the output power is sufficiently high that the additive noise floor remains negligible and sufficiently low that the distortion noises also do, it is the multiplicative noises that limit the ACLR performance due to their scaling with the output power. This behavior is in fact very general as these three groups of noise terms are encountered in any SNR budget in practice, here expressed in terms of ACLR.

The analysis of Figure 7.12 makes apparent the relative importance of various noise contributions, even when belonging to the same group (additive, multiplicative, or distortion). It

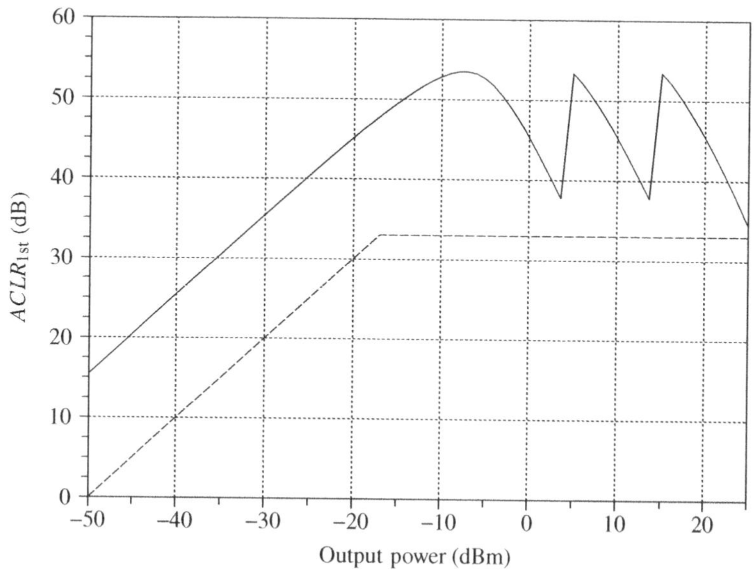

Figure 7.13 ACLR performance achieved for the first adjacent channel as a function of the transmit output power – For the noise contributions shown in Figure 7.12, we achieve the present ACLR performance (solid). As expected, we get an increase in the ACLR of 1 dB per decibel of the transmit output power in the lowest part of its DR due to the additive noise floor, and a degradation by 2 dB per decibel in the highest part of the transmit output power DR due to the distortion terms. Due to this rapid degradation, we need to manage in a consistent way the switching of the CP as a function of the output power in order to fulfill the required mask (dashed).

thus allows us to optimize the characteristics of the line-up in order to achieve a good balance of the constraints between its various blocks.

On top of this balance between the various contributors, we also need to check that the overall ACLR performance conforms to requirements. Figure 7.13 displays the ACLR achieved as a function of the transmit output power when the noise contributions shown in Figure 7.12 are summed together. The requirements are those discussed in "Adjacent channel leakage ratio" (Section 3.2.1), and shown in Figure 3.12(right). We see the direct impact of the different switching points considered for the power consumption reduction of the RF devices. Note that although our present assumptions for the characteristics of the line-up seem to be fine in order to achieve the required ACLR performance, we have to keep in mind that the same characteristics also impact other aspects of the transmitter performance. Thus, in order to conclude on these specifications, we need to check their overall impact through the inspection of the other budgets. This is illustrated for instance by the budget linked to the modulation quality in Section 7.2.4.

Frequency Planning

In order to conclude this section dedicated to the illustration of the budgets linked to respect for the wireless environment, we need to say a few words about frequency planning. For integrated transceivers, this is an important topic to which careful consideration must be given. However, we need to confess that only high level guidelines can be derived, as a case by case analysis often needs to be carried out for a given implementation. This explains why the discussion here is so brief.

Recall that, as detailed in Chapter 6, either due to some feedthrough at the device implementation level, or due to direct electromagnetic coupling between RF paths, a fraction of the RF signals involved in a line-up may be retrieved at an unexpected RF port. Practically speaking, as discussed in Section 6.4, the unexpected RF signals in such a line-up for the most part reduce on the one hand to the LO signals used to drive the analog mixing stages and on the other hand to the clock signals used to drive the digital logic when implemented in a synchronous way. In the latter case, we need to keep in mind that even if a low speed clock is used, due to the square waveform of such signal as required to drive flip-flops, the high order harmonics that lie in the RF world can have a non-negligible magnitude – at least enough to corrupt the transmit spectrum through what is classically a spurious emission.

As a result, even taking all possible care to avoid such coupling, either by careful layout or by direct control of the clocking waveforms, unwanted harmonics of those signals are often retrieved at the output of a transmitter. Thus, it is good practice to carefully select the frequencies of those signals so that we get none of their harmonics falling in the problematic areas of the transmit spectrum. Here, by "problematic" we mean for instance the critical frequency bands often dedicated to other wireless standards, as discussed in "Frequency domain requirements" (Section 3.2.1) and illustrated in Figure 3.10.

However, in light of the above discussion, we may object that in a direct conversion transmitter, the freedom to select at least the LO frequency is somewhat limited. Fortunately, as it is mainly the odd order harmonics of this waveform that are involved, we mostly retrieve unwanted harmonic frequencies that are on the one hand not directly in the frequency bands of interest, and on the other hand can often be handled at the transmitter output, through filtering for instance. This is obviously not the case for the clock signals due to their lower frequency. In that case, we often get that many of their harmonics can lie within the frequency bands of interest. As a result, the handling of the clock spurs can become tricky, as illustrated in Figure 6.43. The frequency planning must then be done with a particular care.

In this section we have mainly discussed the impact of frequency planning on the quality of the out-of-band transmit spectrum. However, this should not be allowed to mask the fact that the frequency planning imposed by the direct conversion, i.e. the fact of having an LO signal at the carrier frequency, can lead to self-degradations in the performance of the transmitter. It can in particular degrade the quality of the modulation through some pulling effects depending on the structure of the synthesizer used for the generation of this LO signal, as discussed in Section 8.1.1.

7.2.4 *Budgets Linked to the Modulation Quality*

Having formulated budgets to check the ability of our transmitter to respect the wireless environment, we need to check the quality of the modulation that it delivers to the air interface in order to ensure the performance in terms of data rate. As discussed in Chapter 3, different metrics can be used. For most of these, however, only a single parameter is involved so a budget is hardly necessary. This is, for instance, the case for the carrier frequency error, which can be considered on its own. All we really need to consider is the EVM budget, and it is to this that we now turn.

EVM Budget

We focus on the EVM recovered at the output of our transmitter – more precisely, the RMS EVM. The RMS EVM budget is sufficient for our purposes, and allows us to keep the derivations simple as it basically reduces to an in-band SNR budget as highlighted in Chapter 3, in particular equation (3.5). That said, we can refine our approach slightly to include additional terms that corrupt the signal and that cannot strictly be considered as noise – in particular, the LO leakage that results in an origin offset in the modulating symbols. Thus, in our RMS EVM approach we deal with the power of the terms involved, but also include deterministic imperfection in our budget.

Let us first list the in-band contributions we have to consider for the derivation of our RMS EVM budget in our direct-conversion transmitter. For that purpose, we again consider the different blocks involved in the line-up and review their potential in-band degradation on the signal being processed. More precisely, we can consider the three high level families of blocks involved in the line-up, i.e. the baseband blocks, those dedicated to the frequency upconversion, and the RF blocks. Looking at Figure 7.4, we can make the following observations:

(i) The baseband blocks have an in-band contribution mainly linked to the intrinsic noise performance of the analog blocks and to the quantization noise of the digital data path. We assume for the sake of simplicity that the linearity of the DAC stage and the analog baseband blocks is good enough so that the corresponding nonlinear EVM generated is negligible. However, as discussed in Section 7.2.2, and illustrated in Figures 7.5 and 7.6, these baseband contributions add to the signal being processed prior to the variable gain in the line-up. As a result, the baseband noise contributions necessarily scale with the signal at the transmitter output. Thus, although we are dealing with additive noise sources, these contributions appear multiplicative when inspected from the transmitter output. Furthermore, we observe that the analog reconstruction filter may generate some linear EVM through its frequency response both in amplitude and phase. However, we assume that this response is either smooth enough in the in-band of the signal being processed, or equalized in the digital domain so that the linear EVM linked to this analog reconstruction filtering stage is negligible in the overall budget.

(ii) The frequency upconversion stage can then have an in-band impact through different contributions. First of all, we have the same contribution as encountered in the ACLR case, i.e. the phase noise of the LO signal used to drive the mixers implemented as choppers. The LO spectrum recovered as transposed around the signal being processed necessarily leads to an in-band SNR limitation. We also have a potential rise in the image signal when dealing with RF impairments between the P and Q branches of the complex upmixer. As detailed in Sections 6.2.2 and 6.3.2, such RF impairments lead to the superposition with the signal being processed of an unwanted component whose complex envelope, when defined as centered around the carrier frequency, is simply the complex conjugate of the one being processed. As a result, the spectrum of this complex envelope is simply the flipped copy around the carrier frequency of that of the complex envelope of the signal being processed, but scaled by some factor. Assuming that we are dealing with a frequency transposition based on the use of the fundamental tone of the LO signal, this scaling factor is simply equal to the IRR. Furthermore, given that the reconstructed complex lowpass modulating waveform is centered around DC at the upconverter input and has a symmetrical spectrum, which is the case for most of the modulating waveforms of interest, as discussed in "Impact of spectral symmetry on top of stationarity" (Section 1.1.3), the spectrum of both the signal being processed and its image have exactly the same shape. This was why there was no additional impact of the image signal on the adjacent channel, as discussed in "ACLR budget" (Section 7.2.3). But it also explains why the resulting image signal directly impacts the in-band performance in our present direct conversion case. Finally, we need to consider in our budget the LO leakage retrieved at the transmitter output either through direct feedthrough at the device implementation level, or due to electromagnetic coupling between RF paths.

(iii) The RF devices can have some in-band contributions that in practice correspond to those encountered in the ACLR case. This means degradations linked to either the intrinsic noise of the blocks or the direct distortion of the signal due to the smooth compression. As discussed in the section "Nonlinear EVM due to odd order nonlinearity" (Section 5.1.3) and "Nonlinear EVM and spectral regrowth due to AM-PM conversion" (Section 5.3.2), such behavior linked to the odd order nonlinear of the line-up directly generates noise terms centered around the carrier frequency due to either AM-AM or AM-PM conversion. Thus, as we assume that we are dealing with an amplitude modulated RF bandpass waveform, we need to consider such nonlinear contributions in our EVM budget.

As during the derivation of the ACLR budget, it is of interest to classify the various noise components depending on whether they can be considered as additive, multiplicative, or distortion noises. This will allow a better interpretation of the root causes of the various limitations that confront us in the final performance of the line-up as discussed at the end of the section. Thus, based on the above review of the main contributions that we can expect in the line-up, the additive noise components we need to consider reduce to the baseband thermal and quantization noises, even though their contributions can appear multiplicative when inspected from the transmitter output as discussed above; the multiplicative noises reduce to the image signal, the LO phase noise, and the LO leakage;[3] and for the distortion noises, we have AM-AM and AM-PM conversion terms, both linked to the smooth compression expected to occur in the RF stages.

Then, we need to determine the power of the in-band noise contributions involved in the RMS EVM budget. For that purpose, in contrast to the ACLR case, we may not need a detailed evaluation of the power spectral densities of those components. As can be seen in Figure 7.14, the noise terms we are dealing with are either wideband noises, assumed white over the frequency band of interest, or bandpass and centered around the carrier frequency. As a result, the power of the former category of contributors can simply be derived as their power spectral density level times the noise bandwidth δf. For the latter category, we observe that due to their bandpass behavior, most of their power is concentrated in-band. Thus, as a first

[3] This can be considered as a multiplicative component at least in the upper part of the output power DR, as discussed in due course.

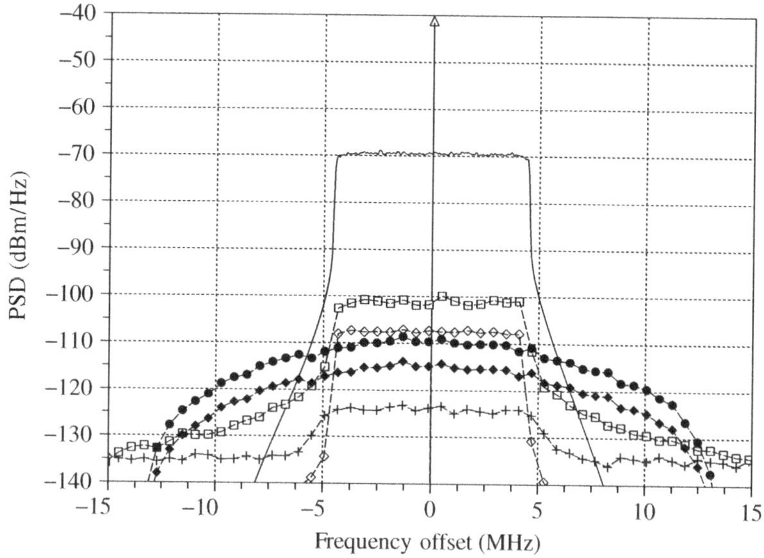

Figure 7.14 Output PSD of the noise components involved in the EVM budget – In the RMS EVM budget, the main noise contributions reduce to the additive noise components composed of both the baseband thermal and quantization noise and of the RF noise floor (pluses); the multiplicative terms composed of the image signal, whose PSD is a scaled flipped copy of that of the wanted signal (diamonds); the LO phase noise (squares) and the LO leakage (triangle); and finally, the distortion noises components that are composed of the AM-AM (dots) and AM-PM terms (filled diamonds).

step it is sufficient to approximate their in-band power as their overall power. This approach allows for the use of analytical formulations more suited to high level budgets than the time domain simulation and spectral estimation approach.

Let us focus first on the derivation of the in-band power of the additive noise contributions, which reduce here to the thermal and quantization noises contributions in the line-up. In fact, this contribution has already been derived in Section 7.2.2 as the additive noise performance of the transmitter clearly depends on the settings of the line-up, mainly through its gain, and thus on the targeted transmit output power. As a result, we thus assume in what follows that we are dealing with the additive noise performance shown in Figure 7.7 for the derivation of our budget. Here, we merely recall that these values are based on a baseband noise performance corresponding to an in-band SNR of 55 dB for a sine wave at the FS and on an RF noise floor of −135 dBm/Hz at the transmitter output.

Then, assuming that we are dealing with a direct frequency transposition using the fundamental tone of the LO waveform, the output power of the image signal can be directly derived from the transmit output power through the IRR value according to the material detailed in Section 6.3.2. In practice, considering reasonable impairments in the generation of the square LO waveforms used to drive mixers implemented as choppers, we can assume as an example a reasonable value of −38 dB for the IRR. Moreover, in our line-up all the variable gain is performed in RF stages, i.e. after the frequency upconversion. As a result, we can assume that the RF impairments at the origin of the rise in the image signal remain constant whatever the settings of the line-up, and the IRR likewise. We can thus simply express the power of the image signal at the transmitter output as $IRR \cdot P_o$, with $IRR = -38$ dB and P_o the transmit output power.

For the LO phase noise, the derivation is again much simpler than in the ACLR case. Indeed, as derived in Section 4.3.3, the SNR achieved through the presence of the LO phase noise centered around its fundamental tone is simply the inverse of the phase noise power, P_{n_ϕ}, as given by equation (4.216). Thus, as we are dealing here with a pure multiplicative noise, we can express the power of the bandpass noise recovered as centered around the carrier frequency due to this phase noise as $P_{n_\phi} P_o$. In the present case, we can assume that the phase noise power P_{n_ϕ} is equal to the realistic value of 1.5° RMS, i.e. $1.5\pi/180$ rad RMS.

For the LO leakage, we may slightly refine our model compared to what might be expected for a simple multiplicative component. We have already said that there are different root causes for such LO leakage recovered at the transmitter output. It may be due either to feedthrough at the mixer implementation level, or to some electromagnetic coupling directly toward the transmitter output. But in practice, these different mechanisms lead to different behaviors of the resulting leakage regarding the transmit signal. Indeed, in the case of a feedthrough at the mixer level, we get a component that is injected in the line-up prior to the variable gain that takes place after the frequency upconverter. As a result, we can expect this component to effectively scale with the signal being processed and thus behave as an exact multiplicative component. But in the case of an electromagnetic coupling directly toward the transmitter output, we get a leakage injected in the line-up after the variable gain. This leakage component thus necessarily behaves as an additive component, i.e. with a level independent of that of the signal being processed. Thus, we can assume that the LO leakage is composed of the superposition of an absolute level linked to the electromagnetic coupling equal to −90 dBm at the transmitter output and a component that scales with the transmitted signal, assumed −40 dBc below this signal.

Finally, we evaluate the power of the distortion noises. For that purpose, we first envisage reusing the analytical expressions for the complex envelopes of those bandpass noises as considered for the derivation of the ACLR budget above. The power we are looking for can be derived from the PSD of those components. But, as highlighted previously, it is often sufficient for a high level budget to approximate the in-band power of such bandpass noise terms as their total power. This offers the possibility of obtaining analytical expressions for those powers that remain more suited to the handling of budgets in practice. In order to do this, we first observe, by looking at equations (5.190) and (5.329), that the output complex envelope of the AM-AM component recovered in the case where both AM-AM and AM-PM conversion exist has the same expression in most practical cases as the one recovered when only AM-AM conversion is present. As a result, the power of the distortion term linked to the AM-AM conversion effect can be evaluated using the analytical derivations in "Nonlinear EVM due to odd order nonlinearity" (Section 5.1.3). Indeed, the average output power $P_{o,\text{AM-AM}}$ of this bandpass noise can simply be derived as half the expectation of the square modulus of any its complex envelopes, as given by equation (1.64). We can thus use equation (5.194) to achieve our aim. Looking at this expression, we see that it is still the equivalent input characteristics of the signal being processed that are involved. Thus, in order to derive our budget as a function of the transmit output power, we can consider the two approaches already discussed for the ACLR budget above. This means on the one hand that we continue to consider such analytical expressions as a function of the equivalent input characteristics of the signal and sweep its equivalent input power P_i. We can then make the link with the transmit output power using the effective gain of the device, G_e, given by equation (5.322). On the other hand, we can transpose these equations as a function of the output characteristics of the signal being processed. For that purpose, assuming that the complex envelopes we are dealing with are defined as centered around the carrier frequency, we can argue that the equivalent output complex envelope of the expected signal, i.e. when undistorted, $\tilde{s}_o(t)$, is related to its input complex envelope through the effective gain of the line-up. Thus, given that $\tilde{s}_o(t) = G_e\tilde{s}_i(t)$, we can write $\rho_o^2(t) = |\tilde{s}_o(t)|^2 = |G_e|^2\rho_i^2(t)$ and $P_o = |G_e|^2 P_i$. As a result, we can transpose equation (5.194) so that we can write

$$P_{\text{o,AM-AM}} = \frac{G^6}{|G_e|^6} P_o \left(\frac{P_o}{G^2 \text{IIP3}}\right)^2 [(\Gamma_6 + 1) - (\Gamma_4 + 1)^2]. \tag{7.6}$$

Whatever the approach chosen, we can express the linearity characteristics of the line-up in terms of the OCP1 as determined in Section 7.2.2. Recall in so doing that G^2IIP3 is the device OIP3. Those quantities are related to each other through the line-up small signal gain G as discussed in Chapter 5. In turn, the device OIP3 can then be expressed as a function of the device OCP1 using equation (5.71). In the above equation, the Γ constants characterize the statistics of the instantaneous amplitude of the bandpass signal being processed in the line-up. Assuming that we are dealing with an OFDM signal, we can approximate these constants with those of the Rayleigh distribution, i.e. $\Gamma_4 = 1$ and $\Gamma_6 = 5$ as given by equations (5.130a) and (5.130b), respectively.

In order to derive the expression for the average output power of the AM-PM term, $P_{\text{o,AM-PM}}$, we can use the same methodology as for the ACLR budget, and simply substitute in the above expression $1/(2\text{IIP3})^2$ by $(|G_e|^2\Theta_o)^2$. We obtain

$$P_{\text{o,AM-PM}} = 4\Theta_o^2 \frac{G^2}{|G_e|^2} P_o^3 [(\Gamma_6 + 1) - (\Gamma_4 + 1)^2]. \tag{7.7}$$

However, as already discussed, the two above expressions are still dependent on the equivalent input power of the signal P_i through the effective gain of the line-up G_e. But, simpler than trying to get rid of it, we can sweep P_i in order to derive an equivalent power transfer function for the line-up according to $P_o = |G_e|^2 P_i$. Then for a given output power, we can derive the effective gain to be used in the above equations by considering the appropriate equivalent input power.

Using the above models, we then achieve the in-band noise power contributions shown in Figure 7.15 as a function of the transmit output power P_o. Looking at this figure, we can identify the different categories of noise components and the associated limitations in the line-up performances. In particular, we can make the following observations:

(i) The additive noises, that reduce here to the sum of the RF thermal noise floor and the fraction of LO leakage directly coupled at the transmitter output, result in a constant output noise contribution independent of the transmit output power. This noise contribution thus leads to an improvement in the EVM of 1 dB per decibel of the transmit output power in the lower part of its DR.

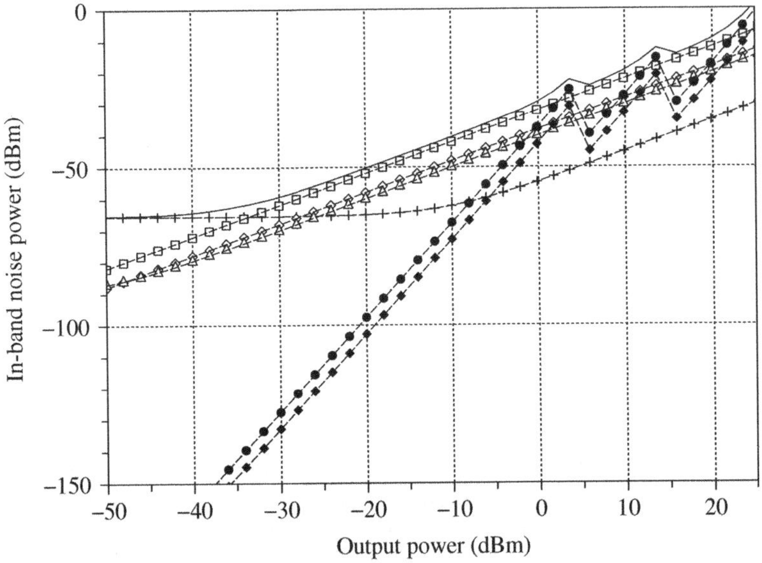

Figure 7.15 Output power of the noise components involved in the EVM budget as a function of the transmit output power – Having the variable gain in the line-up set in RF, the additive RF noise component leads to a constant output noise floor contribution whereas the baseband one scales with the transmit output power (plus). The RF additive noise thus leads to an improvement in the EVM of 1 dB per decibel of the transmit output power in the lowest part of its DR, whereas the baseband component sets a lower bound for it whatever the transmit output power. This is the same for the multiplicative noises, i.e. the image signal (diamonds), the LO phase noise (squares) and the LO leakage in the upper part of the output power DR (triangles). Then the AM-AM (dots) and AM-PM conversion (filled diamonds) terms lead to a degradation in the EVM of 2 dB per decibel of the transmit output power due to their rapid rise. Each kind of limitation can be retrieved on the total noise curve (solid).

(ii) The multiplicative noises, composed of the LO phase noise, the image signal, and the LO leakage component injected in the line-up before the variable gain, lead to an output noise contribution that exactly scales with the transmit output power. As a result, this multiplicative noise contribution leads to a lower bound in the achievable EVM, whatever the considered output power.

(iii) The distortion noises, here composed of the terms linked to AM-AM and AM-PM conversion, lead to an output noise power that increases more quickly than the transmit output power. Based on a simple third order expansion of the nonlinear device transfer function, the power of those noise terms rises three times faster than that of the transmit output power. This results in an EVM term linked to such components that degrades by 2 dB per decibel of the transmit output power.

We thus recover the same behavior as already encountered in the analysis of the ACLR budget, i.e. that it is necessarily the additive noise terms that limit the EVM performance for sufficiently low output power. In contrast, it is the distortion noise terms, whose powers increase more quickly than for any other component, that necessarily limit the performance in the highest part of the transmit output power DR. In between those extremes, i.e. in the range where

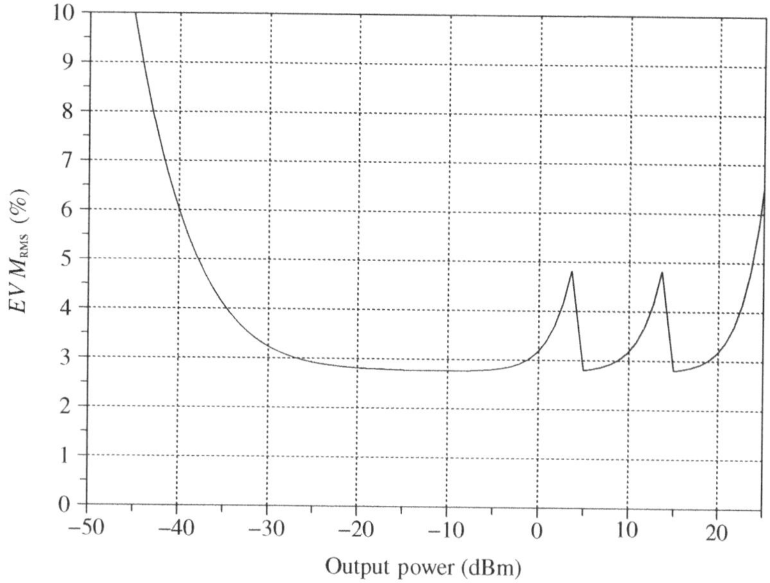

Figure 7.16 RMS EVM performance as a function of the transmit output power – For the noise contributions shown in Figure 7.15, we achieve the RMS EVM performance shown. As expected, we get an improvement of the RMS EVM as a function of the transmit output power in the lowest part of its DR due to the behavior of the additive noise terms. In contrast, we experience a quick degradation in the highest part of the output power DR due to the distortion terms. This degradation needs to be managed in a manner consistent with the switching points set for managing the power consumption of the line-up, but also in a manner consistent with the ACLR performance discussed earlier.

the output power is sufficiently high that the additive noise floor remains negligible but also sufficiently low that the distortion noises also does, it is the multiplicative noises that limit the performance.

Summing together the noise contributions shown in Figure 7.15, we finally achieve the RMS EVM performance shown in Figure 7.16 as a function of the transmit output power. We thus see how such a budget allows us to derive in a straightforward way the impact of the various parameters of the transceiver on its performance. However, as highlighted in "ACLR budget" (Section 7.2.3), this has to be done in a manner consistent with the other budgets.

7.2.5 Conclusion

To conclude this discussion of the transmit side, we first of all observe that the budgets reviewed so far have been derived for a given performance assumed for the blocks that constitute the line-up. But, practically speaking, the performance of at least the RF/analog blocks depends on many parameters, such as their physical temperature or the variations in their power supply. Furthermore, when contemplating mass production of the transceiver we are dimensioning, we have to anticipate some variations in the implementation process used at the factory level so that discrepancies from device to device are taken into account. The end result is that our transmit functionality necessarily achieves variable performance depending on whether we are considering a line-up implemented with a typical process and used in typical temperature and power supply conditions or whether we are dealing with worst case parameters. Thus, we can anticipate variable performance of the line-up depending on what device we are considering and under what conditions of usage. In practice, this may require budgeting to be done with a different set of parameters in order to evaluate the impact of those conditions on the performance.

Furthermore, we need to keep in mind that we have only considered budgets in the strict sense when dealing with the modulation quality for instance. This means budgets that involve many different parameters that need to be managed at the same time for achieving a given performance. However, as highlighted in the introductory part of Section 7.2.4, this should not be allowed to mask the fact that performance is sometimes directly linked to a single implementation limitation and needs to be checked as such. This is for instance the case for the frequency precision achieved for the clocking schemes and the LO signals illustrated in Chapter 9, Section 9.1.5.

In the same way, the purpose of this chapter is only to illustrate how to carry out budgets. As a result, for the sake of simplicity we have considered only the main representative contributors to the signal degradation in the line-up. But additional effects, assumed here to have negligible impacts, such as the DAC or the analog baseband linearity, should also be taken into account if required. Furthermore, we deliberately did not take into account in our budgets potential degradations linked to the structure of the architecture itself. In our present direct conversion case, this could be the pulling effect of the RF synthesizer by some leakage of the transmit signal, as described in "Frequency planning" (Section 7.3.3). Such specific system degradation related to the structure of the architecture itself rather than to a characteristic of one constituent block of the line-up is addressed in Chapter 8 when discussing the main differences between various classical architectures.

7.3 Budgeting a Receiver

Let us now focus on the receive side. For that purpose, as done for the transmit side in the early stages of Section 7.2, we first review the functions embedded in the direct conversion receiver considered as an example in this chapter. This helps us to understand the kind of limitations and degradations we have to deal with in such a line-up while trying to fulfill the requirements. Then we go on to derive some practical budgets.

7.3.1 Review of the ZIF RX Problem

Considering the blocks involved in the implementation of the direct conversion receiver (RX) shown in Figure 7.2, we observe that pollution of the wanted signal can result either from the intrinsic behavior of the signal processing functions embedded in the line-up, at least in the presence of additional unwanted signals, or due to the limitations in the implementation of those functions. We review the theoretical impact of these signal processing functions before moving on to the degradations linked to their physical implementation. This approach highlights the root causes in the limitations we face in the overall performance of a typical direct conversion receiver, as well as how to manage the potential presence of additional unwanted signals like blockers, which is one of the main differences compared to the transmit side.

The Signal Processing Approach

Let us again consider the signal processing functions embedded in the direct conversion receive line-up shown in Figure 7.2. As done for the transmit side, the impact of those functions on the signal being received can be examined in the spectral domain. This approach is even more interesting in the present receive case as it allows a clear representation of the receiver behavior in the presence of unwanted blocking signals.

In the present case where we are considering a receiver embedded in a full duplex transceiver, we are faced with such a blocking signal through the presence of the TX leakage at the input of the line-up, as discussed in "Full duplex vs. half duplex" (Section 3.1.3). Thus, let us consider the situation representative of a reception close to sensitivity where a weak wanted signal is recovered at the input of the receiver in addition to the TX leakage signal. In this configuration, we need to understand that the most likely reason for a weak wanted signal is that the receiver is far from the other device it is connected to, say the base station. In turn, we thus may have our transmitter operating at the maximum output power it can deliver in order to achieve a reasonable uplink quality, still toward the base station. As a result, the reception of a weak wanted signal is almost always associated with the presence of the strongest possible TX leakage at the input of the receiver in a full duplex transceiver. In practice, the difference between the power of the two signals may be up to 80 dB. Thus, assuming such a configuration, we recover the typical spectrum displayed in Figure 7.17 along the receive path. Looking at this figure, the following comments are in order:

(i) Up to the frequency downconversion, we are dealing with blocks in the line-up expected to provide mainly amplification. This means that, when considering an ideal implementation, we recover at the input of the frequency downconversion the same spectral shape as at the input of the receiver, without any particular distortion.

(ii) Then, considering on the one hand a frequency downconversion that uses the fundamental tone of the LO waveform, and on the other hand that this frequency is equal to that of the input wanted signal f_{RX}, we directly recover this wanted signal as centered around DC at the frequency conversion output. Moreover, as we are considering a complex frequency conversion in the present case, when no impairments exist the spectrum of the reconstructed complex output signal $p(t) + jq(t)$ is simply a shift in frequency of the input spectrum, as discussed in Section 6.1. Thus, in addition to the wanted signal that is transposed around DC, we recover the TX leakage signal at the duplex distance $\delta f = |f_{TX} - f_{RX}|$ from it, more precisely around $-\delta f$ in the present case. On top of that, although not represented here due to the frequency planning considered here, we need to highlight that the use of square wave LO signals used to drive mixers implemented as choppers can result in the folding on the wanted signal of blockers lying at the odd order harmonic frequencies $(2l + 1)f_{RX}$ through the harmonic mixing problem. This problem is discussed in "Frequency planning" (Section 7.3.3).

(iii) We then need to provide the signal in a sampled and digitized way to the baseband in order to carry out further signal processing. For that purpose, we may use an ADC at some point. But, as the sampling rate necessarily remains finite, we need to ensure that at least no aliasing of unwanted signals can occur on the wanted signal, as discussed in Section 4.6.1. This means that we often need to use an analog anti-aliasing filter prior to the sampling of the signal in order to sufficiently attenuate the unwanted components.

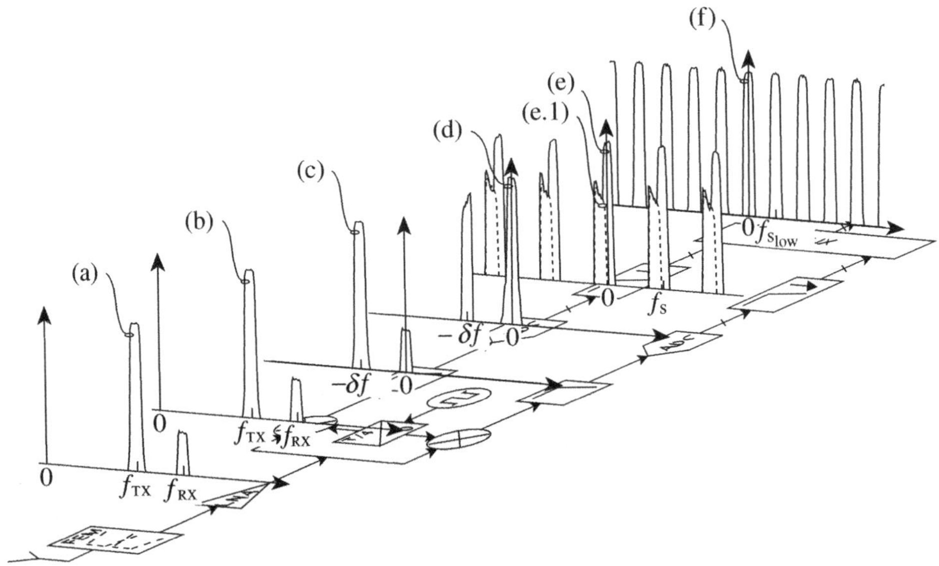

Figure 7.17 Spectrum of the signals recovered at the different stages of a direct conversion receiver: ideal implementation case – Assuming that we are dealing on the one hand with a receiver embedded in a full duplex transceiver and on the other hand with a reception close to sensitivity, we classically are faced with the reception of a strong TX leakage signal, centered around the TX carrier frequency f_{TX}, on top of a weak wanted signal, centered on the receive one f_{RX}, at the input of the active part of the receiver (a). When implemented in an ideal way, the first LNA stage provides only amplification without any distortion or noise addition (b). The amplified signal is then downconverted to baseband through a complex mixer. As we are considering a direct conversion receiver, the wanted signal, originally centered around f_{RX} at the RX input, is downconverted to baseband. In the spectrum of the reconstructed baseband complex signal at the downmixer output, $p(t) + jq(t)$, we thus recover the TX leakage signal at the duplex distance $\delta f = |f_{TX} - f_{RX}|$ from DC, more precisely centered around $-\delta f$ in the present example (c). This TX leakage signal is then attenuated by the anti-aliasing analog filter (d). The filtered signal can then be sampled and digitized by the P and Q ADC blocks. This operation leads to a periodization of the spectrum of the sampled waveform (e). As a side effect, it can result in a potential folding of the TX leakage signal on the wanted signal, depending on the one hand on the duplex distance $-\delta f$ relative to the sampling rate f_s, and on the other hand on the efficiency of the anti-aliasing filtering (e.1). The residual unwanted sidebands that do not yet fold on the wanted signal can then be filtered out by a digital filter before downsampling. This is done in order to implement further advanced signal processing in the baseband modem side at the lowest possible data rate (f).

(iv) However, in practice the sampling at the ADC stage is often done at a higher rate than is strictly required to correctly represent the wanted signal alone. This is obviously done with the aim of reducing the filtering requirements for the analog anti-aliasing filter that traditionally leads to a non-negligible implementation area. As a result, we may need some further decimation stages in the digital domain. But, as we need to avoid any potential aliasing of the residual blocking signals also during this processing, some digital filtering is again required prior to the decimation.

Keeping this signal processing in mind will allow us to understand that even when dealing with theoretical signal processing functions, we already need to take some care in the dimensioning of the line-up, mainly due to the potential presence of unwanted signals. This is obviously the case for the filtering blocks in order to correctly address the aliasing problems as well as the potential harmonic mixing due to the use of mixers implemented as choppers. But we also now need to take into consideration the additional limitations in the physical implementation of those functions.

Impact of Implementation Limitations

Let us now examine the additional degradations that may occur due to the physical implementation of the signal processing functions embedded in our receiver. We continue to look at the spectral shapes recovered along the line-up in order to compare them with those achieved in the ideal case and detailed in the previous section.

Thus, looking at Figure 7.17 and Figure 7.18, we make the following observations:

(i) The early active RF stages of the receiver do not necessarily behave like pure amplifiers as we might expect. We are faced first of all with the degradation of the noise received from the source due to the intrinsic noise contribution of the blocks. In practice, this source noise results from the thermal motion of the electrons in the conductors, the noise signal retrieved from the electromagnetic field by the antenna and the noise component leaking from the transmitter when active in the present full duplex case. We are thus faced with a degradation of the SNR due to this source noise at the input of the receiver when the received signal flows through the receive path. However, as discussed in Section 4.2.4, the main purpose of the early active stages of the receiver is to provide enough amplification so that the additive noise contributions of the following blocks are made negligible. But, on top of those additive noise contributions, we also have the nonlinear behavior of the active RF blocks. As discussed throughout Chapter 5, such nonlinearity can be either of odd order, mainly due to limitations in the power supply of the devices, or even order, mainly due to mismatch in their implementations. But for devices that operate on the received RF bandpass signal, i.e. prior to the frequency downconversion to baseband, only odd order nonlinearity can lead to the generation of distortion noise components that remain centered around the wanted signal in the frequency domain. More precisely, for bandpass signals that are amplitude modulated, those generated terms can be linked to their smooth compression through either AM-AM or AM-PM conversion mechanisms. However, on the receive side we classically deal with signals of lower level than on the transmit side. As a result, often only the AM-AM conversion needs to be considered as in most cases it appears that the equivalent parasitics in the models of the active devices involved in the line-up remain constant over the DR of the received wanted signal. For the sake of simplicity, we only consider this AM-AM conversion in subsequent sections. The associated phenomena to be considered are therefore the nonlinear EVM generated through the self-compression of the wanted signal, or the XM of this signal by strong amplitude modulated blockers. The latter phenomenon still occurs even if the wanted signal remains low enough so that it lies within the linear area of the device transfer

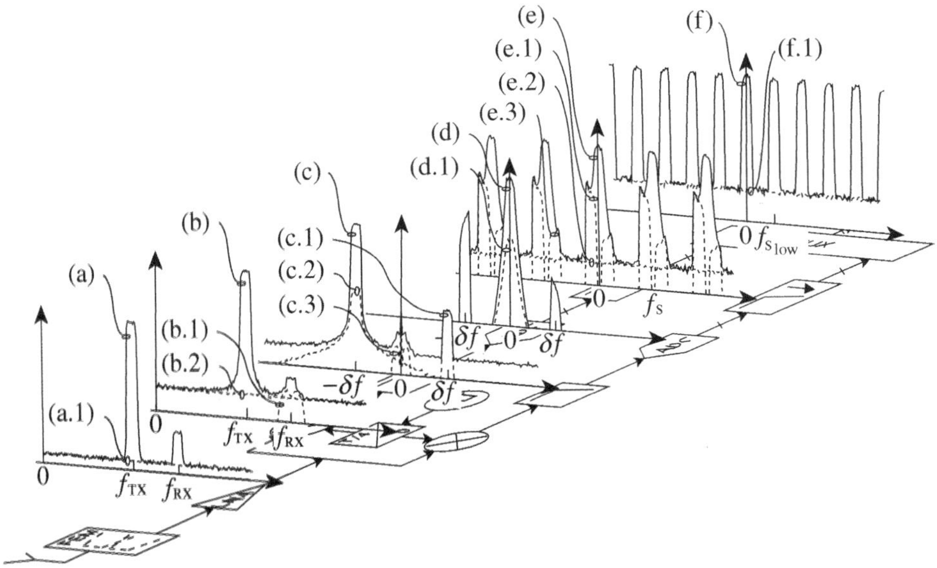

Figure 7.18 Spectrum of the signals recovered at the different stages of a direct conversion receiver: realistic implementation case – In addition to the potential degradations linked to the intrinsic behavior of the ideal signal processing functions embedded in the line-up, as illustrated in Figure 7.17, degradations may occur due to the physical implementation of those functions. At the input of the active part of the receiver, we are faced with the unavoidable noise floor linked to the thermal random motion of the electrons, the noise components captured by the antenna, or the noise leaking from the transmitter (a.1), on top of the received signal (a). Then the first amplification stage necessarily adds some electronic noise as any active device does (b.2). On top of that, even when receiving a weak wanted signal, this RF stage can lead to a degradation due to smooth compression, through the XM of the received signal by the strong TX leakage signal when amplitude modulated (b.1). At the downconversion stage, we then have a rise in the image signal on the reconstructed complex signal $p(t) + jq(t)$ due to the gain and phase imbalance between the P and Q branches of the line-up (c.1). We also have the convolution of the LO phase noise spectrum with that of each input signal (c.2). Finally, we deal with the AM-demodulation of the input signals due to even order nonlinearity (c.3). This last term exhibits a non-negligible DC component that is classically removed by a DC canceler scheme at the PGA stage. This PGA stage also provides amplification and filtering (d), but at the cost of an additional noise contribution also shaped by the filtering effect (d.1). At the output of the ADC blocks we then recover a periodic spectrum (e), with both the presence of the residual TX leakage signal (e.1) and its image (e.3) on the reconstructed complex signal. Furthermore, we have an additional noise floor degradation due to the quantization operation (e.2). Finally, the digital filtering and decimation to a lower sampling rate is necessarily done at the cost of a quantization noise increase (f.1).

function, which corresponds to the situation represented in Figure 7.18 when considering an amplitude modulated TX leakage signal.

(ii) Then, the frequency downconversion stage can impact the received signal in different ways. First of all, we get the degradation linked to the use of a noisy LO signal. As discussed in Chapter 4, this results in the addition of a phase noise component to the signal being processed during the frequency conversion. More precisely, due to the multiplicative behavior between the signal being processed and the LO signal, the phase noise component is recovered as transposed around all the bandpass signals present at the input of the frequency downconversion. This results on the one hand in an in-band SNR limitation due to the phase noise component recovered as centered around the wanted signal and on the other hand in a reciprocal degradation between the bandpass signals being processed due to the frequency extent of the phase noise spectrum. We also have the potential RF impairments between the P and Q branches of the line-up, as discussed in Chapter 6. Such impairments can lead to either a rise in the image signal or a rise in the even order harmonics $2lf_{RX}$ of the square wave LO signals used to drive mixers implemented as choppers, as summarized in "Practical LO spectral content" (Section 6.3.2). The former problem can lead to a direct in-band SNR degradation, while the latter increases the sensitivity of the receiver to the problem of harmonic mixing, as discussed in "Frequency planning" (Section 7.3.3). Finally, we have the linearity of the downmixer, which can lead to additional degradations of performance. But, due to the frequency planning of the direct conversion receiver considered here, it is mainly the even order of the downmixer that leads to degradation due to the AM-demodulation of amplitude modulated blockers, like the TX leakage signal in our example. As discussed in "AM-demodulation due to even order nonlinearity" (Section 5.1.3), whatever the frequency of the considered blocking signal at the input of the nonlinear downmixer, we recover an unwanted term centered around DC at its output. This unwanted term thus remains superposed to the wanted signal that is directly downconverted to baseband in the present case. As discussed in that section, when dealing with an amplitude modulated blocker this unwanted term behaves as an additional noise contribution.

(iii) Then we need to consider the forthcoming ADC stage required to sample and digitize the signal in order to allow for the baseband signal processing. And compared to the ideal case discussed in the previous section where we had to consider only the aliasing problem linked to the limited sampling clock used to drive such analog to digital conversion, we now need to assume that the resolution of the ADC is necessarily finite, as discussed in "Dynamic range and effective number of bits" (Section 4.6.1). Thus, on top of the implementation of an analog anti-alias filter, we also need to consider some additional analog amplification, for the most part set as variable, in order to correctly scale the received signal within the bounds set by the finite DR of the ADC. This results in the implementation of an analog PGA block that provides both filtering and variable gain. But due to its analog implementation, we can thus anticipate a degradation of the received signal through either its own noise contribution, or some nonlinear or linear distortions. However, as highlighted in the introductory part of Chapter 5, for the most part such baseband analog blocks are implemented as feedback systems based on the use of operational amplifiers. As a result, we often get very good linearity for those stages as long as no clipping occurs. Thus, assuming that the signals are scaled correctly in the receiver, as discussed in Section 7.3.2, we can assume for the sake of simplicity that the nonlinear distortion remains negligible compared to the contribution of the RF blocks. However, at the filtering stage, we may also anticipate some linear distortion as detailed in Section 4.4. But, as discussed in "SNR budget" (Section 7.3.4), this last effect can be compensated by proper equalization of the analog filters in the digital domain. Moreover, we recall that some wireless systems use modulating waveforms that are classically equalized in a way that compensates for such distortion. As a result, we also assume in what follows that we do not need to consider this linear EVM contribution in our budgets. Finally, we observe that the PGA stage is often a good place to carry out additional processing such as the cancellation of the DC offset, as discussed in Section 9.2.2. This explains why the DC component retrieved at the downmixer output on the spectrum displayed in Figure 7.18 is no longer present at the PGA output.

(iv) Then we get the ADC stage itself. As highlighted above, this block can first lead to a noise floor degradation due to its finite resolution, but also due to the potential jitter present on its sampling clock, as discussed in more depth in "Clock jitter" (Section 4.6.1). We may have a distortion of the signal being converted due to the nonlinearity of the ADC transfer function, as illustrated in "Linearity" (Section 4.6.1). Lastly, we also need to examine the anti-aliasing problem from the implementation limitation point of view. As the performance of the anti-aliasing analog filter can only be limited in practice, there may be a non-vanishing amount of aliasing of the residual blocking signals.

(v) Finally, we have the remaining digital signal processing functions the aim of which is to filter out the residual unwanted signals in order to allow for the decimation toward a lower sampling rate in a proper way. But, as illustrated in Section 4.5.3, the truncations occurring due to the finite bit resolution used in the fixed point implementation of the signal processing function necessarily lead to a quantization noise floor increase.

We thus see that the degradations linked to the implementation limitations lead to both additional in-band SNR limitations, and to some additional sensitivity to the presence of other unwanted signals.

Bearing in mind the impact of the theoretical functions we expect to implement as well as the impact of the degradations linked to their physical implementation, we can now review of the budgets we have to consider for our receiver.

Budgets to be Considered

As was the case for the transmit side, if we neglect the real time constraints linked to the control of the transceiver, there remain two kinds of budgets to be performed in order to check the performance of the device. As discussed in Chapter 3, we basically need to check on the one hand the quality of the modulation that is delivered to the baseband in order to achieve the data rate, and on the other hand the resistance of the device to its wireless environment. This means that the quality of the modulation must remain consistent even when the receiver has to deal with unwanted signals in addition to the wanted signal. In practice, those unwanted signals can be either adjacent channels that belong to the same network, or blockers that do not.

Recalling the discussion in the previous sections, we observe that an unwanted signal can corrupt the wanted signal in different ways depending on whether or not it lies close to it in the spectral domain. Due to either its convolution with the LO PN spectrum, or its XM by strong blockers like the TX leakage signal, the close in-band part of the spectrum of a signal can be more degraded than the far out-of band part. Thus a blocking signal lying close to the wanted signal can be more of a problem to handle than one lying far away. And we thus anticipate that different terms need to be considered for such budgets aimed at checking the resistance to the wireless environment, depending on the frequency location of the blocking signals, as illustrated in Section 7.3.3.

Following these remarks, we note some commonalities with the transmit side where the budgets also involve different parameters of the line-up depending on which part of the transmit spectrum is being considered. However, this is not to deny that there are fundamental differences between the two line-ups related to the fact that a transmitter processes only the wanted signal while a receiver also has to cope with all the unwanted signals collected by the antenna. This thus leads to situations that are harder to handle as we now need for instance to track the level of the different unwanted signals along the receiver on top of the wanted signal, as discussed in the next section. Moreover, the unwanted signals we need to consider do not necessarily belong to the same network. As a result, their power as recovered at the input of the receiver can have no relationship with the power of the received wanted signal, as discussed in Chapter 3. We can thus anticipate situations where we have to handle strong blocking signals and a weak wanted signal at the same time. This requires a wide enough DR in the line-up in order to handle this kind of situation. Finally, we also mention that in a transmitter, the power to be delivered to the antenna is deterministic. Thus, given the target output power, we can easily scale the signal in the line-up correctly to achieve the best possible performance for a given hardware implementation. This is obviously not the same in a receiver as we do not know the power of the received signal a priori. Thus, we need to consider margins in the line-up to handle some uncertainty in the received power. But more than that, some AGC systems may be required to optimize the setting of the line-up as a function of the received power.

Before discussing the receiver budgets, then, we need to go through the level diagrams associated with such receive line-ups. This will help us to understand how to set the receiver in order to optimally handle the DR associated with the wanted and unwanted signals. As a side effect, it also allows us to highlight why the performance of the receiver is a function of the power of the wanted signal, when considering a classical AGC scheme, exactly in the same way as the performance of the transmitter was a function of the output power. We can then derive typical high level parameters for our receive line-up as a function of the wanted signal input power in order to use them in our receiver budgets.

7.3.2 Level Diagrams and Receiver High Level Parameters

Let us first focus on how to configure our receiver in order to cope in the best possible way with the signals present at its input. For that purpose, we first of all recall the rule of thumb used for the transmit side, i.e. that we should try to set the received signal in the highest part of the receive path DR in order to maximize the signal to additive noise power ratio, while avoiding any compression in order to minimize the distortion.

On the receive side, however, the signal present at the input of the data path may not be composed of the wanted signal alone. What thus needs to be correctly scaled along the path is the total signal present, consisting of the wanted signal and potential unwanted signals. We also need to keep in mind that those unwanted signals are necessarily filtered out along the line-up. They need to be canceled before reaching the signal processing blocks where we aim to recover the data bits. Thus, we should start to consider potential difficulties for setting the receiver in the correct configuration if we only have access to some measurements done at the end of the receive line-up, i.e. after the cancellation of those blockers by the channel filters. This means that for a given level of the wanted signal we need to set the receiver in a configuration that enables us to cope with potential blockers, even if they are not necessarily present in practice.

In this chapter, we do not discuss this topic further as it is the specific problem of the AGC scheme discussed in more depth in Section 9.2.1. Here, we continue to assume for the sake of simplicity that the configuration of the receiver relies on the measurement of the power of the signal present at the end of the line-up, i.e. on the power of the received wanted signal only. We consider separately the case where we have just the wanted signal at the input of the receiver and the case where we have unwanted signals as well. This approach allows us to clearly highlight the impact of those unwanted signals on the configuration of the the receiver, and thus ultimately on its performance.

Wanted Signal Alone: Receiver Max and Min Gains

Let us assume first that we are dealing with a wanted signal alone at the input of our receive line-up. In order to understand our problem, we first need to recall the wide DR for the power of this input signal. As discussed in "Input signal dynamic range" (Section 3.3.1), this may range in practice between −110 dBm and −25 dBm in a wireless system: a DR of more than 80 dB. The problem is that the final signal processing for the purpose of equalizing and determining the data bits often requires us to work on slices of samples with a normalized average power. It is evident from this alone that we need to have a non-negligible amount of variable gain in the receiver in order to compensate for the variations on the received signal power.

We observe that there are major differences in the possibility of physically implementing such variable gain depending on whether we are dealing with an RF, analog baseband or digital implementation. It is quite easy to achieve a wide DR of amplification with a very high resolution in the digital domain, but not particularly so on the analog side, and even less so in the RF world. As a result, we try in practice to minimize as much as possible the use of variable gain stages in the RF/analog world and to implement them on the digital side. An interesting question is then the *minimum* variable gain to be implemented in the RF/analog part of the receiver.

The bottleneck in terms of DR in the receive path is often the ADC core. For power consumption purposes, or even for die area minimization, the DR of such a device is often kept within reasonable limits; we define what we mean by "reasonable" in "Filtering budget vs. ADC dynamic range" (Section 7.3.3). Here we can assume that we are dealing with an ADC that exhibits a dynamic range DR, i.e. a maximum available signal to ADC noise power ratio in the sampling bandwidth as defined through equation (4.285), that is lower than the overall DR of the wanted signal power at the input of the receiver. Under this assumption, the typical level diagrams that can be achieved when considering a wanted signal alone at the input of the receiver are shown in Figure 7.19.

These level diagrams provide explanations for the lower and upper bounds of the maximum and minimum analog gain respectively that we need to implement in the RF/analog part of the receiver. Let us first consider the maximum gain required in our line-up. Obviously, the lower bound for this maximum gain is driven by the configuration in which we are dealing with the minimum level for the received wanted signal, i.e. when we are in a sensitivity configuration. It is in this configuration that we need to provide the greater analog amplification in order to correctly scale the wanted signal at the input of the ADC. In practice, this correct scaling

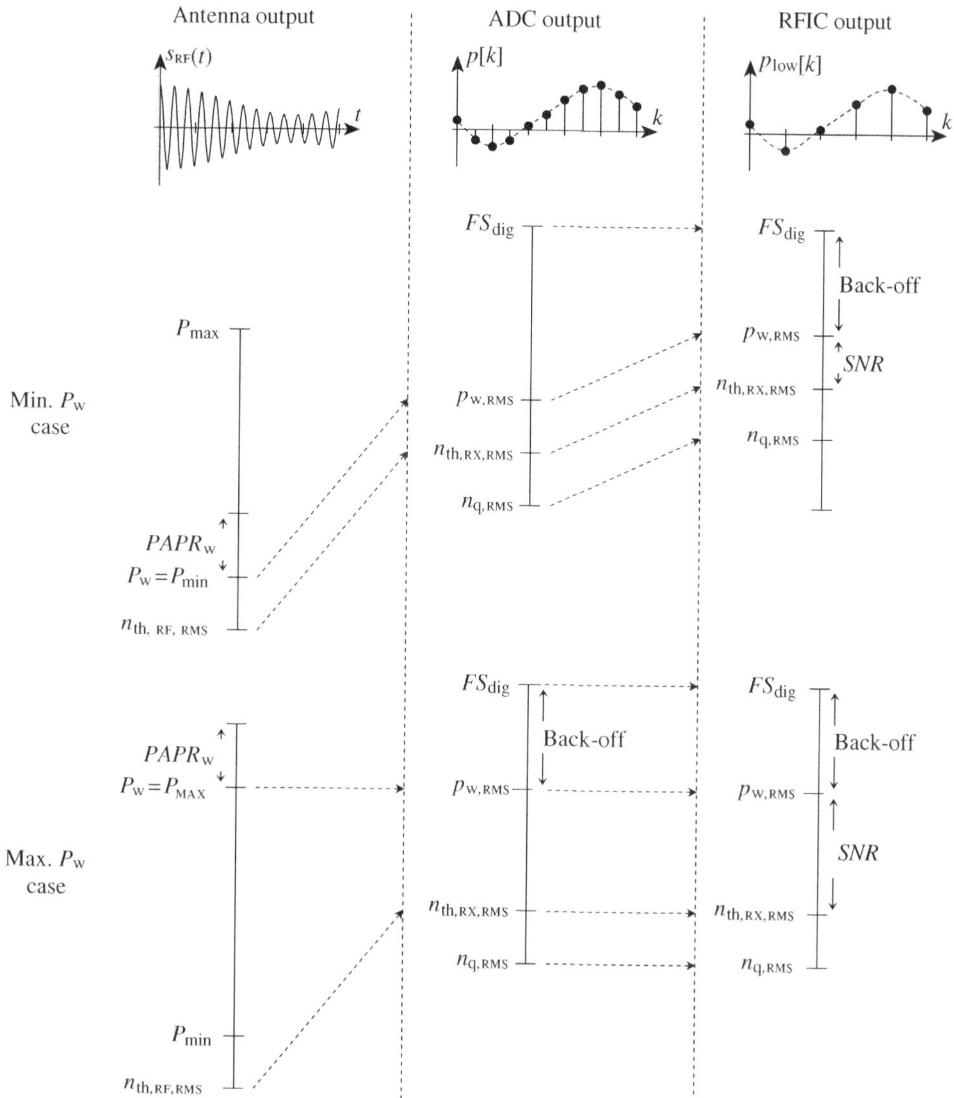

Figure 7.19 Simplified level diagrams along the receive path considering only additive noise components: wanted signal alone case – For the line-up shown in Figure 7.2, the bottleneck in terms of available DR is often the ADC core. In order not to degrade the receiver noise figure when receiving the wanted signal at its minimum input power, P_w, we simply need to amplify the thermal noise of the line-up so that the contribution of the quantization noise remains negligible (top). In the same way, we need to have a sufficiently low receiver gain in order to preserve the target back-off when dealing with the maximum input power (bottom). In both cases, the digital gains need to have enough DR to correctly scale the signal at the RFIC output.

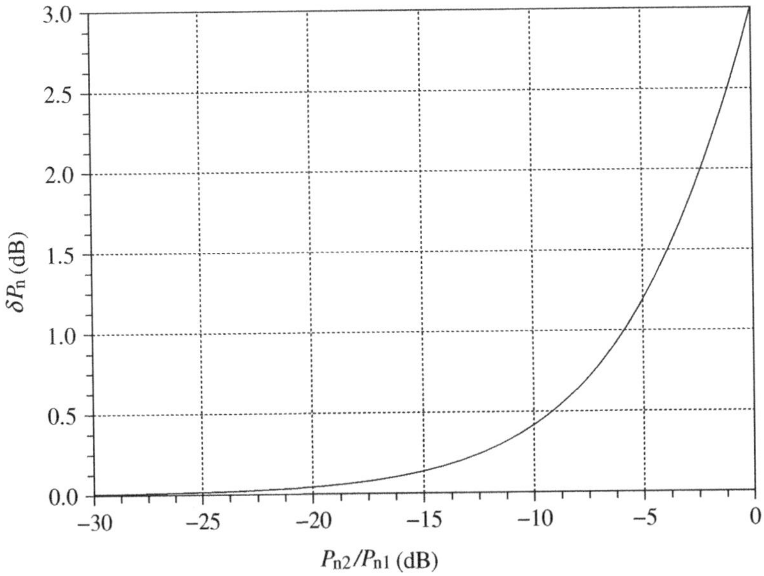

Figure 7.20 Relative noise power degradation due to the superposition of two additive noise terms – The relative noise power degradation, $\delta P_n = (P_{n1} + P_{n2})/P_{n1}$, can reach 3 dB when the power P_{n2} of the additional noise component reaches the power P_{n1} of the originally considered term. As soon as this contribution remains at least 20 dB below the other one, the degradation can be considered as negligible.

simply means that we need to preserve, or at least to minimize the degradation of, the analog SNR. At sensitivity, it is only this parameter that we have to maximize in our receive line-up in order to achieve the best overall performance. For that purpose, we recall that under the assumption that the thermal noise of the receiver is amplified in order to give enough ADC LSB activity, the quantization error can be considered as an additive noise component, as discussed in "Loss below the quantization step" (Section 4.5.3). Given that the receiver thermal noise is amplified so that its power remains above the quantization noise power by, say, 15–20 dB ensures on the one hand that this condition is fulfilled and on the other hand that we have a negligible overall increase in the receiver noise power (see Figure 7.20). As there is no need to add more gain than that, we can derive the maximum required RF/analog gain of the line-up following these guidelines.

Let us, for example, assume that the ADC core involved in the line-up samples voltage waveforms. For the sake of simplicity, let us also assume that the input analog FS of this device is $FS = 1$ V, i.e. 0 dBV. This device exhibits a DR of 52 dB for a sine wave scaled at the FS and in the receiver noise bandwidth δf, assumed for the sake of simplicity here to be equal to the wanted signal bandwidth. As the crest factor CF of the sine wave is $\sqrt{2}$, i.e. 3 dB, we can use equations (4.285) and (4.287) to directly express the equivalent ADC input-referred quantization noise power in the receiver noise bandwidth as

$$\begin{aligned} n_{q,\text{RMS}}|_{\text{dBVrms}} &= 20\log_{10}(FS) - 20\log_{10}(CF) - 10\log_{10}(DR) \\ &= -55\,\text{dBVrms}. \end{aligned} \tag{7.8}$$

If, moreover, we assume that the overall noise figure of the receiver when set with its maximum gain, $F_{\text{RX}(Z_{\text{ant}},T_0)}$, is equal to 4.5 dB in typical conditions, we get that the equivalent overall thermal noise power as referred to the antenna connector is simply given by

$$P_{n_{\text{th}}} = F_{\text{RX}(Z_{\text{ant}},T_0)} k T_0 \delta f \tag{7.9}$$

according to the discussion in Section 4.2. Assuming that the receiver noise bandwidth δf is equal to 9 MHz, i.e. 69.5 dBHz as discussed in "Receiver high level parameters" later in this section, we finally get that

$$P_{n_{\text{th}}}\big|_{\text{dBm}} = 4.5 - 174 + 69.5 = -100\,\text{dBm}. \tag{7.10}$$

Thus, if we want the quantization noise from the ADC to remain negligible in the overall SNR budget at sensitivity, we may amplify the thermal noise as recovered at the antenna connector so that it is at least 20 dB above the quantization noise. This means that its RMS value once amplified by the analog part of the receiver must reach at least $-55 + 20 = -35$ dBVrms at the ADC input. This results in the following lower bound for the maximum gain of the RF/analog part of the receiver:

$$\begin{aligned} G_{\text{RX,ana,max}}\big|_{\text{dBVrms/mW}} &= (n_{\text{q,RMS}}\big|_{\text{dBVrms}} + 20) - P_{n_{\text{th}}}\big|_{\text{dBm}} \\ &= -35 - (-100) = 65\,\text{dBVrms/mW}. \end{aligned} \tag{7.11}$$

This gain has to be understood as a conversion gain from mW at the antenna connector up to Vrms at the ADC input.

In the same way, if we now consider the configuration where we receive the wanted signal with its maximum input level, we can derive the upper bound for the minimum gain in the RF/analog part of the receiver. Staying with Figure 7.19, this can be done by simply considering the margins required at the ADC stage in order to avoid any clipping of the wanted signal. This directly gives the gain to be applied so that the wanted signal attains the correct target level at the ADC input from its maximum power at the receiver input. Anticipating the discussion in "Filtering budget vs. ADC dynamic range" (Section 7.3.3), we observe that in contrast to what we encountered on the transmit side, the margins to be considered to avoid the clipping along the receive path include more than just the CF of the modulating waveform or the RF impairments in the line-up. We need to consider in addition the characteristics of the propagation channel that can impact the statistics of the received signal. Thus, let us assume that we need a consistent back-off of 20 dB along the line-up. We can then deduce the upper bound for the minimum gain in the receiver assuming a maximum input signal power of -25 dBm. In that case, we must be able to set a receiver gain lower than or equal to

$$G_{\text{RX,ana,min}}\big|_{\text{dBVrms/mW}} = (0 - 20) - (-25) = 5\,\text{dBVrms/mW} \tag{7.12}$$

in order to fulfill the minimum target back-off at the ADC input.

We have thus derived lower and upper bounds for the maximum and minimum gains of the RF/analog part of the receiver, respectively. Since we have no particular reason to add more RF/analog gain than $G_{\text{RX,ana,max}}$, or less than $G_{\text{RX,ana,min}}$, we can use those values as specifications in practice. This is what is done in what follows as an example. However, given the overall maximum and minimum gain specified for the RF/analog part of the receiver, we need to distribute the gain among the different blocks. To do so, it is not sufficient to consider the wanted signal alone. We need to discuss the impact of the potential blocking signals present at the input of the receiver, and it is to this that we now turn.

Wanted Signal Plus Unwanted Signals: Gain Split in the Receiver

It is mainly the maximum gain of the receiver that is impacted by the presence of unwanted signals in the line-up, mainly through the potential compression or clipping in the line-up.

These unwanted signals can have high power relative to that of the wanted signal. This is obviously the case when the wanted signal and blocker signal belong to different networks, as discussed in Chapter 3. But it is also the case in practice in a full duplex transceiver, as considered in the present chapter, due to the continuous presence of transmitter leakage. This leakage is at its maximum level when the received wanted signal is close to sensitivity. In both cases, having the wanted signal in the weakest part of the input DR necessarily leads to the use of the maximum RF/analog gain of the receiver in order to preserve the noise performance at, or close to, sensitivity. But, as we are dealing with strong blockers at the same time, we can imagine that the maximum gain to be set in each block is imposed by those signals in order to avoid any compression of the signal.

In contrast, the receiver is set with its minimum RF/analog gain when it processes the wanted signal at its maximum level. We thus expect that in this configuration the adjacent or blocking signals are no longer stronger than the wanted signal. Moreover, these unwanted signals lie outside the wanted signal frequency band in practice. As a result, they are necessarily filtered out along the receive path. We can then imagine that it is more the wanted signal itself that drives the compression performance of the line-up in that maximum input power case.

Thus, let us focus first on how to split the maximum gain of the receiver. In order to optimize the additive noise performance of the line-up, we need to set the maximum possible gain in the early stages of the receiver, as discussed in Section 4.2.4. Thus, having derived the maximum overall gain of the receiver in the previous section, $G_{\text{RX,ana,max}}$, we may want most of it in the early RF stages. In order to determine the limit for it, we then need to consider the potential saturations or clipping we may have in the line-up due to the presence of the blocking signals.

We first of all make some comments on the nature of the signals we process in the physical implementation of our receiver. As discussed in more depth in Chapter 2, in the analog world we process either current or voltage waves carrying the information. But from an implementation point of view there is a major difference between the two kinds of quantity. Due to the classical behavior of transistor based electronic devices, there is more or less always a means of implementing a current source that is able to deliver to a load the amount of current corresponding to the amplitude of a given signal. In contrast, a traditional low frequency voltage source cannot deliver a voltage swing that is higher than its voltage supply. This explains why in modern technologies using low voltage supplies, it is the I/Fs between the blocks using voltages that need to be considered most carefully in order to check that no

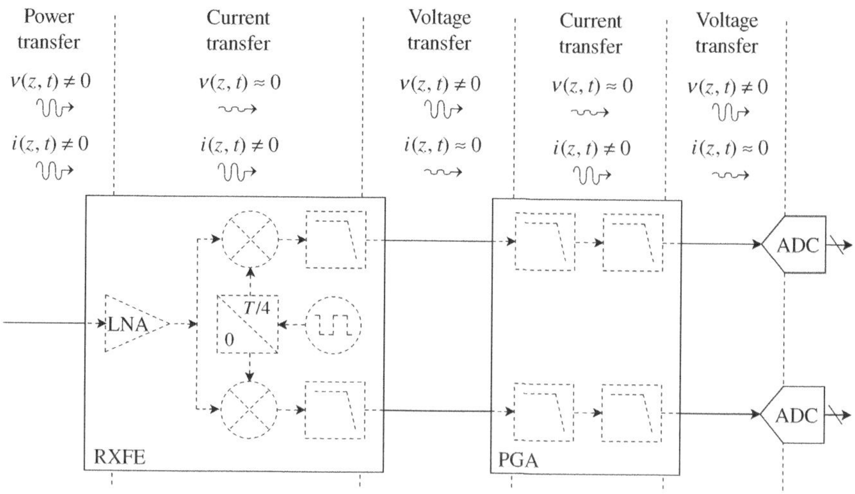

Figure 7.21 Voltage interfaces in the receive line-up – For a physical implementation using transistor devices, the blocks involved in a receive line-up behave either as current or voltage sources with respect to their loads. As a result, at the I/F between them we recover almost exclusively either current or voltage waveforms, but not necessarily both at the same time. But in practice the bottleneck in terms of DR is often the available voltage swing due to the limited voltage supply. We thus need to focus first on those I/Fs to track the level of the signals in the receiver. Looking more deeply into the block content of the receive path shown in Figure 7.2, we can suppose for the sake of simplicity that the main I/Fs in voltage are first that between the RXFE and PGA, and then that between the PGA and ADC.

clipping occurs in practice. In our line-up, for the sake of simplicity we consider only two such I/Fs, one between the RXFE and PGA, and one between the PGA and ADC (see Figure 7.21). As a result, we simply need to focus on the maximum gains of the RXFE and PGA blocks.

Let us now consider the maximum gain we can set in the RXFE block. In line with the above discussion, we need to set the maximum gain so that we can still confidently process the blocking signals. We should obviously consider all the blockers potentially present at the input of the receiver and derive the worst case for our dimensioning. Here, for the sake of simplicity we consider only the TX leakage signal assumed at a power of −25 dBm at the input of the RXFE. Then we need to take into account the potential filtering effects. As illustrated in Figure 7.21, we can assume that we already get some filtering acting on the received signal before the reconstruction of the voltage at the output of the block. Assuming for the sake of simplicity that we are dealing with a single pole that exhibits a cut-off frequency at 3 dB equal to 4.5 MHz, and that the duplex distance is 150 MHz, we get an attenuation of the TX leakage signal of about 30 dB across the block. Furthermore, we can assume that we get the same available voltage FS of 1 V, i.e. 0 dBV, all along the analog part of the receiver path, as assumed in the previous section for the ADC input FS. In the same way, we can assume that we need to achieve the same back-off of 20 dB as considered at the input of the ADC. This should be obviously modified depending on the statistics of the strongest signal present at the I/F of the line-up. However, assuming that we get similar statistics for the different signals considered, the same back-off needs to be considered along the receive path. Under those assumptions, the typical level diagrams that can be achieved in the presence of the transmitter leakage signal are shown in Figure 7.22. We can then derive the maximum gain of the RXFE block:

$$G_{\mathrm{RXFE,max}}|_{\mathrm{dBVrms/mW}} = (0 - 20) - (-25 - 30) = 35\,\mathrm{dBVrms/mW}. \tag{7.13}$$

Thus, given that the maximum gain of the analog part of the receiver, $G_{\mathrm{RX,ana,max}}$, is equal to 65 dBVrms/mW by equation (7.11), the maximum gain of the PGA stage $G_{\mathrm{PGA,max}}$ can be set to

$$\begin{aligned} G_{\mathrm{PGA,max}}|_{\mathrm{dB}} &= G_{\mathrm{RX,ana,max}}|_{\mathrm{dBVrms/mW}} - G_{\mathrm{RXFE,max}}|_{\mathrm{dBVrms/mW}} \\ &= 65 - 35 = 30\,\mathrm{dB}. \end{aligned} \tag{7.14}$$

This gain partitioning thus ensures that we have at least 20 dB of back-off to pass the strongest blocker, here the TX leakage signal, at the critical I/Fs of the receive path. In practice, the same should be done for all the strong unwanted signals potentially present at the receiver input. The maximum gains in the line-up should then be derived as those corresponding to the worst case in terms of available back-off.

Having derived the split for the receiver maximum gain, we need to do the same for the minimum gain. This is straightforward, recalling that we require the same back-off of 20 dB along the receive baseband analog path to avoid any clipping or compression. Furthermore, the minimum gain of the receiver was derived in the previous section for the wanted signal, when received at its maximum input power, to be correctly scaled at the ADC input so as to achieve the 20 dB of back-off. Thus, if we want to achieve the same back-off at the PGA input and output for the wanted signal in such conditions, we simply need to set the PGA voltage gain, $G_{\mathrm{PGA,min}}$, to 0 dB. However, we observe that in practice, the back-off is defined as referred to the FS available for the maximum swing of the signal at the considered node. In the case of different full scales available at the PGA input and output, we should then take this difference into account in the gain of the block. But, supposing for the sake of simplicity that we are dealing with the same analog FS of 1 V, i.e. 0 dBV, along the baseband part of the analog path, we simply get that

$$G_{\mathrm{PGA,min}}|_{\mathrm{dB}} = 0\,\mathrm{dB}. \tag{7.15}$$

Thus, given that the receiver overall minimum gain, $G_{\mathrm{RX,ana,min}}$, is equal to 5 dBVrms/mW by equation (7.12), we can directly deduce that the upper bound for the minimum gain of the RXFE stage $G_{\mathrm{RXFE,min}}$ is equal to

$$G_{\mathrm{RXFE,min}}|_{\mathrm{dBVrms/mW}} = 5\,\mathrm{dBVrms/mW}. \tag{7.16}$$

Finally, we need to derive the intermediate gain values of the receiver between its minimum and maximum values. Given that the wanted signal input power can go continuously from its minimum to its maximum value, we may require a continuous coverage of all the possible receiver RF/analog gain values in order to correctly regulate the level of the signal at the input of the ADC core. In practice, from a system point of view, the continuous coverage of the gain range is not necessarily the easiest way to implement and manage such variable gain. It may

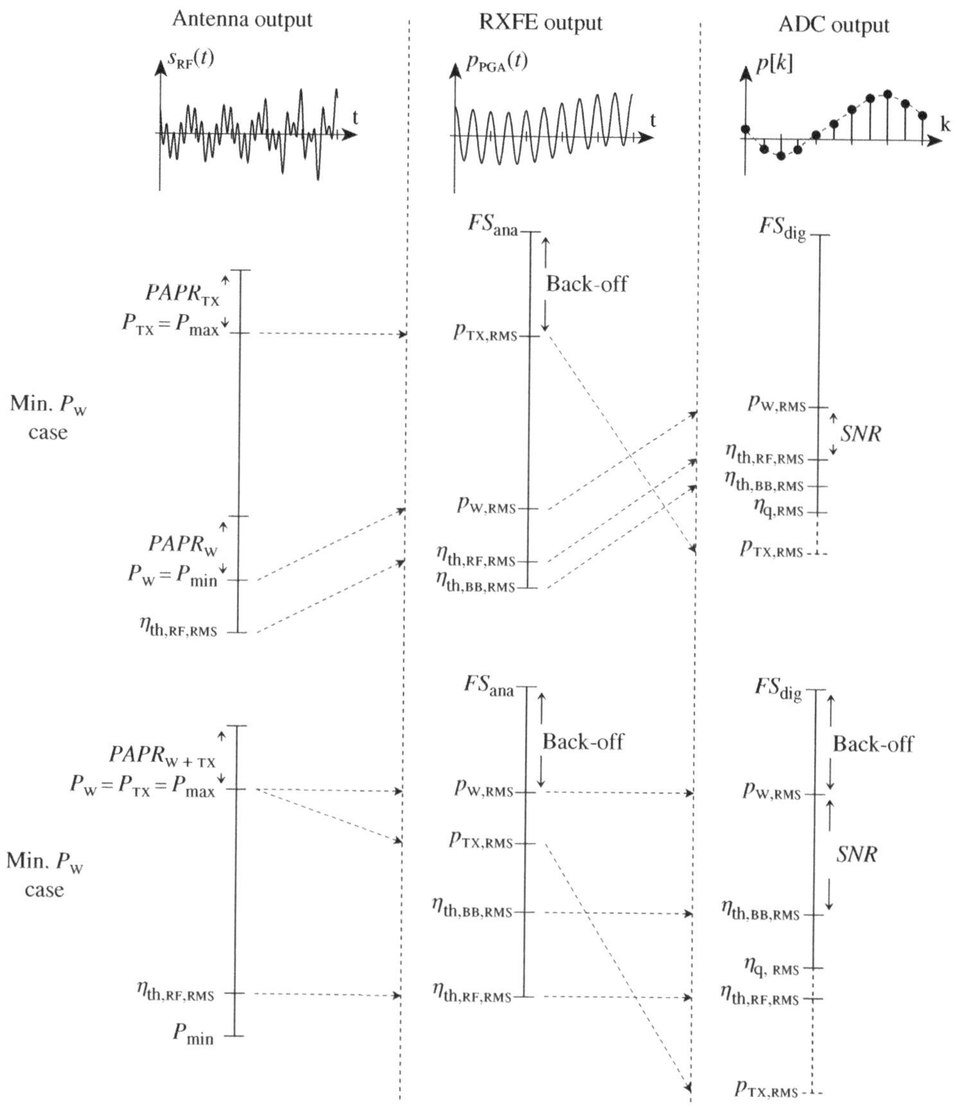

Figure 7.22 Simplified level diagrams along the receive path considering only additive noise components: wanted signal and transmitter leak case – Assuming that the bottlenecks in terms of available DR are the I/Fs between the RXFE and PGA and between the PGA and ADC as illustrated in Figure 7.21, we need to ensure appropriate scaling of the signals at those I/Fs in order to avoid any saturation. This leads to an upper bound for both the maximum RXFE gain in order to pass the blocking signals present when receiving a weak wanted signal (top), and the minimum gain in order to receive the wanted signal at its maximum power (bottom).

be simpler to restrict ourselves to a discrete set of gain steps as this reduces the number of states of the receiver to manage in practice.

We first observe that, due to practical implementation limitations, it is often simpler to achieve a large number of gain steps in the baseband analog world rather than in the RF world. Thus, we may try to put as many gain steps as possible in the PGA stage. Let us assume that the PGA is composed of 11 gain states with 3 dB steps. Thus the PGA gain ranges from 0 dB to 30 dB in 3 dB steps. The value of 3 dB for the gain step may appear arbitrary, but it results in fact from a trade-off between the DR of the ADC and the complexity in the PGA implementation. Indeed, as illustrated in Figure 7.24, the thinner the PGA gain resolution, the more the level at which the signal is regulated at the ADC input can be constant. Thus, the higher the DR of the ADC, the more we can relax the precision on this regulated level and thus the PGA resolution. This simply means that the PGA resolution adds some extra margin to be considered for the minimum DR of the ADC required in order to comply with all the test cases of the wireless standard of interest. In practice, such a margin is taken into account in the back-off as discussed in "Filtering budget vs. ADC dynamic range" (Section 7.3.3).

However, we can see that these PGA gain steps do not allow for a full coverage of the receiver gains between the minimum and maximum value. We thus need to consider the switching of the RXFE gain as well. Although we want to minimize the number of RF gain steps, we should think about how many steps we need for this implementation and what constraints drive their value. Obviously, the constraints do not come from the use cases considered up to now. For instance, we have already derived the maximum gain in the receiver in order to cope with the maximum level of blocking signals. We can thus expect no more constraints from those signals that would require additional gain steps in the RXFE.

But we have just seen that the minimum gain of the RXFE is driven by the maximum input level of the wanted signal through the minimum gain of the receiver. Thus there is an intermediate input power of the wanted signal for which we start to have compression at the RXFE output if we continue to use the maximum gain of this block. We thus see in practice that the wanted signal itself drives the necessity to switch the gain of the RXFE in order to support the wide DR of its input power.

For example, recall the adjacent channel profile discussed in "Adjacent channel selectivity" (Section 3.3.1), and shown in Figure 3.17. Our receiver needs to handle an adjacent channel power increasing linearly from −52 to −25 dBm when the wanted signal power goes from −93 to −66 dBm. This profile, originally taken from the WCDMA standard, is here applied to our receiver, assuming that the frequency offset of the adjacent channel is 10 MHz from the wanted signal carrier frequency. Thus, given on the one hand that the same single pole is embedded in the RXFE as considered above, and on the other hand that the analog FS remains equal to 1 V along the analog baseband receive path, we achieve the back-off, i.e. the ratio between the available FS and the RMS value of the signal, displayed in Figure 7.23 for both the wanted and the adjacent signals. We thus see that in order to avoid any clipping at the RXFE output while not increasing the filtering embedded in this block, we need to decrease the RXFE gain for a low input power of the wanted signal. We observe that if we had to cope with the presence of the wanted signal only, we could keep the highest gain for this block up to a quite high input signal level. But, as our assumption is that the AGC sets the gains based only on the power of the wanted signal, we need to consider a setting of the gains such that the receiver is able to handle the adjacent channel, even if not present in practice.

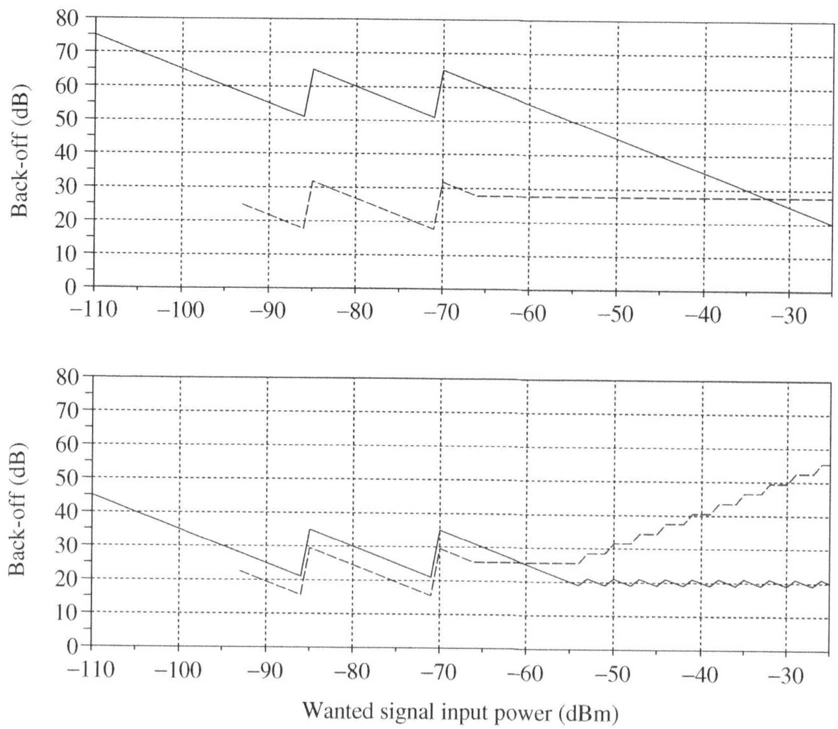

Figure 7.23 Back-off achieved along the receive line-up – In order to avoid the clipping of the adjacent channel (dashed) at the RXFE output (top), we need to switch the RXFE gain for a low power of the received wanted signal. The gain of the PGA stage is then derived in order to regulate the wanted signal (solid) at the ADC input (bottom). For a gain strategy based on the power of the wanted signal only, the same switching points for the gain of the RXFE are necessarily used whether or not an adjacent channel is present. The back-off achieved for the wanted signal is thus higher than would be required if this signal were alone in the line-up.

However, as can be understood from the discussion in "SNR budget" (Section 7.3.4), a direct switching of the RXFE gain from its maximum value down to its minimum value at such low input power can lead to non-negligible degradations of the noise performance of the receiver. This explains why it may be worth using additional gain steps at the early stage of the receiver in order to *smooth* the degradation of the noise performance of the line-up when the input power of the wanted signal increases. For that purpose, in the sequel we assume as an example the use of three gain states for our RXFE stage. More precisely, we assume that the gain steps, classically implemented at the LNA stage in practice, lead to three gain values of 35, 20 and 5 dBVrms/mW for the RXFE, so that we have steps of 15 dB.

Given the above guidelines for the gain split in the receiver, we can now derive the final gain settings to be applied as a function of the input power – more precisely, as a function of the wanted signal input power in accordance with our assumption of gains determined based on this power measurement only. As illustrated in "Filtering budget vs. ADC dynamic range" later in this section, the simplest way to proceed is to suppose that the AGC scheme tries as much as possible to regulate the wanted signal at a constant input level at the ADC input, the AGC set point. Thus, for a given wanted signal input power we can first derive the theoretical overall RF/analog gain to apply in the receive path so that this signal reaches the correct set point at the ADC input. At the same time, given this input power, we can also deduce the RXFE gain, based on the switching points derived previously to avoid the clipping of the potential adjacent channel signal, as shown in Figure 7.23. As a result, given on the one hand the overall RF/analog gain to be used and on the other hand that of the RXFE, we can deduce by subtraction the PGA gain to be applied. However, in practice we may not have sufficient DR for the gain of this PGA block to achieve the theoretical value thus derived. Obviously, when this occurs, only the minimum or maximum PGA gain can be applied in practice. And as a side effect, the wanted signal cannot reach the correct set point at the input of the ADC. This is what happens in particular in the lower part of the input power DR, as can be seen in Figure 7.24. This comes as no great surprise, recalling the derivation of the maximum gain of the receiver in the previous section. The significant criterion for this derivation was

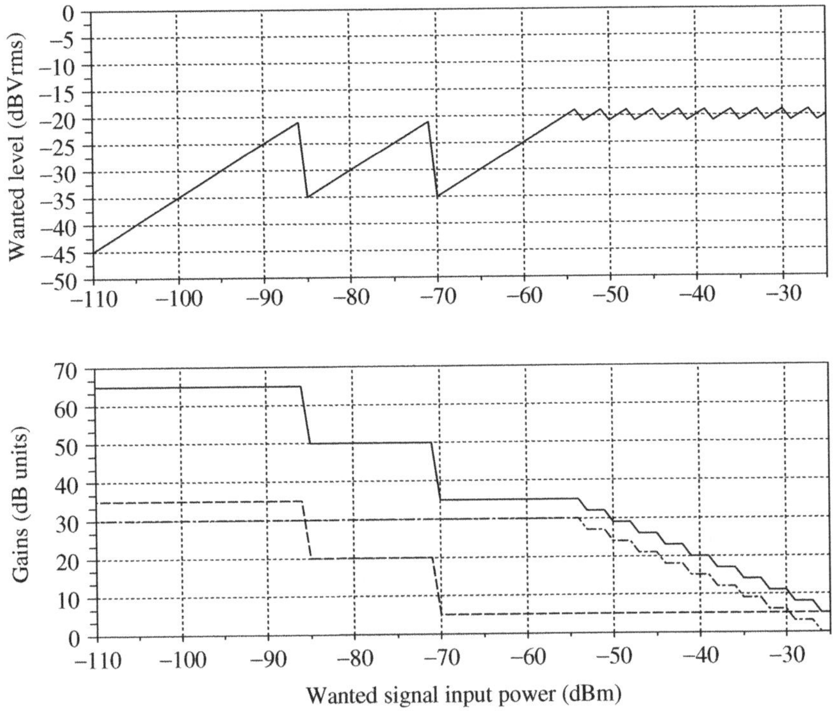

Figure 7.24 Gains applied in the receiver and resulting level of the wanted signal at the ADC input – Assuming a gain strategy based on the power of the wanted signal only and given on the one hand the minimum and maximum gains of the receiver, as derived considering the wanted signal only, and on the other hand the partitioning between the blocks, as derived considering the adjacent and blocking signals, we can derive the receiver gains (bottom, solid for the overall RX in dBVrms/mW, dashed for the RXFE in dBVrms/mW, dot-dashed for the PGA in dB) in order to regulate in the best possible way the wanted signal level at the ADC input (top).

indeed only the preservation of the receiver RF/analog noise performance. And considering only this constraint, the wanted signal does remain in the lower part of the DR when close to sensitivity. However, the preservation of the RF/analog noise performance of the line-up should be checked each time the wanted signal does not reach the correct set point at the ADC input, i.e. each time we switch down the RXFE gain in the present case (see the discussion in "SNR budget" (Section 7.3.4)). Given the overall RXFE and PGA gain strategy as a function of the wanted signal input power shown in Figure 7.24, the digital gain can be set in order to achieve a constant output RMS level toward the forthcoming baseband algorithms.

An unambiguous decoding scheme, that gives the gain to be applied for the different blocks of the line-up as a function of the wanted signal input power, is often referred to as an AGC decoding table. However, this kind of decoding scheme can only be used when the AGC relies on a power measurement at a single stage of the line-up, after the channel filtering in the present case. But more complex AGC schemes are possible in practice in order to overcome some of limitations we have discussed. This is covered in greater depth in Section 9.2.1.

Receiver High Level Parameters

From the discussion in the previous two sections, the setting of the gains in the receiver is necessarily a function of the input power – more precisely, of the wanted signal input power when the AGC scheme is based on the power measurement of this signal only as assumed in this chapter. But, as already discussed for the transmit side, the switching of the gains in the RF/analog blocks of the line-up necessarily leads to a change in their intrinsic performance as well as a change in the overall characteristics of the receiver. This is obviously the case for the additive noise performance of the line-up that is directly impacted by the decrease in the RXFE gain, for instance. As a result, we need to detail what characteristics are effectively impacted by this gain setting and then give illustrative numbers for those characteristics in order to use them later in the illustration of the receiver budgets.

We expect in practice the same kind of impacts on the characteristics of the line-up due to this variable gain setting as encountered on the transmit side. This means up to first order that mainly the additive thermal noise characteristics and the linearity performance of the receiver is expected to be effectively a function of the input power of the wanted signal.

Thus, let us focus first on the additive noise performance of the receive line-up. A convenient way to derive budgets is to work with equivalent input quantities, i.e. referred at the input of the receiver. Indeed, in practice all the use cases of the receiver are defined through the level of a wanted and potential unwanted signals referred at the receiver input. Having the characteristics of the receiver referred to the same node then allows the direct evaluation of the performance for the configuration considered. Dealing with the noise performance of the line-up means that, given a configuration of the receiver, we need to derive the equivalent input noise power that leads to the same SNR as physically recovered at the receiver output when the received signal goes through it.

For that we need to take into account on the one hand the different noise contributions that add to the received signal along the data path and on the other hand the configuration of the gains in the line-up to compose them in the right way as equivalent input quantities. The gains to be used are those derived by our AGC strategy, thus shown in Figures 7.24 and 7.25. Then, it remains to go through the intrinsic noise performance of the different blocks involved in

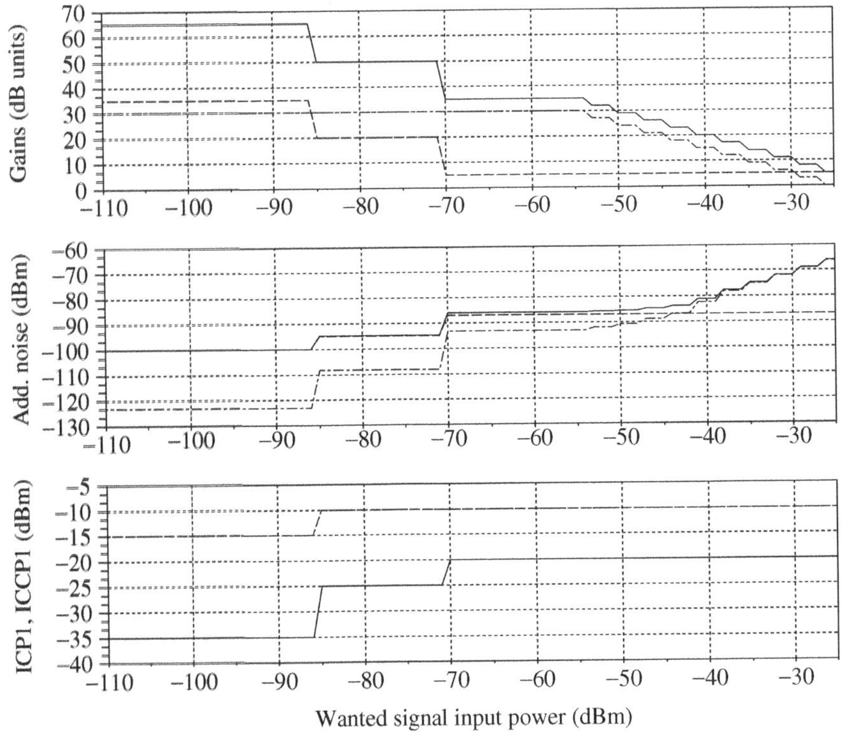

Figure 7.25 Typical receiver configuration and characteristics considered for the derivation of the budgets – Based on the gain strategy derived in the previous sections (top, solid for the overall RX in dBVrms/mW, dashed for the RXFE in dBVrms/mW, dot-dashed for the PGA in dB), we can derive the additive noise performance of the receiver. Considering on the one hand the RXFE and PGA noise characteristics displayed respectively in Tables 7.1 and 7.2, and on the other hand that the insertion loss of the passive FE, assumed to be at T_0, is 2.5 dB, the overall receiver equivalent input additive thermal noise performance (middle, solid) is driven by that of the passive FE plus RXFE in the lower part of the wanted signal input power DR (middle, dashed). But as soon as the gain of the RXFE is low enough and the intrinsic noise contribution of the PGA increases enough due to the reduction of its gain, we get the opposite behavior as experienced in the highest part of the wanted signal input power DR. Here, the RF noise bandwidth δf, is assumed to be equal to the wanted signal bandwidth. It thus corresponds to a 10 MHz LTE waveform, i.e. is equal to $[-4.5, 4.5]$ MHz. Furthermore, the decrease in the gain of the RF stages often results in a better odd order linearity of the receiver. We thus assume here three different values for the ICP1 (bottom, solid) and ICCP1 at the TX frequency offset (bottom, dashed).

the line-up. And the first block to be considered in our line-up comprises the passive devices from the antenna up to the active RF RXFE block input. The noise performance of this passive part of the line-up is obviously a function of the insertion loss and physical temperature of the devices. But, for the sake of simplicity, we assume here that the passive devices we are dealing with are working with a physical temperature equal to the temperature used to refer the noise factor of the receiver to, i.e. T_0. As a result, by the discussion in Section 4.2, the overall noise figure of the line-up simply reduces to the sum of the insertion loss of the passive

Table 7.1 Noise figure assumed for the RXFE.

Gain (dBVrms/mW)	35	20	5
$NF_{(Z_i^*,T_0)}$ (dB)	2	7	15

devices, when expressed in dB, and the noise figure of the active part of the line-up. For our derivations, we can thus just add the insertion loss of the passive FE to the noise figure of the RXFE. Obviously, this is the most favorable case in terms of analytical derivations for our budgets. However, when we need to derive such budgets for devices that have a physical temperature that is not T_0, we need to refine the derivations and use the results presented in Section 4.2.5. In the present case, assuming on the one hand that the insertion loss of the passive devices is of 2.5 dB and on the other hand that the noise performance of the RXFE and PGA is as displayed in Tables 7.1 and 7.2 respectively, we finally achieve the equivalent input additive noise performance shown in Figure 7.25(middle) for the overall receive path.

Let us now focus on the linearity performance of the line-up. Up to first order, it is mainly the odd order linearity performance that is impacted by our gain switch in the line-up. As illustrated in Figure 7.18 and discussed in "Impact of implementation limitations" (Section 7.2.1); up to first order the even order linearity performance of the receiver is driven by the downmixer linearity in the direct conversion receiver. Assuming that no particular gain switching occurs in that stage, there is thus no need to worry about a dependency of the even order performance of the receiver on its internal gain setting. In contrast, the odd order linearity of the line-up, related to the compression behavior, can be impacted by its setting in practice. More precisely, as discussed in the introductory part of Chapter 5, it is often mainly the RF blocks of the line-up that drive the odd order linearity performance. As a result, we can continue to suppose in what follows that the line-up performance is driven by the gain state of the RXFE block. This finally results in the ICP1 and ICCP1 performance shown in Figure 7.25(bottom) following the concepts introduced in "Odd order nonlinearity and IP3, CP1 or Psat" and "Desensitization due to odd order nonlinearity and CCP1" (Section 5.1.2). Here, we observe that the ICCP1 performance has to be understood as that achieved for a blocking signal at the duplex distance as we expect to use this parameter to derive the impact of the TX leakage signal in our budgets.

In conclusion, we observe that we consider as an example for our budgets the same kind of received wanted signal as considered for the derivation of the budgets for the transmit side: an OFDM signal corresponding to the LTE 10 MHz case, at least from the frequency bandwidth point of view. This means a signal with a PSD almost flat over the band [−4.5, 4.5] MHz

Table 7.2 Input-referred noise performances assumed for the PGA.

Gain (dB)	30	27	24	21	18	15	12	9	6	3	0
$n_{\text{PGA,RMS}}$ (nVrms/$\sqrt{\text{Hz}}$)	13	15	17	21	27	44	73	100	150	200	300

as detailed in Section 1.3.3 and shown in Figure 1.19. However, for the sake of simplicity, we assume that this is a pure OFDM signal – more precisely, a signal with a statistical distribution that can be approximated as Gaussian – as it simplifies some of our derivations. We also assume the same statistics for the TX leakage signal. This allows us to highlight some additional phenomena linked to the presence of an amplitude modulated blocker as compared to a simple constant one.

7.3.3 Budgets Linked to the Resistance to the Wireless Environment

As discussed in "Budgets to be considered" (Section 7.2.1), simply considering the resistance of the receiver to its wireless environment, we get that the impact of the unwanted signals depends on their frequency offset relative to the received wanted signal. Different kinds of budgets are required depending on this relative frequency. They involve either the linearity or the noise characteristics of the line-up, as those quantities effectively drive the SNR performance achieved in the presence of those unwanted signals.

As was already the case on the transmit side, there is a particular function embedded in the receiver that is involved in the possibility for a receiver to face strong blockers even if not directly involved in the SNR budgets, the channel filtering. Indeed, we already know that we need to implement enough filtering in our line-up in order to cancel all the unwanted signals and thus deliver a clean wanted signal to the baseband algorithms dedicated to data recovery. But how much of this filtering needs to be done in the RF/analog world and how much in the digital world? Obviously, as discussed when considering the presence of an adjacent channel in "Wanted signal plus unwanted signals: gain split in the receiver" (Section 7.3.2), there is a trade-off between the maximum amplification allowed for the various stages of the line-up, and the filtering required in order to avoid any clipping along the data path. But, on top of that, there is also an impact of the RF/analog filtering on another critical block of the line-up, the ADC through its DR. Indeed, in practice the less the RF/analog filtering, the greater the DR we need to achieve on the ADC in order to cope with the residual blockers present at its input. But, as it is often costly to improve this last parameter in physical ADC implementations, there is often a trade-off with the RF/analog filtering to be implemented.

We first describe this trade-off between the filtering partitioning and the ADC DR before further discussing the budgets needed to check the resistance of the receiver to the wireless environment.

Filtering Budget vs. ADC Dynamic Range

In order to discuss the filtering split in the receive line-up and its associated impact on the ADC DR, we need to keep in mind that we have the same limitation regarding the possibility of implementing such channel filtering in the RF world as encountered on the transmit side when considering direct conversion architectures. Due to the frequency planning associated with such architecture, channel selection is done only during the frequency downconversion down to baseband. As a result, if we want to implement a RF filter whose bandpass remains centered around the wanted signal whatever its carrier frequency within the receive system band, i.e. within the frequency band in which all the users of the same network have their RX signals

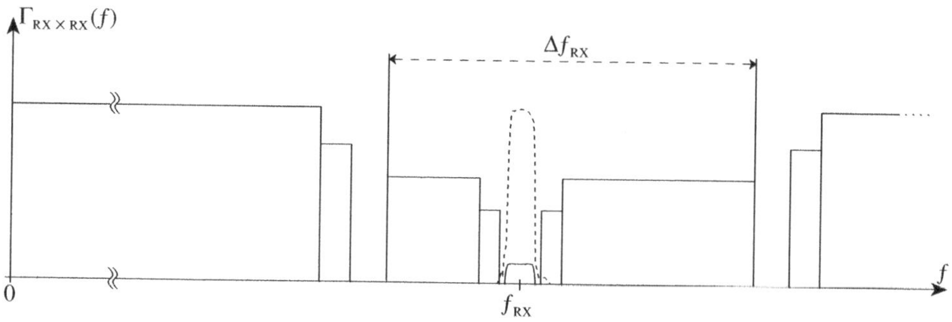

Figure 7.26 Channel filtering problem on the receive side – Given that the carrier frequency, f_{RX}, a receiver has to receive can be anywhere within the total receive system band Δf_{RX}, a channel filter implemented in the RF world would need to be tunable in order to remain bandpass and centered around f_{RX} (dotted). Moreover, given that the power of the close in-band blockers can be much higher than that of the wanted signal, and that the bandwidth of this latter signal can be much smaller than the carrier frequency (solid), channel filtering may require using a sharp filter exhibiting a high quality factor that is hard to achieve in RF. These problems mean that it is preferable to implement the channel filter both after the channel selection, i.e. at a fixed frequency whatever the received carrier frequency f_{RX}, and at a low enough frequency in order to more easily achieve the required selectivity.

multiplexed in frequency, we need to have its center frequency tunable, as illustrated in Figure 7.26. We also need to keep in mind that in practice the unwanted signals to be filtered out can lie close to the wanted signal in the frequency domain, and also that the channel bandwidth can be much smaller than the carrier frequency. As a result, this tunable filter would require a very high quality factor in order to be selective on the unwanted signal. We thus recover the problem symmetrical to that discussed in "Filtering problem" (Section 7.2.3), and illustrated in Figure 7.8. And as was concluded in that section, the implementation of such a RF filter is not really cost effective and leads to some technical issues in practice, even when implemented as active. Thus, at least for the suppression of the close in-band blockers, we continue to rely on a fixed channel filter after the channel selection, i.e. after the downconversion to baseband. However, this does not prevent us from using RF filters expected to be bandpass for the overall receive system band in order to attenuate the out-of-band blockers. These considerations give a new explanation of the structure of the receiver shown in Figure 7.2.

However, to say that most of the filtering, i.e. at least the channel filtering, must take place in baseband does not tell us anything about the split between the analog and digital implementation. Obviously, in practice we may want to put most of it in the digital side. The reasons for this are basically driven by the high cost of the analog filtering implementation, whether through its area or its power consumption. On top of that, the digital implementation also often leads to more flexibility, allowing for instance reconfigurable filter taps. Unfortunately, despite all these apparent benefits, not all of the filtering can be put in the digital side. The bottleneck comes from the ADC core and its limited performance for a given power consumption and area target. In practice, we then need some analog filtering for two reasons.

First, we need to ensure that no residual blocking signal is folded on the wanted receive signal due to the periodization of the spectrum being sampled, as illustrated in Figures 7.17 and 7.18. As a result, an anti-aliasing filter is required prior to the ADC in order to sufficiently attenuate potential blockers, as detailed in Section 4.6.1. In practice, there is a trade-off between the sampling rate f_s, often limited by the technology used for the ADC implementation and by its power consumption, and the required anti-aliasing filtering. Obviously, the higher the sampling frequency, the less filtering required to achieve a given target rejection at the frequency offset equal to f_s.

Second, even if a blocker lies in baseband at a frequency lower than the sampling rate f_s so that we can expect no aliasing problems, we still need to have sufficient DR available at the ADC stage. This DR, defined by equation (4.285) as the ratio between the maximum possible RMS value of the signal being converted and the RMS value of the quantization noise, needs to be sufficiently high to be able to convert confidently the residual blocking signals while having a negligible overall noise degradation. And the less analog filtering we have, the higher the amplitude of the potential blockers at the ADC input, and the higher the DR required to sample those signals correctly. Once again there is a trade-off between the area and power consumption of the analog filtering with regard to the ADC DR.

Obviously, the requirements for the analog filtering due to the anti-aliasing constraints need to be checked in practice. This is often straightforward, though not necessarily for the requirement in terms of ADC DR as a dedicated budget is required.

In order to go into greater depth, we illustrate how to carry out such an RX ADC DR budget. For that purpose, we simply need to consider all the potential configurations of wanted and unwanted signals present at the ADC input and derive the worst case in terms of required DR to pass them. In practice, this has to be done for all the test cases our receiver has to pass in order to be compliant with the considered wireless standard. For the sake of simplicity, we confine ourselves to three of these, representative of what can occur in practice:

(i) We first consider a sensitivity configuration where the received wanted signal is supposed to be alone at the input of the receiver. As a result, we recover only the wanted signal in addition to the RF/analog noises at the input of the ADC core. For the sake of simplicity, we can assume in the present case that we are dealing only with thermal noise in the line-up and that we have an RF/analog SNR of 10 dB. We can also assume that a minimum back-off of 20 dB needs to be achieved in order to avoid any clipping issue. Suppose that we have enough RF analog gain in the line-up in order to amplify the receiver thermal noise up to 20 dB above the ADC quantization noise RMS value, as considered in Section 7.3.2, in order not to degrade the RF/analog noise floor. As a result, it appears that an ADC with a DR of 47 dB is enough for this test case, as illustrated in Figure 7.27. Here, the dynamic range is defined with reference to a sine wave scaled at its FS i.e. for a waveform that exhibits a CF of 3 dB.

(ii) We can then consider a test case corresponding to the achievement of the maximum throughput for the receiver – a configuration where the maximum SNR is expected to be delivered to the baseband algorithms. In that case, we continue to assume that the wanted signal is alone at the receiver input. The only difference compared to the previous case is that we can assume that the noise from the RF/analog part of the receiver is 30 dB below

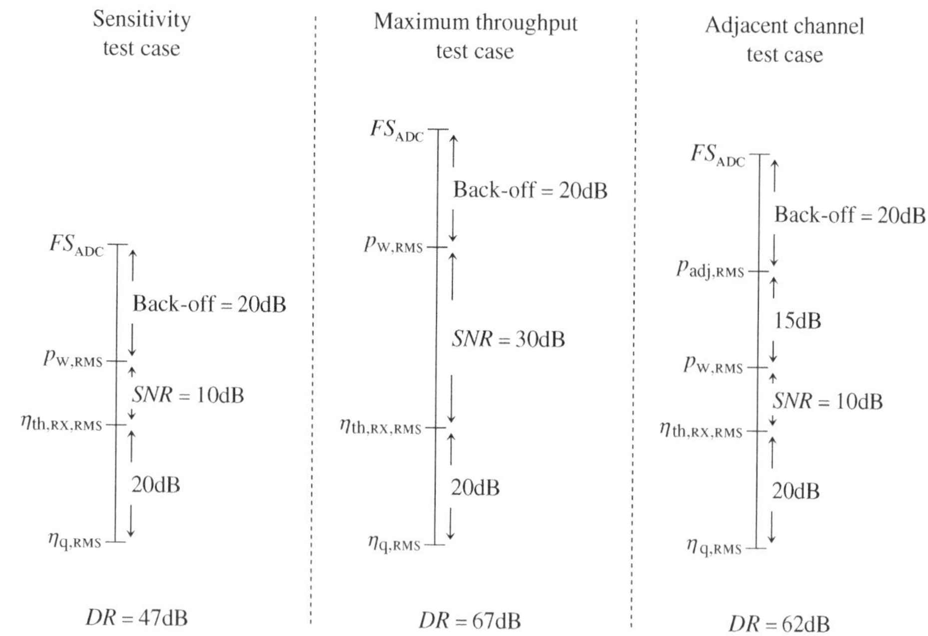

Figure 7.27 RX ADC budget without AGC considerations – The ADC dynamic range, *DR*, defined for a sine wave scaled at the ADC FS, is driven first by the back-off, i.e. the ADC FS divided by the RMS value of the converted signal, taken here as 20 dB. This value covers the CF of the P or Q waveforms, which can be more than 10 dB due to the propagation channel as discussed in "Instantaneous amplitude variations" (Section 2.3.3), some uncertainties in the power measurements as well as in the RF/analog gains; and the headroom burning due to RF impairments like the DC offset as illustrated in Section 9.2.2. Then the signals must be scaled so that the quantization noise, n_q, remains 20 dB below the RF/analog noise, $n_{th,RX}$, as discussed in Section 7.3.2. We then obtain one *DR* requirement per test case.

the wanted signal instead of 10 dB. As a result, the required DR to pass this test is simply 20 dB more than in the previous case and now reaches 67 dB.

(iii) Finally, we can consider a test case where an unwanted signal is present in the adjacent channel. At the input of the receiver, this unwanted signal can have a power somewhat higher than the wanted signal. And due to its frequency location close to the wanted signal, we can expect the attenuation provided by the analog filter to remain quite small in practice. We can then assume that the RMS value of the adjacent channel remains 15 dB above that of the wanted signal at the ADC input. Then, assuming on the one hand that we require the same back-off when it is the adjacent channel that has the highest power at the ADC input, and on the other hand that the RF/analog SNR is still 10 dB as in the sensitivity test case, we finally achieve a required DR of 62 dB for the ADC.

It therefore seems that an ADC with a DR of 67 dB is enough to handle all the test cases considered. This value is indeed the DR required to pass the most demanding test case, i.e. the maximum throughput test here. It is thus natural to suppose that such a DR is appropriate for all the other test cases that require less DR.

Unfortunately, the above reasoning is incorrect in general. To say that the final requirement in terms of ADC DR is simply the maximum of the DR required to pass all the test cases when considered independently is to give only a static view of the problem. We need to take into account the manner in which the signals are effectively scaled at the input of the ADC during the receiver operation. Practically speaking, this means considering how the AGC scheme regulates the signal at the input of the ADC. In the present case where we assume an AGC based only on the power measurement of the wanted signal, it is only the power of the wanted signal that can be regulated at a constant level at the input of the ADC. As discussed in Section 7.3.2 and illustrated in Figure 7.24, this is unfortunately only an approximation in practice as some limitations in the variable gain available in the RF/analog part of the receiver can lead to limitations in this statement. However, for the sake of simplicity, we can assume that the test cases we consider here involve input powers where the wanted signal is indeed regulated at a constant level at the ADC input, the AGC set point. As a result, we then need to assume that all the potential residual signals with a power higher than the wanted signal as well as the margins for clipping need to be taken as an overall margin above this set point. In the same way, all the margins related to the RF/analog SNR preservation must be taken below this point. But, as our simple AGC scheme cannot distinguish between the test cases we consider here, we have to take as the final margin above the AGC set point the worst case margin of all the test cases considered. The same must also be done for the derivation of the margin to be taken below the set point. For the above test cases, we finally achieve a requirement of 82 dB for the ADC DR, still assumed defined for a sine wave scaled at the ADC FS, as illustrated in Figure 7.28, and not 67 dB as derived previously.

From the above discussion, we see that there is a deep relationship between the analog filtering in the line-up, and the ADC characteristics and sampling rate. However, the final conclusion on this trade-off also involves the AGC scheme and the way the signal is effectively scaled at the ADC input. In practice, then, the exact behavior of the AGC scheme must be taken into account for the derivation of the final ADC DR requirement. Anticipating the discussion in Section 9.2.1, we remark that more refined AGC schemes taking into account the residual level of blockers at the ADC input can allow some of this required DR to be saved, and thus some implementation area and current consumption. In contrast, a high DR at the ADC stage can lead to some interesting side effects in addition to a relaxed analog filtering or a potential simplified AGC scheme. Recalling from the discussion in Section 7.3.2 that a higher ADC DR leads to a lower maximum gain being required in the receiver, we then anticipate fewer DC offset problems in the analog part of the line-up and thus simplified schemes to handle this parameter, as discussed in Section 9.2.2.

In any case, we can conclude by saying that once the trade-off between the analog filtering and the ADC DR and sampling frequency has been done, we get that the selected analog filtering is often not efficient enough to totally cancel the unwanted signals. As a result, the additional required filtering is implemented on the digital side. But in practice, such digital filtering can be merged with additional functions like the downsampling toward a more suitable low sampling rate for complex signal processing functions. We can also take advantage of the existence of those filtering stages to perform some static equalization of the transfer function of the analog blocks in order to cancel at least the linear EVM generated through the analog

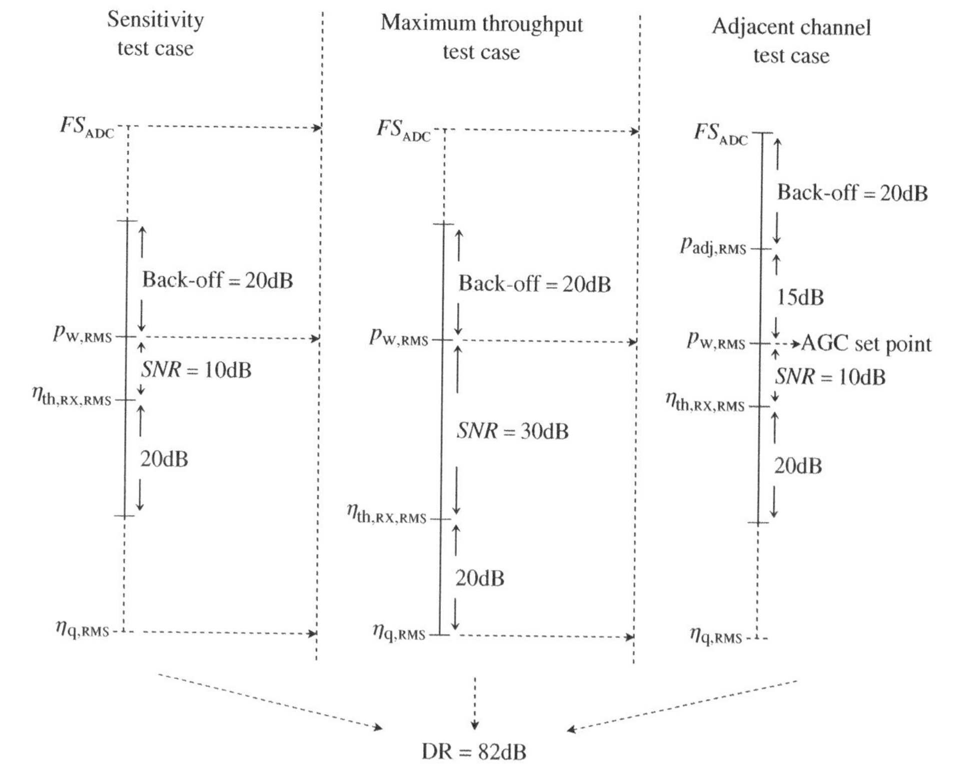

Figure 7.28 RX ADC dynamic range budget with AGC considerations – In contrast to what is shown in Figure 7.27, practical budgets need to take into account the way the AGC sets the signals at the ADC input. We need to consider separately the worst case margins required above and below the AGC set point for all the test cases in order to derive the overall required ADC dynamic range *DR*, defined as before for a sine wave scaled at the ADC FS.

filtering stages, as discussed in Section 4.4. This list is obviously not exhaustive, but only gives an illustration of what can be done in practice.

Budgets for Blocking Test Cases

Having derived a suitable trade-off for the joint optimization of the analog filtering, the ADC DR and the AGC strategy in order to confidently handle the potential unwanted signals along the data path, as detailed in the previous sections, we now need to go through the additional degradations that can result from the limitations in the physical implementation of the line-up.

But first some general preliminary remarks are in order. We recall that there are basically two classes of unwanted signals a receiver can cope with. As discussed in Chapter 3, these are the blocking signals and adjacent channels that are expected to belong to different wireless networks than the received wanted signal. The main consequence is that the power of the blocking signals is expected to be uncorrelated with that of the received wanted signal, in contrast to what we expect for the adjacent channels as illustrated in Section 3.3.1. As a result, a receiver needs to be able to cope with the highest level of such blockers at the same time as receiving a weak wanted signal. Blocker test cases are therefore often written this way, i.e. assuming a constant high blocking signal level. A constant power for the blockers leads to additional in-band noise contributions independent of the wanted signal input power in the corresponding SNR budgets. This is thus the opposite behavior to what occurs with adjacent channels. As a result, it is more consistent to illustrate the impact of the presence of an adjacent channel on the quality of the reception when deriving a full SNR budget as a function of the wanted signal input power. This is what is done in "SNR budget" (Section 7.3.4). For the time being, it is simpler to review the degradations that are likely when considering the reception of high, even if constant, level blocking signals.

Then, we observe that even when dealing with the reception of a wanted signal plus some unwanted blockers, we first of all deal with the reception of a wanted signal. This means that the presence of the unwanted signals has to be seen as the root cause for new degradations, expressed as additional noise contributions in the wanted signal frequency band, on top of those necessarily experienced by the wanted signal as if it were alone in the receiver. We can thus see the budgets to be derived in the presence of such unwanted signals as an incremental step compared to what needs to be done when the wanted signal is alone.

As recalled in "Budgets to be considered" (Section 7.2.1), different kinds of degradations are likely depending on whether the blockers considered lie in the close vicinity of the wanted signal in the frequency domain. Thus, it is worth looking at what happens with blockers that are far from the wanted signal in the frequency domain or in its close vicinity.

Furthermore, we must also remember that we are concerned in this chapter with a full duplex transceiver. As a result, what we call the SNR budget for the reception of the wanted signal alone must be understood as the SNR budget for the reception of the wanted signal alone at the air I/F, but with the TX leakage present at the input of the active part of the receiver. Such a budget necessarily already takes into account the degradations linked to the reception of one single strong blocker, assumed amplitude modulated in the present case, on top of the wanted signal. This means that we have to focus mainly on the additional degradations on the quality of the reception linked to the presence of a second additional blocking signal at the input of the receiver. The noise contributions that need to be considered when dealing with the wanted signal with the TX leakage alone are addressed in the illustration for the SNR budget detailed in "SNR budget" (Section 7.3.4).

Far Out-of-Band Blocker Case

Let us first discuss the additional noise contributions recovered in the wanted signal frequency band when the receiver copes with at least two blockers that are not in the close vicinity of the wanted signal. Assuming a receiver embedded in a full duplex transceiver, one of them is necessarily the TX leakage signal. We thus assume that the other is an additional out-of-band blocker.

Due to this relative frequency location, only two kinds of phenomena can lead to an additional contribution compared to the case where the wanted signal is alone with the TX leakage signal.

First, we get all the bandpass noise contributions that are retrieved as centered around the additional unwanted signals we cope with and that have a sufficient spectral extent so that a fraction of them lie within the wanted signal frequency band. In practice, this cannot be terms related to the distortion of the unwanted signal itself whether through its self-compression or through its XM by the TX leakage signal, here assumed as amplitude modulated. Indeed, as discussed throughout Section 5.1.3, the spectral extent of such distortion noises is related to that of the bandpass signals through the spectral regrowth phenomenon. Thus, considering blockers that are far enough from the wanted signal in the frequency domain, we can assume that no contribution of such distortion terms can reach the wanted signal frequency band. As a result, it remains to consider the bandpass noise contributions retrieved as centered around the blockers that are on the one hand not linked to the receiver linearity and that have on the other hand sufficient spectral extent. For the noise contributions considered for the receiver in Section 7.3.1, we can first of all think about the thermal noise of the line-up that is experienced by all the bandpass signals entering it. However, as this contribution is additive, it results in an in-band SNR limitation independent of the presence or not of the unwanted signals. As a result, only the LO phase noise that is recovered as transposed around all the bandpass signals present at the input of the frequency downconversion can generate such wideband additional degradation. This means that only the reciprocal mixing phenomenon, as detailed in "Reciprocal mixing" (Section 4.3.3), can result in an additional noise contribution with a spectral extent such that a fraction of it can lie in the wanted signal band.

Second, when dealing with multiple unwanted bandpass signals at the input of the receiver, we can cope with intermodulation products that fall back in the wanted signal band, according to the mechanism detailed throughout Section 5.1.2 and in "Revisiting intermodulation and harmonic tones" (Section 5.1.3). In practice, up to first order we can consider mainly the second and third order intermodulation products that involve two unwanted blockers located at particular frequency offsets from the wanted signal, as discussed in "Intermodulation and blocking test case" (Section 3.3.1). Supposing that one of the blockers is the TX leakage signal, the blockers that are involved in the third order intermodulation phenomenon lie at the double duplex and half duplex distance from the wanted signal respectively, as illustrated in Figure 3.19. In both cases, the generated intermodulation tone lying in the wanted signal band can be considered as an additional noise contribution and thus leads to a limitation in the SNR achieved. Practically speaking, the power of these intermodulation tones can be derived using formulas such as (5.159) and (5.164).

As a result, for a wanted signal with TX leakage only, we can carry out budgets for these particular intermodulation configurations considering on the one hand the level of the intermodulation tones and on the other hand the potential contribution of the LO phase noise through the reciprocal mixing with the additional blocker on top of the typical noise contributions, discussed in Section 7.3.4.

Close In-band Blocker Case

Let us now assume that we cope with one of the blockers lying in the close vicinity of the wanted signal in the frequency domain.

However, before going into details about the degradations we are faced with in that case, we observe that due to this particular frequency configuration, our blocker is often an in-band blocker. This means that it lies within the overall receive system band, and thus within the passband of the passive filters present at the input of the receiver. As discussed in "Filtering budget vs. ADC dynamic range" earlier in this section, such RF passive filters are for the most part implemented as non-tunable filters, and are thus passband for the overall receive system

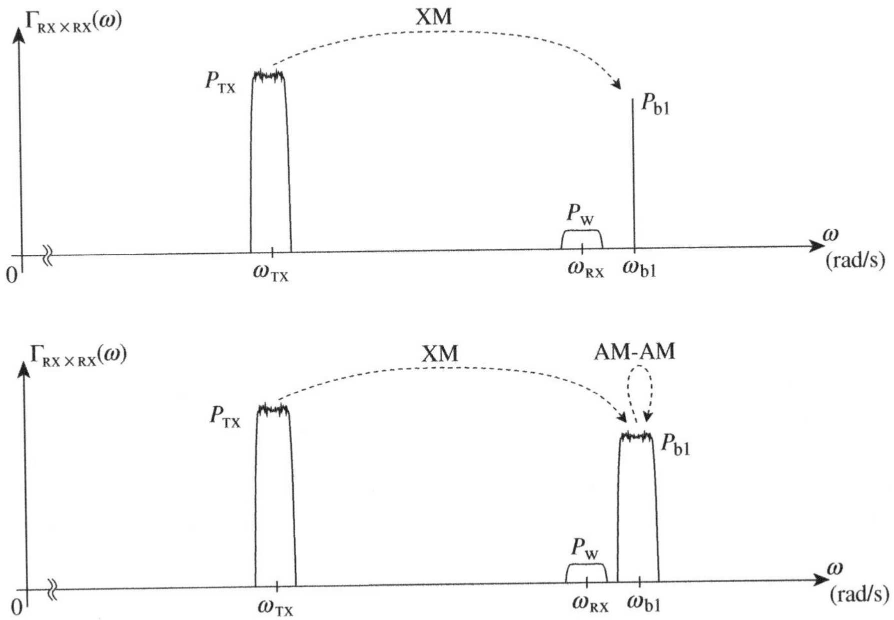

Figure 7.29 Distortion mechanisms involved in the SNR budget when dealing with a close in-band and a far out-of-band blocker – Considering a close-in band blocker, here centered around ω_{bl}, which has a constant instantaneous amplitude, we get that it is insensitive to smooth compression. However, given in addition an amplitude modulated far out-of-band blocker, here assumed to be the TX leakage signal centered around ω_{TX}, we can expect it to cross-modulate the close in-band blocker (top). When this close in-band blocker is also amplitude modulated, it experiences self-distortion due to its compression on top of the XM by the far out-of-band blocker (bottom).

band. As a result, this kind of in-band blocker can reach the input of the active part of the line-up with a quite strong power. Given that such blockers can lie very close to the wanted signal in the frequency domain, they can be tricky to handle in practice.

In order to go further, let us consider the configuration shown in Figure 7.29 where a blocker lies in the close vicinity of the wanted signal in addition to the TX leakage. Looking at this figure, we first see that due to the frequency configuration of the blocking signals, we cannot have any intermodulation products between the blockers that fall in the wanted signal frequency band. As a result, only the bandpass noise terms recovered as centered around the close in-band blocker and having a non-negligible fraction of their power lying in the wanted signal frequency band can lead to an additional in-band SNR reduction.

Obviously, in these contributions we already get those discussed in the previous section dedicated to configurations involving only out-of-band blockers. Indeed, the contributions considered in that case have by definition a wideband structure so that they necessarily still have an impact when the blockers are close in-band ones. Practically speaking, those contributions reduce to the LO phase noise term centered around the blockers and that lead to a reciprocal mixing phenomenon.

Turning to additional bandpass terms and thinking about the various degradations reviewed in Part II of this book and illustrated for our receive line-up in "Impact of implementation limitations" (Section 7.2.1), we can conclude that only the RF compression can lead to such leaking in an adjacent frequency band through the spectral regrowth phenomenon discussed in "Spectral regrowth" (Section 5.1.3) and "Nonlinear EVM and spectral regrowth due to AM-PM conversion" (Section 5.3.2). Thus, even assuming that in our receive case only the AM-AM conversion is of significance due to the lower power of the signals we are dealing with compared to the transmit side, we can cope with both the XM of one bandpass blocker by the other one and with the self-distortion of each bandpass signal. Obviously, these phenomena exist only if the bandpass signals at their origin are amplitude modulated. Assuming this is the case for the TX leakage signal, it is of interest to make the distinction for the close in-band blocker, and to suppose successively that it is amplitude modulated or not, following Figure 7.29.

Supposing that we are dealing with an amplitude modulated TX leakage signal, we get in both cases that the XM of the close in-band blocker by this signal generates a distortion term that is centered around it, as illustrated in Figure 7.30. In practice, due to spectral regrowth,

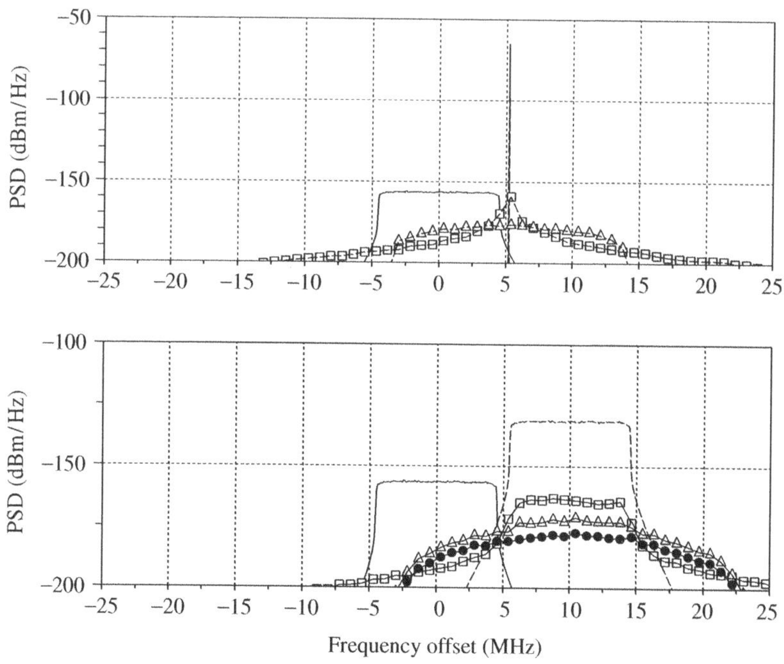

Figure 7.30 Additional noise components PSD involved in the SNR budget when dealing with a close in-band and a far out-of-band blocker – Considering the configurations of blocking signals shown in Figure 7.29, we recover different distortion noises centered around the close in-band blocker due to compression, depending on the nature of the modulations we are dealing with. Given that the far out-of-band blocker is amplitude modulated, when the close in-band blocker is not, here illustrated as a CW (top dashed), only the XM term (triangle) has a contribution in the wanted signal band on top of the LO PN (square). When it is amplitude modulated (bottom), we recover an additional self-distortion term (dot).

the spectral extent of this distortion term is higher than that of the amplitude modulated signal that is at its origin. Thus, assuming in the present case that the TX leakage signal has the same spectral extent as the wanted signal, we can recover a non-negligible amount of unwanted power in the frequency band of the wanted signal. In order to derive our SNR budget, we then need to estimate this fraction of power. This is in fact the same configuration as encountered during the derivation of the ACLR budget on the transmit side. Thus, following the methodology presented in "ACLR budget" (Section 7.2.3), we can reach our goal by means of a spectral analysis based on an analytical expression for this particular distortion term, for instance the first term in equation (5.175).

Supposing now that we are dealing with an amplitude modulated close in-band blocker, we also recover a distortion term linked to the self-compression of this blocker as displayed in Figure 7.30(bottom). In that case we also get that a fraction of the resulting distortion term can lie within the wanted signal band due to the spectral regrowth of the blocker. Here again, the fraction of noise power retrieved in the wanted signal band can be derived using the approach considered for the ACLR budget. This means using a spectral estimation based on an analytical expression for this distortion term, such as equation (5.190).

As a result, we can carry out budgets for such particular configurations that involve at least one close in-band blocker by considering both the reciprocal mixing phenomenon and the additional distortion terms reviewed above in addition to those to be considered when the wanted signal copes with the only TX leakage signal as discussed in Section 7.3.4. However, we observe that given a configuration where the blockers are much stronger than the wanted signal, in real implementations the in-band SNR is often effectively limited by the phenomenon reviewed here rather than by the contributions existing when considering the wanted signal with a TX leakage signal only.

Frequency Planning

In order to conclude this first part dedicated to the illustration of the budgets to be performed to check the resistance of a receiver to its wireless environment, we can say a few words about the frequency planning associated with such a line-up. As stated in Section 7.2.3 concerning the transmit side, the fact that we keep it short does not mean that this is not an important topic. The point is that, unfortunately, we can only give some high level guidelines as this kind of study is highly dependent on the practical implementation considered.

For this topic also the practical orders of magnitude of the phenomenon we are dealing with often lead to a different behavior between the transmit and the receive sides. In the former case, the main purpose is to ensure that no unwanted radiation can occurs in sensitive frequency bands in which strict requirements hold. Then, there may be a need to consider additional phenomena that can degrade the transmitter performance through the degradation of the quality of the modulation. This is the case for instance when the frequency planning leads to some pulling effect of the RF synthesizer used for the generation of the transmitted RF signal. In contrast, it is often the potential degradation of the quality of the reception that drives the frequency planning on the receive side. Then, considerations of potential additional unwanted radiations due to the activity of the receiver can also be required in order to ensure that no pollution of the wireless environment occurs.

However, whatever the consequences in terms of receiver performance, there are basically two types of cause for the degradations. The first type are the mechanisms that lead to a degradation in the performance of the frequency transposition implemented in the RF/analog part of the line-up. As discussed in Section 6.3, this can be for instance related to the use of mixers implemented as choppers. Such implementation results in a harmonic mixing problem due to the presence of odd order harmonics in the spectral content of the reconstructed equivalent complex LO waveform. This behavior, which reduces the resistance of the receiver to blocking signals, is furthermore enhanced in the presence of RF impairments between the P and Q branches of the complex mixing stages, as it results in a rise in the even order harmonics in the spectrum of this LO signal, as detailed in Section 6.4.1. Furthermore, this potential folding of blocking signals can result from the presence of additional unwanted spurious tones in the instantaneous phase spectrum of the LO signal. The presence of these tones can result for instance from the structure of the RF synthesizer used for the LO generation itself. Another cause can be the pulling of the RF oscillator used in this synthesizer by the injection of clock harmonics for instance. In any case, the presence of such spurious tones in the phase spectrum of this LO signal results in the folding on the wanted signal of potential unwanted blockers lying at the corresponding frequency offsets. Such behavior thus leads to the existence of spurious responses, or equivalently to finite ISR capabilities of the receiver that again reduce its resistance to blocking signals.

The second type are the mechanisms that lead to the direct coupling of an unwanted signal along the receive data path. The physical root cause for this can be either some feedthrough at the device implementation level, or the direct electromagnetic coupling between RF paths. In both cases, it can lead to a fraction of the RF signals used in the physical implementation of the line-up being retrieved at an unexpected RF port along the data path. Practically speaking, these unwanted RF signals are often harmonics of the clock signals used to drive the digital logic, the converters, or the frequency mixers, as discussed in Section 6.4. And having harmonics of those signals that are recovered along the data path can result in two kinds of problem. On the one hand, when such RF pollution lies in the wanted signal frequency band, a direct degradation of the quality of the reception can occur, depending on its relative power with respect to that of the wanted signal. On the other hand, if the pollution reaches the antenna connector, there is a potential pollution of the other users. Thus we may also have some spurious emission problems due to the receiver activity. This incidentally explains why a spurious emission requirement can exist, even for receivers as highlighted in "Spurious responses and emissions" (Section 3.3.1).

In practice, we may thus need to select carefully the frequency of all the periodic signals used for the implementation of the line-up in order to try as much as possible to have no harmonics in the problematic frequency bands. Obviously, dealing with a direct conversion receiver, there are not so many degrees of freedom for this with regard to LO frequency planning. This is fortunately not the case for most of the other periodic signals, such as digital clock signals.

In conclusion, we observe that, dealing with a receiver embedded in a full duplex transceiver, such coupling of unwanted tones can also be related to the transmitter activity through the clocks used in the transmitter implementation. As a result, we may also need to take care of the constraints for the receiver in order to carry out the frequency planning of the transmitter.

7.3.4 Budgets Linked to the Modulation Quality

Let us now focus on budgets carried out in order to check the quality of the modulation to be delivered by the RF/analog part of our receiver to the baseband algorithms in charge of the data bit recovery. We recall first that, as discussed in Chapter 3, a wireless standard can hardly do more than require a receiver compliant with it to be able to achieve minimum data rates for given configurations of received signals and propagation channels. Given such throughput requirements and the performance of the baseband algorithms we are dealing with, we can derive requirements for the quality of the modulation to be delivered to those algorithms by the RF/analog part of the receiver.

Practically speaking, these requirements are set for different parameters that impact the quality of the modulation. However, as already encountered on the transmit side, not all of them involve the same level of refinement in the budgets to be carried out. For instance, we can have maximum values that the baseband algorithms can handle for single parameters such as the frequency error or the DC offset. This can result in straightforward specifications for the corresponding blocks in the RF/analog line-up. This is unfortunately not the case for SNR budgets which involve the performance of almost all the blocks present along the data path. Thus, we focus mainly on SNR budgets by way of illustration.

SNR Budget

For the in-band SNR budget of our receiver, we assume that our receiver aims to be compliant with the typical SNR requirements detailed in "Signal to noise power ratio" (Section 3.3.2), and illustrated in Figure 3.20.

Before going any further, we can make some preliminary remarks. As discussed in that section, the SNR requirements derived from the performance of the baseband algorithms are for the most part achieved using pure AWGN simulations, i.e. assuming precise statistical and spectral characteristics for the noise component. We recall that this is a major difference compared to the transmit side as in that case often only a simple RMS EVM performance is required at the transmitter output, regardless of the statistical and spectral characteristics of the noise terms. But on the receive side, obviously many noise contributions cannot be assumed Gaussian or white. This is for instance the case for the distortion noise terms whose characteristics are directly related to those of the bandpass signals being distorted. However, this is often the simplest way to proceed to derive the total noise power at the output of the receiver as the sum of all the contributions, whatever their origin, and thus their statistical or spectral characteristics. Following this methodology, and in order to avoid unpleasant surprises, some margins should be considered with regard to the requirements derived using an AWGN model. We observe that experience shows that for the most part that AWGN is the worst thing that the baseband algorithms can face. As a result, the requirements derived in that way can often be considered as a worst case compared to what we have in practice in a line-up in terms of statistics. However, this does not prevent us from performing some system simulations using realistic behavioral models to check the validity of a budget. In any case, the derivation of an analytical SNR budget as done here is a powerful tool to achieve a deep understanding of the relative importance of the various contributors we are faced with in a line-up. This is the starting point for aiming toward a consistent balance in the constraints between the different blocks of a receiver in order to achieve given SNR requirements.

Recalling the discussion in Section 7.3.2, we also remark that the performance of the receiver, and thus the SNR delivered to the digital algorithms, are necessarily a function of the wanted signal input power. This is due to our assumption of having an AGC scheme based on this single power measurement. However, that performance is also indirectly impacted by the profile we assumed for the adjacent channel signals. This is the profile that drove our choice for the gains switching points as illustrated in "Wanted signal plus unwanted signals: gain split in the receiver" (Section 7.3.2). Moreover, given that their level remains consistent with that of the wanted signal as discussed in Section 3.3.1, we can expect that the absolute level of the in-band additional noise contributions linked to the presence of a signal in the adjacent channel effectively depends on the power of the wanted signal. This is thus a different behavior than what can be expected in the blocker cases as discussed in "Budgets for blocking test cases" (Section 7.3.3). Thus, it is worth completing the examination of the impact of the presence of such a signal in the adjacent channel by deriving the resulting SNR degradation as a function of the wanted signal input power. We consider first the case where the wanted signal is alone with the TX leakage signal and then where there is an additional unwanted signal in the adjacent channel.

Wanted Signal with TX Leakage

We begin with the in-band SNR performance achieved as a function of the wanted signal input power when receiving a single wanted signal present at the antenna connector. Remember that, dealing with a receiver embedded in a full duplex transceiver, we have to consider the unavoidable TX leakage present at the input of its active part. We thus consider here the configuration detailed during the review of the ZIF RX problem in Section 7.3.1.

In order to go further with the corresponding SNR budget, we first need to list the different sources of signal degradation in the line-up. This can be done by reviewing the impact of the different blocks that process the received signal. Practically speaking, three groups of blocks are present in the line-up: RF blocks, the blocks dedicated to the frequency downconversion, and the baseband blocks. Thus, looking at Figure 7.18, we can make the following observations.

The early active RF stages of the receiver have a first direct contribution to the total noise power involved in the line-up by adding their own noise contribution to the noise coming from the outside and recovered at their input. The latter noise is composed of the component retrieved by the antenna from its electromagnetic environment, from the thermal motion of the electrons at the input of the active part of the line-up, and of the noise component leaking from the transmitter up to the receiver input in the present full duplex case.[4] Then, on top of this direct additive noise increase, we often have some smooth compression effect due to the finite power supply of the active RF blocks. As a result, dealing with amplitude modulated bandpass RF signals, we may have to deal with the generation of nonlinear EVM terms. This can be due to either AM-AM or AM-PM conversion mechanisms. However, as highlighted in "Impact of implementation limitations" (Section 7.2.1), on the receive side we often deal with signals of lower level than encountered on the transmit side, even when dealing with their maximum input levels. As a result, the equivalent models for the parasitics of the active devices

[4] Concerning the thermal noise derivations, recall that, as discussed in "Receiver high level parameters" (Section 7.3.2), we assume for the sake of simplicity that both the antenna noise temperature and the line-up physical temperature are equal to T_0, the temperature used for the definition of the noise factor. As discussed in that section, this assumption simplifies the derivation compared to the general case in Section 4.2.5.

involved in the receiver can often be assumed with constant parameters over the dynamics of the signals being processed. This results in no carrier phase shift as a function of the signal instantaneous amplitude. Thus, we can assume here that mainly AM-AM conversion drives this kind of compression performance for our receiver. But even then there may be different in-band contributions, due to either the self-distortion of the received wanted signal or its XM by the TX leakage signal.

The frequency downconversion stage can then impact the quality of the reception toward various degradations. However, due to the present configuration where only the TX leakage is present at the input of the receiver on top of the wanted signal, and considering classical duplex distances between the TX carrier frequency and the RX one, no folding of unwanted signals due to some harmonic mixing for instance, as discussed in "Frequency planning" (Section 6.4), can occur. As a result, only three kinds of degradation in addition to that linked to the obvious intrinsic noise performance of the block have to be considered. First we have the degradation linked to the use of an LO signal that exhibits some phase noise. This results in a direct in-band SNR limitation, as discussed in "SNR limitation" (Section 4.3.3), or a reciprocal mixing phenomenon with the TX leakage signal, as discussed in "Reciprocal mixing" (Section 4.3.3). Second, we have the RF impairments between the P and Q branches of the complex frequency downconversion. As discussed in Chapter 6, such impairments leads to a rise in the image signal. But, due to the frequency planning associated with the direct conversion architecture, the complex envelope of the image signal that is retrieved as superposed with the wanted signal in the frequency domain at the downmixer output is then directly the complex conjugate of that of the wanted signal. As a result, this additional unwanted component behaves as a multiplicative noise term whose power scales with that of the wanted signal according to the discussion in Section 6.2.2. Third, we may have an additional degradation due to the even order nonlinearity of the downmixer. As discussed in "AM-demodulation due to even order nonlinearity" (Section 5.1.3), this kind of nonlinearity leads to the direct AM-demodulation toward baseband of all the bandpass signals present at the input of the nonlinear device. Thus, due to the frequency planning of the direct conversion receiver, such AM-demodulation terms are necessarily superposed on the wanted signal around DC at the output of the downmixer. But, on top of that we need to keep in mind that, as discussed in "Filtering budget vs. ADC dynamic range" (Section 7.3.3), we generally have no channel filtering implemented before the channel selection, i.e. prior to the frequency downconversion down to baseband in the present architecture. It thus follows that we may have high level blocking signals at the input of the downmixer in practice, and thus a non-negligible additional in-band noise contribution.

There is also the contribution of the baseband devices. In practice, such blocks implement amplification, analog channel filtering, and analog to digital conversion. If we consider first the impact of the amplification stages, we can assume here that we are faced mainly with the intrinsic noise contribution of such an active block. This means that we assume that the degradations linked to their linearity are negligible in our budget. Indeed, as highlighted in the introductory part of Chapter 5, the linearity of the baseband blocks is often better than that of the RF devices. We thus assume for the sake of simplicity here that we do not need to consider this kind of contribution any further.[5] Then we need to consider the consequences of the presence of the analog channel filtering. As discussed in Section 4.4, this can result in

[5] Obviously this assumption should be checked on a case by case basis in real life.

the generation of some linear EVM that degrades the quality of the received signal. However, this kind of distortion remains deterministic as long as we know the characteristics of the analog filter implemented. This is in fact more or less true as in practice we necessarily have some spread in the implementation of analog filters, and thus in their characteristics. But, given that those variations remain within acceptable bounds to be determined, we assume that we can rely on an equalization stage present in the line-up that compensates for this linear distortion. Moreover, we recall that there are some classes of modulating waveforms that are classically associated with an equalization in the frequency domain for the compensation of the propagation channel. A side effect of this kind of equalization is that it also naturally compensates for such linear distortion induced by analog filters. This is generally the case for an OFDM signal as considered here. For these two reasons, we thus assume for the sake of simplicity that the linear EVM generated by the analog filtering stages can be neglected for the derivation of our SNR budget. Finally, we need to consider the impact of the limitations in the implementation of the ADC stage. But for that stage also we can assume that its linearity is good enough so that the induced distortion can be neglected. As a result, we consider only its intrinsic noise contribution in what follows through its available DR.

In light of the above observations, we now have a clear overview of the different noise contributions we need to take into account in our in-band SNR budget. Before we go any further, it may of interest to classify those contributions according to whether they are additive, multiplicative or distortion noises terms. As already seen during the derivation of the budgets for the transmit side, such classification allows a straightforward interpretation of the resulting SNR curve and the root cause for the associated limitations. But we also need to keep in mind that we now have an important difference compared to the transmit side due to the presence of TX leakage. In the present case the noise contributions that involve this unwanted signal are necessarily a function of its power. Thus, there may be difficulties classifying the corresponding noise contributions as this refers to the scaling of the noise power with respect to the power of the received wanted signal, and not to that of the TX leakage.

We thus need to make an assumption for the power of this TX leakage. For that purpose, we observe that the power delivered by the transmitter depends on its distance from the device to which it is connected. We thus expect it to be inversely proportional to the power of the received signal. However, for the sake of simplicity, we continue to assume that this TX level is constant over the full range of the wanted signal input power. More precisely, we assume that the power of the TX leakage signal at the input of the active part of the receiver, $P_{\text{i,TX}}$, remains constant and equal to −25 dBm.[6] By thus assumption, we finally get for the additive noise contributions:

(i) the overall additive RF/analog noise contributions in the line-up, at least for a given setting of the gains in the receiver;
(ii) the quantization noise linked to the analog to digital conversion;[7]
(iii) the TX noise leaking in the receive bandwidth;
(iv) the LO phase noise contribution due to the reciprocal mixing with the TX leakage;
(v) the AM-demodulation of the TX leakage.

[6] It may still be necessary to consider a more precise behavior for our unwanted signal in real life.
[7] And to the fixed point implementation of the later digital signal processing.

We can see that at least the last two contributions above behave as additive contributions only due to our assumption of having a constant TX power. Then, for the multiplicative contributions, we get:

(i) the image signal;
(ii) the LO phase noise linked to the transposition of the wanted signal itself;
(iii) the XM with the TX leakage.

As for the additive contributions, we can comment that the generated XM term behaves exactly as a multiplicative noise in the present case due to the assumption of having a constant level for the TX leakage. Last, for the distortion noises we get:

(i) the nonlinear EVM term generated by the self-compression of the wanted signal through the AM-AM conversion.

It then remains to derive the power of these contributions in order to obtain the total noise power. For that purpose, even if the SNR we want to derive is the one that would be achieved at the output of the receive line-up, it is of interest to work with equivalent input quantities for the noise contributions. This has the great advantage of allowing for a direct comparison with the input power of the wanted signal, $P_{\text{i,w}}$. And as discussed in Section 7.3.2, the settings of the receiver and thus its intrinsic performances are indeed a function of $P_{\text{i,w}}$ due to the behavior assumed for the AGC scheme. We can thus anticipate that the SNR performance achieved is in turn a function of $P_{\text{i,w}}$, and thus that directly working on equivalent input noise quantities allows a straightforward interpretation of such performance. Finally, working with equivalent input quantities has the great advantage of clearly highlighting the nature of the noise terms we are dealing with, i.e. additive, multiplicative or distortion, as illustrated at the end of the section.

However, looking at the equivalent input power spectral densities of the noise contributions we are dealing with, shown in Figure 7.31, we see that we have basically two kinds of noise terms: we have flat wideband noise contributions so that their in-band power can be derived in a straightforward way as the noise bandwidth of the receiver, δf, times their PSD level in this frequency band; and bandpass contributions whose spectral extent obviously depends on various parameters, depending on their nature. A spectral evaluation associated with an integration over the frequency band of interest would be required in order to derive the in-band contribution of those terms. However, we observe that unlike in the ACLR case discussed for the transmit side in "ACLR budget" (Section 7.2.3), the power of those bandpass terms is centered around the wanted signal. As a result, up to first order we can approximate the fraction of in-band power of the bandpass terms as their overall power. This has the advantage of allowing us to use simple analytical formulas for the derivation of our budget, as already suggested for the illustration of the EVM budget in "EVM budget" (Section 7.2.4). Such analytical approximation can be seen as an upper bound for the derivation of the total noise power. Experience shows this approach to be useful in practice.

Let us use this methodology to derive the power of the various noise terms we need to consider in our budget. We first focus on the additive noise components. As discussed above, there are two kinds of terms that result in an additive noise contribution. We have on the one

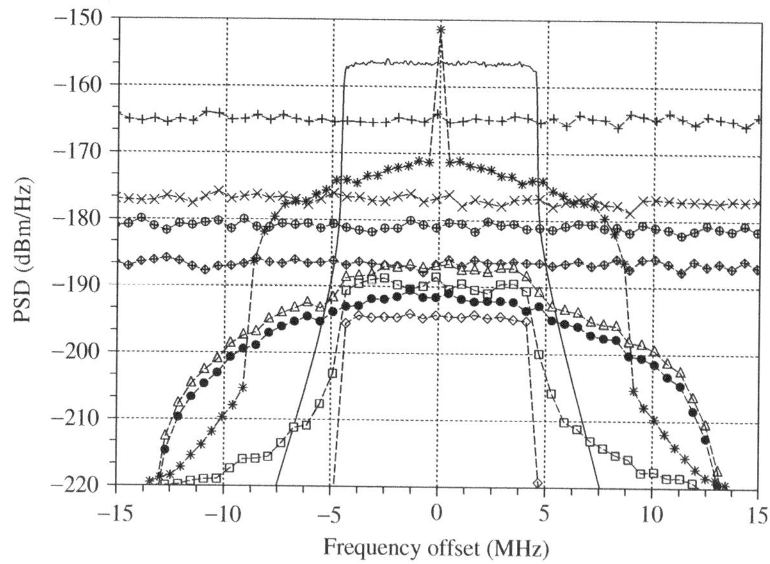

Figure 7.31 Equivalent input PSD of the noise components involved in the SNR budget – The additive noise contributions involved in the SNR budget of our receiver embedded in a full duplex transceiver reduce to additive RF/analog noise (pluses), quantization noise (diamond pluses), TX noise in the receive band (stars), phase noise due to reciprocal mixing with the TX leakage (crosses), and the AM-demodulation component (asterisks). The multiplicative terms are driven by the LO phase noise transposed around the wanted signal (squares), the image signal whose PSD is a scaled flipped copy of that of the wanted signal (diamonds), and the XM with the TX leakage (triangles). Finally, the distortion noise reduces here to the AM-AM term only (dots). The PSDs are plotted as equivalent input quantities, without taking into account the channel filtering effect on them.

hand contributions that are related to the reception of the wanted signal only and on the other hand those related to the presence of the TX leakage.

In the first category, we get the intrinsic RF/analog additive noise performance of the receiver, mainly driven by the thermal noise. This contribution has been discussed in "Receiver high level parameters" (Section 7.3.2). And with our AGC strategy detailed throughout Section 7.3.2, we can finally assume that the intrinsic additive noise performance of the receiver as a function of the wanted signal input power is that shown in Figure 7.25.

Then, we have the contribution of the additive quantization noise. In order to be consistent with the AGC strategy derived in "Wanted signal alone: receiver max and min gains" (Section 7.3.2), we can reuse the ADC DR assumed in that section. This means a DR of 52 dB in the receiver noise bandwidth, assumed equal to the wanted signal frequency band here, and defined for a sine wave set at the ADC FS. If we also continue to assume a FS of 1 V, this results in an equivalent RMS value for the quantization noise of −55 dBVrms at the ADC input as given by equation (7.8). Then, using the gain sets in the RF/analog part of the receiver and derived in Section 7.3.2, we can deduce the equivalent quantization noise power at the receiver input.

It remains now to deal with the additive noise contributions linked to the presence of the TX leakage. In practice, this unwanted term recovered at the input of the receiver is composed of the leakage of the useful part of the TX signal, which can be considered as a blocker in practice, and the noise floor at the transmitter output which lies within the wanted signal frequency band at the receiver input. The latter unwanted term leads to a direct noise contribution that we can assume additive to first order. And as an example, we can assume an in-band power of -130 dBm for this term in our budget.

Then, the useful part of the TX leakage can induce two kinds of additive contributions, one through the reciprocal mixing phenomenon due to the degradation of the signal during the frequency downconversion caused by the LO phase noise, and the other through its AM-demodulation, still at the frequency downconversion stage. Focusing first on the reciprocal mixing with the TX leakage, we can say that considering classical duplex distances, the power of the resulting in-band noise component classically involves mainly the noise floor part of the LO phase noise spectrum. Thus, following the discussion in "Reciprocal mixing" (Section 4.3.3), we can simply evaluate the resulting noise contribution as the product of the TX leakage power, the LO phase noise floor level, and the wanted signal bandwidth, as given by equation (4.241). For the illustration of our budget, let us assume a phase noise floor with a DSB PSD of -155 dBc/Hz. Given a wanted signal bandwidth of 9 MHz and a TX leakage level of -25 dBm, we obtain a noise contribution with an equivalent input power of -110 dBm.

Next, we have to evaluate the power of the TX AM-demodulation term. We first observe that there is an important DC term in this signal that has a constant sign, as can be seen on the power spectral densities shown in Figure 7.31. But in practical receivers implementations this offset is retrieved by the various DC cancellation schemes classically embedded in the receive line-up as discussed in Section 9.2.2. As a result, only the power of the non-constant term in this AM-demodulation signal needs to be considered in our SNR budget. In practice, this power can be evaluated using equation (5.154) for instance. In this expression, the Γ_4 constant, which takes into account the impact of the statistics of the instantaneous amplitude of the blocking signal, as introduced in "Moments of the instantaneous amplitude" (Section 5.1.3), here corresponds to that of the TX leakage signal, $\Gamma_{4,\mathrm{TX}}$. And in the present case where we assume that the TX signal is an OFDM waveform, we can assume that the $\Gamma_{4,\mathrm{TX}}$ constant involved in this equation corresponds to that of the Rayleigh distribution, i.e. is equal to 1 as given by equation (5.130a). The power of this noise contribution can thus be derived accordingly.

Let us now turn to the multiplicative noise components. As listed previously, we consider here on the one hand two components linked to the processing of the wanted signal itself through the receive line-up, namely the image signal and the LO phase noise, and on the other hand one linked to the presence of the TX leakage signal, the XM term. If we first focus on the rise in the image signal, assuming on the one hand that we are dealing with a frequency downconversion that uses mixers implemented as choppers, and on the other hand that this frequency conversion uses the fundamental tone of the LO waveform, the power of the image signal can be directly derived through the IRR value according to the discussion in Section 6.3.2. Considering practical impairments in the generation of the square LO waveforms used to drive the choppers, we can take as an example the same IRR value as used on the transmit side for the EVM budget, i.e. -38 dB. Moreover, assuming that the settings of the mixer itself as well as of the devices used to generate the LO signal remain static whatever the value of $P_{\mathrm{i,w}}$, the IRR value can be taken as constant over the DR of this input power. As a result, the power of the image signal at the transmitter output is simply equal to $IRRP_{\mathrm{i,w}}$, with IRR equal to -38 dB in the present case.

Then the power of the LO phase noise recovered as transposed around the wanted signal during the frequency conversion can be derived in accordance with the discussion in Section 4.3.3. Using equation (4.216), the equivalent input average power of this bandpass noise term is simply equal to $P_{n_\phi} P_{\text{i,w}}$, with P_{n_ϕ} the total power of the phase noise. For illustrative purposes, we can assume that this phase noise power is equal to the realistic value of 1.5° RMS, i.e. $1.5\pi/180$ rad RMS, as already considered for the illustration of the transmit budgets.

It then remains to derive the power of the multiplicative noise term linked to the cross-modulation of the wanted signal by the TX leakage signal. As discussed in "Cross-modulation due to odd order nonlinearity" (Section 5.1.3), the power of the resulting noise term can be evaluated using equation (5.178). But we need to take care that this expression gives the power of the noise term as recovered at the output of the nonlinear device. We thus need to divide it by the device power gain to derive the equivalent input power of the noise term we focus on. However, we have to keep in mind that our purpose is to derive the SNR budget. And recalling the derivation in that section, we need to consider the effective gain of the device G_{e} in order to perform this transposition, as can be understood from equation (5.179). Thus, given that G^2 in this expression is nothing more than the small signal power gain of the device, the equivalent input power of this XM noise term, $P_{\text{i,XM}}$, can be evaluated as

$$P_{\text{i,XM}} = 4\frac{G^2}{G_{\text{e}}^2}\frac{P_{\text{i,w}} P_{\text{i,TX}}^2}{\text{IIP3}^2}\Gamma_{4,\text{TX}}. \tag{7.17}$$

In the present case where only AM-AM conversion is considered, G_{e} is given by equation (5.174), but with the substitution of the blocker by the TX leakage signal to match our configuration. As a side effect, we thus get that G_{e} remains a real quantity. This explains why the power gain $|G_{\text{e}}|^2$, as encountered on the transmit side due to the consideration of the AM-PM conversion, reduces to G_{e}^2 here. We also recover again the $\Gamma_{4,\text{TX}}$ constant that can be approximated here as that of the Rayleigh distribution as discussed above. However, we also remark that we used the ICCP1 of the receiver at the duplex distance to characterize its ability to handle the TX leakage signal. We can thus express the above equation in terms of the device ICCP1 using both equations (5.66) and (5.104).

Finally, we can focus on the self-distortion noise term generated through the AM-AM conversion effect. The power of this term can be evaluated for instance using the analytical derivations in "Nonlinear EVM due to odd order nonlinearity" (Section 5.1.3). For that purpose we can use the fact that the average power of this bandpass noise is simply equal to half the expectation of the square modulus of its complex envelope as given by equation (1.64). Thus, considering again the effective gain of the device G_{e} given by equation (5.174), we can write from equation (5.194) that the equivalent input power, $P_{\text{i,AM-AM}}$, of this distortion term is given by

$$P_{\text{i,AM-AM}} = \frac{G^2}{G_{\text{e}}^2} P_{\text{i,w}} \left(\frac{P_{\text{i,w}}}{\text{IIP3}}\right)^2 [(\Gamma_{6,\text{w}} + 1) - (\Gamma_{4,\text{w}} + 1)^2]. \tag{7.18}$$

In this expression, the Γ constants characterize the statistics of the wanted signal. And here again, considering that we are dealing with an OFDM signal, we can assume that those constants correspond to those of the Rayleigh distribution, i.e. that $\Gamma_{4,\text{w}} = 1$ and $\Gamma_{6,\text{w}} = 5$ as

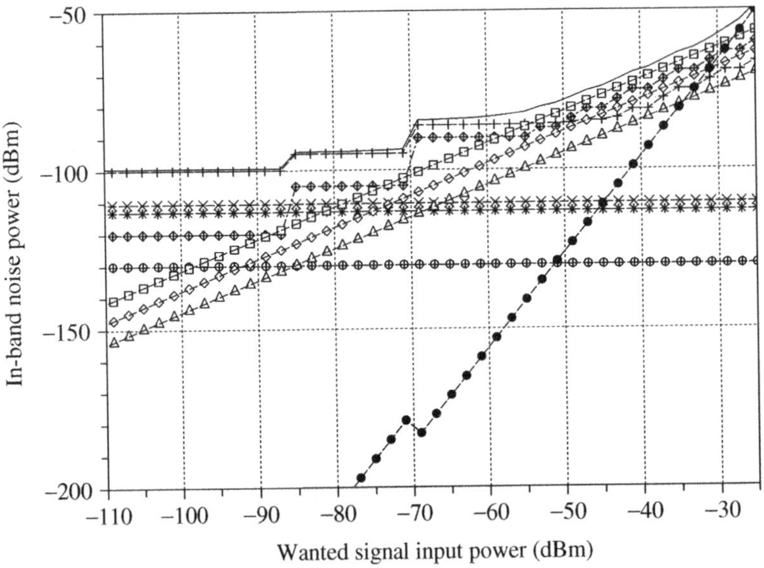

Figure 7.32 Equivalent input power of the noise components involved in the SNR budget as a function of the wanted signal input power: wanted signal plus TX leakage case – Assuming a constant TX leakage power, the thermal noise in the line-up (pluses), the quantization noise (diamond pluses), the TX noise in the receive band (stars), the phase noise due to the reciprocal mixing with the TX leakage (crosses), and the TX AM-demodulation component (asterisks) lead to a constant noise power contribution due to their additive behavior. In contrast, the multiplicative noises composed of the LO phase noise transposed around the wanted signal (squares), the image signal (diamonds), and the XM term due to the compression of the TX leakage (triangles) scale with the received signal power. Finally, the distortion noise due to the AM-AM conversion (dots) has a power that rises more quickly than that of the wanted signal. The total noise is thus dominated by different kinds of contributions depending on the input power area (solid).

given by equations (5.130a) and (5.130b), respectively. We can then use equation (5.66) to formulate this expression as a function of the receiver ICP1, as this is this quantity that has been chosen in "Receiver high level parameters" (Section 7.3.2) to characterize the compression behavior of the receiver on the received wanted signal.

Using the above formulations, we can derive the power of the equivalent input in-band noise contributions experienced by the received wanted signal as a function of its input power. Those contributions, shown in Figure 7.32, then allow the derivation of the achieved in-band SNR when the wanted signal has effectively experienced the degradations linked to the noise contributions, i.e. as recovered at the output of the line-up. Looking at the SNR curve shown in Figure 7.33, we can obviously make the same comments as during the illustration of the budgets for the transmitter about the differences in the way the noise terms impact the overall noise performance depending on whether they are additive, multiplicative, or distortion terms. Indeed, by definition the power of the additive noises depends only on the setting of the line-up, and not on the level of the received signal. In contrast, the power of the multiplicative noises scales with that of the wanted signal, while the distortion terms have a power that rises more quickly than that of the received signal. According to this behavior, there is necessarily a low enough input power of the received signal that leads to additive noise terms that are dominant.

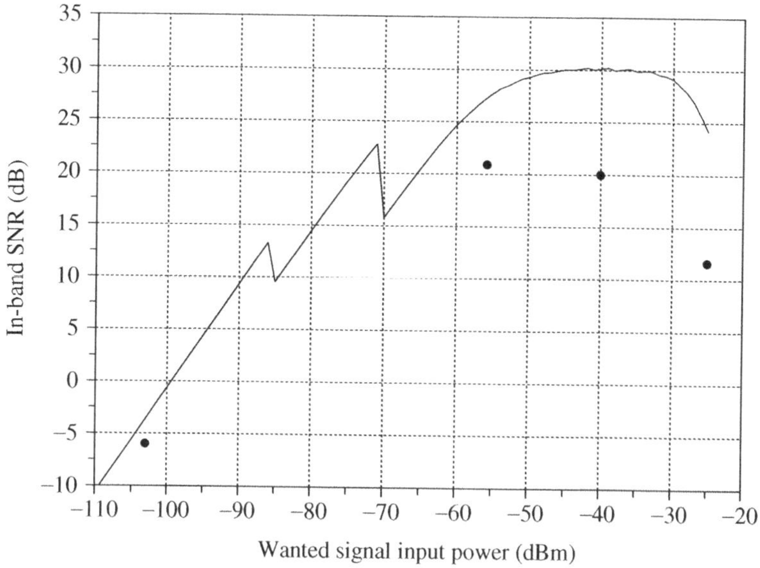

Figure 7.33 Achieved SNR performance as a function of the wanted signal input power: wanted signal plus TX leakage case – Considering the noise contributions shown in Figure 7.32, we achieve an SNR performance (solid) that fulfills the requirements (dots) corresponding to that displayed in Figure 3.20. As expected, as long as the additive noise components are dominant, we get an increase in the SNR performance as a function of the wanted signal input power in the lower part of its DR. In contrast, the rapid rise in the distortion noise leads to a decrease in the SNR performance in the upper part of the input power DR. Between those two extreme situations, the multiplicative noises lead to an upper bound for the achievable SNR performance.

In that range of input power, we thus get a linearly increasing SNR performance as a function of the wanted signal input power. In contrast, as their power rises more quickly than for any other terms, we can expect a sufficiently high input power so that the distortion terms are dominant. For such high input level, we thus get a decrease in the SNR achieved,[8] with a slope −2. In between those two extreme cases, the multiplicative noises lead to an upper bound on the achievable SNR performance, whatever the input power of the received signal. As these considerations are in fact very general, the shape of the SNR curve shown in Figure 7.33 is also very general.

However, we can refine this analysis as we can see in that figure some degradations of the SNR occurring in the area where the additive noise components are dominant. Those degradations, which result in a non-monotonic slope of the SNR curve for low input power, are related to the additive noise performance of the receiver that depends on its setting, mainly through the gain of its early active RF stages in fact. As detailed in Section 7.3.2, the switching of these gains leads to a degradation of the additive noise performance, as illustrated in

[8] Recall that this holds when dealing with an amplitude modulated wanted signal that experiences a smooth RF compression, as assumed here. As detailed in Chapter 5, constant amplitude bandpass signals are insensitive to RF compressions when considering their sideband of interest, i.e. that centered on their fundamental carrier frequency.

Figure 7.25. And as this switching occurs in an area of wanted signal input power where the additive noises remain dominant, we thus recover its impact on the final SNR curve.

At this stage, we can simply recall that due to the AGC strategy considered, based only on the power measurement of the wanted signal, the switching points for those early active RF stages of the line-up, i.e. the RXFE in practice, occur for lower input powers than would be really necessary to avoid the clipping of the only wanted signal. This was done with the aim of passing with enough backoff margins *all* the signals potentially present at the input of the receiver, and in particular the adjacent channels. As a side effect, due to the minimum variable gain considered in the RF/analog part of the line-up, we have that the wanted signal is not perfectly regulated at the ADC input in the lower part of the input power DR as illustrated in Figure 7.24(bottom). However, we can check here that such behavior is not necessarily a problem on its own as the overall noises coming from the RF/analog part of the line-up still remain well scaled at the ADC input. As can be seen in Figure 7.32, the quantization noise remains below the total noise, even when those RXFE gain switchings occur. We thus get a minimum SNR degradation due to the quantization process, which is the best we can do in practice. Then, the final scaling of the wanted signal around a constant target level can be done on the digital side, after the total filtering of the remaining unwanted adjacent or blocking signals.

In conclusion, by using on the one hand a correct setting of the gain switching points and on the other hand a correct specification of the device characteristics, we can finally fulfill the SNR requirements as shown in Figure 7.33. What is important to remark is that that performance depends on the level of the input signals, obviously because of the nature of the various noise components involved in the line-up as discussed above, but also because of the setting of the blocks in the receiver through the AGC that is driven by the input signal level; this setting impacting in turn the intrinsic performance of the blocks in the line-up. There is thus for the most part a deep relationship between the AGC scheme implemented in a receiver and its final performance. This behavior is illustrated in greater depth in Section 9.2.1.

Wanted Signal with TX Leakage and First Adjacent Channel

As already observed, it is of interest to detail the degradation in the quality of the modulation delivered by our line-up to the baseband algorithms when an unwanted signal exists in the adjacent channel. We have already discussed in "Wanted signal plus unwanted signals: gain split in the receiver" (Section 7.3.2) some impacts on the reception linked to the handling of such a signal whose power is expected to scale with that of the wanted signal. These were mainly through the related constraints on the receiver settings based on the assumed AGC behavior. We can now go a step further and examine the degradation of the SNR performance linked to the presence of an adjacent channel signal.

Still considering our receiver embedded in a full duplex transceiver, the presence of the adjacent channel will add some additional contributions to those linked to the reception of the wanted signal plus the TX leakage. Our assumed AGC strategy, detailed in Section 7.3.2, already takes into account the potential presence of an adjacent channel. This means that in our present in-band SNR budget we already consider all the contributions reviewed in the previous section. And given that the same characteristics for our receiver still hold in the present case, we simply need to take into account as a baseline the noise contributions displayed in Figure 7.32. We thus simply need to consider the additional noise contributions linked to the presence of the adjacent channel.

In order to go further and derive an example for our budget, we can suppose for the sake of simplicity that the signal in the adjacent channel is of the same nature as the wanted signal, i.e. is an OFDM-like signal with a useful bandwidth of 9 MHz and a carrier frequency lying at an offset of 10 MHz from the wanted signal. We thus assume here the same spectral configuration as considered in "Close in-band blocker case" (Section 7.3.3) and shown in Figure 7.29(bottom). The direct consequence is that the additional in-band noise contributions that we have to consider due to the presence of the adjacent channel are those already listed in that section and displayed in Figure 7.30(bottom): the reciprocal mixing with the adjacent channel, linked to the LO phase noise spectrum recovered as transposed around this signal during the frequency downconversion; the XM term, linked to the transfer of the amplitude modulation of the TX leakage on the adjacent channel; and the distortion term, due to the self-compression of the adjacent channel. With our present assumptions, this last term reduces to that linked to only the AM-AM conversion as we assume the AM-PM conversion to be negligible in our receiver with the signals input levels considered. These three noise components are the only additional ones linked to the presence of the adjacent channel that leak in the wanted signal bandwidth, whether through its own spectral extent for the LO phase noise spectrum, or through the spectral regrowth phenomenon for the XM and self-compression terms.

The main difference compared the former blocker case detailed in "Close in-band blocker case" (Section 7.3.3), is that we now make the assumption that the adjacent channel is expected to have power that scales with that of the wanted signal, as discussed in Section 3.3.1. For the sake of consistency with the derivation of our receiver gain split and the related AGC strategy performed in Section 7.3.2, we continue to consider the adjacent channel power profile detailed in "Adjacent channel selectivity" (Section 3.3.1), and shown in Figure 3.17(left).

Using this profile, we can now derive the power of the above listed noise contributions in the wanted signal frequency band. However, we can already see that we cannot directly use the same formalism as in the previous section where all the bandpass noise terms of interest for the budget were already centered around the wanted signal. In the present case, the additional contributions we need to take into account are centered around the adjacent channel so that only a small fraction of their total power lies within the wanted signal bandwidth. We thus need to adopt the same approach as considered in "ACLR budget" (Section 7.2.3), to derive the ACLR budget on the transmit side. Practically speaking, this cannot be done in an analytical way for all the noise contributions we are dealing with, in which case a realization of the process considered must be derived based on a time domain analytical expression. Then the fraction of power lying in the wanted signal frequency band can be derived using a spectral estimation of this realization.

Concerning the phase noise contribution, it might occur to us to approximate it using equation (4.241). This approximation simply involves the level of the adjacent channel signal and of the phase noise density at the corresponding frequency offset from the carrier. However, due to the convolution process between the spectrum of the signal being frequency transposed and the spectrum of the LO phase noise, this can lead to inaccuracies when considering on the one hand wideband signals and on the other hand the close in-band part of the resulting spectrum as here. As a result, we continue to keep the above guidelines in mind and employ a spectral estimation. We can use a realistic model of the RF synthesizer as in "ACLR budget" (Section 7.2.3). And in order to keep things simple, we assume here the same model parameters as considered for the ACLR budget, i.e. that $n_\phi(t)$ is generated using a Gaussian process filtered through a first order lowpass filter with a cut-off frequency of 50 kHz. The power of the process is then normalized to achieve an integrated phase noise equal to 1.5° RMS.

For the XM and the self-distortion terms we have fewer degrees of freedom as we can only rely on a time domain simulation to derive a spectral estimation of those components. We can use the discussion in "Cross-modulation due to odd order nonlinearity" (Section 5.1.3) to get time domain expressions for those terms. More precisely, considering the sum of the TX leakage and the adjacent channel at the input of a device that exhibits a smooth compression behavior, the bandpass distortion noise term recovered as centered around the adjacent channel is given in the third order polynomial approximation by equation (5.175). In this expression, to match our present configuration we simply need to substitute the wanted signal with our present adjacent channel, and the blocker with the TX leakage signal. Looking more deeply at this equation, we see that we already get the contribution of the two phenomena of interest, as can indeed be identified by comparing this expression with equation (5.190). Consequently, only the first term in this expression is of interest for the XM phenomenon. However, we also need to take care that this expression gives the complex envelope of the bandpass distortion terms as recovered at the output of the nonlinear device. We thus need to divide it by the device gain to get the equivalent input quantities. For that purpose, as highlighted in the previous section for the derivation of the SNR budget, we need to consider the effective gain of the device G_e. In the present case where only AM-AM conversion is considered, this gain reduces to equation (5.174), but still considering the substitution of the blocker by the TX leakage to match our configuration.

Thus, the equivalent input complex envelope of the XM noise component, defined as centered around the adjacent channel carrier frequency, reduces to

$$\tilde{n}_{\text{i,XM}}(t) = \frac{G}{G_{\text{e}}}\left(\frac{\mathbb{E}\{\rho^2_{\text{i,TX}}\} - \rho^2_{\text{i,TX}}(t)}{\text{IIP3}}\right)\tilde{s}_{\text{i,adj}}(t). \tag{7.19}$$

In this expression, $\rho_{\text{i,TX}}(t)$ represents the modulus of any complex envelope $\tilde{s}_{\text{i,TX}}(t)$ of the TX leakage at the receiver input, and $\tilde{s}_{\text{i,adj}}(t)$ the input one of the adjacent channel, when defined as centered around its own center carrier frequency. We observe that only the second term in this expression can lead to a spectral regrowth phenomenon. It is thus the only one of interest for our present derivation that focuses on the fraction of power leaking in the wanted signal frequency band. We can also express the above equation in terms of the device ICCP1 using equations (5.66) and (5.104).

Turning to the equivalent input AM-AM conversion noise component, we can directly use equation (5.190) to express its equivalent input complex envelope, when defined as centered around the adjacent channel carrier frequency, as

$$\tilde{n}_{\text{i,AM-AM}}(t) = \frac{G}{G_{\text{e}}}\frac{P_{\text{i,adj}}}{\text{IIP3}}\left[(\Gamma_{4,\text{adj}} + 1) - \frac{\rho^2_{\text{i,adj}}(t)}{2P_{\text{i,adj}}}\right]\tilde{s}_{\text{i,adj}}(t), \tag{7.20}$$

with $P_{\text{i,adj}}$ the average input power of the adjacent channel, $\rho_{\text{i,adj}}(t)$ the modulus of any of its complex envelopes at the receiver input, and $\Gamma_{4,\text{adj}}$ the constant that characterizes the statistics of its instantaneous amplitude as defined in "Moments of the instantaneous amplitude"

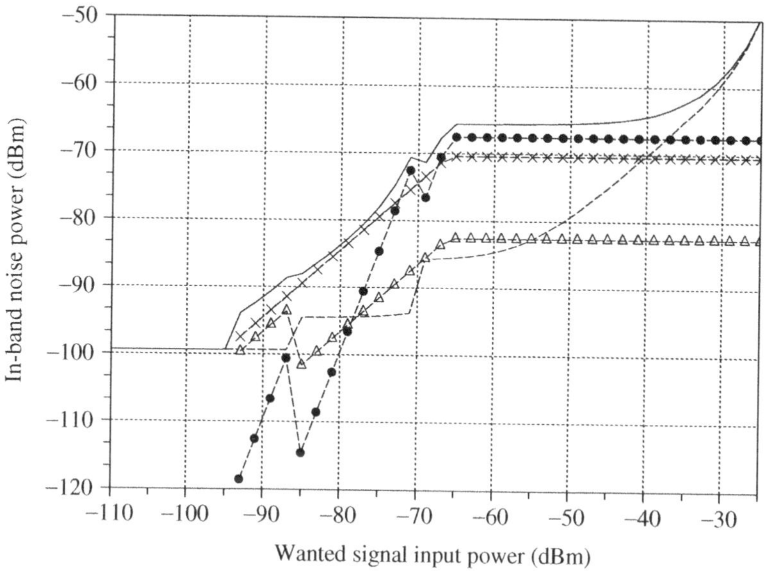

Figure 7.34 Equivalent input power of the noise components involved in the SNR budget as a function of the wanted signal input power: wanted signal plus TX leakage and adjacent channel case – On top of the total noise experienced by the wanted signal when processed alone in the receiver and detailed in Figure 7.32 (dashed), the presence of an adjacent channel leads to additional in-band noise contributions in accordance with the mechanism illustrated in Figure 7.30(bottom). Assuming that the power of the adjacent channel signal scales with that of the wanted signal according to the profile shown in Figure 3.17(left), we can derive the contribution of those additional noise terms as a function of the wanted signal input power, i.e. the LO phase noise due to the reciprocal mixing (crosses), the XM with the TX leakage (triangles), and the self-distortion due to AM-AM compression (dots).

(Section 5.1.3). As for the XM, we observe that only the second term in the above expression leads to a spectral regrowth phenomenon. It is thus the only one of interest for our present derivation in practice. Since we assume that the adjacent channel we are dealing with can also be considered as an OFDM signal, we can continue to approximate the $\Gamma_{4,\mathrm{adj}}$ constant as that of the Rayleigh distribution, i.e. $\Gamma_{4,\mathrm{adj}} = 1$ as given by equation (5.130a). Finally, we can also express this equation in terms of the receiver ICP1 using equation (5.66).

Using the above models, we can then derive the fraction of additional in-band noise power due to the presence of the unwanted adjacent channel. This results in the contributions displayed in Figure 7.34 as a function of the wanted signal input power. Looking at this figure, we see that with our assumptions the degradation due to the LO phase noise remains dominant for most of the DR. However, as its power increases more quickly than the adjacent channel's own power, the contribution of the self-distortion of the adjacent channel becomes non-negligible when the adjacent channel input power reaches its maximum permitted level. In any case, the in-band SNR achieved in the presence of an adjacent channel is poor compared to that achieved when the wanted signal is alone, as illustrated in Figure 7.35. Despite this, the performance still looks good enough to meet the requirements set in the presence of the adjacent channel.

We have thus shown, by appropriate adjustment of the characteristics of the receiver, how to achieve a performance improvement even in the presence of an adjacent channel. In practice,

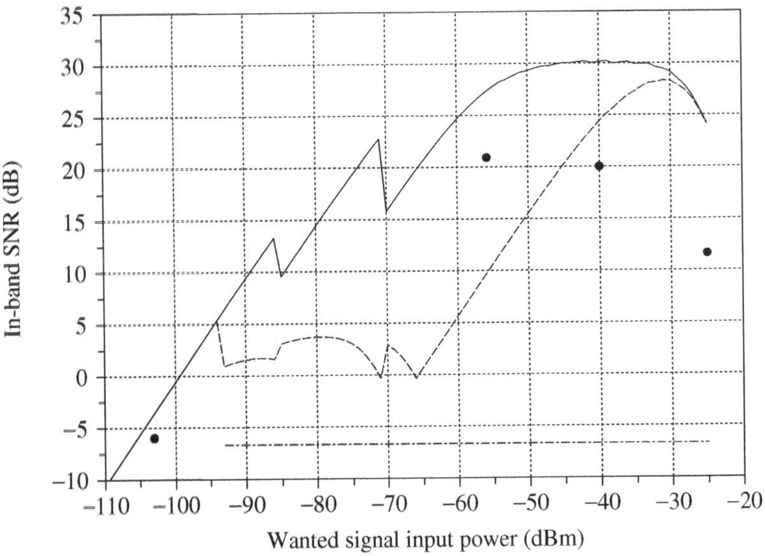

Figure 7.35 SNR performance achieved as a function of the wanted signal input power: wanted signal plus TX leakage and adjacent channel case – The existence of additional in-band noise contributions linked to the presence of the adjacent channel, as shown in Figure 7.34, leads to an in-band SNR degradation (dashed) compared to the performance achieved when the wanted signal is alone (solid). Due to this degraded performance, the existing requirements that hold when the wanted signal is alone in the line-up (dots) are no longer fulfilled. This illustrates why we often have relaxed requirements in the presence of an adjacent channel (dot-dashed).

this has to be balanced with the implementation cost in terms of area and power consumption. However, what is important to keep in mind is that the presence of an unwanted adjacent channel causes additional noise contributions on top of those necessarily existing when the wanted signal alone is processed in the line-up. As a result, the performance in the presence of the adjacent channel can only be *worse* than when the wanted signal is alone. It is up to us to decide what degree of performance degradation we can accept with regard to the associated cost in terms of implementation area and power consumption. Unfortunately, this statement holds for any additional unwanted signal present at the input of the receiver.

7.3.5 Conclusion

We have considered here the derivation of budgets for a given value for each characteristic of the receiver. But as was the case for the transmit side, in real life we may have to deal with variations in those characteristics, whether due to power supply or temperature drifts. There may also be discrepancies due to spreads in the production of the devices so that the intrinsic characteristics may vary from device to device. These variations often need to be taken into account in order to derive the different performance achieved in typical conditions, as compared to worst case conditions.

In the same way, we need to keep in mind that we have considered only budgets in the strict sense. But the performance of receivers, like those of transmitters, is also driven by the accuracy of certain parameters. The example of the frequency accuracy achieved for clocking schemes and LO signals as illustrated in Section 9.2.5 remains valid in the present case.

We can also make the same comments on the fact that our purpose here is to illustrate how to budget a receiver. As a result, only the main contributors to the line-up performance have been considered, and second order effects, such as the ADC or the analog baseband linearity performance, could be considered in practice. Furthermore, structural problems linked to the direct conversion reception should also be considered. This aspect is discussed in more depth in Section 8.2.1.

However, despite these similarities, there is an important difference between the transmit and receive sides. Only the wanted signal is expected to be processed along the data path of a transmitter. This means that we have a deterministic way to set the transmitter for a given target output power. This is not necessarily the case on the receive side due to the possible presence of additional unwanted signals at the input of the receiver for a given wanted signal power. As a result, there are different strategies for the setting of the receiver, mainly through its gains, depending on whether or not we have knowledge of the presence of those unwanted signals. This raises the importance of the AGC strategy on the ultimate performance of the receiver. In the present chapter we have addressed the case where such a setting is based on the knowledge of only the wanted signal power, but other strategies and refinements may be used, as discussed in Section 9.2.1.

8

Transceiver Architectures

The underlying assumption for line-up budgeting, as illustrated in Chapter 7, is that there is enough flexibility in the parameters of the various blocks in the line-up to ensure that the overall requirements can be fulfilled. However, in real life implementations, the tuning range of the parameters is necessarily limited. This may lead to one stage in the line-up behaving as a bottleneck. Changing the architecture of the line-up is one solution to this kind of problem.

We therefore review some classical architectures in order to understand the associated limitations. Obviously, our aim is not to be exhaustive as we can basically imagine as many transceiver architectures as we want. Rather, it is to illustrate through a review of representative structures how to conceive a transceiver at a system level. This means trying to understand how to manage some of the limitations of a given architecture through its modification. In this chapter, the modifications we are referring to have to be understood at the RF/analog implementation level. The same can be said of algorithms, as will be discussed in Chapter 9 through some practical examples.

8.1 Transmitters

Let us start by discussing the transmit side. Basically, from the signal processing point of view, there are two families of architectures, depending on the decomposition of the complex modulating waveform we want to upconvert. More precisely, given the possibility of decomposing this complex signal in terms of either real and imaginary parts or modulus and argument, we can derive a Cartesian or polar line-up, as illustrated at the end of Chapter 1.

We can review the corresponding representative transmit architectures didactically by naturally introducing a given architecture through its potential capability to overcome the limitations associated with another one. To initiate this process, we start by considering the simplest architecture we can think of when dealing with the frequency upconversion of a complex modulating waveform, the direct conversion transmit architecture already taken as an example in Chapter 7.

However, there is another important class of architectures that cannot be considered as driven by the signal processing of the transmit signal as such. Due to the amount of RF

Wireless Transceiver Architecture: Bridging RF and Digital Communications, First Edition. Pierre Baudin.

power that classically needs to be delivered to the radiating element, the transmit line-up is often the most power hungry part of a transceiver. Different approaches have therefore been developed to maximize its electrical efficiency. Some of these will also be discussed in due course.

8.1.1 Direct Conversion Transmitter

One of the simplest architectures we can think of is the direct conversion architecture, also referred to as homodyne or ZIF. As derived in Chapter 1, this architecture embeds the minimum set of signal processing blocks required to deliver an RF signal modulated by a complex waveform represented in Cartesian form. Practically speaking, additional features driven for instance by the theory of electromagnetism as illustrated in Chapter 2, or by the management of wireless networks as discussed in Chapter 3, need to be added to the basic line-up. We need to consider the minimum set of constituent blocks shown in Figure 8.1 and extensively discussed in Chapter 7.

In that chapter we detailed this architecture from the budgeting point of view. We were not unduly concerned about the possibility of physically achieving the characteristics of the different blocks present in the line-up as required to fulfill the overall requirements. But in real life implementations, limitations exist as discussed throughout Part II of this book. And given the ultimate performance of a constituent block, one architecture may be more sensitive to the corresponding parameter than another, depending on the way this block is used.

Take the example of the complex frequency upconversion stage in this direct conversion scheme. As reviewed in Chapter 6, the physical implementation of this function is always associated with some limitations that impact its performance. However, those limitations

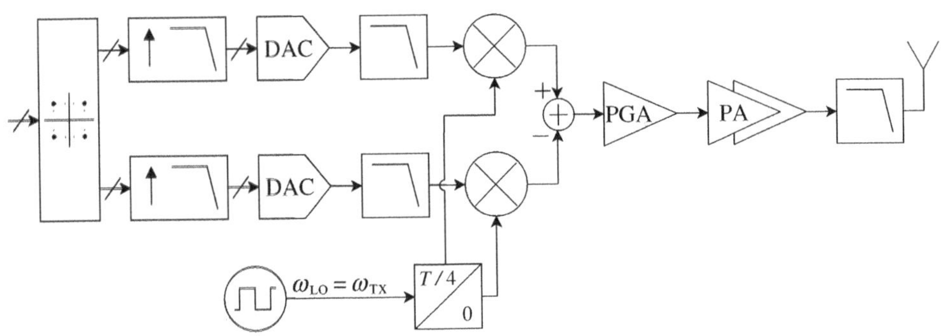

Figure 8.1 Direct conversion transmit architecture – The blocks embedded in a typical direct conversion transmit line-up are driven on the one hand by the minimum set of signal processing functions required to generate a complex modulated RF bandpass signal, to comply with the theory of electromagnetism, or to coexist with other wireless standards. On the other hand, we also need to consider the impact of the limitations linked to the physical implementation of those functions. We thus get the necessary digital signal processing functions dedicated to the generation and upsampling of the digital P and Q modulating waveforms, the digital to analog stages and the associated reconstruction filters, the complex upmixer driven by the quadrature LO signals, the RF variable gain and PAs, and the harmonic filter.

are more problematic to handle in the direct conversion architecture due to the associated frequency planning:

(i) When implemented using choppers, the RF impairments between the P and Q branches of the complex mixing stage can lead to a rise in the even order harmonics of the LO frequency and in the image signal, as discussed in Section 6.3.2. Obviously, the rise in the LO frequency harmonics results in the generation of unwanted sidebands that can degrade the spurious emission capabilities of the line-up. However, as these unwanted components lie outside the wanted signal frequency band at the transmitter output, we can always think of using RF filtering to achieve the required performance. Practically speaking, the filter can take the form of a simple lowpass harmonic filter given that the LO frequency is directly the transmit carrier frequency in this direct conversion scheme. In contrast, a complex modulating waveform centered around DC at the input of the frequency upconversion stage results in an image signal lying in the wanted signal frequency band. This perturbation thus cannot be filtered out and leads to an unavoidable EVM degradation, as illustrated in Section 6.2.2 and Chapter 7.
(ii) Due to electromagnetic coupling or to the direct feedthrough at the device level, we necessarily recover a fraction of the LO signal as a leakage present on the RF ports of the transmitter, as discussed in Section 6.5.1. And here again, due to the particular frequency planning of the direct conversion transmitter, the leakage component recovered at the output of the line-up lies at the transmit carrier frequency, and thus at the middle of the transmit signal frequency band.

Thus, a frequency upconversion of the complex modulating waveform directly from DC up to the transmit carrier angular frequency leads to the presence of unwanted components superposed in the frequency domain on the bandpass RF modulated signal at the transmitter output. Thus we cannot use an RF filter at the line-up output in case we need to cancel them to improve the performance of the transmitter, and would need to use the calibration schemes discussed in Chapter 9 instead.

Another sensitive aspect of this architecture is related to the control of the transmit output power. Some wireless standards may require a wide DR for the control of this average power, as discussed in Chapter 3. Reasons for this may be the need to achieve sharp power transitions, particularly when dealing with burst transmissions, or to regulate the average transmit power over a wide DR in order to achieve accurate compensation of the uplink path loss. In any case, this function is implemented in our architecture through the use of an RF PGA set at the output of the frequency upconverter, as illustrated in Figure 8.1. Practically speaking, mainly for noise purposes, this kind of RF function is often implemented as a variable attenuator, whether active or passive. But, as recalled above for the frequency upconversion stage, we necessarily have some RF coupling between ports in such high frequency implementation. We then necessarily get an upper bound for the achievable DR of this kind of attenuator block due to the direct feedthrough of the input signal toward the output port. This can cause problems, depending on the exact requirements set for this parameter by the considered wireless standard. One might counter that part of the variable gain could take place in another block of the line-up, for instance on the baseband side. But then the problem would be the preservation of the signal to LO leakage and the signal to noise ratios, as discussed in Section 7.2.2. Thus, unless we can ensure appropriate noise performance and solve the LO leakage problem, having the variable gain in the RF world appears to be the safer approach.

A final problem associated with this architecture is linked to its frequency planning. This is the potential frequency pulling of the RF oscillator used to generate the LO signal that physically drives the complex frequency upconverter due to the injection locking of the transmit signal leakage. In order to understand this, we need to keep in mind several preliminary mechanisms:

(i) In realistic physical implementations, we necessarily have a collection of modulated RF bandpass signals in addition to the wanted signal at the transmitter output. As recalled above, when dealing with mixers implemented as choppers, we recover at the output of the frequency upconversion stage both the wanted sideband centered around the transmit carrier angular frequency ω_{TX}, and a collection of unwanted sidebands centered around the odd order harmonics $(2l+1)\omega_{\text{TX}}$. In addition, taking into account the unavoidable RF impairments between the P and Q branches of the line-up, we also have a rise in the sidebands centered around the even order harmonics $2l\omega_{\text{TX}}$. And as in our architecture we also have the LO angular frequency ω_{LO} equal to the transmit carrier angular frequency ω_{TX}, we finally recover at the transmitter output a collection of sidebands of potential strong power centered around the harmonics $k\omega_{\text{LO}}$, as illustrated in Section 6.4.1.

(ii) These sidebands correspond to RF bandpass signals that are complex modulated. More precisely, considering the bandpass signal $s_{\text{H}N}(t)$ that corresponds to the sideband defined as centered around the Nth harmonic angular frequency, $N\omega_{\text{LO}}$, its complex envelope $\tilde{s}_{\text{H}N}(t)$, defined as centered around $N\omega_{\text{LO}}$, is a linear superposition of the transmit complex modulating waveform $\tilde{s}_{\text{TX}}(t)$ and its complex conjugate. Thus $\tilde{s}_{\text{H}N}(t)$ can be written in polar form as

$$\tilde{s}_{\text{H}N}(t) = \rho_{\text{H}N}(t)e^{j\phi_{\text{H}N}(t)}, \tag{8.1}$$

where both $\rho_{\text{H}N}(t)$ and $\phi_{\text{H}N}(t)$ are functions of $\tilde{s}_{\text{TX}}(t)$.

(iii) Due to the unavoidable electromagnetic RF coupling discussed above, a fraction of the modulated RF bandpass signals $s_{\text{H}N}(t)$ can be recovered at any RF port in the line-up, and in particular at the RF oscillator used to generate the LO waveform. In that case, assuming that the coupling factor $\tilde{\alpha} = \rho_{\alpha}e^{j\phi_{\alpha}}$ remains constant over the frequency band of $s_{\text{H}N}(t)$, we necessarily recover a bandpass perturbation in the RF oscillator whose complex envelope, still assumed defined around $N\omega_{\text{LO}}$, is of the form

$$\tilde{\alpha}\tilde{s}_{\text{H}N}(t) = \rho_{\alpha}\rho_{\text{H}N}(t)e^{j\phi_{\text{H}N}(t)+\phi_{\alpha}}. \tag{8.2}$$

Thus, given on the one hand that the oscillator is expected to run at its *constant* free-running angular frequency, $\omega_{\text{osc},0}$, and on the other hand that it faces bandpass pollution centered around $\omega_{\text{osc},0}$ but exhibiting a non-constant instantaneous angular frequency, we may be faced with a pulling issue [44–47]. If this occurs, we may then face a modulation of the instantaneous angular frequency of the bandpass signal standing in the RF oscillator. This phenomenon can then in turn lead to a degradation of the transmit signal modulation due to the use of this non-CW LO signal at the frequency upconversion stage.

Practically speaking, this pulling effect vanishes when the instantaneous angular frequency of the aggressor goes away from $\omega_{\text{osc},0}$. This notion of relative frequency offset should be refined and obviously depends on the quality factor of the resonator. But in practice, for a realistic frequency bandwidth for the modulations we are dealing with compared to the carrier frequency, the pulling effect can only occur in our direct conversion line-up if the free-running oscillator angular frequency $\omega_{\text{osc},0}$, is equal to either the LO angular frequency ω_{LO} or an integer multiple of it. This is the only realistic configuration that allows one of the modulated RF bandpass signals to be recovered at the transmitter output that lies in the vicinity of $\omega_{\text{osc},0}$.

The problem is that in most implementations we effectively need to have this RF oscillator running at an integer multiple of the LO angular frequency. Indeed, even if using devices such as fractional frequency dividers or offset PLL solved the problem, this would lead to degraded noise performance or to the generation of spurs that can make it difficult to fulfill the TX spectrum requirements. In any case we would forfeit much of the benefit of the direct conversion transmit architecture, which theoretically leads to a clean transmit spectrum when associated with an integer ratio between ω_{osc} and ω_{LO}, in contrast to what we can expect with heterodyne approaches as discussed in the next section. Thus, although it should be checked on a case by case basis, stringent wireless systems may require using RF oscillators running at an integer multiple of ω_{LO}, thus corresponding to the illustration shown in Figure 8.2.

The extent of the degradation of the quality of the signal delivered by the oscillator also necessarily depends on the magnitude of the perturbation that is coupled to it. But at the same time, the magnitude of the even order RF modulated sidebands is related to the RF impairments whereas the odd order ones are intrinsic to the chopper behavior of the mixers. For realistic implementation limitations, the level of the even order sidebands is kept much lower than the odd order ones. As a result, if we need to run the oscillator at an integer multiple of the LO angular frequency, we can see the attraction of choosing an even integer multiple rather than an odd order one to allow minimization of the injection locking problem.

There is another side effect of having this pulling issue that vanishes when the angular frequency of the disturbance goes away from that of the oscillator. This behavior means that for a given transmit power, the wider the modulation we are dealing with, the higher the fraction of power of the aggressor that has no impact on the oscillator in terms of injection locking. We may thus conclude that the higher the bandwidth of the modulated signal we are dealing with, the lower the pulling issue we may face. This behavior is enhanced by the fact that the spectrum of the instantaneous frequency of a complex modulated RF bandpass signal is much wider than its own spectrum, as discussed in Section 8.1.6. A direct conversion transmitter using an integer ratio between the RF oscillator angular frequency and the LO angular frequency thus seems more suited to the processing of wideband modulations than to narrowband ones when considering an integrated solution.

The direct conversion transmit architecture is widely used in low cost solutions as it potentially allows the best possible level of integration in silicon, at least when dealing with complex modulations. However, this is achieved at the cost of limited theoretical performance. But as reviewed above, most of those limitations are in fact related to the frequency planning associated with this architecture. This leads for instance to distortions generated through the complex frequency conversion in the frequency band of the transmit signal. It also leads to a potential pulling issue of the RF oscillator by the harmonics of the transmit signal. In order to overcome these limitations it might perhaps be worth considering a different frequency planning.

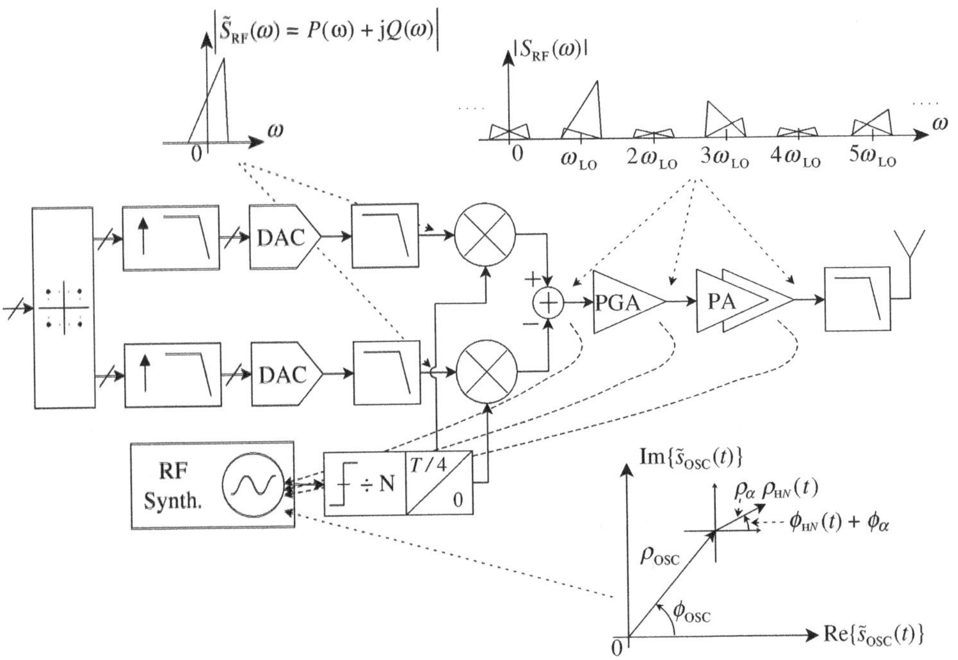

Figure 8.2 RF oscillator pulling problem in the ZIF transmit architecture – For spectral performance, the free-running angular frequency $\omega_{osc,0}$ of the RF oscillator used to generate the LO signal is often set as an integer multiple N of the LO angular frequency ω_{LO}. Given mixers implemented as choppers and faced with RF impairments, we then recover a collection of modulated sidebands centered around the harmonics $k\omega_{LO}$ at the transmitter output, thus with a potential high power. The RF bandpass signal $s_{HN}(t)$, with an instantaneous amplitude $\rho_{HN}(t)$ and phase $N\omega_{LO}t + \phi_{HN}(t)$, which corresponds to the Nth harmonic sideband, can be coupled to the RF oscillator. And having $\phi_{HN}(t) \neq 0$ can result in a pulling effect of the bandpass signal $s_{osc}(t)$ delivered by the RF oscillator.

The simplest course of action would be to consider the possibility of using two frequency conversion stages instead of one. This naturally leads to the heterodyne transmit architecture discussed in the next section.

8.1.2 Heterodyne Transmitter

As highlighted at the end of the previous section, the limitations we face in the homodyne transmit architecture mainly come from its frequency planning, whether directly when considering the LO leakage or the image signal that are necessarily superposed with the wanted signal in the frequency domain, or indirectly when considering the potential pulling issue of the RF oscillator when running at an integer multiple of the LO angular frequency. Thus, it seems natural to consider the possibility of performing the frequency upconversion in two steps in order to overcome those limitations. This leads to the heterodyne transmitter architecture illustrated in Figure 8.3.

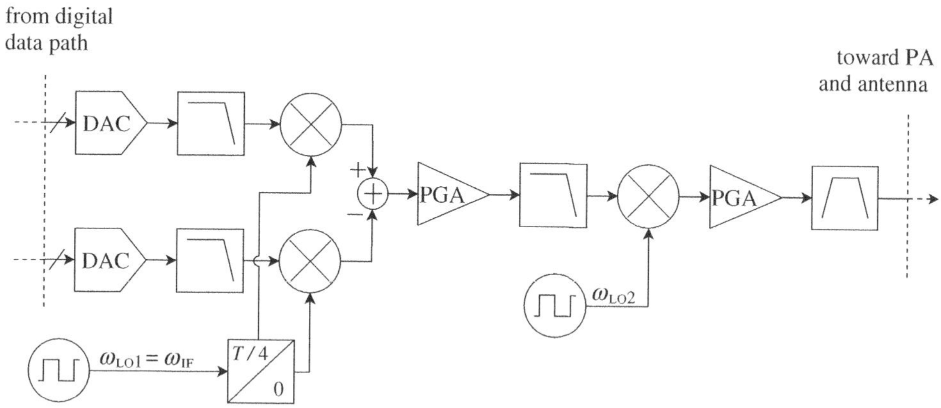

Figure 8.3 Heterodyne transmit architecture – In order to overcome the limitations of the ZIF TX architecture displayed in Figure 8.1, we can try to perform the frequency upconversion in two stages. The first frequency upconversion, which processes the complex lowpass modulating waveform, still needs to be implemented using a complex upmixer, but the second, which processes a modulated IF bandpass signal, can use a simple real upmixer. However, due to the use of the latter structure, and mixers implemented as choppers, we need efficient filtering stages in order to manage the various unwanted sidebands that are generated at the frequency conversion stages, as illustrated in Figure 8.4.

Looking at this figure, we first observe that in the present architecture we are dealing with two different kinds of frequency conversion stages. The first is implemented as a complex upmixing stage. This is required in order to process the complex lowpass modulating waveform decomposed into Cartesian form. The two lowpass signals representing its real and imaginary parts are thus processed along the P and Q paths of the line-up in order to recover a modulated IF bandpass signal whose sideband of interest is centered around the IF angular frequency ω_{IF}, equal to the first LO angular frequency ω_{LO1}. The second frequency upconversion can therefore be implemented as a real frequency upconverter as it transposes one bandpass signal into another. This second stage can be implemented using a single physical mixer. However, as discussed extensively in Chapter 6, we naturally face certain problems in both cases, due to the intrinsic behavior of the signal processing associated with those blocks, or the limitations in their physical implementation.

Beginning with the limitations linked to the signal processing associated with those mixing stages, the real frequency upconversion of a bandpass signal necessarily generates an additional unwanted bandpass image signal lying at the image frequency, as detailed in Section 6.1.2. This means that in order to fulfill the SEM, we often need to use a bandpass RF filter at the output of the transmitter to cancel this signal. However, in order to make the implementation of this filter realistic, we often need to consider it to be bandpass for the overall transmit system band, as discussed in “Filtering problem” (Section 7.2.3). This means that the IF angular frequency, equal to ω_{LO1}, must be chosen sufficiently high that the image signal is generated outside this transmit system band, whatever the selected transmit channel, in order to be canceled by the bandpass RF filter.

Turning to the limitations we face in the implementation of these blocks, there are basically two to consider: the fact that the mixers are for the most part implemented as choppers; and the fact that we face impairments in their implementation. As discussed throughout Section 6.3, as a result of those two phenomena, instead of experiencing the expected theoretical LO waveform – either the single complex exponential during a complex frequency upconversion, or the two complex conjugate complex exponentials during the real frequency upconversion – the signal being processed now experiences a superposition of tones centered around the harmonic angular frequencies of the corresponding LO one. This behavior obviously degrades the quality of the signal recovered at the output of such a mixing stage. But the degradation is now worse in the present case due to the cascade of the two upconversion stages. In order to discuss this point we can detail the mechanism encountered in the line-up when going through the two actual mixing stages.

We first recall that, considering the transmit line-up we are dealing with only up to the complex frequency upconversion stage, we recover nothing more than the signal processing blocks involved in the direct conversion architecture detailed in the previous section. We can therefore expect the same degradation due to the limitation in the physical implementation of this upmixing stage as detailed in that section. On the one hand, this means a rise in the image sideband due to the RF impairments. In this frequency conversion scheme from baseband to the first LO angular frequency ω_{LO1}, the image signal is thus superposed on the wanted signal so that there is no way to filter it out at the output of the upconversion stage. Only the magnitude of the RF impairments can drive the final performance. However, we observe that in realistic physical implementations this magnitude classically increases with the frequency of operation of the mixing stage. Thus, as in the present heterodyne architecture we can set ω_{LO1} to be lower than the angular frequency of the LO waveform used in the direct conversion scheme, we expect fewer problems in this respect. On the other hand, we get the collection of tones related to the chopper behavior of the mixers. Dealing with a single lowpass sideband at the input of the complex upconverter stage, we then recover a collection of unwanted sidebands centered around the harmonic angular frequencies $k\omega_{\text{LO1}}$. We thus recover exactly the same situation as in the ZIF transmitter.

Then we need to consider the incremental impact of the second frequency conversion stage implemented as a real frequency upconverter. Practically speaking, this means that on top of the generation of the additional unwanted bandpass image signal as highlighted previously, we now need to take into account the impact of the harmonic tones present in the equivalent LO waveform of the second upmixer. The situation is thus now that we face a collection of unwanted sidebands lying at the harmonic angular frequencies $k\omega_{\text{LO1}}$ at the input of this stage and an equivalent LO waveform that also exhibits a collection of tones lying at the harmonic angular frequencies $l\omega_{\text{LO2}}$ as illustrated in Figure 8.4. Obviously, we can expect the generation of the intermodulation products corresponding to the transposition of the input sidebands around the angular frequencies of the form $|k\omega_{\text{LO1}} + l\omega_{\text{LO2}}|$ with $k, l \in \mathbb{Z}$. Faced with this problem, a careful choice of ω_{LO1} and ω_{LO2} must be made in order to ensure that none of these intermodulation tones is folded on the wanted sideband of interest. This problem has historically led to the development of various charts to help in the selection of those angular frequencies or in the prediction of the unwanted terms level [72].

Practically speaking, when the frequency planning is well done, we can rely on the RF image reject filter necessarily present at the output of the real mixing stage to reduce the intermodulation tones to an acceptable level. However, when addressing wide RF frequency

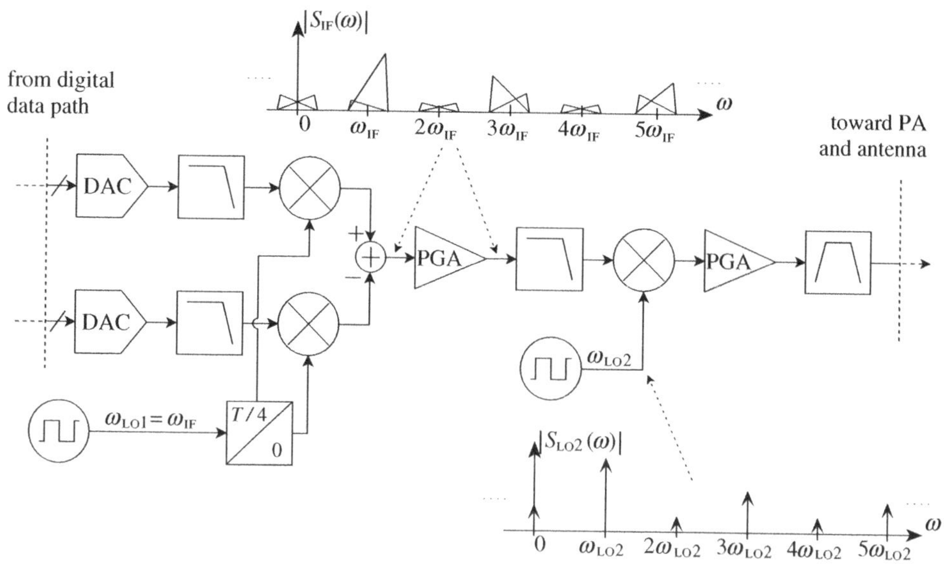

Figure 8.4 Intermodulation product generation in the heterodyne transmit architecture – Due to the use of chopper-like mixers and the RF impairments in the physical implementation of the line-up, we recover a collection of harmonics in the LO spectra. This results in the generation of unwanted sidebands, centered around the harmonic angular frequencies $k\omega_{LO1}$ of the first LO angular frequency, at the complex frequency upconversion stage. At the second upmixing stage, we then deal with the generation of the intermodulation products between those sidebands and the harmonic tones present in the second LO signal, lying at the angular frequencies $l\omega_{LO2}$. A trade-off between an IF lowpass and RF bandpass filter is then required to achieve a sufficiently clean transmit output spectrum.

bands, it can be difficult to achieve the required performance for all the possible configurations of intermodulation. This is for instance the case when the intermodulation tones lie too close to the wanted sideband. It then becomes necessary to filter out the unwanted sidebands generated through the first complex frequency upconversion *before* they reach the input of the second upconversion stage. This explains the presence of the filtering stage in between the two mixing stages as illustrated in Figure 8.3, here implemented as a lowpass filter. However, in order to make this filter as efficient as possible, we observe that it is preferable to select a fixed IF and thus a fixed LO angular frequency ω_{LO1}. As a result, the channel selection is necessarily done using only the second upconversion stage. This allows the filtering of the unwanted sidebands to be optimized while simplifying the implementation and the control of the first RF synthesizer.

When considering a careful trade-off between the frequency planning and the filtering split in the line-up, we can expect to recover a sufficiently clean transmit output spectrum, even if at some additional hardware cost compared to the direct conversion transmitter. Having two successive frequency upconversions in the line-up allows us to overcome some of the intrinsic limitations of this architecture discussed in the previous section.

(i) Careful selection of the frequency planning now leads to the LO leakage systematically lying outside the transmit system frequency band. It can therefore be filtered out by the RF bandpass filter considered up to now for the cancellation of the intermodulation sidebands, thus resulting in no degradation of the modulation quality.

(ii) This flexibility in the frequency planning also allows us to minimize the potential injection locking of the RF oscillators. Given that ω_{TX} is set to $\omega_{LO1} + \omega_{LO2}$ or to $|\omega_{LO1} - \omega_{LO2}|$, depending on whether we expect to use a supradyne or infradyne real frequency upconversion stage, it is quite simple to select the IF so that no RF oscillator runs at an integer multiple of the transmit carrier angular frequency. However, a slight refinement of this may be necessary as we also need to avoid any reciprocal pollution of the two oscillators that can be physically implemented close to each other. Practically speaking, this means that they need to run at angular frequencies that are sufficiently different. Fortunately, offsets in the range of few hundred megahertz are often enough for oscillators that exhibit quality factors as classically encountered in integrated silicon implementations for instance.

(iii) Having two upconversion stages also allows the variable gain to be split into two stages if needed. For example, we could consider one stage processing the IF signal and the other the RF signal, as illustrated in Figure 8.3, while avoiding any limitations in the signal to LO leakage ratio as this last component now lies outside the wanted signal frequency band as highlighted previously. Having a reduced DR in each of those variable gain stages makes their physical implementation more realistic in many cases, while preserving performance.

(iv) Carrying out the complex frequency upconversion at an IF that can be chosen lower than the final RF transmit carrier frequency often enables a better matching between the devices and thus lower RF impairments. We may thus partially ameliorate the problem linked to the generation of the in-band image signal due to the imbalance between the P and Q branches of the line-up.

As a result, we can see that the heterodyne architecture is expected to overcome most of the theoretical intrinsic limitations of the direct conversion transmitter discussed in the previous section. However, this is done at a non-negligible hardware cost as we basically double the number of blocks required to perform the processing. The cost is thus heavy in terms of implementation area and power consumption. In addition to these costly devices, there are the RF synthesizers required to generate the LO waveforms. Perhaps the cost of the solution could be decreased by reducing the number of RF synthesizers while trying to retain the benefits of the double frequency upconversion. This naturally leads to the variable-IF architecture discussed in the next section.

8.1.3 Variable-IF Transmitter

As discussed in the previous section, a two-step frequency upconversion of the modulating waveform leads to the possibility of overcoming some intrinsic limitations of the direct conversion architecture. However, this is obviously achieved at the cost of using additional blocks in the line-up that necessarily increase the implementation area and its power consumption. Among all those additional blocks, perhaps the most critical in respect of those criteria is the RF synthesizer used for the generation of the additional LO signal. Perhaps we could consider using the same RF synthesizer to generate the two LO waveforms. This would simply require

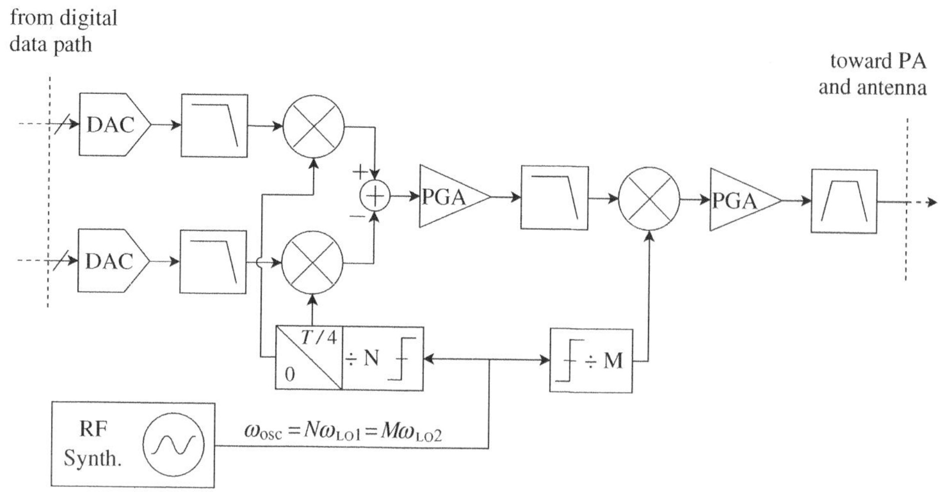

Figure 8.5 Variable-IF transmit architecture – In order to limit the implementation cost of the heterodyne architecture displayed in Figure 8.3 while retaining the benefit of the two-step frequency upconversion, we can use the same RF synthesizer to derive the two LO signals. This requires two frequency dividers to generate the two LO signals according to $\omega_{LO1} = \omega_{osc}/N$ and $\omega_{LO1} = \omega_{osc}/M$, with the transmit carrier angular frequency ω_{TX} equal to $\omega_{LO1} + \omega_{LO2}$ or to $|\omega_{LO1} - \omega_{LO2}|$ depending on whether we are using a supradyne or an infradyne real frequency conversion stage. Having ω_{osc} variable to address the different transmit carrier angular frequencies makes the IF variable.

running the RF oscillator at an angular frequency ω_{osc} sufficiently high that the two LO angular frequencies ω_{LO1} and ω_{LO2} can be generated through appropriate frequency divisions, thus leading to a line-up as shown in Figure 8.5.

This can be illustrated through a practical example. Recall that due to the performance achievable in realistic physical implementations, we often need to suppose that the frequency divisions are performed as integer divisions as discussed in Section 8.1.1. Thus, we can simply assume that $\omega_{LO1} = \omega_{osc}/N$ and $\omega_{LO2} = \omega_{osc}/M$, where N and M are integers and ω_{osc} is the angular frequency of the signal standing in the RF oscillator. Assuming for the sake of simplicity that we are dealing with a supradyne real frequency conversion, i.e. that the LO signal that drives this real mixing stage is chosen with a higher angular frequency than that of the signal being transposed, we thus require

$$\omega_{TX} = \omega_{LO1} + \omega_{LO2} = \omega_{osc}\left(\frac{1}{N} + \frac{1}{M}\right), \tag{8.3}$$

with ω_{TX} the target transmit carrier angular frequency. Thus, rearranging, ω_{osc} must be set to

$$\omega_{osc} = \omega_{TX}\left(\frac{NM}{N+M}\right). \tag{8.4}$$

Thus, by a careful choice of M and N, we have that the RF oscillator does not run at an integer multiple of the transmit carrier angular frequency ω_{TX}. We thus expect no additional pulling issue compared to the heterodyne case. However, this equation also highlights the main characteristic of this architecture, recalling that ω_{TX} can vary over the entire transmit system band. We must then have ω_{osc} variable in order to perform the channel selection and thus address all the possible transmit carrier angular frequencies allowed by the considered wireless standard. But, as the same oscillator is now used to generate the two LO waveforms in the line-up, we necessarily get that ω_{LO1}, and thus the IF is variable. This is thus the main difference compared to the heterodyne architecture discussed in the previous section. In that case, having two separate synthesizers allows us to select a fixed IF and thus optimize in the best possible way the blocks processing the IF signal. This is particularly true for the IF filter that needs to be optimized to minimize the generation of intermodulation products at the second upconversion stage. With a variable IF, it may be difficult to guarantee the overall performance in all the possible situations of frequency planning.

However, except for this problem, which needs to be addressed by both careful frequency planning and optimization of the filtering stages, we can expect the same advantages from the variable-IF transmit architecture as from the pure heterodyne line-up. This is obviously an interesting result, but perhaps we can go further in the integration of the line-up while preserving the benefit of using two upconversion stages. One example of this is the real-IF architecture discussed in the next section.

8.1.4 *Real-IF Transmitter*

As discussed in the previous section, the variable-IF and heterodyne line-ups offer similar advantages in enabling us to overcome the limitations of the direct conversion transmit architecture, both employing a two-step frequency upconversion. The only difference between the two is a single RF synthesizer. Perhaps further simplifications in the implementation of the heterodyne architecture could be made while retaining the benefits of the double frequency upconversion.

Unfortunately, by inspecting the line-up shown in Figure 8.3, we see that the RF synthesizer is indeed the only block that could be easily shared between the two upconversion stages. In order to go further in the integration of the line-up, we thus need to look for a solution other than putting in common additional blocks between the processing functions to be implemented. Practically speaking, this means that we should look for a potentially more efficient way to physically implement those functions, at least in terms of area and cost.

The signal processing functions we are discussing are implemented in the RF/analog world. However, the trend in integrated circuits is for digital implementations to become more and more efficient from one technology node to the next. Perhaps we can implement more processing functions in the digital world than is presently done in the heterodyne architecture. The potential problem with this approach is that we are already implementing all the functions that process the baseband modulating waveform in the digital domain. In order to go further, we thus also need to consider the digital implementation of the functions that process the IF signal. A possible implementation is the line-up shown in Figure 8.6. Obviously, the accuracy associated with digital signal processing provides many improvements.

First, the complex frequency upconversion can now be implemented while avoiding the chopper behavior of most of the RF/analog mixers. In the digital world we can rely on the ideal multiplication of the modulating waveform by pure sine and cosine functions. As a result,

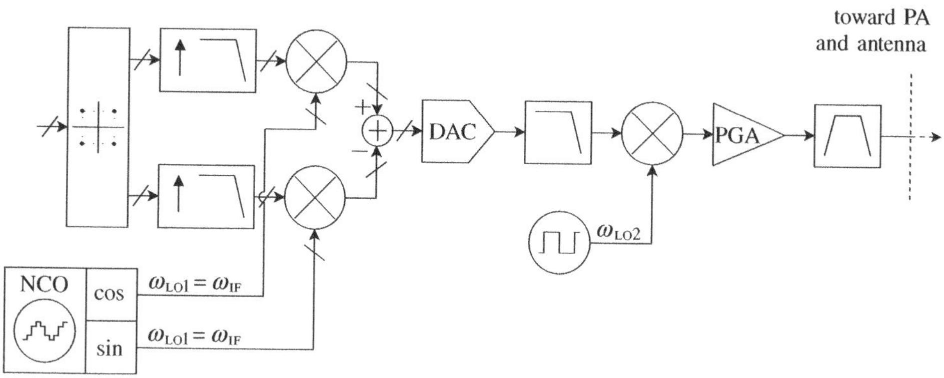

Figure 8.6 Real-IF transmit architecture – In order to go beyond what was achieved with the variable-IF architecture in the integration of the heterodyne line-up displayed in Figure 8.3, we can increase the fraction of signal processing functions implemented in the digital domain. This means the generation of the modulated bandpass IF signal in practice. Implementing this accurately, we can achieve an almost ideal complex frequency upconversion stage. And having the IF signal already generated in its real bandpass form enables us to use only one DAC device for the generation of the analog signal.

we recover the only sideband of interest at the complex frequency upconversion, without a copy centered around the harmonics $k\omega_{IF}$ of the IF angular frequency. In practice, this has to be balanced with the generation of the copy at the digital to analog conversion stage. However, the latter is generated around the harmonics lf_s of the sampling rate, which is necessarily higher than the IF $\omega_{IF}/2\pi$. In addition, it is naturally attenuated by the aperture effect detailed in "Aperture effect" (Section 4.6.2). It thus naturally results in unwanted sidebands more easily handled by the reconstruction filter in the present line-up than than by the IF filter in the heterodyne or variable-IF architectures.

In the same way, the LO signals are now generated in the digital domain using a numerically controlled oscillator (NCO). Practically speaking, this function can be implemented through the use for instance of a look-up table (LUT) that tabulates the sine and cosine functions, or alternatively by using an algorithm such as CORDIC, discussed in Chapter 9. Whatever the chosen implementation, the imbalance between the sine and cosine functions is only a matter of quantization management and can obviously be made negligible compared to what could be achieved in the RF/analog world. We can then expect a negligible in-band image signal increase as well as no LO leakage in practice. Moreover, this digital implementation allows for great agility in phase or frequency jumps of the LO waveforms. This is useful in applications involving frequency hopping for instance.

On top of these benefits linked to the digital implementation, another advantage of this architecture is that we use only one DAC along the data path. This is one motivation for implementing the first complex frequency upconversion in the digital world as the IF signal to be converted is already reconstructed in its real bandpass form, as the name of the architecture suggests. However, this advantage in terms of implementation area must be balanced with the power consumption of this DAC block, which now needs to run at a higher sampling rate, more than twice the IF plus the modulation bandwidth. Practically speaking, this IF can be quite high in order to have the image signal generated through the later real frequency upconversion that lies outside the transmit system band. This is required in order to be able to filter out this unwanted signal by a simple bandpass RF filter, as discussed for the heterodyne architecture in Section 8.1.2. As a result, the possibility of implementing such an architecture depends on the speed capability of the logic available in the chosen implementation technology, relative to the IF we need to generate. The same holds for the DAC that converts the IF bandpass signal into the RF/analog world. In any case, the resulting power consumption must be carefully checked.

So far, then, we have discussed variations on a transmit architecture based on a Cartesian representation of the complex lowpass modulating waveform. None of these is significantly simpler than the direct conversion architecture. This architecture embeds the minimum set of signal processing blocks required to upconvert a complex modulating waveform represented in this Cartesian way. Perhaps we can have still simpler architectures if we consider modulating waveforms that are not complex by nature. An example of practical importance is the phase/frequency modulation, which is still widely used. As illustrated in Section 1.3.1, a real modulating waveform needs only one physical signal to be represented. There is thus no theoretical need to use the two physical signals corresponding to the real and imaginary parts of the modulating complex envelope for the correct representation of the information. We can then surmise that the minimum set of functions to be considered in a transmitter that processes a modulating waveform can be reduced compared to the direct conversion architecture. We discuss this in the next section, where we examine the direct PLL modulator. In so doing, we also introduce the polar transmit architecture more generally.

8.1.5 PLL Modulator

As introduced at the end of the previous section, when considering a pure phase/frequency modulation we can expect some simplifications in the minimum set of processing blocks to be implemented in a transmit line-up compared to a direct conversion architecture. In that case, the RF bandpass signal that needs to be generated has a constant instantaneous amplitude. Thus, rather than thinking about the signal processing associated with the transmit line-up in terms of the upconversion of a modulating waveform as we have so far, we can do so in terms of the direct generation of a constant instantaneous amplitude RF signal, but with an instantaneous frequency that is slowly varying in accordance with the frequency modulation around the transmit carrier angular frequency ω_{TX}.

Obviously, we need to specify what we mean by "slowly varying". Here we observe that in general the modulations we are dealing with are narrowband compared to ω_{TX}. We can thus expect orders of magnitude between the two time constants that characterize the variations of the modulating waveform and the period of the carrier signal. As a result, if we look locally at the resulting RF modulated signal, i.e. over a time frame that is shorter than the time constant that characterizes the modulation, we can see a signal that has all the characteristics of a pure CW signal. With this behavior in mind, perhaps we can get the RF synthesizer, which would be used in other architectures to generate the LO signal, to now track the slow frequency variations of the modulation in order to generate the modulated RF signal. Obviously, if this approach works, it would result in a huge simplification in the transmit architecture as we would simply need to provide the instantaneous angular frequency information to the RF synthesizer and directly recover the modulated RF bandpass signal we are looking for at its output.

In order to explore this possibility in depth, we need to reconsider the architecture of classical RF synthesizers. Recalling the discussion in Section 4.3.1, this block is classically implemented as an RF oscillator embedded in a PLL system. With regard to our application, what is important to remark is that the RF oscillator used to produce the periodic RF bandpass signal is locked in phase and frequency to a stable low frequency reference signal. Perhaps we can modulate the reference signal according to the frequency modulation we are dealing with in order to directly recover the modulated bandpass signal we are looking for at the output of the synthesizer.

An example will serve to illustrate the feasibility of this approach. The problem is that variants exist in the physical implementation of such a synthesizer in practice. For instance, an interesting variant for our application is the all digital PLL (ADPLL). In this implementation, the use of a block that provides a binary word proportional to the delay difference between the signal delivered by the RF oscillator and the reference allows us to have all the functions embedded in the loop implemented as digital blocks [11]. The angular frequency of the oscillator is then tuned through the use of capacitor banks set by digital commands. Such a digitally controlled oscillator thus behaves as a digital to frequency converter from the signal processing point of view. For our application, a digital implementation of this kind is obviously well suited as the digital samples resulting from the generation of the modulating waveform can be directly fed to the ADPLL. The corresponding overall direct PLL modulator transmit line-up reduces in that case to the overview shown in Figure 8.7.

Whatever the physical implementation of the RF synthesizer we are dealing with, we can base our behavioral analysis on the model of the integer PLL used in "Phase locked loop" (Section 4.3.1). The simple model shown in Figure 4.15 is sufficiently representative for us to derive general guidelines. Let us therefore derive the closed loop transfer function linking the instantaneous angular frequency fed at the reference port of the synthesizer to that of the bandpass signal standing in the oscillator and centered around the carrier angular frequency of interest. This can be done quite easily by reusing the results derived in Chapter 4 when considering the instantaneous phase of those signals. The instantaneous angular frequency of

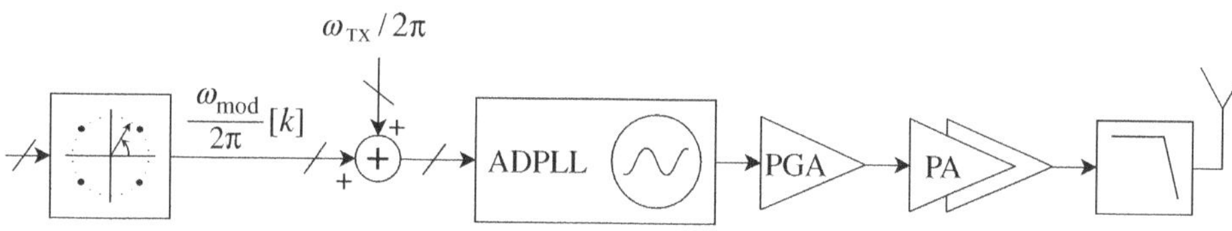

Figure 8.7 Direct PLL modulator transmit architecture – When dealing with a pure phase/frequency modulation, we can simplify the minimum set of signal processing blocks required to upconvert a complex modulating waveform. Indeed, practical RF synthesizers can be designed to track the modulation information provided to them so that they directly generate the expected phase/frequency modulated RF bandpass signal. This constant instantaneous amplitude signal can then experience various amplification stages in order to be delivered with the correct average power to the radiating element.

a bandpass signal is simply the time derivative of its instantaneous phase. We can thus write for instance in the Laplace domain that

$$\Omega_{\text{osc}}(s) = s\Phi_{\text{osc}}(s), \tag{8.5}$$

with $\Omega_{\text{osc}}(s)$ and $\Phi_{\text{osc}}(s)$ the Laplace transform of the instantaneous angular frequency of the fundamental tone delivered by the synthesizer, $\omega_{\text{osc}}(t)$, and of its associated instantaneous phase $\phi_{\text{osc}}(t)$, respectively. Thus, if $\omega_{\text{mod}}(t)$ represents the instantaneous angular frequency of the modulating waveform fed at the reference input of the PLL, we can directly use equation (4.172) to write that

$$\frac{\Omega_{\text{osc}}(s)}{\Omega_{\text{mod}}(s)} = H_{\text{LP}}(s), \tag{8.6}$$

where $H_{\text{LP}}(s)$ is the closed loop transfer function. Obviously, this transfer function depends on the characteristics of the loop filter. Practically speaking, for precision purposes and for the stability of this negative feedback system, this filter necessarily behaves as a lowpass filter. And for the sake of simplicity in our illustration, we continue to assume that we are dealing with a simple integrator as it leads to sufficiently general behavior. Under this assumption, $H_{\text{LP}}(s)$ reduces to the expression given by equation (4.175a).

Looking at this expression, we can see that, as the subscript LP suggests, we are dealing with an overall lowpass transfer function. This behavior leads to a first limitation for our application. When feeding the modulation to the reference port of the loop, only the fraction of the spectrum of the instantaneous angular frequency information that remains within the passband of the closed loop transfer function can be recovered on the signal delivered by the RF oscillator. In other words, the synthesizer cannot track variations of the reference signal that are faster than is allowed by its loop cut-off frequency. The problem is that in practical implementations, mainly to ensure noise performance, this cut-off frequency can be set very low – as low as a few tens of kilohertz. We thus anticipate problems with the present approach as soon as we consider wideband modulating waveforms dedicated to high data rates.

Before tackling possible solutions to this problem, we observe that we are talking here about the spectral extent of the instantaneous angular frequency of the modulated waveform, i.e. of the argument of its complex envelope, and not of the spectrum of the modulated bandpass signal itself. And the fact is that the spectra of these signals can be different in practice. This behavior is discussed in more depth in the next section on polar architecture. Here we merely recall that a relationship indeed exists between the spectrum of a bandpass signal and the spectra of the real and imaginary parts of its complex envelopes. More precisely, as discussed in "Impact of spectral symmetry on top of stationarity" (Section 1.1.3), under common assumptions, the real and imaginary parts of its complex envelopes have the same spectral shape as the complex envelopes themselves, and thus the same as the bandpass signal. This is one of the great benefits of working with Cartesian based architectures, such as direct conversion. In this kind of architecture, we can directly transpose the requirements that hold on the transmit spectrum to the spectrum of the lowpass modulating waveforms and thus directly derive the related baseband signal processing. This is unfortunately not the case when considering the modulus and argument of the complex envelope, which can have spectra that are very different – much wider in some cases – compared to the final modulated bandpass signal. Practically speaking, in our present case of interest this means that the spectrum of

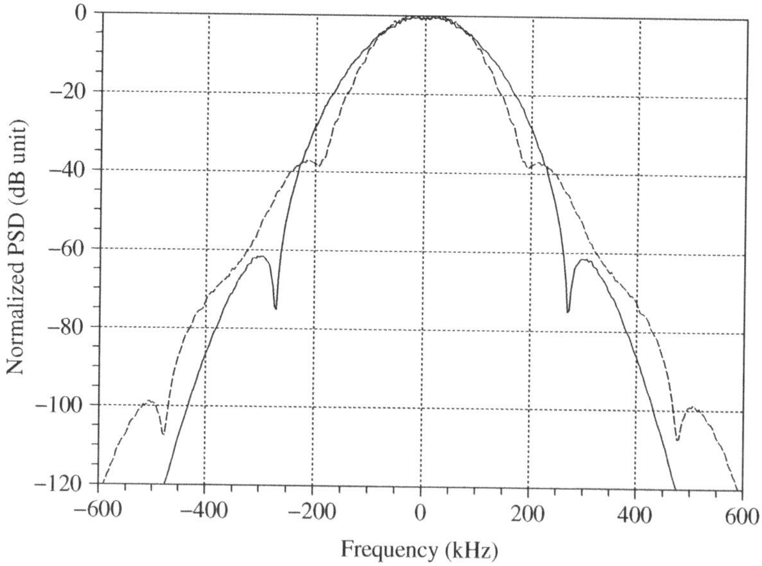

Figure 8.8 PSD of a GMSK modulated bandpass signal and of its instantaneous frequency – The GMSK modulation, used in the GSM standard for instance, and discussed in more depth in Section 1.3.1, is an example of a modulation scheme that leads to a comparable PSD of the modulating instantaneous frequency $f_{mod}(t)$ (solid) and of the corresponding modulated bandpass signal $\cos(2\pi \int f_{mod}(t)dt)$ (dashed).

the instantaneous phase of the modulated RF bandpass signal, equal to $\phi_{osc}(t)$ here, is not necessarily equal to the spectrum of the modulated bandpass signal $\cos(\phi_{osc}(t))$.

This should obviously be qualified on a case by case basis. Some phase/frequency modulating schemes have been conceived in which these two spectra are more or less comparable. This is for instance the case for the GMSK modulation used in the GSM standard. This modulation, detailed in Section 1.3.1, leads to comparable spectra for both the instantaneous angular frequency of the modulated bandpass signal and the modulated signal itself, as illustrated in Figure 8.8. However, even in this particular case optimized for that purpose, we see that the spectrum of the instantaneous frequency spreads over a wider extent than is found for the cut-off frequency of practical PLL implementations, which can be as low as a few tens of kilohertz. We can thus expect difficulties if we want to directly modulate the reference input of an RF synthesizer that has not been optimized for that purpose.

In order to achieve our goal, a workaround must be considered. We briefly review some of the options. One possibility is to implement a predistortion of the modulating instantaneous angular frequency in order to compensate for the impact of the loop filter. Considering a fixed loop transfer function, this can be done through the use of a fixed filter prior to the PLL in order to equalize its transfer function. However, given a lowpass behavior for the loop transfer function, the equalizer necessarily has to be highpass. This can cause problems in the implementation when the filters are steep and we need to carry out the equalization over a wide frequency band. But thinking back the origin of the problem, we also recall that the transfer function experienced by a signal injected in the loop depends on where it is injected. Perhaps we can inject the modulation somewhere else and end up with no distortion of the

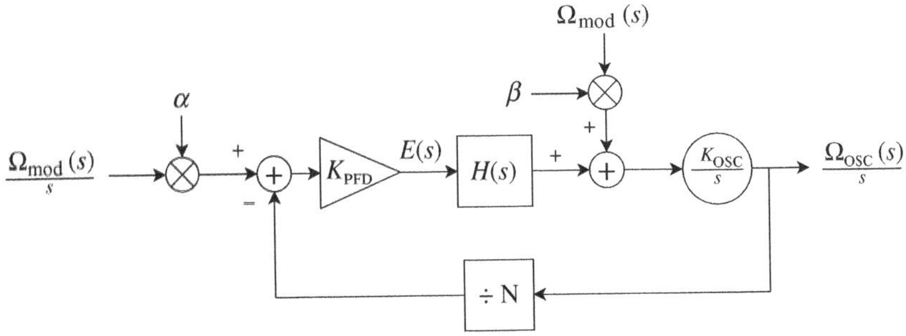

Figure 8.9 Equivalent model for the two-points modulation of a PLL – In practical PLL implementations, the closed loop transfer function experienced by the instantaneous angular frequency from the reference input up to the oscillator output is lowpass, whereas that from the oscillator input up to its output is highpass. Thus, by providing the modulation information at these inputs at the same time, we can achieve some compensation effects that equalize the transfer function.

modulating waveform. This approach is at the origin of the modulation through the frequency divider present in the feedback of the loop for instance. However, this requires a fractional PLL in order to be able to modulate the divider. Another interesting approach would be to inject the modulation at different points in the loop at the same time. Indeed, as discussed in "Phase locked loop" (Section 4.3.1), if the transfer function experienced by a signal injected at the reference input of the PLL is lowpass, that experienced by a signal injected at the RF oscillator stage is highpass. Moreover, looking at the expressions for those two transfer functions (equations (4.175a) and (4.175b)), we see that one is more or less the inverse of the other. By injecting the modulating information simultaneously at those two stages, perhaps it would be possible to have a compensation effect and thus recover finally an overall flat transfer function with respect to the modulation.

In order to investigate this, consider the equivalent model shown in Figure 8.9. As shown in this diagram, we now intend to provide the instantaneous angular frequency information at the input command port of the RF oscillator on top of the reference input. However, assuming that the PFD is sensitive to the phase offset between the signal provided to the reference port of the PLL and the feedback signal, it is still the instantaneous phase of the modulating waveform that appears at the reference port of the PLL on the model. Furthermore, we use simple normalization factors, α and β, in order to match the highpass an lowpass transfer functions experienced by those two modulating signals and thus equalize the overall transfer function linking $\Omega_{mod}(s)$ and $\Omega_{osc}(s)$. Thus, in accordance with this equivalent model, the error signal now takes the form

$$E(s) = K_{PFD}\left(\alpha\frac{\Omega_{mod}(s)}{s} - \frac{1}{N}\frac{\Omega_{osc}(s)}{s}\right). \tag{8.7}$$

At the same time, considering the direct path of the loop, we can write that

$$\frac{\Omega_{osc}(s)}{s} = \frac{K_{osc}}{s}\left(H(s)E(s) + \beta s\frac{\Omega_{mod}(s)}{s}\right). \tag{8.8}$$

Substituting equation (8.7) into equation (8.8) yields

$$\Omega_{\rm osc}(s)\left(\frac{N}{K}s + H(s)\right) = \Omega_{\rm mod}(s)\left(\frac{N}{K}\beta K_{\rm osc}s + \alpha N H(s)\right), \tag{8.9}$$

with $K = K_{\rm PFD}K_{\rm osc}$. Now, if we simply choose α and β such that

$$\alpha = \frac{1}{N}, \tag{8.10a}$$

$$\beta = \frac{1}{K_{\rm osc}}, \tag{8.10b}$$

then we achieve the result we are looking for, i.e. that $\Omega_{\rm osc}(s) = \Omega_{\rm mod}(s)$.

What is interesting is that the PLL transfer function is equalized whatever the loop filter. This is thus a very general approach that theoretically allows the phase/frequency modulation of the RF oscillator whatever the bandwidth of this modulation. However, this simplicity should not blind us to the fact that many relevant parameters impact the final performance in practical implementations. We need to normalize the modulation information injected at the input of the RF oscillator by the factor $\beta = 1/K_{\rm osc}$. But, as with any RF/analog device, the RF oscillator necessarily experiences impairments in its implementation. As a result, its conversion gain $K_{\rm osc}$ can exhibit non-negligible variations, either on a given device when considering different operating conditions, or from device to device in a mass production perspective. It can be difficult to achieve the correct equalization of the PLL transfer function with a fixed predetermined factor β. An oscillator calibration scheme may then be necessary. In the same way, we have to keep in mind that everything that is injected at the input of the RF oscillator experiences a highpass transfer function. This obviously holds for the noise that is injected at that point. Particular care needs to be exercised in generating the modulation information provided to the RF oscillator in order to ensure no degradation of the wideband noise performance of the modulated bandpass signal. In addition, there are constraints linked for instance to potential delay between the two paths providing the modulation, which need to be well aligned in order to achieve appropriate compensation of the transfer functions.

In order to conclude this section, we highlight some advantages of the direct PLL modulator in addition to its simplicity. If the signal to be transmitted is taken as an amplified version of the signal delivered by the RF oscillator, i.e. if we do not use some frequency divider at the synthesizer output, the instantaneous angular frequency of the signal established in the RF oscillator will obviously match that of the transmit signal. As a result, when facing an RF coupling between the oscillator and the high power signal existing in the latter stages of the line-up, only a constant phase offset linked to the delay along the coupling path holds between the instantaneous phase of the signal established in the RF oscillator and that of the potential aggressor. As a result, the pulling issue discussed in Section 8.1.1 when considering the direct conversion architecture should theoretically be avoided in the present case [44–47].

Another advantage comes from the fact that the generated modulated bandpass signal has a truly constant instantaneous amplitude, as does the signal established in the oscillator. This is important in practice as it allows us to use highly efficient PAs that behave as hard limiters with respect to the signal being processed. The problem is that, as discussed in Chapter 5, such devices often exhibit a non-vanishing AM-PM conversion factor. To avoid degrading the phase/frequency modulation of the transmitted signal, we thus need to ensure that no instantaneous amplitude component is present in the signal entering the high power device. The attraction of the direct PLL modulator then becomes obvious as we recover a constant instantaneous amplitude modulated bandpass signal at the output of the PLL. This is not necessarily the case when considering the physical implementation of other architectures that process the P and Q components of the modulating waveform, such as a direct conversion transmitter. In such a line-up, the quality of the generated modulated bandpass signal depends on the imbalance between the P and Q paths, as discussed in Chapter 6. In particular, even when dealing with a pure phase/frequency modulation, such imbalance can lead to the generation of an amplitude modulation component due to the rise in the image signal, as detailed in "Transmit side" (Section 6.2.2) and illustrated in Figure 6.20. This can then lead to potential degradations of the quality of the phase/frequency part of the modulation when facing further AM-PM conversion.

Finally, using a direct PLL modulator, we can expect good noise performance at the output of the transmitter. This is obviously due to the fact that we use the minimum set of processing blocks required to generate a pure phase/frequency modulated waveform. Thus, compared to a more generic architecture such as direct conversion, we necessarily have fewer noise contributors. The noise requirements in such a line-up are thus easier to satisfy.

We have seen that there are many advantages to using the direct PLL modulator. Most (e.g. its noise performance) are related to the small number of processing blocks. But one interesting characteristic is linked to the fact that the instantaneous angular frequency of the signal established in the RF oscillator can exactly match that of the transmit signal. We can thus expect to overcome the pulling issue encountered for instance in the direct conversion architecture. This is obviously a great advantage over the latter architecture. It would therefore be desirable to retain this advantage, even if we are dealing with a complex modulation. With that in mind, we can define the instantaneous angular frequency and amplitude of any modulated RF bandpass signal, those quantities being directly related to the magnitude and phase of the complex envelopes of the signal as discussed in Chapter 1. We can then imagine generating a RF bandpass signal that has the expected instantaneous angular frequency but a constant instantaneous amplitude using a direct PLL modulator scheme. All that would then remain would be to find a way to modulate in turn the instantaneous amplitude of this signal in order to recover the expected overall modulated RF bandpass signal. If we could do this, we would still have the instantaneous angular frequency of the signal established in the RF oscillator matching that of the transmit signal. This approach leads to the polar transmit architecture discussed in the next section.

8.1.6 Polar Transmitter

As introduced at the end of the previous section, on top of the simplicity of the corresponding line-up, one reason for using the direct PLL modulator for the generation of phase/frequency modulated RF bandpass signals is the possibility of the instantaneous angular frequency of the signal established in the RF oscillator matching that of the transmitted signal. The resulting potential robustness to pulling makes this approach attractive for practical integrated solutions in which non-negligible RF coupling can be expected. Even if this is all we achieve, it is sufficient motivation to see if this kind of architecture cannot be extended to complex modulations.

In order to go further, we first quickly review the signal processing functions that would allow such a generalization. This can be done for instance by looking at the mechanisms involved in the generation of the complex modulated RF bandpass signal, $s(t)$, that is complex modulated by the lowpass complex signal $\tilde{s}(t)$. Practically speaking, as the notation suggests, this complex waveform becomes the complex envelope, defined as centered around the transmit carrier angular frequency ω_{TX}, of the bandpass signal $s(t)$. As detailed in Chapter 1, the associated signal processing, corresponding in practice to the frequency upconversion of $\tilde{s}(t)$ around ω_{TX} by using the positive complex exponential $e^{j\omega_{TX}t}$ used in the definition of the complex envelope concept, takes the form

$$s(t) = \mathrm{Re}\left\{\tilde{s}(t)e^{j\omega_{TX}t}\right\}. \tag{8.11}$$

The physical implementation of this processing then depends on how we represent $\tilde{s}(t)$. On the one hand, a Cartesian decomposition such as

$$\tilde{s}(t) = p(t) + jq(t) \tag{8.12}$$

results in the expression for the modulated RF bandpass signal $s(t)$ given by equation (1.28a):

$$\begin{aligned} s(t) &= \mathrm{Re}\left\{(p(t) + jq(t))e^{j\omega_{TX}t}\right\} \\ &= p(t)\cos(\omega_{TX}t) - q(t)\sin(\omega_{TX}t). \end{aligned} \tag{8.13}$$

The physical implementation of the corresponding set of processing blocks then leads to the Cartesian architectures such as direct conversion discussed in Section 8.1.1. On the other hand, a polar decomposition of $\tilde{s}(t)$ such as

$$\tilde{s}(t) = \rho(t)e^{j\phi(t)} \tag{8.14}$$

results in the expression for $s(t)$ given by equation (1.28b):

$$\begin{aligned} s(t) &= \mathrm{Re}\left\{\rho(t)e^{j(\omega_{TX}t+\phi(t))}\right\} \\ &= \rho(t)\cos(\omega_{TX}t + \phi(t)). \end{aligned} \tag{8.15}$$

The physical implementation of this processing leads to the polar architectures. But looking at this expression, we see that rather than thinking of the generated modulated RF bandpass signal $s(t)$ as the result of the frequency upconversion of the modulating complex lowpass waveform $\tilde{s}(t)$, we can interpret it as a two-step process. We can first identify in this expression the constant instantaneous amplitude bandpass signal $\cos(\omega_{TX}t + \phi(t))$. We can thus for instance think of using a direct PLL modulator for the generation of this phase/frequency only modulated bandpass signal as detailed in the previous section. Then we simply need to apply the instantaneous amplitude information $\rho(t)$ to the latter signal. From the signal processing point of view, the latter operation requires multiplication of the phase/frequency modulated RF bandpass signal flowing from the RF synthesizer, and the lowpass instantaneous amplitude information. Practically speaking, this multiplication can be done in different ways depending on its interpretation. We can see this operation as the frequency upconversion of the real lowpass modulating waveform $\rho(t)$; it can thus be implemented through the use of an RF mixer that is driven by the LO signal $\cos(\omega_{TX}t + \phi(t))$ even if phase/frequency modulated. But alternative approaches can also be considered as we can for instance imagine applying the instantaneous amplitude information $\rho(t)$ to the command port of a variable gain device processing the phase/frequency modulated RF signal. Having a variation of the linear gain proportional to $\rho(t)$ applied to this signal would indeed result in the same expected complex modulated bandpass signal $s(t)$. In practice, this approach is of interest in respect of other key performance aspects of the line-up. This is for instance the case if the variations of $\rho(t)$ can be directly applied to the PA in order to vary its own gain through its supply voltage. This strategy may lead to power savings compared to the case where a constant polarization of the device, based on the average transmit output power, is used. This point is discussed in more depth in the next section.

In order to derive a first set of signal processing blocks required for the implementation of such a polar transmitter, we also need to keep in mind that the complex modulating waveforms are for the most part defined through the generation of their real and imaginary parts, i.e. the signals $p(t)$ and $q(t)$, respectively. We thus generally need to consider a Cartesian to polar converter in the digital domain, for instance using a CORDIC algorithm as discussed in Chapter 9, to generate the samples $\rho[k]$ and $\phi[k]$ from the samples of $p[k]$ and $q[k]$. Then, depending on how the RF synthesizer is driven, a time domain derivative function may be required in order to derive the variations of the instantaneous angular frequency of $s(t)$, $\omega[k] = \dot{\phi}[k]$, from the variations of the instantaneous phase data. The resulting line-up takes the general form shown in Figure 8.10.

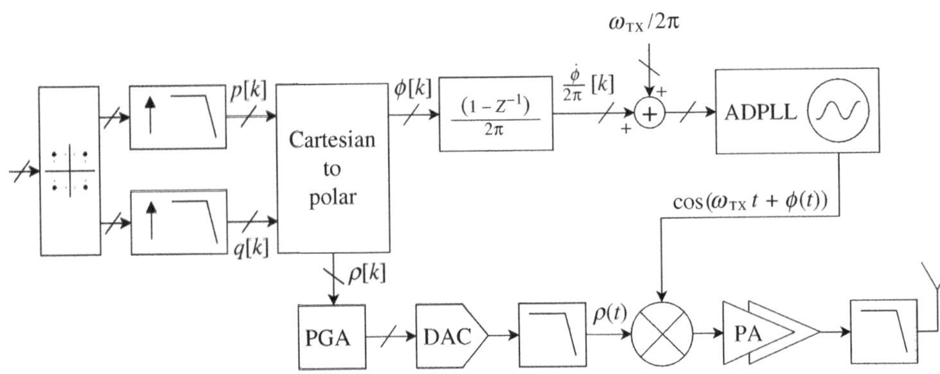

Figure 8.10 Polar transmit architecture – A complex modulated RF bandpass signal that exhibits an instantaneous amplitude and phase, $\rho(t)$ and $\omega_{TX}t + \phi(t)$ respectively, can be generated by first deriving a constant amplitude bandpass signal that has an instantaneous phase varying according to $\phi(t)$ around the central term $\omega_{TX}t$. This can be achieved for instance through a direct PLL modulator. Then the resulting bandpass signal can in turn be modulated in amplitude following $\rho(t)$ through the use of an RF mixer for instance. In addition, the scaling of $\rho(t)$ can allow control of the RF power delivered to the antenna. Finally, with most of the complex modulating waveforms defined in a Cartesian way, the generation of $\rho(t)$ and $\phi(t)$ also requires additional processing to be derived from $p(t)$ and $q(t)$.

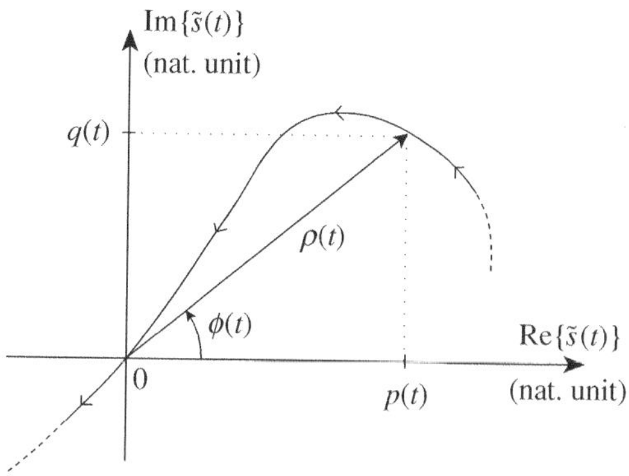

Figure 8.11 Problem linked to the zero crossing of a modulating complex envelope trajectory – In contrast to what occurs for its real and imaginary parts, $p(t)$ and $q(t)$ respectively, the spectral shape of the modulus $\rho(t)$ and argument $\phi(t)$ of a modulating complex envelope $\tilde{s}(t) = p(t) + jq(t) = \rho(t)e^{j\phi(t)}$ can be very different from that of $\tilde{s}(t)$ itself. This is particularly obvious for $\phi(t)$ in the common case where zero crossings occur in the modulating complex envelope trajectory: $\phi(t)$ experiences π phase shifts that necessarily lead to a wideband spectrum for this signal.

In order to go further, we now focus on the potential problems associated with such architecture. We first need to make some general comments on the spectral shape of the signals we are dealing with in our line-up. Indeed, as just highlighted, complex modulated waveforms are mainly defined through the generation of their real and imaginary parts, i.e. through the generation of the signals $p(t)$ and $q(t)$ from the input data bits. As discussed in "Power spectral density" (Section 1.1.3), a good reason for that comes from the fact that under common assumptions, those waveforms have the same spectral shape as the complex envelope $\tilde{s}(t) = p(t) + jq(t)$ and thus the final modulated bandpass signal $s(t)$. It is therefore useful to define and process the $p(t)$ and $q(t)$ waveforms in order to directly control the characteristics of the transmitted output signal, at least in terms of spectral shape. In contrast, the spectral shapes of $\rho(t)$ and $\phi(t)$ can be very different from that of the final modulated bandpass signal $s(t)$. More precisely, those spectra, and particularly that of $\phi(t)$, can be wideband relative to this final transmit spectrum. A good way to understand this behavior is to consider the particular case where the trajectory of the complex envelope $\tilde{s}(t) = p(t) + jq(t) = \rho(t)e^{j\phi(t)}$ exhibits zero crossings. This configuration, which is unfortunately common in most practical modulating schemes, corresponds to the case where both $p(t)$ and $q(t)$ can be null at the same time, resulting in $\rho(t) = 0$. When this occurs, given that $\rho(t)$ is defined as a positive quantity by our definitions, $\phi(t)$ obviously experiences a π phase shift as shown in Figure 8.11. Practically speaking, this means that the analytical expression of the instantaneous phase variations $\phi(t)$ necessarily contains terms $\delta\phi_n$ of the form

$$\delta\phi_n = \pi \mathrm{U}(t - t_n), \tag{8.16}$$

with U(.) the Heaviside unit step function and t_n the instants at which the zero crossings occur. As a result, the expression for $\omega(t)$, i.e. the variations of the instantaneous angular frequency of $s(t)$, which is nothing more than the time domain derivative of $\phi(t)$, contains terms $\delta\omega_n$ of the form [2]

$$\delta\omega_n = \pi\delta(t - t_n). \tag{8.17}$$

Given that such a Dirac delta distribution signal has a constant Fourier transform, the spectrum of $\omega(t)$ theoretically spreads toward infinity.

This behavior can be illustrated for instance by recalling the case of the modified 8PSK modulation used in the GSM/EDGE standard and detailed in Section 1.3.2. In this modulating scheme a continuous rotation of $3\pi/8$ from symbol to symbol has been added to the classical 8PSK modulation so that the modulating trajectory avoids the center of the complex plane, as illustrated in Figure 1.14. The initial motivation for this was the limitation in the DR of the instantaneous amplitude $\rho(t)$ that results in a reduction in the modulation PAPR, as can be understood from the definition of this quantity in equation (1.71). This reduction can indeed in turn lead to a potential reduction of the power consumption of the PA for a given target average transmit output power, as illustrated in the next section. However, from the polar architecture point of view, we observe that the corresponding avoidance of the zero crossings in the trajectory also results in a limitation of the spectrum of the modulating instantaneous angular frequency. This can be understood from Figure 8.12 where the variations in the instantaneous frequency of the resulting modulated bandpass signal and the corresponding spectra are derived when supposing successively that the continuous rotation of $3\pi/8$ from symbol to symbol is applied or not. We see that if the continuous rotation is not used, spikes are effectively present in the instantaneous angular frequency variations due to the zero crossings in the modulating trajectory. As expected, we obtain a spectrum for this instantaneous angular frequency that is wider than when the continuous rotation is used.

In practice, this zero crossing phenomenon can be seen as a limit case. But in general, when the vectorial representation of the modulating complex envelope $\tilde{s}(t) = \rho(t)e^{j\phi(t)}$ experiences a phase shift $\delta\phi$ over a time duration δt, we have an instantaneous angular frequency variation $\delta\omega$ whose magnitude can be estimated as

$$\delta\omega = \frac{\delta\phi}{\delta t}. \tag{8.18}$$

Thus, the closer to the center of the complex plane the trajectory goes, the closer $\delta\phi$ tends to π, and the closer δt tends to 0. We then recover the spikes we were discussing above in the limit case that corresponds to the zero crossing. But in general, δt necessarily remains bounded by the symbol period T_{symb}, which obviously represents the order of magnitude for the maximum transition duration from one symbol representation in the complex plane to the next. As a result, the variations in the modulating instantaneous angular frequency are at least proportional to the inverse of the symbol period, i.e. the symbol rate $R_{\text{symb}} = 1/T_{\text{symb}}$. We thus see that the range of variation of the instantaneous angular frequency of the modulated bandpass signal and its wideband nature are necessarily linked to the data rate we face on top of the shape of the trajectory with respect to the center of the complex plane.

In contrast, if we now focus on the instantaneous amplitude part of the modulated bandpass signal, we find that it has no particular discontinuities. On the other hand, its DR (rather than its spectral shape) is sensitive to the area close to the center of the complex plane. That said, the

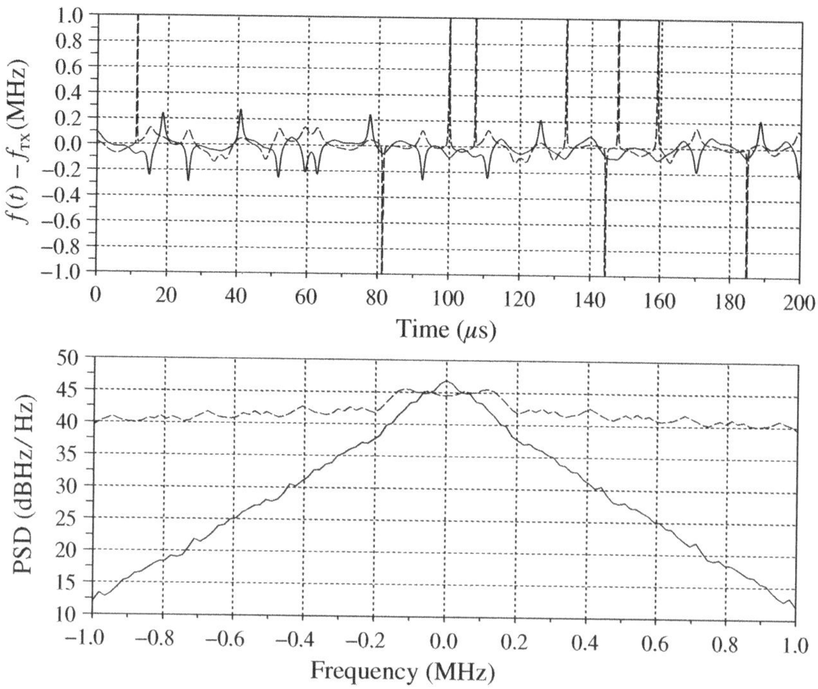

Figure 8.12 Instantaneous frequency variations and corresponding spectra for a GSM/EDGE modified 8PSK modulated waveform with and without $3\pi/8$ continuous rotation – Referring to the trajectories shown in Figure 1.14, the presence of zero crossings when the continuous rotation of $3\pi/8$ is not used leads to π phase shifts and thus spikes in the instantaneous frequency of the modulated bandpass signal (top, dashed). This results in a wideband spectrum for this instantaneous frequency (bottom, dashed). This is not the case when a continuous rotation of $3\pi/8$ is used (solid).

spectrum of $\rho(t)$ can remain different from that of $p(t)$ and $q(t)$, and thus different from the final modulated bandpass signal. Practically speaking, this means that we still need to estimate this particular spectrum in order to correctly dimension the data path used for the processing of $\rho(t)$.

We thus conclude on the one hand that the width of the spectrum of the phase/frequency part of the modulating waveform can be very wide compared to the initial spectrum of the complex envelope we are dealing with, due to the fact that the modulation trajectory can go close to the center of the complex plane. This behavior is obviously enhanced by the data rate of the system we are dealing with. For a given structure of the constellation and modulating waveform, the higher the data rate, the wider the spectrum of the instantaneous angular frequency of the resulting modulated bandpass signal. On the other hand, the DR of the instantaneous amplitude of this modulated signal can be impacted by the shape of the trajectory of the modulating waveform. With such behavior in mind, we can now go through the corresponding impacts on the possibility of physically implementing a polar transmitter.

We first observe that we may experience difficulties in the application of the amplitude part of the modulation to the phase/frequency modulated bandpass signal when $\rho(t)$ has a wide DR. Indeed, in order not to distort the amplitude modulation part, this multiplication must be implemented using an RF device that has a linear behavior over a wide DR with respect to the input port at which $\rho(t)$ is fed. The conclusion must obviously be refined depending on the exact structure of the block that is used for this functionality, but we see the possible benefit of working with a modulation whose trajectory avoids the center of the complex plane in order to limit this DR.

Then, we can anticipate various problems linked to the wideband nature of the signals we are dealing with in this polar decomposition of the complex modulating waveform. The first is linked to the generation of the modulating samples $\rho[k]$ and $\omega[k]$. These samples classically need to be generated from those representing the real and imaginary parts of the expected complex envelope, as illustrated in Figure 8.10. This requires the use of a Cartesian to polar converter. But this block must necessarily run at the sampling rate required to represent the samples $\rho[k]$ and $\omega[k]$ correctly, which may be somewhat higher than is strictly necessary to represent only the $p\,[k]$ and $q\,[k]$ samples correctly. As a side effect, all the subsequent digital signal processing as well as the DACs need to run at the higher sampling rate. We can thus expect a higher power consumption of the digital implementation part of the polar line-up compared to that in a pure Cartesian architecture. On top of that, recalling the discussion in the previous section on the direct PLL modulator, we can anticipate that the wideband nature of the phase/frequency part of the modulation will limit the possibility of using classical structures for the modulation of the RF synthesizer. This may make more complex implementations necessary, such as the two-point modulation PLL discussed in that section.

But on top of that the wideband nature of those signals, and in particular of the instantaneous phase/frequency part of the modulation, leads to stringent requirements in terms of time alignment between the instantaneous amplitude and phase signals in order to achieve a correct reconstruction of the modulated RF bandpass signal. Indeed, we know from the above discussion that the wideband nature of the instantaneous phase variations $\phi(t)$ is linked to the presence of sharp transitions in its time domain expression. But, when such a transition occurs, we necessarily recover the same kind of discontinuity on the phase/frequency only modulated bandpass signal $\cos(\omega_{\text{TX}}t + \phi(t))$ flowing from the RF synthesizer. This is obvious if we recall the didactic configuration where a zero crossing of the modulating trajectory occurs. In that case, we get a π phase shift of $\phi(t)$ that leads to a direct change of sign in this constant instantaneous amplitude signal. Practically speaking, this change of sign means a sharp transition in this signal flowing from the RF synthesizer and in turn a wideband spectrum for this signal. It is only when the multiplication by the instantaneous amplitude part $\rho(t)$ occurs that the final modulated bandpass signal $\rho(t)\cos(\omega_{\text{TX}}t + \phi(t))$ recovers the expected spectral shape. The underlying phenomenon is that the zeros of $\rho(t)$ occur at *the same time* as the change of sign of $\cos(\omega_{\text{TX}}t + \phi(t))$. The corresponding sharp transitions of $\cos(\omega_{\text{TX}}t + \phi(t))$ are thus canceled in the final reconstructed bandpass signal by those zeros, as illustrated in Figure 8.13. We then recover a smooth bandpass signal that exhibits the same spectral shape as the original modulating complex envelope. The problem is that when dealing with sharp transitions, we need to ensure a very good time synchronization between the phase/frequency only modulated bandpass signal $\cos(\omega_{\text{TX}}t + \phi(t))$ flowing from the RF synthesizer and the instantaneous amplitude part $\rho(t)$ in order to ensure that the zeros of the latter signal indeed cancel the transitions linked to the π phase shift of the former. And in practice, the sharper those transitions, the more accurate the time alignment needs to be.

This intuitive conclusion can be confirmed by simple analytical derivations. We simply assume a delay mismatch τ in the recombination of the phase/frequency only modulated

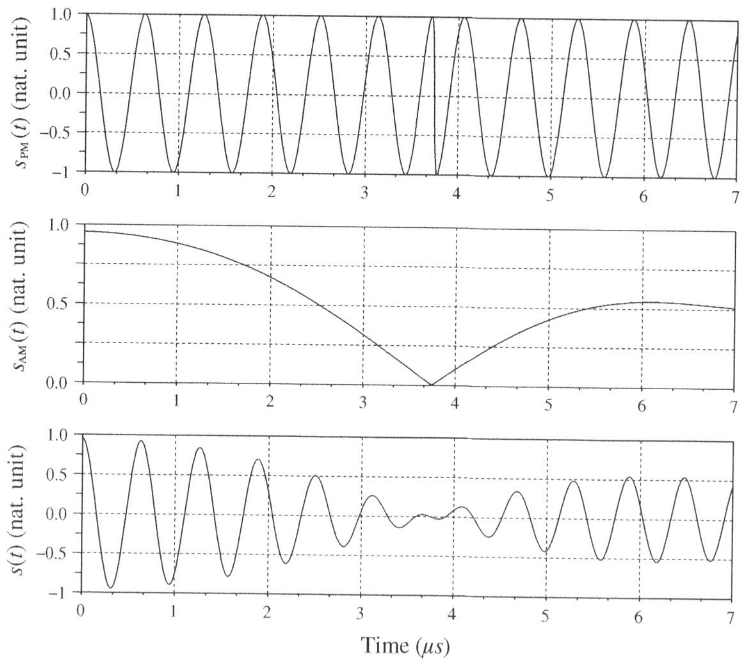

Figure 8.13 Delay alignment problem in the polar decomposition – As illustrated in Figure 8.11, the zero crossings of a complex modulating trajectory lead to π phase shifts in the instantaneous phase variations $\phi(t)$. To recover a clean modulated bandpass signal $s(t) = \rho(t)\cos(\omega_{TX}t + \phi(t))$ (bottom), the corresponding abrupt sign inversions of the constant amplitude signal $s_{PM}(t) = \cos(\omega_{TX}t + \phi(t))$ (top) need to be correctly canceled by the associated zeros of the instantaneous amplitude $s_{AM}(t) = \rho(t)$ (middle). Due to the slope of such inversion, the cancellation requires the correct time alignment between $s_{PM}(t)$ and $s_{AM}(t)$.

bandpass signal $\cos(\omega_{TX}t + \phi(t))$ with the expected instantaneous amplitude $\rho(t)$. As a result, the complex envelope $\tilde{s}_\tau(t)$ of the reconstructed modulated bandpass signal, still defined as centered around the carrier angular frequency ω_{TX}, can now be written in the form

$$\tilde{s}_\tau(t) = \rho(t)e^{j\phi(t-\tau)}. \tag{8.19}$$

Assuming that the delay mismatch remains sufficiently weak, we can thus use a Taylor series expansion of $\phi(t)$ to write, up to first order,

$$\phi(t-\tau) \approx \phi(t) - \tau\dot{\phi}(t), \tag{8.20}$$

with $\dot{\phi}(t)$ the time domain derivative of $\phi(t)$. Now recalling that this derivative is nothing more than the instantaneous angular frequency term $\omega(t)$, we have

$$\phi(t-\tau) \approx \phi(t) - \tau\omega(t). \tag{8.21}$$

Using this expression in equation (8.19), we obtain

$$\tilde{s}_\tau(t) = \rho(t)e^{j\phi(t)}e^{-j\tau\omega(t)} = \tilde{s}(t)e^{-j\tau\omega(t)}. \tag{8.22}$$

As a result, we can deduce the complex envelope $\tilde{e}_\tau(t)$ of the bandpass error signal $e_\tau(t) = s_\tau(t) - s(t)$ recovered at the output of the reconstruction of the modulated RF bandpass signal by simply remarking that

$$\begin{aligned} e_\tau(t) &= \mathrm{Re}\left\{\tilde{s}_\tau(t)e^{j\omega_{TX}t}\right\} - \mathrm{Re}\left\{\tilde{s}(t)e^{j\omega_{TX}t}\right\} \\ &= \mathrm{Re}\left\{(\tilde{s}_\tau(t) - \tilde{s}(t))e^{j\omega_{TX}t}\right\}. \end{aligned} \tag{8.23}$$

Thus, assuming that $\tilde{e}_\tau(t)$ is also defined as centered around ω_{TX}, we can directly write that

$$\tilde{e}_\tau(t) = \tilde{s}_\tau(t) - \tilde{s}(t). \tag{8.24}$$

Finally, using equation (8.22), we have

$$\tilde{e}_\tau(t) = \tilde{s}(t)[e^{-j\tau\omega(t)} - 1]. \tag{8.25}$$

Thus, in order to keep the magnitude of the error signal negligible relative to the modulated signal itself, we need to ensure that

$$|\tilde{e}_\tau(t)| \ll |\tilde{s}(t)|, \tag{8.26}$$

i.e.

$$|e^{-j\tau\omega(t)} - 1| \ll 1. \tag{8.27}$$

We thus need to keep the delay mismatch τ sufficiently low that

$$|\tau\omega(t)| \ll 1, \quad \forall t. \tag{8.28}$$

We thus recover that the wider the difference between the maximum and minimum values of the instantaneous angular frequency of the modulated RF bandpass signal, i.e. the sharper the transition of the instantaneous phase variations, the lower the delay mismatch τ has to be set in order to preserve the performance. Here again, this can be illustrated by recalling the modified 8PSK modulation used in the EDGE standard in the two cases where the continuous rotation of $3\pi/8$ from symbol to symbol is enabled or not. Figure 8.14 shows that when an identical delay mismatch of $T_{symb}/5$ is considered in the two cases, the spectral degradation remains more important in the configuration corresponding to the widest instantaneous angular frequency spectrum, i.e. when the continuous rotation is not used. Obviously, the same behavior can be expected for the other metrics related to the quality of the modulation, such as the EVM.

In practical implementations, it can be tricky to set constraints on the delay mismatch between the paths in order to ensure reasonable performance. Indeed, we have to keep in

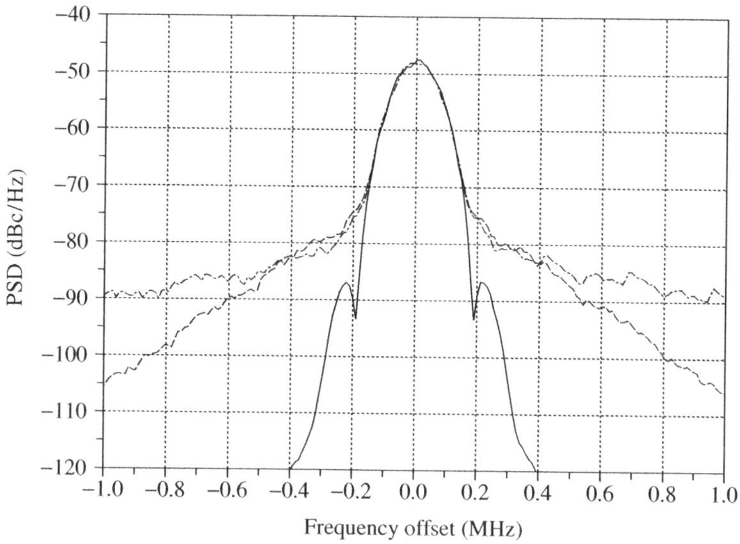

Figure 8.14 PSD of a reconstructed GSM/EDGE modified 8PSK modulated waveform in the presence of delay mismatch with and without $3\pi/8$ continuous rotation – The spectrum of the GSM/EDGE modified 8PSK modulated RF bandpass signal $\rho(t)\cos(\omega_{\text{TX}}t + \phi(t-\tau))$ suffers degradation when a delay $\tau = T_{\text{symb}}/5$ exists between its instantaneous amplitude and phase variations, $\rho(t)$ and $\phi(t)$ respectively, referred to the theoretical one corresponding to $\tau = 0$ (solid). The degradation is more important if the instantaneous angular frequency of the modulated signal exhibits spikes due to the zero crossings in the trajectory (dot-dashed), i.e. when the $3\pi/8$ continuous rotation is not used as illustrated in Figure 8.12, than when the continuous rotation is applied (dashed).

mind that we are necessarily dealing with physical paths that are different for the generation of the phase/frequency modulated RF bandpass signal on the one hand, and the application of the amplitude part of the modulation on the other hand. When a strict requirement holds in terms of delay mismatch, it can therefore be difficult to guarantee the right balance between the delays along those paths in a given operating configuration of the transmitter, and at the same time that those delays are kept equal when this operating configuration changes, e.g. when the temperature of the device or its power supply drifts. Obviously, we do not have the same problem in Cartesian implementations. One reason for that is that we are obviously dealing with waveforms of lower spectral extent than the corresponding instantaneous angular frequency. Having lower slopes in the time domain waveforms we are dealing with leads to less sensitivity in the potential delay mismatch between the $p(t)$ and $q(t)$ waveforms. Then, and most important, in contrast to what occurs in a polar line-up, the P and Q paths of Cartesian architecture can be implemented using strictly the same devices in terms of design. This copy and paste approach in the physical implementation gives the possibility of achieving a natural matching of the line-up, and thus of the delays in practice, whatever the operating conditions.

From the discussion hitherto, the implementation of the polar transmit architecture is fraught with difficulties. Obviously, if we want to avoid tricky calibration schemes or tracking algorithms based on feedback measurements as could be required for the compensation of the delay mismatch for instance [73], we may consider this kind of architecture for appropriate modulating schemes. This means in practice modulating waveforms whose trajectories avoid as much as possible the center of the complex plane and that are used for systems with the lowest possible data rates. This is required in order to limit as much as possible the spectral extent of the instantaneous angular frequency of the modulated bandpass RF signal and the DR of its instantaneous amplitude. That said, we observe a kind of complementarity with the direct conversion architecture discussed in Section 8.1.1. Indeed, one of the main difficulties with this simplest Cartesian transmit architecture when implemented in its classical form, i.e. using an RF oscillator running at an integer multiple of the transmit carrier angular frequency, remains the pulling issue. And as discussed in that section, this problem becomes more and more critical, for a given transmit average output power, as the bandwidth of the modulated RF bandpass signal we are dealing with narrows. This kind of architecture would thus appear better suited to high bandwidth modulation. In contrast, in the polar architecture we can expect to limit this pulling issue as we can run the RF oscillator with an instantaneous angular frequency that matches that of the modulated RF signal flowing from the transmitter. This obviously holds whatever the bandwidth of the modulation we are dealing with. But on top of that, due to the different limitations discussed above and linked to the spectral extent of the instantaneous angular frequency, the polar architecture appears more suited to narrowband modulations, even if it needs to be refined with regard to the zero crossing problem of the modulating complex envelope trajectory.

In conclusion, then, when it is possible to implement it, the polar architecture can also be considered as a way to achieve power savings for the transmit line-up. As previously stated, we can imagine applying the amplitude modulation directly through control of the biasing of the PA. This is an interesting approach that can be seen as part of more general approaches that aim to improve the efficiency of the line-up even when processing a non-constant envelope signal, as discussed in the next section.

8.1.7 Transmitter Architectures for Power Efficiency

While our primary concern is to describe transceiver architectures from a signal processing point of view, there is another aspect to consider if we want a full picture of transmitter architectures, i.e. their power consumption. This topic is important in most wireless applications.

The transmitter is classically the most power hungry line-up in a transceiver due to the RF power that it needs to deliver to the antenna, as detailed in Chapter 2. Obviously, this statement may be qualified depending on the RF power the transmitter needs to deliver to its load. In classical wireless standards the amount of electrical power is generally large, at least considering the maximum transmit power configuration, compared to what is required to supply a receive path. Furthermore, this electrical RF power is necessarily generated through the conversion of the DC power drawn from the power supply of the line-up. But practically speaking, such conversion can only be achieved through a finite efficiency. As a result, a non-negligible amount of electrical power is generally wasted as heat in the final RF stages of a transmit line-up – more precisely, in the final PA stage.

One might object that there are classes of PA that can theoretically achieve 100% efficiency [74]. Unfortunately, they are for the most part switching PAs belonging to non-sinusoidal classes. They thus behave as hard limiters with regard to the signal being amplified. They thus cannot be used as such to process amplitude modulated bandpass waveforms (recall the discussion throughout Chapter 5). And given that such amplitude modulated waveforms are common in present-day wireless standards, we often need to consider the use of a linear PA, resulting in a poor efficiency of the transmit line-up.

To improve the situation, different strategies can be considered. For instance, a first idea might be to try optimizing the design of the PA stage itself. However, this design related topic is outside the scope of this book, and we refer the reader to other sources on such dedicated PA architectures, such as the Doherty structure [74–76].

Given a PA structure, then, we can stay with the system approach and review different ways to improve the global efficiency of the solution while considering a linear transmit line-up in order to support an amplitude modulated waveform. Different approaches can be considered to achieve this [67, 77]. From our system point of view, we can classify these approaches depending on the mode in which the PA is working, as this mode determines the class of solution to consider. We begin with a linear PA that is working in its linear region, then go on to look at one working in its weak nonlinear part, and finally a truly nonlinear device. In the last case, we expect to be able to use a high efficiency PA working in a non-sinusoidal class, even when processing an amplitude modulated signal in the transmit line-up.

PA Working in the Linear Area

As discussed throughout Chapter 5, and especially in Section 5.3.1, through both AM-AM and AM-PM conversion, RF compression can generate bandpass distortion terms centered around the same carrier frequency as the bandpass signal being processed when the latter is amplitude modulated. In order to limit both the in-band and the close in-band degradation of the amplitude modulated signal we thus need to consider a sufficiently linear transmit line-up in order to process it. This means that the back-off in the line-up must be sufficiently high to limit the compression of the signal. And for that, the Psat of the RF stages, and in particular of the PA, must be at least equal to the peak instantaneous power of the waveform it processes plus a given headroom, as illustrated in Figure 8.15.

At the same time, the Psat of the device is directly linked to its power consumption. Thus the power consumption of the PA stage is driven by the instantaneous peak power of the transmit signal to be delivered to the load. But in practice, this instantaneous peak power can be decomposed as the average power of the waveform times its PAPR. So the required Psat, and thus the power consumption of the PA stage, is driven by two parameters that are apparently independent. Indeed, as discussed in Chapter 3, in a classical wireless link the average transmit power has to be set in order to achieve the appropriate uplink quality. On the other hand, the PAPR is driven only by the amplitude modulation part of the transmit waveform. As a result, different strategies can be put in place to limit the unnecessary power consumption of a PA stage with regard to the variations of those two parameters while preserving the linearity of the device.

In light of this remark, we first focus on possible optimization while considering the variations in the average transmit power. For that purpose, we observe that this quantity remains deterministic as the transmitter necessarily knows at what power it is actually delivering the RF signal to the antenna. As a result, we can imagine using different static biasing for the PA corresponding to saturated output power values suited to various ranges of the transmit power. Practically speaking, most existing linear PAs go a step further and switch between different

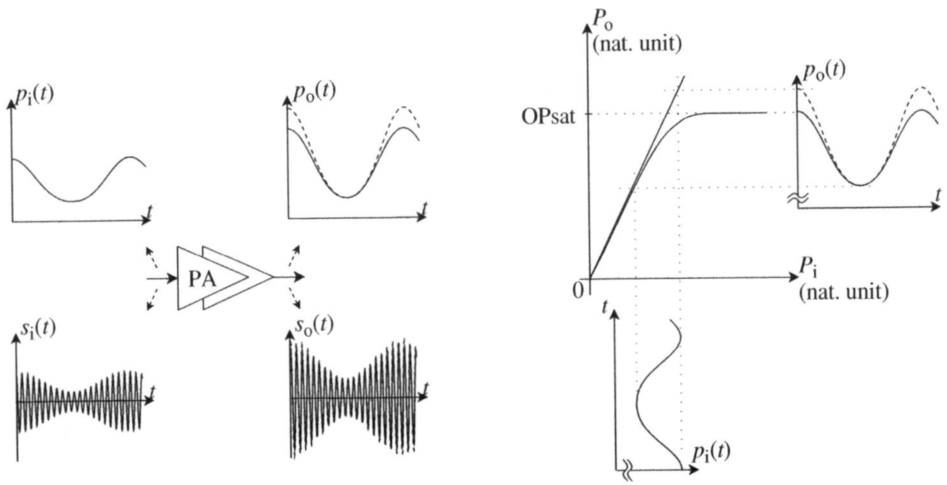

Figure 8.15 Impact of smooth RF compression in a linear PA – Having the Psat of a linear amplifier set too low compared to the instantaneous peak power of the RF modulated bandpass signal being processed leads to the smooth compression of this signal, and thus to its distortion. As a result, the effective output bandpass signal $s_o(t)$ (left, solid) is no longer a scaled version of the input signal of the form $Gs_i(t)$ (left, dashed). The same obviously holds for the instantaneous output power $p_o(t)$ vs. $G^2p_i(t)$. This behavior is explained by the shape of the power transfer function of the device (right).

active devices optimized for given output power DR on top of using different biasing points. From the system point of view, all thus behaves as if we had different physical PAs in parallel in the line-up, as illustrated in Figure 8.16. This kind of strategy in fact corresponds to what we considered in Chapter 7 when illustrating the derivation of performance budgets for a transmit line-up. And as illustrated in that chapter, we must anticipate some side effects due to this change in the PA operating point as various of its characteristics can be impacted on top of its linearity. This is for instance the case for the linear gain, but also for more indirect quantities such as the impedance seen by the RF electromagnetic wave that carries the information. This phenomenon can lead to a problematic carrier phase shift (see the discussion in Section 2.2.2). Depending on the magnitude of the phenomenon we are dealing with, we may need to consider a calibration of the corresponding parameters. This is often the price we pay for using this kind of strategy.

However, such a static biasing strategy does not change the fact that the Psat of the device must be driven by the instantaneous peak power of the transmit waveform, i.e. by its average power times its PAPR, in order to preserve its linear behavior. As a result, for a given average output power we necessarily get a non-negligible fraction of electrical power, proportional to the difference between the Psat and the instantaneous power of the waveform in practice, wasted as heat in the device, as illustrated in Figure 8.17(top). This problem is obviously maximized for high PAPR signals. Indeed, considering for instance an OFDM signal that has a Gaussian-like distribution as discussed in "OFDM" (Section 1.3.3), due to its PAPR of about 7 dB, we need to set the DC voltage supply of a linear device amplifying such signal at a

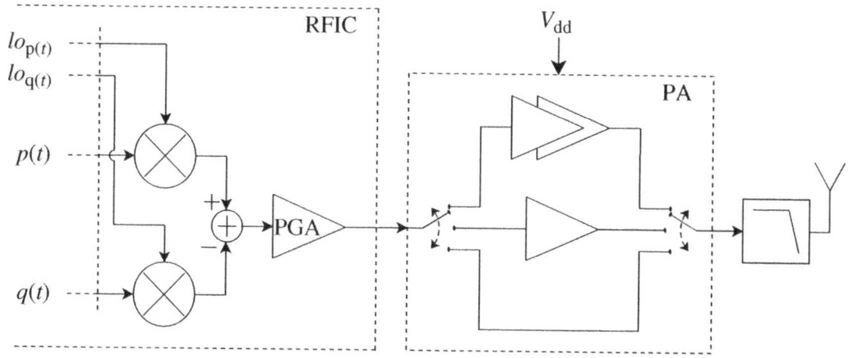

Figure 8.16 Switched gain PA – In order to improve the efficiency of PA stages, we can imagine using different optimized devices depending on the output power DR. However, the switching between these different devices, including their by-pass, can induce discontinuities in the characteristics of the bandpass RF signal being processed, for instance on its instantaneous phase.

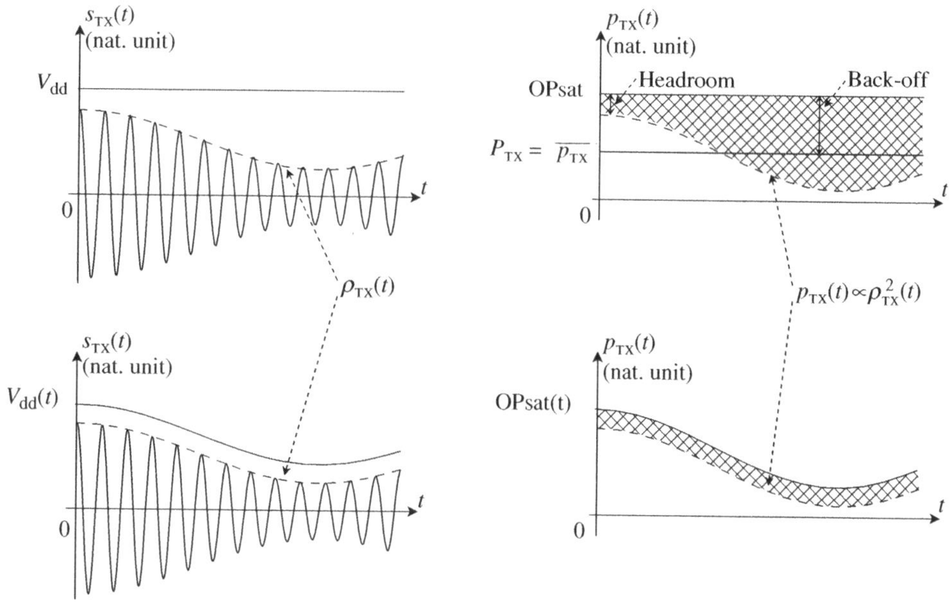

Figure 8.17 Dynamic biasing of a linear amplifier for power efficiency improvement – In order to avoid compression, the static biasing of a linear PA, as done for instance through a static V_{dd} supply voltage (top left), needs to correspond to a Psat, higher than the maximum instantaneous power $p_{TX}(t)$ of the modulated waveform it processes, as illustrated in Figure 8.15. This wastes a non-negligible fraction of electrical power in the amplifier when dealing with a high PAPR waveform (top right, hatched area). In contrast, the adaptation of the supply voltage to the instantaneous amplitude of the signal (bottom left) can minimize this waste (bottom right, hatched area) while keeping the device working in its linear area.

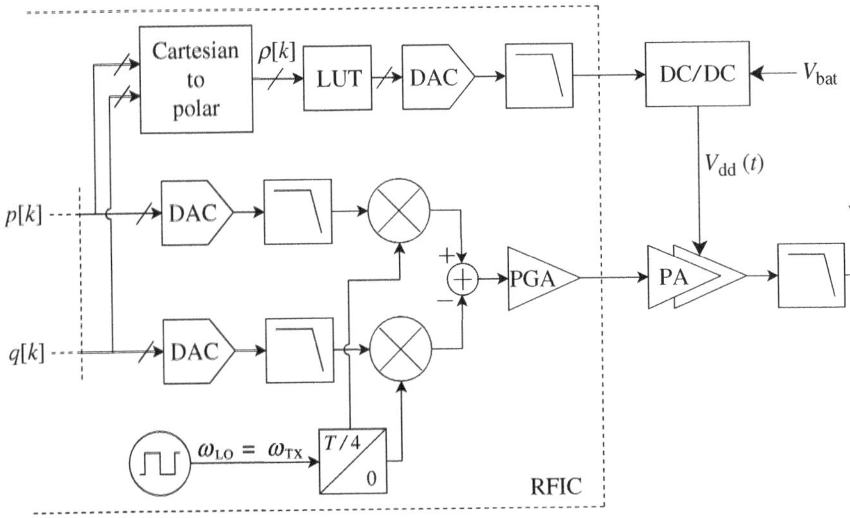

Figure 8.18 Envelope tracking for adaptive PA biasing – Using an efficient DC/DC converter with a sufficiently high bandwidth with regard to its input command port, we can expect to be able to supply a PA with a minimum instantaneous headroom with regard to the envelope of the signal being processed, as illustrated in Figure 8.17. For that purpose, given an analog control of the DC/DC converter, we need to successively reconstruct the instantaneous amplitude information from the P and Q samples of the modulating waveform, and convert those samples in the analog domain through the use of a DAC and a reconstruction filter. The presence of an additional predistortion stage for the supply command, through the use of a LUT for instance, can be assumed to linearize the PA transfer function in addition to improving the efficiency.

level that is about 2.2 times higher than would be required to pass a CW tone with the same average power. We can then expect the same increase in the current drawn from the supply by the device as would be required to deliver the same average power without having amplitude modulation.

In order to improve the situation, we can thus think of using an adaptive biasing that follows the envelope of the transmit signal. Practically speaking, this means a modulation of the voltage supply of the PA according to the variations of the modulated transmit signal envelope as illustrated in Figure 8.17. With such a system, known as envelope tracking (ET), we can expect a non-negligible power saving for high PAPR waveforms. However, there are various implementation limitations that necessarily reduce the gain that can be expected, or even make its implementation infeasible. We first observe that in order to vary the voltage supply with reasonable efficiency, we may need to use a DC/DC converter as illustrated in Figure 8.18. This was in fact already the case when considering different voltages in the static biasing of the PA. But, now considering a dynamic modulation of this biasing, we may be faced with issues as the device is classically not wideband with regard to its input command port. Obviously this has to be compared to the bandwidth of the envelope to track and thus needs to be refined on a case by case basis. However, we may be in for a disappointment as high bandwidth envelopes are for the most part associated with wideband waveforms that exhibit high PAPR.

In the same way, we can mention the sensitivity of the system to the delay mismatch between the modulated signal and the biasing command applied to the device. We need to guarantee that the two signals are sufficiently well aligned to avoid any compression of the modulated signal during the sharpest transitions of its instantaneous amplitude. We thus recover the same kind of constraint as in the polar architecture presented in Section 8.1.6 and reconsidered in "PA working in the truly nonlinear area" earlier in this section. Even if less critical in the present case as the overall complex modulated bandpass signal is already reconstructed, it can still be an issue as we are dealing with physical paths that are necessarily implemented in different ways for the modulated signal and for the voltage biasing. With RF/analog implementations, this means different variations in the delays along each path as a function of the transmitter environment, such as its temperature. As a result, even if we succeed in achieving a correct alignment in the delays at a given temperature for instance, it may be difficult to maintain this behavior when the operating conditions change. The basic solution for that would be to keep sufficient margins between the instantaneous Psat of the device and the instantaneous power of the signal, but obviously at the cost of a degradation in the power saving.

Finally, we mention an additional feature associated with the ET approach that can lead to a modification of the quality of the signal being amplified. This behavior depends on the characteristics of the PA and the strategy adopted. Changes in the supply voltage of some PAs can lead to variations in their intrinsic characteristics, for instance their gain. In the latter case, an ET controlled supply would result in gain variations according to the instantaneous amplitude of the signal being amplified. By definition, this would result in some AM-AM or AM-PM conversion effects, leading to additional distortions of the amplified signal as discussed in Chapter 5. However, relying on a characterization of the transfer function of the PA, we may be able to use this behavior to modulate the supply so as to compensate for this effect, or even to improve the intrinsic characteristics of the PA transfer function. This approach would require using a predistortion of the bias command, as illustrated in Figure 8.18. Obviously, this possibility is related to the effective transfer function of the PA and should be balanced with the efficiency degradation when modifying the supply law with respect to the optimal one for that purpose. This approach may not be appropriate for all devices.

PA Working in the Weak Nonlinear Area

In the previous section, we considered that in order to decrease the power consumption of the PA stage, the back-off, i.e. the ratio between the Psat of the device and the instantaneous power of the signal being processed and delivered to the load, should be minimized. In practice, this strategy leads to either static biasing optimized for given output power ranges or biasing that tracks the variations of the instantaneous power of the signal being processed. In these two cases the underlying assumption is that the back-off is sufficiently high at each instant t that the signal compression is avoided. This is obviously done in order to avoid the deterioration of the amplitude part of the modulation so that we can achieve the performance we expect.

With this approach, the best we can do at any given time is still to use sufficiently high power corresponding at least to the instantaneous peak power of the signal being processed in the PA stage plus a minimum headroom that allows us to minimize the compression. Even if the resulting peak power consumption is limited in time, if we track the variations of the signal envelope, it can still lead to a non-negligible average power consumption, for instance compared to the power consumption required to pass only the average transmit power. As a result, we can wonder if we cannot go a step further and reduce the back-off until we have a reasonable clipping of the waveform, and then compensate for the corresponding distortion induced on the modulation.

This is in fact the idea behind different strategies that try to use the PA, or more generally the RF parts of the transmit line-up, in the weak nonlinear area of their transfer function. There are different techniques for achieving this, which fall under the rubric of feedback systems, feedforward systems, and predistortion systems [67, 74, 77].

Feedback Systems

Let us begin with the feedback approach, or more precisely the negative feedback approach. We know that a negative feedback scheme can linearize a nonlinear device. However, this is obviously achieved at some cost in practice. Consider Figure 8.19, which depicts the application of this principle to our use case. This is obviously a simplified version of our problem as it represents a linearized model of our line-up. That said, it is accurate enough to enable us to understand the limitations associated with our application.

Let us write the closed loop transfer function for the complex envelopes of the bandpass signals of interest in our application, i.e. those corresponding to the sidebands centered around the transmit carrier angular frequency ω_{TX}. Thus, assuming that all the complex envelopes we are dealing with are defined as centered around this angular frequency, the complex envelope $\tilde{s}_o(t)$ of the output bandpass signal is equal to G times the complex envelope $\tilde{\epsilon}(t)$ of the bandpass error signal. So $\tilde{\epsilon}(t) = \tilde{s}_i(t) - G_{FB}\tilde{s}_o(t)$. As a result, we recover the classical form for the closed loop transfer function,

$$\tilde{s}_o(t) = \tilde{s}_i(t)\frac{G}{1 + GG_{FB}}. \tag{8.29}$$

Thus, given sufficient gain in the open loop transfer function, i.e. $|GG_{FB}| \gg 1$,

$$\tilde{s}_o(t) \approx \tilde{s}_i(t)\frac{1}{G_{FB}}. \tag{8.30}$$

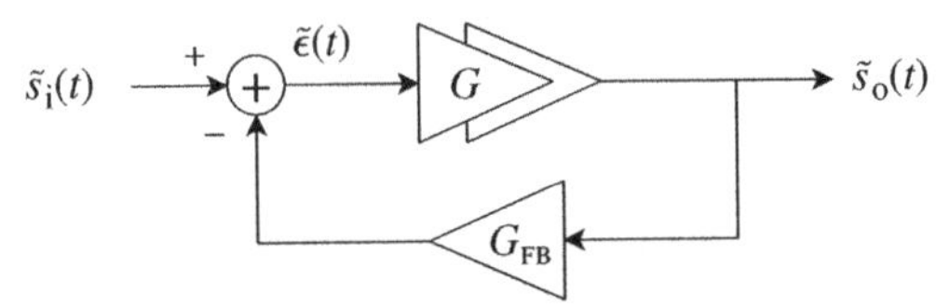

Figure 8.19 Negative feedback principle for the linearization of an RF amplifier – In order to analyze the impact of negative feedback on the transfer function of an amplifier that processes an RF bandpass signal, we can consider an equivalent lowpass model acting on complex envelopes, all assumed defined around the same carrier angular frequency. In the present case, we consider simple linear gains both in the direct and feedback paths, G and G_{FB}, respectively.

We thus have the remarkable result that the closed loop transfer function is independent of G. This result also holds for any nonlinearity. Furthermore, if it is independent of G it is obviously also independent of its potential variations, for instance when operating conditions such as the temperature or power supply drift. This is another important facet of this negative feedback approach. However, if the overall transfer function is indeed independent of G, it now depends on $1/G_{\text{FB}}$. And obviously, if we still want to achieve an overall amplification, we need to have $|G_{\text{FB}}| < 1$. Thus, given that the open loop transfer function needs to satisfy $|GG_{\text{FB}}| \gg 1$ at the same time, we are necessarily faced with a gain loss compared to what we achieved with G. This is a first drawback of this approach, as it is costly to achieve gain in the RF domain. In the same way, with G_{FB} also implemented in the RF domain, we may expect the same kind of limitations in its implementation in terms of linearity or sensitivity to the operating conditions as we face on the direct path. This may lead to a not so different behavior of the transfer function in the closed loop mode from that of the nonlinear RF amplifier alone. A certain skepticism as to the overall efficiency of this approach is justified when dealing with a physical RF implementation.

However, if the direct RF feedback approach seems difficult to achieve, nothing prevents us from considering the implementation of the feedback in the low frequency domain. This approach is appropriate as the information we want to preserve is carried by the modulating part only of our bandpass signal. This information being represented in the physical world through the corresponding lowpass modulating waveforms, our goal can thus be achieved through the comparison of the modulating waveforms representing the modulation of the bandpass signal as effectively recovered at the transmitter output with the theoretical ones. And given that a complex modulation can be physically represented either through a Cartesian or a polar decomposition of the corresponding complex envelope, we can derive the two corresponding implementations as illustrated in Figures 8.20 and 8.21. These obviously correspond to pure analog implementations. Thus, although nowadays we can consider more digitally oriented implementations, such historical implementations are more realistic compared to a pure RF feedback as gain and linearity are more easily achieved with low frequency analog devices. However, such a line-up could indeed be updated by taking advantage of the possibilities offered by the digital signal processing. This would then result in using receivers in the feedback path, including analog to digital conversions and comparison in the digital domain of the recovered modulating waveforms with those expected.

An initial drawback of this approach is that it requires a feedback path to be implemented and thus to be supplied. Obviously the power supply of an additional path depends on its implementation, but when talking about a full receive path, implemented in a coherent way to recover the modulating complex envelope, and potentially with analog to digital conversions to close the loop in the digital domain, there must be a non-negligible amount of additional current consumption. This has to be balanced with the potential gain on the transmit side.

Furthermore, such a solution is likely to be sensitive to one of the main issues that we have to confront in feedback systems, i.e. their stability. Fundamentally, this is because we are trying to correct the cause of an event after this event has occurred. In our case, this means that whereas we expect to compare the effective modulation recovered at the instant t on the RF signal with its theoretical value, in practice we compare it with the theoretical value of the modulating waveform at the time $t + \tau$. This is obviously due to the delay in the detection and feedback path, represented here by the quantity τ. Thus, the higher the rate of change of

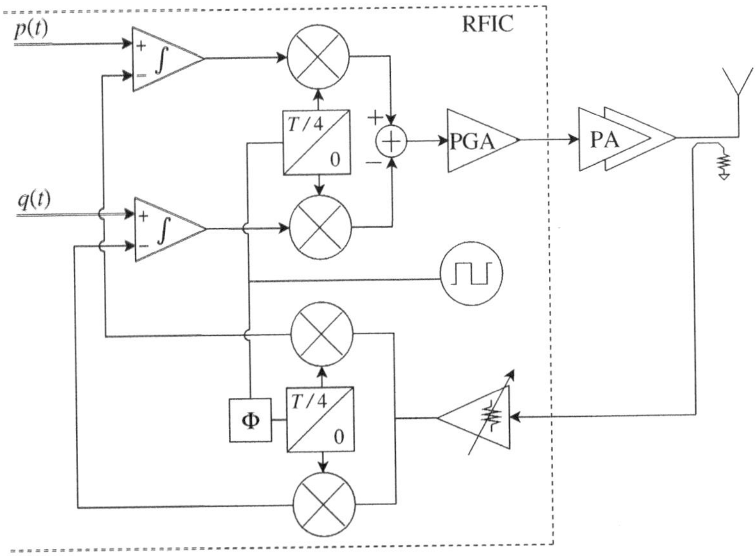

Figure 8.20 Cartesian feedback loop principle – As an alternative to the RF negative feedback approach, we can close the loop in the baseband domain where gain is more easily achieved. In this analog implementation perspective, this can be done by comparing the real and imaginary parts of the transmit complex envelope with the theoretical $p(t)$ and $q(t)$ modulating waveforms. It requires at least a complex demodulation with a phase offset adjustment on the LO signal to compensate for the carrier phase shift along the transmit and coupling path, and error amplifiers to integrate the P and Q error signals.

the modulation during this delay, the less likely we are a stable system as we are comparing apples and pears.

This problem is obviously important for our application as the attraction of using a linearization method to save power consumption increases with the PAPR of the transmit waveform. And in turn, high PAPRs are often encountered with high bandwidth modulated signals, as illustrated in Section 1.3.3. The negative feedback approach is thus likely to cause problems in the practical cases of interest, often involving high bandwidth signals, thus requiring the use of high bandwidth loops hard to stabilize. This issue is obviously even more critical in polar feedback loops due to the higher bandwidth of the instantaneous amplitude and phase components of a modulated bandpass signal compared to its Cartesian counterparts, as discussed in Section 8.1.6. It also needs to be borne in mind in loops involving a digital implementation due to the additional delays in the processing linked to the physical implementation of the logic cells. From the control theory point of view, such pure delay has deep impacts on the stability of the loop as it leads to terms of the form $e^{-j\omega\tau}$ in the open loop transfer function when expressed in the frequency domain. We then get a variation of the phase of the transfer function that increases linearly in frequency without exhibiting any cut-off frequency. This adds difficulties for the stabilization of the loop, especially when dealing with a high cut-off frequency.

Even in cases where a gain in current consumption is demonstrated despite the addition of the feedback path, the physical implementation of such negative feedback can be difficult

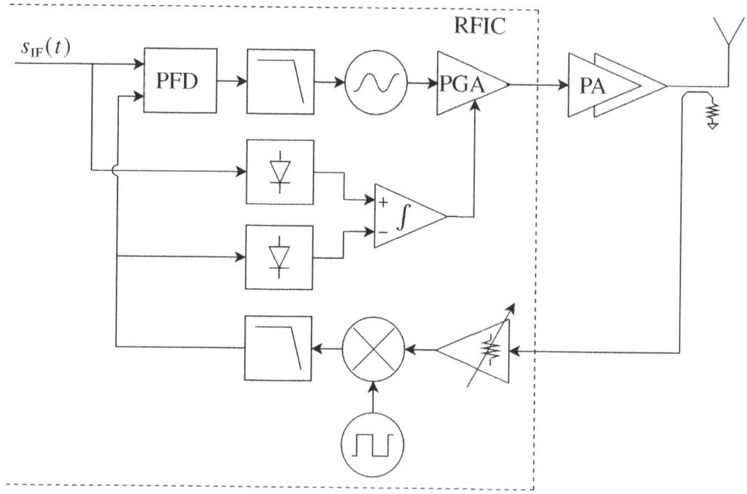

Figure 8.21 Polar feedback loop principle – As an alternative to the Cartesian feedback approach, we can consider a comparison between the instantaneous amplitude and phase of the transmitted modulated bandpass signal on the one hand and the modulus and argument of the theoretical modulating complex envelope on the other hand. In the analog implementation perspective, this can be done through the processing of the corresponding ideal and effective modulating waveforms recovered on equivalent IF bandpass modulated signals. It requires the generation of both the IF signal $s_{IF}(t)$ that carries the ideal modulation, and a copy of the effective transmitted signal downconverted toward this IF. The loop error signals can thus be derived using envelope detectors and a phase comparator.

due to the stability issue for waveforms of interest. This is one of the main motivations for considering the feedforward solution, and it is to this that we now turn.

Feedforward Systems

When a feedback system can be put in place, it theoretically allows very good performance to be achieved. Indeed, given both sufficient open loop gain and a sufficiently high loop bandwidth to detect all the signals of interest, we can expect a cancellation of the error between the output signal and the reference signal. Furthermore, this approach has the great advantage of giving quite stable results whatever the variations in the characteristics of the nonlinear device we are dealing with. This is of particular interest in our application where we consider the compensation of RF devices that often exhibit characteristics sensitive to the operating conditions, e.g. to temperature or power supply drifts.

However, when this cannot be done, mainly due to stability issues in practice, an alternative approach is to try using the error signal in a different manner. This is particularly obvious when reconsidering a feedback loop implemented in the pure RF world, as illustrated in Figure 8.19. In that case, the error signal is nothing more than a metric of the difference between the distorted bandpass signal $s_o(t)$ recovered at the output of the nonlinear device, and the theoretically expected signal $Gs_i(t)$. But, if we know the error on the output signal, all we presumably need to do is simply subtract it from this output signal to obtain the expected result.

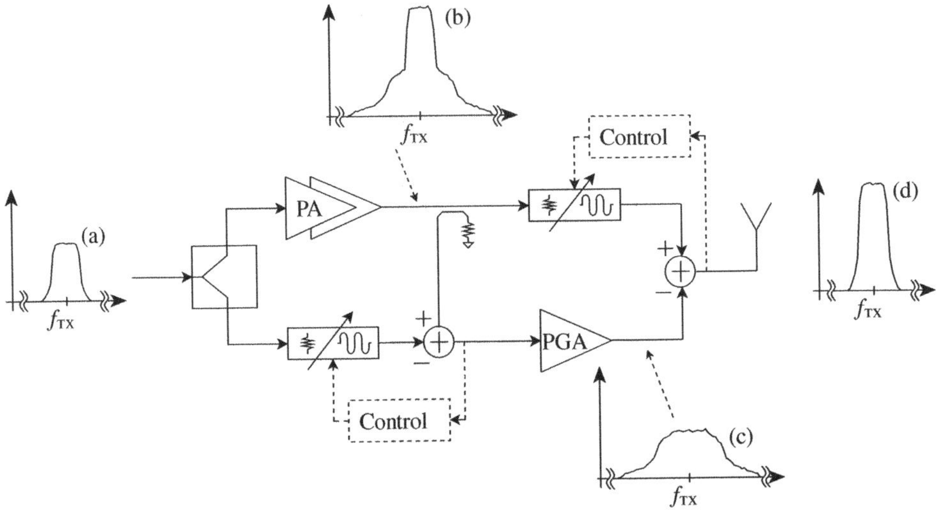

Figure 8.22 Feedforward principle for the linearization of an RF amplifier – In order to linearize a nonlinear RF amplifier, we can use a first loop that estimates the distortion component (c) recovered at the output of this device by subtracting the input signal (a) from the one recovered at its output (b). We can then subtract from this signal the estimated distortion term in order to achieve the expected amplified signal (d). As the correct estimation and cancellation of the distortion term rely on RF/analog subtractions, the performance is limited by the balancing between the paths. Feedback control loops acting on variable gains and delays may be required in order to track and compensate for temperature or supply drifts for instance (dotted).

This simple idea leads to the concept of the feedforward loop illustrated in Figure 8.22 [67, 77]. In practice the error signal is constructed through the detection of the distorted signal recovered at the output of the nonlinear device and by subtracting the theoretical expected signal from it. We observe that in the proposed implementation this subtraction is performed using a small fraction of the distorted signal. One reason for that comes from the desire to preserve the overall efficiency of the system as far as possible by not losing too much RF power through an attenuation of the amplified signal. But another is the desire to avoid amplification of the reference theoretical signal before performing the subtraction. This allows any potential degradation of this reference signal to be minimized, and an exact image of the distortion term to be recovered after subtracting the distorted signal. This reconstructed distortion term needs to be correctly amplified so that its level matches the distortion term present on the main high power path. But once again, as the power of the distortion term is at a lower level than the modulated bandpass signal, the error amplifier embedded in the secondary path of the second loop should cause negligible additional degradations with respect to those induced by the main nonlinear PA. When this holds, the scaled distortion term can be further subtracted from the signal on the main path to recover a clean output signal.

However, even if this scheme seems to work on paper, we can easily anticipate its limitations in practical implementations. Indeed, the generation of the reference distortion and its further subtraction from the high power signal obviously rely on the possibility of achieving good subtractions between RF signals in order to achieve the cancellation of unwanted terms. This means both an appropriate amplitude and a good delay balance between the RF paths. In order to fix the orders of magnitude, we can illustrate the associated constraints by considering the cancellation of an unwanted bandpass signal $s_{\mathrm{u}}(t)$ through its subtraction with its estimated feedforward term $s_{\mathrm{ff}}(t)$ that is nothing more than a scaled and delayed version of it. This means that we assume that $s_{\mathrm{ff}}(t)$ is of the form $gs_{\mathrm{u}}(t-\tau)$, where g and τ are respectively the gain and delay imbalance between the paths. We can use the complex envelopes of these terms for the derivations. Assuming that they are both defined as centered around the same angular frequency, ω_{TX}, we can write for instance that

$$\tilde{s}_{\mathrm{u}}(t) = \rho_{\mathrm{u}}(t)e^{j\phi_{\mathrm{u}}(t)}, \tag{8.31a}$$

$$\tilde{s}_{\mathrm{ff}}(t) = g\rho_{\mathrm{u}}(t-\tau)e^{j[-\omega_{\mathrm{TX}}\tau+\phi_{\mathrm{u}}(t-\tau)]}. \tag{8.31b}$$

As a result, the ratio of the complex envelope $\tilde{s}_{\mathrm{e}}(t)$ of the residual bandpass error signal to that of the initial unwanted signal can be written, at least for $\rho_{\mathrm{u}}(t) \neq 0$, as

$$\frac{\tilde{s}_{\mathrm{e}}(t)}{\tilde{s}_{\mathrm{u}}(t)} = \frac{\tilde{s}_{\mathrm{u}}(t) - \tilde{s}_{\mathrm{ff}}(t)}{\tilde{s}_{\mathrm{u}}(t)} = 1 - g\frac{\rho_{\mathrm{u}}(t-\tau)}{\rho_{\mathrm{u}}(t)}e^{j[-\omega_{\mathrm{TX}}\tau+\phi_{\mathrm{u}}(t-\tau)-\phi_{\mathrm{u}}(t)]}. \tag{8.32}$$

Assuming that the delay mismatch τ remains low enough, we can approximate the terms $\phi_{\mathrm{u}}(t-\tau)$ and $\rho_{\mathrm{u}}(t-\tau)$ through Taylor series expansions up to first order:

$$\phi_{\mathrm{u}}(t-\tau) \approx \phi_{\mathrm{u}}(t) - \tau\dot{\phi}_{\mathrm{u}}(t), \tag{8.33a}$$

$$\rho_{\mathrm{u}}(t-\tau) \approx \rho_{\mathrm{u}}(t) - \tau\dot{\rho}_{\mathrm{u}}(t), \tag{8.33b}$$

with $\dot{\phi}_{\mathrm{u}}(t)$ the time domain derivative of $\phi_{\mathrm{u}}(t)$, and $\dot{\rho}_{\mathrm{u}}(t)$ that of $\rho_{\mathrm{u}}(t)$. We can thus write equation (8.32) as

$$\frac{\tilde{s}_{\mathrm{e}}(t)}{\tilde{s}_{\mathrm{u}}(t)} = 1 - g\left(1 - \tau\frac{\dot{\rho}_{\mathrm{u}}(t)}{\rho_{\mathrm{u}}(t)}\right)e^{-j\tau(\omega_{\mathrm{TX}}+\dot{\phi}_{\mathrm{u}}(t))}, \tag{8.34}$$

Now assume that we are dealing with a sufficiently narrowband modulation. Here, "sufficiently narrowband" would mean that we can assume that the instantaneous angular frequency variations linked to the modulation are much lower than the carrier angular frequency. This common configuration corresponds to $|\dot{\phi}_{\mathrm{u}}(t)| \ll \omega_{\mathrm{TX}}$. Then, assume also that τ is low enough so that the variations of the instantaneous amplitude of $s_{\mathrm{u}}(t)$ during that time are negligible, i.e. that $\tau\dot{\rho}_{\mathrm{u}}(t) \ll \rho_{\mathrm{u}}(t)$. This can be seen as a reasonable assumption at least during the time where $\rho_{\mathrm{u}}(t)$ has a non-negligible magnitude, i.e. when the distortion term $s_{\mathrm{u}}(t)$ has a non-negligible level and that we effectively want to cancel it. We thus get in situations of interest that

$$\frac{\tilde{s}_{\mathrm{e}}(t)}{\tilde{s}_{\mathrm{u}}(t)} \approx 1 - ge^{-j\omega_{\mathrm{TX}}\tau}. \tag{8.35}$$

The magnitude of this term then directly gives the suppression capabilities of the distortion in the presence of gain and delay imbalance between the direct and feedforward paths in practical situations of interest. As illustrated in Figure 8.23, the performance of the system degrades quickly in the face of mismatch values of a few tens of decibels and a few degrees,

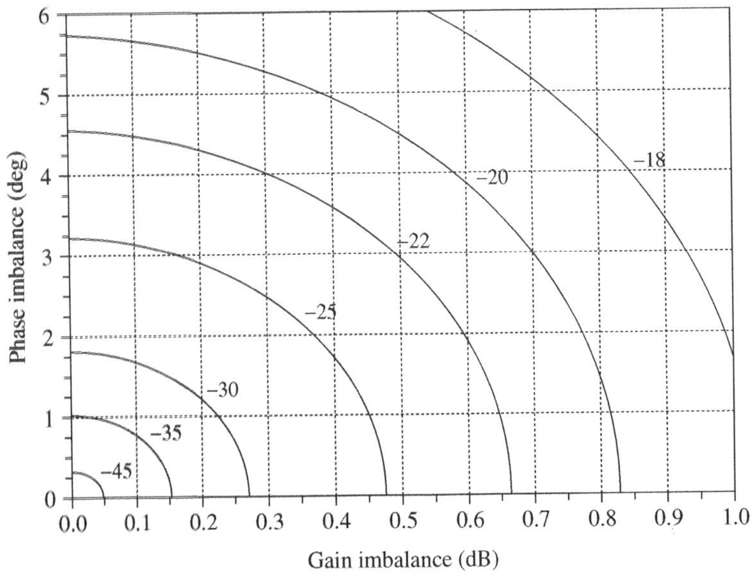

Figure 8.23 Cancellation capability of a narrowband modulated bandpass signal in the presence of gain and phase imbalance – The ability to cancel a narrowband modulated bandpass signal by subtraction of a copy of it is limited by the imbalance between the two signals. Practically speaking, the magnitude of the residual signal relative to the initial signal can be estimated through the modulus of equation (8.35), as a function of both the gain imbalance g and the phase imbalance between the bandpass signals being subtracted. Under the narrowband modulated assumption, this last term has to be understood as the phase rotation of the carrier, $\omega_{TX}\tau$, when a delay mismatch τ exists between the paths.

which are common values in practical RF implementations. Moreover, even if we succeed in achieving a good matching in one configuration of the line-up, it can be difficult to maintain the performance when the operating conditions are changing, for instance when the temperature or the power supply drifts. As a result, we may consider the addition of feedback loops in order to track these variations and scale the parameters of the line-up, as illustrated by the dotted part of Figure 8.22. However, in the present case we observe that we are talking about tracking slow phenomena with very weak time constants. This is therefore a very different situation compared to the feedback schemes considered in the previous section, and results in no particular instability issues in the present case.

Predistortion Systems

The feedforward technique is an alternative to the negative feedback scheme when the latter cannot be implemented. However, despite its poorer performance, the implementation cost of the feedforward system, as depicted in Figure 8.22, is costly as it requires using at least two loops implemented in the RF world, including at least one additional active amplifier. Perhaps another alternative to the negative feedback could be derived with fewer implementation constraints.

Looking back at Figure 8.19, we see that having a bandpass error signal $\epsilon(t)$ that reduces to 0 once the convergence has occurred in the loop means in practice that the bandpass feedback signal $G_{FB}s_o(t)$ that is subtracted from the input signal$s_i(t)$ makes the signal entering the

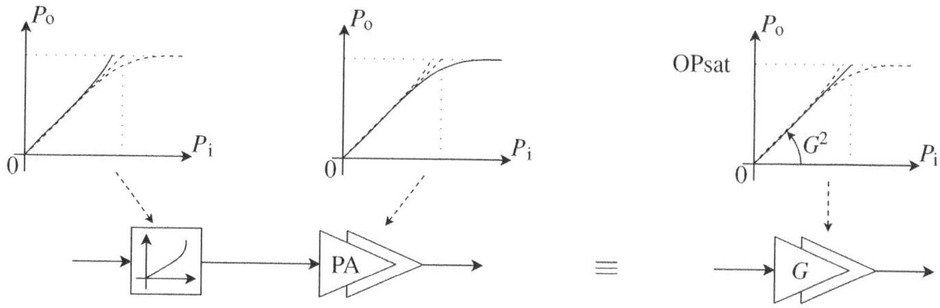

Figure 8.24 Predistortion principle for the linearization of an RF amplifier – In order to linearize a nonlinear RF amplifier we can first process the modulated bandpass signal using a block that exhibits a transfer function (left, solid) that is the inverse of that of the amplifier (center, solid). This results in an overall linear transfer function (right, solid). As seen from the corresponding transfer functions, in this predistortion scheme, as for any other linearization technique, the amplifier cannot deliver an instantaneous RF power higher than its OPsat. Preserving the PAPR of the signal being amplified using a linearization scheme thus means that the average output power may be slightly lower than would be achieved when accepting a smooth compression of the signal.

nonlinear device such that the distortion term that would be generated if $s_i(t)$ had experienced the nonlinearity itself is now canceled. All thus behaves as if the input signal $s_i(t)$ were predistorted by the subtraction of $G_{FB}s_o(t)$ so that the transfer function of the PA now looks linear with respect to the input signal $s_i(t)$.

An alternative to the feedback scheme is thus to perform the predistortion of $s_i(t)$ by going through a nonlinear device that exhibits the inverse transfer function of that of the PA. As a result, by cascading the two transfer functions we obtain the overall linear system illustrated in Figure 8.24.

Such predistortion need not necessarily be performed in the RF world, even if it effectively was in early implementations. It can be implemented as a processing of the baseband lowpass modulating waveform, thus resulting in the possibility of a pure digital implementation taking advantage of modern silicon technologies. This is obviously of interest in terms of integration compared to the feedforward technique, which is mainly an RF approach. However, despite this superiority, we can expect the same kinds of limitations as in the former approach as we are still dealing with an open loop system that cannot track the potential variations of the PA transfer function as such. This means that even if we succeed in the derivation of an exact inverse of the effective PA transfer function in given operating conditions, we may fail when the characteristics of the PA drift. However, as the rate of change of those phenomena is very slow in respect of the modulation time constant for instance, we can still imagine the same approach as in the feedforward case, i.e. performing some periodic updates of the predistortion block to track for those variations. Some examples for this kind of updating and the associated algorithms are further discussed in Section 9.1.4.

PA Working in the Truly Nonlinear Area

Let us now conclude our discussion of the efficiency of transmitters by considering the use of a PA working in a saturated mode, i.e. behaving as a hard limiter with regard to the signal being amplified, from the signal processing point of view. The attraction of such a device obviously lies in the theoretical efficiencies that can be achieved by PAs belonging to non-sinusoidal classes [74]. But, as discussed earlier, the transfer function of such a device necessarily destroys the amplitude modulation of the bandpass signal it processes. This is the main reason why we have focused so far on systems using PAs working in their linear area.

However, although it may seem strange on the face of it, some approaches effectively allow us to use saturated devices even when dealing with amplitude modulated waveforms. But as we know for sure that such devices destroy the information carried by the amplitude modulation, the systems we are looking for will presumably rely on a decomposition of the modulated bandpass signal in order to feed the saturated PA by a constant amplitude signal. And we can already make the link with the polar architecture presented in Section 8.1.6. However, in that section we considered the reconstruction of the complex modulated bandpass signal through the multiplication of the phase/frequency modulated bandpass signal by the amplitude part prior to the PA, as illustrated in Figure 8.10. But nothing prevents us from doing so after the PA, so that this device processes a constant amplitude bandpass signal. In fact, this idea corresponds to the historical approach known as envelope elimination and restoration (EER) [78]. However, in its original form the implementation was somewhat different in practice. For instance the sense of the envelope information to be further applied to the constant amplitude signal was performed in the RF world using an envelope detection. This implies in practice a need to have the complex modulated bandpass signal generated in a classical way before going through its decomposition in terms of an amplitude information on the one hand and a phase/frequency only modulated signal on the other hand. In a modern approach, this can advantageously be replaced by the generation of the amplitude information directly in the digital world by processing the P and Q samples, and of the phase/frequency modulated signal by a direct PLL modulator as considered in the polar architecture approach. In the same way, in the original implementation, the instantaneous amplitude restoration was performed through the use of a variable gain device placed after the high efficiency PA stage. But there is nothing to stop us considering the application of this amplitude information directly on the supply of the PA when possible. This approach can be seen as a limit case of the ET principle discussed previously, but now with the PA operating in its saturated area and with the Psat of the device exactly tracking the variations of the instantaneous power of the waveform as required by its amplitude modulation part. However, whatever the approach, the delay alignment requirement between the amplitude and phase/frequency part of the modulation may lead to problematic implementations in the RF world, mainly due to the variations linked to the changes in the environment, as discussed in Section 8.1.6. In addition, the preservation of the quality of the modulation also requires using a sufficiently linear device with regard to the amplitude modulation part for its application on the phase/frequency modulated signal. This can be difficult to achieve in the RF world and thus problematic when dealing with modulating schemes that exhibit a high instantaneous amplitude DR.

Another interesting approach from the signal processing point of view is the outphasing modulation principle [79]. Practically speaking, this scheme, originally referred to as linear amplification using nonlinear components (LINC), relies on the possibility of decomposing any complex modulated bandpass signal as the sum of two constant amplitude ones. This can be understood by simply considering the trigonometric identity

$$2\cos(\theta_1)\cos(\theta_2) = \cos(\theta_1 + \theta_2) + \cos(\theta_1 - \theta_2). \tag{8.36}$$

If we succeed in decomposing a general complex modulated bandpass signal such as $s(t) = \rho(t)\cos(\omega t + \phi(t))$ in the form corresponding to the left-hand side of this equation, then we automatically get the required decomposition in accordance with the right-hand side. Thus, let us simply set

$$2\rho\cos(\theta_1(t)) = \rho(t), \tag{8.37a}$$

$$\cos(\theta_2(t)) = \cos(\omega t + \phi(t)), \tag{8.37b}$$

where ρ is a constant that reflects the maximum instantaneous amplitude of the modulated bandpass signal $s(t)$. For the sake of simplicity, we can then assume that

$$2\rho = \max_t\{\rho(t)\} \stackrel{\Delta}{=} \rho_{\max}. \tag{8.38}$$

With this definition, we simply need to select $\theta_1(t)$ and $\theta_2(t)$ such that

$$\theta_1(t) = \arccos\left(\frac{\rho(t)}{\rho_{\max}}\right), \tag{8.39a}$$

$$\theta_2(t) = \omega t + \phi(t). \tag{8.39b}$$

Using these expressions in equation (8.36), we finally achieve the decomposition of the complex modulated bandpass signal $s(t) = \rho(t)\cos(\omega t + \phi(t))$ as the sum of two constant amplitude bandpass signals,

$$s_1(t) = \frac{\rho_{\max}}{2}\cos\left[\omega t + \phi(t) + \arccos\left(\frac{\rho(t)}{\rho_{\max}}\right)\right], \tag{8.40a}$$

$$s_2(t) = \frac{\rho_{\max}}{2}\cos\left[\omega t + \phi(t) - \arccos\left(\frac{\rho(t)}{\rho_{\max}}\right)\right]. \tag{8.40b}$$

These expressions explain the terminology of outphasing. Moreover, the instantaneous phase difference reduces to $2\theta_1(t)$ with $\theta_1(t)$ given by equation (8.39a). And given that the complex envelopes are defined in Chapter 1 such that $\rho(t) \geq 0$, $\theta_1(t)$ thus satisfies

$$0 \leq \theta_1(t) \leq \arccos\left(\frac{\rho_{\min}}{\rho_{\max}}\right) \tag{8.41}$$

with $\rho_{\min} = \min_t\{\rho(t)\}$, i.e.

$$0 \leq \theta_1(t) \leq \frac{\pi}{2}. \tag{8.42}$$

We thus get that the maximum relative instantaneous phase offset between $s_1(t)$ and $s_2(t)$ is π rad.

Thus, if we succeed in the generation of $s_1(t)$ and $s_2(t)$ by adequate phase shifts from a general complex modulated bandpass signal $s(t)$, we can target the use of high performance

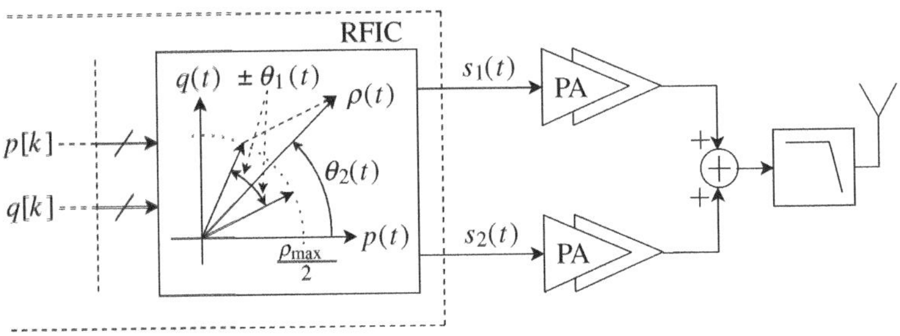

Figure 8.25 Outphasing modulation principle – The decomposition of a complex modulated bandpass signal $s(t)$ as the sum of two constant amplitude signals, $s_1(t)$ and $s_2(t)$, allows us to use two highly nonlinear PAs, theoretically leading to good efficiency overall. This decomposition can be achieved through the setting of a suitable instantaneous phase offset $\pm\theta_1(t)$, with $\theta_1(t)$ given by equation (8.39a), between $s_1(t)$ and $s(t)$ on the one hand and $s_2(t)$ and $s(t)$ on the other hand.

nonlinear PA to amplify them. It then remains to reconstruct the expected amplitude modulated signal through a simple RF summation, thus leading to a line-up as shown in Figure 8.25. However, despite its apparent simplicity from the signal processing point of view, the physical implementation of such a system can face various issues.

First, the generation of the two outphased signals can be difficult at gigahertz frequencies, particularly when employing a historical pure RF/analog approach. However, in a modern implementation we can consider the generation of $s_1(t)$ and $s_2(t)$ through the upconversion of their complex envelopes $\tilde{s}_1(t)$ and $\tilde{s}_2(t)$ that can be derived in a pure digital way from that of $s(t)$, $\tilde{s}(t) = p(t) + jq(t)$. But even then the required processing is not so straightforward as we need to recover the magnitude $\rho(t)$ of $\tilde{s}(t)$ before applying the trigonometric functions to derive first $\theta_1(t)$, using equation (8.39a), and then $\tilde{s}_1(t)$ and $\tilde{s}_2(t)$. This may require efficient algorithms for the computation of the corresponding functions, such as CORDIC (Section 9.1.6). It then remains to perform the frequency upconversion of $\tilde{s}_1(t)$ and $\tilde{s}_2(t)$, thus basically using two transmit line-ups even with an LO synthesizer in common. This already limits the potential gain in the overall power consumption of this approach on top of the implementation cost in terms of die area.

As an alternative approach, we can use the fact that $s_1(t)$ and $s_2(t)$ are constant amplitude bandpass signals to generate them through the direct modulation of two RF PLLs according to the discussion in Section 8.1.5. This is known as the combined analog locked loop universal modulator (CALLUM) [80]. But here again we face a non-negligible additional implementation cost and power consumption in the generation of $s_1(t)$ and $s_2(t)$ due to the presence of two RF synthesizers. Furthermore, given that the two RF signals standing in the RF oscillators have different instantaneous frequencies, we may anticipate potential pulling issues in integrated solutions due to the injection locking at the oscillator stage, as discussed in Section 8.1.1.

Whatever the architecture chosen for the generation of the constant amplitude bandpass signals $s_1(t)$ and $s_2(t)$, a non-negligible amount of additional hardware cost and thus of power consumption for this part of the transceiver needs to be taken into account. This must obviously be balanced with the gain expected at the PA stage. However, the main issue with the outphasing approach is the variable load seen by the two saturated PAs used to amplify

$s_1(t)$ and $s_2(t)$. Indeed, when these constant amplitude signals are physically implemented as voltage waves for instance, we would expect a physical design that leads to an RF current with a constant instantaneous amplitude flowing from each PA, corresponding to the constant load that maximizes their efficiency. But, due to the variable instantaneous phase difference between the two voltages delivered by the two PA devices, we get that the current flowing through the load connected between the nodes that sense those voltages for performing their summation is necessarily a function of this phase offset. Each PA stage thus sees a variable load as a function of this phase offset. This behavior structurally leads to difficulties in the maximization of the overall efficiency of such a system [67, 74, 81]. Thus, there are many difficulties associated with this approach in terms of physical implementation.

8.2 Receivers

Let us now focus on the receive side. As was the case for transmitters, a receiver can be more sensitive to one of the parameters of its constituent blocks, depending on its architecture. This can classically be linked to its frequency planning, which can cause the wanted signal to be superposed or not with some of the perturbations generated through the limitations in the physical implementation of the constituent blocks; or to the associated signal processing itself, whose physical implementation can be harder to achieve depending on the constraints driven by the selected architecture. Thus, depending on the relative importance of the parameter to the final performance we want to achieve, an architecture can be more or less suitable for a given application.

We illustrate this point by reviewing some of the classical receive architectures, proceeding in the same didactic manner as for transmitter architectures. Thus, in reviewing the limitations associated with a given architecture, we try to naturally introduce another architecture which should theoretically be able to overcome them, albeit often at the cost of alternative limitations. To start this process, we consider the simplest architecture we can think of to demodulate a complex modulated bandpass signal, the direct conversion architecture, already extensively discussed in Chapter 7 from the budget perspective.

8.2.1 Direct Conversion Receiver

The direct conversion line-up, also referred to as homodyne or ZIF, is widely used in practice. This is mainly due to the simplicity of its structure and thus to the associated potential high level of integration that makes it suitable for low cost solutions. However, there are performance limitations that are intrinsic to the line-up.

Consider the line-up shown in Figure 8.26. There is obviously a deep symmetry with the direct conversion transmit architecture discussed in Section 8.1.1. We might therefore expect the same root causes for the limitations in the performance of the line-up. Recalling the discussion in that section, we can anticipate that the complex mixing stage is a critical block in this line-up. This is, for instance, the case regarding the image rejection problem which can be more problematic than in other architectures due to the magnitude of the RF impairments between the P and Q paths of the line-up when considering a device working directly at the RF frequency. However, despite this kind of similarity, there are obviously major differences

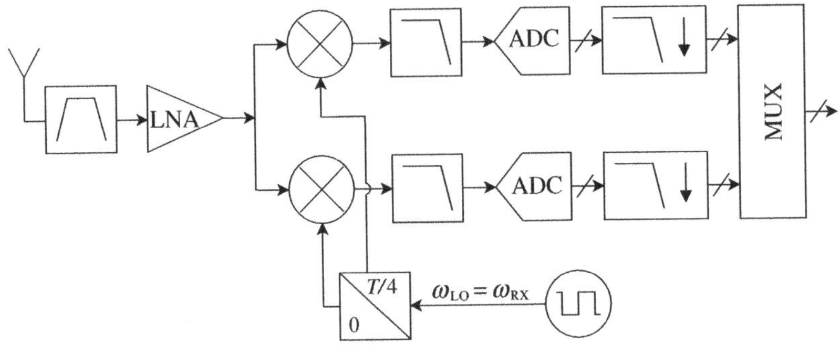

Figure 8.26 Direct conversion receive architecture – The blocks embedded in a typical direct conversion receive line-up are driven by the minimum set of signal processing functions required to demodulate a complex modulated RF bandpass signal, to comply with the electromagnetism theory, or to resist the wireless environment. We also need to consider the impact of the limitations in the physical implementation of those functions. We thus classically need to use a passive RF filter, an LNA stage for the noise performance of the line-up, a complex downmixer driven by the quadrature LO signals, an analog filter acting as an anti-aliasing filter and to limit the DR of the unwanted signals, the analog to digital stages and the associated digital signal processing, i.e. mainly digital filtering and decimation.

we need to keep in mind in the present case. The signals we are dealing with at the input of a receiver are expected to have a much lower power than is encountered on the transmit side. As a result, with special care in the physical implementation of the line-up in order to minimize the RF coupling, the potential pulling issue discussed extensively for integrated direct conversion transmit architecture can be avoided quite easily on the receive side. This is obviously a great advantage that allows us to use a simple frequency planning scheme such as running the oscillator at an integer multiple of the LO frequency, thus leading to a simple and efficient implementation of the direct conversion receiver. However, even if such minimization of the RF coupling can prevent the pulling issue, it can still lead to the presence of a LO leakage recovered at the input of the receive line-up, as discussed in Section 6.5 and illustrated in Figure 6.44. This means that when experiencing an impedance mismatch resulting in a reflected wave back toward the LNA input, we can be faced with the generation of a DC component due to the LO self-mixing. Practically speaking, this additional DC component can be handled by the DC cancellation algorithms classically associated with such receive line-ups, as discussed in Section 9.2.2. However, this CW LO leakage can also be cross-modulated by any other strong amplitude modulated blocking signals present at the receiver input according to the mechanism discussed in "Cross-modulation due to odd order nonlinearity" (Section 5.1.3). This is in particular the case for full-duplex systems due to the transmit signal leakage when the latter is amplitude modulated. In that case, it is no longer a simple DC offset that is recovered at the downmixer output, but also an additional noise component that needs to be considered as a limiting factor in the receiver performance.

We need to keep in mind another major difference with the direct conversion transmit line-up, and that is the fact that, thanks to the natural directivity of the active devices it embeds, a transmitter processes the wanted signal along its line-up almost exclusively. In contrast, a receiver has to cope with all the electromagnetic environment collected by the antenna, finally leading to the presence of additional blocking signals in addition to the wanted signal. This leads to the necessity to embed a deep filtering capability in a receiver in order to cancel those unwanted components. On top of that, we recall that dealing with in-band blocking signals that lie close to the wanted signal in the frequency domain, prior to having performed the channel selection we need to rely on a channel filter of a quality that can hardly be achieved in the RF world, as discussed in Section 7.3.3. Even if this statement needs to be qualified with regard to the possibility of implementing such a tunable RF filter in an active way, it explains the classical structure of the direct conversion line-up of Figure 8.26. But having the channel filtering set in baseband leads to potential limitations. More precisely, having no in-band filtering prior to setting the wanted signal in baseband results in two kinds of problems to be addressed at a system level.

First, there is necessarily an upper bound on the gain that we can set in the early RF stages of the receiver, i.e. in the LNA. This is obviously required in order to avoid any compression in the line-up due to the presence of the strong blockers. However, recalling the discussion in Section 7.3.2 and Figure 7.21, this statement should be qualified when reconsidering the physical nature of the waves carrying the information between the LNA output and the baseband channel filter. But as any RF/analog physical implementation can only be an approximation of the expected signal processing, the impedances we are dealing with are necessarily not the ideal infinite or null quantities we would expect while processing pure voltage or pure current waves, but rather something in between those two extremes. As a result, a voltage swing necessarily exists and thus there is a corresponding limitation in the maximum gain that can be set on the LNA stage in order to avoid compression when using a limited voltage supply. Such behavior can obviously result in various degradations of the performance of the line-up. We can already anticipate a limitation in the overall noise figure of the line-up (see Section 4.2.4). However, this limitation in the LNA gain is linked only to the impossibility of having an efficient RF channel filtering, which holds for many integrated receiver architectures. But in the present direct conversion case, this behavior needs to be examined while considering the joint impact of the frequency planning associated with this architecture. Indeed, having little amplification prior to the downconversion directly toward baseband makes the wanted signal more easily corrupted by the unwanted components present in the low frequency domain at the output of the mixing stage. This is for instance the case for the DC offset that is inherently present in analog implementations as discussed in Section 6.5.2. More precisely, given that such DC is linked mainly to the physical design of the baseband blocks, it thus necessarily behaves as an additive quantity with respect to the signal being processed in the line-up. As a result, for a given amount of offset in the baseband part of the line-up, the lower the gain experienced by the wanted signal prior to its downconversion, the higher the DC to wanted signal ratio at the receiver output. Obviously, some DC cancellation algorithms can be considered in order to solve this issue, as discussed above. However, other potential unwanted components lie in the low frequency domain even if not exactly at DC. This is for instance the case for the flicker noise component when dealing with a CMOS implementation. In that case, only a sufficiently strong amplification of the wanted signal prior to the downconversion toward baseband can prevent an unacceptable SNR degradation.

Second, all the active devices that act before the baseband filter necessarily cope with all the blocking signals present at the input of the receiver. This can result in particular requirements in terms of linearity or of spurious rejection for the corresponding blocks to limit the degradation of the wanted signal. However, we remark that concerning the first

LNA stage, we cannot expect additional requirements in the present direct conversion case compared to other architectures. The LNA stage is the first active device of the line-up in any case, thus having to cope with the blockers flowing from the passive FE regardless of the forthcoming structure of the receiver. This is the same situation for the mixing stage concerning the potential harmonic LO mixing problem discussed in Section 6.4. Indeed, having the first mixing stage of a receiver implemented as a chopper necessarily leads to spurious responses regardless of the frequency planning of the architecture. Only the frequency of interest of those responses is affected by this parameter. However, the situation is obviously different for the linearity aspect of the downconverter. The particular frequency planning associated with the direct conversion architecture makes the line-up sensitive to the IP2 of the mixing stage through the AM-demodulation phenomenon introduced in "AM-demodulation due to even order nonlinearity" (Section 5.1.3). More precisely, given that the wanted signal is directly recovered as centered around DC at the mixing stage output, it is necessarily superposed with the AM-demodulated components of all the signals present at the downmixing stage input. And having no efficient filtering prior to this stage, we see that the potential strong blockers we can cope with at the input of the mixing stage could lead to potential degradations of the signal to noise power ratio. Moreover, by inspection of equation (5.153) we see that the power of the generated noise term is proportional to the square of the blocker power. We thus get the unpleasant surprise that, in the absence of any RF filtering acting on the blocking signal, the AM-demodulated signal over the wanted signal power ratio is proportional to the gain present in the line-up prior to the downmixer stage. This means that, given an IP2 performance of the mixing stage, the higher the LNA gain, the higher the SNR degradation due to this AM-demodulation phenomenon. We thus recover an antagonist requirement in respect of the desire to put the maximum possible gain in the LNA stage in order to achieve the best possible additive noise performance of the line-up. This behavior can thus make the optimization of the line-up not so straightforward.

Thus, carrying out the frequency conversion from the carrier frequency down to baseband in a single stage leads to practical limitations. We thus reach the same conclusion as for the direct conversion transmit architecture even if the mechanisms involved in the performance limitation due to the frequency planning of those line-ups are not the same. As a result, we can surmise that a two-step downconversion would allow us to overcome those limitations, leading to better potential performance. This naturally leads to the heterodyne receive architecture discussed next.

8.2.2 *Heterodyne Receiver*

As highlighted at the end of the previous section, the limitations we face in the homodyne receive architecture are mainly driven by the frequency planning associated with the lack of RF channel filtering. This first results in a limitation of the gain applied on the wanted signal prior to its downconversion to baseband, which makes it sensitive to the various perturbations we inherently find in this low frequency part of the spectrum in the analog domain. Then the unfiltered blockers necessarily reach the mixing stage used for the frequency conversion of the wanted signal down to baseband. As a result, the line-up is necessarily sensitive to the IP2 of this device through the AM-demodulation of the blocking signals, which generates an additional unwanted component superposed on the wanted signal. In order to overcome these

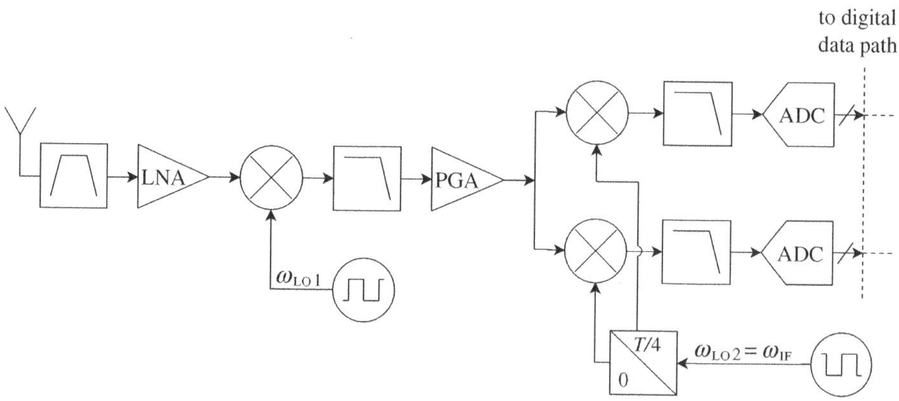

Figure 8.27 Heterodyne receive architecture – In order to overcome the limitations discussed in the previous section for the ZIF RX architecture displayed in Figure 8.26, we can carry out the frequency downconversion in two steps. With the IF sufficiently high so that the wanted signal can be processed as a bandpass signal and the input image signal can be filtered out through a fixed input RF filter, we can then implement the first downconversion stage as a real one. But the second stage, which downconverts this complex modulated IF bandpass signal toward baseband, needs to be implemented as a complex frequency downconversion. The joint use of an IF amplifier and an efficient IF filtering stage allows the sensitivity of the receiver to the nonlinearity of the second mixing stage to be overcome.

limitations, we can carry out the frequency conversion down to baseband in two steps. This naturally leads to the heterodyne receive architecture illustrated in Figure 8.27.

From this figure we can understand that by setting a constant IF whatever that of the received carrier, i.e. by performing the channel selection through the correct setting of the LO angular frequency ω_{LO1} used for the first frequency downconversion stage, we can rely on an optimized IF filtering stage that can already operate as an efficient channel filter. As a result, with fewer potential compression issues due to this filtering, we should be able to set a sufficiently high IF gain to limit the sensitivity of the wanted signal to the perturbations present as centered around DC when downconverted to baseband. At the same time, we can expect a sufficiently strong attenuation of the highest blockers in order to limit the power of the AM-demodulated components recovered at the output of the second mixing stage due its second order nonlinearity. We can thus expect to overcome the limitations of the direct conversion recalled above.

However, the figure also shows that, as for the transmit heterodyne architecture discussed in Section 8.1.2, we are dealing with two kinds of frequency conversion stage. The first is implemented as a real frequency conversion down to an IF. We obviously want a frequency conversion that can be implemented using a single physical mixer, thus allowing a reduced area and power consumption with regard to its complex counterpart. This is theoretically possible as long as the IF remains sufficiently high to allow the correct representation of the modulated wanted IF signal we are dealing with in its real bandpass form. Then we get a second frequency conversion stage implemented as a complex mixer. This is required when dealing with a complex modulation in order to recover the real and imaginary parts of this complex modulating waveform on the P and Q paths of the line-up.

From the signal processing point of view these two kinds of frequency conversion obviously behave differently, as discussed in Chapter 6. Whereas the complex frequency conversion theoretically implements the ideal frequency shift expected during a frequency conversion, its real counterpart leads to the superposition of the double frequency shift of the input spectrum. This results in the folding of the image signal lying such that ω_{LO1} is midway between the received carrier angular frequency and the image angular frequency. As a result, we need to use an image reject filter prior to the real frequency downconversion. And in the present case also, in order to keep this filter as a simple fixed RF filter, we need it to be bandpass with regard to the overall receive system band Δf_{RX}. Thus the IF must be selected sufficiently high that the image signal lies outside this receive system band whatever the carrier frequency of the wanted signal. This is a first constraint for our present architecture as the higher this IF, the more difficult it is to achieve an efficient design of the corresponding IF stages.

Furthermore, we need to keep in mind the limitations in the physical implementation of the frequency downconversion stages. For a start, we need to think about the impact of using mixers implemented as choppers. Obviously, we can anticipate the same difficulties with the induced harmonic LO mixing problem as already discussed in the previous section in the direct conversion architecture context. But in the present case, the problem becomes trickier to handle as folding of the blockers in the wanted signal receive band can result from the succession of the two frequency conversions. We can thus see this problem as symmetric with regard to what we saw on the transmit side when discussing the mechanism involved in the generation of intermodulation sidebands (Section 8.1.2). In order to avoid a multiplicity of spurious responses of the receiver, we thus need to ensure careful frequency planning for selecting the IF, and efficient filtering, set both on the RF side prior to the first frequency conversion and in the IF section [72]. We may also have to deal with an imbalance between sections of the line-up that are expected to behave in exactly the same way. This problem obviously impacts the complex frequency conversion stage through an imbalance between its P and Q paths. And noticing that the remaining part of the heterodyne architecture starting from the IF stage exactly matches a direct conversion receive line-up, we can therefore expect the same kind of problem in the present case as we discussed in the previous section. Practically speaking, this means the same performance limitation in terms of image rejection, reducing in this direct conversion scheme to the imperfect cancellation of the unwanted sideband of the wanted signal, as discussed in Section 6.2.2. However, with this complex frequency conversion stage that is processing an IF signal set at a lower carrier angular frequency than in the direct conversion case, we can expect more easily to achieve an appropriate matching of the devices. As a result, we should see less degradation due to this image signal issue in the present case than in the direct conversion architecture scheme.

Thus, by considering both an optimized frequency planning and efficient RF and IF filtering, we can expect to overcome the limitations of the direct conversion receive architecture discussed in the previous section. However, this is obviously achieved at a non-negligible additional implementation cost which may prove to be prohibitive for low cost integrated solutions. Hence the appeal of a less costly implementation while retaining the advantages of this double frequency conversion. Perhaps we might implement a variable IF scheme in order to avoid one of the two RF synthesizers as was done when considering this problem on the transmit side. This is obviously an interesting approach, but on the receive side an even more efficient approach can be considered. Bear in mind that the constraints linked to the selection of the IF on the receive side are not exactly the same as on the transmit side. In the latter case, the requirements in terms of SEM lead to hard constraints on the generation of the close in-band part of the transmit signal spectrum. When using an IF stage, this results in the classical desire to keep this IF sufficiently high in order to get both the LO leakage and the image signal lying outside the overall transmit system band. Such frequency planning allows us to filter out those components efficiently and recover a sufficiently clean output transmit spectrum. In the same way, the IF may need to be set not too low so that the unwanted sidebands recovered at the output of the transmitter lie at a minimum frequency offset from the RF oscillator free-running frequency. This may obviously be done in order to avoid the potential pulling issue. But on the receive side such requirements do not hold. Rather, the only reason for having an IF is to have the wanted signal that is not superposed on the unwanted components classically recovered in the low frequency part of the spectrum in the analog part of the receiver. However, we might reason that in the heterodyne receive architecture, the IF had been selected sufficiently high to also allow the use of an sufficiently efficient image reject RF filter prior to the first frequency downconversion, implemented as a real one in this architecture. But in practice, nothing prevents us from using a complex mixer as the first frequency conversion stage in order to overcome this image rejection problem. Using this scheme would indeed allow us to select the IF sufficiently high that the wanted signal is not superposed on the unwanted low frequency components, but sufficiently low to allow most of the subsequent signal processing, including the final downconversion of the wanted signal toward baseband, to be implemented in the digital domain, thus allowing a potentially high level of integration. This leads to the low-IF architecture discussed in the next section.

8.2.3 Low-IF Receiver

As discussed in the previous section, the main attraction of using an IF stage on the receive side is the possibility of having the wanted signal that is not directly superposed on the various analog unwanted components lying in the low frequency part of the spectrum, at least prior to having experienced enough amplification. At the same time, by using a complex frequency conversion stage to downconvert the input RF signal toward this IF we can expect to avoid using an image reject filter implemented at the input of this stage as would be required by using a real frequency downconversion. Practically speaking, since we can now neglect the selectivity issue of this filtering stage, we can select the lowest possible IF as long as it allows us to avoid the superposition of the wanted signal on the low frequency unwanted components. Obviously, the exact value of this IF should be determined on a case by case basis as some of the unwanted components can have a non-vanishing spectral extent. This is for instance the case for the AM-demodulated terms that can basically spread over twice the bandwidth of the corresponding input blockers due to the spectral regrowth phenomenon discussed in "Spectral regrowth for AM-demodulation due to even order nonlinearity" (Section 5.1.3). However, if we can select an IF sufficiently low to allow us to implement most of the subsequent signal processing in the digital domain, including the final downconversion of the wanted signal toward baseband, we can target a high level of integration of the receive path. Following this strategy leads to what is known as the low-IF receive architecture shown in Figure 8.28.

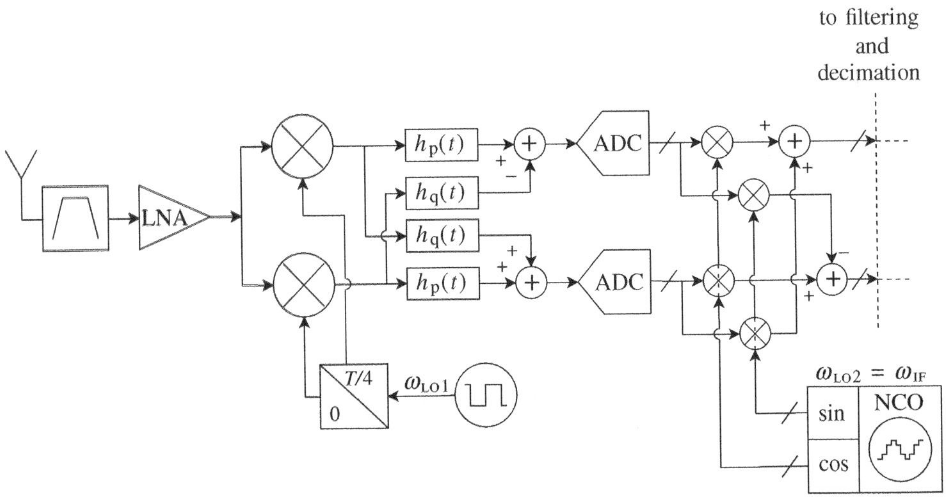

Figure 8.28 Low-IF receive architecture – In order to limit the implementation cost of the heterodyne architecture displayed in Figure 8.27 while retaining the advantages of the two-step frequency downconversion, we can consider an IF angular frequency, ω_{IF}, sufficiently low to be able to implement the second frequency downconversion in the digital domain. This is possible by using a complex frequency conversion in the first place that allows us to avoid the use of an input RF image reject filter. The baseband analog channel filter can be implemented as a complex filter with an impulse response $h_p(t) + jh_q(t)$ that acts on the reconstructed complex signal $p(t) + jq(t)$. This allows attenuation of the signal lying at the symmetric angular frequency $-\omega_{IF}$, as illustrated in Figure 8.29, in order to limit the required ADC DR. The final downconversion can be implemented through the multiplication of the reconstructed complex signal by the ideal exponential $e^{-j\omega_{IF}t}$, before performing the final digital channel filtering and decimation.

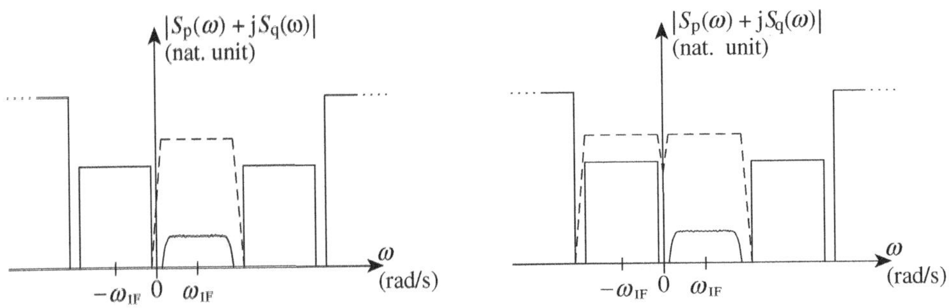

Figure 8.29 Complex analog channel filtering perspective in the low-IF receive architecture – A complex bandpass analog filter centered around the IF angular frequency ω_{IF}, corresponding to the implementation shown in Figure 8.28, can lead to an attenuation of the unwanted signal lying around the angular frequency $-\omega_{IF}$ thanks to its asymmetrical frequency response (left, dashed). In contrast, real filters duplicated independently on the P and Q paths necessarily behave symmetrically, thus providing no attenuation of this unwanted signal (right, dashed). This complex filtering approach provides headroom savings on the ADC stages in the low-IF perspective.

Some comments are in order on the use of the constituent blocks of the line-up in this architecture, in comparison to their counterparts in the direct conversion receive architecture discussed in Section 8.2.1. We might wonder whether the ADC needs to be run at a higher frequency than in the receiver scheme. The main reason for this is that the wanted signal is now centered around an IF rather than at DC, thus necessarily requiring a higher sampling rate to be correctly represented while fulfilling the Nyquist sampling criterion. Recall, however, that in practical receivers the ADC sampling rate is often not only driven by the bandwidth of the wanted signal to be sampled. Rather, it is set at a higher value in order to be able to reduce the costly analog anti-aliasing filtering referred to the digital implementation capabilities of modern silicon technologies. In that case, whether or not we consider a low-IF scheme may make little difference in terms of ADC sampling rate. That said, this conclusion is relevant only for low bandwidth systems as a high OSR may not be realistic when dealing with other kinds of systems, due to practical limitations in the physical implementation of ADC. This constitutes a first argument for claiming that a low-IF receive architecture may not be so well suited to high bandwidth systems.

Next, we observe that the analog baseband filtering stage is implemented as a complex filter in the present case. By the discussion in Section 6.1.4, this kind of filter exhibits a transfer function in the frequency domain that is asymmetric in magnitude with respect to the zero frequency. Of course, this behavior can only be seen on the reconstructed complex signal $p(t) + jq(t)$. On each of the P and Q paths we are dealing with real signals whose spectra are necessarily symmetric. More precisely, they are the sum of the spectrum of this complex reconstructed signal, $p(t) + jq(t)$, and a flipped copy of it according to the mechanism introduced in Chapter 1 and extensively illustrated in Chapter 6. We thus get a scrambling of the wanted signal lying around ω_{IF} with the image signal lying around $-\omega_{\mathrm{IF}}$ on these real signals. As a result, a real filter acting on the real signals cannot provide an attenuation on the unwanted signal that is superposed on the wanted signal in the frequency domain. Only a complex filter acting on the complex reconstructed signal $p(t) + jq(t)$ can do this, as illustrated in Figure 6.8. Thus, we can indeed expect an improvement of the ratio between the wanted signal and the one lying at the symmetric angular frequency $-\omega_{\mathrm{IF}}$ at the output of such a complex analog filter. Moreover, this improvement seen on the complex reconstructed signal at the output of the filter necessarily remains true for each of the real signals $p(t)$ and $q(t)$ due to the relationship between the spectra of these signals recalled above.

Recall that due to the use of a complex frequency conversion stage to downconvert the input RF signal toward the IF, the spectrum of the reconstructed complex signal entering the baseband analog filter is nothing more than a frequency shift of the input RF spectrum. For a practical wireless system, the unwanted signal lying at a frequency offset of $2\omega_{\mathrm{IF}}$ from the wanted signal may be of much higher level than the wanted signal at the input of the receiver as discussed in Section 3.3.1. For the sake of clarity, this behavior can be illustrated by considering the classical configuration where the IF is selected as equal to half the frequency offset between the wanted signal and the adjacent channel. In that case, it is the adjacent channel signal that is recovered as the image signal, i.e. at the angular frequency symmetric to that of the wanted signal at the input of the baseband filter. As a result, without using a complex filter to attenuate this adjacent channel signal, we would recover the same ratio of adjacent channel to wanted signal at the input of the P and Q ADC as at the input of the receiver. And recalling the discussion about the required ADC DR when illustrating the derivation of receiver budgets in Chapter 7, especially in "Filtering budget vs. ADC dynamic range"

(Section 7.3.3), the residual unwanted signals present ahead of the wanted signal at the input of the ADC necessarily increase the DR required for those devices. This should obviously be qualified depending on the AGC strategy and the associated set point at the device input, but this does not change the general principle that the better the rejection of the unwanted signals in the analog domain, the lower the requirements on the ADC stage. In this perspective, we thus see the attraction of using a complex analog filter in a low-IF architecture.

We also need to keep in mind that due to the RF impairments that we necessarily face in the physical implementation of the line-up, only a finite image rejection can be achieved during the complex frequency conversion operation. This is obviously true for all the receive architectures considered so far. But in previous cases, the complex frequency conversion stage was used for the downconversion of the wanted signal down to baseband. Due to this particular frequency planning, the complex envelope of the image signal that is retrieved as superposed on the wanted signal in the frequency domain at the downmixer output is simply the complex conjugate of the wanted signal, as discussed in Section 6.2.2. Thus, given that those two complex conjugate signals have the same power, the achievable IRR directly gives the SNR limitation linked to this impairment problem. This is obviously no longer true in the low-IF configuration as the image signal we are talking about is now the unwanted bandpass signal lying at a frequency offset of $2\omega_{\text{IF}}$ from the wanted signal. Here again, for the sake of clarity we can illustrate this behavior by reconsidering the configuration taken as an example above where the image signal is nothing more than the signal lying in the adjacent channel. In that case, the generated in-band unwanted component is proportional to this adjacent channel signal and not to the wanted signal as in the direct conversion case. This makes the final residual unwanted in-band component behave as an additive noise with respect to the wanted signal. Obviously, this should be qualified on a case by case basis as this conclusion holds only when the power of this adjacent channel signal remains constant at the input of the complex frequency downconversion stage. However, unless the power of the adjacent channel signal is in turn proportional to that of the wanted signal, the residual unwanted in-band component no longer behaves as a multiplicative noise component that sets an upper bound on the achievable in-band SNR in the line-up. Furthermore, classically the signals lying in the adjacent channels can be of much higher power than the wanted signal. As a result, for a given IRR performance we can be faced with a higher degradation of the SNR in the present case than in the direct conversion scheme. In order to preserve the SNR performance of the receiver even in the presence of a strong unwanted signal lying at the image angular frequency, we may need to achieve a better IRR performance than what would be required in a direct conversion receiver. This may make it necessary to use particular calibration schemes or even image cancellation algorithms, as discussed in Section 9.2.3, to compensate for such an imbalance between the P and Q branches of the receiver. However, this should be balanced with the fact that the SNR requirements are often lower when considering the reception of the wanted signal in the presence of unwanted signals, as illustrated in Section 7.3.4. Thus only a deeper analysis on a case by case basis can help us decide whether or not the achievable image rejection is a real problem in the low-IF scheme.

Finally, when the image rejection problem can be handled, this kind of line-up theoretically allows the same level of integration as the direct conversion architecture while retaining the benefit of the two-step frequency downconversion. Only a slight additional hardware cost on the digital side is incurred, which is a great advantage compared to pure heterodyne structures as discussed in the previous section. However, to keep the power consumption reasonable, or even make the design of the ADC stage possible, the IF must be kept very low. And due to the spectral extent of the baseband distortions we want to avoid with such frequency planning, such as the AM-demodulated components of strong blocking signals, we might be inclined to consider this approach mainly for low bandwidth systems.

The various approaches to the direct conversion line-up we have considered so far are able to recover the complex modulating waveform of a received bandpass signal through its real and imaginary parts, $p(t)$ and $q(t)$ respectively. We have thus confined our account to Cartesian receive architectures, the direct conversion being the simplest. But, as already highlighted when discussing transmit architectures, real modulations, in the sense that only one physical signal can represent the modulating waveform, as is the case for pure phase/frequency modulations for instance, are still of interest in practice. In that case, line-ups simpler than Cartesian architectures can be considered, and in the next section we turn our attention to the PLL demodulator.

8.2.4 PLL Demodulator

As stated at the beginning of Section 8.2, the direct conversion receiver is the simplest architecture for recovering a complex modulating waveform. However, this statement fails to hold when dealing with a real modulation. To illustrate this, let us say a few words about the possibility of recovering the information contained in the instantaneous phase or frequency of a bandpass signal by using a PLL demodulator.

A simplified PLL demodulator is shown in Figure 8.30. We examined the behavior of this kind of loop from the modulation perspective when we discussed the PLL as a direct modulator in Section 8.1.5, and from the phase noise perspective in Section 4.3.1. We can therefore use those discussions in our present analysis.

Suppose that we feed the reference port of a PLL with the RF bandpass signal we want to demodulate. By equation (8.6), as long as the cut-off frequency of the closed loop transfer function is sufficiently high with respect to the bandwidth of the instantaneous angular frequency of this signal, $\omega_{\text{ref}}(t)$, we have that

$$\omega_{\text{osc}}(t) = \omega_{\text{ref}}(t), \tag{8.43}$$

where $\omega_{\text{osc}}(t)$ is the instantaneous angular frequency of the bandpass signal standing in the RF oscillator. And assuming for the sake of simplicity that we are dealing with a RF oscillator controlled through an analog voltage command $v_{\text{osc}}(t)$, $\omega_{\text{osc}}(t)$ is simply equal to $K_{\text{osc}} v_{\text{osc}}(t)$, where K_{osc} is the conversion gain of the voltage controlled oscillator. Therefore,

$$v_{\text{osc}}(t) = \frac{\omega_{\text{ref}}(t)}{K_{\text{osc}}}. \tag{8.44}$$

Thus, the lowpass signal recovered as a command of the RF oscillator is nothing more than an image of the instantaneous frequency of the signal fed at the reference input, i.e. precisely the information we were looking for.

However, as highlighted above, for this scheme to work properly, we need to carefully set the cut-off frequency of the closed loop transfer function of the PLL in order to achieve

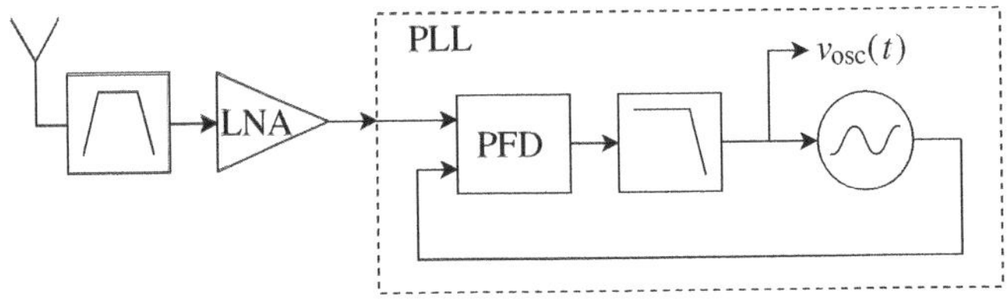

Figure 8.30 PLL demodulator receive architecture – When receiving an RF bandpass signal that is phase/frequency modulated only, the minimum set of signal processing blocks required to recover the modulating waveform can be simplified compared to line-ups designed to recover a complex modulation. For instance, RF synthesizers can be designed to be able to track the instantaneous phase/frequency variations of the bandpass signal fed to their reference port. In that case, the lowpass signal $v_{osc}(t)$ that drives the RF oscillator is an image of those instantaneous phase/frequency variations, thus realizing the demodulation of the signal present at the reference port.

no distortion of the modulation. But, as discussed in Section 8.1.5, the higher this cut-off frequency, the harder it is to achieve stability of the loop. Practically speaking, we may face limitations in the possibility of using the PLL demodulator scheme for the recovery of modulations that exhibit a large difference between the maximum and minimum values of the instantaneous frequency. To get around this problem, we might consider an alternative PLL implementation that allows us to track such wideband phase/frequency modulation as discussed on the transmit side in Section 8.1.5. This could be done by using a two-point PLL for instance. But we need to keep in mind another constraint on the receive side that may prevent us from going down that path: that is the presence of potentially strong unwanted signals at the input of the line-up. There is thus a need to preserve a lowpass effect of the overall receive path, which is not the case in a two-point PLL scheme for instance. However, if this lowpass behavior cannot be achieved directly at the PLL demodulator stage in practice, there is nothing to prevent us from implementing a filtering stage in another part of the receiver. This could be done for instance by considering an IF stage that allows the implementation of a channel filter before going through a wideband PLL demodulator stage. But obviously, this approach would require at least two RF synthesizers, one for the frequency downconversion toward the IF, and another for the demodulation. Moreover, there would necessarily be a certain amount of image rejection to manage at the first frequency downconversion stage. This approach would thus wipe out one of the advantages of this PLL demodulator scheme, i.e. its compactness.

Looking at the simplified PLL structure shown in Figure 8.30, we may notice a kind of similarity with the direct conversion. This is more obvious when considering the PFD device implemented as based on a mixing stage as illustrated in Figure 8.31. Drawn like this, we can interpret the output of the PFD and lowpass filter as the downconversion of the modulated bandpass signal present at the reference input of the PLL. We could thus interpret the present signal processing approach as equivalent to that implemented in the direct conversion receiver except that we now have a single receive branch as when dealing with a real modulation. However, there are fundamental differences as in the present case we recover at the output of the lowpass filter directly the instantaneous frequency information of the wanted signal, and not the wanted signal itself. This is achieved through the feedback toward the RF oscillator in order to make its instantaneous angular frequency equal to that of the wanted signal. We may have at least one integration stage in the loop filter to make this possible. This is a fundamental

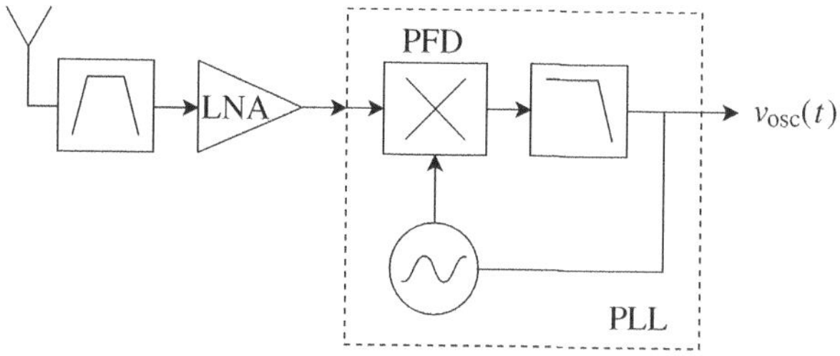

Figure 8.31 PLL demodulator as a real direct conversion path – Considering a PFD based on the use of a multiplier, the structure of the receive line-up shown in Figure 8.30 may resemble a direct conversion line-up at first glance, even if based on a single branch due to the real nature of the modulation we are dealing with. However, the feedback path toward the voltage controlled oscillator as well as the integration functionality necessarily present in the loop filter make the demodulated signal $v_{osc}(t)$ an image of the instantaneous frequency of the input bandpass signal, and not a downconversion copy of it as it would be in a direct conversion scheme.

difference with the signal processing function of the filter embedded in a direct conversion receiver that behaves only as a simple lowpass filter in order to filter out the unwanted signals.

In conclusion, when possible to implement, the PLL demodulation scheme is very efficient in terms of implementation costs as we basically need to use only one RF synthesizer block to achieve the overall signal processing involved in the demodulation. However, the limited bandwidth of such a system may prevent this approach from being used for modulated bandpass signals that exhibit a high bandwidth instantaneous frequency. As a side effect, this can also prevent the generalization of this scheme for complex modulations through the use of some kind of polar demodulator in accordance with the frequency spread of the instantaneous frequency of most of the complex modulated bandpass signals as discussed in Section 8.1.6.

9

Algorithms for Transceivers

In this final chapter, we show how to improve the performance of a given architecture by using a dedicated algorithm. The approach is either to compensate for the distortion experienced by the wanted signal when going through a non-ideal device (e.g. with a predistortion system on the transmit side), or to optimize the usage of the hardware available in the line-up under operating conditions (e.g. with an AGC scheme on the receive side). Such an approach is of interest from the perspective of implementing a line-up using a technology that allows for a high level of digital integration. We examine some of the algorithms classically encountered in wireless transceivers. In keeping with the transceiver architecture point of view, we stay at a system level rather than discussing the design and implementation of the algorithms themselves.

9.1 Transmit Side

Let us begin with algorithms typical of those encountered on the transmit side. We recall a simple fact that drives most of the trade-off involved in the implementation of such algorithms, at least when dealing with the compensation of a default in the line-up. When considering a transmitter as such, unless something special is done, we do not have access to the transmitted signal as effectively recovered at the output of the line-up. This is obviously an important difference compared to a receiver that, by definition, gives access to the received wanted signal. On the transmit side, we thus cannot easily rely on a feedback path in order to adaptively optimize a given compensation for instance.

Obviously, this statement should not mask the fact that in practice it may be necessary to implement this additional measurement path. This may be the case when the desired performance cannot be achieved with an open loop system due to variations in the default to be compensated under operating conditions. However, from an implementation perspective, such a feedback path can be seen as a net cost as it is basically not required in terms of signal processing functions to be theoretically embedded in a transmitter. Moreover, this cost exists in terms of both implementation area and power consumption. Practically speaking, there is thus a desire to minimize the use of an additional path, or at least to try to use the simplest one conceivable for a given application. We can thus expect different strategies for

Wireless Transceiver Architecture: Bridging RF and Digital Communications, First Edition. Pierre Baudin.

a compensation scheme embedded in a transmitter depending on whether or not it requires the use of a feedback path, and what kind of feedback, if any. For a start we can identify the following:

(i) Open loop systems that process the signal with static parameters set for a given configuration of the transmit line-up. Those parameters can basically be determined during the calibration phase of the product or even by design during the conception phase.
(ii) Closed loop systems relying on the sensing of a single RF parameter. As an example, we can mention the sensing of the average power of the transmit signal as required in some power control schemes.
(iii) Closed loop systems relying on the full characterization of the modulated signal being transmitted. If this modulation is complex, it means a full receive path giving for instance access to the P and Q components of the modulating waveform.

Furthermore, we should keep in mind that even if a complex feedback path is required to achieve the required performance in a given line-up, it may be run only periodically over short durations depending on the time constant of the drifts we expect to track using this scheme. This simple consideration illustrates that the power consumption cost of such a solution may not be seen as directly related to the complexity of the sense and feedback path, but rather as related to the real time aspects of the system. However, from this perspective we remark that nothing can perform better than a simple open loop scheme.

We thus anticipate that in order to compensate for an imperfection in a transmitter, different approaches, either based on the use of a feedback path or not, can be considered depending on the trade-off required between the targeted performance and the effective drifts of this imperfection under operating conditions. Thus the types of strategy discussed above may guide our review of the algorithms encountered on the transmit side, at least when dealing with the compensation of an implementation limitation.

9.1.1 Power Control

As introduced in Section 3.2.1, the problem of the power control can be decomposed into two sub-problems. We need to address on the one hand the control of the long-term average power radiated by the line-up during the transmission of the data, and on the other hand the management of its transitions, often over short time frames. The latter problem can be related for instance to transitions from one long-term average output power to another, or to the burst shaping needed in particular schemes such as TDD or TDMA. Even if relying on the control of the same physical quantity, the distinction between the two sub-problems is justified by the difference in the time constants involved. The real time control of the RF power for ensuring correct transitions requires a much more reactive system than the control of its long-term average value. We can thus anticipate differences in the solutions to be considered as well as in their final implementation.

Long-term Average Transmit Power

Consider the problem of control of the long-term average transmit power. Recall that unlike what happens on the receive side, the level of the signal being processed in a transmitter is deterministic. We know how the modulating waveforms are scaled when generated. Thus, knowing the gain of the various stages present in the line-up, we might think that it is quite straightforward to control the level of the signal in the line-up, and thus finally the power effectively delivered to the radiating element.

However, in the present case it is entirely a matter of accuracy. In practical RF/analog physical implementations, we are faced with variations in the characteristics of the devices present in a line-up. This is due either to discrepancies from device to device as a result of mass production or to changes in the operating conditions, such as the temperature or the supply voltage, of a given device. Such variations may be non-negligible compared to the requirement set by classical wireless standards in terms of radiated power. As a result, it is often necessary to improve the intrinsic performance of a transmit line-up in that regard.

For this to be possible, we thus need to incorporate some additional features in the line-up. Depending on the causes and magnitude of the variations, different strategies can be considered. For instance, dealing with variations from part to part in mass production, we can consider a simple calibration scheme at the production stage, as illustrated in Figure 9.1. This would allow us to compensate for those variations by adjusting the gains to be used in order to achieve the expected transmit power for a given target. However, this strategy becomes more difficult to put in place when the inaccuracy in the line-up is related to drifts in temperature and supply. It would require a full characterization of the device under different operating conditions to achieve a reliable compensation in any case. Unless predetermined during the design phase, such characterization often leads to a prohibitive cost for most of the commercial solutions due to its complexity.

A classical alternative to such calibration is to rely on a feedback measurement path that senses the transmit power in operating conditions. The idea is to rely on an embedded measurement capability that allows the implementation of a correction scheme that adjusts the transmit power in real time. Obviously, a simple measurement of the transmit power, and only of the power, is sufficient to achieve the functionality. This allows us to consider a quite simple implementation based on RF couplers and detectors, as illustrated in Figure 9.2, compared to an additional full receive path able to demodulate the transmit signal. However, looking at

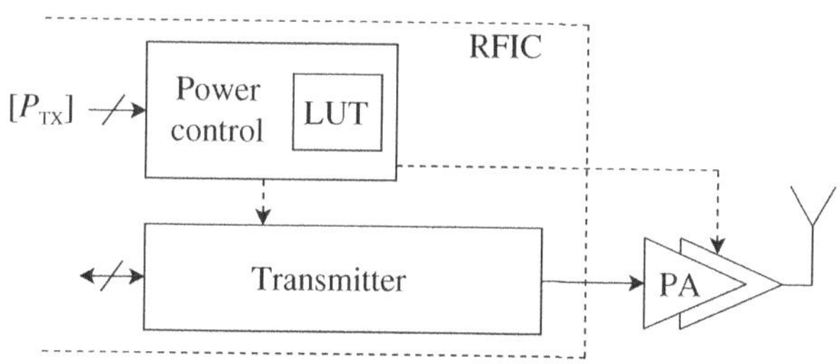

Figure 9.1 Open loop long-term average power control with compensation of variations from part to part – Although the level of the signal being processed in a transmitter is theoretically deterministic, the variations in the gain of the various blocks of the line-up can lead to discrepancies in the actual transmit power P_{TX} compared to the expected power $[P_{TX}]$. To improve the performance, variations from part to part can be compensated by calibrating the line-up at the production level. This allows corrective factors to be applied, here illustrated by through the presence of the lookup table (LUT), to improve the natural accuracy of the line-up.

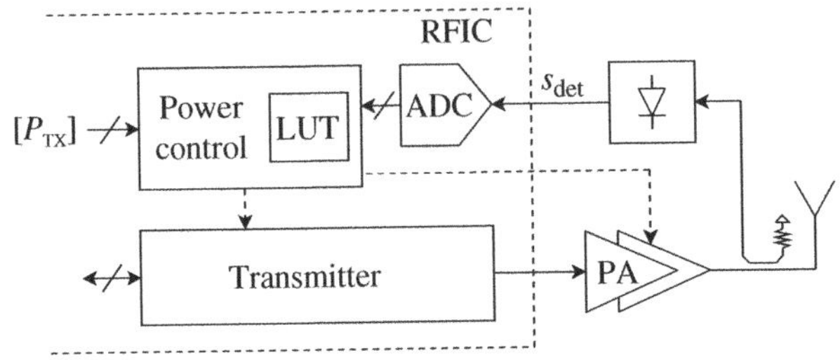

Figure 9.2 Closed loop long-term average power control for handling variations from part to part and in operating conditions – Compared to the open loop configuration shown in Figure 9.1, sensing the actual transmit power P_{TX} allows the implementation of a closed loop system that ensures this quantity matches that corresponding to the command $[P_{TX}]$. However, to be able to predetermine the value of $[P_{TX}]$ that gives the expected output power, we need to rely on a sufficiently accurate feedback information s_{det}. This may make it necessary on the one hand to still perform a calibration in production to cope with the discrepancies of the feedback path from part to part, and on the other hand to minimize variations in operating conditions sufficiently to improve the performance compared to an open loop scheme.

this figure, we observe that in order to improve the situation, this sense path must necessarily provide a sufficiently accurate measurement – by which we mean on the one hand more accurate than the direct path in order to improve the performance compared to the open loop case, and on the other hand more accurate than the required performance of the line-up so that the overall system stands a chance of achieving this target. These preliminary comments prompt us to question why this accuracy might be achieved on the feedback path while apparently not possible on the direct path. Basically, it is indeed possible in physical implementations due to the use of passive devices, such as directive couplers, to perform the sensing of the high power transmit signal. Thus the active devices, such as the detector, used to perform the power measurement in the feedback path, process small signals. This situation finally allows us to achieve weaker operational variations than on the direct path due to the presence of active devices processing high power signals, such as PAs. However, in practical implementations a calibration of the system may still be required in order to compensate for the variations from device to device of the feedback path as depicted in Figure 9.2.

However, this accuracy topic is only one aspect of the problem associated with the control of the long-term average power based on the use of a closed loop scheme. Indeed, due for instance to nonlinearity in the transfer function of the blocks present in the loop, we may also have to address some instability issues. This problem may be illustrated by considering a discontinuity in the transfer function of the gain command path, as may be encountered for instance when dealing with a digital command of those gains. As illustrated in Figure 9.3, due to the resulting incorrect power estimation, we may be faced with a diverging transmit output power. However, in physical implementations we can imagine that at a given power level, the transfer functions involved recover their expected shape so that the divergence stops. This explains why in practice we face an oscillation of the output power rather than a pure divergence. The same behavior obviously holds when the nonlinearity occurs on the power command path of the line-up.

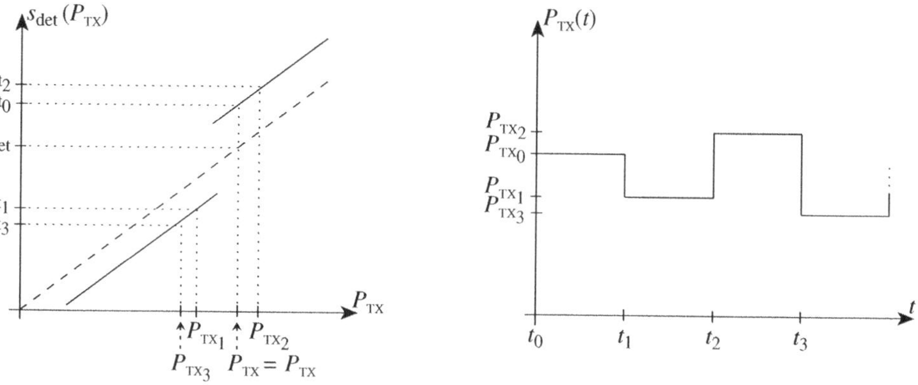

Figure 9.3 Diverging long-term average output power due to a discontinuity in the closed loop control scheme – Given a discontinuity in the actual transfer function of the feedback sense path (left, solid) relative to the expected linear one (left, dashed), we may face a diverging behavior of the control scheme. In the present example, an expected output power P_{TX} leading to a detection level s_{det_0} higher than the theoretical value s_{det} makes the control reduce the power to P_{TX_1}. Due to the shape of the feedback transfer function, the corresponding detection level s_{det_1} is below the expected value s_{det} by an amount that is higher than $|s_{det_0} - s_{det}|$. The closed loop control scheme thus reacts by increasing the output power to a level higher than P_{TX}. Iterating this process may lead to a diverging behavior (right). Here $|t_{n+1} - t_n|$ represents in a simplified way the round trip processing duration in the loop.

As discussed in Section 9.2.1, such oscillating behavior can also be encountered on the receive side, particularly in an AGC scheme. Regulating the received signal at a given level can basically be seen as the same problem as that encountered here, at least from a functional point of view. As a result, we can expect to see the same behavior when dealing with a distortion in the control path of the receiver gains for instance. To overcome the present problem we might think of using a hysteresis scheme to determine the gain switching, as is classically done on the receive side. This is unfortunately not so straightforward on the transmit side. Whereas hysteresis can solve the instability issue discussed above, it cannot do anything for the accuracy of the signal as recovered at the output of the regulating scheme. And whereas we can live with some inaccuracy in the regulating level on the receive side by taking sufficiently high headroom margins along the data path, on the transmit path we need to ensure the accuracy of the transmit power, in terms of absolute level but also in terms of transition steps. This leads to the necessity to get sufficiently linear power commands and feedback sense path in the latter case. In practice, we may need to add LUTs that linearize either the command ports or the feedback path. The content of those tables can be derived during a calibration phase for instance. In any case, whatever the solution chosen to manage this topic, such linearity requirement for the power control on the transmit side must be kept in mind.

Fortunately, this linearity issue is often the only one we need to deal with. For the most part, average power control can be implemented without particular real time issues. Obviously, this notion of real time depends on the constraints set by the network to which the transceiver belongs. But generally speaking, we can anticipate far fewer timing issues than in the control of the power transitions as encountered in burst systems for instance. In the present case, even when dealing with the use of a feedback sense path, some software processing can often be considered, which is hardly possible in the instantaneous power shaping case discussed in the next section.

Short-term Transmit Power

We now turn to the control of the short-term power of the transmitted signal. As highlighted previously, such control needs to be considered either for performing the transitions from one long-term average output power level to another during a continuous transmission, or for shaping a transmit burst, as encountered in some TDD or TDMA systems for instance.

From the implementation perspective those two use cases are not strictly equivalent. In practical configurations, changes in the long-term power during a transmission occur through the application of successive small steps, each less than a few decibels. This is obviously done in order to limit the rate of change of the electromagnetic environment in the network. But the result of this weak magnitude in the power change is that it almost ensures in itself a limited pollution of the other users, thus hopefully leading to no particular additional constraints on the implementation of the functionality. This is obviously not the same in the burst shaping case as we can deal with huge power variations over a short time frame. This behavior can easily lead to the pollution of the other users through some switching transient effects, as introduced in "Time mask and switching transient" (Section 3.2.1). Although this statement may need to be qualified in practice, we can thus focus mainly on this burst shaping problem here. However, the conclusions are obviously valid more generally.

Considering the latter case we remark that, compared to the average power control problem, the burst shaping requires the generation of the waveform to be applied to shape the instantaneous amplitude of the transmit signal according to the expected variations for its power. From a functional perspective, the generation of the burst can thus be seen as the multiplication of a constant long-term average power signal by a window function having the expected shape, as illustrated in Figure 9.4. From the implementation point of view, however, the picture may differ depending on the transmit architecture considered and how the above multiplication can be implemented. For instance, dealing with a transmit signal that is amplitude modulated, the line-up is by definition able to apply the amplitude modulation on the generated RF transmit signal. As a result, such windowing of the signal can be directly superposed on the amplitude modulation information. The multiplication we are talking about can thus be performed in the baseband domain, for instance directly on the $p(t)$ and $q(t)$ signals when dealing with a complex modulated waveform. Such processing can even be implemented in the digital domain for the sake of simplicity. With this approach, the RF bandpass signal generated through the upconversion of the shaped $p(t)$ and $q(t)$ signals directly results in the expected burst form. In contrast, dealing with a phase/frequency only modulated bandpass signal, we may consider using a direct PLL modulator for its generation as discussed in Section 8.1.5. In that case, we necessarily recover a constant average power modulated RF bandpass signal at the PLL output. As a result, the shaping of the waveform must be done in the RF world. The multiplication we want can then be implemented by controlling the gain of a RF device such as a PA.

Although those schemes appear equivalent from the signal processing point of view, they may perform differently in terms of accuracy in the power transitions. For instance, when

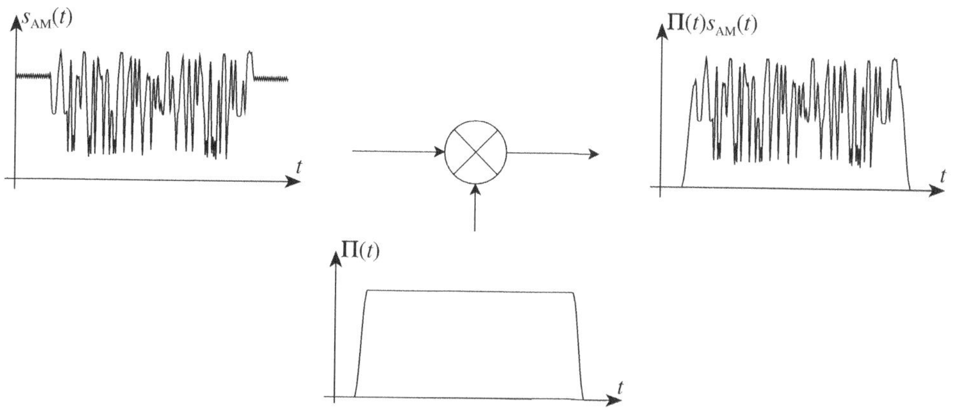

Figure 9.4 Burst shaping as a windowing process – The generation of a RF burst can be seen from the signal processing point of view as the multiplication of a constant long-term average power modulated signal, $s_{AM}(t)$, by a gate function, $\Pi(t)$. The problem is to decide in which physical area to implement this multiplication in practice: digital, analog, or RF.

the shaping is carried out in the digital or even in the low frequency analog domain, we can expect the linearity and accuracy of the devices used to implement the functionality to be good enough that an open loop power control system may be sufficient to achieve the required shaping accuracy. In contrast, dealing with the command of a RF device such as a PA this may not necessarily be the case. We may face both a nonlinear transfer function experienced by the shaping signal and variations in operating conditions. Even when considering a calibration of the transfer function in typical conditions, this can lead to unacceptable degradations under extreme operating conditions.

In the latter case, where we apply the shaping through the command of a RF device (and not through a baseband device), we may need to use a closed loop system as already encountered for the control of the long-term average transmit power. As discussed in the previous section, we can rely for that control on a power sense done through the use of a RF coupler and a detector. Such a feedback path is of the simplest form conceivable to get the information necessary to implement our control scheme. However, we get a major difference in the real time aspect of the control compared to the long-term average transmit power case. Whereas the constraints on this aspect could be considered as relaxed in the case of "long term average power" control, we now need to ensure a fast enough loop processing to control the shaping of the transmit power. Obviously, what we mean by "fast enough" needs to be determined case by case as it basically depends on the rate of change required for this short-term power. But in practical use cases, orders of magnitude in the range of the microseconds are common. This means a loop bandwidth of the order of hundreds of kilohertz. Obviously, such bandwidth limits the choices for the physical implementation, as illustrated in Figure 9.5. Practically speaking, although a digital implementation allows much more flexibility in the control than an analog one, the issue of stability looms larger due to the delay in the processing. For instance, a digital implementation that requires n clock periods, T_{clk}, for the round trip processing leads to a term of the form $e^{-nT_{clk}s}$ in the equivalent open loop transfer function expressed in the Laplace domain. We can thus anticipate potential issues in preserving the phase margin while achieving

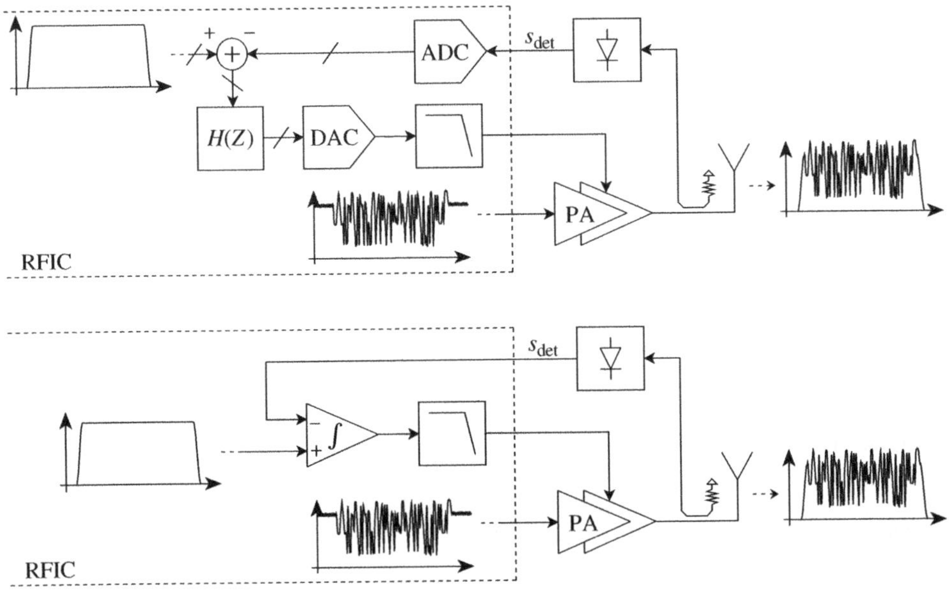

Figure 9.5 Closed loop short-term average power control for RF burst shaping – Due to nonlinearity or variations in the transfer function of an RF device under operating conditions, a closed loop system may be required in order to ensure the shape of the bursts generated through the control of the device, here a PA. But having short durations for the transitions may require high bandwidth loop systems. Thus, although allowing more flexibility, the delay in a digital implementation may limit the achievable cut-off frequency due to instability issues (bottom). An analog implementation may therefore be seen as more suitable for this scheme (top).

a cut-off frequency higher than hundreds of kilohertz. In contrast, an analog implementation as illustrated in the same figure may be more realistic, albeit less flexible. When all is said and done, it is a matter of selecting the best possible implementation options with regard to the timing requirements set by the wireless system.

9.1.2 LO Leakage Cancellation

Another issue that can be improved by a calibration or cancellation scheme on the transmit side is the LO leakage. Such leakage is caused by different phenomena in practice: the direct coupling of the LO signal toward the RF port of the mixer device, or of the transmitter itself when considering an integrated solution; and the presence of DC offset at the low frequency port of the device (see Section 6.5). As discussed in Chapter 6, due to those phenomena we can therefore expect a leakage whose complex envelope, $\tilde{s}_{leak}(t)$, when defined as centered around the LO angular frequency ω_{LO}, is given by equation (6.149) (Cartesian form) or (6.150) (polar form).

From a system point of view, the presence of this leakage impacts the performance of a transmitter in different ways depending on the frequency planning associated with the line-up, as illustrated throughout Chapter 8. For instance, a transmit line-up that uses a two-step frequency upconversion can be optimized in such a way that this LO leakage systematically lies outside the transmit system band. We can then use a simple fixed RF filter at the output of the line-up in order to cancel this signal. This is what is classically done in a heterodyne transmitter, as discussed in Section 8.1.2. In contrast, when dealing with a direct upconversion of the complex lowpass modulating waveform, this leakage is necessarily superposed on the wanted sideband of interest at the output of the line-up, as illustrated in our review of the direct conversion scheme in Section 8.1.1 and Chapter 7. Depending on the structure of the modulating waveform being processed, this can classically result in an origin offset in the symbol constellation, as illustrated in "Origin offset suppression" (Section 3.2.2). With such frequency planning, we thus see that the unwanted leakage component cannot be filtered at the output of the line-up. We need to rely on a sufficient minimization of the LO leakage in order not to degrade the modulation quality of the bandpass signal delivered by the transmitter. But if we have an unacceptable leakage level that prevents us from achieving the required performance, we could consider using a cancellation scheme instead.

In that perspective, we must derive the most suitable structure to enable us to achieve the required functionality with the minimum complexity. Recall equation (6.149), the expression in Cartesian form for the complex envelope $\tilde{s}_{\text{leak}}(t)$ of the leakage we expect to cancel, and suppose for the sake of simplicity that we are dealing with a direct conversion line-up. If we neglect the potential drifts in operating conditions, both the real and imaginary parts of $\tilde{s}_{\text{leak}}(t)$ are constant as they depend only on the coupling factor from the LO port of the mixer up to the RF output port and on the DC offset inherently present at the low frequency input port of this device. Thus $\tilde{s}_{\text{leak}}(t) = \tilde{s}_{\text{leak}}$, as holds for the upconversion of a pure constant DC offset present at the input of the upmixing stage. On the other hand, due to the distributive behavior of the upmixing processing with regard to the signal fed at the input of the mixing stage, the complex envelope of the signal resulting from the upconversion of an intentional lowpass signal that adds to the DC offset inherently present at the input of the upmixer is simply the expected complex envelope plus that achieved without this intentional lowpass signal, i.e. $\tilde{s}_{\text{leak}}$. As a result, applying the values $dc_{\text{comp,p}}$ and $dc_{\text{comp,q}}$ on the P and Q branches of the considered complex upmixer, we get that the complex envelope $\tilde{s}_{\text{TX}}(t)$ of the bandpass signal recovered at the output of the transmitter is of the form

$$\tilde{s}_{\text{TX}}(t) = \tilde{s}_{\text{TX}} = \tilde{s}_{\text{comp}} + \tilde{s}_{\text{leak}}, \tag{9.1}$$

with

$$\tilde{s}_{\text{comp}} = dc_{\text{comp,p}} + j dc_{\text{comp,q}} \tag{9.2}$$

and all the complex envelopes defined as centered around the LO angular frequency ω_{LO}. Since in order to cancel the problematic LO leakage we need $\tilde{s}_{\text{TX}} = 0$, we simply require

$$dc_{\text{comp,p}} = -\text{Re}\{\tilde{s}_{\text{leak}}\}, \tag{9.3a}$$

$$dc_{\text{comp,q}} = -\text{Im}\{\tilde{s}_{\text{leak}}\}. \tag{9.3b}$$

Thus, by simply tuning an additional DC component we can achieve our goal. This is of particular interest from an integrated perspective as we can carry out this addition directly in the digital part of the line-up for instance. This is obviously a great advantage in terms of implementation complexity relative to any attempt to perform compensation directly in the RF world.

Practically speaking, the values of $dc_{\text{comp,p}}$ and $dc_{\text{comp,q}}$ can be derived during line-up calibration. However, as discussed for all the compensation schemes encountered so far, this strategy is valid only if the compensation settings allow the required performance to be achieved whatever the variations of the leakage under operating conditions. But in practice, we can expect a dependency of the DC offset inherently present at the input of the upmixer on both temperature and supply. This may potentially lead to unacceptable variations in the level of the leakage. Although this will vary case by case, it may be necessary to consider a sense and feedback system to track those variations.

In that perspective, rather than potential pure RF tricks that may be impossible to integrate, we can consider using an auxiliary receiver for the sensing of the leakage, as illustrated in Figure 9.6. But as this figure shows, the potential phase offset between the complex envelope demodulated by the auxiliary receiver relative to the TX LO signal on the transmit side may make it necessary to use a dedicated algorithm to derive the compensation values $dc_{\text{comp,p}}$ and $dc_{\text{comp,q}}$ based on the demodulated components $\text{Re}\{\tilde{s}_{\text{leak}}e^{j\phi}\}$ and $\text{Im}\{\tilde{s}_{\text{leak}}e^{j\phi}\}$. However, we observe that in such a compensation scheme, which is of interest mainly when dealing with a direct conversion line-up, we can consider using the same LO signal on the direct and sense paths as highlighted in the figure. But despite this simplification, a quite heavy system may be needed for implementing a tracking of such simple impairment compensation.

For various reasons, we expect to be able to reduce this complexity considerably in many practical configurations of interest. For instance, thinking about the root causes for the

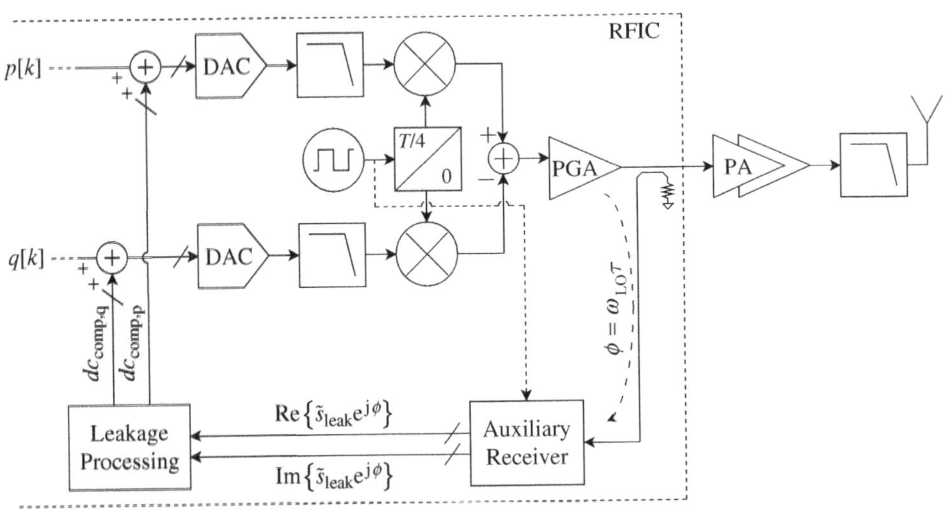

Figure 9.6 Sense and feedback path for the cancellation of LO leakage – In order to track the variations in operating conditions of the LO leakage, it may be necessary to consider a sense and feedback scheme. A coherent receiver may be required due to the phase offset ϕ in the signal effectively sensed relative to the TX LO signal. However, in practical integrated solutions further simplifications can be achieved if ϕ tends toward 0.

generation of the LO leakage, the sensing can be performed quite early in the transmit line-up (practically speaking, done right at the output of the upmixer, or in the case of an integrated solution, at the output of the RFIC when dealing with internal coupling). But even in the latter case there is a big difference compared to other TX related problems that may require a sensing of the output of the total line-up, as for instance in the power control problem discussed in Section 9.1.1. In the present case, this results in the possibility of using a fully integrated sense and feedback system. A side effect of this is that we can expect to face a phase offset ϕ sufficient low that a coherent receiver is no longer of interest with regard to practical orders of magnitude for cancellation purposes. This means that a simple real downconversion can be used and driven successively by the two quadrature TX LO waveforms, $lo_{\mathrm{p}}(t)$ and $lo_{\mathrm{q}}(t)$, in order to recover $\mathrm{Re}\{\tilde{s}_{\mathrm{leak}}\}$ and $\mathrm{Im}\{\tilde{s}_{\mathrm{leak}}\}$. The compensation values can then be derived directly according to equation (9.3), thus avoiding the need for a more complex algorithm for that purpose.

We can go a step further in the simplification if we consider on the one hand the root causes at the origin of the presence of this LO leakage and on the other hand that we are dealing with a CW signal. As recalled above and as can be seen in the complex envelope of this leakage signal given by equation (6.149), this CW leakage is generated first by the DC offset inherently present at the low frequency input port of the upmixer and then by the direct coupling of the LO signal toward the output RF port. But, even in the latter case, we can assume that the main component is the direct feedthrough at the physical device stage rather than a pure electromagnetic coupling toward RF ports. This means that when shutting down one of the P or Q physical mixers, we may cancel the corresponding component of the LO leakage, at least up to first order. As a result, assuming for instance that we are using only the P mixer in the transmit line up while the Q mixer is turned off, we can expect to recover an LO leakage whose complex envelope, when defined as centered around the LO angular frequency ω_{LO}, reduces to $\mathrm{Re}\{\tilde{s}_{\mathrm{leak}}\}$. This means the LO leakage can be given by

$$s_{\mathrm{leak}}(t) = \mathrm{Re}\left\{\mathrm{Re}\{\tilde{s}_{\mathrm{leak}}\}e^{+j\omega_{\mathrm{LO}}t}\right\} = \mathrm{Re}\{\tilde{s}_{\mathrm{leak}}\}\cos(\omega_{\mathrm{LO}}t). \tag{9.4}$$

Now supposing that we inject an additional correction value $dc_{\mathrm{comp,p}}$ at the input of the P branch of the upmixer, we recover a leakage of the form

$$s_{\mathrm{leak}}(t) = (dc_{\mathrm{comp,p}} + \mathrm{Re}\{\tilde{s}_{\mathrm{leak}}\})\cos(\omega_{\mathrm{LO}}t). \tag{9.5}$$

Thus, in order to cancel the leakage in this particular case, we still need to set $dc_{\mathrm{comp,p}}$ such that $dc_{\mathrm{comp,p}} + \mathrm{Re}\{\tilde{s}_{\mathrm{leak}}\} = 0$. But, looking more carefully at the structure of $s_{\mathrm{leak}}(t)$, we see that due to the simplification linked to the fact that we use only one branch of the complex upmixer at a time, a change in the sign of this sum leads to a π phase shift on $s_{\mathrm{leak}}(t)$. As a result, by simply tracking for the phase of this CW signal when changing the value of $dc_{\mathrm{comp,p}}$, we can determine the optimum value by searching for the corresponding π phase shift. This can be done quite easily by using a simple RF phase detector, which is really efficient in terms of implementation complexity. However, for this approach to be possible, it needs to be done without any modulating signal in order to rely on the CW structure of the leakage. As a result, such a simplified scheme is mainly suited to a calibration of the line-up before any transmission.

9.1.3 P/Q Imbalance Compensation

Another classical source of limitations in the performance of transmitters that can be improved by a dedicated digital compensation is the imbalance between the P and Q branches of the line-up when based on a Cartesian representation of the complex modulating waveform. Here, by "imbalance" we mean the equivalent amplitude and phase error between the two LO signals that are used to physically implement a complex frequency upconversion. As extensively discussed in Chapter 6, these two signals should be exactly in quadrature to represent the real and imaginary parts of the complex exponential that is expected to be used to implement such frequency conversion. But due to limitations in the physical implementation, we are necessarily faced with some imbalance in practice. However, we should qualify this slightly as the present problem is not necessarily associated with all the Cartesian transmit architectures. For instance, it is almost non-existent in a real-IF approach, as discussed in Section 8.1.4. Indeed, the complex frequency upconversion is implemented in the digital domain in that case. As a result, only small residual limitations, linked to a fixed point implementation for instance, can limit performance. But, as soon as the complex frequency upconversion is performed in the RF/analog world, we may consider degradations in the overall performance due to such imbalance between the P and Q branches of the line-up.

From the system impact point of view, such imbalance leads to the rise in what we called an image signal. As discussed in Section 6.2.2, in the present complex upconversion case, the image signal recovered at the output of the processing is nothing more than the bandpass signal whose complex envelope, when defined as centered around the LO angular frequency, is the complex conjugate of the expected complex modulating waveform. More precisely, assuming for instance that we rely on the use of the positive complex exponential $e^{j\omega_{LO}t}$ for implementing the frequency upconversion, we can use equation (6.50) to write the complex envelope of the bandpass signal recovered at the output of the processing, $\tilde{s}_{TX}(t)$, as

$$\tilde{s}_{TX}(t) = (\alpha_+^+)\tilde{s}_{mod}(t) + (\alpha_+^-)^* \tilde{s}_{mod}^*(t). \tag{9.6}$$

Here, the complex envelope is defined as centered around the LO angular frequency, and $\tilde{s}_{mod}(t)$ stands for the lowpass modulating waveform present at the input of the complex frequency upconversion. In addition, (α_+^+) and $(\alpha_+^-)^*$ are complex numbers that depend only on the gain and phase imbalance existing between the P and Q LO waveforms in accordance with equation (6.42). Practically speaking, such complex frequency upconversion is of interest mainly when dealing with the frequency upconversion of a waveform that is centered around a sufficiently low frequency that a representation in terms of a real bandpass signal is impossible. This means dealing with a complex modulating waveform centered around DC, or at least around a sufficiently low IF. As a result, in most cases the resulting image signal is either a pure in-band distortion or a distortion lying in the close vicinity of the wanted signal sideband (see Figure 6.19). We thus cannot rely on a potential dedicated filtering stage at the output of the frequency upconversion to cancel this unwanted signal. The only realistic way forward is to find a way to eliminate the term $(\alpha_+^-)^* \tilde{s}_{mod}^*(t)$ in the above expression so that the complex envelope of the bandpass signal of interest $s_{TX}(t)$ indeed matches the complex lowpass modulating waveform $\tilde{s}_{mod}(t)$.

However, this description can be seen as a simplified picture of what happens in a physical implementation. As discussed throughout Section 6.3, most physical RF/analog mixing stages are implemented using chopper-like devices. As a result, rather than a SSB centered around the LO angular frequency ω_{LO} at the output of the processing, we recover a collection of upconverted sidebands centered around all the harmonic angular frequencies $k\omega_{\text{LO}}$. The imbalances we are faced with between the two P and Q LO signals are not directly gain and phase imbalance, but rather DtyCy delay and amplitude imbalance. However, the tones of the same order involved in the series expansion of those LO waveforms do suffer from gain and phase imbalance. But now, the magnitude of this imbalance depends on the harmonic order k. More precisely, the structure of the complex envelope of the bandpass signal centered around the harmonic angular frequency $k\omega_{\text{LO}}$ is of the form given by equation (9.6), as illustrated in Figure 6.40 for instance. But the factors that weight the expected complex envelope, $\tilde{s}_{\text{mod}}(t)$, and that corresponding to the image signal, $\tilde{s}^*_{\text{mod}}(t)$, depend on k according to the material discussed in "Practical LO spectral content" (Section 6.3.2). However, if we refer to the material discussed in that section, we see that only the tones of odd order, i.e. corresponding to k of the form $2l-1$, are of interest for the implementation of the frequency conversion and thus for the definition of the image signal. Consequently, even if the results below could be easily extended to the tones of even order at the cost of a slight modification of the structure of the compensation scheme, we focus here on the use of odd order tones for implementing the frequency conversion. The factors that weight the expected complex envelope and its complex conjugate are then the $(\alpha^+_+)_{2l-1}$ and $(\alpha^-_+)_{2l-1}$ factors given in our general case by equation (6.134). Thus, given the expression of those factors, and relying on the structure of equation (9.6) to derive the compensation to apply, we see that we can address with a single scheme the cancellation of the image signal resulting from the upconversion achieved through the use of any of the odd order harmonics of the LO signal. But, having the factors $(\alpha^+_+)_{2l-1}$ and $(\alpha^-_+)_{2l-1}$ depending on l, the parameters to be used for the compensation also necessarily do. We end up in the same situation as encountered with the digital predistortion discussed in Section 9.1.4: by tuning the lowpass modulating waveform we can expect to clean up the complex envelope of the wanted sideband of interest, but not all the sidebands at the same time. However, this is not an issue as in practical architectures only one sideband is of interest, the others being filtered out through the use of filtering stages present at the output of the upmixing stage. What remains is to improve the in-band performance that cannot be addressed in the same way. For the sake of simplicity, we assume in what follows that this sideband is the one centered around the fundamental LO angular frequency ω_{LO}, i.e. $s_{\text{TX,H1}}(t)$. This sideband corresponds to the one that is of interest in most of the practical applications as the fundamental tone of the LO signal gives the highest conversion gain among all the harmonics present in this waveform. As a result, we then consider the equivalent lowpass model for our study illustrated in Figure 9.7, taking here the simple direct conversion line-up as an example.

In order to derive a structure for the compensation of the corresponding in-band distortion, it is of interest to transpose equation (9.6) into a matrix form that links the real and imaginary parts of the complex envelope of the sideband assumed of interest, i.e. $\tilde{s}_{\text{TX,H1}}(t) = p_{\text{TX,H1}}(t) + \text{j}q_{\text{TX,H1}}(t)$, to those of the complex modulating waveform $\tilde{s}_{\text{mod}}(t) = p_{\text{mod}}(t) + \text{j}q_{\text{mod}}(t)$. For that purpose, we simply reorder equation (9.6) as

$$p_{\text{TX,H1}}(t) + \text{j}q_{\text{TX,H1}}(t) = \left[(\alpha^+_+) + (\alpha^-_+)^*\right]p_{\text{mod}}(t) + \left[(\alpha^+_+) - (\alpha^-_+)^*\right]\text{j}q_{\text{mod}}(t). \tag{9.7}$$

At the same time, we can write from equation (6.42) that

$$(\alpha^+_+) + (\alpha^-_+)^* \propto 1, \tag{9.8a}$$

$$(\alpha^+_+) - (\alpha^-_+)^* \propto g(\cos(\delta\phi) + \text{j}\sin(\delta\phi)), \tag{9.8b}$$

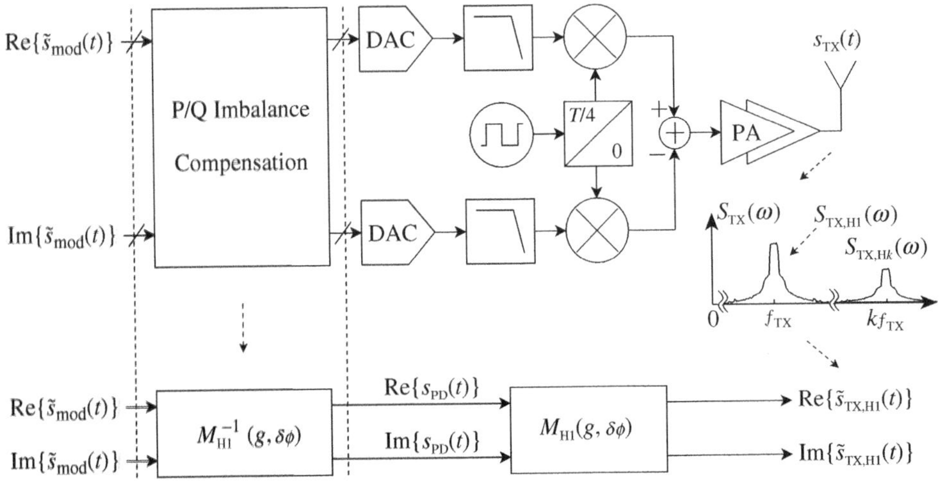

Figure 9.7 Equivalent lowpass model for investigating the digital compensation of the imbalance between the fundamental tones of the P and Q LO signals on the transmit side – With RF/analog mixers implemented as choppers, we recover a collection of sidebands centered around the harmonics of the LO angular frequency, $k\omega_{LO} = k\omega_{TX}$ in the present case. As for any of those sidebands, a linear transformation holds between the real and imaginary parts of the complex envelope $\tilde{s}_{TX,H1}(t)$, defined as centered around ω_{TX}, of the sideband of interest and the real and imaginary parts of the lowpass complex modulating waveform $\tilde{s}_{mod}(t)$. The parameters of the linear transformation, $\boldsymbol{M}_{H1}$, depend on the gain and phase offsets between the fundamental tones of the P and Q LO signals, g and $\delta\phi$ respectively, as defined by equations (6.29) and (6.30). The compensation of the distortion experienced by this sideband can thus be achieved by applying the inverse transformation $\boldsymbol{M}_{H1}^{-1}$.

with the same proportionality factor in the two cases. Here, g and $\delta\phi$ stand for the gain and phase imbalance as existing between the fundamental tones of the P and Q LO signals as defined by equations (6.29) and (6.30). Thus, suppressing this proportionality factor which is of no use in the present discussion, we can write from the two equations above that

$$\begin{pmatrix} p_{TX,H1}(t) \\ q_{TX,H1}(t) \end{pmatrix} = \boldsymbol{M}_{H1}(g, \delta\phi) \begin{pmatrix} p_{mod}(t) \\ q_{mod}(t) \end{pmatrix}, \tag{9.9}$$

with

$$\boldsymbol{M}_{H1}(g, \delta\phi) = \begin{pmatrix} 1 & -g\sin(\delta\phi) \\ 0 & g\cos(\delta\phi) \end{pmatrix}. \tag{9.10}$$

We thus see that a simple linear transformation holds between the real and imaginary parts of $\tilde{s}_{TX,H1}(t)$ and the lowpass complex modulating waveform present at the input of the complex upmixer. As a result, if we feed the complex upconverter stage with a predistorted complex signal $s_{\mathrm{PD}}(t) = p_{\mathrm{PD}}(t) + jq_{\mathrm{PD}}(t)$ such that

$$\begin{pmatrix} p_{\mathrm{PD}}(t) \\ q_{\mathrm{PD}}(t) \end{pmatrix} = \boldsymbol{M}_{\mathrm{H1}}^{-1}(g, \delta\phi) \begin{pmatrix} p_{\mathrm{mod}}(t) \\ q_{\mathrm{mod}}(t) \end{pmatrix}, \tag{9.11}$$

we then achieve our goal, i.e.

$$\begin{aligned} \begin{pmatrix} p_{\mathrm{TX,H1}}(t) \\ q_{\mathrm{TX,H1}}(t) \end{pmatrix} &= \boldsymbol{M}_{\mathrm{H1}}(g, \delta\phi) \begin{pmatrix} p_{\mathrm{PD}}(t) \\ q_{\mathrm{PD}}(t) \end{pmatrix} \\ &= \boldsymbol{M}_{\mathrm{H1}} \boldsymbol{M}_{\mathrm{H1}}^{-1}(g, \delta\phi) \begin{pmatrix} p_{\mathrm{mod}}(t) \\ q_{\mathrm{mod}}(t) \end{pmatrix} = \begin{pmatrix} p_{\mathrm{mod}}(t) \\ q_{\mathrm{mod}}(t) \end{pmatrix}. \end{aligned} \tag{9.12}$$

Thus, in order to compensate for the distortion of the sideband of interest by canceling the corresponding bandpass image signal, we simply need to implement the linear transformation of the complex modulating waveform corresponding to

$$\boldsymbol{M}_{\mathrm{H1}}^{-1}(g, \delta\phi) = \begin{pmatrix} 1 & \theta_{\mathrm{p}}(\delta\phi) \\ 0 & \theta_{\mathrm{q}}(g, \delta\phi) \end{pmatrix}, \tag{9.13}$$

with

$$\theta_{\mathrm{p}}(\delta\phi) = \tan(\delta\phi), \tag{9.14a}$$

$$\theta_{\mathrm{q}}(g, \delta\phi) = 1/[g\cos(\delta\phi)]. \tag{9.14b}$$

Transposed to the digital domain, we have the possible structure for the overall line-up as shown in Figure 9.8. We also observe that given the same structure for the expression for the complex envelopes of all the bandpass signals recovered around any of the odd order LO harmonic angular frequencies, the same structure necessarily holds for the linear transformation linking their real and imaginary parts with those of the complex modulating waveform. The only difference compared to the present case lies in the components of the corresponding matrix that depend on the harmonic order as the gain and phase imbalance do. We thus conclude that the compensation structure shown in Figure 9.8 can be used to clean up any of the odd order sidebands recovered at the output of the upmixing process by proper selection of the parameters θ_{p} and θ_{q}, but not many of them at the same time. However, as highlighted above, only the cleaning of the sideband of interest is important in practice.

However, we still need to find a way to derive the θ_{p} and θ_{q} parameters to be used. In most applications they are unknown a priori. They thus need to be derived, for instance during calibration of the line-up. For that purpose, we observe that the gain and phase imbalance, g and $\delta\phi$ respectively, existing between the two tones of interest in the series expansion of the P

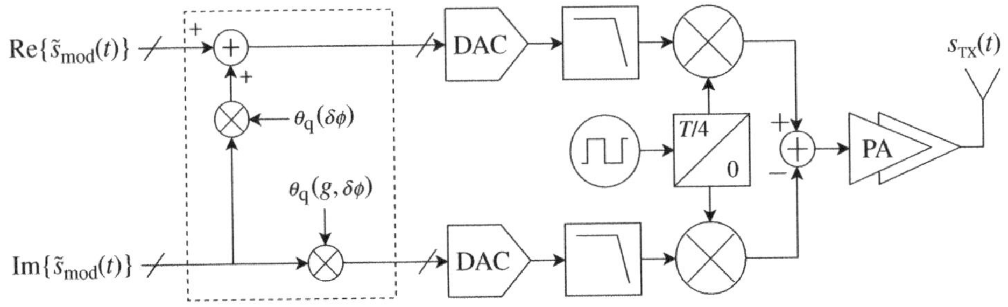

Figure 9.8 Digital compensation of the imbalance between the fundamental tones of the P and Q LO signals on the transmit side – The structure of the complex envelope of the sideband of interest, here assumed to be the one centered around the fundamental tone of the LO signal, in the presence of gain and phase imbalance leads to a possible compensation scheme for the effects of this imbalance based on the linear transformation given by equation (9.13), with $\theta_p(\delta\phi)$ and $\theta_q(g, \delta\phi)$ given by equation (9.14). However, the same compensation structure can be used to compensate for the distortion experienced by any of the odd order sidebands recovered at the output of the upconversion process by proper selection of the θ_p and θ_q parameters.

and Q LO signals cannot be easily evaluated by inspection only of the output bandpass signal of interest. We thus need to rely on indirect measurement, basically through the monitoring of the side effects of the presence of this imbalance, i.e. through the rise in the image signal in practice. However, we see from equation (9.6) that this can be achieved quite easily by proper selection of the complex modulating waveform. By selecting for instance a simple CW test signal lying at a frequency offset $\delta\omega$ from the LO angular frequency, we get

$$\tilde{s}_{mod}(t) = e^{j\delta\omega t}. \tag{9.15}$$

As a result, considering for instance the complex envelope $\tilde{s}_{TX,H1}(t)$ of the sideband centered around the LO fundamental angular frequency as an example, we can write from equation (9.6) that

$$\tilde{s}_{TX,H1}(t) = (\alpha_+^+)e^{j\delta\omega t} + (\alpha_+^-)^* e^{-j\delta\omega t}. \tag{9.16}$$

Thus, by bypassing the compensation block $\boldsymbol{M}_{H1}^{-1}$ during the calibration phase, i.e. by setting $\theta_p(\delta\phi) = 0$ and $\theta_q(g, \delta\phi) = 1$, and by using a pure CW test signal at a given frequency offset, we can have direct access to the parameters (α_+^+) and $(\alpha_+^-)^*$ by measuring the amplitude and phase of the two resulting tones centered around the LO fundamental angular frequency as illustrated in Figure 9.9. This can be done easily through a Fourier transform that gives access to both pieces of information at the same time. Then, given (α_+^+) and $(\alpha_+^-)^*$, we can simply use equation (9.8) to derive

$$ge^{j\delta\phi} = \frac{(\alpha_+^+) - (\alpha_+^-)^*}{(\alpha_+^+) + (\alpha_+^-)^*} = R. \tag{9.17}$$

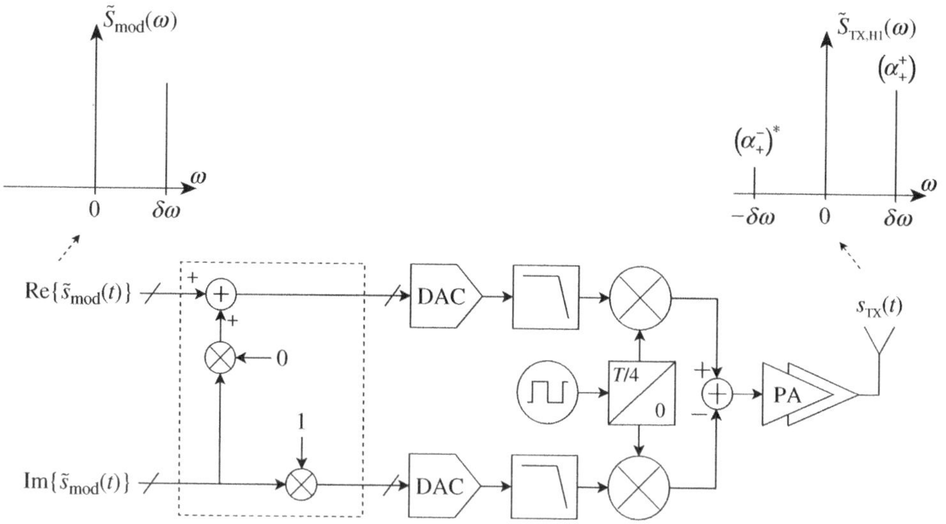

Figure 9.9 Derivation of the parameters for compensating the imbalance between the fundamental tones of the P and Q LO signals on the transmit side by using a CW test signal – In order to derive the parameters of the compensation for the imbalance between the fundamental tones of the P and Q LO signals illustrated in Figure 9.8, a CW test signal can be used. The amplitude and phase of the corresponding RF signals yield the values of (α_+^+) and (α_+^-) and thus finally θ_p and θ_q according to equation (9.18). More generally, the same procedure could be used to derive the $(\alpha_+^+)_{2l-1}$ and $(\alpha_+^-)_{2l-1}$ parameters involved in the distortion of any of the odd order sidebands recovered at the output of the upconversion process.

On the other hand, we have that θ_p and θ_q are functions of g and $\delta\phi$ according to equation (9.14). We can thus deduce the values to be used for our compensation from the measurements of the calibration test tone through

$$\theta_\text{p}(\delta\phi) = \frac{\text{Im}\{R\}}{\text{Re}\{R\}}, \tag{9.18a}$$

$$\theta_\text{q}(g, \delta\phi) = \frac{1}{\text{Re}\{R\}}. \tag{9.18b}$$

Obviously, the same procedure could be considered for any of the odd order sidebands recovered at the output of the upconversion if one aims to compensate for its distortion. The only difference lies in the theoretical expressions for the $(\alpha_+^+)_{2l-1}$ and $(\alpha_+^-)_{2l-1}$ parameters that need to be considered in order to link the amplitude and phase of the measured test tones and the correction parameters θ_p and θ_q.

However, as discussed so far for all the compensation schemes encountered on the transmit side, a static compensation remains efficient enough only as long as the drifts of the parameters to be compensated, g and $\delta\phi$ in the present case, remain low enough in operating conditions to allow the required performance to be achieved. Otherwise, unless we have access to a full characterization of the line-up under operating conditions, we may need to consider an adaptive algorithm to track those drifts. This must obviously rely on an additional sense and feedback path that can be seen as a net cost for a transmit line-up. Moreover, as we are talking here about the compensation of the complex envelope of the transmit sideband of interest, we may need to consider a full complex auxiliary receiver for this feedback, which is the worst configuration imaginable in terms of implementation cost. If this cannot be avoided, the algorithms to be used are quite similar to those encountered for receivers. For those line-ups we indeed naturally have such sensing available to adapt the compensation in real time: see the discussion in Section 9.2.3.

It would thus appear from the above derivation and the resulting structure of the compensation block that we have great flexibility in the imbalance we can compensate. Looking at equation (9.14), unless we have $\delta\phi$ tending toward $\pi/2$, which would result in θ_p and θ_q tending toward infinity, we can compensate for almost any imbalance value. This could be of interest if we recall that the P and Q LO waveforms are often generated through the frequency division of the same high frequency signal delivered by a RF synthesizer. Indeed, depending on the frequency division ratio used, it is not obvious that we can achieve two LO signals of same frequency but in quadrature. We might thus think of using such a quadrature compensation scheme to simplify the frequency planning of the transmitter and rely on LO signals that are not in quadrature. However, this approach is obviously too good to be true. We get two problems associated with the correction of high quadrature errors. First of all, with the parameters θ_p and θ_q increasing, we need to manage signals that need a higher DR to be physically represented in the implementation. This can lead to penalties in the efficiency of the line-up. But, most problematic, the IRR achieved, as given by equation (6.38), varies more and more rapidly as we go away from the exact quadrature corresponding to $g = 1$ and $\delta\phi = 0$, as illustrated in Figure 9.10. As a result, the higher the original quadrature error, the

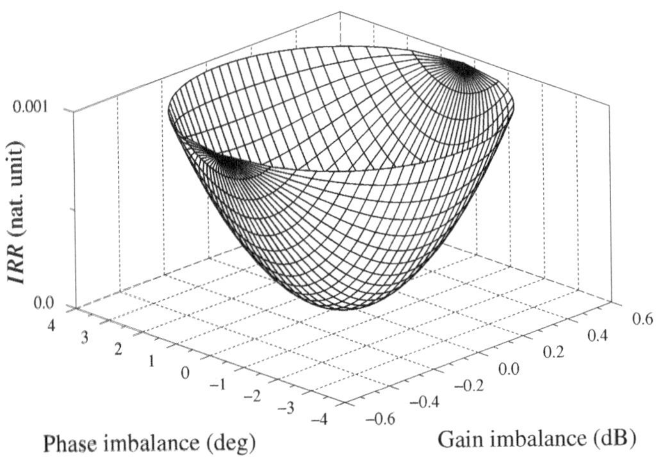

Figure 9.10 IRR as a function of gain and phase imbalance – Given an IRR that is a function of g and $\delta\phi$ according to equation (6.38), $\partial_g IRR(g, \delta\phi) = 0$ and $\partial_{\delta\phi} IRR(g, \delta\phi) = 0$ only when $g = 1$ and $\delta\phi = 0$. Moreover, the higher the quadrature and gain error, the sharper the slopes of the IRR. Thus, the higher a given imbalance, the less stable the performance achievable through a static calibration of the line-up when faced with drifts in operating conditions.

more sensitive the compensation is to any drift in operating conditions. This could make it impossible to rely on a simple calibration of the line-up.

In order to conclude this section, we also observe that the imbalance model used in the present section is by nature independent of the frequency extent of the signal being upconverted. Although this corresponds to the behavior that can be expected from devices behaving exactly as choppers, additional phenomena in the line-up can cause the imbalance experienced by the signal being processed to vary within its frequency bandwidth. This can result for instance from different gain and group delay variations within the signal bandwidth along the P and Q data paths. This problem is more acute when using wide bandwidth modulating waveforms. In that case, the model obviously needs to be refined. Fortunately, the present approach is valid in most practical cases.

9.1.4 Predistortion

Let us now focus on the predistortion of the signal being processed. Here, by "predistortion" we obviously mean a scheme that intentionally distorts the signal so as to nullify the impact of an unwanted distortion experienced at a later stage of the line-up. We can already say that in order to be able to perform such processing, we necessarily talk about the compensation of the impact of a deterministic imperfection in the transfer function experienced by the signal.

Thus, among the various problems discussed throughout Part II of this book, at least two are within our scope here. the distortion induced by the various filtering stages present in the line-up as illustrated in Section 4.4; and the distortion resulting from the nonlinearities as presented throughout Chapter 5. However, recalling the discussion in those chapters, there is a major difference between those two situations related to the fact that we have either a linear or a nonlinear relationship between the signal being processed and the resulting unwanted generated term. As discussed below, this necessarily leads to different compensation schemes.

Linear Distortion Case

As discussed in Section 4.4, a filtering stage can result in the generation of a linear distortion of the signal being processed, thus leading to the creation of what we have called linear EVM. But due to this linear relationship between the signal being processed and the distortion term generated, a linear compensation can succeed in canceling this problematic component. As highlighted in Section 4.4, such compensation can eventually be present on the receive side through channel estimation and equalization. Such processing often reduces to the estimation and compensation of the linear distortions experienced by the received signal, as is the case when going through the propagation channel. Although this may vary from case to case, it explains why we consider the transmit side here for the illustration of a fixed linear distortion compensation. However, equivalent schemes can be considered on the receive path if required.

Practically speaking, such linear compensation means a filtering stage that equalizes the response of the problem filters. This equalization, which can advantageously take place in the digital part of the line-up, has to be understood both in terms of amplitude and group delay variations in the frequency bandwidth of the transmit signal. What we need to know is whether a fixed equalization is sufficient for a targeted performance, or whether an adaptive scheme is required in order to track the drifts in operating conditions of the filters to be compensated. In the former case, a simple derivation of the equalization stage based on the design characteristics of the line-up can be enough. This is obviously not so simple in the latter case, where an adaptive linear equalization scheme may be required. But most problematic, a sense and feedback of the distorted signal is needed for such adaptation to be possible.

As discussed in the introductory part of this chapter, such an additional path is problematic on the transmit side as it can be seen as a net cost from the implementation point of view. However, we observe that in practice the filtering stages we are talking about must have a cut-off frequency close enough to the edge of the spectrum of the signal being processed in order to induce a non-negligible distortion on it. And in most practical transmit architectures, such channel-like filters are located in the baseband part of the line-up, prior to any channel selection, as illustrated throughout Chapter 8. As a result, the feedback that would be required for the derivation of our adaptive scheme can be implemented by sensing the lowpass modulating waveform, for instance through both its real and imaginary parts $p(t)$ and $q(t)$, just at the output of the problem filters, i.e. prior to any frequency upconversion. Such an approach thus limits the implementation cost, both in terms of die area and power consumption, as compared to a full RF sensing path.

We also observe that the only filtering stages from which we can expect variations in operating conditions are the analog filtering stages. But practically speaking, on the transmit side a lowpass analog filter is often used mainly as a reconstuction filter at the DAC output, i.e. in order to attenuate the residual unwanted sideband copies present at their output. This means in the present case that we can set the cut-off frequency of those reconstruction filters sufficiently high so that their contributions are negligible in terms of linear distortion. In that case, only the impact of the deterministic contributions in the line-up, for instance associated with the digital filters or to the DAC aperture effect, are relevant, thus reducing our problem to the derivation of a fixed equalization stage.

Nonlinear Distortion Case

The situation is obviously different in the nonlinear distortion case. On the one hand, a simple linear equalization cannot compensate for the generation of such nonlinear EVM; on the other hand, the problematic device is often located in the RF world. This obviously does not help in reducing the implementation cost of potential systems based on a sense of the distorted signal. However, as pointed out in the discussion in "PA working in the weak nonlinear area" (Section 8.1.7), despite this issue, many schemes have been derived based on such RF sensing. This is for instance the case for the negative feedback based systems that intrinsically lead to robust and performing systems, or to the feedforward techniques that allow the instability issues of the negative feedback schemes to be overcome, albeit at the cost of lower performance. Most of these linearization techniques are based on an extensive use of RF/analog blocks. It may thus seem logical that they have historically been considered first. However, when considering the capabilities of modern silicon technologies in terms of digital signal processing integration, it seems more profitable to consider a predistortion technique whose principles are illustrated in Figure 8.24. Indeed, although in that case again the first historical implementations were pure RF/analog ones [82], nothing prevents us from considering carrying out this predistortion directly on the modulating signal, i.e. still in baseband, and preferably in the digital domain.

As such distortion of the modulation due to RF nonlinearity can occur only for non-constant amplitude schemes, we continue to consider a complex modulation in what follows.

We can already comment that by acting on the lowpass modulating waveform, we can only act in practice on the modulation part of the various sidebands recovered at the output of the transmitter. By "various sidebands" we mean the unwanted bandpass signals generated as centered around the harmonics of the transmit carrier frequency f_{TX} by, on the one hand, the chopper-like behavior of the upmixers or the impairments in their implementation as detailed in Section 6.4.1, and on the other hand, by the nonlinearity in the RF stages as illustrated throughout Chapter 5. As a result, those unwanted signals can only be canceled, or at least attenuated toward an acceptable level, by using an efficient enough filtering in order to recover a clean enough bandpass transmit signal composed of the wanted sideband only. From the system point of view, the interest of a digital predistortion scheme as considered here lies exclusively in the improvement of the characteristics of the wanted bandpass signal centered around f_{TX}, in terms of EVM or ACLR for instance.

Thus, the digital predistortion can be studied through an equivalent lowpass model. For that purpose, we consider the complex gain approach to model the bandpass nonlinearity experienced by the wanted sideband $s_{\text{TX,H1}}(t)$ lying around the transmit carrier frequency. As discussed in Section 5.3, this allows us to take into account both the AM-AM and AM-PM conversion at the same time. As a side effect, this means that we do not consider here the compensation of the memory effect. This is, however, often not the limiting factor in practical wireless devices in respect of the AM-AM and AM-PM effects. As a result, we can consider the equivalent model shown in Figure 9.11 in order to illustrate our discussion.

To discuss further the structure of the transfer function of the digital predistortion block, $\tilde{F}$, which we need to consider for our application, we first recall that the complex gain, $\tilde{G}_{\text{PA}}$, that models the bandpass nonlinearity is a function of the instantaneous power of the bandpass signal entering the nonlinear device, as illustrated by equation (5.308). In our model, assuming that no distortion occurs between the output of the predistortion block and the input of this nonlinear RF device, those instantaneous power variations are proportional to those of the complex modulating signal as recovered at the output of this former block, i.e. of $\tilde{s}_{\text{PD}}(t)$ as illustrated in Figure 9.11. Thus, assuming that all the complex envelopes we are dealing with are defined as centered around the carrier frequency f_{TX}, we can write the complex envelope of the bandpass signal of interest at the transmitter output as

$$\tilde{s}_{\text{TX,H1}}(t) = \tilde{G}_{\text{PA}}(|\tilde{s}_{\text{PD}}(t)|^2)\tilde{s}_{\text{PD}}(t). \tag{9.19}$$

But taking into account the action of the predistortion block, we have

$$\tilde{s}_{\text{PD}}(t) = \tilde{F}(\tilde{s}_{\text{mod}}(t)), \tag{9.20}$$

so that

$$\tilde{s}_{\text{TX,H1}}(t) = \tilde{G}_{\text{PA}}(|\tilde{F}(\tilde{s}_{\text{mod}}(t))|^2)\tilde{F}(\tilde{s}_{\text{mod}}(t)). \tag{9.21}$$

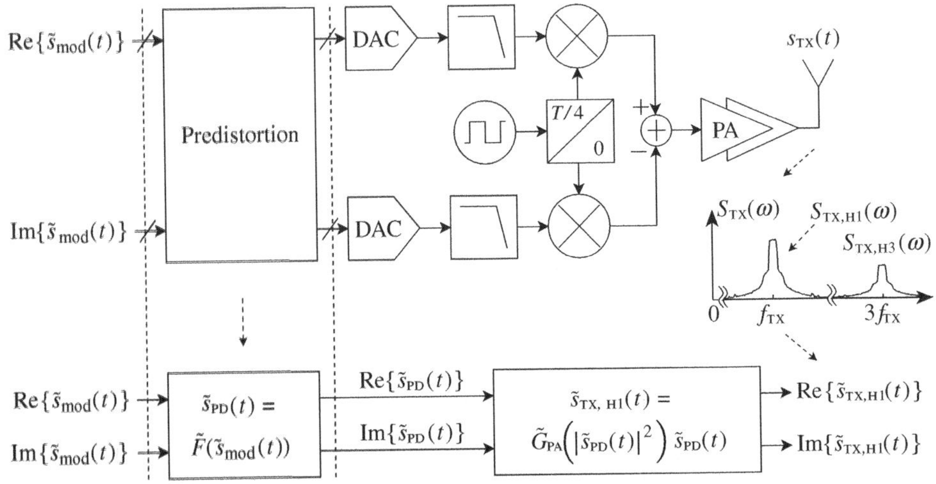

Figure 9.11 Equivalent lowpass model for investigating a digital predistortion scheme expected to compensate a bandpass nonlinearity – Classically, a transmit signal is composed of the bandpass signal $s_{TX,H1}(t)$ corresponding to the sideband of interest, i.e. lying at the carrier frequency f_{TX}, and a collection of unwanted bandpass signals lying at harmonic frequencies, such as $s_{TX,H3}(t)$ at $3f_{TX}$. By acting on the complex lowpass modulating waveform $\tilde{s}_{mod}(t)$, we cannot reduce those unwanted sidebands, but rather compensate for the distortion experienced by the wanted bandpass signal $s_{TX,H1}(t)$. The predistortion scheme dedicated to the improvement of $s_{TX,H1}(t)$ can then be studied by considering a complex gain $\tilde{G}_{PA}$ that models the bandpass nonlinearity experienced by $s_{TX,H1}(t)$.

Thus, if we want the predistortion block to compensate exactly for the bandpass nonlinearity, we must have $\tilde{F}$ such that

$$\tilde{s}_{TX,H1}(t) = \tilde{G}\tilde{s}_{mod}(t), \tag{9.22}$$

with $\tilde{G}$ a complex constant that represents the expected ideal linear gain of the transmit line-up as experienced by the complex modulating waveform. As a result, comparing the two equations above, $\tilde{F}$ must fulfill

$$\tilde{G}_{PA}(|\tilde{F}(\tilde{s}_{mod}(t))|^2)\tilde{F}(\tilde{s}_{mod}(t)) = \tilde{G}\tilde{s}_{mod}(t). \tag{9.23}$$

Looking at this equation, we can surmise that the derivation of $\tilde{F}$ by a direct resolution may be not so straightforward. Moreover, it is unrealistic to expect to have exact analytical expressions for $\tilde{G}_{PA}$ and $\tilde{F}$ in practice. Finally, the direct implementation of a tricky analytical expression may lead to an unacceptable implementation cost in terms of die area and power consumption. For all these reasons, a classical implementation is what is often referred to as a mapping predistortion scheme. As the name suggests, this is based on the direct mapping of $\tilde{s}_{mod}(t)$ onto $\tilde{F}(\tilde{s}_{mod}(t))$, often achieved through the use of a LUT that directly stores the values of $\tilde{F}(\tilde{s}_{mod}(t))$ and that is addressed by the values of $\tilde{s}_{mod}(t)$. This approach is of considerable interest as it basically gives all the flexibility for the shape of $\tilde{F}$ in order to achieve the compensation of $\tilde{G}_{\mathrm{PA}}$. The values required can for instance be derived while calibrating the device. However, in the case of drift of $\tilde{G}_{\mathrm{PA}}$ in operating conditions, such initial calibration may not be sufficient to achieve the overall required performance and an update of the LUT may be required. This point is discussed later on. However, depending on the number of bits on which the modulating waveform $\tilde{s}_{\mathrm{mod}}(t)$ is coded, the memory required may be substantial. Moreover, in the present case the LUT we are talking about is indexed by both the real part $p_{\mathrm{mod}}(t)$ and the imaginary part $q_{\mathrm{mod}}(t)$ of $\tilde{s}_{\mathrm{mod}}(t)$. This may require the storage of an unacceptable number of words, depending on the combinations between the two signals and the number of bits they are coded on. In order to limit this problem, different strategies can be considered. For instance, carrying out the predistortion directly on the modulating constellation can reduce the LUT size as it obviously reduces the number of entries [83]. However, this may lead to other problems as we basically apply the predistortion prior to the pulse shaping filter. Moreover, in some applications such as a RFIC, we may have access only to the modulating waveform in its final form. In this case, we can still consider alternative tricks to reduce the number of LUT entries, e.g. a non-uniform sampling of those entries according to the statistics of the signal, or their subsampling associated with an interpolation function at the LUT output. But the problem with downsampling along the data path is that it necessarily leads to some performance degradation, for instance in terms of EVM as it reduces the number of degrees of freedom in the digital modulating signals [84].

Thus, it may be worth considering the implementation of an analytical formulation for $\tilde{F}$ in order to overcome the drawback of the direct mapping approach, as long as it is simple enough. As highlighted previously, the derivation of such an analytical formulation is not so straightforward in the light of equation (9.23). But, recalling the discussion throughout Chapter 5, we observe that the series expansion of nonlinear transfer functions leads to interesting results, at least from the system point of view so far. Perhaps the same might hold for our present problem. Let us therefore consider a series expansion for the complex gain $\tilde{G}_{\mathrm{PA}}$ in order to see if we cannot derive any interesting information about the structure of $\tilde{F}$. As discussed in Chapter 5, only the odd order part of the nonlinear transfer function can affect the bandpass signal centered around the carrier frequency. This is for instance highlighted by the general expression for the series expansion of the bandpass nonlinear transfer function given by equation (5.307). In order to initiate our investigation, it is relevant to consider the terms up to third order in this expression. Suppose that

$$\tilde{G}_{\mathrm{PA}}(|\tilde{s}_{\mathrm{PD}}(t)|^2) = \tilde{\alpha}_1 + \tilde{\alpha}_3|\tilde{s}_{\mathrm{PD}}(t)|^2. \tag{9.24}$$

In the same way, let us write the output of the predistortion block, $\tilde{s}_{\mathrm{PD}}(t) = \tilde{F}(\tilde{s}_{\mathrm{mod}}(t))$, as

$$\tilde{F}(\tilde{s}_{\mathrm{mod}}(t)) = \tilde{\beta}_1\tilde{s}_{\mathrm{mod}}(t) + \tilde{\delta}, \tag{9.25}$$

where $\tilde{\delta}$ contains terms of higher order in $\tilde{s}_{\mathrm{mod}}(t)$. Thus, by substituting the two expressions above into equation (9.19) and neglecting the terms of order higher than 3, we can write

$$\tilde{s}_{\mathrm{TX,H1}}(t) = \tilde{\alpha}_1\tilde{\beta}_1\tilde{s}_{\mathrm{mod}}(t) + \tilde{\alpha}_1\tilde{\delta} + \tilde{\alpha}_3|\tilde{\beta}_1|^2\tilde{\beta}_1|\tilde{s}_{\mathrm{mod}}(t)|^2\tilde{s}_{\mathrm{mod}}(t). \tag{9.26}$$

In order to cancel the nonlinear term in the expression for the output bandpass signal of interest, we simply need $\tilde{\delta}$ to be proportional to $|\tilde{s}_{\text{mod}}(t)|^2\tilde{s}_{\text{mod}}(t)$. Thus, using this result in equation (9.25), $\tilde{F}(\tilde{s}_{\text{mod}}(t))$ can be expanded up to third order as

$$\tilde{F}(\tilde{s}_{\text{mod}}(t)) = (\tilde{\beta}_1 + \tilde{\beta}_3|\tilde{s}_{\text{mod}}(t)|^2)\tilde{s}_{\text{mod}}(t). \tag{9.27}$$

What is interesting to remark when looking at this equation is that $\tilde{F}(\tilde{s}_{\text{mod}}(t))$ can in turn be expressed in the form of a complex gain $\tilde{G}_{\text{PD}}$ acting on the complex modulating waveform. Furthermore, this complex gain is a function of the instantaneous power of the signal it processes, as is the complex gain that models the bandpass nonlinearity we are considering. Based on the above derivations, the general expression for the transfer function of the predistortion block takes the form

$$\tilde{s}_{\text{PD}}(t) = \tilde{F}(\tilde{s}_{\text{mod}}(t)) = \tilde{G}_{\text{PD}}(|\tilde{s}_{\text{mod}}(t)|^2)\tilde{s}_{\text{mod}}(t), \tag{9.28}$$

with

$$\tilde{G}_{\text{PD}}(|\tilde{s}_{\text{mod}}(t)|^2) = \sum_{l=0}^{\infty} \tilde{\beta}_{2l+1}|\tilde{s}_{\text{mod}}(t)|^{2l}. \tag{9.29}$$

Such decomposition of the transfer function of the predistortion block in terms of complex gain thus leads to the alternative line-up implementation shown in Figure 9.12. Obviously, if

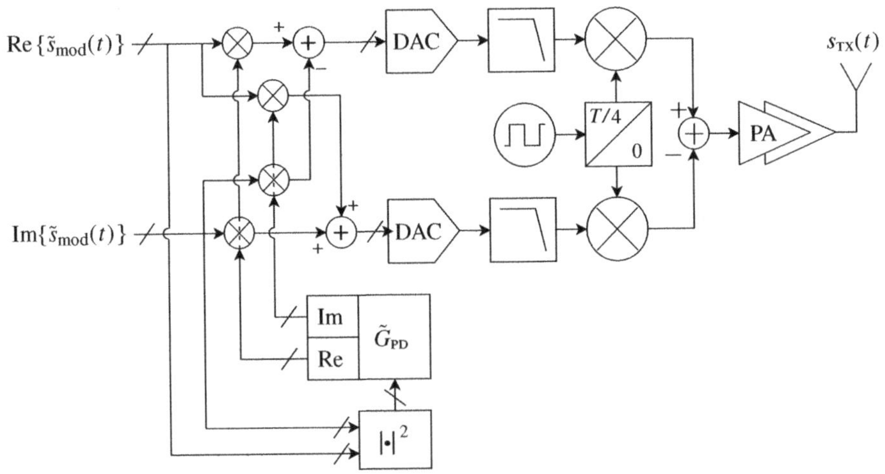

Figure 9.12 Digital predistortion processing as a complex gain – The processing of a predistortion block expected to compensate for a bandpass nonlinearity acting on a sideband of interest can be written as the application of a complex gain, $\tilde{G}_{\text{PD}}$, on the complex waveform $\tilde{s}_{\text{mod}}(t)$ representing the modulation of this bandpass signal. As given by equation (9.28), $\tilde{G}_{\text{PD}}$ is then necessarily a function of the instantaneous power $|\tilde{s}_{\text{mod}}(t)|^2$ of this modulating waveform.

we are dealing with only AM-AM conversion, the α_{2l+1} coefficients in the series expansion of $\tilde{G}_{PA}$ can be made real, as they can for $\tilde{G}_{PD}$. In that case, having $\mathrm{Im}\{\tilde{G}_{PD}\} = 0$ makes the implementation more straightforward but does not change the structure of the compensation. This structure is of interest because the quantity that represents the compensation, $\tilde{G}_{PD}$, is derived outside the direct data path. This means that the accuracy required for this quantity is driven only by the performance we expect in the compensation, and not by the deterioration of the modulating waveform as was the case in the direct mapping approach discussed above. Practically speaking, this means that if we fall back on a LUT based implementation to store the values of $\tilde{G}_{PD}$, we can rely more confidently on a reduced set of words, the exact number being driven only by the dynamic of the AM-AM and AM-PM characteristics that require compensation. Moreover, having the LUT addressed only by the instantaneous power $|\tilde{s}_{mod}(t)|^2$ of the complex modulating waveform instead of its real and imaginary parts as in the mapping approach necessarily helps to reduce the number of entries. However, as expected at the beginning of our derivation, a series expansion based implementation of $\tilde{G}_{PD}$ is also possible [85, 86].

Whatever the scheme used for the implementation of the predistortion, the validity of the compensation when dealing with variations of the nonlinear transfer function in operating conditions is an open question. If the performance degrades too much when such drift occurs, we may consider an adaptive system. For that purpose, unless we have access to a full characterization of the line-up in operating conditions, we can only rely on a sense path to adapt the predistortion block in real time. Here, "real time" simply means that we need to be able to track the temperature or supply drifts. And in practice such drifts may be seen as very slow compared to the time constants involved in the modulating schemes for instance. We thus need not fear any potential instability issue in such a closed loop system here, but can focus on the criteria to be used to update the model. For that purpose, as is the case for all the parameters that we need to track in a transmit line-up, we may wonder what is the simplest sense path that can achieve the adaptation of the block. Obviously, in the present case, our purpose is to compensate for the system impacts of both the AM-AM and the AM-PM conversion. And as discussed in Section 5.3.2, this reduces on the one hand to the generation of nonlinear EVM through an in-band distortion, and on the other hand to the generation of a pollution in the adjacent channels through the spectral regrowth phenomenon.

Basically, dealing with a complex modulation as presently assumed, we need to consider a full demodulator providing both the real and imaginary parts of the complex envelope of the transmit bandpass signal of interest in order to achieve a possible sense of the transmit EVM. This solution is obviously the most costly we can imagine for a transmit line-up, even if we observe that, considering the transmit sideband of interest, the same LO can be used on the direct and feedback paths if we are dealing with a direct conversion transmitter as illustrated in Figure 9.13. However, a sense of the power present in the adjacent channel may require a simpler scheme as it can be based on a real downconversion followed by a bandpass filter and a power detector for instance [86]. In any case, it is interesting to remark that, as illustrated in Section 5.3.2, it is the same generated distortion signal that is the cause for both the in-band EVM and the adjacent channel pollution. This means that whatever the criteria selected for the adaptation of the predistortion block, we can expect to improve the performance of both at the same time. Obviously, this suggests it would be preferable to go through the adjacent channel power measurement in order to keep the implementation cost reasonable if a sense is required

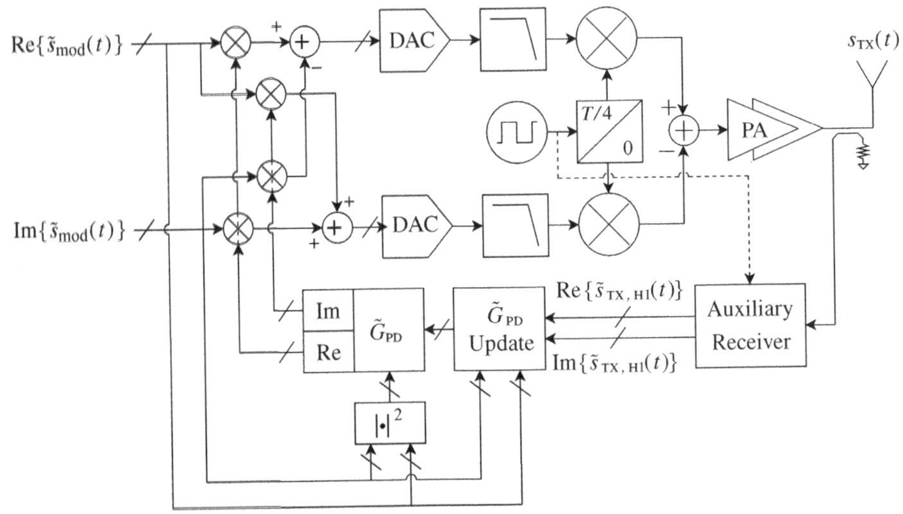

Figure 9.13 Complex gain predistortion block update based on EVM evaluation – When facing drifts in operating conditions of the nonlinear transfer function we are trying to compensate, an update of the predistortion block based on a sense of the transmit signal may be required. As such bandpass nonlinearity leads to both an in-band distortion and an out-of-band pollution, either the EVM or the ACLR metric can be used for the optimization. In the former case, when dealing with a complex modulation a full receiver is required to recover the complex envelope of the transmit bandpass signal of interest.

to track the drifts. In any case, whatever the selected type of feedback, various schemes can be considered for the adaptation of the predistortion block, basically more or less based on the tracking of the solution of equation (9.23) [85, 86].

In conclusion, we observe that, dealing with such a digital predistortion system, if we expect to compensate for a distortion term that leads to an adjacent channel power radiation, the distorted modulating waveform to be upconverted before going through the nonlinear device already necessarily spreads toward a wider bandwidth than the original undistorted signal. Another way to see this is to consider the spectral regrowth that necessarily results from the processing linked to the nonlinear transfer function present in the predistortion block. As a result, the system we are talking about needs to work at a higher sampling rate than is strictly necessary to correctly represent the modulating waveform. We can consider at least 3 times the sampling rate when dealing with a nonlinear term in the form x^3. We can thus expect the same proportionality in the increase of the current consumption of the digital, DAC, and reconstruction filter blocks. This increase in the overall power consumption of the solution is obviously even worse if a feedback path is required. Moreover, as illustrated in Figure 8.24, we know that whatever the approach, we cannot make the instantaneous power delivered by a RF amplifier higher than its Psat. This means that by using a predistortion system, the best we can do is to make the transfer function of the system linear up to its OPsat level. But practically speaking, in order to avoid any clipping of the transmit signal, we still need to keep this Psat higher than the maximum instantaneous power of the signal. As a result, the best we can expect is to save the margins that are taken on the amplifier back-off in the non-predistortion case – a few decibels in practice. As a result, unless dealing with huge RF power, the gain in current drain may remain quite reasonable and then needs to be carefully balanced in most applications with the overall increase in consumption of the system due to the presence of the additional blocks. These simple considerations explain why such a predistortion scheme is mainly encountered in applications where the RF performance cannot be met without it, rather than for improving the overall power consumption of a system that already meets that performance. However, only a deeper case by case analysis can conclude on the real benefits of such an approach for the overall power consumption problem.

9.1.5 Automatic Frequency Correction

Let us now consider the problem linked to the limited accuracy of the reference frequency sources in a transmitter. At least in most low cost consumer products, the reference sources embedded in the system have a limited intrinsic accuracy, say a few parts per million (ppm), in order to fix the orders of magnitude. But at the same time, the clocks to be used to drive the line-up can basically be generated only in two ways. On the one hand, they can be generated through the frequency multiplication of the reference signal using a PLL for instance. This is what is classically done for the generation of the LO signal as discussed in Section 4.3.1. On the other hand, they can be generated through a direct frequency division of this reference signal, for instance in order to derive a low frequency clock as could be used in some areas of the digital part of the line-up. Whatever method is used for the generation of a given clock signal, we get a fixed ratio between the frequency of this clock and that of the reference signal used for its generation. Thus the relative frequency accuracy of the signal generated this way can only be that of the initial reference source, i.e. a few parts per million in most practical cases if nothing is done.

From the system point of view, this inaccuracy has different impacts depending on the use of the clock. In that perspective, we can divide those periodic signals into three groups: clocks used to push the logic when considering a synchronous implementation; clocks used as LO signals to drive the RF/analog frequency converters; and clocks used as sampling clocks for the ADC or DAC blocks. Practically speaking, the system impact of an inaccuracy of a few parts per million will vary depending on the category of the signal. For instance, the synthesis of a synchronous digital block is already performed while taking into account a tolerance in the delays along the clock trees. This is required in order to cope with the variations in the characteristics of the buffering stages in operating conditions. And in practice, this tolerance is far above what is needed to handle the few parts per million we are presently discussing. As a result, this additional inaccuracy in the digital clocks does not lead to implementation issues of great significance. In contrast, there are various system impacts related to the inaccuracy in the LO frequency or in the frequency of the sampling clocks, thus leading to the need to consider a compensation, classically through what is called an automatic frequency correction (AFC) system. We observe that this limitation obviously impacts not only the transmit side, but also the receive side as the same reference frequency source can be used for both sides of a transceiver. This means that most of the concepts discussed in this section can be transposed to the receive side, albeit with some variations in terms of system impacts as discussed in Section 9.2.5.

Before discussing the different strategies for compensating for such frequency inaccuracy, we observe that we obviously need to know the size of the error we are dealing with before deciding how to compensate for it. Although the description of such error recovery is outside the scope of this book, generally speaking, it can only be estimated by comparing the inaccurate frequency we are dealing with, with a source that is accurate enough to behave as a true reference. But, by definition of our present problem, this accurate reference is not available in the physical implementation of the transmit line-up. It thus needs to be provided indirectly. For instance, this can be done through the broadcasting of a reference accurate RF signal by the network our device belongs to. This necessarily means taking a measurement through the receive part of the transceiver that the transmitter belongs to. Alternatively, it can be done through a direct feedback by the network of the error estimated on the carrier frequency delivered by our transmitter. What is interesting to see is that with this kind of strategy, the signal that is evaluated in order to derive the frequency error has necessarily experienced the propagation channel. As discussed in Section 2.3.3, if we are considering a mobile application this means that the frequency error that is evaluated that way also takes the Doppler effect into account. We thus observe that any correction of both the carrier frequency and the sampling rate based on this kind of evaluation also necessarily corrects for the system impacts of the Doppler effect.

Assuming that we know the frequency error we are dealing with, let us now focus on the different ways to manage the system impacts linked to this inaccuracy. For that purpose, we first consider the problem linked to the frequency error of the LO signal. Obviously, if nothing is done to compensate for this error, the first consequence for a transmitter is that the upconverted modulating waveform, originally generated as centered around DC, lies as centered around a carrier frequency that is incorrect by a few parts per million. This is unfortunately not allowed by most wireless networks, which can classically require an accuracy of the order of 0.01 ppm, i.e. some orders of magnitude below. Something therefore needs to be done in order to compensate for that frequency drift.

One possibility is a slight frequency transposition of the complex modulating waveform by an amount that corresponds to the error prior to performing the RF upconversion based on the use of the wrong LO waveform. Practically speaking, assuming a classical order of magnitude for the error of 10 ppm, and that we are dealing with a carrier frequency of 2 GHz, we would have an absolute error of 20 kHz to compensate. This approach thus corresponds to a low-IF transmitter with a very low IF. And dealing with such a very low IF, we can imagine that a digital implementation of the frequency transposition, advantageously located at the early stage of the line-up in order to deal with a low frequency processing, can be considered in practice as illustrated in Figure 9.14(top). However, we need to highlight the potential drawbacks of this approach in addition to the cost linked to the additional block required. The lowpass complex modulating waveform recovered after the frequency transposition is no longer centered around DC. We can thus expect a slight imbalance in the impact of the various filtering stages on the modulation. However, we recall that on the transmit side the filters to be used are often quite wideband due to the fact that we process the wanted signal only. In contrast to what we face on the receive side, as discussed in Section 9.2.5, we can thus be confident that the impact would be negligible in most practical implementations of transmitters. However, we also need to keep in mind that such a frequency transposition block necessarily transposes all the signals present at its input. We might wonder why this is so, as on the transmit side we process the wanted signal only. But referring to potential strategies for the compensation of the LO leakage as discussed in Section 9.1.2, we recall that we may consider using the addition of an offset. In that case, we need to consider adding such compensation after the frequency conversion block in order to keep this quantity at DC at the input of the RF upconverter.

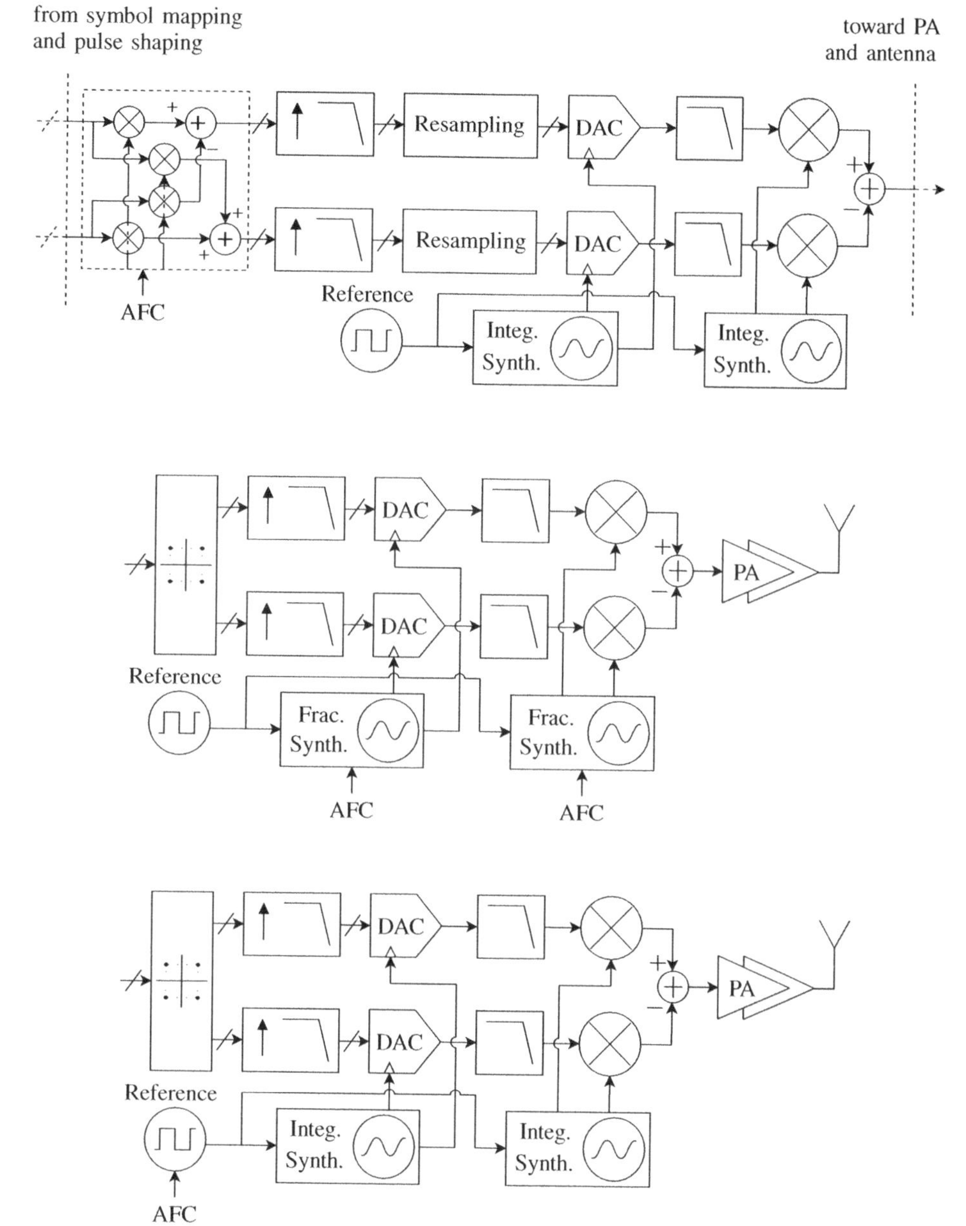

Figure 9.14 Illustration of AFC strategies on the transmit side – Given a frequency error on the reference clock, the main system impacts to be considered come from the resulting error on the LO frequency and on the sampling rate at the DAC stage. Knowing the frequency error we are dealing with, those impacts can be compensated through a resampling and rotation of the complex modulating waveform (top), through the correct regeneration of the LO and DAC clocks even when we have an inappropriate reference (middle), or through the direct correction of the reference source (bottom).

As an alternative approach, we recall that some PLL implementations can generate a wide number of frequencies with a very thin granularity whatever the reference clock used. This is for instance the case for the so-called fractional synthesizers. We can obviously recover the exact carrier frequency for the LO signal even if the reference is wrong. Practically speaking, this means that, at the cost of a trickier synthesizer implementation, we can still process a modulating waveform centered around DC as illustrated in Figure 9.14(middle).

Let us now consider the problem linked to the frequency variations of the clock used to drive the converters, mainly the DAC used to generate the analog version of the modulating waveform in practice. Given that by default the samples of this modulating signal are computed to match an exact sampling rate, we may face distortions in the reconstructed analog signal if those samples are now converted at instants that are not the expected ones due to the drift in the frequency of the DAC clock. In terms of solutions to be considered, we can make a connection with the previous problem of LO frequency error. Indeed, as illustrated in Figure 9.14, we can say the following:

(i) Knowing the sampling rate error, we can consider the resampling of the modulating waveform prior to being fed to the DAC so that the samples match the physical clock frequency. We expect to cancel the distortion that would have occurred otherwise during the digital to analog conversion process. However, the sampling rate at the output of this rate conversion stage is necessarily variable in order to track the drifts of the reference clock in operating conditions. As a result, it would be preferable to put this block close to the DAC so that most of the digital signal processing blocks run at a constant sampling rate. This may be required in order to achieve an overall frequency response of those blocks, typically for the digital filters, that is fixed, thus also achieving a deterministic distortion for the signal. This is generally preferable even though, as stated previously, the filtering stages present in typical transmit line-up are often wideband with regard to the signal being processed. However, despite this we may need to consider the implementation of a resampling block just prior to the DAC block. As a side effect, given that the sampling rate at the DAC input is often highest in the digital part of the line-up, this generally results in the highest penalty imaginable in terms of power consumption of the line-up.

(ii) As an alternative solution, in the present case we can imagine using a synthesizer able to support the generation of the exact sampling frequency. As illustrated in Figure 9.14(middle) this obviously allows us to avoid a potential resampling block even if at the cost of using a more complex synthesizer for the generation of the clock.

In conclusion, we observe that whatever the problem linked to this frequency inaccuracy, the simplest way to manage the issue is to compensate the reference source directly as illustrated in Figure 9.14(bottom). Practically speaking, this can be done by using for this reference an oscillator that can be controlled in frequency. And given that this reference is corrected, all the clocks derived from it necessarily have the expected value, thus canceling the system impacts discussed above.

9.1.6 Cartesian to Polar Conversion

Let us now conclude this review of typical algorithms for transmitters by focusing on the Cartesian to polar conversion. Obviously, in contrast to the algorithms we have considered so far, this kind of scheme is not directly related to the compensation of a degradation linked to a limitation in the physical implementation of the line-up. However, given that most of the complex modulating waveforms are defined in the Cartesian domain, as recalled in Section 8.1.6, a brief discussion of such algorithms will give us a better view of the additional digital implementation cost due to the additional processing that may be required in some alternative transmit architectures such as those that are polar based.

In order to illustrate such conversion when done in the digital domain, let us say a few words about the CORDIC approach [87]. This algorithm is representative in practice of what can be used to efficiently implement Cartesian to polar conversion. Here, "efficient" means using only additions and bit shifts. This is obviously of interest as deriving the modulus of a complex number should require the computation of a square root, and its argument an arctangent. But to achieve this result, CORDIC uses only successive rotations in the complex plane of angles $\epsilon_n\theta_n$, with $\epsilon_n = \pm 1$ and θ_n angles of predetermined positive values. With appropriate selection of the θ_n angles, we can greatly simplify the associated processing.

By way of illustration, let us consider the sample $\tilde{s}[k] = p\,[k] + jq\,[k]$ of a complex modulating waveform $\tilde{s}(t)$, here expressed in Cartesian form. Our goal is thus to derive the magnitude, $\rho[k]$, and argument, $\phi[k]$, of this complex number. For the sake of simplicity in our derivation, we assume that both $p\,[k]$ and $q\,[k]$ are positive quantities so that the vector representation of $\tilde{s}[k]$ lies in the first trigonometric quadrant, as illustrated in Figure 9.15. Let us now suppose that we apply a first rotation of angle $\epsilon_1\theta_1$ to the vector $(p\,[k], q\,[k])^{\mathrm{T}}$. Given that θ_1 is a fixed positive predetermined value, the purpose of the algorithm reduces to the selection of ϵ_1 so that the resulting vector,

$$\begin{pmatrix} p_1[k] \\ q_1[k] \end{pmatrix} = \begin{pmatrix} \cos(\theta_1) & -\epsilon_1 \sin(\theta_1) \\ \epsilon_1 \sin(\theta_1) & \cos(\theta_1) \end{pmatrix} = \cos(\theta_1)\boldsymbol{M}_{\theta_1}(\epsilon_1)\begin{pmatrix} p\,[k] \\ q\,[k] \end{pmatrix}, \tag{9.30}$$

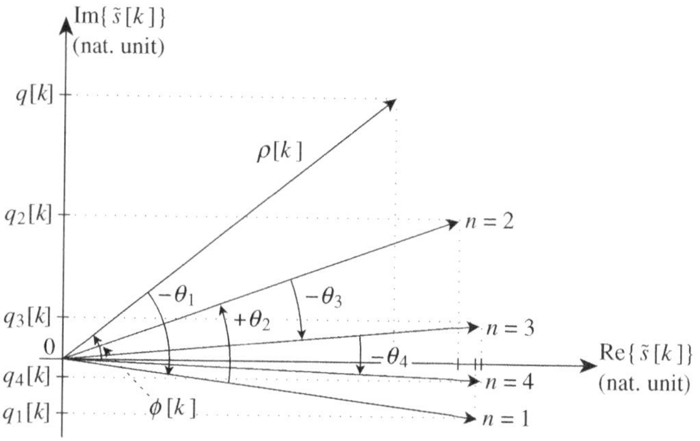

Figure 9.15 CORDIC as a Cartesian to polar converter – Following equation (9.33), we can derive at the same time the magnitude $\rho[k]$ and the argument $\phi[k]$ of a complex number $\tilde{s}[k] = p\,[k] + jq\,[k]$ by using successive rotations of angles $\epsilon_n\theta_n$, with $\epsilon_n = \pm 1$. In this simple scheme, the value of ϵ_n is determined based on the sign of $q_{n-1}[k]$.

with

$$\boldsymbol{M}_{\theta_1}(\epsilon_1) = \begin{pmatrix} 1 & -\epsilon_1 \tan(\theta_1) \\ \epsilon_1 \tan(\theta_1) & 1 \end{pmatrix}, \tag{9.31}$$

is more collinear to the direction vector of the abscissa than the original one. Practically speaking, this means that if $q[k] > 0$, we need to select $\epsilon_1 = -1$ in order to perform a rotation of angle $-\theta_1$, and in contrast, if $q[k] < 0$, we need to select $\epsilon_1 = 1$ in order to perform a rotation of angle θ_1. After N iterations of this process, we obtain the vector

$$\begin{pmatrix} p_N[k] \\ q_N[k] \end{pmatrix} = \prod_{n=1}^{N} \cos(\theta_n)\boldsymbol{M}_{\theta_n}(\epsilon_n) \begin{pmatrix} p[k] \\ q[k] \end{pmatrix}. \tag{9.32}$$

As shown by Figure 9.15, assuming that the θ_n angles are correctly decreasing as a function of n, we obtain

$$\lim_{N\to\infty} q_N[k] = 0, \tag{9.33a}$$

$$\lim_{N\to\infty} p_N[k] = \rho[k], \tag{9.33b}$$

$$\lim_{N\to\infty} \sum_{n=1}^{N} \epsilon_n \theta_n = -\phi[k]. \tag{9.33c}$$

It remains to determine the exact values of the θ_n angles to be used. For that purpose, we simply need to take into consideration the implementation complexity of the algorithm. Considering the processing corresponding to the first iteration as given by equation (9.30), we obseve that the term $\cos(\theta_1)$ remains a constant independent of the quantities ϵ_1 determined during this iteration. Given the same phenomenon at each iteration, we thus see that the term $\prod_{n=1}^{N} \cos(\theta_n)$ involved in equation (9.32) is nothing more than a normalization constant that can be managed independently as such in a fixed point implementation perspective. As a result, the only signal processing we need to consider during one iteration of the process is that which transforms the vector $(x_n, y_n)^{\mathrm{T}}$ into the vector $(x_{n+1}, y_{n+1})^{\mathrm{T}}$ according to

$$\begin{pmatrix} x_{n+1} \\ y_{n+1} \end{pmatrix} = \boldsymbol{M}_{\theta_n}(\epsilon_n) \begin{pmatrix} x_n \\ y_n \end{pmatrix}, \tag{9.34}$$

with $\boldsymbol{M}_{\theta_n}(\epsilon_n)$ of the form given by equation (9.31) for $\boldsymbol{M}_{\theta_1}(\epsilon_1)$. After expansion, this means that we need to consider the iterative processing corresponding to

$$\begin{aligned} x_{n+1} &= x_n - \epsilon_n \tan(\theta_n) y_n, \\ y_{n+1} &= \epsilon_n \tan(\theta_n) x_n + y_n. \end{aligned} \tag{9.35}$$

The implementation of this processing is thus straightforward if we select the angles θ_n such that

$$\theta_n = \arctan\left\{\frac{1}{2^n}\right\} \tag{9.36}$$

when expressed in radians. Hence, only additions and bit shifts are required as claimed at the beginning of the section. Moreover, given that for small angles $\arctan\{1/2^n\} \approx 1/2^n$, we necessarily get the convergence of the process, as illustrated in Figure 9.15. Practically speaking, only the final resolution required on both $\rho[k]$ and $\phi[k]$ drives the number of iterations to be performed for each set of samples.

Finally, when considering a set of input samples $p[k]$ and $q[k]$ that does not correspond to a vector lying in the first trigonometric quadrant, we can always apply the algorithm to $|p[k]|$ and $|q[k]|$. This obviously does not change the value of the magnitude we are looking for. Alternatively, the exact value of $\phi[k]$ can still be determined straightforwardly by adding the appropriate complementary angle based on the signs of those input quantities. We thus conclude that this kind of approach for implementing a Cartesian to polar conversion leads to a quite reasonable processing load compared to other algorithms reviewed so far and required in a classical line-up.

9.2 Receive Side

We now turn our attention to the receive side. Obviously, using the same technology for the physical implementation of a receive line-up as for a transmit line-up, we may consider the same kinds of limitations in the performance of the constituent blocks involved. Thus, even if the final system impacts may be different in the two cases, we may still be able to address the compensation of the impairments in receivers by considering the same kinds of approaches as discussed in Section 9.1 for transmitters.

However, as highlighted in the introductory part of that section, there is an important difference that may impact this statement. In a receiver we necessarily have access to the signal that has experienced the distortion resulting from those impairments, and thus to metrics about the quality of the signal. Practically speaking, this leads to the possibility of going through adaptive schemes for the optimization of the configuration of the line-up. This would obviously allow us to go through more robust compensations in operating conditions compared to simpler open loop systems as encountered in most cases in transmitters. On the other hand, a closed loop system is necessarily associated with some additional cost. Among others, we can mention the increase in the power consumption of the solution if the estimation has to be run continuously. Once again, the trade-off between the performance, the implementation cost and the power consumption of the solution needs to be checked case by case.

9.2.1 Automatic Gain Control

In order to initiate our review of typical algorithms on the receive side, let us consider the problem of the regulation of the level of the received signal along the data path. This problem is quite symmetrical to that linked to the power control on the transmit side as discussed in Section 9.1.1.

In receivers this regulation is achieved through what is classically called automatic gain control. Recall the reasons for putting such regulation in place. This function may be needed regardless of the limitations in the physical implementation of the line-up. Having to determine the data bits based on the shape of the received waveform, we may need up to a point to work on a normalized signal. This is required in order to have a reliable reference for determining the value of those bits. On the other hand, as introduced in Section 3.3.1, the power of the wanted signal at the antenna connector can vary over a wide DR due to the path loss, at least in a mobile application. We thus necessarily need to set some normalization process behaving as a variable gain in such applications, even when dealing with an ideal data path implementation.

However, this is only one aspect of the problem. Considering practical physical implementations, we obviously need to take other constraints into account. We also need to consider the limitation in the DR available all along the receive data path. Here, by "dynamic range" we mean the ratio between the available FS and the noise floor as defined for instance for an ADC through equation (4.285). Practically speaking, as illustrated throughout Section 7.3, this DR not only remains finite along the data path, but also we often try to set it at the minimum value consistent with achieving the required performance. This is generally driven by the cost, typically in terms of power consumption, related to the reduction of the noise floor in a typical physical implementation. However, minimizing this DR is possible only if there is a way to scale the power of the received signal optimally all along the data path. This means keeping on the one hand sufficient back-off with regard to the FS in order to avoid compression, and on the other hand sufficient margin with respect to the noise floor in order to preserve the SNR.

This possibility is particularly well illustrated by recalling the derivation of an ADC DR budget as illustrated in "Filtering budget vs. ADC dynamic range" (Section 7.3.3). In that example, we assumed that the scaling of the signal in the receive line-up was based on the power evaluation of the wanted signal only, regardless the presence of adjacent or blocking signals. This kind of partly blind determination of the gains to be used leads to the necessity to consider an AGC set point at the ADC input corresponding to an increased headroom compared to the case where the wanted signal is effectively alone along the data path, as illustrated in Figure 7.28. Even if this needs to be qualified depending on the analog channel filtering present in the line-up, the root cause for this behavior remains the need for headroom to allow us to deal with potential residual adjacent or blocking signals, whether or not effectively present during the reception. But if we can have access to an evaluation of the power of the total signal present at the ADC input (the sum of the wanted signal plus the residual adjacent or blocking signals), then we can imagine an AGC scheme that regulates this total power at the ADC set point. As a result, reconsidering the simple test cases displayed in Figure 7.28 but with this new strategy, we now achieve a required ADC DR of 70 dB, as illustrated in Figure 9.16. This is 12 dB less than the 82 dB required with the AGC scheme based on the only wanted signal power, i.e. a saving of two equivalent bits on the ADC stage. This is obviously a useful result in terms of power consumption and implementation area of the corresponding block, even if dealing with an ADC with a higher DR can have interesting side effects, for instance on managing the DC offset (see also Section 9.2.2).

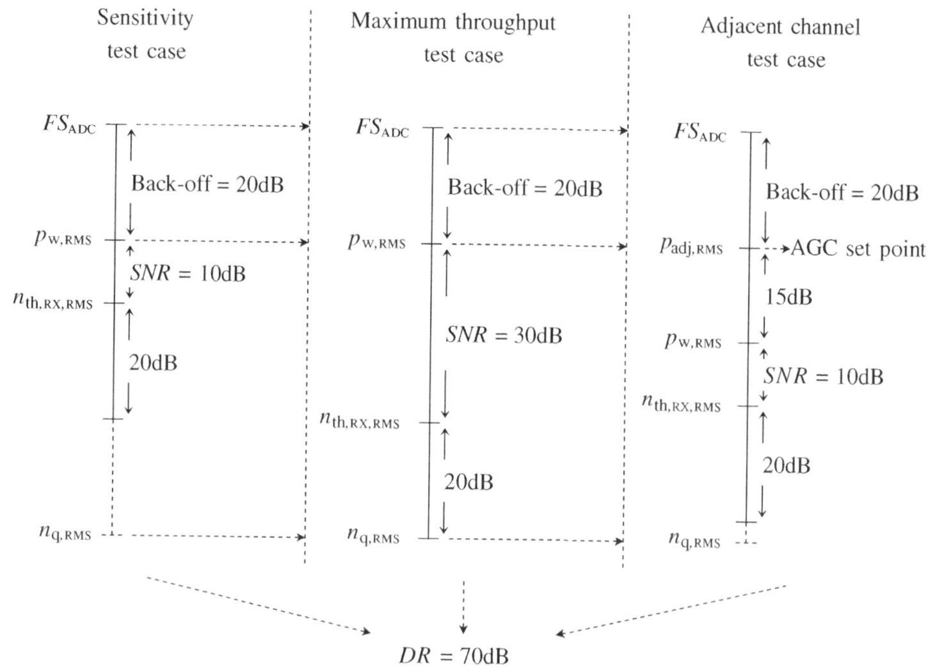

Figure 9.16 Revisited RX ADC dynamic range budget with an alternative AGC scheme – Having an AGC scheme that regulates the power of all the signals present at the input of the ADC, i.e. the sum of the wanted signal and the residual unwanted signals, results in an optimization of the DR of this stage compared to the case where the regulation is based on the power of the wanted signal only. For instance, considering the same test cases as shown in Figure 7.28, we now achieve a required DR of 70 dB, compared to 82 dB in the previous case.

In the discussion above, we considered an ADC stage by way of illustration, but the conclusion remains valid for the various blocks of the line-up as long as we have access to the power measurement of the sum of all the signals present in the data path at this stage. In that perspective, we might wonder which locations in the line-up are of real interest for such sensing. Practically speaking, the problem with a power evaluation located downstream in the data path comes from the filtering stages present in the line-up that make it impossible to evaluate the power of unwanted signals lying outside their passband and present at their input. We thus deduce that the relevant locations for those additional power measurements are dispatched at critical nodes apart from those filtering stages, as illustrated in Figure 9.17. A full sensing of the signals at those filtering stages should theoretically allow an optimum scaling of the signal to be achieved along the data path.

However, this should not mask the fact that there are drawbacks associated with this kind of approach. For instance, it is complex compared to the simple scheme based on the power of the wanted signal only. In this case, we simply need one decoding table which provides the gain split between the blocks of the receiver as a function of the wanted signal power. In contrast, using various power measurements necessarily requires us to determine the gains on the fly

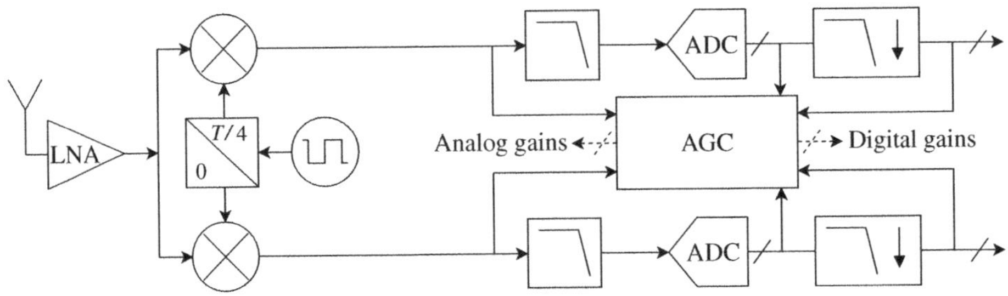

Figure 9.17 AGC scheme based on multiple power measurement evaluation – An evaluation of the power of the total signal, including both the in-band and the out-of-band components, present along the data path, theoretically allows the DR of the line-up to be optimized. This may require the sensing of the power at the critical nodes apart from the filtering stages.

during the reception, potentially through real time processing rather than a simple decoding scheme. This approach may thus also lead to an increase in the power consumption of the overall system on top of the implementation cost linked to this additional complexity. Finally, the unwanted out-of-band signals that we may face in a real life implementation can belong to different wireless systems than that of the wanted signal. This may lead to characteristic time constants in the behavior of those signals that are totally different from those of the wanted signal. It may thus also lead to an increase in the complexity of the control of the solution in order to take those different time constants into account.

Let us now focus on another class of issues related to our regulating scheme. In addition to the problems linked to the system optimization of the line-up discussed above, we need to take into account potential issues related to the physical implementation of such a control system. More precisely, we need to keep in mind that we are talking here about a closed loop system. As a result, we may wonder about the classical issues associated with this kind of scheme, especially its stability. We observe that the characteristic time constants corresponding to a long-term average power evaluation are necessarily very slow compared to those related to the modulating waveforms for instance. Even if this varies case by case, we can thus expect to face a closed loop system with a very low cut-off frequency, thus with no particular stability issues in practice. In contrast, as already highlighted in Section 9.1.1 when discussing the regulation of the transmit power, we may face problems linked to the inaccuracy in the gain of some blocks that are actually switched in the line-up. As illustrated in Figure 9.3, this inaccuracy may lead to some nonlinearity in the transfer function of the loop that may corrupt its good behavior and eventually lead to a continuous switching of the gains of the blocks present in the line-up. This behavior can then be directly transposed on the receive side for our present problem, as illustrated in Figure 9.18(top). Practically speaking, this gain inaccuracy is encountered mainly in RF amplifiers – the main reason being that we can hardly consider using negative feedback techniques for those blocks in a low cost solution. This then often results in non-negligible variations in their characteristics, including their gain, both in mass production and under operating conditions. From the system point of view, this potential continuous switching issue can lead to various problems, typically when the early RF stages of a receiver such as the LNA are involved. The gain switching of such a device necessarily leads to some

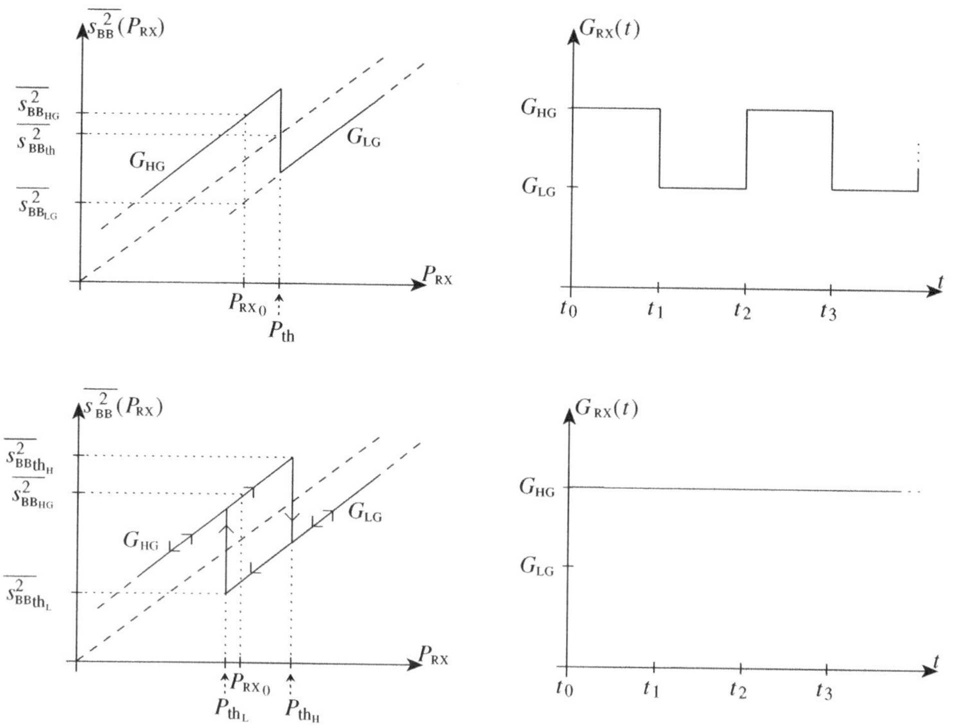

Figure 9.18 Hysteresis mechanism to prevent continuous gain switching in an AGC scheme – Due to inaccuracies in the different receiver gain states, the evaluation of the power of the demodulated baseband signal $\overline{s_{BB}^2}$ can lead to different values above and below the switching threshold $\overline{s_{BBth}^2}$ for a given input power P_{RX_0} (top left). This may lead to a continuous gain switching problem (top right). A hysteresis based on two thresholds, $\overline{s_{BBth_H}^2}$ and $\overline{s_{BBth_L}^2}$, in between which we hold the gain state (bottom left), may solve the problem (bottom right).

degradation of the quality of the reception through various phenomena. This is for instance the case through the introduction of transient responses that may degrade the characteristics of the modulation. It is also the case through the shift of the carrier phase due to the changes in the impedance experienced by the received electromagnetic wave as discussed in Section 2.2.2. Thus, a continuous switching of the state of such a device is highly undesirable if we want to preserve the quality of the radio link.

However, the solution for dealing with the associated problem may be different from what has been discussed on the transmit side. In the transmit side situation, the average power delivered by the line-up to the radiating element needs to be accurate enough with regard to the performance requirements. This is generally the main reason for putting a power control scheme in place in a transmitter. In that case, there is thus a need to linearize the power command law in order to avoid the present problem while achieving at the same time the accuracy of the radiated power. In contrast, we can live with some inaccuracy in the scaling of the signal along the receive data path, at least in its RF/analog part. This obviously needs to be qualified depending on the relative importance of this phenomenon with regard to the available DR. But due to the order of magnitude classically achieved in practice, we often prefer to keep those variations in the line-up and consider a simpler system to prevent instabilities in its regulation. A simple hysteresis mechanism as illustrated in Figure 9.18 may solve the problem.

9.2.2 DC Offset Cancellation

Let us now focus on the DC offset problem as classically encountered in receive line-ups. As introduced in Section 6.5, the reasons for such offset in practical physical implementations are mainly twofold. On the one hand, we have the LO self-mixing resulting from the downconversion of the fraction of LO signal present at the input port of the mixer due to either electromagnetic coupling or a direct feedthrough at the device level. On the other hand, we have the offsets induced by the mismatch in the physical implementation of the analog baseband blocks. As a result, we are basically dealing with the reciprocal situation compared to the LO leakage problem discussed for the transmit side in Section 9.1.2.

From a system point of view, the impacts on performance of the presence of this offset are also twofold. Most obviously, we have the degradation of the performance at the data bit recovery stage. Practically speaking, this degradation can be seen as the direct result of having an origin offset in the trajectory of the received modulating signal. However, we need to keep in mind that the exact degradation on the performance necessarily depends on the system used to physically multiplex the users in the wireless system, as highlighted for instance in "Origin offset suppression" (Section 3.2.2). We also need to keep in mind that the presence of a DC offset can have an additional impact on the good behavior of some additional algorithms present in the line-up. This can be illustrated for instance by the case of the P and Q imbalance compensation schemes discussed in Section 9.2.3.

We also need to consider the reduction in the equivalent FS available to physically represent the signals along the data path of the line-up (see Figure 9.19). At first glance, this phenomenon

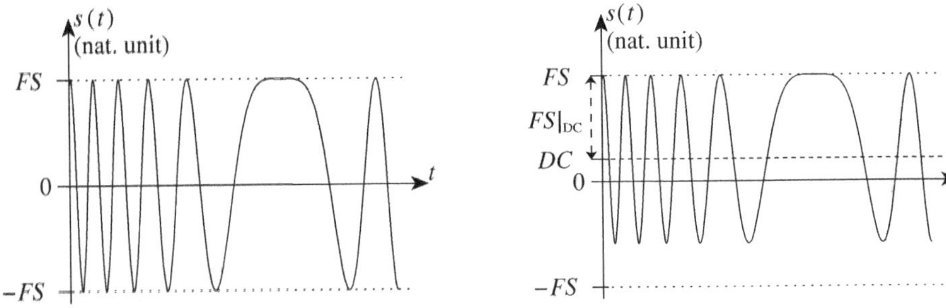

Figure 9.19 DC offset seen as a reduction of the available full scale in a line-up – Given that the FS is defined as the maximum theoretical peak value that a signal can reach without clipping in a physical implementation, the presence of a DC offset DC, thus resulting in a long-term average value on the signal, can be interpreted as a reduction of this quantity (right) compared to the expected situation where the signal being processed is zero-mean (left). In that perspective, the full scale $FS|_{DC}$ in the presence of such an offset reduces to $FS - |DC|$ with FS the full scale in the zero-mean situation.

is not necessarily a limiting factor in the digital part of the line-up as we can always think of adding some spare bits to handle an increase in the dynamic, although this is obviously a sub-optimal approach in terms of implementation cost. In contrast, we necessarily face a maximum available voltage swing in the analog part of it, this maximum swing being driven by the voltage supply that is used. As illustrated in Figure 9.19, we end up with a reduction of the maximum available FS that is directly equal to the DC offset present in the line-up.

These two problems classically yield different strategies in practice. Although we obviously need to cancel efficiently the offset present in the modulating waveform prior to delivering it to the baseband algorithms dedicated to the data bit recovery, this is not necessarily the case in the earlier stages of the line-up, in particular in the analog domain, as long as we can live with the corresponding reduction in the available FS. This is obviously good news as such compensation or cancellation in the analog domain is always trickier and less accurate to implement than in the pure digital domain. It remains to discuss what is really acceptable in terms of residual DC offset in those early analog stages.

It is instructive to first derive orders of magnitude of the DC offsets we may be faced with in practice. Suppose, for instance, that the maximum offset in a differential pair, as classically used as the first stage of a differential amplifier, is of the order of tens of millivolts. Then, recalling the voltage gain of the baseband analog amplifier considered in the illustration of receiver budgets in Section 7.3.2, i.e. 30 dB, we see that the amount of DC can easily exceed hundreds of millivolts at the input port of the ADC stage. This has to be compared with classical maximum voltage swings of the order of volts. Depending on the practical analog voltage gain present in the baseband part of the line-up, however, a range of situations can occur. In applications where only a small analog baseband gain is required, we can have negligible offset levels at the ADC input. In contrast, we can face a saturation in line-ups where huge analog baseband gains are required. Practically speaking, the analog gain required in a line-up is directly related to the DR of the ADC, as discussed in "Filtering budget vs. ADC dynamic range" (Section 7.3.3). We conclude that each situation needs to be considered individually.

Whatever the amount of DC offset in the line-up, we still need to determine whether or not it is acceptable in order to be able to target the right compensation scheme (if any). For that purpose, as stated previously, it is of interest to interpret the offset as a reduction of the available FS in the line-up. This FS drives the available DR for a given noise floor. This is particularly well illustrated by considering the definition of this quantity for an ADC as given by equation (4.285). Recalling the discussion of receiver budgets in Chapter 7, this DR is of importance in passing (i.e. converting correctly and without clipping) the sum of all the signals present in the analog part of the line-up, i.e. both the wanted signal and the residual unwanted signals. As a result, the level of DC offset in the analog part of the line-up can be taken into account through the DR budget of the line-up. This is even more relevant given that the ADC stage is often a bottleneck in terms of DR as illustrated in "Filtering budget vs. ADC dynamic range" (Section 7.3.3). Moreover, due to the cumulative effects of the analog gain stages present in the baseband part of the line-up, it is at the input of the ADC that the impact of the DC offset can be expected to be most important. In any case, we can straightforwardly express the reduction in the available DR due to the presence of the DC offset by simply assuming that this reduction matches that of the available FS for a given noise floor. This FS in the presence of DC offset, $FS|_{\text{DC}}$, is simply related to the original full scale, FS, through

$$FS|_{\text{DC}} = FS - |DC|, \tag{9.37}$$

with DC the DC offset level (see Figure 9.19). As the DR is classically expressed in decibels in order to enable straightforward budget derivations, we can conveniently express the FS loss $FS|_{DC}/FS$ in decibels as a function of the amount of DC relative to the original available full scale FS. The result is displayed in Figure 9.20. By inspecting this figure, we can check that the degradation in the FS and thus in DR increases slowly with the DC level. For instance, with a DC level below 20% of the maximum possible voltage swing we keep the degradation below 2 dB. This obviously has to be compared with orders of magnitude of practical ADC DR but considering classical implementations, the loss remains quite reasonable.

Thus, a residual level of DC offset at the output of the analog part of a line-up is often acceptable. This explains why in practical implementations we can consider only simple DC calibrations or rough feedback loops in order to minimize the amount of DC in compliance with the allowable loss in the ADC DR. Obviously, it is not the same on the digital side as full cancellation is often required in order to achieve the performance at the data bit recovery stage. However, as highlighted previously, dealing with a digital representation of the signal, we can easily consider algorithms compliant with an accurate estimation and cancellation of its average value. Thus we can decompose the problem of DC offset cancellation in a receive line-up into two distinct topics. This classically leads to the implementation of two different schemes, as illustrated in Figure 9.21.

This decomposition of the solution should not mask the potential interaction in terms of control between the two schemes. We have the obvious direct impact of the DC retrieved by the first system on the estimation performed by the second system. But more than that, we also need to take into account potential dynamic effects that can impact the DC offset present in the line-up, and thus the control of those schemes. Although it may seem strange to talk

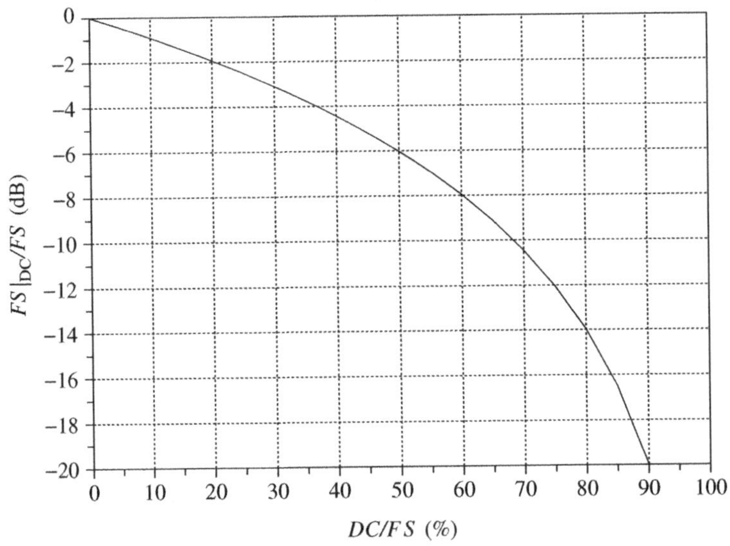

Figure 9.20 Full scale loss due to the presence of DC offset in a line-up – As illustrated in Figure 9.19, the DC offset present in a line-up can be interpreted as a reduction of the available full scale $FS|_{DC}$ for the signal being processed compared to the full scale FS in the zero-mean situation. From the system point of view, this reduction needs to be compared to the available DR in order to determine acceptable levels.

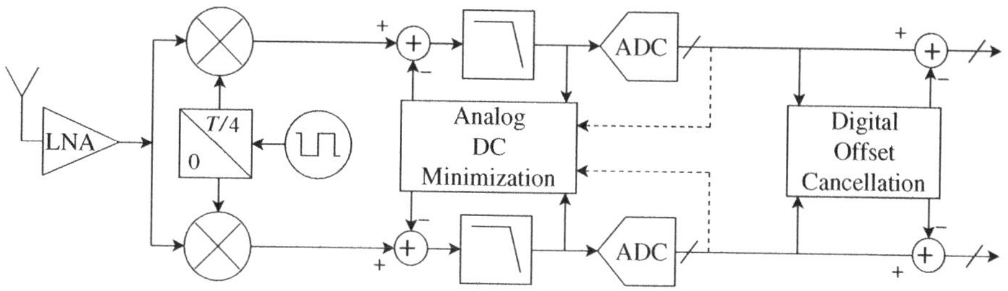

Figure 9.21 DC offset compensation scheme partitioning in a receive line-up – Due to limitations in the estimation and compensation in the analog domain, a minimization of the offset is often the most realistic way forward as long as the reduction in the available DR is reasonable. This can be achieved based on feedback loops that use a sensing of either the output of the analog path or the input of the digital path, depending on their implementation. We can then use an efficient average value cancellation algorithm in the digital domain to eliminate the residual offset.

about variations of a quantity that is centered at zero frequency, we need to keep in mind that from the signal processing perspective, it is the average value of the received signal over the characteristic time constant of the processing used to recover the data bits that is considered as a DC offset in the static sense. But in a real life implementation, there may be variations in this average value from one time frame to the next for various reasons. Obviously, a first reason for such dynamic behavior is the drift in operating conditions of the characteristics of the analog blocks present in the line-up, and their intrinsic offset. However, such drift should be quite slow with respect to the time constants linked to practical modulating schemes. Another source of variation may be changes in the environment of the receiver that may lead to more abrupt transitions. We can for instance mention the changes in the level of the LO leakage reflected back to the receiver due to variations in the input impedance of the line-up through potential antenna detuning effects. This can lead to variations in the level of DC through LO self-mixing. We can also refer to the additional DC generated through the AM-demodulation of strong blockers due to second order nonlinearity as illustrated in "DC offset generation due to even order nonlinearity" (Section 5.1.2). In this case, due to the potential dynamic behavior of the blocking signals, we may have abrupt variations in the level of DC in the line-up. All this may lead to particular measures to control the compensation schemes in order to be able to track those changes in practice.

9.2.3 *P/Q Imbalance Compensation*

Another classical source of degradation in receivers that can be compensated using a dedicated algorithm is the imbalance between the P and Q branches of the line-up when based on a Cartesian representation of the demodulated complex envelope. Practically speaking, such imbalance concerns the equivalent amplitude and phase error between the two LO signals in quadrature that are used to physically implement the complex frequency downconversion. We recover the situation symmetrical to that which we discussed in Section 9.1.3 for the transmit side. Recall from that section that this phenomenon is of importance mainly for complex frequency conversions implemented in the RF/analog world. Indeed, the performance of a digital implementation of this kind of processing is limited only by the fixed point implementation, if any. It can then be made quite negligible in practice. We thus keep in mind the RF/analog implementation perspective in this section.

The system impacts of such imbalance lead to the rise in what we have called an image signal during the frequency conversion. As discussed in Section 6.2.2, in the present complex downconversion case this image signal can be interpreted as the result of the downconversion of an equivalent input bandpass signal whose complex envelope, defined as centered around the LO angular frequency, is proportional to the complex conjugate of the reconstructed complex signal theoretically recovered at the output of the frequency conversion in the absence of any imbalance. Practically speaking, assuming for instance that this conversion is implemented using the negative complex exponential $e^{-j\omega_{LO}t}$, this means that we can write the complex signal $s_{BB}(t) = p_{BB}(t) + jq_{BB}(t)$ reconstructed at the output of the downconversion in the presence of imbalance as

$$s_{BB}(t) = (\alpha_{-}^{-})\tilde{s}_{RX}(t) + (\alpha_{-}^{+})\tilde{s}_{RX}^{*}(t), \tag{9.38}$$

where $\tilde{s}_{RX}(t)$ stands for the reconstructed demodulated waveform expected to be recovered in the absence of such imbalance, as given by equation (6.67). The latter term then represents nothing more than the complex envelope of the expected received signal sideband, when defined as centered around the LO angular frequency ω_{LO}, in a practical implementation. In that expression, (α_{-}^{-}) and (α_{-}^{+}) are complex numbers that depend only on the gain and phase imbalance existing between the P and Q LO waveforms as given by equation (6.37).

Depending on the frequency planning used for the frequency downconversion, we may face a spectral content of $\tilde{s}_{RX}^{*}(t)$ in equation (9.38) that may not necessarily result in an additive unwanted contribution overlapping the wanted signal frequency band. Indeed, from the Fourier transform property given by equation (1.8), $\tilde{s}_{RX}(t)$ and $\tilde{s}_{RX}^{*}(t)$ have spectra that are mutually symmetrical with respect to the zero frequency. As a result, we may be able to cancel the unwanted components in $\tilde{s}_{RX}(t)$ that may be recovered as superposed on the wanted signal through the mechanism of spectrum flip. Practically speaking, this can be achieved by using an image reject filter prior to the frequency downconversion when the frequency planning of the frequency conversion allows, i.e. when the wanted signal lies at a sufficiently high IF at the output of the frequency conversion (see Figure 6.21). Such complex frequency conversion can theoretically implement the ideal frequency shift of the input spectrum. When ideally implemented, it may be possible to avoid the use of costly image reject filters, as discussed in Section 6.1, and thus allow flexibility in the frequency planning for a direct frequency downconversion of the input wanted signal down to baseband, or at least down to a low IF. This is what makes it possible to achieve highly integrated architectures such as direct conversion or low-IF (Section 8.2). But, as in these cases we obviously do not rely on any filtering effect prior to the frequency downconversion, we may face situations where the term $\tilde{s}_{RX}^{*}(t)$ results in a non-vanishing in-band unwanted contribution, thus degrading the quality of the reception. As a side effect, the only realistic way to improve the situation while taking advantage of the complex frequency downconversion by avoiding the use of an RF image reject filter and using an aggressive frequency planning is to find a way to cancel the contribution of the term $(\alpha_{-}^{+})\tilde{s}_{RX}^{*}(t)$ in the above expression.

Before further discussing the strategies for achieving this cancellation, we first need to refine the model as considered hitherto. As already discussed in relation to the transmit side, RF/analog mixers are for the most part implemented as choppers. This means in the present downconversion case that we may recover at baseband not just the input signal as downconverted by the fundamental tone of the LO signal, but the sum of all the signals resulting from its downconversion through all the harmonic tones present in the LO waveform, as illustrated in Figure 6.42. However, even in that case we can interpret the imbalance between the P and Q LO signals in terms of gain and phase imbalance between the tones of the same order involved in their series expansions. It follows that the structure of the reconstructed complex signal resulting from the downconversion through the kth LO harmonic tone is still of the form given by equation (9.38). Only the factors that weight the expected complex envelope $\tilde{s}_{\mathrm{RX}}(t)$ and its complex conjugate $\tilde{s}^*_{\mathrm{RX}}(t)$ now depend on k, according to the discussion in "Practical LO spectral content" (Section 6.3.2). However, if we refer to the material discussed in that section, we see that only the tones of odd order, i.e. corresponding to k of the form $2l - 1$, are of interest for the implementation of the frequency conversion and thus for the definition of the image signal. Consequently, even if the results below could be easily extended to the tones of even order at the cost of a slight modification of the structure of the compensation scheme, we focus here on the use of odd order tones for implementing the frequency conversion of the wanted signal. The factors that weight the expected complex envelope and its complex conjugate are then the $(\alpha^-_-)_{2l-1}$ and $(\alpha^+_-)_{2l-1}$ factors defined through equation (6.141). Thus, given the expression of those factors, and relying on the structure of equation (9.38) to derive the compensation to apply, we see that we can address with a single scheme the cancellation of the image signal rise resulting from the downconversion achieved through the use of any of the odd order harmonics of the LO signal. But, given that the factors $(\alpha^-_-)_{2l-1}$ and $(\alpha^+_-)_{2l-1}$ depend on l, so also do the parameters to be used for the compensation. This means that with the present strategy we cannot compensate for all the image signals resulting from the downconversions through all the odd order LO harmonic tones at the same time. However, in all practical use cases the frequency conversion through only one LO tone is of interest for downconverting the wanted signal. Furthermore, as is the case for the image signal associated with the frequency conversion based on the fundamental tone, the image signal resulting from the frequency conversion through the $(2l - 1)$th harmonic tone can be interpreted as the result of the downconversion of an equivalent input bandpass signal whose complex envelope, when defined as centered around the $(2l - 1)$th LO harmonic angular frequency, is proportional to the complex conjugate of the reconstructed complex signal theoretically recovered through the frequency conversion based on this $(2l - 1)$th harmonic tone in the absence of any imbalance. These image signals are thus proportional to the signals involved in the harmonic LO mixing problem, as illustrated in Figure 6.42. At the same time, referring to the discussion above about the frequency planning of classical receiver architectures for which such complex frequency downconversion is of interest, the fraction of signal downconverted through the $(2l - 1)$th LO harmonic tone and recovered in the frequency band of interest necessarily lies in the close vicinity of the corresponding $(2l - 1)$th LO harmonic angular frequencies at the input of the frequency downconversion (again see Figure 6.42). As a result, an RF filtering stage can often be used prior to the frequency conversion to cancel those problematic signals. This is obviously realistic only when the LO angular frequency is sufficiently high that a low cost wideband RF filter can be used, which is the case in most practical applications. As a side effect, this cancellation necessarily results in preventing the folding of the corresponding image components that are directly proportional to the unwanted signals involved in the harmonic LO mixing problem. Thus, in our present problem, it is meaningful to focus on the compensation of the distortion resulting from the downconversion through only one of the tones present in the LO waveform, i.e. the one used for the downconversion of the wanted signal. For the sake of simplicity, we can assume that this tone is the fundamental one. This is the usual configuration in practice as it provides the highest conversion gain among all the available LO harmonics. Practically speaking, denoting by $\tilde{s}_{\text{RX,H1}}(t)$ the theoretical reconstructed complex signal resulting from the downconversion of the wanted RX bandpass signal using the LO fundamental tone in absence of imbalance, this means that we now focus on how to cancel the image signal term $\tilde{s}^*_{\text{RX,H1}}(t)$ resulting from the presence of imbalance between the P and Q paths of the complex downmixer.

We can proceed in the same way as for the transmit side in Section 9.1.3, and transpose equation (9.38) into a matrix form that links the real and imaginary parts of the reconstructed complex signal recovered at the output of the frequency downconversion in the presence of imbalance, i.e. $s_{\text{BB}}(t) = p_{\text{BB}}(t) + \text{j}q_{\text{BB}}(t)$, to those of the expected complex demodulated waveform $\tilde{s}_{\text{RX,H1}}(t) = p_{\text{RX,H1}}(t) + \text{j}q_{\text{RX,H1}}(t)$. For that, we remark that

$$p_{\text{BB}}(t) + \text{j}q_{\text{BB}}(t) = [(\alpha_-^-) + (\alpha_-^+)]p_{\text{RX,H1}}(t) + [(\alpha_-^-) - (\alpha_-^+)]\text{j}q_{\text{RX,H1}}(t). \tag{9.39}$$

On the other hand, by equation (6.37) we have

$$(\alpha_-^-) + (\alpha_-^+) \propto 1 - jg\sin(\delta\phi), \tag{9.40a}$$

$$(\alpha_-^-) - (\alpha_-^+) \propto g\cos(\delta\phi), \tag{9.40b}$$

with the same proportionality factor in the two equations. Here, g and $\delta\phi$ stand for the gain and phase imbalance that exist between the tones of the same order of the P and Q LO signals, as defined through equations (6.29) and (6.30). As a result, suppressing the proportionality factor that is superfluous to our present discussion, we can write from the two equations above that

$$\begin{pmatrix} p_{\text{BB}}(t) \\ q_{\text{BB}}(t) \end{pmatrix} = \boldsymbol{M}_{\text{H1}}(g, \delta\phi) \begin{pmatrix} p_{\text{RX,H1}}(t) \\ q_{\text{RX,H1}}(t) \end{pmatrix}, \tag{9.41}$$

with

$$\boldsymbol{M}_{\text{H1}}(g, \delta\phi) = \begin{pmatrix} 1 & 0 \\ -g\sin(\delta\phi) & g\cos(\delta\phi) \end{pmatrix}. \tag{9.42}$$

Thus a simple linear transformation holds between the real and imaginary parts of the reconstructed complex signal as effectively recovered at the output of the complex frequency downconversion in the presence of imbalance, $s_{\text{BB}}(t)$, and those of the expected demodulated signal, $\tilde{s}_{\text{RX,H1}}(t)$. Finally, we see that if we apply to $s_{\text{BB}}(t)$ the linear transformation corresponding to

$$\boldsymbol{M}_{\text{H1}}^{-1}(g, \delta\phi) = \begin{pmatrix} 1 & 0 \\ \theta_{\text{p}}(\delta\phi) & \theta_{\text{q}}(g, \delta\phi) \end{pmatrix}, \tag{9.43}$$

with

$$\theta_{\text{p}}(\delta\phi) = \tan(\delta\phi), \tag{9.44a}$$

$$\theta_{\text{q}}(g, \delta\phi) = 1/[g\cos(\delta\phi)], \tag{9.44b}$$

we then achieve our goal. Indeed, the real and imaginary parts of the reconstructed complex signal $\tilde{s}_{\text{mod}}(t) = p_{\text{mod}}(t) + jq_{\text{mod}}(t)$ recovered at the output of the compensation can be written as

$$\begin{pmatrix} p_{\text{mod}}(t) \\ q_{\text{mod}}(t) \end{pmatrix} = \boldsymbol{M}_{\text{H1}}^{-1}(g, \delta\phi) \begin{pmatrix} p_{\text{BB}}(t) \\ q_{\text{BB}}(t) \end{pmatrix}$$

$$= \boldsymbol{M}_{\text{H1}}^{-1}\boldsymbol{M}_{\text{H1}}(g, \delta\phi) \begin{pmatrix} p_{\text{RX,H1}}(t) \\ q_{\text{RX,H1}}(t) \end{pmatrix} = \begin{pmatrix} p_{\text{RX,H1}}(t) \\ q_{\text{RX,H1}}(t) \end{pmatrix}. \tag{9.45}$$

Thus the simple structure shown in Figure 9.22 allows us to cancel the image signal rise as expected. Moreover, as discussed previously, the same structure can be used whatever the odd order LO harmonic tone considered for the downconversion of the wanted signal. Only the $\theta_{\text{p}}(\delta\phi)$ and $\theta_{\text{q}}(g, \delta\phi)$ values would need to be tuned according to the corresponding (α_{-}^{-}) and (α_{-}^{+}) parameters.

It remains to find a way to derive the θ_{p} and θ_{q} parameters to be used. Recall the calibration procedure discussed in Section 9.1.3 for the transmit side. The parameters can be derived quite easily through the use of a CW test tone that gives direct access to the (α_{-}^{-}) and (α_{-}^{+}) quantities, and thus in turn to the parameters we want. However, there are limitations to this approach. The resulting compensation remains efficient enough only as long as the drifts in operating conditions of the parameters to be compensated, g and $\delta\phi$ in the present case, remain low enough to enable the required performance to be achieved. And there is a major difference in the present case compared to the transmit side as we now have direct access to the signal distorted by the impairment we expect to compensate. As a result, we can naturally consider an adaptive scheme that tracks the imbalance in operating conditions and thus improves the robustness of the solution.

In order to be able to implement an adaptive algorithm, we need to find a sufficiently simple scheme with reasonable implementation cost. For that purpose, it is of interest to consider the

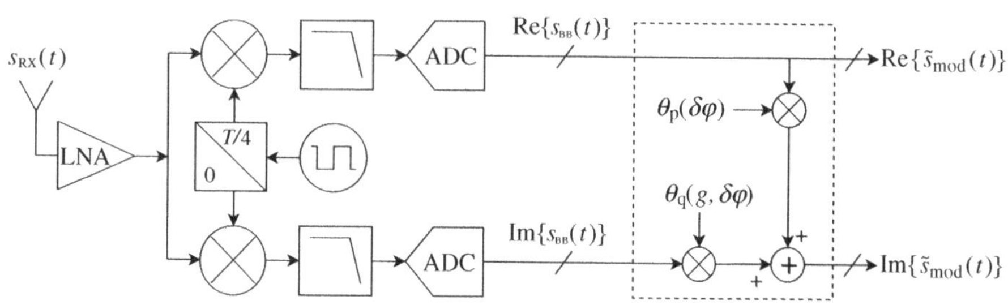

Figure 9.22 Digital compensation of the imbalance between the fundamental tones of the P and Q LO signals on the receive side – The structure of the reconstructed complex signal $s_{\text{BB}}(t)$ recovered at the output of the complex frequency downconversion in the presence of gain and phase imbalance leads to the possibility of using a cancellation scheme for the image signal based on the linear transformation given by equation (9.43), with $\theta_{\text{p}}(\delta\phi)$ and $\theta_{\text{q}}(g, \delta\phi)$ given by equation (9.44). Dealing with mixers implemented as choppers, this structure can be used to compensate for the distortion resulting from the downconversion of the wanted signal using any of the odd order harmonic tones present in the series expansion of the LO waveforms by proper selection of the θ_{p} and θ_{q} parameters.

physical nature of the imbalance we need to estimate and compensate: the gain imbalance g and the phase imbalance $\delta\phi$ that exist between the tones of the P and Q LO signals as defined for the fundamental tone through equations (6.29) and (6.30). Focusing first on g, in practice such imbalance results in a difference in the receiver gain along each of the P and Q paths. At the same time, recalling the discussion in Appendix 2, for most of the modulating schemes encountered in practical wireless systems the P and Q demodulated components have the same power. Thus the gain imbalance in the receiver may directly lead to a difference between the power of the $p_{\mathrm{BB}}(t)$ and $q_{\mathrm{BB}}(t)$ waveforms collected at the output of the line-up. As a result, by evaluating the difference between the power of those signals we can evaluate the gain imbalance g. This can be checked by taking the RMS values of those lowpass waveforms based on equations (9.41) and (9.42). We obtain

$$\begin{aligned}\overline{p_{\mathrm{BB}}^2(t) - q_{\mathrm{BB}}^2(t)} = \overline{p_{\mathrm{RX,H1}}^2(t)} - g^2 \sin^2(\delta\phi)\overline{p_{\mathrm{RX,H1}}^2(t)} - g^2 \cos^2(\delta\phi)\overline{q_{\mathrm{RX,H1}}^2(t)}\\ - 2g^2 \sin(\delta\phi)\cos(\delta\phi)\overline{p_{\mathrm{RX,H1}}(t)q_{\mathrm{RX,H1}}(t)},\end{aligned} \tag{9.46}$$

where $\overline{(.)}$ denotes the time average. Then, assuming that the P and Q components of the modulating waveform we are dealing with have the same power, we can write that $\overline{p_{\mathrm{RX,H1}}^2(t)} = \overline{q_{\mathrm{RX,H1}}^2(t)}$. Moreover, as discussed in Appendix 2, we can also assume that those components are uncorrelated, i.e. that $\overline{p_{\mathrm{RX,H1}}(t)q_{\mathrm{RX,H1}}(t)} = 0$. We can thus simplify the above relationship so that we finally obtain

$$\overline{p_{\mathrm{BB}}^2(t) - q_{\mathrm{BB}}^2(t)} = \overline{p_{\mathrm{RX,H1}}^2(t)}\left(1 - g^2\right) \tag{9.47}$$

as expected. Turning now to the phase imbalance $\delta\phi$ between the P and Q branches of the line-up, we simply recall that a basic phase detector can be implemented by lowpass filtering the product of the corresponding signals. Thus, applying this simple scheme to our present situation, we can directly write from equations (9.41) and (9.42) that

$$\overline{p_{\mathrm{BB}}(t)q_{\mathrm{BB}}(t)} = -g\sin(\delta\phi)\overline{p_{\mathrm{RX,H1}}^2(t)} + g\cos(\delta\phi)\overline{p_{\mathrm{RX,H1}}(t)q_{\mathrm{RX,H1}}(t)}. \tag{9.48}$$

Now, again using the fact that $\overline{p_{\mathrm{RX,H1}}(t)q_{\mathrm{RX,H1}}(t)} = 0$, this equation reduces to

$$\overline{p_{\mathrm{BB}}(t)q_{\mathrm{BB}}(t)} = -g\sin(\delta\phi)\overline{p_{\mathrm{RX,H1}}^2(t)}. \tag{9.49}$$

Finally, supposing for the sake of simplicity that the small imbalance assumption is valid, i.e. that $\delta\phi$ is close to 0 and g close to 1, we can gather equations (9.47) and (9.49) as

$$\overline{p_{\mathrm{BB}}^2(t) - q_{\mathrm{BB}}^2(t)} \propto 1 - g^2, \tag{9.50a}$$

$$\overline{p_{\mathrm{BB}}(t)q_{\mathrm{BB}}(t)} \propto \delta\phi. \tag{9.50b}$$

Practically speaking, the time averaging can be advantageously replaced by a lowpass filtering, thus allowing a straightforward implementation of loops converging toward the imbalance parameters g and $\delta\phi$. However, this illustrative example can also be adapted to derive a scheme that directly converges toward the compensation parameters θ_p and θ_q rather than toward g and $\delta\phi$. This would thus lead to an even simpler implementation that combines the estimation of the imbalance parameters and the compensation of the received signal.

We should point out that the performance of the approach described may be limited in practice by a number of problems. First of all, we need to take into account the potential DC offset present in the line-up, as discussed in Section 9.2.2. The presence of such an offset superposed on the signals $p_{BB}(t)$ and $q_{BB}(t)$ necessarily leads to a correlation term between those signals. Thus, having $\overline{p_{BB}(t)q_{BB}(t)} \neq 0$ prevents the correct estimation of the parameters to compensate. We may also need to take into account the potential variations in the received signal power. This can be the case for instance in a mobile application due to the fading experienced by the received signal, as discussed throughout Chapter 2. As a side effect, we may face variations in the magnitude of the terms $\overline{p_{BB}^2(t) - q_{BB}^2(t)}$ and $\overline{p_{BB}(t)q_{BB}(t)}$, and thus in the proportionality factors involved in the above equations. This may corrupt the stability or the satisfactory convergence of loops based on such equations. There are various ways to deal with this problem, such as using a criterion based only on the sign of those quantities, the price to pay in this case being a longer convergence duration [88].

The above discussion is based on a pragmatic approach to estimating the imbalance to compensate. But we can also consider more refined adaptive schemes, albeit at the cost of more expensive physical implementations [89, 90]. Such methods are for the most part based on the fact that the expected reconstructed complex lowpass demodulated signal is uncorrelated with the reconstructed image term, as discussed in Appendix 2 and illustrated in Section 6.2.2. As a result, compensation algorithms can be derived based on the forcing to zero of the cross-correlation between the reconstructed complex lowpass signal $s_{BB}(t) = p_{BB}(t) + jq_{BB}(t)$ and its image term $s_{BB}^*(t) = p_{BB}(t) - jq_{BB}(t)$. However, it is interesting to note the equivalence with our former approach as can be seen by expanding the cross-correlation function between $s_{BB}(t)$ and $s_{BB}^*(t)$ when considered at the same time. From equation (A1.46b) we have

$$\gamma_{s_{BB} \times s_{BB}^*}(0) = \gamma_{p_{BB} \times p_{BB}}(0) - \gamma_{q_{BB} \times q_{BB}}(0) + 2j\gamma_{p_{BB} \times q_{BB}}(0). \tag{9.51}$$

As a result, forcing $\gamma_{s_{BB} \times s_{BB}^*}(0)$ to zero is equivalent to forcing both $\gamma_{p_{BB} \times p_{BB}}(0) - \gamma_{q_{BB} \times q_{BB}}(0)$ and $\gamma_{p_{BB} \times q_{BB}}(0)$ to zero. This is in fact nothing more than the criterion used in equation (9.50) to estimate the imbalance parameters.

9.2.4 *Linearization Techniques*

Let us now say a few words about the linearization techniques that can be encountered in receivers. As for most of the topics discussed so far, we can again make a link with our discussion on the transmit side in Section 9.1.4, and thus derive solutions of the same kind. This is obviously true for the linear distortion case. Although on the receive side such distortion is linked to the various anti-aliasing or channel filtering stages rather than to the reconstruction filter associated with the DACs as encountered in most classical transmit line-ups, we still have the same structure for the distortion experienced by the wanted signal in both cases. As a result, a simple linear equalization as discussed in that section can also be considered for receivers. However, this should not mask the fact that the ultimate performance to be achieved in terms of equivalent in-band SNR may be different, depending on the line-up. In some low cost applications, we may have tighter requirements for the receiver than for the transmitter. One good reason for this is that on the receive side we have direct access to the signal distorted by the impairment we expect to compensate. As a result, even if the same structure for the correction scheme can be used in both cases, we can naturally consider an adaptive estimation of the parameters to apply on the receive side in order to track the impairments in operating conditions, thus improving the robustness of the solution. Obviously, the need for this approach should be considered on a case by case basis, keeping in mind that some modulating schemes are classically associated with an equalization in the frequency domain for the compensation of the propagation channel, if any. In that case, the linear distortion induced by analog filters is also naturally compensated up to a point. This is generally the case for OFDM based systems, as considered for the illustration of receiver SNR budgets in "SNR budget" (Section 7.3.4).

In contrast, this similarity does not necessarily extend to the nonlinear distortion case. Even if we are dealing with the same kind of limitations at the origin of the nonlinear behavior we face in the physical implementation of the two line-ups, we observe that on the transmit side the signal that experiences the nonlinearity necessarily reduces to the wanted signal only. On top of that, we obviously have access to the characteristics of the lowpass modulating waveform as it is prior to entering the nonlinear part of the line-up. As a result, the unwanted contributions generated through this nonlinearity can theoretically be reconstructed from this ideal baseband signal, and thus further subtracted from it. This is for instance the spirit of the digital predistortion techniques discussed in Section 9.1.4. This approach is obviously of great interest for low cost integrated solutions based on a physical implementation using a process that allows for an intensive digital signal processing. Unfortunately, the problem is less tractable on the receive side. Given in practice that such nonlinearity is encountered mainly in RF amplifiers, the main contributions to the in-band distortion are often generated upstream of any frequency transposition, i.e. upstream of any channel selection in the line-up. Practically speaking, having no channel selection means that we can hardly consider an efficient channel filtering, as discussed in "Filtering budget vs. ADC dynamic range" (Section 7.3.3). As a result, the signal entering the nonlinearity is for the most part composed of both the wanted signal and strong unwanted adjacent or blocking signals. As discussed in "Budgets for blocking test cases" (Section 7.3.3), and illustrated in Figure 7.30, the generated distortion terms that corrupt the wanted signal are also related to the statistics of the corresponding unwanted adjacent or blocking signals. But at the same time, in baseband, and even more in the later stages of the digital part of the line-up, i.e. after efficient channel filtering, we cannot have access to those unwanted signals. Thus we cannot easily rely on the use of the signal present in those later stages of the data path to reconstruct an estimate of the distortion terms experienced by the wanted signal and subtract them from it. In other words, the direct transposition on the receive side of the digital predistortion as considered on the transmit side cannot be done as such. This can be seen as problematic for low cost integrated solutions based on the use of digital signal processing. Indeed, unless going through trickier iterative interference cancellation schemes up to the data bit recovery stage, such cancellation based on the knowledge of the statistics of the signals at the origin of the unwanted contribution are the only realistic way forward in terms of acceptable processing load in most applications.

However, nothing prevents us from considering a mixed approach. Indeed, the limitation in this attempt to transpose the predistortion to the receive side comes from the possibility of having access to the statistics of the RF signals at the origin of the unwanted terms recovered in the wanted signal frequency band when going through the nonlinear device. But, if this seems hardly possible by processing the signal in the digital part of the receive line-up, nothing prevents us from considering a direct RF sensing. This means a sensing of the unwanted signals involved in the distortion process just prior to entering the nonlinearity. Then, by proper processing we can still reconstruct the corresponding distortion terms and their subtraction in the baseband part of the line-up, preferably in the digital domain [91]. This scheme could be seen as a kind of upgrade of the feedforward technique using a downconversion and a nonlinear model in between its input and output. The cost associated with this approach is obviously non-negligible, both in terms of area penalty and power consumption in respect of the main data path of the receiver. The conclusion would be basically the same when considering the transposition for receivers of negative feedback architectures based on frequency transposition techniques such as those introduced in Section 8.1.7.

However, simpler schemes can also be considered in the particular case where a known blocker is the dominant source of distortion experienced by the received wanted signal. This is for instance the case when the receiver is physically close to a transmitter operating at the same time, and possibly also when two transceivers are close to each other but belong to different wireless systems. If we have access to the frequency planning of the transmitter behaving as an aggressor, we can then imagine simplified schemes in terms of sensing of the corresponding blocker statistics in order to reconstruct the distortion terms in a simpler way. This situation can also occur in full duplex systems for which the most problematic blocker is often the transmit signal leakage, as illustrated throughout Section 7.3. In that case we have access not only to the carrier frequency of the aggressor, but also obviously to all the statistics of this signal generated by the transmit part of the transceiver that the receiver belongs to. We can thus envisage even simpler ways to reconstruct the distortion terms linked to this particular blocking signal.

9.2.5 Automatic Frequency Correction

Let us now conclude the review of algorithms typical of those encountered in receivers by recalling the problems associated with the limited accuracy of the reference frequency sources available in such line-ups. Obviously, from the discussion in Section 9.1.5, this problem is common on the transmit side. Thus, we can expect to be faced with the same kind of system impacts as discussed in that section. Practically speaking, this means system impacts linked on the one hand to the inaccurate frequency of the LO signals used to drive the RF/analog mixers, and on the other hand to the inaccurate frequency of the clocks that feed the converters, i.e. mainly the ADCs dedicated to the sampling and quantization of the received signal. However, given different system constraints for receivers and transmitters, it is worth transposing the strategies considered in that section to the receive side. Let us again assume that we have access to the frequency error to be compensated through an indirect estimation.

Consider the problem of the inaccuracy in the LO frequency. If nothing is done to compensate for that, the first obvious consequence for a receiver is that the downconverted signal lies centered around an IF that is incorrect by a few parts per million relative to the expected LO frequency. Suppose for the sake of simplicity that we are dealing with a direct conversion scheme. With classical orders of magnitude, we can say that with a relative error of 10 ppm, and with a carrier frequency of 2 GHz, we would recover the demodulated complex waveform as centered 20 kHz away from DC. Assume for now that such an error is not acceptable from a data bit recovery perspective. As discussed in Section 9.1.5, different strategies can then be considered for handling this error.

One possibility is a slight frequency transposition of the complex modulating waveform by an amount that corresponds to the error. This situation is symmetrical to that discussed for the transmit side. However, in the receive case, additional problems arise. Whereas on the transmit side mainly the center frequency accuracy of the generated modulated transmit signal was of concern, we now need to consider more carefully the distortion of the received signal due to the presence of filtering stages. Indeed, unlike what classically happens on the transmit side, there is a need for narrowband channel filtering stages on the receive side in order to eliminate the unwanted signals. With the orders of magnitude we are talking about, such residual frequency conversion would need to be located prior to the narrower channel filtering stages, often implemented in the digital domain, as illustrated in Figure 9.23. As a side effect, this additional processing block may need to run at a higher sampling rate than that used to represent the wanted signal only, thus potentially leading to a non-negligible power consumption. We also need to keep in mind that a frequency transposition block necessarily transposes all the signals fed to it. This is in particular true for the DC offset inherently present in the analog part of the line-up that would be recovered as centered around the frequency error, assumed equal to 20 kHz in our example above. We thus need to consider implementing the DC evaluation and compensation prior to the final frequency conversion. This is obviously a limitation of the present approach as efficient signal processing, including DC offset management, could be considered in the later stages of the line-up during the processing for data bit recovery.

Alternatively, we can consider using a synthesizer able to support the generation of the exact sampling frequency even when dealing with an incorrect reference. As illustrated in Figure 9.23(middle), this would indeed allow the receive line-up to be kept simple, albeit at the cost of a more complex synthesizer.

Let us now consider the problem of the frequency inaccuracy of the clock used to drive the converters, mainly the ADC used to sample and quantize the received signal. With digital signal processing designed to process samples at a given rate, we may face distortions in the transfer function of the digital data path experienced by the received signal if its samples correspond to incorrect sample time values. This can be particularly problematic for the channel filtering stage that once again may lead to distortions on the wanted signal itself due to its narrowband nature. Various strategies can be considered for this problem.

To deal with incorrect sampling of the received signal, one possibility is direct resampling. As just highlighted, in order to make the transfer function of the remaining digital data path the expected one, it is of interest to put this resampling block closest to the ADC as illustrated in Figure 9.23(top). However, as discussed above for the derotation processing, given that the sampling rate at the ADC output is often highest in the digital part of the line-up, this location for the resampling block necessarily results in the highest penalty conceivable in terms of power consumption of the line-up.

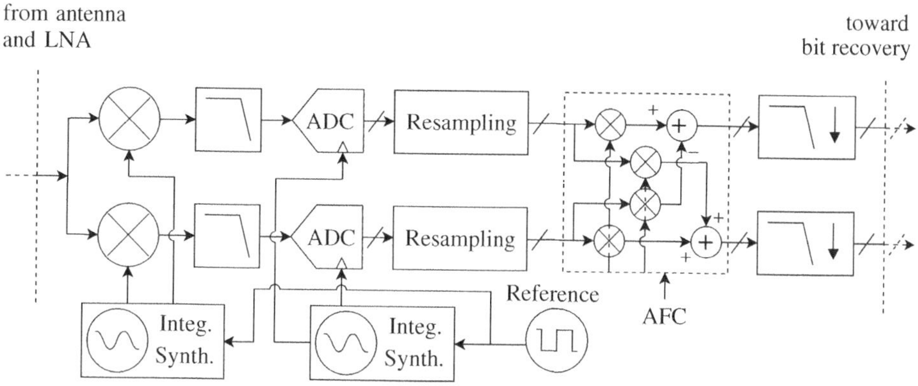

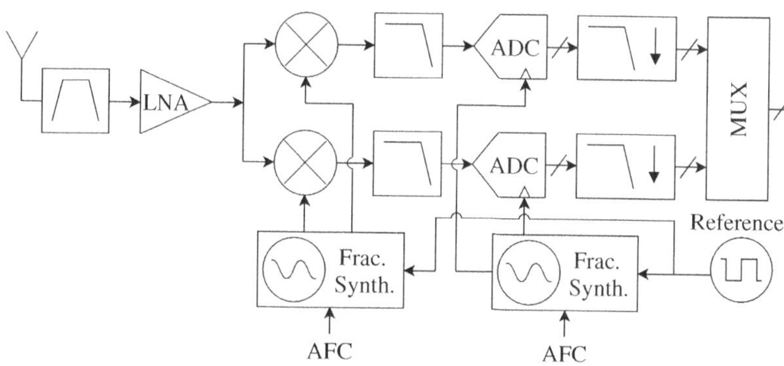

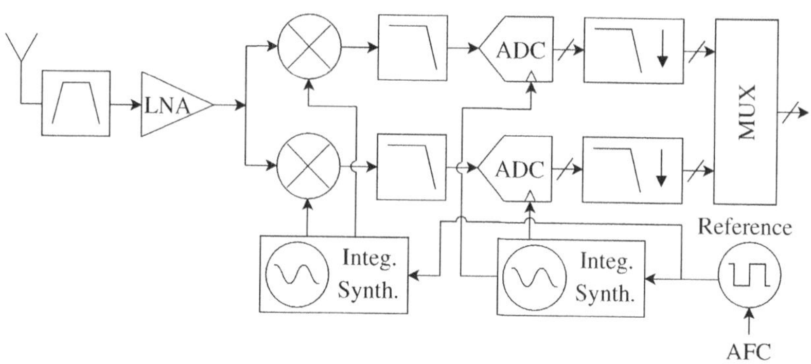

Figure 9.23 Illustration of AFC strategies on the receive side – Given a frequency error on the reference clock, the main system impacts to be considered come from the resulting error on the LO frequency and on the sampling rate at the ADC stage. Knowing the frequency error we are dealing with, these impacts can be compensated through a resampling and derotation of the complex modulating waveform (top), through the correct regeneration of the LO and ADC clocks even with an incorrect reference (middle), or through the direct correction of the reference source (bottom).

Alternatively, we might use a synthesizer able to support the generation of the exact sampling frequency even with an incorrect reference. As illustrated in Figure 9.23(middle), this allows us to avoid a potential resampling block, albeit at the cost of a more complex implementation of the synthesizer used for the generation of the clock.

However, as concluded for the transmit side, the simplest way to manage the issue is the direct compensation of the reference source as illustrated in Figure 9.23(bottom). Given that this reference is corrected, all the clocks derived from it necessarily have the expected value, thus allowing us to keep the line-up as simple as possible.

Appendices

Appendix 1

Correlation

Autocorrelation and cross-correlation functions are often required for analytical derivations throughout this book. For instance, autocorrelation functions are used to derive power spectral densities, while cross-correlation functions are required for checking the non-correlation between two signals. In the latter case the aim may be to check that a given unwanted signal can effectively be considered as an additive noise term for the derivation of an SNR budget, for instance.

Analytical derivations involving RF bandpass signals may be carried out using their lowpass complex envelope representations. It is thus of interest to examine the conditions that such RF bandpass signals need to fulfill for equivalence to hold between their non-correlation and the non-correlation of their corresponding complex envelopes; we do this in the next section. We can then detail some general properties of the cross-correlation and autocorrelation functions that are useful for our analytical derivations.

A1.1 Bandpass Signals Correlations

Let us first focus on the correspondence between the non-correlation of two bandpass signals and the non-correlation of their complex envelopes. We begin with the case of two deterministic bandpass signals. The reasons for this will become apparent later on when making the link with the stochastic case.

Suppose that we want to check the non-correlation between two deterministic bandpass signals, $x(t)$ and $y(t)$, and derive an equivalent condition for their complex envelopes. In order to stay general and make the link with correlation functions, we want to check the non-correlation between $x(t)$ and $y(t-\tau)$, for a given time offset τ that can be zero if necessary. Consequently, let us assume that we want to check that

$$\overline{x(t)y(t-\tau)} = 0, \tag{A1.1}$$

where $\overline{(.)}$ represents the time averaging. At this stage, we observe that checking the correlation or non-correlation between $x(t)$ and $y(t-\tau)$ is meaningful only when those bandpass signals have overlapping spectra. If not, the direct application of the Fourier transform property given

Wireless Transceiver Architecture: Bridging RF and Digital Communications, First Edition. Pierre Baudin.

by equation (1.59) to $x(t)y(t-\tau)$ leads to $\overline{x(t)y(t-\tau)}=0$. Two bandpass signals with non-overlapping spectra are thus necessarily uncorrelated. We therefore assume this is not the case. As a side effect, it is also meaningful to consider for our derivation the complex envelopes $\tilde{x}(t)$ and $\tilde{y}(t)$ of those bandpass signals that are defined as centered around the same center angular frequency, ω_c here. Based on these assumptions, we thus assume that

$$x(t)=\mathrm{Re}\left\{\tilde{x}(t)e^{j\omega_c t}\right\}, \tag{A1.2a}$$

$$y(t)=\mathrm{Re}\left\{\tilde{y}(t)e^{j\omega_c t}\right\}. \tag{A1.2b}$$

Using equation (1.5), we can then rewrite equation (A1.1) as

$$\overline{x(t)y(t-\tau)}=\frac{1}{4}\overline{\left(\tilde{x}(t)e^{j\omega_c t}+\tilde{x}^*(t)e^{-j\omega_c t})(\tilde{y}(t-\tau)e^{j\omega_c(t-\tau)}+\tilde{y}^*(t-\tau)e^{-j\omega_c(t-\tau)}\right)},$$

so that after expansion,

$$\begin{aligned}\overline{x(t)y(t-\tau)}&=\frac{1}{4}\overline{\tilde{x}(t)\tilde{y}^*(t-\tau)e^{j\omega_c\tau}+\tilde{x}^*(t)\tilde{y}(t-\tau)e^{-j\omega_c\tau}}\\&\quad+\frac{1}{4}\overline{\tilde{x}(t)\tilde{y}(t-\tau)e^{j\omega_c(2t-\tau)}+\tilde{x}^*(t)\tilde{y}^*(t-\tau)e^{-j\omega_c(2t-\tau)}}.\end{aligned}$$

Let us now suppose that the spectra of the signals of interest are narrowband relative to ω_c. This argument, extensively used in this book, corresponds to the situation of most practical modulating waveforms of interest in the field of wireless. As a consequence, by the discussion in Section 1.1.3, the time average of the terms centered around $2\omega_c$ in the above equation can be considered negligible compared to that of the lowpass terms. This leads to

$$\begin{aligned}\overline{x(t)y(t-\tau)}&=\frac{1}{4}\left(\overline{\tilde{x}(t)\tilde{y}^*(t-\tau)}e^{j\omega_c\tau}+\overline{\tilde{x}^*(t)\tilde{y}(t-\tau)}e^{-j\omega_c\tau}\right)\\&=\frac{1}{2}\mathrm{Re}\left\{\overline{\tilde{x}(t)\tilde{y}^*(t-\tau)}e^{j\omega_c\tau}\right\}.\end{aligned} \tag{A1.3}$$

Thus the condition for the non-correlation to hold between real bandpass signals can be transposed as a necessary and sufficient condition on their complex envelopes, when defined as centered around the same center frequency, as

$$\overline{x(t)y(t-\tau)}=0 \Leftrightarrow \mathrm{Re}\left\{\overline{\tilde{x}(t)\tilde{y}^*(t-\tau)}e^{j\omega_c\tau}\right\}=0. \tag{A1.4}$$

We then observe that

$$\overline{\tilde{x}(t)\tilde{y}^*(t-\tau)}=0 \Rightarrow \overline{x(t)y(t-\tau)}=0, \tag{A1.5}$$

i.e. that $\overline{\tilde{x}(t)\tilde{y}^*(t-\tau)}=0$ is a sufficient condition for the non-correlation of the corresponding real bandpass signals $x(t)$ and $y(t-\tau)$. However, it is not a necessary condition.

This can be seen by considering the simple case where $x(t)$ and $y(t)$ are two CW signals at the same angular frequency, chosen for the sake of simplicity as equal to ω_c. Following our definitions so far, we can then assume that the two complex envelopes of $x(t)$ and $y(t)$ take the form $\tilde{x} = \rho_\mathrm{x} e^{j\phi_\mathrm{x}}$ and $\tilde{y} = \rho_\mathrm{y} e^{j\phi_\mathrm{y}}$ respectively, both constant in time, i.e. that $x(t)$ and $y(t)$ can be written as

$$x(t) = \mathrm{Re}\left\{\tilde{x} e^{j\omega_\mathrm{c} t}\right\} = \rho_\mathrm{x} \cos(\omega_\mathrm{c} t + \phi_\mathrm{x}), \tag{A1.6a}$$

$$y(t) = \mathrm{Re}\left\{\tilde{y} e^{j\omega_\mathrm{c} t}\right\} = \rho_\mathrm{y} \cos(\omega_\mathrm{c} t + \phi_\mathrm{y}). \tag{A1.6b}$$

To check the non-correlation between $x(t)$ and $y(t-\tau)$, we first follow the direct approach and take the time average of the product of these signals in accordance with equation (A1.1). For that purpose, we first expand $x(t)y(t-\tau)$ using equation (5.19):

$$\begin{aligned} x(t)y(t-\tau) &= \rho_\mathrm{x}\rho_\mathrm{y} \cos(\omega_\mathrm{c} t + \phi_\mathrm{x}) \cos(\omega_\mathrm{c}(t-\tau) + \phi_\mathrm{y}) \\ &= \frac{\rho_\mathrm{x}\rho_\mathrm{y}}{2}[\cos(2\omega_\mathrm{c} t - \omega_\mathrm{c}\tau + \phi_\mathrm{x} + \phi_\mathrm{y}) + \cos(\omega_\mathrm{c}\tau + \phi_\mathrm{x} - \phi_\mathrm{y})]. \end{aligned} \tag{A1.7}$$

Taking the time average of this expression, we deduce that

$$\overline{x(t)y(t-\tau)} = \frac{\rho_\mathrm{x}\rho_\mathrm{y}}{2} \cos(\omega_\mathrm{c}\tau + \phi_\mathrm{x} - \phi_\mathrm{y}). \tag{A1.8}$$

This means that for such CW signals,

$$\overline{x(t)y(t-\tau)} = 0 \Leftrightarrow \omega_\mathrm{c}\tau + \phi_\mathrm{x} - \phi_\mathrm{y} = \pi/2 + k\pi, \quad \text{with } k \in \mathbb{Z}. \tag{A1.9}$$

This result can also be recovered using the equivalent criteria on their complex envelopes. Indeed, following equation (A1.4) we can simply check in the present case that

$$\begin{aligned} \mathrm{Re}\left\{\overline{\tilde{x}(t)\tilde{y}^*(t-\tau)} e^{j\omega_\mathrm{c}\tau}\right\} &= \mathrm{Re}\left\{\tilde{x}\tilde{y}^* e^{j\omega_\mathrm{c}\tau}\right\} \\ &= \rho_\mathrm{x}\rho_\mathrm{y} \cos(\omega_\mathrm{c}\tau + \phi_\mathrm{x} - \phi_\mathrm{y}). \end{aligned} \tag{A1.10}$$

Comparing this expression with equation (A1.9), we thus obtain the same result as the direct approach. However, it is interesting to note that for such simple signals, the sufficient condition given by equation (A1.5) is untrue. We indeed get in the present case that

$$\overline{\tilde{x}(t)\tilde{y}^*(t-\tau)} = \tilde{x}\tilde{y}^* = \rho_\mathrm{x}\rho_\mathrm{y} e^{j(\phi_\mathrm{x}-\phi_\mathrm{y})} = 0. \tag{A1.11}$$

This condition can be fulfilled if and only if $\rho_\mathrm{x} = 0$ or $\rho_\mathrm{y} = 0$, i.e. if at least one of the two CW signals is null. Of course, if one of the two signals is null, the signals are uncorrelated, but as seen above, this is not necessary to get the result. This is a simple example where $\overline{x(t)y(t-\tau)}$ is null whereas only $\mathrm{Re}\left\{\overline{\tilde{x}(t)\tilde{y}^*(t-\tau)} e^{j\omega_\mathrm{c}\tau}\right\}$ is null and not $\overline{\tilde{x}(t)\tilde{y}^*(t-\tau)}$.

The deep reason for this comes from the non-stationarity of the correlation function of the signals. However, we note from our discussion in Appendix 2 that this stationarity, at least up to second order, is a common property of the processes we deal with in transceivers.

It is thus worthwhile to further explore the topic of correlation using this assumption. Let us therefore examine the stochastic case. For that purpose, we assume that $x(t)$ and $y(t)$ are two stationary stochastic processes. Assuming they are ergodic, we can then check the non-correlation between the random variables x_{t_1} and y_{t_2} corresponding to the samples of $x(t)$ and $y(t)$ at times t_1 and t_2 through their stochastic cross-correlation function:

$$\gamma_{x \times y}(t_1, t_2) = \mathbb{E}\{x_{t_1} y_{t_2}\}. \tag{A1.12}$$

Our purpose is thus to find an equivalent condition for

$$\gamma_{x \times y}(t_1, t_2) = 0, \tag{A1.13}$$

using the complex envelopes $\tilde{x}(t)$ and $\tilde{y}(t)$ of $x(t)$ and $y(t)$. We can continue to assume that those complex envelopes are defined as centered around the same center angular frequency ω_c. The random variables corresponding to the samples of $x(t)$ and $y(t)$ at a given time are thus related to those of their complex envelopes at the same time through

$$x_t = \mathrm{Re}\left\{\tilde{x}_t e^{j\omega_c t}\right\}, \tag{A1.14a}$$

$$y_t = \mathrm{Re}\left\{\tilde{y}_t e^{j\omega_c t}\right\}. \tag{A1.14b}$$

Using equation (1.5) we can then rewrite equation (A1.13) as

$$\begin{aligned}\gamma_{x \times y}(t_1, t_2) = \frac{1}{4}\mathbb{E}\Big\{&\tilde{x}_{t_1}\tilde{y}^*_{t_2} e^{j\omega_c(t_1-t_2)} + \tilde{x}^*_{t_1}\tilde{y}_{t_2} e^{-j\omega_c(t_1-t_2)} \\ &+ \tilde{x}_{t_1}\tilde{y}_{t_2} e^{j\omega_c(t_1+t_2)} + \tilde{x}^*_{t_1}\tilde{y}^*_{t_2} e^{-j\omega_c(t_1+t_2)}\Big\}.\end{aligned} \tag{A1.15}$$

Reordering terms yields

$$\begin{aligned}\gamma_{x \times y}(t_1, t_2) = &\frac{1}{2}\mathrm{Re}\left\{\mathbb{E}\left\{\tilde{x}_{t_1}\tilde{y}^*_{t_2}\right\} e^{j\omega_c(t_1-t_2)}\right\} \\ &+ \frac{1}{2}\mathrm{Re}\left\{\mathbb{E}\left\{\tilde{x}_{t_1}\tilde{y}_{t_2}\right\} e^{j\omega_c(t_1+t_2)}\right\}.\end{aligned} \tag{A1.16}$$

But given the stationarity of $x(t)$ and $y(t)$ at least up to second order, $\gamma_{x \times y}(t_1, t_2)$ must depend only on the time difference $\tau = t_1 - t_2$, and thus the right-hand side of this expression also. As a dependency in $t_1 + t_2$ appears in the last term of this equation, we can deduce that

$$\mathrm{Re}\left\{\mathbb{E}\left\{\tilde{x}_{t_1}\tilde{y}_{t_2}\right\} e^{j\omega_c(t_1+t_2)}\right\} = 0. \tag{A1.17}$$

As this relationship needs to be fulfilled for all instants t_1 and t_2, we suspect that $\mathbb{E}\{\tilde{x}_{t_1}\tilde{y}_{t_2}\}$ needs to be null. This is indeed the case even if the direct expansion of the above real part does not lead so obviously to the result for all sample times. However, this result can be achieved using the properties of the Hilbert transform that is involved in the definition of the complex envelopes of bandpass signals, as introduced in Section 1.1.2. To understand this, let us expand

$\tilde{x}(t)$ and $\tilde{y}(t)$ in terms on the one hand of the original bandpass signals $x(t)$ and $y(t)$, and on the other hand of their Hilbert transforms, $\hat{x}(t)$ and $\hat{y}(t)$ respectively. Given that $\tilde{x}(t)$ and $\tilde{y}(t)$ are assumed as defined around the same center angular frequency, we get from equation (1.22) that

$$\tilde{x}(t) = (x(t) + j\hat{x}(t))e^{-j\omega_c t}, \tag{A1.18a}$$

$$\tilde{y}(t) = (y(t) + j\hat{y}(t))e^{-j\omega_c t}. \tag{A1.18b}$$

We can thus write that

$$\mathbb{E}\{\tilde{x}_{t_1}\tilde{y}_{t_2}\} = \mathbb{E}\{x_{t_1}y_{t_2} - \hat{x}_{t_1}\hat{y}_{t_2} + jx_{t_1}\hat{y}_{t_2} + j\hat{x}_{t_1}y_{t_2}\}e^{-j\omega_c(t_1+t_2)}. \tag{A1.19}$$

But, having assumed the stationarity of the real bandpass processes $x(t)$ and $y(t)$, we get that the cross-correlation terms that involve those processes must depend only on $\tau = t_1 - t_2$. The same obviously holds for the terms involving $\hat{x}(t)$ and $\hat{y}(t)$ as those Hilbert transforms can be seen as the result of a linear filtering of $x(t)$ and $y(t)$, by the definition given in Section 1.1.2. We can thus rewrite the above equation as

$$\mathbb{E}\{\tilde{x}_{t_1}\tilde{y}_{t_2}\} = (\gamma_{x\times y}(\tau) - \gamma_{\hat{x}\times\hat{y}}(\tau) + j\gamma_{x\times\hat{y}}(\tau) + j\gamma_{\hat{x}\times y}(\tau))e^{-j\omega_c(t_1+t_2)}. \tag{A1.20}$$

But in the stationary case the terms $\gamma_{\hat{x}\times\hat{y}}(\tau)$, $\gamma_{x\times\hat{y}}(\tau)$ and $\gamma_{\hat{x}\times y}(\tau)$ can be derived from $\gamma_{x\times y}(\tau)$ using the interference formula. This relationship links the cross-correlation functions of two signals considered at the input and output of a filtering stage. More simply, this relationship can be transposed into the frequency domain in order to avoid the convolution operation and thus make the link between the corresponding cross-spectral densities. In that case we have [3]

$$\Gamma_{(h_x\star x)\times(h_y\star y)}(\omega) = H_x(\omega)H_y^*(\omega)\Gamma_{x\times y}(\omega). \tag{A1.21}$$

Here, $H_x(\omega)$ and $H_y(\omega)$ represent the filter transfer functions experienced by the signals $x(t)$ and $y(t)$, respectively. In our present case of interest, the filter that is involved implements the Hilbert transform. Its transfer function in the frequency domain therefore takes the simple form $-j\,\text{sign}\{\omega\}$ (see equation (1.17)). The application of this relationship in our case of interest directly leads to

$$\Gamma_{\hat{x}\times\hat{y}}(\omega) = \Gamma_{x\times y}(\omega), \tag{A1.22a}$$

$$\Gamma_{x\times\hat{y}}(\omega) = j\,\text{sign}\{\omega\}\Gamma_{x\times y}(\omega), \tag{A1.22b}$$

$$\Gamma_{\hat{x}\times y}(\omega) = -j\,\text{sign}\{\omega\}\Gamma_{x\times y}(\omega). \tag{A1.22c}$$

Taking the inverse Fourier transform of those relationships yields

$$\gamma_{\hat{x}\times\hat{y}}(\tau) = \gamma_{x\times y}(\tau), \tag{A1.23a}$$

$$\gamma_{x\times\hat{y}}(\tau) = -\hat{\gamma}_{x\times y}(\tau), \tag{A1.23b}$$

$$\gamma_{\hat{x}\times y}(\tau) = \hat{\gamma}_{x\times y}(\tau), \tag{A1.23c}$$

where $\hat{\gamma}_{x\times y}(\tau)$ stands for the Hilbert transform of $\gamma_{x\times y}(\tau)$. Thus, the terms involved in the right-hand side of equation (A1.20) cancel each other out. We thus have the results expected, i.e. that the complex envelopes $\tilde{x}(t)$ and $\tilde{y}(t)$, defined as centered around the same center frequency, must be such that

$$\mathbb{E}\{\tilde{x}_{t_1}\tilde{y}_{t_2}\} = 0. \tag{A1.24}$$

Now using this result in turn while remembering that $\gamma_{x\times y}(t_1, t_2)$ is a function of $\tau = t_1 - t_2$ only, we get from equation (A1.16) that

$$\begin{aligned}\gamma_{x\times y}(\tau) &= \frac{1}{2}\mathrm{Re}\{\mathbb{E}\{\tilde{x}_{t_1}\tilde{y}^*_{t_2}\}e^{j\omega_c\tau}\}\\ &= \frac{1}{2}\mathrm{Re}\{\gamma_{\tilde{x}\times\tilde{y}}(t_1, t_2)e^{j\omega_c\tau}\}.\end{aligned} \tag{A1.25}$$

We see that $\gamma_{\tilde{x}\times\tilde{y}}(t_1, t_2)$ must depend on τ only as $\gamma_{x\times y}(\tau)$ does. We can therefore write

$$\gamma_{x\times y}(\tau) = \frac{1}{2}\mathrm{Re}\{\gamma_{\tilde{x}\times\tilde{y}}(\tau)e^{j\omega_c\tau}\}. \tag{A1.26}$$

Thus, the condition for the non-correlation between $x(t)$ and $y(t)$ to hold can be transposed onto their complex envelopes defined as centered around the same center angular frequency following

$$\gamma_{x\times y}(\tau) = 0 \Leftrightarrow \mathrm{Re}\{\gamma_{\tilde{x}\times\tilde{y}}(\tau)e^{j\omega_c\tau}\} = 0. \tag{A1.27}$$

We thus obtain the same necessary and sufficient condition on the complex envelopes for the non-correlation to hold as derived in the deterministic case. But here again, the sufficient condition

$$\gamma_{\tilde{x}\times\tilde{y}}(\tau) = 0 \Rightarrow \gamma_{x\times y}(\tau) = 0, \tag{A1.28}$$

alone is enough to get the non-correlation between corresponding bandpass processes. However, in the stationary case, at least up to second order, we can go a step further and show that there is an equivalence between having $\gamma_{x\times y}(\tau) = 0$ and $\gamma_{\tilde{x}\times\tilde{y}}(\tau) = 0$. The deep reason for this to hold comes again from the properties of the Hilbert transform. This transform behaves as an all-pass filter in the frequency domain so that the statistical properties of $\mathrm{Im}\{\gamma_{\tilde{x}\times\tilde{y}}(\tau)\}$ are in fact fully related to those of $\mathrm{Re}\{\gamma_{\tilde{x}\times\tilde{y}}(\tau)\}$. To understand this, let us expand $\gamma_{\tilde{x}\times\tilde{y}}(\tau) = \mathbb{E}\{\tilde{x}_t\tilde{y}^*_{t-\tau}\}$ using equation (A1.18):

$$\begin{aligned}\gamma_{\tilde{x}\times\tilde{y}}(\tau) &= \mathbb{E}\{x_t y_{t-\tau} + \hat{x}_t\hat{y}_{t-\tau} - jx_t\hat{y}_{t-\tau} + j\hat{x}_t y_{t-\tau}\}e^{-j\omega_c\tau}\\ &= (\gamma_{x\times y}(\tau) + \gamma_{\hat{x}\times\hat{y}}(\tau) - j\gamma_{x\times\hat{y}}(\tau) + j\gamma_{\hat{x}\times y}(\tau))e^{-j\omega_c\tau}.\end{aligned} \tag{A1.29}$$

But under our stationarity assumption, we can use equation (A1.23) so that

$$\gamma_{\tilde{x}\times\tilde{y}}(\tau) = 2(\gamma_{x\times y}(\tau) + j\hat{\gamma}_{x\times y}(\tau))e^{-j\omega_c\tau}. \quad \text{(A1.30)}$$

Moreover, given that $\gamma_{x\times y}(\tau) = 0$ for all τ, and calling to mind the definition of the Hilbert transform given by equation (1.20), we get that $\hat{\gamma}_{x\times y}(\tau)$ is also necessarily identically null. It then follows from the above equation that $\gamma_{\tilde{x}\times\tilde{y}}(\tau) = 0$. Given that the reverse relationship holds by equation (A1.28), we then get the result that for stationary bandpass processes, at least up to second order, and for their complex envelopes defined as centered around the same center frequency,

$$\gamma_{\tilde{x}\times\tilde{y}}(\tau) = 0 \Leftrightarrow \gamma_{x\times y}(\tau) = 0. \quad \text{(A1.31)}$$

We can then make the link with the case of the CW signals considered at the beginning of the section. We derived that for such signals only the general condition given by either equation (A1.4) or (A1.27) holds so that the real and imaginary parts of $\gamma_{\tilde{x}\times\tilde{y}}(\tau)$ cannot be null at the same time without one of the signals being null. The deep reason for this comes from the fact that stationarity of the cross-correlation function between the two bandpass signals does not hold. We see from equation (A1.7) that this cross-correlation depends not only on the time difference τ but also on the sample time t. This situation could be overcome if necessary by adding for instance a random phase term uniformly distributed over $[0, 2\pi]$ to each of those sinusoidal signals. This is what is classically done to model noise sources that are for the most part stationary [39]. This approach then allows us to carry out spot noise derivations while taking advantage of the complex notation for managing electrical equations. This approach can be generalized by using general narrowband complex envelopes, as is done in Section 4.2.3.

To conclude, we observe that the expression $\mathbb{E}\{\tilde{x}_t\tilde{y}_t^*\}$ defines a scalar product in the mathematical sense. This is why stationary and independent noise sources that fulfill equation (A1.31) are said to be orthogonal. Generally speaking, it does not mean that an exact instantaneous phase offset of $\pi/2$ stands for all sample times when considering a particular realization of the complex envelopes $\tilde{x}(t)$ and $\tilde{y}(t)$. The situation is obviously different in the deterministic case, as can be seen with the example of the CW signals considered at the beginning of the section. In that particular situation, non-correlation means a $\pi/2$ phase offset between the arguments of the complex envelopes in accordance with equation (A1.9).

A1.2 Properties of Cross-Correlation Functions

Let us now detail some basic properties of cross-correlation functions useful for the analytical derivations in this book. Suppose that $\tilde{x}_{t_1}$ and $\tilde{y}_{t_2}$ represent the two complex random variables that correspond to the two samples of the stochastic processes, $\tilde{x}(t)$ and $\tilde{y}(t)$, considered at times t_1 and t_2 respectively, i.e.

$$\tilde{x}_{t_1} = \tilde{x}(t_1), \quad \text{(A1.32a)}$$

$$\tilde{y}_{t_2} = \tilde{y}(t_2). \quad \text{(A1.32b)}$$

The cross-correlation function, $\gamma_{\tilde{x}\times\tilde{y}}(t_1, t_2)$, of those two stochastic processes is defined in this book as

$$\gamma_{\tilde{x}\times\tilde{y}}(t_1, t_2) = \mathbb{E}\left\{\tilde{x}_{t_1}\tilde{y}^*_{t_2}\right\}, \tag{A1.33}$$

where $\mathbb{E}\{.\}$ stands for the stochastic expectation. In the case of stationarity of the considered random processes, at least up to second order, the cross-correlation function depends only on the time difference between the sample times, $\tau = t_1 - t_2$. Thus, $\gamma_{\tilde{x}\times\tilde{y}}(t_1, t_2)$ can be written as

$$\gamma_{\tilde{x}\times\tilde{y}}(\tau) = \mathbb{E}\{\tilde{x}_t\tilde{y}^*_{t-\tau}\}. \tag{A1.34}$$

From this definition, we can deduce the following properties:

$$\gamma_{\tilde{x}\times\tilde{y}}(t_1, t_2) = \mathbb{E}\left\{\tilde{x}_{t_1}\tilde{y}^*_{t_2}\right\} = \mathbb{E}^*\left\{\tilde{y}_{t_2}\tilde{x}^*_{t_1}\right\} = \gamma^*_{\tilde{y}\times\tilde{x}}(t_2, t_1), \tag{A1.35a}$$

$$\gamma_{\tilde{x}\times\tilde{y}^*}(t_1, t_2) = \mathbb{E}\left\{\tilde{x}_{t_1}\tilde{y}_{t_2}\right\} = \mathbb{E}\left\{\tilde{y}_{t_2}\tilde{x}_{t_1}\right\} = \gamma_{\tilde{y}\times\tilde{x}^*}(t_2, t_1), \tag{A1.35b}$$

$$\gamma_{\tilde{x}^*\times\tilde{y}}(t_1, t_2) = \mathbb{E}\left\{\tilde{x}^*_{t_1}\tilde{y}^*_{t_2}\right\} = \mathbb{E}^*\left\{\tilde{x}_{t_1}\tilde{y}_{t_2}\right\} = \gamma^*_{\tilde{x}\times\tilde{y}^*}(t_1, t_2), \tag{A1.35c}$$

which reduce in the stationary case, at least up to second order, to

$$\gamma_{\tilde{x}\times\tilde{y}}(\tau) = \gamma^*_{\tilde{y}\times\tilde{x}}(-\tau), \tag{A1.36a}$$

$$\gamma_{\tilde{x}\times\tilde{y}^*}(\tau) = \gamma_{\tilde{y}\times\tilde{x}^*}(-\tau), \tag{A1.36b}$$

$$\gamma_{\tilde{x}^*\times\tilde{y}}(\tau) = \gamma^*_{\tilde{x}\times\tilde{y}^*}(\tau). \tag{A1.36c}$$

Of course, these derivations remain valid if the processes concerned are real valued. In that case, all the imaginary parts are null and thus the complex conjugate symbols vanish.

A1.3 Properties of Autocorrelation Functions

Let us now focus on the autocorrelation case. This can simply be derived from the cross-correlation case assuming the same stochastic process, $\tilde{x}(t)$, considered at two different instants. Thus, retaining the notation used for the cross-correlation case, the autocorrelation of $\tilde{x}(t)$ is defined by

$$\gamma_{\tilde{x}\times\tilde{x}}(t_1, t_2) = \mathbb{E}\left\{\tilde{x}_{t_1}\tilde{x}^*_{t_2}\right\}. \tag{A1.37}$$

For a stationary process, at least up to second order, this autocorrelation function reduces to

$$\gamma_{\tilde{x}\times\tilde{x}}(\tau) = \mathbb{E}\{\tilde{x}_t\tilde{x}^*_{t-\tau}\}. \tag{A1.38}$$

Here we observe that in this stationary case, the autocorrelation function considered at $\tau = 0$ simply gives the power $P_{\tilde{x}}$ of the considered process as

$$P_{\tilde{x}} = \mathbb{E}\{\tilde{x}\tilde{x}^*\} = \gamma_{\tilde{x}\times\tilde{x}}(0). \tag{A1.39}$$

In addition, the relationships derived in the cross-correlation case are still valid in our present autocorrelation case considering $\tilde{y} = \tilde{x}$. In particular, equation (A1.36a) reduces to

$$\gamma^*_{\tilde{x}\times\tilde{x}}(\tau) = \gamma_{\tilde{x}\times\tilde{x}}(-\tau). \tag{A1.40}$$

This means that in a general manner, an autocorrelation function fulfills Hermitian symmetry. This is interesting if we consider that in the stationary case, at least up to second order, the PSD of the process $\tilde{x}(t)$ can be derived by taking the Fourier transform of its autocorrelation function [3, 4]. Due to the Fourier transform property given by equation (1.39), we then see that

$$\mathcal{F}_{\{\gamma^*_{\tilde{x}\times\tilde{x}}(\tau)\}}(\omega) = \mathcal{F}_{\{\gamma_{\tilde{x}\times\tilde{x}}(-\tau)\}}(\omega) = \Gamma_{\tilde{x}\times\tilde{x}}(-\omega). \tag{A1.41}$$

Using equation (1.8) in turn, we have

$$\mathcal{F}_{\{\gamma^*_{\tilde{x}\times\tilde{x}}(\tau)\}}(\omega) = \Gamma^*_{\tilde{x}\times\tilde{x}}(-\omega). \tag{A1.42}$$

We thus deduce from the above two equations that

$$\Gamma_{\tilde{x}\times\tilde{x}}(\omega) = \Gamma^*_{\tilde{x}\times\tilde{x}}(\omega). \tag{A1.43}$$

As can be expected for the Fourier transform of a signal with Hermitian symmetry, we see that the PSD is a real quantity. However, we remark that this is true for power spectral densities only, i.e. for Fourier transforms of autocorrelation functions, but not necessarily for power cross-spectral densities, i.e. for Fourier transforms of cross-correlation functions.

To conclude, we can derive the relationships between the autocorrelation function of a complex stochastic process and the correlation functions of its real or imaginary parts. For that purpose, let us expand equation (A1.37), supposing that $\tilde{x}(t) = p(t) + \mathrm{j}q(t)$. We then get that

$$\begin{aligned}
\gamma_{\tilde{x}\times\tilde{x}}(t_1, t_2) &= \mathbb{E}\{(p_{t_1} + \mathrm{j}q_{t_1})(p_{t_2} - \mathrm{j}q_{t_2})\} \\
&= \mathbb{E}\{p_{t_1}p_{t_2}\} + \mathbb{E}\{q_{t_1}q_{t_2}\} + \mathrm{j}\mathbb{E}\{q_{t_1}p_{t_2} - p_{t_1}q_{t_2}\} \\
&= \gamma_{p\times p}(t_1, t_2) + \gamma_{q\times q}(t_1, t_2) + \mathrm{j}(\gamma_{q\times p}(t_1, t_2) - \gamma_{p\times q}(t_1, t_2)).
\end{aligned}$$

Applying equation (A1.35a) and remembering that for real valued variables the cross-correlation functions are real valued, we get

$$\begin{aligned}
\gamma_{\tilde{x}\times\tilde{x}}(t_1, t_2) = \gamma_{p\times p}(t_1, t_2) &+ \gamma_{q\times q}(t_1, t_2) \\
&+ \mathrm{j}(\gamma_{p\times q}(t_2, t_1) - \gamma_{p\times q}(t_1, t_2)).
\end{aligned} \tag{A1.44}$$

In the same way, we have

$$\begin{aligned}
\gamma_{\tilde{x}\times\tilde{x}^*}(t_1, t_2) &= \mathbb{E}\{(p_{t_1} + \mathrm{j}q_{t_1})(p_{t_2} + \mathrm{j}q_{t_2})\} \\
&= \gamma_{p\times p}(t_1, t_2) - \gamma_{q\times q}(t_1, t_2) + \mathrm{j}(\gamma_{q\times p}(t_1, t_2) + \gamma_{p\times q}(t_1, t_2)),
\end{aligned}$$

which can be written using equation (A1.35a) as

$$\gamma_{\tilde{x}\times\tilde{x}^*}(t_1,t_2) = \gamma_{p\times p}(t_1,t_2) - \gamma_{q\times q}(t_1,t_2) \\ + \mathrm{j}(\gamma_{p\times q}(t_2,t_1) + \gamma_{p\times q}(t_1,t_2)). \tag{A1.45}$$

In the stationary case, at least up to second order, those relationships thus reduce to

$$\gamma_{\tilde{x}\times\tilde{x}}(\tau) = \gamma_{p\times p}(\tau) + \gamma_{q\times q}(\tau) + \mathrm{j}(\gamma_{p\times q}(-\tau) - \gamma_{p\times q}(\tau)), \tag{A1.46a}$$

$$\gamma_{\tilde{x}\times\tilde{x}^*}(\tau) = \gamma_{p\times p}(\tau) - \gamma_{q\times q}(\tau) + \mathrm{j}(\gamma_{p\times q}(-\tau) + \gamma_{p\times q}(\tau)). \tag{A1.46b}$$

Appendix 2

Stationarity

It is often convenient to assume that the processes we are dealing with are stationary. In our system design perspective, this is for instance of interest for the derivation of the power spectral densities of stochastic processes that represent either noises or randomly modulated signals.

In wireless transceivers we have to deal with both bandpass signals, which model for instance the RF signals of interest, and lowpass signals, which model for instance the baseband modulating waveforms. But depending on whether we are dealing with RF bandpass processes that have been generated directly in the RF domain, as may be the case for a thermal noise generated at the early stages of a receiver for instance, or by the frequency upconversion of a stationary lowpass process, as may be the case for all the modulated signals we are dealing with, stationarity is not as obvious a property as it might first appear. It is therefore useful to detail the properties and the conditions associated with such stationarity.

A2.1 Stationary Bandpass Signals

Some of the bandpass RF signals we are dealing with in wireless transceivers are inherently stationary, at least over a characteristic time duration for the wireless link considered. Thus we are already dealing with an approximation as stationarity necessarily requires the process concerned to be spread over an infinite duration [3]. However, we can assume that an RF noise source, linked to thermal noise for instance, is stationary in the sense that its statistical characteristics (e.g. variance, PSD) remain constant whatever the observation time. With this definition, we see that by stationarity we mean here strict stationarity, in the sense that all the statistical characteristics of the process are constant in time. This is in particular the case for the PDF if it exists, of the samples of the process. This stationarity can be assumed, for instance, for the RF noise sources we have to deal with in practice even if the second order (or weak, or wide sense) stationarity (WSS), i.e. the stationarity of the moments up to second order, is sufficient for most of our applications, for instance for the derivation of power spectral densities.

Let us suppose for now that we are dealing with a stationary RF bandpass process, at least up to second order, $x(t)$, and let us examine the consequences of this stationarity on its complex

Wireless Transceiver Architecture: Bridging RF and Digital Communications, First Edition. Pierre Baudin.

envelopes. For that purpose, we can focus on the one, $\tilde{x}(t) = p(t) + jq(t)$, that is defined as centered around the angular frequency ω_c. According to equation (1.27), we can then write

$$x(t) = \mathrm{Re}\left\{\tilde{x}(t)e^{j\omega_c t}\right\}. \tag{A2.1}$$

In order to examine the potential stationarity of $\tilde{x}(t)$, we can focus first on the moments, at least up to second order, of this lowpass process.

Let us examine first the characteristics of the first order moment of $\tilde{x}(t)$. Taking expectations in equation (A2.1) yields

$$\mathbb{E}\{x_t\} = \mathrm{Re}\left\{\mathbb{E}\{\tilde{x}_t\}e^{j\omega_c t}\right\} \tag{A2.2}$$

or, as $x(t)$ is assumed stationary,

$$\mathbb{E}\{x\} = \mathrm{Re}\left\{\mathbb{E}\{\tilde{x}_t\}e^{j\omega_c t}\right\}. \tag{A2.3}$$

From this expression, it is not so obvious that having $\mathbb{E}\{x\}$ independent of t leads to having $\mathbb{E}\{\tilde{x}_t\}$ also independent of t. However, in all practical cases of interest, $x(t)$ is a centered process. This is for instance the case for noise processes but also for most of the modulation schemes encountered in the field of wireless as long as the data bits remain equally distributed (see the examples discussed in Chapter 1). With this additional assumption, we then necessarily get that

$$\mathbb{E}\{\tilde{x}_t\} = \mathbb{E}\{\tilde{x}\} = \mathbb{E}\{p\} + j\mathbb{E}\{q\} = 0, \tag{A2.4}$$

and thus the stationarity up to first order of $\tilde{x}(t)$ and of both $p(t)$ and $q(t)$.

Turning now to the characteristics of the second order moment of $\tilde{x}(t)$, we can adopt the same approach as for its first order moment and derive the autocorrelation function of $x(t)$, i.e. $\gamma_{x\times x}(t_1, t_2) = \mathbb{E}\{x_{t_1}x_{t_2}\}$. For that purpose, we can directly rely on the derivations performed in Appendix 1 and write from equation (A.16) that

$$\begin{aligned}\gamma_{x\times x}(t_1, t_2) = {}& \frac{1}{2}\mathrm{Re}\left\{\mathbb{E}\left\{\tilde{x}_{t_1}\tilde{x}^*_{t_2}\right\}e^{j\omega_c(t_1-t_2)}\right\}\\ &+\frac{1}{2}\mathrm{Re}\left\{\mathbb{E}\left\{\tilde{x}_{t_1}\tilde{x}_{t_2}\right\}e^{j\omega_c(t_1+t_2)}\right\}.\end{aligned} \tag{A2.5}$$

But, given that $x(t)$ is a stationary RF bandpass process, at least up to second order, we get that $\gamma_{x\times x}(t_1, t_2)$ is necessarily a function of $\tau = t_1 - t_2$ only. Consequently, the last term of the right-hand side of this equation which depends on $t_1 + t_2$ must be null. By the derivation of equation (A1.24), this holds because we can write

$$\mathbb{E}\{\tilde{x}_{t_1}\tilde{x}_{t_2}\} = \gamma_{\tilde{x}\times\tilde{x}^*}(t_1, t_2) = 0. \tag{A2.6}$$

Then equation (A2.5) reduces to

$$\gamma_{x\times x}(\tau) = \frac{1}{2}\mathrm{Re}\left\{\gamma_{\tilde{x}\times\tilde{x}}(t_1, t_2)e^{j\omega_c\tau}\right\}, \tag{A2.7}$$

and $\gamma_{\tilde{x}\times\tilde{x}}(t_1, t_2)$ must be a function of $\tau = t_1 - t_2$ only as $\gamma_{x\times x}(\tau)$ is. By the discussion so far, for a stationary and centered RF bandpass process $x(t)$, any of its complex envelopes is also stationary, at least up to second order, and with

$$\gamma_{x\times x}(\tau) = \frac{1}{2}\mathrm{Re}\left\{\gamma_{\tilde{x}\times\tilde{x}}(\tau)e^{j\omega_c\tau}\right\}. \tag{A2.8}$$

An alternative approach would have been possible by directly deriving an expression for $\gamma_{\tilde{x}\times\tilde{x}}(t_1, t_2)$. We observe that the stationarity of $x(t)$ involves the stationarity of its Hilbert transform $\hat{x}(t)$ as the latter signal is simply the result of a linear filtering of $x(t)$ (see the definition given for this operation in Section 1.1.2). We can thus directly refer to equation (A1.30) to write in that case that $\gamma_{\tilde{x}\times\tilde{x}}(t_1, t_2) = \gamma_{\tilde{x}\times\tilde{x}}(\tau)$ with

$$\gamma_{\tilde{x}\times\tilde{x}}(\tau) = 2(\gamma_{x\times x}(\tau) + j\hat{\gamma}_{x\times x}(\tau))e^{-j\omega_c\tau}. \tag{A2.9}$$

However, the former approach leading to equation (A2.6) is of particular interest as it allows us to derive interesting properties for the real and imaginary parts of $\tilde{x}(t) = p(t) + jq(t)$. We must first remark that those real and imaginary parts are also stationary up to second order as $\tilde{x}(t)$ is. This can be seen for instance for $p(t)$ by writing $p(t) = (\tilde{x}(t) + \tilde{x}^*(t))/2$ and then expanding $\gamma_{p\times p}(t_1, t_2)$. Using on the one hand that $\tilde{x}(t)$ is stationary up to second order and that equation (A2.6) holds, we obtain $\gamma_{p\times p}(t_1, t_2) = \gamma_{p\times p}(\tau)$. The same property obviously holds for $q(t)$. Consequently, referring to the expansion of $\gamma_{\tilde{x}\times\tilde{x}^*}(t_1, t_2)$ given by equation (A1.45), the stationarity of the bandpass signal $x(t)$ leads to

$$\gamma_{\tilde{x}\times\tilde{x}^*}(t_1, t_2) = \gamma_{\tilde{x}\times\tilde{x}^*}(\tau). \tag{A2.10}$$

As the stationarity of $x(t)$ leads at the same time to equation (A2.6) we can finally write[1] [1, 3]

$$\gamma_{\tilde{x}\times\tilde{x}^*}(\tau) = 0, \tag{A2.11}$$

and thus, from equation (A1.46b),

$$\gamma_{p\times p}(\tau) - \gamma_{q\times q}(\tau) + j(\gamma_{p\times q}(-\tau) + \gamma_{p\times q}(\tau)) = 0. \tag{A2.12}$$

[1] This condition is important in various practical use cases in the field of wireless transceivers. We refer for instance to Chapter 6 where the RF bandpass signal whose complex envelope is the complex conjugate of $\tilde{x}(t)$ is nothing more than the image signal of the bandpass signal $x(t)$ during a complex frequency transposition in the presence of gain and phase imbalance.

Thus for $x(t)$ stationary, at least up to second order, the real and imaginary parts $p(t)$ and $q(t)$ of any of its complex envelopes necessarily fulfill

$$\gamma_{p\times p}(\tau) = \gamma_{q\times q}(\tau), \tag{A2.13a}$$

$$\gamma_{p\times q}(-\tau) = -\gamma_{p\times q}(\tau). \tag{A2.13b}$$

In particular, considering equation (A2.13a) for $\tau = 0$, we see that $p(t)$ and $q(t)$ necessarily have the same power. More generally, taking the Fourier transform of this relationship, $p(t)$ and $q(t)$ also necessarily have the same PSD i.e. the same spectral shape. In the same way, considering equation (A2.13b) for $\tau = 0$, we have the additional property that $p(t)$ and $q(t)$ are necessarily uncorrelated when considered at the same time. All those properties are of importance, for instance when considering the decomposition of an RF bandpass noise in terms of its two quadrature components, as done for instance in Section 1.2.1.

A2.2 Stationary Complex Envelopes

Let us now focus on the reverse situation, where we get an RF bandpass process that is generated through the frequency upconversion of a given stationary lowpass process. This is the case for both the RF signals that carry the information in wireless systems, but also for the RF bandpass noises that result from the frequency upconversion of baseband components on the transmit side for instance. We suppose that we are dealing with two lowpass processes, $p(t)$ and $q(t)$, that correspond respectively to the real and imaginary parts of a modulating complex envelope $\tilde{x}(t) = p(t) + jq(t)$, that is further upconverted to generate the RF bandpass process $x(t)$ according to the processing corresponding to equation (A2.1).

In most applications, this lowpass complex envelope can be considered as a stationary process. This seems as reasonable for baseband noise processes as it was for RF noises considered in the previous section. It also seems reasonable for modulating processes as long as we are dealing with waveforms generated from random bits flowing from a stationary source encoder. However, even if we can assume the stationarity of those lowpass modulating waveforms, the stationarity of the resulting RF bandpass process is not so obvious at first glance. However, it can be derived, at least up to second order, under common assumptions.

Assuming first that we are dealing with centered lowpass processes, the expectations of all the resulting bandpass processes are necessarily null, as can be seen from equation (A2.3). As a side effect, we thus get the independence of this first order moment with respect to the sample time. It then remains to check the dependency of the autocorrelation function of $x(t)$ on the sample times considered. For that purpose, we can recall our discussion in the previous section, and in particular equation (A2.5). According to this equation, to have $\gamma_{x\times x}(t_1, t_2)$ as a function of $\tau = t_1 - t_2$ only, we need to have $\mathbb{E}\{\tilde{x}_{t_1}\tilde{x}_{t_2}\} = \gamma_{\tilde{x}\times\tilde{x}^*}(t_1, t_2) = 0$. Or, using equation (A1.46b) and remembering that the lowpass signals we are dealing with are assumed stationary, we need to check that

$$\gamma_{\tilde{x}\times\tilde{x}^*}(t_1, t_2) = \gamma_{\tilde{x}\times\tilde{x}^*}(\tau) = \gamma_{p\times p}(\tau) - \gamma_{q\times q}(\tau) + j(\gamma_{p\times q}(-\tau) + \gamma_{p\times q}(\tau)) = 0, \tag{A2.14}$$

i.e. that equation (A2.13) holds.

In order to go further, we need to consider the practical characteristics of the lowpass waveforms of interest in transceivers. We make the following observations:

(i) Those waveforms are for the most part generated with the same power spectral densities on the P and Q paths. This leads to the same autocorrelation functions for $p(t)$ and $q(t)$. The same argument can be derived for noise components that are generally generated through the use of identical blocks that are duplicated for the P and Q paths of practical line-ups. In both cases, we can thus assume that

$$\gamma_{p \times p}(\tau) = \gamma_{q \times q}(\tau). \tag{A2.15}$$

(ii) Due to the optimization of the source coding for the efficiency of the transmission, we can assume for the most part that the modulating waveforms are uncorrelated between the P and Q paths. This holds at least during the transmission of the randomly modulated data. In the same way, given noise components generated by different physical blocks on the P and Q paths, the $p(t)$ and $q(t)$ noise processes are independent and thus uncorrelated. In both cases, we can finally assume in practice that

$$\gamma_{p \times q}(\tau) = 0. \tag{A2.16}$$

Consequently, we get that for practical transceiver implementations and for practical modulation schemes, equations (A2.15) and (A2.16) hold so that we finally get from equation (A2.14) that

$$\gamma_{\tilde{x} \times \tilde{x}^*}(t_1, t_2) = \gamma_{\tilde{x} \times \tilde{x}^*}(\tau) = 0. \tag{A2.17}$$

We then recover that the RF bandpass signal resulting from the frequency upconversion of such centered lowpass baseband process is stationary, at least up to second order.

A2.3 Gaussian Case

We remark that additional interesting properties hold when the processes we are dealing with can be assumed Gaussian in addition to being centered and stationary up to second order. In the discussion in Section 1.2.1, $x(t)$, $p(t)$ and $q(t)$ were centered and Gaussian. We can thus also remark that the condition given by equation (A2.11) corresponds to having the complex normal random vector composed of the time samples of $\tilde{x}(t) = p(t) + jq(t)$, which is circular by the definition in Appendix 3. This property is of importance in the analytical derivation of the higher order moments of such process as done in that appendix.

Finally, we observe that second order stationarity is equivalent to strict stationarity for Gaussian processes. Thus $\tilde{x}(t)$ is necessarily stationary in the strict sense when $x(t)$ is stationary, centered and Gaussian. Whereas second order stationarity remains sufficient for most applications in this book, higher order stationarity is of interest in particular situations, such as when examining the spectral regrowth phenomenon involving higher order moments in Chapter 5. It may thus also be of interest to assume signals that can be modeled by Gaussian processes in that case.

Appendix 3

Moments of Normal Random Vectors

Real or complex Gaussian processes are defined so that the vector composed of the random variables corresponding to the time samples of those processes are respectively real or complex normal random vectors. Such processes are of particular importance as they can model both real bandpass signals and lowpass complex envelopes often encountered in wireless transceivers. This applies for instance to most of the analog noises as discussed in Section 1.2, and to most of the wideband modulating waveforms discussed in Section 1.3.3. From the system design point of view, such processes make it possible to carry out analytical derivations with the Gaussian distribution. This is for instance the case when investigating the distortion experienced by a Gaussian signal that goes through a nonlinearity (see Chapter 5).

But to do so, we need to deal with higher order moments of normal random vectors. It is therefore the purpose of this appendix to recall the results on that topic. We start by considering real normal random vectors in order to use the corresponding results to derive moments of complex normal random vectors that are used to represent the time samples of complex envelopes of Gaussian bandpass processes. We also give general results in order to further highlight the simplifications linked to the *stationarity* of the processes as classically encountered in wireless transceivers (see the discussion in Appendix 2).

A3.1 Real Normal Random Vectors

Let us first focus on real Gaussian random vectors. In the wireless transceiver perspective, this means that we consider here a random vector resulting from the time samples either of an RF bandpass Gaussian process or of the real and imaginary parts of the complex envelopes of such bandpass Gaussian process.

For a set of N real random variables $[x_1, \ldots, x_N]$, the vector

$$\mathbf{x} = \begin{pmatrix} x_1 \\ \vdots \\ x_N \end{pmatrix} \tag{A3.1}$$

Wireless Transceiver Architecture: Bridging RF and Digital Communications, First Edition. Pierre Baudin.

is said to be a real normal random vector if its PDF, provided it exists, follows a multivariate normal distribution, i.e. can be expressed as [3]

$$p(x_1, \dots, x_N) = \frac{1}{(2\pi)^{N/2}|\boldsymbol{\Sigma}|^{1/2}} \exp\left(-\frac{1}{2}(\mathbf{x}-\boldsymbol{\mu})^T\boldsymbol{\Sigma}^{-1}(\mathbf{x}-\boldsymbol{\mu})\right), \tag{A3.2}$$

where $.^T$ refers to the transpose operation. Here $\boldsymbol{\mu}$ is the column vector composed of the expectations of the x_j variables,

$$\boldsymbol{\mu} = \begin{pmatrix} \mu_1 \\ \vdots \\ \mu_N \end{pmatrix}, \tag{A3.3}$$

with

$$\mu_j = \mathbb{E}\{x_j\}. \tag{A3.4}$$

In the same way, $\boldsymbol{\Sigma}$ is the covariance matrix defined by

$$\boldsymbol{\Sigma} = \begin{pmatrix} \sigma_{11} & \sigma_{12} & \cdots & \sigma_{1N} \\ \sigma_{21} & \sigma_{22} & \cdots & \sigma_{2N} \\ \cdots & \cdots & \cdots & \cdots \\ \sigma_{N1} & \sigma_{N2} & \cdots & \sigma_{NN} \end{pmatrix}, \tag{A3.5}$$

with

$$\sigma_{jk} = \mathbb{E}\{(x_j - \mu_j)(x_k - \mu_k)\}, \tag{A3.6}$$

and $|\boldsymbol{\Sigma}|$ is the determinant of $\boldsymbol{\Sigma}$. We see that the PDF given by equation (A3.2) may not exist if the covariance matrix is singular. However, we observe that even in that case, the characteristic function of $\mathbf{x}$, $\Phi_{\mathbf{x}}(u_1, \dots, u_N)$, defined by

$$\Phi_{\mathbf{x}}(u_1, \dots, u_N) = \mathbb{E}\left\{e^{j\mathbf{u}^T\mathbf{x}}\right\}, \tag{A3.7}$$

with the vector $\mathbf{u}$ given by

$$\mathbf{u} = \begin{pmatrix} u_1 \\ \vdots \\ u_N \end{pmatrix}, \tag{A3.8}$$

is well defined. This characteristic function can indeed be evaluated as

$$\Phi_{\mathbf{x}}(u_1, \dots, u_N) = \exp\left(j\boldsymbol{\mu}^T\mathbf{u} - \frac{1}{2}\mathbf{u}^T\boldsymbol{\Sigma}\mathbf{u}\right). \tag{A3.9}$$

Moreover, the processes we deal with in the field of wireless transceivers are for the most part centered. This holds for both the noise terms and for the modulation schemes classically encountered in the field of wireless as long as the data bits remain equally distributed (see the examples discussed in Chapter 1). We can thus assume that

$$\boldsymbol{\mu} = 0 \tag{A3.10}$$

so that the characteristic function used in our derivations reduces to

$$\Phi_{\mathbf{x}}(u_1, \dots, u_N) = \exp\left(-\frac{1}{2}\mathbf{u}^{\mathrm{T}}\boldsymbol{\Sigma}\mathbf{u}\right). \tag{A3.11}$$

It is then interesting to remark that this characteristic function allows the derivation of the moments of any order of centered real normal random vectors. For instance, the kth order moment defined as

$$\mathbb{E}\left\{\prod_{l=1}^{k} x_{i_l}\right\}, \tag{A3.12}$$

with indexes i_l in the set $[1, \dots, N]$, can be evaluated by identifying terms of the same order in the Taylor series expansions of the right-hand side of equations (A3.7) and (A3.11). Proceeding in that way, we observe that we cannot have even order terms in u_i in those series. This leads to null odd order moments for our centered real normal vector. Conversely, for $k = 2\lambda$, we have

$$\frac{1}{(2\lambda)!}\mathbb{E}\{(\mathbf{u}^{\mathrm{T}}\mathbf{x})^{2\lambda}\} = \frac{1}{2^{\lambda}\,\lambda!}(\mathbf{u}^{\mathrm{T}}\boldsymbol{\Sigma}\mathbf{u})^{\lambda}. \tag{A3.13}$$

Finally, we get for centered real normal vectors that [3]:

(i) when k is odd, of the form $2\lambda + 1$,

$$\mathbb{E}\left\{\prod_{l=1}^{2\lambda+1} x_{i_l}\right\} = 0; \tag{A3.14}$$

(ii) when k is even, of the form 2λ,

$$\mathbb{E}\left\{\prod_{l=1}^{2\lambda} x_{i_l}\right\} = \sum \mathbb{E}\{x_{i_1}x_{i_2}\}\mathbb{E}\{x_{i_3}x_{i_4}\}\dots\mathbb{E}\{x_{i_{2\lambda-1}}x_{i_{2\lambda}}\}. \tag{A3.15}$$

Here, the sum is taken over all the allocations of the set $[i_1, \dots, i_{2\lambda}]$ into λ unordered pairs. This means for instance that

$$\mathbb{E}\{x_1x_2x_3x_4\} = \mathbb{E}\{x_1x_2\}\mathbb{E}\{x_3x_4\} + \mathbb{E}\{x_1x_3\}\mathbb{E}\{x_2x_4\} + \mathbb{E}\{x_1x_4\}\mathbb{E}\{x_2x_3\}. \tag{A3.16}$$

A3.2 Complex Normal Random Vectors

In order to perform derivations directly on the complex envelopes of Gaussian bandpass processes, we need to generalize the results of the previous section to complex normal random vectors. Let us consider a set of N complex random variables $[\tilde{x}_1, \dots, \tilde{x}_N]$, with $\tilde{x}_j = p_j + \mathrm{j}q_j$. The complex vector

$$\tilde{\mathbf{x}} = \begin{pmatrix} \tilde{x}_1 \\ \vdots \\ \tilde{x}_N \end{pmatrix} = \begin{pmatrix} p_1 \\ \vdots \\ p_N \end{pmatrix} + \mathrm{j} \begin{pmatrix} q_1 \\ \vdots \\ q_N \end{pmatrix} \tag{A3.17}$$

is said to be complex normal if its real and imaginary parts are jointly real normal random vectors, as discussed in the previous section. We again suppose that we are dealing with centered processes and thus with centered complex normal random vectors. However, when focusing on the moments of $\tilde{\mathbf{x}}$, we see a difference compared to the pure real case as a moment of given order can be defined in the present case using either the variable $\tilde{x}_i$ or its complex conjugate. The kth order moment of the complex normal vector can indeed be defined as

$$\mathbb{E}\left\{\prod_{l=1}^{k} \tilde{x}_{i_l}^{\epsilon_l}\right\}, \tag{A3.18}$$

with indexes i_l taken in the set $[1, \dots, N]$, and where ϵ_l can be either $+1$ or $*$. But, considering the product involved in the definition of the moments in terms of real and imaginary parts, we can obtain the moments of the real random vectors corresponding to the real and imaginary part of $\tilde{\mathbf{x}}$. It can thus be shown that for centered variables [3, 92]:

(i) when k is odd, of the form $2\lambda + 1$,

$$\mathbb{E}\left\{\prod_{l=1}^{2\lambda+1} \tilde{x}_{i_l}^{\epsilon_l}\right\} = 0; \tag{A3.19}$$

(ii) when k is even, of the form 2λ,

$$\mathbb{E}\left\{\prod_{l=1}^{2\lambda} \tilde{x}_{i_l}^{\epsilon_l}\right\} = \sum \mathbb{E}\{\tilde{x}_{i_1}^{\epsilon_1}\tilde{x}_{i_2}^{\epsilon_2}\}\mathbb{E}\{\tilde{x}_{i_3}^{\epsilon_3}\tilde{x}_{i_4}^{\epsilon_4}\} \dots \mathbb{E}\{\tilde{x}_{i_{2\lambda-1}}^{\epsilon_{2\lambda-1}}\tilde{x}_{i_{2\lambda}}^{\epsilon_{2\lambda}}\}, \tag{A3.20}$$

with the sum still taken over all the allocations of the set $[i_1, \dots, i_{2\lambda}]$ into λ unordered pairs.

By our discussion in Appendix 2, we can assume *stationarity*, at least up to second order, of the processes of interest in the field of wireless. It follows that the complex envelopes encountered in practice fulfill equation (A2.11). This condition applied to the complex Gaussian random vector composed of the samples of such stationary process leads to $\mathbb{E}\{\tilde{x}_m\tilde{x}_n\} = 0$ for all possible m and n from 1 to N. In matrix notation, this means in particular that

$$\mathbb{E}\{\tilde{\mathbf{x}}\tilde{\mathbf{x}}^{\mathrm{T}}\} = 0. \tag{A3.21}$$

In that case, the complex normal random vector is said to be circular. What is interesting is that this characteristic leads to a simplified version of the moments derived so far. Indeed, we now get that all the terms on the right-hand side of equation (A3.20) that do not involve two different symbols ϵ_j with one equal to $+1$ and the other equal to $*$ vanish. The expression for the moments then simplifies and we see that the moments of order 2λ are non-vanishing if and only if we have the same number λ of symbols equal to $+1$ and equal to $*$. For instance, the general expression for the fourth order moment,

$$\mathbb{E}\left\{\tilde{x}_1\tilde{x}_1^*\tilde{x}_2\tilde{x}_2^*\right\} = \mathbb{E}\left\{\tilde{x}_1\tilde{x}_1^*\right\}\mathbb{E}\left\{\tilde{x}_2\tilde{x}_2^*\right\} + \mathbb{E}\{\tilde{x}_1\tilde{x}_2\}\mathbb{E}\left\{\tilde{x}_1^*\tilde{x}_2^*\right\} + \mathbb{E}\left\{\tilde{x}_1\tilde{x}_2^*\right\}\mathbb{E}\left\{\tilde{x}_1^*\tilde{x}_2\right\}, \quad \text{(A3.22)}$$

reduces under stationarity of the process $\tilde{x}(t)$ to

$$\mathbb{E}\left\{\tilde{x}_1\tilde{x}_1^*\tilde{x}_2\tilde{x}_2^*\right\} = \mathbb{E}\left\{\tilde{x}_1\tilde{x}_1^*\right\}\mathbb{E}\left\{\tilde{x}_2\tilde{x}_2^*\right\} + \mathbb{E}\left\{\tilde{x}_1\tilde{x}_2^*\right\}\mathbb{E}\left\{\tilde{x}_1^*\tilde{x}_2\right\}. \quad \text{(A3.23)}$$

In the same way, the sixth order moment $\mathbb{E}\left\{\tilde{x}_1^2\tilde{x}_1^*\tilde{x}_2^{*2}\tilde{x}_2\right\}$ reduces under stationarity to

$$\mathbb{E}\left\{\tilde{x}_1^2\tilde{x}_1^*\tilde{x}_2^{*2}\tilde{x}_2\right\} = 2\mathbb{E}\left\{\tilde{x}_1\tilde{x}_2^*\right\}\left[2\mathbb{E}\left\{\tilde{x}_1\tilde{x}_1^*\right\}\mathbb{E}\left\{\tilde{x}_2\tilde{x}_2^*\right\} + \left|\mathbb{E}\left\{\tilde{x}_1\tilde{x}_2^*\right\}\right|^2\right]. \quad \text{(A3.24)}$$

The latter two expressions are used for instance in Chapter 5 when performing analytical derivations to examine the spectral regrowth phenomenon.

References

[1] J. G. Proakis, *Digital Communications*, 3rd edn. New York: McGraw-Hill, 1995.

[2] E. Roubine, *Distributions – Signal*, 2nd edn. Paris: Eyrolles, 1990.

[3] B. Picinbono, *Random Signals and Systems*. Upper Saddle River, NJ: Prentice Hall, 1993.

[4] J. W. B. Davenport and W. L. Root, *An Introduction to the Theory of Random Signals and Noise*. New York: McGraw-Hill, 1958.

[5] P. Flandrin, *Time-Frequency/Time-Scale Analysis*. San Diego, CA: Academic Press, 1999.

[6] I. S. Gradshteyn and I. M. Ryzhik, *Table of Integrals, Series, and Products*, 6th edn. San Diego, CA: Academic Press, 2000.

[7] “3rd Generation Partnership Project; Technical Specification Group GSM/EDGE Radio Access Network; Digital Cellular Telecommunications System (phase 2+); Modulation (Release 99)”, 3GPP, TS 05.04 V8.4.0, Nov. 2001.

[8] P. Laurent, “Exact and approximate construction of digital phase modulations by superposition of amplitude modulated pulses (AMP)”, *IEEE Transactions on Communications*, vol. 34, no. 2, pp. 150–160, Feb. 1986.

[9] “3rd Generation Partnership Project; Technical Specification Group Radio Access Network; Spreading and Modulation (FDD) (Release 9)”, 3GPP, TS 25.213 V9.2.0, Sept. 2010.

[10] “3rd Generation Partnership Project; Technical Specification Group Radio Access Network; Spreading and Modulation (TDD) (Release 9)”, 3GPP, TS 25.223 V9.0.0, Dec. 2009.

[11] R. Staszewski, D. Leipold, K. Muhammad, and P. Balsara, “Digitally controlled oscillator (DCO)-based architecture for RF frequency synthesis in a deep-submicrometer CMOS process”, *IEEE Transactions on Circuits and Systems—Part II: Analog and Digital Signal Processing*, vol. 50, no. 11, pp. 815–828, Nov. 2003.

[12] E. Roubine and J. C. Bolomey, *Antennes Volume 1: Introduction générale*, 2nd edn. Paris: Masson, 1986.

[13] M. Hélier, *Techniques micro-ondes: Structures de guidage, dispositifs passifs et tubes micro-ondes*, ser. Technosup. Paris: Ellipses Marketing, 2001.

[14] T. A. Milligan, *Modern Antenna Design*, 2nd edn. New York: McGraw-Hill, 1985.

[15] R. P. Feynman, R. B. Leighton, and M. Sands, *The Feynman Lectures on Physics: The Definitive Edition*, vol. 2, 2nd edn. Reading, MA: Addison-Wesley, 2005.

[16] H. Friis, “A note on a simple transmission formula”, *Proceedings of the IRE*, vol. 34, no. 5, pp. 254–256, May 1946.

[17] D. M. Pozar, *Microwave Engineering*, 3rd edn. Hoboken, NJ: John Wiley & Sons, Inc., 2005.

[18] M. Camus, B. Butaye, L. Garcia, M. Sie, B. Pellat, and T. Parra, “A 5.4 mW/0.07 mm^2 2.4 GHz front-end receiver in 90 nm CMOS for IEEE 802.15.4 WPAN standard”, *IEEE Journal of Solid-State Circuits*, vol. 43, no. 6, pp. 1372–1383, June 2008.

[19] H. W. Bode, *Network Analysis and Feedback Amplifier Design*. New York: Van Nostrand, 1945.

[20] G. L. Matthaei, L. Young, and E. M. T. Jones, *Microwave Filters, Impedance-Matching Networks, and Coupling Structures*. New York: McGraw-Hill, 1964.

[21] K. Kurokawa, “Power waves and the scattering matrix”, *IEEE Transactions on Microwave Theory and Techniques*, vol. 13, no. 2, pp. 194–202, Mar. 1965.

[22] B. Sklar, “Rayleigh fading channels in mobile digital communication systems I. Characterization”, *IEEE Communications Magazine*, vol. 35, no. 7, pp. 90–100, July 1997.

Wireless Transceiver Architecture: Bridging RF and Digital Communications, First Edition. Pierre Baudin.

[23] P. Bello, "Characterization of randomly time-variant linear channels", *IEEE Transactions on Communication Systems*, vol. 11, no. 4, pp. 360–393, Dec. 1963.

[24] J.-C. Pélissolo, "Propagation des ondes radio-électriques 01: Bases théoriques, rôle et influence du sol", École Supérieure d'Électricité, Gif-sur-Yvette, France, Course manual 03057/01, 1969.

[25] W. C. Jakes, *Microwave Mobile Communications*. New York: John Wiley & Sons, Inc., 1974.

[26] R. P. Feynman, R. B. Leighton, and M. Sands, *The Feynman Lectures on Physics: The Definitive Edition*, vol. 1, 2nd edn. Reading, MA: Addison-Wesley, 2005.

[27] M. Gans, "A power-spectral theory of propagation in the mobile-radio environment", *IEEE Transactions on Vehicular Technology*, vol. 21, no. 1, pp. 27–38, Feb. 1972.

[28] P. Dent, G. Bottomley, and T. Croft, "Jakes fading model revisited", *Electronics Letters*, vol. 29, no. 13, pp. 1162–1163, June 1993.

[29] J. Costas, "Synchronous communications", *Proceedings of the IEEE*, Dec. 1956, republished in same journal in Vol 90, no. 8, August 2002 as a classic paper.

[30] "3rd Generation Partnership Project; Technical Specification Group Radio Access Network; Physical Channels and Mapping of Transport Channels onto Physical Channels (FDD) (Release 9)", 3GPP, TS 25.211 V9.2.0, Sept. 2010.

[31] "3rd Generation Partnership Project; Technical Specification Group Radio Access Network; Physical Channels and Mapping of Transport Channels onto Physical Channels (TDD) (Release 9)", 3GPP, TS 25.221 V9.3.0, Sept. 2010.

[32] "3rd Generation Partnership Project; Technical Specification Group Radio Access Network; Evolved Universal Terrestrial Radio Access (E-UTRA); User Equipment (UE) Radio Transmission and Reception (Release 9)", 3GPP, TS 36.101 V9.0.0, June 2009.

[33] "3rd Generation Partnership Project; Technical Specification Group GSM/EDGE Radio Access Network; Radio Transmission and Reception (Release 1999)", 3GPP, TS 05.05 V8.20.0, Nov. 2005.

[34] "3rd Generation Partnership Project; Technical Specification Group Radio Access Network; User Equipment (UE) Radio Transmission and Reception (FDD) (Release 9)", 3GPP, TS 25.101 V9.6.0, Dec. 2010.

[35] P. Baudin and F. Belvèze, "Impact of RF impairments on a DS-CDMA receiver", *IEEE Transactions on Communications*, vol. 52, no. 1, pp. 31–36, Jan. 2004.

[36] A. Van der Ziel, *Noise in Solid State Devices and Circuits*. New York: John Wiley & Sons, Inc., 1986.

[37] C. Cohen-Tannoudji, B. Diu, and F. Laloe, *Quantum Mechanics*, vol. 1. New York: John Wiley & Sons, Inc., June 1977.

[38] K. L. Fong and R. G. Meyer, "Monolithic RF active mixer design", *IEEE Transactions on Circuits and Systems—Part II: Analog and Digital Signal Processing*, vol. 46, no. 3, pp. 231–239, Mar. 1999.

[39] W. R. Bennett, "Methods of solving noise problems", *Proceedings of the IRE*, vol. 44, no. 5, pp. 609–638, May 1956.

[40] P. R. Gray, P. J. Hurst, S. H. Lewis, and R. G. Meyer, *Analysis and Design of Analog Integrated Circuits*, 4th edn. New York: John Wiley & Sons, Inc., 2001.

[41] H. Friis, "Noise figures of radio receivers", *Proceedings of the IRE*, vol. 32, no. 7, pp. 419–422, July 1944.

[42] A. Hajimiri and T. H. Lee, "A general theory of phase noise in electrical oscillators", *IEEE Journal of Solid-State Circuits*, vol. 33, no. 2, pp. 179–194, Feb. 1998.

[43] A. Hajimiri and T. H. Lee, "Corrections to 'A general theory of phase noise in electrical oscillators'", *IEEE Journal of Solid-State Circuits*, vol. 33, no. 6, pp. 928–928, June 1998.

[44] R. Adler, "A study of locking phenomena in oscillators", *Proceedings of the IRE*, vol. 34, no. 6, pp. 351–357, June 1946.

[45] L. Paciorek, "Injection locking of oscillators", *Proceedings of the IEEE*, vol. 53, no. 11, pp. 1723–1727, Nov. 1965.

[46] K. Kurokawa, "Injection locking of microwave solid-state oscillators", *Proceedings of the IEEE*, vol. 61, no. 10, pp. 1386–1410, Oct. 1973.

[47] I. Ali, A. Banerjee, A. Mukherjee, and B. Biswas, "Study of injection locking with amplitude perturbation and its effect on pulling of oscillator", *IEEE Transactions on Circuits and Systems—Part I: Regular Papers*, vol. 59, no. 1, pp. 137–147, Jan. 2012.

[48] F. M. Gardner, *Phaselock Techniques*, 3rd edn. Hoboken, NJ: John Wiley & Sons, Inc., 2005.

[49] J. Chin and A. Cantoni, "Phase jitter ≡ timing jitter?" *IEEE Communications Letters*, vol. 2, no. 2, pp. 54–56, Feb. 1998.

[50] B. Widrow, I. Kollar, and M.-C. Liu, "Statistical theory of quantization", *IEEE Transactions on Instrumentation and Measurement*, vol. 45, no. 2, pp. 353–361, Apr. 1996.

[51] B. Widrow, "A study of rough amplitude quantization by means of Nyquist sampling theory", *IRE Transactions on Circuit Theory*, vol. 3, no. 4, pp. 266–276, Dec. 1956.

[52] A. Sripad and D. Snyder, "A necessary and sufficient condition for quantization errors to be uniform and white", *IEEE Transactions on Acoustics, Speech, and Signal Processing*, vol. 25, no. 5, pp. 442–448, Oct. 1977.

[53] R. M. Gray, "Quantization noise spectra", *IEEE Transactions on Information Theory*, vol. 36, no. 6, pp. 1220–1244, Nov. 1990.

[54] S. R. Norsworthy, R. Schreier, and G. C. Temes, *Delta-Sigma Data Converters: Theory, Design, and Simulation*. New York: IEEE Press, 1996.

[55] L. Schuchman, "Dither signals and their effect on quantization noise", *IEEE Transactions on Communication Technology*, vol. 12, no. 4, pp. 162–165, Dec. 1964.

[56] W. R. Bennett, "Spectra of quantized signals", *Bell System Technical Journal*, vol. 27, no. 4, pp. 446–472, July 1948.

[57] J. Feddeler and B. Lucas, "ADC definitions and specifications", Freescale Semiconductor, Inc., Application Note AN2438, Feb. 2003.

[58] C. Rapp, "Effects of HPA-nonlinearity on a 4-DPSK/OFDM-signal for a digital sound broadcasting signal", *Proceedings of the Second European Conference on Satellite Communications (ECSC-2)*, P. S. Weltevreden, ed., Oct. 1991, pp. 179–184.

[59] R. Meyer and A. Wong, "Blocking and desensitization in RF amplifiers", *IEEE Journal of Solid-State Circuits*, vol. 30, no. 8, pp. 944–946, Aug. 1995.

[60] D. Manstretta, M. Brandolini, and F. Svelto, "Second-order intermodulation mechanisms in CMOS downconverters", *IEEE Journal of Solid-State Circuits*, vol. 38, no. 3, pp. 394–406, Mar. 2003.

[61] D. Schreurs, M. O'Droma, A. A. Goacher, and M. Gadringer, *RF Power Amplifier Behavioral Modeling*. Cambridge: Cambridge University Press, 2009.

[62] J. Jensen, "Sur les fonctions convexes et les inégalités entre les valeurs moyennes", *Acta Mathematica*, vol. 30, no. 1, pp. 175–193, Dec. 1906.

[63] F. Belvèze and P. Baudin, "Specifying receiver IP2 and IP3 based on tolerance to modulated blockers", *IEEE Transactions on Communications*, vol. 56, no. 10, pp. 1677–1682, Oct. 2008.

[64] J. W. B. Davenport, "Signal-to-noise ratios in band-pass limiters", *Journal of Applied Physics*, vol. 24, no. 6, pp. 720–727, 1953.

[65] A. A. M. Saleh, "Frequency-independent and frequency-dependent nonlinear models of TWT amplifiers", *IEEE Transactions on Communications*, vol. 29, no. 11, pp. 1715–1720, Nov. 1981.

[66] A. Ghorbani and M. Sheikhan, "The effect of solid state power amplifiers (SSPAs) nonlinearities on MPSK and M-QAM signal transmission", *Proceedings of the Sixth International Conference on Digital Processing of Signals in Communications*, Sept. 1991, pp. 193–197.

[67] S. C. Cripps, *Advanced Techniques in RF Power Amplifier Design*. Norwood, MA: Artech House, 2002.

[68] J. Pedro and S. Maas, "A comparative overview of microwave and wireless power-amplifier behavioral modeling approaches", *IEEE Transactions on Microwave Theory and Techniques*, vol. 53, no. 4, pp. 1150–1163, Apr. 2005.

[69] E. Ngoya, N. Le Gallou, J. Nebus, H. Buret, and P. Reig, "Accurate RF and microwave system level modeling of wideband nonlinear circuits", in *2000 IEEE MTT-S International Microwave Symposium Digest*, vol. 1, 2000, pp. 79–82.

[70] B. Razavi, *RF microelectronics*. Upper Saddle River, NJ: Prentice Hall, 1998.

[71] T. H. Lee, *The Design of CMOS Radio-Frequency Integrated Circuits*, 2nd edn. Cambridge: Cambridge University Press, 2004.

[72] W. F. Egan, *Practical RF System Design*. Hoboken, NJ: John Wiley & Sons, Inc., 2003.

[73] J.-F. Bercher and C. Berland, "Adaptive time mismatches identification and correction in polar transmitter architecture", in *10th European Conference on Wireless Technologies*, Oct. 2007, pp. 78–81.

[74] S. C. Cripps, *RF Power Amplifiers for Wireless Communications*, 2nd edn. Norwood, MA: Artech House, 2006.

[75] W. Doherty, "A new high efficiency power amplifier for modulated waves", in *Proceedings of the IRE*, vol. 24, no. 9, Sept. 1936, pp. 1163–1182.

[76] F. Raab, "Efficiency of doherty RF power-amplifier systems", *IEEE Transactions on Broadcasting*, vol. BC-33, no. 3, pp. 77–83, Sept. 1987.

[77] F. Raab, P. Asbeck, S. Cripps, P. Kenington, Z. Popovic, N. Pothecary, J. Sevic, and N. Sokal, "Power amplifiers and transmitters for RF and microwave", *IEEE Transactions on Microwave Theory and Techniques*, vol. 50, no. 3, pp. 814–826, Mar. 2002.

[78] L. Kahn, "Single-sideband transmission by envelope elimination and restoration", in *Proceedings of the IRE*, vol. 40, no. 7, July 1952, pp. 803–806.

[79] H. Chireix, "High power outphasing modulation", in *Proceedings of the IRE*, vol. 23, no. 11, Nov. 1935, pp. 1370–1392.

[80] A. Bateman, "The Combined Analogue Locked Loop Universal Modulator (CALLUM)", in *Vehicular Technology Conference, 1992, IEEE 42nd*, vol. 2, May 1992, pp. 759–763.

[81] F. Raab, "Efficiency of outphasing RF power-amplifier systems", *IEEE Transactions on Communications*, vol. 33, no. 10, pp. 1094–1099, Oct. 1985.

[82] M. Nannicini, P. Magni, and F. Oggionni, "Temperature controlled predistortion circuits for 64 QAM microwave power amplifiers", in *Microwave Symposium Digest, 1985 IEEE MTT-S International*, June 1985, pp. 99–102.

[83] A. A. M. Saleh and J. Salz, "Adaptive linearization of power amplifiers in digital radio systems", *Bell System Technical Journal*, vol. 62, no. 4, pp. 1019–1033, Apr. 1983.

[84] K. Muhonen, M. Kavehrad, and R. Krishnamoorthy, "Look-up table techniques for adaptive digital predistortion: a development and comparison", *IEEE Transactions on Vehicular Technology*, vol. 49, no. 5, pp. 1995–2002, Sept. 2000.

[85] J. Cavers, "Amplifier linearization using a digital predistorter with fast adaptation and low memory requirements", *IEEE Transactions on Vehicular Technology*, vol. 39, no. 4, pp. 374–382, Nov. 1990.

[86] S. Stapleton and F. Costescu, "An adaptive predistorter for a power amplifier based on adjacent channel emissions [mobile communications]", *IEEE Transactions on Vehicular Technology*, vol. 41, no. 1, pp. 49–56, Feb. 1992.

[87] J. E. Volder, "The CORDIC trigonometric computing technique", *IRE Transactions on Electronic Computers*, vol. EC-8, no. 3, pp. 330–334, Sept. 1959.

[88] S. Lerstaveesin and B.-S. Song, "A complex image rejection circuit with sign detection only", *IEEE Journal of Solid-State Circuits*, vol. 41, no. 12, pp. 2693–2702, Dec. 2006.

[89] L. Yu and W. Snelgrove, "A novel adaptive mismatch cancellation system for quadrature IF radio receivers", *IEEE Transactions on Circuits and Systems—Part II: Analog and Digital Signal Processing*, vol. 46, no. 6, pp. 789–801, June 1999.

[90] M. Valkama, M. Renfors, and V. Koivunen, "Advanced methods for I/Q imbalance compensation in communication receivers", *IEEE Transactions on Signal Processing*, vol. 49, no. 10, pp. 2335–2344, Oct. 2001.

[91] E. Keehr and A. Hajimiri, "Successive regeneration and adaptive cancellation of higher order intermodulation products in RF receivers", *IEEE Transactions on Microwave Theory and Techniques*, vol. 59, no. 5, pp. 1379–1396, May 2011.

[92] I. Reed, "On a moment theorem for complex Gaussian processes", *IRE Transactions on Information Theory*, vol. 8, no. 3, pp. 194–195, Apr. 1962.

Index

Note: page numbers in **bold** type indicate a page containing important information about an entry, such as a definition or basic usage.

Wireless Transceiver Architecture: Bridging RF and Digital Communications, First Edition. Pierre Baudin.

ted and bound by CPI Group (UK) Ltd, Croydon, CR0 4YY
17/09/2026
14964302-0007